Lecture Notes in Computer Science 10589

Commenced Publication in 1973
Founding and Former Series Editors:
Gerhard Goos, Juris Hartmanis, and Jan van Leeuwen

Frank Nielsen · Frédéric Barbaresco (Eds.)

Geometric Science of Information

Third International Conference, GSI 2017
Paris, France, November 7–9, 2017
Proceedings

Editors
Frank Nielsen
Ecole Polytechnique
Palaiseau
France

and

Sony Computer Science Laboratories, Inc.
Tokyo
Japan

Frédéric Barbaresco
Thales Land and Air Systems
Limours
France

ISSN 0302-9743 ISSN 1611-3349 (electronic)
Lecture Notes in Computer Science
ISBN 978-3-319-68444-4 ISBN 978-3-319-68445-1 (eBook)
https://doi.org/10.1007/978-3-319-68445-1

Library of Congress Control Number: 2017956720

LNCS Sublibrary: SL6 – Image Processing, Computer Vision, Pattern Recognition, and Graphics

Printed on acid-free paper

This Springer imprint is published by Springer Nature
The registered company is Springer International Publishing AG
The registered company address is: Gewerbestrasse 11, 6330 Cham, Switzerland

Preface

GSI 2017 in the Shadow of Blaise Pascal's "Aleae Geometria"

GSI 2017 banner: Euclide, Thales, Clairaut, Legendre, Poncelet, Darboux, Poincaré, Cartan, Fréchet, Libermann, Leray, Koszul, Ferrand, Souriau, Balian, Berger, Choquet-Bruhat, Gromov

On behalf of both the Organizing and the Scientific Committees, it is our pleasure to welcome you to the proceedings of the Third International SEE Conference on Geometric Science of Information (GSI 2017), hosted in Paris, in 2017.

The three-day conference was organized in the framework of the relations set up between SEE and the following scientific institutions or academic laboratories: Ecole Polytechnique, Ecole des Mines ParisTech, Inria, Supélec, Université Paris-Sud, Institut Mathématique de Bordeaux, Sony Computer Science Laboratories, Institut Mines Télécom. GSI 2017 benefited from scientific and financial sponsors.

We would like to express all our thanks to the local organizers for hosting this second scientific event at the interface between geometry, probability, and information geometry.

The GSI conference cycle was initiated by the Brillouin Seminar Team. The GSI 2017 event has been motivated by the continuity of the first initiatives launched in 2013 (https://www.see.asso.fr/gsi2013) and consolidated in 2015 (https://www.see.asso.fr/gsi2015). In 2011, we organized an Indo-French workshop on "Matrix Information Geometry" that yielded an edited book in 2013.

The technical program of GSI 2017 covered all the main topics and highlights in the domain of "geometric science of information" including information geometry manifolds of structured data/information and their advanced applications. These proceedings consist solely of original research papers that have been carefully peer-reviewed by two or three experts before, and revised before acceptance.

The GSI 2017 program included a renowned invited speaker and three distinguished keynote speakers.

Historical Background

Like GSI 2013 and 2015, GSI 2017 addressed the inter-relations between different mathematical domains such as shape spaces (geometric statistics on manifolds and Lie groups, deformations in shape space), probability/optimization and algorithms on manifolds (structured matrix manifold, structured data/Information), relational and discrete metric spaces (graph metrics, distance geometry, relational analysis), computational and Hessian information geometry, algebraic/infinite dimensional/Banach information manifolds, divergence geometry, tensor-valued morphology, optimal transport theory, manifold and topology learning, as well as applications such as geometries of audio-processing, inverse problems, and signal processing. The program was enriched by contributions in the area of (stochastic) geometric mechanics and geometric model of quantum physics. GSI 2017 included new topics such as geometric robotics, information structure on neuroscience, stochastic calculus of variations (Malliavin calculus), and geometric deep learning among others.

At the turn of the century, new and fruitful interactions were discovered between several branches of science: information science (information theory, digital communications, statistical signal processing), mathematics (group theory, geometry and topology, probability, statistics, sheaf theory), and physics (geometric mechanics, thermodynamics, statistical physics, quantum mechanics). The GSI conference cycle aims to discover joint mathematical structures to all these disciplines by elaboration of a "general theory of information" embracing physics science, information science, and cognitive science in a global scheme.

The GSI 2017 program comprised 101 papers presented at 19 sessions:

- Session "Statistics on Non-linear Data" chaired by X. Pennec and S. Sommer
- Session "Shape Space" chaired by S. Allasonnière, S. Durrleman, and A. Trouvé
- Session "Optimal Transport and Applications I" chaired by Q. Merigot, J. Bigot, and B. Maury
- Session "Optimal Transport and Applications II" chaired by J.F. Marcotorchino and A. Galichon
- Session "Statistical Manifold & Hessian Information Geometry" chaired by M. Boyom, A. Matsuzoe, and Hassan Shahid
- Session "Monotone Embedding in Information Geometry" chaired by J. Zhang and J. Naudts
- Session "Information Structure in Neuroscience" chaired by P. Baudot, D. Bennequin, and S. Roy
- Session "Geometric Robotics and Tracking" chaired by S. Bonnabel and A. Barrau
- Session "Geometric Mechanics and Robotics" chaired by G. de Saxcé, J. Bensoam, and J. Lerbet
- Session "Stochastic Geometric Mechanics and Lie Group Thermodynamics" chaired by F. Gay-Balmaz and F. Barbaresco
- Session "Probability on Riemannian Manifolds" chaired by M. Arnaudon and A.-B. Cruzeiro
- Session "Divergence Geometry" chaired by M. Broniatowski and I. Csiszar

- Session "Non-parametric Information Geometry" chaired by N. Ay and J. Armstrong
- Session "Optimization on Manifold" chaired by P.A. Absil and R. Sepulchre
- Session "Computational Information Geometry" chaired by F. Nielsen and O. Schwander
- Session "Probability Density Estimation" chaired by S. Said and E. Chevallier
- Session "Geometry of Tensor-Valued Data" chaired by J. Angulo, Y. Berthoumieu, G. Verdoolaege, and A.M. Djafari
- Session "Geodesic Methods with Constraints" chaired by J.-M. Mirebeau and L. Cohen
- Session "Applications of Distance Geometry" chaired by A. Mucherino and D. Gonçalves

Three keynote speakers' talks opened each day (Prof. A. Trouvé, B. Tumpach, and M. Girolami). An invited honorary speaker (Prof. J.M. Bismut) gave a talk at the end of the first day and a guest Honorary speaker (Prof. D. Bennequin) closed the conference (https://www.see.asso.fr/wiki/18335_invited-keynote-speakers).

Invited Honorary Speaker:

- Jean-Michel Bismut (Paris-Saclay University), "The Hypoelliptic Laplacian"

Guest Honorary Speaker:

- Daniel Bennequin (Paris Diderot University), "Geometry and Vestibular Information"

Keynote Speakers:

- Alain Trouvé (ENS Paris-Saclay), "Hamiltonian Modeling for Shape Evolution and Statistical Modeling of Shapes Variability"
- Barbara Tumpach (Lille University), "Riemannian Metrics on Shape Spaces of Curves and Surfaces"
- Mark Girolami (Imperial College London), "Riemann Manifold Langevin and Hamiltonian Monte Carlo Methods"

GSI 2017 Seeding by Blaise Pascal's Geometry of Chance

Blaise Pascal's ideas are widely debated in 2017, because Pope François decided to initiate a request for his beatification. Among all the genius ideas of Pascal, one was the invention of probability. The "calculation of probabilities" began four years after the death of René Descartes, in 1654, in a correspondence between Blaise Pascal and Pierre Fermat. They exchanged letters on elementary problems of gambling, in this case a problem of dice and a problem of "parties." Pascal and Fermat were particularly interested in this problem and succeeded in the "party rule" by two different methods. One understands the legitimate pride of Pascal in his address of the same year at the Académie Parisienne created by Mersenne, to which he presented, among "the ripe fruit of our geometry" (*"les fruits mûrs de notre Géométrie" in French*), an entirely new treaty, of an absolutely unexplored matter, the distribution of chance in the games. In the same way, Pascal in his introduction to "Les Pensées" wrote that, "under the influence of Méré, given to the game, he throws the bases of the calculation of probabilities and composes the Treatise of the Arithmetical Triangle." If Pascal appears at first sight as the initiator of the calculation of probabilities, watching a little closer, his role in the emergence of this theory is more complex. However, there is no trace of the word "probabilities" in Pascal's work. To designate what might resemble what we now call calculation of probabilities, one does not even find the word in such a context. The only occurrences of probability are found in "Les Provinciales" where he referred to the doctrine of the Jesuits, or in "Les Pensées." We do not find in Pascal's writings the words of "Doctrine des chances" or "Calcul des chances," but only "Géométrie du hasard" (geometry of chance). In 1654, Blaise Pascal submitted a short paper to "Celeberrimae matheseos Academiae Parisiensi" (ancestor of the French Royal Academy of Sciences founded in 1666), with the title "Aleae Geometria" (Geometry of Chance) or "De compositione aleae in ludis ipsi subjectis," which that was the seminal paper founding Probability as a new discipline in Science. In this paper, Pascal said: "… et sic matheseos demonstrationes cum aleae incertitudine jugendo, et quae contraria videntur conciliando, ab utraque nominationem suam accipiens, stupendum hunc titulum jure sibi arrogat: Aleae Geometria," which we can translate as: "By the union thus realized between the demonstrations of mathematics and the uncertainty of chance, and by the conciliation of apparent contradictions, it can derive its name from both sides and arrogate to itself this astonishing title: Geometry of Chance" (« … par l'union ainsi réalisée entre les démonstrations des mathématiques et l'incertitude du hasard, et par la conciliation entre les contraires apparents, elle peut tirer son nom de part et d'autre et s'arroger à bon droit ce titre étonnant: Géométrie du Hasard»). We can observe that Blaise Pascal attached a geometrical sense to probabilities in this seminal paper. Like Jacques Bernoulli, we can also give references to another Blaise Pascal document entitled "Art de penser" (the "Logique" of Port-Royal), published the year of his death (1662), with the last chapters containing elements on the calculus of probabilities applied to history, to medicine, to miracles, to literary criticism, and to events of life, etc.

In "De l'esprit géométrique," the use of reason for knowledge is based on a geometric model. In geometry, the first principles are given by the natural common sense to all men, and there is no need to define them. Other principles are clearly explained by definitions of terms such that it is always possible to mentally substitute the explanation for the defined term [23, 24, 25]. These definitions of terms are completely free, the only condition to be respected is univocity and invariability. Judging his solution as one of his most important contributions to science, Pascal envisioned the drafting of a small treatise entitled "Géométrie du Hasard" (Geometry of Chance). He never wrote it. Inspired by this, Christian Huygens wrote the first treatise on the calculation of chances, the "De ratiociniis in ludo aleae" ("On Calculation in Games of Chance," 1657). We can conclude this preamble by observing that seminal work of Blaise Pascal on probability was inspired by geometry. The objective of the GSI conference is to return to this initial idea that we can geometrize statistics in a rigorous way.

We can also make reference to Blaise Pascal for this GSI conference on computing geometrical statistics because he was the inventor of computers with his "Pascaline" machine. The introduction of Pascaline marks the beginning of the development of mechanical calculus in Europe. This development, which passed from calculating machines to the electrical and electronic calculators of the following centuries, culminated in the invention of the microprocessor. Also, Charles Babbage conceived his analytical machine from 1834 to 1837, a programmable calculating machine that was the ancestor of the computers of the 1940s, combining the inventions of Blaise Pascal and Jacquard's machine, with instructions written on perforated cards, one of the descendants of the Pascaline, the first machine that supplied the intelligence of man. We can observe that these machines were conceived on "mechanical" principles to develop "analytical" computation. GSI could be a source for new HPC machines based on "geometrical" computation. Future machines could be conceived on algorithm (geometrical) structures, with new numerical schemes that will overcome coordinate systems by using an intrinsic approach based on symmetries. We could say that we have to replace René Descartes by Blaise Pascal to build new HPC machines, intrinsically without coordinate systems.

- **Babbage Analytic Machine**

- **Jacquard Loom**

- **Pascaline Machine**

We will conclude with this beautiful citation of Joseph de Maistre comparing geometry and probability:

If we add that the criticism which accustoms the mind, especially in matters of facts, to receive simple probabilities for proofs, is, by this place, less adapted to form it than the geometry which makes it contract the habit of acquiescence only in evidence; We will reply that, strictly speaking, we might conclude from this very difference that criticism gives, on the contrary, more exercise to the mind than geometry: because the evidence, which is one and absolute, first aspect without leaving either the freedom to doubt, or the merit of choosing; Whereas, in order to be in a position to take a decision, it is necessary that they should be compared, discussed, and weighed. A kind of study which, so to speak, breaks the mind to this operation, is certainly of a wider use than that in which everything is subject to the evidence; Because the chances of determining themselves on likelihoods or probabilities are more frequent than those which require that we proceed by demonstrations: why should we not say that they often also hold to much more important objects?
—Joseph de Maistre in L'Esprit de Finesse

Si on ajoute que la critique qui accoutume l'esprit, surtout en matière de faits, à recevoir de simples probabilités pour des preuves, est, par cet endroit, moins propre à le former, que ne le doit être la géométrie qui lui fait contracter l'habitude de n'acquiescer qu'à l'évidence; nous répliquerons qu'à la rigueur on pourrait conclure de cette différence même, que la critique donne, au contraire, plus d'exercice à l'esprit que la géométrie: parce que l'évidence, qui est une et absolue, le fixe au premier aspect sans lui laisser ni la liberté de douter, ni le mérite de choisir; au lieu que les probabilités étant susceptibles du plus et du moins, il faut, pour se mettre en état de prendre un parti, les comparer ensemble, les discuter et les peser. Un genre d'étude qui rompt, pour ainsi dire, l'esprit à cette opération, est certainement d'un usage plus étendu que celui où tout est soumis à l'évidence; parce que les occasions de se déterminer sur des vraisemblances ou probabilités, sont plus fréquentes que celles qui exigent qu'on procède par démonstrations: pourquoi ne dirions –nous pas que souvent elles tiennent aussi à des objets beaucoup plus importants?
—Joseph de Maistre dans L'Esprit de Finesse

October 2017

Frank Nielsen
Frédéric Barbaresco

Organization

Program Chairs

Frédéric Barbaresco	Thales Air Systems, France
Frank Nielsen	Ecole Polytechnique, France
Silvère Bonnabel	Ecole des Mines ParisTech, France

Scientific Committee

Pierre-antoine Absil	University of Louvain, Belgium
Bijan Afsari	Johns Hopkins University, USA
Stéphanie Allassonnière	Ecole Polytechnique, France
Jesus Angulo	Mines ParisTech, France
Jean-pierre Antoine	Université catholique de Louvain, Belgium
John Armstrong	King's College London, UK
Marc Arnaudon	University of Bordeaux, France
Dena Asta	Carnegie Mellon University, USA
Anne Auger	Inria, France
Nihat Ay	MPI, Leipzig, Germany
Roger Balian	Academy of Sciences, France
Frédéric Barbaresco	Thales, France
Michèle Basseville	IRISA, France
Pierre Baudot	Max Planck Institute for Mathematic in the Sciences, Germany
Daniel Bennequin	Paris-Diderot University, France
Joël Bensoam	Ircam, Paris, France
Yannick Berthoumieu	IMS Université de Bordeaux, France
Jérémie Bigot	Université de Bordeaux, France
Silvère Bonnabel	Mines-ParisTech, Paris, France
Michel Boyom	Université de Montpellier, France
Michel Broniatowski	University of Pierre and Marie Curie, France
Alain Chenciner	IMCCE and Université Paris 7, France
Laurent Cohen	University of Paris Dauphine, Paris, France
Patricia Conde Céspedes	Université Pierre et Marie Curie, Paris, France
Ana Cruzeiro	IST, Lisbon University, Portugal
Géry De Saxcé	Université des Sciences et des Technologies de Lille, France
Stanley Durrleman	Inria, France
Diala Ezzeddine	Laboratoire ERIC, Lyon, France
Alfred Galichon	New York University, USA
François Gay-balmaz	CNRS, ENS-ULM, Paris, France
Jean-Pierre Gazeau	Université Paris-Diderot, France

Douglas Gonçalves	Universidade Federal de Santa Catarina, Brazil
Matsuzoe Hiroshi	Nagoya Institute of Technology, Japan
Susan Holmes	Stanford University, USA
Jérémie Jakubowicz	Institut Mines Telecom, France
Nicolas Le bihan	Université de Grenoble, CNRS, France/University of Melbourne, Australia
Alice Le brigant	Université de Bordeaux, France
Jean Lerbet	Université Paris-Saclay, France
Luigi Malagò	Romanian Institute of Science and Technology, Romania
Jonathan Manton	University of Melbourne, Australia
Gaetan Marceau Caron	Thales Theresis, France
Matilde Marcolli	Caltech, USA
Jean-François Marcotorchino	Thales, France
Charles-michel Marle	Université Pierre et Marie Curie, France
Juliette Mattioli	Thales, France
Bertrand Maury	Université Paris-Sud, France
Quentin Mérigot	Université Paris-Dauphine/CNRS, France
Nina Miolane	Inria, France
Jean-marie Mirebeau	University of Paris-Sud, CNRS, University of Paris-Saclay, France
Ali Mohammad-djafari	Supelec, CNRS, France
Antonio Mucherino	IRISA, Université de Rennes 1, France
Florence Nicol	Ecole Nationale de l'Aviation Civile, France
Frank Nielsen	Ecole Polytechnique, France
Richard Nock	Data61, Australia
Steve Oudot	Inria, France
Maks Ovsjanikov	Ecole Polytechnique, France
Pierre Pansu	University of Paris-Sud, France
Xavier Pennec	Inria, France
Michel Petitjean	University of Paris Diderot, CNRS, France
Marion Pilté	Mines ParisTech, France
Giovanni Pistone	Collegio Carlo Alberto, Moncalieri, Italy
Radmila Pribic	Thales, The Netherlands
Olivier Rioul	Telecom ParisTech, France
Salem Said	Laboratoire IMS, Bordeaux, France
Carola-bibiane Schönlieb	University of Cambridge, UK
Olivier Schwander	UPMC, Paris, France
Rodolphe Sepulchre	Cambridge University, UK
Fátima Silva Leite	University of Coimbra, Portugal
Stefan Sommer	University of Copenhagen, Denmark
Dominique Spehner	University of Grenoble Alpes, France
Barbara Trivellato	Politecnico di Torino, Italy
Alain Trouvé	ENS Cachan, France
Barbara Tumpach	Laboratoire Painlevé, Lille, France

Geert Verdoolaege	Ghent University, Belgium
René Vidal	John Hopkins University, USA
Jun Zhang	University of Michigan, Ann Arbor, USA

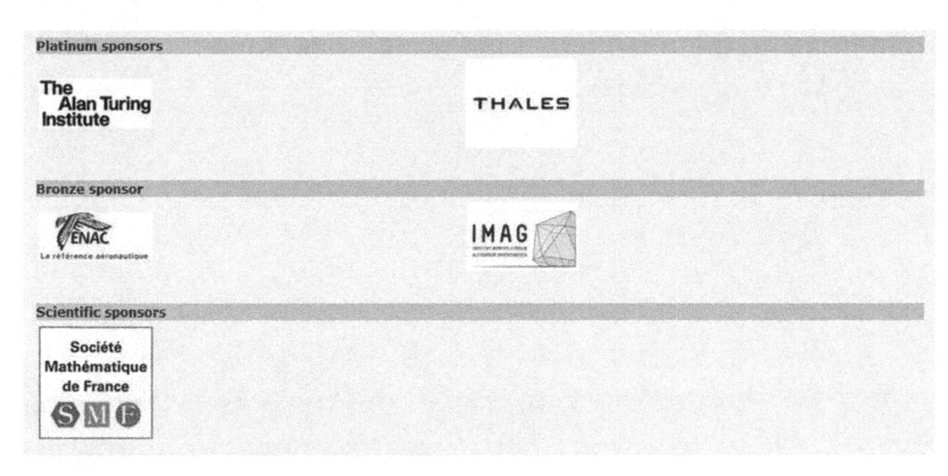

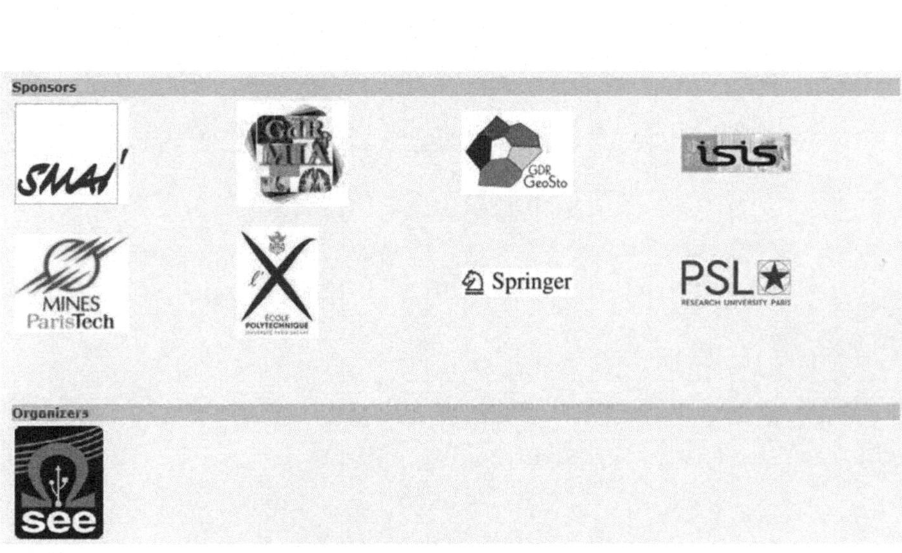

Contents

Geometric Mechanics and Robotics

Stochastic Geometric Mechanics and Lie Group Thermodynamics

Non-parametric Information Geometry

Optimization on Manifold

Computational Information Geometry

Probability Density Estimation

Session Geometry of Tensor-Valued Data

Statistics on Non-linear Data

Stochastic Development Regression Using Method of Moments

Line Kühnel(✉) and Stefan Sommer

Department of Computer Science, University of Copenhagen,
Copenhagen, Denmark
{kuhnel,sommer}@di.ku.dk

Abstract. This paper considers the estimation problem arising when inferring parameters in the stochastic development regression model for manifold valued non-linear data. Stochastic development regression captures the relation between manifold-valued response and Euclidean covariate variables using the stochastic development construction. It is thereby able to incorporate several covariate variables and random effects. The model is intrinsically defined using the connection of the manifold, and the use of stochastic development avoids linearizing the geometry. We propose to infer parameters using the Method of Moments procedure that matches known constraints on moments of the observations conditional on the latent variables. The performance of the model is investigated in a simulation example using data on finite dimensional landmark manifolds.

Keywords: Frame bundle · Non-linear statistics · Regression · Statistics on manifolds · Stochastic development

1 Introduction

There is a growing interest for statistical analysis of non-linear data such as shape data arising in medical imaging and computational anatomy. Non-linear data spaces lack vector space structure, and traditional Euclidean statistical theory is therefore not sufficient to analyze non-linear data. This paper considers parameter inference for the stochastic development regression (SDR) model introduced in [10] that generalizes Euclidean regression models to non-linear spaces. The focus of this paper is to introduce an alternative estimation procedure which is simple and computationally tractable.

Stochastic development regression is used to model the relation between a manifold-valued response and Euclidean covariate variables. Similar to Brownian motions on a manifold, $\mathcal{M}$, defined as the transport of a Euclidean Brownian motion from $\mathbb{R}^n$ to $\mathcal{M}$, the SDR model is defined as the transport of a Euclidean regression model. A Euclidean regression model can be regarded as a time dependent model in which, potentially, several observations have been

F. Nielsen and F. Barbaresco (Eds.): GSI 2017, LNCS 10589, pp. 3–11, 2017.
https://doi.org/10.1007/978-3-319-68445-1_1

observed over time. Given a response variable $y_t \in \mathbb{R}^d$ and covariate vector $\boldsymbol{x}_t = (x_t^1, \ldots, x_t^m) \in \mathbb{R}^m$, the Euclidean regression model can be written as

$$y_t = \alpha_t + \beta_t \boldsymbol{x}_t + \varepsilon_t, \quad t \in [0, 1], \tag{1}$$

where $\alpha_t \in \mathbb{R}^d$ and $\beta_t \in \mathbb{R}^{d \times m}$. A regression model can hence be defined as a stochastic process with drift α_t, covariate dependency through $\beta_t \boldsymbol{x}_t$, and a brownian noise ε_t. The SDR model is then defined as the transport of a regression model of the form (1), from $\mathbb{R}^d$ to the manifold $\mathcal{M}$. The transportation is performed by stochastic development described in Sect. 2. Figure 1 visualizes the idea behind the model.

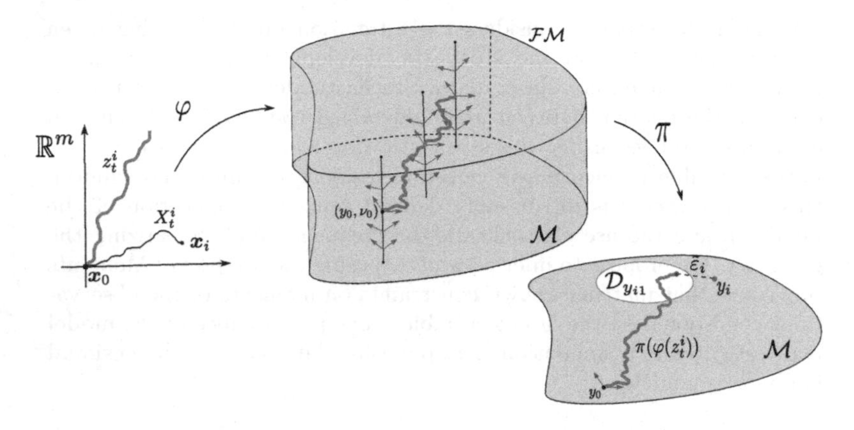

Fig. 1. The idea behind the model. Normal linear regression process z_t^i defined in (1) is transported to the manifold through stochastic development, φ. Here $\mathcal{FM}$ is the frame bundle, π a projection map, and $\mathcal{D}_{y_{i1}}$ the transition distribution of $y_{it} = \pi(\varphi(z_t^i))$. The tangent bundle of $\mathcal{FM}$ can be split in a horizontal and vertical subspace. Changes on $\mathcal{FM}$ in the vertical direction corresponds to fixing a point $y \in \mathcal{M}$ while changing the frame, ν, of the tangent space, $T_y\mathcal{M}$. Changes in the horizontal direction is fixing the frame for the tangent space and changing the point on the manifold. The frame is in this case parallel transported to the new tangent space.

In [10], Laplace approximation was applied for estimation of the parameter vector. However, this method was computational expensive and it was difficult to obtain results for detailed shapes. Alternatively, a Monte Carlo Expectation Maximization (MCEM) method has been considered, but, with this method, high probability samples were hard to obtain, which led to an unstable objective function. As a consequence, this paper examines the Method of Moments (MM) procedure for parameter estimation. The MM procedure is easy to apply and not as computationally expensive as the Laplace approximation. It is a well-known method for estimation in Euclidean statistics (see for example [3,6,14]), where it has been proven in general to provide consistent parameter estimates.

Several versions of the generalized regression model have been proposed in the case of manifold-valued response and Euclidean covariate variables. Local regression is considered in [11,19]. The former defines an intrinsic local regression model, while [11] constructs an extrinsic model. For global regression models, [5,12,16] consider geodesic regression, which is a generalization of the Euclidean linear regression model. There have been several approaches for defining non-geodesic regression models on manifolds. An example is kernel based regression models, in which the model function is estimated by a kernel representation [1,4,13]. In [7,8,17], the non-geodesic relation is modelled by a polynomial or piecewise cubic spline function. Moreover, [2,15] propose estimation of a parametric link function by minimization of the total residual sum of squares and the generalized method of moments procedure respectively.

The paper will be structured as follows. Section 2 gives a brief description of stochastic development and the frame bundle $\mathcal{FM}$. Section 3 introduces the SDR model and Sect. 4 describes the estimation procedure, Method of Moments. At the end, a simulation example is performed in Sect. 5.

2 Stochastic Development

This section gives a brief introduction to frame bundle and stochastic development. For a more detailed description and a reference for the following see [9]. Consider a d-dimensional Riemannian manifold $(\mathcal{M}, g)$ and a probability space $(\Omega, \mathcal{F}, P)$. Stochastic development is a method for transportation of stochastic processes in $\mathbb{R}^d$ to stochastic processes on $\mathcal{M}$. Let $z_t\colon \Omega \to \mathbb{R}^d$ denote a stochastic process for $t \in [0,1]$. In order to define the stochastic development of z_t it is necessary to consider a connection on $\mathcal{M}$. A connection, ∇, defines transportation of vectors along curves on the manifold, such that tangent vectors in different tangent spaces can be compared. A frequently used connection, which will also be used in this paper, is the Levi-Civita connection of a Riemannian metric. Consider a point $q \in \mathcal{M}$ and let ∂_i for $i = 1, \dots, d$ denote a coordinate frame at q, i.e. an ordered basis for $T_q\mathcal{M}$, with dual frame dx^i. A connection ∇ is locally determined by the Christoffel symbols defined by $\nabla_{\partial_i}\partial_j = \Gamma^k_{ij}\partial_k$. The Christoffel symbols for the Levi-Civita connection are given by $\Gamma^k_{ij} = \frac{1}{2}g^{kl}\left(\partial_i g_{jl} + \partial_j g_{il} - \partial_l g_{ij}\right)$, where g_{ij} denotes the coefficients of the metric g in the dual frame dx^i, i.e. $g = g_{ij}dx^i dx^j$, and g^{ij} are the inverse coefficients.

Stochastic development uses the frame bundle, $\mathcal{FM}$, defined as the fiber bundle of tuples (y, ν), $y \in \mathcal{M}$ with $\nu\colon \mathbb{R}^d \to T_y\mathcal{M}$ being a frame for the tangent space $T_y\mathcal{M}$. Given a connection on $\mathcal{FM}$, the tangent bundle of the frame bundle, $T\mathcal{FM}$, can be split into a horizontal, $H\mathcal{FM}$, and vertical, $V\mathcal{FM}$, subspace, i.e. $T\mathcal{FM} = H\mathcal{FM} \oplus V\mathcal{FM}$. Figure 1 shows a visualization of the frame bundle and the horizontal and vertical tangent spaces. The horizontal subspace determines changes in $y \in \mathcal{M}$ while fixing the frame ν, while $V\mathcal{FM}$ fixes $y \in \mathcal{M}$ and describes the change in the frame for $T_y\mathcal{M}$. Given the split of the tangent bundle $T\mathcal{FM}$, an isomorphism $\pi_{\star,(y,\nu)}\colon H_{(y,\nu)}\mathcal{FM} \to T_y\mathcal{M}$ can be defined. The inverse map $\pi^{\star}_{(y,\nu)}$ is called the horizontal lift and pulls a tangent vector in $T_y\mathcal{M}$ to $H_{(y,\nu)}\mathcal{FM}$. The horizontal lift of $v \in T_y\mathcal{M}$ is here denoted $v^\star \in H_{(y,\nu)}\mathcal{FM}$.

Let $e_1, \dots, e_d$ be the canonical basis of $\mathbb{R}^d$ and consider a point $(y, \nu) \in \mathcal{FM}$. Define the horizontal vector fields, $H_1, \dots, H_d$, by $H_i(\nu) = (\nu e_i)^*$. The vector fields $H_1, \dots, H_d$ then form a basis for the subspace $H\mathcal{FM}$. Given this basis for $H\mathcal{FM}$, the stochastic development of a Euclidean stochastic process, z_t, to the frame bundle $\mathcal{FM}$ can be found by the solution to the Stratonovich differential equation $dU_t = H_i(U_t) \circ dz_t^i$, where Einsteins summation notation is used and $\circ$ specifies that it is a Stratonovich differential equation. The stochastic development of a process $z_t \in \mathbb{R}^d$ with reference point (y, ν) will be denoted $\varphi_{(y,\nu)}(z_t)$. A stochastic process on $\mathcal{M}$ can then be obtained by the projection of U_t to $\mathcal{M}$ by the projection map $\pi \colon \mathcal{FM} \to \mathcal{M}$.

3 Model

Consider a d-dimensional manifold $\mathcal{M}$ equipped with a connection ∇ and let $\boldsymbol{y}_1, \dots, \boldsymbol{y}_n$ be n realizations of the response $y \in \mathcal{M}$. Notice that the realizations are assumed to be measured with additive noise, which might pull the observations to an ambient space of $\mathcal{M}$. An example of such additive noise for landmark data is given in Sect. 5. Denote for each observation $i = 1, \dots, n$, $\boldsymbol{x}_i = (x_{i1}, \dots, x_{im}) \in \mathbb{R}^m$ the covariate vector of $m \leq d$ covariate variables. The SDR model is defined as a stochastic process on $\mathcal{M}$ based on the definition of Euclidean regression models regarded as stochastic processes (see (1)). Assume therefore that the response $y \in \mathcal{M}$ is the endpoint of a stochastic process y_t in $\mathcal{M}$ and the covariates, $\boldsymbol{x}_i$, the endpoint of a stochastic process $X_t = (X_{1t}, \dots, X_{mt})$ in $\mathbb{R}^m$. The process X_{jt} is for random covariate variables assumed to be a Brownian motion in $\mathbb{R}$, while for fixed covariate effects it is modelled as a fixed drift. The process y_{it} for each observation $i = 1, \dots, n$ is defined as the stochastic development of a Euclidean model on $\mathbb{R}^m$. Consider the stochastic process, z_{it}, in $\mathbb{R}^m$ defined by the stochastic differential equation equivalent to the Euclidean regression model defined in (1),

$$dz_{it} = \alpha dt + W dX_{it} + d\varepsilon_{it}, \quad t \in [0, 1]. \tag{2}$$

Here αdt is a fixed drift, W the $m \times m$ coefficient matrix and ε_{it} the random error modelled as a Brownian motion in $\mathbb{R}^m$. The response process y_{it} is then given as the stochastic development of z_{it}, i.e. $y_{it} = \varphi_{(\boldsymbol{y}_0, \nu_0)}(z_{it})$ for a reference point $\boldsymbol{y}_0$ and frame $\nu_0 \in T_{\boldsymbol{y}_0}\mathcal{M}$ (see Fig. 1). The realizations are modelled as noisy observations of the endpoints of y_{it}, $\boldsymbol{y}_i = y_{i1} + \tilde{\varepsilon}_i$ in which $\tilde{\varepsilon}_i \sim \mathcal{N}(0, \tau^2 I)$ denotes iid. additive noise. There is a natural relation between W and the frame ν_0. If ν_0 is assumed to be an orthonormal basis and U the $d \times m$-matrix with columns of basis vectors of ν_0, then the matrix $\tilde{W} = UW$ explains the gathered effect of W and ν_0 through U. However, this decomposition is not unique and hence the $\tilde{W}$ matrix is estimated instead of U and W individually.

4 Method of Moments

In this section the MM procedure is introduced for the estimation of the parameters in the regression model. The MM procedure uses known moment conditions to define a set of equations which can be optimized to find the true parameter vector $\theta = (\tau, \alpha, \tilde{W}, \boldsymbol{y}_0)$, see [3,6,14]. Here τ^2 is the additive noise variance, α the drift, $\tilde{W}$ combined effect of covariates and ν_0, and $\boldsymbol{y}_0$ the initial point on $\mathcal{M}$.

In the SDR model the known moment conditions are based on the moments of the additive noise $\tilde{\varepsilon}_i$ and the fact that $\tilde{\varepsilon}_i$ is independent of the covariate variables x_{ik} for each $k = 1, \ldots, m$. Hence, the moment conditions are,

$$\mathbb{E}\left[\tilde{\varepsilon}_{ij}\right] = 0,\ \mathbb{E}\left[\tilde{\varepsilon}_{ij} x_{ik}\right] = 0,\ \mathbb{E}\left[\tilde{\varepsilon}_{ij}^2\right] = \tau^2\ \forall j = 1, \ldots, d, \text{ and } k = 1, \ldots, m.$$

Known consistent estimators for these moments are the sample means. Consider the residuals, $\hat{\varepsilon}_{ij} = y_{ij} - \hat{y}_{ij}$, in which the dependency of the parameter vector, θ, lies in the predictions, $\hat{y}_{ij}$ for $i = 1, \ldots, n,\ j = 1, \ldots, d$. For a proper choice of parameter vector θ, the sample means will approach the true moments. Therefore, the set of equations used to optimize the parameter vector θ are,

$$\frac{1}{n}\sum_{i=1}^{n} \hat{\varepsilon}_{ij} = 0, \quad \frac{1}{n}\sum_{i=1}^{n} x_{ik}\hat{\varepsilon}_{ij} = 0, \quad \text{and} \quad \frac{1}{n-2}\sum_{i=1}^{n} \hat{\varepsilon}_{ij}^2 = \hat{\tau}^2,$$

for all $j = 1, \ldots, d$ and $k = 1, \ldots, m$ and where $\hat{\tau}^2$ is the estimated variance. In Euclidean statistics, the method of moments is known to provide consistent estimators, but these estimators might be biased.

The cost function considered for optimization with respect to θ is,

$$f(\theta) = \frac{1}{d}\sum_{j}\left(\frac{1}{n}\sum_{i=1}^{n}\hat{\varepsilon}_{ij}\right)^2 + \frac{1}{dm}\sum_{j,k}\left(\frac{1}{n}\sum_{i=1}^{n} x_{ik}\hat{\varepsilon}_{ij}\right)^2 + \frac{1}{d}\sum_{j}\left(\frac{1}{n-2}\sum_{i=1}^{n}\hat{\varepsilon}_{ij}^2 - \hat{\tau}^2\right)^2. \tag{3}$$

This cost function depends on predictions from the model based on the given parameter vector in each iteration. In order for the objective function to be stable it has to be evaluated for several predictions. Therefore, the function has been averaged for several predictions to obtain a more stable gradient descent optimization procedure.

The initial value of θ can in practice be chosen as parameters estimated from a Euclidean multivariate linear regression model. Here, the estimated covariance matrix would resemble the $\tilde{W}$ effect and the intercept the initial point $\boldsymbol{y}_0$.

5 Simulation Example

The performance of the estimation procedure will be evaluated using simulated data. We will generate landmark data on Riemannian landmark manifolds as defined in the Large Deformation Diffeomorphic Metric Mapping (LDDMM) framework [18], and use the Levi-Civita connection. Shapes in the landmark manifold $\mathcal{M}$ are defined by a finite landmark representation, i.e. $q \in \mathcal{M}$, $q =$

$(x_1^1, x_1^2, \ldots, x_{n_l}^1, x_{n_l}^2)$, where n_l denotes the number of landmarks. The dimension of $\mathcal{M}$ is hence $d = 2n_l$. Using a kernel K, the Riemannian metric on $\mathcal{M}$ is defined as $g(v, w) = \sum_{i,j}^{n_l} vK^{-1}(\boldsymbol{x}_i, \boldsymbol{x}_j)w$ with K^{-1} denoting the inverse of the kernel matrix. In the following, we use a Gaussian kernel for K with standard deviation $\sigma = 0.1$. We will consider a single covariate variable $x \in \mathbb{R}$ drawn from $\mathcal{N}(0, 36)$ and model the relation to two response variables either with 1 or 3 landmarks. The response variables are simulated from a model with parameters given in Table 1 and Fig. 2 for $n_l = 3$. Examples of simulated data for $n_l = 1$ and 3 are shown in Fig. 2. The additive noise is in this case normally distributed iid. random noise added to each coordinate of landmarks. In this example we consider a simplification of the model, as the random error in z_{it}, given in (2), will be disregarded. Estimation of parameters is examined for three different models: one without additive noise and drift, one without drift, and at last the full model. For $n_l = 3$ only estimation of the two first models is studied, and estimation in the model with no drift has been considered for $n = 70$ and $n = 150$.

Table 1. Parameter estimates found with the MM procedure for 1 landmark. First column shows the true values and each column, estimated parameters in each model.

	True	Excl. τ, α $n = 70$	Excl. α $n = 70$	Excl. α $n = 150$	Full model $n = 150$
τ	0.1	$-(\tau = 0)$	0.256	0.226	0.207
α	40	$-(\alpha = 0)$	$-(\alpha = 0)$	$-(\alpha = 0)$	37.19
$\tilde{W}$	$(0, 2)$	$(0, 2.013)$	$(0.004, 1.996)$	$(0, 2.003)$	$(0, 2.004)$
$\boldsymbol{y}_0$	$(1, 0)$	$(1.064, 0.0438)$	$(1.158, 0.162)$	$(1.026, 0.0227)$	$(1.076, 2.708)$

By the results shown in Table 1 and Fig. 2, the procedure makes a good estimate of the frame matrix $\tilde{W}$ in every situation. For the model with no additive noise and no drift, the procedure finds a reasonable estimate of $\boldsymbol{y}_0$. When noise is added, it is seen that a larger sample size is needed in order to get a good estimate of $\boldsymbol{y}_0$. On the contrary, the variance estimate seems biased in each case. For $n_l = 3$ the variance parameters estimated were $\hat{\tau} = 0.306$ for $n = 70$ and $\hat{\tau} = 0.231$ for $n = 150$. However, when drift is added to the model, the estimation procedure has a hard time recapture the true estimates of $\boldsymbol{y}_0$ and α. This difficulty can be explained by the relation between the variables. In normal linear regression, only one intercept variable is present in the model, but in the SDR this intercept variable is split between α and $\boldsymbol{y}_0$.

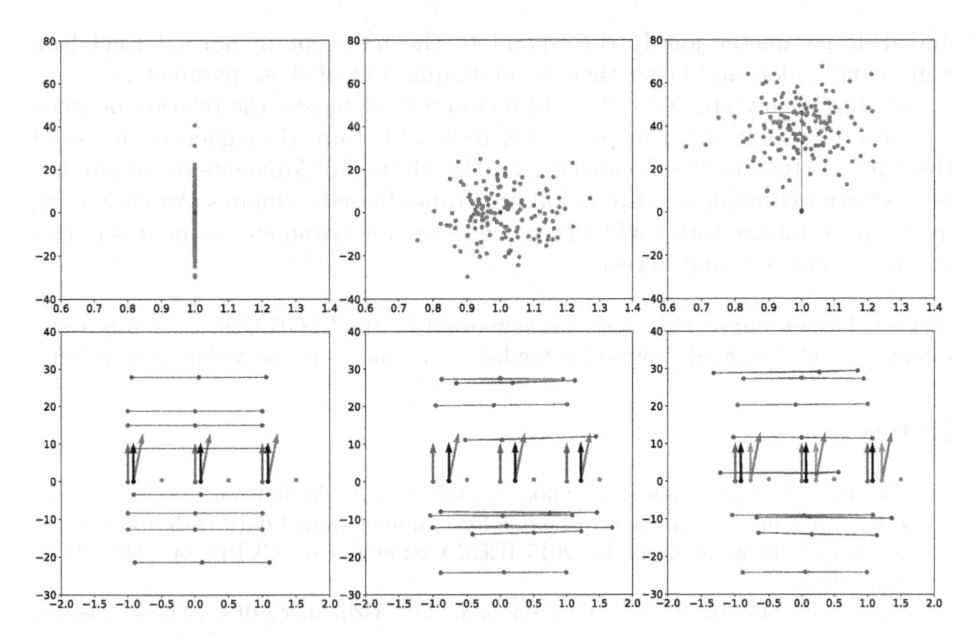

Fig. 2. (upper left) Sample drawn from model without additive noise and drift. (upper center) Sample drawn with additive noise, but no drift. (upper right) Sample drawn from the full model. The vertical lines are the stochastic development of z_{it} and the horizontal corresponds to the additive noise, the blue point is the reference point. (lower left) Model without drift and variance for $n_l = 3$, $n = 70$. (lower center) Model without drift and $n = 70$. (lower right) Model without drift and $n = 150$. These plots show the estimated results. (red) initial, (green) true, and (black) estimated reference point and frame. The gray samples are predicted from the estimated model while the green are a subset of the simulated data. Lower right plot does also show the difference in the estimated parameters for $n = 70$, $n = 150$ for the model with no drift. The magenta parameters in that plot is the estimated parameters for model without drift and $n = 70$, the corresponding black parameters in lower center plot. (Color figure online)

6 Conclusion

Method of Moments procedure has been examined for parameter estimation in the stochastic development regression (SDR) model. The SDR model is a generalization of regression models on Euclidean space to manifold-valued data. This model analyzes the relation between manifold-valued response and Euclidean covariate variables. The performance of the estimation procedure was studied based on a simulation example. The Method of Moments procedure was easier to apply and less computationally expensive than the Laplace approximation considered in [10]. The estimates found for the frame parameters were reasonable, but the procedure had a hard time retrieving the reference point and drift parameter. This is due to a mis-specification of the model as the reference point and drift parameter jointly correspond to the intercept in normal Euclidean regression models and hence there is no unique split of these parameters.

For further investigation, it could be interesting to test the relation between the reference point and drift parameter to be able to retrieve good estimates of these parameters. In the Euclidean case, the Method of Moments procedure has been shown to provide consistent, but sometimes biased estimates. An interesting question for future work could also be, whether the parameter estimates in this model is consistent and biased.

Acknowledgements. This work was supported by the CSGB Centre for Stochastic Geometry and Advanced Bioimaging funded by a grant from the Villum foundation.

References

1. Banerjee, M., Chakraborty, R., Ofori, E., Okun, M.S., Vaillancourt, D.E., Vemuri, B.C.: A nonlinear regression technique for manifold valued data with applications to medical image analysis. In: 2016 IEEE Conference on CVPR, pp. 4424–4432, June 2016
2. Cornea, E., Zhu, H., Kim, P., Ibrahim, J.G.: The Alzheimer's disease neuroimaging initiative. Regression models on Riemannian symmetric spaces. J. Roy. Stat. Soc. B **79**, 463–482 (2017)
3. Cragg, J.G.: Using higher moments to estimate the simple errors-in-variables model. Rand J. Econ. **28**, S71–S91 (1997)
4. Davis, B.C., Fletcher, P.T., Bullitt, E., Joshi, S.: Population shape regression from random design data. In: 2007 IEEE 11th International Conference on Computer Vision, pp. 1–7, October 2007
5. Fletcher, P.T.: Geodesic regression and the theory of least squares on Riemannian manifolds. Int. J. Comput. Vision **105**, 171–185 (2012)
6. Hazelton, M.L.: Methods of moments estimation. In: Lovric, M. (ed.) International Encyclopedia of Statistical Science, pp. 816–817. Springer, Heidelberg (2011). doi:10.1007/978-3-642-04898-2_364
7. Hinkle, J., Muralidharan, P., Fletcher, P.T., Joshi, S.: Polynomial Regression on Riemannian Manifolds. arXiv:1201.2395 (2012)
8. Hong, Y., Kwitt, R., Singh, N., Vasconcelos, N., Niethammer, M.: Parametric regression on the Grassmannian. IEEE Trans. Pattern Anal. Mach. Intell. **38**(11), 2284–2297 (2016)
9. Hsu, E.P.: Stochastic Analysis on Manifolds. American Mathematical Society, Providence (2002)
10. Kühnel, L., Sommer, S.: Stochastic development regression on non-linear manifolds. In: Niethammer, M., Styner, M., Aylward, S., Zhu, H., Oguz, I., Yap, P.-T., Shen, D. (eds.) IPMI 2017. LNCS, vol. 10265, pp. 53–64. Springer, Cham (2017). doi:10.1007/978-3-319-59050-9_5
11. Lin, L., St Thomas, B., Zhu, H., Dunson, D.B.: Extrinsic local regression on manifold-valued data. arXiv:1508.02201, August 2015
12. Niethammer, M., Huang, Y., Vialard, F.-X.: Geodesic regression for image time-series. In: Fichtinger, G., Martel, A., Peters, T. (eds.) MICCAI 2011. LNCS, vol. 6892, pp. 655–662. Springer, Heidelberg (2011). doi:10.1007/978-3-642-23629-7_80
13. Nilsson, J., Sha, F., Jordan, M.I.: Regression on manifolds using kernel dimension reduction. In: Proceedings of the 24th ICML, pp. 697–704. ACM (2007)

14. Pal, M.: Consistent moment estimators of regression coefficients in the presence of errors in variables. J. Econom. **14**(3), 349–364 (1980)
15. Niethammer, M., Huang, Y., Vialard, F.-X.: Geodesic regression for image time-series. In: Fichtinger, G., Martel, A., Peters, T. (eds.) MICCAI 2011. LNCS, vol. 6892, pp. 655–662. Springer, Heidelberg (2011). doi:10.1007/978-3-642-23629-7_80
16. Singh, N., Hinkle, J., Joshi, S., Fletcher, P.T.: Hierarchical geodesic models in diffeomorphisms. Int. J. Comput. Vision **117**, 70–92 (2016)
17. Singh, N., Vialard, F.-X., Niethammer, M.: Splines for diffeomorphisms. Med. Image Anal. **25**(1), 56–71 (2015)
18. Younes, L.: Shapes and Diffeomorphisms. Springer, Heidelberg (2010)
19. Yuan, Y., Zhu, H., Lin, W., Marron, J.S.: Local polynomial regression for symmetric positive definite matrices. J. Roy. Stat. Soc. Ser. B Stat. Methodol. **74**(4), 697–719 (2012)

Bootstrapping Descriptors for Non-Euclidean Data

Benjamin Eltzner(✉) and Stephan Huckemann

Felix-Bernstein-Institute for Mathematical Statistics in the Biosciences, University of Goettingen, Goettingen, Germany
beltzne@uni-goettingen.de

Abstract. For data carrying a non-Euclidean geometric structure it is natural to perform statistics via geometric descriptors. Typical candidates are means, geodesics, or more generally, lower dimensional subspaces, which carry specific structure. Asymptotic theory for such descriptors is slowly unfolding and its application to statistical testing usually requires one more step: Assessing the distribution of such descriptors. To this end, one may use the bootstrap that has proven to be a very successful tool to extract inferential information from small samples. In this communication we review asymptotics for descriptors of manifold valued data and study a non-parametric bootstrap test that aims at a high power, also under the alternative.

1 Introduction

In recent years, the study of data on non-Euclidean spaces has found increasing attention in statistics. Non-Euclidean data spaces have lead to a surge of specialized fields: directional statistics is concerned with data on spheres of different dimensions (e.g. [15]); shape analysis studies lead to data on quotient spaces (e.g. [6]), some of which are manifolds and some of which are non-manifold stratified spaces; and applications in population genetics have lead to increasing interest in data on non-manifold phylogenetic tree spaces (e.g. [4]) and to graph data in general.

As a basis for statistics on these spaces, it is important to investigate asymptotic consistency of estimators, as has been done for intrinsic and extrinsic Fréchet means on manifolds by [3,8], and more generally for a class of descriptors called *generalized Fréchet means* by [11,12]. Examples of such generalized Fréchet means are not only Procrustes means on non-manifold shape spaces ([6,11]) but also geodesic principal components on such spaces (cf. [10]), or more generally, barycentric subspaces by [17], see also [16] for a similar approach on phylogenetic tree spaces, or more specifically, small and great subspheres for spherical data by [14,18].

B. Eltzner and S. Huckemann—Acknowledging the Niedersachsen Vorab of the Volkswagen Foundation.

F. Nielsen and F. Barbaresco (Eds.): GSI 2017, LNCS 10589, pp. 12–19, 2017.
https://doi.org/10.1007/978-3-319-68445-1_2

In particular, the question of asymptotic consistency and normality of principal nested spheres analysis [14], say, goes beyond generalized Fréchet means analysis. In all *nested* schemes, several estimators are determined sequentially, where each estimation depends on all previous ones. Recently, asymptotic consistency of *nested generalized Fréchet means* was introduced in [13], as a generalization of classical PCA's asymptotics, e.g. by [1], where nestedness of approximating subspaces is not an issue because it is trivially given.

Based on asymptotic consistency of nested and non-nested descriptors, hypothesis tests, like the two-sample test can be considered. Since by construction, every sample determines only one single descriptor and not its distribution, resampling techniques like the bootstrap are necessary to produce confidence sets. Notably, this is a very generic technique independent of specific sample spaces and descriptors. In the following, after introducing non-nested and nested generalized Fréchet means, we will elaborate on bootstrapping quantiles for a two-sample test. We will show that a *separated* approach in general leads to greatly increased power of the test in comparison to a *pooled* approach, both with correct asymptotic size. Also, we illustrate the benefit of *nested* over non-nested descriptors.

2 Descriptors for Manifold Valued Data

2.1 Single Descriptors

With a silently underlying probability space $(\Omega, \mathfrak{A}, \mathbb{P})$, *random elements* on a topological space Q are mappings $X : \Omega \to Q$ that are measurable with respect to the Borel σ-algebra of Q.

For a topological space Q we say that a continuous function $d : Q \times Q \to [0, \infty)$ is a *loss function* if $d(q, q') = 0$ if and only if $q = q'$.

Definition 1 (Generalized Fréchet Means [11]**).** *Let Q be a separable topological space, called the* data space, *and P a separable topological space, called the* descriptor space, *with loss function $d : P \times P \to [0, \infty)$ and a continuous map $\rho : Q \times P \to [0, \infty)$. Random elements $X_1, \ldots, X_n \overset{\text{i.i.d.}}{\sim} X$ on Q give rise to* population *and* sample descriptors

$$\mu \in \underset{p \in P}{\operatorname{argmin}}\, \mathbb{E}[\rho(X, p)^2], \qquad \mu_n \in \underset{p \in P}{\operatorname{argmin}} \sum_{j=1}^{n} \rho(X_j, p)^2.$$

The descriptors are also called generalized ρ-Fréchet means. *The sample descriptor is a* least squares M-estimator.

Asymptotic theory for generalized ρ-Fréchet means under additional assumptions, among them that the means be unique and attained on a twice differentiable manifold part of P has been established by [11,12].

2.2 Nested Descriptors

For nested descriptors, we need to establish a notion of nestedness and the relations between the successive descriptor spaces.

Definition 2 ([13]). *A separable topological data space Q admits* backward nested families of descriptors *(BNFDs) if*

(i) there is a collection P_j $(j = 0, \dots, m)$ of topological separable spaces with loss functions $d_j : P_j \times P_j \to [0, \infty)$;
(ii) $P_m = \{Q\}$;
(iii) every $p \in P_j$ $(j = 1, \dots, m)$ is itself a topological space and gives rise to a topological space $\emptyset \neq S_p \subset P_{j-1}$ which comes with a continuous map

$$\rho_p : p \times S_p \to [0, \infty);$$

(iv) for every pair $p \in P_j$ $(j = 1, \dots, m)$ and $s \in S_p$ there is a measurable projection *map*

$$\pi_{p,s} : p \to s.$$

For $j \in \{1, \dots, m-2\}$ call a family

$$f = \{p^j, \dots, p^{m-1}\}, \text{ with } p^{k-1} \in S_{p^k}, k = j+1, \dots, m$$

a backward nested family of descriptors (BNFD) ending in *P_j, where we ignore the unique $p^m = Q \in P_m$. The space of all BNFDs ending in P_j is given by*

$$T_j = \left\{ f = \{p^k\}_{k=j}^{m-1} : p^{k-1} \in S_{p^k}, k = j+1, \dots, m \right\} \subseteq \prod_{k=j}^{m-1} P_k.$$

For $j \in \{1, \dots, m\}$, given a BNFD $f = \{p^k\}_{k=j}^{m-1}$ set

$$\pi_f = \pi_{p^{j+1}, p^j} \circ \ldots \circ \pi_{p^m, p^{m-1}} : p^m \to p^j$$

which projects along each descriptor. For another BNFD $f' = \{p'^k\}_{k=j}^{m-1} \in T_j$ set

$$d^j(f, f') = \sqrt{\sum_{k=j}^{m-1} d_k(p^k, p'^k)^2}.$$

Building on this notion, we can now define nested population and sample descriptors similar to Definition 1.

Definition 3 (Nested Generalized Fréchet Means [13]**).** *Random elements $X_1, \dots, X_n \overset{\text{i.i.d.}}{\sim} X$ on a data space Q admitting BNFDs give rise to* backward nested population *and* sample descriptors *(abbreviated as BN descriptors)*

$$\{E^{f^j} : j = m-1, \dots, 0\}, \quad \{E_n^{f_n^j} : j = m-1, \dots, 0\}$$

recursively defined using $p^m = Q = p_n^m$ *via*

$$E^{f^j} = \operatorname*{argmin}_{s\in S_{p^{j+1}}} \mathbb{E}[\rho_{p^{j+1}}(\pi_{f^{j+1}} \circ X, s)^2], \qquad f^j = \{p^k\}_{k=j}^{m-1}$$

$$E_n^{f_n^j} = \operatorname*{argmin}_{s\in S_{p_n^{j+1}}} \sum_{i=1}^{n} \rho_{p_n^{j+1}}(\pi_{f_n^{j+1}} \circ X_i, s)^2, \qquad f_n^j = \{p_n^k\}_{k=j}^{m-1}.$$

where $p^j \in E^{f^j}$ *and* $p_n^j \in E^{f_n^j}$ *is a measurable choice for* $j = 1, \ldots, m-1$.

We say that a BNFD $f = \{p^k\}_{k=0}^{m-1}$ *gives* unique *BN population descriptors if* $E^{f^j} = \{p^j\}$ *with* $f^j = \{p^k\}_{k=j}^{m-1}$ *for all* $j = 0, \ldots, m-1$.

Each of the E^{f^j} *and* $E_n^{f_n^j}$ *is called a* nested generalized Fréchet mean *and* $E_n^{f_n^j}$ *can be viewed as* nested least squares M-estimator.

Asymptotic theory for such backward nested families of descriptors, again under additional assumptions, among them being assumed on twice-differentiable manifold parts, has been established in [13].

In order to asses asymptotics of single elements in a family of nested generalized ρ-Fréchet means, the last element, say, a key ingredient is the following definition from [13].

Definition 4 (Factoring Charts [13]**).** *Let* $W \subset T_j$, $U \subset P^j$ *open subsets with* C^2 *manifold structure,* $f' = (p'^{m-1}, \ldots, p'^j) \in W$ *and* $p'^j \in U$, *and with local chart*

$$\psi : W \to \psi(W) \subset \mathbb{R}^{\dim(W)}, \qquad f = (p^{m-1}, \ldots, p^j) \mapsto \eta = (\theta, \xi)$$

the chart ψ factors, *if there is a chart* ϕ *and projections* π^U, $\pi^{\phi(U)}$

$$\phi : U \to \phi(U) \subset \mathbb{R}^{\dim(U)}, \; p^j \mapsto \theta$$

$$\pi^U : W \to U, \; f \mapsto p^j, \qquad \pi^{\phi(U)} : \psi(W) \to \phi(U), \; (\theta, \xi) \mapsto \theta$$

such that the following diagram commutes

$$\begin{array}{ccc} W & \xrightarrow{\quad\psi\quad} & \psi(W) \\ \Big\downarrow{\scriptstyle \pi^U} & & \Big\downarrow{\scriptstyle \pi^{\phi(U)}} \\ U & \xrightarrow{\quad\phi\quad} & \phi(U) \end{array} \tag{1}$$

In case that factoring charts exist, from the asymptotics of an entire backward nested descriptor family it is possible to project to a chart, describing the last element descriptor only, and such a projection preserves asymptotic Gaussianity, cf. [13].

3 Bootstrap Testing

Based on the central limit theorems proved in [11,13], it is possible to introduce a T^2-like two-sample test for non-nested descriptors, BNFDs and single nested descriptors.

3.1 The Test Statistic

Suppose that we have two independent i.i.d. samples $X_1, \ldots, X_n \sim X \in Q$, $Y_1, \ldots, Y_m \sim Y \in Q$ in a data space Q admitting non-nested descriptors, BNFDs and single nested descriptors in P and we want to test

$$H_0 : X \sim Y \quad \text{versus} \quad H_1 : X \not\sim Y$$

using descriptors in $p \in P$. Here, $p \in P$ stands either for a single $p_k \in P_k$ or for a suitable sequence $f \in T_j$. We assume that the first sample gives rise to $\hat{p}_n^X \in P$, the second to $\hat{p}_m^Y \in P$, and that these are unique. We introduce shorthand notation to simplify the following complex expressions

$$d_{n,b}^{X,*} = \phi(\hat{p}_{n,b}^{X,*}) - \phi(\hat{p}_n^X) \qquad d_{m,b}^{Y,*} = \phi(\hat{p}_{m,b}^{Y,*}) - \phi(\hat{p}_m^Y)$$

$$\Sigma_{\phi,n}^{X,*} := \frac{1}{B} \sum_{b=1}^{B} d_{n,b}^{X,*} d_{n,b}^{X,*T} \qquad \Sigma_{\phi,m}^{Y,*} := \frac{1}{B} \sum_{b=1}^{B} d_{m,b}^{Y,*} d_{m,b}^{Y,*T}.$$

Define the statistic

$$T^2 := \left(\phi(\hat{p}_n^X) - \phi(\hat{p}_m^Y)\right)^T \left(\Sigma_{\phi,n}^{X,*} + \Sigma_{\phi,m}^{Y,*}\right)^{-1} \left(\phi(\hat{p}_n^X) - \phi(\hat{p}_m^Y)\right). \tag{2}$$

Under H_0 and the assumptions of the CLTs shown in [11,13], this is asymptotically Hotelling T^2 distributed if the corresponding bootstrapped covariance matrices exist. Notably, under slightly stronger regularity assumptions, which are needed for the bootstrap, this estimator is asymptotically consistent, cf. [5, Corollary 1].

3.2 Pooled Bootstrapped Quantiles

Since the test statistic (2) is only asymptotically T^2 distributed and especially deeply nested estimators may have sizable bias for finite sample size, it can be advantageous to use the bootstrap to simulate quantiles, whose covering rate usually has better convergence properties, cf. [7]. A pooled approach to simulated quantiles runs as follows. From $X_1, \ldots, X_n, Y_1 \ldots, Y_m$, sample $Z_{1,b}, \ldots, Z_{n+m,b}$ and compute the corresponding T^{*2}_b ($b = 1, \ldots, B$) following (2) from $X_{i,b}^* = Z_{i,b}, Y_{j,b}^* = Z_{n+j,b}$ ($i = 1, \ldots, n$, $j = 1, \ldots, m$). From these, for a given level $\alpha \in (0,1)$ we compute the empirical quantile $c_{1-\alpha}^*$ such that

$$\mathbb{P}\{T^{*2} \le c_{1-\alpha}^* | X_1, \ldots, X_n, Y_1, \ldots, Y_m\} = 1 - \alpha.$$

We have then under H_0 that $c_{1-\alpha}^*$ gives an asymptotic coverage of $1 - \alpha$ for T^2, i. e. $\mathbb{P}\{T^2 \le c_{1-\alpha}^*\} \to 1 - \alpha$ as $n, m \to \infty$ if $n/m \to c$ with a fixed $c \in (0, \infty)$. Under H_1, however, the bootstrap samples $X_{i,b}^*$ and $Y_{j,b}^*$ have substantially higher variance than both the original X_i and Y_j. This leads to a large spread between the values of the quantiles and thus to diminished power of the test. This will be exemplified in the simulations below.

3.3 Separated Bootstrapped Quantiles

To improve the power of the test while still achieving the asymptotic size, we simulate a slightly changed statistic under H_0, by again bootstrapping, but now separately, from $X_1, \ldots, X_n$ and $Y_1 \ldots, Y_m$ (for $b = 1, \ldots, B$),

$$T^{*2} = \left(d_{n,b}^{X,*} - d_{m,b}^{Y,*}\right)^T \left(\Sigma_{\phi,n}^{X,*} + \Sigma_{\phi,m}^{Y,*}\right)^{-1} \left(d_{n,b}^{X,*} - d_{m,b}^{Y,*}\right). \tag{3}$$

From these values, for a given level $\alpha \in (0,1)$ we compute the empirical quantile $c_{1-\alpha}^*$ such that

$$\mathbb{P}\{T^{*2}(A) \leq c_{1-\alpha}^* | X_1, \ldots, X_n, Y_1, \ldots, Y_m\} = 1 - \alpha.$$

Then, in consequence of [2, Theorems 3.2 and 3.5], asymptotic normality of $\sqrt{n}((\phi(\hat{p}_n^X) - \phi(\hat{p}^X))$, and $\sqrt{m}((\phi(\hat{p}_m^Y) - \phi(\hat{p}^Y))$, guaranteed by the CLT in [13], extends to the same asymptotic normality for $\sqrt{n}\, d_{n\,b}^{X\,*}$, and $\sqrt{m}\, d_{m\,b}^{Y\,*}$, respectively. We have then under H_0 that $c_{1-\alpha}^*$ gives an asymptotic coverage of $1 - \alpha$ for T^2 from Eq. (2), i. e. $\mathbb{P}\{T^{*2} \leq c_{1-\alpha}^*\} \to 1 - \alpha$ as $B, n, m \to \infty$ if $n/m \to c$ with a fixed $c \in (0, \infty)$.

We note that also the argument from [3, Corollary 2.3 and Remark 2.6] extends at once to our setup, as we assume that the corresponding population covariance matrix Σ_ψ or Σ_ϕ, respectively, is invertible.

4 Simulations

We perform simulations to illustrate two important points. For our simulations we use the nested descriptors of Principal Nested Great Spheres (PNGS) analysis [14] and the intrinsic Fréchet mean [3]. In all tests and simulated quantiles we use $B = 1000$ bootstrap samples for each data set.

4.1 Differences Between Pooled and Separated Bootstrap

The first simulated example uses the nested mean and first geodesic principal component (GPC) to compare the two different bootstrapped quantiles with T^2-distribution quantiles in order to illustrate the benefits provided by separated quantiles. The two data sets we use are concentrated along two great circle arcs on an $\mathbb{S}^2$ which are perpendicular to each other. The data sets are normally distributed along these clearly different great circles with common nested mean and have sample size of 60 and 50 points, respectively, cf. Fig. 1a.

We simulate 100 samples from the two distributions and compare the p-values for the different quantiles. By design, we expect a roughly uniform distribution of p-values for the nested mean, indicating correct size of the test, and a clear rejection of the null for the first GPC, showing the power of the test. Both is satisfied for the separated quantiles and T^2-quantiles but not for the pooled quantiles, leading to diminished power under the alternative, cf. Fig. 1c. Under closer inspection, Fig. 1b shows that separated quantile p-values are closer to T^2-quantile p-values than pooled quantile p-values, which are systematically higher due to the different covariance structures rendering the test too conservative.

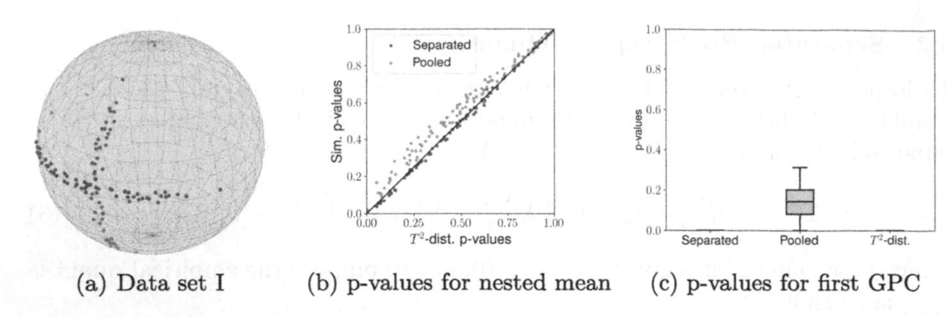

(a) Data set I (b) p-values for nested mean (c) p-values for first GPC

Fig. 1. Simulated data set I on $\mathbb{S}^2$ (a) with correct size under the null hypothesis of equal nested means (b) and power under the alternative of different first GPCs (c). The red sample has 50 points, the blue 60 points; we use p-values for 100 simulations each. (Color figure online)

4.2 Nested Descriptors May Outperform Non-nested Descriptors

The second point we highlight is that the nested mean of PNGS analysis is generically much closer to the data than the ordinary intrinsic mean and can thus, in specific situations, be more suitable to distinguish two populations. The same may also hold true for other nested estimators in comparison with their non-nested kin. The data set II considered here provides an example for such a situation. It consists of two samples of 300 and 100 points, respectively, on an $\mathbb{S}^2$ with coinciding intrinsic mean but different nested mean.

Here we only consider separated simulated quantiles, for both nested and intrinsic means. For the intrinsic mean two-sample test, we also use the bootstrap to estimate covariances for simplicity as outlined by [3], although closed forms for variance estimates exist, cf. [9]. Data set II and the distribution of resulting

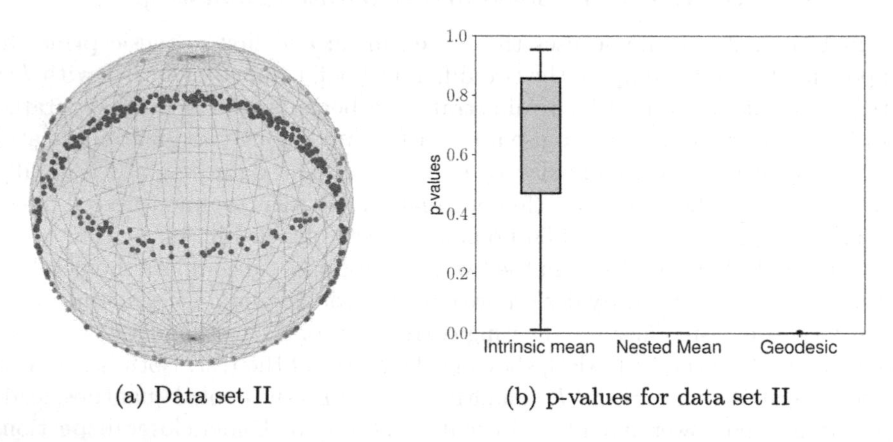

(a) Data set II (b) p-values for data set II

Fig. 2. Simulated data set II (red: 100 points, blue: 300 points) on $\mathbb{S}^2$ (left), and box plots displaying the distribution of 100 p-values for PNGS nested mean and intrinsic mean (right) from the two-sample test. (Color figure online)

p-values are displayed in Fig. 2. These values are in perfect agreement with the intuition guiding the design of the data showing that the nested mean is suited to distinguish the data sets where the intrinsic mean fails to do so.

References

1. Anderson, T.: Asymptotic theory for principal component analysis. Ann. Math. Stat. **34**(1), 122–148 (1963)
2. Arcones, M.A., Giné, E.: On the bootstrap of m-estimators and other statistical functionals. In: LePage, R., Billard, L. (eds.) Exploring the Limits of Bootstrap, pp. 13–47. Wiley (1992)
3. Bhattacharya, R.N., Patrangenaru, V.: Large sample theory of intrinsic and extrinsic sample means on manifolds II. Ann. Stat. **33**(3), 1225–1259 (2005)
4. Billera, L., Holmes, S., Vogtmann, K.: Geometry of the space of phylogenetic trees. Adv. Appl. Math. **27**(4), 733–767 (2001)
5. Cheng, G.: Moment consistency of the exchangeably weighted bootstrap for semiparametric m-estimation. Scand. J. Stat. **42**(3), 665–684 (2015)
6. Dryden, I.L., Mardia, K.V.: Statistical Shape Analysis. Wiley, Chichester (1998)
7. Fisher, N.I., Hall, P., Jing, B.Y., Wood, A.T.: Improved pivotal methods for constructing confidence regions with directional data. J. Am. Stat. Assoc. **91**(435), 1062–1070 (1996)
8. Hendriks, H., Landsman, Z.: Asymptotic behaviour of sample mean location for manifolds. Stat. Probab. Lett. **26**, 169–178 (1996)
9. Huckemann, S., Hotz, T., Munk, A.: Intrinsic MANOVA for Riemannian manifolds with an application to Kendall's space of planar shapes. IEEE Trans. Pattern Anal. Mach. Intell. **32**(4), 593–603 (2010)
10. Huckemann, S., Hotz, T., Munk, A.: Intrinsic shape analysis: geodesic principal component analysis for Riemannian manifolds modulo Lie group actions (with discussion). Stat. Sin. **20**(1), 1–100 (2010)
11. Huckemann, S.: Inference on 3D procrustes means: tree boles growth, rank-deficient diffusion tensors and perturbation models. Scand. J. Stat. **38**(3), 424–446 (2011)
12. Huckemann, S.: Intrinsic inference on the mean geodesic of planar shapes and tree discrimination by leaf growth. Ann. Stat. **39**(2), 1098–1124 (2011)
13. Huckemann, S.F., Eltzner, B.: Backward nested descriptors asymptotics with inference on stem cell differentiation (2017). arXiv:1609.00814
14. Jung, S., Dryden, I.L., Marron, J.S.: Analysis of principal nested spheres. Biometrika **99**(3), 551–568 (2012)
15. Mardia, K.V., Jupp, P.E.: Directional Statistics. Wiley, New York (2000)
16. Nye, T., Tang, X., d Weyenberg, G., Yoshida, R.: Principal component analysis and the locus of the Fréchet mean in the space of phylogenetic trees. arXiv:1609.03045 (2016)
17. Pennec, X.: Barycentric subspace analysis on manifolds. arXiv preprint arXiv:1607.02833 (2016)
18. Schulz, J., Jung, S., Huckemann, S., Pierrynowski, M., Marron, J., Pizer, S.M.: Analysis of rotational deformations from directional data. J. Comput. Graph. Stat. **24**(2), 539–560 (2015)

Sample-Limited L_p Barycentric Subspace Analysis on Constant Curvature Spaces

Xavier Pennec(✉)

Côte d'Azur University (UCA), Inria, Sophia-Antipolis, France
xavier.pennec@inria.fr

Abstract. Generalizing Principal Component Analysis (PCA) to manifolds is pivotal for many statistical applications on geometric data. We rely in this paper on *barycentric subspaces*, implicitly defined as the locus of points which are weighted means of $k+1$ reference points [8,9]. Barycentric subspaces can naturally be nested and allow the construction of inductive forward or backward nested subspaces approximating data points. We can also consider the whole hierarchy of embedded barycentric subspaces defined by an ordered series of points in the manifold (a flag of affine spans): optimizing the accumulated unexplained variance (AUV) over all the subspaces actually generalizes PCA to non Euclidean spaces, a procedure named Barycentric Subspaces Analysis (BSA).

In this paper, we first investigate sample-limited inference algorithms where the optimization is limited to the actual data points: this transforms a general optimization into a simple enumeration problem. Second, we propose to robustify the criterion by considering the unexplained p-variance of the residuals instead of the classical 2-variance. This construction is very natural with barycentric subspaces since the affine span is stable under the choice of the value of p. The proposed algorithms are illustrated on examples in constant curvature spaces: optimizing the (accumulated) unexplained p-variance (L_p PBS and BSA) for $0 < p \leq 1$ can identify reference points in clusters of a few points within a large number of random points in spheres and hyperbolic spaces.

1 Introduction

Principal Component Analysis (PCA) is the ubiquitous tool to obtain low dimensional representation of the data in linear spaces. To generalize PCA to Riemannian manifolds, one can analyze the covariance matrix of the data in the tangent space at the Fréchet mean (Tangent PCA). This is often sufficient when data are sufficiently centered around a central value (unimodal or Gaussian-like data), but generally fails for multimodal or distributions with a large variability with respect to the curvature. Instead of maximizing the explained variance, methods minimizing the unexplained variance were proposed: Principal Geodesic Analysis (PGA) [4] and Geodesic PCA (GPCA) [5] minimize the distance to a Geodesic Subspace (GS) spanned by the geodesics going through a point with tangent vector in a linear subspace of the tangent space.

F. Nielsen and F. Barbaresco (Eds.): GSI 2017, LNCS 10589, pp. 20–28, 2017.
https://doi.org/10.1007/978-3-319-68445-1_3

Barycentric subspaces are a new type of subspaces in manifolds recently introduced which are implicitly defined as the locus of weighted means of k+1 reference points (with positive or negative weights) [8,9]. Depending on the definition of the mean, we obtain the Fréchet, Karcher or Exponential Barycentric subspaces (FBS/KBS/EBS). The Fréchet (resp. Karcher) barycentric subspace of the points $(x_0, \ldots x_k) \in \mathcal{M}^{k+1}$ is the locus of weighted Fréchet (resp. Karcher) means of these points, i.e. the set of global (resp. local) minima of the weighted variance: $\sigma^2(x, \lambda) = \frac{1}{2} \sum_{i=0}^{k} \underline{\lambda}_i \operatorname{dist}^2(x, x_i)$, where $\underline{\lambda} = \lambda / (\sum_{j=0}^{k} \lambda_j)$:

$$\operatorname{FBS}(x_0, \ldots x_k) = \left\{ \arg\min_{x \in \mathcal{M}} \sigma^2(x, \lambda), \lambda \in \mathcal{P}_k^* = \left\{ \lambda \in \mathbb{R}P^n / \mathbb{1}^\top \lambda \neq 0 \right\} \right\}.$$

The EBS is the locus of weighted exponential barycenters of the reference points (critical points of the weighted variance) defined outside their cut-locus by:

$$\operatorname{EBS}(x_0, \ldots x_k) = \{ x \in \mathcal{M} \setminus C(x_0, \ldots x_k) | \exists \lambda \in \mathcal{P}_k^* : \textstyle\sum_i \lambda_i \log_x(x_i) = 0 \}.$$

Thus, we clearly see the inclusion $FBS \subset KBS \subset EBS$. The metric completion of the the EBS is called the *affine span* $\operatorname{Aff}(x_0, \ldots x_k)$. Its completeness allows ensuring that a closest point exists on the subspace, which is fundamental in practice for optimizing the subspaces by minimizing the residuals of the data to their projection. This definition works on metric spaces more general than Riemannian manifolds. In stratified metric spaces, the barycentric subspace spanned by points belonging to different strata naturally maps over several strata.

Barycentric subspaces can be characterized using the matrix field $Z(x) = [\log_x(x_0), \ldots \log_x(x_k)]$ of the log of the reference points x_i. This is a smooth field outside the cut locus of the reference points. The EBS is the zero level-set of the smallest singular value of $Z(x)$. The associated right singular vector gives the weights λ that satisfy the barycentric equation $\sum_i \lambda_i \log_x(x_i) = 0$. This simple equation generates a very rich geometry: at regular points where the Hessian of the weighted distance to the reference points is not degenerate, the EBS is a stratified space of maximal dimension k. In general, the largest stratum defines locally a submanifold of dimension k.

From PCA to Barycentric Subspace Analysis. The nestedness of approximation spaces is one of the most important characteristics for generalizing PCA to more general spaces [1]. Barycentric subspaces can easily be nested by adding or removing one or several points at a time, which corresponds to put the barycentric weight of this (or these) point(s) to zero. This gives a family of embedded submanifolds called a flag because this generalizes flags of vector spaces [9].

With a forward analysis, we compute iteratively the flag of affine spans by adding one point at a time keeping the previous ones fixed. Thus, we begin by computing the optimal barycentric subspace $\operatorname{Aff}(x_0) = \{x_0\}$, which may be a Karcher mean or more generally a stationary value of the unexplained variance, i.e. a Karcher mean. Adding a second point amounts to computing the geodesic passing through the mean that best approximates the data. Adding a third point now generally differs from PGA. In practice, the forward analysis should be stopped at a fixed number or when the variance of the residuals reaches a threshold (typically 5% of the original variance). We call this method the forward barycentric subspace (FBS) decomposition. Due to the greedy nature of this forward method, the affine span of dimension k defined by the first $k+1$ points is not in general the optimal one minimizing the unexplained variance.

The backward analysis consists in iteratively removing one dimension. As the affine span of $n+1$ linearly independent points generate the full manifold, the optimization really begins with n points. Once they are fixed, the optimization boils down to test which point should be removed. In practice, we may rather optimize $k+1$ points to find the optimal k-dimensional affine span, and then reorder the points using a backward sweep to find inductively the one that least increases the unexplained variance. We call this method the k-dimensional pure barycentric subspace with backward ordering (k-PBS). With this method, the k-dimensional affine span is optimizing the unexplained variance, but there is no reason why any of the lower dimensional ones should do.

In order to obtain optimal subspaces which are embedded, it is necessary to define a criterion which depends on the whole flag of subspaces and not on each of the subspaces independently. In PCA, one often plots the unexplained variance as a function of the number of modes used to approximate the data. This curve should decreases as fast as possible from the variance of the data (for 0 modes) to 0 (for n modes). A standard way to quantify the decrease consists in summing the values at all steps. This idea gives the Accumulated Unexplained Variances (AUV) criterion [9], which is analogous to the Area-Under-the-Curve (AUC) in Receiver Operating Characteristic (ROC) curves. This leads to an interesting generalization of PCA on manifolds called Barycentric Subspaces Analysis (BSA). In practice, one can stop at a maximal dimension k like for the forward analysis in order to limit the computational complexity. This analysis limited to a flag defined by $k+1$ points is denoted k-BSA.

2 Sample-Limited L_p Barycentric Subspace Inference

This paper investigates variants of the three above barycentric subspace analysis algorithms (FBS, k-PBS and k-BSA) along two main directions. First, we limit in Sect. 2 the optimization of flags of barycentric subspaces to the sample points of the data: this transforms a general optimization into a very simple enumeration problem. Second, we robustify in Sect. 2 the optimized criteria by considering the unexplained p-variance of the residuals instead of the classical 2-variance. This construction is very natural with barycentric subspaces since the affine span is stable under the choice of the value of p.

Sample-Limited Barycentric Subspace Inference. In several domains, it has been proposed to limit the inference of the Fréchet mean to the data-points only. In neuroimaging studies, the individual image minimizing the sum of square deformation distance to other subject images is a good alternative to the mean template (a Fréchet mean in deformation and intensity space) because it conserves the original characteristics of a real subject image [7]. Beyond the Fréchet mean, [3] proposed to define the first principal component mode as the unexplained variance minimizing geodesic going through two of the data points. The method named *set statistics* was aiming to accelerate the computation of statistics on tree spaces. [11] further explored this idea under the name of *sample-limited geodesics* in the context of PCA in phylogenetic tree space. In both cases, defining higher order principal modes was seen as a challenging research topic.

With barycentric subspaces, the idea of sample-limited statistics naturally extends to any dimension by restricting the search to the (flag of) affine spans that are parametrized by points sampled form the data. The implementation boils down to an enumeration problem. With this technique, the reference points are never interpolated as they are by definition sampled from the data. This is a important advantage for interpreting the modes of variation since we may go back to other information about the samples like the medical history and disease type. The search can be done exhaustively for a small number of reference points. The main drawback is the combinatorial explosion of the computational complexity with the dimension for the optimal order-k flag of affine spans, which is involving $O(N^{k+1})$ operations, where N is the number of data points. In this paper we perform an exhaustive search, but approximate optima can be sought using a limited number of randomly sampled points [3].

Stability of Barycentric Subspaces by L_p Norms. Since barycentric subspaces minimize the weighted variance, one could think of taking a power p of the metric to define the p-variance $\sigma^p(x) = \frac{1}{p}\sum_{i=0}^k \text{dist}^p(x, x_i)$. The global minima of this p-variance defines the Fréchet median for $p = 1$, the Fréchet mean for $p = 2$ and the barycenter of the support of the distribution for $p = \infty$. This suggests to further generalize barycentric subspaces by taking the locus of the minima of the weighted p-variance $\sigma^p(x, \lambda) = \frac{1}{p}\sum_{i=0}^k \underline{\lambda}_i \text{ dist}^p(x, x_i)$. However, it turns out that the critical points of the weighted p-variance are necessarily included in the EBS: the gradient of the p-variance at a non-reference point is

$$\nabla_x \sigma^p(x, \lambda) = \nabla_x \frac{1}{p}\sum_{i=0}^k \lambda_i(\text{ dist}^2(x, x_i))^{p/2} = -\sum_{i=0}^k \underline{\lambda}_i \text{ dist}^{p-2}(x, x_i) \log_x(x_i).$$

Thus, we see that the critical points of the p-variance satisfy the equation $\sum_{i=0}^k \lambda'_i \log_x(x_i) = 0$ for the new weights $\lambda'_i = \lambda_i \text{ dist}^{p-2}(x, x_i)$. Thus, they are also elements of the EBS and changing the power of the metric just amounts to a reparametrization of the barycentric weights. This stability of the EBS/affine span with respect to the power of the metric p shows that the affine span is really a central notion.

L_p Barycentric Subspaces Fitting and Analysis. While changing the power does not change the subspace definition, it has a drastic impact on its estimation: minimizing the sum of L_p distance to the subspace for non-vanishing residuals obviously changes the relative influence of points. It is well known that medians are more robust than least-squares estimators: the intuitive idea is to minimize the power of residuals with $1 \leq p \leq 2$ to minimize the influence of outliers.

For $0 < p < 1$, the influence of the closest points becomes predominant, at the cost of non-convexity. In general, this is a problem for optimization. However, since we perform an exhaustive search in our sample-limited setting, this is not a problem here. At the limit of $p = 0$, all the barycentric subspaces containing $k+1$ points (i.e. all the sample-limited barycentric subspaces of dimension k that we consider) have the same L_0 sum of residuals, which is a bit less interesting.

For a Euclidean space, minimizing the sum L_p norm of residuals under a rank k constraint is essentially the idea of the robust R1-PCA [2]. However, as noted in [6], an optimal rank k subspace is not in general a subspace of the optimal subspace of larger ranks: we loose the nestedness property. In this paper, we do not follow the PCA-L1 approach they propose, which maximizes the L_1 dispersion within the subspace. On manifolds, this would lead to a generalization of tangent-PCA maximizing the explained p-variance. In contrast, we solve this problem by minimizing the Accumulated Unexplained p-Variance (L_p AUV) over all the subspaces of the flag which is considered. Since the subspaces definition is not impacted by the power p, we can compare the subspaces' parameters (the reference points) for different powers. It also allows to simplify the algorithms: as the (positive) power of a (positive) distance is monotonic, the closest point to an affine span for the 2-distance remains the closest point for the p-distance. This give rise to three variations of our previous estimation algorithms:

- The Forward Barycentric Subspace decomposition (L_p k-FBS) iteratively adds the point that minimizes the unexplained p-variance up to $k+1$ points.
- The optimal Pure Barycentric Subspace with backward reordering (L_p k-PBS) estimates the $k+1$ points that minimize the unexplained p-variance, and then reorders the points accordingly for lower dimensions.
- The Barycentric Subspace Analysis of order k (L_p k-BSA) looks for the flag of affine spans defined by $k+1$ ordered points that optimized the L_p AUV.

3 Examples on Constant Curvature Spaces

We consider here the exhaustive sample-limited version of the three above algorithms and we illustrate some of their properties on spheres and hyperbolic spaces. Affine spans in spheres are simply lower dimensional great subspheres [8,9]. The projection of a point of a sphere on a subsphere is almost always unique (with respect to the spherical measure) and corresponds to the renormalization of the projection on the Euclidean subspace containing the subsphere. The same property can be established for hyperbolic spaces, which can be viewed as pseudo-spheres embedded in a Minkowski space. Affine spans are great pseudo-spheres (hyperboloids) generated by the intersection of the plane containing the reference points with the pseudo-sphere, and the closest point on the affine span is the renormalization of the unique Minkowski projection on that plane. In both cases, implementing the Riemannian norm of the residuals is very easy and the difficulty of sample-limited barycentric subspace algorithms analysis resides in the computational complexity of the exhaustive enumeration of tuples of points.

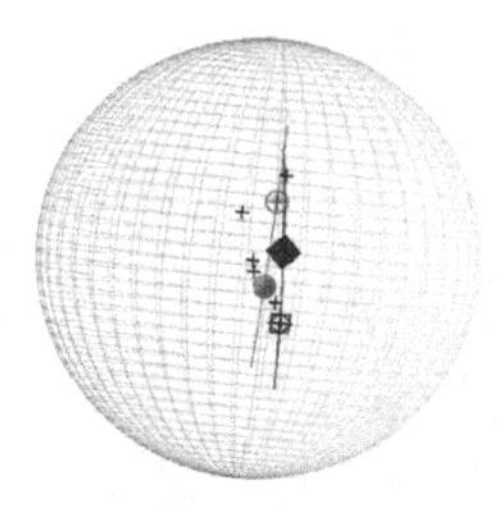
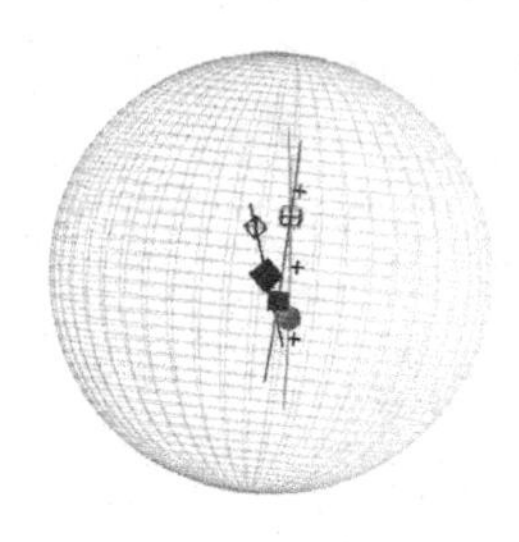
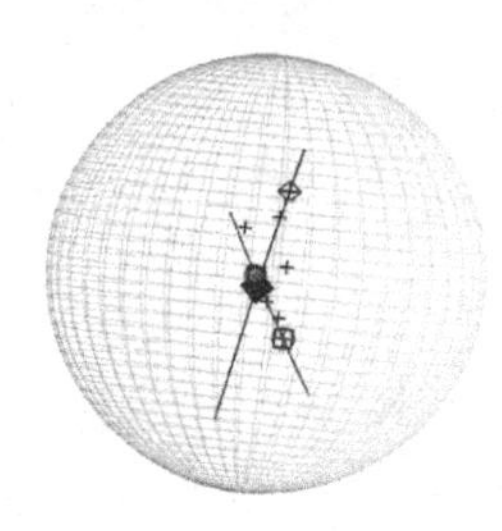

Fig. 1. Mount Tom Dinosaur trackway 1 data (symbol +), with $p = 2$ (**left**), $p = 1$ (**middle**) and $p = 0.1$ (**right**). For each method (FBS in blue, 1-PBS in green and 1-BSA in red), the first reference point has a solid symbol. The 1D mode is the geodesic joining this point to the second reference point (empty symbols). (Color figure online)

Example on Real Shape Data. For planar triangles, the shape space (quotient of the triad by similarities) boils down to the sphere of radius 1/2. The shape of three successive footprints of Mount Tom Dinosaur trackway 1 described in [10, p.181] is displayed on Fig. 1 (sample of 9 shapes). In this example, the reference points of the L_2 BSA stay the same from $k = 0$ to 3 and identical to the ones of the L_2 FBS. This is a behavior that we have observed for simulated examples when the variance of each mode is sufficiently different. The optimal L_2 1-PBS (the best geodesic approximation) picks up different reference points. For $p = 1$, the L_1 FBS is highly influenced by the location of the Fréchet median (solid blue symbol at the center Fig. 1) and we see that the optimal L_1 1-PBS and 1-BSA pick-up a different zero-th order mode (solid green and red symbols at the center). For a very low value $p = 0.1$, the optimal 1D subspace $L_{0.1}$ 1-PBS and the 1-BSA agree on points defining a geodesic excluding the 3 points located on the top right while the forward method gives something less intuitive.

3 Clusters on a 5D Sphere. In this synthetic dataset, we consider three clusters of 10, 9 and 8 points around the axes e_1, e_2 and e_3 (the vertices of an equilateral triangle of side length $\pi/2$) on a 5-dimensional sphere (embedded in 6D) with an error of standard deviation $\sigma = 6°$. We add 30 points uniformly sampled on the sphere to simulate three clusters on a 2-sphere with 50% of outliers. The ideal flag of subspaces is a pure 2D subspace spanning the first three coordinates with points at the cluster centers (Fig. 2).

For the L_2 metric, one first observes that at zero-th and first order, FBS, PBS and BSA estimate the same reference points which do not fall into any of the three clusters (blue point and geodesic on Fig. 2, left). For the second order approximation, which should cover the ideal 2-sphere, 2-BSA and 2-FBS continue to agree on the previous reference points and pick-up a third reference point within the smallest cluster (dark green circle on top of the sphere). Thus, we get at most one of the reference point in one of the clusters, except for the optimal 2-subspace (2-PBS) which makes a remarkable job by picking one point in each cluster (dark green point on Fig. 2, left).

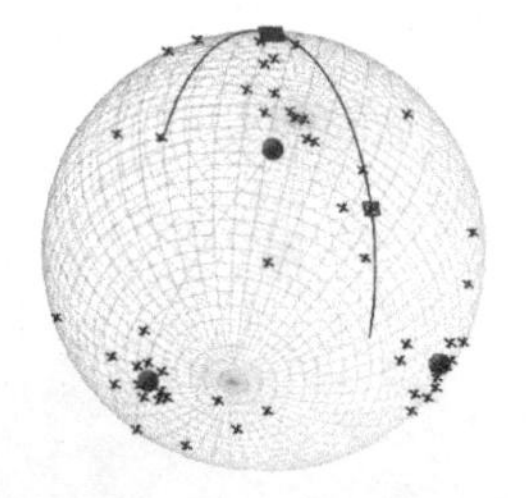
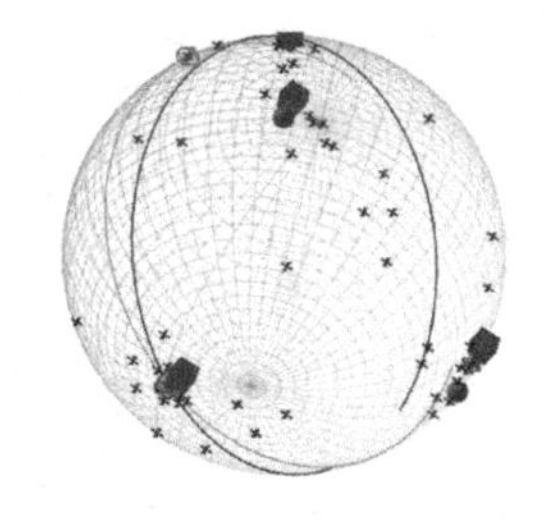
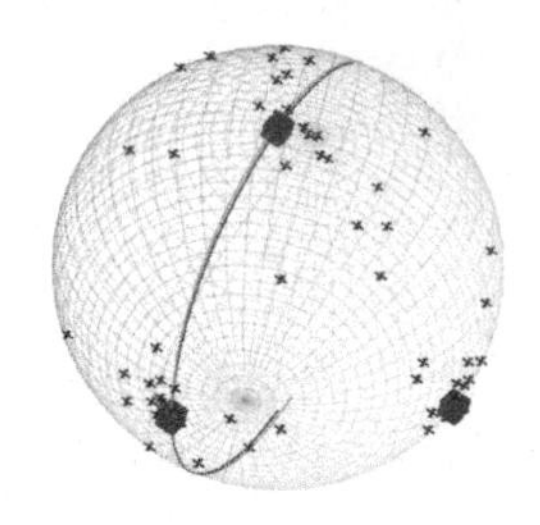

Fig. 2. Analysis of 3 clusters on a 5D sphere, projected to the expected 2-sphere, with $p = 2$ (**left**), $p = 1$ (**middle**) and $p = 0.1$ (**right**). For each method (FBS in blue, 1-PBS in green, 1-BSA in red), the 1D mode is a geodesic joining the two reference point. The three reference points of 2-PBS are represented with dark green solid circles, and the ones of 2-BSA with deep pink solid boxes. (Color figure online)

With the L_1 metric, we first observe that the FBS is fooled by the geometric median, which is not in any of the three clusters. The two other reference points successively added fall in two of the clusters. The L_1 optimal subspace (1-PBS) and 1-BSA find one of their reference points in a cluster, but the second point is still an outlier. When we come to 2D subspaces, both the 2-PBS and 2-BSA algorithms pick-up reference points in the three clusters, although they are not the same (dark green circles and deep pink solid boxes on Fig. 2, center). For a lower value of the power $p = 0.1$, all three reference points of the FBS are identical and within the three clusters, demonstrating the decrease in sensibility of the method to the outliers. The first and second order PBS and BSA consistently find very similar reference points within the clusters (Fig. 2, right).

3 Clusters on a 5D Hyperbolic Space. This example emulates the same example as above but on the 5D hyperbolic space: we draw 5 random points (tangent Gaussian with variance 0.015) around each vertex of an equilateral triangle of length 1.57 centered at the bottom of the 5D hyperboloid embedded in the (1,5)-Minkowski space. As outliers, we add 15 points drawn according to a tangent Gaussian of variance 1.0 truncated at a maximum distance of 1.5 around the bottom of the 5D hyperboloid. This simulates three clusters on a 2-pseudo-sphere with 50% of outliers (Fig. 3). With the L_2 hyperbolic distance, the 1-FBS and 1-BSA methods select outliers for their two reference points. 1-PBS manages to get one point in a cluster. For the two dimensional approximation, the 2-FBS and the 2-PBS select only one reference points within the clusters while 2-BSA correctly finds the clusters (Fig. 3 left, dark green points). With the L_1 distance, FBS, PBS and BSA select 3 very close points within the three clusters (Fig. 3 center). Lowering the power to $p = 0.5$ leads to selecting exactly the same points optimally centered within the 3 clusters for all the methods (Fig. 3 right).

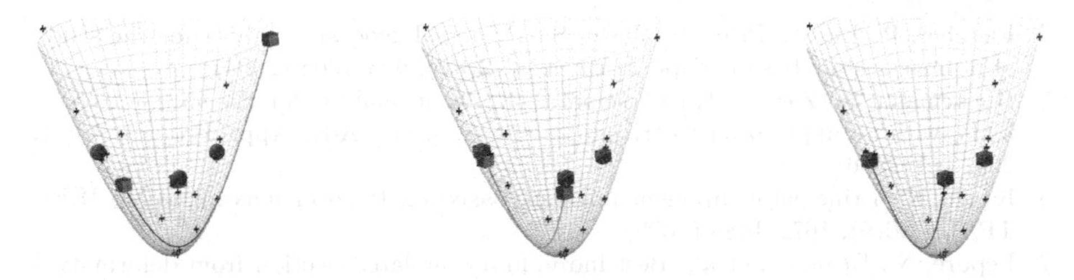

Fig. 3. Analysis of 3 clusters on a 5D hyperbolic space, projected to the expected 2-pseudo-sphere, with $p = 2$ (**left**), $p = 1$ (**middle**) and $p = 0.5$ (**right**). For each method (FBS in blue, 1-PBS in green, 1-BSA in red), the 1d mode is figured as a geodesic joining the two reference point. The three reference points of 2-PBS are represented with dark green solid circles, and the ones of 2-BSA with deep pink solid boxes. (Color figure online)

4 Conclusion

We have presented in this paper the extension of the barycentric subspace analysis approach to L_p norms and developed sample-limited inference algorithms which are quite naturally suited to barycentric subspaces, thanks to their definition using points rather than vectors as in more classical extensions of PCA. Experimental results on spheres and hyperbolic spaces demonstrate that the forward and optimal estimations of a k-subspace may differ from the barycentric subspace analysis optimizing the full flag of embedded subspaces together, even with the L_2 norm on residuals. This behavior differs from the one in Euclidean space where all methods are identical. Experiments also demonstrate that taking the L_p norm for $p < 2$ improves the robustness. Combined with the sample-limited estimation technique, we can even go well below $p = 1$ using exhaustive optimization. The main limitation of the optimal pure barycentric subspace (PBS) and barycentric subspace analysis (BSA) algorithms is their computational complexity which is exponential in the number of reference points. Thus, in order to increase the dimensionality, we now nee to develop efficient stochastic sampling techniques which allow to quickly pick up good reference points.

References

1. Damon, J., Marron, J.S.: Backwards principal component analysis and principal nested relations. J. Math. Imaging Vis. **50**(1–2), 107–114 (2013)
2. Ding, C., Zhou, D., He, X., Zha, H.: R1-PCA: Rotational invariant L1-norm principal component analysis for robust subspace factorization. In: Proceedings of ICML 2006, New York, NY, USA, pp. 281–288 (2006)
3. Feragen, A., Owen, M., Petersen, J., Wille, M.M.W., Thomsen, L.H., Dirksen, A., de Bruijne, M.: Tree-space statistics and approximations for large-scale analysis of anatomical trees. In: Gee, J.C., Joshi, S., Pohl, K.M., Wells, W.M., Zöllei, L. (eds.) IPMI 2013. LNCS, vol. 7917, pp. 74–85. Springer, Heidelberg (2013). doi:10.1007/978-3-642-38868-2_7

4. Fletcher, P., Lu, C., Pizer, S., Joshi, S.: Principal geodesic analysis for the study of nonlinear statistics of shape. IEEE TMI **23**(8), 995–1005 (2004)
5. Huckemann, S., Ziezold, H.: Principal component analysis for Riemannian manifolds, with an application to triangular shape spaces. Adv. Appl. Probab. **38**(2), 299–319 (2006)
6. Kwak, N.: Principal component analysis based on l1-norm maximization. IEEE TPAMI **30**(9), 1672–1680 (2008)
7. Leporé, N., Brun, C., et al.: Best individual template selection from deformation tensor minimization. In Proceedings of ISBI 2008, France, pp. 460–463 (2008)
8. Pennec, X.: Barycentric subspaces and affine spans in manifolds. In: Nielsen, F., Barbaresco, F. (eds.) GSI 2015. LNCS, vol. 9389, pp. 12–21. Springer, Cham (2015). doi:10.1007/978-3-319-25040-3_2
9. Pennec, X.: Barycentric subspace analysis on manifolds. Ann. Stat. (2017, to appear). arXiv:1607.02833
10. Small, C.: The Statistical Theory of Shapes. Springer Series in Statistics. Springer, New York (1996)
11. Zhai, H.: Principal component analysis in phylogenetic tree space. Ph.D. thesis, University of North Carolina at Chapel Hill (2016)

Parallel Transport in Shape Analysis: A Scalable Numerical Scheme

Maxime Louis[1,2(✉)], Alexandre Bône[1,2], Benjamin Charlier[2,3], Stanley Durrleman[1,2], and The Alzheimer's Disease Neuroimaging Initiative

[1] Sorbonne Universités, UPMC Université Paris 06, Inserm, CNRS, Institut du Cerveau et de la Moelle (ICM) – Hôpital Pitié-Salpêtrière, 75013 Paris, France
maxime.louis.x2012@gmail.com
[2] Aramis Project-Team, Inria Paris, 75013 Paris, France
[3] Université de Montpellier, Montpellier, France

Abstract. The analysis of manifold-valued data requires efficient tools from Riemannian geometry to cope with the computational complexity at stake. This complexity arises from the always-increasing dimension of the data, and the absence of closed-form expressions to basic operations such as the Riemannian logarithm. In this paper, we adapt a generic numerical scheme recently introduced for computing parallel transport along geodesics in a Riemannian manifold to finite-dimensional manifolds of diffeomorphisms. We provide a qualitative and quantitative analysis of its behavior on high-dimensional manifolds, and investigate an application with the prediction of brain structures progression.

1 Introduction

Riemannian geometry is increasingly meeting applications in statistical learning. Indeed, working in flat space amounts to neglecting the underlying geometry of the laws which have produced the considered data. In other words, such a simplifying assumption ignores the intrinsic constraints on the observations. When prior knowledge is available, top-down methods can express invariance properties as group actions or smooth constraints and model the data as points in quotient spaces, as for Kendall shape space. In other situations, manifold learning can be used to find a low-dimensional hypersurface best describing a set of observations.

Once the geometry has been modeled, classical statistical approaches for constrained inference or prediction must be adapted to deal with structured data, as it is done in [4,5,11,13]. Being an isometry, the parallel transport arises as a natural tool to compare features defined at different tangent spaces.

In a system of coordinates, the parallel transport is defined as the solution to an ordinary differential equation. The integration of this equation requires to compute the Christoffel symbols, which are in general hard to compute – e.g. in the case of the Levi-Civita connection – and whose number is cubic in

M. Louis and A. Bône—Equal contributions.

F. Nielsen and F. Barbaresco (Eds.): GSI 2017, LNCS 10589, pp. 29–37, 2017.
https://doi.org/10.1007/978-3-319-68445-1_4

the dimension. The Schild's ladder [5], later improved into the Pole ladder [7] when transporting along geodesics, is a more geometrical approach which only requires the computation of Riemannian exponentials and logarithms. When the geodesic equation is autonomous, the scaling and squaring procedure [6] allows to compute exponentials very efficiently. In Lie groups, the Baker-Campbell Haussdorff formula allows fast computations of logarithms with a controlled precision. In such settings, the Schild's ladder is computationally tractable. However, no theoretical study has studied the numerical approximations or has provided a convergence result. In addition, in the more general case of Riemannian manifolds, the needed logarithm operators are often computationally intractable.

The Large Deformation Diffeomorphic Metric Mapping (LDDMM) framework [1] focuses on groups of diffeomorphisms, for shape analysis. Geodesic trajectories can be computed by integrating the Hamiltonian equations, which makes the exponential operator computationally tractable, when the logarithm remains costly and hard to control in its accuracy. In [12] is suggested a numerical scheme which approximates the parallel transport along geodesics using only the Riemannian exponential and the metric. The convergence is proved in [8].

In this paper, we translate this so-called *fanning sheme* to finite-dimensional manifolds of diffeomorphisms built within the LDDMM framework [2]. We provide a qualitative and quantitative analysis of its behavior, and investigate a high-dimensional application with the prediction of brain structures progression. Section 2 gives the theoretical background and the detailed steps of the algorithm, in the LDDMM context. Section 3 describes the considered application and discusses the obtained results. Section 4 concludes.

2 Theoretical Background and Practical Description

2.1 Notations and Assumptions

Let $\mathcal{M}$ be a finite-dimensional Riemannian manifold with metric g and tangent space norm $\|\cdot\|_g$. Let $\gamma : t \to [0,1]$ be a geodesic whose coordinates are known at all time. Given $t_0, t \in [0,1]$, the parallel transport of a vector $w \in T_{\gamma(s)}\mathcal{M}$ from $\gamma(t_0)$ to $\gamma(t)$ along γ will be noted $\mathrm{P}_{t_0,t}(w) \in T_{\gamma(t)}\mathcal{M}$. We recall that this mapping is uniquely defined by the integration from $u = t_0$ to t of the differential equation $\nabla_{\dot\gamma(u)}\mathrm{P}_{t_0,u}(w) = 0$ with $\mathrm{P}_{t_0,t_0}(w) = w$ where ∇ is the Levi-Civita covariant derivative.

We denote Exp the exponential map, and for h small enough we define $\mathrm{J}^w_{\gamma(t)}(h)$, the Jacobi Field emerging from $\gamma(t)$ in the direction $w \in T_{\gamma(t)}\mathcal{M}$ by:

$$\mathrm{J}^w_{\gamma(t)}(h) = \left.\frac{\partial}{\partial\varepsilon}\right|_{\varepsilon=0} \mathrm{Exp}_{\gamma(t)}\big(h\,[\dot\gamma(t) + \varepsilon w]\,\big) \in T_{\gamma(t+h)}\mathcal{M}. \tag{1}$$

2.2 The Key Identity

The following proposition relates the parallel transport to a Jacobi field [12]:

Proposition. *For all $t > 0$ small enough and $w \in T_{\gamma(0)}\mathcal{M}$, we have:*

$$\mathrm{P}_{0,t}(w) = \frac{\mathrm{J}^{w}_{\gamma(0)}(t)}{t} + \mathrm{O}(t^2). \tag{2}$$

Proof. Let $X(t)$ be the time-varying vector field corresponding to the parallel transport of w, i.e. such that $\dot{X}^i + \Gamma^i_{kl} X^l \dot{\gamma}^k = 0$ with $X(0) = w$. At $t = 0$, in normal coordinates the differential equation simplifies into $\dot{X}^i(0) = 0$. Besides, near $t = 0$ in the same local chart, the Taylor expansion of $X(t)$ writes $X^i(t) = w^i + \mathrm{O}(t^2)$. Noticing that the ith normal coordinate of $\mathrm{Exp}_{\gamma(0)}(t[\dot{\gamma}(t) + \varepsilon w])$ is $t(v_0^i + \varepsilon w^i)$, the ith coordinate of $\mathrm{J}^{w}_{\gamma(0)}(t) = \frac{\partial}{\partial \varepsilon}|_{\varepsilon=0} \mathrm{Exp}_{\gamma(0)}(t[\dot{\gamma}(0) + \varepsilon w])$ is therefore tw^i, and we thus obtain the desired result. □

Subdividing $[0, 1]$ into N intervals and iteratively computing the Jacobi fields $\frac{1}{N}\mathrm{J}^{w}_{\gamma(k/N)}(\frac{1}{N})$ should therefore approach the parallel transport $\mathrm{P}_{0,1}(w)$. With an error in $\mathrm{O}(\frac{1}{N^2})$ at each step, a global error in $\mathrm{O}(\frac{1}{N})$ can be expected. We propose below an implementation of this scheme in the context of a manifold of diffeomorphisms parametrized by control points and momenta. Its convergence with a rate of $\mathrm{O}(\frac{1}{N})$ is proved in [8].

2.3 The Chosen Manifold of Diffeomorphisms

The LDDMM-derived construction proposed in [2] provides an effective way to build a finite-dimensional manifold of diffeomorphims acting on the d-dimensional ambient space $\mathbb{R}^d$. Time-varying vector fields $v_t(.)$ are generated by the convolution of a Gaussian kernel $k(x, y) = \exp\left[-\|x - y\|^2/2\sigma^2\right]$ over n_{cp} time-varying control points $c(t) = [c_i(t)]_i$, weighted by n_{cp} associated momenta $\alpha(t) = [\alpha_i(t)]_i$, i.e. $v_t(.) = \sum_{i=1}^{n_{cp}} k[\,.\,, c_i(t)]\,\alpha_i(t)$. The set of such vector fields forms a Reproducible Kernel Hilbert Space (RKHS).

Those vector fields are then integrated along $\partial_t \phi_t(.) = v_t[\phi(.)]$ from $\phi_0 = \mathrm{Id}$ into a flow of diffeomorphisms. In [10], the authors showed that the kernel-induced distance between ϕ_0 and ϕ_1 – which can be seen as the deformation kinetic energy – is minimal i.e. the obtained flow is geodesic when the control points and momenta satisfy the Hamiltonian equations:

$$\dot{c}(t) = \mathrm{K}_{c(t)}\alpha(t), \quad \dot{\alpha}(t) = -\frac{1}{2}\ \mathrm{grad}_{c(t)}\left\{\alpha(t)^T\,\mathrm{K}_{c(t)}\,\alpha(t)\right\}, \tag{3}$$

where $K_{c(t)}$ is the kernel matrix. A diffeomorphism is therefore fully parametrized by its initial control points c and momenta α.

Those Hamiltonian equations can be integrated with a Runge-Kutta scheme without computing the Christoffel symbols, thus avoiding the associated curse of dimensionality. The obtained diffeomorphisms then act on shapes embedded in $\mathbb{R}^d$, such as images or meshes.

For any set of control points $c = (c_i)_{i\in\{1,..,n\}}$, we define the finite-dimensional subspace $V_c = \text{span}\left\{k(.,c_i)\xi \,|\, \xi \in \mathbb{R}^d,\ i \in \{1,..,n\}\right\}$ of the vector fields' RKHS. We fix an initial set $c = (c_i)_{i\in\{1,..,n\}}$ of distinct control points and define the set $\mathcal{G}_c = \{\phi_1 \,|\, \partial_t\phi_t = v_t \circ \phi_t,\ v_0 \in V_c, \phi_0 = \text{Id}\}$. Equipped with $\text{K}_{c(t)}$ as – inverse – metric, $\mathcal{G}_c$ is a Riemannian manifold such that $T_{\phi_1}\mathcal{G}_c = V_{c(1)}$, where for all t in $[0,1]$, $c(t)$ is obtained from $c(0) = c$ through the Hamiltonian equations (3) [9].

2.4 Summary of the Algorithm

We are now ready to describe the algorithm on the Riemannian manifold $\mathcal{G}_c$.

Algorithm. Divide $[0,1]$ into N intervals of length $h = \frac{1}{N}$ where $N \in \mathbb{N}$. We note ω_k the momenta of the transported diffeomorphism, c_k the control points and α_k the momenta of the geodesic γ at time $\frac{k}{N}$. Iteratively:

(i) Compute the main geodesic control points c_{k+1} and momenta α_{k+1}, using a Runge-Kutta 2 method.
(ii) Compute the control points $c_{k+1}^{\pm h}$ of the perturbed geodesics $\gamma_{\pm h}$ with initial momenta and control points $(\alpha_k \pm h\omega_k,\, c_k)$, using a Runge-Kutta 2 method.
(iii) Approximate the Jacobi field J_{k+1} by central finite difference:

$$\text{J}_{k+1} = \frac{c_{k+1}^{+h} - c_{k+1}^{-h}}{2h}. \tag{4}$$

(iv) Compute the transported momenta $\tilde{\omega}_{k+1}$ according to Eq. (2):

$$\text{K}_{c_{k+1}}\tilde{\omega}_{k+1} = \frac{\text{J}_{k+1}}{h}. \tag{5}$$

(v) Correct this value with $\omega_{k+1} = \beta_{k+1}\tilde{\omega}_{k+1} + \delta_{k+1}\alpha_{k+1}$, where β_{k+1} and δ_{k+1} are normalization factors ensuring the conservation of $\|\omega\|_{V_c} = \omega_k^T K_{c_k}\omega_k$ and of $\langle\alpha_k, \omega_k\rangle_{c_k} = \alpha_k^T K_{c_k}\omega_k$.

As step of the scheme is illustrated in Fig. 1. The Jacobi field is computed with only four calls to the Hamiltonian equations. This operation scales quadratically with the dimension of the manifold, which makes this algorithm practical in high dimension, unlike Christoffel-symbol-based solutions. Step (iv) – solving a linear system of size n_{cp} – is the most expensive one, but remained within reasonable computational time in the investigated examples.

In [8], the authors prove the convergence of this scheme, and show that the error increases linearly with the size of the step used. The convergence is guaranteed as long as the step (ii) is performed with a method of order at least two. A first order method in step (iii) is also theoretically sufficient to guarantee convergence. Those variations will be studied in Sect. 3.3.

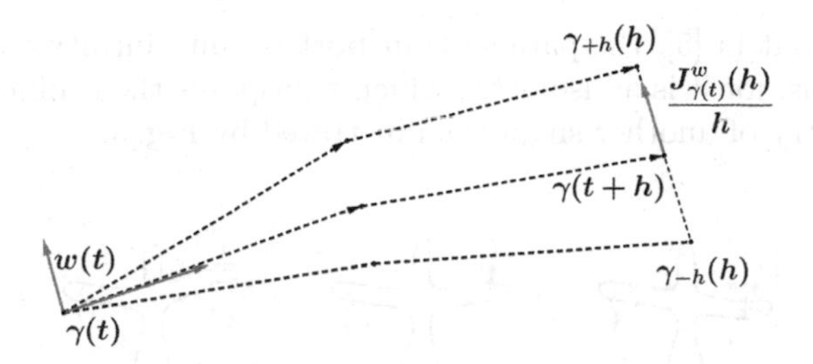

Fig. 1. Step of the parallel transport of the vector w (blue arrow) along the geodesic γ. J^w_γ is computed by central finite difference with the perturbed geodesics γ_h and γ_{-h}, integrated with a second-order Runge-Kutta scheme (dotted black arrows). A fan of geodesics is formed. (Color figure online)

3 Application to the Prediction of Brain Structures

3.1 Introducing the Exp-parallelization Concept

Exploiting the fanning scheme described in Sect. 2.4, we can parallel-transport any set of momenta along any given reference geodesic. Figure 2 illustrates the procedure. The target shape is first registered to the reference geodesic: the diffeomorphism that best transforms the chosen reference shape into the target one is estimated with a gradient descent algorithm on the initial control points and momenta [2]. Such a procedure can be applied generically to images or meshes. Once this geodesic is obtained, its initial set of momenta is parallel-transported along the reference geodesic. Taking the Riemannian exponential of the transported vector at each point of the geodesic defines a new trajectory, which we will call *exp-parallel* to the reference one.

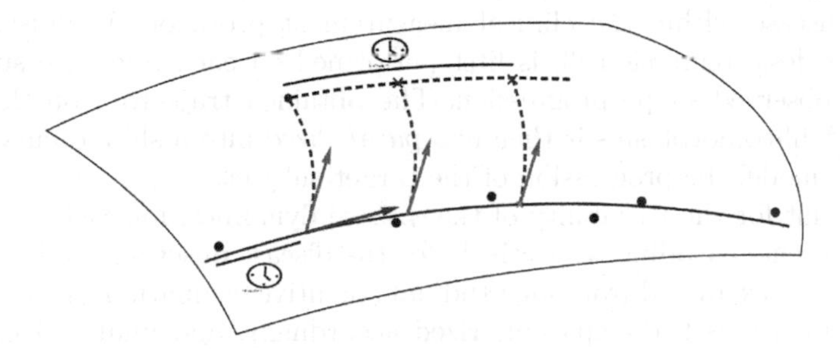

Fig. 2. Time-reparametrized *exp-parallelization* of a reference geodesic model. The black dots are the observations, on which are fitted a geodesic regression (solid black curve, parametrized by the blue arrow) and a matching (leftmost red arrow). The red arrow is then parallel-transported along the geodesic, and exponentiated to define the *exp-parallel* curve (black dashes). (Color figure online)

As pointed out in [5], the parallel transport is quite intuitive in the context of shape analysis, for it is an isometry which transposes the evolution of a shape into the geometry of another shape, as illustrated by Fig. 3.

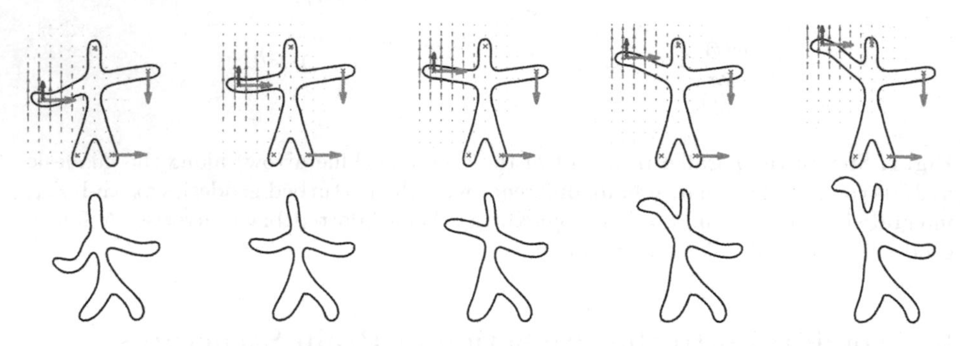

Fig. 3. Illustration of the *exp-parallelization* concept. Top row: the reference geodesic at successive times. Bottom row: the exp-parallel curve. Blue arrows: the geodesic momenta and velocity field. Red arrows: the momenta describing the initial registration with the target shape and its transport along the geodesic. (Color figure online)

3.2 Data and Experimental Protocol

Repeated Magnetic Resonance Imaging (MRI) measurements from 71 subjects are extracted from the ADNI database and preprocessed through standard pipelines into affinely co-registered surface meshes of hippocampi, caudates and putamina. The geometries of those brain sub-cortical structures are altered along the Alzheimer's disease course, which all considered subjects finally convert to.

Two subjects are successively chosen as references, for they have fully developed the disease within the clinical measurement protocol. As illustrated on Fig. 2, a geodesic regression [3] is first performed on each reference subject to model the observed shape progression. The obtained trajectory on the chosen manifold of diffeomorphisms is then *exp-parallelized* into a shifted curve, which is hoped to model the progression of the target subject.

To account for the variability of the disease dynamics, for each subject two scalar coefficients encoding respectively for the disease onset age and the rate of progression are extracted from longitudinal cognitive evaluations as in [11]. The exp-parallel curve is time-reparametrized accordingly, and finally gives predictions for the brain structures. In the proposed experiment, the registrations and geodesic regressions typically feature around 3000 control points in $\mathbb{R}^3$, so that the deformation can be seen as an element of a manifold of dimension 9000.

3.3 Estimating the Error Associated to a Single Parallel Transport

To study the error in this high-dimensional setting, we compute the parallel transport for a varying number of discretization steps N, thus obtaining

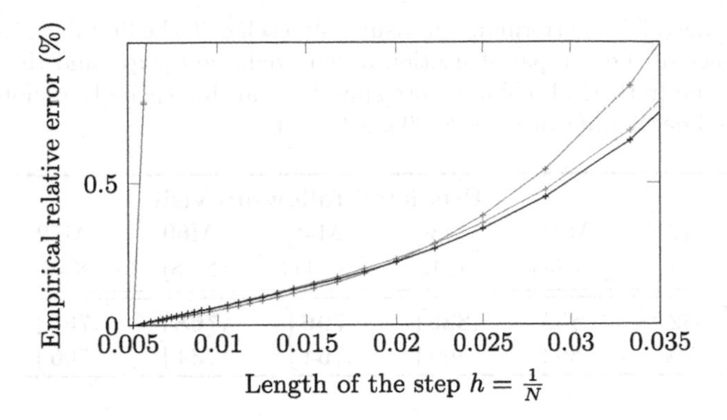

Fig. 4. Empirical relative error of the parallel transport in a high-dimensional setting. In black the proposed algorithm, in green the WEC variant, in red the RK4 variant, and in blue the SPG one. (Color figure online)

increasingly accurate estimations. We then compute the empirical relative errors, taking the most accurate computation as reference.

Arbitrary reference and target subjects being chosen, Fig. 4 gives the results for the proposed algorithm and three variations: without enforcing the conservations at step (v) [WEC], using a Runge-Kutta of order 4 at step (ii) [RK4], and using a single perturbed geodesic to compute J at step (iii) [SPG]. We recover a linear behavior with the length of the step $\frac{1}{N}$ in all cases. The SPG variant converges much slower, and is excluded from the following considerations.

For the other algorithms, the empirical relative error remains below 5% with 15 steps or more, and below 1% with 25 steps or more. The slopes of the asymptotic linear behaviors, estimated with the last 10 experimental measurements, range from 0.10 for the RK4 method to 0.13 for the WEC one. Finally, an iteration takes respectively 4.26, 4.24 and 8.64 s for the proposed algorithm, the WEC variant and the RK4 one. Therefore the initially detailed algorithm in Sect. 2.4 seems to achieve the best tradeoff between accuracy and speed in the considered experimental setting.

3.4 Prediction Performance

Table 1 gathers the predictive performance of the proposed exp-parallelization method. The performance metric is the Dice coefficient, which ranges from 0 for disjoint structures to 1 for a perfect match. A Mann-Witney test is performed to quantify the significance of the results in comparison to a naive methodology, which keeps constant the baseline structures over time. Considering the very high dimension of the manifold, failing to accurately capture the disease progression trend can quickly translates into unnatural predictions, much worse than the naive approach.

Table 1. Averaged Dice performance measures. In each cell, the first line gives the average performance of the exp-parallelization-based prediction [exp], and the second line the reference one [ref]. Each column corresponds to an increasingly remote predicted visit from baseline. Significance levels [.05, .01, .001].

Method	Predicted follow-up visit						
	M12	M24	M36	M48	M60	M72	M96
	N=140	N=134	N=123	N=113	N=81	N=62	N=17
[exp]	.882	.852	**.825** (**)	**.796** (***)	**.768** (**)	**.756** (***)	**.730** (**)
[ref]	**.884**	.852	.809	.764	.734	.706	.636

The proposed paradigm significantly outperforms the naive prediction three years or later from the baseline, thus demonstrating the relevance of the exp-parallelization concept for disease progression modeling, made computationally tractable thanks to the operational qualities of the fanning scheme for high-dimensional applications.

4 Conclusion

We detailed the fanning scheme for parallel transport on a high-dimensional manifold of diffeomorphisms, in the shape analysis context. Our analysis unveiled the operational qualities and computational efficiency of the scheme in high dimensions, with a empirical relative error below 1% for 25 steps only. We then took advantage of the parallel transport for accurately predicting the progression of brain structures in a personalized way, from previously acquired knowledge.

References

1. Beg, M.F., Miller, M.I., Trouvé, A., Younes, L.: Computing large deformation metric mappings via geodesic flows of diffeomorphisms. Int. J. Comput. Vis. **61**, 139–157 (2005)
2. Durrleman, S., Allassonnière, S., Joshi, S.: Sparse adaptive parameterization of variability in image ensembles. Int. J. Comput. Vis. **101**(1), 161–183 (2013)
3. Fletcher, T.: Geodesic regression and the theory of least squares on Riemannian manifolds. Int. J. Comput. Vis. **105**(2), 171–185 (2013)
4. Lenglet, C., Rousson, M., Deriche, R., Faugeras, O.: Statistics on the manifold of multivariate normal distributions: theory and application to diffusion tensor MRI processing. J. Math. Imaging Vis. **25**(3), 423–444 (2006)
5. Lorenzi, M., Ayache, N., Pennec, X.: Schild's ladder for the parallel transport of deformations in time series of images. In: Székely, G., Hahn, H.K. (eds.) IPMI 2011. LNCS, vol. 6801, pp. 463–474. Springer, Heidelberg (2011). doi:10.1007/978-3-642-22092-0_38
6. Lorenzi, M., Pennec, X.: Geodesics, parallel transport & one-parameter subgroups for diffeomorphic image registration. Int. J. Comput. Vis. **105**(2), 111–127 (2013)

7. Marco, L., Pennec, X.: Parallel transport with pole ladder: application to deformations of time series of images. In: Nielsen, F., Barbaresco, F. (eds.) GSI 2013. LNCS, vol. 8085, pp. 68–75. Springer, Heidelberg (2013). doi:10.1007/978-3-642-40020-9_6
8. Louis, M., Charlier, B., Jusselin, P., Pal, S., Durrleman, S.: A fanning scheme for the parallel transport along geodesics on Riemannian manifolds, July 2017. https://hal.archives-ouvertes.fr/hal-01560787
9. Micheli, M.: The differential geometry of landmark shape manifolds: metrics, geodesics, and curvature. Ph.D. thesis, Providence, RI, USA (2008). aAI3335682
10. Miller, M.I., Trouvé, A., Younes, L.: Geodesic shooting for computational anatomy. J. Math. Imaging Vis. **24**(2), 209–228 (2006)
11. Schiratti, J.B., Allassonnière, S., Colliot, O., Durrleman, S.: Learning spatiotemporal trajectories from manifold-valued longitudinal data. In: NIPS (2015)
12. Younes, L.: Jacobi fields in groups of diffeomorphisms and applications. Q. Appl. Math. **65**(1), 113–134 (2007)
13. Zhang, M., Fletcher, P.: Probabilistic principal geodesic analysis. In: Advances in Neural Information Processing Systems, vol. 26, pp. 1178–1186 (2013)

Maximum Likelihood Estimation of Riemannian Metrics from Euclidean Data

Georgios Arvanitidis(✉), Lars Kai Hansen, and Søren Hauberg

Section for Cognitive Systems, Technical University of Denmark, Richard Petersens Plads, B321, 2800 Kongens Lyngby, Denmark
{gear,lkai,sohau}@dtu.dk

Abstract. Euclidean data often exhibit a nonlinear behavior, which may be modeled by assuming the data is distributed near a nonlinear submanifold in the data space. One approach to find such a manifold is to estimate a Riemannian metric that locally models the given data. Data distributions with respect to this metric will then tend to follow the nonlinear structure of the data. In practice, the learned metric rely on parameters that are hand-tuned for a given task. We propose to estimate such parameters by maximizing the data likelihood under the assumed distribution. This is complicated by two issues: (1) a change of parameters imply a change of measure such that different likelihoods are incomparable; (2) some choice of parameters renders the numerical calculation of distances and geodesics unstable such that likelihoods cannot be evaluated. As a practical solution, we propose to (1) re-normalize likelihoods with respect to the usual Lebesgue measure of the data space, and (2) to bound the likelihood when its exact value is unattainable. We provide practical algorithms for these ideas and illustrate their use on synthetic data, images of digits and faces, as well as signals extracted from EEG scalp measurements.

Keywords: Manifold learning · Metric learning · Statistics on manifolds

1 Introduction

The *"manifold assumption"* is often applied in machine learning research to express that data is believed to lie near a (nonlinear) submanifold embedded in the data space. Such an assumption finds uses e.g. in dynamical or periodic systems, and in many problems with a smooth behavior. When the manifold structure is known a priori it can be incorporated into the problem specification, but unfortunately such structure is often not known. In these cases it is necessary to estimate the manifold structure from the observed data, a process known as *manifold learning*. In this work, we approach manifold learning *geometrically* by estimating a Riemannian metric that captures local behavior of the data, and *probabilistically* by estimating unknown parameters of the metric using maximum likelihood. First we set the stage with background information on manifold learning (Sect. 1.1) and geometry (Sect. 1.2), followed by an

F. Nielsen and F. Barbaresco (Eds.): GSI 2017, LNCS 10589, pp. 38–46, 2017.
https://doi.org/10.1007/978-3-319-68445-1_5

exposition of our model (Sect. 2) and the proposed maximum likelihood scheme (Sect. 3). Finally results are presented (Sect. 4) and discussed (Sect. 5).

1.1 Background and Related Work

Given observations $\mathbf{x}_{1:N} = \{\mathbf{x}_1, \dots, \mathbf{x}_N\}$ in $\mathbb{R}^D$, the key task in manifold learning is to estimate a data representation that reflect the nonlinear structure of the original data. The intuition behind most methods for this was phrased by Saul and Roweis [18] as *"Think Globally, Fit Locally"*, practically meaning that locally linear models are fitted to all points in data space and these then are merged to a global representation (details depend on the method).

The *Isomap* method [20] famously replace Euclidean distances with geodesic distances defined on a neighborhood graph and then embed the data in a lower dimensional space where Euclidean distances approximate the geodesic counterparts. While this approach is popular, its discrete nature only describes the observed data points and consequently cannot be used to develop probabilistic generative models. Similar comments hold for other graph-based methods [2,18].

As a smooth alternative, Lawrence [16] proposed a probabilistic extension of standard surface models by assuming that each dimension of the data is described as $x_d = f_d(\mathbf{z})$, where $\mathbf{z}$ is a low-dimensional latent variable and f_d is a Gaussian process. The latent variables then provide a low-dimensional parametrization that capture the manifold structure. Tosi et al. [21] give this a geometric interpretation by deriving the distribution of the induced Riemannian pull-back metric and show how geodesics can be computed under this uncertain metric.

Often manifold learning is viewed as a form of dimensionality reduction, but this need not be the case. Hauberg et al. [12] suggest to model the local behavior of the data manifold via a locally-defined Riemannian metric, which is constructed by interpolating a set of pre-trained metric tensors at a few select points in data space. Once a Riemannian metric is available existing tools can be used for dimensionality reduction [8,11,22], mixture modeling [1,19], tracking [13,14], hypothesis testing [17], transfer learning [9] and more. Our approach follow this line of work.

1.2 The Basics of Riemannian Geometry

For completeness we start with an informal review of Riemannian manifolds, but refer the reader to standard text books [5] for a more detailed exposition.

Definition 1. *A smooth manifold $\mathcal{M}$ together with a Riemannian metric $\mathbf{M} : \mathcal{M} \to \mathbb{R}^{D \times D}$ and $\mathbf{M} \succ 0$ is called a Riemannian manifold. The Riemannian metric $\mathbf{M}$ encodes a smoothly changing inner product $\langle \mathbf{u}, \mathbf{M}(\mathbf{x})\mathbf{v} \rangle$ on the tangent space $\mathbf{u}, \mathbf{v} \in \mathcal{T}_{\mathbf{x}}\mathcal{M}$ of each point $\mathbf{x} \in \mathcal{M}$.*

Since the Riemannian metric $\mathbf{M}(\mathbf{x})$ acts on tangent vectors it may be interpreted as a standard Mahalanobis metric restricted to an infinitesimal region around $\mathbf{x}$. This local inner product is a suitable model for capturing local behavior of data, i.e. *manifold learning*. Shortest paths (*geodesics*) are then length-minimizing curves connecting two points $\mathbf{x}, \mathbf{y} \in \mathcal{M}$, i.e.

$$\hat{\gamma} = \underset{\gamma}{\operatorname{argmin}} \int_0^1 \sqrt{\langle \gamma'(t), \mathbf{M}(\gamma(t))\gamma'(t) \rangle} dt, \quad \text{s.t.} \quad \gamma(0) = \mathbf{x}, \; \gamma(1) = \mathbf{y}. \tag{1}$$

Here $\mathbf{M}(\gamma(t))$ is the metric tensor at $\gamma(t)$, and the tangent vector γ' denotes the derivative (velocity) of γ. The distance between $\mathbf{x}$ and $\mathbf{y}$ is defined as the length of the geodesic. Geodesic can be found as the solution to a system of 2nd order ordinary differential equations (ODEs):

$$\gamma''(t) = -\frac{1}{2}\mathbf{M}^{-1}(\gamma(t)) \left[\frac{\partial \text{vec}[\mathbf{M}(\gamma(t))]}{\partial \gamma(t)} \right]^{\mathsf{T}} (\gamma'(t) \otimes \gamma'(t)) \tag{2}$$

subject to $\gamma(0) = \mathbf{x}, \; \gamma(1) = \mathbf{y}$. Here vec$[\cdot]$ stacks the columns of a matrix into a vector and $\otimes$ is the Kronecker product.

This differential equation allows us to define basic operations on the manifold. The *exponential map* at a point $\mathbf{x}$ takes a tangent vector $\mathbf{v} \in \mathcal{T}_\mathbf{x}\mathcal{M}$ to $\mathbf{y} = \text{Exp}_\mathbf{x}(\mathbf{v}) \in \mathcal{M}$ such that the curve $\gamma(t) = \text{Exp}_\mathbf{x}(t \cdot \mathbf{v})$ is a geodesic originating at $\mathbf{x}$ with initial velocity $\mathbf{v}$ and length $\|\mathbf{v}\|$. The inverse mapping, which takes $\mathbf{y}$ to $\mathcal{T}_\mathbf{x}\mathcal{M}$ is known as the *logarithm map* and is denoted $\text{Log}_\mathbf{x}(\mathbf{y})$. By definition $\|\text{Log}_\mathbf{x}(\mathbf{y})\|$ corresponds to the geodesic distance from $\mathbf{x}$ to $\mathbf{y}$. The exponential and the logarithmic map can be computed by solving Eq. 2 numerically, as an *initial value problem* or a *boundary value problem* respectively.

2 A Locally Adaptive Normal Distribution

We have previously provided a simple nonparametric manifold learning scheme that conceptually mimics a local principal component analysis [1]. At each point $\mathbf{x} \in \mathbb{R}^D$ a local covariance matrix is computed and its inverse then specify a local metric. For computational efficiency and to prevent overfitting we restrict ourselves to diagonal covariances

$$M_{dd}(\mathbf{x}) = \left(\sum_{n=1}^{N} w_n(\mathbf{x})(x_{nd} - x_d)^2 + \rho \right)^{-1}, \tag{3}$$

$$w_n(\mathbf{x}) = \exp\left(-\frac{\|\mathbf{x}_n - \mathbf{x}\|_2^2}{2\sigma^2} \right). \tag{4}$$

Here the subscript d is the dimension, n corresponds to the given data, and ρ is a regularization parameter to avoid singular covariances. The weight-function $w_n(\mathbf{x})$ changes smoothly such that the resulting metric is Riemannian. It is easy to see that if $\mathbf{x}$ is outside of the support of the data, then the metric tensor is large. Thus, geodesics are "pulled" towards the data where the metric is small (see Fig. 1).

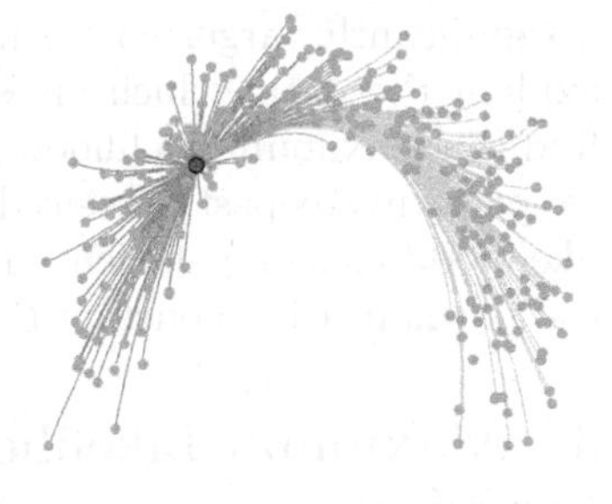

Fig. 1. Example geodesics.

The weight-function $w_n(\mathbf{x})$ depends on a parameter σ that effectively determine the size of the neighborhood used to define the data manifold. Small values of σ gives a manifold with high curvature, while a large σ gives an almost flat manifold. The main contribution of this paper is a systematic approach to determine this parameter.

For a given metric (and hence σ), we can estimate data distributions with respect to this metric. We consider Riemannian normal distributions [17]

$$p_{\mathcal{M}}(\mathbf{x} \mid \boldsymbol{\mu}, \boldsymbol{\Sigma}) = \frac{1}{\mathcal{C}} \exp\left(-\frac{1}{2} d^2_{\boldsymbol{\Sigma}}(\mathbf{x}, \boldsymbol{\mu})\right), \quad \mathbf{x} \in \mathcal{M} \tag{5}$$

and mixtures thereof. Here $\mathcal{M}$ denote the manifold induced by the learned metric, $\boldsymbol{\mu}$ and $\boldsymbol{\Sigma}$ are the mean and covariance, and $d^2_{\boldsymbol{\Sigma}}(\mathbf{x}, \boldsymbol{\mu}) = \langle \mathrm{Log}_{\boldsymbol{\mu}}(\mathbf{x}), \boldsymbol{\Sigma}^{-1} \mathrm{Log}_{\boldsymbol{\mu}}(\mathbf{x}) \rangle$. The normalization constant $\mathcal{C}$ is by definition

$$\mathcal{C}(\boldsymbol{\mu}, \boldsymbol{\Sigma}) = \int_{\mathcal{M}} \exp\left(-\frac{1}{2} d^2_{\boldsymbol{\Sigma}}(\mathbf{x}, \boldsymbol{\mu})\right) \mathrm{d}\mathcal{M}(\mathbf{x}), \tag{6}$$

where $\mathrm{d}\mathcal{M}(\mathbf{x})$ denotes the measure induced by the Riemannian metric. Note that this measure depends in σ. Figure 2 show an example of the resulting distribution under the proposed metric. As the distribution adapts locally to the data we coin it a *locally adaptive normal distribution (LAND).*

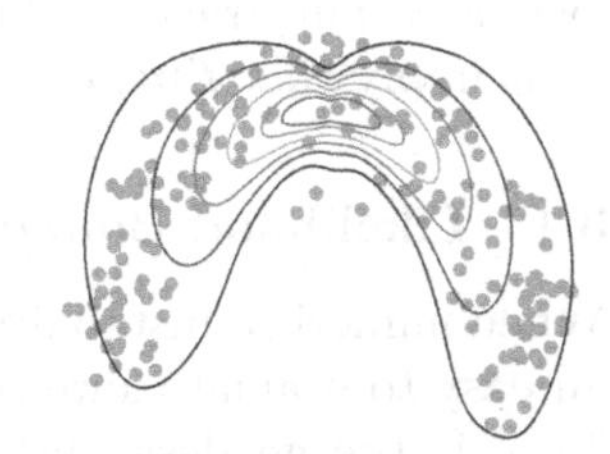

Fig. 2. Example of the *locally adaptive normal distribution (LAND).*

Assuming that the data are generated from a distribution $q_{\mathcal{M}}$ then commonly the mean $\boldsymbol{\mu}$ and covariance $\boldsymbol{\Sigma}$ are estimated with intrinsic least squares (ILS)

$$\hat{\boldsymbol{\mu}} = \underset{\boldsymbol{\mu} \in \mathcal{M}}{\mathrm{argmin}} \int_{\mathcal{M}} d^2(\boldsymbol{\mu}, \mathbf{x}) q_{\mathcal{M}}(\mathbf{x}) \mathrm{d}\mathcal{M}(\mathbf{x}), \tag{7}$$

$$\hat{\boldsymbol{\Sigma}} = \int_{\mathcal{T}_{\hat{\boldsymbol{\mu}}}\mathcal{M}} \mathrm{Log}_{\hat{\boldsymbol{\mu}}}(\mathbf{x}) \mathrm{Log}_{\hat{\boldsymbol{\mu}}}(\mathbf{x})^{\intercal} p_{\mathcal{M}}(\mathbf{x}) \mathrm{d}\mathcal{M}(\mathbf{x}), \tag{8}$$

where $d^2(\cdot, \cdot)$ denotes the squared geodesic distance. These parameter estimates naturally generalize their Euclidean counterparts, and they can be further shown to have maximal likelihood when the manifold is also a symmetric space [7]. For more general manifolds, like the ones under consideration in this paper, these estimates do not attain maximal likelihood. Figure 3 show both the ILS estimate of $\boldsymbol{\mu}$ and the maximum likelihood (ML) estimate. Since the ILS estimate falls outside the support of the data, a significantly larger covariance matrix is needed to explain the data, which gives a poor likelihood. To find the maximum likelihood parameters of $\boldsymbol{\mu}$ and $\boldsymbol{\Sigma}$ we perform steepest descent directly on the data log-likelihood using an efficient Monte Carlo estimator of the normalization constant $\mathcal{C}$ [1].

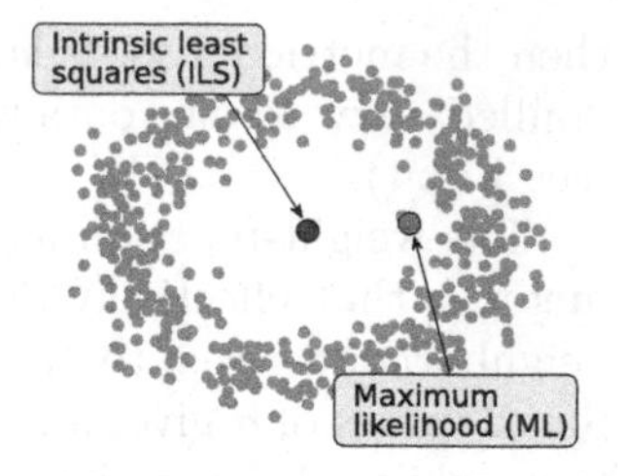

Fig. 3. ML and ILS means.

3 Maximum Likelihood Metric Learning

Determining the optimal metric (parametrized by σ) is an open question. Since the LAND is a parametric probabilistic model it is natural to perform this model selection using maximum likelihood. The complete data log-likelihood is

$$\mathcal{L}(\sigma) = -\frac{1}{2}\sum_{n=1}^{N} d^2_{\boldsymbol{\Sigma}}(\mathbf{x}_n, \boldsymbol{\mu}) - N \log \mathcal{C}(\boldsymbol{\mu}, \boldsymbol{\Sigma}). \tag{9}$$

It is tempting to evaluate $\mathcal{L}(\sigma)$ for several values of σ and pick the one with maximal likelihood. This, however, is both theoretically and practically flawed.

The first issue is that the measure $\mathrm{d}\mathcal{M}(\cdot)$ used to define the LAND depends on σ. This imply that $\mathcal{L}(\sigma)$ cannot be compared for different values of σ as they do not rely on the same measure. The second issue is that $\mathrm{Log}_{\boldsymbol{\mu}}(\mathbf{x}_n)$ must be evaluated numerically, which can become unstable when $\mathcal{M}$ has high curvature. This imply that $\mathcal{L}(\sigma)$ can often not be fully evaluated when σ is small.

3.1 Likelihood Bounds to Cope with Numerical Instabilities

When numerical instabilities prevent us from evaluating $\mathcal{L}(\sigma)$ we instead rely on an easy-to-evaluate lower bound $\overline{\mathcal{L}(\sigma)}$. To derive this, let $\mathbf{v}_n = \mathrm{Log}_{\boldsymbol{\mu}}(\mathbf{x}_n)$. Then $\|\mathbf{v}_n\|$ is the geodesic distance between $\boldsymbol{\mu}$ and $\mathbf{x}_n$, while $\mathbf{v}_n/\|\mathbf{v}_n\|$ is the initial direction of the connecting geodesic. It is easy to provide an upper bound on the geodesic distance by taking the length of a non-geodesic connecting curve, here chosen as the straight line connecting $\boldsymbol{\mu}$ and $\mathbf{x}_n$. The bound then becomes

$$\|\mathbf{v}_n\| \leq \tilde{d}_n = \int_0^1 \sqrt{\langle (\mathbf{x}_n - \boldsymbol{\mu}), \mathbf{M}(t\mathbf{x}_n + (1-t)\boldsymbol{\mu})(\mathbf{x}_n - \boldsymbol{\mu})\rangle}\mathrm{d}t. \tag{10}$$

The initial orientation $\mathbf{v}_n/\|\mathbf{v}_n\|$ influence the log-likelihood as the covariance $\boldsymbol{\Sigma}$ is generally anisotropic. This is, however, easily bounded by picking the initial direction as the eigenvector of $\boldsymbol{\Sigma}$ corresponding to the smallest eigenvalue $\lambda_{\min}$. This then gives the final lower bound

$$\overline{\mathcal{L}(\sigma)} = -\frac{1}{2}\sum_{n=1}^{N} \frac{\tilde{d}_n^2}{\lambda_{\min}} - N \log \mathcal{C}(\boldsymbol{\mu}, \boldsymbol{\Sigma}). \tag{11}$$

In practice, we only use the bound for data points $\mathbf{x}$ where the logarithm map cannot be evaluated, and otherwise use the correct log-likelihood.

3.2 Comparing Likelihoods

Since the measure $\mathrm{d}\mathcal{M}(\cdot)$ changes with σ we cannot directly compare $\mathcal{L}(\sigma)$ across inputs. In order to make this comparison feasible, we propose to re-normalize the LAND with respect to the usual Lebesgue measure of the data space $\mathbb{R}^D$. This amount to changing the applied measure in Eq. 6. As we lack closed-form expressions, we perform this re-normalization using importance sampling [4]

$$\tilde{\mathcal{C}}(\boldsymbol{\mu}, \boldsymbol{\Sigma}) = \int_{\mathbb{R}^D} \exp\left(-\frac{1}{2} d^2_{\boldsymbol{\Sigma}}(\boldsymbol{\mu}, \mathbf{x})\right) \mathrm{d}\mathbf{x} = \int_{\mathbb{R}^D} \frac{\exp\left(-\frac{1}{2} d^2_{\boldsymbol{\Sigma}}(\boldsymbol{\mu}, \mathbf{x})\right)}{q(\mathbf{x})} q(\mathbf{x}) \mathrm{d}\mathbf{x} \quad (12)$$

$$\approx \frac{1}{S} \sum_{s=1}^{S} w_s \exp\left(-\frac{1}{2} d^2_{\boldsymbol{\Sigma}}(\boldsymbol{\mu}, \mathbf{x}_s)\right), \quad \mathbf{x}_s \sim q(\mathbf{x}), \; w_s = \frac{1}{q(\mathbf{x}_s)}, \quad (13)$$

where $q(\mathbf{x})$ is the proposal distribution from which we draw S samples. In our experiments we choose $q(\mathbf{x}) = \mathcal{N}(\mathbf{x}|\boldsymbol{\mu}, \boldsymbol{\Sigma})$ with the linear mean and covariance of the data. Thus, we ensure that the support of the proposal captures the data manifold, but any other distribution with the desired properties can be used.

4 Results

Experimental setup: We evaluate the proposed method on both synthetic and real data. The two-dimensional synthetic data is drawn from an arc-shaped distribution (see Fig. 4c) [1]. We further consider features extracted from EEG measurements during human sleep [1]; the digit "1" from MNIST; and the "Frey faces"[1]. Both image modalities are projected onto their first two principal components, and are separated into 10 and 5 folds respectively. To each data modality, we fit a mixture of LANDs with K components.

Verification: First, we validate the importance sampling scheme in Fig. 4b where we compare with an expensive numerical integration scheme on a predefined grid. It is evident that importance sampling quickly gives a good approximation to the true normalization constant. However, choosing the correct proposal distribution is usually crucial for the success of the approximation [4]. Then, in Fig. 4c we show the impact of σ on the geodesic solution. When σ is small (0.01) the true geodesic cannot be computed numerically and a straight line is used to bound the likelihood (Sect. 3). For larger values of σ the geodesic can be computed. Note that the geodesic becomes increasingly "straight" for large values of σ.

Model selection: Figures 4d–g show the log-likelihood bound proposed in Sect. 3 for all data sets. In particular, we can distinguish three different regions for the σ parameter. (1) For small values of σ the manifold has high curvature and some geodesics cannot be computed, such that the bound penalizes the data log-likelihood. (2) There is a range of σ values where the construction of the

[1] http://www.cs.nyu.edu/~roweis/data.html.

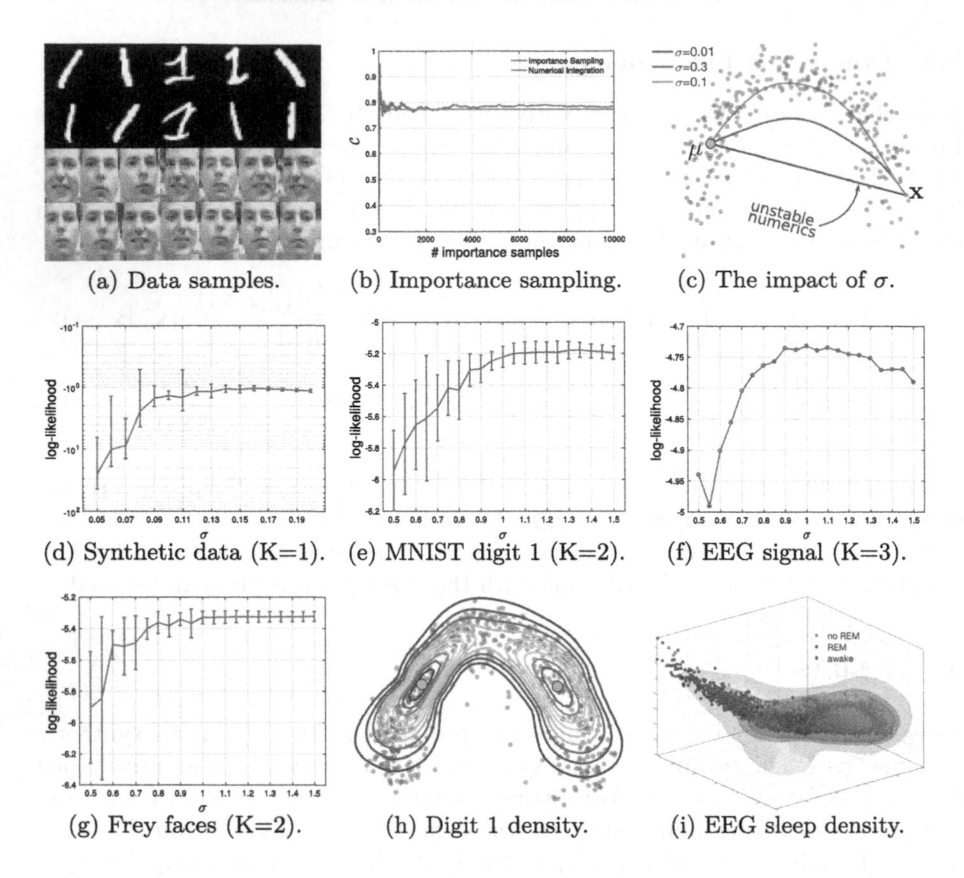

(a) Data samples. (b) Importance sampling. (c) The impact of σ.

(d) Synthetic data (K=1). (e) MNIST digit 1 (K=2). (f) EEG signal (K=3).

(g) Frey faces (K=2). (h) Digit 1 density. (i) EEG sleep density.

Fig. 4. Experimental results on various data sets; see text for details.

manifold captures the actual underlying data structure, and in those cases we achieve the best log-likelihood. (3) For larger values of σ the manifold becomes flat, and even if we are able compute all the geodesics the likelihood is reduced. The reason is that when the manifolds becomes flat, significant probability mass is assigned to regions outside of the data support, while in the other case all the probability mass is concentrated near the data resulting to higher likelihood.

5 Discussion

Probability density estimation in non-linear spaces is essential in data analysis [6]. With the current work, we have proposed practical tools for model selection of the metric underlying the *locally adaptive normal distribution (LAND)* [1]. The basic idea amounts to picking the metric that maximize the data likelihood. A *theoretical* concern is that different metrics gives different measures implying that likelihoods are not comparable. We have proposed to solve this by re-normalizing according to the Lebesgue measure associated with the data space. *Practically* our idea face numerical challenges when the metric has high curvature as geodesics then become unstable to compute. Here we have proposed an easy-to-compute bound on the data likelihood, which has the added benefit that metrics giving rise to numerical instabilities are penalized. Experimental results on diverse data sets indicate that the approach is suitable for model selection.

In this paper we have considered maximum likelihood estimation on the training data, which can potentially overfit [10]. While we did not observe such behavior in our experiments it is still worth investigating model selection on a held-out test set or to put a prior on σ and pick the value that maximize the posterior probability. Both choices are straight-forward.

An interesting alternative to bounding the likelihood appears when considering probabilistic solvers [15] for the geodesic equations (2). These represent the numerical estimate of geodesics with a Gaussian process whose uncertainty captures numerical approximation errors. Efficient algorithms then exist for estimating the distribution of the geodesic arc length [3]. With these solvers, hard-to-estimate geodesics will be associated with high variance, such that the now-stochastic data log-likelihood also has high variance. Model selection should then take this variance into account.

Acknowledgements. SH was supported by a research grant (15334) from VILLUM FONDEN. LKH was supported by the Novo Nordisk Foundation Interdisciplinary Synergy Program 2014, 'Biophysically adjusted state-informed cortex stimulation (BASICS)'.

References

1. Arvanitidis, G., Hansen, L.K., Hauberg, S.: A locally adaptive normal distribution. In: Advances in Neural Information Processing Systems (NIPS) (2016)
2. Belkin, M., Niyogi, P.: Laplacian eigenmaps for dimensionality reduction and data representation. Neural Comput. **15**(6), 1373–1396 (2003)
3. Bewsher, J., Tosi, A., Osborne, M., Roberts, S.: Distribution of Gaussian process arc lengths. In: AISTATS (2017)
4. Bishop, C.M.: Pattern Recognition and Machine Learning. Information Science and Statistics. Springer, New York (2006)
5. Carmo, M.: Riemannian Geometry. Birkhäuser, Boston (1992)
6. Chevallier, E., Barbaresco, F., Angulo, J.: Probability density estimation on the hyperbolic space applied to radar processing. In: Nielsen, F., Barbaresco, F. (eds.) GSI 2015. LNCS, vol. 9389, pp. 753–761. Springer, Cham (2015). doi:10.1007/978-3-319-25040-3_80
7. Fletcher, P.T.: Geodesic regression and the theory of least squares on Riemannian manifolds. IJCV **105**(2), 171–185 (2013)
8. Fletcher, P.T., Lu, C., Pizer, S.M., Joshi, S.: Principal geodesic analysis for the study of nonlinear statistics of shape. IEEE TMI **23**(8), 995–1005 (2004)
9. Freifeld, O., Hauberg, S., Black, M.J.: Model transport: towards scalable transfer learning on manifolds. In: CVPR (2014)
10. Hansen, L.K., Larsen, J.: Unsupervised learning and generalization. In: IEEE International Conference on Neural Networks, vol. 1, pp. 25–30. IEEE (1996)

11. Hauberg, S.: Principal curves on Riemannian manifolds. IEEE Trans. Pattern Anal. Mach. Intell. (TPAMI) **38**, 1915–1921 (2016)
12. Hauberg, S., Freifeld, O., Black, M.J.: A geometric take on metric learning. In: Advances in Neural Information Processing Systems (NIPS), pp. 2033–2041 (2012)
13. Hauberg, S., Lauze, F., Pedersen, K.S.: Unscented Kalman filtering on Riemannian manifolds. J. Math. Imaging Vis. **46**(1), 103–120 (2013)
14. Hauberg, S., Pedersen, K.S.: Stick It! articulated tracking using spatial rigid object priors. In: Kimmel, R., Klette, R., Sugimoto, A. (eds.) ACCV 2010. LNCS, vol. 6494, pp. 758–769. Springer, Heidelberg (2011). doi:10.1007/978-3-642-19318-7_59
15. Hennig, P., Hauberg, S.: Probabilistic solutions to differential equations and their application to Riemannian statistics. In: AISTATS, vol. 33 (2014)
16. Lawrence, N.: Probabilistic non-linear principal component analysis with Gaussian process latent variable models. J. Mach. Learn. Res. **6**, 1783–1816 (2005)
17. Pennec, X.: Intrinsic statistics on Riemannian manifolds: basic tools for geometric measurements. J. Math. Imaging Vis. **25**(1), 127–154 (2006)
18. Saul, L.K., Roweis, S.T.: Think globally, fit locally: unsupervised learning of low dimensional manifolds. J. Mach. Learn. Res. **4**, 119–155 (2003)
19. Straub, J., Chang, J., Freifeld, O., Fisher III, J.W.: A Dirichlet process mixture model for spherical data. In: AISTATS (2015)
20. Tenenbaum, J.B., de Silva, V., Langford, J.C.: A global geometric framework for nonlinear dimensionality reduction. Science **290**(5500), 2319 (2000)
21. Tosi, A., Hauberg, S., Vellido, A., Lawrence, N.D.: Metrics for probabilistic geometries. In: The Conference on Uncertainty in Artificial Intelligence (UAI) (2014)
22. Zhang, M., Fletcher, P.T.: Probabilistic principal geodesic analysis. In: Advances in Neural Information Processing Systems (NIPS), vol. 26, pp. 1178–1186 (2013)

Shape Space

Shape Analysis on Lie Groups and Homogeneous Spaces

Elena Celledoni, Sølve Eidnes, Markus Eslitzbichler, and Alexander Schmeding(✉)

NTNU Trondheim, Trondheim, Norway
{elena.celledoni,solve.eidnes,alexander.schmeding}@ntnu.no,
m.eslitzbichler@gmail.com

Abstract. In this paper we are concerned with the approach to shape analysis based on the so called Square Root Velocity Transform (SRVT). We propose a generalisation of the SRVT from Euclidean spaces to shape spaces of curves on Lie groups and on homogeneous manifolds. The main idea behind our approach is to exploit the geometry of the natural Lie group actions on these spaces.

Keywords: Shape analysis · Lie group · Homogeneous spaces · SRVT

Shape analysis methods have significantly increased in popularity in the last decade. Advances in this field have been made both in the theoretical foundations and in the extension of the methods to new areas of application. Originally developed for planar curves, the techniques of shape analysis have been successfully extended to higher dimensional curves, surfaces, activities, character motions and a number of different types of digitalized objects.

In the present paper, shapes are unparametrized curves, evolving on a vector space, on a Lie group, or on a manifold. Shape spaces and spaces of curves are infinite-dimensional Riemannian manifolds, whose Riemannian metrics are the crucial tool to compare and analyse shapes.

We are concerned with one particular approach to shape analysis, which is based on the Square Root Velocity Transform (SRVT) [10]. On vector spaces, the SRVT maps parametrized curves (i.e. smooth immersions) to appropriately scaled tangent vector fields along them via

$$\mathcal{R}\colon \operatorname{Imm}([0,1],\mathbb{R}^d) \to C^\infty([0,1],\mathbb{R}^d \setminus \{0\}), \quad c \mapsto \frac{\dot{c}}{\sqrt{\|\dot{c}\|}}. \tag{1}$$

The transformed curves are then compared computing geodesics in the L^2 metric, and the scaling induces reparametrization invariance of the pullback metric. Note that it is quite natural to consider an L^2 metric directly on the original parametrized curves. Constructing the L^2 metric with respect to integration by arc-length, one obtains a reparametrisation invariant metric. However, this metric is unsuitable for our purpose as it leads to vanishing geodesic distance on

F. Nielsen and F. Barbaresco (Eds.): GSI 2017, LNCS 10589, pp. 49–56, 2017.
https://doi.org/10.1007/978-3-319-68445-1_6

the quotient shape space [6] and consequently also on the space of parametrised curves [1]. This infinite-dimensional phenomenon prompted the investigation of alternative, higher order Sobolev type metrics [7], which however can be computationally demanding. Since it allows geodesic computations via the L^2 metric on the transformed curves, the SRVT technique is computationally attractive. It is also possible to prove that this algorithmic approach corresponds, at least locally, to a particular Sobolev type metric, see [2,4].

We propose a generalisation of the SRVT to construct well-behaved Riemannian metrics on shape spaces with values in Lie groups and homogeneous manifolds. Our methodology is alternative to what was earlier proposed in [5,11] and the main idea is, following [4], to take advantage of the Lie group acting transitively on the homogeneous manifold. Since we want to compare curves, the main tool here is an SRVT which transports the manifold valued curves into the Lie algebra or a subspace of the Lie algebra.

1 SRVT for Lie Group Valued Shape Spaces

In the Lie group case, the obvious choice for this tangent space is of course the Lie algebra $\mathfrak{g}$ of the Lie group G. The idea is to use the derivative T_eR_g of the right translation for the transport and measure with respect to a right-invariant Riemannian metric.[1] Instead of the ordinary derivative, one thus works with the right-logarithmic derivative $\delta^r(c)(t) = T_eR_{c(t)^{-1}}(\dot{c}(t))$ (here e is the identity element of G) and defines an SRVT for Lie group valued curves as (see [4]):

$$\mathcal{R}\colon \mathrm{Imm}([0,1],G) \to C^\infty([0,1],\mathfrak{g}\setminus\{0\}), \quad c \mapsto \frac{\delta^r(c)}{\sqrt{\|\dot{c}\|}}. \tag{2}$$

We will use the short notation $I = [0,1]$ in what follows. Using tools from Lie theory, we are then able to describe the resulting pullback metric on the space $\mathcal{P}_*$ of immersions $c\colon [0,1] \to G$ which satisfy $c(0) = e$:

Theorem 1 (The Elastic metric on Lie group valued shape spaces [4]). *Let $c \in \mathcal{P}_*$ and consider $v, w \in T_c\mathcal{P}_*$. The pullback of the L^2-metric on $C^\infty(I,\mathfrak{g}\setminus\{0\})$ under the SRVT (2) to $\mathcal{P}_*$ is given by the first order Sobolev metric:*

$$\begin{aligned} G_c(v,w) = \int_I \frac{1}{4}\langle D_sv, u_c\rangle\langle D_sw, u_c\rangle \\ + \langle D_sv - u_c\langle D_sv,u_c\rangle, D_sw - u_c\langle D_sw,u_c\rangle\rangle\, ds, \end{aligned} \tag{3}$$

where $D_sv := T_c\delta^r(v)/\|\dot{c}\|$, $u_c := \delta^r(c)/\|\delta^r(c)\|$ is the unit tangent vector of $\delta^r(c)$ and $ds = \|\dot{c}(t)\|\,dt$.

The geodesic distance of this metric descends to a nonvanishing metric on the space of unparametrized curves. In particular, this distance is easy to compute as one can prove [4, Theorem 3.16] that

[1] Equivalently one could instead use left translations and a left-invariant metric here.

Theorem 2. *If* $\dim \mathfrak{g} > 2$*, then the geodesic distance of* $C^\infty(I, \mathfrak{g}\backslash\{0\})$ *is globally given by the* L^2*-distance. In particular, in this case the geodesic distance of the pullback metric* (3) *on* $\mathcal{P}_*$ *is given by*

$$d_{\mathcal{P}_*}(c_0, c_1) := \sqrt{\int_I \|\mathcal{R}(c_0)(t) - \mathcal{R}(c_1)(t)\|^2 dt}.$$

These tools give rise to algorithms which can be used in, among other things, tasks related to computer animation and blending of curves, as shown in [4]. The blending $c(t,s)$ of two curves $c_0(t)$ and $c_1(t)$, $t \in I$, amounts simply to a convex linear convex combination of their SRV transforms:

$$c(t,s) = \mathcal{R}^{-1}\left(s\,\mathcal{R}(c_0(t)) + (1-s)\mathcal{R}(c_1(t))\right), \qquad s \in [0,1].$$

Using the transformation of the curves to the Lie algebra by the SRVT, we also propose a curve closing algorithm allowing one to remove discontinuities from motion capturing data while preserving the general structure of the movement. (See Fig. 1.)

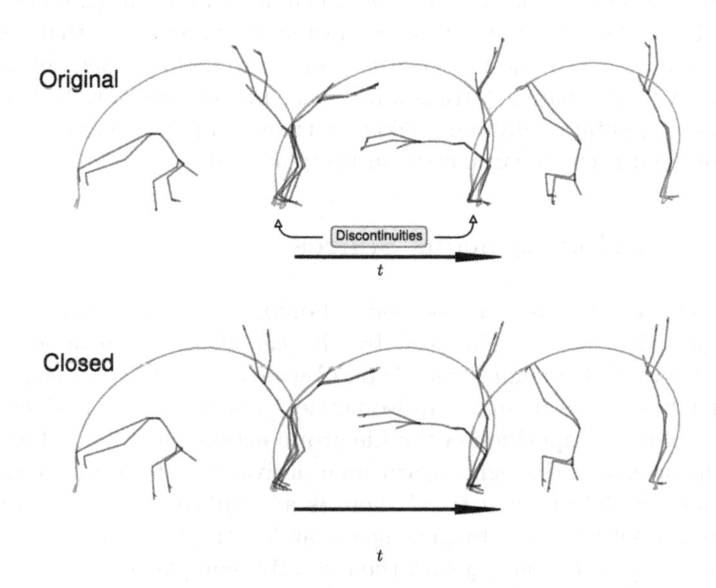

Fig. 1. Application of closing algorithm to a cartwheel animation. Note the large difference between start and end poses, on the right and the left respectively. The motion is repeated once and suffers from a strong jerk when it repeats, especially in the left hand. In the second row, the curve closing method has been used to alleviate this discontinuity.

2 The Structure of the SRVT

Analysing the constructions for the square root velocity transform, e.g. (1) and (2) or the generalisations proposed in the literature, every SRVT is composed of three distinct building blocks. While two of these blocks can not be changed, there are many choices for the second one (transport) in constructing an SRVT:

- **Differentiation:** The basic building block of every SRVT, taking a curve to its derivative.
- **Transport:** Bringing a curve into a common space of reference. In general there are many choices for this transport[2] (in our approach we use the Lie group action to transport data into the Lie algebra of the acting group).
- **Scaling:** The second basic building block, assures reparametrization invariance of the metrics obtained.

In constructing the SRVT, we advocate the use of Lie group actions for the transport. This action allows us to transport derivatives of curves to our choice of base point and to lift this information to a curve in the Lie algebra.

Other common choices for the transport usually arise from parallel transport (cf. e.g. [5,11]). The advantage of using the Lie group action is that we obtain a global transport, i.e. we do not need to restrict to certain open submanifolds to make sense of the (parallel) transport.[3] Last but not least, right translation is in general computationally more efficient than computing parallel transport using the original Riemannian metric on the manifold.

3 SRVT on Homogeneous Spaces

Our approach [3] for shape analysis on a homogeneous manifold $\mathcal{M} = G/H$ exploits again the geometry induced by the canonical group action $\Lambda\colon G \times \mathcal{M} \to \mathcal{M}$. We fix a Riemannian metric on G which is right H-invariant, i.e. the maps R_h for $h \in H$ are Riemannian isometries. The SRVT is obtained using a right inverse of the composition of the Lie group action with the evolution operator (i.e. the inverse of the right-logarithmic derivative) of the Lie group. If the homogeneous manifold is reductive,[4] there is an explicit way to construct this right inverse. Identifying the tangent space at $[e]$, the equivalence class of the identity, via $\omega_e\colon T_{[e]}\mathcal{M} \to \mathfrak{m} \subseteq \mathfrak{g}$ with the reductive complement. Then we define

[2] In the literature, e.g. [11], a common choice is parallel transport with respect to the Riemannian structure.

[3] The problem in these approaches arises from choosing curves along which the parallel transport is conducted. Typically, one wants to transport along geodesics to a reference point and this is only well-defined outside of the cut locus (also cf. [8]).

[4] Recall that a homogeneous space G/H is reductive if the Lie subalgebra $\mathfrak{h}$ of $H \subseteq G$ admits a reductive complement, i.e. $\mathfrak{g} = \mathfrak{h} \oplus \mathfrak{m}$, where $\mathfrak{m}$ is a subvector space invariant under the adjoint action of H.

the map $\omega([g]) = \mathrm{Ad}(g).\omega_e(T\Lambda(g^{-1},\cdot)[g])$ (which is well-defined by reductivity) and obtain a square root velocity transform for reductive homogeneous spaces as

$$\mathcal{R}\colon \mathrm{Imm}([0,1],\mathcal{M}) \to C^\infty([0,1],\mathfrak{g}\setminus\{0\}), \quad c \mapsto \frac{\omega \circ \dot{c}}{\sqrt{\|\omega \circ \dot{c}\|}} \tag{4}$$

Conceptually this SRVT is somewhat different from the one for Lie groups, as it does not establish a bijection between the manifolds of smooth mappings. However, one can still use (4) to construct a pullback metric on the manifold of curves to the homogeneous space by pulling back the L^2 inner product of curves on the Lie algebra through the SRVT. Different choices of Lie group actions will give rise to different Riemannian metrics (with different properties).

4 Numerical Experiments

We present some results about the realisation of this metric through the SRVT framework in the case of reductive homogeneous spaces. Further, our results are illustrated in a concrete example. We compare the new methods for curves into the sphere $\mathrm{SO}(3)/\mathrm{SO}(2)$ with results derived from the Lie group case.

In the following, we use the Rodrigues' formula for the Lie group exponential $\exp\colon \mathfrak{so}(3) \to \mathrm{SO}(3)$,

$$\exp(\hat{x}) = I + \frac{\sin(\alpha)}{\alpha}\hat{x} + \frac{1-\cos(\alpha)}{\alpha^2}\hat{x}^2, \qquad \alpha = \|x\|_2$$

and the corresponding formula for the logarithm $\log\colon \mathrm{SO}(3) \to \mathfrak{so}(3)$,

$$\log(X) = \frac{\sin^{-1}(\|y\|)}{\|y\|}\hat{y}, \quad X \neq I, \quad X \text{ close to } I,$$

are used, where $\hat{y} = \frac{1}{2}(X - X^{\mathrm{T}})$, and the relationship between x and $\hat{x}$ is given by the isomorphism between $\mathbb{R}^3$ and $\mathfrak{so}(3)$ known as the hat map

$$x = \begin{pmatrix} x_1 \\ x_2 \\ x_3 \end{pmatrix} \mapsto \hat{x} = \begin{pmatrix} 0 & -x_3 & x_2 \\ x_3 & 0 & -x_1 \\ -x_2 & x_1 & 0 \end{pmatrix}.$$

4.1 Lie Group Case

Consider a continuous curve $z(t), t \in [t_0, t_N]$, in $\mathrm{SO}(3)$. We approximate it by $\bar{z}(t)$, interpolating between $N+1$ values $\bar{z}_i = z(t_i)$, with $t_0 < t_1 < \ldots < t_N$, as:

$$\bar{z}(t) := \sum_{i=0}^{N-1} \chi_{[t_i,t_{i+1})}(t) \exp\left(\frac{t-t_i}{t_{i+1}-t_i}\log\left(\bar{z}_{i+1}\bar{z}_i^{\mathrm{T}}\right)\right)\bar{z}_i, \tag{5}$$

where χ is the characteristic function.

The SRVT (2) of $\bar{z}(t)$ is a piecewise constant function $\bar{p}(t)$ in $\mathfrak{so}(3)$ with values $\bar{p}_i = \bar{p}(t_i)$, $i = 0, ..., N-1$, found by

$$\bar{p}_i = \frac{\eta_i}{\sqrt{\|\eta_i\|}}, \qquad \eta_i = \frac{\log(\bar{z}_{i+1}\bar{z}_i^{\mathrm{T}})}{t_{i+1} - t_i}.$$

The inverse $\mathcal{R}^{-1} : \mathfrak{so}(3) \to \mathrm{SO}(3)$ is then given by (5), with the discrete points

$$\bar{z}_{i+1} = \exp\left(\|\bar{p}_i\|\bar{p}_i\right)\bar{z}_i, \quad i = 1, ..., N-1, \quad \bar{z}_0 = z(t_0).$$

4.2 Homogeneous Manifold Case

As an example of the homogeneous space case, consider the curve $c(t)$ on the sphere SO(3)/SO(2) (i.e. S^2), which we approximate by $\bar{c}(t)$, interpolating between the $N+1$ values $\bar{c}_i = c(t_i)$:

$$\bar{c}(t) := \sum_{i=0}^{N-1} \chi_{[t_i,t_{i+1})}(t) \exp\left(\frac{t - t_i}{t_{i+1} - t_i}\left(v_i\bar{c}_i^{\mathrm{T}} - \bar{c}_i v_i^{\mathrm{T}}\right)\right)\bar{c}_i, \tag{6}$$

where v_i are approximations to $\frac{d}{dt}\Big|_{t=t_i} c(t)$ found by solving the equations

$$\bar{c}_{i+1} = \exp\left(v_i\bar{c}_i^{\mathrm{T}} - \bar{c}_i v_i^{\mathrm{T}}\right)\bar{c}_i, \tag{7}$$

$$\text{constrained by} \quad v_i^{\mathrm{T}}\bar{c}_i = 0. \tag{8}$$

Observing that if $\kappa = \bar{c}_i \times v_i$, then $\hat{\kappa} = v_i\bar{c}_i^{\mathrm{T}} - \bar{c}_i v_i^{\mathrm{T}}$, and assuming that the sphere has radius 1, we have by (8) that $\|\bar{c}_i \times v_i\|_2 = \|\bar{c}_i\|_2\|v_i\|_2 = \|v_i\|_2$. By (7) we get

$$\bar{c}_{i+1} = \frac{\sin\left(\|v_i\|_2\right)}{\|v_i\|_2} v_i + \cos\left(\|v_i\|_2\right)\bar{c}_i.$$

Calculations give $\bar{c}_i^{\mathrm{T}}\bar{c}_{i+1} = 1 - \cos\left(\|v_i\|_2\right)$ and $\|v_i\|_2 = \arccos\left(\bar{c}_i^{\mathrm{T}}\bar{c}_{i+1}\right)$, leading to $v_i = \left(\bar{c}_{i+1} - \bar{c}_i^{\mathrm{T}}\bar{c}_{i+1}\bar{c}_i\right)\frac{\arccos\left(\bar{c}_i^{\mathrm{T}}\bar{c}_{i+1}\right)}{\sqrt{1-\left(\bar{c}_i^{\mathrm{T}}\bar{c}_{i+1}\right)^2}}$, which we insert into (6) to get

$$\bar{c}(t) = \sum_{i=0}^{N-1} \chi_{[t_i,t_{i+1})}(t) \exp\left(\frac{t - t_i}{t_{i+1} - t_i}\frac{\arccos\left(\bar{c}_i^{\mathrm{T}}\bar{c}_{i+1}\right)}{\sqrt{1 - \left(\bar{c}_i^{\mathrm{T}}\bar{c}_{i+1}\right)^2}}\left(\bar{c}_{i+1}\bar{c}_i^{\mathrm{T}} - \bar{c}_i\bar{c}_{i+1}^{\mathrm{T}}\right)\right)\bar{c}_i. \tag{9}$$

The SRVT (4) of $\bar{c}(t)$ is a piecewise constant function $\bar{q}(t)$ in $\mathfrak{so}(3)$, taking values $\bar{q}_i = \bar{q}(t_i)$, $i = 0, ..., N-1$, where

$$\begin{aligned}
\bar{q}_i = \mathcal{R}(\bar{c}_i) &= \frac{a_{\bar{c}_i}(v_i)}{\|a_{\bar{c}_i}(v_i)\|^{\frac{1}{2}}} = \frac{v_i\bar{c}_i^{\mathrm{T}} - \bar{c}_i v_i^{\mathrm{T}}}{\|v_i\bar{c}_i^{\mathrm{T}} - \bar{c}_i v_i^{\mathrm{T}}\|^{\frac{1}{2}}} \\
&= \frac{\arccos^{\frac{1}{2}}\left(\bar{c}_i^{\mathrm{T}}\bar{c}_{i+1}\right)}{\left(1 - \left(\bar{c}_i^{\mathrm{T}}\bar{c}_{i+1}\right)^2\right)^{\frac{1}{4}}\|\bar{c}_{i+1}\bar{c}_i^{\mathrm{T}} - \bar{c}_i\bar{c}_{i+1}^{\mathrm{T}}\|^{\frac{1}{2}}}\left(\bar{c}_{i+1}\bar{c}_i^{\mathrm{T}} - \bar{c}_i\bar{c}_{i+1}^{\mathrm{T}}\right)
\end{aligned}$$

The inverse of this SRVT is given by (9), with the discrete points found as in the Lie group case by $\bar{c}_{i+1} = \exp(\|\bar{q}_i\|\bar{q}_i)\bar{c}_i$ and $\bar{c}_0 = c(t_0)$.

As an alternative, we define the reductive SRVT [3] by

$$\mathcal{R}_{\mathfrak{m}}(\bar{c}_i) := \mathcal{R}([U, U^{\perp}]_i^{\mathrm{T}} \bar{c}_i),$$

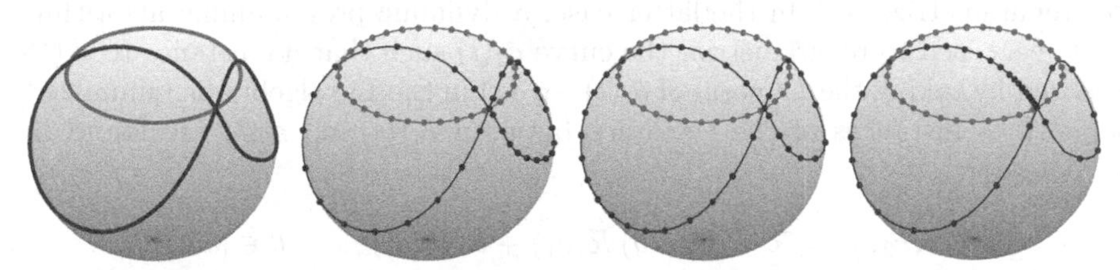

(a) From left to right: Two curves on the sphere, their original parametrizations, the reparametrization minimizing the distance in SO(3) and the reparametrization minimizing the distance in S^2, using the reductive SRVT.

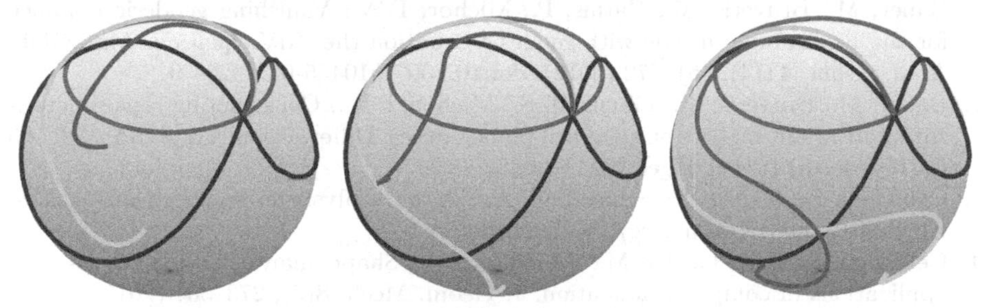

(b) The interpolated curves at times $\theta = \left\{\frac{1}{4}, \frac{1}{2}, \frac{3}{4}\right\}$, from left to right, before reparametrization, on S^2 (blue line) and SO(3) (yellow line).

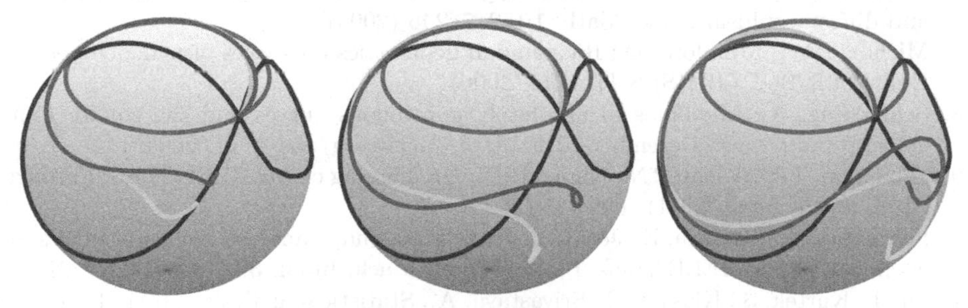

(c) The interpolated curves at times $\theta = \left\{\frac{1}{4}, \frac{1}{2}, \frac{3}{4}\right\}$, from left to right, after reparametrization, on S^2 (blue line) and SO(3) (yellow line).

Fig. 2. Interpolation between two curves on S^2, with and without reparametrization, obtained by the reductive SRVT. The results are compared to the corresponding SRVT interpolation between curves on SO(3). The SO(3) curves are mapped to S^2 by multiplying with the vector $(0, 1, 1)^{\mathrm{T}}/\sqrt{2}$. (Colour figure online)

where $[U, U^{\perp}]_{i+1} = \exp(a_{\bar{c}_i}(v_i))[U, U^{\perp}]_i$ for $i = 0, ..., N-1$, and $[U, U^{\perp}]_0$ can be found e.g. by QR-factorization of $c(t_0)$.

In Fig. 2 we show instants of the computed geodesic in the shape space of curves on the sphere between two curves $\bar{c}_1$ and $\bar{c}_2$, using the reductive SRVT. We compare this to the geodesic between the curves $\bar{z}_1$ and $\bar{z}_2$ in SO(3) which when mapped to S^2 gives $\bar{c}_1$ and $\bar{c}_2$. We show the results obtained before and after reparametrization. In the latter case, a dynamic programming algorithm, see [9], was used to reparametrize the curve $\bar{c}_2(t)$ such that its distance to $\bar{c}_1(t)$, measured by taking the L^2 norm of $\bar{q}_1(t) - \bar{q}_2(t)$ in the Lie algebra, is minimized. The various instances of the geodesics between $\bar{c}_1(t)$ and $\bar{c}_2(t)$ are found by interpolation,

$$\bar{c}_{\text{int}}(\bar{c}_1, \bar{c}_2, \theta) = \mathcal{R}^{-1}\left((1-\theta)\,\mathcal{R}(\bar{c}_1) + \theta\,\mathcal{R}(\bar{c}_2)\right), \qquad \theta \in [0, 1].$$

References

1. Bauer, M., Bruveris, M., Harms, P., Michor, P.W.: Vanishing geodesic distance for the Riemannian metric with geodesic equation the KdV-equation. Ann. Glob. Anal. Geom. **41**(4), 461–472 (2012). doi:10.1007/s10455-011-9294-9
2. Bauer, M., Bruveris, M., Marsland, S., Michor, P.W.: Constructing reparameterization invariant metrics on spaces of plane curves. Differ. Geom. Appl. **34**, 139–165 (2014). doi:10.1016/j.difgeo.2014.04.008
3. Celledoni, E., Eidnes, S., Schmeding, A.: Shape analysis on homogeneous spaces, April 2017. http://arxiv.org/abs/1704.01471v1
4. Celledoni, E., Eslitzbichler, M., Schmeding, A.: Shape analysis on Lie groups with applications in computer animation. J. Geom. Mech. **8**(3), 273–304 (2016)
5. Le Brigant, A.: Computing distances and geodesics between manifold-valued curves in the SRV framework (2016). https://arxiv.org/abs/1601.02358
6. Michor, P.W., Mumford, D.: Vanishing geodesic distance on spaces of submanifolds and diffeomorphisms. Doc. Math. **10**, 217–245 (2005)
7. Michor, P.W., Mumford, D.: Riemannian geometries on spaces of plane curves. J. Eur. Math. Soc. (JEMS) **8**(1), 1–48 (2006)
8. Schmeding, A.: Manifolds of absolutely continuous curves and the square root velocity framework, December 2016. http://arxiv.org/abs/1612.02604v1
9. Sebastian, T.B., Klein, P.N., Kimia, B.B.: On aligning curves. IEEE Trans. Pattern Anal. Mach. Intell. **25**(1), 116–125 (2003)
10. Srivastava, A., Klassen, E., Joshi, S., Jermyn, I.: Shape analysis of elastic curves in euclidean spaces. IEEE Trans. Pattern Anal. Mach. Intell. **33**, 1415–1428 (2011)
11. Su, J., Kurtek, S., Klassen, E., Srivastava, A.: Statistical analysis of trajectories on Riemmannian manifolds: bird migration, hurricane tracking and video surveillance. Ann. Appl. Stat. **8**(2), 530–552 (2014)

Optimal Matching Between Curves in a Manifold

Alice Le Brigant[1,2(✉)], Marc Arnaudon[1], and Frédéric Barbaresco[2]

[1] Institut Mathématique de Bordeaux, UMR 5251, Université de Bordeaux and CNRS, Bordeaux, France
alice.lebrigant@gmail.com

[2] Thales Air System, Surface Radar Domain, Technical Directorate, Voie Pierre-Gilles de Gennes, 91470 Limours, France

Abstract. This paper is concerned with the computation of an optimal matching between two manifold-valued curves. Curves are seen as elements of an infinite-dimensional manifold and compared using a Riemannian metric that is invariant under the action of the reparameterization group. This group induces a quotient structure classically interpreted as the "shape space". We introduce a simple algorithm allowing to compute geodesics of the quotient shape space using a canonical decomposition of a path in the associated principal bundle. We consider the particular case of elastic metrics and show simulations for open curves in the plane, the hyperbolic plane and the sphere.

Keywords: Optimal matching · Manifold-valued curves · Elastic metric

1 Introduction

A popular way to compare shapes of curves is through a Riemannian framework. The set of curves is seen as an infinite-dimensional manifold on which acts the group of reparameterizations, and is equipped with a Riemannian metric G that is invariant with respect to the action of that group. Here we consider the set of open oriented curves in a Riemannian manifold $(M, \langle\cdot,\cdot\rangle)$ with velocity that never vanishes, i.e. smooth immersions,

$$\mathcal{M} = \mathrm{Imm}([0,1], M) = \{c \in C^\infty([0,1], M) : c'(t) \neq 0 \ \forall t \in [0,1]\}.$$

It is an open submanifold of the Fréchet manifold $C^\infty([0,1], M)$ and its tangent space at a point c is the set of infinitesimal vector fields along the curve c in M,

$$T_c\mathcal{M} = \{w \in C^\infty([0,1], TM) : w(t) \in T_{c(t)}M \ \forall t \in [0,1]\}.$$

A curve c can be reparametrized by right composition $c \circ \varphi$ with an increasing diffeomorphism $\varphi : [0,1] \to [0,1]$, the set of which is denoted by $\mathrm{Diff}^+([0,1])$. We consider the quotient space $\mathcal{S} = \mathcal{M}/\mathrm{Diff}^+([0,1], M)$, interpreted as the space of "shapes" or "unparameterized curves". If we restrict ourselves to elements of $\mathcal{M}$ on which the diffeomorphism group acts freely, then we obtain a principal bundle

F. Nielsen and F. Barbaresco (Eds.): GSI 2017, LNCS 10589, pp. 57–64, 2017.
https://doi.org/10.1007/978-3-319-68445-1_7

$\pi : \mathcal{M} \rightarrow \mathcal{S}$, the fibers of which are the sets of all the curves that are identical modulo reparameterization, i.e. that project on the same "shape" (Fig. 1). We denote by $\bar{c} := \pi(c) \in \mathcal{S}$ the shape of a curve $c \in \mathcal{M}$. Any tangent vector $w \in T_c\mathcal{M}$ can then be decomposed as the sum of a vertical part $w^{ver} \in \mathrm{Ver}_c$, that has an action of reparameterizing the curve without changing its shape, and a horizontal part $w^{hor} \in \mathrm{Hor}_c = (\mathrm{Ver}_c)^{\perp_G}$, G-orthogonal to the fiber,

$$T_c\mathcal{M} \ni w = w^{ver} + w^{hor} \in \mathrm{Ver}_c \oplus \mathrm{Hor}_c,$$
$$\mathrm{Ver}_c = \ker T_c\pi = \{mv := mc'/|c'| : \; m \in C^\infty([0,1],\mathbb{R}), m(0) = m(1) = 0\},$$
$$\mathrm{Hor}_c = \{h \in T_c\mathcal{M} : \; G_c(h, mv) = 0, \; \forall m \in C^\infty([0,1],\mathbb{R}), m(0) = m(1) = 0\}.$$

If we equip $\mathcal{M}$ with a Riemannian metric $G_c : T_c\mathcal{M} \times T_c\mathcal{M} \rightarrow \mathbb{R}$, $c \in \mathcal{M}$, that is constant along the fibers, i.e. such that

$$G_{c\circ\varphi}(w \circ \varphi, z \circ \varphi) = G_c(w, z), \quad \forall \varphi \in \mathrm{Diff}^+([0,1]), \tag{1}$$

then there exists a Riemannian metric $\bar{G}$ on the shape space $\mathcal{S}$ such that π is a Riemannian submersion from $(\mathcal{M}, G)$ to $(\mathcal{S}, \bar{G})$, i.e.

$$G_c(w^{hor}, z^{hor}) = \bar{G}_{\pi(c)}\left(T_c\pi(w), T_c\pi(z)\right), \quad \forall w, z \in T_c\mathcal{M}.$$

This expression defines $\bar{G}$ in the sense that it does not depend on the choice of the representatives c, w and z ([4], Sect. 29.21). If a geodesic for G has a horizontal initial speed, then its speed vector stays horizontal at all times - we say it is a horizontal geodesic - and projects on a geodesic of the shape space for $\bar{G}$ ([4], Sect. 26.12). The distance between two shapes for $\bar{G}$ is given by

$$\bar{d}\left(\overline{c_0}, \overline{c_1}\right) = \inf\left\{\, d\left(c_0, c_1 \circ \varphi\right) \mid \varphi \in \mathrm{Diff}^+([0,1]) \,\right\}.$$

Solving the boundary value problem in the shape space can therefore be achieved either through the construction of horizontal geodesics e.g. by minimizing the horizontal path energy [1,7], or by incorporating the optimal reparameterization of one of the boundary curves as a parameter in the optimization problem [2,6,8]. Here we introduce a simple algorithm that computes the horizontal geodesic linking an initial curve with fixed parameterization c_0 to the closest reparameterization $c_1 \circ \varphi$ of the target curve c_1. The optimal reparameterization φ yields what we will call an *optimal matching* between the curves c_0 and c_1.

2 The Optimal Matching Algorithm

We want to compute the geodesic path $s \mapsto \bar{c}(s)$ between the shapes of two curves c_0 and c_1, that is the projection $\bar{c} = \pi(c_h)$ of the horizontal geodesic $s \mapsto c_h(s)$ - if it exists - linking c_0 to the fiber of c_1 in $\mathcal{M}$, see Fig. 1. This horizontal path verifies $c_h(0) = c_0$, $c_h(1) \in \pi^{-1}(\overline{c_1})$ and $\partial c_h/\partial s(s) \in \mathrm{Hor}_{c_h(s)}$ for all $s \in [0,1]$. Its end point gives the optimal reparameterization $c_1 \circ \varphi$ of the target curve c_1 with respect to the initial curve c_0, i.e. such that

$$\bar{d}(\overline{c_0}, \overline{c_1}) = d(c_0, c_1 \circ \varphi) = d(c_0, c_h(1)).$$

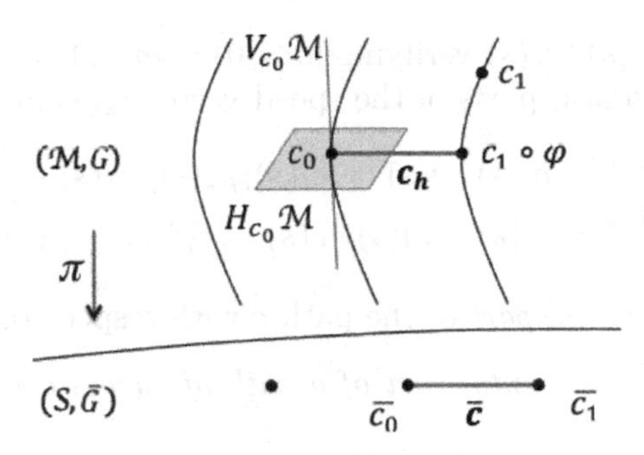

Fig. 1. Schematic representation of the shape bundle.

In all that follows we identify a path of curves $[0,1] \ni s \mapsto c(s) \in \mathcal{M}$ with the function of two variables $[0,1] \times [0,1] \ni (s,t) \mapsto c(s,t) \in M$ and denote by $c_s := \partial c/\partial s$ and $c_t := \partial c/\partial t$ its partial derivatives with respect to s and t. We decompose any path of curves $s \mapsto c(s)$ in $\mathcal{M}$ into a horizontal path reparameterized by a path of diffeomorphisms, i.e. $c(s) = c^{hor}(s) \circ \varphi(s)$ where $c_s^{hor}(s) \in \mathrm{Hor}_{c^{hor}(s)}$ and $\varphi(s) \in \mathrm{Diff}^+([0,1])$ for all $s \in [0,1]$. That is,

$$c(s,t) = c^{hor}(s, \varphi(s,t)) \quad \forall s,t \in [0,1]. \tag{2}$$

The horizontal and vertical parts of the speed vector of c can be expressed in terms of this decomposition. Indeed, by taking the derivative of (2) with respect to s and t we obtain

$$c_s(s) = c_s^{hor}(s) \circ \varphi(s) + \varphi_s(s) \cdot c_t^{hor}(s) \circ \varphi(s), \tag{3a}$$

$$c_t(s) = \varphi_t(s) \cdot c_t^{hor}(s) \circ \varphi(s), \tag{3b}$$

and so if $v^{hor}(s,t) := c_t^{hor}(s,t)/|c_t^{hor}(s,t)|$ denotes the normalized speed vector of c^{hor}, (3b) gives since $\varphi_t > 0$, $v(s) = v^{hor}(s) \circ \varphi(s)$. We can see that the first term on the right-hand side of Eq. (3a) is horizontal. Indeed, for any $m : [0,1] \to C^\infty([0,1], \mathbb{R})$ such that $m(s,0) = m(s,1) = 0$ for all s, since G is reparameterization invariant we have

$$\begin{aligned} G\left(c_s^{hor}(s) \circ \varphi(s),\, m(s) \cdot v(s)\right) &= G\left(c_s^{hor}(s) \circ \varphi(s),\, m(s) \cdot v^{hor}(s) \circ \varphi(s)\right) \\ &= G\left(c_s^{hor}(s),\, m(s) \circ \varphi(s)^{-1} \cdot v^{hor}(s)\right) \\ &= G\left(c_s^{hor}(s),\, \tilde{m}(s) \cdot v^{hor}(s)\right), \end{aligned}$$

with $\tilde{m}(s) = m(s) \circ \varphi(s)^{-1}$. Since $\tilde{m}(s,0) = \tilde{m}(s,1) = 0$ for all s, the vector $\tilde{m}(s) \cdot v^{hor}(s)$ is vertical and its scalar product with the horizontal vector $c_s^{hor}(s)$ vanishes. On the other hand, the second term on the right hand-side of Eq. (3a) is vertical, since it can be written

$$\varphi_s(s) \cdot c_t^{hor} \circ \varphi(s) = m(s) \cdot v(s),$$

with $m(s) = |c_t(s)|\varphi_s(s)/\varphi_t(s)$ verifying $m(s,0) = m(s,1) = 0$ for all s. Finally, the vertical and horizontal parts of the speed vector $c_s(s)$ are given by

$$c_s(s)^{ver} = m(s) \cdot v(s) = |c_t(s)|\varphi_s(s)/\varphi_t(s) \cdot v(s), \tag{4a}$$

$$c_s(s)^{hor} = c_s(s) - m(s) \cdot v(s) = c_s^{hor}(s) \circ \varphi(s). \tag{4b}$$

We call c^{hor} the *horizontal part* of the path c with respect to G.

Proposition 1. *The horizontal part of a path of curves c is at most the same length as c*

$$L_G(c^{hor}) \le L_G(c).$$

Proof. Since the metric G is reparameterization invariant, the squared norm of the speed vector of the path c at time $s \in [0,1]$ is given by, if $\|\cdot\|_G^2 := G(\cdot,\cdot)$,

$$\begin{aligned}\|c_s(s,\cdot)\|_G^2 &= \|c_s^{hor}(s,\varphi(s,\cdot))\|_G^2 + |\varphi_s(s,\cdot)|^2 \|c_t^{hor}(s,\varphi(s,\cdot)\|_G^2 \\ &= \|c_s^{hor}(s,\cdot)\|_G^2 + |\varphi_s(s,\cdot)|^2 \|c_t^{hor}(s,\cdot)\|_G^2,\end{aligned}$$

This gives $\|c_s^{hor}(s)\|_G \le \|c_s(s)\|$ for all s and so $L_G(c^{hor}) \le L_G(c)$.

Now we will see how the horizontal part of a path of curves can be computed.

Proposition 2 (Horizontal part of a path). *Let $s \mapsto c(s)$ be a path in $\mathcal{M}$. Then its horizontal part is given by $c^{hor}(s,t) = c(s, \varphi(s)^{-1}(t))$, where the path of diffeomorphisms $s \mapsto \varphi(s)$ is solution of the PDE*

$$\varphi_s(s,t) = m(s,t)/|c_t(s,t)| \cdot \varphi_t(s,t), \tag{5}$$

with initial condition $\varphi(0,\cdot) = Id$, and where $m(s) : [0,1] \to \mathbb{R}$, $t \mapsto m(s,t) := |c_s^{ver}(s,t)|$ is the vertical component of $c_s(s)$.

Proof. This is a direct consequence of Eq. (4a), which states that the vertical part of $c_s(s)$ is $m(s) \cdot v(s)$ where $m(s) = |c_t(s)|\varphi_s(s)/\varphi_t(s)$.

If we take the horizontal part of the geodesic linking two curves c_0 and c_1, we will obtain a horizontal path linking c_0 to the fiber of c_1 which will no longer be a geodesic path. However this path reduces the distance between c_0 and the fiber of c_1, and gives a "better" representative $\tilde{c}_1 = c_1 \circ \varphi(1)$ of the target curve. By computing the geodesic between c_0 and this new representative $\tilde{c}_1$, we are guaranteed to reduce once more the distance to the fiber. The algorithm that we propose simply iterates these two steps and is detailed in Algorithm 1.

3 Example: Elastic Metrics

In this section we consider the particular case of the two-parameter family of elastic metrics, introduced for plane curves by Mio et al. in [5]. We denote by ∇ the Levi-Civita connection of the Riemannian manifold M, and by $\nabla_t w := \nabla_{c_t} w$, $\nabla_t^2 w := \nabla_{c_t}\nabla_{c_t} w$ the first and second order covariant derivatives of a vector field

Data: $c_0, c_1 \in \mathcal{M}$
Result: $\tilde{c}_1$
Set $\tilde{c}_1 \leftarrow c_1$ and Gap $\leftarrow 2 \times$ Threshold;
while *Gap* $>$ *Threshold* **do**
  construct the geodesic $s \mapsto c(s)$ between c_0 and $\tilde{c}_1$;
  compute the horizontal part $s \mapsto c^{hor}(s)$ of c;
  set Gap $\leftarrow \text{dist}_{L^2}\left(c^{hor}(1), \tilde{c}_1\right)$ and $\tilde{c}_1 \leftarrow c^{hor}(1)$;
end

Algorithm 1. Optimal matching.

w along a curve c of parameter t. For manifold-valued curves, elastic metrics can be defined for any $c \in T_c\mathcal{M}$ and $w, z \in T_c\mathcal{M}$ by

$$G_c^{a,b}(w, z) = \langle w(0), z(0) \rangle + \int_0^1 \left(a^2 \langle \nabla_\ell w^N, \nabla_\ell z^N \rangle + b^2 \langle \nabla_\ell w^T, \nabla_\ell z^T \rangle\right) \mathrm{d}\ell, \quad (6)$$

where $\mathrm{d}\ell = |c'(t)|\mathrm{d}t$ and $\nabla_\ell = \frac{1}{|c'(t)|}\nabla_t$ respectively denote integration and covariant derivation according to arc length. In the following section, we will show simulations for the special case $a = 1$ and $b = 1/2$: for this choice of coefficients, the geodesic equations are easily numerically solved [3] if we adopt the so-called square root velocity representation [6], in which each curve is represented by the pair formed by its starting point and speed vector renormalized by the square root of its norm. Let us characterize the horizontal subspace for $G^{a,b}$, and give the decomposition of a tangent vector.

Proposition 3 (Horizontal part of a vector for an elastic metric). *Let $c \in \mathcal{M}$ be a smooth immersion. A tangent vector $h \in T_c\mathcal{M}$ is horizontal for the elastic metric* (6) *if and only if it verifies the ordinary differential equation*

$$\left((a/b)^2 - 1\right) \langle \nabla_t h, \nabla_t v \rangle - \langle \nabla_t^2 h, v \rangle + |c'|^{-1} \langle \nabla_t c', v \rangle \langle \nabla_t h, v \rangle = 0. \quad (7)$$

The vertical and horizontal parts of a tangent vector $w \in T_c\mathcal{M}$ are given by

$$w^{ver} = mv, \quad w^{hor} = w - mv,$$

where the real function $m \in C^\infty([0,1], \mathbb{R})$ verifies $m(0) = m(1) = 0$ and

$$\begin{aligned} &m'' - \langle \nabla_t c'/|c'|, v \rangle m' - (a/b)^2 |\nabla_t v|^2 m \quad (8) \\ &\quad = \langle \nabla_t \nabla_t w, v \rangle - \left((a/b)^2 - 1\right) \langle \nabla_t w, \nabla_t v \rangle - \langle \nabla_t c'/|c'|, v \rangle \langle \nabla_t w, v \rangle. \end{aligned}$$

Proof. Let $h \in T_c\mathcal{M}$ be a tangent vector. It is horizontal if and only if it is orthogonal to any vertical vector, that is any vector of the form mv with $m \in C^\infty([0,1], \mathbb{R})$ such that $m(0) = m(1) = 0$. We have $\nabla_t(mv) = m'v + m\nabla_t v$ and since $\langle \nabla_t v, v \rangle = 0$ we get $\nabla_t(mv)^N = m\nabla_t v$ and $\nabla_t(mv)^T = m'v$. Since

$m(0) = 0$ the non integral part vanishes and the scalar product is written

$$G_c^{a,b}(h, mv) = \int_0^1 \left(a^2 m\langle\nabla_t h, \nabla_t v\rangle + b^2 m'\langle\nabla_t h, v\rangle\right) |c'|^{-1}\mathrm{d}t$$

$$= \int_0^1 a^2 m\langle\nabla_t h, \nabla_t v\rangle|c'|^{-1}\mathrm{d}t - \int_0^1 b^2 m \frac{d}{dt}(\langle\nabla_t h, v\rangle|c'|^{-1})\mathrm{d}t$$

$$= \int_0^1 m/|c'|\Big((a^2 - b^2)\langle\nabla_t h, \nabla_t v\rangle - b^2\langle\nabla_t\nabla_t h, v\rangle + b^2\langle\nabla_t c', v\rangle\langle\nabla_t h, v\rangle|c'|^{-1}\Big)\mathrm{d}t,$$

where we used integration by parts. The vector h is horizontal if and only if $G_c^{a,b}(h, mv) = 0$ for all such m, and so we obtain the desired equation. Now consider a tangent vector w and a real function $m : [0,1] \to \mathbb{R}$ such that $m(0) = m(1) = 0$. Then $w - mv$ is horizontal if and only if it verifies the ODE (7). Noticing that $\langle\nabla_t v, v\rangle = 0$, $\langle\nabla_t\nabla_t v, v\rangle = -|\nabla_t v|^2$ and $\nabla_t\nabla_t(mv) = m''v + 2m'\nabla_t v + m\nabla_t\nabla_t v$, we easily get the desired equation.

This allows us to characterize the horizontal part of a path of curves for $G^{a,b}$.

Proposition 4 (Horizontal part of a path for an elastic metric). *Let $s \mapsto c(s)$ be a path in $\mathcal{M}$. Then its horizontal part is given by $c^{hor}(s,t) = c(s, \varphi(s)^{-1}(t))$, where the path of diffeomorphisms $s \mapsto \varphi(s)$ is solution of the PDE*

$$\varphi_s(s,t) = m(s,t)/|c_t(s,t)| \cdot \varphi_t(s,t), \tag{9}$$

with initial condition $\varphi(0,\cdot) = Id$, and where $m(s) : [0,1] \to \mathbb{R}$, $t \mapsto m(s,t)$ is solution for all s of the ODE

$$\begin{aligned} &m_{tt} - \langle\nabla_t c_t/|c_t|, v\rangle m_t - (a/b)^2|\nabla_t v|^2 m \\ &\quad = \langle\nabla_t\nabla_t c_s, v\rangle - \left((a/b)^2 - 1\right)\langle\nabla_t c_s, \nabla_t v\rangle - \langle\nabla_t c_t/|c_t|, v\rangle\langle\nabla_t c_s, v\rangle. \end{aligned} \tag{10}$$

Proof. This is a direct consequence of Propositions 2 and 3.

We numerically solve the PDE of Proposition 4 using Algorithm 2.

Data: path of curves $s \mapsto c(s)$
Result: path of diffeomorphisms $s \mapsto \varphi(s)$
for $k = 1$ **to** n **do**
  estimate the derivative $\varphi_t(\frac{k}{n}, \cdot)$;
  solve ODE (10) using a finite difference method to obtain $m(\frac{k}{n}, \cdot)$;
  set $\varphi_s(\frac{k}{n}, t) \leftarrow m(\frac{k}{n}, t)/|c_t(\frac{k}{n}, t)| \cdot \varphi_t(\frac{k}{n}, t)$ for all t;
  propagate $\varphi(\frac{k+1}{n}, t) \leftarrow \varphi(\frac{k}{n}, t) + \frac{1}{n}\varphi_s(\frac{k}{n}, t)$ for all t;
end

Algorithm 2. Decomposition of a path of curves.

4 Simulations

We test the optimal matching algorithm for the elastic metric with parameters $a = 2b = 1$ - for which all the formulas and tools to compute geodesics are available [3] - and for curves in the plane, the hyperbolic half-plane $\mathbb{H}^2$ and the sphere $\mathbb{S}^2$. The curves are discretized and geodesics are computed using a discrete geodesic shooting method presented in detail in [3]. Useful formulas and algorithms in $\mathbb{H}^2$ and $\mathbb{S}^2$ are available in [3] and [8] respectively. Figure 2 shows results of the optimal matching algorithm for a pair of segments in $\mathbb{H}^2$. We consider 5 different combinations of parameterizations of the two curves, always fixing the parameterization of the curve on the left-hand side while searching for the optimal reparameterization of the curve on the right-hand side. On the top row, the points are "evenly distributed" along the latter, and on the bottom row, along the former. For each set of parameterizations, the geodesic between the initial parameterized curves (more precisely, the trajectories taken by each point) is shown in blue, and the horizontal geodesic obtained as output of the optimal matching algorithm is shown in red. The two images on the bottom right corner show their superpositions, and their lengths are displayed in Table 1, in the same order as the corresponding images of Fig. 2. We can see that the horizontal geodesics redistribute the points along the right-hand side curve in a way that seems natural: similarly to the distribution of the points on the left curve. Their superposition shows that the underlying shapes of the horizontal geodesics are very similar, which is not the case of the initial geodesics. The horizontal geodesics are always shorter than the initial geodesics, as expected, and have always approximatively the same length. This common length is the distance between the shapes of the two curves. The same exercise can be carried out on spherical curves (Fig. 3) and on plane curves, for which we show the superposition of the geodesics and horizontal geodesics between different parameterizations in Fig. 4. The execution time varies from a few seconds to a few minutes, depending on the curves and the ambient space: the geodesics between plane curves are computed using explicit equations whereas for curves in a nonlinear manifold, we use a time-consuming geodesic shooting algorithm.

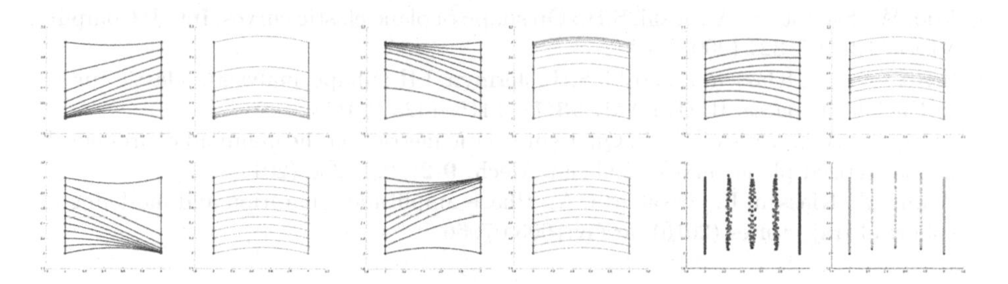

Fig. 2. Geodesics between parameterized curves (blue) and corresponding horizontal geodesics (red) in the hyperbolic half-plane, and their superpositions. (Color figure online)

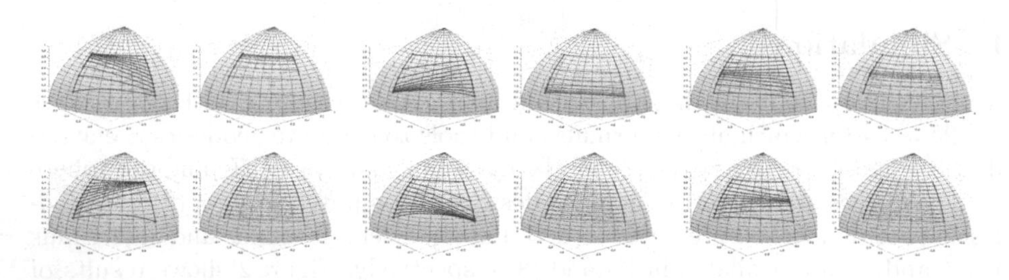

Fig. 3. Initial and horizontal geodesics between spherical parameterized curves.

Fig. 4. Superposition of the initial (blue) and horizontal (red) geodesics obtained for different sets of parameterizations of three pairs of plane curves. (Color figure online)

Table 1. Length of the geodesics of the hyperbolic half-plane shown in Fig. 2.

0.6287	0.5611	0.6249	0.5633	0.5798	0.5608
0.7161	0.5601	0.7051	0.5601		

References

1. Bauer, M., Harms, P., Michor, P.W.: Almost local metrics on shape space of hypersurfaces in n-space. SIAM J. Imaging Sci. **5**(1), 244–310 (2012)
2. Bauer, M., Bruveris, M., Harms, P., Møller-Andersen, J.: A numerical framework for sobolev metrics on the space of curves. SIAM J. Imaging Sci. **10**, 47–73 (2017)
3. Le Brigant, A.: Computing distances and geodesics between manifold-valued curves in the SRV framework. J. Geom. Mech. **9**(2), 131–156 (2017)
4. Michor, P.W.: Topics in differential geometry. In: Graduate Studies in Mathematics, vol. 93. American Mathematical Society, Providence (2008)
5. Mio, W., Srivastava, A., Joshi, S.H.: On shape of plane elastic curves. Int. J. Comput. Vision **73**, 307–324 (2007)
6. Srivastava, A., Klassen, E., Joshi, S.H., Jermyn, I.H.: Shape analysis of elastic curves in Euclidean spaces. IEEE PAMI **33**(7), 1415–1428 (2011)
7. Tumpach, A.B., Preston, S.C.: Quotient elastic metrics on the manifold of arc-length parameterized plane curves. J. Geom. Mech. **9**(2), 227–256 (2017)
8. Zhang, Z., Klassen, E., Srivastava, A.: Phase-amplitude separation and modeling of spherical trajectories (2016). arXiv:1603.07066

Optimization in the Space of Smooth Shapes

Kathrin Welker(✉)

Department of Mathematics, Trier University, 54286 Trier, Germany
welker@uni-trier.de

Abstract. The theory of shape optimization problems constrained by partial differential equations is connected with the differential-geometric structure of the space of smooth shapes.

1 Introduction

A lot of real world problems can be reformulated as shape optimization problems which are constrained by partial differential equations (PDEs), see, for instance, [6,16]. The subject of shape optimization is covered by several fundamental monographs (cf. [5,20]). In recent work, PDE constrained shape optimization problems are embedded in the framework of optimization on shape spaces. Finding a shape space and an associated metric is a challenging task and different approaches lead to various models. One possible approach is to define shapes as elements of a Riemannian manifold as proposed in [11]. In [12], a survey of various suitable inner products is given, e.g., the curvature weighted metric and the Sobolev metric. From a theoretical point of view this is attractive because algorithmic ideas from [1] can be combined with approaches from differential geometry. In [17], shape optimization is considered as optimization on a Riemannian shape manifold which contains smooth shapes, i.e., shapes with infinitely differentiable boundaries. We consider exactly this manifold in the following.

A well-established approach in shape optimization is to deal with shape derivatives in a so-called Hadamard form, i.e., in the form of integrals over the surface, as well as intrinsic shape metrics (cf. [13,20]). Major effort in shape calculus has been devoted towards such surface expressions (cf. [5,20]), which are often very tedious to derive. Along the way, volume formulations appear as an intermediate step. Recently, it has been shown that this intermediate formulation has numerical advantages, see, for instance, [3,6,14]. In [9], also practical advantages of volume shape formulations have been demonstrated. E.g., they require less smoothness assumptions. Furthermore, the derivation as well as the implementation of volume formulations require less manual and programming work. However, volume integral forms of shape derivatives require an outer metric on the domain surrounding the shape boundary. In [18], both points of view are harmonized by deriving a metric from an outer metric. Efficient shape optimization algorithms based on this metric, which reduce the analytical effort so far involved in the derivation of shape derivatives, are proposed in [18,19].

F. Nielsen and F. Barbaresco (Eds.): GSI 2017, LNCS 10589, pp. 65–72, 2017.
https://doi.org/10.1007/978-3-319-68445-1_8

The main aim of this paper is to explain how shape calculus can be combined with geometric concepts of the space of smooth shapes and to outline how this combination results in efficient optimization techniques. This paper reports on ongoing work and has the following structure. A short overview of basics concepts in shape optimization is given in Sect. 2. Afterwards, in Sect. 3, we do not only introduce the space of smooth shapes, but we also consider the surface and volume form of shape derivatives and summarize the way from shape derivatives to entire optimization algorithms in this shape space for each formulation.

2 A Brief Introduction in Shape Optimization

In this section, we set up notation and terminology of basic shape optimization concepts. For a detailed introduction into shape calculus, we refer to [5,20].

Shape optimization deals with shape functionals, which are defined as a functions $J\colon \mathcal{A} \to \mathbb{R}$, $\Omega \mapsto J(\Omega)$ with $\mathcal{A} \subset \{\Omega\colon \Omega \subset D\}$, where D denotes a non-empty subset of $\mathbb{R}^d$. One of the main focuses of shape optimization is to solve shape optimization problems. A shape optimization problem is given by $\min_\Omega J(\Omega)$, where J is a shape functional. When J depends on a solution of a PDE, we call the shape optimization problem PDE constrained. To solve PDE constrained shape optimization problems, we need their shape derivatives:

Let D be as above. Moreover, let $\{F_t\}_{t\in[0,T]}$ be a family of mappings $F_t\colon \overline{D} \to \mathbb{R}^d$ such that $F_0 = id$, where $\overline{D}$ denotes the closure of D and $T > 0$. This family transforms the domain Ω into new *perturbed domains* $\Omega_t := F_t(\Omega) = \{F_t(x)\colon x \in \Omega\}$ with $\Omega_0 = \Omega$ and the boundary Γ of Ω into new *perturbed boundaries* $\Gamma_t := F_t(\Gamma) = \{F_t(x)\colon x \in \Gamma\}$ with $\Gamma_0 = \Gamma$. Such a transformation can be described, e.g., by the *perturbation of identity*, which is defined by $F_t(x) := x + tV(x)$, where V denotes a sufficiently smooth vector field.

Definition 1. *Let $D \subset \mathbb{R}^d$ be open, where $d \geq 2$ is a natural number. Moreover, let $k \in \mathbb{N} \cup \{\infty\}$, let $\Omega \subset D$ be measurable and let Ω_t denote the perturbed domains defined above. The Eulerian derivative of a shape functional J at Ω in direction $V \in \mathcal{C}_0^k(D,\mathbb{R}^d)$ is defined by*

$$DJ(\Omega)[V] := \lim_{t\to 0^+} \frac{J(\Omega_t) - J(\Omega)}{t}. \quad (1)$$

If for all directions $V \in \mathcal{C}_0^k(D,\mathbb{R}^d)$ the Eulerian derivative (1) exists and the mapping $G(\Omega)\colon \mathcal{C}_0^k(D,\mathbb{R}^d) \to \mathbb{R}$, $V \mapsto DJ(\Omega)[V]$ is linear and continuous, the expression $DJ(\Omega)[V]$ is called the shape derivative of J at Ω in direction $V \in \mathcal{C}_0^k(D,\mathbb{R}^d)$. In this case, J is called shape differentiable of class $\mathcal{C}^k$ at Ω.

There are a lot of options to prove shape differentiability of shape functionals, e.g., the min-max approach [5], the chain rule approach [20], the Lagrange method of Céa [4] and the rearrangement method [7]. Note that there are cases where the method of Céa fails (cf. [15]). A nice overview about these approaches is given in [21].

In many cases, the shape derivative arises in two equivalent notational forms:

$$DJ_\Omega[V] := \int_\Omega RV(x)\,dx \qquad \text{(volume formulation)} \tag{2}$$

$$DJ_\Gamma[V] := \int_\Gamma r(s)\langle V(s), n(s)\rangle ds \qquad \text{(surface formulation)} \tag{3}$$

Here $r \in L^1(\Gamma)$ and R is a differential operator acting linearly on the vector field V with $DJ_\Omega[V] = DJ(\Omega)[V] = DJ_\Gamma[V]$. Surface expressions of shape derivatives are often very tedious to derive. Along the way, volume formulations appear as an intermediate step. These volume expressions are preferable over surface forms. This is not only because of saving analytical effort, but also due to additional regularity assumptions, which usually have to be required in order to transform volume into surface forms, as well as because of saving programming effort. In the next section, it is outlined how shape calculus and in particular surface as well as volume shape derivatives can be combined with geometric concepts of the space of smooth shapes.

3 Shape Calculus Combined with Geometric Concepts of the Space of Smooth Shapes

In this section, we analyze the connection of Riemannian geometry on the space of smooth shapes to shape optimization. Moreover, we summarize the way from shape derivatives to entire optimization algorithms in the space of smooth shapes for both, surface and volume shape derivative formulations.

First, we introduce the space of smooth shapes. In [11], the *set of all two-dimensional smooth shapes* is characterized by

$$B_e(S^1, \mathbb{R}^2) := \mathrm{Emb}(S^1, \mathbb{R}^2)/\mathrm{Diff}(S^1),$$

i.e., the orbit space of $\mathrm{Emb}(S^1, \mathbb{R}^2)$ under the action by composition from the right by the Lie group $\mathrm{Diff}(S^1)$. Here $\mathrm{Emb}(S^1, \mathbb{R}^2)$ denotes set of all embeddings from the unit circle S^1 into the plane $\mathbb{R}^2$ and $\mathrm{Diff}(S^1)$ is the set of all diffeomorphisms from S^1 into itself. In [8], it is proven that $B_e(S^1, \mathbb{R}^2)$ is a smooth manifold. For the sake of completeness it should be mentioned that the shape space $B_e(S^1, \mathbb{R}^2)$ together with appropriate inner products is even a Riemannian manifold. In [12], a survey of various suitable inner products is given. Note that the shape space $B_e(S^1, \mathbb{R}^2)$ and its theoretical results can be generalized to higher dimensions (cf. [10]). The tangent space $T_cB_e(S^1, \mathbb{R}^2)$ is isomorphic to the set of all smooth normal vector fields along $c \in B_e(S^1, \mathbb{R}^2)$, i.e.,

$$T_cB_e(S^1, \mathbb{R}^2) \cong \{h\colon h = \alpha n,\, \alpha \in \mathcal{C}^\infty(S^1)\}, \tag{4}$$

where n denotes the exterior unit normal field to the shape boundary c such that $n(\theta) \perp c_\theta(\theta)$ for all $\theta \in S^1$, where $c_\theta = \frac{\partial c}{\partial \theta}$ denotes the circumferential derivative.

Due to the Hadamard Structure Theorem given in [20, Theorem 2.27], there exists a scalar distribution r on the boundary Γ of the domain Ω under consideration. If we assume $r \in L^1(\Gamma)$, the shape derivative can be expressed on the boundary Γ of Ω (cf. (3)). The distribution r is often called the *shape gradient.* However, note that gradients depend always on chosen scalar products defined on the space under consideration. Thus, it rather means that r is the usual L^2-shape gradient. If we want to optimize on the shape manifold B_e, we have to find a representation of the shape gradient with respect to an appropriate inner product. This representation is called the *Riemannian shape gradient* and required to formulate optimization methods in B_e.

In order to deal with surface formulations of shape derivatives in optimization techniques, e.g., the Sobolev metric is an appropriate inner product. Of course, there are a lot of further metrics on B_e (cf. [12]), but the Sobolev metric is the most suitable choice for our applications. One reason for this is that the Riemannian shape gradient with respect to g^1 acts as a Laplace-Beltrami smoothing of the usual L^2-shape gradient (cf. Definition 2). Thus, in the following, we consider the first Sobolev metric g^1 on the shape space B_e. It is given by

$$g^1 \colon T_cB_e(S^1,\mathbb{R}^2) \times T_cB_e(S^1,\mathbb{R}^2) \to \mathbb{R},\ (h,k) \mapsto \int_{S^1} \langle (I - A\triangle_c)\alpha, \beta \rangle \, ds,$$

where $h = \alpha n$, $k = \beta n$ denote elements of the tangent space $T_cB_e(S^1,\mathbb{R}^2)$, $A > 0$ and $\triangle_c$ denotes the Laplace-Beltrami operator on the surface c. For the definition of the Sobolev metric g^1 in higher dimensions we refer to [2].

Now, we have to detail the Riemannian shape gradient with respect to g^1. The shape derivative can be expressed as

$$DJ_\Gamma[V] = \int_\Gamma \alpha r \, ds \tag{5}$$

if $V\big|_{\partial\Omega} = \alpha n$. In order to get an expression of the Riemannian shape gradient with respect to the Sobolev metric g^1, we look at the isomorphism (4). Due to this isomorphism, a tangent vector $h \in T_\Gamma B_e$ is given by $h = \alpha n$ with $\alpha \in \mathcal{C}^\infty(\Gamma)$. This leads to the following definition.

Definition 2. *The Riemannian shape gradient of a shape differentiable objective function J in terms of the Sobolev metric g^1 is given by*

$$\text{grad}(J) = qn \quad \textit{with} \quad (I - A\triangle_\Gamma)q = r,$$

where $\Gamma \in B_e$, $A > 0$, $q \in \mathcal{C}^\infty(\Gamma)$ and r is the L^2-shape gradient given in (3).

The Riemannian shape gradient is required to formulate optimization methods in the shape space B_e. In the setting of PDE constrained shape optimization problems, a Lagrange-Newton method is obtained by applying a Newton method to find stationary points of the Lagrangian of the optimization problem. In contrast to this method, which requires the Hessian in each iteration, quasi-Newton methods only need an approximation of the Hessian. Such an approximation is realized, e.g., by a limited memory Broyden-Fletcher-Goldfarb-Shanno (BFGS) update. In a limited-memory BFGS method, a representation of the shape gradient with respect to the Sobolev metric g^1 has to be computed and applied as a Dirichlet boundary condition in the linear elasticity mesh deformation. We refer to [17] for the limited-memory BFGS method in B_e. In Fig. 1, the entire optimization algorithm for the limited-memory BFGS case is summarized. Note that this method boils down to a steepest descent method by omitting the computation of the BFGS-update. This method only needs the gradient—but not the Hessian—in each iteration.

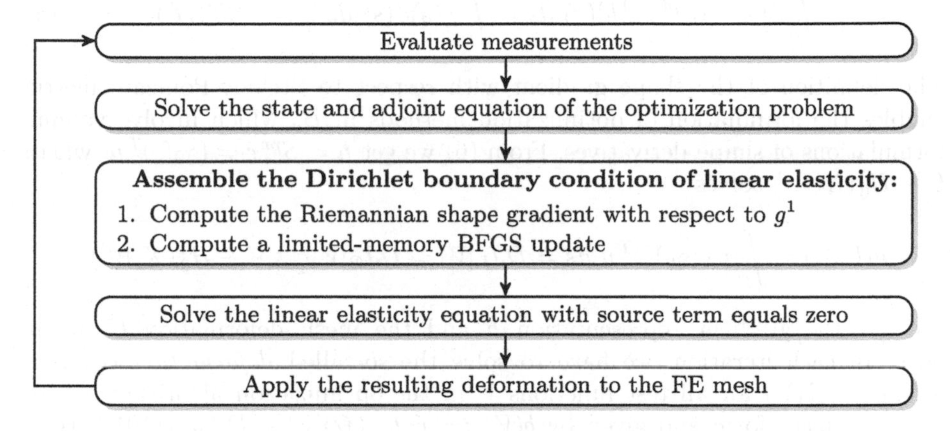

Fig. 1. Entire optimization algorithm based on surface expressions and g^1.

One possible approach to use volume formulations of shape derivatives is to consider Steklov-Poincaré metrics. In order to define these metrics, let us consider a compact domain $\Omega \subset X \subset \mathbb{R}^d$ with $\Omega \neq \emptyset$ and $\mathcal{C}^\infty$-boundary $\Gamma := \partial\Omega$, where X denotes a bounded domain with Lipschitz-boundary $\Gamma_{\text{out}} := \partial X$. In particular, this means $\Gamma \in B_e(S^{d-1}, \mathbb{R}^d)$. In this setting, the Steklov-Poincaré metric is defined by

$$g^S : H^{1/2}(\Gamma) \times H^{1/2}(\Gamma) \to \mathbb{R}, (\alpha, \beta) \mapsto \int_\Gamma \alpha(s) \cdot [(S^{pr})^{-1}\beta](s)\, ds.$$

Here S^{pr} denotes the projected Poincaré-Steklov operator which is given by

$$S^{pr} : H^{-1/2}(\Gamma_{\text{out}}) \to H^{1/2}(\Gamma_{\text{out}}),\ \alpha \mapsto (\gamma_0 U)^T n,$$

where $\gamma_0 : H_0^1(X, \mathbb{R}^d) \to H^{1/2}(\Gamma_{\text{out}}, \mathbb{R}^d)$, $U \mapsto U\big|_{\Gamma_{\text{out}}}$ and $U \in H_0^1(X, \mathbb{R}^d)$ solves the Neumann problem

$$a(U, V) = \int_{\Gamma_{\text{out}}} \alpha \cdot (\gamma_0 V)^T n\, ds \quad \forall V \in H_0^1(X, \mathbb{R}^d)$$

with $a(\cdot, \cdot)$ being a symmetric and coercive bilinear form.

Due to isomorphism (4) and expression (5), we can state the connection of B_e with respect to g^S to shape calculus:

Definition 3. *Let r denote the L^2-shape gradient given in (3). Moreover, let S^{pr} and γ_0 be as above. A representation $h \in T_\Gamma B_e \cong \mathcal{C}^\infty(\Gamma)$ of the shape gradient in terms of g^S is determined by*

$$g^S(\phi, h) = (r, \phi)_{L^2(\Gamma)} \quad \forall \phi \in \mathcal{C}^\infty(\Gamma),$$

which is equivalent to

$$\int_\Gamma \phi(s) \cdot [(S^{pr})^{-1}h](s)\, ds = \int_\Gamma r(s)\phi(s)\, ds \quad \forall \phi \in \mathcal{C}^\infty(\Gamma). \tag{6}$$

The definition of the shape gradient with respect to Steklov-Poincaré metric enables the formulation of optimization methods in B_e which involve volume formulations of shape derivatives. From (6) we get $h = S^{pr} r = (\gamma_0 U)^T n$, where $U \in H_0^1(X, \mathbb{R}^d)$ solves

$$a(U, V) = \int_\Gamma r \cdot (\gamma_0 V)^T n\, ds = DJ_\Gamma[V] = DJ_\Omega[V] \quad \forall V \in H_0^1(X, \mathbb{R}^d).$$

We get the gradient representation h and the mesh deformation U all at once. In each iteration, we have to solve the so-called *deformation equation* $a(U, V) = b(V)$ for all test functions V in the optimization algorithm, where $b(\cdot)$ is a linear form and given by $b(V) := DJ_{\text{vol}}(\Omega)[V] + DJ_{\text{surf}}(\Omega)[V]$. Here $J_{\text{surf}}(\Omega)$ denotes parts of the objective function leading to surface shape derivative expressions. It is incorporated as a Neumann boundary condition. Parts of the objective function leading to volume shape derivative expressions are denoted by $J_{\text{vol}}(\Omega)$. Note that from a theoretical point of view the volume and surface shape derivative formulations have to be equal to each other for all test functions. Thus, $DJ_{\text{vol}}[V]$ is assembled only for test functions V whose support includes Γ.

Figure 2 summarizes the entire optimization algorithm in the setting of the Steklov-Poincaré metric and, thus, in the case of volume shape derivative expressions. This algorithm is very attractive from a computational point of view. The computation of a representation of the shape gradient with respect to the chosen inner product of the tangent space is moved into the mesh deformation itself. The elliptic operator is used as an inner product and a mesh deformation. This leads to only one linear system, which has to be solved. In shape optimization one usually computes a descent direction as a deformation of the variable boundary. Note that this method also boils down to a steepest descent method by omitting the computation of the BFGS-update. In contrast to the algorithm based on the Sobolev metric (cf. Fig. 1), the metric used here interprets the descent direction as a volumetric force within the FE grid. For more details about this approach and in particular the implementation details we refer to [18,19].

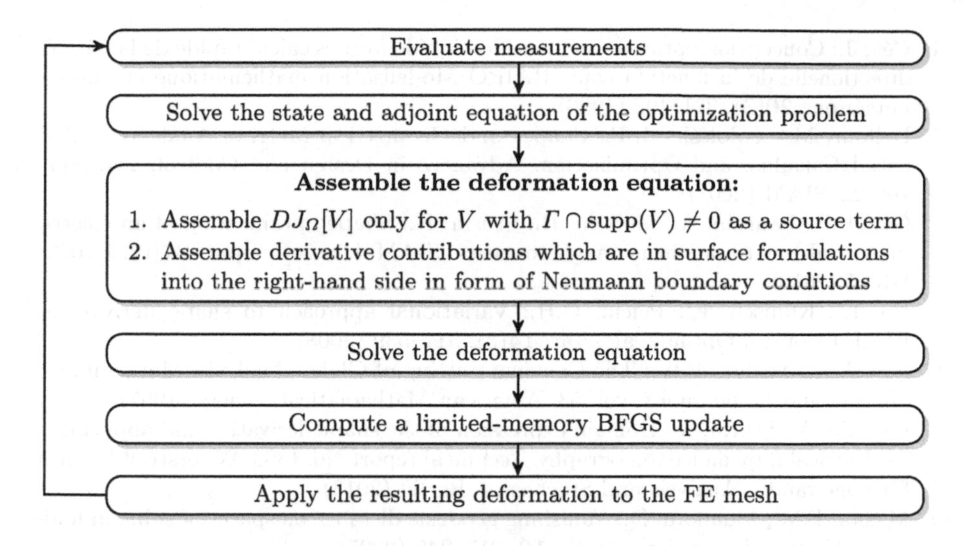

Fig. 2. Entire optimization algorithm based on volume formulations and g^S.

4 Conclusion

The differential-geometric structure of the space of smooth shapes is applied to the theory of PDE constrained shape optimization problems. In particular, a Riemannian shape gradient with respect to the Sobolev metric and the Steklov-Poincaré metric are defined. If we consider Sobolev metrics, we have to deal with surface formulations of shape derivatives. An intermediate and equivalent result in the process of deriving surface forms is the volume expression, which is preferable over surface forms. One possible approach to use volume forms is to consider Steklov-Poincaré metrics. The gradients with respect to both, g^1 and g^S, open the door to formulate optimization algorithms in the space of smooth shapes.

Acknowledgement. This work has been partly supported by the German Research Foundation within the priority program SPP 1962 under contract number Schu804/15-1.

References

1. Absil, P.A., Mahony, R., Sepulchre, R.: Optimization Algorithms on Matrix Manifolds. Princeton University Press, Princeton (2008)
2. Bauer, M., Harms, P., Michor, P.M.: Sobolev metrics on shape space of surfaces. J. Geom. Mech. **3**(4), 389–438 (2011)
3. Berggren, M.: A unified discrete–continuous sensitivity analysis method for shape optimization. In: Fitzgibbon, W., Kuznetsov, Y., Neittaanmäki, P., Périaux, J., Pironneau, O. (eds.) Applied and Numerical Partial Differential Equations. Computational Methods in Applied Sciences, vol. 15, pp. 25–39. Springer, Dordrecht (2010). doi:10.1007/978-90-481-3239-3_4

4. Céa, J.: Conception optimale ou identification de formes calcul rapide de la dérivée directionelle de la fonction coût. RAIRO Modelisation mathématique et analyse numérique **20**(3), 371–402 (1986)
5. Delfour, M.C., Zolésio, J.-P.: Shapes and Geometries: Metrics, Analysis, Differential Calculus, and Optimization. Advances in Design and Control, 2nd edn., vol. 22. SIAM (2001)
6. Gangl, P., Laurain, A., Meftahi, H., Sturm, K.: Shape optimization of an electric motor subject to nonlinear magnetostatics. SIAM J. Sci. Comput. **37**(6), B1002–B1025 (2015)
7. Ito, K., Kunisch, K., Peichl, G.H.: Variational approach to shape derivatives. ESAIM Control Optim. Calc. Var. **14**(3), 517–539 (2008)
8. Kriegl, A., Michor, P.W.: The Convient Setting of Global Analysis. Mathematical Surveys and Monographs, vol. 53. American Mathematical Society (1997)
9. Laurain, A., Sturm, K.: Domain expression of the shape derivative and application to electrical impedance tomography. Technical report No. 1863, Weierstraß-Institut für angewandte Analysis und Stochastik, Berlin (2013)
10. Michor, P.M., Mumford, D.: Vanishing geodesic distance on spaces of submanifolds and diffeomorphisms. Doc. Math. **10**, 217–245 (2005)
11. Michor, P.M., Mumford, D.: Riemannian geometries on spaces of plane curves. J. Eur. Math. Soc. **8**(1), 1–48 (2006)
12. Michor, P.M., Mumford, D.: An overview of the Riemannian metrics on spaces of curves using the Hamiltonian approach. Appl. Comput. Harmon. Anal. **23**(1), 74–113 (2007)
13. Mohammadi, B., Pironneau, O.: Applied Shape Optimization for Fluids. Oxford University Press, Oxford (2001)
14. Paganini, A.: Approximate shape gradients for interface problems. In: Pratelli, A., Leugering, G. (eds.) New Trends in Shape Optimization. ISNM, vol. 166, pp. 217–227. Springer, Cham (2015). doi:10.1007/978-3-319-17563-8_9
15. Pantz, O.: Sensibilitè de l'èquation de la chaleur aux sauts de conductivitè. Comptes Rendus Mathematique de l'Académie des Sciences **341**(5), 333–337 (2005)
16. Schmidt, S., Wadbro, E., Berggren, M.: Large-scale three-dimensional acoustic horn optimization. SIAM J. Sci. Comput. **38**(6), B917–B940 (2016)
17. Schulz, V.H., Siebenborn, M., Welker, K.: Structured inverse modeling in parabolic diffusion problems. SIAM J. Control Optim. **53**(6), 3319–3338 (2015)
18. Schulz, V.H., Siebenborn, M., Welker, K.: Efficient PDE constrained shape optimization based on Steklov-Poincaré type metrics. SIAM J. Optim. **26**(4), 2800–2819 (2016)
19. Siebenborn, M., Welker, K.: Computational aspects of multigrid methods for optimization in shape spaces. SIAM J. Sci. Comput. (2016). Submitted
20. Sokolowski, J., Zolésio, J.-P.: Introduction to Shape Optimization. Computational Mathematics, vol. 16. Springer, Heidelberg (1992)
21. Sturm, K.: Shape differentiability under non-linear PDE constraints. In: Pratelli, A., Leugering, G. (eds.) New Trends in Shape Optimization. ISNM, vol. 166, pp. 271–300. Springer, Cham (2015). doi:10.1007/978-3-319-17563-8_12

Surface Matching Using Normal Cycles

Pierre Roussillon(✉) and Joan Alexis Glaunès

MAP5, UMR 8145 CNRS,
Université Paris Descartes, Sorbonne Paris Cité, Paris, France
{pierre.roussillon,alexis.glaunes}@parisdescartes.fr

Abstract. In this article we develop in the case of triangulated meshes the notion of normal cycle as a dissimilarity measure introduced in [13]. Our construction is based on the definition of kernel metrics on the space of normal cycles which take explicit expressions in a discrete setting. We derive the computational setting for discrete surfaces, using the Large Deformation Diffeomorphic Metric Mapping framework as model for deformations. We present experiments on real data and compare with the varifolds approach.

1 Introduction

The field of computational anatomy focuses on the analysis of datasets composed of anatomical shapes through the action of deformations on these shapes. The key algorithm in this framework is the estimation of an optimal deformation which matches any two given shapes. This problem is most of the time formulated as the minimization of a functional composed of two terms. The first one is an energy term which enforces the regularity of the deformation. The second one is a data-fidelity term which measures a remaining distance between the deformed shape and the target. This data attachment term is of importance since it drives the registration and relaxes the constraint of exact matching.

In the case of shapes given as curves or surfaces, a framework based on currents have been developed in [8, 14] to provide a satisfying data attachment term, which does not necessitate point correspondences. However, currents are sensitive to orientation, and consequently insensitive to high curvature points of the shapes, which can lead to incorrect matchings of these points, or boundaries of shapes. To overcome this drawback, the *varifold* representation of shapes was introduced in [4]. Such a representation is orientation-free, and thus overcomes the difficulties experienced with currents. In [13], we developed a new data-attachment term using the theory of normal cycles. The normal cycle of a shape is the current associated with its normal bundle. It is orientation-free and encodes curvature information of the shape. The general framework have been set in [13] as well as the application to three dimensional curves.

In this article, we extend this framework to the case of surfaces. Section 2 focuses on the description of the normal bundle for a triangulation. When this description is done, we will introduce kernel metrics on normal cycles, so that we

F. Nielsen and F. Barbaresco (Eds.): GSI 2017, LNCS 10589, pp. 73–80, 2017.
https://doi.org/10.1007/978-3-319-68445-1_9

have an explicit distance between shapes represented as normal cycles (Sect. 3). In Sect. 4 we present some results of surface matching using the Large Deformation Diffeomorphic Metric Mapping (LDDMM) framework and kernel metrics on normal cycles. We illustrate the properties of a matching with normal cycles, as well as some limitations. Using parallel computations, we are able to provide examples on real data with a large number of points (around 6000 for each shape).

2 Normal Cycle of a Triangulated Mesh

Normal Cycles. This section requires basics knowledge about currents. The interested reader can see [5,7] for an approach in the field of computational anatomy. Moreover, we only very briefly remind the mathematical notion of normal cycle in this section, one should refer to [13] for a more extensive presentation.

Normal cycles are defined for sets with *positive reach* (see [6] for the original definition). For such a set $X \in \mathbb{R}^d$, one can consider its normal bundle $\mathcal{N}_X$, which is the set of all (x, n), $x \in X$, with n unit normal vector at point x. Here the notion of normal vector is considered in a generalized sense (see again [6]). For a given point $x \in X$, we denote $\mathrm{Nor}^u(X, x)$ all the unit normal vectors of the set X at this point. In the following, we denote $\Lambda^{d-1}(\mathbb{R}^d \times \mathbb{R}^d)$ the space of $(d-1)$-vectors in $\mathbb{R}^d \times \mathbb{R}^d$ and $\Omega_0^{d-1}(\mathbb{R}^d \times S^{d-1}) := \mathcal{C}^0(\mathbb{R}^d \times S^{d-1}, \Lambda^{d-1}(\mathbb{R}^d \times \mathbb{R}^d)^*)$ the space of continuous $(d-1)$-differential forms of $\mathbb{R}^d \times S^{d-1}$ vanishing at infinity, endowed with the supremum norm.

Definition 1 (Normal cycle). *The normal cycle of a positive reach set $X \subset \mathbb{R}^d$ is the $(d-1)$-current associated with $\mathcal{N}_X$ with its canonical orientation (independent of any orientation of X). For any differential form $\omega \in \Omega_0^{d-1}(\mathbb{R}^d \times S^{d-1})$, one has:*

$$N(X)(\omega) := [\mathcal{N}_X](\omega) = \int_{\mathcal{N}_X} \omega_{(x,n)}(\tau_{\mathcal{N}_X}(x, n)) d\mathcal{H}^{d-1}(x, n) \tag{1}$$

The theory of normal cycles can be extended to the case of finite unions of sets with positive reach, as done in [15], with the use of the following additive property:

$$N(C \cup S) := N(C) + N(S) - N(C \cap S) \tag{2}$$

This allows to define normal cycles for a very large class of subsets, in particular for unions of triangles, which will be used in our discrete model.

Normal Cycle of a Triangle. Consider a single triangle T, with vertices x_1, x_2, x_3 and edges: $f_1 = x_2 - x_1$, $f_2 = x_3 - x_2$, $f_3 = x_1 - x_3$. The normal vectors of the face are: $n_T = \frac{f_1 \times f_2}{|f_1 \times f_2|}$ and $-n_T$. The description of the normal bundle of a triangle is quite straightforward. As illustrated in Fig. 1, it can be decomposed into a planar part, composed of two triangles, a cylindrical part, composed of three "half" cylinders located at the edges, and a spherical part, composed of three portions of sphere located at the vertices:

$$\mathcal{N}_T^p := \cup_{x \in T \setminus \partial T} \mathrm{Nor}^u(T, x) = T \times \{-n_T, n_t\},$$
$$\mathcal{N}_T^c := \cup_{i=1}^3 [x_i, x_{i+1}] \times S^{\perp +}_{f_i, f_i \times n_T},$$
$$\mathcal{N}_T^s := \cup_{i=1}^3 \{x_i\} \times S^{+}_{f_{i-1}, -f_{i+1}},$$

where for any non zero vectors $\alpha, \beta \in \mathbb{R}^3$, we denote the semicircle $S^{\perp +}_{\alpha,\beta} = (S^2 \cap \alpha^\perp) \cap \{u | \langle u, \beta \rangle \geq 0\}$, and the portion of sphere $S^{+}_{\alpha,\beta} := \{u \in S^2, \langle u, \alpha \rangle \geq 0, \langle u, \beta \rangle \geq 0\}$.

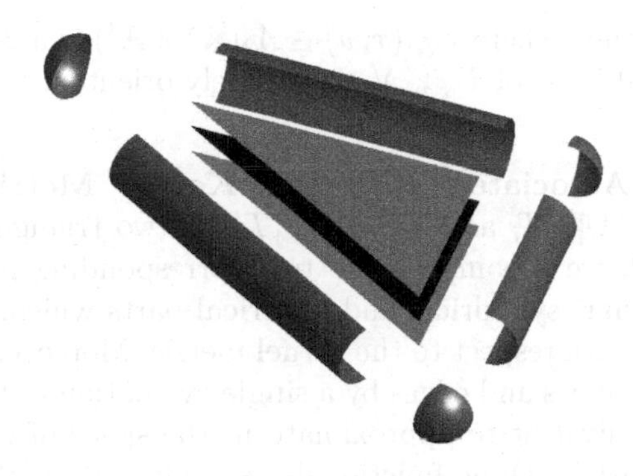

Fig. 1. Illustration of the decomposition of the normal bundle of a triangle into a planar (in purple), a cylindrical (in blue) and a spherical (in yellow) parts. Note that the actual normal bundle lives in $\mathbb{R}^3 \times S^2$ (Color figure online)

Normal Cycle of a Triangulated Mesh. Let $\mathcal{T}$ be a triangulated mesh, which we define as a finite union of triangles $\mathcal{T} = \cup_{i=1}^{n_T} T_i$ such that the intersection of any two triangles is either empty or a common edge. The normal cycle of a triangulated mesh is defined using the additive formula (2) as a combination of normal cycles of its faces (triangles), edges (segments) and vertices (points). All these elements further decompose into planar, cylindrical and spherical parts.

3 Kernel Metrics on Normal Cycles with Constant Normal Kernel

Construction of the Kernel Metric. As detailed in [13], we use the framework of Reproducing Kernel Hilbert Spaces (RKHS) to define a metric between normal cycles. The kernel has the form

$$\begin{aligned} K_W :\ & (\mathbb{R}^3 \times S^2)^2 \ \to \mathcal{L}\big(\Lambda_2(\mathbb{R}^3 \times \mathbb{R}^3)\big) \\ & ((x,u),(y,v)) \mapsto k_p(x,y) k_n(u,v) \mathrm{Id}_{\Lambda_2(\mathbb{R}^3 \times \mathbb{R}^3)}, \end{aligned}$$

This defines a RKHS W, and under some regularity conditions on the kernels ([13], Proposition 25), we have $W \hookrightarrow \Omega_0^2(\mathbb{R}^3 \times S^2)$, and thus, $\Omega_0^2(\mathbb{R}^3 \times S^2)' \subset W'$. The corresponding metric on W' can be used as a data attachment term for shapes represented as normal cycles. In this work, for simplicity and efficiency reasons, we consider the following normal kernel: $k_n(u,v) = 1$ (constant kernel). Other simple and interesting choices will be $k_n(u,v) = \langle u, v\rangle$ (linear kernel) or $k_n(u,v) = 1 + \langle u, v\rangle$, but we keep them for future work.

The expression of scalar product between two normal cycles $N(C)$ and $N(S)$, associated with shapes S and C is then:

$$\langle N(C), N(S)\rangle_{W'} = \int_{\mathcal{N}_C}\int_{\mathcal{N}_S} k_p(x,y)\, \langle \tau_{\mathcal{N}_C}(x,u), \tau_{\mathcal{N}_S}(y,v)\rangle \, d\mathcal{H}^2(x,u) d\mathcal{H}^2(y,v), \quad (3)$$

for the constant kernel, where $\tau_{\mathcal{N}_C}(x,u) \in \Lambda_2(\mathbb{R}^3 \times \mathbb{R}^3)$ is a 2-vector associated with an orthonormal basis of $T_{(x,u)}\mathcal{N}_C$, positively oriented.

Scalar Product Associated with the Kernel Metric for Discrete Surfaces. Let $\mathcal{T} = \cup_{i=1}^n T_i$ and $\mathcal{T}' = \cup_{i=1}^m T_i'$ be two triangulated meshes. As explained in Sect. 2, we decompose the two corresponding normal cycles into combinations of planar, cylindrical and spherical parts which, as was proven in [13], are orthogonal with respect to the kernel metric. Moreover, we approximate integrations over triangles and edges by a single evaluation at the center of these elements. This is equivalent to approximate in the space of currents the cylindrical and planar part by Dirac functionals as explained in [13], Sect. 3.2.2. For integrations over the sphere however, we can get simple analytic formulas for the integrations with our choice of kernel. The new approximations in the space of currents are denoted $\tilde{N}(\mathcal{T})$ and $\tilde{N}(\mathcal{T}')$. We do not further detail the calculus of the different integrations over the normal bundle and express only the result obtained:

$$\begin{aligned}
&\left\langle \tilde{N}(\mathcal{T}), \tilde{N}(\mathcal{T}')\right\rangle_{W'} \\
&= \sum_{\substack{f \text{ edge of the} \\ \text{border of } \mathcal{T}}} \sum_{\substack{g \text{ edge of the} \\ \text{border of } \mathcal{T}'}} \frac{\pi^2}{4}\left(k_p(x_f^1, y_g^1) + k_p(x_f^2, y_g^2) - k_p(x_f^2, y_g^1) - k_p(x_f^1, y_g^2)\right) \left\langle \frac{f}{|f|}, \frac{g}{|g|}\right\rangle \\
&+ \sum_{i=1}^{F}\sum_{j=1}^{G} 4k_p(c_i, d_j)\, \langle f_i, g_j\rangle \left\langle \sum_{\{T | f_i \text{ edge of } T\}} n_{T,f_i}, \sum_{\{T' | g_j \text{ edge of } T'\}} n_{T',g_j} \right\rangle,
\end{aligned} \quad (4)$$

where

- x_f^1 and x_f^2 are the two vertices of f with $f = x_f^2 - x_f^1$.
- c_i (resp. d_j) is the middle of the edge f_i (resp. g_j).
- n_{T,f_i} is the normal vector of the triangle T such that $n_{T,f_i} \times f_i$ is oriented inward for the triangle T.

Let us make a few remarks here. First, we recall that the previous expression does not necessitate a coherent orientation for the mesh.

Secondly, even with a constant kernel k_n for the normal part, the metric is sensitive to curvature. Indeed, for an edge f, the cylindrical part of the scalar product involves scalar products between normal vectors of the adjacent triangles which are required quantities to compute the discrete mean curvature.

Another interesting feature to notice is that the scalar product involves a specific term for the boundary which will enforce the matching of the boundaries of the shapes. The fact that the boundary has a special behaviour for the normal cycle metric is not surprising. Indeed a normal cycle encodes generalized curvature information of the shape. Hence, the boundary corresponds to a singularity of the curvature and has a specific behaviour in the kernel metric. We will see in Sect. 4 that this feature is of interest for a matching purpose.

4 Results

We used the Large Deformation Diffeomorphic Metric Mapping (LDDMM) framework (see for example [1,2,12]) in our experiments to model deformations of the ambient space. We emphasize here that this choice of framework is not mandatory, and that other registration models could be used together with our normal cycle data attachment term. We used the discrete formulation of LDDMM via initial momentum parametrization and a geodesic shooting algorithm [1,12]. For the optimization of the functional we used a quasi Newton Broyden Fletcher Goldfarb Shanno algorithm with limited memory (L-BFGS) [11]. The step in the descent direction is fixed by a Wolfe line search. For the numerical integrations of geodesic and backward equations, a Runge-Kutta (4,5) scheme is used (function ode45 in Matlab). In order to improve the computational cost, the convolution operations involved are done with parallel computing on a graphic card. The CUDA mex files using GPU are included in the MATLAB body program. The algorithm is run until convergence with a stopping criterion on the norm of the successive iterations, with a tolerance of 10^{-6}.

For all the following matchings, the geometric kernel k_p is a Gaussian kernel of width σ_W, and k_n is a constant kernel. The kernel K_V is a sum of 4 Gaussian kernels of decreasing sizes, in order to capture different features of the deformation (see [3]). The trade-off parameter γ is fixed at 0.1 for all the experiments.

The first example is a matching of two human hippocampi. Each shape has around 7000 vertices. Three runs at different geometric kernel sizes are performed (see Fig. 2). We can see that the final deformation matches well the two hippocampi, even the high curved regions of the shape.

The second data set was provided by B. Charlier, N. Charon and M.F. Beg. It is a set of retina layers from different subjects [10], which has been already used for computational anatomy studies [9]. The retinas are surfaces of typical size 8 mm^2. Each retina is sampled with approximately 5000 points. All the details of the matching are in Fig. 3. The retinas have a boundary which will be seen as a region with singularities for the kernel metric on normal cycles. This is not the case for the varifolds metric which makes the matching of the corresponding corners harder. The matching of the boundaries is better with normal cycles, and provides a much more regular deformation (see Fig. 3).

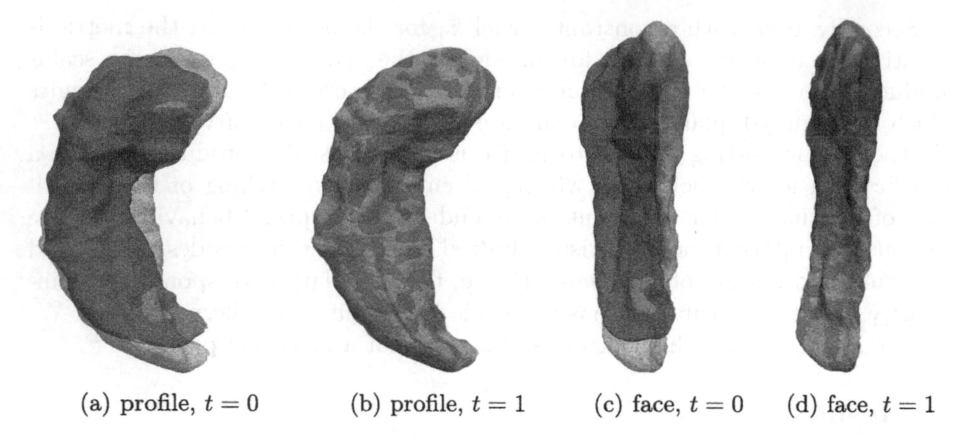

(a) profile, $t = 0$ (b) profile, $t = 1$ (c) face, $t = 0$ (d) face, $t = 1$

Fig. 2. Two views (profile and face) at times $t = 0$ and $t = 1$ of the matching of two hippocampi with normal cycles. The target shape is in orange and the source in blue. Each shape has 6600 points. Three runs at different geometric kernel sizes are performed ($\sigma_W = 25, 10, 5$) and the kernel of deformation is a sum of Gaussian kernels with $\sigma_V = 10, 5, 2.5, 1.25$ (the diameter of hippocampus is about 40 mm). Each run ended respectively at 62, 66 and 48 iterations for a total time of 4076 s (23 s per iteration). (Color figure online)

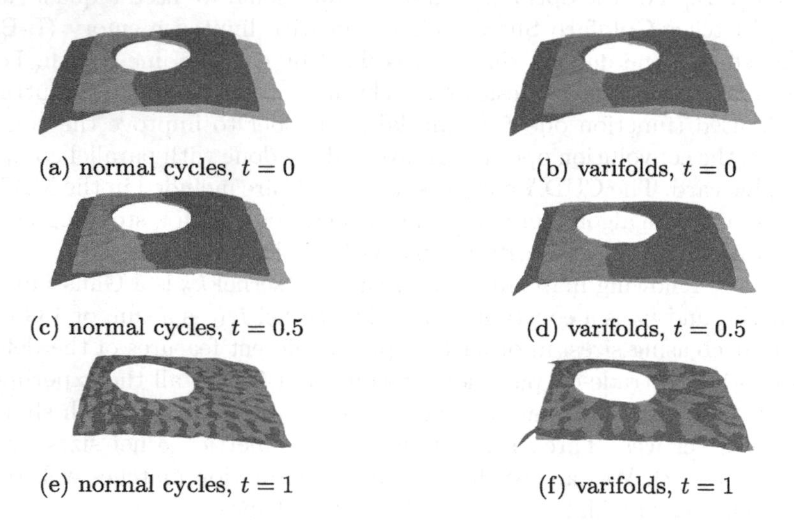

(a) normal cycles, $t = 0$ (b) varifolds, $t = 0$

(c) normal cycles, $t = 0.5$ (d) varifolds, $t = 0.5$

(e) normal cycles, $t = 1$ (f) varifolds, $t = 1$

Fig. 3. Matching of two retinas with kernel metric on normal cycles (left) and varifolds (right). The target shape is in orange and the source shape is in blue. Each shape has 5000 points. For the varifolds metric, the geometric kernel is Gaussian and the kernel on the Grassmanian is chosen linear. The same parameters are used for each data attachment term. Three runs at different geometric kernel sizes are performed ($\sigma_W = 0.8, 0.4, 0.2$). K_V is a sum of Gaussian kernels with $\sigma_V = 2.4, 1.2, 0.6, 0.3$. For normal cycles, each run ended respectively at 88, 297 and 5 iterations for a total time of 5487 s (14 s/it). For varifolds, it was 55, 1 and 1 iterations for a total time of 2051 s (35 s/it). (Color figure online)

In the last example (Fig. 4), the two retinas are the result of an unsatisfactory segmentation. This leads to artifacts in each retina: two triangles for the source retina (in blue, Fig. 4) and only one for the target, in orange. These are regions of high curvature and as we could expect, the kernel metric on normal cycles will make a correspondence between those points. As we can see in the second row of Fig. 4, the two triangles are crushed together, into one triangle, even though the cost of the resulting deformation is high. This example shows how sensitive to noise or artifacts normal cycles are.

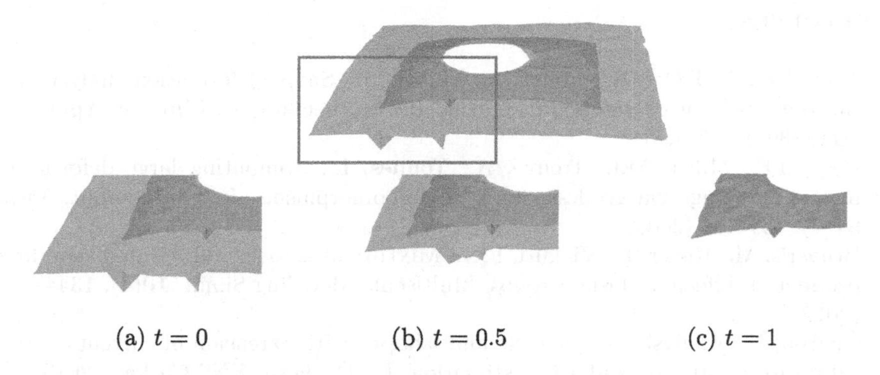

(a) $t = 0$ (b) $t = 0.5$ (c) $t = 1$

Fig. 4. Matching of two retinas with normal cycles: the target (in orange) and the source (in blue). Three runs at different geometric kernel sizes are performed ($\sigma_W = 0.8, 0.4, 0.2$). K_V is a sum of Gaussian kernels with $\sigma_V = 2.4, 1.2, 0.6, 0.3$. The first row shows the initial configuration. The second row shows the matching in the specific zone delimited by the red rectangle. The metric on normal cycles enforces the matching of corresponding high curvature points, which leads to the alignment of the two triangles into the single one of the target. Each run ended respectively at 211, 90 and 202 iterations for a total time of 8114 s (16 s/it). (Color figure online)

5 Conclusion

In this article we extended to the case of surfaces the methodology introduced in [13] for curve matching, based on the notion of normal cycle. Compared to the representation of shapes with currents or varifolds, this model encodes the curvature information of the surfaces. We define a scalar product between two triangulations represented as normal cycles using the theory of reproducing kernels. The intrinsic complexity of the model is simplified by using a constant normal kernel for the metric. Even though it may seem rough, we do not get rid of all curvature information of the shape, as it can be seen in Eq. (4) or in the example showed in Figs. 3 and 4. Using parallel computing on GPU, we are able to match two surfaces with a large number of points in a reasonable time, with a descent algorithm that is run until convergence.

The examples of this article show promising first results. For the retinas data set, the weighting of the boundaries and corner points provided by the metric on normal cycles allows a much more precise and regular deformation than with varifolds. As a future work, it will be interesting to study more complex normal kernels k_n, as the linear kernel or a combination of a linear kernel and a constant kernel. The exact type of curvatures (mean, Gaussian) that we are able to retrieve with such kernels is not clear yet and should be investigated. We also would like to work on data sets were the refinement of normal cycles is relevant.

References

1. Arguillère, S., Trélat, E., Trouvé, A., Younès, L.: Shape deformation analysis from the optimal control viewpoint. Journal de Mathématiques Pures et Appliquées **104**, 139–178 (2015)
2. Beg, M.F., Miller, M.I., Trouvé, A., Younes, L.: Computing large deformation metric mappings via geodesic flows of diffeomorphisms. Int. J. Comput. Vision **61**(2), 139–157 (2005)
3. Bruveris, M., Risser, L., Vialard, F.X.: Mixture of kernels and iterated semidirect product of diffeomorphisms groups. Multiscale Modeling Simul. **10**(4), 1344–1368 (2012)
4. Charon, N.: Analysis of geometric and fshapes with extension of currents. Application to registration and atlas estimation. Ph.D. thesis, ÉNS Cachan (2013)
5. Durrleman, S.: Statistical models of currents for measuring the variability of anatomical curves, surfaces and their evolution. Ph.D. thesis, Université Nice, Sophia Antipolis (2010)
6. Federer, H.: Curvature measures. Trans. Amer. Maths. Soc. **93**, 418–491 (1959)
7. Glaunès, J.: Transport par difféomorphismes de points, de mesures et de courants pour la comparaison de formes et l'anatomie numérique. Ph.D. thesis, Université Paris 13 (2005)
8. Glaunès, J., Qiu, A., Miller, M., Younes, L.: Large deformation diffeomorphic metric curve mapping. Int. J. Comput. Vision **80**(3), 317–336 (2008)
9. Lee, S., Charon, N., Charlier, B., Popuri, K., Lebed, E., Sarunic, M., Trouvé, A., Beg, M.: Atlas-based shape analysis and classification of retinal optical coherence tomography images using the fshape framework. Med. Image Anal. **35**, 570–581 (2016)
10. Lee, S., Han, S.X., Young, M., Beg, M.F., Sarunic, M.V., Mackenzie, P.J.: Optic nerve head and peripapillary morphometrics in myopic glaucoma. Invest. Ophthalmol. Vis. Sci. **55**(7), 4378 (2014)
11. Liu, D.C., Nocedal, J.: On the limited memory BFGS method for large scale optimization. Math. Program. **45**(1–3), 503–528 (1989)
12. Miller, M.I., Trouvé, A., Younes, L.: Geodesic shooting for computational anatomy. J. Math. Imaging Vis. **24**(2), 209–228 (2006)
13. Roussillon, P., Glaunès, J.: Kernel metrics on normal cycles and application to curve matching. SIAM J. Imaging Sci. **9**, 1991–2038 (2016)
14. Vaillant, M., Glaunès, J.: Surface matching via currents. In: Christensen, G.E., Sonka, M. (eds.) IPMI 2005. LNCS, vol. 3565, pp. 381–392. Springer, Heidelberg (2005). doi:10.1007/11505730_32
15. Zähle, M.: Curvatures and currents for unions of set with positive reach. Geom. Dedicata. **23**, 155–171 (1987)

Optimal Transport and Applications: Image Processing

Regularized Barycenters in the Wasserstein Space

Elsa Cazelles(✉), Jérémie Bigot, and Nicolas Papadakis

Université de Bordeaux, CNRS, Institut de Mathématiques de Bordeaux, UMR 5251, Talence, France
elsa.cazelles@u-bordeaux.fr

Abstract. This paper is an overview of results that have been obtain in [2] on the convex regularization of Wasserstein barycenters for random measures supported on $\mathbb{R}^d$. We discuss the existence and uniqueness of such barycenters for a large class of regularizing functions. A stability result of regularized barycenters in terms of Bregman distance associated to the convex regularization term is also given. Additionally we discuss the convergence of the regularized empirical barycenter of a set of n iid random probability measures towards its population counterpart in the real line case, and we discuss its rate of convergence. This approach is shown to be appropriate for the statistical analysis of discrete or absolutely continuous random measures. In this setting, we propose an efficient minimization algorithm based on accelerated gradient descent for the computation of regularized Wasserstein barycenters.

Keywords: Wasserstein space · Fréchet mean · Barycenter of probability measures · Convex regularization · Bregman divergence

1 Introduction

This paper is concerned by the statistical analysis of data sets whose elements may be modeled as random probability measures supported on $\mathbb{R}^d$. It is an overview of results that have been obtain in [2]. In the special case of one dimension ($d = 1$), we are able to provide refined results on the study of a sequence of discrete measures or probability density functions (e.g. histograms) that can be viewed as random probability measures. Such data sets appear in various research fields. Examples can be found in neuroscience [10], biodemographic and genomics studies [11], economics [7], as well as in biomedical imaging [9]. In this paper, we focus on first-order statistics methods for the purpose of estimating, from such data, a population mean measure or density function.

The notion of averaging depends on the metric that is chosen to compare elements in a given data set. In this work, we consider the Wasserstein distance W_2 associated to the quadratic cost for the comparison of probability measures. Let Ω be a subset of $\mathbb{R}^d$ and $\mathcal{P}_2(\Omega)$ be the set of probability measures supported on Ω with finite order second moment.

F. Nielsen and F. Barbaresco (Eds.): GSI 2017, LNCS 10589, pp. 83–90, 2017.
https://doi.org/10.1007/978-3-319-68445-1_10

Definition 1. *As introduced in* [1], *an empirical Wasserstein barycenter* $\bar{\nu}_n$ *of a set of n probability measures* $\nu_1, \ldots, \nu_n$ *(not necessarily random) in* $\mathcal{P}_2(\Omega)$ *is defined as a minimizer of*

$$\mu \mapsto \frac{1}{n}\sum_{i=1}^{n} W_2^2(\mu, \nu_i), \;\; over\; \mu \in \mathcal{P}_2(\Omega). \tag{1}$$

The Wasserstein barycenter corresponds to the notion of empirical Fréchet mean [6] that is an extension of the usual Euclidean barycenter to nonlinear metric spaces.

However, depending on the data at hand, such a barycenter may be irregular. As an example let us consider a real data set of neural spike trains which is publicly available from the MBI website[1]. During a squared-path task, the spiking activity of a movement-encoded neuron of a monkey has been recorded during 5 s over $n = 60$ repeated trials. Each spike train is then smoothed using a Gaussian kernel (further details on the data collection can be found in [10]). For each trial $1 \leq i \leq n$, we let ν_i be the measure with probability density function (pdf) proportional to the sum of these Gaussian kernels centered at the times of spikes. The resulting data are displayed in Fig. 1(a). For probability measures supported on the real line, computing a Wasserstein barycenter simply amounts to averaging the quantile functions of the ν_i's (see e.g. Sect. 6.1 in [1]). The pdf of the Wasserstein barycenter $\bar{\nu}_n$ is displayed in Fig. 1(b). This approach clearly leads to the estimation of a very irregular mean template density of spiking activity.

In this paper, we thus introduce a convex regularization of the optimization problem (1) for the purpose of obtaining a regularized Wasserstein barycenter. In this way, by choosing an appropriate regularizing function (e.g. the negative entropy in Subsect. 2.1), it is of possible to enforce this barycenter to be absolutely continuous with respect to the Lebesgue measure on $\mathbb{R}^d$.

2 Regularization of Barycenters

We choose to add a penalty directly into the computation of the Wasserstein barycenter in order to smooth the Fréchet mean and to remove the influence of noise in the data.

Definition 2. *Let* $\mathbb{P}_n^\nu = \frac{1}{n}\sum_{i=1}^n \delta_{\nu_i}$ *where* δ_{ν_i} *is the dirac distribution at* ν_i*. We define a regularized empirical barycenter* $\mu_{\mathbb{P}_n^\nu}^\gamma$ *of the discrete measure* $\mathbb{P}_n^\nu$ *as a minimizer of*

$$\mu \mapsto \frac{1}{n}\sum_{i=1}^{n} W_2^2(\mu, \nu_i) + \gamma E(\mu) \;\; over\; \mu \in \mathcal{P}_2(\Omega), \tag{2}$$

where $\mathcal{P}_2(\Omega)$ *is the space of probability measures on* Ω *with finite second order moment,* $E : \mathcal{P}_2(\Omega) \to \mathbb{R}_+$ *is a smooth convex penalty function, and* $\gamma > 0$ *is a regularization parameter.*

[1] http://mbi.osu.edu/2012/stwdescription.html.

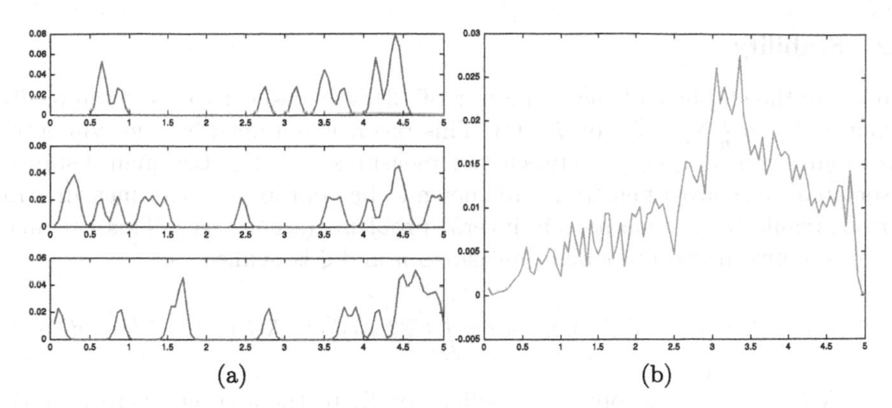

Fig. 1. (a) A subset of 3 smoothed neural spike trains out of $n = 60$. Each row represents one trial and the pdf obtained by smoothing each spike train with a Gaussian kernel of width 50 ms. (b) Probability density function of the empirical Wasserstein barycenter $\bar{\nu}_n$ for this data set.

In what follows, we present the main properties on the regularized empirical Wasserstein barycenter $\mu^{\gamma}_{\mathbb{P}^{\nu}_n}$.

2.1 Existence and Uniqueness

We consider the wider problem of

$$\min_{\mu \in \mathcal{P}_2(\Omega)} J^{\gamma}_{\mathbb{P}}(\mu) = \int W_2^2(\mu, \nu) d\mathbb{P}(\nu) + \gamma E(\mu). \tag{3}$$

Hence, (2) corresponds to the minimization problem (3) where $\mathbb{P}$ is discrete i.e. $\mathbb{P} = \mathbb{P}_n = \frac{1}{n} \sum_{i=1}^{n} \delta_{\nu_i}$.

Theorem 1 (Theorem 3.2 in [2]**).** *Let $E : \mathcal{P}_2(\Omega) \to \mathbb{R}_+$ be a proper, lower semicontinuous and differentiable function that is strictly convex on its domain $\mathcal{D}(E) = \{\mu \in \mathcal{P}_2(\Omega)$ such that $E(\mu) < +\infty\}$. Then, the functional $J^{\gamma}_{\mathbb{P}}$ define by* (3) *admits a unique minimizer.*

Such assumptions on E are supposed to be always satisfied throughout the paper. A typical example of regularization function satisfying such assumptions is the negative entropy defined as

$$E(\mu) = \begin{cases} \int_{\mathbb{R}^d} f(x) \log(f(x)) dx, & \text{if } \mu \text{ admits a density } f \text{ with respect to} \\ & \text{the Lebesgue measure } dx \text{ on } \Omega, \\ +\infty & \text{otherwise.} \end{cases}$$

2.2 Stability

We study the stability of the minimizer of (3) with respect to the discrete distribution $\mathbb{P}_n^{\nu} = \frac{1}{n}\sum_{i=1}^{n}\delta_{\nu_i}$ on $\mathcal{P}_2(\Omega)$. This result is obtained for the symmetric Bregman distance $d_E(\mu,\zeta)$ between two measures μ and ζ. Bregman distances associated to a convex penalty E are known to be appropriate error measures for various regularization methods in inverse problems (see e.g. [4]). This Bregman distance between two probability measures μ and ζ is defined as

$$d_E(\mu,\zeta) := \langle \nabla E(\mu) - \nabla E(\zeta), \mu - \zeta\rangle = \int_{\Omega} (\nabla E(\mu)(x) - \nabla E(\zeta)(x))(d\mu - d\zeta)(x),$$

where $\nabla E : \Omega \to \mathbb{R}$ denotes the gradient of E. In the setting where E is the negative entropy and $\mu = \mu_f$ (resp. $\zeta = \zeta_g$) admits a density f (resp. g) with respect to the Lebesgue measure, then d_E is the symmetrised Kullback-Leibler divergence

$$d_E(\mu_f,\zeta_g) = \int (f(x) - g(x)) \log\left(\frac{f(x)}{g(x)}\right) dx.$$

The stability result of the regularized empirical barycenter can then be stated as follows.

Theorem 2 (Theorem 3.3 in [2]). *Let $\nu_1,\ldots,\nu_n$ and $\eta_1,\ldots,\eta_n$ be two sequences of probability measures in $\mathcal{P}_2(\Omega)$. If we denote by $\mu^{\gamma}_{\mathbb{P}_n^{\nu}}$ and $\mu^{\gamma}_{\mathbb{P}_n^{\eta}}$ the regularized empirical barycenter associated to the discrete measures $\mathbb{P}_n^{\nu}$ and $\mathbb{P}_n^{\eta}$, then the symmetric Bregman distance (associated to E) between these two barycenters is bounded as follows*

$$d_E\left(\mu^{\gamma}_{\mathbb{P}_n^{\nu}}, \mu^{\gamma}_{\mathbb{P}_n^{\eta}}\right) \leq \frac{2}{\gamma n} \inf_{\sigma\in\mathcal{S}_n} \sum_{i=1}^{n} W_2(\nu_i, \eta_{\sigma(i)}), \tag{4}$$

where $\mathcal{S}_n$ denotes the permutation group of the set of indices $\{1,\ldots,n\}$.

In particular, inequality (4) allows to compare the case of data made of n absolutely continuous probability measures $\nu_1,\ldots,\nu_n$, with the more realistic setting where we have only access to a dataset of random variables $\boldsymbol{X} = (\boldsymbol{X}_{i,j})_{1\leq i\leq n;\ 1\leq j\leq p_i}$ organized in the form of n experimental units, such that $\boldsymbol{X}_{i,1},\ldots,\boldsymbol{X}_{i,p_i}$ are iid observations in $\mathbb{R}^d$ sampled from the measure ν_i for each $1 \leq i \leq n$. If we denote by $\boldsymbol{\nu}_{p_i} = \frac{1}{p_i}\sum_{j=1}^{p_i}\delta_{\boldsymbol{X}_{i,j}}$ the usual empirical measure associated to ν_i, it follows from inequality (4) that

$$\mathbb{E}\left(d_E^2\left(\mu^{\gamma}_{\mathbb{P}_n^{\nu}}, \boldsymbol{\mu}^{\gamma}_{\boldsymbol{X}}\right)\right) \leq \frac{4}{\gamma^2 n}\sum_{i=1}^{n}\mathbb{E}\left(W_2^2(\nu_i,\boldsymbol{\nu}_{p_i})\right),$$

where $\boldsymbol{\mu}^{\gamma}_{\boldsymbol{X}}$ is given by $\boldsymbol{\mu}^{\gamma}_{\boldsymbol{X}} = \underset{\mu\in\mathcal{P}_2(\Omega)}{\operatorname{argmin}} \frac{1}{n}\sum_{i=1}^{n} W_2^2(\mu, \frac{1}{p_i}\sum_{j=1}^{p_i}\delta_{\boldsymbol{X}_{i,j}}) + \gamma E(\mu)$.
This result allows to discuss the rate of convergence (for the symmetric squared Bregman distance) of $\boldsymbol{\mu}^{\gamma}_{\boldsymbol{X}}$ to $\mu^{\gamma}_{\mathbb{P}_n^{\nu}}$ as a function of the rate of convergence (for the

squared Wasserstein distance) of the empirical measure $\boldsymbol{\nu}_{p_i}$ to ν_i for each $1 \leq i \leq n$ (in the asymptotic setting where $p = \min_{1 \leq i \leq n} p_i$ is let going to infinity). As an illustrative example, in the one-dimensional case (that is $d = 1$), one may use the work in [3] on a detailed study of the variety of rates of convergence of an empirical measure on the real line toward its population counterpart for the expected squared Wasserstein distance. For example, by Theorem 5.1 in [3], it follows that

$$\mathbb{E}\left(W_2^2(\nu_i, \boldsymbol{\nu}_{p_i})\right) \leq \frac{2}{p_i + 1} J_2(\nu_i), \text{ with } J_2(\nu_i) = \int_{\Omega} \frac{F_i(x)(1 - F_i(x))}{f_i(x)} dx,$$

where f_i is the pdf of ν_i, and F_i denotes its cumulative distribution function. Therefore, provided that $J_2(\nu_i)$ is finite for each $1 \leq i \leq n$, one obtains the following rate of convergence of $\boldsymbol{\mu}_{\mathbf{X}}^{\gamma}$ to $\mu_{\mathbb{P}_n^{\nu}}^{\gamma}$ (in the case of measures ν_i supported on an interval Ω of $\mathbb{R}$)

$$\mathbb{E}\left(d_E^2\left(\mu_{\mathbb{P}_n^{\nu}}^{\gamma}, \boldsymbol{\mu}_{\mathbf{X}}^{\gamma}\right)\right) \leq \frac{8}{\gamma^2 n} \sum_{i=1}^{n} \frac{J_2(\nu_i)}{p_i + 1} \leq \frac{8}{\gamma^2} \left(\frac{1}{n} \sum_{i=1}^{n} J_2(\nu_i)\right) p^{-1}. \quad (5)$$

Note that by the results in Appendix A in [3], a necessary condition for $J_2(\nu_i)$ to be finite is to assume that f_i is almost everywhere positive on the interval Ω.

2.3 Convergence to a Population Wasserstein Barycenter

Introducing this symmetric Bregman distance also allows to analyze the consistency of the regularized barycenter $\mu_{\mathbb{P}_n}^{\gamma}$ as the number of observations n tends to infinity and the parameter γ is let going to zero. When $\boldsymbol{\nu}_1, \ldots, \boldsymbol{\nu}_n$ are supposed to be independent and identically distributed (iid) random measures in $\mathcal{P}_2(\Omega)$ sampled from a distribution $\mathbb{P}$, we analyze the convergence of $\boldsymbol{\mu}_{\mathbb{P}_n^{\nu}}^{\gamma}$ with respect to the population Wasserstein barycenter defined as

$$\mu_{\mathbb{P}}^0 \in \underset{\mu \in \mathcal{P}_2(\Omega)}{\operatorname{argmin}} \int W_2^2(\mu, \nu) d\mathbb{P}(\nu),$$

and its regularized version

$$\mu_{\mathbb{P}}^{\gamma} = \underset{\mu \in \mathcal{P}_2(\Omega)}{\operatorname{argmin}} \int W_2^2(\mu, \nu) d\mathbb{P}(\nu) + \gamma E(f).$$

In the case where Ω is a compact of $\mathbb{R}^d$ and $\nabla E(\mu_{\mathbb{P}}^0)$ is bounded, we prove that $\mu_{\mathbb{P}}^{\gamma}$ converges to $\mu_{\mathbb{P}}^0$ as $\gamma \to 0$ for the Bregman divergence associated to E. This result corresponds to showing that the bias term (as classically referred to in nonparametric statistics) converges to zero when $\gamma \to 0$. We also analyze the rate of convergence of the variance term when Ω is a compact of $\mathbb{R}$:

Theorem 3 (Theorem 4.5 in [2]). *For Ω compact included in $\mathbb{R}$, there exists a constant $C > 0$ (not depending on n and γ) such that*

$$\mathbb{E}\left(d_E^2\left(\boldsymbol{\mu}_{\mathbb{P}_n^{\nu}}^{\gamma}, \mu_{\mathbb{P}}^{\gamma}\right)\right) \leq \frac{C}{\gamma^2 n}.$$

Therefore, when $\boldsymbol{\nu}_1, \ldots, \boldsymbol{\nu}_n$ are iid random measures with support included in a compact interval Ω, it follows that if $\gamma = \gamma_n$ is such that $\lim_{n\to\infty} \gamma_n^2 n = +\infty$ then

$$\lim_{n\to\infty} \mathbb{E}(d_E^2\left(\boldsymbol{\mu}_{\mathbb{P}_n^{\boldsymbol{\nu}}}^{\gamma}, \mu_{\mathbb{P}}^0)\right) = 0.$$

3 Numerical Experiments

We consider a simulated example where the measures $\boldsymbol{\nu}_i$ are discrete and supported on a small number p_i of data points ($5 \leq p_i \leq 10$). To this end, for each $i = 1, \ldots, n$, we simulate a sequence $(\boldsymbol{X}_{ij})_{1\leq j \leq p_i}$ of iid random variables sampled from a Gaussian distribution $\mathcal{N}(\boldsymbol{\mu}_i, \boldsymbol{\sigma}_i^2)$, and the $\boldsymbol{\mu}_i$'s (resp. $\boldsymbol{\sigma}_i$) are iid random variables such that $-2 \leq \boldsymbol{\mu}_i \leq 2$ and $0 \leq \boldsymbol{\sigma}_i \leq 1$ with $\mathbb{E}(\boldsymbol{\mu}_i) = 0$ and $\mathbb{E}(\boldsymbol{\sigma}_i) = 1/2$. The target measure that we wish to estimate in these simulations is the population (or true) Wasserstein barycenter of the random distribution $\mathcal{N}(\boldsymbol{\mu}_1, \boldsymbol{\sigma}_1^2)$ which is $\mathcal{N}(0, 1/4)$ thanks to the assumptions $\mathbb{E}(\boldsymbol{\mu}_1) = 0$ and $\mathbb{E}(\boldsymbol{\sigma}_1) = 1/2$. Then, let $\boldsymbol{\nu}_i = \frac{1}{p_i}\sum_{j=1}^{p_i} \delta_{\boldsymbol{X}_{ij}}$, where δ_x is the Dirac measure at x.

In order to compute the regularized barycenter, we solve (3) with an efficient minimization algorithm based on accelerated gradient descent (see [5]) for the computation of regularized barycenters in 1-D (see Appendix C in [2]).

To illustrate the benefits of regularizing the Wasserstein barycenter of the $\boldsymbol{\nu}_i$'s, we compare our estimator with the one obtained by the following procedure which we refer to as the kernel method. In a preliminary step, each measure $\boldsymbol{\nu}_i$ is smoothed using a standard kernel density estimator to obtain

$$\hat{\boldsymbol{f}}_{i,h_i}(x) = \frac{1}{p_i h_i}\sum_{j=1}^{p_i} K\left(\frac{x - \boldsymbol{X}_{ij}}{h_i}\right), \; x \in \Omega,$$

where K is a Gaussian kernel. The bandwidth h_i is chosen by cross-validation. An alternative estimator is then defined as the Wasserstein barycenter of the smoothed measures with density $\hat{\boldsymbol{f}}_{1,h_1}, \ldots, \hat{\boldsymbol{f}}_{n,h_n}$. Thanks, to the well-know quantile averaging formula, the quantile function $\bar{F}_n^{-1}$ of this smoothed Wasserstein barycenter is given by $\bar{F}_n^{-1} = \frac{1}{n}\sum_{i=1}^{n} F_{\hat{\boldsymbol{f}}_{i,h_i}}^{-1}$ where F_g^{-1} denotes the quantile function of a given pdf g. The estimator $\bar{F}_n^{-1}$ corresponds to the notion of smoothed Wasserstein barycenter of multiple point processes as considered in [8]. The density of $\bar{F}_n^{-1}$ is denoted by $\hat{\boldsymbol{f}}_n$, and it is displayed in Fig. 2. Hence, it seems that a preliminary smoothing of the $\boldsymbol{\nu}_i$ followed quantile averaging is not sufficient to recover a satisfactory Gaussian shape when the number p_i of observations per unit is small.

Alternatively, we have applied our algorithm directly on the (non-smoothed) discrete measures $\boldsymbol{\nu}_i$ to obtain the regularized barycenter $\boldsymbol{f}_{\mathbb{P}_n}^{\gamma}$ defined as the minimizer of (2). For the penalty function E, we took either the negative entropy or a Dirichlet regularization. The densities of the penalized Wasserstein barycenters associated to these two choices for E and for different values of γ are displayed as solid curves in warm colors in Fig. 2. For both penalty functions and despite a

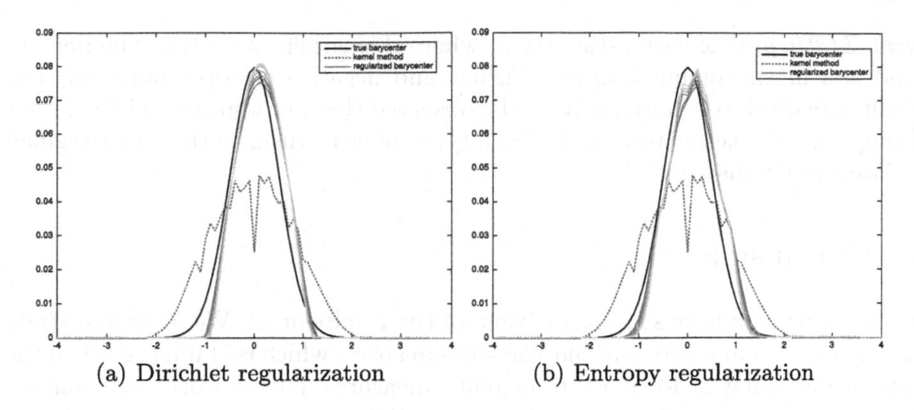

(a) Dirichlet regularization (b) Entropy regularization

Fig. 2. Simulated data from Gaussian distributions with random means and variances. In all the figures, the black curve is the density of the true Wasserstein barycenter. The blue and dotted curve represents the pdf of the smoothed Wasserstein barycenter obtained by a preliminary kernel smoothing step. Pdf of the regularized Wasserstein barycenter $\mu^{\gamma}_{\mathbb{P}_n}$ (a) for $20 \leq \gamma \leq 50$ with $E(f) = ||f'||^2$ (Dirichlet), and (b) for $0.08 \leq \gamma \leq 14$ with $E(f) = \int f \log(f)$ (negative entropy). (Color figure online)

small number of observations per experimental units, the shape of these densities better reflects the fact that the population Wasserstein barycenter is a Gaussian distribution.

Finally, we provide Monte-Carlo simulations to illustrate the influence of the number $n = 100$ of observed measures on the convergence of these estimators. For a given $10 \leq n_0 \leq n$, we randomly draw n_0 measures $\boldsymbol{\nu}_i$ from the whole sample, and we compute a smoothed barycenter via the kernel method and a regularized barycenter for a chosen γ. For given value of n_0, this procedure is repeated 200 times, which allows to obtain an approximation of the expected

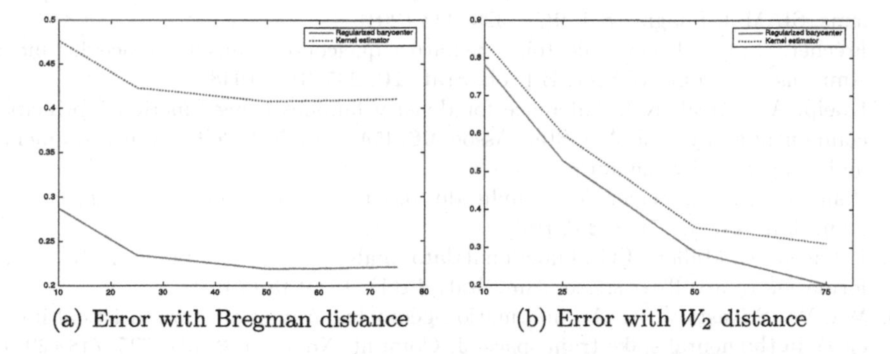

(a) Error with Bregman distance (b) Error with W_2 distance

Fig. 3. Errors in terms of expected Bregman and Wasserstein distances between the population barycenter and the estimated barycenters (kernel method in dashed blue, regularized barycenter in red) for a sample of size $n_0 = 10, 25, 50$ and 75. (Color figure online)

error $\mathbb{E}\left(d(\hat{\mu}, \mu_{\mathbb{P}})\right)$ of each estimator $\hat{\mu}$, where d is either d_E or W_2. The penalty used is a linear combinaison of Dirichlet and negative entropy functions. The results are displayed in Fig. 3. It can be observed that our approach yields better results than the kernel method for both types of error (using either the Bregman or Wasserstein distance).

4 Conclusion

In this paper, we have summarize some of the results of [2]. We provide a study on regularized barycenters in the Wasserstein space, which is of interest when the data are irregular or for noisy probability measures. Future works will concern the numerical computation and the study of the convergence to a population Wasserstein barycenter for $\Omega \subset \mathbb{R}^d$.

Acknowledgment. This work has been carried out with financial support from the French State, managed by the French National Research Agency (ANR) in the frame of the GOTMI project (ANR-16-CE33-0010-01).

References

1. Agueh, M., Carlier, G.: Barycenters in the Wasserstein space. SIAM J. Math. Anal. **43**(2), 904–924 (2011)
2. Bigot, J., Cazelles, E., Papadakis, N.: Penalized barycenters in the Wasserstein space. Submitted. https://128.84.21.199/abs/1606.01025
3. Bobkov, S., Ledoux, M.: One-dimensional empirical measures, order statistics and Kantorovich transport distances (2014). Book in preparation. http://perso.math.univ-toulouse.fr/ledoux/files/2013/11/Order.statistics.10.pdf
4. Burger, M., Osher, S.: Convergence rates of convex variational regularization. Inverse Prob. **20**(5), 1411 (2004)
5. Cuturi, M., Peyré, G.: A smoothed dual approach for variational Wasserstein problems. SIAM J. Imaging Sci. **9**(1), 320–343 (2016)
6. Fréchet, M.: Les éléments aléatoires de nature quelconque dans un espace distancié. Ann. Inst. H. Poincaré Sect. B Prob. Stat. **10**, 235–310 (1948)
7. Kneip, A., Utikal, K.J.: Inference for density families using functional principal component analysis. J. Am. Stat. Assoc. **96**(454), 519–542 (2001). With comments and a rejoinder by the authors
8. Panaretos, V.M., Zemel, Y.: Amplitude and phase variation of point processes. Ann. Stat. **44**(2), 771–812 (2016)
9. Petersen, A., Müller, H.G.: Functional data analysis for density functions by transformation to a Hilbert space. Ann. Stat. **44**(1), 183–218 (2016)
10. Wu, W., Srivastava, A.: An information-geometric framework for statistical inferences in the neural spike train space. J. Comput. Neurosci. **31**(3), 725–748 (2011)
11. Zhang, Z., Müller, H.-G.: Functional density synchronization. Comput. Stat. Data Anal. **55**(7), 2234–2249 (2011)

Extremal Curves in Wasserstein Space

Giovanni Conforti[1] and Michele Pavon[2(✉)]

[1] Laboratoire Paul Painlevé, Université des Sciences et Technologies de Lille 1, 59655 Villeneuve d'Ascq Cedex, France
giovanniconfort@gmail.com

[2] Dipartimento di Matematica "Tullio Levi-Civita", Università di Padova, via Trieste 63, 35121 Padova, Italy
pavon@math.unipd.it

Abstract. We show that known Newton-type laws for Optimal Mass Transport, Schrödinger Bridges and the classic Madelung fluid can be derived from variational principles on Wasserstein space. The second order differential equations are accordingly obtained by annihilating the first variation of a suitable action.

Keywords: Kantorovich-Rubinstein metric · Calculus of variations · Displacement interpolation · Entropic interpolation · Schrödinger bridge · Madelung fluid

1 Introduction

Continuous random evolutions which are critical for some suitable action occur in many diverse fields of science. We have in mind, in particular, the following three famous problems: The Benamou-Brenier formulation of the *Optimal Mass Transport* (OMT) problem with quadratic cost [3], the *Schrödinger Bridge Problem* (SBP) [44,45] and the quantum evolution of a nonrelativistic particle in Madelung's fluid (NSM) [24,31]. All three problems are considered in their *fluid-dynamic* form. The flow of one-time marginals $\{\mu_t; 0 \leq t \leq 1\}$ of each solution may be thought of as a curve in *Wasserstein space*.

It is known that SBP may be viewed as a "regularization" of OMT, the latter problem being recovered through a "zero-noise limit" [10,11,15,29,30,33–35]. Recently, it was shown by von Renesse [48], for the quantum mechanical *Madelung fluid* [24], and by Conforti [13], for the Schrödinger bridge, that their flows satisfy suitable Newton-like laws in Wasserstein space. In [22], the foundations of a Hamilton-Jacobi theory in Wasserstein space were laid.

In this paper, we outline some of the results of [14], where we show that the solution flows of OMT, SBP and NSM may all be seen as extremal curves in Wasserstein space of a suitable action. The actions only differ by the presence or the sign of a (relative) *Fisher information functional* besides the kinetic energy term. The solution marginals flows correspond to critical points, i.e. annihilate the first variation, of the respective functionals. The extremality conditions

F. Nielsen and F. Barbaresco (Eds.): GSI 2017, LNCS 10589, pp. 91–99, 2017.
https://doi.org/10.1007/978-3-319-68445-1_11

imply indeed the local form of Newton-type second laws in analogy to classical mechanics [25, p. 1777]. These are then interpreted in the frame of Otto's formal Riemannian calculus for optimal transport of probability measures as second-order differential equations in Wasserstein space involving the covariant derivative. Although some of these results are present in some form in the cited literature, our goal here is to develop a coherent framework where an actual calculus of variations on Wasserstein space can be developed for various significant problems.

The paper is outlined as follows. In Sects. 2 and 3, we provide some essential background on OMT and SBP, respectively. In Sect. 4, we obtain the second-order differential equation from an extremality condition for the fluid-dynamic version of the Schrödinger problem. The same is then accomplished for the Madelung fluid - Nelson's stochastic mechanics in Sect. 5.

2 Background on Optimal Mass Transport

The literature on this problem is by now vast. We refer the reader to the following monographs and survey papers [1,2,18,39,41,43,46,47]. We shall only briefly review some concepts and results which are relevant for the topics of this paper.

The optimal transport problem may be used to introduce a useful distance between probability measures. Indeed, let $\mathcal{P}_2(\mathbb{R}^N)$ be the set of probability measures μ on $\mathbb{R}^N$ with finite second moment. For $\nu_0, \nu_1 \in \mathcal{P}_2(\mathbb{R}^N)$, the Kantorovich-Rubinstein (Wasserstein) quadratic distance, is defined by

$$W_2(\nu_0, \nu_1) = \left(\inf_{\pi \in \Pi(\nu_0, \nu_1)} \int_{\mathbb{R}^N \times \mathbb{R}^N} \|x - y\|^2 d\pi(x, y) \right)^{1/2}, \tag{1}$$

where $\Pi(\nu_0, \nu_1)$ are "couplings" of ν_0 and ν_1, namely probability distributions on $\mathbb{R}^N \times \mathbb{R}^N$ with marginals ν_0 and ν_1. As is well known [46, Theorem 7.3], W_2 is a *bona fide* distance. Moreover, it provides a most natural way to "metrize" weak convergence[1] in $\mathcal{P}_2(\mathbb{R}^N)$ [46, Theorem 7.12], [1, Proposition 7.1.5] (the same applies to the case $p \geq 1$ replacing 2 with p everywhere). The *Wasserstein space* $\mathcal{W}_2$ is defined as the metric space $\left(\mathcal{P}_2(\mathbb{R}^N), W_2\right)$. It is a *Polish space*, namely a separable, complete metric space. A *dynamic* version of the OMT problem was elegantly accomplished by Benamou and Brenier in [3] by showing that

$$W_2^2(\nu_0, \nu_1) = \inf_{(\mu, v)} \int_0^1 \int_{\mathbb{R}^N} \|v(x, t)\|^2 \mu_t(dx) dt, \tag{2a}$$

$$\frac{\partial \mu}{\partial t} + \nabla \cdot (v\mu) = 0, \tag{2b}$$

$$\mu_0 = \nu_0, \quad \mu_1 = \nu_1. \tag{2c}$$

Here the flow $\{\mu_t; 0 \leq t \leq 1\}$ varies over continuous maps from $[0,1]$ to $\mathcal{P}_2(\mathbb{R}^N)$ and v over smooth fields. In [47, Chap. 7], Villani provides some motivation to

[1] μ_k converges weakly to μ if $\int_{\mathbb{R}^N} f d\mu_k \to \int_{\mathbb{R}^N} f d\mu$ for every continuous, bounded function f.

study the time-dependent version of OMT. Further reasons are the following. It allows to view the optimal transport problem as an (atypical) optimal control problem [7–11]. It provides a ground on which the Schrödinger bridge problem appears as a regularization of the former [10,11,15,29,30,33–35]. Similarly with the Madelung fluid, see below. In some applications, such as interpolation of images [12] or spectral morphing [27], the interpolating flow is crucial.

Let $\{\mu_t^*; 0 \leq t \leq 1\}$ and $\{v^*(x,t); (x,t) \in \mathbb{R}^N \times [0,1]\}$ be optimal for (2). Then

$$\mu_t^* = [(1-t)I + t\nabla\varphi]\,\#\nu_0,$$

with $T = \nabla\varphi$ solving Monge's problem, provides, in McCann's language, the *displacement interpolation* between ν_0 and ν_1 (# denotes "push-forward"). Then $\{\mu_t^*; 0 \leq t \leq 1\}$ may be viewed as a constant-speed geodesic joining ν_0 and ν_1 in Wasserstein space. This formally endows $\mathcal{W}_2$ with a kind of Riemannian structure. McCann discovered [32] that certain functionals are *displacement convex*, namely convex along Wasserstein geodesics. This has led to a variety of applications. Following one of Otto's main discoveries [28,38], it turns out that a large class of PDE's may be viewed as *gradient flows* on the Wasserstein space $\mathcal{W}_2$. This interpretation, because of the displacement convexity of the functionals, is well suited to establish uniqueness and to study energy dissipation and convergence to equilibrium. A rigorous setting in which to make sense of the Otto calculus has been developed by Ambrosio, Gigli and Savaré [1] for a suitable class of functionals. Convexity along geodesics in $\mathcal{W}_2$ also leads to new proofs of various geometric and functional inequalities [32], [46, Chap. 9]. The *tangent space* of $\mathcal{P}_2(\mathbb{R}^N)$ at a probability measure μ, denoted by $T_\mu\mathcal{P}_2(\mathbb{R}^N)$ [1] may be identified with the closure in L^2_μ of the span of $\{\nabla\varphi : \varphi \in C_c^\infty\}$, where C_c^∞ is the family of smooth functions with compact support. It is equipped with the scalar product of L^2_μ.

3 Schrödinger Bridges and Entropic Interpolation

Let $\Omega = C([t_0,t_1];\mathbb{R}^N)$ be the space of $\mathbb{R}^N$ valued continuous functions. Let $W_x^{\sigma^2}$ denote Wiener measure on Ω with variance $\sigma^2 I_N$ starting at the point x at time t_0. If, instead of a Dirac measure concentrated at x, we give the volume measure as initial condition, we get the unbounded measure on path space $W^{\sigma^2} = \int_{\mathbb{R}^N} W_x^{\sigma^2} dx$. It is useful to introduce the family of distributions $\mathcal{P}$ on Ω which are equivalent to it. Let $\mathcal{P}(\rho_0,\rho_1)$ denote the set of distributions in $\mathcal{P}$ having the prescribed marginal marginals densities at $t = 0$ and $t = 1$, respectively. Then, the *Schrödinger bridge problem* (SBP) with W^{σ^2} as "prior" is the maximum entropy problem

$$\text{Minimize} \quad H(P|W^{\sigma^2}) = \mathbb{E}_P\left[\log\frac{dP}{dW^{\sigma^2}}\right] \quad \text{over} \quad P \in \mathcal{P}(\rho_0,\rho_1). \tag{3}$$

Conditions for existence and uniqueness for this problem and properties of the minimizing measure (with general Markovian prior) have been studied by many authors, most noticeably by Fortet, Beurlin, Jamison and Föllmer [4,20,21,26,29],[30, Proposition 2.5]. The solution P^* is called *the Schrödinger bridge* from ρ_0 to ρ_1 over P [20]. We shall tacitly assume henceforth that they are satisfied so that P^* is well defined. In view of Sanov's theorem [42], solving the maximum entropy problem is equivalent to a problem of large deviations of the empirical distribution as showed by Föllmer [20] recovering Schrödinger's original motivation [44,45]. It has been observed since the early nineties that SBP can be turned, thanks to *Girsanov's theorem*, into a stochastic control problem with atypical boundary constraints, see [5,16,17,19,40]. The latter has a fluid dynamic counterpart: When prior is W^{σ^2} stationary Wiener measure with variance σ^2 [2], the solution of the SBP with marginal densities ρ_0 and ρ_1 can be characterized as the solution of the fluid-dynamic problem [10, p. 683], [23, Corollary 5.8]:

$$\inf_{(\rho,v)} \int_{\mathbb{R}^n}\int_0^1 \left[\frac{1}{2\sigma^2}\|v(x,t)\|^2 + \frac{\sigma^2}{8}\|\nabla \log \rho(x,t)\|^2\right]\rho(x,t)dtdx, \tag{4a}$$

$$\frac{\partial \rho}{\partial t} + \nabla\cdot(v\rho) = 0, \tag{4b}$$

$$\rho(0,x) = \rho_0(x), \quad \rho(1,y) = \rho_1(y), \tag{4c}$$

Notice that the only difference from the Benamou-Brenier problem (2) is given by an extra term in the action with the form of a *Fisher Information functional*

$$I(\rho) = \int_{\mathbb{R}^n} \frac{\|\nabla\rho\|^2}{\rho}dx. \tag{5}$$

4 Variational Analysis for the Fluid-Dynamic SBP

Let $\mathcal{P}_{\rho_0\rho_1}$ be the family of continuous flows of probability densities $\rho = \{\rho(\cdot,t); 0\le t\le 1\}$ satisfying (4c) and let $\mathcal{V}$ be the family of continuous feedback control laws $v(\cdot,\cdot)$. Consider the unconstrained minimization over $\mathcal{P}_{\rho_0\rho_1}\times\mathcal{V}$ of the Lagrangian

$$\begin{aligned}\mathcal{L}(\rho,v;\lambda) = \int_{\mathbb{R}^n}\int_0^1 \Bigg\{&\left[\frac{1}{2\sigma^2}\|v(x,t)\|^2 + \frac{\sigma^2}{8}\|\nabla\log\rho(x,t)\|^2\right]\rho(x,t)\\ &+\lambda(x,t)\left(\frac{\partial\rho}{\partial t}+\nabla\cdot(v\rho)\right)\Bigg\}dtdx,\end{aligned}$$

where λ is a C^1 Lagrange multiplier. After integration by parts and discarding the constant boundary terms, we get the problem of minimizing over $\mathcal{P}_{\rho_0\rho_1}\times\mathcal{V}$

$$\int_{\mathbb{R}^n}\int_0^1\left[\frac{1}{2\sigma^2}\|v(x,t)\|^2 + \frac{\sigma^2}{8}\|\nabla\log\rho(x,t)\|^2 + \left(-\frac{\partial\lambda}{\partial t} - \nabla\lambda\cdot v\right)\right]\rho(x,t)dtdx. \tag{6}$$

[2] The case of a general reversible Markovian prior is treated in [14] where all the details of the variational analysis may also be found.

Pointwise minimization with respect to v for a fixed flow in $\mathcal{P}_{\rho_0\rho_1}$ gives

$$v_\rho^*(x,t) = \sigma^2 \nabla\lambda(x,t), \tag{7}$$

which is continuous. Plugging this into (6), we get to minimize over $\mathcal{P}_{\rho_0\rho_1}$

$$J(\rho) = -\int_{\mathbb{R}^n}\int_0^1 \left[\frac{\partial\lambda}{\partial t} + \frac{\sigma^2}{2}\|\nabla\lambda\|^2 - \frac{\sigma^2}{8}\|\nabla\log\rho\|^2\right]\rho dt dx. \tag{8}$$

Setting the first variation of J in direction $\delta\rho$ equal to zero for all smooth variations vanishing at times $t=0$ and $t=1$, we get the *extremality condition*

$$\frac{\partial\lambda}{\partial t} + \frac{\sigma^2}{2}\|\nabla\lambda\|^2 + \frac{\sigma^2}{8}\|\nabla\log\rho\|^2 + \frac{\sigma^2}{4}\Delta\log\rho = 0. \tag{9}$$

By (9), the convective derivative of v^* in (7) yields the acceleration field

$$a^*(x,t) = \left(\frac{\partial}{\partial t} + v^*\cdot\nabla\right)(v^*)(x,t) = -\sigma^2\nabla\left(\frac{\sigma^2}{8}\|\nabla\log\rho\|^2 + \frac{\sigma^2}{4}\Delta\log\rho\right). \tag{10}$$

The term appearing in the right-hand side of (10) may be viewed as a gradient in Wasserstein space of the entropic part of the Lagrangian (4a), namely

$$-\nabla\left(\frac{\sigma^2}{8}\|\nabla\log\rho\|^2 + \frac{\sigma^2}{4}\Delta\log\rho\right) = \nabla^{\mathcal{W}_2}\frac{\sigma^2}{8}\mathcal{I}(\rho). \tag{11}$$

This relation can be found in [48, A.2] and [13]. Indeed, a calculation shows

$$\frac{d}{dt}\int_{\mathbb{R}^n}\|\nabla\log\rho(x,t)\|^2\rho(x,t)dx = -\int_{\mathbb{R}^n}\nabla\left[\|\nabla\log\rho\|^2 + 2\Delta\log\rho\right]\cdot v(x,t)\rho(x,t)dx.$$

Finally, since v^* is a gradient, it follows that the convective derivative a^* is in fact the *covariant derivative* $\nabla_{\dot\mu}^{\mathcal{W}_2}\dot\mu$ for the smooth curve $t\to\mu_t=\rho(x,t)dx$, see [48, A.3] and [2, Chap. 6]. Thus, (10) takes on the form of a Newton-type law [13] on $\mathcal{W}_2$

$$\nabla_{\dot\mu}^{\mathcal{W}_2}\dot\mu = \frac{\sigma^4}{8}\nabla^{\mathcal{W}_2}\mathcal{I}(\mu). \tag{12}$$

In the case when $\sigma^2\searrow 0$, we recover in the limit that displacement interpolation provides constant speed geodesics for which $\nabla_{\dot\mu}^{\mathcal{W}_2}\dot\mu\equiv 0$.

5 Optimal Transport and Nelson's Stochastic Mechanics

There has been some interest in connecting optimal transport with Nelson's stochastic mechanics [6], [47, p. 707] or directly with the Madelung fluid [48]. Consider the case of a free non-relativistic particle of mass m (the case of a particle in a force field is treated similarly). Then, a variational principle leading to the Schrödinger equation, can be based on the Guerra-Morato action functional [25] which, in fluid dynamic form, is

$$\mathcal{A}_{GM} = \int_{t_0}^{t_1} \left[\int_{\mathbb{R}^n} \frac{m}{2} \|v(x,t)\|^2 \rho_t(x) dx - \frac{\hbar^2}{8m} I(\rho_t) \right] dt. \tag{13}$$

For the Nelson process, we have $\sigma^2 = \hbar/m$. It can be derived directly from the classical action [37]. Instead, the Yasue action [49] in fluid-dynamic form is

$$\mathcal{A}_Y(t_0, t_1) = \int_{t_0}^{t_1} \left[\int_{\mathbb{R}^N} \frac{m}{2} \|v(x,t)\|^2 \rho_t(x) dx + \frac{\hbar^2}{8m} I(\rho_t) \right] dt. \tag{14}$$

In [6, p. 131], Eric Carlen poses the question of minimizing the Yasue action subject to the continuity equation (4b) for given initial and final marginals (4c). In view of the formulation (4), we already know the solution: It is provided by Nelson's current velocity [36] and the flow of one-time densities of the Schrödinger bridge with (4c) and stationary Wiener measure as a prior. Finally observe that the action in (4a) and $\frac{1}{\hbar}\mathcal{A}_{GM}$ only differ by the sign in front of the Fisher information functional! Thus the variational analysis can be carried out as in the previous section obtaining eventually the Newton-type law as in [48]

$$\nabla^{\mathcal{W}_2}_{\dot{\mu}} \dot{\mu} = -\frac{\hbar^2}{m^2 8} \nabla^{\mathcal{W}_2} \mathcal{I}(\mu). \tag{15}$$

6 Conclusion

We have outlined some of the results of [14] where the optimal evolutions for OMT, SBP and NSM are shown to be critical curves for suitable actions. These extremality conditions lead to second-order differential equations on Wasserstein space. A number of piecemeal results available in various fields of science such as Optimal Mass Transport, Statistical Mechanics and Quantum Mechanics can all be cast in this coherent picture.

References

1. Ambrosio, L., Gigli, N., Savaré, G.: Gradient Flows in Metric Spaces and in the Space of Probability Measures. Lectures in Mathematics ETH Zürich, 2nd edn. Birkhäuser Verlag, Basel (2008)
2. Ambrosio, L., Gigli, N.: A user's guide to optimal transport. In: Piccoli, B., Rascle, M. (eds.) Modeling and optimisation of flows on networks. Lecture Notes in Mathematics, vol. 2062, pp. 1–155. Springer, Heidelberg (2013). doi:10.1007/978-3-642-32160-3_1
3. Benamou, J., Brenier, Y.: A computational fluid mechanics solution to the Monge-Kantorovich mass transfer problem. Numer. Math. **84**, 375–393 (2000)
4. Beurling, A.: An automorphism of product measures. Ann. Math. **72**, 189–200 (1960)

5. Blaquière, A.: Controllability of a Fokker-Planck equation, the Schrödinger system, and a related stochastic optimal control. Dyn. Control **2**(3), 235–253 (1992)
6. Carlen, E.: Stochastic mechanics: a look back and a look ahead. In: Faris, W.G. (ed.) Diffusion, Quantum Theory and Radically Elementary Mathematics. Mathematical Notes, vol. 47, pp. 117–139. Princeton University Press (2006)
7. Chen, Y., Georgiou, T.T., Pavon, M.: Optimal steering of a linear stochastic system to a final probability distribution. Part I. IEEE Trans. Aut. Control **61**(5), 1158–1169 (2016). arXiv:1408.2222v1
8. Chen, Y., Georgiou, T.T., Pavon, M.: Optimal steering of a linear stochastic system to a final probability distribution. Part II. IEEE Trans. Aut. Control **61**(5), 1170–1180 (2016). arXiv:1410.3447v1
9. Chen, Y., Georgiou, T.T., Pavon, M.: Fast cooling for a system of stochastic oscillators. J. Math. Phys. **56**(11), 113302 (2015). arXiv:1411.1323v2
10. Chen, Y., Georgiou, T.T., Pavon, M.: On the relation between optimal transport and Schrödinger bridges: a stochastic control viewpoint. J. Optim. Theory Appl. **169**(2), 671–691 (2016). doi:10.1007/s10957-015-0803-z. Published Online 2015. arXiv:1412.4430v1
11. Chen, Y., Georgiou, T.T., Pavon, M.: Optimal transport over a linear dynamical system. IEEE Trans. Autom. Control (2017, to appear). arXiv:1502.01265v1
12. Chen, Y., Georgiou, T.T., Pavon, M.: Entropic and displacement interpolation: a computational approach using the Hilbert metric. SIAM J. Appl. Math. **76**(6), 2375–2396 (2016). arXiv:1506.04255v1
13. Conforti, G.: A second order equation for Schrödinger bridges with applications to the hot gas experiment and entropic transportation cost. (2017, preprint). arXiv:1704.04821v2
14. Conforti, G., Pavon, M.: Extremal flows on Wasserstein space (2017, under preparation)
15. Cuturi, M.: Sinkhorn distances: lightspeed computation of optimal transport. In: Advances in Neural Information Processing Systems, pp. 2292–2300 (2013)
16. Dai Pra, P.: A stochastic control approach to reciprocal diffusion processes. Appl. Math. Optim. **23**(1), 313–329 (1991)
17. Dai Pra, P., Pavon, M.: On the Markov processes of Schroedinger, the Feynman-Kac formula and stochastic control. In: Kaashoek, M.A., van Schuppen, J.H., Ran, A.C.M. (eds.) Proceedings of MTNS Conference on Realization and Modeling in System Theory, pp. 497–504. Birkaeuser, Boston (1990)
18. Evans, L.C.: Partial differential equations and Monge-Kantorovich mass transfer. Curr. Dev. Math. **1977**, 65–126 (1999). International Press, Boston
19. Fillieger, R., Hongler, M.-O., Streit, L.: Connection between an exactly solvable stochastic optimal control problem and a nonlinear reaction-diffusion equation. J. Optimiz. Theory Appl. **137**, 497–505 (2008)
20. Föllmer, H.: Random fields and diffusion processes. In: Hennequin, P.-L. (ed.) École d'Été de Probabilités de Saint-Flour XV–XVII, 1985–1987. LNM, vol. 1362, pp. 101–203. Springer, Heidelberg (1988). doi:10.1007/BFb0086180
21. Fortet, R.: Résolution d'un système d'equations de M. Schrödinger. J. Math. Pure Appl. **IX**, 83–105 (1940)
22. Gangbo, W., Nguyen, T., Tudorascu, A.: Hamilton-Jacobi equations in the Wasserstein space. Methods Appl. Anal. **15**(2), 155–183 (2008)
23. Gentil, I., Lonard, C., Ripani, L.: About the analogy between optimal transport and minimal entropy. Ann. Fac. Toulouse (2017, to appear). arXiv: 1510.08230
24. Guerra, F.: Structural aspects of stochastic mechanics and stochastic field theory. Phys. Rep. **77**, 263–312 (1981)

25. Guerra, F., Morato, L.: Quantization of dynamical systems and stochastic control theory. Phys. Rev. D **27**(8), 1774–1786 (1983)
26. Jamison, B.: The Markov processes of Schrödinger. Z. Wahrscheinlichkeitstheorie verw. Gebiete **32**, 323–331 (1975)
27. Jiang, X., Luo, Z., Georgiou, T.: Geometric methods for spectral analysis. IEEE Trans. Signal Process. **60**(3), 106471074 (2012)
28. Jordan, R., Kinderlehrer, D., Otto, F.: The variational formulation of the Fokker-Planck equation. SIAM J. Math. Anal. **29**, 1–17 (1998)
29. Léonard, C.: A survey of the Schroedinger problem and some of its connections with optimal transport. Discrete Contin. Dyn. Syst. A **34**(4), 1533–1574 (2014)
30. Léonard, C.: From the Schrödinger problem to the Monge-Kantorovich problem. J. Funct. Anal. **262**, 1879–1920 (2012)
31. Madelung, E.: Quantentheorie in hydrodynamischer Form. Z. Phys. **40**, 322–326 (1926)
32. McCann, R.: A convexity principle for interacting gases. Adv. Math. **128**(1), 153–179 (1997)
33. Mikami, T.: Monge's problem with a quadratic cost by the zero-noise limit of h-path processes. Probab. Theory Relat. Fields **129**, 245–260 (2004)
34. Mikami, T., Thieullen, M.: Duality theorem for the stochastic optimal control problem. Stoch. Proc. Appl. **116**, 1815–1835 (2006)
35. Mikami, T., Thieullen, M.: Optimal transportation problem by stochastic optimal control. SIAM J. Control Optim. **47**(3), 1127–1139 (2008)
36. Nelson, E.: Dynamical Theories of Brownian Motion. Princeton University Press, Princeton (1967)
37. Nelson, E.: Stochastic mechanics and random fields. In: Hennequin, P.-L. (ed.) École d'Été de Probabilités de Saint-Flour XV–XVII, 1985–87. LNM, vol. 1362, pp. 427–459. Springer, Heidelberg (1988). doi:10.1007/BFb0086184
38. Otto, F.: The geometry of dissipative evolution equations: the porous medium equation. Commun. Part. Differ. Equ. **26**(1–2), 101–174 (2001)
39. Ollivier, Y., Pajot, H., Villani, C.: Optimal Transportation. Theory and Applications, London Mathematical Society. Lecture Notes Series, vol. 413. Cambridge University Press (2014)
40. Pavon, M., Wakolbinger, A.: On free energy, stochastic control, and Schroedinger processes. In: Di Masi, G.B., Gombani, A., Kurzhanski, A. (eds.) Modeling, Estimation and Control of Systems with Uncertainty, pp. 334–348. Birkauser, Boston (1991)
41. Rachev, T., Rüschendorf, L.: Mass Transportation Problems, Volume I: Theory, Volume II: Applications. Probability and Its Applications. Springer, New York (1998)
42. Sanov, I.S.: On the probability of large deviations of random magnitudes (in Russian). Mat. Sb. N. S. **42**(84) 111744 (1957). Select. Transl. Math. Stat. Probab. **1**, 213–244 (1961)
43. Santambrogio, F.: Optimal Transport for Applied Mathematicians. Birkhäuser (2015)
44. Schrödinger, E.: Über die Umkehrung der Naturgesetze. Sitzungsberichte der Preuss Akad. Wissen. Berlin Phys. Math. Klasse, 144–153 (1931)
45. Schrödinger, E.: Sur la théorie relativiste de l'electron et l'interpretation de la mécanique quantique. Ann. Inst. H. Poincaré **2**, 269 (1932)
46. Villani, C.: Topics in Optimal Transportation, vol. 58. AMS (2003)
47. Villani, C.: Optimal Transport. Old and New. Grundlehren der Mathematischen Wissenschaften, vol. 338. Springer, Heidelberg (2009)

48. von Renesse, M.-K.: An optimal transport view of Schrödinger's equation. Canad. Math. Bull. **55**, 858–869 (2012)
49. Yasue, K.: Stochastic calculus of variations. J. Funct. Anal. **41**, 327–340 (1981)

Semi-discrete Optimal Transport in Patch Space for Enriching Gaussian Textures

Bruno Galerne[1], Arthur Leclaire[2](✉), and Julien Rabin[3]

[1] Laboratoire MAP5, Université Paris Descartes and CNRS, Sorbonne Paris Cité, Paris, France

[2] CMLA, ENS Cachan, CNRS, Université Paris-Saclay, 94235 Cachan, France
Arthur.Leclaire@cmla.ens-cachan.fr

[3] Normandie Univ, ENSICAEN, CNRS, GREYC, 14000 Caen, France

Abstract. A bilevel texture model is proposed, based on a local transform of a Gaussian random field. The core of this method relies on the optimal transport of a continuous Gaussian distribution towards the discrete exemplar patch distribution. The synthesis then simply consists in a fast post-processing of a Gaussian texture sample, boiling down to an improved nearest-neighbor patch matching, while offering theoretical guarantees on statistical compliancy.

Keywords: Optimal transport · Texture synthesis · Patch distribution

1 Introduction

Designing models for realistic and fast rendering of structured textures is a challenging research topic. In the past, many models have been proposed for exemplar-based synthesis, which consists in synthesizing a piece of texture having the same perceptual characteristics as an observed texture sample with still some innovative content. Several authors [2,24] have proposed non-parametric methods based on progressive sampling of the texture using a copy-paste principle. This paved the way to many other successful synthesis methods relying on patch-based sampling [3,11–13,16,19] (see [23] for a detailed review). Even if these methods can be applied very efficiently [13], they do not offer much mathematical guarantees (except asymptotic results [14]) which reflects for example in the growing garbage effect described in [2].

In contrast Gaussian texture models [4] are stationary and inherently respect the frequency content of the input texture. They can be efficiently simulated on the discrete plane $\mathbb{Z}^2$ [6] and even generalized to the framework of procedural noises defined in the continuous domain $\mathbb{R}^2$ (see e.g. [7] and references therein). They also allow for dynamic texture synthesis and mixing [25] and inpainting [5]. However the Gaussian model is intrinsically limited since the color distribution of the output is always symmetric around the mean color and the local texture patterns cannot contain salient features such as contrasted contours.

F. Nielsen and F. Barbaresco (Eds.): GSI 2017, LNCS 10589, pp. 100–108, 2017.
https://doi.org/10.1007/978-3-319-68445-1_12

The main purpose of this work is to propose a theoretically sound post-processing of the Gaussian model to cope with some of these limitations. Since the color and local pattern information is a part of the patch distribution, we propose to apply a local operation that will transform the patch distribution of the Gaussian texture into the patch distribution of the input image. This can naturally be addressed using a semi-discrete optimal transport plan.

Several tools from optimal transport (OT) have already been applied to texture synthesis: Rabin et al. [20] and Xia et al. [25] formulate texture mixing via Wasserstein barycenters, Tartavel et al. [21] use Wasserstein distances in a variational formulation of texture synthesis, and Gutierrez et al. [9] apply discrete OT in order to specify a global patch distribution. Here, we suggest to locally apply a semi-discrete transport plan which can be seen as a tweaked nearest-neighbor projection.

Indeed, semi-discrete OT corresponds to assign the centers of Laguerre cells to each point of a well-designed Laguerre partition [1,17]. So far, deterministic methods for solving semi-discrete methods have been limited to dimensions $D = 2$ [17] and $D = 3$ [15] using explicit geometric construction of the Laguerre tessellation. Recently, Genevay et al. [8] proposed several stochastic optimization schemes for solving entropy-regularized OT problems.

In Sect. 2 we summarize the non-regularized semi-discrete OT framework and the numerical solution given in [8]. In Sect. 3 we show how to use this algorithm to transport patch distributions (in dimensions $D = 27$ or higher) thus enriching the Gaussian texture model. The resulting bilevel algorithm, can be seen as a non-iterative version of [11] with a global statistical constraint, which is confirmed by the experiments of Sect. 4.

2 Semi-discrete Optimal Transport

In this section, we recall the framework for semi-discrete optimal transport and the numerical solution given by Genevay et al. [8].

2.1 The Optimal Transport Problem and Its Dual Formulation

Let μ, ν be two probability measures on $\mathbb{R}^D$. We assume that μ has a bounded probability density function ρ and that ν is a discrete measure $\nu = \sum_{y \in S} \nu_y \delta_y$ with finite support S. Let us denote by $\Pi(\mu, \nu)$ the set of probability measures on $\mathbb{R}^D \times \mathbb{R}^D$ having marginal distributions μ, ν. If T is a measurable map, we denote by $T_\sharp \mu$ the push-forward measure defined as $T_\sharp \mu(A) = \mu(T^{-1}(A))$. If $v \in \mathbb{R}^S$, we define the c-transform of v with respect to the cost $c(x, y) = \|x - y\|^2$ as $v^c(x) = \min_{y \in S} \|x - y\|^2 - v(y)$, and T_v the corresponding assignment uniquely defined almost everywhere by

$$T_v(x) = \operatorname*{argmin}_{y \in S} \|x - y\|^2 - v(y). \tag{1}$$

When $v = 0$, we get the nearest neighbor (NN) projection which assigns to x the closest point in S (unique for almost all x).

The preimages of T_v define a partition of $\mathbb{R}^D$ (up to a negligible set), called the power diagram or Laguerre tessellation, the cells of which are defined as

$$\mathrm{Pow}_v(y) = \{x \in \mathbb{R}^D \mid \forall z \in S \setminus \{y\},\ \|x-y\|^2 - v(y) < \|x-z\|^2 - v(z)\}. \quad (2)$$

The Kantorovich formulation of optimal transport consists in solving

$$\min_{\pi \in \Pi(\mu,\nu)} \int_{\mathbb{R}^D \times \mathbb{R}^D} \|x-y\|^2 d\pi(x,y). \quad (3)$$

It has been shown [1,10,15,22] that this problem admits solutions of the form $(\mathrm{Id} \times T_v)_\sharp \mu$ where v solves the dual problem

$$\operatorname*{argmax}_{v \in \mathbb{R}^S} H(v) \quad \text{where} \quad H(v) = \int_{\mathbb{R}^D} v^c(x)\rho(x)dx + \sum_{y \in S} v(y)\nu_y. \quad (4)$$

The same authors have shown that the function H is concave, $\mathcal{C}^1$-smooth, and that its gradient is given by $\frac{\partial H}{\partial v(y)} = -\mu(\mathrm{Pow}_v(y)) + \nu_y$. Thus, v is a critical point of H if and only if $\mu(\mathrm{Pow}_v(y)) = \nu_y$ for all y, which means that $(T_v)_\sharp \mu = \nu$.

2.2 Stochastic Optimization

Genevay et al. [8] have suggested to address the maximization of (4) by using a stochastic gradient ascent, which is made possible by writing

$$H(v) = \mathbb{E}[h(X,v)] \quad \text{where} \quad h(x,v) = v^c(x) + \sum_{y \in S} v(y)\nu_y \quad (5)$$

and where X is a random variable with distribution μ. Notice that for $x \in \mathrm{Pow}_v(y)$, $v \mapsto v^c(x)$ is smooth with gradient $-e_y$ (where $\{e_y\}_{y \in S}$ is the canonical basis of $\mathbb{R}^S$). Therefore, for any $w \in \mathbb{R}^S$, for almost all $x \in \mathbb{R}^D$, $v \mapsto h(x,v)$ is differentiable at w and $\nabla_v h(x,w) = -e_{T_w(x)} + \nu$. (abusing notation $(\nu_y) \in \mathbb{R}^S$).

In order to minimize $-H$, Genevay et al. propose the following averaged stochastic gradient descent (ASGD) initialized with $\tilde{v}_1 = 0$

$$\begin{cases} \tilde{v}_k = \tilde{v}_{k-1} + \frac{C}{\sqrt{k}} \nabla_v h(x_k, \tilde{v}_{k-1}) & \text{where } x_k \sim \mu \\ v_k = \frac{1}{k}(\tilde{v}_1 + \ldots + \tilde{v}_k). \end{cases} \quad (6)$$

Since $\nabla_v h(x, \tilde{v}_{k-1})$ exists x-a.s. and is bounded, the convergence of this algorithm is ensured by [18, Theorem 7], in the sense $\max(H) - \mathbb{E}[H(v_k)] = \mathcal{O}(\frac{\log k}{\sqrt{k}})$. The authors of [8] also proposed to address a regularized transport problem with a similar method but we do not discuss it here due to lack of space.

3 A Bilevel Model for Texture Synthesis

In this section, we introduce a bilevel model for texture synthesis that consists in first synthesizing a Gaussian version of the texture (with long range correlations but no geometric structures) and next transforming each patch of the synthesis with an optimal assignment (in order to enforce the patch distribution of the exemplar texture).

To be more precise, let us denote by $u : \Omega \to \mathbb{R}^d$ the exemplar texture defined on a discrete domain $\Omega \subset \mathbb{Z}^2$. Let $U : \mathbb{Z}^2 \to \mathbb{R}^d$ be the asymptotic discrete spot noise (ADSN) [4,6] associated with u, which is defined as

$$\forall x \in \mathbb{Z}^2,\ U(x) = \bar{u} + \sum_{y \in \mathbb{Z}^2} t_u(y) W(x-y) \quad \text{where} \begin{cases} \bar{u} = \frac{1}{|\Omega|} \sum_{x \in \Omega} u(x), \\ t_u = \frac{1}{\sqrt{|\Omega|}} (u - \bar{u}) \mathbf{1}_\Omega \end{cases} \tag{7}$$

where $\mathbf{1}_\Omega$ is the indicator function, and where W is a normalized Gaussian white noise on $\mathbb{Z}^2$ (the convolution between t_u and W is computed in Fourier domain). U is a stationary Gaussian random field whose first and second order moments are the empirical mean and covariance of the exemplar texture. In particular

$$\mathbb{E}[(U(x) - \bar{u})(U(y) - \bar{u})] = a_u(y-x) \quad \text{where} \quad a_u(z) = \sum_{x \in \mathbb{Z}^2} t_u(z) t_u(x+z)^T. \tag{8}$$

Thus, U can be considered as a "Gaussianized" version of u, that has the correct correlations but no salient structures.

The second step consists in a patchwise operation. Let $\omega = \{-r, \dots, r\}^2$ be the patch domain with $r \in \mathbb{N}$. Let us denote by μ the distribution of patches of U, that is $\mu = \mathcal{N}(\bar{u}, C)$ with $C(x, y) = a_u(y - x)$, $x, y \in \omega$. Let us denote by ν the empirical distribution of patches of the exemplar texture u, that is $\nu = \frac{1}{|S|} \sum_{p \in S} \delta_p$, where $S = \{u_{|x+\omega} \mid x + \omega \subset \Omega\}$. Actually, in practice we approximate it with $\nu = \frac{1}{J} \sum_{j=1}^{J} \delta_{p_j}$ where $p_1, \dots, p_J$ are $J = 1000$ patches randomly drawn from the exemplar texture. Thus, μ and ν are two probability measures on $\mathbb{R}^D$ with $D = d(2r+1)^2$. Besides, μ is a Gaussian distribution that, except in degenerate cases, admits a probability density function ρ with respect to the Lebesgue measure. Using the algorithm explained in Sect. 2.2, we compute the optimal assignment T_v that realizes the semi-discrete OT from μ to ν. We then apply this mapping T_v to each patch of the Gaussian synthesis U, and we recompose an image V by averaging the "transported" patches: the value at pixel x is the average of values of x in all overlapping patches. More formally,

$$\forall x \in \mathbb{Z}^2, \quad P_x = T_v(U_{|x+\omega}), \tag{9}$$

$$\forall x \in \mathbb{Z}^2, \quad V(x) = \frac{1}{|\omega|} \sum_{h \in \omega} P_{x-h}(h). \tag{10}$$

Proposition 1. *V is a stationary random field on $\mathbb{Z}^2$ and satisfies the following long-range independence property: if Γ denotes the finite support of the auto-correlation function a_u defined in* (8)*, then for every $A, B \subset \mathbb{Z}^2$ such that $(A - B) \cap (\Gamma + 4\omega) = \varnothing$ the restrictions $V_{|A}, V_{|B}$ are independent.*

Proof. Since U is Gaussian, $U(x) \perp\!\!\!\perp U(y)$ as soon as $x - y \notin \Gamma$. Therefore, if $x - y \notin \Gamma + 2\omega$, then $U_{|x+\omega} \perp\!\!\!\perp U_{|y+\omega}$ and thus $P_x \perp\!\!\!\perp P_y$. After averaging we get $V(x) \perp\!\!\!\perp V(y)$ as soon as $x - y \notin \Gamma + 4\omega$. The generalization to subsets A, B is straightforward.

This property is a guarantee of spatial stability for synthesis, meaning that the synthesis algorithm will not start to "grow garbage" as may do the method of [2]. We also have a guarantee on the patch distribution. Indeed, if T_v is the true solution to the semi-discrete optimal-transport problem, then any patch P_x is exactly distributed according to ν. After recomposition, the distribution of $V_{|\omega}$ may not be exactly ν but is expected to be not too far away (as will be confirmed in Fig. 4). In this sense, V still respects the long range correlations (inherited from U) while better preserving the local structures.

4 Results and Discussion

In this section, we discuss experimental results of texture synthesis obtained with the proposed bilevel model. Figure 1 compares the synthesis results obtained with the Gaussian model before and after local transformation. One clearly observes the benefit of applying the optimal transport in the patch domain: it restores the color distribution of the exemplar and also creates salient features from the Gaussian content. In particular (Rows 1 and 3), it is able to break the symmetry of the color distribution which is a strong restriction of the Gaussian model.

Figures 1 and 2 demonstrate that the optimal assignment (OT) is better suited than the simple NN projection, as illustrated in Fig. 3(a). While these two operators project on the exemplar patches, the assignment is optimized to globally respect a statistical constraint. On Fig. 2 we also question the use of different patch sizes. The simple 1×1 case (which only amounts to apply a color map to the Gaussian field) is already interesting because it precisely respects the color distribution of the exemplar. Thus the bilevel model with 1×1 patch can be seen as an extension of the ADSN model with prescribed marginal distributions. For a larger patch size, we observe only minor improvements of the model with slightly cleaner geometric structures.

One possible explanation could be the slow convergence speed in very high dimensions of the stochastic gradient descent scheme. In this non-smooth setting ($\nabla v(x, \cdot)$ is not Lipschitz continuous), the convergence rate given in [18] is only $\mathcal{O}(\frac{\log k}{\sqrt{k}})$. Besides, in our setting, we apply this algorithm for 10^6 iterations with gradient step parameter $C = 1$ in a very high-dimensional space $\mathbb{R}^D$ ($D = 27$ for 3×3 RGB patches), and with discrete target distributions ν having large

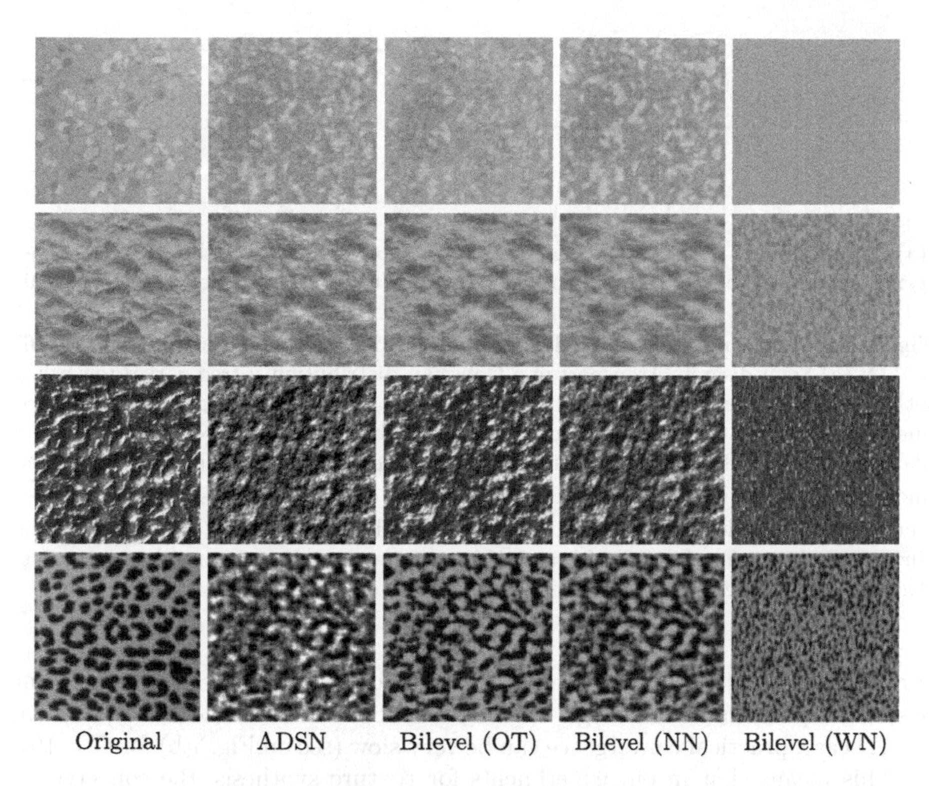

Fig. 1. Bilevel synthesis. Each row displays an exemplar texture, a sample of the associated ADSN model, and samples of the bilevel models obtained with 3×3 patch optimal assignment (OT) or patch nearest neighbor projection (NN), and the one with white noise initialization (WN). The OT assignment better preserves patch statistics than the NN projection. Besides, the last column illustrates the importance to start from a spatially correlated Gaussian model at the first level.

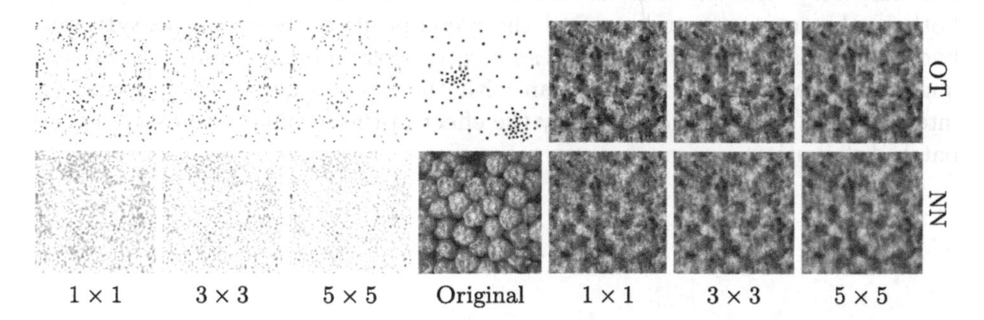

Fig. 2. Influence of the patch size. On the middle column we display two original textures and on the other columns we display samples of the bilevel models with varying patch size using the patch optimal assignment (OT, first row) and the patch nearest neighbor projection (NN, second row). The OT performs in general better than a NN (it better preserves the color/patch statistics) but fails to reproduce complex geometry (like in the right example). (Color figure online)

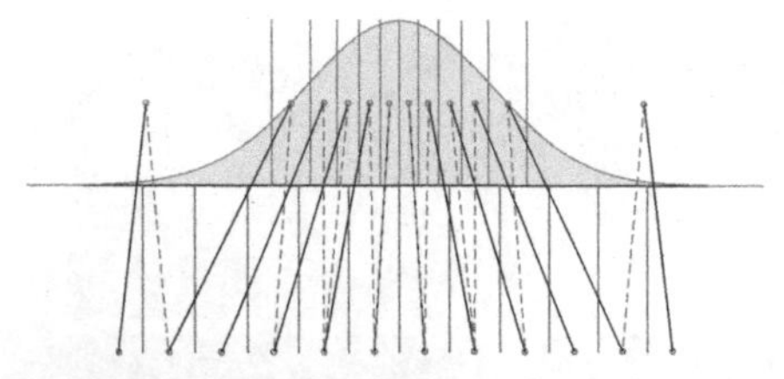

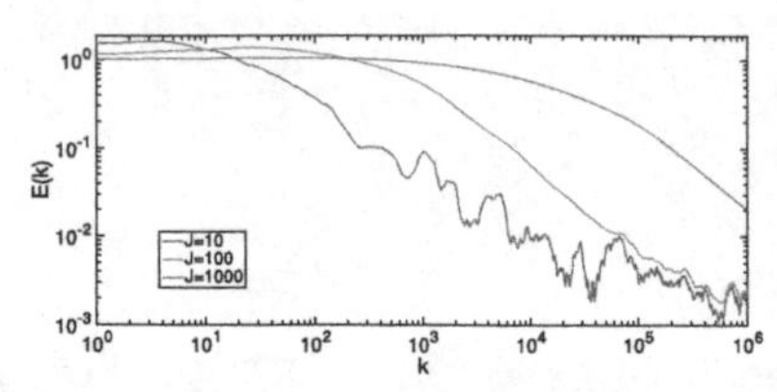

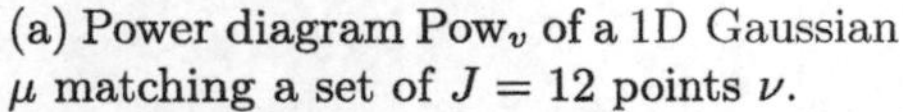

(a) Power diagram Pow_v of a 1D Gaussian μ matching a set of $J = 12$ points ν.

(b) Convergence of the ASGD algorithm (6) in 1D for J points, and $C = 10$

Fig. 3. Illustration of the semi-discrete OT and the convergence of ASGD (6) **in 1D.** (a) Semi-discrete Transport of a Gaussian distribution μ (red curve) towards a set of points ν (blue dots with $J = 12$). The corresponding power diagram Pow_v (in red lines) is compared to the Voronoi diagram Pow_0 (blue lines). The optimal transport plan T_v (black lines) is compared to the nearest-neighbor matching T_0 (grey dotted lines). (b) ASGD in 1D. Evolution of the relative error, defined as $E(k) = \frac{\|v_k - v^\star\|}{\|v^\star\|}$ where $v^\star$ is the (closed-form) optimal solution and k the number of iterations. The curves are shown for $J = 10$, 10^2, and 10^3 points, using the same random sequence. (Color figure online)

support ($J = |S| = 10^3$). Updating all coordinates of v during the gradient descent requires to visit enough all the power cells. As a consequence, even in the 1D case, practical convergence can be very slow (see on Fig. 3(b) for $C = 10$).

This means that in our experiments for texture synthesis, the convergence is not reached, and yet the OT solution provides better synthesis results than a simple NN approach, which is confirmed by examining the output patch distribution (Fig. 4). Interestingly, it is also true after patch recomposition (which remains to be properly justified).

The results obtained with this bilevel model raise several questions. First, one may very well question the use of the ℓ^2-distance for patch comparison, both in the transport problem and the recomposition step. It is already well known [11] that using other distances for recomposition may improve the visual quality of the results (less blur than with the ℓ^2 average). Also, it would be interesting to analyze more precisely the effect of the recomposition step on the patch distribution.

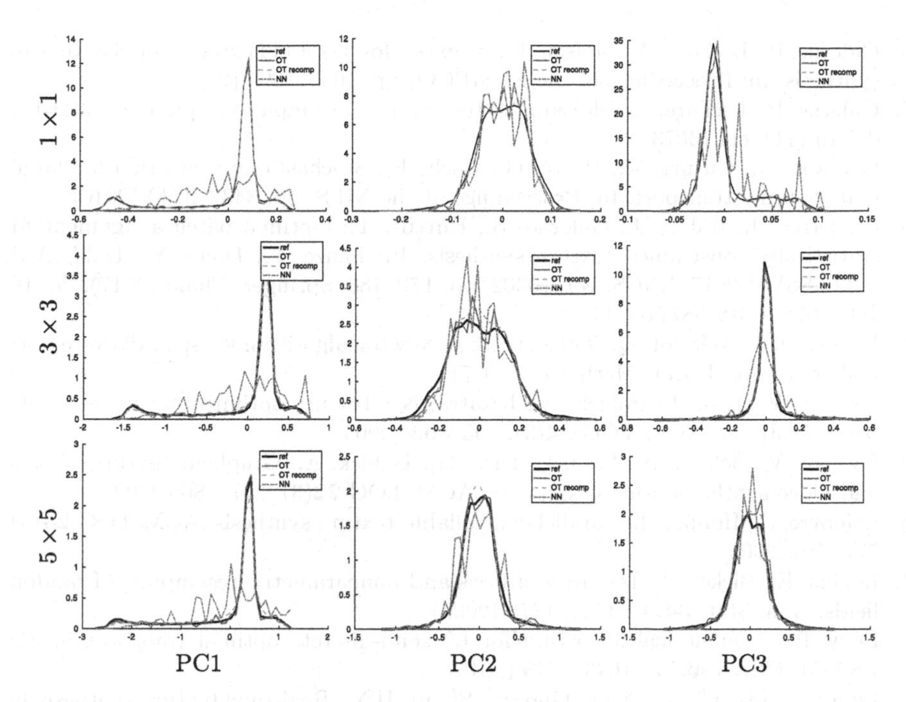

Fig. 4. Patch distribution (three first principal components). For the first image of Fig. 1 we plot the estimated distribution of patches in the three first principal components (columns) for different patch sizes (rows). The PCA transform is obtained on the exemplar patch distribution. We compare the patch distributions of the exemplar image (legend "ref"), of the synthesized image before patch recomposition (legend "OT") and after (legend "OT recomp"), and of the transformed patch with nearest-neighbor projection (legend "NN"). Even if we only approximate the optimal transport mapping, it suffices to reproduce the reference patch distribution better than the NN projection.

Acknowledgments. This work has been partially funded by Project Texto (Projet Jeunes Chercheurs du GdR Isis).

References

1. Aurenhammer, F., Hoffmann, F., Aronov, B.: Minkowski-type theorems and least-squares clustering. Algorithmica **20**(1), 61–76 (1998)
2. Efros, A.A., Leung, T.K.: Texture synthesis by non-parametric sampling. In: Proceedings of the ICCV 1999, p. 1033 (1999)
3. Efros, A., Freeman, W.: Image quilting for texture synthesis and transfer. In: ACM TOG, pp. 341–346, August 2001
4. Galerne, B., Gousseau, Y., Morel, J.M.: Random phase textures: theory and synthesis. IEEE Trans. Image Process. **20**(1), 257–267 (2011)
5. Galerne, B., Leclaire, A.: Texture inpainting using efficient Gaussian conditional simulation. SIIMS **10**(3), 1446–1474 (2017)

6. Galerne, B., Leclaire, A., Moisan, L.: A texton for fast and flexible Gaussian texture synthesis. In: Proceedings of the EUSIPCO, pp. 1686–1690 (2014)
7. Galerne, B., Leclaire, A., Moisan, L.: Texton noise. Comput. Graph. Forum (2017). doi:10.1111/cgf.13073
8. Genevay, A., Cuturi, M., Peyré, G., Bach, F.: Stochastic optimization for large-scale optimal transport. In: Proceedings of the NIPS, pp. 3432–3440 (2016)
9. Gutierrez, J., Rabin, J., Galerne, B., Hurtut, T.: Optimal patch assignment for statistically constrained texture synthesis. In: Lauze, F., Dong, Y., Dahl, A.B. (eds.) SSVM 2017. LNCS, vol. 10302, pp. 172–183. Springer, Cham (2017). doi:10.1007/978-3-319-58771-4_14
10. Kitagawa, J., Mérigot, Q., Thibert, B.: A Newton algorithm for semi-discrete optimal transport. J. Eur. Math Soc. (2017)
11. Kwatra, V., Essa, I., Bobick, A., Kwatra, N.: Texture optimization for example-based synthesis. ACM TOG **24**(3), 795–802 (2005)
12. Kwatra, V., Schödl, A., Essa, I., Turk, G., Bobick, A.: Graphcut textures: image and video synthesis using graph cuts. ACM TOG **22**(3), 277–286 (2003)
13. Lefebvre, S., Hoppe, H.: Parallel controllable texture synthesis. ACM TOG **24**(3), 777–786 (2005)
14. Levina, E., Bickel, P.: Texture synthesis and nonparametric resampling of random fields. Ann. Stat. **34**(4), 1751–1773 (2006)
15. Lévy, B.: A numerical algorithm for L2 semi-discrete optimal transport in 3D. ESAIM: M2AN **49**(6), 1693–1715 (2015)
16. Liang, L., Liu, C., Xu, Y.Q., Guo, B., Shum, H.Y.: Real-time texture synthesis by patch-based sampling. ACM TOG **20**(3), 127–150 (2001)
17. Mérigot, Q.: A multiscale approach to optimal transport. Comput. Graph. Forum **30**(5), 1583–1592 (2011)
18. Moulines, E., Bach, F.: Non-asymptotic analysis of stochastic approximation algorithms for machine learning. In: Proceedings of the NIPS, pp. 451–459 (2011)
19. Raad, L., Desolneux, A., Morel, J.: A conditional multiscale locally Gaussian texture synthesis algorithm. J. Math. Imaging Vis. **56**(2), 260–279 (2016)
20. Rabin, J., Peyré, G., Delon, J., Bernot, M.: Wasserstein barycenter and its application to texture mixing. In: Bruckstein, A.M., ter Haar Romeny, B.M., Bronstein, A.M., Bronstein, M.M. (eds.) SSVM 2011. LNCS, vol. 6667, pp. 435–446. Springer, Heidelberg (2012). doi:10.1007/978-3-642-24785-9_37
21. Tartavel, G., Peyré, G., Gousseau, Y.: Wasserstein loss for image synthesis and restoration. SIAM J. Imaging Sci. **9**(4), 1726–1755 (2016)
22. Villani, C.: Topics in Optimal Transportation. American Mathematical Society, Providence (2003)
23. Wei, L.Y., Lefebvre, S., Kwatra, V., Turk, G.: State of the art in example-based texture synthesis. In: Eurographics, State of the Art Reports, pp. 93–117 (2009)
24. Wei, L.Y., Levoy, M.: Fast texture synthesis using tree-structured vector quantization. In: Proceedings of the SIGGRAPH 2000, pp. 479–488 (2000)
25. Xia, G., Ferradans, S., Peyré, G., Aujol, J.: Synthesizing and mixing stationary gaussian texture models. SIAM J. Imaging Sci. **7**(1), 476–508 (2014)

Analysis of Optimal Transport Related Misfit Functions in Seismic Imaging

Yunan Yang(✉) and Björn Engquist

Department of Mathematics, The University of Texas at Austin, University Station C1200, Austin, TX 78712, USA
yunanyang@math.utexas.edu

Abstract. We analyze different misfit functions for comparing synthetic and observed data in seismic imaging, for example, the Wasserstein metric and the conventional least-squares norm. We revisit the convexity and insensitivity to noise of the Wasserstein metric which demonstrate the robustness of the metric in seismic inversion. Numerical results illustrate that full waveform inversion with quadratic Wasserstein metric can often effectively overcome the risk of local minimum trapping in the optimization part of the algorithm. A mathematical study on Fréchet derivative with respect to the model parameters of the objective functions further illustrates the role of optimal transport maps in this iterative approach. In this context we refer to the objective function as misfit. A realistic numerical example is presented.

Keywords: Full waveform inversion · Optimal transport · Seismic imaging · Optimization · Inverse problem

1 Introduction

Seismic data contains interpretable information about subsurface properties. Imaging predicts the spatial locations as well as properties that are useful in exploration seismology. The inverse method in the imaging predicts more physical properties if a full wave equation is employed instead of an asymptotic far-field approximation to it [9].

This, so called full waveform inversion (FWI) is a data-driven method to obtain high resolution subsurface properties by minimizing the difference or misfit between observed and synthetic seismic waveforms [12]. In the past three decades, the least-squares norm (L^2) has been widely used as a misfit function [10], which is known to suffer from cycle skipping issues (local minimum trapping) and sensitivity to noise [12].

Optimal transport has become a well developed topic in mathematics since it was first proposed by Gaspard Monge in 1781. The idea of using optimal transport for seismic inversion was first proposed in 2014 [3]. A useful tool from the theory of optimal transport, the Wasserstein metric computes the optimal cost of rearranging one distribution into another given a cost function. In computer

F. Nielsen and F. Barbaresco (Eds.): GSI 2017, LNCS 10589, pp. 109–116, 2017.
https://doi.org/10.1007/978-3-319-68445-1_13

science the metric is often called the "Earth Mover's Distance" (EMD). Here we will focus on the quadratic Wasserstein metric (W_2).

In this paper, we briefly review the theory of optimal transport and revisit the convexity and noise insensitivity of W_2 that were proved in [4]. The properties come from the analysis of the objective function. Next, we compare the Fréchet derivative with respect to the model parameters in different misfit functions using the adjoint-state method [8]. Discussions and comparisons between large scale inversion results using W_2 and L^2 metrics illustrate that the W_2 metric is very promising for overcoming the cycle skipping issue in seismic inversion.

2 Theory

2.1 Full Waveform Inversion and the Least Squares Functional

Full waveform inversion is a PDE-constrained optimization problem, minimizing the data misfit $J(f,g)$ by updating the model m, i.e.:

$$m^{\star} = \underset{m}{\operatorname{argmin}} \ J(f(x_r,t;m), g(x_r,t)), \tag{1}$$

where g is observed data, f is simulated data, x_r are receiver locations. We get the modeled data $f(x,t;m)$ by numerically solving in both the space and time domain [1].

Generalized least squares functional is a weighted sum of the squared errors and hence a generalized version of the standard least-squares misfit function. The formulation is

$$J_1(m) = \sum_r \int |W(f(x_r,t;m)) - W(g(x_r,t))|^2 \, dt. \tag{2}$$

In the conventional L^2 misfit, the weighting operator W is the identity I.

The integral wavefields misfit functional [5] is a generalized least squares functional applied on full-waveform inversion (FWI) with weighting operator $W(u) = \int_0^t u(x,\tau)d\tau$. If we define the integral wavefields $U(x,t) = \int_0^t u(x,\tau)d\tau$, then misfit function becomes the ordinary least squares difference between $\int_0^t g(x_r,\tau)d\tau$ and $\int_0^t f(x_r,\tau;m)d\tau$. The integral wavefields still satisfy the original acoustic wave equation with a different source term: $\delta(\boldsymbol{x}-\boldsymbol{x}_s)H(t) * S(t)$, where S is the original source term and $H(t)$ is the Heaviside step function [5]. We will refer this misfit function as H^{-1} norm in this paper.

Normalized Integration Method (NIM) is another generalized least squares functional, with an additional normalization step than integral wavefields misfit functional [6]. The weighting operator is

$$W(u)(x_r,t) = \frac{\int_0^t P(u)(x_r,\tau)d\tau}{\int_0^T P(u)(x_r,\tau)d\tau}, \tag{3}$$

where function P is included to make the data nonnegative. Three common choices are $P_1(u) = |u|$, $P_2(u) = u^2$ and $P_3 = E(u)$, which correspond to the absolute value, the square and the envelop of the signal [6].

2.2 Optimal Transport

Optimal transport is a problem that seeks the minimum cost required to transport mass of one distribution into another given a cost function, e.g. $|x-y|^2$. The mathematical definition of the distance between the distributions $f : X \to \mathbb{R}^+$ and $g : Y \to \mathbb{R}^+$ can then be formulated as

$$W_2^2(f,g) = \inf_{T_{f,g} \in \mathcal{M}} \int_X |x - T_{f,g}(x)|^2 f(x)\, dx \tag{4}$$

where $\mathcal{M}$ is the set of all maps $T_{f,g}$ that rearrange the distribution f into g [11].

The Wasserstein metric is an alternative misfit function for FWI to measure the difference between synthetic data f and observed data g. We can compare the data trace by trace and use the Wasserstein metric (W_p) in 1D to measure the misfit. The overall misfit is then

$$J_2(m) = \sum_{r=1}^{R} W_p^p(f(x_r, t; m), g(x_r, t)), \tag{5}$$

where R is the total number of traces. In this paper, we mainly discuss about quadratic Wasserstein metric (W_2) when $p = 2$ in (4) and (5).

Here we consider $f_0(t)$ and $g_0(t)$ as synthetic data and observed data from one single trace. After proper scaling with operator P, we get preconditioned data $f = P(f_0)$ and $g = P(g_0)$ which are positive and having total sum one. If we consider they are probability density functions (pdf), then after integrating once, we get the cumulative distribution function (cdf) $F(t)$ and $G(t)$.

If f is continuous we can write the explicit formulation for the 1D Wasserstein metric as:

$$W_2^2(f,g) = \int_0^1 |F^{-1}(t) - G^{-1}(t)|^2 dt = \int_0^T (G^{-1}F(t) - t)^2 f(t) dt. \tag{6}$$

The interesting fact is that W_2 computes the L^2 misfit between F^{-1} and G^{-1} (Fig. 1), while the objective function of NIM measures the L^2 misfit between F and G (Fig. 1). This is identical to the mathematical norm of Sobolev space H^{-1}, $||f-g||^2_{H^{-1}}$, given f and g are nonnegative and sharing equal mass.

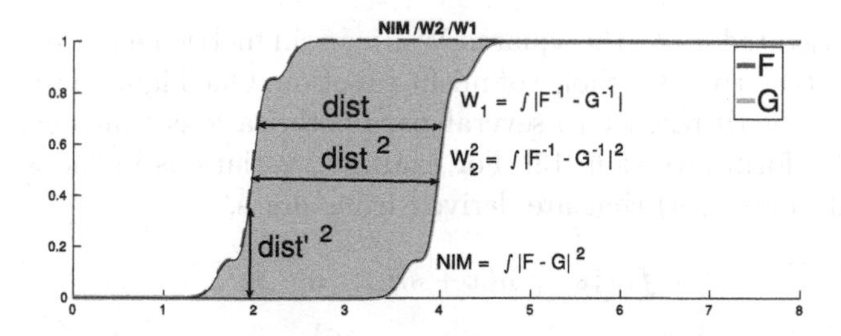

Fig. 1. After data normalization NIM measures $\int(F-G)^2 dt$, while W_2 considers $\int(F^{-1} - G^{-1})^2 dt$ and W_1 considers $\int|F^{-1} - G^{-1}| dt$.

3 Properties

Figure 2a shows two signals f and its shift g, both of which contain two ricker wavelets. Shift of signals are common in seismic data when we have incorrect velocity. We compute the L^2 norm and W_2 norm between f and g, and plot the misfit curves in terms of s in Fig. 2b and c. The L^2 difference between two signals has many local minima and maxima as s changes. It is a clear demonstration of the cycle skipping issue of L^2 norm. The global convexity of Fig. 2c is a motivation to further study the ideal properties of W_2 norm.

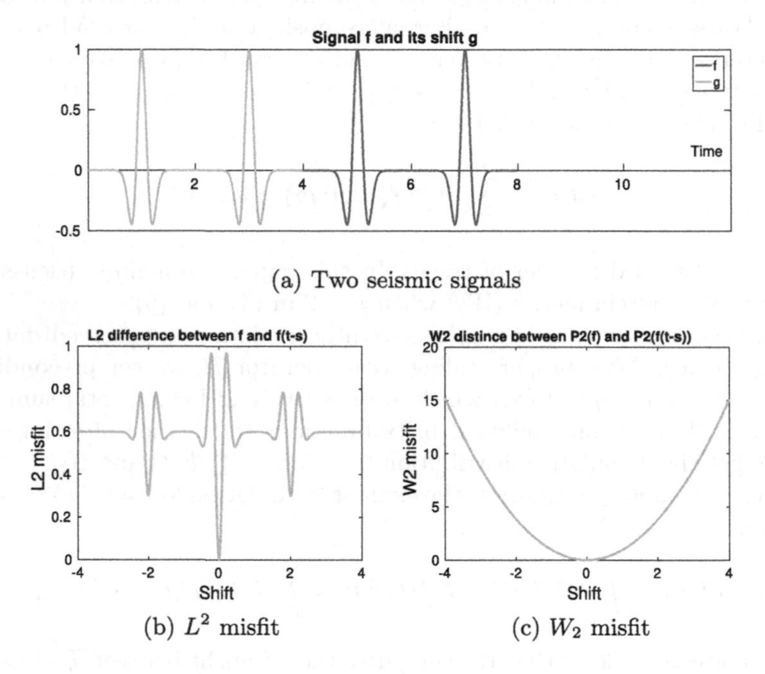

(a) Two seismic signals

(b) L^2 misfit

(c) W_2 misfit

Fig. 2. (a) A signal consisting two Ricker wavelets (blue) and its shift (red). (b) L^2 norm between f and g which is a shift of f. (c) W_2 norm between $P_2(f)$ and $P_2(g)$ in terms of different shift s. (Color figure online)

As demonstrated in [4], the squared Wasserstein metric has several properties that make it attractive as a choice of misfit function. One highly desirable feature is its convexity with respect to several parameterizations that occur naturally in seismic waveform inversion [13]. For example, variations in the wave velocity lead to simulations $f(m)$ that are derived from shifts,

$$f(x; s) = g(x + s\eta), \quad \eta \in \mathbb{R}^n, \tag{7}$$

or dilations,

$$f(x; A) = g(Ax), \quad A^T = A, A > 0, \tag{8}$$

applied to the observation g. Variations in the strength of a reflecting surface or the focusing of seismic waves can also lead to local rescalings of the form

$$f(x;\beta) = \begin{cases} \beta g(x), & x \in E \\ g(x), & x \in \mathbb{R}^n \backslash E. \end{cases} \tag{9}$$

In Theorem 1, f and g are assumed to be nonnegative with identical integrals.

Theorem 1 (Convexity of squared Wasserstein metric [4]**).** *The squared Wasserstein metric $W_2^2(f(m), g)$ is convex with respect to the model parameters m corresponding to a shift s in* (7)*, the eigenvalues of a dilation matrix A in* (8)*, or the local rescaling parameter β in* (9).

The Fig. 2c numerically exemplifies Theorem 1. Even if the scaling $P(u) = u^2$ perfectly fits the theorem it has turned out not to work well in generating an adjoint source that works well in inversion. The linear scaling, $P(u) = au + b$, on the other hand works very well even if the related misfit lacks strict convexity with respect to shifts. The two-variable example described below and Fig. 3 are based on the linear scaling. It gives the convexity with respect to other variables in velocity than a simple shift in the data.

The example from [7] shows a convexity result in higher dimensional model domain. The model velocity is increasing linearly in depth as $v(x, z) = v_{p,0} + \alpha z$, where $v_{p,0}$ is the starting velocity on the surface, α is vertical gradient and z is depth. The reference for $(v_{p,0}, \alpha)$ is $(2\,\text{km/s},\ 0.7\,\text{s}^{-1})$, and we plot the misfit curves with $\alpha \in [0.4, 1]$ and $v_0 \in [1.75,\ 2.25]$ on 41×45 grid in Fig. 3. We observe many local minima and maxima in Fig. 3a. The curve for W_2 (Fig. 3b) is globally convex in model parameters $v_{p,0}$ and α. It demonstrates the capacity of W_2 in mitigating cycle skipping issues.

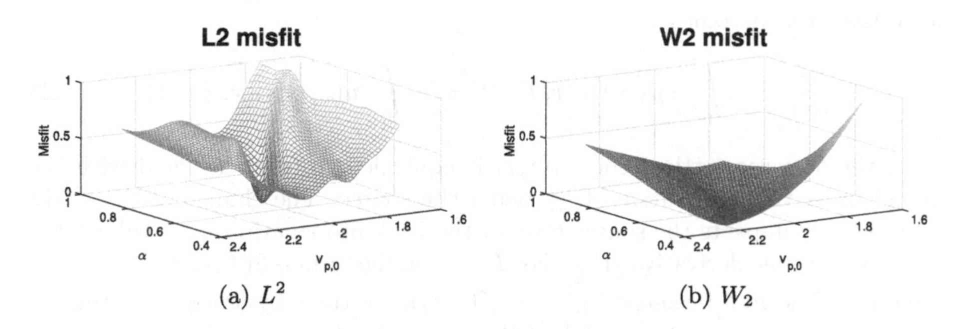

(a) L^2 (b) W_2

Fig. 3. (a) Conventional L^2 misfit function (b) W_2 misfit function trace-by-trace

Another ideal property of optimal transport is the insensitivity to noise. All seismic data contains either natural or experimental noise. For example, the ocean waves lead to extremely low frequency data in marine acquisition. Wind and cable motions also generate random noise.

The L^2 norm is known to be sensitive to noise since the misfit between clean and noisy data is calculated as the sum of squared noise amplitude at each sampling point. In [4] W_2 norm is proved to be insensitive to mean-zero noise and the property apply for any dimension of the data. This is a natural result from optimal transport theory since the W_2 metric defines a global comparison that not only considers the change in signal intensity but also the phase difference.

Theorem 2 (Insensitivity to noise [4]). *Let f_{ns} be f with a piecewise constant additive noise of mean zero uniform distribution. The squared Wasserstein metric $W_2^2(f, f_{ns})$ is of $\mathcal{O}(\frac{1}{N})$ where N is the number of pieces of the additive noise in f_{ns}.*

4 Discussions

Typically we solve the linearized problem iteratively to approximate the solution in FWI. This approach requires the Fréchet derivatives of the misfit function $J(m)$ which is expensive to compute directly. The adjoint-state method [8] provides an efficient way of computing the gradient. This approach requires the Fréchet derivative $\frac{\partial J}{\partial f}$ and two modelings by solving the wave equations. Here we will only discuss about the acoustics wave Eq. (10).

$$m\frac{\partial^2 u(x,t)}{\partial t^2} - \Delta u(x,t) = S(x,t) \tag{10}$$

In the adjoint-state method, we first forward propagate the source wavelet with zero initial conditions. The simulated data f is the source wavefield u recorded on the boundary. Next we back propagate the Fréchet derivative $\frac{\partial J}{\partial f}$ as the source with zero final conditions and get the receiver wavefield v.

With both the forward wavefield u and backward wavefield v, the Fréchet derivative of m becomes

$$\frac{\partial J}{\partial m} = -\int_0^T u_{tt}(x,t)v(x,T-t) = -\int_0^T u(x,t)v_{tt}(x,T-t) \tag{11}$$

In the acoustic setting, the $v_{tt}(x,t)$ is equivalent to the wavefield with the second order time derivative of $\frac{\partial J}{\partial f}$ being the source. The change of the misfit function only impacts the source term of the back propagation, particularly the second order time derivative of $\frac{\partial J}{\partial f}$. For L^2 norm, the term is $2(f_{tt}(x,t) - g_{tt}(x,t))$, and for H^{-1} norm it becomes $2(g(x,t) - f(x,t))$. For trace-by-trace W_2 norm, the second order time derivative of $\frac{\partial W_2^2(f,g)}{\partial f}$ is $2\left(\frac{g(x,t')-f(x,t)}{g(x,t')}\right)$ where $t' = G^{-1}F(t)$, the optimal coupling of t for each trace.

Compared with L^2 norm, the source term of H^{-1} does not has the two time derivatives and therefore has more of a focus on the lower frequency part of the data. Lower frequency components normally provide a wider basin of attraction in optimization. The source term of W_2 is similar to the one of H^{-1} norm, but the order of signal g in time has changed with the optimal map for each trace at receiver x. The optimal couplings often change the location of the wavefront. For example, if g is a shift of f, then the wavefront of g will be mapped to the wavefront of f even if two wavefronts do not match in time. The change of time order in g also helps generate a better image under the reflectors when we back propagate the source and compute the gradient as in (11).

5 Numerical Example

In this section, we use a part of the BP 2004 benchmark velocity model [2] (Fig. 4a) and a highly smoothed initial model without the upper salt part (Fig. 4b) to do inversion with W_2 and L^2 norm respectively. A fixed-spread surface acquisition is used, involving 11 shots located every 1.6 Km on top. A Ricker wavelet centered on 5 Hz is used to generate the synthetic data with a bandpass filter only keeping 3 to 9 Hz components. We stopped the inversion after 300 L-BFGS iterations.

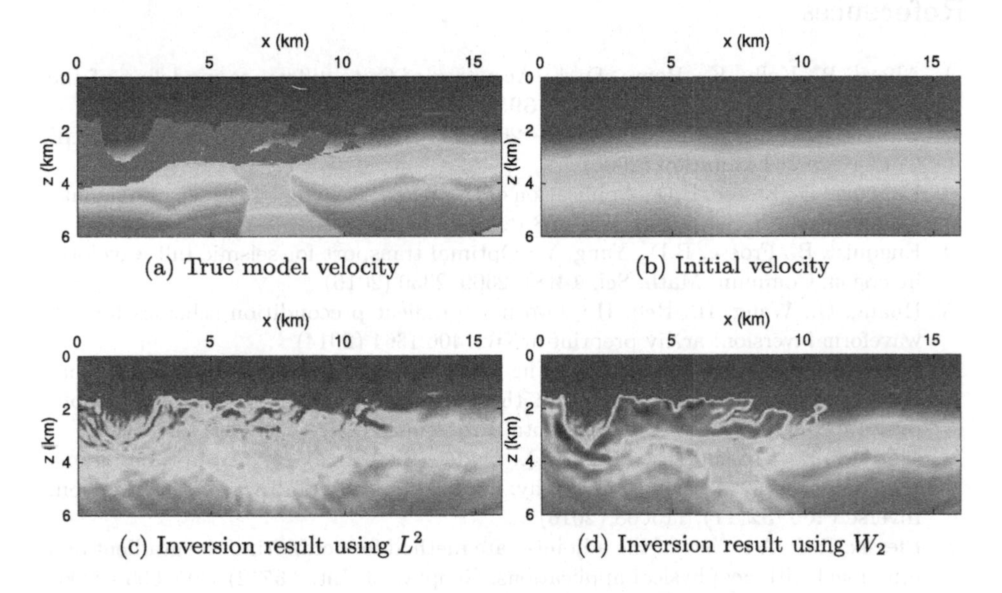

(a) True model velocity (b) Initial velocity

(c) Inversion result using L^2 (d) Inversion result using W_2

Fig. 4. Large scale FWI example

Here we precondition the data with function $P(f) = a \cdot f + b$ to satisfy the nonnegativity and mass balance in optimal transport. Inversion with trace-by-trace W_2 norm successfully construct the shape of the salt bodies (Fig. 4d), while FWI with the conventional L^2 failed to recover boundaries of the salt bodies as shown by Fig. 4c.

6 Conclusion

In this paper, we revisited the quadratic Wasserstein metric from the optimal transport theory in the application of seismic inversion. The desirable properties of convexity and insensitivity to noise make it a promising alternative misfit function in FWI. We also analyze the conventional least-squares inversion (L^2 norm), the integral wavefields misfit function (H^{-1} norm) and the quadratic Wasserstein metric (W_2) in terms of the model parameter gradient using the adjoint-state method. The analysis further demonstrate the effectiveness of optimal transport ideas in dealing with cycle skipping.

Acknowledgments. We thank Sergey Fomel, Junzhe Sun and Zhiguang Xue for helpful discussions, and thank the sponsors of the Texas Consortium for Computational Seismology (TCCS) for financial support. This work was also partially supported by NSF DMS-1620396.

References

1. Alford, R., Kelly, K., Boore, D.M.: Accuracy of finite-difference modeling of the acoustic wave equation. Geophysics **39**(6), 834–842 (1974)
2. Billette, F., Brandsberg-Dahl, S.: The 2004 BP velocity benchmark. In: 67th EAGE Conference & Exhibition (2005)
3. Engquist, B., Froese, B.D.: Application of the Wasserstein metric to seismic signals. Commun. Math. Sci. **12**(5), 979–988 (2014)
4. Engquist, B., Froese, B.D., Yang, Y.: Optimal transport for seismic full waveform inversion. Commun. Math. Sci. **14**(8), 2309–2330 (2016)
5. Huang, G., Wang, H., Ren, H.: Two new gradient precondition schemes for full waveform inversion. arXiv preprint arXiv:1406.1864 (2014)
6. Liu, J., Chauris, H., Calandra, H.: The normalized integration method-an alternative to full waveform inversion? In: 25th Symposium on the Application of Geophpysics to Engineering & Environmental Problems (2012)
7. Métivier, L., Brossier, R., Mrigot, Q., Oudet, E., Virieux, J.: An optimal transport approach for seismic tomography: application to 3D full waveform inversion. Inverse Prob. **32**(11), 115008 (2016)
8. Plessix, R.E.: A review of the adjoint-state method for computing the gradient of a functional with geophysical applications. Geophys. J. Int. **167**(2), 495–503 (2006)
9. Stolt, R.H., Weglein, A.B.: Seismic Imaging and Inversion: Volume 1: Application of Linear Inverse Theory, vol. 1. Cambridge University Press, Cambridge (2012)
10. Tarantola, A., Valette, B.: Generalized nonlinear inverse problems solved using the least squares criterion. Rev. Geophys. **20**(2), 219–232 (1982)
11. Villani, C.: Topics in Optimal Transportation. Graduate Studies in Mathematics, vol. 58. American Mathematical Society, Providence (2003)
12. Virieux, J., Operto, S.: An overview of full-waveform inversion in exploration geophysics. Geophysics **74**(6), WCC1–WCC26 (2009)
13. Yang, Y., Engquist, B., Sun, J., Froese, B.D.: Application of optimal transport and the quadratic wasserstein metric to full-waveform inversion. arXiv preprint arXiv:1612.05075 (2016)

Optimal Transport and Applications: Signal Processing

Information Geometry of Wasserstein Divergence

Ryo Karakida[1(✉)] and Shun-ichi Amari[2]

[1] National Institute of Advanced Industrial Science and Technology, Koto-ku, Tokyo 135-0064, Japan
karakida.ryo@aist.go.jp

[2] RIKEN Brain Science Institute, Wako-shi, Saitama 351-0198, Japan
amari@brain.riken.jp

Abstract. There are two geometrical structures in a manifold of probability distributions. One is invariant, based on the Fisher information, and the other is based on the Wasserstein distance of optimal transportation. We propose a unified framework which connects the Wasserstein distance and the Kullback-Leibler (KL) divergence to give a new information-geometrical theory. We consider the discrete case consisting of n elements and study the geometry of the probability simplex S_{n-1}, the set of all probability distributions over n atoms. The Wasserstein distance is introduced in S_{n-1} by the optimal transportation of commodities from distribution $\boldsymbol{p} \in S_{n-1}$ to $\boldsymbol{q} \in S_{n-1}$. We relax the optimal transportation by using entropy, introduced by Cuturi (2013) and show that the entropy-relaxed transportation plan naturally defines the exponential family and the dually flat structure of information geometry. Although the optimal cost does not define a distance function, we introduce a novel divergence function in S_{n-1}, which connects the relaxed Wasserstein distance to the KL-divergence by one parameter.

1 Introduction

Information geometry studies invariant properties of a manifold of probability distributions, which are useful for various applications in statistics, machine learning, signal processing, optimization and others. Two geometrical structures have been introduced from two different backgrounds. One is constructed based on the invariance principle: The geometry is invariant under reversible transformations of random variables. We then have the Fisher information matrix as the unique invariant Riemannian metric (Rao 1945; Chentsov 1982; Amari 2016). Moreover, two dually coupled affine connections are given as invariant connections. These structures are useful for various applications. Another geometrical structure is introduced through the transportation problem. A distribution of commodities in a manifold is transported to another distribution. The transportation with the minimal cost defines a distance between the two distributions, called the Monge-Kantorovich-Wasserstein distance or earth-mover distance. This gives a tool to study the geometry of distributions taking the metric of the supporting manifold into account.

F. Nielsen and F. Barbaresco (Eds.): GSI 2017, LNCS 10589, pp. 119–126, 2017.
https://doi.org/10.1007/978-3-319-68445-1_14

Let $X = \{1, \cdots, n\}$ be the support of a probability measure $\boldsymbol{p}$. The invariant geometry gives a structure which is invariant under permutations of elements of X. It leads to an efficient estimator in statistical estimation. On the other hand, when we consider a picture over n^2 pixels $X = \{(ij); i, j = 1, \cdots, n\}$, neighboring pixels are close. A permutation of X destroys such a neighboring structure, so the invariance should not be required. The Wasserstein distance is responsible for such a structure. Therefore, it is useful for problems having neighboring structure in support X.

An interesting question arises how these two geometrical structures are related. They are useful structures in their own right, but it is intriguing to find a unified framework to include the two. For this purpose in mind, the present paper treats the discrete case over n elements, such that a probability distribution is given by a probability vector $\boldsymbol{p} = (p_1, \cdots, p_n)$ in the probability simplex S_{n-1}, letting a general case of continuous distributions over a manifold to be studied in future.

Cuturi (2013) modified the transportation problem such that the cost is minimized under the entropy constraint. This is called the entropy-relaxed optimal translation problem. In many applications, his group showed the quasi-distance defined by the entropy-constrained optimal solution gives superior properties to the information-geometric distance such as the KL divergence or the Hellinger distance. As an application, consider a set of normalized histograms over X. A clustering problem categorizes them in some classes such that a class consists of similar histograms. Since a histogram is regarded as an empirical probability distribution, the problem is formulated within the probability simplex S_{n-1} in the discrete case and the distances among supporting pixels play a fundamental role.

We follow the entropy-relaxed framework of Cuturi (2013), Cuturi and Avis (2014), Cuturi and Peyré (2016), etc. and introduce a Lagrangian function which is a linear combination of the transportation cost and the entropy. Given distribution $\boldsymbol{p}$ of commodities at the sender and $\boldsymbol{q}$ at the receiver, the optimal transportation plan is the minimizer of the Lagrangian function. We reveal that it is a convex function of $\boldsymbol{p}$ and $\boldsymbol{q}$ so it defines a dually flat geometric structure in $S_{n-1} \times S_{n-1}$. The m-flat coordinates are $(\boldsymbol{p}, \boldsymbol{q})$ and their dual, e-flat coordinates $(\boldsymbol{\alpha}, \boldsymbol{\beta})$ are given from the Lagrangian duality of nonlinear optimization problems. The set of the optimal transportation plans is an exponential family with the canonical parameters $(\boldsymbol{\alpha}, \boldsymbol{\beta})$, where the expectation parameters are $(\boldsymbol{p}, \boldsymbol{q})$. Furthermore, we introduce a novel divergence between $\boldsymbol{p}$ and $\boldsymbol{q}$ in S_{n-1}. It connects the relaxed Wasserstein distance to the KL-divergence by a one parameter family. Our divergence will be expected to be useful for practical applications, because a divergence is a general concept including the square of a distance and more flexible admitting non-symmetricity between $\boldsymbol{p}$ and $\boldsymbol{q}$.

2 Entropy-Constrained Transportation Problem

Let us consider n terminals $X = (X_1, \cdots, X_n)$ at which amounts $p_1, \cdots, p_n$ of commodities are stocked. We transport them within X such that amounts

$q_1, \cdots, q_n$ are newly stored at $X_1, \cdots, X_n$. We normalize the total amount to be equal to 1, so $\boldsymbol{p} = (p_1, \cdots, p_n)$ and $\boldsymbol{q} = (q_1, \cdots, q_n)$ are regarded as probability distributions in the probability simplex S_{n-1},

$$\sum p_i = 1, \quad \sum q_i = 1, \quad p_i > 0, \quad q_i > 0. \tag{1}$$

We consider a transportation plan $\mathbf{P} = (P_{ij})$, where P_{ij} is the amount of commodity transported from X_i to X_j. A plan $\mathbf{P}$ is regarded as a joint probability distribution of commodities flowing from X_i to X_j, satisfying the sender and receiver constraints,

$$\sum_j P_{ij} = p_i, \quad \sum_i P_{ij} = q_j. \tag{2}$$

The set of $\mathbf{P}$'s satisfying (2) is denoted by $U(\boldsymbol{p}, \boldsymbol{q})$.

Let $\mathbf{c} = (c_{ij})$ be the cost matrix, where c_{ij} denotes the cost of transporting one unit of commodities from X_i to X_j.

The transportation cost is defined by

$$c(\mathbf{P}) = \langle \mathbf{c}, \mathbf{P} \rangle = \sum c_{ij} P_{ij}. \tag{3}$$

The Wasserstein distance is defined by the minimal cost of transporting distribution $\boldsymbol{p}$ at the senders to $\boldsymbol{q}$ at the receivers,

$$c(\boldsymbol{p}, \boldsymbol{q}) = \min_{\mathbf{P} \subset U(\boldsymbol{p}, \boldsymbol{q})} \langle \mathbf{c}, \mathbf{P} \rangle, \tag{4}$$

where min is taken over all $\mathbf{P}$ satisfying constraints (2). See e.g., Villani (2013).

Given $\boldsymbol{p}$ and $\boldsymbol{q}$, let us consider a special transportation plan $\mathbf{P}_D$ defined by the direct product of $\boldsymbol{p}$ and $\boldsymbol{q}$,

$$\mathbf{P}_D = \boldsymbol{p} \otimes \boldsymbol{q} = (p_i q_j). \tag{5}$$

This plan transports commodities from each sender to the receivers according to the receiver distribution $\boldsymbol{q}$, irrespective of $\mathbf{c}$. The entropy of $\mathbf{P}_D$,

$$H(\mathbf{P}_D) = -\sum \mathrm{P}_{Dij} \log \mathrm{P}_{Dij} = H(\boldsymbol{p}) + H(\boldsymbol{q}), \tag{6}$$

is the minimum among all $\mathbf{P}$'s belonging to $U(\boldsymbol{p}, \boldsymbol{q})$, because of $H(\mathbf{P}) \leq H(\boldsymbol{p}) + H(\boldsymbol{q})$, where $H(\mathbf{P})$, $H(\boldsymbol{p})$ and $H(\boldsymbol{q})$ are the respective entropies and the equality holds for $\mathbf{P} = \mathbf{P}_D$.

We consider a constrained problem of searching for $\mathbf{P}$ that minimizes $\langle \mathbf{c}, \mathbf{P} \rangle$ within a KL-divergence ball centered at $\mathbf{P}_D$,

$$KL\left[\mathbf{P} : \mathbf{P}_D\right] \leq d \tag{7}$$

for constant d. As d increases, the entropy of $\mathbf{P}$ increases within the ball. This is equivalent to the entropy constrained problem that minimizes a linear combination of the transportation cost $\langle \mathbf{c}, \mathbf{P} \rangle$ and entropy $H(\mathbf{P})$,

$$F_\lambda(\mathbf{P}) = \frac{1}{1+\lambda} \langle \mathbf{c}, \mathbf{P} \rangle - \frac{\lambda}{1+\lambda} H(\mathbf{P}) \tag{8}$$

for constant λ (Cuturi 2013). Here, λ is a Lagrangian multiplier and λ becomes smaller as d becomes larger.

3 Solution of Entropy-Constrained Problem

Since $\mathbf{P}$ satisfies constraints (2), by using Lagrange multipliers α_i, β_j, minimization of (8) is formulated in the Lagrangian form,

$$L_\lambda(\mathbf{P}) = \frac{1}{1+\lambda}\langle \mathbf{c}, \mathbf{P}\rangle - \frac{\lambda}{1+\lambda} H(\mathbf{P}) - \sum_{i,j} (\alpha_i + \beta_j) P_{ij}. \tag{9}$$

Let us fix λ, considering it as a parameter controlling the magnitude of the entropy or the size of the KL-ball. By differentiating (9) with respect to P_{ij}, we have the following solution,

$$P_{ij} = \exp\left\{-\frac{c_{ij}}{\lambda} + \frac{1+\lambda}{\lambda}(\alpha_i + \beta_j) - 1\right\}. \tag{10}$$

Let us put

$$a_i = \exp\left(\frac{1+\lambda}{\lambda}\alpha_i\right) \quad b_j = \exp\left(\frac{1+\lambda}{\lambda}\beta_j\right), \quad K_{ij} = \exp\left\{-\frac{c_{ij}}{\lambda}\right\}, \tag{11}$$

and the optimal solution is written as

$$P^*_{\lambda ij} \propto a_i b_j K_{ij}, \tag{12}$$

where a_i and b_j correspond to the Lagrange multipliers α_i and β_j to be determined from the constraints (2). Note that $2n$ constraints (2) are not independent. Because of $\sum p_i = 1$, we can obtain a_n by $a_n = 1 - \sum_{i \neq n} a_i$. Further, we note that $\mu \boldsymbol{a}$ and $\boldsymbol{b}/\mu$ give the same answer for any $\mu > 0$, where $\boldsymbol{a} = (a_i)$ and $\boldsymbol{b} = (b_j)$. Therefore, the degrees of freedom of $\boldsymbol{a}$ and $\boldsymbol{b}$ are $2(n-1)$, which are to be determined from $\boldsymbol{p}$ and $\boldsymbol{q}$ of which degrees of freedom are also $2(n-1)$. Therefore, we may choose $\boldsymbol{a}$ and $\boldsymbol{b}$ such that they satisfy

$$\sum a_i = 1, \quad \sum b_j = 1. \tag{13}$$

Then, $\boldsymbol{a}, \boldsymbol{b} \in S_{n-1}$ and we have the following theorem.

Theorem 1. The optimal transportation plan $\mathbf{P}^*_\lambda$ is given by

$$P^*_{\lambda ij} = c a_i b_j K_{ij}, \tag{14}$$

$$c = \frac{1}{\sum a_i b_j K_{ij}}, \tag{15}$$

where two vectors $\boldsymbol{a}$ and $\boldsymbol{b}$ are determined from $\boldsymbol{p}$ and $\boldsymbol{q}$.

Cuturi (2013) obtained the above $P^*_{\lambda ij}$ and applied the Sinkhorn-Knopp algorithm to iteratively compute $\boldsymbol{a}$ and $\boldsymbol{b}$.

The following lemma is useful for later calculations.

Lemma 1. The optimal value

$$\varphi_\lambda(\boldsymbol{p}, \boldsymbol{q}) = \min F_\lambda(\mathbf{P}) \tag{16}$$

is given by

$$\varphi_\lambda(\boldsymbol{p}, \boldsymbol{q}) = \frac{\lambda}{1+\lambda}\left(\sum p_i \log a_i + \sum q_j \log b_j + \log c\right). \tag{17}$$

Proof. We first calculate $H\left(\mathbf{P}^*_\lambda\right)$. Substituting (15) in $H\left(\mathbf{P}^*_\lambda\right)$, we have

$$H\left(\mathbf{P}^*_\lambda\right) = -\sum_{ij} \mathrm{P}^*_{\lambda ij}\left(-\frac{c_{ij}}{\lambda} + \log c a_i b_j\right) \tag{18}$$

$$= \frac{1}{\lambda}\langle \mathbf{c}, \mathbf{P}^*_\lambda\rangle - \sum p_i \log a_i - \sum q_j \log b_j - \log c. \tag{19}$$

Hence, (17) follows.

4 Exponential Family of Optimal Transportation Plans

A transportation plan $\mathbf{P}$ is a probability distribution over branches (i, j) connecting terminals i and j. Let x denote branches and $\delta_{ij}(x) = 1$ when x is (i, j) and 0 otherwise. Then $\mathbf{P}$ is a probability distribution of random variable x,

$$P(x) = \sum_{i,j=1}^{n} P_{ij}\delta_{ij}(x). \tag{20}$$

By introducing new parameters

$$\theta^{ij} = \log\frac{P_{ij}}{P_{nn}}, \quad \boldsymbol{\theta} = \left(\theta^{ij}\right), \tag{21}$$

it is rewritten in a parameterized form as

$$P(x, \boldsymbol{\theta}) = \exp\left\{\sum_{i,j}\theta^{ij}\delta_{ij}(x) + \log P_{nn}\right\}. \tag{22}$$

This shows that the set of transportation plans is an exponential family, where θ^{ij} are the canonical parameters and $\eta_{ij} = P_{ij}$ the expectation parameters. They form an $\left(n^2 - 1\right)$-dimensional manifold denoted by S_{TP}, because $\theta^{nn} = 0$.

An optimal transportation plan is specified by $(\boldsymbol{\alpha}, \boldsymbol{\beta})$ in (10), $\boldsymbol{\alpha} = (\alpha_i)$, $\boldsymbol{\beta} = (\beta_j)$ which are determined from $(\boldsymbol{p}, \boldsymbol{q})$. It is written as

$$P(x, \boldsymbol{\alpha}, \boldsymbol{\beta}) = \exp\left[\sum_{i,j}\left\{\frac{\lambda+1}{\lambda}(\alpha_i + \beta_j) - \frac{c_{ij}}{\lambda}\right\}\delta_{ij}(x) - \frac{\lambda+1}{\lambda}\psi\right], \tag{23}$$

where

$$\psi(\boldsymbol{\alpha}, \boldsymbol{\beta}) = \frac{\lambda}{1+\lambda} \log \sum_{i,j} \exp\left\{\frac{\lambda+1}{\lambda}(\alpha_i + \beta_j) - \frac{c_{ij}}{\lambda}\right\} \tag{24}$$

is the normalization factor. By putting

$$\theta^{ij} = \frac{1+\lambda}{\lambda}(\alpha_i + \beta_j) - \frac{c_{ij}}{\lambda}, \tag{25}$$

we see that the set S_{OTP} of optimal transformation plans is a submanifold of S_{TP}. Because (25) is linear in $\boldsymbol{\alpha}$ and $\boldsymbol{\beta}$, S_{OTP} itself is an exponential family, where the canonical parameters are $(1+\lambda)/\lambda$ times $(\boldsymbol{\alpha}, \boldsymbol{\beta})$ and the expectation parameters are $(\boldsymbol{p}, \boldsymbol{q}) \in S_{n-1} \times S_{n-1}$, since

$$E\left[\sum_j \delta_{ij}(x)\right] = p_i, \tag{26}$$

$$E\left[\sum_i \delta_{ij}(x)\right] = q_j. \tag{27}$$

Since each of $\boldsymbol{p}$, $\boldsymbol{q} \in S_{n-1}$ has $n-1$ degrees of freedom, S_{OPT} is a $2(n-1)$-dimensional dually flat manifold. We may put $\alpha_n = \beta_n = 0$ without loss of generality, which correspond to putting $a_n = b_n = 1$ instead of $\sum a_i = \sum b_j = 1$.

We calculate the relaxed cost function $\varphi_\lambda(\boldsymbol{p}, \boldsymbol{q})$ corresponding to $\mathbf{P}(\boldsymbol{\alpha}, \boldsymbol{\beta})$. We then have

$$\varphi_\lambda(\boldsymbol{p}, \boldsymbol{q}) = \frac{1}{1+\lambda}\langle \mathbf{c}, \mathbf{P}\rangle + \frac{\lambda}{1+\lambda}\sum_{i,j} P_{ij}\left\{\frac{1+\lambda}{\lambda}(\alpha_i + \beta_j) - \frac{c_{ij}}{\lambda} - \frac{1+\lambda}{\lambda}\psi_\lambda\right\} \tag{28}$$

$$= \boldsymbol{p} \cdot \boldsymbol{\alpha} + \boldsymbol{q} \cdot \boldsymbol{\beta} - \psi_\lambda(\boldsymbol{\alpha}, \boldsymbol{\beta}). \tag{29}$$

When we use new notations $\boldsymbol{\eta} = (\boldsymbol{p}, \boldsymbol{q})^T$, $\boldsymbol{\theta} = (\boldsymbol{\alpha}, \boldsymbol{\beta})^T$, we have

$$\psi_\lambda(\boldsymbol{\theta}) + \varphi_\lambda(\boldsymbol{\eta}) = \boldsymbol{\theta} \cdot \boldsymbol{\eta}, \tag{30}$$

which is the Legendre relation between $\boldsymbol{\theta}$ and $\boldsymbol{\eta}$. Thus, we have the following theorem.

Theorem 2. The relaxed cost function φ_λ and the free energy (cumulant generating function) ψ_λ of the exponential family are both convex, connected by the Legendre transformation,

$$\boldsymbol{\theta} = \nabla_{\boldsymbol{\eta}} \varphi_\lambda(\boldsymbol{\eta}), \quad \boldsymbol{\eta} = \nabla_{\boldsymbol{\theta}} \psi_\lambda(\boldsymbol{\theta}), \tag{31}$$

$$\boldsymbol{\alpha} = \nabla_{\boldsymbol{p}} \varphi_\lambda(\boldsymbol{p}, \boldsymbol{q}),\ \boldsymbol{\beta} = \nabla_{\boldsymbol{q}} \varphi_\lambda(\boldsymbol{p}, \boldsymbol{q}), \tag{32}$$

$$\boldsymbol{p} = \nabla_{\boldsymbol{\alpha}} \psi_\lambda(\boldsymbol{\alpha}, \boldsymbol{\beta}),\ \boldsymbol{q} = \nabla_{\boldsymbol{p}} \psi_\lambda(\boldsymbol{\alpha}, \boldsymbol{\beta}). \tag{33}$$

The Riemannian metric $\mathbf{G}_\lambda$ is given to $S_{n-1} \times S_{n-1}$ by

$$\mathbf{G}_\lambda = \nabla_{\boldsymbol{\eta}} \nabla_{\boldsymbol{\eta}} \varphi_\lambda(\boldsymbol{\eta}) \tag{34}$$

in the $\boldsymbol{\eta}$-coordinate system $(\boldsymbol{p}, \boldsymbol{q})$. Its inverse is

$$\mathbf{G}_\lambda^{-1} = \nabla_{\boldsymbol{\theta}} \nabla_{\boldsymbol{\theta}} \psi_\lambda(\boldsymbol{\theta}). \tag{35}$$

In addition, we can calculate $\mathbf{G}_\lambda^{-1}$ explicitly from (24).

Theorem 3. The Fisher information matrix $\mathbf{G}_\lambda^{-1}$ in the $\boldsymbol{\theta}$-coordinate system is given by

$$\mathbf{G}_\lambda^{-1} = \frac{1+\lambda}{\lambda} \left\{ \begin{bmatrix} \operatorname{diag}(\boldsymbol{p}) & \mathbf{P} \\ \mathbf{P}^{\mathrm{T}} & \operatorname{diag}(\boldsymbol{q}) \end{bmatrix} - \begin{bmatrix} \boldsymbol{p} \\ \boldsymbol{q} \end{bmatrix} \begin{bmatrix} \boldsymbol{p}^T & \boldsymbol{q}^T \end{bmatrix} \right\}, \tag{36}$$

or in the component form as

$$\mathbf{G}_\lambda^{-1} = \frac{1+\lambda}{\lambda} \left[\begin{array}{c|c} p_i\delta_{ij} - p_i p_j & P_{ij} - p_i q_j \\ \hline P_{ij} - p_i q_j & q_i \delta_{ij} - q_i q_j \end{array} \right]. \tag{37}$$

Remark 1. The $\boldsymbol{p}$-part of $\mathbf{G}_\lambda^{-1}$ is a scalar multiple of the Fisher information of $\boldsymbol{p}$ in S_{n-1} in the e-coordinate system. So is the $\boldsymbol{q}$-part. They are independent of the cost matrix c_{ij}, but the off-diagonal blocks of $\mathbf{G}_\lambda^{-1}$ depend on it.

Remark 2. The $\boldsymbol{p}$-part of $\mathbf{G}_\lambda$ is not equal to the Fisher information of $\boldsymbol{p}$ in the m-coordinate system. It is the $\boldsymbol{p}$-part of the inverse of $\mathbf{G}_\lambda^{-1}$, depending on $\boldsymbol{q}$, too.

5 λ-Divergence in S_{n-1}

The relaxed Wasserstein distance $\varphi_\lambda(\boldsymbol{p} : \boldsymbol{q})$ does not satisfy a criterion of divergence, i.e. $\varphi_\lambda(\boldsymbol{p} : \boldsymbol{p}) \neq 0$, because $\varphi_\lambda(\boldsymbol{p} : \boldsymbol{q})$ is minimalized at $\boldsymbol{q} \neq \boldsymbol{p}$ in general. In contrast, the original Wasserstein distance literally satisfies the criteria of distance and those of divergence. To recover the property of divergence in the relaxed form, we introduce a canonical divergence between two transportation plans $(\boldsymbol{p}, \boldsymbol{p})$ and $(\boldsymbol{p}, \boldsymbol{q})$, which is composed of the Legendre pair of the convex functions φ_λ and ψ_λ (Amari 2016):

$$D_\lambda[\boldsymbol{p} : \boldsymbol{q}] = \psi_\lambda(\boldsymbol{\alpha}, \boldsymbol{\beta}) + \varphi_\lambda(\boldsymbol{p}, \boldsymbol{p}) - \boldsymbol{\alpha} \cdot \boldsymbol{p} - \boldsymbol{\beta} \cdot \boldsymbol{p}, \tag{38}$$

where $(\boldsymbol{\alpha}, \boldsymbol{\beta})$ corresponds to $(\boldsymbol{p}, \boldsymbol{q})$. We call this a λ-divergence $D_\lambda[\boldsymbol{p} : \boldsymbol{q}]$ in S_{n-1} from $\boldsymbol{p}$ to $\boldsymbol{q}$. It connects the Wasserstein distance and the KL-divergence in the following way.

This λ-divergence can be transformed into a Bregman-like divergence with the relaxed cost function φ_λ (not a Bregman divergence constructed from a convex function of a single variable $\boldsymbol{q}$):

$$D_\lambda[\boldsymbol{p} : \boldsymbol{q}] = \varphi_\lambda(\boldsymbol{p}, \boldsymbol{p}) - \varphi_\lambda(\boldsymbol{p}, \boldsymbol{q}) - \langle \nabla_{\boldsymbol{q}} \varphi_\lambda(\boldsymbol{p}, \boldsymbol{q}), \boldsymbol{p} - \boldsymbol{q} \rangle. \tag{39}$$

As easily confirmed by substituting (14) to (38), the λ-divergence is equivalent to the KL-divergence between the two transportation plans, up to a constant factor:

$$D_\lambda[\boldsymbol{p} : \boldsymbol{q}] = \frac{\lambda}{1+\lambda} KL\left[\mathbf{P}' : \mathbf{P}\right], \tag{40}$$

where $\mathbf{P}'$ and $\mathbf{P}$ are the optimal plans from $\boldsymbol{p}$ to $\boldsymbol{p}$ and $\boldsymbol{p}$ to $\boldsymbol{q}$, respectively. It is easy to see that $D_\lambda[\boldsymbol{p} : \boldsymbol{q}]$ satisfies the criteria of divergence. However, it is not dually flat in general.

Let us consider the case of $\lambda \to \infty$. Then,

$$\mathbf{P}' = (p_i p_j), \quad \mathbf{P} = (p_i q_j), \tag{41}$$

and hence

$$D_\lambda[\boldsymbol{p} : \boldsymbol{q}] = KL[\boldsymbol{p} : \boldsymbol{q}], \tag{42}$$

converging to the KL-divergence of S_{n-1}.

6 Conclusions

We have opened a new way of studying the geometry of probability distributions. We showed that the entropy-relaxed transportation plan in a probability simplex naturally defines the exponential family and the dually flat structure of information geometry. We also introduced a one-parameter family which connects the relaxed Wasserstein distance to the KL-divergence.

It remains as future problems to extend the information geometry of the relaxed Wasserstain distance into a general case of continuous distributions on a metric manifold. Another direction of research is to study the geometrical properties of the manifold through the new family of λ-divergence and to apply it to various practical applications, where some modifications of D_λ might be useful.

References

Amari, S.: Information Geometry and Its Applications. Springer, Tokyo (2016)

Chentsov, N.N.: Statistical Decision Rules and Optimal Inference. Nauka (1972). Translated in English. AMS (1982)

Cuturi, M.: Sinkhorn distances: light speed computation of optimal transport. In: Advances in Neural Information Processing Systems, pp. 2292–2300 (2013)

Cuturi, M., Avis, D.: Ground metric learning. J. Mach. Learn. Res. **15**, 533–564 (2014)

Cuturi, M., Peyré, G.: A smoothed dual formulation for variational Wasserstein problems. SIAM J. Imaging Sci. **9**, 320–343 (2016)

Rao, C.R.: Information and accuracy attainable in the estimation of statistical parameters. Bull. Calcutta Math. Soc. **37**, 81–91 (1945)

Villani, C.: Topics in Optimal Transportation. Graduate Studies in Math. AMS (2013)

Anomaly Detection in Network Traffic with a Relationnal Clustering Criterion

Damien Nogues[1,2(✉)]

[1] Thales, Paris, France
damien.nogues@thalesdigital.io
[2] Complex Networks, Université Pierre et Marie Curie, Paris, France

Abstract. Unsupervised anomaly detection is a very promising technique for intrusion detection. Among many other approaches, clustering algorithms have often been used to perform this task. However, to describe network traffic, both numerical and categorical variables are commonly used. So most clustering algorithms are not very well-suited to such data. Few clustering algorithms have been proposed for such heterogeneous data. Many approaches do not possess suitable complexity. In this article, using Relational Analysis, we propose a new, unified clustering criterion. This criterion is based on a new similarity function for values in a lattice, which can then be applied to both numerical and categorical variables. Finally we propose an optimisation heuristic of this criterion and an anomaly score which outperforms many state of the art solutions.

1 Anomaly Detection Using Clustering: Motivations

To detect breaches in information systems, most organisations rely on traditional approaches such as anti-viruses and firewalls. However, these approaches are too often too straightforward to evade. Many techniques designed to circumvent detection have been proposed in the literature [8,17].

For this reason, new security systems have been developed: Intrusion Detection Systems (IDS). At their core, these systems act as a system wide alarm system. But most IDS use a misuse detection technique. This means, they rely on some sort of signature database to perform the detection.

Unfortunately, signatures do not permit to detect new or modified attacks. Hence, such systems are often not sufficient to detect advanced, targeted attacks.

This is the reason why anomaly based IDS have been proposed. Furthermore, due to the prohibitive cost of labelling a training dataset of sufficient size, unsupervised anomaly detection is often preferred in intrusion detection. In unsupervised anomaly detection, some hypothesis are necessary. First, normal behaviour must be far more frequent than anomalies, second anomalies and normal events must be structurally different.

Many anomaly detection techniques have been proposed. Interested readers are encouraged to consult this very complete survey [4] and reference therein.

F. Nielsen and F. Barbaresco (Eds.): GSI 2017, LNCS 10589, pp. 127–134, 2017.
https://doi.org/10.1007/978-3-319-68445-1_15

In this paper, we will focus solely on clustering based anomaly detection. These techniques usually work in the following way: first the data is partitioned using a clustering algorithm, then large clusters are used to define what is the normal state of the system, allowing to decide which small clusters are composed of anomalies.

However, most clustering algorithms are not well suited for intrusion detection. Some do not offer sufficient computational performances. Many algorithms are unable to create very small clusters which contain the anomalies, especially when the number of clusters is a given parameter. And most are not able to handle both qualitative and numerical data. To adress those issues, we developed a new clustering algorithm, able to handle heterogeneous data, and particularly efficient for anomaly detection.

2 A New Clustering Criterion

Events used in intrusion detection are often described using both qualitative and numerical data. For example, network flows are described with numerical features such as the duration of the communication and the quantity of exchanged information, and qualitative features such as the protocol and port numbers used. But, most clustering algorithms are not designed to handle heterogeneous data.

Most algorithms, starting with the most common *k-means* [15], rely on distances. Hence, they are only able to handle numerical data.

Some approaches are designed for categorical data, but relatively few approaches tackle heterogeneous data. Most of the time, either numerical data will be discretized, or numerical values will be assigned to categorical data modalities. But these approaches will lead to either loss or addition of information to the data, making them not really satisfactory.

Our approach however rely on an unified notion of similarity and dissimilarity, defined on values of lattices, which can be applied to both types of variables. In a way, it is similar to the approach introduced in [3] called triordonnances. But unfortunately, this approach has a cubic complexity, making it unpractical.

2.1 Intuitive Definition

Defining a clustering criterion using similarities and dissimilarities is quite straightforward. Often, the intuitive definition given to define the goal of clustering is the following: clustering is the task of partitioning a set of objects so that objects within the same cluster are more similar to each other than they are to objects belonging to different clusters.

To define our clustering criterion, we will use the theory of Relational Analysis. This theory consists in representing relations between objects with a binary coding. Readers unfamiliar with this theory can refer to the following [11,12]. A clustering, being nothing more than a partition of the object set, can be thought of as the equivalence relation "being in the same cluster". This will be represented with the following matrix:

$$X_{i,j} = \begin{cases} 1 \text{ if } o_i \text{ and } o_j \text{ belong to the same cluster} \\ 0 \qquad\qquad \text{otherwise} \end{cases}$$

Using the formalism of relational analysis this intuitive definition can easily be translated into a mathematical formula. Let us assume we have n objects $o_1, \ldots, o_n$ and let us consider two functions s and $\overline{s}$ which measure the similarity (resp. the dissimilarity) between pairs of objects. Let us also define the similarity matrix S (resp. the dissimilarity matrix $\overline{S}$) with:

$$S_{i,j} = s(o_i, o_j)$$
$$\overline{S}_{i,j} = \overline{s}(o_i, o_j)$$

We can now consider the following score for any given value of α:

$$Q_\alpha(X) = \sum_{1 \leq i,j \leq n} \alpha S_{i,j} X_{i,j} + (1 - \alpha)\overline{S}_{i,j}\overline{X}_{i,j}, \text{ with } 0 \leq \alpha \leq 1 \quad (1)$$

assuming the following are satisfied:

- $\forall i, j,\ X_{i,j} \in \{0; 1\}$
- $\forall i, j,\ \overline{X}_{i,j} = 1 - X_{i,j}$
- $\forall i,\ X_{i,i} = 1$ (reflexive relation)
- $\forall i, j,\ X_{i,j} = X_{j,i}$ (symmetrical relation)
- $\forall i, j, k,\ X_{i,j} + X_{j,k} - X_{i,k} \leq 1$ (transitive relation)

These constraints ensure we obtain an equivalence relation.

This criterion is quite straightforward to analyse: $\sum_{1 \leq i,j \leq n} S_{i,j} X_{i,j}$ is the sum of similarities of objects belonging to the same clusters and $\sum_{1 \leq i,j \leq n} \overline{S}_{i,j} \overline{X}_{i,j}$ is the sum of dissimilarities between objects from distinct clusters, which are the values we wish to maximize.

Our clustering algorithm will then consist of a maximisation heuristic of $Q_\alpha(X)$.

We need only to define the similarity and dissimilarity functions to obtain our criterion. This will be done with a single definition of similarity.

The main difference between categorical and numerical values is the notion of order. There is a natural order on numerical values, but it makes no sense to compare two modalities of a categorical variable. This is why we will use a special case of partially ordered set: lattices, to define our criterion.

2.2 Similarity and Dissimilarity on a Lattice

To obtain the similarity between objects described with both numerical and categorical values, we used an unified definition for values belonging to a lattice.

A lattice is a partially ordered set of objects $(L, \leq)$ where every pair of elements (x, y) have a unique least upper bound $x \vee y$ and a unique greatest lower bound $x \wedge y$. Meaning $\forall x, y \in L$ there exists $x \wedge y \in L$ and $x \vee y \in L$ such as:

- $x \wedge y \leq x \leq x \vee y$
- $x \wedge y \leq y \leq x \vee y$
- if $z \leq x$ and $z \leq y$ then $z \leq x \wedge y$ (resp. if $x \leq z$ and $y \leq z$ then $x \vee y \leq z$)

Using these notions we can generalise the notion of interval. Let us consider $I_{a,b}$, the interval defined by two elements of a lattice L as $I_{a,b} = \{x \in L, a \wedge b \leq x \leq a \vee b\}$.

And let us consider S a finite sequence which takes its values in L. The number of elements of S belonging to the interval $I_{a,b}$ will be noted $n_{a,b}$.

Using these notations we can now define the similarity between two elements a and b of a sequence of size n of elements belonging to a lattice as:

$$S_{a,b} = 1 - \frac{n_{a,b}}{n}$$

This similarity function verifies intuitive properties of similarities. It takes its values between 0 and 1, and any object is more similar to itself, than it is to other objects: $\forall a \neq b, S_{a,a} > S_{a,b}$.

This function is also particularly interesting for anomaly detection. It gives a greater importance to values which define a sparse interval.

Using this similarity we can now also define a dissimilarity function. The less two objects are similar, the more they will be dissimilar. The dissimilarity should also be positive. And we can also wish that the dissimilarity of an object with itself be 0, so that the dissimilarity is a semi metric function.

For those reasons $\overline{S}_{a,b} = \frac{S_{a,a}+S_{b,b}}{2} - S_{a,b}$ has been chosen as a dissimilarity function.

2.3 Similarity on Heterogeneous Data

With these notions we can now define a similarity and a dissimilarity function for objects represented by heterogeneous variables.

Each object is defined by several variables. So the similarity (and dissimilarity) between two objects will simply be the sum of similarity (resp.dissimilarity) of each variables.

Let us now see how our definition of similarity on lattices can be applied to qualitative and numerical values. To that end, we will see how we can create lattices from our variables.

Similarity on numerical variables. For numerical variables, the values of the variables and their natural order define a lattice. With this order, we simply obtain $a \vee b = max(a, b)$ and $a \wedge b = min(a, b)$.

It follows that if a variable takes the values a and b with $a \leq b$, then $n_{a,b}$ is the number of objects which take its value in $[a, b]$ for this variable. It follows that $S_{a,b}$ is the proportion of objects which does not take their values in $[a, b]$.

This similarity offers several advantages. We do not use the actual values of the variables, only their relative orders, hence normalisation of the values is unnecessary. Also, unlike distance based similarities, our similarity will not be dominated by a few abhorrent values. Finally, it also allows us to consider "ordinal" variables, such as a variable which takes the values "low", "medium", and "high", exactly the same way we consider a numerical variable.

Similarity on categorical variables. Comparing modalities of a categorical variable is senseless. This is reflected in the lattice we construct.

If we have a variable whose set of modalities is U, let us consider the following lattice $(U \cup \{\top, \bot\}, \leq)$ where the values of U are not comparable, i.e. $\forall a, b \in U$, if $a \neq b$ then $a \nleq b$ and $b \nleq a$, and $\forall a \in U$ $a \leq \top$ and $\bot \leq a$.

A graphical representation for such a lattice is given Fig. 1.

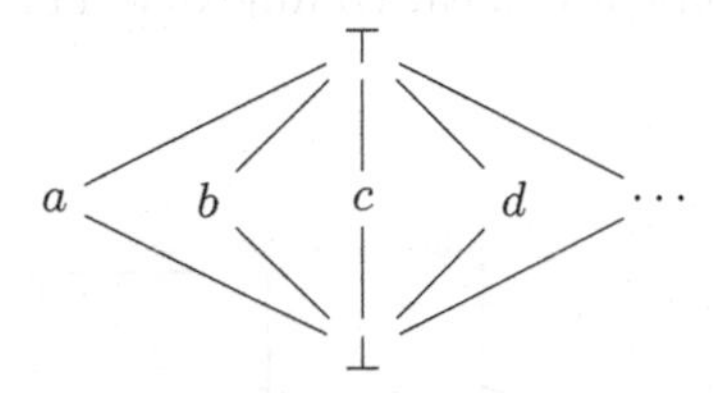

Fig. 1. Graphical representation of the lattice associated to a categorical variable whose modalities are $a, b, c, d, \ldots$.

It follows, if $a \neq b$, $I_{a,b} = I_{\bot,\top}$, hence $n_{a,b} = n$ the number of objects. Also, $n_{a,a} = n_a$ is the number of objects whose variable takes the modality a. The similarity function becomes:

$$S_{a,b} = \begin{cases} 1 - \frac{n_a}{n} & \text{if } a = b \\ 0 & \text{otherwise} \end{cases}$$

3 Optimisation of the Criterion and Anomaly Detection

Having defined a new clustering criterion, we will now see an optimisation heuristic we developed to optimize it. As the finality of this work is anomaly detection, we will propose an anomaly score for the obtained clusters which offers good performances.

3.1 The Optimisation Procedure, a Gradient Descent

Computing the best clustering for a given criterion is a difficult task. An exhaustive approach swiftly becomes infeasible. The number of partitions of a set of n elements is B_n the nth Bell number:

$$B_n = e^{-1} \sum_{k=1}^{\infty} \frac{k^n}{k!}$$

This value is gigantic for relatively small values of n. For example, $B_{10} = 115975$ and $B_{75} > 10^{80}$ the estimated number of atoms in the observable universe. That explains why optimisations heuristics are usually employed.

Here, we proposed a version of a gradient descent [2] in the lattice of sub-partitions. Without going into detail, our approach will look for a good clustering by either merging two clusters or splitting a cluster into two sub-clusters. These operations will be performed when they improve our criterion until a local maximum has been found. Details about the algorithm can be found in [13].

This simple heuristic gives good performances. On small sets where the optimum solution is known such as the "felin" or "iris" datasets, it gives the best solution. Also, by using small optimisations and an equivalent formula for the criterion it also gives empirically linear complexity. The computations times are provided Fig. 2.

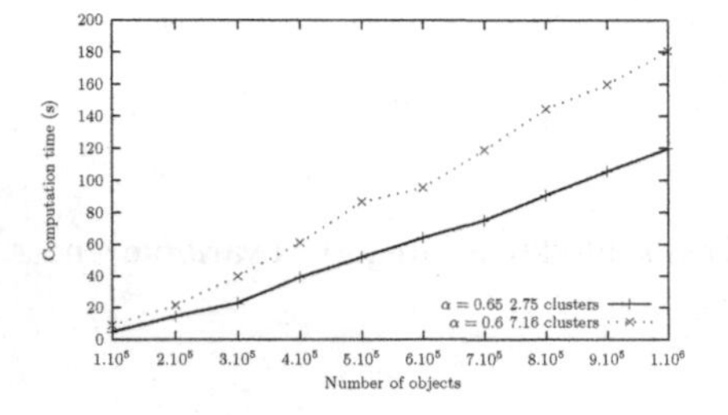

(a) Mean computation time for the clustering algorithm (on 100 runs) as a function of the number of elements for $\alpha = 0.6$ and $\alpha = 0.65$. The computation time depends on the computed clusters structure. So, to maintain a coherent structure, the clustered data is multiple copies of the "German Credit" dataset [10]. This dataset contains 1000 elements. This dataset has been copied from 100 times (10^5 elements) to 1000 times (10^6 elements).

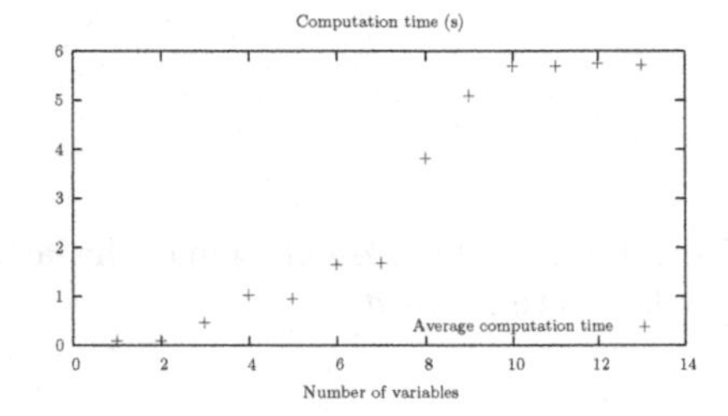

(b) Mean computation time for the clustering algorithm (on 100 runs) as a function of the number of variables for $\alpha = 0.5$. The data used to has been randomly generated.

Fig. 2. Computational performances of our algorithm.

3.2 Anomaly Score and Performances

Our goal is to perform anomaly detection on network traffic using the computed clustering. As mentioned previously, we assume that normal traffic represents a broad majority of our data. This will lead us to suppose that large clusters are composed of normal events.

So, to detect anomalous clusters we will consider clusters that are highly dissimilar to large clusters. The anomaly score we used to assess whether a cluster is anomalous uses the weighted sum of mean dissimilarity between the cluster and the others. Formally, if $C = C_1, \dots, C_p$ is the computed clustering, and $n_1, \dots n_p$ are the sizes of those clusters, the mean dissimilarity between C_i and C_j is $\frac{1}{n_i n_j} \sum_{a \in C_i} \sum_{b \in C_j} \overline{S}_{a,b}$.

The anomaly score for cluster C_i will be

$$A_i = \sum_{j \neq i} n_j \frac{1}{n_i n_j} \sum_{a \in C_i} \sum_{b \in C_j} \overline{S}_{a,b} \tag{2}$$

$$= \frac{1}{n_i} \sum_{j \neq i} \sum_{a \in C_i} \sum_{b \in C_j} \overline{S}_{a,b} \tag{3}$$

Using this anomaly score we obtain excellent performances. To assess the performances of our approach we used subsets of the KDD 99 dataset. It is necessary to select only a subset of the dataset because Denial Of Service (DOS) attacks represent roughly 80% of the records.

For comparison, the performances of our approach are detailled in Table 1a and b.

Table 1. Performances of our approach compared to state of the art.

Algorithm	AUC
our method	**0.990**
Modified clustering-TV [14]	0.973
SVM [5]	0.949
Fixed-width clustering [5]	0.940
K-NN [5]	0.895
fpMafia [9]	0.867

(a) Comparison of the Areas Under the ROC Curve (AUC) obtained with various methods. These values and the method to select the dataset have been taken from [9]. This dataset consists of the KDD 99 dataset, from which the two most frequent denial of service attacks : "smurf" and "neptune" have been removed.

Algorithm	AUC
our method	**0.994**
MLKR [6]	0.983
LPF [16]	0.98
RKOF [7]	≤ 0.962
Active learning [1]	0.935

(b) Comparison of the Areas Under the ROC Curve (AUC) obtained with various methods. These values and the method to select the dataset have been taken from [6]. This dataset consists of the KDD 99 dataset, from which all but the *User to root* types of attacks have been removed from the test dataset. It consists of 60593 benign events and 228 attacks.

To measure these performances, we used the area under the ROC curve (AUC). The ROC curve is the curve created by plotting the true positive rate against the false positive rate for various anomaly score cut-off values. A perfect classifier would yield an AUC of 1. On two different subsets of the dataset used in the literature, thanks to its ability to create small clusters of attacks, our method outperforms the state of the art techniques.

References

1. Abe, N., Zadrozny, B., Langford, J.: Outlier detection by active learning. In: Proceedings of the 12th ACM SIGKDD International Conference on Knowledge Discovery and Data Mining, pp. 504–509. ACM (2006)
2. Cauchy, A.: Méthode générale pour la résolution des systemes d'équations simultanées. Comptes rendus hebdomadaires des séances de l'Académie des sciences **25**(1847), 536–538 (1847)
3. Chah, S.: Comparaisons par triplets en classification automatique. Revue de statistique appliquée **34**(1), 61–79 (1986)
4. Chandola, V., Banerjee, A., Kumar, V.: Anomaly detection: a survey. ACM Comput. Surv. (CSUR) **41**(3), 15 (2009)
5. Eskin, E., Arnold, A., Prerau, M., Portnoy, L., Stolfo, S.: A geometric framework for unsupervised anomaly detection. In: Barbará, D., Jajodia, S. (eds.) Applications of Data Mining in Computer Security. ADIS, vol. 6, pp. 77–101. Springer, Heidelberg (2002). doi:10.1007/978-1-4615-0953-0_4
6. Gao, J., Hu, W., Li, W., Zhang, Z., Wu, O.: Local outlier detection based on kernel regression. In: 2010 20th International Conference on Pattern Recognition (ICPR), pp. 585–588. IEEE (2010)
7. Gao, J., Hu, W., Zhang, Z.M., Zhang, X., Wu, O.: RKOF: robust kernel-based local outlier detection. In: Huang, J.Z., Cao, L., Srivastava, J. (eds.) PAKDD 2011. LNCS, vol. 6635, pp. 270–283. Springer, Heidelberg (2011). doi:10.1007/978-3-642-20847-8_23
8. Handley, M., Paxson, V., Kreibich, C.: Network intrusion detection: evasion, traffic normalization, and end-to-end protocol semantics. In: USENIX Security Symposium, pp. 115–131 (2001)
9. Leung, K., Leckie, C.: Unsupervised anomaly detection in network intrusion detection using clusters. In: Proceedings of the Twenty-Eighth Australasian Conference on Computer Science, vol. 38, pp. 333–342. Australian Computer Society Inc. (2005)
10. Lichman, M.: UCI machine learning repository (2013)
11. Marcotorchino, J.-F., Michaud, P.: Optimisation en analyse ordinale des données. Masson, Paris (1979)
12. Marcotorchino, J.-F., Michaud, P.: Heuristic approach of the similarity aggregation problem. Methods Oper. Res. **43**, 395–404 (1981)
13. Nogues, D.: Method for unsupervised classification of a plurality of objects and device for unsupervised classification associated with said method, EP Patent App. EP20,140,200,529 (2015)
14. Oldmeadow, J., Ravinutala, S., Leckie, C.: Adaptive clustering for network intrusion detection. In: Dai, H., Srikant, R., Zhang, C. (eds.) PAKDD 2004. LNCS, vol. 3056, pp. 255–259. Springer, Heidelberg (2004). doi:10.1007/978-3-540-24775-3_33
15. Steinhaus, H.: Sur la division des corps matériels en parties. Bull. Acad. Polon. Sci. Cl. III **4**, 801–804 (1956)
16. Yang, J., Zhong, N., Yao, Y., Wang, J.: Local peculiarity factor and its application in outlier detection. In: Proceedings of the 14th ACM SIGKDD International Conference on Knowledge Discovery and Data Mining, KDD 2008, New York, NY, USA, pp. 776–784. ACM (2008)
17. You, I., Yim, K.: Malware obfuscation techniques: a brief survey. In: 2010 International Conference on Broadband, Wireless Computing, Communication and Applications, pp. 297–300. IEEE (2010)

Diffeomorphic Random Sampling Using Optimal Information Transport

Martin Bauer[1], Sarang Joshi[2], and Klas Modin[3](✉)

[1] Department of Mathematics, Florida State University, Tallahassee, USA
bauer@math.fsu.edu
[2] Department of Bioengineering, Scientific Computing and Imaging Institute, University of Utah, Salt Lake City, USA
sjoshi@sci.utah.edu
[3] Mathematical Sciences, Chalmers and University of Gothenburg, Gothenburg, Sweden
klas.modin@chalmers.se

Abstract. In this article we explore an algorithm for diffeomorphic random sampling of nonuniform probability distributions on Riemannian manifolds. The algorithm is based on *optimal information transport* (OIT)—an analogue of optimal mass transport (OMT). Our framework uses the deep geometric connections between the Fisher-Rao metric on the space of probability densities and the right-invariant *information metric* on the group of diffeomorphisms. The resulting sampling algorithm is a promising alternative to OMT, in particular as our formulation is semi-explicit, free of the nonlinear Monge–Ampere equation. Compared to Markov Chain Monte Carlo methods, we expect our algorithm to stand up well when a large number of samples from a low dimensional nonuniform distribution is needed.

Keywords: Density matching · Information geometry · Fisher–Rao metric · Optimal transport · Image registration · Diffeomorphism groups · Random sampling

MSC2010: 58E50· 49Q10· 58E10

1 Introduction

We construct algorithms for random sampling, addressing the following problem.

Problem 1. Let μ be a probability distribution on a manifold M. Generate N random samples from μ.

The classic approach to sample from a probability distribution on a higher dimensional space is to use Markov Chain Monte Carlo (MCMC) methods, for example the Metropolis–Hastings algorithm [6]. An alternative idea is to use diffeomorphic density matching between the density μ and a standard density μ_0 from

F. Nielsen and F. Barbaresco (Eds.): GSI 2017, LNCS 10589, pp. 135–142, 2017.
https://doi.org/10.1007/978-3-319-68445-1_16

which samples can be drawn easily. Standard samples are then transformed by the diffeomorphism to generate non-uniform samples. In Bayesian inference, for example, the distribution μ would be the posterior distribution and μ_0 would be the prior distribution. In case the prior itself is hard to sample from the uniform distribution can be used. For M being a subset of the real line, the standard approach is to use the cumulative distribution function to define the diffeomorphic transformation. If, however, the dimension of M is greater then one there is no obvious change of variables to transform the samples to the distribution of the prior. We are thus led to the following matching problem.

Problem 2. Given a probability distribution μ on M, find a diffeomorpism φ such that

$$\varphi_* \mu_0 = \mu.$$

Here, μ_0 denotes a standard distribution on M from which samples can be drawn, and φ_* is the push-forward of φ acting on densities, *i.e.*,

$$\varphi_* \mu_0 = |D\varphi| \mu_0 \circ \varphi,$$

where $|D\varphi|$ is the Jacobian determinant.

A benefit of transport-based methods over traditional MCMC methods is cheap computation of additional samples; it amounts to drawing uniform samples and then evaluating the transformation. On the other hand, transport-based methods scale poorly with increasing dimensionality of M, contrary to MCMC.

The action of the diffeomorphism group on the space of smooth probability densities is transitive (Moser's lemma [13]), so existence of a solution to Problem 2 is guaranteed. However, if the dimension of M is greater then one, there is an infinite-dimensional space of solutions. Thus, one needs to select a specific diffeomorphism within the set of all solutions. Moselhy and Marzouk [12] and Reich [15] proposed to use optimal mass transport (OMT) to construct the desired diffeomorphism φ, thereby enforcing $\varphi = \nabla c$ for some convex function c. The OMT approach implies solving, in one form or another, the heavily nonlinear Monge–Ampere equation for c. A survey of the OMT approach to random sampling is given by Marzouk *et al.* [9].

In this article we pursue an alternative approach for diffeomorphic based random sampling, replacing OMT by *optimal information transport* (OIT), which is diffeomorphic transport based on the Fisher–Rao geometry [11]. Building on deep geometric connections between the Fisher–Rao metric on the space of probability densities and the right-invariant *information metric* on the group of diffeomorphisms [7,11], we developed in [3] an efficient numerical method for density matching. The efficiency stems from a solution formula for φ that is explicit up to inversion of the Laplace operator, thus avoiding the solution of nonlinear PDE such as Monge–Ampere. In this paper we explore this method for random sampling (the initial motivation in [3] is medical imaging, although other applications, including random sampling, are also suggested). The resulting algorithm is implemented in a short MATLAB code, available under MIT license at https://github.com/kmodin/oit-random.

2 Density Transport Problems

Let M be an d–dimensional orientable, compact manifold equipped with a Riemannian metric $g = \langle .,. \rangle$. The volume density induced by g is denoted μ_0 and without loss of generality we assume that the total volume of M with respect to μ_0 is one, *i.e.*, $\int_M \mu_0 = 1$. Furthermore, the space of smooth probability densities on M is given by

$$\mathrm{Prob}(M) = \{\mu \in \Omega^d(M) \mid \mu > 0, \quad \int_M \mu = 1\}, \tag{1}$$

where $\Omega^d(M)$ denotes the space of smooth d-forms. The group of smooth diffeomorphisms $\mathrm{Diff}(M)$ acts on the space of probability densities via push-forward:

$$\mathrm{Diff}(M) \times \mathrm{Prob}(M) \mapsto \mathrm{Prob}(M) \tag{2}$$

$$(\varphi, \mu) \to \varphi_* \mu. \tag{3}$$

By a result of Moser [13] this action is transitive.

We introduce the subgroup of volume preserving diffeomorphisms

$$\mathrm{SDiff}(M) = \{\varphi \in \mathrm{Diff}(M) \mid \varphi_* \mu_0 = \mu_0\}. \tag{4}$$

Note that $\mathrm{SDiff}(M)$ is the isotropy group of μ_0 with respect to the action of $\mathrm{Diff}(M)$. The spaces $\mathrm{Prob}(M)$, $\mathrm{Diff}(M)$, and $\mathrm{SDiff}(M)$ all have the structure of smooth, infinite dimensional Fréchet manifold. Furthermore, $\mathrm{Diff}(M)$ and $\mathrm{SDiff}(M)$ are infinite dimensional Fréchet Lie groups. A careful treatment of these Fréchet topologies can be found in the work by Hamilton [5].

In the following we will focus our attention on the diffeomorphic density matching problem (Problem 2). A common approach to overcome the non-uniqueness in the solution is to add a regularization term to the problem. That is, to search for a minimum energy solution that has the required matching property, for some energy functional E on the diffeomorphism group. Following ideas from mathematical shape analysis [10] it is a natural approach to define this energy functional using the geodesic distance function dist of a Riemannian metric on the diffeomorphism group. Then the regularized diffeomorphic matching problem can be written as follows.

Problem 3. Given a probability density $\mu \in \mathrm{Prob}(M)$ we want to find the diffeomorphism $\varphi \in \mathrm{Diff}(M)$ that minimizes the energy functional

$$E(\varphi) = \mathrm{dist}^2(\mathrm{id}, \varphi) \tag{5}$$

over all diffeomorphisms φ with $\varphi_* \mu_0 = \mu$.

The free variable in the above matching problem is the choice of Riemannian metric—thus distance function—on the group of diffeomorphisms. Although not formulated as here, Moselhy and Marzouk [12] proposed to use the L^2 metric on $\mathrm{Diff}(M)$

$$G_\varphi(u \circ \varphi, v \circ \varphi) = \int_M \langle u \circ \varphi, v \circ \varphi \rangle \, \mu_0 \tag{6}$$

for $u \circ \varphi, v \circ \varphi \in T_\varphi \mathrm{Diff}(M)$. This corresponds to distance-squared *optimal mass transport* (OMT), which induces the Wasserstein L^2 distance on $\mathrm{Prob}(M)$, see, for example, [8,14,16].

In this article we use the right-invariant H^1-type metric

$$G^I_\varphi(u \circ \varphi, v \circ \varphi) = -\int_M \langle \Delta u, v \rangle \mu_0 + \lambda \sum_{i=1}^{k} \int_M \langle u, \xi_i \rangle \mu_0 \int_M \langle v, \xi_i \rangle \mu_0, \quad (7)$$

where $\lambda > 0$, Δ is the Laplace–de Rham operator lifted to vector fields, and $\xi_1, \dots, \xi_k$ is an orthonormal basis of the harmonic 1-forms on M. Because of the Hodge decomposition theorem, G^I is independent of the choice of orthonormal basis $\xi_1, \dots, \xi_k$ for the harmonic vector fields. This construction is related to the Fisher-Rao metric on the space of probability density [2,4], which is predominant in the field of information geometry [1]. We call G^I the *information metric*. See [3,7,11] for more information on the underlying geometry.

The connection between the information metric and the Fisher-Rao metric allows us to construct almost explicit solutions formulas for Problem 2 using the explicit formulas for the geodesics of the Fisher-Rao metric.

Theorem 1 [3,11]. *Let $\mu \in \mathrm{Prob}(M)$ be a smooth probability density. The diffeomorphism $\varphi \in \mathrm{Diff}(M)$ minimizing $\mathrm{dist}_{G^I}(\mathrm{id}, \varphi)$ under the constraint $\varphi_* \mu_0 = \mu$ is given by $\varphi(1)$, where $\varphi(t)$ is obtained as the solution to the problem*

$$\begin{aligned} \Delta f(t) &= \frac{\dot{\mu}(t)}{\mu(t)} \circ \varphi(t), \\ v(t) &= \nabla(f(t)), \\ \frac{d}{dt}\varphi(t)^{-1} &= v(t) \circ \varphi(t)^{-1}, \quad \varphi(0) = \mathrm{id} \end{aligned} \quad (8)$$

where $\mu(t)$ is the (unique) Fisher-Rao geodesic connecting μ_0 and μ

$$\mu(t) = \left(\frac{\sin((1-t)\theta)}{\sin\theta} + \frac{\sin(t\theta)}{\sin\theta} \sqrt{\frac{\mu}{\mu_0}} \right)^2 \mu_0, \quad \cos\theta = \int_M \sqrt{\frac{\mu}{\mu_0}}\, \mu_0. \quad (9)$$

The algorithm for diffeomorphic random sampling, described in the following section, is directly based on solving the Eq. (8).

3 Numerical Algorithm

In this section we explain the algorithm for random sampling using optimal information transport. It is a direct adaptation of [3, Algorithm 1] (If needed, one may also compute the inverse by $\varphi_{k+1}^{-1} = \varphi_k^{-1} + \varepsilon v \circ \varphi_k^{-1}$.).

The algorithm generates N random samples $y_1, \dots, y_N$ from the distribution μ. One can save φ_K and repeat 8–9 whenever additional samples are needed.

Algorithm 1. (OIT based random sampling)

Assume we have a numerical way to represent functions, vector fields, and diffeomorphisms on M, and numerical methods for

- composing functions and vector fields with diffeomorphisms,
- computing the gradient of functions,
- computing solutions to Poisson's equation on M,
- sampling from the standard distribution μ_0 on M, and
- evaluating diffeomorphisms.

An OIT based algorithm for Problem 1 is then given as follows:

1. Choose a step size $\varepsilon = 1/K$ for some positive integer K and calculate the Fisher-Rao geodesic $\mu(t)$ and its derivative $\dot{\mu}(t)$ at all time points $t_k = \frac{k}{K}$ using equation (9).
2. Initialize $\varphi_0 = \mathrm{id}$. Set $k \leftarrow 0$.
3. Compute $s_k = \frac{\dot{\mu}(t_k)}{\mu(t_k)} \circ \varphi_k$ and solve the Poisson equation

$$\Delta f_k = s_k. \tag{10}$$

4. Compute the gradient vector field $v_k = \nabla f_k$.
5. Construct approximations ψ_k to $\exp(-\varepsilon v_k)$, for example

$$\psi_k = \mathrm{id} - \varepsilon v_k. \tag{11}$$

6. Update the diffeomorphism

$$\varphi_{k+1} = \varphi_k \circ \psi_k. \tag{12}$$

7. Set $k \leftarrow k+1$ and continue from step 3 unless $k = K$.
8. Draw N random samples $x_1, \ldots x_N$ from the uniform distribution μ_0.
9. Set $y_n = \varphi_K(x_n)$, $n \in \{1, \ldots N\}$.

The computationally most intensive part of the algorithm is the solution of Poisson's equation at each time step. Notice, however, that we do not need to solve nonlinear equations, such as Monge–Ampere, as is necessary in OMT.

4 Example

In this example we consider $M = \mathbb{T}^2 \simeq (\mathbb{R}/2\pi\mathbb{Z})^2$ with distribution defined in Cartesian coordinates $x, y \in [-\pi, \pi)$ by

$$\mu \sim 3\exp(-x^2 - 10(y - x^2/2 + 1)^2) + 2\exp(-(x+1)^2 - y^2) + 1/10, \tag{13}$$

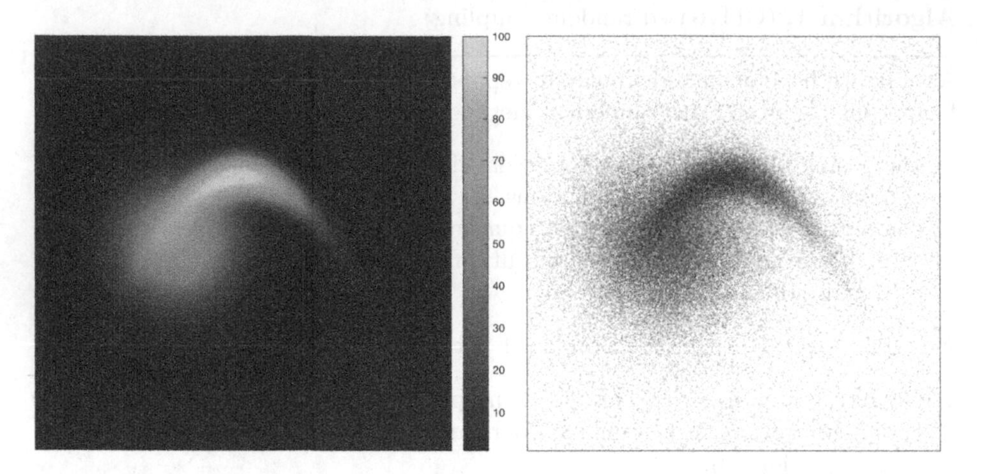

Fig. 1. (left) The probability density μ of (13). The maximal density ratio is 100. (right) 10^5 samples from μ calculated using our OIT based random sampling algorithm.

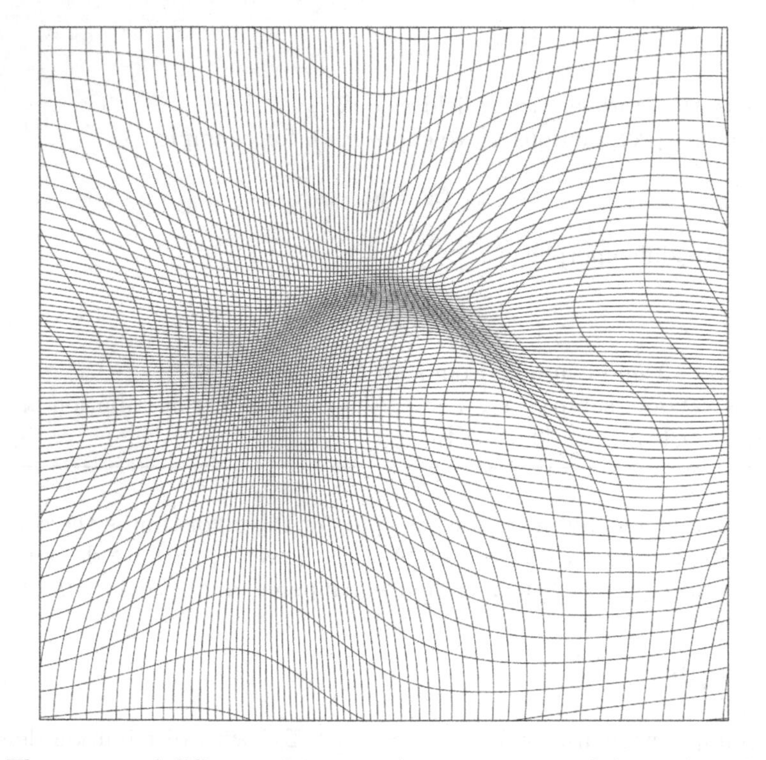

Fig. 2. The computed diffeomorphism φ_K shown as a warp of the uniform 256×256 mesh (every 4th mesh-line is shown). Notice that the warp is periodic. It satisfies $\varphi_* \mu_0 = \mu$ and solves Problem 3 by minimizing the information metric (7). The ratio between the largest and smallest warped volumes is 100.

normalized so that the ratio between the maximum and mimimum of μ is 100. The resulting density is depicted in Fig. 1(left).

We draw 10^5 samples from this distribution using a MATLAB implementation of our algorithm, available under MIT license at https://github.com/kmodin/oit-random.

The implementation can be summarized as follows. To solve the lifting Eq. (8) we discretize the torus by a 256×256 mesh and use the fast Fourier transform (FFT) to invert the Laplacian. We use 100 time steps. The resulting diffeomorphism is shown as a mesh warp in Fig. 2. We then draw 10^5 uniform samples on $[-\pi, \pi]^2$ and apply the diffeomorphism on each sample (applying the diffeomorphism corresponds to interpolation on the warped mesh). The resulting random samples are depicted in Fig. 1(right). To draw new samples is very efficient. For example, another 10^7 samples can be drawn in less than a second on a standard laptop.

5 Conclusions

In this paper we explore random sampling based on the optimal information transport algorithm developed in [3]. Given the semi-explicit nature of the algorithm, we expect it to be an efficient competitor to existing methods, especially for drawing a large number of samples from a low dimensional manifold. However, a detailed comparison with other methods, including MCMC methods, is outside the scope of this paper and left for future work.

We provide an example of a complicated distribution on the flat 2-torus. It is straighforward to extended the method to more elaborate manifolds, *e.g.*, by using finite element methods for Poisson's equation on manifolds. For non-compact manifolds, most importantly $\mathbb{R}^n$, one might use standard techniques, such as Box–Muller, to first transform the required distribution to a compact domain.

References

1. Amari, S., Nagaoka, H.: Methods of information geometry. American Mathematical Society, Providence (2000)
2. Bauer, M., Bruveris, M., Michor, P.W.: Uniqueness of the Fisher-Rao metric on the space of smooth densities. Bull. Lond. Math. Soc. **48**(3), 499–506 (2016)
3. Bauer, M., Joshi, S., Modin, K.: Diffeomorphic density matching by optimal information transport. SIAM J. Imaging Sci. **8**(3), 1718–1751 (2015)
4. Friedrich, T.: Die Fisher-information und symplektische strukturen. Math. Nachr. **153**(1), 273–296 (1991)
5. Hamilton, R.S.: The inverse function theorem of Nash and Moser. Bull. Am. Math. Soc. (N.S.) **7**(1), 65–222 (1982)
6. Hastings, W.K.: Monte Carlo sampling methods using Markov chains and their applications. Biometrika **57**(1), 97–109 (1970)
7. Khesin, B., Lenells, J., Misiołek, G., Preston, S.C.: Geometry of diffeomorphism groups, complete integrability and geometric statistics. Geom. Funct. Anal. **23**(1), 334–366 (2013)

8. Khesin, B., Wendt, R.: The Geometry of Infinite-dimensional Groups. A Series of Modern Surveys in Mathematics, vol. 51. Springer, Berlin (2009)
9. Marzouk, Y., Moselhy, T., Parno, M., Spantini, A.: Sampling via measure transport: An introduction. In: Ghanem, R., Higdon, D., Owhadi, H. (eds.) Handbook of Uncertainty Quantification, pp. 1–14. Springer International Publishing, Cham (2016). doi:10.1007/978-3-319-11259-6_23-1
10. Miller, M.I., Trouvé, A., Younes, L.: On the metrics and euler-lagrange equations of computational anatomy. Annu. Rev. Biomed. Eng. **4**, 375–405 (2002)
11. Modin, K.: Generalized Hunter-Saxton equations, optimal information transport, and factorization of diffeomorphisms. J. Geom. Anal. **25**(2), 1306–1334 (2015)
12. Moselhy, T.A.E., Marzouk, Y.M.: Bayesian inference with optimal maps. J. Comput. Phys. **231**(23), 7815–7850 (2012)
13. Moser, J.: On the volume elements on a manifold. Trans. Am. Math. Soc. **120**, 286–294 (1965)
14. Otto, F.: The geometry of dissipative evolution equations: the porous medium equation. Comm. Partial Diff. Eqn. **26**(1–2), 101–174 (2001)
15. Reich, S.: A nonparametric ensemble transform method for Bayesian inference. SIAM J. Sci. Comput. **35**(4), A2013–A2024 (2013)
16. Villani, C.: Optimal Transport: Old and New, Grundlehren der Mathematischen Wissenschaften, vol. 338. Springer, Berlin (2009)

Optimal Transport to Rényi Entropies

Olivier Rioul(✉)

LTCI, Télécom ParisTech, Université Paris-Saclay, 75013 Paris, France
olivier.rioul@telecom-paristech.fr
http://perso.telecom-paristech.fr/rioul/

Abstract. Recently, an optimal transportation argument was proposed by the author to provide a simple proof of Shannon's entropy-power inequality. Interestingly, such a proof could have been given by Shannon himself in his 1948 seminal paper. In fact, by 1948 Shannon established all the ingredients necessary for the proof and the transport argument takes the form of a simple change of variables.

In this paper, the optimal transportation argument is extended to Rényi entropies in relation to Shannon's entropy-power inequality and to a reverse version involving a certain conditional entropy. The transportation argument turns out to coincide with Barthe's proof of sharp direct and reverse Young's convolutional inequalities and can be applied to derive recent Rényi entropy-power inequalities.

Keywords: Rényi entropy · Entropy-power inequality · Optimal transport

1 Introduction: A Proof that Shannon Missed

2016 was the Shannon Centenary which marked the life and influence of Claude E. Shannon on the 100th anniversary of his birth. On this occasion many scientific events were organized throughout the world in honor of his achievements—on top of which his 1948 seminal paper [1] which developed the mathematical foundations of communication. The French edition of the book re-edition of Shannon's paper [2] has recently been published.

Remarkably, Shannon's revolutionary work, in a single publication [1], established the *fully* formed field of information theory, with all insights and mathematical proofs, albeit in sketched form. There seems to be only one exception in which Shannon's proof turned out to be flawed: the celebrated *entropy-power inequality* (EPI).

The EPI can be described as follows. Letting $P(X) = \frac{1}{n}\mathbb{E}\{\|X\|^2\}$ be the average power of a random vector X taking values in $\mathbb{R}^n$, Shannon defined the *entropy-power* $N(X)$ as the *power* of a zero-mean white Gaussian random vector X^* having the same *entropy* as X. He argued [1, Sect. 21] that for continuous random vectors it is more convenient to work with the entropy-power $N(X)$ than with the differential entropy $h(X)$. By Shannon's formula [1, Sect. 20.6]

F. Nielsen and F. Barbaresco (Eds.): GSI 2017, LNCS 10589, pp. 143–150, 2017.
https://doi.org/10.1007/978-3-319-68445-1_17

$h(X^*) = \frac{n}{2}\log\bigl(2\pi e P(X^*)\bigr)$ for the entropy of the white Gaussian X^*, the closed-form expression of $N(X) = P(X^*)$ when $h(X^*) = h(X)$ is

$$N(X) = \frac{\exp\bigl(\frac{2}{n}h(X)\bigr)}{2\pi e} \tag{1}$$

which is essentially e to the *power* a multiple of the *entropy* of X, also recognized as the "entropy power" of X in this sense. Since the Gaussian maximizes entropy for a given power [1, Sect. 20.5]: $h(X) \leq \frac{n}{2}\log\bigl(2\pi e P(X)\bigr)$, the entropy-power does not exceed the actual power: $N(X) \leq P(X)$ with equality if and only if X is white Gaussian. The power of a *scaled* random vector is given by $P(aX) = a^2 P(X)$, and the same property holds for the entropy-power:

$$N(aX) = a^2 N(X) \tag{2}$$

thanks to the well-known scaling property of the entropy [1, Sect. 20.9]:

$$h(aX) = h(X) + n\log|a| \tag{3}$$

Now for any two *independent* continuous random vectors X and Y, the power of the sum equals the sum of the individual powers: $P(X+Y) = P(X) + P(Y)$ and clearly the same relation holds for the entropy-power in the case of white Gaussian vectors (or Gaussian vectors with proportional covariances). In general, however, the entropy-power of the sum exceeds the sum of the individual entropy-powers:

$$N(X+Y) \geq N(X) + N(Y) \tag{4}$$

where equality holds only if X and Y are Gaussian with proportional covariances. This is the celebrated entropy-power inequality (EPI) as stated by Shannon.

It is remarkable that Shannon had the intuition of this inequality since it turns out to be quite difficult to prove. Shannon's proof [1, Appendix 6] is an incomplete variational argument which shows that Gaussian densities yield a stationary point for $N(X+Y)$ with fixed $N(X)$ and $N(Y)$ but this does not exclude the possibility that the stationary point is not a global minimum.

The first actual proof of the EPI occurred more than ten years later and was quite involved; subsequent proofs used either integration over a path of a continuous Gaussian perturbation or the sharp version of Young's inequality where the EPI is obtained as a limit (which precludes to settle the equality condition in this case). We refer to [3] for a comprehensive list of references and a detailed history.

Recently, an optimal transportation argument was proposed by the author [4,5] to provide a simple proof of the entropy-power inequality, including the equality condition. Interestingly, as we shall now demonstrate, such a proof, appropriately rephrased, could have been given by Shannon himself in his 1948 seminal paper. In fact, by 1948 Shannon established all the ingredients necessary for the proof. As in Shannon's paper [1], to simplify the presentation we assume, without loss of generality, that all considered random vectors have *zero mean* and we here restrict ourselves to real-valued random *variables* in one dimension $n = 1$.

The optimal transport argument takes the form of a simple change of variables: if e.g., X^* is Gaussian, then there exists a (possibly nonlinear) nondecreasing transformation T such that $T(X^*)$ is identically distributed as X—so that one would take $X = T(X^*)$ in what follows. Similarly if Y^* is Gaussian on can take $Y = U(Y^*)$. A detailed proof of this change of variable is given in [4,5] but this is easily seen as a generalization of the inverse c.d.f. method used e.g., for sampling random variables.

Theorem 1 (Shannon's Entropy-Power Inequality). *Let X, Y be independent zero-mean random variables with continuous densities. Then $N(X+Y) \geq N(X) + N(Y)$.*

Proof. The proof is in several steps, each being a direct consequence of Shannon's basic results established in [1].

1. We first proceed to prove the apparently more general inequality
$$N(aX + bY) \geq a^2 N(X) + b^2 N(Y) \tag{5}$$
for any real-valued coefficients a, b. By the scaling property of the entropy-power (2), this is in fact equivalent to the original EPI (4).
2. We can always assume that X and Y have the same entropy-power $N(X) = N(Y)$, or equivalently, have the same entropy $h(X) = h(Y)$. Otherwise, one could find constants c, d such that cX and dY have equal entropy-power (e.g., $c = \exp(-h(X))$ and $d = \exp(-h(Y))$) and applying (5) to cX and dY yields the general case, again thanks to the scaling property of the entropy-power.
3. Let X^*, Y^* be independent zero-mean Gaussian variables with the same entropy as X, Y. Since the entropies of X^* and Y^* are equal they have the same variance and are, therefore, identically distributed. Since equality holds in (5) for X^*, Y^*, we have $a^2 N(X) + b^2 N(Y) = a^2 N(X^*) + b^2 N(Y^*) = N(aX^* + bY^*)$ so that (5) is equivalent to $N(aX + bY) \geq N(aX^* + bY^*)$ or (taking the logarithm)
$$h(aX + bY) \geq h(aX^* + bY^*) \tag{6}$$
4. To prove (6) we may always assume the change of variables $X = T(X^*)$, $Y = U(Y^*)$ as explained above. One is led to prove that
$$h(aT(X^*) + bU(Y^*)) \geq h(aX^* + bY^*) \tag{7}$$
which is written only in terms of the Gaussian variables.
5. Since X^* and Y^* are i.i.d. Gaussian, the Gaussian variables $\tilde{X} = aX^* + bY^*$ and $\tilde{Y} = -bX^* + aY^*$ are uncorrelated and, therefore, independent. Letting $\Delta = a^2 + b^2$ we can write $X^* = (a\tilde{X} - b\tilde{Y})/\Delta$ and $Y^* = (b\tilde{X} + a\tilde{Y})/\Delta$. Since conditioning reduces entropy [1, Sect. 20.4],
$$\begin{aligned} h(aT(X^*) + bU(Y^*)) &= h(aT(\tfrac{a\tilde{X}-b\tilde{Y}}{\Delta}) + bU(\tfrac{b\tilde{X}+a\tilde{Y}}{\Delta})) \\ &\geq h(aT(\tfrac{a\tilde{X}-b\tilde{Y}}{\Delta}) + bU(\tfrac{b\tilde{X}+a\tilde{Y}}{\Delta})|\tilde{Y}) \end{aligned} \tag{8}$$

6. By the change of variable in the entropy [1, Sect. 20.8], for any transformation T, $h(T(X)) = h(X) + \mathbb{E}\log T'(X)$ where $T'(X) > 0$ is the jacobian of the transformation. Applying the transformation in $\tilde{X}$ for fixed $\tilde{Y}$ in the right-hand side of (8) we obtain

$$h(aT(\tfrac{a\tilde{X}-b\tilde{Y}}{\Delta}) + bU(\tfrac{b\tilde{X}+a\tilde{Y}}{\Delta})|\tilde{Y}) = h(\tilde{X}|\tilde{Y}) + \mathbb{E}\log\big(\tfrac{a^2}{\Delta}T'(\tfrac{a\tilde{X}-b\tilde{Y}}{\Delta}) + \tfrac{b^2}{\Delta}U'(\tfrac{b\tilde{X}+a\tilde{Y}}{\Delta})\big) \quad (9)$$

7. By the concavity of the logarithm,

$$\begin{aligned}\log\big(\tfrac{a^2}{\Delta}T'(\tfrac{a\tilde{X}-b\tilde{Y}}{\Delta}) + \tfrac{b^2}{\Delta}U'(\tfrac{b\tilde{X}+a\tilde{Y}}{\Delta})\big) &= \log\big(\tfrac{a^2}{\Delta}T'(X^*) + \tfrac{b^2}{\Delta}U'(Y^*)\big)\\ &\geq \tfrac{a^2}{\Delta}\log T'(X^*) + \tfrac{b^2}{\Delta}\log U'(Y^*) \quad (10)\end{aligned}$$

   but again from change of variable in the entropy [1, Sect. 20.8], $\mathbb{E}\log T'(X^*) = h(T(X^*)) - h(X^*) = h(X) - h(X^*) = 0$ and similarly $\mathbb{E}\log U'(Y^*) = 0$. Thus the second term in the right-hand side of (9) is ≥ 0.
8. Since $\tilde{X}, \tilde{Y}$ are independent, one has [1, Sect. 20.2] $h(\tilde{X}|\tilde{Y}) = h(\tilde{X}) = h(aX^* + bY^*)$, which is the right-hand side of (7). Combining the established inequalities this proves the EPI. □

Remark 1. The case of equality can easily be settled by noting that equality holds in (10) only if $T'(X) = U'(Y)$ a.e., which since X and Y are independent implies that $T' = U'$ is constant, hence transformations T, U are linear and X, Y are Gaussian (see [4] for details).

Going back to the proof it is interesting to note that the only place where the gaussianity of X^*, Y^* is used is for the simplification $h(\tilde{X}|\tilde{Y}) = h(\tilde{X})$. If we drop this assumption we obtain the more general statement:

Corollary 1. *Let X, Y be independent zero-mean random variables with continuous densities, and similarly let X^*, Y^* be independent zero-mean random variables with continuous densities, all of equal entropies. Then for any real a, b,*

$$h(aX + bY) \geq h(aX^* + bY^* | -bX^* + aY^*) \quad (11)$$

If in addition we drop the assumption of equal entropies than letting $\lambda = a^2/\Delta$, $1 - \lambda = b^2/\Delta$ we obtain

Corollary 2. *Let X, Y be independent zero-mean random variables with continuous densities, and similarly let X^*, Y^* be independent zero-mean random variables with continuous densities. Then for any $0 < \lambda < 1$,*

$$\begin{aligned}&h(\sqrt{\lambda}X + \sqrt{1-\lambda}Y) - \lambda h(X) - (1-\lambda)h(Y) \quad (12)\\ &\geq h(\sqrt{\lambda}X^* + \sqrt{1-\lambda}Y^* | -\sqrt{1-\lambda}X^* + \sqrt{\lambda}Y^*) - \lambda h(X^*) - (1-\lambda)h(Y^*)\end{aligned}$$

In fact since the choice of (X, Y) and (X^*, Y^*) is arbitrary the latter inequality can be split into two inequalities [5], the EPI and a *reverse* EPI:

$$\begin{aligned}h(\sqrt{\lambda}X + \sqrt{1-\lambda}Y) &\geq \lambda h(X) + (1-\lambda)h(Y)\\ h(\sqrt{\lambda}X^* + \sqrt{1-\lambda}Y^* | -\sqrt{1-\lambda}X^* + \sqrt{\lambda}Y^*) &\leq \lambda h(X^*) + (1-\lambda)h(Y^*).\end{aligned} \quad (13)$$

2 Generalization to Rényi Entropies

We now extend the same argument to Rényi entropies.

Definition 1 (Hölder Conjugate). *Let $p > 0$, its Hölder conjugate is p' such that $\frac{1}{p} + \frac{1}{p'} = 1$. We write $p' = \infty$ if $p = 1$; note that p' can be negative if $p < 1$.*

Definition 2 (Rényi Entropy). *The Rényi entropy of order p of a random vector X with density $f \in L^p(\mathbb{R}^n)$ is defined by*

$$h_p(X) = -p' \log \|f\|_p = \frac{1}{1-p} \log \int_{\mathbb{R}^n} f^p. \tag{14}$$

As is well known, $h_p(X)$ is non-increasing in p and we recover Shannon's entropy by letting $p \to 1$ from above or below: $h(X) = \lim_{p\to 1} h_p(X)$. We also make the following definitions.

Definition 3 (Power Transformation). *Given a random vector X with density $f \in L^\alpha$, we define X_α as the random vector with density*

$$f_\alpha = \frac{f^\alpha}{\int f^\alpha}. \tag{15}$$

Definition 4 (Young's Triple). *A* Young triple *(p, q, r) consists of three positive real numbers such that p', q', r' are of the same sign and*

$$\frac{1}{p'} + \frac{1}{q'} = \frac{1}{r'}. \tag{16}$$

The triple rate *λ associated to (p, q, r) is the ratio of $1/p'$ in $1/r'$:*

$$\lambda = \frac{1/p'}{1/r'} = \frac{r'}{p'} \qquad 1 - \lambda = \frac{1/q'}{1/r'} = \frac{r'}{q'}. \tag{17}$$

In other words $1/p + 1/q = 1 + 1/r$ as in the classical Young's inequality. If all p', q', r' are > 0 then $p, q, r > 1$; otherwise $p', q', r' < 0$ and $p, q, r < 1$. Thus we always have $0 < \lambda < 1$.

Definition 5 (Dual Young's Triple). *A Young triple (p^*, q^*, r^*) (with rate λ^*) is* dual *to (p, q, r) if it satisfies $r^* = \frac{1}{r}$ and $\lambda^* = 1 - \lambda$.*

From the definition we have $p, q, r > 1 \iff p^*, q^*, r^* < 1$ and *vice versa*. Since $\frac{1}{p^{*\prime}} = \lambda^* \frac{1}{r^{*\prime}} = \frac{1/r' - 1/p'}{1/r'}(1 - r) = \frac{1/r - 1/p}{1/r}$ and similarly for $q^{*\prime}$, the definition fully determines (p^*, q^*, r^*) as

$$(p^* = \frac{p}{r}, q^* = \frac{q}{r}, r^* = \frac{1}{r}) \tag{18}$$

We observe from the definition that the dual of (p^*, q^*, r^*) is the original triple (p, q, r).

We can now state the following

Theorem 2. *Let X, Y be independent zero-mean random variables with continuous densities, and similarly let X^*, Y^* be independent zero-mean random variables with continuous densities. Then for any Young's triple (p, q, r) with dual (p^*, q^*, r^*),*

$$\begin{aligned} h_r(\sqrt{\lambda}X_{1/p} + \sqrt{1-\lambda}Y_{1/q}) - \lambda h_p(X_{1/p}) - (1-\lambda)h_q(Y_{1/q}) \qquad (19)\\ \geq \lambda^* h_{p^*}(X^*_{1/p^*}) + (1-\lambda^*)h_{q^*}(Y^*_{1/q^*}) - h_{r^*}(-\sqrt{\lambda^*}X^*_{1/p^*} + \sqrt{1-\lambda^*}Y^*_{1/q^*}) \end{aligned}$$

Proof. The proof uses the same transportation argument $X = T(X^*)$, $Y = U(Y^*)$ as above, combined with an application of Hölder's inequality. It is omitted due to lack of space (but see Sect. 3.2 below). □

Remark 2. In (19) terms like $h_p(X_{1/p})$ may be simplified since

$$h_p(X_{1/p}) = \frac{1}{1-p}\log\frac{\int f}{(\int f^{1/p})^p} = \frac{1}{1-1/p}\log\int f^{1/p} = h_{1/p}(X). \qquad (20)$$

The above form was chosen to stress the similarity with (12).

Remark 3. The inequality (19) is invariant by duality, in the sense that if we permute the roles of all variables (p, q, r, λ, X, Y) and starred variables $(p^*, q^*, r^*, \lambda^*, X^*, Y^*)$ we obtain the exact same inequality.

Remark 4. The case of equality can be determined as in the proof of Theorem 1: this is the case where $T' = U'$ is constant, hence transformations T, U are linear. Hence equality holds in (19) if and only if there exists a constant $c > 0$ such that X has the same distribution as cX^* and Y has the same distribution as cY^*.

3 Some Applications

3.1 Back to Shannon's Entropy-Power Inequality

There is a striking similarity between Theorem 2 and Corollary 2. In fact for fixed $\lambda = 1 - \lambda^*$, we can let $p, q, r \to 1$ from above (or below) so that $p^*, q^*, r^* \to 1$ from below (or above) to obtain

$$\begin{aligned} h(\sqrt{\lambda}X + \sqrt{1-\lambda}Y) - \lambda h(X) - (1-\lambda)h(Y) \qquad (21)\\ \geq (1-\lambda)h(X^*) + \lambda h(Y^*) - h(-\sqrt{1-\lambda}X^* + \sqrt{\lambda}Y^*). \end{aligned}$$

This is exactly (12) in Corollary 2 because the right-hand side can be rewritten as

$$\begin{aligned} &(1-\lambda)h(X^*) + \lambda h(Y^*) - h(-\sqrt{1-\lambda}X^* + \sqrt{\lambda}Y^*)\\ &= h(X^*) + h(Y^*) - h(-\sqrt{1-\lambda}X^* + \sqrt{\lambda}Y^*) - \lambda h(X^*) - (1-\lambda)h(Y^*)\\ &= h(\sqrt{\lambda}X^* + \sqrt{1-\lambda}Y^*, -\sqrt{1-\lambda}X^* + \sqrt{\lambda}Y^*)\\ &\qquad - h(-\sqrt{1-\lambda}X^* + \sqrt{\lambda}Y^*) - \lambda h(X^*) - (1-\lambda)h(Y^*) \qquad (22)\\ &= h(\sqrt{\lambda}X^* + \sqrt{1-\lambda}Y^* | -\sqrt{1-\lambda}X^* + \sqrt{\lambda}Y^*) - \lambda h(X^*) - (1-\lambda)h(Y^*) \qquad (23) \end{aligned}$$

where (22) holds because the entropy is invariant by rotation.

Thus, Theorem 2 implies the classical Shannon's entropy-power inequality. It is the natural generalization to Rényi entropies using optimal transport arguments.

Remark 5. The above calculation (22)–(23) also shows that the EPI and the "reverse EPI" (13) are in fact equivalent, as already noted in [5]. This is due to the fact that Theorem 2 is invariant by duality (Remark 3).

3.2 Relation to Sharp Young Direct and Reverse Inequalities

To simplify the presentation we stay with one-dimensional random variables. As in Corollary 2, since the choice of (X, Y) and (X^*, Y^*) is arbitrary, (19) can be simplified. If we let $X_{1/p}, Y_{1/q}$ be i.i.d. centered Gaussian, $\sqrt{\lambda}X_{1/p} + \sqrt{1-\lambda}Y_{1/q}$ also has the same Gaussian distribution, and since the Rényi entropy of a Gaussian variable $X \sim \mathcal{N}(m, \sigma^2)$ is easily found to be

$$h_p(X) = -\frac{p' \log p}{2p} + \log\sqrt{2\pi\sigma^2}, \tag{24}$$

the l.h.s. of (19) is equal to $-\frac{r'}{2}(\frac{\log r}{r} - \frac{\log p}{p} - \frac{\log q}{q})$. By the equality case (Remark 4) this expression is also the value taken by the r.h.s. of (19) when $X^*_{1/p^*}, Y^*_{1/q^*}$ are i.i.d. Gaussian (this can also be checked directly from the above definition of the dual Young's triple). Therefore, the expression $-\frac{r'}{2}(\frac{\log r}{r} - \frac{\log p}{p} - \frac{\log q}{q})$ can be inserted between the two sides of (19) in Theorem 2. In other words, (19) is split into two equivalent inequalities which can be rewritten as

$$h_r(\sqrt{\lambda}X + \sqrt{1-\lambda}Y) - \lambda h_p(X) - (1-\lambda)h_q(Y) \geq -\frac{r'}{2}(\frac{\log r}{r} - \frac{\log p}{p} - \frac{\log q}{q}) \tag{25}$$

with equality if and only if X and Y are i.i.d. Gaussian. Plugging the definition (14) of Rényi entropies and dividing by r' (which can be positive of negative), it is easily found [5] that (25) yields the optimal Young's direct and reverse inequalities:

$$\sqrt{\frac{r^{1/r}}{|r'|^{1/r'}}}\|f * g\|_r \leq \sqrt{\frac{p^{1/p}}{|p'|^{1/p'}}}\|f\|_p \cdot \sqrt{\frac{q^{1/q}}{|q'|^{1/q'}}}\|g\|_q. \tag{26}$$

for $p, q, r > 1$ $(r' > 0)$ and the reverse inequality for $0 < p, q, r < 1$ $(r' < 0)$, where f and g denote the densities of $\sqrt{\lambda}X$ and $\sqrt{1-\lambda}Y$. Equality holds if and only if $X/\sqrt{p'}$ and $Y/\sqrt{q'}$ are i.i.d. Gaussian. In fact, a closer look at (19) shows that it coincide with Barthe's transportation proof of sharp Young's inequalities [6, Lemma 1] which uses the same change of variables $X = T(X^*)$, $Y = U(Y^*)$ as above.

3.3 Rényi Entropy-Power Inequalities

Again to simplify the presentation we stay with two one-dimensional independent random variables X, Y. By analogy with the entropy-power (1), the Rényi entropy-power of order p is defined by

$$N_p(X) = \frac{\exp\big(\frac{2}{n}h_p(X)\big)}{2\pi e} \tag{27}$$

We have the following characterization which is an immediate generalization of the classical case $r = c = 1$:

Lemma 1. *Let $r > 0$, $c > 0$. The Renyi entropy-power inequality*

$$N_r(X+Y) \geq c\big(N_r(X) + N_r(Y)\big) \tag{28}$$

is equivalent to

$$h_r(\sqrt{\lambda}X + \sqrt{1-\lambda}Y) - \lambda h_r(X) - (1-\lambda)h_r(Y) \geq \frac{1}{2}\log c \quad \big(\forall \lambda \in (0,1)\big). \tag{29}$$

Now suppose $p^*, q^*, r^* < 1$ so that $r > 1$ is greater than p and q. Since $h_p(X)$ is non-increasing in p, one has $h_p(X) \geq h_r(X)$ and $h_q(Y) \geq h_r(Y)$, hence Theorem 2 in the form (25) implies (29) for any $\lambda \in (0,1)$ provided that $\frac{1}{2}\log c$ is taken as the minimum of the r.h.s. of (25) taken over all p, q such that $1/p + 1/q = 1 + 1/r$.

The method can easily be generalized to more than two independent random variables. In this way we obtain the recent Renyi entropy-power inequalities obtained by Bobkov and Chistyakov [7] and by Ram and Sason [8].

References

1. Shannon, C.E.: A mathematical theory of communication. Bell Syst. Tech. J. **27**, 379–423, 623–656 (1948)
2. Shannon, C.E., Weaver, W.: La théorie mathématique de la communication. Cassini, Paris (2017)
3. Rioul, O.: Information theoretic proofs of entropy power inequalities. IEEE Trans. Inf. Theory **57**(1), 33–55 (2011)
4. Rioul, O.: Yet another proof of the entropy power inequality. IEEE Trans. Inf. Theory **63**(6), 3595–3599 (2017)
5. Rioul, O.: Optimal transportation to the entropy-power inequality. In: IEEE Information Theory and Applications Workshop (ITA 2017), San Diego, USA, February 2017
6. Barthe, F.: Optimal Young's inequality and its converse: a simple proof. GAFA Geom. Funct. Anal. **8**(2), 234–242 (1998)
7. Bobkov, S.G., Chistyakov, G.P.: Entropy power inequality for the Rényi entropy. IEEE Trans. Inf. Theory **61**(2), 708–714 (2015)
8. Ram, E., Sason, I.: On Rényi entropy power inequalities. IEEE Trans. Inf. Theory **62**(12), 6800–6815 (2016)

Statistical Manifold and Hessian Information Geometry

Statistical Manifolds Admitting Torsion, Pre-contrast Functions and Estimating Functions

Masayuki Henmi(✉)

The Institute of Statistical Mathematics,
10-3 Midori-cho, Tachikawa, Tokyo 190-8562, Japan
henmi@ism.ac.jp

Abstract. It is well-known that a contrast function defined on a product manifold $M \times M$ induces a Riemannian metric and a pair of dual torsion-free affine connections on the manifold M. This geometrical structure is called a statistical manifold and plays a central role in information geometry. Recently, the notion of pre-contrast function has been introduced and shown to induce a similar differential geometrical structure on M, but one of the two dual affine connections is not necessarily torsion-free. This structure is called a statistical manifold admitting torsion. This paper summarizes such previous results including the fact that an estimating function on a parametric statistical model naturally defines a pre-contrast function to induce a statistical manifold admitting torsion and provides some new insights on this geometrical structure. That is, we show that the canonical pre-contrast function can be defined on a partially flat space, which is a flat manifold with respect to only one of the dual connections, and discuss a generalized projection theorem in terms of the canonical pre-contrast function.

Keywords: Statistical manifold · Torsion · Contrast function · Estimating function · Standardization · Godambe information matrix · Information geometry

1 Introduction

In information geometry, a central role is played by a statistical manifold, which is a Riemannian manifold with a pair of two dual torsion-free affine connections. This geometrical structure is induced from an asymmetric (squared) distance-like smooth function called a contrast function by taking its second and third derivatives [1,2]. The Kullback-Leibler divergence on a regular parametric statistical model is a typical example of contrast functions and its induced geometrical objects are the Fisher metric, the exponential and mixture connections. The structure determined by these objects play an important role in the geometry of statistical inference, as is widely known [3,4].

M. Henmi—This work was supported by JSPS KAKENHI Grant Number 15K00064.

F. Nielsen and F. Barbaresco (Eds.): GSI 2017, LNCS 10589, pp. 153–161, 2017.
https://doi.org/10.1007/978-3-319-68445-1_18

A statistical manifold admitting torsion (SMAT) is a Riemannian manifold with a pair of two dual affine connections, where only one of them must be torsion-free but the other is necessarily *not*. This geometrical structure naturally appears in a quantum statistical model (i.e. a set of density matrices representing quantum states) [3] and the notion of SMAT was originally introduced to study such a geometrical structure from a mathematical point of view [5]. A pre-contrast function was subsequently introduced as a generalization for the first derivative of a contrast function and it was shown that an pre-contrast function induces a SMAT by taking its first and second derivatives [6].

Henmi and Matsuzoe [7] showed that a SMAT also appears in "classical" statistics through an estimating function. More precisely, an estimating function naturally defines a pre-contrast function on a parametric statistical model and a SMAT is induced from it.

This paper summarizes such previous results and provides some new insights for this geometrical structure. That is, we show that the canonical pre-contrast function can be defined on a partially flat space, which is a SMAT where only one of its dual connections is flat, and discuss a generalized projection theorem in a partially flat space. This theorem relates orthogonal projection of the geodesic with respect to the flat connection to the canonical pre-contrast function.

2 Statistical Manifolds and Contrast Functions

In this paper, we assume that all geometrical objects on differentiable manifolds are smooth and restrict our attention to Riemannian manifolds, although the most of the concepts can be defined for semi-Riemannian manifolds.

Let (M, g) be a Riemannian manifold and ∇ be an affine connection on M. The *dual connection* ∇^* of ∇ with respect to g is defined by

$$Xg(Y, Z) = g(\nabla_X Y, Z) + g(Y, \nabla^*_X Z) \quad (\forall X, \forall Y, \forall Z \in \mathcal{X}(M))$$

where $\mathcal{X}(M)$ is the set of all vector fields on M.

For a affine connection ∇ on M, its curvature tensor field R and torsion tensor field T are defined by the following equations as usual:

$$R(X, Y)Z := \nabla_X \nabla_Y Z - \nabla_Y \nabla_X Z - \nabla_{[X,Y]} Z \quad (\forall X, \forall Y, \forall Z \in \mathcal{X}(M)),$$
$$T(X, Y) := \nabla_X Y - \nabla_Y X - [X, Y] \quad (\forall X, \forall Y \in \mathcal{X}(M)).$$

It is said that an affine connection ∇ is *torsion-free* if $T = 0$. Note that for a torsion-free affine connection ∇, $\nabla^* = \nabla$ implies that ∇ is the Levi-Civita connection with respect to g. Let R^* and T^* be the curvature and torsion tensor fields of ∇^*, respectively. It is easy to see that $R = 0$ always implies $R^* = 0$, but $T = 0$ does not necessarily implies $T^* = 0$.

Let ∇ be a torsion-free affine connection on a Riemannian manifold (M, g). Following [8], we say that (M, g, ∇) is a *statistical manifold* if and only if ∇g is a symmetric $(0, 3)$-tensor field, that is

$$(\nabla_X g)(Y, Z) = (\nabla_Y g)(X, Z) \quad (\forall X, \forall Y, \forall Z \in \mathcal{X}(M)). \tag{1}$$

This condition is equivalent to $T^* = 0$ under the condition that ∇ is a torsion-free. If (M, g, ∇) is a statistical manifold, so is (M, g, ∇^*) and it is called the *dual statistical manifold* of (M, g, ∇). Since ∇ and ∇^* are both torsion-free for a statistical manifold (M, g, ∇), $R = 0$ implies that ∇ and ∇^* are both flat. In this case, (M, g, ∇, ∇^*) is called a *dually flat space.*

Let ϕ be a real-valued function on the direct product $M \times M$ of a manifold M and $X_1, ..., X_i, Y_1, ..., Y_j$ be vector fields on M. The functions $\phi[X_1, ..., X_i|Y_1, ..., Y_j]$, $\phi[X_1, ..., X_i|\ ]$ and $\phi[\ |Y_1, ..., Y_j]$ on M are defined by the equations

$$\phi[X_1, \ldots, X_i|Y_1, \ldots, Y_j](r) := (X_1)_p \cdots (X_i)_p(Y_1)_q \cdots (Y_j)_q \phi(p, q)|_{p=r,q=r}, \quad (2)$$

$$\phi[X_1, \ldots, X_i|\ ](r) := (X_1)_p \cdots (X_i)_p \phi(p, r)|_{p=r}, \quad (3)$$

$$\phi[\ |Y_1, \ldots, Y_j](r) := (Y_1)_q \cdots (Y_j)_q \phi(r, q)|_{q=r} \quad (4)$$

for any $r \in M$, respectively [1]. Using these notations, a *contrast function* ϕ is defined to be a real-valued function which satisfies the following conditions on M [1,2]:

(*a*) $\phi(p, p) = 0 \quad (\forall p \in M)$
(*b*) $\phi[X|\ ] = \phi[\ |X] = 0 \quad (\forall X \in \mathcal{X}(M))$
(*c*) $g(X, Y) := -\phi[X|Y] \ (\forall X, \forall Y \in \mathcal{X}(M))$ is a Riemannian metric on M.

Note that these conditions imply that in some neighborhood of the diagonal set $\{(r, r)|r \in M\}$ in $M \times M$,

$$\phi(p, q) \geq 0, \quad \phi(p, q) = 0 \Longleftrightarrow p = q.$$

Although a contrast function is not necessarily symmetric, this inequality means that a contrast function measures some discrepancy between two points on M (at least locally). For a given contrast function ϕ, the two affine connections ∇ and ∇^* are defined by

$$g(\nabla_X Y, Z) = -\phi[XY|Z], \ g(Y, \nabla^*_X Z) = -\phi[Y|XZ] \quad (\forall X, \forall Y, \forall Z \in \mathcal{X}(M)).$$

In this case, ∇ and ∇^* are both torsion-free and dual to each other with respect to g, which means that both of (M, g, ∇) and (M, g, ∇^*) are statistical manifolds. In particular, (M, g, ∇) is called the statistical manifold induced from the contrast function ϕ.

Now we briefly mention a typical example of contrast functions. Let $S = \{p(\boldsymbol{x}; \boldsymbol{\theta}) \mid \boldsymbol{\theta} = (\theta^1, ..., \theta^d) \in \Theta \subset \boldsymbol{R}^d\}$ be a regular parametric statistical model, which is a set of probability density functions with respect to a dominating measure ν on a sample space $\mathcal{X}$. Each element is indexed by a parameter (vector) $\boldsymbol{\theta}$ in an open subset Θ of $\boldsymbol{R}^d$ and the set S satisfies some regularity conditions, under which S can be seen as a differentiable manifold. The Kullback-Leibler divergence of the two density functions $p_1(\boldsymbol{x}) = p(\boldsymbol{x}; \boldsymbol{\theta}_1)$ and $p_2(\boldsymbol{x}) = p(\boldsymbol{x}; \boldsymbol{\theta}_2)$ in S is defined to be

$$\phi_{KL}(p_1, p_2) := \int_{\mathcal{X}} p_2(\boldsymbol{x}) \log \frac{p_2(\boldsymbol{x})}{p_1(\boldsymbol{x})} \nu(d\boldsymbol{x}).$$

It is easy to see that the Kullback-Leibler divergence satisfies the conditions (a), (b) and (c), and so it is a contrast function on S. Its induced Riemannian metric and dual connections are Fisher metric, the exponential an mixture connections, respectively, and given as follows:

$$g_{jk}(\boldsymbol{\theta}) := g(\partial_j, \partial_k) = E_{\boldsymbol{\theta}}\{s^j(\boldsymbol{x}, \boldsymbol{\theta})s^k(\boldsymbol{x}, \boldsymbol{\theta})\},$$

$$\begin{cases} \Gamma_{ij,k}(\boldsymbol{\theta}) := g(\nabla_{\partial_i}\partial_j, \partial_k) = E_{\boldsymbol{\theta}}[\{\partial_i s^j(\boldsymbol{x}, \boldsymbol{\theta})\}s^k(\boldsymbol{x}, \boldsymbol{\theta})] \\ \Gamma^*_{ik,j}(\boldsymbol{\theta}) := g(\partial_j, \nabla^*_{\partial_i}\partial_k) = \int_{\mathcal{X}} s^j(\boldsymbol{x}, \boldsymbol{\theta})\partial_i\partial_k p(\boldsymbol{x}; \boldsymbol{\theta})\nu(d\boldsymbol{x}) \end{cases},$$

where $E_{\boldsymbol{\theta}}$ indicates that the expectation is taken with respect to $p(\boldsymbol{x}; \boldsymbol{\theta})$, $\partial_i = \frac{\partial}{\partial \theta^i}$ and $s^i(\boldsymbol{x}; \boldsymbol{\theta}) = \partial_i \log p(\boldsymbol{x}; \boldsymbol{\theta})$ $(i = 1, \ldots, d)$. As is widely known, this geometrical structure plays the most fundamental and important role in the differential geometry of statistical inference [3,4].

3 Statistical Manifolds Admitting Torsion and Pre-contrast Functions

A statistical manifold admitting torsion is an abstract notion for the geometrical structure where only one of the dual connections is allow to have torsion, which naturally appears in a quantum statistical model [3]. The definition is obtained by generalizing (1) in the definition of statistical manifold as follows [5].

Let (M, g) be a Riemannian manifold and ∇ be an affine connection on M. We say that (M, g, ∇) is a *statistical manifold admitting torsion* (SMAT for short) if and only if

$$(\nabla_X g)(Y, Z) - (\nabla_Y g)(X, Z) = -g(T(X, Y), Z) \quad (\forall X, \forall Y, \forall Z \in \mathcal{X}(M)). \quad (5)$$

This condition is equivalent to $T^* = 0$ in the case where ∇ possibly has torsion. Note that the condition (5) reduces to (1) if ∇ is torsion-free and that (M, g, ∇^*) is not necessarily a statistical manifold although ∇^* is torsion-free. It should be also noted that (M, g, ∇^*) is a SMAT whenever a torsion-free affine connection ∇ is given on a Riemannian manifold (M, g).

For a SMAT (M, g, ∇), $R = 0$ does not necessarily imply that ∇ is flat, but it implies that ∇^* is flat since $R^* = 0$ and $T^* = 0$. In this case, we call (M, g, ∇, ∇^*) a *partially flat space*.

Let ρ be a real-valued function on the direct product $TM \times M$ of a manifold M and its tangent bundle TM, and $X_1, ..., X_i, Y_1, ..., Y_j, Z$ be vector fields on M. The function $\rho[X_1, ..., X_i Z | Y_1, ..., Y_j]$ is defined by

$$\rho[X_1, \ldots, X_i Z | Y_1, \ldots, Y_j](r) := (X_1)_p \cdots (X_i)_p (Y_1)_q \cdots (Y_j)_q \rho(Z_p, q)|_{p=r, q=r}$$

for any $r \in M$. Note that the role of Z is different from vector fields in the notation of (2). The functions $\rho[X_1, ..., X_i Z| \ ]$ and $\rho[\ |Y_1, ..., Y_j]$ are also defined in the similar way to (3) and (4).

We say that ϕ is a *pre-contrast function* on M if and only if the following conditions are satisfied [6,7]:

(a) $\rho(f_1X_1 + f_2X_2, q) = 0 \ \ (\forall f_i \in C^\infty(M), \forall X_i \in \mathcal{X}(M) \ (i = 1, 2), \forall q \in M)$
(b) $\rho[X|\,] = 0 \ \ (\forall X \in \mathcal{X}(M)) \ \ i.e. \ \rho(X_p, p) = 0 \ \ (\forall p \in M)$
(c) $g(X, Y) := -\rho[X|Y] \ (\forall X, \forall Y \in \mathcal{X}(M))$ is a Riemannian metric on M.

Note that for any contrast function ϕ, the function ρ_ϕ which is defined by $\rho_\phi(X_p, q) := X_p\phi(p, q) \ \ (\forall p, \forall q \in M, \ \forall X_p \in T_p(M))$ is a pre-contrast function on M. The notion of pre-contrast function is obtained by taking the fundamental properties of the first derivative of a contrast function as axioms. For a given pre-contrast function, two affine connections ∇ and ∇^* are defined by the following equations in the same way as a contrast function:

$$g(\nabla_X Y, Z) = -\rho[XY|Z], \ g(Y, \nabla^*_X Z) = -\rho[Y|XZ] \ \ (\forall X, \forall Y, \forall Z \in \mathcal{X}(M)).$$

In this case, ∇ and ∇^* are dual to each other with respect to g and ∇^* is torsion-free. However, the affine connection ∇ possibly has torsion. This means that (M, g, ∇) is a SMAT and it is called the SMAT induced from the pre-contrast function ρ.

4 Generalized Projection Theorem in Partially Flat Spaces

In a dually flat space (M, g, ∇, ∇^*), it is well-known that the canonical contrast functions (called ∇- and ∇^*- divergences) are naturally defined, and the Pythagorean theorem and the projection theorem are stated in terms of the ∇ and ∇^* geodesics and the canonical contrast functions [3,4]. In a partially flat space (M, g, ∇, ∇^*), where $R = R^* = 0$ and $T^* = 0$, a pre-contrast function which seems to be canonical can be defined and a projection theorem holds on the "canonical" pre-contrast function and the ∇^*- geodesic.

Proposition 1 (Canonical Pre-contrast Functions). *Let (M, g, ∇, ∇^*) be a partially flat space (i.e. (M, g, ∇) is a SMAT with $R = R^* = 0$ and $T^* = 0$) and (U, η_i) be an affine coordinate neighborhood with respect to ∇^* in M. The function ρ on $TU \times U$ defined by the following equation is a pre-contrast function on U which induces the SMAT (U, g, ∇):*

$$\rho(Z_p, q) := -g_p(Z_p, \dot{\gamma}^*(0)) \ \ (\forall p, \forall q \in U, \forall Z_p \in T_p(U)), \tag{6}$$

where $\gamma^* : [0, 1] \to U$ is the ∇^*-geodesic such that $\gamma^*(0) = p, \gamma^*(1) = q$ and $\dot{\gamma}^*(0)$ is the tangent vector of γ^* on p.

Proof. For the function ρ defined as (6), the condition (a) in the definition of pre-contrast functions follows from the bilinearity of the inner product g_p. The condition (b) immediately follows from $\dot{\gamma}^*(0) = 0$ when $p = q$. By calculating the derivatives of ρ with the affine coordinate system (η_i), it can be shown that the condition (c) holds and that the induced Riemannian metric and dual affine connections coincide with the original g, ∇ and ∇^*. □

In particular, if (M, g, ∇, ∇^*) is a dually flat space, the pre-contrast function ρ defined in (6) coincides with the directional derivative of ∇^*-divergence $\phi^*(\cdot, q)$ with respect to Z_p (cf. [9,10]). Hence, the definition of (6) seems to be natural one and we call the function ρ in (6) the *canonical pre-contrast function* in a partially flat space (U, g, ∇, ∇^*).

From the definition of the canonical pre-contrast function, we can immediately obtain the following theorem.

Corollary 1 (Generalized Projection Theorem). *Let U be an affine coordinate neighborhood and ρ be the canonical pre-contrast function defined in Proposition 1. For any submanifold N in U, the following conditions are equivalent:*

(i) The ∇^*-geodesic starting at $q \in U$ is perpendicular to N at $p \in N$
(ii) $\rho(Z_p, q) = 0$ for any Z_p in $T_p(N)$.

In the case where (U, g, ∇, ∇^*) is a dually flat space, the projection theorem states that the minimum of the ∇^*-divergence $\phi^*(\cdot, q) : N \to \boldsymbol{R}$ should attain at the point $p \in N$ where the ∇^*-geodesic starting at q is perpendicular to N. It immediately follows from the generalized projection theorem, since the directional derivative of $\phi^*(\cdot, q)$ is the canonical pre-contrast function.

5 Statistical Manifolds Admitting Torsion Induced from Estimating Functions

As we mentioned in Introduction, a SMAT naturally appears through estimating functions in a "classical" statistical model as well as in a quantum statistical model. In this section, we briefly explain how a SMAT is induced on S from an estimating function. See [7] for more details including a concrete example.

Let $S = \{p(\boldsymbol{x}; \boldsymbol{\theta}) \mid \boldsymbol{\theta} = (\theta^1, ..., \theta^d) \in \Theta \subset \boldsymbol{R}^d\}$ be a regular parametric statistical model. An estimating function on S, which we consider here, is a $\boldsymbol{R}^d$-valued function $\boldsymbol{u}(\boldsymbol{x}, \boldsymbol{\theta})$ satisfying the following conditions:

$$E_{\boldsymbol{\theta}}\{\boldsymbol{u}(\boldsymbol{x}, \boldsymbol{\theta})\} = \boldsymbol{0}, \ E_{\boldsymbol{\theta}}\{\|\boldsymbol{u}(\boldsymbol{x}, \boldsymbol{\theta})\|^2\} < \infty, \ \det\left[E_{\boldsymbol{\theta}}\left\{\frac{\partial \boldsymbol{u}}{\partial \boldsymbol{\theta}}(\boldsymbol{x}, \boldsymbol{\theta})\right\}\right] \neq 0 \ (\forall \boldsymbol{\theta} \in \Theta).$$

The first condition is called the unbiasedness of estimating functions, which is important to ensure the consistency of the estimator obtained from an estimating function. Let $X_1, \ldots, X_n$ be a random sample from an unknown probability distribution $p(\boldsymbol{x}; \boldsymbol{\theta}_0)$ in S. The estimator $\hat{\boldsymbol{\theta}}$ for $\boldsymbol{\theta}_0$, which is obtained as a solution to the estimating equation $\sum_{i=1}^n \boldsymbol{u}(\boldsymbol{X}_i, \boldsymbol{\theta}) = \boldsymbol{0}$, is called an M-estimator. The M-estimator $\hat{\boldsymbol{\theta}}$ has the consistency $\hat{\boldsymbol{\theta}} \to \boldsymbol{\theta}_0$ (in probability as $n \to \infty$) and the asymptotic normality $\sqrt{n}(\hat{\boldsymbol{\theta}} - \boldsymbol{\theta}_0) \to N(\boldsymbol{0}, \mathrm{Avar}(\hat{\boldsymbol{\theta}}))$ (in distribution as $n \to \infty$) under some additional regularity conditions [11], where $\mathrm{Avar}(\hat{\boldsymbol{\theta}})$ is an asymptotic variance-covariance matrix of $\hat{\boldsymbol{\theta}}$ and is given by

$\mathrm{Avar}(\hat{\boldsymbol{\theta}}) = \{A(\boldsymbol{\theta}_0)\}^{-1}B(\boldsymbol{\theta}_0)\{A(\boldsymbol{\theta}_0)\}^{-T}$ with $A(\boldsymbol{\theta}) := E_{\boldsymbol{\theta}}\{(\partial \boldsymbol{u}/\partial\boldsymbol{\theta})(\boldsymbol{x},\boldsymbol{\theta})\}$ and $B(\boldsymbol{\theta}) := E_{\boldsymbol{\theta}}\left\{\boldsymbol{u}(\boldsymbol{x},\boldsymbol{\theta})\boldsymbol{u}(\boldsymbol{x},\boldsymbol{\theta})^T\right\}$.

In order to induce the structure of SMAT on S from an estimating function, we consider the notion of *standardization* of estimating functions. For an estimating function $\boldsymbol{u}(\boldsymbol{x},\boldsymbol{\theta})$, its standardization (or *standardized estimating function*) is defined by

$$\boldsymbol{u}_*(\boldsymbol{x},\boldsymbol{\theta}) := E_{\boldsymbol{\theta}}\left\{\boldsymbol{s}(\boldsymbol{x},\boldsymbol{\theta})\boldsymbol{u}(\boldsymbol{x},\boldsymbol{\theta})^T\right\}\left[E_{\boldsymbol{\theta}}\left\{\boldsymbol{u}(\boldsymbol{x},\boldsymbol{\theta})\boldsymbol{u}(\boldsymbol{x},\boldsymbol{\theta})^T\right\}\right]^{-1}\boldsymbol{u}(\boldsymbol{x},\boldsymbol{\theta}),$$

where $\boldsymbol{s}(\boldsymbol{x},\boldsymbol{\theta}) = (\partial/\partial\boldsymbol{\theta})\log p(\boldsymbol{x};\boldsymbol{\theta})$ is the score function [12]. Geometrically, the ith component of the standardized estimating function $\boldsymbol{u}_*(\boldsymbol{x},\boldsymbol{\theta})$ is the orthogonal projection of the ith component of the score function $\boldsymbol{s}(\boldsymbol{x},\boldsymbol{\theta})$ onto the linear space spanned by all components of the estimating function $\boldsymbol{u}(\boldsymbol{x},\boldsymbol{\theta})$ in the Hilbert space

$$\mathcal{H}_{\boldsymbol{\theta}} := \{a(\boldsymbol{x}) \mid E_{\boldsymbol{\theta}}\{a(\boldsymbol{x})\} = 0,\ E_{\boldsymbol{\theta}}\{a(\boldsymbol{x})^2\} < \infty\}$$

with the inner product $< a(\boldsymbol{x}), b(\boldsymbol{x}) >_{\boldsymbol{\theta}} := E_{\boldsymbol{\theta}}\{a(\boldsymbol{x})b(\boldsymbol{x})\}$ $(\forall a(\boldsymbol{x}), \forall b(\boldsymbol{x}) \in \mathcal{H}_{\boldsymbol{\theta}})$. In terms of the standardization, the asymptotic variance-covariance matrix can be rewritten as $\mathrm{Avar}(\hat{\boldsymbol{\theta}}) = \{G(\boldsymbol{\theta}_0)\}^{-1}$, where $G(\boldsymbol{\theta}) := E_{\boldsymbol{\theta}}\left\{\boldsymbol{u}_*(\boldsymbol{x},\boldsymbol{\theta})\boldsymbol{u}_*(\boldsymbol{x},\boldsymbol{\theta})^T\right\}$. The matrix $G(\boldsymbol{\theta})$ is called a Godambe information matrix [13], which is a generalization of the Fisher information matrix.

As we have seen in Sect. 2, the Kullback-Leibler divergence ϕ_{KL} is a contrast function on S. Hence, the first derivative of ϕ_{KL} is a pre-contrast function on S and given by

$$\rho_{KL}((\partial_j)_{p_1}, p_2) := (\partial_j)_{p_1}\phi_{KL}(p_1,p_2) = -\int_{\mathcal{X}} s^j(\boldsymbol{x},\boldsymbol{\theta}_1)p(\boldsymbol{x};\boldsymbol{\theta}_2)\nu(d\boldsymbol{x})$$

for any two probability distributions $p_1(\boldsymbol{x}) = p(\boldsymbol{x};\boldsymbol{\theta}_1)$, $p_2(\boldsymbol{x}) = p(\boldsymbol{x};\boldsymbol{\theta}_2)$ in S and $j = 1,\ldots,d$. This observation leads to the following proposition.

Proposition 2 (Pre-contrast Functions from Estimating Functions). *For an estimating function $\boldsymbol{u}(\boldsymbol{x},\boldsymbol{\theta})$ on the parametric model S, a pre-contrast function $\rho_{\boldsymbol{u}} : TS \times S \to \boldsymbol{R}$ is defined by*

$$\rho_{\boldsymbol{u}}((\partial_j)_{p_1}, p_2) := -\int_{\mathcal{X}} u_*^j(\boldsymbol{x},\boldsymbol{\theta}_1)p(\boldsymbol{x};\boldsymbol{\theta}_2)\nu(d\boldsymbol{x})$$

for any two probability distributions $p_1(\boldsymbol{x}) = p(\boldsymbol{x};\boldsymbol{\theta}_1)$, $p_2(\boldsymbol{x}) = p(\boldsymbol{x};\boldsymbol{\theta}_2)$ in S and $j = 1,\ldots,d$, where $u_*^j(\boldsymbol{x},\boldsymbol{\theta})$ is the jth component of the standardization $\boldsymbol{u}_*(\boldsymbol{x},\boldsymbol{\theta})$ of $\boldsymbol{u}(\boldsymbol{x},\boldsymbol{\theta})$.

The use of the standardization $\boldsymbol{u}_*(\boldsymbol{x},\boldsymbol{\theta})$ instead of $\boldsymbol{u}(\boldsymbol{x},\boldsymbol{\theta})$ ensures that the definition of the function $\rho_{\boldsymbol{u}}$ does not depend on the choice of coordinate system (parameter) of S. In fact, for a coordinate transformation (parameter transformation) $\boldsymbol{\eta} = \Phi(\boldsymbol{\theta})$, the estimating function $\boldsymbol{u}(\boldsymbol{x},\boldsymbol{\theta})$ is changed into $\boldsymbol{v}(\boldsymbol{x},\boldsymbol{\eta}) = \boldsymbol{u}(\boldsymbol{x},\Phi^{-1}(\boldsymbol{\eta}))$ and we have $\boldsymbol{v}_*(\boldsymbol{x},\boldsymbol{\eta}) = (\partial\boldsymbol{\theta}/\partial\boldsymbol{\eta})^T\,\boldsymbol{u}_*(\boldsymbol{x},\boldsymbol{\theta})$. The proof of

Proposition 2 is straightforward. In particular, the condition (b) in the definition of pre-contrast function follows from the unbiasedness of the (standardized) estimating function. The Riemannian metric g, dual connections ∇ and ∇^* induced from the pre-contrast function $\rho_{\boldsymbol{u}}$ are given as follows:

$$g_{jk}(\boldsymbol{\theta}) := g(\partial_j, \partial_k) = E_{\boldsymbol{\theta}}\{u_*^j(\boldsymbol{x}, \boldsymbol{\theta}) u_*^k(\boldsymbol{x}, \boldsymbol{\theta})\} = G(\boldsymbol{\theta})_{jk},$$

$$\begin{cases} \Gamma_{ij,k}(\boldsymbol{\theta}) := g(\nabla_{\partial_i} \partial_j, \partial_k) = E_{\boldsymbol{\theta}}[\{\partial_i u_*^j(\boldsymbol{x}, \boldsymbol{\theta})\} s^k(\boldsymbol{x}, \boldsymbol{\theta})] \\ \Gamma_{ik,j}^*(\boldsymbol{\theta}) := g(\partial_j, \nabla_{\partial_i}^* \partial_k) = \int_{\mathcal{X}} u_*^j(\boldsymbol{x}, \boldsymbol{\theta}) \partial_i \partial_k p(\boldsymbol{x}; \boldsymbol{\theta}) \nu(d\boldsymbol{x}) \end{cases},$$

where $G(\boldsymbol{\theta})_{jk}$ is the (j, k) component of the Godambe information matrix $G(\boldsymbol{\theta})$. Note that ∇^* is always torsion-free since $\Gamma_{ik,j}^* = \Gamma_{ki,j}^*$, whereas ∇ is not necessarily torsion-free unless $\boldsymbol{u}_*(\boldsymbol{x}, \boldsymbol{\theta})$ is integrable with respect to $\boldsymbol{\theta}$.

Henmi and Matsuzoe [7] discussed the quasi score function in [14], which is a well-known example of non-integrable estimating functions. They showed that one of the induced affine connections actually has torsion and the other connection is flat, that is, a partially flat space is induced. The pre-contrast function defined from the estimating function coincides with the canonical pre-contrast function and the generalized projection theorem can be applied. However, its statistical meaning has not been clarified yet. Although it is expected that the SMAT induced from an estimating function has something to do with statistical inference based on the estimating function, the clarification on it is a future problem.

References

1. Eguchi, S.: Geometry of minimum contrast. Hiroshima Math. J. **22**, 631–647 (1992)
2. Matsuzoe, H.: Geometry of contrast functions and conformal geometry. Hiroshima Math. J. **29**, 175–191 (1999)
3. Amari, S., Nagaoka, H.: Method of Information Geometry. Amer. Math. Soc., Providence, Oxford University Press, Oxford (2000)
4. Amari, S.: Information Geometry and Its Applications. AMS, vol. 194. Springer, Tokyo (2016). doi:10.1007/978-4-431-55978-8
5. Kurose, T.: Statistical manifolds admitting torsion. Geometry and Something, Fukuoka University (2007)
6. Matsuzoe, H.: Statistical manifolds admitting torsion and pre-contrast functions. Information Geometry and Its Related Fields, Osaka City University (2010)
7. Henmi, M., Matsuzoe, H.: Geometry of pre-contrast functions and non-conservative estimating functions. In: AIP Conference Proceedings, vol. 1340, pp. 32–41 (2011)
8. Kurose, T.: On the divergences of 1-conformally flat statistical manifolds. Tohoku Math. J. **46**, 427–433 (1994)
9. Henmi, M., Kobayashi, R.: Hooke's law in statistical manifolds and divergences. Nagoya Math. J. **159**, 1–24 (2000)
10. Ay, N., Amari, S.: A novel approach to canonical divergences within information geometry. Entropy **17**, 8111–8129 (2015)
11. van der Vaart, A.W.: Asymptotic Statistics. Cambridge University Press, Cambridge (2000)

12. Heyde, C.C.: Quasi-Likelihood and Its Application. Springer, New York (1997). doi:10.1007/b98823
13. Godambe, V.: An optimum property of regular maximum likelihood estimation. Ann. Math. Statist. **31**, 1208–1211 (1960)
14. McCullagh, P., Nelder, J.A.: Generalized Linear Models, 2nd edn. Chapman and Hall, Boca Raton (1989)

Generalized Wintegen Type Inequality for Lagrangian Submanifolds in Holomorphic Statistical Space Forms

Michel Nguiffo Boyom[1], Mohd. Aquib[2(✉)], Mohammad Hasan Shahid[3], and Mohammed Jamali[4]

[1] IMAG, Alexander Grothendieck Research Institute, Université of Montpellier,Montpellier, France
nguiffo.boyom@gmail.com
[2] Department of Mathematics, Jamia Millia Islamia University, New Delhi, India
aquib80@gmail.com
[3] Department of Mathematics, Jamia Millia Islamia University, New Delhi, India
hasan_jmi@yahoo.com
[4] Department De Mathematiques, Al-Falah University, Faridabad, Haryana, India
jamali_dbd@yahoo.co.in

Abstract. Statistical manifolds are abstract generalizations of statistical models introduced by Amari [1] in 1985. Such manifolds have been studied in terms of information geometry which includes the notion of dual connections, called conjugate connection in affine geometry. Recently, Furuhata [5] defined and studied the properties of holomorphic statistical space forms.

In this paper, we obtain the generalized Wintgen type inequality for Lagrangian submanifolds in holomorphic statistical space forms. We also obtain condition under which the submanifold becomes minimal or H is some scalar multiple of H^*.

Keywords: Wintgen inequality · Lagrangian submanifold · Holomorphic statistical space forms

1 Introduction

The history of statistical manifold was started from investigations of geometric structures on sets of certain probability distributions. In fact, statistical manifolds introduced, in 1985, by Amari [1] have been studied in terms of information geometry and such manifolds include the notion of dual connections, called conjugate connection in affine geometry, closely related to affine differential geometry and which has application in various fields of science and engineering such as

F. Nielsen and F. Barbaresco (Eds.): GSI 2017, LNCS 10589, pp. 162–169, 2017.
https://doi.org/10.1007/978-3-319-68445-1_19

string theory, robot control, digital signal processing etc. The geometry of submanifolds of statistical manifolds is still a young geometry, therefore it attracts our attention.

Moreover, the Wintgen inequality is a sharp geometric inequality for surface in 4-dimensional Euclidean space involving Gauss curvature (intrinsic invariant), normal curvature and square mean curvature (extrinsic invariant). The generalized Wintgen inequality was conjectured by De Smet, Dillen, Verstraelen and Vrancken in 1999 for the submanifolds in real space forms also known as DDVV conjecture.

In present article, we will prove the generalized Wintgen type inequalities for Lagrangian submanifolds in statistical holomorphic space forms some of its applications.

2 Statistical Manifolds and Submanifolds

A statistical manifold is a Riemannian manifold (M, g) endowed with a pair of torsion-free affine connections $\overline{\nabla}$ and $\overline{\nabla}^*$ satisfying

$$Zg(X,Y) = g(\overline{\nabla}_Z X, Y) + g(X, \overline{\nabla}^*_Z Y), \tag{1}$$

for $X, Y, Z \in \Gamma(TM)$. It is denoted by $(M, g, \overline{\nabla}, \overline{\nabla}^*)$. The connections $\overline{\nabla}$ and $\overline{\nabla}^*$ are called dual connections and it is easily shown that $(\overline{\nabla}^*)^* = \overline{\nabla}$. The pair $(\overline{\nabla}, g)$ is said to be a statistical structure. If $(\overline{\nabla}, g)$ is a statistical structure on $\overline{M}$, then $(\overline{\nabla}^*, g)$ is also statistical structure on $\overline{M}$. Denote by $\overline{R}$ and $\overline{R}^*$ the curvature tensor fields of $\overline{\nabla}$ and $\overline{\nabla}^*$, respectively. Then the curvature tensor fields $\overline{R}$ and $\overline{R}^*$ satisfies

$$g(\overline{R}^*(X,Y)Z, W) = -g(Z, \overline{R}(X,Y)W). \tag{2}$$

Let $\overline{M}$ be a $2m$-dimensional manifold and let M be a n-dimensional submanifolds of $\overline{M}$. Then, the corresponding Gauss formulas according to [7] are:

$$\overline{\nabla}_X Y = \nabla_X Y + h(X,Y) \tag{3}$$

$$\overline{\nabla}^*_X Y = \nabla^*_X Y + h^*(X,Y) \tag{4}$$

where h and h^* are symmetric and bilinear, called imbedding curvature tensor of M in $\overline{M}$ for $\overline{\nabla}$ and the imbedding curvature tensor of M in $\overline{M}$ for $\overline{\nabla}^*$, respectively. Let us denote the normal bundle of M by $\Gamma(TM^{\perp})$. Since h and h^* are bilinear, we have the linear transformations A_ξ and A^*_ξ defined by

$$g(A_\xi X, Y) = g(h(X,Y), \xi), \tag{5}$$

$$g(A^*_\xi X, Y) = g(h^*(X,Y), \xi), \tag{6}$$

for any $\xi \in \Gamma(TM^{\perp})$ and $X, Y \in \Gamma(TM)$. The corresponding Weingarten formulas [7] are:

$$\overline{\nabla}_X \xi = -A^*_\xi X + \nabla^{\perp}_X \xi, \tag{7}$$

$$\overline{\nabla}^*_X \xi = -A_\xi X + \nabla^{*\perp}_X \xi, \tag{8}$$

for any $\xi \in \Gamma(TM^{\perp})$ and $X \in \Gamma(TM)$. The connections $\nabla^{\perp}_X$ and $\nabla^{*\perp}_X$ given in the above equations are Riemannian dual connections with respect to the induced metric on $\Gamma(TM^{\perp})$.

The corresponding Gauss, Codazzi and Ricci equations are given by the following results.

Proposition 1 ([7]). *Let $\overline{\nabla}$ be a dual connection on $\overline{M}$ and ∇ the induced connection on M. Let $\overline{R}$ and R be the Riemannian curvature tensors of $\overline{\nabla}$ and ∇, respectively. Then,*

$$\begin{aligned} g(\overline{R}(X,Y)Z,W) = g(R(X,Y)Z,W) + g(h(X,Z),h^*(Y,W)) \\ - g(h^*(X,W),h(Y,Z)), \end{aligned} \tag{9}$$

$$\begin{aligned} (\overline{R}(X,Y)Z)^{\perp} = \nabla^{\perp}_X h(Y,Z) - h(\nabla_X Y, Z) - h(Y, \nabla_X Z) \\ - \{\nabla^{\perp}_Y h(X,Z) - h(\nabla_Y X, Z) - h(X, \nabla_Y Z)\}, \end{aligned} \tag{10}$$

$$g(R^{\perp}(X,y)\xi,\eta) = g(\overline{R}(X,y)\xi,\eta) + g([A^*_\xi, A_\eta]X,Y), \tag{11}$$

where $R^{\perp}$ is the Riemannian curvature tensor on $TM^{\perp}$, $\xi, \eta \in \Gamma(TM^{\perp})$ and $[A^*_\xi, A_\eta] = A^*_\xi A_\eta - A_\eta A^*_\xi$.

Similarly, for the dual connection $\overline{\nabla}^*$ on $\overline{M}$, we have

Proposition 2 ([7]). *Let $\overline{\nabla}^*$ be a dual connection on $\overline{M}$ and ∇^* the induced connection on M. Let $\overline{R}^*$ and R^* be the Riemannian curvature tensors of $\overline{\nabla}^*$ and ∇^*, respectively. Then,*

$$\begin{aligned} g(\overline{R}^*(X,Y)Z,W) = g(R^*(X,Y)Z,W) + g(h^*(X,Z),h(Y,W)) \\ - g(h(X,W),h^*(Y,Z)), \end{aligned} \tag{12}$$

$$\begin{aligned} (\overline{R}^*(X,Y)Z)^{\perp} = \nabla^{*\perp}_X h^*(Y,Z) - h^*(\nabla^*_X Y, Z) - h^*(Y, \nabla^*_X Z) \\ - \{\nabla^{*\perp}_Y h^*(X,Z) - h^*(\nabla^*_Y X, Z) - h^*(X, \nabla^*_Y Z)\}, \end{aligned} \tag{13}$$

$$g(R^{*\perp}(X,y)\xi,\eta) = g(\overline{R}^*(X,y)\xi,\eta) + g([A_\xi, A^*_\eta]X,Y), \tag{14}$$

where $R^{*\perp}$ is the Riemannian curvature tensor for $\nabla^{\perp *}$ on $TM^{\perp}$, $\xi, \eta \in \Gamma(TM^{\perp})$ and $[A_\xi, A^*_\eta] = A_\xi A^*_\eta - A^*_\eta A_\xi$.

Definition 1 ([5]). *A $2m$-dimensional statistical manifold M is said to be a holomorphic statistical manifold if it admits an endomorphism over the tangent bundle $\Gamma(M)$ and a metric g and a fundamental form ω given by $\omega(X,Y) = g(X,JY)$ such that*

$$J^2 = -Id; \qquad \overline{\nabla}\omega = 0, \tag{15}$$

for any vector fields $X, Y \in \Gamma(M)$. Since ω is skew-symmetric, we have $g(X,JY) = -g(JX,Y)$.

Definition 2 ([5]). *A holomorphic statistical manifold M is said to be of constant holomorphic curvature $c \in R$ if the following curvature equation holds:*

$$\begin{aligned}\overline{R}(X,Y)Z = \frac{c}{4}\{&g(Y,Z)X - g(X,Z)Y + g(X,JZ)JY\\ &- g(Y,JZ)JX + 2g(X,JY)JZ\}.\end{aligned} \tag{16}$$

According to the behavior of the tangent space under the action of J, submanifolds in a Hermitian manifold is divided into two fundamental classes namely: *Invariant submanifold* and *totally real submanifold.*

Definition 3. *A totally real submanifold of maximal dimension is called Lagrangian submanifold.*

Let $\{e_1, \ldots, e_n\}$ and $\{e_{n+1}, \ldots, e_{2m}\}$ be tangent orthonormal frame and normal orthonormal frame, respectively, on M. The mean curvature vector field is given by

$$H = \frac{1}{n}\sum_{i=1}^{n} h(e_i, e_i) \tag{17}$$

and

$$H^* = \frac{1}{n}\sum_{i=1}^{n} h^*(e_i, e_i). \tag{18}$$

We also set

$$\|h\|^2 = \sum_{i,j=1}^{n} g(h(e_i, e_j), h(e_i, e_j)) \tag{19}$$

and

$$\|h^*\|^2 = \sum_{i,j=1}^{n} g(h^*(e_i, e_j), h^*(e_i, e_j)). \tag{20}$$

3 Generalized Wintgen Type Inequality

We denote by K and $R^{\perp}$ the sectional curvature function and the normal curvature tensor on M, respectively. Then the normalized scalar curvature ρ is given by [7]

$$\rho = \frac{2\tau}{n(n-1)} = \frac{2}{n(n-1)} \sum_{1 \le i < j \le n} K(e_i \wedge e_j), \tag{21}$$

where τ is scalar curvature, and the normalized normal scalar curvature by [2]

$$\rho^{\perp} = \frac{2\tau^{\perp}}{n(n-1)} = \frac{2}{n(n-1)} \sqrt{\sum_{1 \le i < j \le n} \sum_{1 \le \alpha < \beta \le 2m} (R^{\perp}(e_i, e_j, \xi_\alpha, \xi_\beta))^2}. \tag{22}$$

Following [8] we put

$$K_N = \frac{1}{4} \sum_{r,s=1}^{2m-n} Trace[A_r^*, A_s]^2 \tag{23}$$

and called it the scalar normal curvature of M. The normalized scalar normal curvature is given by [6] $\rho_N = \frac{2}{n(n-1)} \sqrt{K_N}$.

Obviously

$$\begin{aligned} K_N &= \frac{1}{2} \sum_{1 \le r < s \le 2m-n} Trace[A_r^*, A_s]^2 \\ &= \sum_{1 \le r < s \le 2m-n} \sum_{1 \le i < j \le n} g([A_r^*, A_s] e_i . e_j)^2, \end{aligned} \tag{24}$$

for $i, j \in \{1, \ldots, n\}$ and $r, s \in \{1, \ldots, 2m-n\}$.

In term of the components of the second fundamental form, we can express K_N by the formula [6]

$$K_N = \sum_{1 \le r < s \le 2m-n} \sum_{1 \le i < j \le n} \Big(\sum_{k=1}^{n} h_{jk}^{*r} h_{ik}^{s} - h_{jk}^{r} h_{ik}^{*s} \Big)^2. \tag{25}$$

We prove the following.

Theorem 1. *Let M be a Lagrangian submanifold of a holomorphic statistical space form $\overline{M}$. Then*

$$(\rho^{\perp})^2 \ge \frac{c}{n(n-1)} (\rho - \frac{c}{4}) + \frac{c}{(n-1)^2} \big[g(H^*, H) - \|H\| \|H^*\| \big]. \tag{26}$$

Proof. Let M be a Lagrangian submanifold of a holomorphic statistical space form $\overline{M}$ and $\{e_1, \ldots, e_n\}$ an orthonormal frame on M; then $\{\xi_1 = Je_1, \ldots, \xi_n = Je_n\}$ is the orthonormal frame in the normal bundle $\Gamma(TM^{\perp})$. Putting $X = W = e_i$, $Y = Z = e_j$, $i \neq j$ from (16), we have

$$\begin{aligned}\overline{R}(e_i, e_j, e_j, e_i) = \frac{c}{4}\{&g(e_j, e_j)g(e_i, e_i) - g(e_i, e_j)g(e_j, e_i)\\ &+ g(e_i, Je_j)g(Je_j, e_i) - g(e_j, Je_j)g(Je_i, e_i)\\ &+ 2g(e_i, Je_j)g(Je_j, e_i)\}.\end{aligned} \tag{27}$$

Combining Eqs. (9) and (27), we obtain

$$\begin{aligned}R(e_i, e_j, e_j, e_i) = \frac{c}{4}\{&g(e_j, e_j)g(e_i, e_i) - g(e_i, e_j)g(e_j, e_i)\\ &+ g(e_i, Je_j)g(Je_j, e_i) - g(e_j, Je_j)g(Je_i, e_i)\\ &+ 2g(e_i, Je_j)g(Je_j, e_i)\} + g(h(e_i, e_i), h^*(e_j, e_j))\\ &- g(h^*(e_i, e_j), h(e_i, e_j)).\end{aligned} \tag{28}$$

By taking summation $1 \leq i, j \leq n$ and using (17), (18) in (28), we derive

$$2\tau = n(n-1)\frac{c}{4} + n^2 g(H, H^*) - g(h^*(e_i, e_j), h(e_i, e_j)). \tag{29}$$

Using (21) in (29), we get

$$\rho = \frac{c}{4} + \frac{n}{n-1}g(H, H^*) - \frac{1}{n(n-1)}g(h^*(e_i, e_j), h(e_i, e_j)). \tag{30}$$

Now, using Cauchy-Schwarz inequality, (19) and (20) in the above equation, we find

$$\rho \leq \frac{c}{4} + \frac{n}{n-1}g(H, H^*) - \frac{1}{n(n-1)}\|h^*\|\|h\|, \tag{31}$$

which imply

$$\|h^*\|\|h\| \leq n(n-1)(\frac{c}{4} - \rho) + n^2\|H\|\|H^*\|. \tag{32}$$

Further, Eq. (11) implies

$$R^{\perp}(e_i, e_j, \xi_r, \xi_s) = \frac{c}{4}\{-(\delta_{ir}\delta_{js} - \delta_{jr}\delta_{is})\} + g([A^*_{\xi_r}, A_{\xi_s}]e_i, e_j), \tag{33}$$

for all $i, j \in \{1, \ldots, n\}$ and $r, s \in \{1, \ldots, n\}$.

Then we have

$$\begin{aligned}(\tau^{\perp})^2 &= (R^{\perp}(e_i, e_j, \xi_r, \xi_s))^2\\ &= (\frac{c}{4}\{(\delta_{ir}\delta_{js} - \delta_{jr}\delta_{is})\} - g([A^*_{\xi_r}, A_{\xi_s}]e_i, e_j))^2\\ &= \frac{c^2}{16}\frac{n(n-1)}{2} + K_N - \frac{c}{4}g(h(e_i, e_j), h^*(e_i, e_j))\\ &\quad + \frac{c}{4}g(h^*(e_i, e_i), h(e_j, e_j))\end{aligned} \tag{34}$$

Above equation can be re-written as

$$\begin{aligned}(\rho^{\perp})^2 &= \frac{c^2}{8n(n-1)} + \rho_N^2 - \frac{c}{n^2(n-1)^2} g(h(e_i, e_j), h^*(e_i, e_j)) \\ &+ \frac{c}{n^2(n-1)^2} g(h^*(e_i, e_i), h(e_j, e_j)) \\ &\geq \frac{c^2}{8n(n-1)} + \rho_N^2 - \frac{c}{n^2(n-1)^2} \|h\| \|h^*\| \\ &+ \frac{c}{n^2(n-1)^2} g(h^*(e_i, e_i), h(e_j, e_j)). \end{aligned} \tag{35}$$

Now, from (32) and (35), we have

$$\begin{aligned}(\rho^{\perp})^2 &\geq \frac{c^2}{8n(n-1)} + \rho_N^2 - \frac{c}{n^2(n-1)^2} \big[n(n-1)(\frac{c}{4} - \rho) \\ &+ n^2 \|H\| \|H^*\|\big] + \frac{c}{n^2(n-1)^2} g(h^*(e_i, e_i), h(e_j, e_j)) \\ &\geq \frac{c^2}{8n(n-1)} + \rho_N^2 + \frac{c}{n(n-1)}(\rho - \frac{c}{4}) \\ &+ \frac{c}{(n-1)^2}\big[g(H, H^*) - \|H\| \|H^*\|\big] \\ &\geq \frac{c}{n(n-1)}(\rho - \frac{c}{4}) + \frac{c}{(n-1)^2}\big[g(H, H^*) - \|H\| \|H^*\|\big]\end{aligned}$$

An immediate consequence of the Theorem 1 yields the following.

Corollary 1. *Let M be a Lagrangian submanifold of negatively curved holomorphic space form with flat normal bundle. If $\rho = \frac{c}{4}$, then M is either minimal or H is some scalar multiple of H^*.*

Further, we observe

Proposition 3. *Let M be a Lagrangian submanifold of a holomorphic statistical space form $\overline{M}$. If θ be the angle between H and H^*, then*

$$(\rho^{\perp})^2 \geq \frac{c}{n(n-1)}(\rho - \frac{c}{4}) + \frac{c}{(n-1)^2}\big[\|H\| \|H^*\|(\cos\theta - 1)\big].$$

Corollary 2. *Let M be a Lagrangian submanifold of a holomorphic statistical space form $\overline{M}$. If H and H^* are parallel, then*

$$(\rho^{\perp})^2 \geq \frac{c}{n(n-1)}(\rho - \frac{c}{4}).$$

Corollary 3. *Let M be a Lagrangian submanifold of a holomorphic statistical space form $\overline{M}$. If H and H^* are perpendicular, then*

$$(\rho^{\perp})^2 \geq \frac{c}{n(n-1)}(\rho - \frac{c}{4}) - \frac{c}{(n-1)^2} \|H\| \|H^*\|.$$

Notes and Comments

1. The above results are verified for Lagrangian submanifold in complex space form, which is ordinary case of Lagrangian submanifold in holomorphic statistical space form when H and H^* coincides.
2. Please notice that giving a lagrangian submanifold M is a singular foliation with a unique positive dimensional leaf M, the 0-dimensional leaves are singletons $\{x\}$. In the future, we may plan investigating the case of lagrangian foliations whose leaves are subjects the requirements which are assumed in this paper. Therefore, what about the orthogonal lagrangian distribution? That would be our forthcoming challenge.

References

1. Amari, S.: Differential Geometric Methods in Statistics. LNS. Springer, Heidelberg (1985). doi:10.1007/978-1-4612-5056-2
2. Aydin, M.E., Mihai, I.: Wintgen inequality for statistical surfaces (2015). arXiv:1511.04987 [math.DG]
3. Boyom, M.N.: Foliations-webs, Hessian geometry, information geometry, entropy and co-homology. Entropy **18**(12), 433 (2016)
4. Boyom, M.N., Wolak, R.: Transversely Hessian foliations and information geometry. Int. J. Math. **27**(11), 1650092 (2016)
5. Furuhata, H.: Hypersurfaces in statistical manifolds. Diff. Geom. Appl. **27**, 420–429 (2009)
6. Mihai, I.: On the generalized Wintgen inequality for lagrangian submanifolds in complex space form. Nonlinear Anal. **95**, 714–720 (2014)
7. Vos, P.W.: Fundamental equations for statistical submanifolds with applications to the Bartlett correction. Ann. Inst. Stat. Math. **41**(3), 429–450 (1989)
8. Yano, K., Kon, M.: Anti-invariant Submanifolds. M. Dekker, New York (1976)

Amari Functors and Dynamics in Gauge Structures

Michel Nguiffo Boyom[1] and Ahmed Zeglaoui[1,2(✉)]

[1] IMAG, Institut Montpelliérain Alexander Grothendieck, Université de Montpellier, CC051, Place Eugńe Bataillon, 34095 Montpellier, France
boyom@math.univ-montp2.fr, ahmed.zeglaoui@umontpellier.fr

[2] Laboratory of Algebra and Number Theory, Faculté de Mathématiques, University of Science and Technology Houari Boumediene (USTHB), 16111 Bab Ezzouar, Algeria
ahmed.zeglaoui@gmail.com

Abstract. We deal with finite dimensional differentiable manifolds. All items are concerned with are differentiable as well. The class of differentiability is C^{∞}. A metric structure in a vector bundle E is a constant rank symmetric bilinear vector bundle homomorphism of $E \times E$ in the trivial bundle line bundle. We address the question whether a given gauge structure in E is metric. That is the main concerns. We use generalized Amari functors of the information geometry for introducing two index functions defined in the moduli space of gauge structures in E. Beside we introduce a differential equation whose analysis allows to link the new index functions just mentioned with the main concerns. We sketch applications in the differential geometry theory of statistics.

Keywords: Gauge structure · Metric structure · Amari functor · Index functions · Metric dynamic

1 Introduction

A metric structure $(\mathbf{E}, \mathbf{g})$ in a vector bundle $\mathbf{E}$ assigns to every fiber $\mathbf{E}_x$ a symmetric bilinear form $\mathbf{g}_x : \mathbf{E}_x \times \mathbf{E}_x \to \mathbb{R}$. Every finite rank vector bundle admits nondegenerate positive metric structures. One uses the paracompacity for constructing those positive regular metric structures. At another side every nondegenerate metric vector bundle $(\mathbf{E}, \mathbf{g})$ admits metric gauge structures, viz gauge structures $(\mathbf{E}, \nabla)$ subject to the requirement $\nabla \mathbf{g} = 0$. In a nondegenerate structure the values of the curvature tensor of a metric gauge structure $(\mathbf{E}, \nabla)$ belong to the orthogonal sub-algebra $o(\mathbf{E}, \nabla)$ of the Lie algebra $\mathbb{G}(\mathbf{E})$. Arises the question whether a gauge structure $(\mathbf{E}, \nabla)$ is a metric gauge structure in $\mathbf{E}$.

Our concern is to relate this existence question with some methods of the information geometry. In fact in the family ∇^{α} of α-connections in a non singular statistical model $[\mathbf{E}, \pi, M, D, p]$ the 0-connection yields a metric gauge structure in (TM, g). Here g in the Fisher information of the statistical model

F. Nielsen and F. Barbaresco (Eds.): GSI 2017, LNCS 10589, pp. 170–178, 2017.
https://doi.org/10.1007/978-3-319-68445-1_20

as in [2,4]. The question what about the cases $\alpha \neq 0$ deserves the attention. More generally arises the question when the pair $(\nabla, \nabla^\star)$ in a statistical manifold $(M, g, \nabla, \nabla^\star)$ is a pair of a metric gauge structures? Our aim is to address those questions in the general framework of finite rank real vector bundle over finite dimensional smooth manifolds. Our investigation involve two dynamics in the category $\mathfrak{Ga}(\mathbf{E})$. The first dynamic is the natural action of the gauge group $\mathcal{G}(\mathbf{E})$. The second is the action of the infinitely generated Coxeter group generated by the family $\mathfrak{Me}(\mathbf{E})$ of regular metric structures $(\mathbf{E}, \mathbf{g})$. This second dynamic is derived from Amari functors.

2 The Gauge Dynamic in $\mathfrak{Ga}$(E)

2.1 The Gauge Group of a Vector Bundle

Let $\mathbf{E}^\star$ be the dual vector bundle of $\mathbf{E}$. Throughout this Sect. 2 we go to identify the vector bundles $\mathbf{E}^\star \otimes \mathbf{E}$ and $Hom(\mathbf{E}, \mathbf{E})$. Actually $Hom(\mathbf{E}, \mathbf{E})$ is the vector bundle of vector bundle homomorphisms from $\mathbf{E}$ to $\mathbf{E}$. The sheaf of sections of $\mathbf{E}^\star \otimes \mathbf{E}$ is denoted by $\mathbb{G}(\mathbf{E})$. This $\mathbb{G}(\mathbf{E})$ is a Lie algebra sheaf bracket is defined by

$$(\phi, \psi) \longmapsto [\phi, \psi] = \phi \circ \psi - \psi \circ \phi.$$

Actually $\mathbf{E}^\star \otimes \mathbf{E}$ is a Lie algebras bundle. It is called the Lie algebra of infinitesimals gauge transformations. The sheaf of inversible sections of $\mathbf{E}^\star \otimes \mathbf{E}$ is denoted by $\mathcal{G}(\mathbf{E})$. This $\mathcal{G}(\mathbf{E})$ is a Lie groups sheaf whose composition is the composition of applications of $\mathbf{E}$ in $\mathbf{E}$. Elements of $\mathcal{G}(\mathbf{E})$ are called gauge transformations of the vector bundle $\mathbf{E}$. Consequently the set $\mathcal{G}_x(\mathbf{E}) \subset Hom(\mathbf{E}_x, \mathbf{E}_x)$ is nothing but the Lie group $GL(\mathbf{E}_x)$. This $\mathcal{G}(\mathbf{E})$ is the seheaf of sections of the Lie groups bundle $\mathbf{E}^\star \tilde{\otimes} \mathbf{E} \subset \mathbf{E}^\star \otimes \mathbf{E}$. We abuse by calling $\mathcal{G}(\mathbf{E})$ and $\mathbf{E}^\star \tilde{\otimes} \mathbf{E}$ the gauge group of the vector bundle $\mathbf{E}$.

2.2 Gauge Structures in a Vector Bundle E

A gauge structure in a vector bundle $\mathbf{E}$ is a pair $(\mathbf{E}, \nabla)$ where ∇ is a Koszul connection in $\mathbf{E}$. The set of gauge structures is denoted by $\mathfrak{Ga}(\mathbf{E})$. We define the action of the gauge group in $\mathfrak{Ga}(\mathbf{E})$ as it follows

$$\mathcal{G}(\mathbf{E}) \times \mathfrak{Ga}(\mathbf{E}) \longrightarrow \mathfrak{Ga}(\mathbf{E}),$$

$$\phi^\star(\mathbf{E}, \nabla) = (\mathbf{E}, \phi^\star\nabla).$$

The Koszul connection $\phi^\star\nabla$ is defined by

$$(\phi^\star\nabla)_X s = \phi(\nabla_X \phi^{-1} s)$$

for all $s \in \mathfrak{G}a(\mathbf{E})$ and all vectors field X on M. We denoted the gauge moduli space by $Ga(\mathbf{E})$, viz

$$Ga(\mathbf{E}) = \frac{\mathfrak{Ga}(\mathbf{E})}{\mathcal{G}(\mathbf{E})}$$

2.3 The Equation $FE(\nabla\nabla^\star)$

Inspired by the appendix to [4] and by [6] and by we define a map from pairs of gauge structures in the space of differential operators $DO(Eo^* \otimes E, T^*M \otimes E^* \otimes E)$.

To every pair of gauge structures $[(\mathbf{E}, \nabla), (\mathbf{E}, \nabla^\star)]$ we introduce the first order differential operator $D^{\nabla\nabla^*}$ of $\mathbf{E}^\star \otimes \mathbf{E}$ in $T^\star M \otimes \mathbf{E}^\star \otimes \mathbf{E}$ which is defined as it follows

$$D^{\nabla\nabla^*}(\phi)(X, s) = \nabla^*_X(\phi(s)) - \phi(\nabla_X s)$$

for all s and for all vector fields X.

Assume the rank of $\mathbf{E}$ is equal to r and the dimension of M is equal to m. Assume (x_i) is a system of local coordinate functions defined in an open subset $U \subset M$ and (s_α) is a basis of local sections of $\mathbf{E}$ defined in U. We set

$$\nabla_{\partial_{x_i}} s_\alpha = \sum_\beta \Gamma^\beta_{i:\alpha} s_\beta,\ \nabla^\star_{\partial_{x_i}} s_\alpha = \sum_\beta \Gamma^{\star\beta}_{i:\alpha} s_\beta \text{ and } \phi\left(s_\alpha\right) = \sum_\beta \phi^\beta_\alpha s_\beta.$$

Our concern is the analysis of system of partial derivative equations

$$[FE\left(\nabla\nabla^\star\right)]^\gamma_{i:\alpha} : \frac{\partial \phi^\beta_\alpha}{\partial x_i} + \sum_{\beta=1}^{r} \left\{\phi^\beta_\alpha \Gamma^{\star\gamma}_{i:\beta} - \phi^\gamma_\beta \Gamma^\beta_{i:\alpha}\right\} = 0.$$

When we deal with the vector tangent bundles the differential operator $D^{\nabla\nabla^*}$ plays many outstanding roles in the global analysis of the base manifold [6]. In general though every vector bundle admits positive metric structures this same claim is far from being true for symplectic structure and for positive signature metric structures. We aim at linking those open problems with the differential equation $FE(\nabla\nabla^*)$.

The sheaf of germs of solutions to $FE(\nabla\nabla^\star)$ is denoted by $\mathcal{J}_{\nabla\nabla^\star}(\mathbf{E})$.

3 The Metric Dynamics in $\mathfrak{Ga}(\mathrm{E})$

3.1 The Amari Functors in the Category $\mathfrak{Ga}(\mathrm{E})$

Without the express statement of the contrary a metric structure in a vector bundle $\mathbf{E}$ is a constant rank symmetric bilinear vector bundle homomorphism $\mathbf{g}$ of $\mathbf{E} \otimes \mathbf{E}$ in $\tilde{\mathbb{R}}$. Such a metric structure is denoted by $(\mathbf{E}, \mathbf{g})$. A nondegenerate metric structure si called regular, otherwise it is called singular. The category of regular metric structures in $\mathbf{E}$ is denoted by $\mathfrak{Me}(\mathbf{E})$.

Henceforth the concern is the dynamic

$$\begin{aligned} \mathcal{G}(\mathbf{E}) \times \mathfrak{Me}(\mathbf{E}) &\longrightarrow \ \mathfrak{Me}(\mathbf{E}) \\ (\phi, (\mathbf{E}, \mathbf{g})) \ &\longmapsto (E, \phi_\star \mathbf{g}) . \end{aligned}$$

Here the metric $\phi_\star \mathbf{g}$ is defined by

$$\phi_\star \mathbf{g}(s, s') = \mathbf{g}(\phi^{-1}(s), \phi^{-1}(s')).$$

This leads to the moduli space of regular metric structures in a vector bundle $\mathbf{E}$

$$Me(\mathbf{E}) = \frac{\mathfrak{Me}(\mathbf{E})}{\mathcal{G}(\mathbf{E})}.$$

A gauge structure $(\mathbf{E}, \nabla)$ is called metric if there exist a metric structure $(\mathbf{E}, \mathbf{g})$ subject to the requirement $\nabla \mathbf{g} = 0$.

We consider the functor $\mathfrak{Me}(\mathbf{E}) \times \mathfrak{Ga}(\mathbf{E}) \to \mathfrak{Ga}(\mathbf{E})$ which is defined by

$$[(\mathbf{E}, \mathbf{g}), (\mathbf{E}, \nabla) \longmapsto (\mathbf{E}, \mathbf{g}.\nabla)].$$

Here the Koszul connection $\mathbf{g}.\nabla$ is defined by

$$\mathbf{g}(\mathbf{g}.\nabla_X s, s') = X(\mathbf{g}(s, s')) - \mathbf{g}(s, \nabla_X s').$$

The functor just mentioned is called the general Amari functor of the vector bundle $\mathbf{E}$. According to [6], the general Amari functor yield two restrictions:

$$\begin{array}{ccc} \{\mathbf{g}\} \times \mathfrak{Ga}(\mathbf{E},) & \longrightarrow & \mathfrak{Ga}(\mathbf{E}) \\ \nabla & \longmapsto & \mathbf{g}.\nabla \end{array} \tag{1}$$

$$\begin{array}{ccc} \mathfrak{Me}(\mathbf{E}) \times \{\nabla\} & \longrightarrow & \mathfrak{Ga}(\mathbf{E}) \\ \mathbf{g} & \longmapsto & \mathbf{g}.\nabla \end{array} \tag{2}$$

The restriction (1) is called the metric Amari functor of the gauge structure $(\mathbf{E}, \nabla)$. The restriction (2) is called the gauge Amari functor of the metric vector bundle $(\mathbf{E}, \mathbf{g})$.

We observe that $\nabla \mathbf{g} = 0$ if and only if $\mathbf{g}.\nabla = \nabla$. The restriction (1) gives rise to the involution of $\mathfrak{Ga}(\mathbf{E}) : \nabla \to \mathbf{g}.\nabla$. In other words $\mathbf{g}.(\mathbf{g}.\nabla) = \nabla$ for all $(\mathbf{E}, \nabla) \in \mathfrak{Ga}(\mathbf{E})$. In general the question whether an involution admits fixed points has negative answers. In the framework we are concerned with every involution defined by a regular metric structure has fixed points formed by metric gauge structures in $(\mathbf{E}, \mathbf{g})$.

The dynamics

$$\begin{array}{ccc} \mathcal{G}(\mathbf{E}) \times \mathfrak{Ga}(\mathbf{E}) & \longrightarrow & \mathfrak{Ga}(\mathbf{E}) \\ (\phi, \nabla) & \longmapsto & \phi^\star \nabla \end{array}$$

$$\begin{array}{ccc} \mathcal{G}(\mathbf{E}) \times \mathfrak{Me}(\mathbf{E}) & \longrightarrow & \mathfrak{Me}(\mathbf{E}) \\ (\phi, \mathbf{g}) & \longmapsto & \phi_\star \mathbf{g} \end{array}$$

are linked with the metric Amari functor by the formula

$$\phi^\star \mathbf{g}.\nabla = \phi_\star \mathbf{g}.\phi^\star \nabla.$$

We go to introduce the metric dynamics in $\mathfrak{Ga}(\mathbf{E})$. The abstract group of all isomorphisms of $\mathfrak{Ga}(\mathbf{E})$ is denoted by $ISO(\mathfrak{Ga}(\mathbf{E}))$. By the metric Amari functor every regular metric structure $(\mathbf{E}, \mathbf{g})$ yields the involution $(\mathbf{E}, \nabla) \to (\mathbf{E}, \mathbf{g}.\nabla)$.

The subgroup of $ISO(\mathfrak{Ga}(\mathbf{E}))$ which is generated by all regular metric structures in $\mathbf{E}$ is denoted by $\mathcal{G}m(\mathbf{E})$. This group $\mathcal{G}m(\mathbf{E})$ looks like an infinitely generated Coxeter group. Using this analogy we call $\mathcal{G}m(\mathbf{E})$ the metric Coxeter group of $\mathfrak{Ga}(\mathbf{E})$. For instance every metric structure $(\mathbf{E}, \mathbf{g})$ generates a dihedral group of order 2.

3.2 The Quasi-commutativity Property of the Metric Dynamic and the Gauge Dynamic

At the present step we are dealing with both the gauge dynamic

$$\begin{array}{ccc} \mathcal{G}(\mathbf{E}) \times \mathfrak{Ga}(\mathbf{E}) & \longrightarrow & \mathfrak{Ga}(\mathbf{E}) \\ (\phi, \nabla) & \longmapsto & \phi^\star \nabla \end{array} \tag{3}$$

and the metric dynamic

$$\begin{array}{ccc} \mathcal{G}m(\mathbf{E}) \times \mathfrak{Ga}(\mathbf{E}) & \longrightarrow & \mathfrak{Ga}(\mathbf{E}) \\ (\gamma, \nabla) & \longmapsto & \gamma.\nabla \end{array} \tag{4}$$

What we call the quasi commutativity property of (1) and (2) is the link

$$\phi^\star \mathbf{g}.\nabla = \phi_\star \mathbf{g}.\phi^\star \nabla.$$

We consider two regular metric structures $(\mathbf{E}, \mathbf{g}^0)$ and $(\mathbf{E}, \mathbf{g})$. There exists a unique $\phi \in \mathcal{G}(\mathbf{E})$ subject to the requirement

$$\mathbf{g}^0(s, s') = \mathbf{g}(\phi(s), s').$$

By direct calculations one sees that for every gauge structure $(\mathbf{E}, \nabla)$ one has

$$\mathbf{g}.\nabla = \phi^\star(\mathbf{g}^0.\nabla).$$

The quasi-commutativity property shows that every regular metric structure acts in the moduli space $Ga(\mathbf{E})$. Further the gauge orbit $\mathcal{G}(\mathbf{E})(\mathbf{g}.\nabla)$ does not depend on the choice of the regular metric structure $(\mathbf{E}, \mathbf{g})$. Thus the metric Coxeter group $\mathcal{G}m(\mathbf{E})$ acts in the moduli space $Ga(\mathbf{E})$. When there is no risk of confusion the orbit of $[\nabla] \in Ga(\mathbf{E})$ is denoted by $\mathcal{G}m.[\nabla]$ while its stabilizer subgroup is denoted by $\mathcal{G}m_{[\nabla]}$. Consequently one has

Proposition 1. *The index of every stabilizer subgroup $\mathcal{G}m_{[\nabla]} \subset \mathcal{G}m(\mathbf{E})$ is equal to* 1 *or to* 2.

We go to rephrase Proposition 1 versus the orbits of the metric Coxeter group in the moduli space $Ga(\mathbf{E})$.

Proposition 2. *For every orbit $\mathcal{G}m(\mathbf{E}).[\nabla]$ cardinal $\sharp(\mathcal{G}m(\mathbf{E}).[\nabla]) \in \{1, 2\}$.*

3.3 The Metric Index Function

The concern is the metric dynamic

$$\begin{array}{ccc} \mathcal{G}m(\mathbf{E}) \times \mathfrak{Ga}(\mathbf{E}) & \longrightarrow & \mathfrak{Ga}(\mathbf{E}) \\ (\gamma, \nabla) & \longmapsto & \gamma.\nabla \end{array} \tag{5}$$

The length of $\gamma \in \mathcal{G}m(\mathbf{E})$ is denoted by $\mathfrak{l}(\gamma)$. It is defined as it follows

$$\mathfrak{l}(\gamma) = \min\{p \in \mathbb{N} : \gamma = \mathbf{g}_1\mathbf{g}_2 \ldots \mathbf{g}_p, \quad \mathbf{g}_j \in \mathfrak{Me}(\mathbf{E})\}$$

For every gauge structure $(\mathbf{E}, \nabla)$ the metric index of $(\mathbf{E}, \nabla)$ is defined by

$$ind(\nabla) = \min_{\gamma \in \mathcal{G}^* m_\nabla} \{\mathfrak{l}(\gamma) - 1\}.$$

Here $\mathcal{G}^* m_\nabla$ stands for the subset formed of elements of the isotropy subgroup that differ from the unit element. The flowing statement is a straightforward consequence of the quasi-commutativity property.

Lemma 1. *The non negative integer $ind(\nabla)$ is a gauge invariant.*

Consequently we go to encode every orbit $[\nabla] = \mathcal{G}(\mathbf{E})^\star \nabla$ with metric index $ind([\nabla]) = ind(\nabla)$.

Definition 1. *By Lemma 1 we get the metric index function*

$$\mathcal{G}a(\mathbf{E}) \ni [\nabla] \to ind([\nabla]) \in \mathbb{Z}$$

3.4 The Gauge Index Function

We consider the general Amari functor

$$\begin{array}{ccc} \mathfrak{Me}\,(\mathbf{E}) \times \mathfrak{Ga}\,(\mathbf{E}) & \longrightarrow & \mathfrak{Ga}\,(\mathbf{E}) \\ (\mathbf{g}, \nabla) & \longmapsto & \mathbf{g}.\nabla \end{array}$$

For convenience we set $\nabla^{\mathbf{g}} = \mathbf{g}.\nabla$. Therefore to a pair $[(\mathbf{E}, \mathbf{g}), (\mathbf{E}, \nabla)]$ we assign the differential equation $FE(\nabla\nabla^{\mathbf{g}})$. The sheaf of solutions to $FE(\nabla\nabla^{\mathbf{g}})$ is denoted by $\mathcal{J}_{\nabla\nabla^{\mathbf{g}}}(\mathbf{E})$. We go to perform a formalism which is developed in [5]. See also [6] for the case of tangent bundles of a manifolds.

The concerns are metric structures in vector bundles. We recall that a singular metric structure in $\mathbf{E}$ is a constant rank degenerate symmetric bilinear vector bundles homomorphism $\mathbf{g} : \mathbf{E} \times \mathbf{E} \to \tilde{\mathbb{R}}$. Let $(\mathbf{E}, \mathbf{g})$ be a regular metric structure. We pose $\nabla^{\mathbf{g}} = \mathbf{g}.\nabla$.

For every $\phi \in \mathbb{G}(\mathbf{E})$ there exists a unique pair $(\Phi, \Phi^*) \subset \mathbb{G}(\mathbf{E})$ subject to the following requirements

$$\mathbf{g}(\Phi(s), s') = \frac{1}{2}\left[\mathbf{g}(\phi(s), s') + \mathbf{g}(s, \phi(s'))\right], \tag{6}$$

$$\mathbf{g}(\Phi^\star(s), s') = \frac{1}{2}\left[\mathbf{g}(\phi(s), s') - \mathbf{g}(s, \phi(s'))\right] \tag{7}$$

We put $q(s, s') = \mathbf{g}(\Phi(s), s')$ and $\omega(s, s') = \mathbf{g}(\Phi^\star(s), s')$.

Proposition 3 ([5]). *If ϕ is a solution to $FE(\nabla\nabla^\star)$ then Φ and $\Phi^\star$ are solutions to $FE(\nabla\nabla^\star)$. Furthermore,*

$$\nabla q = 0,$$
$$\nabla \omega = 0.$$

By the virtue of the Proposition 3 one has $rank(\Phi) = Constant$ and $rank(\Phi^\star) = Constant$.

Corollary 1. *We assume that the regular metric structure* $(\mathbf{E}, \mathbf{g})$ *is positive definite then we have*

$$\mathbf{E} = \ker(\Phi) \oplus Im(\Phi) \tag{8}$$

$$\mathbf{E} = \ker(\Phi^\star) \oplus Im(\Phi^\star) \tag{9}$$

Further one has the following gauge reductions

$$(\ker(\Phi), \nabla) \subset (\mathbf{E}, \nabla), \tag{10}$$

$$(Im(\Phi), \nabla^{\mathbf{g}}) \subset (\mathbf{E}, \nabla^{\mathbf{g}}), \tag{11}$$

$$(\ker(\Phi^\star), \nabla) \subset (\mathbf{E}, \nabla), \tag{12}$$

$$(Im(\Phi^\star), \nabla^{\mathbf{g}}) \subset (\mathbf{E}, \nabla^{\mathbf{g}}). \tag{13}$$

Corollary 2. *Assume that* $(\mathbf{E}, \mathbf{g}, \nabla, \nabla^{\mathbf{g}})$ *is the vector bundle versus of a statistical manifold* $(M, \mathbf{g}, \nabla, \nabla^{\mathbf{g}})$ *here* $\nabla =$. *Then* $(10, 11, 12, 13)$ *is (quasi)* 4*-web in the base manifold* M.

Given a metric vector bundle $(\mathbf{E}, \mathbf{g})$ and gauge structure $(\mathbf{E}, \nabla)$. The triple $(\mathbf{E}, \mathbf{g}, \nabla)$ is called special if the differential equation $FE(\nabla\nabla^{\mathbf{g}})$ has non trivial solutions. We deduce from Corollary 2 that every special statistical manifold supports a canonical (quasi) 4-web, viz 4 foliations in (quasi) general position.

Before pursing we remark that among formalisms introduce in [6], many (of them) walk in the category of vector bundles. We go to perform this remark. To every special triple $(\mathbf{E}, \mathbf{g}, \nabla)$ we assign the function

$$\begin{array}{ccc} \mathcal{J}_{\nabla\nabla^{\mathbf{g}}}(\mathbf{E}) & \longrightarrow & \mathbb{Z} \\ \phi & \longmapsto & rank(\Phi) \end{array}$$

Reminder: The map Φ is the solution to $FE(\nabla\nabla^g)$ given by $\mathbf{g}(\Phi(s), s') = \frac{1}{2}\left[\mathbf{g}(\phi(s), s') + \mathbf{g}(s, \phi(s'))\right]$.

We define the following non negatives integers

$$s^b(\nabla, \mathbf{g}) = \min_{\phi \in \mathcal{J}_{\nabla\nabla^{\mathbf{g}}}(\mathbf{E})} corank(\Phi), \tag{14}$$

$$s^b(\nabla) = \min_{(\mathbf{E},\mathbf{g}) \in \mathfrak{Me}(\mathbf{E})} s^b(\nabla, \mathbf{g}). \tag{15}$$

Proposition 4. *The non negative integer* $s^b(\nabla)$ *is a gauge invariant* $\mathfrak{Ga}(\mathbf{E})$*, viz* $s^b(\nabla) = s^b(\Phi^\star\nabla)$ *for all gauge transformation* Φ.

Definition 2. *By Proposition 4 we get the gauge index function*

$$\mathcal{G}a(\mathbf{E}) \ni [\nabla] \to s^b([\nabla]) \in \mathbb{Z}$$

4 The Topological Nature of the Index Functions

4.1 Index Functions as Characteristic Obstruction

According to [1], every positive Riemannian foliation (nice singular metric in the tangent bundle of a smooth manifold) admits a unique symmetric metric connection. A combination of [1,4] shows that all those metrics are constructed using methods of the information geometry as in [4] (see the exact sequence (16) below). Remind that we are concerned with the question whether a gauge structure $(\mathbf{E}, \nabla)$ is metric. By the virtue of [1,5] one has

Theorem 1. *In a finite rank vector bundle* $\mathbf{E}$ *a gauge structure* $(\mathbf{E}, \nabla)$ *is metric if and only if for some regular metric structure* $(\mathbf{E}, \mathbf{g})$ *the differential equation* $FE(\nabla\nabla^{\mathbf{g}})$ *admits non trivial solutions.*

Remark 1. If for some metric structure $(\mathbf{E}, \mathbf{g}^0)$ the differential equation $FE(\nabla\nabla^{g^0})$ admits non trivial solutions then for every regular metric structure $(\mathbf{E}, \mathbf{g})$ the differential equation $FE(\nabla\nabla^{g})$ admits non trivial solutions.

Hint: use the following the short exact sequence as in [5]

$$0 \longrightarrow \Omega_2^{\nabla}(TM) \longrightarrow \mathcal{J}_{\nabla\nabla^{\mathbf{g}}}(TM) \longrightarrow S_2^{\nabla}(TM) \longrightarrow 0. \tag{16}$$

We recall that the concern is the question whether a gauge structure $(\mathbf{E}, \nabla)$ is metric. By the remark raised above, this question is linked with the solvability of differential equations $FE(\nabla\nabla^{\mathbf{g}})$ which locally is a system of linear PDE with non constant coefficients. Theorem 1 highlights the links of its solvability with the theory of Riemannian foliations which are objects of the differential topology. The key of those links are items of the information geometry. So giving $(\mathbf{E}, \nabla)$, the property of $(\mathbf{E}, \nabla)$ to be metric is equivalent to the property of $FE(\nabla\nabla^{\mathbf{g}})$ to admit non trivial solutions. Henceforth, our aim is to relate the question just mentioned and the invariants $ind(\nabla)$ and $s^b(\nabla)$. We assume that $(\mathbf{E}, \nabla)$ is special.

Theorem 2. *In a gauge structure* $(\mathbf{E}, \nabla)$, *the following assertions are equivalent*

1. *The gauge structure* $(\mathbf{E}, \nabla)$ *is regularly special.*
2. *The metric index function vanishes at* $[\nabla] \in \mathcal{G}a(\mathbf{E})$ *i.e.* $ind([\nabla]) = 0$.
3. *The gauge index function vanishes at* $[\nabla] \in \mathcal{G}a(\mathbf{E})$ *i.e.* $s^b([\nabla]) = 0$.

Theorem 3. *A gauge structure* $(\mathbf{E}, \nabla)$ *is regularly metric if and only if* $(\mathbf{E}, \mathbf{g}.\nabla)$ *is regularly metric for all regular metric structure* $(\mathbf{E}, \mathbf{g})$.

By Theorem 1 both $ind(\nabla)$ and $s^b(\nabla)$ are characteristic obstructions to $(\mathbf{E}, \nabla)$ being regularly special. We have no relevant interpretation of the case $ind(\nabla) \neq 0$. Regarding the case $s^b(\nabla) \neq 0$, we have

Proposition 5. *Let* $(\mathbf{E}, \nabla)$ *be a gauge structure with* $s^b(\nabla) \neq 0$. *Then there exists a metric structure* $(\mathbf{E}, \mathbf{g})$ *such subject to the following requirement:* $rank(\mathbf{g}) = s^b(\nabla)$, *further* $\mathbf{g}$ *is optimal for those requirement, viz every* ∇*-parallel metric structure* $(\mathbf{E}, \mathbf{g})$ *has rank smaller than* $s^b(\nabla)$.

4.2 Applications to the Statistical Geometry

Theorem 4. *Let $\{\nabla^\alpha\}$ be the family of α-connections of a statistical manifold. If ∇^α is regularly metric for all of the positive real numbers α then all of the α-connections are regularly metric.*

References

1. Affane, A., Chergui, A.: Quasi-connections on degenerate semi-riemanniann manifolds. Mediterr. J. Math. **14**(3), 1–15 (2017). Springer
2. Amari, S.I., Nagaoka, H.: Methods of Information Geometry. Translations of Mathematical Monographs, vol. 191. AMS-OXFORD, Oxford (2007)
3. Bel'ko, I.V.: Degenerate Riemannian metric. Math. Notes. Acad. Sci. USSR **18**(5), 1046–1049 (1975)
4. Nguiffo Boyom, M.: Foliations-Webs-Hessian geometry-information geometry and cohomology. Entropy **18**, 433 (2016)
5. Nguiffo Boyom, M.: Analytic anchored Victor bundle, metric algebroids and stratified Riemannian foliation. In: Naseem, A., Shehzad, H., Arshad, K., Yaya, A. (eds.) Algebra, Geometry, Analysis and their Applications, pp. 1–23. Narosa Publishing House, New Delhi (2016)
6. Nguiffo Boyom, M.: Numerical properties of Koszul connections (to appear)
7. Nguiffo Boyom, M., Wolak, R.A.: Transversely Hessian foliations and information geometry. Int. J. Math. Word Sci. **27**(11) (2016). 17 p
8. Kozlov, S.E.: Levi-Civita connections on degenerate pseudo-Riemannian manifolds. J. Math. Sci. **104**(4), 1338–1342 (2001)

Sasakian Statistical Manifolds II

Hitoshi Furuhata(✉)

Department of Mathematics, Hokkaido University, Sapporo 060-0810, Japan
furuhata@math.sci.hokudai.ac.jp

1 Introduction

This article is a digest of [2,3] with additional remarks on invariant submanifolds of Sasakian statistical manifolds.

We set $\Omega = \{1, \ldots, n+1\}$ as a sample space, and denote by $\mathcal{P}^+(\Omega)$ the set of positive probability densities, that is, $\mathcal{P}^+(\Omega) = \{p : \Omega \to \mathbb{R}_+ \mid \sum_{x \in \Omega} p(x) = 1\}$, where $\mathbb{R}_+$ is the set of positive real numbers. Let M be a smooth manifold as a parameter space, and $s : M \ni u \mapsto p(\cdot, u) \in \mathcal{P}^+(\Omega)$ an injection with the property that $p(x, \cdot) : M \to \mathbb{R}_+$ is smooth for each $x \in \Omega$. Consider a family of positive probability densities on Ω parametrized by M in this manner. We define a $(0,2)$-tensor field on M by

$$g_u(X, Y) = \sum_{x \in \Omega} \{X \log p(x, \cdot)\}\{Y \log p(x, \cdot)\} p(x, u)$$

for tangent vectors $X, Y \in T_u M$. We say that an injection $s : M \to \mathcal{P}^+(\Omega)$ is a *statistical model* if g_u is nondegenerate for each $u \in M$, namely, if g is a Riemannian metric on M, which is called the *Fisher information metric* for s. Define $\varphi : M \to \mathbb{R}^{n+1}$ for a statistical model s by $\varphi(u) = {}^t[2\sqrt{p(1,u)}, \ldots, 2\sqrt{p(n+1,u)}]$. It is known that the metric on M induced by φ from the Euclidean metric on $\mathbb{R}^{n+1}$ coincides with the Fisher information metric g. Since the image $\varphi(M)$ lies on the n-dimensional hypersphere $S^n(2)$ of radius 2, the Fisher information metric is considered as the Riemannian metric induced from the standard metric of the hypersphere. For example, we set

$$M = \{u = {}^t[u^1, \ldots, u^n] \in \mathbb{R}^n \mid u^j > 0, \sum_{l=1}^n u^l < 1\},$$

$$s : M \ni u \mapsto p(x, u) = \begin{cases} u^k, & x = k \in \{1, \ldots, n\}, \\ 1 - \sum_{l=1}^n u^l, & x = n+1. \end{cases}$$

Then $\varphi(M) = S^n(2) \cap (\mathbb{R}_+)^{n+1}$ and the Fisher information metric is the restriction of the standard metric of $S^n(2)$. It shows that a hypersphere with the standard metric plays an important role in information geometry. It is an interesting question whether a *whole* hypersphere plays another part there.

In this article, we give a certain statistical structure on an odd-dimensional hypersphere, and explain its background.

F. Nielsen and F. Barbaresco (Eds.): GSI 2017, LNCS 10589, pp. 179–185, 2017.
https://doi.org/10.1007/978-3-319-68445-1_21

2 Sasakian Statistical Structures

Throughout this paper, M denotes a smooth manifold, and $\Gamma(E)$ denotes the set of sections of a vector bundle $E \to M$. All the objects are assumed to be smooth. For example, $\Gamma(TM^{(p,q)})$ means the set of all the C^∞ tensor fields on M of type (p, q).

At first, we will review the basic notion of Sasakian manifolds, which is a classical topic in differential geometry (See [5] for example). Let $g \in \Gamma(TM^{(0,2)})$ be a Riemannian metric, and denote by ∇^g the Levi-Civita connection of g. Take $\phi \in \Gamma(TM^{(1,1)})$ and $\xi \in \Gamma(TM)$.

A triple (g, ϕ, ξ) is called an *almost contact metric structure* on M if the following equations hold for any $X, Y \in \Gamma(TM)$:

$$\begin{aligned}
&\phi\,\xi = 0, \quad g(\xi, \xi) = 1,\\
&\phi^2 X = -X + g(X, \xi)\xi,\\
&g(\phi X, Y) + g(X, \phi Y) = 0.
\end{aligned}$$

An almost contact metric structure on M is called a *Sasakian structure* if

$$(\nabla^g_X \phi)Y = g(Y, \xi)X - g(Y, X)\xi \tag{1}$$

holds for any $X, Y \in \Gamma(TM)$. We call a manifold equipped with a Sasakian structure a *Sasakian manifold*.

It is known that on a Sasakian manifold the formula

$$\nabla^g_X \xi = \phi X \tag{2}$$

holds for $X \in \Gamma(TM)$. A typical example of a Sasakian manifold is a hypersphere of odd dimension as mentioned below.

We now review the basic notion of statistical manifolds to fix the notation (See [1] and references therein). Let ∇ be an affine connection of M, and $g \in \Gamma(TM^{(0,2)})$ a Riemannian metric. The pair (∇, g) is called a *statistical structure* on M if (i) $\nabla_X Y - \nabla_Y X - [X, Y] = 0$ and (ii) $(\nabla_X g)(Y, Z) = (\nabla_Y g)(X, Z)$ hold for any $X, Y, Z \in \Gamma(TM)$. By definition, (∇^g, g) is a statistical structure on M.

We denote by R^∇ the curvature tensor field of ∇, and by ∇^* the dual connection of ∇ with respect to g, and set $S = S^{(\nabla,g)} \in \Gamma(TM^{(1,3)})$ as the mean of the curvature tensor fields of ∇ and of ∇^*, that is, for $X, Y, Z \in \Gamma(TM)$,

$$\begin{aligned}
&R^\nabla(X, Y)Z = \nabla_X \nabla_Y Z - \nabla_Y \nabla_X Z - \nabla_{[X,Y]} Z,\\
&Xg(Y, Z) = g(\nabla_X Y, Z) + g(Y, \nabla^*_X Z),\\
&S(X, Y)Z = \frac{1}{2}\{R^\nabla(X, Y)Z + R^{\nabla^*}(X, Y)Z\}.
\end{aligned} \tag{3}$$

A statistical manifold (M, ∇, g) is called a *Hessian manifold* if $R^\nabla = 0$. If so, we have $R^{\nabla^*} = S = 0$ automatically.

For a statistical structure (∇, g) on M, we set $K = \nabla - \nabla^g$. Then the following hold:

$$K \in \Gamma(TM^{(1,2)}),\\ K_X Y = K_Y X, \quad g(K_X Y, Z) = g(Y, K_X Z) \tag{4}$$

for any $X, Y, Z \in \Gamma(TM)$. Conversely, if K satisfies (4), the pair $(\nabla = \nabla^g + K, g)$ is a statistical structure on M.

The formula

$$S(X,Y)Z = R^g(X,Y)Z + [K_X, K_Y]Z \tag{5}$$

holds, where $R^g = R^{\nabla^g}$ is the curvature tensor field of the Levi-Civita connection of g.

For a statistical structure (∇, g), we often use the expression like $(\nabla = \nabla^g + K, g)$, and write $K_X Y$ by $K(X,Y)$.

Definition 1. A quadruplet $(\nabla = \nabla^g + K, g, \phi, \xi)$ is called a *Sasakian statistical structure* on M if (i) (g, ϕ, ξ) is a Sasakian structure and (ii) (∇, g) is a statistical structure on M, and (iii) $K \in \Gamma(TM^{(1,2)})$ for (∇, g) satisfies

$$K(X, \phi Y) + \phi K(X,Y) = 0 \quad \text{for } X, Y \in \Gamma(TM). \tag{6}$$

These three conditions are paraphrased in the following three conditions ([3, Theorem 2.17]: (i') (g, ϕ, ξ) is an almost contact metric structure and (ii) (∇, g) is a statistical structure on M, and (iii') they satisfy

$$\nabla_X(\phi Y) - \phi \nabla^*_X Y = g(\xi, Y)X - g(X,Y)\xi, \tag{7}$$

$$\nabla_X \xi = \phi X + g(\nabla_X \xi, \xi)\xi. \tag{8}$$

We get the following formulas for a Sasakian statistical manifold:

$$K(X,\xi) = \lambda g(X,\xi)\xi, \quad g(K(X,Y),\xi) = \lambda g(X,\xi) g(Y,\xi), \tag{9}$$

where

$$\lambda = g(K(\xi,\xi),\xi). \tag{10}$$

Proposition 2. *For a Sasakian statistical manifold $(M, \nabla, g, \phi, \xi)$,*

$$S(X,Y)\xi = g(Y,\xi)X - g(X,\xi)Y \tag{11}$$

holds for $X, Y \in \Gamma(TM)$.

Proof. By (9), we have $[K_X, K_Y]\xi = 0$, from which (5) implies $S = R^g$. It is known that R^g is written as the right hand side of (11) (See [5]). □

A quadruplet $(\widetilde{M}, \widetilde{\nabla} = \nabla^{\widetilde{g}} + \widetilde{K}, \widetilde{g}, \widetilde{J})$ is called a *holomorphic statistical manifold* if $(\widetilde{g}, \widetilde{J})$ is a Kähler structure, $(\widetilde{\nabla}, \widetilde{g})$ is a statistical structure on $\widetilde{M}$, and

$$\widetilde{K}(X, \widetilde{J}Y) + \widetilde{J}\widetilde{K}(X,Y) = 0 \tag{12}$$

holds for $X, Y \in \Gamma(T\widetilde{M})$. The notion of Sasakian statistical manifold can be also expressed in the following: The *cone* over M defined below is a holomorphic statistical manifold. Let $(M, \nabla = \nabla^g + K, g, \phi, \xi)$ be a statistical manifold with an almost contact metric structure. Set $\widetilde{M}$ as $M \times \mathbb{R}_+$, and define a Riemannian metric $\widetilde{g} = r^2 g + (dr)^2$ on $\widetilde{M}$. Take a vector field $\Psi = r\dfrac{\partial}{\partial r} \in \Gamma(T\widetilde{M})$, and define $\widetilde{J} \in \Gamma(T\widetilde{M}^{(1,1)})$ by $\widetilde{J}\Psi = \xi$ and $\widetilde{J}X = \phi X - g(X,\xi)\Psi$ for any $X \in \Gamma(TM)$. Then, $(\widetilde{g}, \widetilde{J})$ is an almost Hermitian structure on $\widetilde{M}$, and furthermore, (g, ϕ, ξ) is a Sasakian structure on M if and only if $(\widetilde{g}, \widetilde{J})$ is a Kähler structure on $\widetilde{M}$. We construct connection $\widetilde{\nabla}$ on $\widetilde{M}$ by

$$\begin{cases} \widetilde{\nabla}_\Psi \Psi = -\lambda\xi + \Psi, \\ \widetilde{\nabla}_X \Psi = \widetilde{\nabla}_\Psi X = X - \lambda g(X,\xi)\Psi, \\ \widetilde{\nabla}_X Y = \nabla_X Y - g(X,Y)\Psi, \end{cases}$$

that is,

$$\widetilde{K}(\Psi,\Psi) = -\lambda\xi, \quad \widetilde{K}(X,\Psi) = -\lambda g(X,\xi)\Psi, \quad \widetilde{K}(X,Y) = K(X,Y)$$

for $X, Y \in \Gamma(TM)$, where λ is in (10). We then have that $(M, \nabla, g, \phi, \xi)$ is a Sasakian statistical manifold if and only if $(\widetilde{M}, \widetilde{\nabla}, \widetilde{g}, \widetilde{J})$ is a holomorphic statistical manifold (A general statement is given as [2, Proposition 4.8 and Theorem 4.10]). It is derived from the fact that the formula (12) holds if and only if both (6) and (9) hold.

Example 3. Let S^{2n-1} be a unit hypersphere in the Euclidean space $\mathbb{R}^{2n}$. Let J be a standard almost complex structure on $\mathbb{R}^{2n}$ considered as $\mathbb{C}^n$, and set $\xi = -JN$, where N is a unit normal vector field of S^{2n-1}. Define $\phi \in \Gamma(T(S^{2n-1})^{(1,1)})$ by $\phi(X) = JX - \langle JX, N\rangle N$. Denote by g the standard metric of the hypersphere. Then such a (g, ϕ, ξ) is known as a standard Sasakian structure on S^{2n-1}. We set

$$K(X,Y) = g(X,\xi)g(Y,\xi)\xi \tag{13}$$

for any $X, Y \in \Gamma(TS^{2n-1})$. Since K satisfies (4) and (6), we have a Sasakian statistical structure $(\nabla = \nabla^g + K, g, \phi, \xi)$ on S^{2n-1}.

Proposition 4. *Let (M, g, ϕ, ξ) be a Sasakian manifold. Set ∇ as $\nabla^g + fK$ for $f \in C^\infty(M)$, where K is given in (13). Then (∇, g, ϕ, ξ) is a Sasakian statistical structure on M. Conversely, we define $\nabla_X Y = \nabla^g_X Y + L(X,Y)V$ for some unit vector field V and $L \in \Gamma(TM^{(0,2)})$. If (∇, g, ϕ, ξ) is a Sasakian statistical structure, then $L \otimes V$ is written as $L(X,Y)V = fg(X,\xi)g(Y,\xi)\xi$ for some $f \in C^\infty(M)$, as above.*

Proof. The first half is obtained by direct calculation. To get the second half, we have by (4),

$$\begin{aligned} 0 &= L(X,Y)V - L(Y,X)V = \{L(X,Y) - L(Y,X)\}V, \\ 0 &= g(L(X,Y)V, Z) - g(Y, L(X,Z)V) = g(L(X,Y)Z - L(X,Z)Y, V). \end{aligned} \tag{14}$$

Substituting V for Z in (14), we have

$$L(X,Y) = L(V,V)g(X,V)g(Y,V).$$

Accordingly, we get by (6),

$$0 = L(X,\phi Y)V + \phi\{L(X,Y)V\} = L(V,V)g(X,V)\{-g(Y,\phi V)V + g(Y,V)\phi V\},$$

which implies that $\phi V = 0$ if $L(V,V) \neq 0$, and hence $V = \pm\xi$. □

3 Invariant Submanifolds

Let $(\widetilde{M},\widetilde{g},\widetilde{\phi},\widetilde{\xi})$ be a Sasakian manifold, and M a submanifold of $\widetilde{M}$. We say that M is an *invariant submanifold* of $\widetilde{M}$ if (i) $\widetilde{\xi}_u \in T_uM$, (ii) $\widetilde{\phi}X \in T_uM$ for any $X \in T_uM$ and $u \in M$. Let $g \in \Gamma(TM^{(0,2)})$, $\phi \in \Gamma(TM^{(1,1)})$ and $\xi \in \Gamma(TM)$ be the restriction of $\widetilde{g}$, $\widetilde{\phi}$ and $\widetilde{\xi}$, respectively. Then it is shown that (g,ϕ,ξ) is a Sasakian structure on M.

A typical example of an invariant submanifold of a Sasakian manifold S^{2n-1} in Example 3 is an odd dimensional unit sphere. Furthermore, we have the following example. Let $\iota : Q \to \mathbb{C}P^{n-1}$ be a complex hyperquadric in the complex projective space, and $\widetilde{Q}$ the principal fiber bundle over Q induced by ι from the Hopf fibration $\pi : S^{2n-1} \to \mathbb{C}P^{n-1}$. We denote the induced homomorphism by $\widetilde{\iota} : \widetilde{Q} \to S^{2n-1}$. Then it is known that $\widetilde{\iota}(\widetilde{Q})$ is an invariant submanifold (See [4], [5]).

We briefly review the statistical submanifold theory to study invariant submanifolds of a Sasakian statistical manifold. Let $(\widetilde{M},\widetilde{\nabla},\widetilde{g})$ be a statistical manifold, and M a submanifold of $\widetilde{M}$. Let g be the metric on M induced from $\widetilde{g}$, and consider the orthogonal decomposition with respect to $\widetilde{g}$: $T_u\widetilde{M} = T_uM \oplus T_uM^{\perp}$. According to this decomposition, we define an affine connection ∇ on M, $B \in \Gamma(TM^{\perp} \otimes TM^{(0,2)})$, $A \in \Gamma((TM^{\perp})^{(0,1)} \otimes TM^{(1,1)})$, and a connection $\nabla^{\perp}$ of the vector bundle $TM^{\perp}$ by

$$\widetilde{\nabla}_XY = \nabla_XY + B(X,Y), \quad \widetilde{\nabla}_XN = -A_NX + \nabla^{\perp}_XN \tag{15}$$

for $X,Y \in \Gamma(TM)$ and $N \in \Gamma(TM^{\perp})$. Then (∇,g) is a statistical structure on M. In the same fashion, we define an affine connection ∇^* on M, $B^* \in \Gamma(TM^{\perp} \otimes TM^{(0,2)})$, $A^* \in \Gamma((TM^{\perp})^{(0,1)} \otimes TM^{(1,1)})$, and a connection $(\nabla^{\perp})^*$ of $TM^{\perp}$ by using th dual connection $\widetilde{\nabla}^*$ instead of $\widetilde{\nabla}$ in (15).

We remark that $\widetilde{g}(B(X,Y),N) = g(A^*_NX,Y)$ for $X,Y \in \Gamma(TM)$ and $N \in \Gamma(TM^{\perp})$, and remark that ∇^* coincides with the dual connection of ∇ with respect to g. See [1] for example.

Theorem 5. *Let $(\widetilde{M},\widetilde{\nabla},\widetilde{g},\widetilde{\phi},\widetilde{\xi})$ be a Sasakian statistical manifold, and M an invariant submanifold of $\widetilde{M}$ with $g,\phi,\xi,\nabla,B,A,\nabla^{\perp},\nabla^*,B^*,A^*,(\nabla^{\perp})^*$ defined as above. Then the following hold:*

(i) *A quintuplet (M,∇,g,ϕ,ξ) is a Sasakian statistical manifold.*
(ii) *$B(X,\xi) = B^*(X,\xi) = 0$ for any $X \in \Gamma(TM)$.*

(iii) $B(X, \phi Y) = B(\phi X, Y) = \widetilde{\phi}B^*(X, Y)$ *for any* $X, Y \in \Gamma(TM)$. *In particular,* $\mathrm{tr}_g B = \mathrm{tr}_g B^* = 0$.
(iv) *If* B *is parallel with respect to the Van der Weaden-Bortolotti connection* $\widetilde{\nabla}'$ *for* $\widetilde{\nabla}$, *then* B *and* B^* *vanish. Namely, if* $(\widetilde{\nabla}'_X B)(Y, Z) = \nabla^{\perp}_X B(Y, Z) - B(\nabla_X Y, Z) - B(Y, \nabla_X Z) = 0$ *for* $Z \in \Gamma(TM)$, *then* $B^*(X, Y) = 0$.
(v) $\widetilde{g}(\widetilde{S}(X, \widetilde{\phi}X)\widetilde{\phi}X - S(X, \phi X)\phi X, X) = 2\widetilde{g}(B^*(X, X), B(X, X))$ *for* $X \in \Gamma(TM)$, *where* $S = S^{(\nabla, g)}$ *and* $\widetilde{S} = S^{(\widetilde{\nabla}, \widetilde{g})}$ *as in (3).*

Corollary 6. *Let* $(\widetilde{M}, \widetilde{\nabla}, \widetilde{g}, \widetilde{\phi}, \widetilde{\xi})$ *be a Sasakian statistical manifold of constant* $\widetilde{\phi}$*-sectional curvature* c, *and* M *an invariant submanifold of* $\widetilde{M}$. *The induced Sasakian statistical structure on* M *has constant* ϕ*-sectional curvature* c *if and only if* $\widetilde{g}(B^*(X, X), B(X, X)) = 0$ *for any* $X \in \Gamma(TM)$ *orthogonal to* ξ.

If we take the Levi-Civita connection as $\widetilde{\nabla}$, the properties above reduce to the ones for an invariant submanifold of a Sasakian manifold. It is known that an invariant submanifold of a Sasakian manifold of constant $\widetilde{\phi}$-sectional curvature c is of constant ϕ-sectional curvature c if and only if it is totally geodesic. It is obtained by setting $B = B^*$ in Corollary 6. It is an interesting question whether there is an interesting invariant submanifold having nonvanishing B with the above property.

Outline of Proof of Theorem 5. The proof of (i) can be omitted.

By (i) and (8), we calculate that $\nabla_X \xi + B(X, \xi) = \widetilde{\nabla}_X \xi = \widetilde{\phi}X + \widetilde{g}(\widetilde{\nabla}_X \xi, \widetilde{\xi})\widetilde{\xi} = \phi X + g(\nabla_X \xi, \xi)\xi$. Comparing the normal components, we have (ii).

By (7), we have $\widetilde{g}(Y, \widetilde{\xi})X - \widetilde{g}(Y, X)\widetilde{\xi} = \widetilde{\nabla}_X(\widetilde{\phi}Y) - \widetilde{\phi}\widetilde{\nabla}^*_X Y = \nabla_X(\phi Y) + B(X, \phi Y) - \widetilde{\phi}(\nabla^*_X Y + B^*(X, Y)) = g(Y, \xi)X - g(Y, X)\xi + B(X, \phi Y) - \widetilde{\phi}B^*(X, Y)$. Comparing the normal components, we have (iii).

By (i) and (ii), we get that $0 = \nabla^{\perp}_X B(Y, \xi) - B(\nabla_X Y, \xi) - B(Y, \nabla_X \xi) = -B(Y, \phi X) = -\widetilde{\phi}B^*(X, Y)$, which implies (iv).

To get (v), we use the Gauss equation in the submanifold theory. The tangential component of $R^{\widetilde{\nabla}}(X, Y)Z$ is given as

$$R^{\nabla}(X, Y)Z - A_{B(Y,Z)}X + A_{B(X,Z)}Y,$$

for $X, Y, Z \in \Gamma(TM)$, which implies that

$$\begin{aligned} 2\widetilde{g}(\widetilde{S}(X, Y)Z, W) &= 2g(S(X, Y)Z, W) \\ &\quad - \widetilde{g}(B^*(X, W), B(Y, Z)) + \widetilde{g}(B^*(Y, W), B(X, Z)) \\ &\quad - \widetilde{g}(B(X, W), B^*(Y, Z)) + \widetilde{g}(B(Y, W), B^*(X, Z)). \end{aligned}$$

Therefore, we prove (v) from (iii). □

To get Corollary 6, we have only to review the definition. A Sasakian statistical structure (∇, g, ϕ, ξ) is said to be of constant ϕ-sectional curvature c if the sectional curvature defined by using S equals c for each ϕ-section, the plane spanned by X and ϕX for a unit vector X orthogonal to ξ: $g(S(X, \phi X)\phi X, X) = cg(X, X)^2$ for $X \in \Gamma(TM)$ such that $g(X, \xi) = 0$.

Acknowledgments. The author thanks the anonymous reviewers for their careful reading of the manuscript. This work was supported by JSPS KAKENHI Grant Number JP26400058.

References

1. Furuhata, H., Hasegawa, I.: Submanifold theory in holomorphic statistical manifolds. In: Dragomir, S., Shahid, M.H., Al-Solamy, F.R. (eds.) Geometry of Cauchy-Riemann Submanifolds, pp. 179–215. Springer, Singapore (2016). doi:10.1007/978-981-10-0916-7_7
2. Furuhata, H., Hasegawa, I., Okuyama, Y., Sato, K.: Kenmotsu statistical manifolds and warped product. J. Geom. To appear. doi:10.1007/s00022-017-0403-1
3. Furuhata, H., Hasegawa, I., Okuyama, Y., Sato, K., Shahid, M.H.: Sasakian statistical manifolds. J. Geom. Phys. **117**, 179–186 (2017). doi:10.1016/j.geomphys.2017.03.010
4. Kenmotsu, K.: Invariant submanifolds in a Sasakian manifold. Tohoku Math. J. **21**, 495–500 (1969)
5. Yano, K., Kon, M.: Structures on Manifolds. World Scientific, Singapore (1984)

(Para-)Holomorphic Connections for Information Geometry

Sergey Grigorian[1] and Jun Zhang[2(✉)]

[1] University of Texas Rio Grande Valley, Edinburg, TX 78539, USA
sergey.grigorian@utrgv.edu
[2] University of Michigan, Ann Arbor, MI 48109, USA
junz@umich.edu

Abstract. On a statistical manifold (M, g, ∇), the Riemannian metric g is coupled to an (torsion-free) affine connection ∇, such that ∇g is totally symmetric; $\{\nabla, g\}$ is said to form "Codazzi coupling". This leads ∇^*, the g -conjugate of ∇, to have same torsion as that of ∇. In this paper, we investigate how statistical structure interacts with L in an almost Hermitian and almost para-Hermitian manifold (M, g, L), where L denotes, respectively, an almost complex structure J with $J^2 = -\text{id}$ or an almost para-complex structure K with $K^2 = \text{id}$. Starting with ∇^L, the L -conjugate of ∇, we investigate the interaction of (generally torsion-admitting) ∇ with L, and derive a necessary and sufficient condition (called "Torsion Balancing" condition) for L to be integrable, hence making (M, g, L) (para-)Hermitian, and for ∇ to be (para-)holomorphic. We further derive that ∇^L is (para-)holomorphic if and only if ∇ is, and that ∇^* is (para-)holomorphic if and only if ∇ is (para-)holomorphic and Codazzi coupled to g. Our investigations provide concise conditions to extend statistical manifolds to (para-)Hermitian manifolds.

1 Introduction

On the tangent bundle TM of a differentiable manifold M, one can introduce two separate structures: affine connection ∇ and pseudo-Riemannian metric g. A manifold M equipped with a g and a torsion-free connection ∇ is called a *statistical manifold* if (g, ∇) is Codazzi-coupled [Lau87]. This is the setting of "classical" information geometry, where the (g, ∇) pair arises from a general construction of divergence ("contrast") functions. To accommodate for torsions in affine connections, the concept of pre-contrast functions was introduced [HM11]. Codazzi coupling has been traditionally studied by affine geometers [NS94, Sim00]. The robustness of Codazzi coupling was investigated by perturbing both the metric and the affine connection [SSS09] and by its interaction with other transformations of connection [TZ16]. Below, we provide a succinct overview.

1.1 *g*-conjugate Connection, Cubic Form, and Codazzi Coupling

Given the pair (g, ∇), we construct the (0, 3)-tensor C by

$$C(X, Y, Z) := (\nabla_Z g)(X, Y) = Zg(X, Y) - g(\nabla_Z X, Y) - g(X, \nabla_Z Y). \tag{1}$$

F. Nielsen and F. Barbaresco (Eds.): GSI 2017, LNCS 10589, pp. 186–194, 2017.
https://doi.org/10.1007/978-3-319-68445-1_22

The tensor C is sometimes referred to as the *cubic form* associated to the pair (∇, g). When $C = 0$, we say g is parallel under ∇.

Given the pair (g, ∇), we can also construct ∇^*, called g -conjugate connection, by

$$Zg(X,Y) = g(\nabla_Z X, Y) + g(X, \nabla^*_Z Y). \tag{2}$$

It can be checked easily that (i) ∇^* is indeed a connection and (ii) g-conjugation of a connection is involutive, i.e., $(\nabla^*)^* = \nabla$.

These two constructions from an arbitrary (g, ∇) pair are related via

$$C(X,Y,Z) = g(X, (\nabla^* - \nabla)_Z Y), \tag{3}$$

so that

$$C^*(X,Y,Z) := (\nabla^*_Z g)(X,Y) = -C(X,Y,Z).$$

Therefore $C(X,Y,Z) = C^*(X,Y,Z) = 0$ if and only if $\nabla^* = \nabla$, that is, ∇ is g-self-conjugate. A connection is both g -self-conjugate and torsion-free defines what is called the Levi-Civita connection ∇^{LC} associated to g.

Simple calculation reveals that

$$\begin{aligned} C(X,Y,Z) - C(Z,Y,X) &= (\nabla_Z g)(X,Y) - (\nabla_X g)(Z,Y), \\ C(X,Y,Z) - C(X,Z,Y) &= g(X, T^{\nabla^*}(Z,Y) - T^{\nabla}(Z,Y)), \end{aligned} \tag{4}$$

where T^{∇} denotes the torsion of ∇

$$T^{\nabla}(X,Y) = \nabla_X Y - \nabla_Y X - [X,Y].$$

Note that $C(X,Y,Z) = C(Y,X,Z)$ always holds, due to $g(X,Y) = g(Y,X)$. Therefore, imposing either of the following is equivalent:

1. $C(X,Y,Z) = C(Z,Y,X)$,
2. $C(X,Y,Z) = C(X,Z,Y)$;

this is because either (i) or (ii) will make C totally symmetric in all of its indices. In the case of (i), we say that g and ∇ are *Codazzi-coupled*:

$$(\nabla_Z g)(X,Y) = (\nabla_X g)(Z,Y). \tag{5}$$

In the case of (ii), ∇ and ∇^* have same torsion. These well-known facts are summarized in the following Lemma.

Lemma 1. *Let g be a pseudo-Riemannian metric, ∇ an arbitrary affine connection, and ∇^* be the g-conjugate connection of ∇. Then the following statements are equivalent:*

1. *(∇, g) is Codazzi-coupled;*
2. *(∇^*, g) is Codazzi-coupled;*
3. *C is totally symmetric;*
4. *C^* is totally symmetric;*
5. *$T^{\nabla} = T^{\nabla^*}$.*

In the above case, (g, ∇, ∇^*) is called a *Codazzi triple.* Codazzi-coupling between g and ∇ or, equivalently, the existence of Codazzi triple (g, ∇, ∇^*) is the key feature of a statistical manifold. In "quantum" information geometry, ∇ is allowed to carry torsion, and [Mat13] introduced *Statistical Manifold Admitting Torsion (SMAT)* as a manifold (M, g, ∇) satisfying

$$(\nabla_Y g)(X, Z) - (\nabla_X g)(Y, Z) = g(T^\nabla(X, Y), Z).$$

Note that ∇^* is torsion-free if and only if (M, g, ∇) is a SMAT. However, in a SMAT, neither ∇ nor ∇^* is Codazzi coupled to g; the deviation from Codazzi coupling is measured by the torsion T^∇ of ∇.

2 Structure of TM Arising from L

A tangent bundle isomorphism L may induce a splitting of TM, corresponding to the eigenbundles associated with the eigenvalues of L. How the action of an arbitrary connection ∇ respects such splitting is the focus of our current paper.

2.1 Splitting of TM by L

For a smooth manifold M, an isomorphism L of the tangent bundle TM is a smooth section of the bundle $\mathrm{End}(TM)$ such that it is invertible everywhere. By definition, L is called an *almost complex structure* if $L^2 = -\mathrm{id}$, or an *almost para-complex structure* if $L^2 = \mathrm{id}$ and the multiplicities of the eigenvalues ± 1 are equal. We will use J and K to denote almost complex structures and almost para-complex structures, respectively, and use L when these two structures can be treated in a unified way. It is clear from our definition that such structures exist only when M is of even dimension.

Denote eigenvalues of L as $\pm\alpha$, where $\alpha = 1$ for $L = K$ and $\alpha = i$ for $L = J$, respectively. Following the standard procedure, we (para-)complexify TM by tensoring with $\mathbb{C}$ or para-complex (also known as split-complex) field $\mathbb{D}$, and use $T^L M$ to denote the resulting $TM \otimes \mathbb{C}$ or $TM \otimes \mathbb{D}$, depending on the type of L. In analogy with standard notation in the complex case, let $T^{(1,0)}M$ and $T^{(0,1)}M$ be the eigenbundles of L corresponding to the eigenvalues $\pm\alpha$, i.e., at each point $p \in M$, the fiber is defined by

$$\begin{aligned}
T^{(1,0)}(p) &:= \{X \in T^L_p M : L_p(X) = \alpha X\} \ , \\
T^{(0,1)}(p) &:= \{X \in T^L_p M : L_p(X) = -\alpha X\} \ .
\end{aligned}$$

As sub-bundles of the (para-)complexified tangent bundle $T^L M$, $T^{(1,0)}M$ and $T^{(0,1)}M$ are distributions. A distribution is called a foliation if it is closed under the bracket $[\cdot,\cdot]$. We will refer to vectors to be of type $(1,0)$ and $(0,1)$ if they take values in $T^{(1,0)}M$ and $T^{(0,1)}M$ respectively. Moreover, define $\pi^{(1,0)}$ and $\pi^{(0,1)}$ to be the projections of a vector field to $T^{(1,0)}M$ and $T^{(0,1)}M$ respectively.

The Nijenhuis tensor N_L associated with L is defined as

$$N_L(X, Y) = -L^2[X, Y] + L[X, LY] + L[LX, Y] - [LX, LY]. \tag{6}$$

When $N_L = 0$, the operator L is said to be integrable. It is well-known that both $T^{(1,0)}M$ and $T^{(0,1)}M$ are foliations if and only if L is integrable, i.e., the integrability condition $N_L = 0$ is satisfied.

2.2 L-conjugate of ∇

Starting from a (not necessarily torsion-free) connection ∇ operating on sections of TM, we can apply an L-conjugate transformation to obtain a new connection $\nabla^L := L^{-1}\nabla L$, or

$$\nabla^L_X Y = L^{-1}(\nabla_X(LY)) \tag{7}$$

for any vector fields X and Y; here L^{-1} denotes the inverse isomorphism of L. It can be verified that indeed ∇^L is an affine connection.

Define a (1, 2)-tensor (vector-valued bilinear form) S via the expression

$$S(X,Y) = (\nabla_X L)Y - (\nabla_Y L)X, \tag{8}$$

where

$$(\nabla_X L)Y = \nabla_X(LY) - L(\nabla_X Y).$$

We say that L and ∇ are *Codazzi-coupled* if $S = 0$. The following is known.

Lemma 2 *(e.g., [SSS09]). Let ∇ be an affine connection, and let L be an arbitrary tangent bundle isomorphism. Then the following statements are equivalent:*

(i) (∇, L) is Codazzi-coupled.
(ii) $T^\nabla(X,Y) = T^{\nabla^L}(X,Y)$.
(iii) (∇^L, L^{-1}) is Codazzi-coupled.

Lemma 3. *For the special case of (para-)complex operators $L^2 = \pm\text{id}$,*

1. *$\nabla^L = \nabla^{L^{-1}}$, i.e., L-conjugate transformation is involutive, $(\nabla^L)^L = \nabla$.*
2. *(∇, L) is Codazzi-coupled if and only if (∇^L, L) is Codazzi-coupled.*

As an affine connection, ∇ gives rise to a map

$$\nabla : \Omega^0(TM) \to \Omega^1(TM),$$

where $\Omega^i(TM)$ is the space of smooth i-forms with value in TM. We may extend this to a map

$$d^\nabla : \Omega^i(TM) \to \Omega^{i+1}(TM)$$

by

$$d^\nabla(\alpha \otimes v) = d\alpha \times v + (-1)^i \alpha \wedge \nabla v$$

for any i-form α and vector field v. In the case that ∇ is flat, then $(d^\nabla)^2 = 0$ and we get a chain complex whose cohomology is the de Rham cohomology twisted by the local system determined by ∇. Regarding L as an element of $\Omega^1(TM)$, it is easy to check using local coordinates that

$$(d^\nabla L)(X,Y) = (\nabla_X L)Y - (\nabla_Y L)X + LT^\nabla(X,Y). \tag{9}$$

Therefore, Codazzi coupling of ∇ and L can also be expressed as

$$(d^\nabla L)(X,Y) = T^\nabla(LX,Y). \tag{10}$$

2.3 Integrability of L

In [FZ17, Lemma 2.5] an expression for $N_L(X,Y)$ in terms of T^{∇} has been derived assuming $S=0$. Using exactly the same procedure, we can write down $N_L(X,Y)$ for an arbitrary S.

Lemma 4. *Given a connection ∇ with torsion T^{∇}, the Nijenhuis tensor N_L of a (para-)complex operator L is given by*

$$N_L(X,Y) = L^2 T^{\nabla}(X,Y) - LT^{\nabla}(X,LY) - LT^{\nabla}(LX,Y) + T^{\nabla}(LX,LY) \\ + LS(X,Y) - L^{-1}S(LY,LX).$$

Now, define θ to be

$$\theta(X,Y) = \frac{1}{2}(\nabla^L_X Y - \nabla_X Y) = \frac{1}{2}L^{-1}(\nabla_X L)Y. \tag{11}$$

with

$$L\theta(X,Y) + \theta(X,LY) = 0. \tag{12}$$

In particular, we see that

$$\frac{1}{2}L^{-1}(S(X,Y)) = \theta(X,Y) - \theta(Y,X),$$

and therefore, θ is symmetric if and only if L and ∇ are Codazzi-coupled. Introduce

$$\tilde{\nabla} = \frac{1}{2}(\nabla + \nabla^L),$$

which satisfies

$$\tilde{\nabla} L \equiv 0.$$

A connection with respect to which L is parallel is called *(para-)complex connection*, and in particular, such a connection preserves the decomposition $T^L M \cong T^{(1,0)}M \oplus T^{(0,1)}M$. So starting from any connection ∇, we can construct its conjugate ∇^L, the average of which is the (para-)complex connection $\tilde{\nabla}$. This situation mirrors the relationship between Levi-Civita connection and the pair of g-conjugate connections ∇, ∇^*. Note that we can also write $\nabla = \tilde{\nabla} - \theta$ and $\nabla^L = \tilde{\nabla} + \theta$, so the quantity θ measures the failure of both ∇ and ∇^L to be a (para-)complex connection.

3 (Para-)Holomorphicity of ∇ Associated to L

3.1 (Para-)Holomorphic Connections

The (para-)Dolbeault operator $\bar{\partial}$ for a given L on $T^L M$ is defined as

$$\bar{\partial}_X Y = \frac{1}{4}\left([X,Y] - L^2[LX,LY] - L^{-1}[LX,Y] + L^{-1}[X,LY]\right) \tag{13}$$

for any vector fields X and Y. It can be checked easily that this expression is tensorial in X, that is $\bar{\partial}_{fX}Y = f\left(\bar{\partial}_X Y\right)$ and is a derivation. In the case when $L = J$, this defines the *holomorphic structure* on $T^{\mathbb{C}}M$ and locally defines the differentiation of vector fields of type $(1,0)$ with respect to the anti-holomorphic coordinates $\frac{\partial}{\partial \bar{z}^i}$. Similarly for *para-holomorphic structure* on $T^{\mathbb{D}}M$ when $L = K$.

From (13) we obtain that if X and Y are of the same type, then $\bar{\partial}_X Y = 0$. However, if $Y \in T^{(1,0)}M$ and $X \in T^{(0,1)}M$, then

$$\bar{\partial}_X Y = \pi^{(1,0)}\left[X, Y\right] \tag{14}$$

and similarly $\bar{\partial}_X Y = \pi^{(0,1)}\left[X, Y\right]$ if $Y \in T^{(0,1)}M$ and $X \in T^{(1,0)}M$. Equivalently, note that if $X \in T^{(1,0)}M$, then $\bar{\partial}X$ is a vector-valued 1-form, of type $(1,0)$ as a vector and type $(0,1)$ as a 1-form, and conversely if $X \in T^{(0,1)}M$.

Given a connection ∇ operating on $T^L M$, we can ask the question whether ∇ is compatible with $\bar{\partial}$. To understand this we may define an alternative operator $\bar{\partial}^{\nabla}$, which for $Y \in T^{(1,0)}M$ is defined as taking the $(0,1)$-part of the vector-valued 1-form ∇Y (and conversely on $T^{(0,1)}M$). This can be expressed as

$$\bar{\partial}^{\nabla}_X Y = \frac{1}{2}\left(\nabla_X Y - \nabla_{LX}\left(L^{-1}Y\right)\right) \tag{15}$$

for any vector fields X and Y in $T^L M$. Clearly, $\bar{\partial}^{\nabla}_X Y = 0$ if X and Y are of the same type and is just $\nabla_X Y$ if X and Y are of opposite type. On a (para-)holomorphic vector bundle, a connection is said to be *(para)-holomorphic* if these two Dolbeault operators coincide. We extend this notion to arbitrary connections on $T^L M \cong T^{(1,0)}M \oplus T^{(0,1)}M$ (that do not necessarily preserve $T^{(1,0)}M$ and $T^{(0,1)}M$) – we say a connection ∇ is *(para-)holomorphic* if $\bar{\partial}^{\nabla}_X Y = \bar{\partial}_X Y$ for any vector fields X and Y.

It can be readily shown that

Theorem 1. *∇^L is (para-)holomorphic if and only if ∇ is (para-)holomorphic.*

Theorem 2. *When ∇ is (para-)holomorphic, the quantity $\theta\left(X, Y\right)$ satisfies:*

$$L\theta\left(X, Y\right) = -\theta\left(X, LY\right) = -\theta\left(LX, Y\right) = L^{-1}\theta\left(LX, LY\right). \tag{16}$$

Theorem 2 shows that $\theta\left(X, Y\right)$ vanishes whenever X and Y are of different types. Moreover, if X and Y are both of type $(1,0)$, $\theta\left(X, Y\right)$ is of type $(0,1)$, and vice versa.

Using (13) and (15), we can also prove

Lemma 5. *Given an arbitrary connection ∇ and an L on a manifold, the connection ∇ is (para-)holomorphic if and only if*

$$S(X, Y) = T^{\nabla}(LX, Y) - LT^{\nabla}(X, Y) - \frac{1}{2}L^2 N_L(LX, Y). \tag{17}$$

From this, we prove the main theorem of our paper.

Theorem 3. *Given the an arbitrary pair* (∇, L) *on a manifold, the connection* ∇ *is (para-)holomorphic and* L *is integrable if and only if*

$$S(X,Y) = T^{\nabla}(LX,Y) - LT^{\nabla}(X,Y). \tag{18}$$

The significance of Theorem 3 is that this gives us a generalization of the Codazzi coupling condition for L that was used in [FZ17] in the case $T^{\nabla} = 0$. In fact, it follows immediately that if $T^{\nabla} = 0$ then Codazzi coupling of ∇ with L makes L integrable *and* makes ∇ (para-)holomorphic.

The condition (18) can be recast in another form to reveal its meaning:

Theorem 4. *Given* ∇ *and* L *on a manifold, then* ∇ *is (para-)holomorphic and* L *is integrable if and only if*

$$T^{\nabla}(LX,Y) = L(T^{\nabla^L}(X,Y)). \tag{19}$$

Theorem 4 shows that the (para-)holomorphicity condition on ∇ can be thought of as requiring "Torsion-Balancing" between ∇ and ∇^L.

3.2 Almost (Para-)Hermitian Structure

The compatibility condition between g and an almost (para-)complex structure $J(K)$ is well-known. We say that g is compatible with J if J is orthogonal, i.e.

$$g(JX,JY) = g(X,Y) \tag{20}$$

holds for any vector fields X and Y. Similarly we say that g is compatible with K if

$$g(KX,KY) = -g(X,Y) \tag{21}$$

is always satisfied, which implies that g must be of split signature. When expressed using L, (20) and (21) have the same form

$$g(X,LY) + g(LX,Y) = 0. \tag{22}$$

When specified in terms of compatible g and L, the manifold (M,g,L) is said to be almost (para-)Hermitian, and (para-)Hermitian manifold if L is integrable.

For any almost (para)-Hermitian manifold, we can define the 2-form $\omega(X,Y) = g(LX,Y)$, called the *fundamental form*, which turns out to satisfy $\omega(X,LY) + \omega(LX,Y) = 0$. The three structures, a pseudo-Riemannian metric g, a nondegenerate 2-form ω, and a tangent bundle isomorphism $L : TM \to TM$ forms a "compatible triple" such that given any two, the third one is uniquely specified; the triple is rigidly "interlocked".

It can be shown that for almost (para-)Hermitian manifolds,

$$(\nabla^L_X g)(LY,Z) + (\nabla_X g)(Y,LZ) = 0. \tag{23}$$

3.3 (Para-)Holomorphicity of ∇^*

We have seen in Theorem 1 that ∇ is (para-)holomorphic if and only if ∇^L is also (para-)holomorphic. We now investigate conditions under which ∇^* is also (para-)holomorphic whenever ∇ is.

Lemma 6. *Given arbitrary g and L on a manifold, with a (para-)holomorphic connection ∇. Then ∇^* is also (para-)holomorphic if and only if*

$$C(LX, Y, Z) = C(X, Y, LZ) \tag{24}$$

for any vector fields X, Y, Z. If moreover, g and L are compatible, i.e., (22) holds, then (24) is equivalent to

$$C(X, Y, Z) = g(\theta(Z, X), Y) + g(X, \theta(Z, Y)). \tag{25}$$

The condition that ∇^* is (para-)holomorphic is a very strong one as the theorem below shows.

Theorem 5. *Let ∇ be a (para-)holomorphic connection ∇ on an almost (para-) Hermitian manifold (M, g, L). Then, the connection $\tilde{\nabla} = \frac{1}{2}(\nabla + \nabla^L)$ is metric-compatible if and only if ∇^* is also (para-)holomorphic.*

In fact, since we already know that $\tilde{\nabla}$ is a (para-)complex connection, i.e. it preserves L, the condition of ∇^* being (para-)holomorphic is then equivalent to $\tilde{\nabla}$ being an almost (para-)Hermitian connection. Moreover, if we assume L to be integrable, since $\tilde{\nabla}$ is also (para-)holomorphic, we can conclude that when restricted to bundle $T^{(1,0)}M$, it must be equal to the (para-)Chern connection. In the theory of holomorphic vector bundles, Chern connection is the unique Hermitian holomorphic connection on a holomorphic vector bundle, and in particular on $T^{(1,0)}M$ on complex manifolds [Mor07]. In general, the Chern connection has torsion, however it is torsion-free on $T^{(1,0)}M$ if and only if (g, J) define a Kähler structure.

It is significant that if g is Codazzi-coupled to a (para-)holomorphic connection ∇, then ∇^* is (para-)holomorphic, and hence $\tilde{\nabla}$ is (para-)Hermitian.

Theorem 6. *Let (M, g, L) be a (para-)Hermitian manifold and let (∇, ∇^*, g) be a Codazzi triple. Then (∇^*, g) is (para-)holomorphic if and only if (∇, g) is (para-)holomorphic.*

This generalizes the results on a Codazzi-(para-)Kähler manifold [FZ17] which admit a pair of torsion-free connections to a (para-)Hermitian manifold which admits holomorphic connections with torsion. The Torsion-Balancing condition, while breaking the requirements of (para-)Kähler structure by possibly violating $d\omega = 0$, still preserves the integrability of L.

4 Summary and Discussions

(Para-)holomorphic connections have hardly been systematically studied in information geometry except in restricted setting of flat connections (see [Fur09]). Connections investigated in this paper are neither curvature-free nor torsion-free. We gave a necessary and sufficient condition("Torsion Balance") of a ∇ to be (para-)holomorphic in the presence of a (para-)complex structure L on the manifold. Given a (para-)holomorphic connection ∇, we then showed that (i) ∇^L, its L-conjugate, is also (para-)holomorphic; (ii) ∇^*, its g -conjugate, is (para-)holomorphic if and only if g and ∇ are Codazzi coupled. These concise characterizations allow us to enhance a statistical structure to a (para-)Hermitian structure, as well as understand the properties of L-conjugaty and g-conjugacy of a connection of a (para-)Hermitian manifold.

Acknowledgement. This research is supported by DARPA/ARO Grant W911NF-16-1-0383 to the University of Michigan (PI: Jun Zhang).

References

[Fur09] Furuhata, H.: Hypersurfaces in statistical manifolds. Differ. Geom. Appl. **27**(3), 420–429 (2009)

[HM11] Henmi, M., Matsuzoe, H.: Geometry of pre-contrast functions and non-conservative estimating functions. In: International Workshop on Complex Structures, Integrability and Vector Fields, vol. 1340, no. 1. AIP Publishing (2011)

[Lau87] Lauritzen, S.L.: Statistical manifolds. In: Differential Geometry in Statistical Inference. IMS Lecture Notes Monograph Series, vol. 10, pp. 163–216. Institute of Mathematical Statistics (1987)

[Mat13] Matsuzoe, H.: Statistical manifolds and geometry of estimating functions. In: Recent Progress in Differential Geometry and Its Related Field, pp. 187–202. World Sci. Publ. (2013)

[Mor07] Moroianu, A.: Lectures on Kähler Geometry. London Mathematical Society Student Texts, vol. 69. Cambridge University Press, Cambridge (2007)

[NS94] Nomizu, K., Sasaki, T.: Affine Differential Geometry: Geometry of Affine Immersions. Cambridge Tracts in Mathematics, vol. 111. Cambridge University Press, Cambridge (1994)

[Sim00] Simon, U.: Affine differential geometry. In: Handbook of Differential Geometry, vol. 1, pp. 905–961. North-Holland (2000)

[SSS09] Schwenk-Schellschmidt, A., Simon, U.: Codazzi-equivalent affine connections. RM **56**(1–4), 211–229 (2009)

[TZ16] Tao, J., Zhang, J.: Transformations and coupling relations for affine connections. Differ. Geom. Appl. **49**, 111–130 (2016)

[FZ17] Fei, T., Zhang, J.: Interaction of Codazzi couplings with (para-)Kahler geometry. RM, 05 July 2017. doi:10.1007/s0025-017-0711-7

Matrix Realization of a Homogeneous Hessian Domain

Hideyuki Ishi[1,2(✉)]

[1] Graduate School of Mathematics, Nagoya University,
Furo-cho, Nagoya 464-8602, Japan
hideyuki@math.nagoya-u.ac.jp

[2] JST, PRESTO, 4-1-8, Honcho, Kawaguchi 332-0012, Japan

Abstract. Extending previous results about matrix realization of a homogeneous cone by the author, we realize any homogeneous Hessian domain as a set of symmetric matrices with a specific block decomposition. A global potential function as well as a transitive affine group action preserving the Hessian structure is also expressed in terms of the matrix realization.

Keywords: Homogeneous Hessian domain · Left-symmetric algebra · Normal Hessian algebra

1 Introduction

A Riemannian manifold with a flat connection is called *a Hessian manifold* if the metric is locally expressed as the Hessian matrix of a smooth function with respect to affine coordinates. Such Hessian structure is very important in Information Geometry [9]. A Hessian manifold is said to be *homogeneous* if its automorphism group acts on the manifold transitively, where the automorphism group is defined as the set of all diffeomorphisms preserving both the flat connection and the metric. Shima [8] established a basic theory of homogeneous Hessian manifolds. He showed that the universal covering space of a homogeneous Hessian manifold is a convex domain equipped with a homogeneous Hessian metric. Furthermore, the convex domain is shown to be the direct product of an affine homogeneous convex domain (containing no straight line) and a vector space. A normal Hessian algebra, which is a left-symmetric algebra with a compatible inner product, plays an important role in Shima's theory as a convenient algebraic tool.

In this paper, combining the theory of normal Hessian algebras with the matrix realization method developed by the author [3–5], we realize any homogeneous Hessian domain as a set of symmetric matrices with a specific block decomposition (Theorem 4). A global potential function as well as a transitive affine group action preserving the Hessian structure is also expressed in terms of the matrix realization (see (8) and (9) respectively). The details with complete proofs of statements will be published elsewhere. This research was supported by JST PRESTO and JSPS KAKENHI Grant Number 16K05174.

F. Nielsen and F. Barbaresco (Eds.): GSI 2017, LNCS 10589, pp. 195–202, 2017.
https://doi.org/10.1007/978-3-319-68445-1_23

2 Normal Hessian Algebras

Let V be a finite dimensional real vector space, $\mathcal{D}$ a domain in V, and g a Hessian metric on $\mathcal{D}$. By definition, g is locally expressed as the Hessian matrix of a convex function. If there exists a smooth function $\varphi : \mathcal{D} \to \mathbb{R}$ whose Hessian matrix equals g at every point in $\mathcal{D}$, we call φ *a global potential* of the Hessian metric g. The automorphism group $\mathrm{Aut}(\mathcal{D}, g)$ of the Hessian domain $(\mathcal{D}, g)$ is defined by $\mathrm{Aut}(\mathcal{D}, g) := \{ \alpha \in \mathrm{Aff}(V) ; \alpha(D) = D, \ \alpha^* g = g \}$. A Hessian domain $(\mathcal{D}, g)$ is said to be homogeneous if $\mathrm{Aut}(\mathcal{D}, g)$ acts transitively on $\mathcal{D}$.

Let $(\mathcal{D}, g)$ be a homogeneous Hessian domain in what follows. Shima [8] showed that there exists a triangular solvable Lie subgroup $H \subset \mathrm{Aut}(\mathcal{D}, g)$ acting simply transitively on $\mathcal{D}$. Here the word *triangular* means that there exists a basis of V such that the linear part of every affine transformation $h \in H$ is expressed as an upper triangular matrix with respect to the basis. Let us take and fix a point $p_0 \in \mathcal{D}$. Then we have a diffeomorphism $H \ni h \mapsto h \cdot p_0 \in \mathcal{D}$. Differentiating the diffeomorphism, we have a linear isomorphism $\mathfrak{h} \ni X \mapsto X(p_0) \in V \equiv T_{p_0}\mathcal{D}$, where $\mathfrak{h}$ is the Lie algebra of H, and $\mathfrak{h}$ is identified with a set of affine vector fields. For $v \in V$, we denote by X_v a unique vector field belonging to $\mathfrak{h}$ such that $X_v(p_0) = v$, and by L_v the linear part of X_v. Then we have $X_v(p) = L_v(p - p_0) + v$. Now we introduce a bilinear product $\triangle$ on V by

$$x \triangle y := L_x y \in V \qquad (x, y \in V).$$

The algebra $(V, \triangle)$ is not commutative nor associative in general. Instead, we have the following equality

$$[L_x, L_y] = L_{x\triangle y - y \triangle x} \qquad (x, y \in V),$$

which is equivalent to

$$x\triangle(y\triangle z) - (x\triangle y)\triangle z = y\triangle(x\triangle z) - (y\triangle x)\triangle z \quad (x, y, z \in V). \tag{1}$$

The algebra satisfying the equality above is called *a left-symmetric algebra* (*Koszul-Vinberg algebra* [1] or *pre-Lie algebra* [7]). Since $\mathfrak{h}$ is triangular, all eigenvalues of the left-multiplication operator L_x are real for every $x \in V$. A left-symmetric algebra is said to be *normal* if this eigenvalue condition of L_x is satisfied.

By means of the identification $V \equiv T_{p_0}\mathcal{D}$, we transfer an inner product $(\cdot|\cdot)$ on V from the metric g on $T_{p_0}\mathcal{D}$. Then we have

$$(x|y\triangle z) - (x\triangle y|z) = (y|x\triangle z) - (y\triangle x|z) \quad (x, y, z \in \mathcal{D}). \tag{2}$$

Shima named a normal left-symmetric algebra with an inner product satisfying (2) *a normal Hessian algebra*, and we call the inner product *a Hessian inner product* on the algebra. We have seen that a homogeneous Hessian domain gives rise to a normal Hessian algebra. The converse is also true. When a normal Hessian algebra $(V, \triangle)$ is given, we take any point $p_0 \in V$, and consider the set $\mathfrak{h}$

of affine vector fields X_v $(v \in V)$ defined by $X_v(p) := L_v(p - p_0) + v$. Thanks to the left-symmetry, $\mathfrak{h}$ forms a Lie algebra. Let H be the Lie subgroup of $\mathrm{Aff}(V)$ whose Lie algebra is $\mathfrak{h}$. It is shown that the H-orbit $\mathcal{D} = H \cdot p_0$ through p_0 is a convex domain in V. Let g be the H-invariant metric on $\mathcal{D}$ which coincides with the Hessian inner product on V at the tangent space $T_{p_0}\mathcal{D} \equiv V$. Then $(\mathcal{D}, g)$ is a homogeneous Hessian domain. In this way, we have a one-to-one correspondence between homogeneous Hessian domains and normal Hessian algebras up to isomorphisms.

A left-symmetric algebra is said to be *compact* if there exists a linear form $\xi \in V^*$ such that $(x|y)_\xi := \xi(x \triangle y)$ gives a positive inner product on V. In view of (1), we see that the inner product $(\cdot|\cdot)_\xi$ satisfies (2). Namely, a compact normal left-symmetric algebra (*clan*) is a normal Hessian algebra. The clans are studied by Koszul [6] and Vinberg [10]. Vinberg showed that clans are in one-to-one correspondence with affine homogeneous convex domains, and that a clan has a unit element if and only if the corresponding domain is a homogeneous cone. Using the latter fact, the author realized all homogeneous cones as well as clans with a unit element as the set of symmetric matrices with certain block decompositions in [5] (similar results are obtained by many researchers, e.g. [2,11,12]). We recall the results of [5] briefly in the next section.

3 Matrix Realization of a Clan with a Unit Element

Let V_n be the vector space of real symmetric matrices of size n. We define a bilinear product $\triangle$ on V_n by

$$x \triangle y := \underset{\sim}{x} y + y\,{}^{\mathrm{t}}(\underset{\sim}{x}) \quad (x, y \in V_n),$$

where $\underset{\sim}{x}$ is a lower triangular matrix defined by

$$(\underset{\sim}{x})_{ij} := \begin{cases} 0 & (i < j), \\ x_{ii}/2 & (i = j), \\ x_{ij} & (i > j). \end{cases}$$

Then $(V_n, \triangle)$ is a clan with a unit element E_n. Let $n = n_1 + n_2 + \cdots + n_r$ be a partition, and let $\mathcal{V}_{lk} \subset \mathrm{Mat}(n_l, n_k; \mathbb{R})$ $(1 \le k < l \le r)$ be vector spaces satisfying

(V1) $A \in \mathcal{V}_{lk},\ B \in \mathcal{V}_{ki} \Rightarrow AB \in \mathcal{V}_{li}$ for $1 \le i < k < l \le r$,
(V2) $A \in \mathcal{V}_{li},\ B \in \mathcal{V}_{ki} \Rightarrow A\,{}^{\mathrm{t}}B \in \mathcal{V}_{lk}$ for $1 \le i < k < l \le r$,
(V3) $A \in \mathcal{V}_{lk} \Rightarrow A\,{}^{\mathrm{t}}A \in \mathbb{R}E_{n_l}$ for $1 \le k < l \le r$.

Let $\mathcal{Z}_{\mathcal{V}}$ be the subspace of V_n consisting of symmetric matrices x of the form

$$x = \begin{pmatrix} X_{11} & {}^{\mathrm{t}}X_{21} & \cdots & {}^{\mathrm{t}}X_{r1} \\ X_{21} & X_{22} & & {}^{\mathrm{t}}X_{r2} \\ \vdots & & \ddots & \\ X_{r1} & X_{r2} & & X_{rr} \end{pmatrix} \qquad \begin{pmatrix} X_{ll} = x_{ll}E_{n_l},\ x_{ll} \in \mathbb{R}\ (l = 1, \dots, r) \\ X_{lk} \in \mathcal{V}_{lk}\ (1 \le k < l \le r) \end{pmatrix}. \tag{3}$$

Thanks to (V1)–(V3), $\mathcal{Z}_{\mathcal{V}}$ is a subalgebra of the clan $(V_n, \triangle)$.

Theorem 1 ([5]). *Every clan with a unit element is isomorphic to the algebra* $(\mathcal{Z}_\mathcal{V}, \triangle)$ *with appropriate* $\{\mathcal{V}_{lk}\}_{1\le k<l\le r}$.

We define

$$\Omega_\mathcal{V} := \{\, x \in \mathcal{Z}_\mathcal{V}\,;\, x \text{ is positive definite} \,\},$$

$$\mathfrak{h}_\mathcal{V} := \left\{\, \underset{\sim}{x}\,;\, x \in \mathcal{Z}_\mathcal{V} \,\right\}$$

$$= \left\{ T = \begin{pmatrix} T_{11} & & & \\ T_{21} & T_{22} & & \\ \vdots & & \ddots & \\ T_{r1} & T_{r2} & & T_{rr} \end{pmatrix}\,;\, \begin{array}{l} T_{ll} = t_{ll}E_{n_l},\ t_{ll} \in \mathbb{R}\ (l = 1, \dots, r) \\ T_{lk} \in \mathcal{V}_{lk}\ (1 \le k < l \le r) \end{array} \right\},$$

$$H_\mathcal{V} := \{\, T \in \mathfrak{h}_\mathcal{V}\,;\, t_{ll} > 0\ (l = 1, \dots, r) \,\}.$$

Then $\mathfrak{h}_\mathcal{V}$ forms a linear Lie algebra, and $H_\mathcal{V}$ equals the corresponding Lie group $\exp \mathfrak{h}_\mathcal{V}$. The domain $\Omega_\mathcal{V}$ is a homogeneous cone on which $H_\mathcal{V}$ acts simply transitively by the action $\rho(T)x := Tx{}^{\mathrm{t}}T$ $(x \in \mathcal{Z}_\mathcal{V},\ T \in H_\mathcal{V})$. In fact, the transformation group $\rho(H_\mathcal{V})$ on $\Omega_\mathcal{V}$ coincides with the integration of the vector fields X_v $(v \in \mathcal{Z}_\mathcal{V})$ given by

$$X_v(p) := v\triangle(p - E_n) + v \qquad (p \in \Omega_\mathcal{V}).$$

By (V3), we can define an inner product $(\cdot|\cdot)_{\mathcal{V}_{lk}}$ on each $\mathcal{V}_{lk}$ $(1 \le k < l \le r)$ in such a way that

$$X_{lk}{}^{\mathrm{t}}Y_{lk} + Y_{lk}{}^{\mathrm{t}}X_{lk} = 2(X_{lk}|Y_{lk})_{\mathcal{V}_{lk}} E_{n_l} \qquad (X_{lk}, Y_{lk} \in \mathcal{V}_{lk}).$$

For $\underline{s} = (s_1, \dots, s_r) \in \mathbb{R}^r_{>0}$, define a linear form $\xi_{\underline{s}}$ on $\mathcal{Z}_\mathcal{V}$ by

$$\xi_{\underline{s}}(x) := \sum_{k=1}^{r} s_k x_{kk} \quad (x \in \mathcal{Z}_\mathcal{V}).$$

Then

$$(x|y)_{\underline{s}} := \xi_{\underline{s}}(x\triangle y) = \sum_{l=1}^{r} s_l\Big(x_{ll}y_{ll} + \sum_{k<l}(X_{lk}|Y_{lk})_{\mathcal{V}_{lk}}\Big) \quad (x, y \in \mathcal{Z}_\mathcal{V})$$

exhausts all the Hessian inner products on $\mathcal{Z}_\mathcal{V}$. We set

$$N_1 := 1, \quad N_k := n_1 + \cdots + n_{k-1} + 1\ (k = 2, \dots, r), \tag{4}$$

and

$$\sigma_r := s_r, \quad \sigma_k := s_k - n_k \sum_{l=k+1}^{r} \sigma_l\ (k = 1, \dots, r-1). \tag{5}$$

Then, for $x \in \mathcal{Z}_\mathcal{V}$, we have $\xi_{\underline{s}}(x) = \sum_{k=1}^{r} \sigma_k \mathrm{tr} x^{[N_k]}$, where $x^{[N]}$ denotes the left-top corner submatrix of x of size N. By [5, Sect. 4], a global potential function $\phi_{\underline{s}}$ of the Hessian metric $g_{\underline{s}}$ corresponding to $(\cdot|\cdot)_{\underline{s}}$ is given by

$$\phi_{\underline{s}}(x) := -\log\Big(\prod_{k=1}^{r}(\det x^{[N_k]})^{\sigma_k}\Big) \qquad (x \in \Omega_\mathcal{V}). \tag{6}$$

4 Matrix Realization of Normal Hessian Algebras

Let us assume $r \geq 2$ in this section. For $i = 1, \ldots, r$, let e_i be the element of $\mathcal{Z}_\mathcal{V}$ whose x_{ii}-component is 1 and all the other components are zero. We define

$$\mathcal{Z}_\mathcal{V}^\vee := \{ x \in \mathcal{Z}_\mathcal{V} ; x_{11} = 0 \}, \quad \mathcal{Z}_\mathcal{V}^\star := \{ x \in \mathcal{Z}_\mathcal{V} ; x_{11} = x_{rr} = 0 \}.$$

Then we have $\mathcal{Z}_\mathcal{V} = \mathbb{R}e_1 \oplus \mathcal{Z}_\mathcal{V}^\vee = \mathbb{R}e_1 \oplus \mathbb{R}e_r \oplus \mathcal{Z}_\mathcal{V}^\star$. We note that $\mathcal{Z}_\mathcal{V}^\vee$ is a subalgebra of $\mathcal{Z}_\mathcal{V}$, but $\mathcal{Z}_\mathcal{V}^\star$ is not in general. Let us introduce a linear map $\pi_\mathcal{V} : \mathcal{Z}_\mathcal{V} \ni x \mapsto x - x_{rr}e_r \in \mathcal{Z}_\mathcal{V}$. Then $\pi_\mathcal{V}$ gives a projection from $\mathcal{Z}_\mathcal{V}^\vee$ onto $\mathcal{Z}_\mathcal{V}^\star$. We define a bilinear product $\triangle^\star$ on $\mathcal{Z}_\mathcal{V}^\star$ by

$$x \triangle^\star y := \pi_\mathcal{V}(x \triangle y) \in \mathcal{Z}_\mathcal{V}^\star \qquad (x, y \in \mathcal{Z}_\mathcal{V}^\star).$$

Theorem 2. (i) *The algebra $(\mathcal{Z}_\mathcal{V}^\star, \triangle^\star)$ with the inner product $(\cdot|\cdot)_{\underline{s}}$ on $\mathcal{Z}_\mathcal{V}^\star$ forms a normal Hessian algebra.*
(ii) *Every normal Hessian algebra is isomorphic to the algebra $(\mathcal{Z}_\mathcal{V}^\star, \triangle^\star)$ with appropriate $\{\mathcal{V}_{lk}\}_{1 \leq k < l \leq r}$.*

Define $\mathfrak{h}_\mathcal{V}^\star := \left\{ \underline{x} ; x \in \mathcal{Z}_\mathcal{V}^\star \right\}$. Then $\mathfrak{h}_\mathcal{V}^\star$ is a Lie subalgebra of $\mathfrak{h}_\mathcal{V}$, even though $(\mathcal{Z}_\mathcal{V}^\star, \triangle^\star)$ is not a subalgebra of $(\mathcal{Z}, \triangle)$. Let $H_\mathcal{V}^\star$ be the subgroup $\exp \mathfrak{h}_\mathcal{V}^\star$ of $H_\mathcal{V}$. Then $H_\mathcal{V}^\star = \{ T \in H_\mathcal{V} ; t_{11} = t_{rr} = 1 \}$. We define a group action $\sigma : H_\mathcal{V}^\star \to \mathrm{Aff}(\mathcal{Z}_\mathcal{V}^\star)$ in such a way that

$$\pi_\mathcal{V}(\rho(T)(e_1 + x)) = e_1 + \sigma(T)x \qquad (x \in \mathcal{Z}_\mathcal{V}^\star, \ T \in H_\mathcal{V}^\star).$$

Then $\sigma(H_\mathcal{V}^\star)$ is exactly the integration of the vector fields $X_v^\star$ $(v \in \mathcal{Z}_\mathcal{V}^\star)$ given by

$$X_v(x) = v \triangle^\star (x - E^\star) + v \qquad (x \in \mathcal{Z}_\mathcal{V}^\star),$$

where $E^\star := E_n - e_1 - e_r \in \mathcal{Z}_\mathcal{V}^\star$. Let $\mathcal{D}_\mathcal{V}^\star$ be the $\sigma(H_\mathcal{V}^\star)$-orbit through $E^\star$ in $\mathcal{Z}_\mathcal{V}^\star$. Then $\sigma(H_\mathcal{V}^\star)$ acts on $\mathcal{D}_\mathcal{V}^\star$ simply transitively, so that a $\sigma(H_\mathcal{V}^\star)$-invariant Riemannian metric $g_{\underline{s}}^\star$ is uniquely determined in such a way that $g_{\underline{s}}^\star$ on $T_{E^\star}\mathcal{D}_\mathcal{V}^\star = \mathcal{Z}_\mathcal{V}^\star$ coincides with $(\cdot|\cdot)_{\underline{s}}$. Recalling the correspondence between normal Hessian algebras and homogeneous Hessian domains, we deduce the following Theorem from Theorem 2.

Theorem 3. *The Riemannian metric $g_{\underline{s}}^\star$ on the domain $\mathcal{D}_\mathcal{V}^\star \subset \mathcal{Z}_\mathcal{V}^\star$ is Hessian. Every homogeneous Hessian domain is isomorphic to some $(\mathcal{D}_\mathcal{V}^\star, g_{\underline{s}}^\star)$ with appropriate $\{\mathcal{V}_{lk}\}_{1 \leq l < l \leq r}$ and $\underline{s} \in \mathbb{R}_{>0}^r$.*

5 Description of Homogeneous Hessian Domains

We continue using the same notation as in the previous section. If $r = 2$, then $\mathcal{Z}_\mathcal{V}^\star = \left\{ \begin{pmatrix} 0 & {}^t X_{21} \\ X_{21} & 0 \end{pmatrix} ; X_{21} \in \mathcal{V}_{21} \right\} \simeq \mathcal{V}_{21}$. In this case, it is easy to check that $x \triangle^\star y = 0$ for all $x, y \in \mathcal{Z}_\mathcal{V}^\star$, and that $\mathcal{D}_\mathcal{V}^\star$ equals the vector space $\mathcal{Z}_\mathcal{V}^\star$ with the Euclidean metric given by $g_{\underline{s}}(y, y')_x = s_2 (Y_{21}|Y'_{21})_{\mathcal{V}_{21}}$ $(x, y, y' \in \mathcal{Z}_\mathcal{V}^\star)$.

Now we assume $r \geq 3$. Let us consider the subfamilies $\mathcal{V}' := \{\mathcal{V}_{lk}\}_{1\leq k<l\leq r-1}$ and $\mathcal{V}'' := \{\mathcal{V}_{lk}\}_{2\leq k<l\leq r-1}$ of $\{\mathcal{V}_{lk}\}_{1\leq k<l\leq r}$. Then both $\mathcal{V}'$ and $\mathcal{V}''$ satisfy the axioms (V1) – (V3), so that we have clans $\mathcal{Z}_{\mathcal{V}'} \subset V_{n'}$ and $\mathcal{Z}_{\mathcal{V}''} \subset V_{n''}$, where $n' := n_1+\cdots+n_{r-1}$ and $n'' := n_2+\cdots+n_{r-1}$. Recall that $\mathcal{Z}^{\vee}_{\mathcal{V}'} = \{\, x \in \mathcal{Z}_{\mathcal{V}'}\,;\, x_{11} = 0\,\}$ is a subalgebra of $\mathcal{Z}_{\mathcal{V}'}$. We define

$$\mathcal{U}_{\mathcal{V}} := \left\{ u = \begin{pmatrix} X_{21} \\ \vdots \\ X_{r-1,1} \end{pmatrix} ; \; X_{k1} \in \mathcal{V}_{k1} \; (2 \leq k \leq r-1) \right\} \subset \mathrm{Mat}(n'', n_1; \mathbb{R}),$$
$$\mathcal{W}_{\mathcal{V}} := \left\{ w = \begin{pmatrix} X_{r1} \; X_{r2} \; \ldots \; X_{r,r-1} \end{pmatrix} ; \; X_{rk} \in \mathcal{V}_{rk} \; (1 \leq k \leq r-1) \right\} \subset \mathrm{Mat}(n_r, n'; \mathbb{R}),$$
$$\mathcal{W}_{\mathcal{V}''} := \left\{ w = \begin{pmatrix} X_{r2} \; \ldots \; X_{r,r-1} \end{pmatrix} ; \; X_{rk} \in \mathcal{V}_{rk} \; (2 \leq k \leq r-1) \right\} \subset \mathrm{Mat}(n_r, n''; \mathbb{R}).$$

Then every $x \in \mathcal{Z}^{\star}_{\mathcal{V}}$ is written as

$$x = \begin{pmatrix} x' & {}^{\mathrm{t}}w \\ w & 0 \end{pmatrix} = \begin{pmatrix} 0 & {}^{\mathrm{t}}u & {}^{\mathrm{t}}w' \\ u & x'' & w'' \\ w' & w'' & 0 \end{pmatrix} \left(=: x(u, w', x'', w'') \right) \tag{7}$$
$$(x' \in \mathcal{Z}^{\vee}_{\mathcal{V}'},\; w \in \mathcal{W}_{\mathcal{V}},\; x'' \in \mathcal{Z}_{\mathcal{V}''}, u \in \mathcal{U}_{\mathcal{V}},\; w' \in \mathcal{V}_{r1},\; w'' \in \mathcal{W}_{\mathcal{V}''}).$$

Theorem 4. *The homogeneous Hessian domain $\mathcal{D}^{\star}_{\mathcal{V}} \subset \mathcal{Z}^{\star}_{\mathcal{V}}$ is described as*

$$\mathcal{D}^{\star}_{\mathcal{V}} = \left\{ x \in \mathcal{Z}^{\star}_{\mathcal{V}}\,;\, e_1 + x' \in \Omega_{\mathcal{V}'} \right\} = \left\{ x(u, w', x'', w'')\,;\, x'' - u{}^{\mathrm{t}}u \in \Omega_{\mathcal{V}''} \right\}.$$

In particular, $\mathcal{D}^{\star}_{\mathcal{V}}$ is linearly isomorphic to the direct product of a real Siegel domain $\mathcal{S}_{\mathcal{V}'} := \left\{ x' = \begin{pmatrix} 0 & {}^{\mathrm{t}}u \\ u & x'' \end{pmatrix} \in \mathcal{Z}^{\vee}_{\mathcal{V}}\,;\, x'' - u{}^{\mathrm{t}}u \in \Omega_{\mathcal{V}''} \right\}$ and the vector space $\mathcal{W}_{\mathcal{V}}$.

We remark that the real Siegel domain $\mathcal{S}_{\mathcal{V}'} \subset \mathcal{Z}^{\vee}_{\mathcal{V}'}$ is the affine homogeneous convex domain corresponding to the clan $\mathcal{Z}^{\vee}_{\mathcal{V}'}$ (see Vinberg [10, Chap. 2]).

Theorem 5. *A global potential function $\psi_{\underline{s}}$ on $\mathcal{D}^{\star}_{\mathcal{V}}$ of the Hessian metric $g^{\star}_{\underline{s}}$ is given by*

$$\psi_{\underline{s}}(x) := -\log\Bigl(\prod_{k=2}^{r-1} (\det(e_1 + x')^{[N_k]})^{\sigma_k} \Bigr) + \frac{s_r}{n_r} \mathrm{tr}\, {}^{\mathrm{t}}w (e_1 + x')^{-1} w \tag{8}$$

for $x \in \mathcal{D}^{\star}_{\mathcal{V}}$ of the form (7)*, where σ_k and N_k are defined in* (4) *and* (5) *respectively.*

Let us describe the affine action $\sigma(H^{\star}_{\mathcal{V}})$ on $\mathcal{Z}^{\star}_{\mathcal{V}}$ explicitly. Every element of $H^{\star}_{\mathcal{V}}$ is written as

$$T(u, w', T'', w'') = \begin{pmatrix} E_{n_1} & 0 & 0 \\ u & T'' & 0 \\ w' & w'' & E_{n_r} \end{pmatrix}$$
$$(T'' \in H_{\mathcal{V}''}, u \in \mathcal{U}_{\mathcal{V}},\; w' \in \mathcal{V}_{r1},\; w'' \in \mathcal{W}_{\mathcal{V}''}),$$

which is factorized as

$$T(u, 0, E_{n''}, 0)\, T(0, w', E_{n''}, 0)\, T(0, 0, T'', 0)\, T(0, 0, E_{n''}, w'').$$

Now take $x_0 = x(u_0, w_0', x_0'', w_0'') \in \mathcal{Z}_{\mathcal{V}}^\star$ with $u_0 \in \mathcal{U}_{\mathcal{V}}$, $w_0' \in \mathcal{V}_{r1}$, $x_0'' \in \mathcal{Z}_{\mathcal{V}''}$, $w_0'' \in \mathcal{W}_{\mathcal{V}''}$. Then we get

$$\sigma(T)x_0 = \begin{cases} x(u_0, w_0' + w''u_0, x_0'', w_0'' + w''x_0'') & (T = T(0,0,E_{n''}, w'')), \\ x(T''u_0, w_0', \rho(T'')x_0'', w_0'{}^{t}T'') & (T = T(0,0,T'',0)), \\ x(u_0, w_0' + w', x_0'', w_0'' + w'^{t}u_0) & (T = T(0,w',E_{n''},0)), \\ x(u_0 + u, w_0', x_0'' + 2u^{t}u_0 + u^{t}u, w_0'' + w_0'{}^{t}u) & (T = T(u,0,E_{n''},0)). \end{cases} \tag{9}$$

Finally, we present two examples of homogeneous Hessian domains of dimension 2.

(1) When

$$\mathcal{Z}_{\mathcal{V}} = \left\{ \begin{pmatrix} x_{11} & x_{21} & 0 \\ x_{21} & x_{22} & 0 \\ 0 & 0 & x_{33} \end{pmatrix} ; x_{ij} \in \mathbb{R} \right\},$$

we have

$$\mathcal{D}_{\mathcal{V}}^\star = \left\{ x = \begin{pmatrix} 0 & u & 0 \\ u & \xi & 0 \\ 0 & 0 & 0 \end{pmatrix} ; \xi - u^2 > 0 \right\}.$$

A global potential of a homogeneous Hessian metric is parametrized by $s > 0$ as $\psi_s(x) = -s \log(\xi - u^2)$. Moreover, we have

$$\sigma(T) \begin{pmatrix} 0 & u & 0 \\ u & \xi & 0 \\ 0 & 0 & 0 \end{pmatrix} = \begin{pmatrix} 0 & au + b & 0 \\ au + b & a^2\xi + 2abu + b^2 & 0 \\ 0 & 0 & 0 \end{pmatrix} \quad \left(T = \begin{pmatrix} 1 & 0 & 0 \\ b & a & 0 \\ 0 & 0 & 1 \end{pmatrix} \in H_{\mathcal{V}}^\star \right).$$

(2) When

$$\mathcal{Z}_{\mathcal{V}} = \left\{ \begin{pmatrix} x_{11} & 0 & 0 \\ 0 & x_{22} & x_{32} \\ 0 & x_{32} & x_{33} \end{pmatrix} ; x_{ij} \in \mathbb{R} \right\},$$

we have

$$\mathcal{D}_{\mathcal{V}}^\star = \left\{ x = \begin{pmatrix} 0 & 0 & 0 \\ 0 & \xi & w \\ 0 & w & 0 \end{pmatrix} ; \xi > 0,\ w \in \mathbb{R} \right\}.$$

A global potential is parametrized by $\underline{s} = (s_2, s_3) \in \mathbb{R}^2_{>0}$ as $\psi_{\underline{s}}(x) = -s_2 \log \xi + s_3 w^2/\xi$. Moreover, we have

$$\sigma(T) \begin{pmatrix} 0 & 0 & 0 \\ 0 & \xi & w \\ 0 & w & 0 \end{pmatrix} = \begin{pmatrix} 0 & 0 & 0 \\ 0 & a^2\xi & a(w + b\xi) \\ 0 & a(w + b\xi) & 0 \end{pmatrix} \quad \left(T = \begin{pmatrix} 1 & 0 & 0 \\ 0 & a & 0 \\ 0 & b & 1 \end{pmatrix} \in H_{\mathcal{V}}^\star \right).$$

As a matter of fact, any 2-dimensional homogeneous Hessian domain is isomorphic to the Euclidean vector space $\mathbb{R}^2$, a homogeneous cone $\mathbb{R}^2_{>0}$, or either of the two domains $\mathcal{D}_{\mathcal{V}}^\star$ above.

References

1. Boyom, M.N.: The cohomology of Koszul-Vinberg algebra and related topics. Afr. Diaspora J. Math. (N.S.) **9**, 53–65 (2010)
2. Chua, C.B.: Relating homogeneous cones and positive definite cones via T-algebras. SIAM J. Optim. **14**, 500–506 (2003)
3. Ishi, H.: Representation of clans and homogeneous cones. Vestnik Tambov Univ. **16**, 1669–1675 (2011)
4. Ishi, H.: Homogeneous cones and their applications to statistics. In: Modern Methods of Multivariate Statistics, pp. 135–154, Travaux en Cours 82, Hermann, Paris (2014)
5. Ishi, H.: Matrix realization of a homogeneous cone. In: Nielsen, F., Barbaresco, F. (eds.) GSI 2015. LNCS, vol. 9389, pp. 248–256. Springer, Cham (2015). doi:10.1007/978-3-319-25040-3_28
6. Koszul, J.L.: Sur la forme hermitienne canonique des espaces homogènes complexes. Canad. J. Math. **7**, 562–576 (1955)
7. Manchon, D.: A short survey on pre-Lie algebras. In: Noncommutative Geometry and Physics: Renormalisation, Motives, Index Theory, pp. 89–102. ESI Lect. Math. Phys., Eur. Math. Soc., Zurich (2011)
8. Shima, H.: Homogeneous Hessian manifolds. Ann. Inst. Fourier Grenoble **30**, 91–128 (1980)
9. Shima, H.: The Geometry of Hessian Structures. World Scientific Publishing, Hackensack (2007)
10. Vinberg, E.B.: The theory of convex homogeneous cones. Trans. Moscow Math. Soc. **12**, 340–403 (1963)
11. Xu, Y.-C.: Theory of Complex Homogeneous Bounded Domains. Kluwer, Dordrecht (2005)
12. Yamasaki, T., Nomura, T.: Realization of homogeneous cones through oriented graphs. Kyushu J. Math. **69**, 11–48 (2015)

Monotone Embedding in Information Geometry

Information Geometry Under Monotone Embedding. Part I: Divergence Functions

Jun Zhang[1(✉)] and Jan Naudts[2]

[1] University of Michigan, Ann Arbor, MI, USA
junz@umich.edu
[2] Universiteit Antwerpen, Antwerpen, Belgium
Jan.Naudts@uantwerpen.be

Abstract. The standard model of information geometry, expressed as Fisher-Rao metric and Amari-Chensov tensor, reflects an embedding of probability density by log-transform. The standard embedding was generalized by one-parametric families of embedding function, such as α-embedding, q-embedding, κ-embedding. Further generalizations using arbitrary monotone functions (or positive functions as derivatives) include the deformed-log embedding (Naudts), U-embedding (Eguchi), and rho-tau dual embedding (Zhang). Here we demonstrate that the divergence function under the rho-tau dual embedding degenerates, upon taking $\rho = id$, to that under either deformed-log embedding or U-embedding; hence the latter two give an identical divergence function. While the rho-tau embedding gives rise to the most general form of cross-entropy with two free functions, its entropy reduces to that of deformed entropy of Naudts with only one free function. Fixing the gauge freedom in rho-tau embedding through normalization of dual-entropy function renders rho-tau cross-entropy to degenerate to U cross-entropy of Eguchi, which has the simpler property, not true for general rho-tau cross-entropy, of reducing to the deformed entropy upon setting the two pdfs to be equal. In Part I, we investigate monotone embedding in divergence function, entropy and cross-entropy, whereas in the sequel (Part II), in induced geometries and probability families.

1 Introduction: A Plethora of Probability Embeddings

One motivation to study probability embedding functions is to extend the framework of information geometry beyond the now-classic expressions of Fisher-Rao metric and Amari-Chensov tensor. Realizing that the standard α-geometry is based on log-embedding of probability functions, various approaches have been proposed to generalize such probability embedding, using a one-parameter family of specific functions at the first level of generality, and using arbitrarily chosen (monotone or positive) functions at the second level of generality.

J. Zhang and J. Naudts contributed equally to this paper.

F. Nielsen and F. Barbaresco (Eds.): GSI 2017, LNCS 10589, pp. 205–214, 2017.
https://doi.org/10.1007/978-3-319-68445-1_24

(i). α-embedding. It was Amari [1] who first investigated the one-parameter family of embeddings $\log_\alpha : \mathbb{R}^+ \to \mathbb{R}$ defined by

$$\log_\alpha(u) = \begin{cases} \log u & \alpha = 1 \\ \frac{2}{1-\alpha} u^{(1-\alpha)/2} & \alpha \neq 1 \end{cases} \tag{1}$$

Under this α-embedding, α-divergence becomes canonical divergence, and α-connections have a simple Γ^1, Γ^{-1}-like characteristics [2].

(ii). q-exponential embedding. Tsallis [3], in investigating the equilibrium distribution of statistical physics which maximizes the Boltzmann-Gibbs-Shannon entropy under constraints, replaced the entropy function by a q-dependent entropy, resulting in a deformed version of statistical physics; here, $q \in \mathbb{R}$. The q-logarithmic/exponential functions were introduced [4]:

$$\log_q(u) = \frac{1}{1-q}\left(u^{1-q} - 1\right), \quad \exp_q(u) = [1 + (1-q)u]^{1/(1-q)}, \quad q \neq 1.$$

Note that q-embedding and α-embedding functions are different: $\log_q(\cdot) \neq \log_\alpha(\cdot)$, even after the identification $\alpha = 2q - 1$. Like α-embedding, q-embedding reduces to the standard logarithm as $\lim_{q\to 1}$.

(iii). κ-exponential embedding. An alternative to the q-deformed exponential model for statistical physics is the κ-model [5], where

$$\log_\kappa(u) = \frac{1}{2\kappa}\left(u^\kappa - u^{-\kappa}\right), \quad \exp_\kappa(u) = \left(\kappa u + \sqrt{1 + \kappa^2 u^2}\right)^{\frac{1}{\kappa}}, \quad \kappa \neq 0;$$

the case of $\lim_{\kappa\to 0}$ corresponds to the standard exponential/logarithm.

(iv). ϕ-, U-, and(ρ, τ)-embedding. Generalizing any parametric forms of embedding functions further leads to the consideration of probability embedding using arbitrary monotone (or after taking derivative, positive) functions. The prominent inventions are Naudts' phi-embedding [7], Eguchi's U-embedding [8], and Zhang's rho-tau embedding [6], though they have been re-invented/renamed by later authors, causing confusion and distraction. We discuss these in the next section.

Below we first review the deformed logarithm, $\log_\phi$, and deformed exponential, $\exp_\phi$, functions. Then we point out that $\log_\phi$ and $\exp_\phi$ are nothing but an arbitrary pair of mutually inverse monotone functions, and are representable as derivatives of a pair of conjugate convex functions f, f^*. The deformed divergence $D_\phi(p, q)$ is then precisely the Bregman divergence $D_f(p, q)$ associated with f. The construction of entropy and cross-entropy from this deformed approach is reviewed, as well as their construction from the U-embedding. Then, we review the rho-tau embedding, which provides two independently chosen embedding functions, and explicitly identify its entropy and cross-entropy. Our Main Theorem shows that the divergence function and entropy function of the rho-tau embedding reduce as a special case to those given by the phi-embedding and U-embedding, while the rho-tau cross-entropy reduces as another special case to the U cross-entropy.

2 Deformation Versus Embedding

2.1 "Deforming" Exponential and Logarithmic Functions

Naudts [7,9] defines the phi-deformed logarithm

$$\log_\phi(u) = \int_1^u \frac{1}{\phi(v)} \mathrm{d}v.$$

Here, $\phi(v)$ is a strictly positive function. In the context of discrete probabilities it suffices that it is strictly positive on the open interval $(0, 1)$, possibly vanishing at the end points. In the case of a probability density function it is assumed to be strictly positive on the interval $(0, +\infty)$. Note that by construction one has $\log_\phi(1) = 0$. The inverse of the phi-logarithm is denoted $\exp_\phi(u)$, and called phi-exponential function:

$$\exp_\phi(\log_\phi(u)) = \log_\phi(\exp_\phi(u)) = u.$$

The phi-exponential has an integral expression

$$\exp_\phi(u) = 1 + \int_0^u \mathrm{d}v\, \psi(v),$$

where the function $\psi(u)$ is given by

$$\psi(u) = \frac{\mathrm{d}}{\mathrm{d}u} \exp_\phi(u) = \frac{\mathrm{d}}{\mathrm{d}u} (\log_\phi)^{-1}(u).$$

In terms of ϕ, ψ, we have the following relations:

$$\psi(u) = \phi(\exp_\phi(u)),$$
$$\phi(u) = \psi(\log_\phi(u)).$$

We want to stress that all four functions, $\phi, \psi, \log_\phi, \exp_\phi$, arise out of choosing one positive-valued function ϕ.

As examples, $\phi(v) = v$ gives rise to the classic natural logarithm and exponential. Taking $\phi(u) = \frac{u}{1+u}$ in [13] leads to $\log_\phi(u) = u - 1 + \log(u)$. Taking $\phi(u) = u(1 + \epsilon u)$ in [14] leads to

$$\log\left(\frac{(1+\epsilon)u}{1+\epsilon u}\right), \quad \exp_\phi(u) = \frac{1}{(1+\epsilon)e^{-v} - \epsilon}.$$

2.2 Deformed Entropy and Deformed Divergence Functions

The phi-entropy of the probability distribution p is defined by [9]

$$S_\phi(p) = -\mathrm{E}_p \log_\phi p + \int_{\mathcal{X}} \mathrm{d}x \int_0^{p(x)} \mathrm{d}u \frac{u}{\phi(u)} + \text{ constant.} \tag{2}$$

By partial integration one obtains an equivalent expression

$$S_\phi(p) = -\int_{\mathcal{X}} \mathrm{d}x \int_1^{p(x)} \mathrm{d}u \, \log_\phi(u) + \text{ constant.} \tag{3}$$

For standard logarithm $\phi(u) = u$, the above expression is the well-known entropy of Boltzmann-Gibbs-Shannon

$$S(p) = -\mathrm{E}_p \log p.$$

The phi-divergence of two probability functions p and q is defined by [9]

$$D_\phi(p,q) = \int_{\mathcal{X}} \mathrm{d}x \int_{q(x)}^{p(x)} \mathrm{d}v \, \left[\log_\phi(v) - \log_\phi(q(x))\right], \tag{4}$$

which has another equivalent expression

$$D_\phi(p,q) = S_\phi(q) - S_\phi(p) - \int_{\mathcal{X}} \mathrm{d}x \, [p(x) - q(x)] \log_\phi(q(x)). \tag{5}$$

Now let us express these quantities in terms of a strictly convex function f, satisfying $f'(u) = \log_\phi(u)$. We have:

$$S_\phi(p) = -\int_{\mathcal{X}} \mathrm{d}x \, f(p(x)) + \text{ constant,} \tag{6}$$

$$D_\phi(p,q) = \int_{\mathcal{X}} \mathrm{d}x \, \{f(p(x)) - f(q(x)) - [p(x) - q(x)] f'(q(x))\}. \tag{7}$$

One can readily recognize that $D_\phi(p,q)$ is nothing but the Bregman divergence, whereas the function f itself determines the deformed entropy $S_\phi(p)$. Note that $p \mapsto S_\phi(p)$ is strictly concave while the map $p \mapsto D_\phi(p,q)$ is strictly convex.

2.3 U-embedding

Eguchi [8] introduces the U-embedding, which is essentially the Bregman divergence under a strictly convex function U coupled with an embedding using $\psi = (U')^{-1}$. The U cross-entropy $C_U(p,q)$ is defined as:

$$C_U(p,q) = \int_{\mathcal{X}} \mathrm{d}x \, \{U(\psi(q(x))) - p(x) \cdot \psi(q(x))\}, \tag{8}$$

whereas the U entropy H_U is defined as $H_U(p) = C_U(p,p)$. The U-divergence is

$$\begin{aligned} D_U(p,q) &= C_U(p,q) - H_U(p,p) \\ &= \int_{\mathcal{X}} \mathrm{d}x \left\{ U\left(\psi(q(x))\right) - U\left(\psi(p(x))\right) - p(x)\left[\left(\psi(q(x)) - \psi(p(x))\right]\right. \right\}. \end{aligned} \tag{9}$$

Note that the U-embedding only has one arbitrarily chosen function, as does phi-embedding.

2.4 Dual rho-tau Embedding

In contrast with the "single function" embedding of the phi-model and the U-model, Zhang's (2004) rho-tau framework uses *two* arbitrarily and independently chosen monotone functions. He starts with the observation that a pair of mutually inverse functions occurs naturally in the context of convex duality. Indeed, if f is strictly convex and f^* is its convex dual then the derivatives f' and $(f^*)'$ are inverse functions of each other:

$$f' \circ (f^*)'(u) = (f^*)' \circ f'(u) = u.$$

Here the definition of the convex dual f^* of f is:

$$f^*(u) = \sup\{uv - f(v)\}.$$

For u in the range of f' it is given by

$$f^*(u) = u(f')^{-1}(u) - f \circ (f')^{-1}(u).$$

Take the derivative of this expression to find $(f^*)' \circ f'(u) = u$. By convex duality then follows that also $f' \circ (f^*)'(u) = u$. Take an additional derivative to obtain

$$f''((f^*)'(u)) \cdot (f^*)''(u) = (f^*)''(f'(u)) \cdot f''(u) = 1. \tag{10}$$

This identity will be used further on.

Consider now a pair $(\rho(\cdot), \tau(\cdot))$ of strictly increasing functions. Then there exists a strictly convex function $f(\cdot)$ satisfying $f'(u) = \tau \circ \rho^{-1}(u)$. This is because the family of strictly increasing functions form a group, with function composition as group operation, an observation made in [6,12]. In terms of the conjugate function f^*, the relation is $(f^*)'(u) = \rho \circ \tau^{-1}(u)$. The derivatives of $f(u)$ and of its conjugate $f^*(u)$ have the property that

$$f'(\rho(u)) = \tau(u) \quad \text{and} \quad (f^*)'(\tau(u)) = \rho(u). \tag{11}$$

Among the triple (f, ρ, τ), given any two functions, the third is specified. When we arbitrarily choose two strictly increasing functions ρ and τ as embedding functions, then they are automatically linked by a pair of conjugated convex functions f, f^*. On the other hand, we may also independently choose to specify $(\rho, f), (\rho, f^*), (\tau, f)$, or (τ, f^*), with the others being fixed. Therefore, rho-tau embedding is a mechanism with *two* independently chosen functions; this differs from both the phi-embedding and the U-embedding. The following identities will be useful:

$$f''(\rho(u))\, \rho'(u) = \tau'(u)\ , \qquad (f^*)''(\tau(u))\, \tau'(u) = \rho'(u)\ , \tag{12}$$

$$f''(\rho(u))\, (\rho'(u))^2 = (f^*)''(\tau(u))\, (\tau'(u))^2\ , \tag{13}$$

$$f''(\rho(u))\, (f^*)''(\tau(u)) = 1. \tag{14}$$

2.5 Divergence of the rho-tau Embedding

Zhang (2004) introduces[1] the rho-tau divergence (see Proposition 6 of [6])

$$D_{\rho,\tau}(p,q) = \int_{\mathcal{X}} \mathrm{d}x \, \{f(\rho(p(x))) + f^*(\tau(q(x))) - \rho(p(x))\tau(q(x))\}, \tag{15}$$

where f is a strictly convex function satisfying $f'(\rho(u)) = \tau(u)$.

Lemma 1. *Expression (15) can be written as*

$$\begin{aligned} D_{\rho,\tau}(p,q) &= \int_{\mathcal{X}} \mathrm{d}x \left\{ f(\rho(p(x))) - f(\rho(q(x))) - [\rho(p(x)) - \rho(q(x))]\tau(q(x)) \right\} \\ &= \int_{\mathcal{X}} \mathrm{d}x \int_{q(x)}^{p(x)} [\tau(v) - \tau(q(x))] \, \mathrm{d}\rho(v) \\ &= \int_{\mathcal{X}} \mathrm{d}x \int_{\rho(q(x))}^{\rho(p(x))} \mathrm{d}u \, [f'(u) - f'(\rho(q(x)))] \, . \end{aligned} \tag{16}$$

In particular this implies that $D_{\rho,\tau}(p,q) \geq 0$, with equality if and only if $p = q$. We note the following identity:

$$f(\rho(p(x))) - \rho(p(x))\tau(p(x)) + f^*(\tau(p(x))) = 0. \tag{17}$$

The "reference-representation biduality" [6,10,12] reveals as

$$D_{\rho,\tau}(p,q) = D_{\tau,\rho}(q,p).$$

2.6 Entropy and Cross-Entropy of rho-tau Embedding

It is now obvious to give the following definition of the rho-tau entropy

$$S_{\rho,\tau}(p) = -\int_{\mathcal{X}} \mathrm{d}x \, f(\rho(p(x))), \tag{18}$$

where $f(u)$ is a strictly convex function satisfying $f'(u) = \tau \circ \rho^{-1}(u)$. This can be written as

$$\begin{aligned} S_{\rho,\tau}(p) &= -\int_{\mathcal{X}} \mathrm{d}x \int^{\rho(p(x))} f'(v)\mathrm{d}v + \text{ constant} \\ &= -\int_{\mathcal{X}} \mathrm{d}x \int^{p(x)} \tau(u)\mathrm{d}\rho(u) + \text{ constant}. \end{aligned} \tag{19}$$

[1] The original definition as found in [6,12] uses the notation $D_{f,\rho}(p,q)$ and treats f and ρ as independent. In the present definition $D_{\rho,\tau}(p,q)$ the definition of f depends on ρ, τ. The difference in only notational and inconsequential.

Note that the rho-tau entropy $S_{\rho,\tau}(p)$ is concave in $\rho(p)$, but not necessarily in p. This has consequences further on. We likewise define rho-tau cross-entropy

$$C_{\rho,\tau}(p,q) = -\int_{\mathcal{X}} \mathrm{d}x\, \rho(p(x))\tau(q(x))$$

with $C_{\rho,\tau}(p,q) = C_{\tau,\rho}(q,p)$.

The rho-tau divergence can then be given by

$$\begin{aligned} D_{\rho,\tau}(p,q) &= S_{\rho,\tau}(q) - S_{\rho,\tau}(p) - \int_{\mathcal{X}} \mathrm{d}x\, [\rho(p(x)) - \rho(q(x))]\tau(q(x)). \\ &= [S_{\rho,\tau}(q) - C_{\rho,\tau}(q,q)] - [S_{\rho,\tau}(p) - C_{\rho,\tau}(p,q)] \end{aligned}$$

Note that in general $S_{\rho,\tau}(q) \neq C_{\rho,\tau}(q,q)$; this is because

$$S_{\rho,\tau}(p) - C_{\rho,\tau}(p,p) = \int_{\mathcal{X}} \mathrm{d}x\, f^*(\tau(p(x))).$$

So unless $f(u) = cu$ for constant c, f^* would not vanish. In fact, denote

$$S^*_{\rho,\tau}(p) = -\int_{\mathcal{X}} \mathrm{d}x\, f^*(\tau(p(x))). \tag{20}$$

Then $S^*_{\rho,\tau}(p) = S_{\tau,\rho}(p)$, and

$$S_{\rho,\tau}(p) - C_{\rho,\tau}(p,p) + S^*_{\rho,\tau}(p) = 0 \tag{21}$$

which is, after integrating $\int_{\mathcal{X}} \mathrm{d}x$, a re-write of (17). Therefore,

$$D_{\rho,\tau}(p,q) = S_{\rho,\tau}(p) - C_{\rho,\tau}(p,q) + S^*_{\rho,\tau}(q). \tag{22}$$

Because $D_{\rho,\tau}(p,q)$ is non-negative and vanishes if and only if $p = q$, the function $p \mapsto S_{\rho,\tau}(p) - C_{\rho,\tau}(p,q)$ has its unique maximum at $p = q$. Therefore, minimizing $p \mapsto D_{\rho,\tau}(p,q)$ is equivalent with maximizing $p \mapsto S_{\rho,\tau}(p) - C_{\rho,\tau}(p,q)$.

2.7 Gauge Freedom of the rho-tau Embedding

Because rho-tau embedding has the freedom of two functions, it reduces to the single-function embeddings (either phi- or U-embedding) upon fixing one embedding function.

Divergence. In the phi-embedding, Expression (15) of $D_{\rho,\tau}(p,q)$ reduces to the phi-divergence $D_\phi(p,q)$ for instance if $\rho = id$, the identity function; in this case, $\tau(u) = \log_\phi(u) = f'(u)$.

The U-embedding is also a special case of the rho-tau embedding, with $\rho = id$ identification: $U = f^*$, $\tau = (U')^{-1} = f'$. So phi-divergence (7) and U-divergence (9) are identical. U- and phi-embedding are the same, with $U' = \exp_\phi$, as noted in [11].

Entropy. By virtue of gauge selection $\rho = id$ in the rho-tau embedding, any phi-deformed entropy (3) is a special case of rho-tau entropy (18)

$$S_{\rho,\tau}(p) = S_\phi(\rho(p)).$$

On the other hand, though the rho-tau entropy (18) has two free functions in appearance, it is the result of their function composition that matters. So any rho-tau entropy is also a phi-entropy for a well-chosen ϕ.

The situation with the U-embedding is the same, because U-entropy is identical with phi-entropy:

$$\begin{aligned} H_U(p) &= \int_{\mathcal{X}} \mathrm{d}x \, \left[U((U')^{-1}(p(x))) - p(x) \cdot (U')^{-1}(p(x))\right] \\ &= \int_{\mathcal{X}} \mathrm{d}x \, \left[f^*(f'(p(x))) - p(x) \cdot f'(p(x))\right] = -\int_{\mathcal{X}} \mathrm{d}x f(p(x)) = S_\phi(p). \end{aligned}$$

Cross-entropy. The rho-tau embedding identifies $C_{\rho,\tau}(p,q)$ as the cross-entropy with a dual embedding mechanism, one free function for each of the p, q. In this most general form, however, we do not require that $C_{\rho,\tau}(p,q)$ reduce to either $S_{\rho,\tau}(p)$ or $S^*_{\rho,\tau}(p) \equiv S_{\tau,\rho}(p)$ when $p = q$. This is *different* from the approach of the U-embedding, where its cross-entropy $C_U(p,q)$ is such that $C_U(p,p) = H_U(p)$. It turns out that $C_U(p,q)$ given by (8) equals the rho-tau cross-entropy minus the dual rho-tau entropy (after adopting the $\rho = id$ gauge):

$$C_{\rho,\tau}(p,q) - S^*_{\rho,\tau}(q) = C_U(p,q). \tag{23}$$

Below, we *extend* Eguchi's definition of U cross-entropy by removing the $\rho = id$ restriction. In other words, we can call the left-hand side of (23) U cross-entropy, which depends on two free functions ρ, τ, and obtain from (22)

$$D_{\rho,\tau} = C_U(p,q) - C_U(p,p).$$

2.8 The Normalization Gauge

Let us fix the gauge by $f^* = \tau^{-1}$. In this case, $\int_{\mathcal{X}} \mathrm{d}x \, f^*(\tau(p(x))) = \int_{\mathcal{X}} p(x)\mathrm{d}x = 1$, so $S^*_{\rho,\tau}(p) = S^*_{\rho,\tau}(q) = -1$.

Adopting the $f^* = \tau^{-1}$ gauge (we call this "normalization gauge") implies that

$$\rho(p) = (f^*)'(\tau(p)) = (\tau^{-1})'(\tau(p)) = \frac{1}{\tau'(p)}.$$

So the transformation

$$\lambda : \tau(\cdot) \longrightarrow \frac{1}{\tau'(\cdot)} \equiv (\tau^{-1})'(\tau(\cdot))$$

reflects a transformation of embedding functions. In the phi-embedding language, $\tau \rightarrow \rho$ is simply $\log_\phi \rightarrow \phi$, or the phi-exponentiation operation.

This transformation is important in studying phi-exponential family of pdfs (Part II).

Fixing the gauge freedom by normalization simplifies the form of $D_{\rho,\tau}$. Making use of (21), with $S^* = const$, implies that the rho-tau cross-entropy $C_{\rho,\tau}$ and U cross-entropy $C_U(\rho,\tau)$, as given by left-hand side of (23), are equal and are denoted C_0:

$$C_0(p,q) = -\int_{\mathcal{X}} \mathrm{d}x\, \rho(p(x)) \cdot \tau(q(x)) = -\int_{\mathcal{X}} \mathrm{d}x\, (\tau^{-1})'(\tau(p(x))) \cdot \tau(q(x))$$

or, in terms of deformed-logarithm notation,

$$C_0(p,q) = -\int_{\mathcal{X}} \mathrm{d}x\, \rho(p(x)) \log_\rho(q(x)).$$

Then

$$H_0(p) \equiv C_0(p,p) = -\int_{\mathcal{X}} \mathrm{d}x\, \rho(p(x)) \log_\rho(p(x)),$$

with

$$\begin{aligned} D_0(p,q) &= C_0(p,q) - C_0(p,p) \\ &= \int_{\mathcal{X}} \mathrm{d}x\, \rho(p(x)) \cdot (\log_\rho(p(x)) - \log_\rho(q(x))) \\ &= \int_{\mathcal{X}} \mathrm{d}x\, \frac{1}{\tau'(p(x))} (\tau(p(x)) - \tau(q(x))). \end{aligned} \tag{24}$$

Note that $D_0 \neq D_\phi$; they both degenerate from $D_{\rho,\tau}$ under different gauges.

We summarize the above conclusions in the following theorem:

Theorem 1. *The (ρ,τ) embedding reduces to special cases upon fixing the gauge as:*

(i) $\rho = id$*: rho-tau divergence $D_{\rho,\tau}$ reduces to deformed phi-divergence D_ϕ with $\tau = f' = \log_\phi$, and to U-divergence D_U with $U = f^*$ and $\tau = f' = (U')^{-1}$;*

(ii) $f^* = \tau^{-1}$*: rho-tau cross-entropy $C_{\rho,\tau}$ reduces to U-cross-entropy as redefined in (23). In this case, $\rho = \phi, \tau = \log_\phi$, i.e., $\tau \to \rho = (\tau^{-1})' \circ \tau \equiv 1/\tau'$ is taking phi-exponentiation operation;*

(iii) $\rho = \tau$*: rho-tau divergence $D_{\rho,\tau}$ becomes $\int dx(\rho(p(x)) - \rho(q(x)))^2/2$.*

3 Discussion

The main thesis of our paper is that the divergence function $D_{\rho,\tau}$ constructed from (ρ,τ)-embedding subsumes both the phi-divergence D_ϕ constructed from the deformed-log embedding and the U-divergence constructed from the U-embedding. A highlight of our analysis is that the rho-tau divergence $D_{\rho,\tau}$ provides a clear distinction between entropy and cross-entropy as *two* distinct quantities *without* requiring the latter to degenerate to the former. This is significant in terms of the resulting geometry generated by these two quantities (see Part II).

On the other hand, upon fixing the gauge $f^* = \tau^{-1}$ (normalization gauge) renders the rho-tau cross-entropy to be U cross-entropy, where the dual-entropy is constant. In this case, $\tau \leftrightarrow \rho$ is akin to $\log_\phi \leftrightarrow \phi$ transformation encountered in studying normalization of phi-exponential family. A thorough discussion of the geometries induced from the rho-tau divergence and from the phi-exponential family will be given in Part II.

Acknowledgement. The first author is supported by DARPA/ARO Grant W911NF-16-1-0383.

References

1. Amari, S.: Differential-Geometric Methods in Statistics. LNS, vol. 28. Springer, Heidelberg (1985). doi:10.1007/978-1-4612-5056-2
2. Amari, S., Nagaoka, H.: Methods of Information Geometry. Translations of Mathematical Monographs, vol. 191. Am. Math. Soc., Oxford University Press, Oxford (2000). Originally in Japanese (Iwanami Shoten, Tokyo, 1993)
3. Tsallis, C.: Possible generalization of Boltzmann-Gibbs statistics. J. Stat. Phys. **52**, 479–487 (1988)
4. Tsallis, C.: What are the numbers that experiments provide? Quim. Nova **17**, 468 (1994)
5. Kaniadakis, G.: Non-linear kinetics underlying generalized statistics. Phys. A Stat. Mech. Appl. **296**(3), 405–425 (2001)
6. Zhang, J.: Divergence function, duality, and convex analysis. Neural Comput. **16**, 159–195 (2004)
7. Naudts, J.: Estimators, escort probabilities, and phi-exponential families in statistical physics. J. Ineq. Pure Appl. Math. **5**, 102 (2004). arXiv:math-ph/0402005
8. Eguchi, S.: Information geometry and statistical pattern recognition. Sugaku Expo. (Amer. Math. Soc.) **19**, 197–216 (2006). Originally Sūgaku 56 (2004) 380 in Japanese
9. Naudts, J.: Generalised Thermostatistics. Springer, London (2011). ISBN: 978-0-85729-354-1
10. Zhang, J.: Nonparametric information geometry: from divergence function to referential-representational biduality on statistical manifolds. Entropy **15**, 5384–5418 (2013)
11. Naudts, J., Anthonis, B.: The exponential family in abstract information theory. In: Nielsen, F., Barbaresco, F. (eds.) GSI 2013. LNCS, vol. 8085, pp. 265–272. Springer, Heidelberg (2013). doi:10.1007/978-3-642-40020-9_28
12. Zhang, J.: On monotone embedding in information geometry. Entropy **17**, 4485–4499 (2015)
13. Newton, N.J.: Information geometric nonlinear filtering. Inf. Dim. Anal. Quantum Prob. Rel. Topics **18**, 1550014 (2015)
14. Zhou, J.: Information theory and statistical mechanics revisited, arXiv:1604.08739

Information Geometry Under Monotone Embedding. Part II: Geometry

Jan Naudts[1(✉)] and Jun Zhang[2]

[1] Physics Department, Universiteit Antwerpen, Universiteitsplein 1, 2610 Antwerpen, Belgium
jan.naudts@uantwerpen.be
[2] University of Michigan, Ann Arbor, MI, USA
junz@umich.edu

Abstract. The rho-tau embedding of a parametric statistical model defines both a Riemannian metric, called "rho-tau metric", and an alpha family of rho-tau connections. We give a set of equivalent conditions for such a metric to become Hessian and for the ± 1-connections to be dually flat. Next we argue that for any choice of strictly increasing functions $\rho(u)$ and $\tau(u)$ one can construct a statistical model which is Hessian and phi-exponential. The metric derived from the escort expectations is conformally equivalent with the rho-tau metric.

Keywords: Hessian geometry · Dually-flat · rho-tau embedding · phi-exponential family · Escort probability

1 Introduction

Amari [1,2] introduced the alpha family of connections $\Gamma^{(\alpha)}$ for a statistical model belonging to the exponential family. He showed that $\Gamma^{(\alpha)}$ and $\Gamma^{(-\alpha)}$ are each others dual and that for $\alpha = \pm 1$ the corresponding geometries are flat. Both the notions of an alpha family of connections and that of an exponential family of statistical models have been generalized. The present paper combines two general settings, that of the alpha family of connections determined by rho-tau embeddings [3] and that of phi-deformed exponential families [4].

Let $\mathcal{M}$ denote the space of probability density functions over the measure space $(\mathcal{X}, \mathrm{d}x)$. A parametric model p^θ is a map from some open domain in $\mathbb{R}^n$ into $\mathcal{M}$. It becomes a parametric statistical model if $\theta \to p^\theta$ is a Riemannian manifold with metric tensor $g(\theta)$.

Throughout the paper it is assumed that two strictly increasing functions ρ and τ are given. The rho-tau divergence (see Part I) induces a metric tensor g on finite-dimensional manifolds of probability distributions and makes them into Riemannian manifolds.

J. Naudts and J. Zhang contributed equally to this paper.

F. Nielsen and F. Barbaresco (Eds.): GSI 2017, LNCS 10589, pp. 215–222, 2017.
https://doi.org/10.1007/978-3-319-68445-1_25

2 The Metric Tensor

The rho-tau divergence $D_{\rho,\tau}(p,q)$ can be used [3,5,6] to define a metric tensor $g(\theta)$ by

$$g_{i,j}(\theta) = \partial_j \partial_i D_{\rho,\tau}(p, p^\theta)\Big|_{p=p^\theta}, \tag{1}$$

with $\partial_i = \partial/\partial\theta^i$. A short calculation gives

$$g_{ij}(\theta) = \int_{\mathcal{X}} \mathrm{d}x \, \left[\partial_i \tau(p^\theta(x))\right] \left[\partial_j \rho(p^\theta(x))\right]. \tag{2}$$

Because $\tau = f' \circ \rho$, the rho-tau metric $g(\theta)$ also takes the form:

$$\begin{aligned} g_{ij}(\theta) &= \int_{\mathcal{X}} \mathrm{d}x \, \left[\partial_i f'(\rho(p^\theta(x)))\right] \left[\partial_j \rho(p^\theta(x))\right] \\ &= \int_{\mathcal{X}} \mathrm{d}x \, f''(\rho(p^\theta(x))) \left[\partial_i \rho(p^\theta(x))\right] \left[\partial_j \rho(p^\theta(x))\right]. \end{aligned}$$

This shows that the matrix $g(\theta)$ is symmetric. Moreover, it is positive-definite, because the derivatives ρ' and f'' are strictly positive and the matrix with components $\left(\partial_j p^\theta(x)\right)\left(\partial_i p^\theta(x)\right)$ has eigenvalues 0 and 1 (assuming $\theta \to p^\theta$ has no stationary points). Finally, $g(\theta)$ is covariant, so g is indeed a metric tensor on the Riemannian manifold p^θ. From (2) follows that it is invariant under the exchange of ρ and τ.

The rho-tau entropy $S_{\rho,\tau}$ of the parametric family p^θ can be written as

$$S_{\rho,\tau}(p^\theta) = -\int_{\mathcal{X}} \mathrm{d}x \, f(\rho(p^\theta(x))). \tag{3}$$

So its second derivative

$$h_{ij}(\theta) = -\partial_i \partial_j S_{\rho,\tau}(p^\theta)$$

is symmetric in i, j. When positive-definite, $h(\theta)$ can also serve as a metric tensor as is found sometimes in the Physics literature.

Note that $h(\theta)$ differs from $g(\theta)$ in general: the former is induced by the entropy function $S_{\rho,\tau}(p)$, whose definition depends on the single function $f \circ \rho$, the latter is derived from the function $D_{\rho,\tau}(p,q)$.

3 Gauge Freedom

Write the rho-tau metric g_{ij} as

$$g_{ij}(\theta) = \int_{\mathcal{X}} \mathrm{d}x \, \frac{1}{\phi(p^\theta)} \left[\partial_i p^\theta(x)\right] \left[\partial_j p^\theta(x)\right], \tag{4}$$

where $\phi(u) = 1/(\rho'(u)\tau'(u))$. So despite of the two independent choices of embedding functions ρ and τ, the metric tensor g_{ij} is determined by one function ϕ only. More remarkably,

$$\begin{aligned} g_{ij}(\theta) &= \int_{\mathcal{X}} \mathrm{d}x\, f''(\rho(p^\theta(x))) \left[\partial_i(\rho(p^\theta(x))\right] \left[\partial_j \rho(p^\theta(x))\right] \\ &= \int_{\mathcal{X}} \mathrm{d}x\, (f^*)''(\tau(p^\theta(x))) \left[\partial_i \tau(p^\theta(x))\right] \left[\partial_j \tau(p^\theta(x))\right], \end{aligned}$$

so the gauge freedom in g_{ij} exists independent of the embedding – there is freedom in choosing an arbitrary function f in the case of the ρ-embedding and an arbitrary function f^* in the case of the τ-embedding of p^θ.

Without loss of generality, we choose τ-embedding and denote $X^\theta(x) = \tau(p^\theta(x))$. From the form of the rho-tau metric

$$g_{ij}(\theta) = \int_{\mathcal{X}} \mathrm{d}x\, \frac{\rho'(p^\theta(x))}{\tau'(p^\theta(x))} \left[\partial_i \tau(p^\theta(x))\right] \left[\partial_j \tau(p^\theta(x))\right],$$

we introduce a bilinear form $\langle \cdot, \cdot \rangle$ defined on pairs of random variables $u(x), v(x)$

$$\langle u, v \rangle_\theta = \int_{\mathcal{X}} \mathrm{d}x\, \frac{\rho'(p^\theta(x))}{\tau'(p^\theta(x))} u(x)\, v(x).$$

For any random variable u it holds that

$$\partial_j \int_{\mathcal{X}} \mathrm{d}x\, \rho(p^\theta(x)) u(x) = \int_{\mathcal{X}} \mathrm{d}x\, \frac{\rho'(p^\theta(x))}{\tau'(p^\theta(x))} \partial_j \tau(p^\theta(x)) u(x) = \langle \partial_j X^\theta, u \rangle_\theta$$

Following [2], $\partial_j X^\theta$ is then, by definition, tangent to the rho-representation $\rho(p^\theta)$ of the model p^θ. We also have

$$-\partial_j S_{\rho,\tau}(p^\theta) = \langle \partial_j X^\theta, X^\theta \rangle_\theta. \tag{5}$$

The difference of the metrics $g(\theta)$ and $h(\theta)$ can be readily appreciated:

$$g_{ij}(\theta) = \langle \partial_j X^\theta, \partial_i X^\theta \rangle_\theta$$

whereas

$$\begin{aligned} h_{ij}(\theta) = -\partial_i \partial_j S_{\rho,\tau}(p^\theta) &= \partial_i \langle \partial_j X^\theta, X^\theta \rangle_\theta \\ &= g_{ij}(\theta) + \int_{\mathcal{X}} \mathrm{d}x\, \tau(p^\theta(x)) \partial_i \partial_j \rho(p^\theta(x)). \end{aligned} \tag{6}$$

4 The Hessian Case

We now consider the condition under which the rho-tau metric g becomes Hessian.

Theorem 1. *Let be given a C^∞-manifold of probability distributions p^θ. For fixed strictly increasing functions ρ and τ, let the metric tensor $g(\theta)$ be given by (2). Then the following statements are equivalent:*

(i) g is Hessian, i.e., there exists $\Phi(\theta)$ such that

$$g_{ij}(\theta) = \partial_i\partial_j\Phi(\theta).$$

(ii) There exists a function $V(\theta)$ such that

$$\frac{\partial^2 V}{\partial\theta^i\partial\theta^j} = -\int_{\mathcal{X}} \mathrm{d}x\, \tau(p^\theta(x))\partial_i\partial_j\rho(p^\theta(x)). \tag{7}$$

(iii) There exists a function $W(\theta)$ such that

$$\frac{\partial^2 W}{\partial\theta^i\partial\theta^j} = -\int_{\mathcal{X}} \mathrm{d}x\, \rho(p^\theta(x))\partial_i\partial_j\tau(p^\theta(x)). \tag{8}$$

(iv) There exist coordinates $\eta_i(\theta)$ for which

$$g_{ij}(\theta) = \partial_j\eta_i.$$

(v) There exist coordinates ξ_i such that

$$\partial_j\xi_i(\theta) = -\int_{\mathcal{X}} \mathrm{d}x\, \tau(p^\theta(x))\partial_i\partial_j\rho(p^\theta(x)). \tag{9}$$

(vi) There exist coordinates ζ_i such that

$$\partial_j\zeta_i(\theta) = -\int_{\mathcal{X}} \mathrm{d}x\, \rho(p^\theta(x))\partial_i\partial_j\tau(p^\theta(x)). \tag{10}$$

Proof.
(i) $\longleftrightarrow$ (iv) This is well-known: the existence of a strictly convex function Φ is equivalent to the existence of dual coordinates η_i.
(ii) $\longleftrightarrow$ (v) From (ii) to (v): Given the existence of $V(\theta)$ satisfying (7), choose $\xi_i = \partial_i V$, and (9) is satisfied. From (v) to (ii): Since the right-hand side of (9) is symmetric with respect to i, j, we have $\partial_j\xi_i = \partial_i\xi_j$. Hence there exists a function $V(\theta)$ such that $\xi_i = \partial_i V$; this is the V function satisfying (7).
(iii) $\longleftrightarrow$ (vi) The proof is similar to the previous paragraph, by simply changing V to W and ξ to ζ.
(i) $\longleftrightarrow$ (ii) From the identity (6), the existence of $\Phi(\theta)$ to represent g_{ij} as its second derivatives allows us to choose the function V as $V = \Phi + S$. So from (i) we obtain (ii). Conversely when the integral term can be represented by the second derivative of $V(\theta)$, we can choose $\Phi = V - S$ that would satisfy (6). This yields (i) from (ii).
(i) $\longleftrightarrow$ (iii) The proof is similar to that of the previous paragraph, except that we will invoke the following identity instead of (6):

$$-\partial_i\partial_j S^*_{\rho,\tau}(p^\theta) = g_{ij}(\theta) + \int_{\mathcal{X}} \mathrm{d}x\, \rho(p^\theta(x))\partial_i\partial_j\tau(p^\theta(x)).$$

□

The case when g is Hessian is very special, because of the existence of various bi-orthogonal coordinates.

The η_i are the dual coordinates of the θ^i. The ζ_i are called *escort coordinates.* They are linked to η_i by

$$\zeta_i = -\int_{\mathcal{X}} \mathrm{d}x\, \rho(p^\theta(x))\partial_i\tau(p^\theta(x)) + \eta_i = \partial_i S^*_{\rho,\tau}(p^\theta) + \eta_i. \tag{11}$$

They satisfy

$$\partial_j\partial_k\zeta_i = -\langle\partial_k X^\theta, \partial_i\partial_j X^\theta\rangle.$$

The *dual escort coordinates* ξ_i are given by

$$\xi_j(\theta) = \partial_j S_{\rho,\tau}(p^\theta) + \eta_j. \tag{12}$$

The Hessian of the function $V(\theta)$, when it does not vanish, causes a discrepancy between a metric tensor h defined as minus the Hessian of the entropy and the metric tensor g as defined by (2).

5 Zhang's rho-tau Connections

Given a pair of strictly increasing functions ρ and τ and a model p^θ, Zhang introduced the following connections [3]

$$\begin{aligned}\Gamma^{(\alpha)}_{ij,k} &= \frac{1+\alpha}{2}\int_{\mathcal{X}} \mathrm{d}x\, \left[\partial_i\partial_j\rho(p^\theta(x))\right]\left[\partial_k\tau(p^\theta(x))\right]\\ &+\frac{1-\alpha}{2}\int_{\mathcal{X}} \mathrm{d}x\, \left[\partial_i\partial_j\tau(p^\theta(x))\right]\left[\partial_k\rho(p^\theta(x))\right],\end{aligned} \tag{13}$$

where $\Gamma^{(\alpha)}_{ij,k} \equiv (\Gamma^{(\alpha)})^l_{ij} g_{lk}$. One readily verifies

$$\Gamma^{(\alpha)}_{ij,k} + \Gamma^{(-\alpha)}_{jk,i} = \partial_i g_{jk}(\theta). \tag{14}$$

This shows that, by definition, $\Gamma^{(-\alpha)}$ is the dual connection of $\Gamma^{(\alpha)}$.

The coefficients of the connection $\Gamma^{(-1)}$ vanish identically if

$$\int_{\mathcal{X}} \mathrm{d}x\, \left[\partial_i\partial_j\tau(p^\theta(x))\right]\left[\partial_k\rho(p^\theta(x))\right] = 0. \tag{15}$$

This condition can be written as

$$\partial_j\partial_k\zeta_i = -\langle\partial_i\partial_j X^\theta, \partial_k X^\theta\rangle_\theta = 0. \tag{16}$$

It states that the escort coordinates are affine functions of θ and expresses that the second derivatives $\partial_i\partial_j X^\theta$ are orthogonal to the tangent plane of the statistical manifold. If satisfied then the dual of $\Gamma^{(-1)}$ satisfies

$$\Gamma^{(1)}_{ij,k} = \partial_i g_{jk}(\theta). \tag{17}$$

Likewise, the coefficients of the connection $\Gamma^{(1)}$ vanish identically if

$$\int_{\mathcal{X}} \mathrm{d}x \, \left[\partial_i\partial_j\rho(p^\theta(x))\right] \left[\partial_k\tau(p^\theta(x))\right] = 0. \tag{18}$$

Proposition 1. *With respect to conditions (15) and (18),*

1. *When (15) holds, the coordinates θ^i are affine coordinates for $\Gamma^{(-1)}$; the dual coordinates η_i are affine coordinates for $\Gamma^{(1)}$;*
2. *When (18) holds, the coordinates θ^i are affine coordinates for $\Gamma^{(1)}$; the dual coordinates η_i are affine coordinates for $\Gamma^{(-1)}$;*
3. *In either case above, $g(\theta)$ is Hessian.*

Proof.

One recalls that when $\Gamma = 0$ under a coordinate system θ, then θ^i's are affine coordinates – the geodesics are straight lines:

$$\theta(t) = (1-t)\theta_{(t=1)} + t\theta_{(t=0)}.$$

The geodesics of the dual connection Γ^* satisfies the Euler-Lagrange equations

$$\frac{\mathrm{d}^2}{\mathrm{d}t^2}\theta^i + \Gamma^i_{km}\left(\frac{\mathrm{d}}{\mathrm{d}t}\theta^k\right)\left(\frac{\mathrm{d}}{\mathrm{d}t}\theta^m\right) = 0. \tag{19}$$

Its solution is such that the dual coordinates η are affine coordinates:

$$\eta(t) = (1-t)\eta_{(t=1)} + t\eta_{(t=0)}.$$

For Statement 1, we apply the above knowledge, taking $\Gamma = \Gamma^{(-1)}$ and $\Gamma^* = \Gamma^{(1)}$; for Statement 2, taking $\Gamma = \Gamma^{(1)}$ and $\Gamma^* = \Gamma^{(-1)}$.

To prove Statement 3 observe that

$$\partial_k g_{ij}(\theta) = \int_{\mathcal{X}} \mathrm{d}x \, \left[\partial_i\tau(p^\theta(x))\right] \partial_j\partial_k\rho(p^\theta(x)) + \int_{\mathcal{X}} \mathrm{d}x \, \left[\partial_j\rho(p^\theta(x))\right] \partial_i\partial_k\tau(p^\theta(x)).$$

So the vanishing of either term, i.e., either (15) or (18) holding, will lead $\partial_k g_{ij}(\theta)$ to be symmetric in j, k or in i, k, respectively. This, in conjunction with the fact that g_{ij} is symmetric in i, j, leads to the conclusion that $\partial_k g_{ij}(\theta)$ is totally symmetric in an exchange of any two of the three indices i, j, k. This implies that η_i exist for which $g_{ij}(\theta) = \partial_j\eta_i$. That g is Hessian follows now from Theorem 1.

□

6 Rho-tau Embedding of phi-exponential Models

Let $\phi(u) = 1/(\rho'(u)\tau'(u))$ as before and fix real random variables $F_1, F_2, \cdots, F_n$. These functions determine a phi-exponential family $\theta \to p^\theta$ by the relation (see [4,7,8])

$$p^\theta(x) = \exp_\phi\left[\theta^k F_k(x) - \alpha(\theta)\right]. \tag{20}$$

The function $\alpha(\theta)$ is determined by the requirement that p^θ is a probability distribution and must be normalized to 1.

Assume that the integral

$$z(\theta) = \int_{\mathcal{X}} \mathrm{d}x\, \phi(p^\theta(x))$$

converges. Then the *escort family* of probability distributions $\tilde{p}^\theta$ is defined by

$$\tilde{p}^\theta(x) = \frac{1}{z(\theta)}\phi(p^\theta(x)).$$

The corresponding escort expectation is denoted $\tilde{\mathbb{E}}_\theta$. From the normalization of the p^θ follows that $\partial_i\alpha(\theta) = \tilde{\mathbb{E}}_\theta F_i$. Now calculate, starting from (4),

$$\begin{aligned} g_{ij}(\theta) &= \int_{\mathcal{X}} \mathrm{d}x\, \frac{1}{\phi(p^\theta(x))}\left[\partial_i p^\theta(x)\right]\left[\partial_j p^\theta(x)\right] \\ &= \int_{\mathcal{X}} \mathrm{d}x\, \phi(p^\theta(x))\left[F_i - \partial_i\alpha(\theta)\right]\left[F_j - \partial_j\alpha(\theta)\right] \\ &= z(\theta)\left[\tilde{\mathbb{E}}_\theta F_i F_j - \tilde{\mathbb{E}}_\theta F_i \tilde{\mathbb{E}}_\theta F_j\right]. \end{aligned} \tag{21}$$

The latter expression is the metric tensor of the phi-exponential model as introduced in [4]. It implies that the rho-tau metric tensor is conformally equivalent with the metric tensor as derived from the escort expectation of the random variables F_i.

Finally, let $\eta_i = \mathbb{E}_\theta F_i$. A short calculation shows that

$$\begin{aligned} \partial_j\eta_i &= \int_{\mathcal{X}} \mathrm{d}x\, \phi(p^\theta(x))\left[F_j - \partial_j\alpha(\theta)\right] F_i \\ &= z(\theta)\left[\tilde{\mathbb{E}}_\theta F_i F_j - \tilde{\mathbb{E}}_\theta F_i \tilde{\mathbb{E}}_\theta F_j\right] \\ &= g_{ij}(\theta). \end{aligned} \tag{22}$$

By (iv) of Theorem 1 this implies that the metric tensor g_{ij} is Hessian. Note that the η_i are dual coordinates. As defined here, they only depend on ϕ and not on the particular choice of embeddings ρ and τ. In particular, also the dually flat geometry does not depend on it.

One concludes that for any choice of strictly increasing functions $\rho(u)$ and $\tau(u)$ one can always construct statistical models for which the rho-tau metric is Hessian. These are phi-exponential models, with ϕ given by $\phi(u) = 1/\rho'(u)\tau'(u)$.

Conversely, given a phi-exponential model, its metric tensor is always a rho-tau metric tensor, with ρ, τ subject to the condition that $\rho'(u)\tau'(u) = 1/\phi(u)$. Two special cases are that either ρ or τ is the identity map, with the other being identified as the $\log_\phi$ function.

In the terminology of Zhang [3] the models of the phi-exponential family are called ρ-affine models where the normalization condition is, however, not imposed.

7 Discussion

This paper studies parametrized statistical models p^θ and the geometry induced on them by the choice of a pair of strictly increasing functions ρ and τ.

Theorem 1 gives equivalent conditions for the metric to be Hessian. It is shown that for the existence of a dually flat geometry the metric has to be Hessian.

The rho-tau metric tensor depends on a single function ϕ which is defined by $\phi(u) = 1/(\rho'(u)\tau'(u))$. If the model is phi-exponential for the same function ϕ then the rho-tau metric coincides with the metric used in the context of phi-exponential families and in particular the metric is Hessian. This shows that it is always possible to construct models which are Hessian for the given rho-tau metric.

Acknowledgement. The second author is supported by DARPA/ARO Grant W911NF-16-1-0383.

References

1. Amari, S.: Differential-Geometric Methods in Statistics. LNS, vol. 28. Springer, Heidelberg (1985). doi:10.1007/978-1-4612-5056-2
2. Amari, S., Nagaoka, H.: Methods of Information Geometry. Translations of Mathematical Monographs, vol. 191. Am. Math. Soc., Oxford University Press (2000). Originally in Japanese (Iwanami Shoten, Tokyo, 1993)
3. Zhang, J.: Divergence function, duality, and convex analysis. Neural Comput. **16**, 159–195 (2004)
4. Naudts, J.: Estimators, escort probabilities, and phi-exponential families in statistical physics. J. Ineq. Pure Appl. Math. **5**, 102 (2004)
5. Zhang, J.: Nonparametric information geometry: from divergence function to referential-representational biduality on statistical manifolds. Entropy **15**, 5384–5418 (2013)
6. Zhang, J.: On monotone embedding in information geometry. Entropy **17**, 4485–4499 (2015)
7. Naudts, J.: Generalised exponential families and associated entropy functions. Entropy **10**, 131–149 (2008)
8. Naudts, J.: Generalised Thermostatistics. Springer, London (2011). doi:10.1007/978-0-85729-355-8

A Sequential Structure of Statistical Manifolds on Deformed Exponential Family

Hiroshi Matsuzoe[1(✉)], Antonio M. Scarfone[2], and Tatsuaki Wada[3]

[1] Department of Computer Science and Engineering, Graduate School of Engineering, Nagoya Institute of Technology, Gokiso-cho, Showa-ku, Nagoya 466-8555, Japan
matsuzoe@nitech.ac.jp
[2] Istituto dei Sistemi Complessi (ISC-CNR) c/o, Politecnico di Torino, Corso Duca degli Abruzzi 24, 10129 Torino, Italy
[3] Department of Electrical and Electronic Engineering, Ibaraki University, Nakanarusawa-cho, Hitachi 316-8511, Japan

Abstract. Heavily tailed probability distributions are important objects in anomalous statistical physics. For such probability distributions, expectations do not exist in general. Therefore, an escort distribution and an escort expectation have been introduced. In this paper, by generalizing such escort distributions, a sequence of escort distributions is introduced. For a deformed exponential family, we study the fundamental properties of statistical manifold structures derived from the sequence of escort expectations.

Keywords: Statistical manifold · Escort distribution · Escort expectation · Deformed exponential family · Information geometry

1 Introduction

Heavily tailed probability distributions are important objects in anomalous statistical physics (cf. [11,15]). Such probability distributions do not have expectations in general. Therefore the notion of escort distribution has been introduced [4] in order to give a suitable down weight for heavy tail probability. Consequently, there exists a modified expectation for such a probability distributions.

For a deformed exponential family, an escort distribution is given by the differential of a deformed exponential function. Therefore, the first named author considered further generalizations of escort distributions In q-exponential case, he introduced a sequential structure of escort distributions [7].

In this paper, we consider a sequential structure of escort distributions on a deformed exponential family. It is known that a deformed exponential family naturally has at least three kinds of different statistical manifold structures [8]. We elucidate relations between these statistical manifold structures and the

H. Matsuzoe—This research was partially supported by JSPS (Japan Society for the Promotion of Science), KAKENHI (Grants-in-Aid for Scientific Research) Grant Numbers JP26108003, JP15K04842 and JP16KT0132.

F. Nielsen and F. Barbaresco (Eds.): GSI 2017, LNCS 10589, pp. 223–230, 2017.
https://doi.org/10.1007/978-3-319-68445-1_26

structures derived from the sequence of escort expectations. Consequently, we find that dually flat structures and generalized conformal structures for statistical manifolds naturally arise in this framework.

2 Deformed Exponential Families

Throughout this paper, we assume that all the objects are smooth. In this section, we summarize foundations of deformed exponential functions and deformed exponential families. For further details, see [11].

Let χ be a strictly increasing function from $\mathbf{R}_{++}$ to $\mathbf{R}_{++}$. We call this function χ a *deformation function*. By use of a deformation function, we define a *χ-exponential function* $\exp_\chi t$ (or a *deformed exponential function*) by the eigenfunction of the following non-linear differential equation

$$\frac{d}{dt}\exp_\chi t = \chi(\exp_\chi t).$$

The inverse of a χ-exponential function is called a *χ-logarithm function* or a *deformed logarithm function*, and it is given by

$$\ln_\chi s := \int_1^s \frac{1}{\chi(t)}dt.$$

If the deformation function is a power function $\chi(t) = t^q$ $(q > 0, q \neq 1)$, the deformed exponential and the deformed logarithm are given by

$$\exp_q t := (1 + (1-q)t)^{\frac{1}{1-q}}, \qquad (1 + (1-q)t > 0),$$

$$\ln_q s := \frac{s^{1-q} - 1}{1-q}, \qquad (s > 0),$$

and they are called a *q-exponential* and a *q-logarithm*, respectively.

We suppose that a statistical model S_χ has the following expression

$$S_\chi = \left\{ p(x,\theta) \;\middle|\; p(x;\theta) = \exp_\chi\left[\sum_{i=1}^n \theta^i F_i(x) - \psi(\theta)\right],\ \theta \in \Theta \subset \mathbf{R}^n \right\},$$

where $F_1(x), \ldots, F_n(x)$ are functions on the sample space Ω, $\theta = {}^t(\theta^1, \ldots, \theta^n)$ is a parameter, and $\psi(\theta)$ is the normalization defined by $\int_\Omega p(x;\theta)dx = 1$. We call the statistical model S_χ a *χ-exponential family* or a *deformed exponential family*. Under suitable conditions, S_χ is regarded as a manifold with coordinate system $\theta = (\theta^1, \ldots, \theta^n)$. When the deformed exponential function is a q-exponential, we denote the statistical model by S_q and call it a *q-exponential family*.

We remark that the regularity conditions for S_χ is very difficult. To elucidate such conditions is quite an open problem. For example, regularity conditions for a statistical model (see Chap. 2 in [1]) and the well-definedness of a deformed exponential function should be satisfied simultaneously. A few arguments of this problem is given in the first and the third named author's previous work [9].

3 A Sequential Structure of Expectations

In this section we consider a sequential structure of expectations. As we will see later, statistical manifold structures are defined from this sequence.

Let $S_\chi = \{p_\theta\} = \{p(x;\theta)\}$ be a χ-exponential family. We say that $P_\chi(x;\theta)$ is an *escort distribution* of $p_\theta \in S_\chi$ if

$$P_\chi(x;\theta) := P_{\chi,(1)}(x;\theta) := \chi(p_\theta).$$

We say that $P_\chi^{esc}(x;\theta)$ is a *normalized escort distribution* of p_θ if

$$P_\chi^{esc}(x;\theta) := P_{\chi,(1)}^{esc}(x;\theta) := \frac{\chi(p_\theta)}{Z_\chi(p_\theta)},$$

$$\text{where} \quad Z_\chi(p_\theta) := Z_{\chi,(1)}(p_\theta) := \int_\Omega \chi(p_\theta)dx.$$

We generalize the escort distribution by use of higher-order differentials.

Definition 1. *Let S_χ be a χ-exponential family. Denote by $\exp_\chi^{(n)} x$ the n-th differential of the χ-exponential function. For $p_\theta \in S_\chi$, we define the* n-th escort distribution $P_{\chi,(n)}(x;\theta)$ *by*

$$P_{\chi,(n)}(x;\theta) := \exp_\chi^{(n)}(\ln_\chi p_\theta) = \exp_\chi^{(n)}\left(\sum_{i=1}^n \theta^i F_i(x) - \psi(\theta)\right),$$

and the normalized n-th escort distribution $P_{\chi,(n)}^{esc}(x;\theta)$ *by*

$$P_{\chi,(n)}^{esc}(x;\theta) := \frac{P_{\chi,(n)}(x;\theta)}{Z_{\chi,(n)}(p_\theta)}, \quad \textit{where} \quad Z_{\chi,(n)}(p_\theta) = \int_\Omega P_{\chi,(n)}(x;\theta)dx.$$

For a given function $f(x)$ on Ω, we define the n-th escort expectation *of $f(x)$ and the* normalized n-th escort expectation *of $f(x)$ by*

$$E_{\chi,(n),p}[f(x)] := \int_\Omega f(x)P_{\chi,(n)}(x;\theta)dx,$$

$$E_{\chi,(n),p}^{esc}[f(x)] := \int_\Omega f(x)P_{\chi,(n)}^{esc}(x;\theta)dx,$$

respectively.

For example, in the case of q-exponential family S_q, the n-th escort distribution of $p_q(x;\theta)$ is given by

$$P_{q,(n)}(x;\theta) := \{q(2q-1)\cdots((n-1)q-(n-2))\}\{p_q(x;\theta)\}^{nq-(n-1)}.$$

When we consider geometric structure determined from the unbiasedness of generalized score function, that is,

$$E_{\chi,(1),p}[\partial_i \ln_\chi p(x;\theta)] = 0,$$

a sequential structure of expectations naturally arises. This is one of our motivations to study sequential expectations. When we consider correlations of random variables, another kinds of sequence of expectations will be required.

4 Geometry of Statistical Models

Let (M, g) be a Riemannian manifold, and C be a totally symmetric $(0,3)$-tensor field on M. We call the triplet (M, g, C) a *statistical manifold* [6]. In this case, the tensor field C is called a *cubic form*. For a given statistical manifold (M, g, C), we can define one parameter family of affine connections by

$$g(\nabla_X^{(\alpha)} Y, Z) := g(\nabla_X^{(0)} Y, Z) - \frac{\alpha}{2} C(X, Y, Z), \tag{1}$$

where $\alpha \in \mathbf{R}$ and $\nabla^{(0)}$ is the Levi-Civita connection with respect to g. It is easy to check that $\nabla^{(\alpha)}$ and $\nabla^{(-\alpha)}$ are mutually dual with respect to g, that is,

$$Xg(Y, Z) = g(\nabla_X^{(\alpha)} Y, Z) + g(Y, \nabla_X^{(-\alpha)} Z).$$

We say that S is a *statistical model* if S is a set of probability density functions on Ω with parameter $\xi \in \Xi$ such that

$$S = \left\{ p(x;\xi) \,\middle|\, \int_\Omega p(x;\xi)dx = 1,\ p(x;\xi) > 0,\ \xi = (\xi^1, \ldots, \xi^n) \in \Xi \subset \mathbf{R}^n \right\}.$$

Under suitable conditions, we can define a *Fisher metric* g^F on S by

$$\begin{aligned} g_{ij}^F(\xi) &= \int_\Omega \left(\frac{\partial}{\partial \xi^i} \ln p(x;\xi) \right) \left(\frac{\partial}{\partial \xi^j} \ln p(x;\xi) \right) p(x;\xi)\, dx && (2) \\ &= \int_\Omega \left(\frac{\partial}{\partial \xi^i} \ln p(x;\xi) \right) \left(\frac{\partial}{\partial \xi^j} p(x;\xi) \right) dx && (3) \\ &= E_p[\partial_i l_\xi \partial_j l_\xi], \end{aligned}$$

where $\partial_i = \partial / \partial \xi^i$, $l_\xi = l(x;\xi) = \ln p(x;\xi)$, and $E_p[f]$ is the standard expectation of $f(x)$ with respect to $p(x;\xi)$.

Next, we define a totally symmetric $(0,3)$-tensor field C^F by

$$C_{ijk}^F(\xi) = E_p\left[(\partial_i l_\xi)(\partial_j l_\xi)(\partial_k l_\xi) \right].$$

From Eq. (1), we can define one parameter family of affine connections. In particular, the connection $\nabla^{(e)} = \nabla^{(1)}$ is called the*exponential connection* and $\nabla^{(m)} = \nabla^{(-1)}$ is called the *mixture connection*. These connections are given by

$$\Gamma_{ij,k}^{(e)}(\xi) = \int_\Omega (\partial_i \partial_j \ln p_\xi)(\partial_k p_\xi) dx,$$
$$\Gamma_{ij,k}^{(m)}(\xi) = \int_\Omega (\partial_k \ln p_\xi)(\partial_i \partial_j p_\xi) dx.$$

It is known that g^F and C^F are independent of the choice of reference measure on Ω. Therefore, the triplet (S, g^F, C^F) is called an *invariant statistical manifold*. If a statistical model S is an exponential family, then the invariant statistical manifold (S, g^F, C^F) determines a dually flat structure on S.

(See [1,13].) However, this fact may not be held for a deformed exponential family S_χ and an invariant structure may not be important for S_χ. Therefore, we consider another statistical manifold structures.

We summarize statistical manifold structures for S_χ based on [8].

Let S_χ be a χ-exponential family. We define a Riemannian metric g^M by

$$g_{ij}^M(\theta) := \int_\Omega (\partial_i \ln_\chi p_\theta)(\partial_j p_\theta)\, dx,$$

where $\partial_i = \partial/\partial\theta^i$. The Riemannian metric g^M is a generalization of the representation of Fisher metric (3). A pair of dual affine connections are given by

$$\Gamma_{ij,k}^{M(e)}(\theta) = \int_\Omega (\partial_i\partial_j \ln_\chi p_\theta)(\partial_k p_\theta) dx,$$
$$\Gamma_{ij,k}^{M(m)}(\theta) = \int_\Omega (\partial_k \ln_\chi p_\theta)(\partial_i\partial_j p_\theta) dx.$$

The difference of two affine connections $C_{ijk}^M = \Gamma_{ij,k}^{M(m)} - \Gamma_{ij,k}^{M(e)}$ determines a cubic form. In addition, from the definition of the deformed exponential family S_χ, $\Gamma_{ij,k}^{M(e)}(\theta)$ always vanishes. Therefore, we have the following proposition.

Proposition 1. *For a χ-exponential family S_χ, the triplet (S_χ, g^M, C^M) is a statistical manifold. In particular, $(S_\chi, g^M, \nabla^{M(e)}, \nabla^{M(m)})$ is a dually flat space.*

By setting

$$U_\chi(s) := \int_0^s (\exp_\chi t)\, dt,$$

we define a *U-divergence* [10] by

$$D_\chi(p||r) = \int_\Omega \{U_\chi(\ln_\chi r(x)) - U_\chi(\ln_\chi p(x)) - p(x)(\ln_\chi r(x) - \ln_\chi p(x))\} dx.$$

It is known that the U-divergence $D_\chi(p||r)$ on S_χ coincides with the canonical divergence for $(S_\chi, g^M, \nabla^{M(m)}, \nabla^{M(e)})$ (See [8,10]).

Next, we define another statistical manifold structure from the viewpoint of Hessian geometry.

For a χ-exponential family S_χ, suppose that the normalization ψ is strictly convex. Then we can define a *χ-Fisher metric* g^χ and a *χ-cubic form* C^χ [3] by

$$g_{ij}^\chi(\theta) := \partial_i\partial_j\psi(\theta),$$
$$C_{ijk}^\chi(\theta) := \partial_i\partial_j\partial_k\psi(\theta).$$

Obviously, the triplet (S_χ, g^χ, C^χ) is a statistical manifold. From Eq. (1), we can define a torsion-free affine connection $\nabla^{\chi(\alpha)}$ by

$$g^\chi(\nabla_X^{\chi(\alpha)}Y, Z) := g^\chi(\nabla_X^{\chi(0)}Y, Z) - \frac{\alpha}{2}C^\chi(X, Y, Z),$$

where $\nabla^{\chi(0)}$ is the Levi-Civita connection with respect to g^χ. By standard arguments in Hessian geometry [13], $(S_\chi, g^\chi, \nabla^{\chi(1)}, \nabla^{\chi(-1)})$ is a dually flat space. The canonical divergence for $(S_\chi, g^\chi, \nabla^{\chi(-1)}, \nabla^{\chi(1)})$ is given by

$$D^\chi(p||r) = E_{\chi,r}^{esc}[\ln_\chi r(x) - \ln_\chi p(x)].$$

5 Statistical Manifolds Determined from Sequential Escort Expectations

In this section, we consider statistical manifold structures determined from sequential escort expectations.

For a χ-exponential family S_χ, we define $g^{(n)}$ and $C^{(n)}$ by

$$g_{ij}^{(n)}(\theta) := \int_\Omega (\partial_i \ln_\chi p_\theta)(\partial_j \ln_\chi p_\theta) P_{\chi,(n)}(x;\theta)dx,$$

$$C_{ijk}^{(n)}(\theta) := \int_\Omega (\partial_i \ln_\chi p_\theta)(\partial_j \ln_\chi p_\theta)(\partial_k \ln_\chi p_\theta) P_{\chi,(n+1)}(x;\theta)dx.$$

We suppose that $g^{(n)}$ is a Riemannian metric on S_χ. Then we obtain a sequence of statistical manifolds:

$$(S_\chi, g^{(1)}, C^{(1)}) \rightarrow (S_\chi, g^{(2)}, C^{(2)}) \rightarrow \cdots \rightarrow (S_\chi, g^{(n)}, C^{(n)}) \rightarrow \cdots.$$

The limit of this sequence is not clear at this moment. In the q-Gaussian case, the sequence of normalized escort distributions $\{P_{q,(n)}^{esc}(x;\theta)\}$ converges to the Dirac's delta function $\delta(x-\mu)$ (cf. [14]).

Theorem 1. *Let $S_q = \{p(x;\theta)\}$ be a χ-exponential family. Then $(S_\chi, g^{(1)}, C^{(1)})$ coincides with (S_χ, g^M, C^M).*

Proof. From the definition of χ-logarithm and $P_\chi(x;\theta) = P_{\chi,(1)}(x;\theta) = \chi(p_\theta)$, we obtain

$$(\partial_i \ln_\chi p_\theta) P_{\chi,(1)}(x;\theta) = \frac{\partial_i p_\theta}{\chi(p_\theta)} \chi(p_\theta) \;=\; \partial_i p_\theta.$$

Therefore, we obtain

$$\begin{aligned} g_{ij}^M(\theta) &= \int_\Omega (\partial_i \ln_\chi p_\theta)(\partial_j p_\theta)dx \;=\; \int_\Omega (\partial_i \ln_\chi p_\theta)(\partial_j \ln_\chi p_\theta) P_{\chi,(1)}(x;\theta)dx \\ &= g^{(1)}(\theta). \end{aligned}$$

Recall that $\{\theta^i\}$ is a $\nabla^{M(e)}$-affine coordinate system [8]. In addition, the generalized score function $\partial_i \ln_\chi p_\theta$ is unbiased with respect to the escort expectation, that is,

$$E_{\chi,p}[\partial_i \ln_\chi p_\theta] = \int_\Omega (\partial_i \ln_\chi p_\theta) P_{\chi,(1)}(x;\theta)dx = \int_\Omega \partial_i p_\theta dx = 0.$$

Therefore we obtain

$$\begin{aligned} C_{ijk}^M(\theta) = \Gamma_{ij,k}^{M(m)}(\theta) &= \int_\Omega (\partial_k \ln_\chi p_\theta)(\partial_i \partial_j p_\theta)dx \\ &= \int_\Omega (\partial_k \ln_\chi p_\theta)\partial_i\{(\partial_j \ln_\chi p_\theta) P_{\chi,(1)}(x;\theta)\}dx \\ &= 0 + \int_\Omega (\partial_k \ln_\chi p_\theta)(\partial_j \ln_\chi p_\theta)(\partial_i \ln_\chi p_\theta) P_{\chi,(2)}(x;\theta)dx \\ &= C_{ijk}^{(1)}(\theta). \end{aligned}$$

From the second escort expectation, we have the following theorem.

Theorem 2. *Let $S_q = \{p(x;\theta)\}$ be a χ-exponential family. Then $(S_\chi, g^{(2)}, C^{(2)})$ and (S_χ, g^χ, C^χ) have the following relations:*

$$g^{(2)}_{ij}(x;\theta) = Z_\chi(p_\theta) g^\chi_{ij}(\theta),$$
$$C^{(2)}_{ijk}(x;\theta) = Z_\chi(p_\theta) C^\chi_{ij}(\theta) + g^\chi_{ij}(\theta)\partial_k Z_\chi(p_\theta) + g^\chi_{jk}(\theta)\partial_i Z_\chi(p_\theta) + g^\chi_{ki}(\theta)\partial_j Z_\chi(p_\theta).$$

Proof. Set $u(x) = (\exp_q x)'$. Then we have

$$\begin{aligned}
\partial_i p(x;\theta) &= u\left(\sum \theta^k F_k(x) - \psi(\theta)\right)(F_i(x) - \partial_i\psi(\theta)) \\
\partial_i\partial_j p(x;\theta) &= u'\left(\sum \theta^k F_k(x) - \psi(\theta)\right)(F_i(x) - \partial_i\psi(\theta))(F_j(x) - \partial_j\psi(\theta)) \\
&\quad - u\left(\sum \theta^k F_k(x) - \psi(\theta)\right)\partial_i\partial_j\psi(\theta) \\
&= P_{\chi,(2)}(x;\theta)(\partial_i \ln_\chi p_\theta)(\partial_j \ln_\chi p_\theta) - P_{\chi,(1)}(x;\theta)\partial_i\partial_j\psi(\theta).
\end{aligned}$$

Since $\int_\Omega \partial_i p(x;\theta)dx = \int_\Omega \partial_i\partial_j p(x;\theta)dx = 0$ and $Z_\chi(p) = \int_\Omega \chi(p(x;\theta))dx = \int_\Omega P_{\chi,(1)}(x;\theta)dx$, we obtain

$$g^{(2)}_{ij}(\theta) = Z_\chi(p_\theta) g^\chi_{ij}(\theta).$$

From a straight forward calculation, we have

$$\begin{aligned}
\partial_i\partial_j\partial_k p(x;\theta) &= u''\left(\sum \theta^l F_l(x) - \psi(\theta)\right) \\
&\quad \times (F_i(x) - \partial_i\psi(\theta))(F_j(x) - \partial_j\psi(\theta))(F_k(x) - \partial_k\psi(\theta)) \\
&\quad - u'\left(\sum \theta^l F_l(x) - \psi(\theta)\right)(F_k(x) - \partial_k\psi(\theta))\partial_i\partial_j\psi(\theta) \\
&\quad - u'\left(\sum \theta^l F_l(x) - \psi(\theta)\right)(F_i(x) - \partial_i\psi(\theta))\partial_j\partial_k\psi(\theta) \\
&\quad - u'\left(\sum \theta^l F_l(x) - \psi(\theta)\right)(F_j(x) - \partial_j\psi(\theta))\partial_k\partial_i\psi(\theta) \\
&\quad - u\left(\sum \theta^l F_l(x) - \psi(\theta)\right)\partial_i\partial_j\partial_k\psi(\theta), \qquad (4)
\end{aligned}$$

$$\begin{aligned}
\partial_i Z_\chi(p_\theta) &= \int_\Omega \partial_i P_{\chi,(1)}(x;\theta)dx \\
&= \int_\Omega u\left(\sum \theta^l F_l(x) - \psi(\theta)\right)(F_i(x) - \partial_i\psi(\theta))dx.
\end{aligned}$$

By integrating (4), we obtain the relation $C^{(2)}$ and C^χ.

We remark that the statistical manifold $(S_\chi, g^{(2)}, C^{(2)})$ cannot determine a dually flat structure in general whereas (S_χ, g^χ, C^χ) determines a dually flat structure. The relations in Theorem 2 imply that two statistical manifolds have a generalized conformal equivalence relation in the sense of Kurose [5].

6 Concluding Remarks

In this paper, we considered a sequential structure of escort expectations and statistical manifold structures that are defined from the sequence of escort expectations. Further geometric properties of the sequence $\{(S_\chi, g^{(n)}, C^{(n)})\}_{n\in\mathbf{N}}$ are not clear at this moment. However. the sequential structure will be important in the geometric theory of non-exponential type statistical models. Actually, in the case of q-exponential family, $(S_q, g^{(1)}, C^{(1)})$ is induced from a β-divergence. In addition, $(S_q, g^{(2)}, C^{(2)})$ are essentially equivalent to the invariant statistical manifold structure $(S_q, g^F.C^F)$, which are induced from an α-divergence [7].

The authors would like to express their sincere gratitude to the referees for giving helpful comments to improve this paper.

References

1. Amari, S., Nagaoka, H.: Method of Information Geometry. Amer. Math. Soc., Providence, Oxford University Press, Oxford (2000)
2. Amari, S.: Information Geometry and Its Applications. AMS, vol. 194. Springer, Tokyo (2016). doi:10.1007/978-4-431-55978-8
3. Amari, S., Ohara, A., Matsuzoe, H.: Geometry of deformed exponential families: invariant, dually-flat and conformal geometry. Phys. A **391**, 4308–4319 (2012)
4. Beck, C., Schlögl, F.: Thermodynamics of Chaotic Systems: An Introduction. Cambridge University Press, Cambridge (1993)
5. Kurose, T.: On the divergences of 1-conformally flat statistical manifolds. Tôhoku Math. J. **46**, 427–433 (1994)
6. Lauritzen, S.L.: Statistical manifolds. In: Differential Geometry in Statistical Inferences. IMS Lecture Notes Monograph Series, vol. 10, pp. 96–163. Hayward California, Institute of Mathematical Statistics (1987)
7. Matsuzoe, H.: A sequence of escort distributions and generalizations of expectations on q-exponential family. Entropy **19**(1), 7 (2017)
8. Matsuzoe, H., Henmi, M.: Hessian structures and divergence functions on deformed exponential families. In: Nielsen, F. (ed.) Geometric Theory of Information. SCT, pp. 57–80. Springer, Cham (2014). doi:10.1007/978-3-319-05317-2_3
9. Matsuzoe, H., Wada, T.: Deformed algebras and generalizations of independence on deformed exponential families. Entropy **17**(8), 5729–5751 (2015)
10. Murata, N., Takenouchi, T., Kanamori, T., Eguchi, S.: Information geometry of U-boost and Bregman divergence. Neural Comput. **16**, 1437–1481 (2004)
11. Naudts, J.: Generalised Thermostatistics. Springer, London (2011). doi:10.1007/978-0-85729-355-8
12. Sakamoto, M., Matsuzoe, H.: A generalization of independence and multivariate student's t-distributions. In: Nielsen, F., Barbaresco, F. (eds.) GSI 2015. LNCS, vol. 9389, pp. 740–749. Springer, Cham (2015). doi:10.1007/978-3-319-25040-3_79
13. Shima, H.: The Geometry of Hessian Structures. World Scientific, Singapore (2007)
14. Tanaka, M.: Meaning of an escort distribution and τ-transformation. J. Phys: Conf. Ser. **201**, 012007 (2010)
15. Tsallis, C.: Introduction to Nonextensive Statistical Mechanics: Approaching a Complex World. Springer, New York (2009). doi:10.1007/978-0-387-85359-8

Normalization and φ-function: Definition and Consequences

Luiza H.F. de Andrade[1(✉)], Rui F. Vigelis[2], Francisca L.J. Vieira[3], and Charles C. Cavalcante[4]

[1] Center of Exact and Natural Sciences,
Federal Rural University of Semi-Arid Region, Mossoró, RN, Brazil
luizafelix@ufersa.edu.br

[2] Computer Engineering, Campus Sobral,
Federal University of Ceará, Sobral, CE, Brazil
rfvigelis@ufc.br

[3] Department of Mathematics, Regional University of Cariri,
Juazeiro do Norte, CE, Brazil
leidmar.vieira@urca.br

[4] Department of Teleinformatics Engineering, Federal University of Ceará,
Fortaleza, CE, Brazil
charles@ufc.br

Abstract. It is known from the literature that a φ-function may be used to construct the φ-families of probability distributions. In this paper, we assume that one of the properties in the definition of φ-function is not satisfied and we analyze the behavior of the normalizing function near the boundary of its domain. As a consequence, we find a measurable function that does not belong to the Musielak–Orlicz class, but the normalizing function applied to this found function converges to a finite value near the boundary of its domain. We conclude showing that this change in the definition of φ-function affects the behavior of the normalizing function.

1 Introduction

In [7] was obtained a generalization of exponential families of probability distributions [4,5], called φ-families. The construction of these families is based on Musielak–Orlicz spaces [2] and on a function, called φ-function, which satisfies some properties. Another generalization of exponential families of probability distributions in infinite-dimensional setting was studied in [8]. In [6], it was studied the Δ_2-condition and its consequences on φ-families of probability distributions, we explain briefly this condition in Sect. 3.1. More specifically, the behavior of the normalizing function near the boundary of its domain was analyzed, considering that the Musielak–Orlicz function Φ_c does not satisfies the Δ_2-condition. In [1] the authors found an example that has the same form of a φ-function, but does not satisfy all the properties of a φ-function. Our aim in this paper is to analyze the behavior of the normalizing function near the boundary of its domain, considering functions, as the one found in [1], which do not satisfy all the properties of the definition of a φ-function.

F. Nielsen and F. Barbaresco (Eds.): GSI 2017, LNCS 10589, pp. 231–238, 2017.
https://doi.org/10.1007/978-3-319-68445-1_27

2 Preliminary Considerations

In this section we provide an introduction to φ-families of probability distributions. Let (T, Σ, μ) be a σ-finite, non-atomic measure space on which probability distributions are defined. In this space, T may be thought of as the set of real numbers $\mathbb{R}$. These families are based on the replacement of the exponential function by a φ-function $\varphi\colon \mathbb{R} \to (0,\infty)$ that satisfies the following properties [7]:

(a1) $\varphi(\cdot)$ is convex and injective;
(a2) $\lim_{u\to-\infty}\varphi(u) = 0$ and $\lim_{u\to\infty}\varphi(u) = \infty$;
(a3) There exists a measurable function $u_0\colon T \to (0,\infty)$ such that

$$\int_T \varphi(c + \lambda u_0)d\mu < \infty, \qquad \text{for all } \lambda > 0,$$

for every measurable function $c\colon T \to \mathbb{R}$ for which $\int \varphi(c)d\mu = 1$.

There are many examples of φ-functions that satisfy (a1)–(a3) [7].

In the definition of φ-function, the constraint $\int_T \varphi(c)d\mu = 1$ can be replaced by $\int_T \varphi(c)d\mu < \infty$ since this fact was shown in [1, Lemma 1]. Thus, condition (a3) can be rewritten as:

(a3') There exists a measurable function $u_0\colon T \to (0,\infty)$ such that

$$\int_T \varphi(c + \lambda u_0)d\mu < \infty, \qquad \text{for all } \lambda > 0,$$

for every measurable function $c\colon T \to \mathbb{R}$ for which $\int_T \varphi(c)d\mu < \infty$.

Thus (a3) and (a3') are equivalent.

Also, there are functions that satisfy (a1)–(a2) but do not satisfy (a3') and an example was given in [1, Example 2]:

$$\varphi(u) = \begin{cases} e^{(u+1)^2/2}, & u \geq 0, \\ e^{(u+1/2)}, & u \leq 0. \end{cases} \tag{1}$$

Clearly, $\lim_{u\to\infty}\varphi(u) = \infty$ and $\lim_{u\to-\infty}\varphi(u) = 0$. It was shown in [1] that for the function (1) there exists a measurable function $c\colon T \to \mathbb{R}$ and another function u_0, both functions were defined in [1], such that $\int_T \varphi(c)d\mu < \infty$ but $\int_T \varphi(c + u_0)d\mu = \infty$. This function in (1) is a deformed exponential function up to a trivial multiplicative factor, as discussed in [3].

The φ-families of probability distributions were built based on the Musielak–Orlicz spaces [2]. Let φ be a φ-function. The Musielak–Orlicz function was defined in [7] by

$$\Phi_c(t, u) = \varphi(c(t) + u) - \varphi(c(t)),$$

where $c\colon T \to \mathbb{R}$ is a measurable function such that $\varphi(c)$ is μ-integrable. Then we have, the Musielak-Orlicz space L^{Φ_c} and the Musielak–Orlicz class $\tilde{L}^{\Phi_c}$, denoted by L_c^φ and $\tilde{L}_c^\varphi$, respectively.

Let $\mathcal{K}_c^\varphi$ be the set of all the functions $u \in L_c^\varphi$ such that $\varphi(c+\lambda u)$ is μ-integrable for every λ in a neighborhood of $[0,1]$. We know that $\mathcal{K}_c^\varphi$ is an open set in L_c^φ [7, Lemma 2] and, for $u \in \mathcal{K}_c^\varphi$, the function $\varphi(c+u)$ is not necessarily in $\mathcal{P}_\mu$, so the normalizing function $\psi\colon \mathcal{K}_c^\varphi \to \mathbb{R}$ is introduced in order to make the density $\varphi(c+u-\psi(u)u_0)$ is in $\mathcal{P}_\mu$ [7]. For any $u \in \mathcal{K}_c^\varphi$, $\psi(u) \in \mathbb{R}$ is the unique function which $\varphi(c+u-\psi(u)u_0)$ is in $\mathcal{P}_\mu$ [7, Proposition 3]. Let

$$B_c^\varphi = \left\{u \in L_c^\varphi : \int_T u\varphi'_+(t,c(t))d\mu = 0\right\}$$

be a closed subspace of L_c^φ, thus for every $u \in \mathcal{B}_c^\varphi = B_c^\varphi \cap \mathcal{K}_c^\varphi$, by the convexity of φ, one has $\psi(u) \geq 0$ and $\varphi(c+u-\psi(u)u_0) \in \mathcal{P}_\mu$.

For each measurable function $c\colon T \to \mathbb{R}$ such that $p = \varphi(c) \in \mathcal{P}_\mu$ is associated a parametrization $\boldsymbol{\varphi}_c\colon \mathcal{B}_c^\varphi \to \mathcal{F}_c^\varphi$, given by

$$\boldsymbol{\varphi}_c(u) = \boldsymbol{\varphi}(c+u-\psi(u)u_0),$$

where the operator $\boldsymbol{\varphi}$ acts on the set of real-value functions $u\colon T \to \mathbb{R}$ given by $\boldsymbol{\varphi}(u)(t) = \varphi(u(t))$ and the set $\mathcal{F}_c^\varphi = \boldsymbol{\varphi}_c(\mathcal{B}_c^\varphi) \subseteq \mathcal{P}_\mu$ where $\mathcal{P}_\mu = \bigcup\{\mathcal{F}_c^\varphi : \varphi(c) \in \mathcal{P}_\mu\}$ and the map $\boldsymbol{\varphi}_c$ is a bijection from $\mathcal{B}_c^\varphi$ to $\mathcal{F}_c^\varphi$. In the following section we will study the behavior of the normalizing function ψ near the boundary of $\mathcal{B}_c^\varphi$.

3 The Behavior of ψ Near the Boundary of $\mathcal{B}_c^\varphi$

Let us suppose that the condition (a3) (or the equivalent (a3')) on φ-function definition is not fulfilled. In others words, it is possible to find functions $\widetilde{c}\colon T \to \mathbb{R}$ such that $\int_T \varphi(\widetilde{c})d\mu < \infty$ but $\int_T \varphi(\widetilde{c}+\lambda u_0)d\mu = \infty$ for some $\lambda > 0$. Now, for u being a function in $\partial\mathcal{B}_c^\varphi$ we want to know whether $\psi(\alpha u)$ converges to a finite value as $\alpha \uparrow 1$ or not. First, let us remember how the normalizing function ψ behaves near $\partial\mathcal{B}_c^\varphi$, assuming that the condition (a3') is satisfied and the Musielak–Orlicz function Φ_c does not satisfies the Δ_2-condition [6].

3.1 Δ_2-Condition and the Normalizing Function

The Δ_2-condition of Musielak–Orlicz functions and φ-families of probability distributions was studied in [6], where the behavior of the normalizing function ψ near the boundary of $\mathcal{B}_c^\varphi$ was discussed. Remember that the set $\mathcal{B}_c^\varphi = \mathcal{K}_c^\varphi \cap B_c^\varphi$ is open in B_c^φ, then a function $u \in B_c^\varphi$ belongs to $\partial\mathcal{B}_c^\varphi$, the boundary of $\mathcal{B}_c^\varphi$, if and only if $\int_T \varphi(c+\lambda u)d\mu < \infty$ for all $\lambda \in (0,1)$ and $\int_T \varphi(c+\lambda u)d\mu = \infty$ for all $\lambda > 1$.

We say that a Musielak–Orlicz function satisfies the Δ_2-condition, if one can find a constant $K > 0$ and a non-negative function $f \in \tilde{L}_c^\varphi$ such that

$$\Phi(t,2u) \leq K\Phi(t,u), \qquad \text{for all } u \geq f(t), \quad \text{and } \mu\text{-a.e. } t \in T.$$

If the Musielak–Orlicz function $\Phi_c(u) = \varphi(c(t)+u) - \varphi(c(t))$ satisfies the Δ_2-condition, then $\int_T \varphi(c+u)d\mu < \infty$ for all $u \in L_c^\varphi$ and $\partial\mathcal{B}_c^\varphi$ is empty.

Assuming that the Musielak–Orlicz function Φ_c does not satisfies the Δ_2-condition, the boundary of $\mathcal{B}_c^\varphi$ is non-empty. Let u be a function in $\partial\mathcal{B}_c^\varphi$, for $\alpha \in [0,1)$. It was shown in [6, Proposition 6] that if $\int_T \varphi(c+u)d\mu < \infty$ then the normalizing function $\psi_u(\alpha) = \psi(\alpha u) \to \beta$, with $\beta \in (0,\infty)$ as $\alpha \uparrow 1$, and if $\int_T \varphi(c+u)d\mu = \infty$ then $(\psi_u)'_+(\alpha) \to \infty$ as $\alpha \uparrow 1$. Now, it follows our first result, which states that it is possible to show that $\psi(\alpha u) \to \infty$ as $\alpha \uparrow 1$, when $\int_T \varphi(c+u)d\mu = \infty$, with $u \in \partial\mathcal{B}_c^\varphi$, as in the following proposition:

Proposition 1. *For a function $u \in \partial\mathcal{B}_c^\varphi$ such that $\int_T \varphi(c+u)d\mu = \infty$. Then $\psi(\alpha u) \to \infty$ as $\alpha \uparrow 1$.*

Proof. Suppose that, for some $\overline{\lambda} > 0$, the function u satisfies $\psi(\alpha u) \leq \overline{\lambda}$ for all $\alpha \in [0,1)$. Denote $A = \{u \geq 0\}$. Observing that

$$\int_A \varphi(c+\alpha u - \overline{\lambda}u_0)d\mu \leq \int_T \varphi(c+\alpha u - \overline{\lambda}u_0)d\mu \leq \int_T \varphi(c+\alpha u - \psi(\alpha u)u_0)d\mu = 1,$$

we obtain that $\int_A \varphi(c+u-\overline{\lambda}u_0)d\mu < \infty$. In addition, it is clear that

$$\int_{T\setminus A} \varphi(c+u-\overline{\lambda}u_0)d\mu \leq \int_{T\setminus A} \varphi(c)d\mu \leq 1.$$

As a result, we have $\int_T \varphi(c+u-\overline{\lambda}u_0)d\mu < \infty$. From the condition (a3'), it follows that $\int_T \varphi(c+u)d\mu < \infty$, which is a contradiction.

From this result we can investigate the behavior of ψ near the boundary of $\mathcal{B}_c^\varphi$ in terms of whether the condition (a3') is satisfied or not. We will discuss about this in the following section.

3.2 The Definition of φ-function and Its Consequences

We know there are functions that satisfy conditions (a1)–(a2) in the definition of φ-function but do not satisfy (a3') as seen in (1). In this section we discuss about the behavior of the normalizing function ψ near the boundary of $\mathcal{B}_c^\varphi$ in cases where the condition (a3') is not satisfied. To begin with, let us prove that the condition (a3') is equivalent to the existence of constants $\overline{\lambda}, \alpha > 0$ and a non-negative function $f \in \tilde{L}_c^\varphi$ such that

$$\alpha\Phi_c(t,u) \leq \Phi_{c-\overline{\lambda}u_0}(t,u), \qquad \text{for all } u > f(t). \tag{2}$$

For this we need the following lemma.

Lemma 1 [2, Theorem 8.4]. *Let Ψ and Φ be finite-value Musielak–Orlicz functions. Then the inclusion $\tilde{L}^\Phi \subset \tilde{L}^\Psi$ is satisfied if and only if there exist $\alpha > 0$ and a non-negative function $f \in \tilde{L}^\Phi$ such that*

$$\alpha\Psi(t,u) \leq \Phi(t,u), \textit{ for all } u > f(t).$$

Using the above lemma we can prove the equivalence between the condition (a3') in the definition of φ-function and the inequality (2).

Proposition 2. *A measurable function u_0 satisfies (a3') in the definition of φ-function if and only if for some measurable function $c\colon T \to \mathbb{R}$ such that $\varphi(c)$ is μ-integrable, we can find constants $\overline{\lambda}, \alpha > 0$ and a non-negative function $f \in \tilde{L}^{\Phi_c}$ such that*

$$\alpha\Phi_c(t,u) \leq \Phi_{c-\overline{\lambda}u_0}(t,u), \qquad \textit{for all}\, u > f(t). \tag{3}$$

Proof. Suppose that u_0 satisfies condition (a3'). Let $c\colon T \to \mathbb{R}$ be any measurable function such that $\int_T \varphi(c)d\mu < \infty$. As u is a measurable function with $\int_T \varphi(c - \overline{\lambda}u_0 + u)d\mu < \infty$ then

$$\int_T \varphi(c+u)d\mu = \int_T \varphi(c - \overline{\lambda}u_0 + u + \overline{\lambda}u_0)d\mu < \infty.$$

This result implies $\tilde{L}^{\Phi_{c-\overline{\lambda}u_0}} \subset \tilde{L}^{\Phi_c}$. Inequality (3) follows from Lemma 1.

Now suppose that inequality (3) is satisfied. By Lemma 1 we have $\tilde{L}^{\Phi_{c-\lambda u_0}} \subset \tilde{L}^{\Phi_c}$. Therefore, $u \in \tilde{L}^{\Phi_c}$ implies $u + \overline{\lambda}u_0 \in \tilde{L}^{\Phi_{c-\overline{\lambda}u_0}} \subset \tilde{L}^{\Phi_c}$. Or, equivalently, if u is a measurable function such that $\varphi(c+u)$ is μ-integrable, then $\varphi(c+u+\overline{\lambda}u_0)$ is μ-integrable. As a result, we conclude that $\int_T \varphi(c+u+\lambda u_0)d\mu < \infty$ for all $\lambda > 0$. Let $\widetilde{c}\colon T \to \mathbb{R}$ be any measurable function satisfying $\int_T \varphi(\widetilde{c})d\mu < \infty$. Denote $A = \{\widetilde{c} > c\}$. Thus, for each $\lambda > 0$, it follows that

$$\int_T \varphi(\widetilde{c}+\lambda u_0)d\mu = \int_T \varphi(c+(\widetilde{c}-c)+\lambda u_0)d\mu \leq \int_T \varphi(c+(\widetilde{c}-c)\chi_A+\lambda u_0)d\mu < \infty,$$

which shows that u_0 is stated in the definition of φ-functions.

From this proposition we have that the condition (a3') is satisfied if, and only if, there exists a measurable function $u\colon T \to \mathbb{R}$ such that $\int_T \varphi(c+u)d\mu = \infty$ but $\int_T \varphi(c+u-\overline{\lambda}u_0)d\mu < \infty$ for some $\overline{\lambda} > 0$. For our main result we make use of the lemmas below.

Lemma 2 [2, Lemma 8.3]. *Consider a non-atomic and σ-finite measure μ. If $\{u_n\}$ is a sequence of finite-value, non-negative, measurable functions, and $\{\alpha_n\}$ is a sequence of positive, real numbers, such that*

$$\int_T u_n d\mu \geq 2^n \alpha_n, \qquad \textit{for all}\ n \geq 1,$$

then an increasing sequence $\{n_i\}$ of natural numbers and a sequence $\{A_i\}$ of pairwise disjoint, measurable sets can be found, such that

$$\int_{A_i} u_{n_i} d\mu = \alpha_{n_i}, \qquad \textit{for all}\ i \geq 1.$$

For the next lemma we denote the functional $I_{\Phi_c} = \int_T \Phi_c(t, \mid u(t) \mid)d\mu$ for any $u \in L^0$.

Lemma 3. *Consider* $c\colon T \to [0,\infty)$ *a measurable function such that* $\int_T \varphi(c)d\mu < \infty$. *Suppose that, for each* $\lambda > 0$, *we cannot find* $\alpha > 0$ *and* $f \in \tilde{L}^{\Phi_c}$ *such that*

$$\alpha\Phi_c(t,u) \le \Phi_{c-\lambda u_0}(t,u), \qquad \text{for all } u > f(t). \tag{4}$$

Then a strictly decreasing sequence $0 < \lambda_n \downarrow 0$, *and sequences* $\{u_n\}$ *and* $\{A_n\}$ *of finite-value, measurable functions, and pairwise disjoint, measurable sets, respectively, can be found such that*

$$I_{\Phi_c}(u_n\chi_{A_n}) = 1, \quad \text{and} \quad I_{\Phi_{c-\lambda_n u_0}}(u_n\chi_{A_n}) \le 2^{-n}, \qquad \text{for all } n \ge 1. \tag{5}$$

Proof. Let $\{\lambda_m\}$ be a strictly decreasing sequence such that $0 < \lambda_m \downarrow 0$. Define the non-negative functions

$$f_m(t) = \sup\{u > 0 : 2^{-m}\Phi_c(t,u) > \Phi_{c-\lambda_m u_0}(t,u)\}, \text{ for all } m \ge 1,$$

where we adopt the convention that $\sup\emptyset = 0$. Since (4) is not satisfied, we have that $I_{\Phi_c}(f_m) = \infty$ for each $m \ge 1$. For every rational number $r > 0$, define the measurable sets

$$A_{m,r} = \{t \in T : 2^{-m}\Phi_c(t,r) > \Phi_{c-\lambda_m u_0}(t,r)\},$$

and the simple functions $u_{m,r} = r\chi_{A_{m,r}}$. For $r = 0$, set $u_{m,r} = 0$. Let $\{r_i\}$ be an enumeration of the non-negative rational numbers with $r_1 = 0$. Define the non-negative, simple functions $v_{m,k} = \max_{1\le i\le k} u_{m,r_i}$, for each $m, k \ge 1$. By continuity of $\Phi_c(t,\cdot)$ and $\Phi_{c-\lambda_m u_0}(t,\cdot)$, it follows that $v_{m,k} \uparrow f_m$ as $k \to \infty$. In virtue of the Monotone Convergence Theorem, for each $m \ge 1$, we can find some $k_m \ge 1$ such that the function $v_m = v_{m,k_m}$ satisfies $I_{\Phi_c}(v_m) \ge 2^m$. Clearly, we have that $\Phi_c(t, v_m(t)) < \infty$ and $2^{-m}\Phi_c(t, v_m(t)) \ge \Phi_{c-\lambda_m u_0}(t, v_m(t))$. By Lemma 2, there exist an increasing sequence $\{m_n\}$ of indices and a sequence $\{A_n\}$ of pairwise disjoint, measurable sets such that $I_{\Phi_c}(v_{m_n}\chi_{A_n}) = 1$. Taking $\lambda_n = \lambda_m$, $u_n = v_{m_n}$ and A_n, we obtain (5).

Finally, our main result follows.

Proposition 3. *Assuming that the condition (a3') is not satisfied in the definition of* φ*-function, then there exists* $u \in \partial\mathcal{B}_c^{\varphi}$ *such that* $\int_T \varphi(c+u)d\mu = \infty$ *but* $\psi(\alpha u) \to \beta$, *with* $\beta \in (0,\infty)$, *as* $\alpha \uparrow 1$.

Proof. Let $\{\lambda_n\}$, $\{u_n\}$ and $\{A_n\}$ as in Lemma 3. Given any $\lambda > 0$, take $n_0 \ge 1$ such that $\lambda \ge \lambda_n$ for all $n \ge n_0$. Denote $B = T \setminus \bigcup_{n=n_0}^{\infty} A_n$, then we define $u = \sum_{n=n_0}^{\infty} u_n\chi_{A_n}$. From (5), it follows that

$$\begin{aligned}\int_T \varphi(c+u-\lambda u_0)d\mu &= \int_B \varphi(c-\lambda u_0)d\mu + \sum_{n=n_0}^{\infty}\int_{A_n}\varphi((c-\lambda u_0)+u_n)d\mu \\ &= \int_B \varphi(c-\lambda u_0)d\mu\end{aligned}$$

$$+\sum_{n=n_0}^{\infty}\left\{\int_{A_n}\varphi(c-\lambda u_0)d\mu+I_{\Phi_{c-\lambda u_0}}(u_n\chi_{A_n})\right\}$$

$$\leq\int_T\varphi(c-\lambda u_0)d\mu+\sum_{n=n_0}^{\infty}2^{-n}<\infty.$$

Consequently, for $\alpha\in(0,1)$, we can write

$$\begin{aligned}\int_T\varphi(c+\alpha u)d\mu&=\int_T\varphi\Big(c+\alpha(u-\lambda u_0)+(1-\alpha)\frac{\alpha\lambda}{1-\alpha}u_0\Big)d\mu\\&\leq\alpha\int_T\varphi(c+u-\lambda u_0)d\mu+(1-\alpha)\int_T\varphi\Big(c+\frac{\alpha\lambda}{1-\alpha}u_0\Big)d\mu\\&<\infty.\end{aligned}$$

On the other hand, for $\alpha\geq 1$, it follows that

$$\begin{aligned}\int_T\varphi(c+\alpha u)d\mu&\geq\int_B\varphi(c)d\mu+\sum_{n=n_0}^{\infty}\int_{A_n}\varphi(c+u_n)d\mu\\&\geq\int_B\varphi(c)d\mu+\sum_{n=n_0}^{\infty}\left\{\int_{A_n}\varphi(c)d\mu+I_{\Phi_c}(u_n\chi_{A_n})\right\}\\&=\int_T\varphi(c)d\mu+\sum_{n=n_0}^{\infty}1=\infty.\end{aligned}$$

We can choose $\lambda'<0$ such that

$$w=\lambda' u_0\chi_B+\sum_{n=n_0}^{\infty}u_n\chi_{A_n}$$

satisfies $\int_T w\varphi'_+(c)d\mu=0$. Clearly, $\int_T\varphi(c+w)d\mu=\infty$, $\int_T\varphi(c+\alpha w)d\mu<\infty$ for $\alpha\in(0,1)$ and $\int_T\varphi(c+\alpha w)d\mu=\infty$ for $\alpha>1$, that is, $w\in\partial\mathcal{B}_c^{\varphi}$ and $\int_T\varphi(c+w-\lambda u_0)d\mu<\infty$ for some fixed $\lambda>0$. Suppose that $\psi(\alpha w)\uparrow\infty$, then for all $K>0$, there exists $\delta>0$ such that $0<|\alpha-1|<\delta$ implies that $\psi(\alpha w)>K$. Let $\lambda''>\lambda$ be such that $\int_T\varphi(c+w-\lambda'' u_0)d\mu<1$, taking $K=\lambda''$ we have $\varphi(c+\alpha w-\psi(w)u_0)<\varphi(c+\alpha w_{\{w>0\}}-\lambda'' u_0)<\varphi(c+w_{\{w>0\}}-\lambda'' u_0)$, that is a μ-integrable function. Therefore by the Dominated Convergence Theorem we have

$$\lim_{\alpha\uparrow 1}\int_T\varphi(c+\alpha w-\lambda'' u_0)d\mu=\int_T\varphi(c+w-\lambda'' u_0)d\mu,$$

then

$$\begin{aligned}1&=\lim_{\alpha\uparrow 1}\int_T\varphi(c+\alpha w-\psi(\alpha w)u_0)d\mu\\&\leq\lim_{\alpha\uparrow 1}\int_T\varphi(c+\alpha w-\lambda'' u_0)d\mu=\int_T\varphi(c+w-\lambda'' u_0)d\mu<1,\end{aligned}$$

which is a contradiction.

4 Conclusions

This paper focused on the behavior of the normalizing function ψ near the boundary of its domain. Assuming that the condition (a3') is satisfied, it has been shown in [6] that for all measurable function u in the boundary of the normalizing function domain such that $E[\varphi(c+u)] = \infty$, $\psi(\alpha u)$ converges to infinity as α approaches 1. Now, whereas that the condition (a3') in φ-function definition is not satisfied, we found a measurable function $w\colon T \to \mathbb{R}$ in the boundary of normalizing function domain such that $E[\varphi(c+w)] = \infty$ but $\psi(\alpha w)$ converges to a finite value as α approaches 1. We conclude that the condition (a3') in the definition of φ-function affects the behavior of the normalizing function near the boundary of its domain. A perspective for future works is to investigate the behavior of normalizing function considering that the φ-function is not necessarily injective.

Acknowledgement. The authors would like to thank CAPES and CNPq (Proc. 309055/2014-8) for partial funding of this research.

References

1. de Souza, D.C., Vigelis, R.F., Cavalcante, C.C.: Geometry induced by a generalization of rényi divergence. Entropy **18**(11), 407 (2016)
2. Musielak, J.: Orlicz Spaces and Modular Spaces. LNM, vol. 1034. Springer, Heidelberg (1983). doi:10.1007/BFb0072210
3. Naudts, J.: Generalised Thermostatistics. Springer, London (2011). doi:10.1007/978-0-85729-355-8
4. Pistone, G., Rogantin, M.P., et al.: The exponential statistical manifold: mean parameters, orthogonality and space transformations. Bernoulli **5**(4), 721–760 (1999)
5. Pistone, G., Sempi, C.: An infinite-dimensional geometric structure on the space of all the probability measures equivalent to a given one. Ann. Stat. **23**, 1543–1561 (1995)
6. Vigelis, R.F., Cavalcante, C.C.: The Δ_2-condition and Φ-families of probability distributions. In: Nielsen, F., Barbaresco, F. (eds.) GSI 2013. LNCS, vol. 8085, pp. 729–736. Springer, Heidelberg (2013). doi:10.1007/978-3-642-40020-9_81
7. Vigelis, R.F., Cavalcante, C.C.: On φ-families of probability distributions. J. Theor. Probab. **26**(3), 870–884 (2013)
8. Zhang, J., Hästö, P.: Statistical manifold as an affine space: a functional equation approach. J. Math. Psych. **50**(1), 60–65 (2006)

Deformed Exponential Bundle: The Linear Growth Case

Luigi Montrucchio[1] and Giovanni Pistone[2(✉)]

[1] Collegio Carlo Alberto, Torino, Italy
luigi.montrucchio@unito.it
[2] de Castro Statistics, Collegio Carlo Alberto, Torino, Italy
giovanni.pistone@carloalberto.org
http://www.giannidiorestino.it

Abstract. Vigelis and Cavalcante extended the Naudts' deformed exponential families to a generic reference density. Here, the special case of Newton's deformed logarithm is used to construct an Hilbert statistical bundle for an infinite dimensional class of probability densities.

1 Introduction

Let $\mathcal{P}$ be a family of positive probability densities on the probability space $(\mathbb{X}, \mathcal{X}, \mu)$. At each $p \in \mathcal{P}$ we have the Hilbert space of square-integrable random variables $L^2(p \cdot \mu)$ so that we can define the *Hilbert bundle* consisting of $\mathcal{P}$ with linear fibers $L^2(p{\cdot}\mu)$. Such a bundle supports most of the structure of Information Geometry, cf. [1] and the non-parametric version in [6,7].

If $\mathcal{P}$ is an exponential manifold, there exists a splitting of each fiber $L(p{\cdot}\mu) = H_p \oplus H_p^\perp$, such that H_p is equal or contains as a dense subset, the tangent space of the manifold at p. Moreover, the geometry on $\mathcal{P}$ is affine and, as a consequence, there are natural transport mappings on the Hilbert bundle.

We shall study a similar set-up when the manifold is defined by charts based on mapping other than the exponential, while retaining an affine structure, see e.g. [10]. Here, we use $p = \exp_A(v)$, where $\exp_A$ is exponential-like function with linear growth at $+\infty$. In such a case, the Hilbert bundle has fibers which are all sub-spaces of the same $L^2(\mu)$ space.

The formalism of deformed exponentials by Naudts [4] is reviewed and adapted in Sect. 2. The following Sect. 3 is devoted to the adaptation of that formalism to the non-parametric case. Our construction is based on the work of Vigelis and Cavalcante [9], and we add a few more details about the infinite-dimensional case. Section 4 discusses the construction of the Hilbert statistical bundle in our case.

2 Background

We recall a special case of a nice and useful formalism introduced by Naudts [4]. Let $A\colon [0, +\infty[\to [0, 1[$ be an increasing, concave and differentiable function

F. Nielsen and F. Barbaresco (Eds.): GSI 2017, LNCS 10589, pp. 239–246, 2017.
https://doi.org/10.1007/978-3-319-68445-1_28

with $A(0) = 0$, $A(+\infty) = 1$ and $A'(0+) = 1$. We focus on the case $A(x) = 1 - 1/(1+x) = x/(1+x)$ that has been firstly discussed by Newton [5]. The deformed A-logarithm is the function $\log_A(x) = \int_1^x A(\xi)^{-1}\, d\xi = x - 1 + \log x$, $x \in]0, +\infty[$. The deformed A-exponential is $\exp_A = \log_A^{-1}$ which turns out to be the solution to the Cauchy problem $e'(y) = A(e(y)) = 1 + 1/(1+e(y))$, $e(0) = 1$.

In the spirit of [8,9] we consider the curve in the space of positive measures on $(\mathbb{X}, \mathcal{X})$ given by $t \mapsto \mu_t = \exp_A(tu + \log_A p) \cdot \mu$, where $u \in L^2(\mu)$. As $\exp_A(a+b) \leq a^+ + \exp_A(b)$, each μ_t is a finite measure, $\mu_t(\mathbb{X}) \leq \int (tu)^+\, d\mu + 1$, with $\mu_0 = p \cdot \mu$. The curve is actually continuous and differentiable because the pointwise derivative of the density $p_t = \exp_A(tu + \log_A(p))$ is $\dot{p}_t = A(p_t)u$ so that $|\dot{p}_t| \leq |u|$. In conclusion $\mu_0 = p$ and $\dot{\mu}_0 = u$.

Notice that there are two ways to normalize the density p_t, either dividing by a normalizing constant $Z(t)$ to get the statistical model $t \mapsto \exp_A(tu - \log_A p)/Z(t)$ or, subtracting a constant $\psi(t)$ from the argument to get the model $t \mapsto \exp_A(tu - \psi(t) + \log_A(p))$. In the standard exponential case the two methods lead to the same result, which is not the case for deformed exponentials where $\exp_A(\alpha + \beta) \neq \exp_A(\alpha)\exp_A(\beta)$. We choose in the present paper the latter option.

3 Deformed Exponential Family Based on $\exp_A$

Here we use the ideas of [4,8,9] to construct deformed non-parametric exponential families. Recall that we are given: the measure space $(\mathbb{X}, \mathcal{X}, \mu)$; the set $\mathcal{P}$ of probability densities; the function $A(x) = x/(1+x)$. Throughout this section, the density $p \in \mathcal{P}$ will be fixed.

Proposition 1. *1. The mapping $L^1(\mu) \ni u \mapsto \exp_A(u + \log_A p) \in L^1(\mu)$ has full domain and is 1-Lipschitz. Consequently, the mapping*

$$u \mapsto \int g \exp_A(u + \log_A p)\, d\mu$$

is $\|g\|_\infty$-Lipschitz for each bounded function g.

2. For each $u \in L^1(\mu)$ there exists a unique constant $K(u) \in \mathbb{R}$ such that $\exp_A(u - K(u) + \log_A p) \cdot \mu$ is a probability.

3. It holds $K(u) = u$ if, and only if, u is constant. In such a case,

$$\exp_A(u - K(u) + \log_A p) \cdot \mu = p \cdot \mu \ .$$

Otherwise, $\exp_A(u - K(u) + \log_A p) \cdot \mu \neq p \cdot \mu$.

4. A density q takes the form $q = \exp_A(u - K(u) + \log_A p)$, with $u \in L^1(\mu)$ if, and only if, $\log_A q - \log_A p \in L^1(\mu)$.

5. If $u, v \in L^1(\mu)$

$$\exp_A(u - K(u) + \log_A p) = \exp_A(v - K(v) + \log_A p) \ ,$$

then $u - v$ is constant.

6. *The functional* $K \colon L^1(\mu) \to \mathbb{R}$ *is translation invariant. More specifically,* $c \in \mathbb{R}$ *implies* $K(u+c) = K(u) + cK(1)$.
7. *The functional* $K \colon L^1(\mu) \to \mathbb{R}$ *is continuous and quasi-convex, namely all its sub-levels* $L_\alpha = \left\{u \in L^1(\mu) \middle| K(u) \leq \alpha\right\}$ *are convex.*
8. $K : L^1(\mu) \to \mathbb{R}$ *is convex.*

Proof. 1. As $\exp_A(u + \log_A p) \leq u^+ + p$ and so $\exp_A(u + \log_A p) \in L^1(\mu)$ for all $u \in L^1(\mu)$. The estimate $|\exp_A(u + \log_A p) - \exp_A(v + \log_A p)| \leq |u - v|$ leads to the desired result.
2. For all $\kappa \in \mathbb{R}$ the integral $I(\kappa) = \int \exp_A(u - \kappa + \log_A p)\, d\mu$ is bounded by $1 + \int (u - \kappa)^+\, d\mu < \infty$ and the function $\kappa \mapsto I(k)$ is continuous and strictly decreasing. Convexity of $\exp_A$ together with the equation for its derivative imply $\exp_A(u - \kappa + \log_A p) \geq \exp_A(u + \log_A p) - A(\exp_A(u + \log_A p))\kappa$, so that $\int \exp_A(u - \kappa + \log_A p)\, d\mu \geq \int \exp_A(u + \log_A p)\, d\mu - \kappa \int A(\exp_A(u + \log_A p))\, d\mu$, where the coefficient of κ is positive. Hence $\lim_{\kappa \to -\infty} \int \exp_A(u - \kappa + \log_A p)\, d\mu = +\infty$. For each $\kappa \geq 0$, we have $\exp_A(u - \kappa + \log_A p) \leq \exp_A(u + \log_A p) \leq p + u^+$ so that by dominated convergence we get $\lim_{\kappa \to \infty} I(\kappa) = 0$. Therefore $K(u)$ will be the unique value for which $\int \exp_A(u - \kappa + \log_A p)\, d\mu = 1$.
3. If the function u is a constant, then $\int \exp_A(u - u + \log_A p)\, d\mu = \int p\, d\mu = 1$ and so $K(u) = u$. The converse implication is trivial. The equality $\exp_A(u - K(u) + \log_A p) = p$ holds if, and only if, $u - K(u) = 0$.
4. If $\log_A q = u - K(u) + \log_A p$, then $\log_A q - \log_A p = u - K(u) \in L^1(\mu)$. Conversely, if $\log_A q - \log_A p = v \in L^1(\mu)$, then $q = \exp_A(v + \log_A p)$. As q is a density, then $K(v) = 0$.
5. If $u - K(u) + \log_A p = v - K(v) + \log_A p$, then $u - v = K(u) - K(v)$.
6. Clearly, $K(c) = c = cK(1)$ and $K(u + c) = K(u) + c$.
7. Observe that $\int \exp_A(u + \log_A p)\, d\mu \leq 1$ if, and only if, $K(u) \leq 0$. Hence $u_1, u_2 \in L_0$, implies $\int \exp_A(u_i + \log_A p)\, d\mu \leq 1$, $i = 1, 2$. Thanks to the convexity of the function $\exp_A$, we have $\int \exp_A((1-\alpha)u_1 + \alpha u_2) + \log_A p\, d\mu \leq (1 - \alpha) \int \exp_A(u_1 + \log_A p)\, d\mu + \alpha \int \exp_A(u_2 + \log_A p)\, d\mu \leq 1$, that provides $K((1-\alpha)u_1 + \alpha u_2) \leq 0$. Hence the sub-level L_0 is convex. Notice that all the other sub-levels are convex since they are obtained by translation of L_0. More precisely, $L_\alpha = L_0 + \alpha$. Clearly both the sets $\left\{\int \exp_A(u + \log_A p)\, d\mu \leq 1\right\}$ and $\left\{\int \exp_A(u + \log_A p)\, d\mu \geq 1\right\}$ are closed in $L^1(\mu)$, since the functional $u \to \int \exp_A(u)\, d\mu$ is continuous. Hence $u \to K(u)$ is continuous as well.
8. A functional which is translation invariant and quasiconvex is necessarily convex. Though this property is more or less known, a proof is gathered below.

Lemma 1. *A translation invariant functional on a vector space* V*, namely* $I \colon V \to \mathbb{R}$ *such that for some* $v \in V$ *one has* $I(x + \lambda v) = I(x) + \lambda I(v)$ *for all* $x \in V$ *and* $\lambda \in \mathbb{R}$*, is convex if and only if* I *is quasiconvex, namely all level sets are convex, provided* $I(v) \neq 0$.

Proof. Let I be quasiconvex, then the sublevel $L_0(I) = \{x \in V : I(x) \leq 0\}$ is nonempty and convex. Clearly, $L_\lambda(I) = L_0(I) + (\lambda / I(v))v$ holds for every $\lambda \in \mathbb{R}$.

Hence, if λ and μ are any pair of assigned real numbers and $\alpha \in (0,1)$, $\bar{\alpha} = 1-\alpha$, then

$$\alpha L_\lambda (I) + \bar{\alpha} L_\mu (I) = \alpha L_0 (I) + \bar{\alpha} L_0 (I) + \frac{\alpha\lambda + \bar{\alpha}\mu}{I(v)} v$$
$$= L_0 (I) + \frac{\alpha\lambda + \bar{\alpha}\mu}{I(v)} v = L_{\alpha\lambda+\bar{\alpha}\mu} (I) \; .$$

Therefore, if for any pair of points $x, y \in V$, we set $I(x) = \lambda$ and $I(y) = \mu$, then $x \in L_\lambda (I)$ and $y \in L_\mu (I)$. Consequently $\alpha x + \bar{\alpha} y \in \alpha L_\lambda (I) + \bar{\alpha} L_\mu (I) = L_{\alpha\lambda+\bar{\alpha}\mu}(I)$. That is, $I(\alpha x + \bar{\alpha} y) \leq \alpha\lambda + \bar{\alpha}\mu = \alpha I(x) + \bar{\alpha} I(y)$ that shows the convexity of I. Of course the converse holds in that a convex function is quasi-convex.

For each positive density q, define its *escort density* to be $\widetilde{q} = A(q)/\int A(q)\, d\mu$, see [4]. Notice that $0 < A(q) < 1$. The next proposition provides a subgradient of the convex function K.

Proposition 2. *Let $v \in L^1(\mu)$ and $q(v) = \exp_A(v - K(v) + \log_A p)$. For every $u \in L^1(\mu)$, the inequality $K(u+v) - K(v) \geq \int u\widetilde{q}(v)\, d\mu$ holds i.e., the density $\widetilde{q}(v) \in L^\infty(\mu)$ is a subgradient of K at v.*

Proof. Thanks to convexity of $\exp_A$ and the derivation formula, we have

$$\exp_A(u + v - K(u+v) + \log_A p) - q \geq A(q)(u - K(u+v) + K(v)) \; .$$

If we take μ-integral of both sides,

$$0 \geq \int uA(q)\, d\mu - (K(u+v) - K(v)) \int A(q)\, d\mu \; .$$

Isolating the increment $K(u+v) - K(v)$, the desired inequality obtains.

By Proposition 2, if the functional K were differentiable, the gradient mapping would be $v \mapsto \widetilde{q}(v)$, whose strong continuity requires additional assumptions. We would like to show that K is differentiable by means of the Implicit Function Theorem. That too, would require specific assumptions. In fact, it is in general not true that a superposition operator such as $L^1(\mu) \ni u \mapsto \exp_A(u + \log_A p) \in L^1(\mu)$ is differentiable, cf. [2, Sect. 1.2]. In this perspective, we prove the following.

Proposition 3. *1. The superposition operator $L^2(\mu) \ni v \mapsto \exp_A(v + \log_A p) \in L^1(\mu)$ is continuously Fréchet differentiable with derivative*

$$d\exp_A(v) = (h \mapsto A(\exp_A(v + \log_A p))h) \in \mathcal{L}(L^2(\mu), L^1(\mu)) \; .$$

2. The functional $K : L^2(\mu) \to \mathbb{R}$, implicitly defined by the equation

$$\int \exp_A(v - K(v) + \log_A p)\, d\mu = 1, \quad v \in L^2(\mu)$$

is continuously Fréchet differentiable with derivative

$$dK(v) = (h \mapsto \int h\tilde{q}(v)\, d\mu), \quad q(v) = \exp_A(v - K(v))$$

where

$$\tilde{q}(v) = \frac{A \circ q(v)}{\int A \circ q(v)\, d\mu}$$

is the escort density of p.

Proof. 1. It is easily seen that

$$\exp_A(v + h + \log_A p) - \exp_A(v + \exp_A p) - A[\exp_A(v + \log_A p)]h = R_2(h),$$

with the bound $|R_2(h)| \leq (1/2)\,|h|^2$. It follows

$$\frac{\int |R_2(h)|\, d\mu}{\left(\int |h|^2\, d\mu\right)^{\frac{1}{2}}} \leq \frac{\frac{1}{2}\int |h|^2\, d\mu}{\left(\int |h|^2\, d\mu\right)^{\frac{1}{2}}} = \frac{1}{2}\left(\int |h|^2\, d\mu\right)^{\frac{1}{2}}.$$

Therefore $\|R_2(h)\|_{L^1(\mu)} = o\left(\|h\|_{L^2(\mu)}\right)$ and so the operator $v \mapsto \exp_A(v + \log_A p)$ is Fréchet-differentiable with derivative $h \mapsto A(\exp_A(v + \log_A p))h$ at v. Let us show that the F-derivative is a continuous map $L^2(\mu) \to \mathcal{L}(L^2(\mu), L^1(\mu))$. If $\|h\|_{L^2(\mu)} \leq 1$ and $v, w \in L^2(\mu)$ we have

$$\int |(A[\exp_A(v + \log_A p)] - A[\exp_A(w + \log_A p)])h|\, d\mu$$
$$\leq \|A[\exp_A(v + \log_A p) - A[\exp_A(w + \log_A p)]\|_{L^2(\mu)} \leq \|v - w\|_{L^2(\mu)},$$

hence the derivative is 1-Lipschitz.

2. Frechét differentiability of K is a consequence of the Implicit Function Theorem in Banach spaces, see [3], applied to the C^1-mapping

$$L^2(\mu) \times \mathbb{R} \ni (v, \kappa) \mapsto \int \exp_A(v - \kappa + \log_A p)\, d\mu.$$

The derivative can be easily obtained from the computation of the subgradient.

In the expression $q(u) = \exp_A(u - K(u) + \log_A p)$, $u \in L^1(\mu)$, the random variable u is identified up to a constant. We can choose in the class a unique representative, by assuming $\int u\tilde{p}\, d\mu = 0$, the expected value being well defined as the escort density is bounded. In this case we can solve for u and get

$$u = \log_A q - \log_A p - \mathrm{E}_{\tilde{p}}[\log_A p - \log_A q]$$

In analogy with the exponential case, we can express the functional K as a divergence associated to the N.J. Newton logarithm:

$$K(u) = \mathrm{E}_{\tilde{p}}[\log_A p - \log_A q(u)] = D_A(p\|q(u)).$$

It would be interesting to proceed with the study of the convex conjugation of K and the related properties of the divergence, but do not do that here.

4 Hilbert Bundle Based on $\exp_A$

In this section $A(x) = x/(1+x)$ and $\mathcal{P}(\mu)$ denotes the set of all μ-densities on the probability space $(\mathbb{X}, \mathcal{X}, \mu)$ of the form $q = \exp_A(u - K(u))$ with $u \in L^2(\mu)$ and $\mathrm{E}_\mu[u] = 0$, cf. [5]. Notice that $1 \in \mathcal{P}(\mu)$ because we can take $u = 0$. Equivalently, $\mathcal{P}(\mu)$ is the set of all densities q such that $\log_A q \in L^2(\mu)$ because in such a case we can take $u = \log_A q - \mathrm{E}_\mu[\log_A q]$. The condition for $q \in \mathcal{P}(\mu)$ can be expressed by saying that both q and $\log q$ are in $L^2(\mu)$. In fact, as $\exp_A$ is 1-Lipschitz, we have $\|q-1\|_\mu \le \|u - K(u)\|_\mu$ and the other inclusion follows from $\log q = \log_A q + 1 - q$. An easy but important consequence of such a characterization is the compatibility of the class $\mathcal{P}(\mu)$ with the product of measures. If $q_i = \exp_A(u_i - K_1(u_i)) \in \mathcal{P}(\mu_i)$, $i = 1, 2$, the product is $(q_1 \cdot \mu_1) \otimes (q_2 \cdot \mu_2) = (q_2 \otimes q_2) \cdot (\mu_1 \otimes \mu_2)$, hence $q_2 \otimes q_2 \in \mathcal{P}(\mu_1 \otimes \mu_2)$ since $\|q_1 \otimes q_2\|_{\mu_1 \otimes \mu_2} = \|q_1\|_{\mu_1} \|q_2\|_{\mu_2}$. Moreover $\log(q_1 \otimes q_2) = \log q_1 + \log q_2$, hence $\|\log(q_1 \otimes q_2)\|_{\mu_1 \otimes \mu_2} \le \|\log q_1\|_{\mu_1} + \|\log q_2\|_{\mu_2}$.

We proceed now to define an Hilbert bundle with base $\mathcal{P}(\mu)$. For each $p \in \mathcal{P}(\mu)$ consider the Hilbert spaces $H_p = \left\{u \in L^2(\mu) \middle| \mathrm{E}_{\widetilde{p}}[u] = 0\right\}$ with scalar product $\langle u, v\rangle_p = \int uv \, d\mu$ and form the Hilbert bundle

$$H\mathcal{P}(\mu) = \{(p, u) | p \in \mathcal{P}(\mu), u \in H_p\} .$$

For each $p, q \in \mathcal{P}(\mu)$ the mapping $\mathbb{U}_p^q u = u - \mathrm{E}_{\widetilde{q}}[u]$ is a continuous linear mapping from H_p to H_q. We have $\mathbb{U}_q^r \mathbb{U}_p^q = \mathbb{U}_p^r$. In particular, $\mathbb{U}_q^p \mathbb{U}_p^q$ is the identity on H_p, hence $\mathbb{U}_p^q$ is an isomorphism of H_p onto H_q. In the next proposition we construct an atlas of charts for which $\mathcal{P}(\mu)$ is a Riemannian manifold and $H\mathcal{P}(\mu)$ is an expression of the tangent bundle.

In the following proposition we introduce an affine atlas of charts and use it to define our Hilbert bundle which is an expression of the tangent bundle. The velocity of a curve $t \mapsto p(t) \in \mathcal{P}(\mu)$ is expressed in the Hilbert bundle by the so called A-score that, in our case, takes the form $A(p(t))^{-1}\dot{p}(t)$, with $\dot{p}(t)$ computed in $L^1(\mu)$.

Proposition 4. *1. $q \in \mathcal{P}(\mu)$ if, and only if, both q and $\log q$ are in $L^2(\mu)$.*

2. Fix $p \in \mathcal{P}(\mu)$. Then a positive density q can be written as

$$q = \exp_A(v - K_p(v) + \log_A p), \quad \text{with } v \in L^2(\mu) \text{ and } \mathrm{E}_{\widetilde{p}}[v] = 0,$$

if, and only if, $q \in \mathcal{P}(\mu)$.

3. For each $p \in \mathcal{P}(\mu)$ the mapping

$$s_p \colon \mathcal{P}(\mu) \ni q \mapsto \log_A q - \log_A p - \mathrm{E}_{\widetilde{p}}[\log_A q - \log_A p] \in H_p$$

is injective and surjective, with inverse $e_p(u) = \exp_A(u - K_p(u) + \log_A p)$.

4. The atlas $\{s_p | p \in \mathcal{P}(\mu)\}$ is affine with transitions

$$s_q \circ e_p(u) = \mathbb{U}_p^q u + s_p(q) .$$

5. *The expression of the velocity of the differentiable curve* $t \mapsto p(t) \in \mathcal{P}(\mu)$ *in the chart* s_p *is* $ds_p(p(t))/dt \in H_p$. *Conversely, given any* $u \in H_p$, *the curve* $p\colon t \mapsto \exp_A(tu - K_p(tu) + \log_A p)$ *has* $p(0) = p$ *and has velocity at* $t = 0$ *expressed in the chart* s_p *by* u. *If the velocity of a curve is expressed in the chart* s_p *by* $t \mapsto \dot{u}(t)$, *then its expression in the chart* s_q *is* $\mathbb{U}_p^q \dot{u}(t)$.
6. *If* $t \mapsto p(t) \in \mathcal{P}(\mu)$ *is differentiable with respect to the atlas then it is differentiable as a mapping in* $L^1(\mu)$. *It follows that the A-score is well-defined and is the expression of the velocity of the curve* $t \mapsto p(t)$ *in the moving chart* $t \mapsto s_{p(t)}$.

Proof. 1. Assume $q = \exp_A(u - K(u))$ with $u \in L_0^2(\mu)$. It follows $u - K(u) \in L^2(\mu)$ hence $q \in L^2(\mu)$ because $\exp_A$ is 1-Lipschitz. As moreover $q + \log q - 1 = u - K(u) \in L^2(\mu)$, then $\log q \in L^2(\mu)$. Conversely, $\log_a q = q - 1 + \log q = v \in L^2(\mu)$ and we can write $q = \exp_A v = \exp_A((v - \mathrm{E}_p[v]) + \mathrm{E}_p[v])$ and we can take $u = v - \mathrm{E}_\mu[v]$.

2. The assumption $p, q \in \mathcal{P}(\mu)$ is equivalent to $\log_A p, \log_A q \in L^2(\mu)$. Define $u = \log_A q - \log_A p - \mathrm{E}_{\widetilde{p}}[\log_A q - \log_A p]$ and $D_A(p\|q) = \mathrm{E}_{\widetilde{p}}[\log_A p - \log_A q]$. It follows $u \in L^2(\mu)$, $\mathrm{E}_{\widetilde{p}}[u] = 0$, and $\exp_A(u - D_A(p\|q) + \log_A p) = q$. Conversely, $\log_A q = u - K_p(u) + \log_A p \in L^2(\mu)$.
3. This has been already proved.
4. All simple computations.
5. If $p(t) = \exp_A(u(t) - K_p(u(t)) + \log_A p)$, with $u(t) = s_p(u(t))$ then in that chart the velocity is $\dot{u}(t) \in H_p$. When $u(t) = tu$ the expression of the velocity will be u. The proof of the second part follows from the fact that $\mathbb{U}_p^q$ is the linear part of the affine change of coordinates $s_q \circ e_p$.
6. Choose a chart s_p and express the curve as $t \mapsto s_p(p(t)) = u(t)$ so that $p(t) = \exp_A(u(t) - K_p(u(t)) + \log_A p)$. It follows that the derivative of $t \mapsto p(t)$ exists in $L^1(\mu)$ by derivation of the composite function and it is given by $\dot{p}(t) = A(p(t))\mathbb{U}_p^{p(t)}\dot{u}(t)$, hence $A(p(t))^{-1}\dot{p}(t) = \mathbb{U}_p^{p(t)}\dot{u}(t)$. If the velocity at t is expressed in the chart centered at $p(t)$, then its expression is the score.

5 Conclusions

We have constructed an Hilbert statistical bundle using an affine atlas of charts based on the A-logarithm with $A(x) = x/(1 + x)$. In particular, this entails a Riemannian manifold of densities. On the other end, our bundle structure could be useful in certain contexts. The general structure of the argument mimics the standard case of the exponential manifold. We would like to explicit some, hopefully new, features of our set-up.

The proof of the convexity and continuity of the functional K when defined on $L^1(\mu)$ relies on the property of translation invariance. Whenever K is restricted to $L^2(\mu)$, it is shown to be differentiable along with the deformed exponential and this, in turn, provides a rigorous construction of the A-score.

The gradient mapping of K is continuous and 1-to-1, but its inverse cannot be continuous as it takes values which are bounded functions. It would be interesting to analyze the analytic properties of the convex conjugate of K^*, as both K and K^* are the coordinate expression of relevant divergences.

If F is a section of the Hilbert bundle namely, $F: \mathcal{P}(\mu) \to L^2(\mu)$ with $E_{\widetilde{p}}[F(p)] = 0$ for all p, differential equations take the form $A(p(t))\dot{p}(t) = F(p(t))$ in the atlas, which in turn implies $\dot{p}(t) = A(p(t))F(p(t))$ in $L^1(\mu)$. This is important for some applications e.g., when the section F is the gradient with respect to the Hilbert bundle of a real function. Namely, the gradient, $\operatorname{grad}\phi$, of a smooth function $\phi: \mathcal{P}(\mu) \to \mathbb{R}$ is a section of the Hilbert bundle such that

$$\frac{d}{dt}\phi(p(t)) = \langle \operatorname{grad}\phi(p(t)), A(p(t))\dot{p}(t)\rangle_\mu$$

for each differentiable curve $t \mapsto p(t) \in \mathcal{P}(\mu)$.

Acknowledgments. L. Montrucchio acknowledges the support of Collegio Carlo Alberto Foundation. G. Pistone is a member of GNAFA-INDAM and acknowledges the support of de Castro Statistics Foundation and Collegio Carlo Alberto Foundation.

References

1. Amari, S.: Dual connections on the Hilbert bundles of statistical models. In: Geometrization of Statistical Theory (Lancaster, 1987), pp. 123–151. ULDM Publ., Lancaster (1987)
2. Ambrosetti, A., Prodi, G.: A Primer of Nonlinear Analysis, Cambridge Studies in Advanced Mathematics, vol. 34. Cambridge University Press, Cambridge (1993)
3. Dieudonné, J.: Foundations of Modern Analysis. Academic Press, New York (1960)
4. Naudts, J.: Generalised Thermostatistics. Springer, London (2011). doi:10.1007/978-0-85729-355-8
5. Newton, N.J.: An infinite-dimensional statistical manifold modelled on Hilbert space. J. Funct. Anal. **263**(6), 1661–1681 (2012)
6. Pistone, G.: Nonparametric information geometry. In: Nielsen, F., Barbaresco, F. (eds.) GSI 2013. LNCS, vol. 8085, pp. 5–36. Springer, Heidelberg (2013). doi:10.1007/978-3-642-40020-9_3
7. Pistone, G., Sempi, C.: An infinite-dimensional geometric structure on the space of all the probability measures equivalent to a given one. Ann. Statist. **23**(5), 1543–1561 (1995)
8. Schwachhöfer, L., Ay, N., Jost, J., Lê, H.V.: Parametrized measure models. Bernoulli (online-to appear)
9. Vigelis, R.F., Cavalcante, C.C.: On ϕ-families of probability distributions. J. Theor. Probab. **26**, 870–884 (2013)
10. Zhang, J., Hästö, P.: Statistical manifold as an affine space: a functional equation approach. J. Math. Psychol. **50**(1), 60–65 (2006)

On Affine Immersions of the Probability Simplex and Their Conformal Flattening

Atsumi Ohara(✉)

University of Fukui, Fukui 910-8507, Japan
ohara@fuee.u-fukui.ac.jp

Abstract. Embedding or representing functions play important roles in order to produce various information geometric structure. This paper investigates them from a viewpoint of affine differential geometry [2]. By restricting affine immersions to a certain class, the probability simplex is realized to be 1-conformally flat [3] statistical manifolds immersed in $\mathbf{R}^{n+1}$. Using this fact, we introduce a concept of conformal flattening of such manifolds to obtain dually flat statistical (Hessian) ones with conformal divergences, and show explicit forms of potential functions, dual coordinates. Finally, we demonstrate applications of the conformal flattening to nonextensive statistical physics and certain replicator equations on the probability simplex.

Keywords: Conformal flattening · Affine differential geometry

1 Introduction

In the theory of information geometry for statistical models, the logarithmic function is crucially significant to give a standard information geometric structure for exponential family [1]. By changing the logarithmic function to the other ones we can deform the standard structure to new one keeping its basic property as a statistical manifold, which consists of a pair of mutually dual affine connections (∇, ∇^*) with respect to Riemannian metric g. There exists several ways [4–6] to introduce such freedom of functions to deform statistical manifold structure and the functions are sometimes called *embedding* or *representing functions.*

In this paper we elucidate common geometrical properties of statistical manifolds defined by representing functions, using concepts from affine differential geometry [2,3].

2 Affine Immersion of the Probability Simplex

Let $\mathcal{S}^n$ be the probability simplex defined by

$$\mathcal{S}^n := \left\{ p = (p_i) \,\middle|\, p_i \in \mathbf{R}_+,\ \sum_{i=1}^{n+1} p_i = 1 \right\},$$

where $\mathbf{R}_+$ denotes the set of positive numbers.

F. Nielsen and F. Barbaresco (Eds.): GSI 2017, LNCS 10589, pp. 247–254, 2017.
https://doi.org/10.1007/978-3-319-68445-1_29

Consider an affine immersion [2] (f, ξ) of the simplex $\mathcal{S}^n$. Let D be the canonical flat affine connection on $\mathbf{R}^{n+1}$. Further, let f be an immersion of $\mathcal{S}^n$ into $\mathbf{R}^{n+1}$ and ξ be a transversal vector field on $\mathcal{S}^n$. For a given *affine immersion* (f, ξ) of $\mathcal{S}^n$, the induced torsion-free connection ∇ and the affine fundamental form h are defined from the Gauss formula by

$$D_X f_*(Y) = f_*(\nabla_X Y) + h(X,Y)\xi, \quad X, Y \in \mathcal{X}(\mathcal{S}^n), \tag{1}$$

where $\mathcal{X}(\mathcal{S}^n)$ is the set of vector fields on $\mathcal{S}^n$.

It is well known [2,3] that the realized geometric structure $(\mathcal{S}^n, \nabla, h)$ is a statistical manifold if and only if (f, ξ) is non-degenerate and equiaffine, i.e., h is non-degenerate and ∇h is symmetric. Further, a statistical manifold $(\mathcal{S}^n, \nabla, h)$ is 1-conformally flat [3] (but not necessarily dually flat nor of constant curvature).

Now we consider the affine immersion with the following assumptions.

Assumptions:

1. The affine immersion (f, ξ) is nondegenerate and equiaffine,
2. The immersion f is given by the component-by-component and common representing function L, i.e.,

$$f : \mathcal{S}^n \ni p = (p_i) \mapsto x = (x^i) \in \mathbf{R}^{n+1}, \quad x^i = L(p_i),\ i = 1, \cdots, n+1,$$

3. The representing function $L : \mathbf{R}_+ \to \mathbf{R}$ is concave with $L'' < 0$ and strictly increasing, i.e., $L' > 0$. Hence, the inverse of L denoted by E exists, i.e., $E \circ L = \mathrm{id}$.
4. Each component of ξ satisfies $\xi^i < 0,\ i = 1, \cdots, n+1$ on $\mathcal{S}^n$.

Remark 1. From the third assumption, it follows that $L'E' = 1$, $E' > 0$ and $E'' > 0$. Note that L is concave with $L'' < 0$ or convex $L'' > 0$ if and only if there exists ξ for h to be positive definite. Hence, we can regard h as a Riemannian metric on $\mathcal{S}^n$. The details are described later.

2.1 Conormal Vector and the Geometric Divergence

Define a function Ψ on $\mathbf{R}^{n+1}$ by

$$\Psi(x) := \sum_{i=1}^{n+1} E(x^i),$$

then $f(\mathcal{S}^n)$ immersed in $\mathbf{R}^{n+1}$ is expressed as a level surface of $\Psi(x) = 1$. Denote by $\mathbf{R}_{n+1}$ the dual space of $\mathbf{R}^{n+1}$ and by $\langle \nu, x \rangle$ the pairing of $x \in \mathbf{R}^{n+1}$ and $\nu \in \mathbf{R}_{n+1}$. The conormal vector [2] $\nu : \mathcal{S}^n \to \mathbf{R}_{n+1}$ for the affine immersion (f, ξ) is defined by

$$\langle \nu(p), f_*(X) \rangle = 0,\ \forall X \in T_p\mathcal{S}^n, \qquad \langle \nu(p), \xi(p) \rangle = 1 \tag{2}$$

for $p \in \mathcal{S}^n$. Using the assumptions and noting the relations:

$$\frac{\partial \Psi}{\partial x^i} = E'(x^i) = \frac{1}{L'(p_i)} > 0, \quad i = 1, \cdots, n+1,$$

we have

$$\nu_i(p) := \frac{1}{\Lambda} \frac{\partial \Psi}{\partial x^i} = \frac{1}{\Lambda(p)} E'(x^i) = \frac{1}{\Lambda(p)} \frac{1}{L'(p_i)}, \quad i = 1, \cdots, n+1, \tag{3}$$

where Λ is a normalizing factor defined by

$$\Lambda(p) := \sum_{i=1}^{n+1} \frac{\partial \Psi}{\partial x^i} \xi^i = \sum_{i=1}^{n+1} \frac{1}{L'(p_i)} \xi^i(p). \tag{4}$$

Then we can confirm (2) using the relation $\sum_{i=1}^{n+1} X^i = 0$ for $X = (X^i) \in \mathcal{X}(\mathcal{S}^n)$. Note that $v : \mathcal{S}^n \to \mathbf{R}_{n+1}$ defined by

$$v_i(p) = \Lambda(p)\nu_i(p) = \frac{1}{L'(p_i)}, \quad i = 1, \cdots, n+1,$$

also satisfies

$$\langle v(p), f_*(X) \rangle = 0, \ \forall X \in T_p\mathcal{S}^n. \tag{5}$$

Further, it follows, from (3), (4) and the assumption 4, that

$$\Lambda(p) < 0, \quad \nu_i(p) < 0, \quad i = 1, \cdots, n+1,$$

for all $p \in \mathcal{S}^n$.

It is known [2] that the affine fundamental form h can be represented by

$$h(X, Y) = -\langle \nu_*(X), f_*(Y) \rangle, \quad X, Y \in T_p\mathcal{S}^n.$$

In our case, it is calculated via (5) as

$$\begin{aligned} h(X, Y) &= -\Lambda^{-1} \langle v_*(X), f_*(Y) \rangle - (X\Lambda^{-1}) \langle v, f_*(Y) \rangle \\ &= -\frac{1}{\Lambda} \sum_{i=1}^{n+1} \left(\frac{1}{L'(p_i)} \right)' L'(p_i) X^i Y^i = \frac{1}{\Lambda} \sum_{i=1}^{n+1} \frac{L''(p_i)}{L'(p_i)} X^i Y^i. \end{aligned}$$

Since h is positive definite from the assumptions 3 and 4, we can regard it as a Riemannian metric.

Utilizing these notions from affine differential geometry, we can introduce the function ρ on $\mathcal{S}^n \times \mathcal{S}^n$, which is called a *geometric divergence* [3], as follows:

$$\begin{aligned} \rho(p, r) &= \langle \nu(r), f(p) - f(r) \rangle = \sum_{i=1}^{n+1} \nu_i(r)(L(p_i) - L(r_i)) \\ &= \frac{1}{\Lambda(r)} \sum_{i=1}^{n+1} \frac{L(p_i) - L(r_i)}{L'(r_i)}, \qquad p, r \in \mathcal{S}^n. \end{aligned} \tag{6}$$

We can easily see that ρ is a contrast function [1,7] of the geometric structure $(\mathcal{S}^n, \nabla, h)$ because it holds that

$$\rho[X|] = 0, \quad h(X,Y) = -\rho[X|Y], \tag{7}$$
$$h(\nabla_X Y, Z) = -\rho[XY|Z], \quad h(Y, \nabla^*_X Z) = -\rho[Y|XZ], \tag{8}$$

where $\rho[X_1 \cdots X_k | Y_1 \cdots Y_l]$ stands for

$$\rho[X_1 \cdots X_k | Y_1 \cdots Y_l](p) := (X_1)_p \cdots (X_k)_p (Y_1)_r \cdots (Y_l)_r \rho(p,r)|_{p=r}$$

for $p, r \in \mathcal{S}^n$ and $X_i, Y_j \in \mathcal{X}(\mathcal{S}^n)$.

2.2 Conformal Divergence and Conformal Transformation

Let σ be a positive function on $\mathcal{S}^n$. Associated with the geometric divergence ρ, the *conformal divergence* [3] of ρ with respect to a conformal factor $\sigma(r)$ is defined by

$$\tilde{\rho}(p,r) = \sigma(r)\rho(p,r), \qquad p, r \in \mathcal{S}^n.$$

The divergence $\tilde{\rho}$ can be proved to be a contrast function for $(\mathcal{S}^n, \tilde{\nabla}, \tilde{h})$, which is conformally transformed geometric structure from $(\mathcal{S}^n, \nabla, h)$, where $\tilde{h}$ and $\tilde{\nabla}$ are given by

$$\tilde{h} = \sigma h, \tag{9}$$
$$h(\tilde{\nabla}_X Y, Z) = h(\nabla_X Y, Z) - d(\ln \sigma)(Z) h(X,Y). \tag{10}$$

When there exists such a positive function σ that relates $(\mathcal{S}^n, \nabla, h)$ with $(\mathcal{S}^n, \tilde{\nabla}, \tilde{h})$ as in (9) and (10), they are said 1-conformally equivalent and $(\mathcal{S}^n, \tilde{\nabla}, \tilde{h})$ is also a statistical manifold [3].

2.3 A Main Result

Generally, the induced structure $(\mathcal{S}^n, \tilde{\nabla}, \tilde{h})$ from the conformal divergence $\tilde{\rho}$ is not also dually flat, which is the most abundant structure in information geometry. However, by choosing the conformal factor σ carefully, we can demonstrate $(\mathcal{S}^n, \tilde{\nabla}, \tilde{h})$ is dually flat. Hereafter, we call such a transformation as *conformal flattening.*

Define

$$Z(p) := \sum_{i=1}^{n+1} \nu_i(p) = \frac{1}{\Lambda(p)} \sum_{i=1}^{n+1} \frac{1}{L'(p_i)},$$

then it is negative because each $\nu_i(p)$ is. The conformal divergence to ρ with respect to the conformal factor $\sigma(r) := -1/Z(r)$ is

$$\tilde{\rho}(p,r) = -\frac{1}{Z(r)} \rho(p,r).$$

Proposition 1. *If the conformal factor is given by $\sigma = -1/Z$, then statistical manifold $(\mathcal{S}^n, \tilde{\nabla}, \tilde{h})$ that is 1-conformally transformed from $(\mathcal{S}^n, \nabla, h)$ is dully flat and $\tilde{\rho}$ is canonical where mutually dual potential functions and coordinate systems are explicitly given by*

$$\theta^i(p) = x^i(p) - x^{n+1}(p) = L(p_i) - L(p_{n+1}), \quad i = 1, \cdots, n \tag{11}$$

$$\eta_i(p) = P_i(p) := \frac{\nu_i(p)}{Z(p)}, \quad i = 1, \cdots, n, \tag{12}$$

$$\psi(p) = -x_{n+1}(p) = -L(p_{n+1}), \tag{13}$$

$$\varphi(p) = \frac{1}{Z(p)} \sum_{i=1}^{n+1} \nu_i(p) x^i(p) = \sum_{i=1}^{n+1} P_i(p) L(p_i). \tag{14}$$

Proof. Using given relations, we first show that the conformal divergence $\tilde{\rho}$ is the canonical divergence for $(\mathcal{S}^n, \tilde{\nabla}, \tilde{h})$:

$$\begin{aligned}
\tilde{\rho}(p, r) &= -\frac{1}{Z(r)} \langle \nu(r), f(p) - f(r) \rangle = \langle P(r), f(r) - f(p) \rangle \\
&= \sum_{i=1}^{n+1} P_i(r)(x^i(r) - x^i(p)) \\
&= \sum_{i=1}^{n+1} P_i(r) x^i(r) - \sum_{i=1}^{n} P_i(r)(x^i(p) - x^{n+1}(p)) - \left(\sum_{i=1}^{n+1} P_i(r) \right) x^{n+1}(p) \\
&= \varphi(r) - \sum_{i=1}^{n} \eta_i(r) \theta^i(p) + \psi(p).
\end{aligned} \tag{15}$$

Next, let us confirm that $\partial \psi / \partial \theta^i = \eta_i$. Since $\theta^i(p) = L(p_i) + \psi(p)$, $i = 1, \cdots, n$, we have

$$p_i = E(\theta^i - \psi), \quad i = 1, \cdots, n+1,$$

by setting $\theta^{n+1} := 0$. Hence, we have

$$1 = \sum_{i=1}^{n+1} E(\theta^i - \psi).$$

Differentiating by θ^j, we have

$$\begin{aligned}
0 &= \frac{\partial}{\partial \theta^j} \sum_{i=1}^{n+1} E(\theta^i - \psi) = \sum_{i=1}^{n+1} E'(\theta^i - \psi) \left(\delta^i_j - \frac{\partial \psi}{\partial \theta^j} \right) \\
&= E'(x^j) - \left(\sum_{i=1}^{n+1} E'(x^i) \right) \frac{\partial \psi}{\partial \theta^j}.
\end{aligned}$$

This implies that

$$\frac{\partial \psi}{\partial \theta^j} = \frac{E'(x^j)}{\sum_{i=1}^{n+1} E'(x^i)} = P_j = \eta_j.$$

Together with (15) and this relation, φ is confirmed to be the Legendre transform of ψ.

The dual relation $\partial\varphi/\partial\eta_i = \theta^i$ follows automatically from the property of the Legendre transform. Q.E.D.

Remark 2. Note that $\tilde{h} = -h/Z$ and the dual affine connections ∇^* and $\tilde{\nabla}^*$ are projectively equivalent [3]. The form of $\eta_i(p) = P_i(p)$ can be interpreted as generalization of the *escort probability* [10] (See the following example).

Corollary 1. *The choice of ξ does not affect on the obtained dually flat structure $(\mathcal{S}^n, \tilde{\nabla}, \tilde{h})$.*

Proof. We have the following alternative expressions of $\eta_i = P_i$ with respect to L and E:

$$P_i(p) = \frac{1}{L'(p_i)\sum_{k=1}^{n+1} 1/L'(p_k)} = \frac{E'(x_i)}{\sum_{i=1}^{n+1} E'(x_i)}.$$

Hence, all the expressions in proposition 1 does not depend on ξ, and the statement follows. Q.E.D.

2.4 Examples

If we take L to be the logarithmic function $L(t) = \ln(t)$, we immediately have the standard dually flat structure [1] $(g^F, \nabla^{(1)}, \nabla^{(-1)})$ on the simplex $\mathcal{S}^n$, where g^F denotes the Fisher metric.

Next let the affine immersion (f, ξ) be defined by the following L and ξ:

$$L(t) := \frac{1}{1-q}t^{1-q}, \quad x^i(p) = \frac{1}{1-q}(p_i)^{1-q},$$

and

$$\xi^i(p) = -q(1-q)x^i(p),$$

with $0 < q$ and $q \neq 1$, then it realizes the alpha-geometry [1] $(\mathcal{S}^n, \nabla^{(\alpha)}, g^F)$ with $q = (1+\alpha)/2$. Following the procedure of conformally flattening described in the above, we have [8]

$$\Psi(x) = \sum_{i=1}^{n+1}((1-q)x^i)^{1/1-q}, \quad \Lambda(p) = -q, \ (constant)$$

$$\nu_i(p) = -\frac{1}{q}(p_i)^q, \quad -\frac{1}{Z(p)} = \frac{q}{\sum_{k=1}^{n+1}(p_i)^q},$$

and obtain dually flat structure $(\tilde{h}, \tilde{\nabla}, \tilde{\nabla}^*)$ via the formulas in proposition 1:

$$\eta_i = \frac{(p_i)^q}{\sum_{k=1}^{n+1}(p_k)^q}, \quad \theta^i = \frac{1}{1-q}(p_i)^{1-q} - \frac{1}{1-q}(p_{n+1})^{1-q} = \ln_q(p_i) - \psi(p),$$

$$\psi(p) = -\ln_q(p_{n+1}), \quad \varphi(p) = \ln_q\left(\frac{1}{\exp_q(S_q(p))}\right).$$

Here, $\ln_q$ and $S_q(p)$ are the *q-logarithmic function* and the *Tsallis entropy* [10], respectively defined by

$$\ln_q(t) = \frac{t^{1-q}-1}{1-q}, \quad S_q(p) = \frac{\sum_{i=1}^{n+1}(p_i)^q - 1}{1-q}.$$

3 An Application to Gradient Flows on $(\mathcal{S}^n, \tilde{\nabla}, \tilde{h})$

Recall the *replicator system* on the simplex $\mathcal{S}^n$ for given functions $f_i(p)$ defined by

$$\dot{p}_i = p_i(f_i(p) - \bar{f}(p)),\ i = 1, \cdots, n+1, \quad \bar{f}(p) := \sum_{i=1}^{n+1} p_i f_i(p), \tag{16}$$

which is extensively studied in evolutionary game theory. It is known [11] that

(i) the solution of (16) is the gradient flow that maximizes a function $V(p)$ satisfying
$$f_i = \frac{\partial V}{\partial p_i},\ i = 1, \cdots, n+1, \tag{17}$$
with respect to the *Shahshahani metric* g^S (See below),

(ii) the KL divergence is a local Lyapunov function for an equilibrium called the *evolutionary stable state (ESS)*.

The Shahshahani metric g^S is defined on the positive orthant $\mathbf{R}_+^{n+1}$ by

$$g_{ij}^S(p) = \frac{\sum_{k=1}^{n+1} p_k}{p_i}\delta_{ij}, \quad i, j = 1, \cdots, n+1.$$

Note that the Shahshahani metric induces the Fisher metric g^F on $\mathcal{S}^n$. Further, the KL divergence is the canonical divergence [1] of $(g^F, \nabla^{(1)}, \nabla^{(-1)})$. Thus, the replicator dynamics (16) are closely related with the standard dually flat structure $(g^F, \nabla^{(1)}, \nabla^{(-1)})$, which associates with exponential and mixture families of probability distributions.

Similarly it would be of interest to investigate gradient flows for dually flat geometry $(\mathcal{S}^n, \tilde{\nabla}, \tilde{h})$ (or $(\mathcal{S}^n, \nabla, h)$). Since $\tilde{h}$ can be naturally extended to $\mathbf{R}_+^{n+1}$ as a diagonal form:

$$\tilde{h}_{ij}(p) = \sigma(p)h_{ij}(p) = -\frac{1}{Z(p)\Lambda(p)}\frac{L''(p_i)}{L'(p_i)}\delta_{ij}, \qquad i, j = 1, \cdots, n+1,$$

we can define the gradient flow for $V(p)$ on $\mathcal{S}^n$ as

$$\dot{p}_i = \tilde{h}_{ii}^{-1}(f_i - \bar{f}^H), \quad \bar{f}^H(p) := \sum_{k=1}^{n+1} H_k(p) f_k(p), \quad H_i(p) := \frac{\tilde{h}_{ii}^{-1}(p)}{\sum_{k=1}^{n+1} \tilde{h}_{kk}^{-1}(p)}. \tag{18}$$

We can verify that $\dot{p} \in T_p\mathcal{S}^n$ and

$$\tilde{h}(X,\dot{p}) = \sum_{i=1}^{n+1} f_i X^i - \bar{f}^H \sum_{i=1}^{n+1} X^i = \sum_{i=1}^{n+1} \frac{\partial V}{\partial p_i} X^i, \quad \forall X = (X^i) \in \mathcal{X}(\mathcal{S}^n).$$

For the flow (18) of special case: $L(t) = t^{1-q}/(1-q)$, we have shown the following result [9]:

Proposition 2. *The trajectories of gradient flow (18) with respect to the conformal metric $\tilde{h}$ for $L(t) = t^{1-q}/(1-q)$ coincide with those of (16) while velocities of time-evolutions are different by the factor $-Z(p)$.*

On the other hand, we here demonstrate another aspect of the flow (18). Let us consider the following f_i:

$$f_i(p) := \frac{L''(p_i)}{(L'(p_i))^2} \sum_{j=1}^{n+1} a_{ij}(p) P_j(p), \quad a_{ij}(p) = -a_{ji}(p), \quad i,j = 1,\cdots,n+1. \tag{19}$$

Note that f_i's are not integrable, i.e., non-trivial V satisfying (17) does not exist because of anti-symmetry of a_{ij}. Hence, for this case, (18) is no longer a gradient flow. However, we can prove the following result:

Theorem 1. *Assume that there exists an equilibrium $r \in \mathcal{S}^n$ for the flow (18) on $(\mathcal{S}^n, \tilde{\nabla}, \tilde{h})$ with the functions f_i defined by (19). Then, $\tilde{\rho}(r,p)$ is the first integral (conserved quantity) of the flow.*

References

1. Amari, S-I., Nagaoka, H.: Methods of Information Geometry. Translations of Mathematical Monographs, vol. 191. American Mathematical Society and Oxford University Press (2000)
2. Nomizu, K., Sasaki, T.: Affine Differential Geometry. Cambridge University Press, Cambridge (1993)
3. Kurose, T.: Tohoku Math. J. **46**, 427–433 (1994)
4. Zhang, J.: Divergence function, duality, and convex analysis. Neural Comput. **16**, 159–195 (2004)
5. Naudts, J.: Continuity of a class of entropies and relative entropies. Rev. Math. Phys. **16**(6), 809–822 (2004)
6. Eguchi, S.: Information geometry and statistical pattern recognition. Sugaku Expositions **19**, 197–216 (2006). (originally 2004 Sugaku 56 380–399 in Japanese)
7. Eguchi, S.: Geometry of minimum contrast. Hiroshima Math. J. **22**, 631–647 (1992)
8. Ohara, A., Matsuzoe, H., Amari, S.-I.: A dually flat structure on the space of escort distributions. J. Phys: Conf. Ser. **201**, 012012 (2010)
9. Ohara, A., Matsuzoe, H., Amari, S.: Information geometry and statistical pattern recognition. Mod. Phys. Lett. B **26**(10), 1250063 (2012). (14 pages)
10. Tsallis, C.: Introduction to Nonextensive Statistical Mechanics: Approaching a Complex World. Springer, New York (2009)
11. Hofbauer, J., Sigmund, K.: The Theory of Evolution and Dynamical Systems: Mathematical Aspects of Selection. Cambridge University Press, Cambridge (1988)

Information Structure in Neuroscience

Pairwise Ising Model Analysis of Human Cortical Neuron Recordings

Trang-Anh Nghiem[1,3], Olivier Marre[2], Alain Destexhe[1], and Ulisse Ferrari[2(✉)]

[1] Laboratory of Computational Neuroscience, Unité de Neurosciences, Information et Complexité, CNRS, Gif-Sur-Yvette, France

[2] Institut de la Vision, INSERM and UMPC, 17 Rue Moreau, 75012 Paris, France
ulisse.ferrari@gmail.com

[3] European Institute for Theoretical Neuroscience (EITN), Paris, France

Abstract. During wakefulness and deep sleep brain states, cortical neural networks show a different behavior, with the second characterized by transients of high network activity. To investigate their impact on neuronal behavior, we apply a pairwise Ising model analysis by inferring the maximum entropy model that reproduces single and pairwise moments of the neuron's spiking activity. In this work we first review the inference algorithm introduced in Ferrari, *Phys. Rev. E* (2016) [1]. We then succeed in applying the algorithm to infer the model from a large ensemble of neurons recorded by multi-electrode array in human temporal cortex. We compare the Ising model performance in capturing the statistical properties of the network activity during wakefulness and deep sleep. For the latter, the pairwise model misses relevant transients of high network activity, suggesting that additional constraints are necessary to accurately model the data.

Keywords: Ising model · Maximum entropy principle · Natural gradient · Human temporal cortex · Multielectrode array recording · Brain states

Advances in experimental techniques have recently enabled the recording of the activity of tens to hundreds of neurons simultaneously [2] and has spurred the interest in modeling their collective behavior [3–9]. To this purpose, the pairwise Ising model has been introduced as the maximum entropy (most generic [10]) model able to reproduce the first and second empirical moments of the recorded neurons. Moreover it has already been applied to different brain regions in different animals [3,5,6,9] and shown to work efficiently [11].

The inference problem for a pairwise Ising model is a computationally challenging task [12], that requires devoted algorithms [13–15]. Recently, we proposed a *data-driven* algorithm and applied it on rat retinal recordings [1]. In the present work we first review the algorithm structure and then describe our successful application to a recording in the human temporal cortex [4].

We use the inferred Ising model to test if a model that reproduces empirical pairwise covariances without assuming any other additional information, also

F. Nielsen and F. Barbaresco (Eds.): GSI 2017, LNCS 10589, pp. 257–264, 2017.
https://doi.org/10.1007/978-3-319-68445-1_30

predicts empirical higher-order statistics. We apply this strategy separately to brain states of wakefulness (Awake) and Slow-Wave Sleep (SWS). In contrast to the former, the latter is known to be characterized by transients of high activity that modulate the whole population behavior [16]. Consistently, we found that the Ising model does not account for such global oscillations of the network dynamics. We do not address Rapid-Eye Movement (REM) sleep.

1 The Model and the Geometry of the Parameter Space

The pairwise Ising model is a fully connected Boltzmann machine without hidden units. Consequently it belongs to the exponential family and has probability distribution:

$$P_\eta(\,\boldsymbol{X}\,) = \exp\left(\,\boldsymbol{T}(X)\cdot\boldsymbol{\eta}\; - \;\log Z[\boldsymbol{\eta}]\,\right), \tag{1}$$

where $\boldsymbol{X} \in [0,1]^N$ is the row vector of the N system's free variables and $Z[\boldsymbol{\eta}]$ is the normalization. $\boldsymbol{\eta} \in \mathcal{R}^D$ is the column vector of model parameters, with $D = N(N+1)/2$ and $\boldsymbol{T}(X) \in [0,1]^D$ is the vector of model sufficient statistics. For the fully-connected pairwise Ising model the latter is composed of the list of free variables $\boldsymbol{X}$ and their pairwise products:

$$\{T_a(\boldsymbol{X})\}_{a=1}^D = \{\;\{X_i\}_{i=1}^N, \{X_iX_j\}_{i=1,j=i+1}^N\;\} \in [0,1]^D. \tag{2}$$

A dataset Ω for the inference problem is composed by a set of τ_Ω *i.i.d.* empirical configurations $\boldsymbol{X}$: $\Omega = \{\boldsymbol{X}(t)\}_{t=1}^{\tau_\Omega}$. We cast the inference problem as a log-likelihood maximization task, which for the model (1) takes the shape:

$$\boldsymbol{\eta}^* \equiv \underset{\boldsymbol{\eta}}{\text{argmax}}\,\ell[\boldsymbol{\eta}]; \quad \ell[\boldsymbol{\eta}] \equiv \mathbf{T}_{\boldsymbol{\Omega}}\cdot\boldsymbol{\eta} - \log Z[\boldsymbol{\eta}], \tag{3}$$

where $\mathbf{T}_{\boldsymbol{\Omega}} \equiv \mathbf{E}[\,\boldsymbol{T}(\boldsymbol{X}) \mid \Omega\,]$ is the empirical mean of the sufficient statistics. As a consequence of the exponential family properties, the log-likelihood gradient may be written as:

$$\boldsymbol{\nabla}\,\ell[\boldsymbol{\eta}] = \mathbf{T}_{\boldsymbol{\Omega}} - \mathbf{T}_\eta, \tag{4}$$

where $\mathbf{T}_\eta = \mathbf{E}[\,\boldsymbol{T}(\boldsymbol{X}) \mid \boldsymbol{\eta}\,]$ is the mean of $\boldsymbol{T}(\boldsymbol{X})$ under the model distribution (1) with parameters $\boldsymbol{\eta}$. Maximizing the log-likelihood is then equivalent to imposing $\mathbf{T}_{\boldsymbol{\Omega}} = \mathbf{T}_\eta$: the inferred model then reproduces the empirical averages.

Parameter Space Geometry. In order to characterize the geometry of the model parameter space, we define the minus log-likelihood Hessian $\mathcal{H}[\boldsymbol{\eta}]$, the model Fisher matrix $J[\boldsymbol{\eta}]$ and the model susceptibility matrix $\chi[\boldsymbol{\eta}]$ as:

$$\chi_{ab}[\boldsymbol{\eta}] \equiv \mathbf{E}[\,T_aT_b \mid \boldsymbol{\eta}\,] - \mathbf{E}[\,T_a \mid \boldsymbol{\eta}\,]\mathbf{E}[\,T_b \mid \boldsymbol{\eta}\,] \tag{5}$$

$$J_{ab}[\boldsymbol{\eta}] \equiv \mathbf{E}[\,\nabla_a \log P_\eta(\mathbf{X})\;\nabla_b \log P_\eta(\mathbf{X}) \mid \boldsymbol{\eta}\,], \tag{6}$$

$$\mathcal{H}_{ab}[\boldsymbol{\eta}] \equiv -\nabla_a\nabla_b l[\,\boldsymbol{\eta}\,], \tag{7}$$

As a property inherited from the exponential family, for the Ising model (1):

$$\chi_{ab}[\boldsymbol{\eta}] = J_{ab}[\boldsymbol{\eta}] = \mathcal{H}_{ab}[\boldsymbol{\eta}]. \tag{8}$$

This last property is the keystone of the present algorithm.

Moreover, the fact that the log-likelihood Hessian can be expressed as a covariance matrix ensures its non-negativity. Some zero Eigenvalues can be present, but they can easily be addressed by $L2$-regularization [1,14]. The inference problem is indeed convex and consequently the solution of (3) exists and is unique.

2 Inference Algorithm

The inference task (3) is an hard problem because the partition function $Z[\boldsymbol{\eta}]$ cannot be computed analytically. Ref. [1] suggests applying an approximated natural gradient method to numerically address the problem. After an initialization of the parameters to some initial value $\boldsymbol{\eta}_0$, the natural gradient [17,18] iteratively updates their values with:

$$\eta_{n+1} = \eta_n - \alpha J^{-1}[\boldsymbol{\eta}_n] \cdot \boldsymbol{\nabla}\, \ell[\eta_n]. \tag{9}$$

For sufficiently small α, the convexity of the problem and the positiveness of the Fisher matrix ensure the convergence of the dynamics to the solution $\boldsymbol{\eta}^*$.

As computing $J[\boldsymbol{\eta}_n]$ at each n is computationally expensive, we use (8) to approximate the Fisher with an empirical estimate of the susceptibility [1]:

$$J[\boldsymbol{\eta}] = \chi[\boldsymbol{\eta}] \approx \chi[\boldsymbol{\eta}^*] \approx \chi_\Omega \equiv \mathrm{Cov}[\ \boldsymbol{T} \mid \Omega\ ]. \tag{10}$$

The first approximation becomes exact upon convergence of the dynamics, $\boldsymbol{\eta}_n \rightarrow \boldsymbol{\eta}^*$. The second assumes that (i) the distribution underlying the data belongs to the family (1), and that (ii) the error in the estimate of χ_Ω, arising from the dataset's finite size, is small.

We compute χ_Ω of Eq. (10) only once, and then we run the inference algorithm that performs the following approximated natural gradient:

$$\eta_{n+1} = \eta_n - \alpha \chi_\Omega^{-1} \cdot \boldsymbol{\nabla}\, \ell[\eta_n]. \tag{11}$$

Stochastic Dynamics.[1] The dynamics (11) require estimating $\boldsymbol{\nabla}\, \ell[\eta]$ and thus of $\mathbf{T}_\eta$ at each iteration. This is accounted by a Metropolis Markov-Chain Monte Carlo (MC), which collects Γ_η, a sequence of τ_Γ *i.i.d.* samples of the distribution (1) with parameters $\boldsymbol{\eta}$ and therefore estimates:

$$\mathbf{T}_\eta^{\mathrm{MC}} \equiv \mathbf{E}[\ \boldsymbol{T}(\boldsymbol{X}) \mid \Gamma_\eta\ ]. \tag{12}$$

This estimate itself is a random variable with mean and covariance given by:

$$\mathbf{E}[\ \mathbf{T}_\eta^{\mathrm{MC}} \mid \{\Gamma_\eta\}\ ] = \mathbf{T}_\eta; \quad \mathrm{Cov}[\ \mathbf{T}_\eta^{\mathrm{MC}} \mid \{\Gamma_\eta\}\ ] = \frac{J[\boldsymbol{\eta}]}{\tau_\Gamma}, \tag{13}$$

where $\mathbf{E}[\ \cdot \mid \{\Gamma_\eta\}\ ]$ means expectation with respect to the possible realizations Γ_η of the configuration sequence.

[1] The results of this section are grounded on the repeated use of central limit theorem. See [1] for more detail.

Data: $\mathbf{T}_{\boldsymbol{\Omega}}$,$\chi_{\Omega}$
Result: $\boldsymbol{\eta}^*$,$\mathbf{T}^{\text{MC}}_{\boldsymbol{\eta}^*}$
Initialization: set $\tau_{\Gamma} = \tau_{\Omega}$,$\alpha = 1$ and $\boldsymbol{\eta}_0$; estimate $\mathbf{T}^{\text{MC}}_{\eta_0}$ and compute ϵ_0 ;
while $\epsilon > 1$ **do**
  $\boldsymbol{\eta}_{n+1} \leftarrow \boldsymbol{\eta}_n - \alpha\chi_{\Omega}^{-1} \cdot \boldsymbol{\nabla}\, l[\eta_n]$;
  estimate $\mathbf{T}^{\text{MC}}_{\eta_{n+1}}$ and compute ϵ_{n+1};
  if $\epsilon_{n+1} < \epsilon_n$ **then**
    increase α, keeping $\alpha \leq 1$;
  else
    decrease α and set $\boldsymbol{\eta}_{n+1} = \boldsymbol{\eta}_n$;
  end
  $n \leftarrow n + 1$;
end
Fix $\alpha < 1$ and perform several iterations.

Algorithm 1. Algorithm pseudocode for the ising model inference.

For $\boldsymbol{\eta}$ sufficiently close to $\boldsymbol{\eta}^*$, after enough iterations, this last result allows us to compute the first two moments of $\nabla\ell^{\text{MC}}_{\eta} \equiv \mathbf{T}_{\boldsymbol{\Omega}} - \mathbf{T}^{\text{MC}}_{\eta}$, using a second order expansion of the log-likelihood (3):

$$\mathbf{E}\big[\ \nabla\ell^{\text{MC}}_{\eta}\ \big|\ \{\Gamma_{\eta}\}\ \big] = \mathcal{H}[\boldsymbol{\eta}]\cdot(\boldsymbol{\eta} - \boldsymbol{\eta}^*);\ \text{Cov}\big[\ \nabla\ell^{\text{MC}}_{\eta}\ \big|\ \{\Gamma_{\eta}\}\ \big] = \frac{J[\boldsymbol{\eta}^*]}{\tau_{\Gamma}}. \tag{14}$$

In this framework, the learning dynamics becomes stochastic and ruled by the master equation:

$$P_{n+1}(\boldsymbol{\eta}') = \int d\boldsymbol{\eta}\ P_n(\boldsymbol{\eta})\ W_{\boldsymbol{\eta}\to\boldsymbol{\eta}'}[\boldsymbol{\eta}];\quad W_{\boldsymbol{\eta}\to\boldsymbol{\eta}'}[\boldsymbol{\eta}] = \text{Prob}\big(\nabla\ell^{\text{MC}}_{\eta} = \boldsymbol{\eta}' - \boldsymbol{\eta}\big), \tag{15}$$

where $W_{\boldsymbol{\eta}\to\boldsymbol{\eta}}[\boldsymbol{\eta}]$ is the probability of transition from $\boldsymbol{\eta}$ to $\boldsymbol{\eta}'$. For sufficiently large τ_{Γ} and thanks to the equalities (8), the central limit theorem ensures that the unique stationary solution of (15) is a Normal Distribution with moments:

$$\mathbf{E}\big[\ \boldsymbol{\eta}\ \big|\ P_{\infty}(\boldsymbol{\eta})\ \big] = \boldsymbol{\eta}^*\ ;\quad \mathbf{Cov}\big[\ \boldsymbol{\eta}\ \big|\ P_{\infty}(\boldsymbol{\eta})\ \big] = \frac{\alpha}{(2-\alpha)\tau_{\Gamma}}\chi^{-1}[\boldsymbol{\eta}^*]\,. \tag{16}$$

Algorithm. Thanks to (8) one may compute the mean and covariance of the model posterior distribution (with flat prior):

$$\mathbf{E}\big[\ \boldsymbol{\eta}\ \big|\ P^{\text{Post}}(\boldsymbol{\eta})\ \big] = \boldsymbol{\eta}^*\ ;\quad \mathbf{Cov}\big[\ \boldsymbol{\eta}\ \big|\ P^{\text{Post}}(\boldsymbol{\eta})\ \big] = \frac{1}{\tau_{\Omega}}\chi^{-1}[\boldsymbol{\eta}^*] \tag{17}$$

where τ_{Ω} is the size of the training dataset. From (14), if $\boldsymbol{\eta} \sim P^{\text{Post}}$ we have:

$$\mathbf{E}\big[\ \nabla\ell^{\text{MC}}_{\eta}\ \big|\ \{\Gamma_{\eta\sim P^{\text{Post}}}\}\ \big] = 0;\ \text{Cov}\big[\ \nabla\ell^{\text{MC}}_{\eta}\ \big|\ \{\Gamma_{\eta\sim P^{\text{Post}}}\}\ \big] = \frac{2\chi[\boldsymbol{\eta}^*]}{\tau_{\Gamma}}. \tag{18}$$

Interestingly, by imposing:

$$\frac{1}{\tau_\Omega} = \frac{\alpha}{(2-\alpha)\tau_\Gamma} \tag{19}$$

the moments (16) equal (17) [1]. To evaluate the inference error at each iteration we define:

$$\epsilon_n = \left\|\nabla \ell_{\eta_n}^{\mathrm{MC}}\right\|_{\chi_\Omega} = \sqrt{\frac{\tau_\Omega}{2D}\nabla \ell_{\eta_n}^{\mathrm{MC}} \cdot \chi_\Omega^{-1} \cdot \nabla \ell_{\eta_n}^{\mathrm{MC}}}. \tag{20}$$

Averaging ϵ over the posterior distribution, see (18), gives $\epsilon = 1$. Consequently, if $\boldsymbol{\eta}_n \neq \boldsymbol{\eta}^*$ implies $\epsilon_n > 1$ with high probability, for $\boldsymbol{\eta}_n \to \boldsymbol{\eta}^*$ thanks to (19) we expect $\epsilon_n = \sqrt{\tau_\Omega/\tau_\Gamma/(2-\alpha)}$ [1]. As sketched in pseudocode 1, we iteratively update $\boldsymbol{\eta}_n$ through (11) with $\tau_\Gamma = \tau_\Omega$ and $\alpha < 1$ until $\epsilon_n < 1$ is reached.

3 Analysis of Cortical Recording

As in [4,7], we analyze $\sim$ 12 h of intracranial multi-electrode array recording of neurons in the temporal cortex of a single human patient. The dataset is composed of the spike times of $N = 59$ neurons, including $N^{\mathrm{I}} = 16$ inhibitory neurons and $N^{\mathrm{E}} = 43$ excitatory neurons. During the recording session, the subject alternates between different brain states [4]. Here we focused on wakefulness (Awake) and Slow-Wave Sleep (SWS) periods. First, we divided each recording into τ_Ω short 50 ms-long time bins and encoded the activity of each neuron $i = 1, \ldots, N$ in each time bin $t = 1, \ldots, \tau_\Omega$ as a binary variable $X_i(t) \in [0, 1]$ depending on whether the cell i was silent ($X_i(t) = 0$) or emitted at least one spike ($X_i(t) = 1$) in the time window t. We thus obtain one training dataset $\Omega = \{\{X_i(t)\}_{i=1}^N\}_{t=1}^{\tau_\Omega}$ per brain state of interest. To apply the Ising model we assume that this binary representation of the spiking activity is representative of the neural dynamics and that subsequent time-bins can be considered as independent. We then run the inference algorithm on the two datasets separately to obtain two sets of Ising model parameters $\boldsymbol{\eta}_{\mathrm{Awake}}$ and $\boldsymbol{\eta}_{\mathrm{SWS}}$.

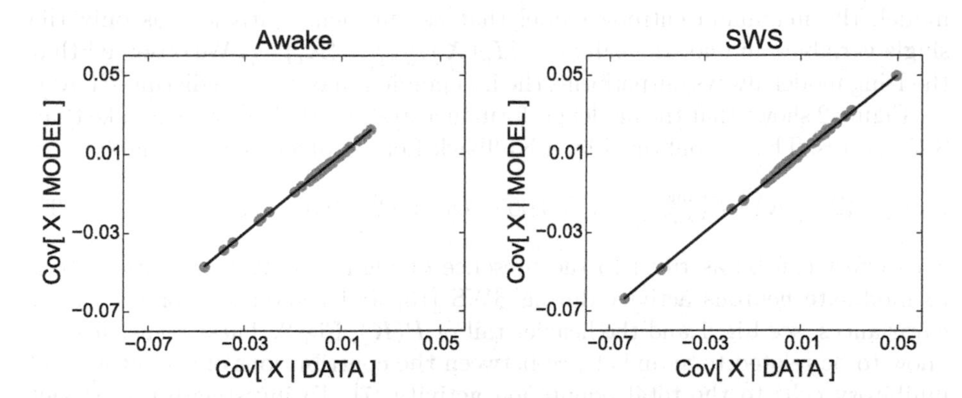

Fig. 1. Empirical pairwise covariances against their model prediction for Awake and SWS. The goodness of the match implies that the inference task was successfully completed. Note the larger values in SWS than Awake

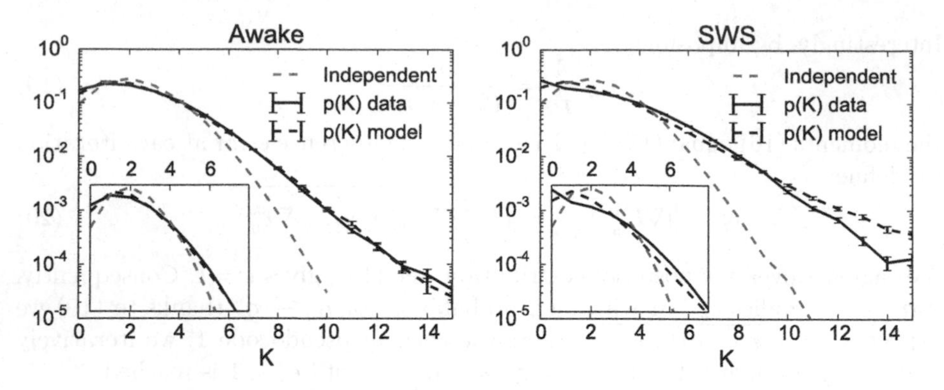

Fig. 2. Empirical and predicted distributions of the whole population activity $K = \sum_i X_i$. For both Awake and SWS periods the pairwise Ising model outperforms the independent model (see text). However, Ising is more efficient at capturing the population statistics during Awake than SWS, expecially for medium and large K values. This is consistent with the presence of transients of high activity during SWS.

Thanks to (4), when the log-likelihood is maximized, the pairwise Ising model reproduces the covariances $\mathbf{E}[\, X_i X_j \mid \Omega \,]$ for all pairs $i \neq j$. To validate the inference method, in Fig. 1 we compare the empirical and model-predicted pairwise covariances and found that the first were always accurately predicted by the second in both Awake and SWS periods.

This shows that the inference method is successful. Now we will test if this model can describe well the statistics of the population activity. In particular, synchronous events involving many neurons may not be well accounted by the pairwise nature of the Ising model interactions. To test this, as introduced in Ref. [6], we quantify the empirical probability of having K neurons active in the same time window: $K = \sum_i X_i$. In Fig. 2 we compare empirical and model prediction for $P(K)$ alongside with the prediction from an *independent* neurons model, the maximum entropy model that as sufficient statistics has only the single variables and not the pairwise: $\{T_a(\boldsymbol{X})\}_{a=1}^{N} = \{X_i\}_{i=1}^{N}$. We observed that the Ising model always outperforms the independent model in predicting $P(\,K\,)$.

Figure 2 shows that the model performance are slightly better for Awake than SWS states. This is confirmed by a Kullback-Leibler divergence estimate:

$$D_{\mathrm{KL}}\big(\, P^{\mathrm{Data}}_{\mathrm{Awake}}(K) \mid P^{\mathrm{Ising}}_{\mathrm{Awake}}(K) \,\big) = 0.005;\; D_{\mathrm{KL}}\big(\, P^{\mathrm{Data}}_{\mathrm{SWS}}(K) \mid P^{\mathrm{Ising}}_{\mathrm{SWS}}(K) \,\big) = 0.030.$$

This effect can be ascribed to the presence of high activity transients, known to modulate neurons activity during SWS [16] and responsible for the larger covariances, see Fig. 1 and the heavier tail of $P(K)$, Fig. 2. These transients are know to be related to an unbalance between the contributions of excitatory and inhibitory cells to the total population activity [7]. To investigate the impact of these transients, in Fig. 3 we compare $P(K)$ for the two populations with the corresponding Ising model predictions. For the Awake state, the two contributions are very similar, probably in consequence of the excitatory/inhibitory

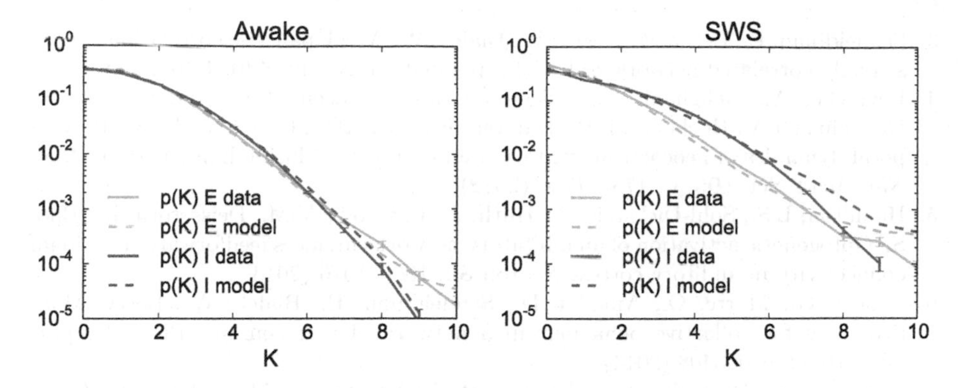

Fig. 3. Empirical and predicted distributions of excitatory (red) and inhibitory (blue) population activity. During SWS, the pairwise Ising model fails at reproducing high activity transients, especially for inhibitory cells. (Color figure online)

balance [7]. Moreover the model is able to reproduce both behaviors. For SWS periods, instead, the two populations are less balanced [7], with the inhibitory (blue line) showing a much heavier tail. Moreover, the model partially fails in reproducing this behavior, notably strongly overestimating large K probabilities.

4 Conclusions

(i) The pairwise Ising model offers a good description of the neural network activity observed during wakefulness. (ii) By contrast, taking into account pairwise correlations is not sufficient to describe the statistics of the ensemble activity during SWS, where (iii) alternating periods of high and low network activity introduce high order correlations among neurons, especially for inhibitory cells [16]. (iv) This suggests that neural interactions during wakefulness are more local and short-range, whereas (v) these in SWS are partially modulated by internally-generated activity, synchronizing neural activity across long distances [4,16,19].

Acknowledgments. We thank B. Telenczuk, G. Tkacik and M. Di Volo for useful discussion. Research funded by European Community (Human Brain Project, H2020-720270), ANR TRAJECTORY, ANR OPTIMA, French State program Investissements dAvenir managed by the Agence Nationale de la Recherche [LIFESENSES: ANR-10-LABX-65] and NIH grant U01NS09050.

References

1. Ferrari, U.: Learning maximum entropy models from finite-size data sets: a fast data-driven algorithm allows sampling from the posterior distribution. Phys. Rev. E **94**, 023301 (2016)
2. Stevenson, I.H., Kording, K.P.: How advances in neural recording affect data analysis. Nat. Neurosci. **14**(2), 139–142 (2011)

3. Schneidman, E., Berry, M., Segev, R., Bialek, W.: Weak pairwise correlations imply strongly correlated network states in a population. Nature **440**, 1007 (2006)
4. Peyrache, A., Dehghani, N., Eskandar, E.N., Madsen, J.R., Anderson, W.S., Donoghue, J.A., Hochberg, L.R., Halgren, E., Cash, S.S., Destexhe, A.: Spatiotemporal dynamics of neocortical excitation and inhibition during human sleep. Proc. Nat. Acad. Sci. **109**(5), 1731–1736 (2012)
5. Hamilton, L.S., Sohl-Dickstein, J., Huth, A.G., Carels, V.M., Deisseroth, K., Bao, S.: Optogenetic activation of an inhibitory network enhances feedforward functional connectivity in auditory cortex. Neuron **80**, 1066–1076 (2013)
6. Tkacik, G., Marre, O., Amodei, D., Schneidman, E., Bialek, W., Berry, M.J.: Searching for collective behaviour in a network of real neurons. PloS Comput. Biol. **10**(1), e1003408 (2014)
7. Dehghani, N., Peyrache, A., Telenczuk, B., Le Van Quyen, M., Halgren, E., Cash, S.S., Hatsopoulos, N.G., Destexhe, A.: Dynamic balance of excitation and inhibition in human and monkey neocortex. Sci. Rep. **6** (2016). Article no: 23176. doi:10.1038/srep23176
8. Gardella, C., Marre, O., Mora, T.: A tractable method for describing complex couplings between neurons and population rate. Eneuro **3**(4), 0160 (2016)
9. Tavoni, G., Ferrari, U., Cocco, S., Battaglia, F.P., Monasson, R.: Functional coupling networks inferred from prefrontal cortex activity show experience-related effective plasticity. Netw. Neurosci. **0**(0), 1–27 (2017). doi:10.1162/NETN_a_00014
10. Jaynes, E.T.: On The rationale of maximum-entropy method. Proc. IEEE **70**, 939 (1982)
11. Ferrari, U., Obuchi, T., Mora, T.: Random versus maximum entropy models of neural population activity. Phys. Rev. E **95**, 042321 (2017)
12. Ackley, D.H., Hinton, G.E., Sejnowski, T.J.: A learning algorithm for Boltzmann machines. Cogn. Sci. **9**, 147–169 (1985)
13. Broderick, T., Dudik, M., Tkacik, G., Schapire, R.E., Bialek, W.: Faster solutions to the inverse pairwise Ising problem. arXiv:0712.2437 (2007)
14. Cocco, S., Monasson, R.: Adaptive cluster expansion for inferring Boltzmann machines with noisy data. Phys. Rev. Lett. **106**, 090601 (2011)
15. Sohl-Dickstein, J., Battaglino, P.B., DeWeese, M.R.: New method for parameter estimation in probabilistic models: minimum probability flow. Phys. Rev. Lett. **107**, 220601 (2011)
16. Renart, A., De La Rocha, J., Bartho, P., Hollender, L., Parga, N., Reyes, A., Harris, K.D.: The asynchronous state in cortical circuits. Science **327**(5965), 587–590 (2010)
17. Amari, S.: Natural gradient works efficiently in learning. Neural Comput. **10**, 251–276 (1998)
18. Amari, S., Douglas, S.C.: Why natural gradient? Proc. IEEE **2**, 1213–1216 (1998)
19. Le Van Quyen, M., Muller, L.E., Telenczuk, B., Halgren, E., Cash, S., Hatsopoulos, N.G., Dehghani, N., Destexhe, A.: High-frequency oscillations in human and monkey neocortex during the wake-sleep cycle. Proc. Nat. Acad. Sci. **113**(33), 9363–93680 (2016)

Prevalence and Recoverability of Syntactic Parameters in Sparse Distributed Memories

Jeong Joon Park, Ronnel Boettcher, Andrew Zhao, Alex Mun, Kevin Yuh, Vibhor Kumar, and Matilde Marcolli(✉)

California Institute of Technology,
1200 E. California Blvd, Pasadena, CA 91125, USA
matilde@caltech.edu

Abstract. We propose a new method, based on sparse distributed memory, for studying dependence relations between syntactic parameters in the Principles and Parameters model of Syntax. By storing data of syntactic structures of world languages in a Kanerva network and checking recoverability of corrupted data from the network, we identify two different effects: an overall underlying relation between the prevalence of parameters across languages and their degree of recoverability, and a finer effect that makes some parameters more easily recoverable beyond what their prevalence would indicate. The latter can be seen as an indication of the existence of dependence relations, through which a given parameter can be determined using the remaining uncorrupted data.

Keywords: Syntactic structures · Principles and parameters · Kanerva networks

1 Introduction

The general idea behind the Principles and Parameters approach to Syntax, [2], is the encoding of syntactic properties of natural languages as a vector of binary variables, referred to as *syntactic parameters*. (For an expository introduction, see [1].) While this model has controversial aspects, syntactic parameters are especially suitable from the point of view of a mathematical approach to understanding the geometry of the syntactic parameters space and the distribution of features across language families, with geometric methods of modern data analysis, see [15,17,19–21]. Among the shortcomings ascribed to the Principles and Parameters model (see for instance [6]) is the lack of a complete set of such variable, the unclear nature of the dependence relations between them, and the lack of a good set of independent coordinates.

In this paper we rely on data of syntactic structures collected in the "Syntactic Structures of the World's Languages" (SSWL) database [22]. We selected a list of 21 syntactic parameters (numbered 1 to 20 and A01 in [22]), which mostly describe word order relations[1], and a list of 166 languages, selected

[1] A detailed description of the properties described by these syntactic features can be found at http://sswl.railsplayground.net/browse/properties.

F. Nielsen and F. Barbaresco (Eds.): GSI 2017, LNCS 10589, pp. 265–272, 2017.
https://doi.org/10.1007/978-3-319-68445-1_31

so that they cut across a broad range of different linguistic families, for which the values of these 21 parameters are fully recorded in the SSWL database. By storing these data of syntactic parameters in a Kanerva Network, we test for recoverability when one of the binary variables is corrupted.

We find an overall relation between recoverability and prevalence across languages, which depends on the functioning of the sparse distributed memory. Moreover, we also see a further effect, which deviates from a simple relation with the overall prevalence of a parameter. This shows that certain syntactic parameters have a higher degree of recoverability in a Kanerva Network. This property can be interpreted as a consequence of existing underlying dependence relations between different parameters. With this interpretation, one can envision a broader use of Kanerva Networks as a method to identify further, and less clearly visible, dependence relations between other groups of syntactic parameters. Another reason why it is interesting to analyze syntactic parameters using Kanerva Networks is the widespread use of the latter as models of human memory, [5,9,11]. In view of the problem of understanding mechanism of language acquisition, and how the syntactic structure of language may be stored in the human brain, sparse distributed memories appear to be a promising candidate for the construction of effective computational models.

2 Sparse Distributed Memory

Kanerva Networks were developed by Pentti Kanerva in 1988, [8,9], as a mathematical model of human long term memory. The model allows for approximate accuracy storage and recall of data at any point in a high dimensional space, using fixed hard locations distributed randomly throughout the space. During storage of a datum, hard locations close to the datum encode information about the data point. Retrieval of information at a location in the space is performed by pooling nearby hard locations and aggregating their encoded data. The mechanism allows for memory addressability of a large memory space with reasonable accuracy in a sparse representation. Kanerva Networks model human memory in the following way: a human thought, perception, or experience is represented as an (input) feature vector – a point in a high dimensional space. Concepts stored by the brain are also represented as feature vectors, and are usually stored relatively far from each other in the high dimensional space (the mind). Thus, addressing the location represented by the input vector will yield, to a reasonable degree of accuracy, the concept stored near that location. Thus, Kanerva Networks model the fault tolerance of the human mind – the mind is capable of mapping imprecise input experiences to well defined concepts. For a short introduction to Kanerva Networks aimed at a general public, see Sect. 13 of [4].

The functioning of Kanerva Network models can be summarized as follows. Over the field $\mathbb{F}_2 = \{0, 1\}$, consider a vector space (Boolean space) $\mathbb{F}_2^N$ of sufficiently large dimension N. Inside $\mathbb{F}_2^N$, choose a uniform random sample of 2^k hard locations, with $2^k << 2^N$ (a precise estimate is derived in Sect. 6 of [8]). Compute the median Hamming distance between hard locations. The *access sphere* of a point in the space $\mathbb{F}_2^N$ is a Hamming sphere of radius slightly larger than this median value (see Sect. 6 of [8] for some precise estimates). When writing to the network at some location X in the space $\mathbb{F}_2^N$, data is distributively stored by writing to all hard locations within the access sphere of that point X. Namely, each hard location stores N counters (initialized to 0), and all hard locations within the access sphere of X have their i-th counter incremented or decremented by 1, depending on the value of the i-th bit of X, see Sect. 3.3.1 of [9]. When the operation is performed for a set of locations, each hard location stores a datum whose i-th entry is determined by the majority rule of the corresponding i-th entries for all the stored data. One reads at a location Y in the network a new datum, whose i-th entry is determined by comparing 0 to the i-th counters of all the hard locations that fall within the access sphere of Y, that is, the i-th entry read at Y is itself given by the majority rule on the i-th entries of all the data stored at all the hard locations accessible from Y. The network is typically successful in reconstructing stored data, because intersections between access spheres are infrequent and small. Thus, copies of corrupted data in hard locations within the access sphere of a stored datum X are in the minority with respect to hard locations faithful to X's data. When a datum is corrupted by noise (*i.e.* flipping bit values randomly), the network is sometimes capable of correctly reconstructing these corrupted bits. The ability to reconstruct certain bits hints that these bits are derived from the remaining, uncorrupted bits in the data. Thus, Kanerva networks are a valuable general tool for detecting dependencies in a high-dimensional data sets, see [7].

3 Recoverability of Syntactic Features

The 21 SSWL syntactic features and 166 languages considered provide 166 data points in a Kanerva Network with Boolean space $\mathbb{F}_2^{21}$, where each data point is a concatenated binary string of all the values, for that particular language, of the 21 syntactic parameters considered. The Kanerva network was initialized with an access sphere of $n/4$, with n the median Hamming distance between items. This was the optimal value we could work with, because larger values resulted in an excessive number of hard locations being in the sphere, which became computationally unfeasible with the Python SDM library.

Three different methods of corruption were tested. First, the correct data was written to the Kanerva network, then reads at corrupted locations were tested. A known language bit-string, with a single corrupted bit, was used as the read location, and the result of the read was compared to the original bit-string in order to test bit recovery. The average Hamming distance resulting from the corruption of a given bit, corresponding to a particular syntactic parameter, was calculated across all languages. In order to test for relationships independent of the prevalence of the features, another test was run that normalized for this. For each feature, a subset of languages of fixed size was chosen randomly such that half of the languages had that feature. Features that had too few languages with or without the feature to reach the chosen fixed size were ignored for this purpose.

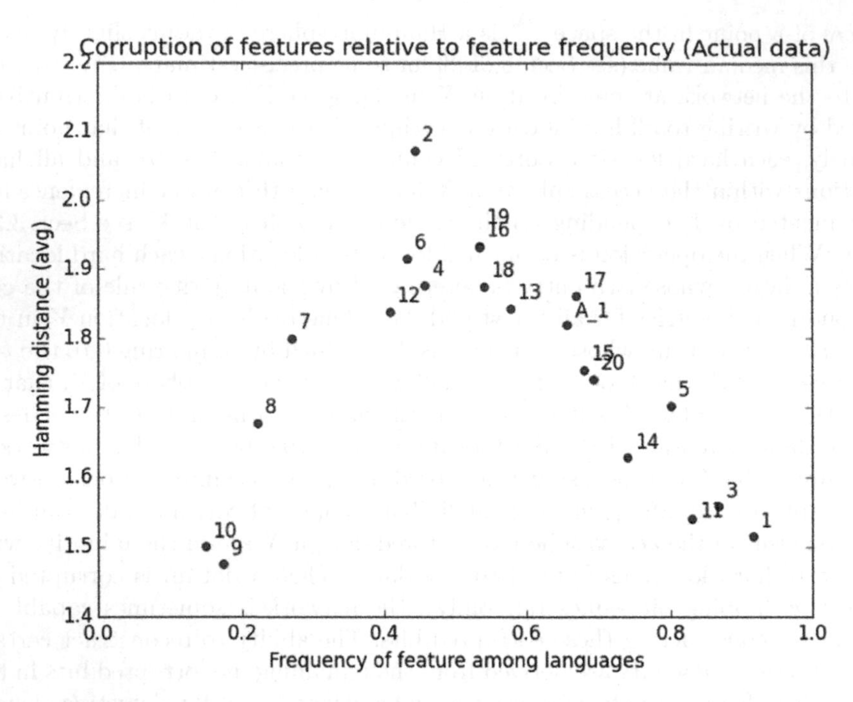

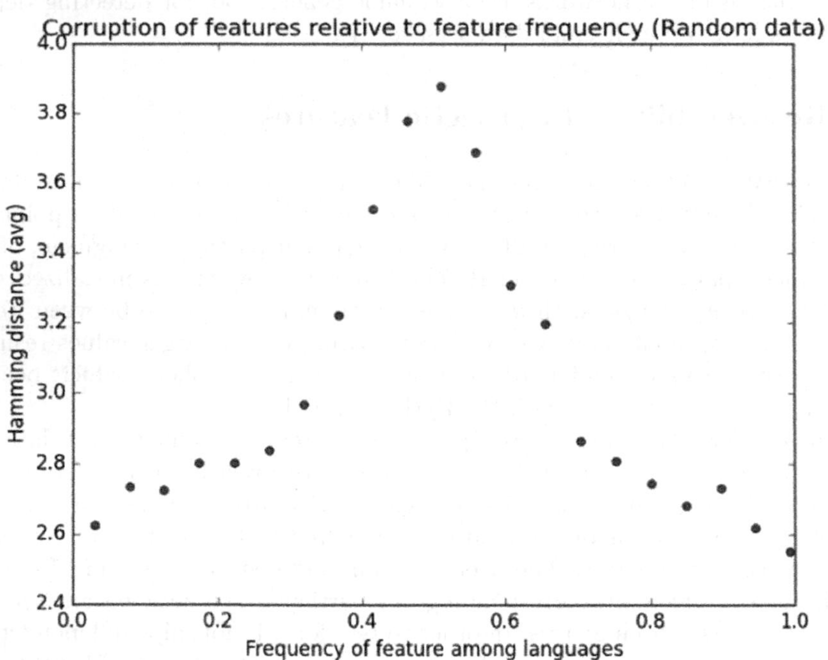

Fig. 1. Prevalence and recoverability for syntactic parameters in a Kanerva Network (actual data compared with random data).

For this test, a fixed size of 95 languages was chosen, as smaller sizes would yield less significant results, and larger sizes would result in too many languages being skipped. The languages were then written to the Kanerva network and the recoverability of that feature was measured. Finally, to check whether the different recovery rates we obtained for different syntactic parameters were really a property of the language data, rather than of the Kanerva network itself, the test was run again with random data generated with an approximately similar distribution of bits. The results for the actual data and for random data are reported in Fig. 1.

The random data show an overall general shape of the curve that reflects a property of the Kanerva network relating frequency of occurrence and recoverability. This overall effect, relating frequencies and recoverability, seen in random data with the same frequencies as the chosen set of parameters, seems in itself interesting, given ongoing investigations on how prevalence rates of different syntactic parameters may correlate to neuroscience models, see for instance [12].

The magnitude of the values for the actual data, however, differs significantly from the random data curve. This indicates that the recoverability rates observed for the syntactic parameters are also being influenced by the existence of dependence relations between different syntactic parameters. The normalized test indicates a smaller but still significant variation in feature recoverability even when all features considered had the same prevalence among the dataset.

3.1 Recoverability Scores

To each parameter we assign a score, obtained by computing the average Hamming distance between the resulting bit-vector in the corruption experiment and the original one. The lower the score, the more easily recoverable a parameter is from the uncorrupted data, hence from the other parameters.

The resulting levels of recoverability of the syntactic parameters are listed in the table below along with the frequency of expression among the given set of languages. The results of the normalized test are given, for a selection of parameters, in Fig. 2.

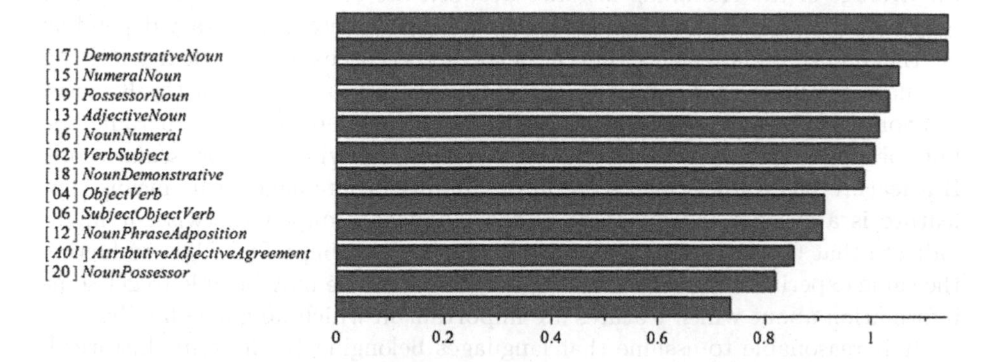

Fig. 2. Corruption (normalized test) of some syntactic parameters.

Parameter	Frequency	Corruption (non-normalized)
[01] Subject–Verb	0.64957267	1.50385541439
[02] Verb–Subject	0.31623933	2.03638553143
[03] Verb–Object	0.61538464	1.56180722713
[04] Object–Verb	0.32478634	1.86186747789
[05] Subject–Verb–Object	0.56837606	1.6709036088
[06] Subject–Object–Verb	0.30769232	1.88596384645
[07] Verb–Subject–Object	0.1923077	1.7879518199
[08] Verb–Object–Subject	0.15811966	1.66993976116
[09] Object–Subject–Verb	0.12393162	1.46596385241
[10] Object–Verb–Subject	0.10683761	1.4907228899
[11] Adposition–Noun–Phrase	0.58974361	1.52427710056
[12] Noun–Phrase–Adposition	0.2905983	1.81512048125
[13] Adjective–Noun	0.41025642	1.82927711248
[14] Noun–Adjective	0.52564102	1.6037349391
[15] Numeral–Noun	0.48290598	1.74969880581
[16] Noun–Numeral	0.38034189	1.94036144018
[17] Demonstrative–Noun	0.47435898	1.87596385121
[18] Noun–Demonstrative	0.38461539	1.87463855147
[19] Possessor–Noun	0.38034189	1.91487951279
[20] Noun–Possessor	0.49145299	1.74102410674
[A 01] Attributive–Adjective–Agreement	0.46581197	1.79102409244

4 Further Questions and Directions

We outline here some possible directions in which we plan to expand the present work on an approach to the study of syntactic parameters using Kanerva Networks.

One limitation of our result is that this scalar score is simply computed as the average of the Hamming distance between the resultant bit-vector and the original bit-vector. The derivability of a certain parameter might vary depending on the family of languages that it belongs to. For example, when a certain language feature is not robust to corruption in certain regions of the Kanerva Network, which means the parameter is not dependent on other parameters, but robust to corruption in all the other regions, we will get a low scalar score. If a feature has a low scalar score in one family of languages, this means that feature is a sharing characteristic of the language group. Otherwise, it might indicate that the feature is a changeable one in the group. Thus, by conducting the same experiments grouped by language families, we may be able to get some information about which features are important in which language family.

It is reasonable to assume that languages belonging to the same historical-linguistic family are located near each other in the Kanerva Network. However, a more detailed study where data are broken down by different language families will be needed to confirm whether syntactic proximity as detected by a Kanerva network corresponds to historical poximity. Under the assumption that closely related languages remain near in the Kanerva Network, the average of dependencies of a given parameter over the whole space might be less informative globally, because there is no guarantee that the dependencies would hold throughout all regions of the Kanerva Network. However, this technique may help identifying specific relations between syntactic parameters that hold within specific language families, rather than universally across all languages. The existence of such relations is consistent with the topological features identified in [17] which also vary across language families.

One of the main open frontiers in understanding human language is relating the structure of natural languages to the neuroscience of the human brain. In an idealized vision, one could imagine a Universal Grammar being hard wired in the human brain, with syntactic parameters being set during the process of language acquisition (see [1] for an expository account). This view is inspired by Chomsky's original proposals about Universal Grammar. A serious difficulty lies in the fact that there is, at present, no compelling evidence from the neuroscience perspective that would confirm this elegant idea. Some advances in the direction of linking a Universal Grammar model of human language to neurobiological data have been obtained in recent years: for example, some studies have suggested Broca's area as a biological substrate for Universal Grammar, [16]. Recent studies like [12] found indication of possible links between cross linguistic prevalence of syntactic parameters relating to word order structure and neuroscience models of how action is represented in Broca's area of the human brain. This type of results seems to cast a more positive light on the possibility of relating syntactic parameters to computational neuroscience models. Universal Grammar should be seen in the plasticity adaptive rules (storing algorithms) that shape the network structure and that are known to be universal across cortical areas and neural networks. Models of language acquisition based on neural networks have been previously developed, see for example the survey [18]. Various results, [3,7,10,11,13], have shown advantages of Kanerva's sparse distributed memories over other models of memory based on neural networks. To our knowledge, Kanerva Networks have not yet been systematically used in models of language acquisition, although the use of Kanerva Networks is considered in the work [14] on emergence of language. Thus, a possible way to extend the present model will be storing data of syntactic parameters in Kanerva Network, with locations representing (instead of different world languages) events in a language acquisition process that contain parameter-setting cues. In this way, one can try to create a model of parameter setting in language acquisition, based on sparse distributed memories as a model of human memory. We will return to this approach in future work.

Acknowledgment. This work was performed in the last author's Mathematical and Computational Linguistics lab and CS101/Ma191 class at Caltech. The last author was partially supported by NSF grants DMS-1201512 and PHY-1205440.

References

1. Baker, M.: The Atoms of Language. Basic Books, New York City (2001)
2. Chomsky, N., Lasnik, H.: The theory of principles and parameters. In: Syntax: An International Handbook of Contemporary Research, pp. 506–569. de Gruyter (1993)
3. Chou, Ph.A.: The capacity of the Kanerva associative memory. IEEE Trans. Inform. Theory **35**(2), 281–298 (1989)
4. Franklin, S.: Artificial Minds. MIT Press, Cambridge (2001)
5. Furber, S.B., Brown, G., Bose, J., Cumpstey, J.M., Marshall, P., Shapiro, J.L.: Sparse distributed memory using rank-order neural codes. IEEE Trans. Neural Netw. **18**(3), 648–659 (2007)
6. Haspelmath, M.: Parametric versus functional explanations of syntactic universals. In: The Limits of Syntactic Variation, pp. 75–107. John Benjamins (2008)
7. Hely, T.A., Willshaw, D.J., Hayes, G.M.: A new approach to Kanerva's sparse distributed memory. IEEE Trans. Neural Netw. **8**(3), 791–794 (1997)
8. Kanerva, P.: Sparse Distributed Memory. MIT Press, Cambridge (1988)
9. Kanerva, P.: Sparse distributed memory and related models. In: Hassoun, M.H. (eds.) Associative Neural Memories: Theory and Implementation, pp. 50–76. Oxford University Press (1993)
10. Kanerva, P.: Encoding structure in Boolean space. In: Niklasson, L., Boden, M., Ziemke, T. (eds.) ICANN 98. Perspectives in Neural Computing, pp. 387–392. Springer, Heidelberg (1998). doi:10.1007/978-1-4471-1599-1_57
11. Keeler, J.D.: Capacity for patterns and sequences in Kanerva's SDM as compared to other associative memory models. In: Anderson, D.Z. (ed.) Neural Information Processing Systems, pp. 412–421. American Institute of Physics (1988)
12. Kemmerer, D.: The cross-linguistic prevalence of SOV and SVO word orders reflects the sequential and hierarchical representation of action in Broca's area. Lang. Linguist. Compass **6**(1), 50–66 (2012)
13. Knoblauch, A., Palm, G., Sommer, F.T.: Memory capacities for synaptic and structural plasticity. Neural Comput. **22**, 289–341 (2010)
14. MacWhinney, B.: Models of the emergence of language. Annu. Rev. Psychol. **49**, 199–227 (1998)
15. Marcolli, M.: Syntactic parameters and a coding theory perspective on entropy and complexity of language families. Entropy **18**(4), 17 (2016). Paper No. 110
16. Marcus, G.F., Vouloumanos, A., Sag, I.A.: Does Broca's play by the rules? Nat. Neurosci. **6**(7), 651–652 (2003)
17. Port, A., Gheorghita, I., Guth, D., Clark, J.M., Liang, C., Dasu, S., Marcolli, M.: Persistent topology of syntax. arXiv:1507.05134 [cs.CL]
18. Poveda, J., Vellido, A.: Neural network models for language acquisition: a brief survey. In: Corchado, E., Yin, H., Botti, V., Fyfe, C. (eds.) IDEAL 2006. LNCS, vol. 4224, pp. 1346–1357. Springer, Heidelberg (2006). doi:10.1007/11875581_160
19. Siva, K., Tao, J., Marcolli, M.: Spin glass models of syntax and language evolution. arXiv:1508.00504 [cs.CL]
20. Shu, K., Marcolli, M.: Syntactic structures and code parameters. Math. Comp. Sci. **11**(1), 79–90 (2017)
21. Shu, K., Aziz, S., Huynh, V.L., Warrick, D., Marcolli, M.: Syntactic phylogenetic trees. arXiv:1607.02791 [cs.CL]
22. Syntactic Structures of World Languages. http://sswl.railsplayground.net/

Multi-scale Activity Estimation with Spatial Abstractions

Majd Hawasly[1(✉)], Florian T. Pokorny[2], and Subramanian Ramamoorthy[3]

[1] FiveAI Inc., Edinburgh, UK
m.hawasly@five.ai
[2] KTH Royal Institute of Technology, Stockholm, Sweden
[3] School of Informatics, The University of Edinburgh, Edinburgh, UK

Abstract. Estimation and forecasting of dynamic state are fundamental to the design of autonomous systems such as intelligent robots. State-of-the-art algorithms, such as the particle filter, face computational limitations when needing to maintain beliefs over a hypothesis space that is made large by the dynamic nature of the environment. We propose an algorithm that utilises a hierarchy of such filters, exploiting a filtration arising from the geometry of the underlying hypothesis space. In addition to computational savings, such a method can accommodate the availability of evidence at varying degrees of coarseness. We show, using synthetic trajectory datasets, that our method achieves a better normalised error in prediction and better time to convergence to a true class when compared against baselines that do not similarly exploit geometric structure.

1 Introduction

Autonomous agents acting in dynamic environments need the capacity to make predictions about the environment within which they are acting, so as to take actions that are suited to the present world state. Traditionally, tools for *state estimation* are geared to the case wherein uncertainty arises from noise in the dynamics or sensorimotor processes. For example, the particle filter is a state estimation method utilising a nonparametric representation of beliefs over the state space, used extensively in robotics. However, in problems involving spatial activity, e.g., robot navigation, the underlying dynamics are best described in a hierarchical fashion, as movement is not just determined by local physical laws and noise characteristics, but also by longer-term *goals* and *preferences*. This has a few implications for predictive models: we require techniques that (1) accept evidence at varying scales - from very precise position measurements to coarser forms of knowledge, e.g. human feedback, and (2) make predictions at multiple scales to support decision making. These form the primary focus of this paper.

Early models of large-scale spatial navigation [5] considered ways in which multiple representations, ranging from coarse and intuitive topological notions of connectivity between landmarks to a more detailed metrical and control level

At the time of this study, the first author was with The University of Edinburgh.

F. Nielsen and F. Barbaresco (Eds.): GSI 2017, LNCS 10589, pp. 273–281, 2017.
https://doi.org/10.1007/978-3-319-68445-1_32

description of action selection, could be brought together in a coherent framework and implemented on robots. Other recent methods, e.g. [1,3], propose ways in which control vector fields could be abstracted so as to support reasoning about larger-scale tasks. While these works provide useful inspiration, the hierarchy in these methods is often statically defined by the designer, while in many applications it is of interest to learn it directly from data, e.g. to enable continual adaptation over time. Also, these approaches are often silent on how best to integrate tightly with Bayesian belief estimates, such as within a particle filter.

There is indeed prior work on the notion of hierarchy in state estimation with particle filters. For instance, Verma et al. [11] define a variable resolution particle filter for operation in large state spaces, where chosen states are aggregated to reduce the complexity of the filter. Brandao et al. [2] devise a subspace hierarchical particle filter wherein state estimation can be run in parallel with factored parallel computation. Other ways to factoring computation exist, e.g. [6,10], and a hierarchy of feature encodings can be used [13]. However, to the best of our knowledge, no prior method allows tracking a process on multiple scales at once and accepts evidence with variable resolutions.

In this paper, we learn a spatial hierarchy directly from input trajectories, using which we devise a novel construction of a bank of particle filters - one at each scale in a geometric filtration - which maintain consistent beliefs over the trajectories as a whole and, through that, over the state space. We present an agglomerative clustering scheme [7] using the Fréchet distance between trajectories [4] to compute a tree-structured representation of trajectory classes that correspond to incrementally-coarser partitions of the underlying space. This is inspired by persistent homology on trajectories [8,9] whose output is also such a hierarchical representation. We then define a *linear dynamics* model at each of the levels of the hierarchy based on the subset of trajectories they represent, and show how that can be used with a stream of observations to provide updates to the probability that the system is following the dynamics associated with each of the abstracted trajectory classes. This construction of the filter allows us to fluently incorporate readings of varying resolution if they were accompanied by an indication of the coarseness with which the observation is to be interpreted.

We show that our proposed method performs better than baselines both in terms of normalised error in prediction with respect to the ground truth, and in terms of the time taken for the belief to converge to the true trajectory of a class (where convergence is defined with respect to the resolution of the prediction being considered). We perform experiments with synthetic datasets which brings out the qualitative behaviour of the procedure in a visually intuitive manner.

2 Multiscale Hierarchy of Particle Filters

The *Multiscale Hierarchy of Particle Filters* (MHPF) is a bank of *consistent* particle filters defined over *abstractions* of the state space induced by example trajectories. The lowest level of this hierarchy consists of the complete set of trajectories with cardinality equal to the size of the trajectory dataset, while each other abstract level has coarser descriptions of the trajectory shape defined by equivalence class of similar trajectories for increasing thresholds. With a particle filter defined at each level, this *inclusion* property of the representation allows evidence at various degrees of coarseness to be incorporated into the full bank of filters while maintaining consistency across all levels, see Fig. 2.

Construction. To create a filtration of spatial abstractions from trajectories we consider agglomerative hierarchical clustering [12] by means of a trajectory distance measure. In this paper, we use the discrete Fréchet distance [4]: for two discretised d-dimensional trajectories $\tau_1 : [0, m] \to \mathbb{R}^d$ and $\tau_2 : [0, n] \to \mathbb{R}^d$, the distance $\delta_F(\tau_1, \tau_2) = \inf_{\alpha,\beta} \max_{j \le m+n} \delta_E(\tau_1(\alpha(j)), \tau_2(\beta(j)))$, where α and β are discrete, monotonic re-parametrisations $\alpha : [1 : m+n] \to [0 : m]$, $\beta : [1 : m+n] \to [0 : n]$ which align the trajectories to each other point-wise, and $\delta_E(.,.)$ is Euclidean distance. Thus, δ_F corresponds to the maximal point-wise distance between optimal reparameterisations of τ_1 and τ_2, which can be computed efficiently using dynamic programming in $O(mn)$ time [4].

Let D be the distance matrix of the input trajectories, $D_{i,j} = D_{j,i} = \delta_F(\tau_i, \tau_j)$. A *single-linkage* hierarchical agglomerative clustering of D results in a *tree* $\mathcal{T}$ of trajectory clusters in which the leaves are the single trajectories, while every other tree layer is created when the pair with the smallest distance from the previous layer combine together (Fig. 2). If τ_i and τ_j are such a pair, we call the new cluster τ_{ij} a *parent* to its constituents and write $\tau_{ij} = \rho(\tau_i) = \rho(\tau_j)$. Let the *birth index* b be that minimum distance that indexes the creation of a layer (e.g., $b_{ij} = D_{i,j}$ for τ_{ij}), and the *death index* d be the distance at which a cluster is subsumed to its parent (e.g., $d_i = d_j = D_{i,j}$ for τ_i and τ_j). Let $\mathcal{C}$ be the set of all clusters in $\mathcal{T}$. A class $c_i \in \mathcal{C}$ is *alive* at some index x if $b_i \le x < d_i$. A *level* in the tree $\mathcal{C}_x \subseteq \mathcal{C}$ at index x contains all the classes that are alive at x. Figure 1 (Left) illustrates an example clustering.

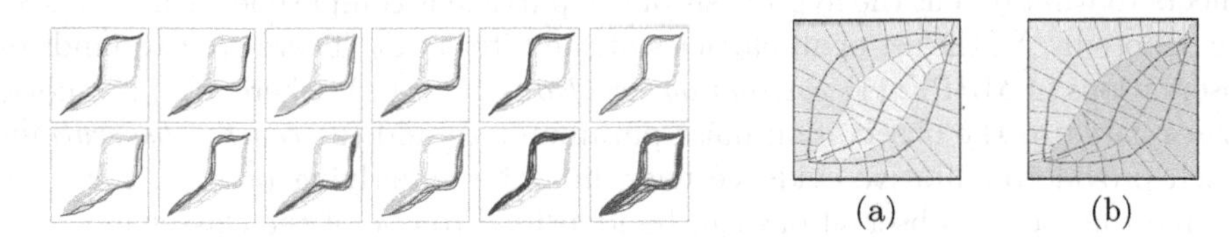

Fig. 1. (Left) Trajectory clusters with increasing birth indices of a tree of 14 trajectories using hierarchical single-linkage agglomerative clustering with Fréchet distance. (Right) The intuition behind the probability operations in a toy example 2D domain, where (a) three classes merge into two (b).

Thus, a cluster $c \in \mathcal{C}$ is a collection of qualitatively *similar* trajectories at some *level of resolution* (index) b. The *class* of behaviour that c represents could be modelled as a generative model $\mathbf{P}(z'|z, c)$, $z, z' \in \mathbb{R}^d$. With the assumption of no self-intersecting trajectories, we can approximate the dynamics of c at some

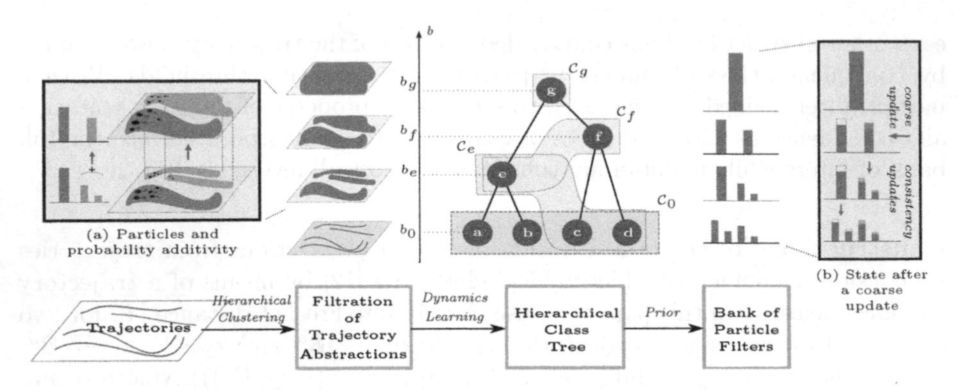

Fig. 2. An overview of the approach. Trajectories are hierarchically clustered into a filtration of spatial abstractions (*classes*), organised in a tree structure by birth indices. Shaded areas on the tree show *levels* of the hierarchy, with $\mathcal{C}_0$ being the finest level with single trajectory classes. Inset (a) shows an example particle set and how the tree structure enforces the consistency of class probabilities. The distributions on the right show an example of a consistent estimate across the tree maintained by a bank of particle filters. Inset (b) shows an example of a coarse observation received at one level, and how updates propagate throughout the tree to maintain consistency.

arbitrary point $z \in \mathbb{R}^d$ using a weighted average of velocity at local points of c in an ϵ-ball around z: $B_\epsilon(z) = \{z' \in c : \delta_E(z', z) < \epsilon\}$, where ϵ relates to the density or sparsity of the trajectories. Hence, $\dot{z} = \frac{1}{\eta} \sum_{z' \in B_\epsilon(z)} \frac{\dot{z}'}{\delta_E(z,z')}$, with normalisation $\eta = \sum_{z' \in B_\epsilon(z)} \frac{1}{\delta_E(z,z')}$. Thus, $z' \sim z + \dot{z} + \gamma(\kappa)$, where $\gamma(\kappa)$ is a noise term related to dynamics noise κ.

At each level of the tree $\mathcal{C}_b$ we define a particle filter where a *particle* x^t represents a weighted hypothesis of *both* the class of behaviour $c \in \mathcal{C}_b$ and the position $z^t \in \mathbb{R}^d$ at time t. We write $(x^t(z^t, c), w^t)$ where w^t is a weight that reflects to what extent the hypothesis of the particle is compatible with evidence. We denote by X_b the set of all particles of the filter at $\mathcal{C}_b$. There are two kinds of observations in MHPF: (1) *position observations* $z^t + \gamma(\psi)$, where $\gamma(\psi)$ is a noise term related to the observation noise parameter ψ; and (2) *coarse observations* which provide qualitative evidence regarding the underlying process. Here, we assume that coarse observations can be identified to one of the classes in $\mathcal{C}$.

Algorithm 1 presents the full MHPF procedure. First, the particle set X_0 of the filter at $\mathcal{C}_0$ is created by sampling N particles from a prior over initial positions and class assignment from $\mathcal{C}_0$ (individual trajectories) with uniform weights. Denote by N_i the number of particles of class c_i, such that $\sum_{c_i \in \mathcal{C}_0} N_i = N$. The prior probabilities of the classes $c_i \in \mathcal{C}_0$ can be computed as N_i/N. These probabilities propagate recursively upwards in the tree by additivity: a parent's probability is the sum of its children's probabilities, $\mathbf{P}^t(\bar{c}) = \sum_{\underline{c}=\rho^{-1}(\bar{c})} \mathbf{P}^t(\underline{c})$. By the same principle, the children of a class proportionally inherit their parent's probability when moving down the tree. For the sake of intuition, consider the simple example in Fig. 1 (Right), where a cluster

Algorithm 1. Multiscale hierarchy of particle filters

Require: Prior over particles, number of basic particles N, the depletion parameter v, tree structure $\mathcal{T}$

1: Create X_0: sample N particles from a prior over $\mathcal{C}_0 \times \mathbb{R}^d$ with equal weights.
2: **for** each time step $t > 0$ **do**
3: Build the tree probabilities up from X_0 and $\mathcal{C}_0$ (Algorithm 2).
4: **for** parents $\bar{c}$ of $\mathcal{C}_0$ classes recursively **to** the root of $\mathcal{T}$ **do**
5: Sample $N_{\bar{c}}$ particles; $N_{\bar{c}} = \sum_{\underline{c}=\rho^{-1}(\bar{c})} N_{\underline{c}}$, with equal weights
6: **end for**
7: Sample a new position per particle, $z^t \sim \mathbf{P}(z|z^{t-1}, c)$
8: Receive observation ξ^t.
9: **if** fine observation **then**
10: $\mathcal{C}_\xi = \mathcal{C}_0$.
11: update X_0 weights with Euclidean distance to ξ^t: $w^t \propto -\log(\delta_E(., \xi^t))$.
12: **else if** coarse observation at tree level b_ξ **then**
13: Find all alive classes at b_ξ: $\mathcal{C}_\xi = \{c_i \in \mathcal{C} : \ b_i \le b_\xi < d_i\}$.
14: Compute tree distance $\delta_\mathcal{T}(c, \xi^t)$, for all $c \in \mathcal{C}_\xi$.
15: Update weights in X_c, $c \in \mathcal{C}_\xi$ relative to distance: $w^t \propto -\log(\delta_\mathcal{T}(c, \xi^t))$.
16: **end if**
17: Rebuild the tree probabilities from $\mathcal{C}_\xi$ and X_ξ (Algorithm 2).
18: Update particle weights in $X \setminus X_\xi$: $w^t = w^{t-1} \frac{\mathbf{P}^t(c)}{\mathbf{P}^{t-1}(c)}$
19: Update X_0: resample $N\,(1-v)$ particles from X_0 based on new weights w^t, and $N\,v$ particles uniformly randomly from $\mathcal{C}_0$.
20: **end for**

Algorithm 2. Tree probability rebuild

Require: Tree structure $\mathcal{T}$, tree level $\mathcal{C}_b$, particle set X_b

1: Update the probabilities of $c_i \in \mathcal{C}_b$ from X_b weights: $\mathbf{P}^t(c_i) = \frac{\sum_{c(x)=c_i} w^t(x)}{\sum_{x \in X_b} w^t(x)}$
2: **for** children of $\mathcal{C}_b$ classes recursively **to** the leaves of $\mathcal{T}$ **do**
3: Update child $\underline{c}$ probability relative to its parent $\bar{c}$: $\mathbf{P}^t(\underline{c}) = \mathbf{P}^{t-1}(\underline{c}) \frac{\mathbf{P}^t(\bar{c})}{\mathbf{P}^{t-1}(\bar{c})}$
4: **end for**
5: **for** parents of $\mathcal{C}_0$ classes recursively **to** the root of $\mathcal{T}$ **do**
6: Update parent $\bar{c}$ probability relative to its children $\underline{c}$: $\mathbf{P}^t(\bar{c}) = \sum_{\underline{c}=\rho^{-1}(\bar{c})} \mathbf{P}^t(\underline{c})$
7: **end for**

of trajectories can be understood spatially as the union of Voronoi cells of trajectory discretisation. The corresponding probability of this class is the probability of the agent being in that region. Then, when classes merge at some level of resolution their corresponding regions merge, and thus their probabilities are added up. This simple technique guarantees consistency of the filters by design.

With the class probabilities specified, the same number of particles as assigned to the children are sampled for parents, $N_{\bar{c}} = \sum_{\underline{c}=\rho^{-1}(\bar{c})} N_{\underline{c}}$, and this is repeated recursively to the top of the tree. Note that, any arbitrary level $\mathcal{C}_b$ of the tree would have exactly N particles with a proper probability distribution. The last stage of the tree construction is to sample new positions for the particles. Note that the class assignment of a particle does not change due to sampling.

Updates. A coarse observation $\xi \in \mathcal{C}$ with resolution b_ξ targets all the particles from classes that are *alive* at $\mathcal{C}_\xi = \{c_i \in \mathcal{C} | \ b_i \leq b_\xi < d_i\}$. To update these particles, we use the *tree distance* between classes $\delta_\mathcal{T}(.,.)$ which we define as the birth index of the *youngest* shared parent of the two classes in the tree. This measures how large the ϵ-balls around the points of one class need to be to include the other. For example, in Fig. 2, $\delta_\mathcal{T}(e, g) = \delta_\mathcal{T}(c, e) = b_g$. The weight of a particle is updated relative to the distance of the observation from the particle's class c, $w \propto -\log(\delta_\mathcal{T}(\xi, c))$. A position observation $\xi \in \mathbb{R}^d$, on the other hand, updates a particle relative to the Euclidean distance between the observation and the particle's position z, $w \propto -\log(\delta_E(\xi, z))$.

After updating all particles in X_ξ, the probabilities of the corresponding classes in $\mathcal{C}_\xi$ are recomputed as the sum of their particles' normalised weights, then propagate to the rest of the tree as in Algorithm 2. Here, children classes of $\mathcal{C}_\xi$ are updated first recursively relative to their parents' new probabilities, $\mathbf{P}^t(\underline{c}) = \mathbf{P}^{t-1}(\underline{c})\frac{\mathbf{P}^t(\bar{c})}{\mathbf{P}^{t-1}(\bar{c})}$, $\forall \underline{c} = \rho^{-1}(\bar{c})$, then the updates propagate upwards to update all the remaining parents $\mathbf{P}^t(\bar{c}) = \sum_{\underline{c}=\rho^{-1}(\bar{c})} \mathbf{P}^t(\underline{c})$. Then, particle weights are updated to reflect the updated class probabilities, $w^t = w^{t-1}\frac{\mathbf{P}^t(c)}{\mathbf{P}^{t-1}(c)}$, $\forall x \in X \setminus X_\xi$. The final step is to sample N particles from X_0 with uniform weights to get the posterior particle set after incorporating the evidence ξ. To guard against particle depletion, we replace the classes of a small percentage v of all particles uniformly randomly to classes from $\mathcal{C}_0$.

3 Experiments

We evaluate the performance of **MHPF** with $N = 100$ particles in two synthetic 2-dimensional navigation domains, one representing a 2-dimensional configuration space with 33 trajectories, and the other with 13 trajectories (Fig. 3 (Left)). We compare the performance to particle filters without access to the hierarchical structure: **BL1** is a basic particle filter with $N = 100$ particles, each follows the dynamics of a single trajectory (classes $c \in \mathcal{C}_0$); and **BL2** is a particle filter with $N = 100$ particles which all follow the averaging dynamics of the trajectories together with κ noise (Note that **BL1** is equivalent to the filter at the bottom layer of MHPF stack, and **BL2** is equivalent to the filter at the top layer.) We use as metrics: (1) the mean squared error of the filter's point prediction, (2) the tree distance of the filter's predicted class to the ground truth, and (3) the time to convergence to the true class. Each experiment is run with 10 randomly-selected ground truth trajectories, reporting averaged scores of 25 repetitions. Trajectories are uniformly discretised, and the length of a trial depends on the number of trajectory points. Observation at time t is generated from the discretised ground truth $z^t \in \mathbb{R}^2$ and the observation noise ψ. A fine observation is defined as $z^t + \gamma$ where $\gamma \in [0, \psi] \times [0, \psi]$, while a coarse observation is selected by sampling $n = 10$ points from $\mathcal{N}(z^t, \psi^2)$ then finding the class that is most likely to generate these samples. We use the localised dynamics model as in Sect. 2 with $\epsilon = b_c$ for some coarse class c and the noise parameter κ. At the end of every step, $v = 1\%$ of the particles is changed randomly.

In the configuration space dataset, we compute the filter's predicted position at time t as the w-weighted average of the particle positions when using fine observations only, and report the average mean squared error (MSE) of the ground truth over time. **MHPF** achieved a mean of 0.27 (standard deviation of 0.04), beating **BL1** 0.38(0.14) and **BL2** 0.53(0.13). Figure 3 (Right) illustrates

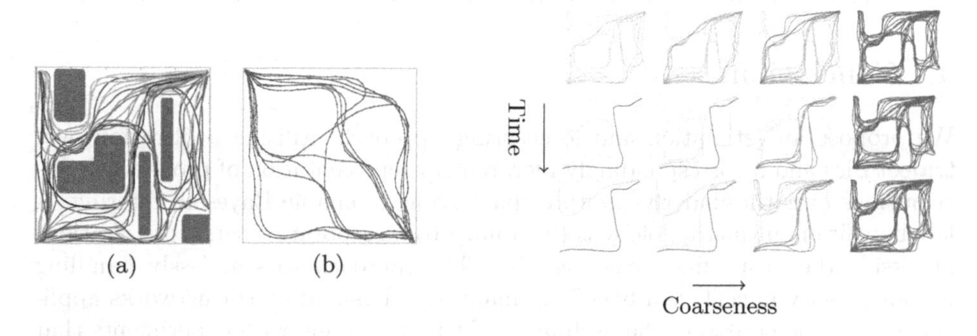

Fig. 3. (Left) Datasets used. (Right) Evolution of MHPF prediction. The columns show levels across the tree with the finest at the left, and rows show time steps with the first at the top. Each panel shows the trajectories of the alive classes. The opacity of the line reflects the probability of the class.

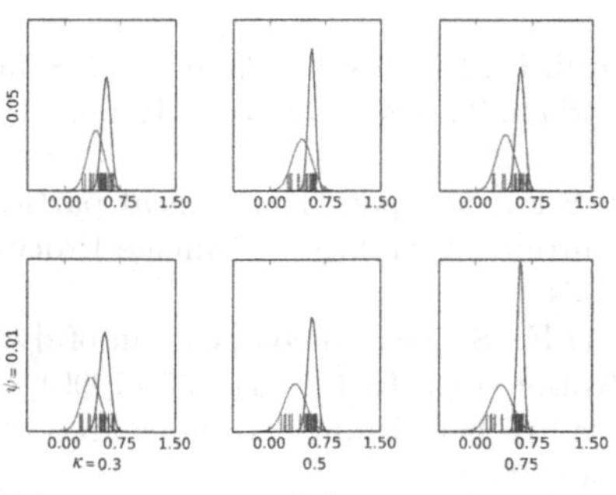

(a) Tree distance of the ground truth to MAP prediction. Coarse instructions were provided stochastically 50% of the time. The plot shows robustness against noise as dynamics noise varies between 30% (Left) to 75% (Right) of range, and observation noise ranges between 1% (Bottom) to 5% (Top). MHPF converges to a better solution than the baseline (statistically significant at p-value= 0.004).

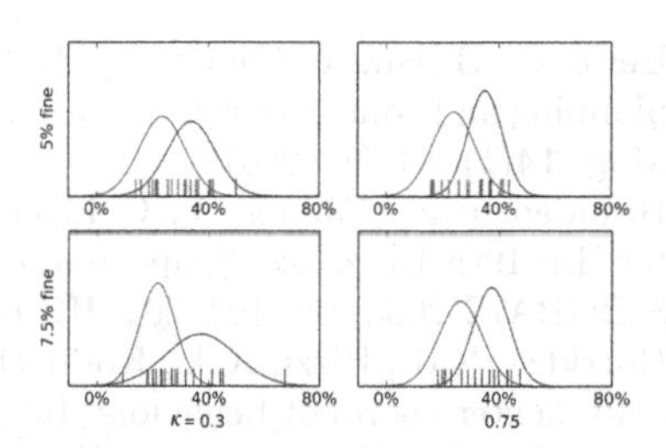

(b) Time needed to reach within 33% of convergence. Fine observations are provided for a lead-in period (5% (Top) and 7.5% (bottom) of trial time). The plot shows the benefit of coarse observations to convergence time. Dynamics noise ranges from 30% (Left) to 75% (Right), and observation noise is set to 1%. MHPF converges faster to the correct solution than the baseline (statistically significant at a p-value = 0.02).

Fig. 4. Performance results comparing MHPF (red) and BL1 (blue) - lower is better. (Color figure online)

the kind of multi-resolution output MHPF can produce, showing the *maximum a posteriori* (MAP) class in time at different levels of the tree.

Next, we compare **MHPF** with **BL1** using the 13 trajectory dataset in a situation where fine observations are consistently generated, but coarse observations are produced stochastically 50% of the time. We analyse the benefit of this additional knowledge by plotting, in Fig. 4a, the average tree distance of the MAP prediction to the ground truth, with noise parameters ($\kappa = 30\%, 50\%, 75\%$) and ($\psi = 1\%, 5\%$). Finally, when fine observations are only provided for a lead-in period of 5%/ 7.5% of trial length followed by only coarse observations, we show in Fig. 4b the time needed for the tree distance to converge within the 33%-ball of the ground truth with noise parameters $\psi = 1\%$ and $\kappa = 30\%, 75\%$.

4 Conclusion

We propose an estimation and forecasting approach utilising a filtration over trajectories and a correspondingly hierarchical representation of probability distributions over the underlying state space so as to enable Bayesian filtering. A key benefit of our methodology is the ability to incorporate 'coarse' observations alongside the basic 'fine' scale signals. This approach to seamlessly handling inhomogeneity in scale is a benefit in many robotics and sensor networks applications. We demonstrate the usefulness of this technique with experiments that show performance gains over a conventional particle filtering scheme that does not similarly exploit the geometric structure in the hypothesis space.

References

1. Belta, C., Bicchi, A., Egerstedt, M., Frazzoli, E., Klavins, E., Pappas, G.: Symbolic planning and control of robot motion [grand challenges of robotics]. Robot. Autom. Mag. **14**(1), 61–70 (2007)
2. Brandao, B.C., Wainer, J., Goldenstein, S.K.: Subspace hierarchical particle filter. In: 19th Brazilian Symposium on Computer Graphics and Image Processing, SIBGRAPI 2006, pp. 194–204. IEEE (2006)
3. Burridge, R.R., Rizzi, A.A., Koditschek, D.E.: Sequential composition of dynamically dexterous robot behaviors. Int. J. Robot. Res. **18**(6), 534–555 (1999)
4. Eiter, T., Mannila, H.: Computing discrete Fréchet distance. Technical report CD-TR 94/64. Technical University of Vienna (1994)
5. Kuipers, B.: The spatial semantic hierarchy. Artif. Intell. **119**(1), 191–233 (2000)
6. MacCormick, J., Isard, M.: Partitioned sampling, articulated objects, and interface-quality hand tracking. In: Vernon, D. (ed.) ECCV 2000. LNCS, vol. 1843, pp. 3–19. Springer, Heidelberg (2000). doi:10.1007/3-540-45053-X_1
7. Müllner, D.: Modern hierarchical, agglomerative clustering algorithms. arXiv preprint arXiv:1109.2378 (2011)
8. Pokorny, F.T., Hawasly, M., Ramamoorthy, S.: Multiscale topological trajectory classification with persistent homology. In: Robotics: Science and Systems (2014)
9. Pokorny, F.T., Hawasly, M., Ramamoorthy, S.: Topological trajectory classification with filtrations of simplicial complexes and persistent homology. Int. J. Robot. Res. **35**, 201–223 (2015)

10. Shabat, G., Shmueli, Y., Bermanis, A., Averbuch, A.: Accelerating particle filter using randomized multiscale and fast multipole type methods. IEEE Trans. Pattern Anal. Mach. Intell. **PP**(99), 1 (2015)
11. Verma, V., Thrun, S., Simmons, R.: Variable resolution particle filter. In: International Joint Conference on Artificial Intelligence (IJCAI), pp. 976–984 (2003)
12. Xu, R., Wunsch, D.I.: Survey of clustering algorithms. IEEE Trans. Neural Netw. **16**(3), 645–678 (2005)
13. Yang, C., Duraiswami, R., Davis, L.: Fast multiple object tracking via a hierarchical particle filter. In: Tenth IEEE International Conference on Computer Vision (ICCV), vol. 1, pp. 212–219 (2005)

Geometry of Policy Improvement

Guido Montúfar[1,2(✉)] and Johannes Rauh[1]

[1] Max Planck Institute for Mathematics in the Sciences, InselstraßE 22, 04103 Leipzig, Germany
montufar@mis.mpg.de
[2] Departments of Mathematics and Statistics, UCLA, Los Angeles, CA 90095-1555, USA

Abstract. We investigate the geometry of optimal memoryless time independent decision making in relation to the amount of information that the acting agent has about the state of the system. We show that the expected long term reward, discounted or per time step, is maximized by policies that randomize among at most k actions whenever at most k world states are consistent with the agent's observation. Moreover, we show that the expected reward per time step can be studied in terms of the expected discounted reward. Our main tool is a geometric version of the policy improvement lemma, which identifies a polyhedral cone of policy changes in which the state value function increases for all states.

Keywords: Partially Observable Markov Decision Process · Reinforcement learning · Memoryless stochastic policy · Policy gradient theorem

1 Introduction

We are interested in the amount of randomization that is needed in action selection mechanisms in order to maximize the expected value of a long term reward, depending on the uncertainty of the acting agent about the system state.

It is known that in a Markov Decision Process (MDP), the optimal policy may always be chosen deterministic (see, e.g., [5]), in the sense that the action a that the agent chooses is a deterministic function of the world state w the agent observes. This is no longer true in a Partially Observable MDP (POMDP), where the agent does not observe w directly, but only the value s of a sensor. In general, optimal memoryless policies for POMDPs are stochastic. However, the more information the agent has about w, the less stochastic an optimal policy needs to be. As shown in [4], if a particular sensor value s uniquely identifies w, then the optimal policy may be chosen such that, on observing s, the agent always chooses the same action. We generalize this as follows: The agent may choose an optimal policy such that, if a given sensor value s can be observed from at most k world states, then the agent chooses an action probabilistically among a set of at most k actions.

Such characterizations can be used to restrict the search space when searching for an optimal policy. In [1], it was proposed to construct a low-dimensional

F. Nielsen and F. Barbaresco (Eds.): GSI 2017, LNCS 10589, pp. 282–290, 2017.
https://doi.org/10.1007/978-3-319-68445-1_33

manifold of policies that contains an optimal policy in its closure and to restrict the learning algorithm to this manifold. In [4], it was shown how to do this in the POMDP setting when it is known that the optimal policy can be chosen deterministic in certain sensor states. This construction can be generalized and gives manifolds of even smaller dimension when the randomization of the policy can be further restricted.

As in [4], we study the case where at each time step the agent receives a reward that depends on the world state w and the chosen action a. We are interested in the long term reward in either the average or the discounted sense [6]. Discounted rewards are often preferred in theoretical analysis, because of the properties of the dynamic programming operators. In [4], the analysis of average rewards was much more involved than the analysis of discounted rewards. While the case of discounted rewards follows from a policy improvement argument, an elaborate geometric analysis was needed for the case of average rewards.

Various works have compared average and discounted rewards [2,3,8]. Here, we develop a tool that allows us to transfer properties of optimal policies from the discounted case to the average case. Namely, the average case can be seen as the limit of the discounted case when the discount factor γ approaches 1. If the Markov chain is irreducible and aperiodic, this limit is uniform, and the optimal policies of the discounted case converge to optimal policies of the average case.

2 Optimal Policies for POMDPs

A (discrete time) partially observable Markov decision process (POMDP) is defined by a tuple $(W, S, A, \alpha, \beta, R)$, where W, S, A are finite sets of world states, sensor states, and actions, $\beta\colon W \to \Delta_S$ and $\alpha\colon W \times A \to \Delta_W$ are Markov kernels describing sensor measurements and world state transitions, and $R\colon W \times A \to \mathbb{R}$ is a reward signal. We consider stationary (memoryless and time independent) action selection mechanisms, described by Markov kernels of the form $\pi\colon S \to \Delta_A$. We denote the set of stationary policies by $\Delta_{S,A}$. We write $p^\pi(a|w) = \sum_s \beta(s|w)\pi(a|s)$ for the effective world state policy. Standard reference texts are [5,6].

We assume that the Markov chain starts with a distribution $\mu \in \Delta_W$ and then progresses according to α, β and a fixed policy π. We denote by $\mu_\pi^t \in \Delta_W$ the distribution of the world state at time t. It is well known that the limit $p_\mu^\pi := \lim_{T\to\infty} \frac{1}{T}\sum_{t=0}^{T-1} \mu_\pi^t$ exists and is a stationary distribution of the Markov chain. The following technical assumption is commonly made:

$(*)$ For all π, the Markov chain over world states is aperiodic and irreducible.

The most important implication of irreducibility is that the limit distribution p_μ^π is independent of μ. If the chain has period s, then $p_\mu^\pi = \lim_{T\to\infty} \frac{1}{s}\sum_{t=1}^{s} \mu_\pi^{T+t}$. In particular, under assumption $(*)$, $\mu_\pi^t \to p_\mu^\pi$ for any μ. (Since we assume finite sets, all notions of convergence of probability distributions are equivalent.)

The objective of learning is to maximize the expected value of a long term reward. The (normalized) discounted reward with discount factor $\gamma \in [0, 1)$ is

$$\mathcal{R}^{\gamma}_{\mu}(\pi) = (1-\gamma)\sum_{t=0}^{\infty}\gamma^t\sum_{w}\mu^t_{\pi}(w)\sum_{a}p^{\pi}(a|w)R(w,a) = (1-\gamma)\mathbb{E}_{\pi,\mu}\Big[\sum_{t=0}^{\infty}\gamma^t R(w_t,a_t)\Big].$$

The average reward is

$$\mathcal{R}_{\mu}(\pi) = \sum_{w}p^{\pi}_{\mu}(w)\sum_{a}p^{\pi}(a|w)R(w,a).$$

Under assumption $(*)$, $\mathcal{R}_{\mu}$ is independent of the choice of μ and depends continuously on π, as we show next. Since $\Delta_{S,A}$ is compact, the existence of optimal policies is guaranteed. Without assumption $(*)$, optimal policies for $\mathcal{R}_{\mu}$ need not exist. On the other hand, the expected discounted reward $\mathcal{R}^{\gamma}_{\mu}$ is always continuous, so that, for this, optimal policies always exist.

Lemma 1. *Under assumption $(*)$, $\mathcal{R}_{\mu}(\pi)$ is continuous as a function of π.*

Proof. By $(*)$, p^{π}_{μ} is the unique solution to a linear system of equations that smoothly depends on π. Thus, $\mathcal{R}_{\mu}$ is continuous as a function of π. □

Lemma 2. *For fixed μ and $\gamma \in [0,1)$, $\mathcal{R}^{\gamma}_{\mu}(\pi)$ is continuous as a function of π.*

Proof. Fix $\epsilon > 0$. There exists $l > 0$ such that $(1-\gamma)\sum_{t=l}^{\infty}\gamma^t R \leqslant \epsilon/4$, where $R = \max_{w,a}|R(w,a)|$. For each t, the distribution μ^t_{π} depends continuously on π. For fixed π, let U be a neighborhood of π such that $|\mu^t_{\pi}(w) - \mu^t_{\pi'}(w)| \leqslant \frac{1}{2|W|R}\epsilon$ for $t = 0,\ldots,l-1$, $w \in W$ and $\pi' \in U$. Then, for all $\pi' \in U$,

$$|\mathcal{R}^{\gamma}_{\mu}(\pi) - \mathcal{R}^{\gamma}_{\mu}(\pi')| \leqslant \frac{\epsilon}{2} + (1-\gamma)\sum_{t=0}^{l-1}\gamma^t\sum_{w}|\mu^t_{\pi}(w) - \mu^t_{\pi'}(w)|R \leqslant \frac{\epsilon}{2} + \frac{|W|}{2|W|R}\epsilon R = \epsilon.$$

□

The following refinement of the analysis of [4] is our main result.

Theorem 1. *Consider a POMDP (W,S,A,α,β,R), and let $\mu \in \Delta_W$ and $\gamma \in [0,1)$. There is a stationary (memoryless, time independent) policy $\pi^* \in \Delta_{S,A}$ with $|\operatorname{supp}(\pi^*(\cdot|s))| \leqslant |\operatorname{supp}(\beta(s|\cdot))|$ for all $s \in S$ and $\mathcal{R}^{\gamma}_{\mu}(\pi^*) \geqslant \mathcal{R}^{\gamma}_{\mu}(\pi)$ for all $\pi \in \Delta_{S,A}$. Under assumption $(*)$, the same holds true for $\mathcal{R}_{\mu}$ in place of $\mathcal{R}^{\gamma}_{\mu}$.*

We prove the discounted case in Sect. 3 and the average case in Sect. 4.

3 Discounted Rewards from Policy Improvement

The state value function V^{π} of a policy π is defined as the unique solution of the Bellman equation

$$V^{\pi}(w) = \sum_{a}p^{\pi}(a|w)\Big[R(w,a) + \gamma\sum_{w'}\alpha(w'|w,a)V^{\pi}(w')\Big], \quad w \in W.$$

It is useful to write $V^\pi(w) = \sum_a p^\pi(a|w) Q^\pi(w,a)$, where

$$Q^\pi(w,a) = R(w,a) + \gamma \sum_{w'} \alpha(w'|w,a) V^\pi(w'), \quad w \in W, a \in A,$$

is the state action value function. Observe that $\mathcal{R}^\gamma_\mu(\pi) = (1-\gamma)\sum_w \mu(w) V^\pi(w)$. If two policies π, π' satisfy $V^{\pi'}(w) \geqslant V^\pi(w)$ for all w, then $\mathcal{R}^\gamma_\mu(\pi') \geqslant \mathcal{R}^\gamma_\mu(\pi)$ for all μ. The following is a more explicit version of a lemma from [4]:

Lemma 3 (Policy improvement lemma). *Let $\pi, \pi' \in \Delta_{S,A}$ and $\epsilon(w) = \sum_a p^{\pi'}(a|w) Q^\pi(w,a) - V^\pi(w)$ for all $w \in W$. Then*

$$V^{\pi'}(w) = V^\pi(w) + \mathbb{E}_{\pi', w_0 = w}\Big[\sum_{t=0}^\infty \gamma^t \epsilon(w_t)\Big] \quad \textit{for all } w \in W.$$

If $\epsilon(w) \geqslant 0$ for all $w \in W$, then

$$V^{\pi'}(w) \geqslant V^\pi(w) + d^{\pi'}(w)\epsilon(w) \quad \textit{for all} w \in W,$$

where $d^{\pi'}(w) = \sum_{t=0}^\infty \gamma^t \Pr(w_t = w|\pi', w_0 = w) \geqslant 1$ is the discounted expected number of visits to w.

Proof. $V^\pi(w) = \sum_a p^{\pi'}(a|w) Q^\pi(w,a) - \epsilon(w)$

$$\begin{aligned}
&= \mathbb{E}_{\pi', w_0 = w}\Big[\Big(R(w_0, a_0) - \epsilon(w_0)\Big) + \gamma V^\pi(w_1)\Big] \\
&= \mathbb{E}_{\pi', w_0 = w}\Big[\Big(R(w_0, a_0) - \epsilon(w_0)\Big) + \gamma\Big(\sum_a p^{\pi'}(a|w_1) Q^\pi(w_1, a) - \epsilon(w_1)\Big)\Big] \\
&= \mathbb{E}_{\pi', w_0 = w}\Big[\sum_{t=0}^\infty \gamma^t \Big(R(w_t, a_t) - \epsilon(w_t)\Big)\Big] = V^{\pi'}(w) - \mathbb{E}_{\pi', w_0 = w}\Big[\sum_{t=0}^\infty \gamma^t \epsilon(w_t)\Big].
\end{aligned}$$

□

Lemma 3 allows us to find policy changes that increase $V^\pi(w)$ for all $w \in W$ and thereby $\mathcal{R}^\gamma_\mu(\pi)$ for any μ.

Definition 1. *Fix a policy $\pi \in \Delta_{S,A}$. For each sensor state $s \in S$ consider the set* $\operatorname{supp}(\beta(s|\cdot)) = \{w \in W : \beta(s|w) > 0\} = \{w^s_1, \ldots, w^s_{k_s}\}$, *and define the linear forms*

$$l^{\pi,s}_i : \Delta_A \to \mathbb{R};\ q \mapsto \sum_a q(a) Q^\pi(w^s_i, a), \quad i = 1, \ldots, k_s.$$

The policy improvement cone *at policy π and sensation s is*

$$L^{\pi,s} = \{q \in \Delta_A : l^{\pi,s}_i(q) \geqslant l^{\pi,s}_i(\pi(\cdot|s)) \textit{ for all } i = 1, \ldots, k_s\}.$$

The (total) policy improvement cone *at policy π is*

$$L^\pi = \{\pi' \in \Delta_{S,A} : \pi'(\cdot|s) \in L^{\pi,s} \textit{ for all } s \in S\}.$$

$L^{\pi,s}$ and L^{π} are intersections of Δ_A and $\Delta_{S,A}$ with intersections of affine half-spaces (see Fig. 1). Since $\pi \in L^{\pi}$, the policy improvement cones are never empty.

Lemma 4. *Let $\pi \in \Delta_{S,A}$ and $\pi' \in L^{\pi}$. Then, for all w,*

$$V^{\pi'}(w) - V^{\pi}(w) \geqslant d^{\pi'}(w) \sum_s \beta(s|w) \sum_a (\pi'(a|s) - \pi(a|s)) Q^{\pi}(w,a) \geqslant 0.$$

Proof. Fix $w \in W$. In the notation from Lemma 3, suppose that $\operatorname{supp}(\beta(\cdot|w)) = \{s_1, \ldots, s_l\}$ and that $w = w_{i_j}^{s_j}$ for $j = 1 \ldots, l$. Then

$$\begin{aligned}\epsilon(w) &= \sum_a p^{\pi'}(a|w) Q^{\pi}(w,a) - \sum_a p^{\pi}(a|w) Q^{\pi}(w,a) \\ &= \sum_{j=1}^{l} \beta(s_j|w) l_{i_j}^{\pi,s_j}(\pi'(\cdot|s_j) - \pi(\cdot|s_j)) \geqslant 0,\end{aligned}$$

since $\pi' \in L^{\pi}$. The statement now follows from Lemma 3. □

Remark 1. Lemma 4 relates to the policy gradient theorem [7], which says that

$$\frac{\partial V^{\pi}(w)}{\partial \pi(a'|s')} = d^{\pi}(w) \sum_s \beta(s|w) \sum_a \frac{\partial \pi(a|s)}{\partial \pi(a'|s')} Q^{\pi}(w,a). \quad (1)$$

Our result adds that, for each w, the value function $V^{\pi'}(w)$ is bounded from below by a linear function of π' that takes value at least $V^{\pi}(w)$ within the entire policy improvement cone L^{π}. See Fig. 1.

Now we show that there is an optimal policy with small support.

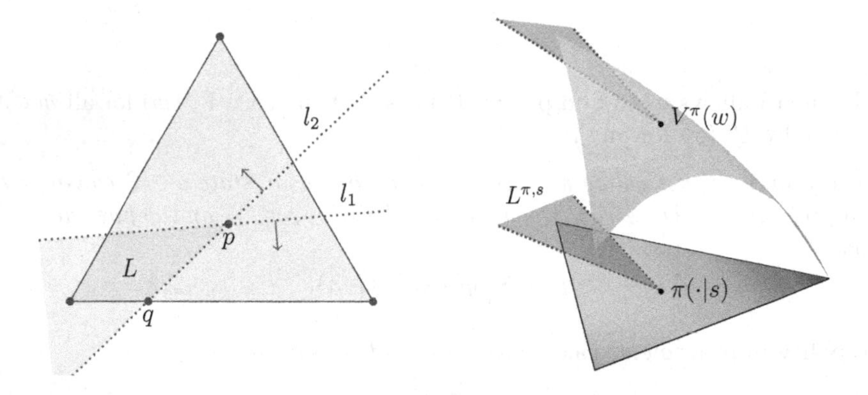

Fig. 1. Left: illustration of the policy improvement cone. Right: illustration of the state value function $V^{\pi}(w)$ for some fixed w, showing the linear lower bound over the policy improvement cone $L^{\pi,s}$. This numerical example is discussed further in Sect. 5. (Color figure online)

Lemma 5. *Let P be a polytope, and let $l_1, \ldots, l_k$ be linear forms on P. For any $p \in P$, let $L_{i,+} = \{q \in P : l_i(q) \geqslant l_i(p)\}$. Then $\bigcap_{i=1}^{k} L_{i,+}$ contains an element q that belongs to a face of P of dimension at most $k-1$.*

Proof. The argument is by induction. For $k = 1$, the maximum of l_1 on P is attained at a vertex q of P. Clearly, $l_1(q) \geqslant l_1(p)$, and so $q \in L_{1,+}$.

Now suppose that $k > 1$. Let $P' := P \cap L_{k,+}$. Each face of P' is a subset of a face of P of at most one more dimension. By induction, $\bigcap_{i=1}^{k-1} L_{i,+} \cap P'$ contains an element q that belongs to a face of P' of dimension at most $k-2$. □

Proof (of Theorem 1 *for discounted rewards).* By Lemma 5, each policy improvement cone $L^{\pi,s}$ contains an element q that belongs to a face of Δ_A of dimension at most $(k-1)$ (that is, the support of q has cardinality at most k), where $k = |\operatorname{supp}(\beta(s|\cdot))|$. Putting these together, we find a policy π' in the total policy improvement cone that satisfies $|\operatorname{supp}(\pi(\cdot|s))| \leqslant |\operatorname{supp}(\beta(s|\cdot))|$ for all s. By Lemma 4, $V^{\pi'}(w) \geqslant V^{\pi}(w)$ for all w, and so $\mathcal{R}_\mu^\gamma(\pi') \geqslant \mathcal{R}_\mu^\gamma(\pi)$. □

Remark 2. The $|\operatorname{supp}\beta(s|\cdot)|$ positive probability actions at sensation s do not necessarily correspond to the actions that the agent would choose if she knew the identity of the world state, as our example in Sect. 5 shows.

4 Average Rewards from Discounted Rewards

The average reward per time step can be written in terms of the discounted reward as $\mathcal{R}(\pi) = \mathcal{R}^\gamma_{p^\pi_\mu}$. However, the hypothesis $V^{\pi'}(w) \geqslant V^\pi(w)$ for all w, does not directly imply any relation between $\mathcal{R}(\pi')$ and $\mathcal{R}(\pi)$, since they compare the value function against different stationary distributions. We show that results for discounted rewards translate nonetheless to results for average rewards.

Lemma 6. *Let μ be fixed, and assume $(*)$. For any $\epsilon > 0$ there exists $l > 0$ such that for all π and all $t \geqslant l$, $|\mu^t_\pi(w) - p^\pi_\mu(w)| \leqslant \epsilon$ for all w.*

Proof. By $(*)$, the transition matrix of the Markov chain has the eigenvalue one with multiplicity one, with left eigenvector denoted by p^π_μ. Let $p_2, \ldots, p_{|W|}$ be orthonormal left eigenvectors to the other eigenvalues $\lambda_2, \ldots, \lambda_{|W|}$, ordered such that λ_2 has the largest absolute value. There is a unique expansion $\mu = c_1 p^\pi_\mu + c_2 p_2 + \cdots + c_{|W|} p_{|W|}$. Then $\mu^t_\pi = c_1 p^\pi_\mu + \sum_{i=2}^{|W|} c_i \lambda_i^t p_i$. Letting $t \to \infty$, it follows that $c_1 = 1$. By orthonormality, $|c_i|^2 \leqslant \sum_{i=2}^{|W|} c_i^2 \leqslant \|\mu\|_2^2 \leqslant 1$ and $|p_i(w)| \leqslant 1$ for $i = 2, \ldots, |W|$. Therefore, $|\mu^t_\pi(w) - p^\pi_\mu(w)| = |\sum_{i=2}^{|W|} c_i \lambda_i^t p_{|W|}(w)| \leqslant |W||\lambda_2|^t$.

$|\lambda_2|$ depends continuously on the transition matrix, which itself depends continuously on π. Since $\Delta_{S,A}$ is compact, $|\lambda_2| = |\lambda_2(\pi)|$ has a maximum d, and $d < 1$ due to $(*)$. Therefore, $|\mu^t_\pi(w) - p^\pi_\mu(w)| \leqslant |W| d^t$ for all π. The statement follows from this. □

Proposition 1. *For fixed μ, under assumption $(*)$, $\mathcal{R}^\gamma_\mu(\pi) \to \mathcal{R}_\mu(\pi)$ uniformly in π as $\gamma \to 1$.*

Proof. For fixed μ and ϵ, let l be as in Lemma 6. Let $R = \max_{w,a} |R(w,a)|$. Then

$$\begin{aligned}\mathcal{R}^\gamma_\mu(\pi) &= (1-\gamma)\sum_{k=0}^{l-1}\gamma^k \sum_w \mu^k_\pi(w)\sum_a \pi(a|w)R(w,a) \\ &\quad + (1-\gamma)\gamma^l \sum_{k=0}^{\infty}\gamma^k \sum_w p^\pi_\mu(w)\sum_a \pi(a|w)R(w,a) + O(\epsilon R)(1-\gamma)\sum_{k=0}^{\infty}\gamma^k \\ &= O((1-\gamma)lR) + O(\epsilon R) + \gamma^l \mathcal{R}_\mu(\pi)\end{aligned}$$

for all π. For given $\delta > 0$, we can choose $\epsilon > 0$ such that the term $O(\epsilon R)$ is smaller in absolute value than $\delta/3$. This also fixes $l = l(\epsilon)$. Then, for any $\gamma < 1$ large enough, the term $O((1-\gamma)lR)$ is smaller than $\delta/3$, and also $|(\gamma^l - 1)\mathcal{R}_\mu(\pi)| \leqslant \delta/3$. This shows that for $\gamma < 1$ large enough, $|\mathcal{R}^\gamma_\mu(\pi) - \mathcal{R}_\mu(\pi)| \leqslant \delta$, independent of π. The statement follows since $\delta > 0$ was arbitrary. □

Theorem 2. *For any $\gamma \in [0,1)$, let $\hat{\pi}_\gamma$ be a policy that maximizes $\mathcal{R}^\gamma_\mu$. Let $\hat{\pi}$ be a limit point of a convergent subsequence as $\gamma \to 1$. Then $\hat{\pi}$ maximizes $\mathcal{R}_\mu$, and $\lim_{\gamma\to 1}\mathcal{R}^\gamma_\mu(\hat{\pi}_\gamma) = \mathcal{R}_\mu(\hat{\pi})$.*

Proof. For any $\epsilon > 0$, there is $\delta > 0$ such that $\gamma \geqslant 1-\delta$ implies $|\mathcal{R}_\mu(\pi) - \mathcal{R}^\gamma_\mu(\pi)| \leqslant \epsilon$ for all π. Thus $|\max_\pi \mathcal{R}_\mu(\pi) - \max_\pi \mathcal{R}^\gamma_\mu| \leqslant \epsilon$, whence $\lim_{\gamma\to 1}\max_\pi \mathcal{R}^\gamma_\mu(\pi) = \max_\pi \mathcal{R}_\mu(\pi)$. Moreover, $|\max_\pi \mathcal{R}_\mu(\pi) - \mathcal{R}_\mu(\hat{\pi}_\gamma)| \leqslant 2\epsilon + |\max_\pi \mathcal{R}^\gamma_\mu(\pi) - \mathcal{R}^\gamma_\mu(\hat{\pi}_\gamma)| = 2\epsilon$. By continuity, the limit value of $\mathcal{R}_\mu$ applied to a convergent subsequence of the $\hat{\pi}_\gamma$ is the maximum of $\mathcal{R}_\mu$. □

Corollary 1. *Fix a world state w, and let $r \geqslant 0$. If there exists for each $\gamma \in [0,1)$ a policy $\hat{\pi}_\gamma$ that is optimal for $\mathcal{R}^\gamma_\mu$ with $|\operatorname{supp}(\pi(\cdot|s))| \leqslant r$, then there exists a policy $\hat{\pi}$ with $|\operatorname{supp}(\pi(\cdot|s))| \leqslant r$ that is optimal for $\mathcal{R}_\mu$.*

Proof. Take a limit point of the family $\hat{\pi}_\gamma$ as $\gamma \to 1$ and apply Theorem 2. □

Remark 3. Without (*), one can show that $\mathcal{R}^\gamma_\mu(\pi)$ still converges to $\mathcal{R}_\mu(\pi)$ for each fixed π, but convergence is no longer uniform. Also, $\mathcal{R}_\mu$ need not be continuous in π, and so an optimal policy need not exist.

5 Example

We illustrate our results on an example from [4]. Consider an agent with sensor states $S = \{1,2,3\}$ and actions $A = \{1,2,3\}$. The system has world states $W = \{1,2,3,4\}$ with the transitions and rewards illustrated in Fig. 2. At $w = 1,4$ all actions produce the same outcomes. States $w = 2,3$ are observed as $s = 2$. Hence we can focus on $\pi(\cdot|s=2) \in \Delta_A$. We evaluate 861 evenly spaced policies in this 2-simplex. Figure 2 shows color maps of the expected reward (interpolated between evaluations), with lighter colors corresponding to higher values. As in Fig. 1, red vectors are the gradients of the linear forms (corresponding to $Q^\pi(w,\cdot), w = 2,3$), and dashed blue lines limit the policy improvement cones $L^{\pi,s=2}$. Stepping into the improvement cone always increases $V^{\pi}(w) = \mathcal{R}^{\gamma}_{\mu=\delta_w}(\pi)$ for all $w \in W$. Note that each cone contains a policy at an edge of the simplex, i.e., assigning positive probability to at most two actions. The convergence of $\mathcal{R}^{\gamma}_{\mu}$ to $\mathcal{R}_{\mu}$ as $\gamma \to 1$ is visible. Note also that for $\gamma = 0.6$ the optimal policy requires two positive probability actions, so that our upper bound $|\operatorname{supp}(\pi(\cdot|s))| \leqslant |\operatorname{supp}(\beta(s|\cdot))|$ is attained.

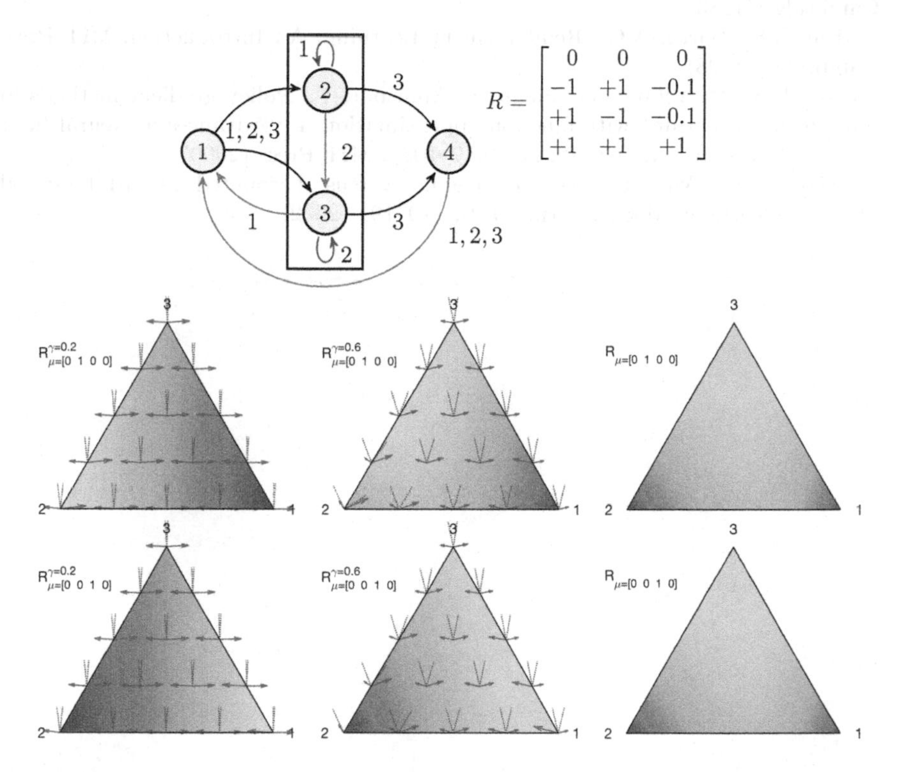

Fig. 2. Illustration of the example form Sect. 5. Top: state transitions and reward signal. Bottom: numerical evaluation of the expected long term reward. (Color figure online)

Acknowledgment. We thank Nihat Ay for support and insightful comments.

References

1. Ay, N., Montúfar, G., Rauh, J.: Selection criteria for neuromanifolds of stochastic dynamics. In: Yamaguchi, Y. (ed.) Advances in Cognitive Neurodynamics (III), pp. 147–154. Springer, Dordrecht (2013). doi:10.1007/978-94-007-4792-0_20
2. Hutter, M.: General discounting versus average reward. In: Balcázar, J.L., Long, P.M., Stephan, F. (eds.) ALT 2006. LNCS, vol. 4264, pp. 244–258. Springer, Heidelberg (2006). doi:10.1007/11894841_21

3. Kakade, S.: Optimizing average reward using discounted rewards. In: Helmbold, D., Williamson, B. (eds.) COLT 2001. LNCS, vol. 2111, pp. 605–615. Springer, Heidelberg (2001). doi:10.1007/3-540-44581-1_40
4. Montúfar, G., Ghazi-Zahedi, K., Ay, N.: Geometry and determinism of optimal stationary control in partially observable Markov decision processes. arXiv:1503.07206 (2015)
5. Ross, S.M.: Introduction to Stochastic Dynamic Programming. Academic Press Inc., Cambridge (1983)
6. Sutton, R.S., Barto, A.G.: Reinforcement Learning: An Introduction. MIT Press, Cambridge (1998)
7. Sutton, R.S., McAllester, D., Singh, S., Mansour, Y.: Policy gradient methods for reinforcement learning with function approximation. In: Advances in Neural Information Processing Systems 12, pp. 1057–1063. MIT Press (2000)
8. Tsitsiklis, J.N., Van Roy, B.: On average versus discounted reward temporal-difference learning. Mach. Learn. **49**(2), 179–191 (2002)

Joint Geometric and Photometric Visual Tracking Based on Lie Group

Chenxi Li[1,2,3,4](✉), Zelin Shi[1,3,4], Yunpeng Liu[1,3,4], and Tianci Liu[1,2,3,4]

[1] Shenyang Institute of Automation, Chinese Academy of Sciences, Shenyang 110016, Liaoning, China
lichenxi@sia.cn
[2] University of Chinese Academy of Sciences, Beijing 100049, China
[3] Key Laboratory of Opto-electronic Information Processing, Chinese Academy of Sciences, Shenyang 110016, Liaoning, China
[4] The Key Lab of Image Understanding and Computer Vision, Shenyang 110016, Liaoning Province, China

Abstract. This paper presents a novel efficient and robust direct visual tracking method under illumination variations. In our approach, non-Euclidean Lie group characteristics of both geometric and photometric transformations are exploited. These transformations form Lie groups and are parameterized by their corresponding Lie algebras. By applying the efficient second-order minimization trick, we derive an efficient second-order optimization technique for jointly solving the geometric and photometric parameters. Our approach has a high convergence rate and low iterations. Moreover, our approach is almost not affected by linear illumination variations. The superiority of our proposed method over the existing direct methods, in terms of efficiency and robustness is demonstrated through experiments on synthetic and real data.

Keywords: Visual tracking · Illumination variations · Lie algebra · Efficient second-order minimization · Lie group

1 Introduction

Direct visual tracking can be formulated as finding the incremental transformations between a reference image and successive frames of a video sequence. As utilizing all information of pixels of interest to estimate the transformation, it can give sub-pixel accuracy, which is necessary for certain applications, e.g., augmented reality, vision-based robot control [1], medical image analysis. Direct visual tracking problem can be made as complex as possible by considering illumination variations, occlusions and multiple-modality. In this paper, we focus on direct visual tracking which is formulated as iterative registration problem under global illumination variations.

Traditional direct visual tracking methods often assume intensity constancy under Lucas-Kanade framework, where the sum of squared differences (SSD) is used as similarity metric. The inverse compositional (IC) method [2] and the efficient

F. Nielsen and F. Barbaresco (Eds.): GSI 2017, LNCS 10589, pp. 291–298, 2017.
https://doi.org/10.1007/978-3-319-68445-1_34

second-order minimization (ESM) method [1] are two of the most efficient method. The drawback of these methods is their sensitivity to illumination variations.

Two different strategies have been employed to improve the robustness to illumination variations. First, robust similarity metrics are used, such as the normalized correlation coefficient (NCC), the mutual information (MI) [3], the enhanced correlation coefficient (ECC) [4], and the sum of conditional variance (SCV) [5]. Recently, robust multi-dimensional features are used [6]. These methods have superior robustness to illumination variations, even multi-modality. However, these advantages come either at a high computational cost, or at low convergent radius. The second approach relies on modeling the illumination variations [7–12]. The affine photometric model is often used to compensate for illumination variations, either in a global way [7, 12] or in a local way [8–11]. In these approaches, all of them but DIC algorithm [12] used additive rule to update the photometric parameters in the optimization process.

In this paper, we propose a very efficient and robust direct image registration approach that jointly performs geometric and photometric registration by extending the efficient second-order minimization method. We also use the affine transformation to model illumination variations. Different from [11] where the photometric parameters were updated using additive rule, we employ the compositional rule to update both the geometric and photometric parameters, similar to [12]. Based on the joint Lie algebra parameterization of geometric and photometric transformation we derive a second-order optimization technique for image registration. Our approach preserves the advantages of the original ESM with low iteration number, high convergence frequency.

The rest of the paper is organized as follows. In Sect. 2, we give the necessary theoretical background of our work. The details of our algorithm are given in Sect. 3. Experimental results are presented in Sect. 4. A conclusion is provided in Sect. 5.

2 Theoretical Background

2.1 Lie Algebra Parameterization of Geometric Transformations

We consider homography as the geometric transformations as it is the most general cases for planar objects. The coordinates of a pixel $\mathbf{q}^*$ in the interest region $\mathcal{R}^*$ of reference image $\mathcal{I}^*$ are related to its corresponding $\mathbf{q}$ in the current image $\mathcal{I}$ by a projective homography $\mathbf{G}$ from which a warp $\mathbf{w}(\mathbf{q}^*;\mathbf{G})$ can be induced.

We employ the same parameterization way of homographies as in [1, 11]. The set of homographies is identified with the 3-D special linear group defined as $\mathbb{SL}(3)$. The Lie group $\mathbb{SL}(3)$ and its Lie algebra $sl(3)$ are related via exponential map, then a homography $\mathbf{G}(\mathbf{x}) \in \mathbb{SL}(3)$ can be parameterized as follows:

$$\mathbf{G}(\mathbf{x}) = \exp(\mathrm{A}(\mathbf{x})) = \sum\nolimits_{i=1}^{\infty} \frac{1}{i!}(\mathrm{A}(\mathbf{x}))^i. \tag{1}$$

where $\mathrm{A}(\mathbf{x})$ is the Lie algebra element of $\mathbf{G}(\mathbf{x})$, $\mathbf{x} = [x_1, \cdots, x_8]^T \in \mathbb{R}^8$ the geometric parameters vector [1, 11].

2.2 Lie Algebra Parameterization of Photometric Transformation

For gray-level images, we model the global illumination variations as affine photometric transformation, which is also referred as the gain and bias model. Based on this model, the reference image $\mathcal{I}^*$ and the current image $\mathcal{I}$ are related as $\mathcal{I}^* = \alpha\mathcal{I} + \beta$. where α is the gain and β the bias. We rewrite this relation as matrix form $\begin{bmatrix} \mathcal{I}^* \\ 1 \end{bmatrix} = \begin{bmatrix} \alpha & \beta \\ 0 & 1 \end{bmatrix} \begin{bmatrix} \mathcal{I} \\ 1 \end{bmatrix}$. Then the set of photometric transformations can be identified with 1-D affine Lie group $\mathbb{GA}(1) = \left\{ \mathbf{P}(\alpha, \beta) = \begin{bmatrix} \alpha & \beta \\ 0 & 1 \end{bmatrix} \middle| \alpha, \beta \in \mathbb{R} \right\}$. In our problem, $\alpha > 0$. The Lie algebra associated with $\mathbb{GA}(1)$ is $ga(1) = \left\{ \begin{bmatrix} t_1 & t_2 \\ 0 & 0 \end{bmatrix} \middle| t_1, t_2 \in \mathbb{R} \right\}$. Let $\{\mathbf{B}_1, \mathbf{B}_2\}$ be a basis of Lie algebra $ga(1)$. Each element $\mathbf{B} \in ga(1)$ can be written as a combination of $\mathbf{B}_i$, $\mathbf{B}(\mathbf{t}) = \sum_{i=1}^{2} t_i \mathbf{B}_\mathbf{i}$, with $\mathbf{t} = [t_1, t_2]^T \in \mathbb{R}^2$ the photometric parameters vector. In this paper, we choose $\mathbf{B}_1 = \begin{bmatrix} 1 & 0 \\ 0 & 0 \end{bmatrix}$,$\mathbf{B}_2 = \begin{bmatrix} 0 & 1 \\ 0 & 0 \end{bmatrix}$. For each photometric transformation $\mathbf{P}(\alpha, \beta) \in \mathbb{GA}(1)$, we can parameterize it using its Lie algebra via exponential map like geometric transformations as follows:

$$\mathbf{P}(\alpha(\mathbf{t}), \beta(\mathbf{t})) = \exp(\mathbf{B}(\mathbf{t})) = \sum\nolimits_{i=1}^{\infty} \frac{1}{i!} \left(\begin{bmatrix} t_1 & t_2 \\ 0 & 0 \end{bmatrix} \right)^i \tag{2}$$

We remake that this parameterization is smooth and one-to-one onto, with a smooth inverse, for all $\alpha, \beta \in \mathbb{R}, \alpha > 0$. In the following, if we emphasis on the photometric parameters $\mathbf{t}$, then the Lie algebra parameterized photometric transformation $\mathbf{P}(\alpha(\mathbf{t}), \beta(\mathbf{t}))$ simply denoted as $\mathbf{P}(\mathbf{t})$. For convenience, we define the group action of the photometric transformation $\mathbf{P}(\alpha, \beta)$ on the intensity $\mathcal{I}$ as $\mathbf{P}(\alpha, \beta) \circ \mathcal{I} = \alpha\mathcal{I} + \beta$. It satisfies

$$\mathbf{P}(\alpha_2, \beta_2) \circ \mathbf{P}(\alpha_1, \beta_1) \circ \mathcal{I} = \mathbf{P}(\alpha_2, \beta_2) \circ (\mathbf{P}(\alpha_1, \beta_1) \circ \mathcal{I}) = (\mathbf{P}(\alpha_2, \beta_2)\mathbf{P}(\alpha_1, \beta_1)) \circ \mathcal{I} \tag{3}$$

according to the group properties of Lie group $\mathbb{GA}(1)$.

3 The Proposed Visual Tracking Method

Considering illumination variations, the visual tracking problem can be formulated as a search for the optimal geometric and photometric transformations between two the reference image and the current frame. Given an estimated geometric transformation $\hat{\mathbf{G}}$ and an estimated photometric transformation $\hat{\mathbf{P}}(\hat{\alpha}, \hat{\beta})$ which are often given by previous frame, our considered problem can be formulated as

$$\min_{\mathbf{x}\in\mathbb{R}^8,\mathbf{t}\in\mathbb{R}^2} \frac{1}{2} \sum_{\mathbf{q}_i^*\in\mathcal{R}^*} [\mathbf{P}(\mathbf{t}) \circ \hat{\mathbf{P}} \circ \mathcal{I}\left(\mathbf{w}\left(\mathbf{q}_i^*; \hat{\mathbf{G}}\mathbf{G}(\mathbf{x})\right)\right) - \mathcal{I}^*(\mathbf{q}_i^*)]^2 \tag{4}$$

where $\mathbf{G}(\mathbf{x})$ and $\mathbf{P}(\mathbf{t})$ are the incremental geometric and photometric transformation. Problem (4) can be explicitly written as

$$\min_{\mathbf{x}\in\mathbb{R}^8,\mathbf{t}\in\mathbb{R}^2} \sum_{\mathbf{q}_i^*\in\mathcal{R}^*} \frac{1}{2} [(\hat{\alpha}\mathcal{I}\left(\mathbf{w}\left(\mathbf{q}_i^*; \hat{\mathbf{G}}\mathbf{G}(\mathbf{x})\right)\right) + \hat{\beta})\alpha(\mathbf{t}) + \beta(\mathbf{t}) - \mathcal{I}^*(\mathbf{q}_i^*)]^2 \tag{5}$$

Note that the geometric and photometric transformation in problem (4) can be defined over the joint Lie group $\mathbb{SL}(3) \times \mathbb{GA}(1)$ whose corresponding Lie algebra is $sl(3) \oplus ga(1)$. If we denote $\boldsymbol{\theta}= [\mathbf{x}^T,\ \mathbf{t}^T]^T$ to be the joint parameters, then $\boldsymbol{\theta} \in sl(3) \oplus ga(1) = \mathbb{R}^{10}$. Let the error function in problem (5) be denoted as $D_{\mathbf{q}_i^*}(\boldsymbol{\theta}) = (\hat{\alpha}\mathcal{I}\left(\mathbf{w}\left(\mathbf{q}_i^*; \hat{\mathbf{G}}\mathbf{G}(\mathbf{x})\right)\right) + \hat{\beta})\alpha(\mathbf{t}) + \beta(\mathbf{t}) - \mathcal{I}^*(\mathbf{q}_i^*)$, It can be shown that the second-order approximation of $D_{\mathbf{q}_i^*}(\boldsymbol{\theta})$ around $\boldsymbol{\theta} = \mathbf{0}$ is given by

$$D_{\mathbf{q}_i^*}(\boldsymbol{\theta}) = D_{\mathbf{q}_i^*}(\mathbf{0}) + \frac{1}{2}(\mathbf{J}_{\mathbf{q}_i^*}(\mathbf{0}) + \mathbf{J}_{\mathbf{q}_i^*}(\boldsymbol{\theta}))\boldsymbol{\theta} + \mathbf{O}(\|\boldsymbol{\theta}\|^3) \tag{6}$$

where $\mathbf{J}_{\mathbf{q}_i^*}(\mathbf{0}) = [\nabla_{\mathbf{x}} D_{\mathbf{q}_i^*}(\mathbf{0}), \nabla_{\mathbf{t}} D_{\mathbf{q}_i^*}(\mathbf{0})]$ and $\mathbf{J}_{\mathbf{q}_i^*}(\boldsymbol{\theta}) = [\nabla_{\mathbf{x}} D_{\mathbf{q}_i^*}(\boldsymbol{\theta}), \nabla_{\mathbf{t}} D_{\mathbf{q}_i^*}(\boldsymbol{\theta})]$ are the Jacobians evaluated in $\mathbf{0}$ and $\boldsymbol{\theta}$ respectively. The expressions for $\nabla_{\mathbf{x}} D_{\mathbf{q}_i^*}(\mathbf{0})$ and $\nabla_{\mathbf{t}} D_{\mathbf{q}_i^*}(\mathbf{0})$ can be directly computed as $\nabla_{\mathbf{x}} D_{\mathbf{q}_i^*}(\mathbf{0}) = \hat{\alpha}\mathbf{J}_{\mathcal{I}}\mathbf{J}_{\mathbf{w}}\mathbf{J}_{\mathbf{G}}$, $\nabla_{\mathbf{t}} D_{\mathbf{q}_i^*}(\mathbf{0}) = [\hat{\alpha}\mathcal{I}(\mathbf{w}(\mathbf{q}_i^*; \hat{\mathbf{G}})) + \hat{\beta},\ 1]$. The detailed computation of the derivatives $\mathbf{J}_{\mathcal{I}}$, $\mathbf{J}_{\mathbf{w}}$ and $\mathbf{J}_{\mathbf{G}}$ can be found in [1]. $\nabla_{\mathbf{x}} D_{\mathbf{q}_i^*}(\boldsymbol{\theta})$ and $\nabla_{\mathbf{t}} D_{\mathbf{q}_i^*}(\boldsymbol{\theta})$ depend on the unknown parameters $\boldsymbol{\theta}$, therefore are not easy to compute. However, suppose that $\boldsymbol{\theta}^*= [\mathbf{x}^{*T},\ \mathbf{t}^{*T}]^T$ are the solution of problem (5), based on the Lie algebra parameterization of both geometric transformations and photometric transformations, we can get that $\nabla_{\mathbf{x}} D_{\mathbf{q}_i^*}(\boldsymbol{\theta}^*)= \mathbf{J}_{\mathcal{I}^*}\mathbf{J}_{\mathbf{w}}\mathbf{J}_{\mathbf{G}}$ [1] and $\nabla_{\mathbf{t}} D_{\mathbf{q}_i^*}(\boldsymbol{\theta}^*) = [\mathcal{I}^*(\mathbf{q}_i^*),\ 1]$. Let $\mathbf{J}_{esm}^{gp}$ be the following (1×10) matrix:

$$\begin{aligned} \mathbf{J}_{esm}^{gp}(\mathbf{q}_i^*) &= \frac{1}{2}(\mathbf{J}_{\mathbf{q}_i^*}(\mathbf{0}) + \mathbf{J}_{\mathbf{q}_i^*}(\boldsymbol{\theta}^*)) \\ &= \frac{1}{2}[(\hat{\alpha}\mathbf{J}_{\mathcal{I}} + \mathbf{J}_{\mathcal{I}^*})\mathbf{J}_{\mathbf{w}}\mathbf{J}_{\mathbf{G}}, \hat{\alpha}\mathcal{I}\left(\mathbf{w}(\mathbf{q}_i^*; \hat{\mathbf{G}})\right) + \hat{\beta} + \mathcal{I}^*(\mathbf{q}_i^*),\ 2] \end{aligned} \tag{7}$$

Then the problem (5) can be approximated as a linear least squares problem:

$$\min_{\boldsymbol{\theta}\in\mathbb{R}^{10}} \frac{1}{2} \sum_{\mathbf{q}_i^*\in\mathcal{R}^*} [D_{\mathbf{q}_i^*}(\mathbf{0}) + \mathbf{J}_{esm}^{gp}(\mathbf{q}_i^*)\boldsymbol{\theta}]^2 \tag{8}$$

The solution of problem (8) is given by

$$\boldsymbol{\theta}_0 = [\mathbf{x}_0^T,\ \mathbf{t}_0^T]^T = -\Big(\sum_{\mathbf{q}_i^* \in \mathcal{R}^*} \mathbf{J}_{esm}^{gp}(\mathbf{q}_i^*)^T \mathbf{J}_{esm}^{gp}(\mathbf{q}_i^*)\Big)^{-1} \cdot \Big(\sum_{\mathbf{q}_i^* \in \mathcal{R}^*} \mathbf{J}_{esm}^{gp}(\mathbf{q}_i^*)^T D_{\mathbf{q}_i^*}(\mathbf{0})\Big) \tag{9}$$

The geometric transformation $\hat{\mathbf{G}}$ and photometric transformation $\hat{\mathbf{P}}$ are simultaneously updated as follows:

$$\hat{\mathbf{G}} \leftarrow \hat{\mathbf{G}}\mathbf{G}(\mathbf{x}_0) = \hat{\mathbf{G}}\exp(\mathrm{A}(\mathbf{x}_0))\ \&\ \hat{\mathbf{P}} \leftarrow \mathbf{P}(\mathbf{t}_0)\hat{\mathbf{P}} = \exp(\mathbf{B}(\mathbf{t}_0))\hat{\mathbf{P}} \tag{10}$$

The process is iterated until $\|\boldsymbol{\theta}_0\| < \varepsilon$, where $\varepsilon = 1 \times 10^{-2}$ in our experiments.

4 Experimental Results

In this section, we compare our algorithm with four different algorithms which are also designed for template tracking under illumination variations. They are DIC [12]; the algorithm proposed in [8], which we terms as ESM-PA because the *photometric* parameters are updated using *additive* rule; (ECC) [4]; SCV [5]. Our implementation uses a PC equipped with Intel® Core™ i5-3470 CPU at 3.20 GHz and 4G RAM.

4.1 Convergence Comparison

Figure 1 shows the set of images used as the reference images whose gray-lever versions were used. Two templates with size of 100 × 100 pixels were cropped from each of the reference image. As in [2, 4], a homography was simulated by adding Gaussian noise with standard deviation γ (γ captures the magnitude of geometric deformations) to the coordinates of the four corners of the template. The current image was generated by the simulated homography in conjunction with a gain α and a bias β. The initial values for geometric and photometric transformations were set to identity maps. The max number of iterations was set to 50.

Fig. 1. The selected images used for synthetic experiments

We reported the performance comparison in terms of number of iterations and convergence frequency in Fig. 2. The results were averaged over 500 trials according to 500 random geometric transformations.

As shown in Fig. 2a, our proposed algorithm always has the lowest number of iterations under different magnitude of the geometric transformation. Here the gain and

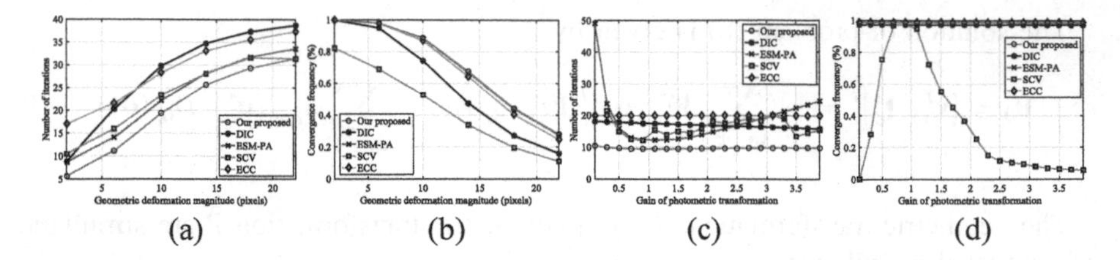

Fig. 2. Performance comparison on image registration task.

bias are fixed to 1.3 and 15 respectively. One trial is considered to be converged if the difference between the estimated coordinates of the corners of the template and the ground truth is below 2 pixels. The convergence frequency is the percentage of the convergent trials over the whole 500 trials. As shown in Fig. 2b, our proposed algorithm always has the highest convergence frequency.

From Fig. 2c, we can see that our proposed algorithm always has the lowest number of iterations under different gain values. Here the geometric deformation magnitude was fixed to 5 pixels. Note that our proposed algorithm is almost free of the influence of linear illumination variations, similar to ECC and DIC. Obviously, ESM-PA which use additive update rule of photometric parameters is heavily affected by illumination variations. Figure 2d presents the convergence frequencies versus the gain. Our proposed algorithm has almost the same best results with ESM-PA and ECC. DIC performs slightly worse. SCV is affected heavily by variations of the gain.

4.2 Tests on Template Tracking

We selected three videos from (www.cs.cmu.edu/~halismai/bitplanes). These videos contain sudden illumination variations and low light, therefore are challenging for direct visual tracking. While the original videos are recorded at 120 Hz, we extracted images from the videos at 40 Hz resulting in three image sequences where large inter-frame displacements are induced. In this experiment, we fixed the max number of iterations for each algorithm to 30 and each frame was resized to 180×320.

Figure 3 plots the number of iterations for each frame during tracking. Figure 4 presents some examples of tracking results. The legends in Fig. 4 correspond to those in Fig. 3. Sever illumination variation occurs at frame #172 and #362 in the first image sequence, at frame #116 in the second image sequence and at frame #414 in the third image sequence. Note that the number of iterations of ESM-PA increases dramatically when sever illumination variation occurs as shown in Fig. 3. In the first and third image sequences, SCV is affected heavily by the sever illumination variations as well. In fact, SCV failed at frame #433 in the first sequence, as shown in Fig. 4a. DIC failed in all the three sequences as shown in Fig. 4. ECC failed in the second image sequence at frame #281. Our proposed algorithm can successfully track the template in all of these image sequences. Table 1 shows the average number of iterations and runtime per frame. We can see that in all of the three image sequences our proposed algorithm needs the lowest iterations and runtime.

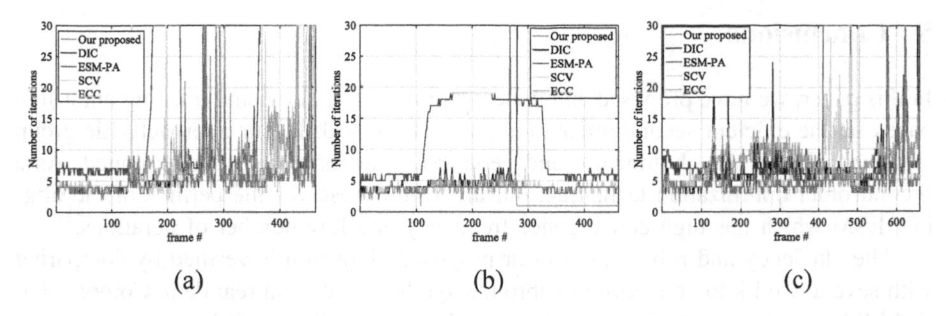

Fig. 3. The number of iterations for each algorithm when tracking in the three image sequences. (a) Sequence 1 (b) Sequence 2 (c) Sequence 3.

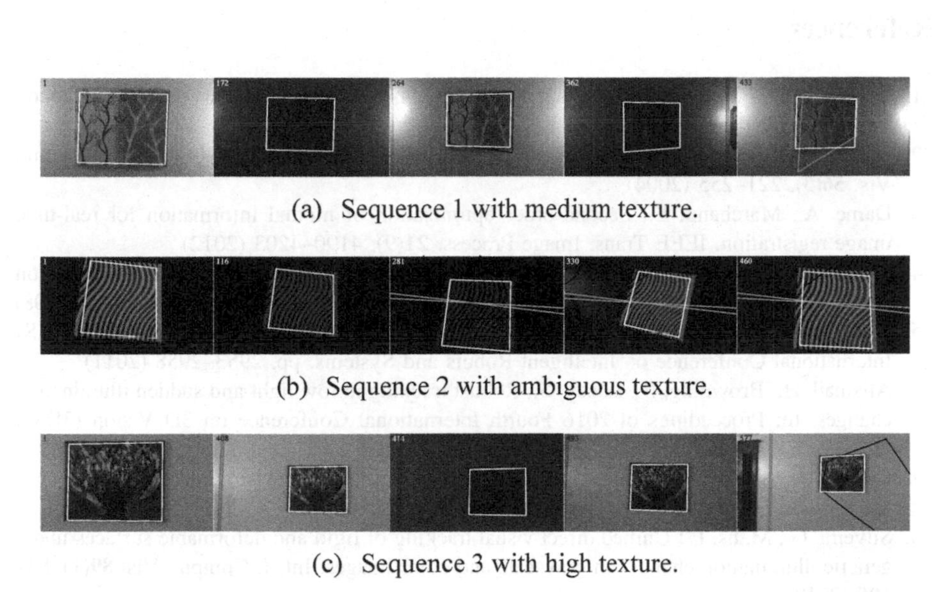

(a) Sequence 1 with medium texture.

(b) Sequence 2 with ambiguous texture.

(c) Sequence 3 with high texture.

Fig. 4. Tracking results in the three image sequences with gray-level intensities. (a) Sequence 1 with medium texture. (b) Sequence 2 with ambiguous texture. (c) Sequence 3 with high texture.

Table 1. Template tracking average number of iterations per frame. In parenthesis we show the average runtime (seconds). N/A stands for tracking failure.

Sequence (template size)	Our proposed	DIC	ESM-PA	SCV	ECC
Sequence 1 (128 × 163)	**5.834 (0.0446)**	N/A	14.50 (0.1139)	N/A	7.567 (0.1183)
Sequence 2 (133 × 140)	**3.833 (0.0305)**	N/A	11.45 (0.0873)	3.903 (0.0351)	N/A
Sequence 3 (145 × 175)	**4.900 (0.0481)**	N/A	9.675 (0.0949)	5.740 (0.0646)	8.102 (0.1523)

5 Conclusions

In this paper, we have proposed an efficient and robust direct visual tracking algorithm based on the efficient second-order minimization method. In our approach, Lie group structure of both the photometric and geometric transformations are exploited. As a second-order optimization technique, our algorithm preserves the permits of the original ESM which has high convergence frequency and low number of iterations.

The efficiency and robustness of our proposed algorithm is verified by comparing with several well-known algorithms through synthetic data and real data. Compared to ESM-PA, our algorithm is more efficient under illumination variations.

References

1. Benhimane, S., Malis, E.: Homography-based 2D visual tracking and servoing. Int. J. Robot. Res. **26**(7), 661–676 (2007)
2. Baker, S., Matthews, I.: Lucas-Kanade 20 years on: a unifying framework. Int. J. Comput. Vis. **56**(3), 221–255 (2004)
3. Dame, A., Marchand, E.: Second-order optimization of mutual information for real-time image registration. IEEE Trans. Image Process. **21**(9), 4190–4203 (2012)
4. Evangelidis, G.D., Psarakis, E.Z.: Parametric image alignment using enhanced correlation coefficient maximization. IEEE Trans. Pattern Anal. Mach. Intell. **30**(10), 1858–1865 (2008)
5. Richa, R., et al.: Visual tracking using the sum of conditional variance. In: 2011 IEEE/RSJ International Conference on Intelligent Robots and Systems, pp. 2953–2958 (2011)
6. Alismail, H., Browning, B., Lucey, S.: Robust tracking in low light and sudden illumination changes. In: Proceedings of 2016 Fourth International Conference on 3D Vision (3DV), pp. 389–398 (2016)
7. Luong, H.Q., et al.: Joint photometric and geometric image registration in the total least square sense. Pattern Recogn. Lett. **32**(15), 2061–2067 (2011)
8. Silveira, G., Malis, E.: Unified direct visual tracking of rigid and deformable surfaces under generic illumination changes in grayscale and color images. Int. J. Comput. Vis. **89**(1), 84–105 (2010)
9. Gouiffes, M., et al.: A study on local photometric models and their application to robust tracking. Comput. Vis. Image Underst. **116**(8), 896–907 (2012)
10. Fouad, M.M., Dansereau, R.M., Whitehead, A.D.: Image registration under illumination variations using region-based confidence weighted m-estimators. IEEE Trans. Image Process. **21**(3), 1046–1060 (2012)
11. Silveira, G., Malis, E.: Real-time visual tracking under arbitrary illumination changes. In: 2007 IEEE Conference on Computer Vision and Pattern Recognition, vol. 1–8, pp. 1–6 (2007)
12. Bartoli, A.: Groupwise geometric and photometric direct image registration. IEEE Trans. Pattern Anal. Mach. Intell. **30**(12), 2098–2108 (2008)

Geometric Robotics and Tracking

Drone Tracking Using an Innovative UKF

Marion Pilté[1,2](✉), Silvère Bonnabel[1], and Frédéric Barbaresco[2]

[1] Mines ParisTech, PSL Research University, Center for Robotics, Paris, France
{marion.pilte,silvere.bonnabel}@mines-paristech.fr
[2] Thales Air Systems, Hameau de Roussigny, Limours, France
frederic.barbaresco@thalesgroup.com

Abstract. This paper addresses the drone tracking problem, using a model based on the Frenet-Serret frame. A kinematic model in 2D, representing intrinsic coordinates of the drone is used. The tracking problem is tackled using two recent filtering methods. On the one hand, the Invariant Extended Kalman Filter (IEKF), introduced in [1] is tested, and on the other hand, the second step of the filtering algorithm, *i.e.* the update step of the IEKF is replaced by the update step of the Unscented Kalman Filter (UKF), introduced in [2]. These two filters are compared to the well known Extended Kalman Filter. The estimation precision of all three algorithms are computed on a real drone tracking problem.

Keywords: Tracking · Geometric estimation · Kalman Filtering

1 Introduction

Very few works have been done on drone tracking using radars rather than computer vision technologies such as the use of cameras. In this paper, we will apply algorithms devoted to more usual targets for radars, such as planes or missiles, to the problem of drone tracking. Indeed, more and more drones are used, for military applications as well as for civilian applications, and it is crucial to track them so that they do not interfere with regular air traffic operations, especially when they are close to an airport. The challenge is different from that of regular target tracking. The drones are much smaller and behave differently as aircrafts, they fly slower, which is also a challenge for radars. The filtering algorithms used for aircrafts thus have to be robustified.

The model chosen in this paper is based on the Frenet-Serret frame in 2D, which is attached to the drone, and which represents some intrinsic parameters of the motion, such as the curvature of the trajectory (through the angular velocity of the target). The use of such intrinsic models has already been addressed in [3] and is applied here to drone tracking.

There are a large variety of filters designed to perform state estimation, the most well-known being the Kalman Filter [4], and its most widespread extension to nonlinear models, the Extended Kalman Filter (EKF), presented in [5]. However, the EKF is unstable when confronted to large initial errors and highly nonlinear evolution or measurement functions, so we opt here for more evolved

F. Nielsen and F. Barbaresco (Eds.): GSI 2017, LNCS 10589, pp. 301–309, 2017.
https://doi.org/10.1007/978-3-319-68445-1_35

filtering techniques, such as the Invariant Extended Kalman Filter (IEKF) and the Unscented Kalman Filter (UKF). These filters are much more stable than the EKF, and more appropriate to the model formulation we have chosen. However, contrary to previous use of Kalman filtering on Lie groups to perform robot localization, we do not have access to any odometer measurements, and we have to extend the theory presented in [6], to the case when the angular and tangential velocities are unknown. Another type of filters used to perform estimation are the particle filters, as in [7], or the Rao-Blackwell particle filter, see [8], however, we do not want to use any particles for this study, due to the computational cost they induce.

This paper is organized as follows. In Sect. 2 the kinematic model is presented. In Sect. 3 we recall the IEKF equations for this model, as described in [9]. In Sect. 4 we develop the UKF update step and adapt it to fit our IEKF propagation step, the filter obtained will be called the left-UKF. Finally, in Sect. 5 we compare the precision of these two filters, and of the Extended Kalman Filter when applied to some real drone tracking problems.

2 Kinematic Model

A drone is controlled by some commands activated either automatically or by a human being. It seems thus natural to consider these control commands piecewise constant. These commands are expressed in a frame attached to the drone, and are called intrinsic coordinates. This was already proposed for instance in [3]. Drone positions are known only in range and bearing coordinates (the radar does not give accurate altitude measurements for this type of target). We thus need to use a 2D model to derive the evolution equation of the drone. They are presented in [9] for instance, and they read:

$$\begin{aligned} &\frac{d}{dt}\theta_t = \omega_t + w_t^\theta, \frac{d}{dt}x_t^{(1)} = (u_t + w_t^x)\cos(\theta_t), \frac{d}{dt}x_t^{(2)} = (u_t + w_t^x)\sin(\theta_t) \\ &\frac{d}{dt}\omega_t = 0 + w_t^\omega, \frac{d}{dt}u_t = 0 + w_t^u \end{aligned} \tag{1}$$

where θ_t is the direction of the drone, $\left(x_t^{(1)}, x_t^{(2)}\right)$ is its cartesian position, ω_t is the angular velocity and u_t is the tangential velocity (also called the norm of the velocity). All these parameters form the state vector of the drone. $w_t^\theta, w_t^x, w_t^\omega, w_t^u$ are white gaussian noises. The measurement equation writes:

$$Y_n = (r_n, \alpha_n) + v_n = h(x_{t_n}^{(1)}, x_{t_n}^{(2)}) + v_n = \left(\sqrt{(x_{t_n}^{(1)})^2 + (x_{t_n}^{(2)})^2}, \arctan\left(\frac{x_{t_n}^{(2)}}{x_{t_n}^{(1)}}\right)\right) + v_n \tag{2}$$

r_n is called the range coordinate and α_n the bearing coordinate. v_n is a white Gaussian noise, with covariance N.

We cast the angle θ_t in a rotation matrix $R(\theta_t) = \begin{pmatrix} \cos\theta_t & -\sin\theta_t \\ \sin\theta_t & \cos\theta_t \end{pmatrix}$, which enables us to work on the matrix Lie group $SE(2)$ with the partial state matrix

χ_t and the evolution matrix ν_t, as in (3).

$$\chi_t = \begin{pmatrix} \cos\theta_t & -\sin\theta_t & x_t^{(1)} \\ \sin\theta_t & \cos\theta_t & x_t^{(2)} \\ 0 & 0 & 1 \end{pmatrix}, \nu_t = \begin{pmatrix} 0 & -\omega_t & u_t \\ \omega_t & 0 & 0 \\ 0 & 0 & 0 \end{pmatrix} \tag{3}$$

The model evolution thus writes in a more compact way:

$$\frac{d}{dt}\chi_t = \chi_t(\nu_t + w_t^\chi), \frac{d}{dt}\omega_t = 0 + w_t^\omega, \frac{d}{dt}u_t = 0 + w_t^u \tag{4}$$

This kinematic model is used to design two different filters, the Invariant Extended Kalman Filter, presented in the next Section, and an innovative UKF, called the left-UKF, explained in Sect. 4.

3 Invariant Extended Kalman Filter Equations

We apply to this model the methodology of the IEKF, as explained in [1,10] for instance. The method for the particular model (1) is also developed in [9].

We call exp the exponential of the Lie group $SE(2)$, so we have exp: $\mathfrak{se}(2) \to SE(2)$, with $\mathfrak{se}(2)$ the Lie algebra of SE(2), for more precision on Lie groups, see [11]. We also need to define the matrices (5).

$$A_t = \begin{pmatrix} 0 & 0 & 0 & 1 & 0 \\ 0 & 0 & \hat{\omega}_t & 0 & 1 \\ \hat{u}_t & -\hat{\omega}_t & 0 & 0 & 0 \\ 0 & 0 & 0 & 0 & 0 \\ 0 & 0 & 0 & 0 & 0 \end{pmatrix}, H_n = \nabla h_{\hat{x}_{t_n}} R(\hat{\theta}_{t_n}) \begin{pmatrix} 0 & 1 & 0 & 0 & 0 \\ 0 & 0 & 1 & 0 & 0 \end{pmatrix} \tag{5}$$

The IEKF equations are summarized below.

1. Propagation step:

$$\begin{aligned} &\frac{d}{dt}\hat{\theta}_t = \hat{\omega}_t, \frac{d}{dt}\hat{x}_t = \begin{pmatrix} \cos\hat{\theta}_t \\ \sin\hat{\theta}_t \end{pmatrix} \hat{u}_t, \frac{d}{dt}\hat{\omega}_t = 0, \frac{d}{dt}\hat{u}_t = 0 \\ &\frac{d}{dt}P_t = A_t P_t + P_t A_t + Q_t \end{aligned} \tag{6}$$

2. Update step:

$$\begin{aligned} K_n &= P_{t_n} H_n (H_n P_{t_n} H_n^T + N)^{-1} \\ z_n &= R(\hat{\theta}_{t_n})^T (Y_n - \hat{x}_{t_n}) \\ e &= K_n z_n, \text{ let us call } e = (e_1, e_2, e_3, e_4, e_5)^T \\ \hat{\chi}_{t_n}^+ &= \hat{\chi}_{t_n} \exp(e_1, e_2, e_3), \hat{\omega}_{t_n}^+ = \hat{\omega}_{t_n} + e_4, \hat{u}_{t_n}^+ = \hat{u}_{t_n} + e_5 \end{aligned} \tag{7}$$

The strength of the IEKF is that in a perfect theoretical setting, the linearizations (they intervene in the equations as A_t for the propagation step and as H_n for the update step) do not depend on the predicted state $(\hat{\chi}_t, \hat{\omega}_t, \hat{u}_t)$. In the previous equations however, we see that with our model, the matrices A_t and H_n depend on the predicted state. For the propagation step, this does not seem too preoccupying, since A_t only depends on $\hat{\omega}_t$ and $\hat{u}_t$, and not directly on the position. However, for H_n the problem is different, since it depends on $(\hat{x}_t^{(1)}, \hat{x}_t^{(2)})$, and we have the same approximation and stability problems as for the EKF. We then need to find another method to avoid computing the Jacobian of h. The UKF update step seems appropriate for this (see [12]), we present it in the next section.

4 Left-UKF Filter

The Unscented Kalman Filter (UKF), see for instance [2], allows to approximate the posterior (Gaussian) distribution $p(X|Y)$ thanks to the use of so-called sigma-points. This UKF is adapted here as in [6] to suit the model formulation, this adaptation is called the left-UKF (l-UKF). We combine the prediction step of the IEKF with the update step of the left-UKF.

Instead of performing a linearization of the nonlinear model, the unscented transform is used to pick a minimal set of sigma points around the mean state. These sigma points are updated through the nonlinear function h, and a new mean and covariance are derived from this update.

The idea is to increase the dimension of the state and of its covariance. Let us call $\bar{\chi}$ the mean of the whole state put in matrix form, that is:

$$\bar{\chi} = \begin{pmatrix} \cos\bar{\theta} & -\sin\bar{\theta} & \bar{x}^{(1)} & 0 & \bar{u} \\ \sin\bar{\theta} & \cos\bar{\theta} & \bar{x}^{(2)} & \bar{\omega} & 0 \\ 0 & 0 & 1 & 0 & 0 \\ 0 & 0 & 0 & 1 & 0 \\ 0 & 0 & 0 & 0 & 1 \end{pmatrix} \tag{8}$$

Let us define the augmented covariance as $P_n^a = diag(P_n, N)$.

We then construct a set of $2L+1$ sigma points (in our model $L = 7$, it is the dimension of the augmented state) as in (9), and where λ is a scaling factor.

$$\begin{aligned} \bar{\alpha} = [0^T, v^T], \quad \alpha_n^0 = \bar{\alpha}, \quad \alpha_n^i &= \bar{\alpha} + \left(\sqrt{(L+\lambda)P_n^a}\right)_i, i = 1, \ldots, L \\ \alpha_n^i &= \bar{\alpha} - \left(\sqrt{(L+\lambda)P_n^a}\right)_i, i = L+1, \ldots, 2L \end{aligned} \tag{9}$$

We denote $[\xi_i, v_i] = \alpha_n^i$, and our state at time n is $\bar{\chi}$. Then these sigma points go through the measurement function h:

$$y_i = h(\bar{\chi} \exp \xi_i) + v_i, i = 0, \ldots, 2L$$

The measure is thus $\bar{y} = \sum_{i=0}^{2L} W_s^i y_i$. The values for the weights W_s^i can be found explicitly in [6].

The state and covariance are then updated as:

$$
\begin{aligned}
P_{yy} &= \sum_{i=0}^{2L} W_c^i (y_i - \bar{y})(y_i - \bar{y})^T, \quad P_{\alpha y} = \sum_{i=0}^{2L} W_c^i (\alpha_i - \bar{\alpha})(y_i - \bar{y})^T \\
[\bar{\xi}^T, *]^T &= P_{\alpha y} P_{yy}^{-1} (y - \bar{y}), \quad \chi^+ = \bar{\chi} \exp(\bar{\xi}) \\
P^+ &= P - P_{\alpha y} (P_{\alpha y} P_{yy}^{-1})^T
\end{aligned}
\tag{10}
$$

The final filter, that we call the l-UKF (left-Unscented Kalman Filter), is composed of the propagation step of the IEKF (Eq. (6)) and of the left-UKF update step (Eqs. (9) to (10)). This does not interfere with the consistency properties of the IEKF in the optimal setting, but this allows to get around the approximations of the measurement functions linearization.

5 Application on Real Drone Flights

In this section, we present results obtained on real drone flights. The data come from GPS measurements. We have thus added noise by hand, with amplitude similar to that of real radar noise. The drone positions are only known in 2D, so our 2D model is well suited for these positions. The IEKF and l-UKF algorithms can be adapted to 3D range, bearing and elevation measurements. The 3D IEKF for the target tracking problem is for example presented in [13]. In the model (1), the tangential and angular velocities (u and ω) were supposed constant. However, it is not exactly the case in practice, as they are only piecewise constant. The process noise tuning thus has to be adapted to the amplitude of the variations of these parameters.

We have compared the EKF with the IEKF and the l-UKF on three different drone trajectories. The trajectories are presented on Fig. 1, without measurement noise for better readability. The trajectories were obtained with different types of drones: a quadcopter drone, a hexacopter drone, and a flying wing drone. Position estimations and RMSE for the EKF, the IEKF and the l-UKF respectively are presented on Figs. 2, 3 and 4. The results for the IEKF and the l-UKF are more precise than that of the EKF. It is mostly visible on the RMSE figures.

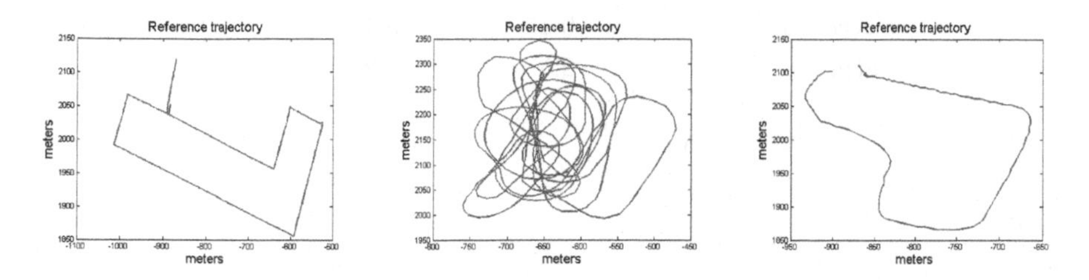

Fig. 1. Three different drone trajectories

We have computed the Root Mean Squared Errors (RMSE) of each parameter for each trajectory, with the same initialization for an EKF, an IEKF and a l-UKF. The process noises used for each filtering algorithm and each trajectory were optimzed by maximising the measurement likelihood, as in [14]. The same measurement noises are used for all three trajectories. These RMSE results are presented in Table 1. As we have already seen with the position RMSE plots, the position estimation precision is better for the IEKF and the l-UKF than for the EKF. But what is more remarquable is the orientation θ precision. Indeed, it is notably better for the IEKF and the l-UKF filters. We can also notice that the l-UKF performs overall slightly better than the IEKF, especially on orientation, angular and tangential velocities. For the radar application, the orientation precision is very important, indeed, the orientation parameter gives the direction of the velocity of the target, and this is needed to refresh the beam of the radar. This estimation is thus of great impact, and it is very valuable to have a precise orientation estimation.

Table 1. RMSE for each parameter on 100 Monte Carlo, for each one of the three trajectories, and for the three algorithms

Algorithm	Parameter	Trajectory 1	Trajectory 2	Trajectory 3
EKF	$x^{(1)}(m)$	4.6	11	6.0
	$x^{(2)}(m)$	1.9	3.4	2.3
	θ (RMSE for $1 - \cos\theta$)	0.45	0.22	0.38
	$\omega(rad/s)$	0.34	0.45	0.96
	$u(m/s)$	3.6	3.6	1.7
IEKF	$x^{(1)}(m)$	4.5	7.2	5.3
	$x^{(2)}(m)$	1.9	2.3	2.2
	θ (RMSE for $1 - \cos\theta$)	0.34	0.17	0.21
	$\omega(rad/s)$	0.30	0.43	0.95
	$u(m/s)$	2.7	3.6	1.4
l-UKF	$x^{(1)}(m)$	4.1	7.6	6.8
	$x^{(2)}(m)$	2.2	2.9	2.8
	θ (RMSE for $1 - \cos\theta$)	0.29	0.17	0.23
	$\omega(rad/s)$	0.25	0.42	0.95
	$u(m/s)$	2.5	3.2	1.2

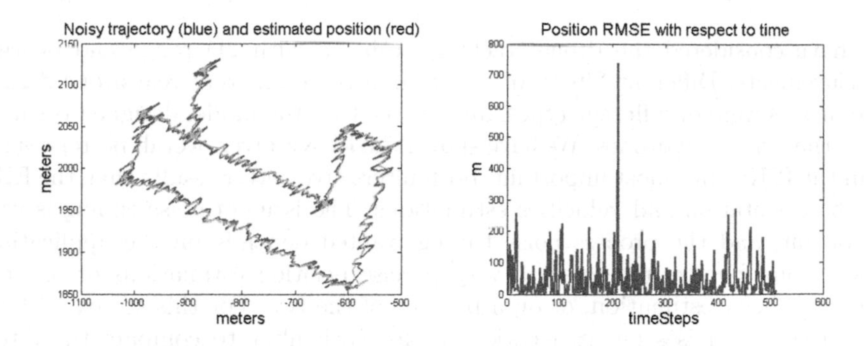

Fig. 2. Estimation and RMSE of the position for the EKF algorithm

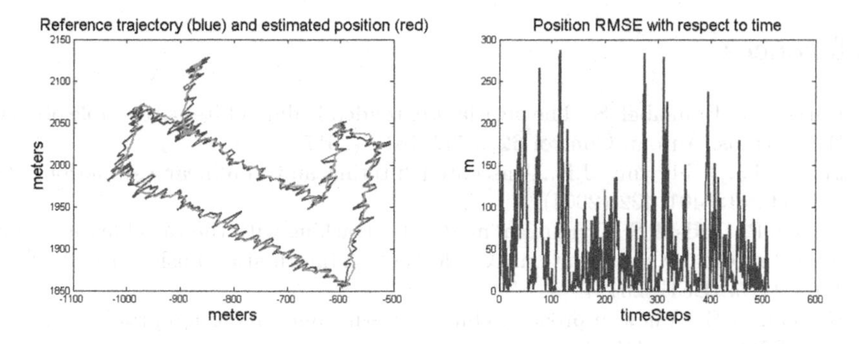

Fig. 3. Estimation and RMSE of the position for the IEKF algorithm

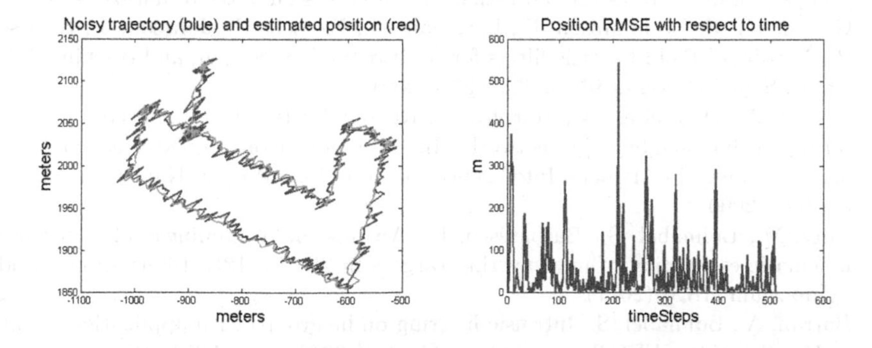

Fig. 4. Estimation and RMSE of the position for the l-UKF algorithm

6 Conclusion

We have considered the drone tracking problem, with 2D range and bearing measurements. Different filters were tested on three different real drone flights. The drones were of different types, and we see that the model designed is suited to all these types of drones. We have shown the l-UKF gives overall better results than the IEKF, but most important, both filters give better results than the EKF for the orientation and velocities estimations. The issue of noise tuning is very important, and the process noise tuning wanted depends on the application. Indeed, one can be interested in very precise position estimations or on very precise velocity estimation, or on a balance of the two. For this study, we have optimized the noises on each trajectory for each filter to compare the filters with equal treatment. A more robust solution for the noise tuning is to use the Castella method, see [15], which can be used for all kind of filters. This method is used to adapt the process noise in real time. The position estimation is thus more precise, however, this is at the cost of a lesser velocity estimation precision.

References

1. Barrau, A., Bonnabel, S.: The invariant extended Kalman filter as a stable observer. IEEE Trans. Autom. Control **62**, 1797–1812 (2017)
2. Julier, S.J., Uhlmann, J.K.: Unscented filtering and nonlinear estimation. Proc. IEEE **92**(3), 401–422 (2004)
3. Bunch, P., Godsill, S.: Dynamical models for tracking with the variable rate particle filter. In: 2012 15th International Conference on Information Fusion (FUSION), pp. 1769–1775. IEEE (2012)
4. Kalman, R.E.: A new approach to linear filtering and prediction problems. J. Basic Eng. **82**(1), 35–45 (1960)
5. Bar-Shalom, Y., Li, X., Kirubarajan, T.: Estimation with Applications to Tracking and Navigation: Theory Algorithms and Software. Wiley, New York (2004). https://books.google.fr/books?id=xz9nQ4wdXG4C
6. Brossard, M., Bonnabel, S., Condomines, J.-P.: Unscented Kalman filtering on lie groups. soumis à IROS (2017). https://hal.archives-ouvertes.fr/hal-01489204
7. Gustafsson, F., Gunnarsson, F., Bergman, N., Forssell, U., Jansson, J., Karlsson, R., Nordlund, P.-J.: Particle filters for positioning, navigation, and tracking. IEEE Trans. Signal Process. **50**(2), 425–437 (2002)
8. Doucet, A., De Freitas, N., Murphy, K., Russell, S.: Rao-blackwellised particle filtering for dynamic bayesian networks. In: Proceedings of the Sixteenth conference on Uncertainty in Artificial Intelligence, pp. 176–183. Morgan Kaufmann Publishers Inc. (2000)
9. Pilté, M., Bonnabel, S., Barbaresco, F.: An innovative nonlinear filter for radar kinematic estimation of maneuvering targets in 2D. In: 18th International Radar Symposium (IRS) (2017)
10. Barrau, A., Bonnabel, S.: Intrinsic filtering on lie groups with applications to attitude estimation. IEEE Trans. Autom. Control **60**(2), 436–449 (2015)
11. Eade, E.: Lie groups for 2d and 3d transformations. http://ethaneade.com/lie.pdf. Accessed Dec 2013

12. Julier, S.J., Uhlmann, J.K.: New extension of the kalman filter to nonlinear systems. In: AeroSense 1997, pp. 182–193. International Society for Optics and Photonics (1997)
13. Pilté, M., Bonnabel, S., Barbaresco, F.: Tracking the Frenet-Serret frame associated to a highly maneuvering target in 3D. working paper or preprint. https://hal-mines-paristech.archives-ouvertes.fr/hal-01568908
14. Abbeel, P., Coates, A., Montemerlo, M., Ng, A.Y., Thrun, S.: Discriminative training of Kalman filters. In: Robotics: Science and systems, vol. 2, p. 1 (2005)
15. Castella, F.R.: An adaptive two-dimensional Kalman tracking filter. IEEE Trans. Aerosp. Electron. Syst. **AES–16**(6), 822–829 (1980)

Particle Observers for Contracting Dynamical Systems

Silvère Bonnabel[1(✉)] and Jean-Jacques Slotine[2]

[1] MINES ParisTech, PSL Reasearch University, Centre for Robotics, 60 Bd Saint-Michel, 75006 Paris, France
silvere.bonnabel@mines-paristech.fr
[2] Department of Mechanical Engineering, Massachusetts Institute of Technology, Cambridge, MA 02139, USA
jjs@mit.edu

Abstract. In the present paper we consider a class of partially observed dynamical systems. As in the Rao-Blackwellized particle filter (RBPF) paradigm (see e.g., Doucet et al. 2000), we assume the state x can be broken into two sets of variables $x = (z, r)$ and has the property that conditionally on z the system's dynamics possess geometrical contraction properties, or is amenable to such a system by using a nonlinear observer whose dynamics possess contraction properties. Inspired by the RBPF we propose to use particles to approximate the r variable and to use a simple copy of the dynamics (or an observer) to estimate the rest of the state. This has the benefits of 1- reducing the computational burden (a particle filter would sample the variable x also), which is akin to the interest of the RBPF, 2- coming with some indication of stability stemming from contraction (actual proofs of stability seem difficult), and 3- the obtained filter is well suited to systems where the dynamics of x conditionally on z is precisely known and the dynamics governing the evolution of z is quite uncertain.

1 A Primer on Contraction Theory

1.1 Background on Contraction Theory

Consider a Riemannian manifold $(\mathcal{M}, g)$, where g denotes the metric. Consider local coordinates. In the present paper, we will simplify the exposure by systematically assuming that $\mathcal{M} = \mathbb{R}^n$. The squared infinitesimal length is given by the quadratic form:

$$\|dx\|^2 = \sum_{1\leq i,j\leq n} g_{ij}(x)dx_i dx_j$$

The matrix $G = (g_{ij})_{1\leq i,j\leq n}$ is called the Riemannian metric tensor and it generally depends on x. Now, consider the continuous time deterministic system described by the following ordinary differential equation (ODE) on $\mathbb{R}^n$:

$$\frac{d}{dt}x = f(x), \tag{1}$$

F. Nielsen and F. Barbaresco (Eds.): GSI 2017, LNCS 10589, pp. 310–317, 2017.
https://doi.org/10.1007/978-3-319-68445-1_36

with f a smooth nonlinear function satisfying the usual conditions for global existence and unicity of the solution. For a detailed proof of the following theorem, see e.g., Pham and Slotine (2013).

Theorem 1 (Lohmiller and Slotine (1998)). *Let $J_f(x)$ denote the Jacobian matrix of $f(x)$. Assume that $M(x) = G^T(x)G(x)$ is uniformly positive definite, and that $G(x)J_f(x)G^{-1}(x)$ is uniformly negative definite, then all trajectories exponentially converge to a single trajectory. Moreover, the convergence rate is equal to $\lambda > 0$ which is the supremum over x of the largest eigenvalue of $G(x)J_f(x)G^{-1}(x)$. More precisely, if $a(t)$ and $b(t)$ are two trajectories of* (1)*, we have:*

$$d_g(a(t), b(t)) \leq d_g(a(0), b(0))e^{-2\lambda t},$$

where d_g denotes the Riemannian distance associated to metric g.

1.2 Nonlinear Observers for Contracting Systems

Consider the system (1) where $x(t) \in \mathbb{R}^N$, with partial observations

$$y(t) = h(x(t)), \tag{2}$$

The goal of observer design, is to estimate in real time the *unknown* quantity $x(t)$ with the greatest possible accuracy given all the measurements up to current time t. Assume that for a class of functions $y(t)$, the dynamics

$$\frac{d}{dt}z = f(z) + K(z, y)(y - h(z)) \tag{3}$$

can be proved to be contractive with rate $\lambda > 0$. Then, the observer for the system (1) and (2) defined by

$$\frac{d}{dt}\hat{x} = f(\hat{x}) + K(\hat{x}, y)(y - h(\hat{x})), \tag{4}$$

possesses convergence properties. Indeed, as the simulated $\hat{x}(t)$ and the true trajectory $x(t)$ are both solutions of Eq. (3), Theorem 1 applies and we have:

$$d_g(\hat{x}(t), x(t)) \leq d_g(\hat{x}(0), x(0))e^{-2\lambda t}.$$

2 The Basic Particle Observer

2.1 The Rao-Blackwellized Particle Filter (RBPF)

Consider a (discrete) Markov process r_t of initial distribution $p(r_0)$ and transition equation $p(r_t \mid r_{t-1})$. The variable r_t is hidden, and assume we have as observation a random variable y_t at time t, which is correlated with r_t. The observations are assumed to be conditionally independent given the process r_t.

The goal of discrete time filtering is to infer online the hidden variables from the observations, that is, to compute:

$$p(r_t \mid y_{1:t}), \quad \text{where} \quad y_{1:t} = \{y_1, \cdots, y_t\},$$

or more generally $p(r_{1:t} \mid y_{1:t})$. Assume now, that we also want to infer another related process z_t, such that $p(z_t \mid y_{1:t}, r_{1:t})$ can be analytically evaluated. This is typically the case using a Kalman filter when conditionally on r the system is linear and Gaussian. A simple version of the RBPF is given by Algorithm 1.

Algorithm 1. RBPF with prior sampling (see e.g., Doucet et al. 2000)

Draw N particles from the prior initial distribution $p(r_0)$
loop
Sample from the prior

$$r_t^{(i)} \sim p(r_t \mid r_{t-1}^{(i)}), \quad \text{and let} \quad r_{1:t}^{(i)} = \left(r_t^{(i)}, r_{1:t-1}^{(i)}\right)$$

Evaluate and update weights

$$w_t^{(i)} = p(y_t \mid y_{1:t-1}, r_{1:t}^{(i)})\; w_{t-1}^{(i)}$$

Normalize weights

$$\tilde{w}_t^{(i)} = \frac{w_t^{(i)}}{[\sum_j w_t^{(j)}]^{-1}}$$

The estimate of the expected value $\mathbb{E}\,(F(z,r))$ of any function F is

$$\sum_i \tilde{w}_t^{(i)} \mathbb{E}_{p(z_t \mid y_{1:t}, r_{1:t}^{(i)})} \left(F(r_t^{(i)}, z_t)\right)$$

Resample if necessary, i.e., duplicate and suppress particles to obtain N random samples with equal weights (i.e., equal to $1/N$).
end loop

2.2 The Particle Observer for Conditonnally Contracting Systems

Consider a noisy dynamical system of the form

$$\frac{d}{dt}z = f(z,r) \tag{5}$$

$$\frac{d}{dt}r = g(r) + w(t) \tag{6}$$

where f, g are smooth maps, $w(t)$ is a process noise, and we have an initial prior distribution $\pi_0(z,r)$ at time $t=0$. Assume one has access to discrete time uncertain measurements $y_n = h(z_{t_n}, r_{t_n}) + V_n$ at times $t_0 < t_1 < t_2 < \cdots$, and where V_n are unknown independent identically distributed random variables with known density l, that is, $p(y \mid z, r) = l(y - h(z,r))$. We introduce the following definition.

Definition 1. *The system* (5) *and* (6) *is said to be a contraction conditionally on z if Eq.* (5) *is a contraction when $r(t)$ is considered as a known input.*

The rationale of our particle observer is as follows. If $r(t)$ were known, then, all trajectories of the system (5) would converge to each other due to the conditional contraction properties we assume. Thus, if we call $\hat{z}(t)$ a solution of (5) associated to some trajectory $\{r(t)\}_{t\geq 0}$, then asymptotically we have $p(z(t) \mid \{r(s)\}_{0\leq s\leq t}) \approx \delta(z(t) - \hat{z}(t))$, which means that contrarily to the RBPF paradigm we can not compute the conditional distributions in closed form but we have access to relevant approximations to them. Thus, letting $(\hat{z}_t^{(i)}, r_t^{(i)})$ be a solution to the stochastic differential Eqs. (5) and (6), we have the following approximations that stem from the partial contraction properties of the system:

$$p(z_t \mid y_{1:t}, r_{1:t_n}^{(i)}) \approx \delta(z_t - \hat{z}_t^{(i)}), \quad p(y_{t_n} \mid y_{1:t_{n-1}}, r_{1:t_n}^{(i)}) \approx l(y_{t_n} - h(\hat{z}_{t_n}^{(i)}, r_{t_n}^{(i)})).$$

Thus, resorting to those approximation, and applying the RBPF methodology to the above system (5) and (6) we propose the following Algorithm 2.

Algorithm 2. The PO with prior sampling

Draw N particles $(z_{t_0}^{(1)}, r_{t_0}^{(1)}), \cdots, (z_{t_0}^{(N)}, r_{t_0}^{(N)})$ from the prior initial distribution $\pi_0(z, r)$
loop

Sample $(z_{t_n}^{(i)}, r_{t_n}^{(i)})$ from the prior by numerically integrating the stochastic differential Eqs. (5) and (6) from time t_{n-1} to t_n.
Evaluate and update weights

$$w_t^{(i)} = l(y_{t_n} - h(z_{t_n}^{(i)}, r_{t_n}^{(i)}))\ w_{t-1}^{(i)}.$$

Numerically enforce that at least one weight is not equal to zero (i.e., if all weights are zero, set e.g. $w_t^{(1)} = 1$).
Normalize weights

$$\tilde{w}_t^{(i)} = \frac{w_t^{(i)}}{[\sum_j w_t^{(j)}]^{-1}}.$$

The estimate of the expected value $\mathbb{E}\left(F(z,r)\right)$ of any function F is approximated by

$$\sum_i \tilde{w}_t^{(i)} F(r_t^{(i)}, z_t^{(i)})$$

In particular the state is approximated by

$$\sum_i \tilde{w}_t^{(i)} (r_t^{(i)}, z_t^{(i)})$$

Resample if necessary, i.e., duplicate and suppress particles to obtain N random samples with equal weights (i.e., equal to $1/N$).
end loop

2.3 Some Comments on the Choice of Model

The relevance of system (5) and (6) is debatable, for the two following reasons. First, it might be surprising that the dynamics of z conditionally on r be deterministic, whereas the dynamics of r be noisy. Second, because it is rare to find systems that are naturally (conditionally) contracting. Both issues will be partly addressed in the extensions outlined in the sequel. At this stage, we can make the following comments regarding the first issue. Assume both Eqs. (5) and (6) to be noisy. Then, thanks to the contraction property, the asymptotic distribution of $z(t)$ conditionally on $r(t)$ is not very dispersed if the process noise is moderate, see Pham et al. (2009). So the method may yield good results in practice. Assume on the other hand, both Eqs. (5) and (6) to be deterministic. Then, it is hopeless to estimate and track efficiently the state with a (RB) particle filter, as the state space will not be explored adequately. Indeed, because multiple copies are produced after each resampling step, the diversity of the particle system decreases to a few points, which can be very different from the true state. To solve this degeneracy problem, the regularized particle filter was proposed in Musso and Oujdane (1998). Albeit debatable, this technique may yield good results in practice. Following this route, we can postulate noisy Eq. (6) to implement our particle filter.

Remark 1. *Note that, here we do* not *deal with parameter identification, as in e.g., Saccomani et. al (2003). Although this might look similar, $r(t)$ is not a parameter, preventing us to directly apply the results of e.g., Wills et. al (2008)*

3 A Chemical Reactor Example

3.1 Retained Model

Consider the exothermic chemical reactor of Adebekun and Schork (1989). It was shown in Lohmiller and Slotine (1998) that, if the temperature T is known, and thus can be considered as an input, then the system is a contraction. But to achieve best performance, and filter the noise out of the temperature measurements, the temperature should be considered as a (measured) part of the state as in Adebekun and Schork (1989). This leads to a system that is not a contraction. To make our point, we even propose to slightly modify the temperature dynamics to make it clearly unstable, yielding the more challenging following system:

$$\frac{d}{dt}I = \frac{q(t)}{V}(I_f - I) - k_d e^{-\frac{E_d}{RT(t)}} I \tag{7}$$

$$\frac{d}{dt}M = \frac{q(t)}{V}(M_f - M) - 2k_p e^{-\frac{E_d}{RT(t)}} M^2 I \tag{8}$$

$$\frac{d}{dt}P = \frac{q(t)}{V}(P_f - P) + k_p e^{-\frac{E_d}{RT(t)}} M^2 I \tag{9}$$

$$\frac{d}{dt}T = \beta T + \sigma_2 w(t) \tag{10}$$

where $w(t)$ is a white Gaussian standard noise, and $\sigma_2 > 0$ a parameter encoding the noise amplitude. Letting $V_n \sim \mathcal{N}(0,1)$ a random standard centered Gaussian, we assume discrete temperature measurements of the form:

$$y_n = T(t_n) + \sigma_1 V_n. \tag{11}$$

Lohmiller and Slotine (1998) already proved the system is contracting conditionally on $T(t)$. Thus, we can use the method described in Algorithm 2.

3.2 Simulation Results

The true system is simulated according to the Eqs. (7), (8), (9) and (10) where we turned the noise off in Eq. (10) (this means we started from a noise-free system for which the RBPF would not work properly, and used the regularization technique discussed at Sect. 2.3). The noisy output (11) was also simulated, where an observation is made every 5 steps. We chose $\tilde{\sigma}_2 = 0.1$. Density l is dictated by the observation noise, that is, $l(u) = \frac{1}{\sigma_1\sqrt{2\pi}}\exp(-\frac{u^2}{2\sigma_1^2})$ with $\sigma_1 = 1K$.

To apply our methodology, we assume that we have plausible physical upper bounds on the concentrations inside, and denote them by $I_{\max}, M_{\max}, P_{\max}$ and we let π_0 be the uniform distribution on the hyperrectangle $[0, I_{\max}] \times [0, M_{\max}] \times [0, P_{\max}]$ with a Dirac on the measured initial temperature. In the simulation, all those upper bounds are set equal to 4 mol. $\frac{q(t)}{V}$ and $T(t)$ are slowly oscillating around $1\,\text{min}^{-1}$ and $300K$, we have $k_d e^{+\frac{E_d}{RT(t)}} \approx 0.8\,\text{min}^{-1}$ and $k_p e^{+\frac{E_d}{RT(t)}} \approx 0.2L\ \text{mol}^{-1}\,\text{min}^{-1}$. We also let $\beta = 0.01\,\text{min}^{-1}$.

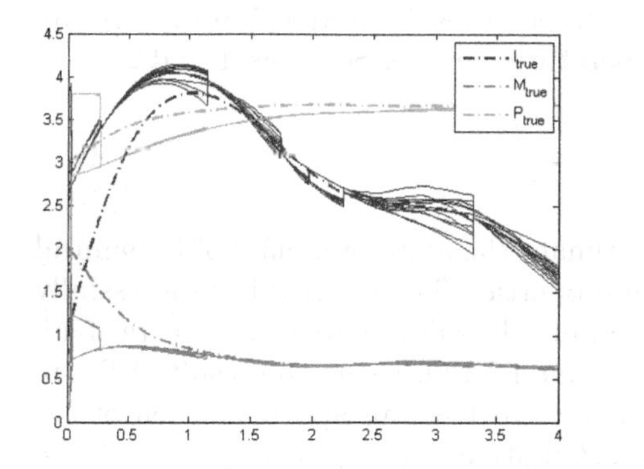

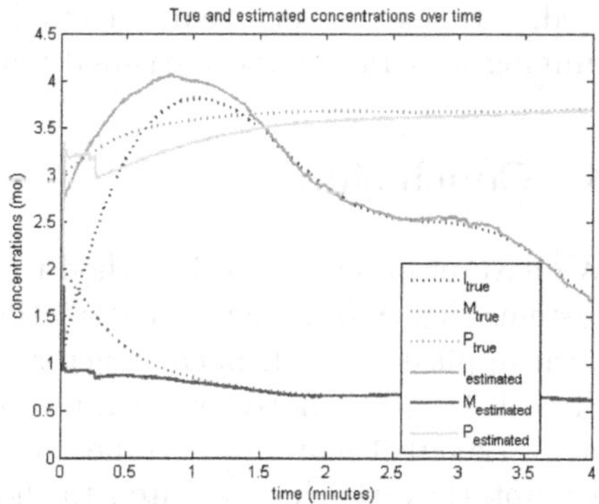

Fig. 1. Left: True concentrations (dashed lines) and trajectories of the 15 particles. We see the effect of resampling, that refocuses the bundle of trajectories on the fittest ones, when too many become unlikely. Right: True concentrations (dashed lines), and estimates of the particle observer (solid lines).

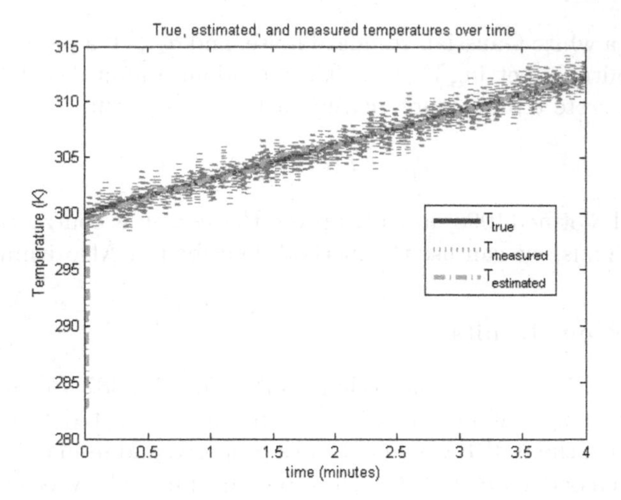

Fig. 2. True (solid line), measured (noisy line), estimated (dashed line, output by the RBPO) temperatures.

N=15 particles are used (which results in a very cheap to implement particle filter, as each particle is associated only to a naive observer). We resampled[1] each time the number of effective particles $1/(\sum_1^M w_j^2)$ drops below $N/4$, i.e., 25% of the total population. The resampling step is *necessary*, so that all particles gradually improve their estimation of the temperature, allowing the concentrations to be well estimated in turn.

The noise is efficiently filtered and all values asymptotically very well recovered, although a very reduced number of particles is used (15 observers are running in parallel) and measured temperature is noisy. See Figs. 1 and 2.

4 Conclusion

We have proposed a novel method to estimate the state of a class of dynamical systems that possess partial contraction properties. The method has successfully been applied to a chemical reactor example. Possible extensions are twofold. First, if Eq. (5) is noisy, one can use the same RBPO. Using the result of Pham et al. (2009), Pham and Slotine (2013), we can have an approximation of the asymptotic variance associated to the distribution $p(z(t) \mid \{r(s)\}_{0\leq s\leq t})$. Thus, a Gaussian approximation to this distribution can be leveraged to implement a RBPF. Second, if f is not contracting conditionally on r, *but*, is amenable to it

[1] i.e., draw N particles from the current particle set with probabilities proportional to their weights; replace the current particle set with this new one. Instead of setting the weights of the new particles equal to $1/N$ as in the standard methodology, we preferred in the simulations to assign them their former weight and then normalize.

using an observer of the form

$$\frac{d}{dt}z = f(z) + K(z, y, r)(y - h(z)),$$

then the method may still be applied.

The ideas introduced in this short paper might also be applied to differentially positive systems Forni and Sepulchre (2016), Bonnabel et al. (2011). In the future, we would also like to study the behavior of particle filters for systems with contraction properties. A starting point could be to seek how to use the recent results of Pham et al. (2009), Pham and Slotine (2013), Tabareau et. al (2010) on stochastic contraction.

References

Adebekun, D.K., Schork, F.J.: Continuous solution polymerization reactor control. 1. Nonlinear reference control of methyl methacrylate polymerization. Ind. Eng. Chem. Res. **28**(9), 1308–1324 (1989)

Bonnabel, S., Astolfi, A., Sepulchre, R.: Contraction and observer design on cones. In: IEEE Conference on Decision and Control (CDC 2011) (2011)

Doucet, A., De Freitas, N., Murphy, K.: Rao-blackwellised particle filtering for dynamic Bayesian networks. In: Proceedings of the Sixteenth Conference on Uncertainty in Artificial Intelligence, pp. 176–183 (2000)

Forni, F., Sepulchre, R.: Differentially positive systems. IEEE Trans. Autom. Control **61**(2), 346–359 (2016)

Lohmiller, W., Slotine, J.: On contraction analysis for nonlinear systems. Automatica **34**(6), 683–696 (1998)

Musso, C., Oudjane, N.: Regularization schemes for branching particle systems as a numerical solving method of the nonlinear filtering problem. In: Proceedings of the Irish Signals and Systems Conference (1998)

Pham, Q.C., Tabareau, N., Slotine, J.J.: A contraction theory approach to stochastic incremental stability. IEEE Trans. Autom. Control **54**(4), 816–820 (2009)

Pham, Q.C., Slotine, J.J.: Stochastic contraction in Riemannian metrics. arXiv preprint arXiv:1304.0340 (2013)

Saccomani, M.P., Audoly, S., D'Angió, L.: Parameter identifiability of nonlinear systems: the role of initial conditions. Automatica **39**(4), 619–632 (2003)

Tabareau, N., Slotine, J.J., Pham, Q.C.: How synchronization protects from noise. PLoS Comput. Biol. **6**(1), e1000637 (2010)

Wills, A., Schön, T.B., Ninness, B.: Parameter estimation for discrete-time nonlinear systems using EM. IFAC Proc. Vol. **41**(2), 4012–4017 (2008)

Sigma Point Kalman Filtering on Matrix Lie Groups Applied to the SLAM Problem

James Richard Forbes[1(✉)] and David Evan Zlotnik[2]

[1] Department of Mechanical Engineering, McGill University, Montréal, QC H3A 0C3, Canada
james.richard.forbes@mcgill.ca

[2] Department of Aerospace Engineering, University of Michigan, Ann Arbor, MI 48109, USA
dzlotnik@umich.edu

Abstract. This paper considers sigma point Kalman filtering on matrix Lie groups. Sigma points that are elements of a matrix Lie group are generated using the matrix exponential. Computing the mean and covariance using the sigma points via weighted averaging and effective use of the matrix natural logarithm, respectively, is discussed. The specific details of estimating landmark locations, and the position and attitude of a vehicle relative to the estimated landmark locations, is considered.

Keywords: Estimation · Sigma-point Kalman filtering · Matrix lie group · Simultaneous localization and mapping

1 Introduction

The extended Kalman filter (EKF) [1, pp. 54–64], sigma point Kalman filter (SPKF) [2], particle filter [1, pp. 96–113], and other approximations of the Bayes filter [1, pp. 26–33], assume the system state is an element of a vector space, such as $\mathbb{R}^{n_x}$, where n_x is the state dimension. When the system state space is an element a matrix Lie group, such as the special orthogonal group, denoted $SO(3)$, the EKF, SPKF, etc., are not directly applicable. This paper investigates SPKFing on matrix Lie groups. After reviewing the sigma point transformation (SPT) it is generalized and used to estimate means and covariances of nonlinear functions with states that are elements of a matrix Lie group. Next, the matrix Lie group SPT is used to construct a SPKF leading to the matrix Lie group SPKF (MLG SPKF). For simplicity of exposition, the unscented transformation (UT) [3] is the specific SPT used in all derivations, although the term SPT is retained throughout the paper. Particular attention is paid to how to compute the weighted average of sigma points when the underlying matrix Lie group is $SO(3)$. This is particularly relevant in robotics applications where vehicles can rotate in space. Two methods to compute a weighted mean on $SO(3)$ are discussed, one that employs a singular value decomposition, and another based on [4]. The method of [4] is popular in the aerospace community and, perhaps more importantly, the weighted averaging employed in [4] is computationally simple.

F. Nielsen and F. Barbaresco (Eds.): GSI 2017, LNCS 10589, pp. 318–328, 2017.
https://doi.org/10.1007/978-3-319-68445-1_37

The papers [4–8] are closest to the present work. The work of [4] employes a parameterization of $SO(3)$, but in doing so, a simple way to compute the mean of a set of sigma points is derived. The papers [5,6] also present SPKFs (and, to be more specific, UKFs) for systems with states that are elements of nonlinear manifolds. The formulations presented in [5,6] are similar to the formulation in the present paper, but with some differences, such as how the correction step is realized, and how the weighted mean of sigma points is computed in the special case that the matrix Lie group is $SO(3)$. The papers [7,8] share similarities to the present work also. In particular, although [7] is specific to the special Euclidian group, denoted $SE(3)$, the way uncertainty is propagated and the way sigma points are generated is the same to what is presented in this paper. Additionally, [7] proposes an iterative method to fuse multiple poses in order to compute a mean pose, while computing the mean attitude, which is an element of $SO(3)$, using alternative methods is discussed in this paper. In [8] SPKFing on matrix Lie groups is also considered. In [8] both left and right Gaussian distributions on the matrix Lie group are considered, while [7] and the present work only consider right Gaussian distributions. The most significant difference between [8] and the present work is the way the mean is computed in the prediction step of the filter. In [8] the mean of the previous step is propagated using the process model input in a noise-free manner, while herein the mean is computed using sigma points generated using the prior state covariance and process model noise covariance. Finally, the proposed MLG SPKF is applied to the problem of simultaneous localization and mapping (SLAM), a challenging problem not considered in [4–8]. This paper essentially combines and applies the results of [4–8] to the SLAM problem with particular attention being paid to computing the mean attitude from a set of sigma points.

2 The Sigma Point Transformation

The SPKF can be best understood in terms of approximating [2, p. 81] [9, p. 128]

$$E[\mathbf{f}(\mathbf{x}_k)] = \int_{-\infty}^{\infty} \mathbf{f}(\mathbf{x}_k)p(\mathbf{x}_k)\mathrm{d}\mathbf{x}_k, \tag{1}$$

$$E\left[(\mathbf{f}(\mathbf{x}_k) - E[\mathbf{f}(\mathbf{x}_k)])(\mathbf{f}(\mathbf{x}_k) - E[\mathbf{f}(\mathbf{x}_k)])^\mathsf{T}\right] = \int_{-\infty}^{\infty} (\mathbf{f}(\mathbf{x}_k) - E[\mathbf{f}(\mathbf{x}_k)])(\mathbf{f}(\mathbf{x}_k) - E[\mathbf{f}(\mathbf{x}_k)])^\mathsf{T} p(\mathbf{x}_k)\mathrm{d}\mathbf{x}_k, \tag{2}$$

using a sigma point transformation. The probability density function $p(\mathbf{x}_k)$ is assumed to be Gaussian, denoted $\mathcal{N}(\hat{\mathbf{x}}_k, \mathbf{P}_k)$, where $\hat{\mathbf{x}}_k \in \mathbb{R}^{n_x}$ is the mean, $\mathbf{P}_k \in \mathbb{R}^{n_x \times n_x}$ is the covariance, and n_x is the dimension of $\mathbf{x}_k$. Using a Cholesky decomposition, $\mathbf{P}_k = \mathbf{S}_k\mathbf{S}_k^\mathsf{T}$ where $\mathbf{S}_k$ is lower-triangular and $\mathbf{S}_k = [\mathbf{s}_{1,k} \cdots \mathbf{s}_{i,k} \cdots \mathbf{s}_{L,k}]$, a set of sigma points are computed as $\boldsymbol{\mathcal{X}}_{0,k} = \hat{\mathbf{x}}_k$, $\boldsymbol{\mathcal{X}}_{i,k} = \hat{\mathbf{x}}_k + \sqrt{L+\kappa}\ \mathbf{s}_{i,k}$, and $\boldsymbol{\mathcal{X}}_{i+L,k} = \hat{\mathbf{x}}_k - \sqrt{L+\kappa}\ \mathbf{s}_{i,k}$ where $i = 1, \ldots, L$ and $L = n_x$. Passing the sigma points through the nonlinear function $\mathbf{f}(\cdot)$ results in $\boldsymbol{\mathcal{X}}_{i,k}^{+} = \mathbf{f}(\boldsymbol{\mathcal{X}}_{i,k})$, $i = 0, \ldots, 2L$. The mean, $\hat{\mathbf{x}}_k^{+}$, and covariance, $\mathbf{P}_k^{+}$, are then approximated as

$$\hat{\mathbf{x}}_k^{+} = \sum_{i=0}^{2L} w_i \boldsymbol{\mathcal{X}}_{i,k}^{+}, \quad \mathbf{P}_k^{+} = \sum_{i=0}^{2L} w_i \left(\boldsymbol{\mathcal{X}}_{i,k}^{+} - \hat{\mathbf{x}}_k^{+}\right)\left(\boldsymbol{\mathcal{X}}_{i,k}^{+} - \hat{\mathbf{x}}_k^{+}\right)^\mathsf{T}, \tag{3}$$

where w_i are weights of the form $w_0 = \kappa/(\kappa + L)$ and $w_i = 1/(2\kappa + 2L)$. The unscented transformation [3] is the specific sigma point transformation used in this paper, but there are other sigma point transformations. See [2, Chap. 5], [9, Chap. 5], or [10].

3 Matrix Lie Groups

A matrix Lie group, denoted G, is composed of full rank, and therefore invertible, $n \times n$ matrices that is closed under matrix multiplication [11, p. 98]. The matrix Lie algebra associated with a matrix Lie group, denoted $\mathfrak{g}$, is the tangent space of G at the identity, denoted $T_{\mathbf{1}}G$. The tangent space of G at a point $\mathbf{X} \in G$ is denoted $T_{\mathbf{X}}G$. The matrix Lie algebra is a vector space closed under the operation of the matrix Lie bracket defined by $[\mathbf{A}, \mathbf{B}] = \mathbf{AB} - \mathbf{BA}$, $\forall \mathbf{A}, \mathbf{B} \in \mathfrak{g}$. Moreover, $\mathbf{XAX}^{-1} \in \mathfrak{g}\ \forall \mathbf{X} \in G$, $\forall \mathbf{A} \in \mathfrak{g}$ [11, p. 98]. Let $\{\mathbf{B}_1, \dots, \mathbf{B}_n\}$ be a basis for the matrix Lie algebra, called the generators, so that any $\mathbf{A} \in \mathfrak{g}$ can be written $\mathbf{A} = \mathbf{S}(\mathbf{a}) = \sum_{i=1}^{n} a_i \mathbf{B}_i$ where $\mathbf{a} = [a_1, \ \dots, \ a_n]^{\mathsf{T}} \in \mathbb{R}^n$ is the column matrix of coefficients associated with $\mathbf{A}$ [12]. The definition of $\mathbf{S} : \mathbb{R}^n \to \mathfrak{g}$ naturally leads to the definition of the inverse operation $\mathbf{S}^{-1} : \mathfrak{g} \to \mathbb{R}^n$. The matrix Lie group G and its associated matrix Lie algebra are related through the matrix exponential and matrix natural logarithm. Specifically, $\exp(\cdot) : \mathfrak{g} \to G$, $\ln(\cdot) : G \to \mathfrak{g}$ so that $\exp(\ln(\mathbf{X})) = \mathbf{X}$, $\forall \mathbf{X} \in G$ [13, p. 19].

4 The Sigma Point Transformation on Matrix Lie Groups

Let $\mathcal{N}(\hat{\mathbf{X}}_k, \mathbf{P}_k)$ denote a Gaussian distribution with mean $\hat{\mathbf{X}}_k \in G$ and covariance $\mathbf{P}_k \in \mathbb{R}^{n_x \times n_x}$ where n_x is the dimension of $\mathfrak{g}$. A realization, also referred to as a sample, from $\mathcal{N}(\hat{\mathbf{X}}_k, \mathbf{P}_k)$ must respect the group structure of G. Herein realizations are generated using the matrix exponential [7,14]. Specifically, using $\mathbf{S}(\boldsymbol{\xi}_k) \in \mathfrak{g}$ where $\boldsymbol{\xi}_i \in \mathbb{R}^{n_x}$ and $\boldsymbol{\xi}_i \sim \mathcal{N}(\mathbf{0}, \mathbf{P}_k)$, a realization is generated via $\mathbf{X}_k \leftarrow \exp(\mathbf{S}(\boldsymbol{\xi}_k))\hat{\mathbf{X}}_k$.

The task at hand is to compute the mean and covariance of a nonlinear function $\mathbf{f} : G \to G$ where G is a matrix Lie Group. Starting with $\mathcal{N}(\hat{\mathbf{X}}_k, \mathbf{P}_k)$, a Cholesky decomposition $\mathbf{P}_k = \mathbf{S}_k \mathbf{S}_k^{\mathsf{T}}$ where $\mathbf{S}_k$ is lower-triangular and $\mathbf{S}_k = [\mathbf{s}_{1,k} \ \cdots \ \mathbf{s}_{i,k} \ \cdots \ \mathbf{s}_{L,k}]$, is used to generate sigma points. In particular, $\boldsymbol{\mathcal{X}}_{0,k} = \hat{\mathbf{X}}_k$, $\boldsymbol{\mathcal{X}}_{i,k} = \exp\left(\sqrt{L+\kappa}\ \mathbf{S}(\mathbf{s}_{i,k})\right)\hat{\mathbf{X}}_k$, $\boldsymbol{\mathcal{X}}_{i+L,k} = \exp\left(-\sqrt{L+\kappa}\ \mathbf{S}(\mathbf{s}_{i,k})\right)\hat{\mathbf{X}}_k$, where $i = 1, \dots, L$ and $L = n_x$. The sigma points are passed through the nonlinear function $\mathbf{f}(\cdot)$ resulting in $\boldsymbol{\mathcal{X}}_{i,k}^{+} = \mathbf{f}(\boldsymbol{\mathcal{X}}_{i,k})$, $i = 0, \dots, 2L$. The mean and covariance, $\hat{\mathbf{X}}_k^{+}$ and $\mathbf{P}_k^{+}$, respectively, cannot be computed using (3) because adding or subtracting elements of G does not yield an element of G. The mean can be computing by solving

$$\hat{\mathbf{X}}_k^{+} = \arg \min_{\mathbf{X}_k \in G} \sum_{i=0}^{2L} w_i d^2(\boldsymbol{\mathcal{X}}_{i,k}^{+}, \mathbf{X}_k), \tag{4}$$

where $d(\cdot,\cdot)$ denotes distance on G. Usually $d(\cdot,\cdot)$ is the geodesic distance but, as discussed in [15,16], it is possible to employ alternative distance measures. For instance, when the matrix Lie group is $SO(3)$ the chordal distance may be employed over the geodesic distance. Although [5] suggest solving (4) as well, a detailed discussion on how to go do so is not provided. On the other hand, [6,7,14] proposes finding a solution to (4) in a similar manner to [17,18], that is, in an iterative manner.

The covariance is computed by first defining $\exp(\mathbf{S}(\mathbf{e}_{i,k})) \in G$ where $\mathbf{S}(\mathbf{e}_{i,k}) \in \mathfrak{g}$ and $\mathbf{e}_{i,k} \in \mathbb{R}^{n_x}$ such that $\boldsymbol{\mathcal{X}}^{+}_{i,k} = \exp(\mathbf{S}(\mathbf{e}_{i,k}))\hat{\mathbf{X}}^{+}_{k}$, where $\hat{\mathbf{X}}^{+}_{k}$ is the solution to (4). The term $\exp(\mathbf{S}(\mathbf{e}_{i,k}))$ can be thought of as the difference or error between sigma point $\boldsymbol{\mathcal{X}}^{+}_{i,k}$ and the mean $\hat{\mathbf{X}}^{+}_{k}$. Using the matrix logarithm and $\mathbf{S}^{-1}$, each column matrix $\mathbf{e}_{i,k}$, $i = 0,\ldots,2L$, can be computed via $\exp(\mathbf{S}(\mathbf{e}_{i,k})) = \boldsymbol{\mathcal{X}}^{+}_{i,k}(\hat{\mathbf{X}}^{+}_{k})^{-1}$, $\mathbf{S}(\mathbf{e}_{i,k}) = \ln(\boldsymbol{\mathcal{X}}^{+}_{i,k}(\hat{\mathbf{X}}^{+}_{k})^{-1})$, $\mathbf{e}_{i,k} = \mathbf{S}^{-1}(\ln(\boldsymbol{\mathcal{X}}^{+}_{i,k}(\hat{\mathbf{X}}^{+}_{k})^{-1}))$. It follows that the covariance can be approximated as $\mathbf{P}^{+}_{k} = \sum_{i=0}^{2L} w_i \mathbf{e}_{i,k}\mathbf{e}^{\mathsf{T}}_{i,k}$.

5 Specific Use of the Sigma Point Transformation for State Estimation on Matrix Lie Groups

Consider the nonlinear process and measurement models

$$\mathbf{X}_k = \mathbf{f}_{k-1}(\mathbf{X}_k, \mathbf{u}_{k-1}, \mathbf{w}_{k-1}), \qquad \mathbf{w}_k \sim \mathcal{N}(\mathbf{0}, \mathbf{Q}_k), \tag{5}$$

$$\mathbf{y}_k = \mathbf{g}_k(\mathbf{X}_k, \mathbf{v}_k), \qquad \mathbf{v}_k \sim \mathcal{N}(\mathbf{0}, \mathbf{R}_k), \tag{6}$$

where $\mathbf{X}_k \in G$, $\mathbf{u}_k \in \mathbb{R}^{n_u}$, $\mathbf{w}_k \in \mathbb{R}^{n_w}$, $\mathbf{y}_k \in \mathbb{R}^{n_y}$ and $\mathbf{v}_k \in \mathbb{R}^{n_y}$. The process and measurement models could alternatively have $\mathbf{u}_k \in G$, $\mathbf{w}_k \in G$, $\mathbf{y}_k \in G$, and $\mathbf{v}_k \in G$.

To estimate $\mathbf{X}_k$ using the SPKF, the properties of the group G must be respected. The state estimation procedure starts with definition of $\mathbf{z}_{k-1} = \left[(\mathbf{S}^{-1}(\ln(\hat{\mathbf{X}}_{k-1})))^{\mathsf{T}}\ \mathbf{0}^{\mathsf{T}}\right]^{\mathsf{T}}$, and $\mathbf{Y}_{k-1} = \operatorname{diag}\{\mathbf{P}_{k-1}, \mathbf{Q}_{k-1}\}$. Using a Cholesky decomposition, let $\mathbf{Y}_{k-1} = \mathbf{S}_{k-1}\mathbf{S}^{\mathsf{T}}_{k-1}$ where $\mathbf{S}_{k-1} = [\mathbf{s}_{1,k-1} \cdots \mathbf{s}_{L,k-1}]$ is lower triangular. Further partition each $\mathbf{s}_{1,k-1}$ as $\mathbf{s}_{i,k-1} = \left[\mathbf{s}^{\mathsf{T}}_{i,k-1,x}\ \mathbf{s}^{\mathsf{T}}_{i,k-1,w}\right]^{\mathsf{T}}$ where $\mathbf{s}_{i,k-1,x} \in \mathbb{R}^{n_x}$ and $\mathbf{s}_{i,k-1,w} \in \mathbb{R}^{n_w}$. The sigma points are then computed as

$$\boldsymbol{\mathcal{Z}}_{0,k-1} = \begin{bmatrix} \mathbf{S}^{-1}(\ln(\hat{\mathbf{X}}_{k-1})) \\ \boldsymbol{\mathcal{W}}_{0,k-1} \end{bmatrix}, \qquad \boldsymbol{\mathcal{Z}}_{i,k-1} = \begin{bmatrix} \mathbf{S}^{-1}(\ln(\exp(\sqrt{L+\kappa}\ \mathbf{S}(\mathbf{s}_{i,k-1,x}))\hat{\mathbf{X}}_{k-1})) \\ \boldsymbol{\mathcal{W}}_{0,k-1} + \sqrt{L+\kappa}\mathbf{s}_{i,k-1,w} \end{bmatrix},$$

$$\boldsymbol{\mathcal{Z}}_{i+L,k-1} = \begin{bmatrix} \mathbf{S}^{-1}(\ln(\exp(-\sqrt{L+\kappa}\ \mathbf{S}(\mathbf{s}_{i,k-1,x}))\hat{\mathbf{X}}_{k-1})) \\ \boldsymbol{\mathcal{W}}_{0,k-1} - \sqrt{L+\kappa}\mathbf{s}_{i,k-1,w} \end{bmatrix}.$$

Partitioning the $i = 0,\ldots,2L$ sigma points, $\boldsymbol{\mathcal{Z}}_{i,k-1} = [(\mathbf{S}^{-1}(\ln(\boldsymbol{\mathcal{X}}_{i,k-1})))^{\mathsf{T}}\ \boldsymbol{\mathcal{W}}^{\mathsf{T}}_{i,k-1}]^{\mathsf{T}}$, where $\boldsymbol{\mathcal{X}}_{i,k-1} = \exp(\sqrt{L+\kappa}\ \ \mathbf{S}(\mathbf{s}_{i,k-1,x}))\hat{\mathbf{X}}_{k-1}$ for $i = 1,\ldots,L$ and $\boldsymbol{\mathcal{X}}_{i,k-1} = \exp(-\sqrt{L+\kappa}\ \ \mathbf{S}(\mathbf{s}_{i,k-1,x}))\hat{\mathbf{X}}_{k-1}$ for $i = L+1,\ldots,2L$, and passing them through the nonlinear motion model in conjunction with $\mathbf{u}_{k-1}$ gives $\boldsymbol{\mathcal{X}}^{-}_{i,k} = \mathbf{f}_{k-1}(\boldsymbol{\mathcal{X}}_{i,k-1}, \mathbf{u}_{k-1}, \boldsymbol{\mathcal{W}}_{i,k-1})$, $i = 0,\ldots,2L$. The prediction step is completed by computing the predicted state and covariance in a weighted fashion that respects the group structure of G. As discussed in Sect. 4,

$$\hat{\mathbf{X}}_k^- = \arg\min_{\mathbf{X}_k \in G} \sum_{i=0}^{2L} w_i d^2(\boldsymbol{\mathcal{X}}_{i,k}^-, \mathbf{X}_k), \quad \mathbf{P}_k^- = \sum_{i=0}^{2L} w_i \mathbf{e}_{i,k}^- \mathbf{e}_{i,k}^{-\mathsf{T}}, \quad \text{where } \mathbf{e}_{i,k}^- = \mathbf{S}^{-1}(\ln(\boldsymbol{\mathcal{X}}_{i,k}^-(\hat{\mathbf{X}}_k^-)^{-1})).$$

To execute the correction step, define $\mathbf{z}_k^- = \left[(\mathbf{S}^{-1}(\ln(\hat{\mathbf{X}}_k^-)))^\mathsf{T} \; \mathbf{0}^\mathsf{T} \right]^\mathsf{T}$, and $\mathbf{Y}_k^- = \text{diag}\left\{\mathbf{P}_k^-, \mathbf{R}_k\right\}$, where $L = n_x + n_y$. Partition $\mathbf{Y}_k^-$ using Cholesky decomposition, $\mathbf{Y}_k^- = \mathbf{S}_k^- \mathbf{S}_k^{-\mathsf{T}}$, where $\mathbf{S}_k^-$ is lower triangular, and $\mathbf{s}_{j,k}^-$ is the jth column of $\mathbf{S}_k^-$. Moreover, partition the columns of $\mathbf{S}_k^-$ as $\mathbf{s}_{j,k}^- = \left[(\mathbf{s}_{j,k,x}^-)^\mathsf{T} \; (\mathbf{s}_{j,k,y}^-)^\mathsf{T} \right]^\mathsf{T}$. Compute a set of sigma points via

$$\boldsymbol{\mathcal{Z}}_{0,k} = \begin{bmatrix} \mathbf{S}^{-1}(\ln(\hat{\mathbf{X}}_k^-)) \\ \boldsymbol{\mathcal{V}}_{0,k} \end{bmatrix}, \qquad \boldsymbol{\mathcal{Z}}_{j,k}^- = \begin{bmatrix} \mathbf{S}^{-1}(\ln(\exp(\sqrt{L+\kappa}\; \mathbf{S}(\mathbf{s}_{j,k,x}^-))\hat{\mathbf{X}}_k^-)) \\ \boldsymbol{\mathcal{V}}_{0,k} + \sqrt{L+\kappa}\mathbf{s}_{j,k,y}^- \end{bmatrix},$$

$$\boldsymbol{\mathcal{Z}}_{j+L,k} = \begin{bmatrix} \mathbf{S}^{-1}(\ln(\exp(-\sqrt{L+\kappa}\; \mathbf{S}(\mathbf{s}_{j,k}^-))\hat{\mathbf{X}}_k^-)) \\ \boldsymbol{\mathcal{V}}_{0,k} - \sqrt{L+\kappa}\mathbf{s}_{j,k,y}^- \end{bmatrix}.$$

Partitioning the sigma points as $\boldsymbol{\mathcal{Z}}_{j,k}^- = \left[(\mathbf{S}^{-1}(\ln(\boldsymbol{\mathcal{X}}_{j,k}^-)))^\mathsf{T} \; \boldsymbol{\mathcal{V}}_{j,k}^\mathsf{T} \right]^\mathsf{T}$ where $\boldsymbol{\mathcal{X}}_{j,k}^- = \exp(\sqrt{L+\kappa}\; \mathbf{S}(\mathbf{s}_{j,k,x}^-))\hat{\mathbf{X}}_k^-$ for $j = 1, \ldots, L$ and $\boldsymbol{\mathcal{X}}_{j,k}^- = \exp(-\sqrt{L+\kappa}\; \mathbf{S}(\mathbf{s}_{j,k}^-))\hat{\mathbf{X}}_k^-$ for $j = L+1, \ldots, 2L$, and passing the sigma points through the nonlinear observation model yields $\boldsymbol{\mathcal{Y}}_{j,k} = \mathbf{g}_k(\boldsymbol{\mathcal{X}}_{j,k}^-, \boldsymbol{\mathcal{V}}_{j,k})$, $j = 0, \ldots, 2L$. The predicted measurement and its covariance are $\hat{\mathbf{y}}_k = \sum_{j=0}^{2L} w_j \boldsymbol{\mathcal{Y}}_{j,k}$ and $\mathbf{V}_k = \sum_{j=0}^{2L} w_j \left(\boldsymbol{\mathcal{Y}}_{j,k} - \hat{\mathbf{y}}_k\right)\left(\boldsymbol{\mathcal{Y}}_{j,k} - \hat{\mathbf{y}}_k\right)^\mathsf{T}$, where w_j are weights. The matrix $\mathbf{U}_k$ is computed as $\mathbf{U}_k = \sum_{j=0}^{2L} w_j \mathbf{e}_{j,k} \left(\boldsymbol{\mathcal{Y}}_{j,k} - \hat{\mathbf{y}}_k\right)^\mathsf{T}$, where $\mathbf{e}_{j,k} = \mathbf{S}^{-1}(\ln(\boldsymbol{\mathcal{X}}_{j,k}^-(\hat{\mathbf{X}}_k^-)^{-1}))$, and $\mathbf{K}_k = \mathbf{U}_k \mathbf{V}_k^{-1}$. Defining $\delta\boldsymbol{\xi}_k = \mathbf{K}_k(\mathbf{y}_k - \hat{\mathbf{y}}_k)$ so that $\mathbf{S}(\delta\boldsymbol{\xi}_k) \in \mathfrak{g}$, it follows that

$$\hat{\mathbf{X}}_k = \exp(\delta\boldsymbol{\xi}_k)\hat{\mathbf{X}}_k^-, \qquad \mathbf{P}_k = \mathbf{P}_k^- - \mathbf{U}_k \mathbf{V}_k^{-1} \mathbf{U}_k^\mathsf{T}. \tag{7}$$

In [5] the correction is not computed on G, but rather in $T_{\hat{\mathbf{X}}_k^-} G$, while [6,8] corrects in the same way as given in (7).

6 Application to Simultaneous Localization and Mapping

The MLG SPKF will now be applied to a popular problem in robotics, namely the SLAM problem. Estimating attitude and position of a vehicle relative to known landmarks is referred to as localization. When landmarks are not *a priori* known, creating a map of observed landmarks and, at the same time, localizing the vehicle relative to the map created, is referred to as SLAM. Consider a rigid-body vehicle endowed with a body-fixed frame, denoted $\mathcal{F}_b$, able to rotate relative to a datum frame, denoted $\mathcal{F}_a$. The matrix $\mathbf{C}_{ba} \in SO(3)$ describes the attitude of $\mathcal{F}_b$ relative to $\mathcal{F}_a$, where $SO(3) = \left\{\mathbf{C} \in \mathbb{R}^{3\times 3} \mid \mathbf{C}^\mathsf{T}\mathbf{C} = \mathbf{1},\ \det \mathbf{C} = +1\right\}$ is the special orthogonal group [19]. In the aerospace community the matrix $\mathbf{C}_{ba} \in$

$SO(3)$ is referred to as the direction cosine matrix (DCM) [20, p. 8]. The matrix Lie algebra associated with $SO(3)$, denoted $\mathfrak{so}(3)$, is the vectors space of 3×3 skew-symmetric matrices $\mathfrak{so}(3) = \{\boldsymbol{S} \in \mathbb{R}^{3\times 3} \mid \boldsymbol{S} = -\boldsymbol{S}^\mathsf{T}\}$. In particular, define $(\cdot)^\times : \mathbb{R}^3 \to \mathfrak{so}(3)$ where $\mathbf{y}^\times \mathbf{z} = -\mathbf{z}^\times \mathbf{y}$, $\forall \mathbf{y}, \mathbf{z} \in \mathbb{R}^3$, defines the typical cross product operation, and $(\cdot)^\vee : \mathfrak{so}(3) \to \mathbb{R}^3$ such that $(\mathbf{z}^\times)^\vee = \mathbf{z}$, $\forall \mathbf{z} \in \mathbb{R}^3$. The DCM $\mathbf{C}_{ba}$ can be written using axis/angle parameters as $\mathbf{C}_{ba} = \exp(-\phi \mathbf{a}^\times) = \cos\phi \mathbf{1} + (1 - \cos\phi)\mathbf{a}\mathbf{a}^\mathsf{T} - \sin\phi \mathbf{a}^\times$, where $\phi \in \mathbb{R}$ and $\mathbf{a} \in \mathbb{R}^3$ where $\mathbf{a}^\mathsf{T}\mathbf{a} = 1$. The matrix exponential and matrix logarithm are defined as $\exp(-\boldsymbol{\phi}^\times) = \mathbf{C}_{ba}$ and $\ln(\mathbf{C}_{ba}) = -\boldsymbol{\phi}^\times$ so that $\exp(\ln(\mathbf{C}_{ba})) = \mathbf{C}_{ba}$ where $\boldsymbol{\phi} = \phi\mathbf{a}$ is the rotation vector (or, strictly speaking, the rotation column matrix). The negative sign in the matrix exponential, and the"$-\sin\phi\mathbf{a}^\times$" term in $\mathbf{C}_{ba}$, follows the aerospace convention of describing attitude in terms of the $\mathcal{F}_b$ relative to $\mathcal{F}_a$ [20, p. 12, p. 32]. Consider two arbitrary points, points z and w. Let the position of point z relative to point w resolved in $\mathcal{F}_a$ be denoted $\mathbf{r}_a^{zw}$. The relationship between $\mathbf{r}_a^{zw} \in \mathbb{R}^3$ and $\mathbf{r}_b^{zw} \in \mathbb{R}^3$ is given by $\mathbf{r}_b^{zw} = \mathbf{C}_{ba}\mathbf{r}_a^{zw}$.

6.1 The Process and Measurement Models

Consider a robotic vehicle that can rotate and translate in space. Denote the body-fixed frame associated with the vehicle by $\mathcal{F}_b$, and the datum frame by $\mathcal{F}_a$. Assume the vehicle is equipped with two interoceptive sensors, an accelerometer and rate gyro, located at point z on the vehicle. Assume also the vehicle is equipped with an exteroceptive sensor located at point z on the vehicle that is able to observe and identify landmarks $p_1, \ldots, p_\ell$ where $\ell \in \mathbb{N}$ relative to point z, such as a camera. Let point w be another point. Point z can be thought of as the "origin" of $\mathcal{F}_b$ while point w can be though of as the "origin" of $\mathcal{F}_a$.

Accelerometer, Rate Gyro, and the Process Model — The discrete-time relationship between the position, velocity, and the accelerometer measurement $\mathbf{u}_{b_k}^1 \in \mathbb{R}^3$ is

$$\mathbf{r}_a^{z_k w} = \mathbf{r}_a^{z_{k-1} w} + T\mathbf{v}_a^{z_{k-1} w/a}, \qquad \mathbf{v}_a^{z_k w/a} = \mathbf{v}_a^{z_{k-1} w/a} + T\mathbf{C}_{b_{k-1}a}^\mathsf{T}(\mathbf{u}_{b_{k-1}}^1 + \mathbf{w}_{b_{k-1}}^1) - T\mathbf{g}_a, \quad (8)$$

where $T = t_k - t_{k-1}$, $\mathbf{r}_a^{z_k w}$ is the position of point z relative to point w resolved in $\mathcal{F}_a$ at time t_k, $\mathbf{v}_a^{z_k w/a} \in \mathbb{R}^3$ is the velocity of point z relative to point w with respect to $\mathcal{F}_a$ resolved in $\mathcal{F}_a$ at time t_k, $\mathbf{g}_a \in \mathbb{R}^3$ is Earth's gravitational acceleration resolved in $\mathcal{F}_a$, and $\mathbf{w}_{b_k}^1 \in \mathbb{R}^3$ is noise.

The discrete-time relationship between attitude and the rate-gyro measurement $\mathbf{u}_{b_k}^2 \in \mathbb{R}^3$ is

$$\mathbf{C}_{b_k a} = \exp(-\boldsymbol{\psi}_{k-1}^\times)\mathbf{C}_{b_{k-1}a}, \quad (9)$$

where $\boldsymbol{\psi}_{k-1} = T\mathbf{u}_{b_{k-1}}^2 + T\mathbf{w}_{b_{k-1}}^2$, $\mathbf{w}_{b_k}^2$ is noise.

Let $\mathbf{r}_a^{p_{j_k} w} \in \mathbb{R}^3$ denote the position of landmark $j \in \{1, \cdots, \ell\}$ relative to point w, resolved in $\mathcal{F}_a$. The process model associated with $\mathbf{r}_a^{p_{j_k} w}$ is

$$\mathbf{r}_a^{p_{j_k} w} = \mathbf{r}_a^{p_{j_{k-1}} w} + T\mathbf{w}_k^{p_j}, \quad \forall j \in \{1, \cdots, \ell\}, \quad (10)$$

where $\mathbf{w}_k^{p_j}$ is noise. This random-walk process model is used to ensure the prediction step of the filter is not overconfident. An interpretation of such a process model is that the landmarks are assumed to be moving very slowly. A similar approach is taken when estimating bias in sensors [4].

Combining (8), (9), and (10) results in

$$\begin{bmatrix} \mathbf{r}_a^{z_k w} \\ \mathbf{v}_a^{z_k w/a} \\ -(\ln(\mathbf{C}_{b_k a}))^\vee \\ \mathbf{r}_a^{p_{1_k} w} \\ \vdots \\ \mathbf{r}_a^{p_{\ell_k} w} \end{bmatrix} = \begin{bmatrix} \mathbf{r}_a^{z_{k-1} w} + T\mathbf{v}_a^{z_{k-1} w/a} \\ \mathbf{v}_a^{z_{k-1} w/a} + T\mathbf{C}_{b_{k-1} a}^\mathsf{T}(\mathbf{u}_{b_{k-1}}^1 + \mathbf{w}_{b_{k-1}}^1) - T\mathbf{g}_a \\ -(\ln(\exp(-\boldsymbol{\psi}_{k-1}^\times)\mathbf{C}_{b_{k-1} a}))^\vee \\ \mathbf{r}_a^{p_{1_{k-1}} w} + T\mathbf{w}_k^{p_1} \\ \vdots \\ \mathbf{r}_a^{p_{\ell_{k-1}} w} + T\mathbf{w}_k^{p_\ell} \end{bmatrix}, \tag{11}$$

the discrete-time process model.

Camera and the Measurement Model — Let $\mathcal{A} = \{1, \cdots, \ell\}$ denote the set of all landmark indices and let $\mathcal{I}_k = \{1, \cdots, q\} \subseteq \mathcal{A}$, $q \leq \ell$, denote the set of all landmarks observed from time $t = 0$ to time $t = kT$. Further, let $\mathcal{O}_k = \{\alpha, \cdots, \zeta\} \subseteq \mathcal{A}$ denote the set of landmarks observed at time $t = kT$. The exteroceptive sensor measures $\mathbf{y}_{b_k}^i = \mathbf{C}_{b_k a}(\mathbf{r}_a^{p_i w} - \mathbf{r}_a^{z_k w}) + \mathbf{v}_{b_k}^i$ where $\mathbf{r}_a^{p_i w} \in \mathbb{R}^3$ is unknown but constant and $\mathbf{v}_{b_k}^i \in \mathbb{R}^3$ is noise. Combining all exteroceptive measurements together to yields $\mathbf{y}_k = \mathbf{g}_k(\mathbf{r}_a^{z_k w}, \mathbf{C}_{b_k a}, \mathbf{r}_a^{p_\alpha w}, \cdots, \mathbf{r}_a^{p_\zeta w}, \mathbf{v}_{b_k}^\alpha, \cdots, \mathbf{v}_{b_k}^\zeta)$.

6.2 MLG SPKF Applied to the SLAM Problem

To be concise, let $\hat{\mathbf{r}}_{k-1} = \hat{\mathbf{r}}_a^{z_{k-1} w}$, $\hat{\mathbf{v}}_{k-1} = \mathbf{v}_a^{z_{k-1} w/a}$, $\hat{\mathbf{C}}_{k-1} = \hat{\mathbf{C}}_{b_{k-1} a}$, and $\hat{\mathbf{m}}_{k-1}^j = \hat{\mathbf{r}}_a^{p_{j,k-1} w}$ $\forall j \in \mathcal{I}_{k-1}$. As discussed in Sect. 5, to execute the prediction step, first define $\mathbf{z}_{k-1}$ and $\mathbf{Y}_{k-1}$ where $L = 15 + 6q$. Decomposing $\mathbf{Y}_{k-1}$ using a Cholesky decomposition, forming sigma points, and passing the sigma points through the nonlinear process model in conjunction with the measurements $\mathbf{u}_{b_{k-1}}^1$ and $\mathbf{u}_{b_{k-1}}^2$ gives

$$\begin{aligned} \boldsymbol{\mathcal{R}}_{i,k}^- &= \boldsymbol{\mathcal{R}}_{i,k-1} + T\boldsymbol{\mathcal{V}}_{i,k-1}, \quad \boldsymbol{\mathcal{V}}_{i,k}^- = \boldsymbol{\mathcal{V}}_{i,k-1} + T\boldsymbol{\mathcal{C}}_{i,k-1}^\mathsf{T}(\mathbf{u}_{b_{k-1}}^1 + \boldsymbol{\mathcal{W}}_{i,k-1}^1) - T\mathbf{g}_a, \\ \boldsymbol{\mathcal{C}}_{i,k}^- &= \exp(-\boldsymbol{\psi}_{k-1}^\times)\boldsymbol{\mathcal{C}}_{i,k-1}, \qquad \text{where} \qquad \boldsymbol{\psi}_{k-1} = T\mathbf{u}_{b_{k-1}}^2 + T\boldsymbol{\mathcal{W}}_{i,k-1}^2, \\ \boldsymbol{\mathcal{M}}_{i,k}^{j-} &= \boldsymbol{\mathcal{M}}_{i,k-1}^j + T\boldsymbol{\mathcal{W}}_{i,k-1}^{p_j}, \ \forall j \in \mathcal{I}_{k-1}. \end{aligned}$$

Using weights w_i it follows that $\hat{\mathbf{r}}_k^- = \sum_{i=0}^{2L} w_i \boldsymbol{\mathcal{R}}_{i,k}^-$, $\hat{\mathbf{v}}_k^- = \sum_{i=0}^{2L} w_i \boldsymbol{\mathcal{V}}_{i,k}^-$, $\hat{\mathbf{m}}_k^{j-} = \sum_{i=0}^{2L} w_i \boldsymbol{\mathcal{M}}_{i,k}^{j-}$, $\forall j \in \mathcal{I}_{k-1}$. On the other hand, computing $\hat{\mathbf{C}}_k^-$ must be done with care. One way to compute $\hat{\mathbf{C}}_k^-$ is to solve

$$\hat{\mathbf{C}}_k^- = \arg \min_{\mathbf{C}_k \in SO(3)} \sum_{i=0}^{2L} w_i d^2(\boldsymbol{\mathcal{C}}_{i,k}^-, \mathbf{C}_k) \tag{12}$$

in an iterative manner [17,18]. Details of alternative methods to compute $\hat{\mathbf{C}}_k^-$ are discussed in Sect. 6.3. The covariance $\mathbf{P}_k^-$ is

$$\mathbf{P}_k^- = \sum_{i=0}^{2L} w_i \boldsymbol{\mathcal{E}}_{i,k}^- \boldsymbol{\mathcal{E}}_{i,k}^{-\mathsf{T}}, \text{ where } \boldsymbol{\mathcal{E}}_{i,k}^- = \begin{bmatrix} \boldsymbol{\mathcal{R}}_{i,k}^- - \hat{\mathbf{r}}_k^- \\ \boldsymbol{\mathcal{V}}_{i,k}^- - \hat{\mathbf{v}}_k^- \\ \boldsymbol{\xi}_{i,k}^- \\ \boldsymbol{\mathcal{M}}_{i,k}^{1-} - \hat{\mathbf{m}}_k^{1-} \\ \vdots \\ \boldsymbol{\mathcal{M}}_{i,k}^{q-} - \hat{\mathbf{m}}_k^{q-} \end{bmatrix}, \text{ and } \boldsymbol{\xi}_{i,k}^- = -(\ln(\boldsymbol{\mathcal{C}}_{i,k}^- \hat{\mathbf{C}}_k^{-\mathsf{T}}))^{\vee}.$$

When a landmark, such as $\mathbf{m}_k^j$, is first observed, it is initialized as $\hat{\mathbf{m}}_k^{j-} = \hat{\mathbf{C}}_k^{-\mathsf{T}} \mathbf{y}_{b_k}^j + \hat{\mathbf{r}}_k^-$, $\forall j \in \mathcal{O}_k$, and the covariance is augmented as $\mathbf{P}_k^- = \begin{bmatrix} \mathbf{P}_k^- & \mathbf{P}_{\mathbf{x}_k \mathbf{m}_k^j}^- \\ \mathbf{P}_{\mathbf{x}_k \mathbf{m}_k^j}^{-\mathsf{T}} & \mathbf{P}_{\mathbf{m}_k^j \mathbf{m}_k^j}^- \end{bmatrix}$, where $\mathbf{P}_{\mathbf{m}_k^j \mathbf{m}_k^j}^- = \sum_{i=0}^{2L} w_i (\boldsymbol{\mathcal{M}}_{i,k}^j - \hat{\mathbf{m}}_k^{j-})(\boldsymbol{\mathcal{M}}_{i,k}^j - \hat{\mathbf{m}}_k^{j-})^{\mathsf{T}}$, $\mathbf{P}_{\mathbf{x}_k \mathbf{m}_k^j}^- = \sum_{i=0}^{2L} w_i \boldsymbol{\mathcal{E}}_{i,k}^- (\boldsymbol{\mathcal{M}}_{i,k}^j - \hat{\mathbf{m}}_k^{j-})^{\mathsf{T}}$, and $\boldsymbol{\mathcal{M}}_{i,k}^j = \boldsymbol{\mathcal{C}}_{i,k}^{-\mathsf{T}} \mathbf{y}_{b_k}^j + \boldsymbol{\mathcal{R}}_{i,k}^-$.

Following Sect. 5, to execute the correction step, define $\mathbf{z}_k$ and $\mathbf{Y}_k^-$ where $L = 9 + 3q + 3o$, and o is the cardinality of $\mathcal{O}_k$. Recall that the discretized exteroceptive measurement model is $\mathbf{y}_k = \mathbf{g}_k(\mathbf{r}_a^{z_k w}, \mathbf{C}_{b_k a}, \mathbf{r}_a^{p_\alpha w}, \cdots, \mathbf{r}_a^{p_\zeta w}, \mathbf{v}_{b_k}^{\alpha}, \cdots, \mathbf{v}_{b_k}^{\zeta})$. Decomposing $\mathbf{Y}_k^-$ using a Cholesky decomposition, forming sigma points, and passing the sigma points through exteroceptive measurement model results in $\boldsymbol{\mathcal{Y}}_{j,k} = \mathbf{g}_k(\boldsymbol{\mathcal{R}}_{j,k}^-, \boldsymbol{\mathcal{C}}_{j,k}^-, \boldsymbol{\mathcal{M}}_{j,k}^{\alpha-}, \cdots, \boldsymbol{\mathcal{M}}_{j,k}^{\zeta-}, \boldsymbol{\mathcal{N}}_{j,k}^-)$, $j = 0, \ldots, 2L$. The predicted measurement and associated covariance are then $\hat{\mathbf{y}}_k = \sum_{j=0}^{2L} w_j \boldsymbol{\mathcal{Y}}_{j,k}$, and $\mathbf{V}_k = \sum_{j=0}^{2L} w_j (\boldsymbol{\mathcal{Y}}_{j,k} - \hat{\mathbf{y}}_k)(\boldsymbol{\mathcal{Y}}_{j,k} - \hat{\mathbf{y}}_k)^{\mathsf{T}}$. The matrix $\mathbf{U}_k$ is computed in a similar way to $\mathbf{P}_k^-$, that is

$$\mathbf{U}_k = \sum_{j=0}^{2L} w_j \boldsymbol{\mathcal{E}}_{j,k}^- (\boldsymbol{\mathcal{Y}}_{j,k} - \hat{\mathbf{y}}_k)^{\mathsf{T}} \text{ where } \boldsymbol{\mathcal{E}}_{j,k}^- = \begin{bmatrix} \boldsymbol{\mathcal{R}}_{j,k}^- - \hat{\mathbf{r}}_k^- \\ \boldsymbol{\mathcal{V}}_{j,k}^- - \hat{\mathbf{v}}_k^- \\ \boldsymbol{\xi}_{j,k}^- \\ \boldsymbol{\mathcal{M}}_{j,k}^{1-} - \hat{\mathbf{m}}_k^{1-} \\ \vdots \\ \boldsymbol{\mathcal{M}}_{j,k}^{q-} - \hat{\mathbf{m}}_k^{q-} \end{bmatrix} \text{ where } \boldsymbol{\xi}_{j,k}^- = -\ln(\boldsymbol{\mathcal{C}}_{j,k}^- \hat{\mathbf{C}}_k^{-\mathsf{T}})^{\vee}.$$

It follows that the Kalman gain is $\mathbf{K}_k = \mathbf{U}_k \mathbf{V}_k^{-1}$. To correct the state estimate first define $\delta\boldsymbol{\chi}_k = \mathbf{K}_k(\mathbf{y}_k - \hat{\mathbf{y}}_k^-)$ where

$$\delta\boldsymbol{\chi}_k = \begin{bmatrix} \delta\boldsymbol{\chi}_{k,r} \\ \delta\boldsymbol{\chi}_{r,v} \\ \delta\boldsymbol{\xi}_k \\ \delta\boldsymbol{\xi}_{k,m_1} \\ \vdots \\ \delta\boldsymbol{\xi}_{k,m_q} \end{bmatrix} = \begin{bmatrix} \hat{\mathbf{r}}_a^{z_k w} - \hat{\mathbf{r}}_k^- \\ \hat{\mathbf{v}}_a^{z_k w/a} - \hat{\mathbf{v}}_k^- \\ \delta\boldsymbol{\xi}_k \\ \hat{\mathbf{r}}_a^{p_{1,k} w} - \hat{\mathbf{m}}_k^{1-} \\ \vdots \\ \hat{\mathbf{r}}_a^{p_{q,k} w} - \hat{\mathbf{m}}_k^{q-} \end{bmatrix}.$$

Then, to compute $\hat{\mathbf{r}}_a^{z_k w}$, $\hat{\mathbf{v}}_a^{z_k w/a}$, $\mathbf{C}_{b_k a}$, and $\hat{\mathbf{r}}_a^{p_{j,k} w}$, $j \in \mathcal{I}_k$, $\hat{\mathbf{r}}_a^{z_k w} = \hat{\mathbf{r}}_k^- + \delta\boldsymbol{\chi}_{k,r}$, $\hat{\mathbf{v}}_a^{z_k w/a} = \hat{\mathbf{v}}_k^- + \delta\boldsymbol{\chi}_{k,v}$, $\hat{\mathbf{C}}_{b_k a} = \exp(-\delta\boldsymbol{\xi}_k^{\times})\hat{\mathbf{C}}_k^-$, $\hat{\mathbf{r}}_a^{p_{j,k} w} = \hat{\mathbf{m}}_k^{j-} + \delta\boldsymbol{\xi}_{k,m_j}$. The estimated covariance is updated as $\mathbf{P}_k = \mathbf{P}_k^- - \mathbf{U}_k \mathbf{V}_k^{-1} \mathbf{U}_k^{\mathsf{T}}$.

6.3 Weighted Averaging

To solve (12), namely $\hat{\mathbf{C}}_k^- = \arg\min_{\mathbf{C}_k \in SO(3)} \sum_{i=0}^{2L} w_i d^2(\mathcal{C}_{i,k}^-, \mathbf{C}_k)$, the iterative method of [17,18] can be employed when $d(\cdot,\cdot)$ is the geodesic distance. A similar approach is proposed in [6]. The $SE(3)$ case is considered in [7,14]. Alternatively, when $d(\cdot,\cdot)$ is a chordal distance [15,16], namely $d^2(\mathbf{C}_1, \mathbf{C}_0) = \|\mathbf{C}_1 - \mathbf{C}_0\|_{\mathsf{F}}^2$, a singular value decomposition (SVD) can be used to solve for $\hat{\mathbf{C}}_k^-$ in the following way. Consider the objective function $J_k(\mathbf{C}_k) = \sum_{i=0}^{2L} w_i \left\|\mathcal{C}_{i,k}^- - \mathbf{C}_k\right\|_{\mathsf{F}}^2$, where $\mathbf{C}_k \in SO(3)$ [15]. The objective function can be written as

$$J_k(\mathbf{C}_k) = \sum_{i=0}^{2L} w_i \mathrm{tr}\left(\mathcal{C}_{i,k}^- - \mathbf{C}_k\right)^{\mathsf{T}}\left(\mathcal{C}_{i,k}^- - \mathbf{C}_k\right) = 6\sum_{i=0}^{2L} w_i - 2\sum_{i=0}^{2L} \mathrm{tr}\left(w_i \mathcal{C}_{i,k}^{-\mathsf{T}} \mathbf{C}_k\right).$$

Minimizing $J_k(\cdot)$ as a function of $\mathbf{C}_k$ is equivalent to maximizing $\mathcal{J}_k(\mathbf{C}_k) = \mathrm{tr}\left(\mathbf{B}^{\mathsf{T}}\mathbf{C}_k\right)$ where $\mathbf{B}^{\mathsf{T}} = \sum_{i=0}^{2L} w_i \mathcal{C}_{i,k}^{-\mathsf{T}}$. The maximizing solution, $\hat{\mathbf{C}}_k^-$, is $\hat{\mathbf{C}}_k^- = \mathbf{V}_B \,\mathrm{diag}\left\{1, 1, \det \mathbf{V}_B \det \mathbf{U}_B\right\} \mathbf{U}_B^{\mathsf{T}}$, where the SVD $\mathbf{B} = \mathbf{V}_B \boldsymbol{\Sigma}_B \mathbf{U}_B^{\mathsf{T}}$, $\mathbf{V}_B^{\mathsf{T}}\mathbf{V}_B = \mathbf{1}$, $\mathbf{U}_B^{\mathsf{T}}\mathbf{U}_B = \mathbf{1}$, and $\boldsymbol{\Sigma}_B = \mathrm{diag}\left\{\sigma_1, \sigma_2, \sigma_3\right\}$ where $\sigma_1 \geq \sigma_2 \geq \sigma_3 \geq 0$ has been employed [21,22].

An alternative way to compute $\hat{\mathbf{C}}_k^-$ using generalized Rodrigues parameters is explored in [4]. Herein an exposition of something equivalent, although different, to [4] is presented that is computationally simpler than both the iterative method of [18] and the SVD method previously discussed. Consider $\mathcal{C}_{i,k}^- = \mathbf{E}_{i0}\mathcal{C}_{0,k}^-$, $i = 1, \ldots, 2L$, where $\mathbf{E}_{i0} \in SO(3)$ represents the attitude of $\mathcal{F}_{b_{k_i}}$ relative to $\mathcal{F}_{b_{k_0}}$. Write $\mathbf{E}_{i0}$ as $\mathbf{E}_{i0} = \mathcal{C}_{i,k}^-\mathcal{C}_{0,k}^{-\mathsf{T}} = \cos\phi^{i0}\mathbf{1} + (1 - \cos\phi^{i0})\mathbf{a}^{i0}\mathbf{a}^{i0^{\mathsf{T}}} - \sin\phi^{i0}\mathbf{a}^{i0^{\times}}$, where $\mathbf{a}^{i0}$ can be interpreted as the axis of rotation resolved in $\mathcal{F}_{b_{k_0}}$. Owing to the fact that $\mathbf{E}_{i0}\mathbf{a}^{i0} = \mathbf{a}^{i0}$ it is also correct to state that $\mathbf{a}^{i0}$ can be interpreted as the axis of rotation resolved in $\mathcal{F}_{b_{k_i}}$. However, by resolving all $i = 1, \ldots, 2L$ axes of rotation in $\mathcal{F}_{b_{k_0}}$, they can be added in a weighted fashion. Similarly, by resolving all $i = 1, \ldots, 2L$ rotation vectors $\boldsymbol{\phi}_{b_{k_0}}^{i0} = \phi^{i0}\mathbf{a}^{i0}$ in $\mathcal{F}_{b_{k_0}}$, their weighted sum can be added. Said another way, each $\boldsymbol{\phi}_{b_{k_0}}^{i0^{\times}}$ are being expressed in the same tangent space, and therefore can be added. In particular, $\boldsymbol{\phi}_{b_{k_0}} = \sum_{i=1}^{2L} w_i \boldsymbol{\phi}_{b_{k_0}}^{i0}$, where $w_i = \frac{1}{2(L+\kappa)}$. It follows then that $\hat{\mathbf{C}}_k^- = \exp(-\boldsymbol{\phi}_{b_{k_0}}^{\times})\hat{\mathbf{C}}_{b_{k-1}a}$.

7 Conclusion

This paper formulates a sigma point Kalman filter (SPKF) specifically for systems with states that are elements of a matrix Lie group. First, how the the sigma point transformation is generalized and used to compute means and covariances of nonlinear functions with states that are elements of matrix Lie groups is discussed. The matrix Lie Group SPKF (MLG SPKF) is then formulated in general. The MLG SPKF is applied to the problem of simultaneous localization and mapping (SLAM) where the position and attitude of a vehicle are estimated relative to a set of estimated landmark locations and, simultaneously, the estimates of the landmark locations are refined. Approaches to averaging elements of $SO(3)$ are discussed, including using the a chordal distance measure, as well as a method adopted from [4]. Using the chordal distance measure allows for the used of an SVD solution to the averaging of elements of $SO(3)$.

References

1. Thrun, S., Burgard, W., Fox, D.: Probabilistic Robotics. MIT Press, Cambridge (2006)
2. Särkkä, S.: Bayesian Filtering and Smoothing. Cambridge University Press, Cambridge (2013)
3. Julier, S.J., Uhlmann, J.K., Durrant-Whyte, H.F.: A new method for the nonlinear transformation of means and covariances in filters and estimators. IEEE Trans. Autom. Control **45**(3), 477–482 (2000)
4. Crassidis, J.L., Markley, F.L.: Unscented filtering for spacecraft attitude estimation. J. Guidance Control Dyn. **26**(4), 536–542 (2003). AIAA
5. Hauberg, S., Lauze, F., Pedersen, K.S.: Unscented Kalman filtering on Riemannian manifolds. J. Math. Imaging Vis. **46**(1), 103–120 (2013)
6. Hertzberg, C., Wagner, R., Frese, U., Schröder, L.: Integrating generic sensor fusion algorithms with sound state representations through encapsulation of manifolds. Inf. Fusion **14**(1), 57–77 (2013)
7. Barfoot, T.D., Furgale, P.T.: Associating uncertainty with three-dimensional poses for use in estimation problems. IEEE Trans. Robot. **30**(3), 679–693 (2014)
8. Brossard, M., Bonnabel, S., Condomines, J.-P.: Unscented Kalman Filtering on Lie Groups. Soumis à IROS 2017 (2017). <hal-01489204v2>. https://hal.archives-ouvertes.fr/hal-01489204v2
9. Haug, A.J.: Bayesian Estimation and Tracking - A Practical Guide. Wiley, Hoboken (2012)
10. Wu, Y., Hu, D., Wu, M., Hu, X.: A numerical-integration perspective on gaussian filters. IEEE Trans. Signal Process. **54**, 2910–2921 (2006)
11. Bloch, A., Baillieul, J., Crouch, P., Marsden, J.E., Krishnaprasad, P.S., Murray, R., Zenkov, D.: Nonholonomic Mechanics and Control, vol. 24. Springer, New York (2003)
12. Eade, E.: Lie Groups for Computer Vision. http://ethaneade.com/lie_groups.pdf. Accessed 13 Apr 2017
13. Chirikjian, G.S.: Stochastic Models, Information Theory, and Lie Groups, vol. 2. Birkhauser, Springer Science + Business Media Inc., Boston (2011)
14. Barfoot, T.D.: State Estimation for Robotics. Cambridge University Press, Cambridge (2017). In preparation (draft compiled February 14, 2017)
15. Moakher, M.: Means and averaging in the group of rotations. SIAM J. Matrix Anal. Appl. **24**(1), 1–16 (2002)
16. Hartley, R., Trumpf, J., Dai, Y., Li, H.: Rotation averaging. Int. J. Comput. Vis. **103**(3), 267–305 (2013)
17. Manton, J.H.: A globally convergent numerical algorithm for computing the centre of mass on compact lie groups. In: 8th Control, Automation, Robotics and Vision Conference, vol. 3, pp. 2211–2216, December 2004
18. Fiori, S., Tanaka, T.: An algorithm to compute averages on matrix lie groups. IEEE Trans. Signal Process. **57**, 4734–4743 (2009)

19. Murray, R.N., Li, Z., Sastry, S.S.: A Mathematical Introduction to Robotic Manipulation. CRC Press Inc., Boca Raton (1993)
20. Hughes, P.C.: Spacecraft Attitude Dynamics, 2nd edn. Dover, Mineola (2004)
21. Markley, F.L.: Attitude determination using vector observations and the singular value decomposition. J. Astronaut. Sci. **36**(3), 245–258 (1988)
22. de Ruiter, A., Forbes, J.R.: On the solution of Wahba's problem on $SO(n)$. J. Astronaut. Sci. **60**(1), 1–31 (2015)

A Topological View on Forced Oscillations and Control of an Inverted Pendulum

Ivan Polekhin(✉)

Steklov Mathematical Institute of Russian Academy of Sciences,
8 Gubkina Str., 119991 Moscow, Russia
ivanpolekhin@mi.ras.ru

Abstract. We consider a system of a planar inverted pendulum in a gravitational field. First, we assume that the pivot point of the pendulum is moving along a horizontal line with a given law of motion. We prove that, if the law of motion is periodic, then there always exists a periodic solution along which the pendulum never becomes horizontal (never falls). We also consider the case when the pendulum with a moving pivot point is a control system, in which the mass point is constrained to be strictly above the pivot point (the rod cannot fall 'below the horizon'). We show that global stabilization of the vertical upward position of the pendulum cannot be obtained for any smooth control law, provided some natural assumptions.

Keywords: Inverted pendulum · Forced oscillations · Global stabilization · Control design

1 Introduction

Below we consider the following system

$$\begin{aligned} \dot{q} &= p, \\ \dot{p} &= u(q,p,t)\cdot \sin q - \cos q. \end{aligned} \tag{1}$$

Here u is a smooth function. When $u = \ddot{\xi}(t)$ is a function of time, the system describes the motion of an inverted pendulum in a gravitational field with a moving pivot point (without loss of generality, we assume that the mass and the length of the pendulum equals 1 and the gravity acceleration is also 1). In this case, the law of motion of the pivot point is defined by the smooth function $\xi(t)$. We show that for any T-periodic function $\ddot{\xi}(t)$, there always exists a T-periodic solution such that $q(t) \in (0, \pi)$ for all t, i.e., the pendulum never falls.

When u is a smooth function from $\mathbb{R}^3$ to $\mathbb{R}$, we have a control system. When u is periodic in q and autonomous, it can be shown that the problem of stabilization of the vertical position of an inverted pendulum does not allow continuous control which would asymptotically lead the pendulum to the vertical from any initial position. This follows from the fact that a continuous function on a circle, which

F. Nielsen and F. Barbaresco (Eds.): GSI 2017, LNCS 10589, pp. 329–335, 2017.
https://doi.org/10.1007/978-3-319-68445-1_38

takes values of opposite sign, has at least two zeros, i.e., system (1) has at least two equilibria.

The following questions naturally arise. First, do the above statements remain true if we consider the pendulum only in the positions where its mass point is above the pivot point (often there exists a physical constraint in the system which do not allow the rod to be below the plane of support and it is meaningless to consider the pendulum in such positions). Second, is it true that global stabilization cannot be obtained when the control law is a time-dependent function and it is also a non-periodic function of the position of the pendulum? For a relatively broad class of problems, which may appear in practice, we show that the answers are positive for the both questions.

The main idea of both proofs goes back to the topological method of Ważewski [3,4] and its developments, which we shortly consider below.

2 Forced Oscillations

In this section we will discuss the existence of a forced oscillation in the system (1) when $u = \ddot{\xi}(t)$ is a smooth T-periodic function. For brevity of exposition, we will proof the main result of this section for a slightly modified system

$$\begin{aligned} \dot{q} &= p, \\ \dot{p} &= \ddot{\xi}(t) \cdot \sin q - \cos q - \mu \dot{q}. \end{aligned} \tag{2}$$

This system differs from (1) by the term $-\mu\dot{q}$, which describes an arbitrary small viscous friction (we assume that $\mu > 0$ can be arbitrarily small).

First, we introduce some definitions and a result from [5,6] which we slightly modify for our use.

Let $v \colon \mathbb{R} \times M \to TM$ be a time-dependent vector field on a manifold M

$$\dot{x} = v(t, x). \tag{3}$$

For $t_0 \in \mathbb{R}$ and $x_0 \in M$, the map $t \mapsto x(t, t_0, x_0)$ is the solution for the initial value problem for the system (3), such that $x(0, t_0, x_0) = x_0$. If $W \subset \mathbb{R} \times M$, $t \in \mathbb{R}$, then we denote

$$W_t = \{x \in M \colon (t, x) \in W\}.$$

Definition 1. *Let $W \subset \mathbb{R} \times M$. Define the exit set W^- as follows. A point (t_0, x_0) is in W^- if there exists $\delta > 0$ such that $(t + t_0, x(t, t_0, x_0)) \notin W$ for all $t \in (0, \delta)$.*

Definition 2. *We call $W \subset \mathbb{R} \times M$ a Ważewski block for the system (3) if W and W^- are compact.*

Definition 3. *A set $W \subset [a, b] \times M$ is called a simple periodic segment over $[a, b]$ if it is a Ważewski block with respect to the system (3), $W = [a, b] \times Z$, where $Z \subset M$, and $W^-_{t_1} = W^-_{t_2}$ for any $t_1, t_2 \in [a, b)$.*

Definition 4. *Let W be a simple periodic segment over $[a,b]$. The set $W^{--} = [a,b] \times W_a^-$ is called the essential exit set for W.*

In our case, the result from [5,6] can be presented as follows.

Theorem 1. *Let W be a simple periodic segment over $[a,b]$. Then the set*

$$U = \{x_0 \in W_a : x(t-a,a,x_0) \in W_t \setminus W_t^{--} \text{ for all } t \in [a,b]\}$$

is open in W_a and the set of fixed points of the restriction $x(b-a,a,\cdot)|_U : U \to W_a$ is compact. Moreover, the fixed point index of $x(b-a,a,\cdot)|_U$ can be calculated by means of the Euler-Poincaré characteristics of W and W_a^- as follows

$$\operatorname{ind}(x(b-a,a,\cdot)|_U) = \chi(W_a) - \chi(W_a^-).$$

In particular, if $\chi(W_a) - \chi(W_a^-) \neq 0$ then $x(b-a,a,\cdot)|_U$ has a fixed point in W_a.

Theorem 2. *Suppose that the function $\ddot{\xi} : \mathbb{R} \to \mathbb{R}$ in (2) is T-periodic, then for any $\mu > 0$ there exists q_0 and p_0 such that for all $t \in \mathbb{R}$*

1. *$q(t,0,q_0,p_0) = q(t+T,0,q_0,p_0)$ and $p(t,0,q_0,p_0) = p(t+T,0,q_0,p_0)$,*
2. *$q(t,0,q_0,p_0) \in (0,\pi)$.*

Proof. First, in order to apply Theorem 1, we show that a periodic Ważewski segment for our system can be defined as follows

$$W = \{(t,q,p) \in [0,T] \times \mathbb{R}/2\pi\mathbb{Z} \times \mathbb{R} : 0 \leqslant q \leqslant \pi, -p' \leqslant p \leqslant p'\},$$

where p' satisfies

$$p' > \frac{1}{\mu} \sup_{t \in [0,T]} (|\ddot{\xi}| + 1). \tag{4}$$

It is clear that W is compact. Let us show that W^{--} is compact as well and

$$\begin{aligned} W^{--} = &\{(t,q,p) \in [0,T] \times \mathbb{R}/2\pi\mathbb{Z} \times \mathbb{R} : q = 0, -p' \leqslant p \leqslant 0\} \\ &\cup \{(t,q,p) \in [0,T] \times \mathbb{R}/2\pi\mathbb{Z} \times \mathbb{R} : q = \pi, 0 \leqslant p \leqslant p'\}. \end{aligned}$$

If $p = p'$, then from (2) and (4) we have $\dot{p} < 0$. Therefore, $(t,q,p) \notin W^{--}$ for $0 < q < \pi$, $t \in [0,T]$, $p = p'$. When $q = 0$ and $t \in [0,T]$, we have $(t,q,p) \in W^{--}$ for $-p' \leqslant p < 0$ and $(t,q,p) \notin W^{--}$ for any $p > 0$. Moreover, it can be proved that $(t,q,p) \in W^{--}$ when $p = 0$. Indeed, for $q = p = 0$, we have

$$\ddot{q} = \dot{p} = -1.$$

Therefore, any solution starting at $q = p = 0$ leaves W. The cases $p = -p'$ and $q = \pi$ can be considered in a similar way. Finally, we obtain $\chi(W_0) - \chi(W_0^-) = -1$ and Theorem 1 can be applied.

As we said above, this result can be proved without the assumption of the presence of friction. To be more precise, the following holds

Theorem 3. *Suppose that the function $u = \ddot{\xi}\colon \mathbb{R} \to \mathbb{R}$ in (1) is T-periodic, then there exists q_0 and p_0 such that for all $t \in \mathbb{R}$*

1. $q(t, 0, q_0, p_0) = q(t + T, 0, q_0, p_0)$ *and* $p(t, 0, q_0, p_0) = p(t + T, 0, q_0, p_0)$,
2. $q(t, 0, q_0, p_0) \in (0, \pi)$.

In this case, the proof is basically the same, but the Ważewski block W has a different form. However, it can be continuously deformed to a simple periodic segment and an extended version of Theorem 1 can be applied (Fig. 1).

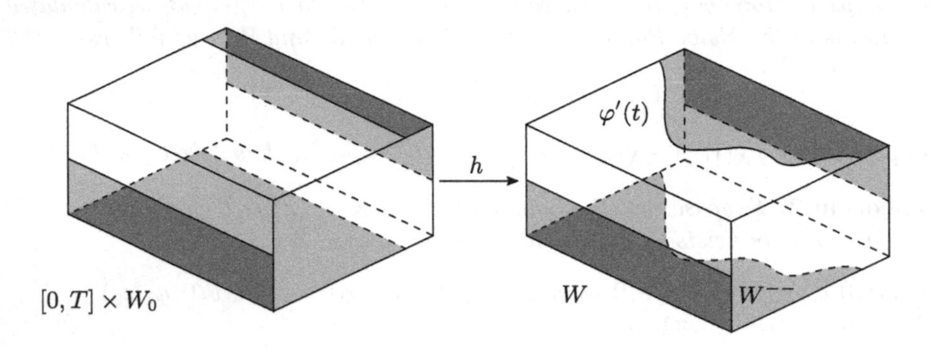

Fig. 1. Ważewski block W. W^{--} is in gray.

3 Global Stabilization

The key observation in the proof of Theorem 2 was that the solutions, that reach the points where $q = 0$ or $q = \pi$ and $p = 0$, are externally tangent to W. From this we obtain that if a solution leaves W, then all solutions with close initial data also leave W at close points. It can be seen that this property also holds for system (1) for any u. From this we immediately obtain

Theorem 4. *For any smooth control function $u\colon \mathbb{R}^3 \to \mathbb{R}$ in (1), there exists q_0 and p_0 such that $q(t, 0, q_0, p_0) \in (0, \pi)$ on the interval of existence of the solution.*

Proof. Consider an arbitrary line segment γ in the hyperplane $t = 0$ which connects the set $\{q, p, t\colon q = 0,\ p \leqslant 0\}$ with the set $\{q, p, t\colon q = \pi,\ p \geqslant 0\}$. Suppose that all solutions starting at γ reach the hyperplane $q = 0$ or the hyperplane $q = \pi$. As it was mentioned before, if some solution reaches the set $q = 0$, then all solutions with close initial data also reach this set at close points. In other words, we have a continuous map from γ to the above hyperplanes. Therefore, we can construct a continuous map from the line segment to its boundary points. This contradiction proves the theorem.

Similar arguments can be applied to the case when we try to stabilize our system in a vicinity of the vertical upward position. Suppose that we are looking for a control that would stabilize system (1) in a vicinity of a certain equilibrium position in the following sense. Let M be a subset of the phase space of the system such that the points of M correspond to the positions of the pendulum in which the rod is above the horizontal line (in our case, $M = \{0 < q < \pi\}$) and $\mu = (\pi/2, 0) \in M$ is the equilibrium for a given control u. We assume that the control function u is chosen in such a way that there exists a compact subset $U \subset M$, $\mu \in U \setminus \partial U$ and a C^1-function $V\colon U \to \mathbb{R}$) with the following properties

L1. $V(\mu) = 0$ and $V > 0$ in $U \setminus \mu$.
L2. Derivative $\dot{V}$ with respect to system (1) is negative in $U \setminus \mu$ for all t.

Since the function V can be considered as a Lyapunov function for our system, the equilibrium μ is stable. For instance, such a function exists in the following case. Suppose that for a given u, system (1) can be written as follows in a vicinity of μ

$$\dot{x} = Ax + f(x, t),$$

where $x = (q, p)$, A is a constant matrix and its eigenvalues have negative real parts, f is a continuous function and $f(t, x) = o(\|x\|)$ uniformly in t. Then there exists [7] a function V satisfying properties L1, L2 (in this case, μ is asymptotically stable).

Theorem 5. *For a given control $u(q, p, t)$, suppose there exists a function V satisfying L1 and L2 for system (1). Then there exists an initial condition (q_0, p_0) and a neighborhood $B \subset M$ of μ such that on the interval of existence the solution $(q(t, 0, q_0, p_0), p(t, 0, q_0, p_0))$ stays in $M \setminus B$.*

Proof. The proof is similar to the one in Theorem 4. It can be shown that for $\varepsilon > 0$ small enough, the level set $V = \varepsilon$ is a circle (topologically). Let $B = \{q, p\colon V(q, p) \leqslant \varepsilon\}$. Let γ_1 and γ_2 be two line segments in the plane $t = 0$ connecting the sets $\{q, p, t\colon q = 0, p \leqslant 0\}$ and $\{q, p, t\colon q = \pi, p \geqslant 0\}$ with boundary ∂B, correspondingly. Suppose that any solution starting at $\gamma_1 \cup \gamma_2 \cup \partial B$ leaves $M \setminus B$. From the same arguments as in Theorem 4, we conclude that if the considered solution leaves this set, then all solutions with close initial data also leave it. Therefore, we obtain a continuous map between $\gamma_1 \cup \gamma_2 \cup \partial B$ and a disconnected set (∂B and two boundary points of γ_1 and γ_2). The contradiction proves the theorem.

Remark 1. Actually, as it can be seen from the proofs of Theorems 4 and 5, we obtain not a single solution, that does not leave the considered sets, but a one-parameter family of such solutions. This family can be constructed by varying the line segments considered in the proofs.

4 Conclusion

In this note, we have presented topological ideas which can be used for studying forced oscillations and global stabilization of an inverted pendulum with a moving pivot point. In both cases, the results can be generalized and extended to a broader class of systems.

For instance, the result on the existence of a forced oscillation can be proved for the case when we consider a mass point on a manifold with a boundary in a periodic external field [8,9]. In particular, similar result holds for the spherical inverted pendulum [1] with a moving pivot point (here manifold with a boundary is the upper semi-sphere). Moreover, similar result can be proved for groups of interacting nonlinear systems [10]. As an illustration, we can consider the following system. Let us have a finite number of planar pendulums moving with viscous friction (can be arbitrarily small) in a gravitational field (Fig. 2). Let r_i be a radius-vector of the massive point of the i-th pendulum. Suppose that their pivot points are moving along a horizontal line in accordance with a T-periodic law of motion $h\colon \mathbb{R}/T\mathbb{Z} \to \mathbb{R}$, which is the same for all pendulums. Let us also assume the following: for any two pendulums there is a repelling force F_{ij} acting on the mass point of the i-th pendulum from the j-th pendulum (F_{ij} is parallel to $r_i - r_j$). It is possible to prove that in this system with non-local interaction (each pendulum is influenced by all other pendulums), there always exists a forced oscillation and along this solution the pendulums never become horizontal.

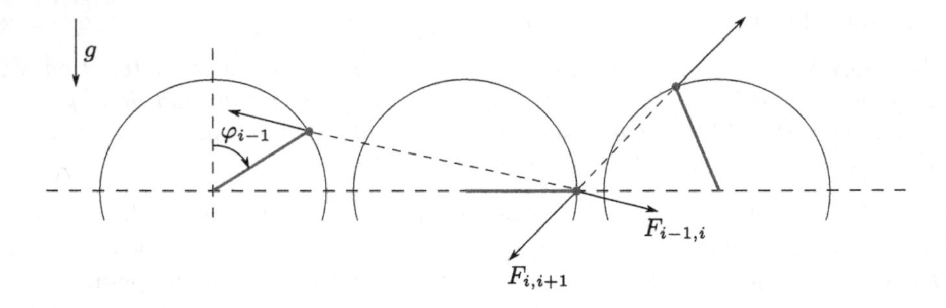

Fig. 2. When the i-th pendulum is horizontal, the repelling forces acting on it are directed downward.

Our simple results on global stabilization can also be proved for various similar systems. One of the main possible generalizations is the system of an inverted pendulum on a cart, which is more correct from the physical point of view [2]. It is also possible to consider multidimensional systems or systems with friction. We can also omit the requirement of the existence of a Lyapunov-type function V satisfying L1 and L2 (we just need the existence of a 'capturing' set containing the point μ). Moreover, systems without the assumption on the uniqueness of the solutions can also be considered, since only the right-uniqueness is important for our considerations above. For instance, we can consider systems with set-valued right-hand sides (Filippov-type systems), including systems with dry friction.

References

1. Srzednicki, R.: On periodic solutions in the Whitney's inverted pendulum problem. arXiv:1709.08254v1
2. Polekhin, I.: On topological obstructions to global stabilization of an inverted pendulum. arXiv:1704.03698v2
3. Ważewski, T.: Sur un principe topologique de l'examen de l'allure asymptotique des intégrales des équations différentielles ordinaires. Ann. Soc. Polon. Math. **20**, 279–313 (1947)
4. Reissig, R., Sansone, G., Conti, R.: Qualitative Theorie nichtlinearer Differentialgleichungen. Edizioni Cremonese, Rome (1963)
5. Srzednicki, R.: Periodic and bounded solutions in blocks for time-periodic nonautonomous ordinary differential equations. Nonlinear Anal. Theor. Methods Appl. **22**, 707–737 (1994)
6. Srzednicki, R., Wójcik, K., Zgliczyński, P.: Fixed point results based on the Ważewski method. In: Brown, R.F., Furi, M., Górniewicz, L., Jiang, B. (eds.) Handbook of Topological Fixed Point Theory, pp. 905–943. Springer, Dordrecht (2005). doi:10.1007/1-4020-3222-6_23
7. Demidovich, B.P.: Lectures on the Mathematical Theory of Stability. Nauka, Moscow (1967)
8. Polekhin, I.: Forced oscillations of a massive point on a compact surface with a boundary. Nonlinear Anal. Theor. Methods Appl. **128**, 100–105 (2015)
9. Bolotin, S.V., Kozlov, V.V.: Calculus of variations in the large, existence of trajectories in a domain with boundary, and Whitney's inverted pendulum problem. Izv. Math. **79**(5), 894–901 (2015)
10. Polekhin, I.: On forced oscillations in groups of interacting nonlinear systems. Nonlinear Anal. Theor. Methods Appl. **135**, 120–128 (2016)

Uniform Observability of Linear Time-Varying Systems and Application to Robotics Problems

Pascal Morin[1(✉)], Alexandre Eudes[2], and Glauco Scandaroli[1]

[1] ISIR, Sorbonne Universités, UPMC Univ. Paris 06, CNRS UMR 7222, Paris, France
morin@isir.upmc.fr, scandaroli@upmc.fr
[2] ONERA - The French Aerospace Lab, Palaiseau, France
alexandre.eudes@onera.fr

Abstract. Many methods have been proposed to estimate the state of a nonlinear dynamical system from uncomplete measurements. This paper concerns an approach that consists in lifting the estimation problem into a higher-dimensional state-space so as to transform an original nonlinear problem into a linear problem. Although the associated linear system is usually time-varying, one can then rely on Kalman's linear filtering theory to achieve strong convergence and optimality properties. In this paper, we first present a technical result on the uniform observability of linear time-varying systems. Then, we illustrate through a problem arising in robotics how this result and the lifting method evoked above lead to explicit observability conditions and linear observers.

Keywords: Observability · Observer design · Filtering

1 Introduction

The general problem of observability and observer design concerns the reconstruction of the state x of a dynamical system $\dot{x} = f(x, u, t)$ from the knowledge of the input u and an output function $y = h(x, u, t)$. There is a vast control litterature on this topic for both linear [11,12] and nonlinear systems [9]. For linear systems, i.e. $f(x, u, t) = A(t)x + B(t)u,\ h(x, u, t) = C(t)x + D(t)u$, observability is independent of the input u and is directly related to the system's Grammian, which only depends on the matrices A and C. By contrast, observability of a nonlinear system usually depends on the input. This dependence is a source of difficulty for both observability analysis and observer design. As a matter of fact, the main observability characterization result for nonlinear systems [9] only ensures a "weak" form of observability, i.e. existence of control inputs for which the system is observable. Neither the characterization of these "good inputs", nor the characterization of the associated observability property (e.g. uniform versus non-uniform) is provided in [9]. Concerning observer design, in many application fields involving nonlinear systems state estimation relies on Extended

These results were obtained while all authors were with ISIR. This work was supported by the "Chaire d'excellence en Robotique RTE-UPMC".

F. Nielsen and F. Barbaresco (Eds.): GSI 2017, LNCS 10589, pp. 336–344, 2017.
https://doi.org/10.1007/978-3-319-68445-1_39

Kalman Filters (EKFs), which often perform well but also, sometimes, yield divergent estimation errors or/and unconsistent results. One of the main progress achieved in recent years on the topic of nonlinear observers concerns systems with symmetries [2]. When both the system's dynamics and output functions are invariant under a transformation group, so-called "invariant observers" can be built with improved convergence properties w.r.t. EKFs. In particular, if the system's state space is a Lie group and the system's dynamics and output functions are invariant w.r.t. the Lie group operation, observers with error dynamics independent of the trajectory can be obtained [3]. W.r.t. EKFs this implies stronger convergence results, as demonstrated for several applications [4,15]. As a remaining difficulty, the error dynamics is still nonlinear. This can make global or semi-global convergence properties difficult to achieve. This paper concerns a different approach, which consists in lifting the estimation problem into a higher-dimensional state-space so as to transform the original nonlinear estimation problem into a linear estimation problem in higher-dimension. Like for invariant observers, the objective is still to simplify the observability analysis and observer design. In this case the goal is fully achieved thanks to the strong observability and observer design results for linear time-varying systems.

The paper is organized as follows. The main technical result of this paper is given in Sect. 2. Then, a robotics application example is treated in Sect. 3.

2 Observability of Linear Time-Varying Systems

Consider a general linear time-varying (LTV) system

$$\begin{cases} \dot{x} = A(t)x + B(t)u \\ y = C(t)x \end{cases} \tag{1}$$

There exist different types of observability properties for LTV systems, like e.g., differential, instantaneous, or uniform observability (see e.g. [6, Chap. 5] for more details). Here we focus on *uniform observability*, which ensures that the state estimation process is well-conditionned and can be solved via the design of exponentially stable observers. The following assumption will be used.

Assumption 1. *The matrix-valued functions A, B, and C of the LTV system* (1) *are continuous and bounded on* $[0, +\infty)$.

Definition 1. *A LTV system* (1) *satisfying Assumption 1 is* uniformly observable *if there exist $\tau, \delta > 0$ such that*

$$\forall t \geq 0, \qquad 0 < \delta I \leq W(t, t+\tau) \triangleq \int_t^{t+\tau} \Psi(s,t)^T C^T(s) C(s) \Psi(s,t)\, ds \tag{2}$$

with $\Psi(s,t)$ the state transition matrix of $\dot{x} = A(t)x$ and I the identity matrix. The matrix W is called observability Grammian *of System* (1).

From this definition uniform observability is independent of B. Thus, we say without distinction that System (1) or the pair (A, C) is uniformly observable. Note also, as a consequence of Assumption 1, that $W(t, t+\tau)$ is upper bounded by some $\bar{\delta} I$ for any $t \geq 0$.

The following theorem recalls two properties of uniformly observable systems. The first property follows from [1, Lemma 3] and the duality principle (see, e.g. [6, Theorem 5–10]). This principle, together with [10, Theorem 3], imply the second property.

Theorem 1. *For a LTV system (1) satisfying Assumption 1 the following properties hold.*

1. *The pair (A, C) is uniformly observable iff the pair $(A - LC, C)$ is uniformly observable, with $L(.)$ any bounded matrix-valued time-function.*
2. *If the pair (A, C) is uniformly observable, then for any $a > 0$ there exists a bounded matrix $L_a(t)$ such that the linear observer*

$$\dot{\hat{x}} = A(t)\hat{x} + L_a(t)(y - C(t)\hat{x})$$

is uniformly exponentially stable with convergence rate given by a, i.e. there exists $c_a > 0$ such that $\|\hat{x}(t) - x(t)\| \leq c_a e^{-a(t-t_0)} \|\hat{x}(t_0) - x(t_0)\|$ for any $t \geq t_0$ and any $x(0), \hat{x}(0)$.

Main technical result: Checking uniform observability of a LTV system can be difficult since calculation of the Grammian requires integration of the solutions of $\dot{x} = A(t)x$. It is well known that observability properties of LTV systems are related to properties of the observability space $\mathcal{O}(t)$ defined by [6, Chap. 5]:

$$\mathcal{O}(t) \triangleq \begin{pmatrix} N_0(t) \\ N_1(t) \\ \vdots \end{pmatrix}, \qquad N_0 \triangleq C, \quad N_{k+1} \triangleq N_k A + \dot{N}_k \quad \text{for } k = 1, \dots \tag{3}$$

For example, instantaneous observability at t is guaranteed if $\text{Rank}(\mathcal{O}_{n-1}(t)) = n$. For general LTV systems, however, uniform observability cannot be characterized in term of rank conditions only. We propose below a sufficient condition for uniform observability.

Proposition 1. *Consider a LTV system (1) satisfying Assumption 1. Assume that there exists a positive integer K such that:*

1. *The k-th order derivative of A (resp. C) is well defined and bounded on $[0, +\infty)$ up to $k = K$ (resp. up to $k = K + 1$).*
2. *There exist a $n \times n$ matrix M composed of row vectors of $N_0, \dots, N_K$, and two scalars $\bar{\delta}, \bar{\tau} > 0$ such that*

$$\forall t \geq 0, \qquad 0 < \bar{\delta} \leq \int_t^{t+\bar{\tau}} |det(M(s))| \, ds \tag{4}$$

with $det(M)$ the determinant of M.

Then, System (1) *is uniformly observable.*

The proof of this result is given in the appendix. As illustrated in the following section, this result leads to very explicit uniform observability conditions.

Remark: In [5] a sufficient condition for *uniform complete observability* (a property equivalent to uniform observability under Assumption 1) is provided. There are similarities between that condition and (4) but the latter is less demanding as it only requires positivity "in average" while the positivity condition in [5] must hold for any time-instant.

3 Application to a Robotics Estimation Problem

A classical robotics problem consists in recovering the motion of a robot from measurements given by a vision system. For compacity reasons, monocular vision can be preferred to stereo vision but then, full 3D motion estimation cannot be performed due to depth ambiguity. To remedy this difficuty, vision data can be fused with measurements provided by an IMU (Inertial Measurement Unit). This is called visuo-inertial fusion. In this section, we describe how the problem of monocular visuo-inertial fusion is commonly posed as a nonlinear estimation problem, and we show how it can be transformed by lifting into a linear estimation problem in higher dimension. The material presented in this section is based on [7] to which we refer the reader for more details.

The monocular visuo-inertial problem. Consider two images $\mathbf{I}_A$ and $\mathbf{I}_B$ of a planar scene taken by a monocular camera. Each image $\mathbf{I}_*$ ($* \in \{A, B\}$) is taken from a specific pose of the camera and we denote by $\mathcal{F}_*(* \in \{A, B\})$ an associated camera frame with origin corresponding to the optical center of the camera and third basis vector aligned with the optical axis. We also denote by d_* and n_* respectively the distance from the origin of $\mathcal{F}_*$ to the planar scene and the normal to the scene expressed in $\mathcal{F}_*$. Let R denote the rotation matrix from $\mathcal{F}_B$ to $\mathcal{F}_A$ and $p \in \mathbb{R}^3$ the coordinate vector of the origin of $\mathcal{F}_B$ expressed in $\mathcal{F}_A$. The problem here considered consists in estimating R and p. From $\mathbf{I}_A$ and $\mathbf{I}_B$ one can compute (see, e.g., [13]) the so-called "homography matrix"

$$H = R^T - \frac{1}{d_A} R^T p n_A^T \tag{5}$$

Considering also a (strapped-down) IMU, we obtain as additional measurements ω, the angular velocity vector of the sensor w.r.t. the inertial frame expressed in body frame, and a_s, the so-called specific acceleration. Assuming that $\mathcal{F}_A$ is an inertial frame and $\mathcal{F}_B$ is the body frame[1], ω and a_s are defined by:

$$\dot{R} = RS(\omega)\,, \quad \ddot{p} = g_A + Ra_s \tag{6}$$

[1] For simplicity we assume that the camera frame and IMU frame coincide.

where g_A denotes the gravitational acceleration field expressed in $\mathcal{F}_A$ and $S(x)$ is the skew-symmetric matrix associated with the cross product by x, i.e. $S(x)y = x \times y$ with $\times$ the cross product. Note that a_s can also be defined by the relation

$$\dot{v} = -S(\omega)v + a_s + g_B \tag{7}$$

where $v = R^T\dot{p}$ denotes the velocity of $\mathcal{F}_B$ w.r.t. $\mathcal{F}_A$ expressed in $\mathcal{F}_B$, and $g_B = R^T g_A$ is the gravitational acceleration field expressed in $\mathcal{F}_B$.

Visuo-inertial fusion: Diverse estimation algorithms have been proposed. In [16] the state is defined as $x = (R, p, v, n_A, d_A)$, with measurement $y = (H, \omega, a_s)$. Since H is a nonlinear function of x, the estimation problem is nonlinear and an EKF is used. In [8,14] the state is defined as $x = (\bar{H}, M, n_A)$ where $\bar{H} = \det(H)^{-\frac{1}{3}}H$ and $M = v\frac{n_A^T}{d_A}$, with measurement $y = (\bar{H}, \omega, a_s)$. The measurement then becomes a linear function of x, but the dynamics of x is nonlinear. Using the fact that $\bar{H}$ belongs to the Special Linear group $SL(3)$, nonlinear observers with convergence guarantees are proposed in [8,14], but under restrictive motion assumptions.

As an alternative solution, define the state as $x = (H, M, n_s, Q)$ with H defined by (5) and $M = vn_s^T$, $n_s = \frac{n_A}{d_A}$, $Q = g_B n_s^T$. Since n_A and d_A are constant quantities, one verifies from (6) and (7) that:

$$\begin{cases} \dot{H} = -S(\omega)H - M\,, & \dot{n}_s = 0 \\ \dot{M} = -S(\omega)M + Q + a_s n_s^T\,, & \dot{Q} = -S(\omega)Q \end{cases} \tag{8}$$

Since ω and a_s are known time-functions, the above system is a linear time-varying system in x. In other words, the estimation problem has been transformed into a linear estimation problem by lifting to a higher-dimensional state space. One verifies (see [7] for details) that R and p can be extracted from x. From this point, one can make use of existing tools of linear estimation theory. Proposition 1 provides the following characterization of uniform observability in term of the IMU data. It was initially obtained in [7].

Proposition 2. *System* (8) *with measurement* $y = (H, \omega, a_s)$ *is uniformly observable if*

(i) ω *and* a_s *are continuous and bounded on* $[0, +\infty)$, *and their first, second, and third-order time-derivatives are well defined and bounded on* $[0, +\infty)$*;*
(ii) there exists two scalars $\delta, \sigma > 0$ *such that*

$$\forall t \geq 0, \quad 0 < \delta \leq \int_t^{t+\sigma} \|\dot{a}_s(\tau) + \omega(\tau) \times a_s(\tau)\|\, d\tau \tag{9}$$

4 Conclusion

We have provided a technical result on uniform observability of linear time-varying system and we have illustrated the application of this result to an estimation problem. We have also shown through this problem how an original nonlinear estimation problem could be transformed into a linear estimation problem through lifting of the state space. The main open problem is to characterize systems for which such a lifting exists.

Appendix: Proof of Proposition 1: We must show the existence of constants $\tau, \delta > 0$ such that (2) is satisfied. The inequality in (2) is equivalent to $x^T W(t, t+\tau)x \geq \delta\|x\|^2$ for any vector $x \in D = \{x \in \mathbb{R}^n : \|x\| = 1\}$. Thus, the proof consists in showing the existence of constants $\tau, \delta > 0$ such that

$$\forall t \geq 0, \qquad 0 < \delta \leq \inf_{x \in D} \int_t^{t+\tau} \|C(s)\Psi(s,t)x\|^2 \, ds$$

We proceed by contradiction. Assume that

$$\forall \tau > 0, \forall \delta > 0, \; \exists t(\tau,\delta): \quad \inf_{x \in D} \int_{t(\tau,\delta)}^{t(\tau,\delta)+\tau} \|C(s)\Psi(s,t(\tau,\delta))x\|^2 \, ds < \delta \tag{10}$$

Take $\tau = \bar{\tau}$ with $\bar{\tau}$ the constant in (4), and consider the sequence $(\delta_p = 1/p)$. Thus, for any $p \in \mathbb{N}$, there exists t_p such that

$$\inf_{x \in D} \int_{t_p}^{t_p+\bar{\tau}} \|C(s)\Psi(s,t_p)x\|^2 \, ds < \frac{1}{p}$$

so that there exists $x_p \in D$ such that

$$\int_{t_p}^{t_p+\bar{\tau}} \|C(s)\Psi(s,t_p)x_p\|^2 \, ds < \frac{1}{p} \tag{11}$$

Since D is compact, a sub-sequence of the sequence (x_p) converges to some $\bar{x} \in D$. From Assumption 1, A is bounded on $[0, +\infty)$. Therefore,

$$\forall x \in \mathbb{R}^n, \; \forall t \leq s, \qquad e^{-(s-t)\|A\|_\infty}\|x\| \leq \|\Psi(s,t)x\| \leq e^{(s-t)\|A\|_\infty}\|x\| \tag{12}$$

with $\|A\|_\infty = \sup_{t \geq 0} \|A(t)\|$. Since C is also bounded (from Assumption 1) and the interval of integration in (11) is of fixed length $\bar{\tau}$, it follows that

$$\lim_{p \to +\infty} \int_{t_p}^{t_p+\bar{\tau}} \|C(s)\Psi(s,t_p)\bar{x}\|^2 \, ds = 0$$

By a change of integration variable, this equation can be written as

$$\lim_{p \to +\infty} \int_0^{\bar{\tau}} \|f_p(s)\|^2 \, ds = 0 \tag{13}$$

with $f_p(t) = C(t+t_p)\Psi(t+t_p,t_p)\bar{x}$. It is well known, and easy to verify, that

$$f_p^{(k)}(t) = N_k(t+t_p)\Psi(t+t_p,t_p)\bar{x} \tag{14}$$

with $f_p^{(k)}$ the k-th order derivative of f_p and N_k defined by (3). The existence of $f_p^{(k)}$, for any $k = 0, \cdots, K+1$, follows by Assumption 1 of Proposition 1. The end of the proof relies on the following lemma, proved further.

Lemma 1. *Assume that* (10) *is satisfied. Then,* $\forall k = 0, \ldots, K$,

$$\lim_{p \to +\infty} \int_0^{\bar{\tau}} \|f_p^{(k)}(s)\|^2 \, ds = 0 \tag{15}$$

Since the matrix M in (4) is composed of row vectors of $N_0, \ldots, N_K$, it follows from (14) that

$$\int_0^{\bar{\tau}} \|M(s+t,p)\Psi(s+t_p,t_p)\bar{x}\|^2 \, ds \le \sum_{k=0}^{K} \int_0^{\bar{\tau}} \|f_p^{(k)}(s)\|^2 \, ds$$

Therefore, from Lemma 1,

$$\lim_{p \to +\infty} \int_{t_p}^{t_p+\bar{\tau}} \|M(s)\Psi(s,t_p)\bar{x}\|^2 \, ds = \lim_{p \to +\infty} \int_0^{\bar{\tau}} \|M(s+t_p)\Psi(s+t_p,t_p)\bar{x}\|^2 \, ds = 0 \tag{16}$$

Then, for any $\xi \in \mathbb{R}^n$

$$\|M(s)\xi\|^2 = \xi^T M^T(s) M(s) \xi \ge \|\xi\|^2 \min_i \lambda_i(M^T(s)M(s)) = \|\xi\|^2 \lambda_1(M^T(s)M(s)) \tag{17}$$

with $\lambda_1(M^T(s)M(s)) \le \cdots \le \lambda_n(M^T(s)M(s))$ the eigenvalues of $M^T(s)M(s)$ in increasing order. Furthermore, since M is bounded on $[0, +\infty)$ (as a consequence of Assumption 1 and the definition of M), there exists a constant $c > 0$ such that $\max_i \lambda_i(M^T(s)M(s)) \le c$ for all s. Thus

$$\lambda_1(M^T(s)M(s)) = \frac{\det(M^T(s)M(s))}{\prod_{j>1} \lambda_j(M^T(s)M(s))} \ge \frac{\det(M^T(s)M(s))}{c^{n-1}} \ge \frac{(\det(M(s)))^2}{c^{n-1}}$$

It follows from this inequality, (12) and (17), and the fact that $\|\bar{x}\| = 1$ that

$$\forall p \in \mathbb{N}, \quad \int_{t_p}^{t_p+\bar{\tau}} \|M(s)\Psi(s,t_p)\bar{x}\|^2 \, ds \ge \bar{c} \int_{t_p}^{t_p+\bar{\tau}} (\det(M(s)))^2 \, ds \tag{18}$$

with $\bar{c} = e^{-2\bar{\tau}\|A\|_\infty}/c^{n-1} > 0$. Furthermore, Schwarz inequality implies that

$$\int_{t_p}^{t_p+\bar{\tau}} |\det(M(s))| \, ds \le \left(\int_{t_p}^{t_p+\bar{\tau}} 1 \, ds \right)^{1/2} \left(\int_{t_p}^{t_p+\bar{\tau}} (\det(M(s)))^2 \, ds \right)^{1/2}$$

Thus, it follows from (4) and (18) that

$$\forall p \in \mathbb{N}, \quad \int_{t_p}^{t_p+\bar{\tau}} \|M(s)\Psi(s,t_p)\bar{x}\|^2 \, ds \ge \bar{c}\bar{\delta}^2/\bar{\tau} > 0$$

which contradicts (16). To complete the proof, we must prove Lemma 1.

Proof of Lemma 1: We proceed by induction. By assumption (10) is satisfied, which implies that (13) holds true. Thus, (15) holds true for $k = 0$. Assuming now that (15) holds true for $k = 0, \ldots, \bar{k} < K$, we show that it holds true for $k = \bar{k}+1$

too. From Assumption 1 of Proposition 1, recall that $\forall j = 1, \dots K+1$, $f_p^{(j)}$ is well defined and bounded on $[0, \bar{\tau}]$, uniformly w.r.t. p. We claim that $f_p^{(\bar{k})}(0)$ tends to zero as p tends to $+\infty$. Assume on the contrary that $f_p^{(\bar{k})}(0)$ does not tend to zero. Then, $\exists \varepsilon > 0$ and a subsequence $(f_{p_j}^{(\bar{k})})$ of $(f_p^{(\bar{k})})$ such that $\|f_{p_j}^{(\bar{k})}(0)\| > \varepsilon$, $\forall j \in \mathbb{N}$. Since $\|f_{p_j}^{(\bar{k}+1)}(0)\|$ is bounded uniformly w.r.t. j (because $f_p^{(\bar{k}+1)}$ is bounded on $[0, \bar{\tau}]$ uniformly w.r.t. p), $\exists t' > 0$ such that $\forall j \in \mathbb{N}$, $\forall t \in [0, t']$, $\|f_{p_j}^{(\bar{k})}(t)\| > \varepsilon/2$. By (15), this contradicts the induction hypothesis. Therefore, $f_p^{(\bar{k})}(0)$ tends to zero as p tends to $+\infty$. By a similar argument, one can show that $f_p^{(\bar{k})}(\bar{\tau})$ tends to zero as p tends to $+\infty$. Now,

$$\begin{aligned}
\int_0^{\bar{\tau}} \|f_p^{(\bar{k}+1)}(s)\|^2 \, ds &= \sum_{i=1}^{n} \int_0^{\bar{\tau}} \left(f_{p,i}^{(\bar{k}+1)}(s)\right)^2 ds \\
&= -\sum_{i=1}^{n} \int_0^{\bar{\tau}} f_{p,i}^{(\bar{k})}(s) f_{p,i}^{(\bar{k}+2)}(s) \, ds + \sum_{i=1}^{n} \left[f_{p,i}^{(\bar{k})}(s) f_{p,i}^{(\bar{k}+1)}(s)\right]_0^{\bar{\tau}} \\
\leq \sum_{i=1}^{n} \left(\int_0^{\bar{\tau}} \left(f_{p,i}^{(\bar{k})}(s)\right)^2 ds\right)^{1/2} & \left(\int_0^{\bar{\tau}} \left(f_{p,i}^{(\bar{k}+2)}(s)\right)^2 ds\right)^{1/2} + \sum_{i=1}^{n} \left[f_{p,i}^{(\bar{k})}(s) f_{p,i}^{(\bar{k}+1)}(s)\right]_0^{\bar{\tau}}
\end{aligned}$$

Each term

$$\left(\int_0^{\bar{\tau}} \left(f_{p,i}^{(\bar{k})}(s)\right)^2 ds\right)^{1/2} \left(\int_0^{\bar{\tau}} \left(f_{p,i}^{(\bar{k}+2)}(s)\right)^2 ds\right)^{1/2}$$

in the first sum tends to zero as p tends to infinity due to (15) for $k = \bar{k}$ and the fact that $f_p^{(\bar{k}+2)}$ is bounded uniformly w.r.t. p. Boundary terms in the second sum also tend to zero as p tends to infinity since $f_p^{(\bar{k})}(0)$ and $f_p^{(\bar{k})}(\bar{\tau})$ tend to zero, and $f_p^{(\bar{k}+1)}$ is bounded. As a result, (15) is satisfied for $k = \bar{k} + 1$. ■

References

1. Anderson, B.D.O., Moore, J.B.: New results in linear system stability. SIAM J. Control **7**(3), 398–414 (1969)
2. Bonnabel, S., Martin, P., Rouchon, P.: Symmetry-preserving observers. IEEE Trans. Autom. Control **53**(11), 2514–2526 (2008)
3. Bonnabel, S., Martin, P., Rouchon, P.: Non-linear symmetry-preserving observers on lie groups. IEEE Trans. Autom. Control **54**(7), 1709–1713 (2009)
4. Bonnabel, S., Martin, P., Salaün, E.: Invariant extended Kalman filter: theory and application to a velocity-aided attitude estimation problem. In: IEEE Conference on Decision and Control, pp. 1297–1304 (2009)
5. Bristeau, P.J., Petit, N., Praly, L.: Design of a navigation filter by analysis of local observability. In: IEEE Conference on Decision and Control, pp. 1298–1305 (2010)
6. Chen, C.-T.: Linear System Theory and Design. Oxford University Press, New York (1984)
7. Eudes, A., Morin, P.: A linear approach to visuo-inertial fusion for homography-based filtering and estimation. In: IEEE/RSJ International Conference on Intelligent Robots and Systems (IROS), pp. 3095–3101 (2014)

8. Eudes, A., Morin, P., Mahony, R., Hamel, T.: Visuao-inertial fusion for homography-based filtering and estimation. In: IEEE/RSJ International Conference on Intelligent Robots and Systems (IROS), pp. 5186–5192 (2013)
9. Hermann, R., Krener, A.J.: Nonlinear controllability and observability. IEEE Trans. Autom. Control **22**, 728–740 (1977)
10. Ikeda, M., Maeda, H., Kodama, S.: Stabilization of linear systems. SIAM J. Control **4**, 716–729 (1972)
11. Kalman, R.E.: Contributions to the theory of optimal control. Bol. Soc. Math. Mex **5**, 102–119 (1960)
12. Kalman, R.E.: On the general theory of control systems. In: First International Congress of Automatic Control (IFAC), pp. 481–493 (1960)
13. Ma, Y., Soatto, S., Kosecka, J., Sastry, S.S.: An Invitation to 3-D Vision: From Images to Geometric Models. Springer, New York (2003)
14. Mahony, R., Hamel, T., Morin, P., Malis, E.: Nonlinear complementary filters on the special linear group. Int. J. Control **85**, 1557–1573 (2012)
15. Mahony, R., Hamel, T., Pflimlin, J.-M.: Nonlinear complementary filters on the special orthogonal group. IEEE Trans. Autom. Control **53**(5), 1203–1218 (2008)
16. Servant, F., Houlier, P., Marchand, E.: Improving monocular plane-based slam with inertial measures. In: IEEE/RSJ International Conference on Intelligent Robots and Systems (IROS), pp. 3810–3815 (2010)

Global Exponential Attitude and Gyro Bias Estimation from Vector Measurements

Philippe Martin[1(✉)] and Ioannis Sarras[2]

[1] Centre Automatique et Systèmes, MINES ParisTech, PSL Research University, 75006 Paris, France
philippe.martin@mines-paristech.fr
[2] ONERA – The French Aerospace Lab, 91123 Palaiseau, France
ioannis.sarras@onera.fr

Abstract. We consider the classical problem of estimating the attitude and gyro biases of a rigid body from at least two vector measurements and a triaxial rate gyro. We propose a solution based on a dynamic nonlinear estimator designed without respecting the geometry of SO(3), which achieves uniform global exponential convergence. The convergence is established thanks to a dynamically scaled Lyapunov function.

1 Introduction

Estimating the attitude of a rigid body from vector measurements (obtained for instance from accelerometers, magnetometers, sun sensors, etc.) has been for decades a problem of interest, because of its importance for a variety of technological applications such as satellites or unmanned aerial vehicles. The attitude of the body can be described by the rotation matrix $R \in \mathrm{SO}(3)$ from body to inertial axes. On the other hand, the (time-varying) measurement vectors $u_1, \cdots, u_n \in \mathbb{R}^3$ correspond to the expression in body axes of known and not all collinear vectors $U_1, \cdots, U_n \in \mathbb{R}^3$ which are constant in inertial axes, i.e., $u_k(t) = R^T(t)U_k$. The goal then is to reconstruct the attitude at time t using only the knowledge of the measurement vectors until t. The solution to the problem would be very easy if the vector measurements were perfect and two of them were linearly independent: indeed, using for instance only the two vectors $u_1(t)$ and $u_2(t)$ and noticing that $R^T(x \times y) = R^T x \times R^T y$ since R is a rotation matrix, we readily find

$$\begin{aligned} R^T(t) &= R^T(t) \cdot \big(U_1\, U_2\, U_1 \times U_2\big) \cdot \big(U_1\, U_2\, U_1 \times U_2\big)^{-1} \\ &= \big(u_1(t)\, u_2(t)\, u_1(t) \times u_2(t)\big) \cdot \big(U_1\, U_2\, U_1 \times U_2\big)^{-1}. \end{aligned}$$

But in real situations, the measurement vectors are always corrupted at least by noise. Moreover, the U_k's may possibly be not strictly constant: for instance a triaxial magnetometer measures the (locally) constant Earth magnetic field, but is easily perturbed by ferromagnetic masses and electromagnetic perturbations; similarly, a triaxial accelerometer can be considered as measuring the direction

F. Nielsen and F. Barbaresco (Eds.): GSI 2017, LNCS 10589, pp. 345–351, 2017.
https://doi.org/10.1007/978-3-319-68445-1_40

of gravity provided it is not undergoing a substantial acceleration (see e.g. [13] for a detailed discussion of this assumption and its consequences in the framework of quadrotor UAVs). That is why, despite the additional cost, it may be interesting to use a triaxial rate gyro to supplement the possibly deteriorated vector measurements.

The current literature on attitude estimation from vector measurements can be broadly divided into three categories: (i) optimization-based methods; (ii) stochastic filtering; (iii) nonlinear observers. Details on the various approaches can be found e.g. in the surveys [6,17] and the references therein. The first category sets the problem as the minimization of a cost function, and is usually referred to as Wahba's problem. The attitude is algebraically recovered at time t using only the measurements at time t. No filtering is performed, and possibly available velocity information from rate gyros is not exploited. The second category mainly hinges on Kalman filtering and its variants. Despite their many qualities, the drawback of those designs is that convergence cannot in general be guaranteed except for mild trajectories. Moreover the tuning is not completely obvious, and the computational cost may be too high for small embedded processors. The third, and more recent, approach proposes nonlinear observers with a large guaranteed domain of convergence and a rather simple tuning through a few constant gains. These observers can be designed: (a) directly on SO(3) (or the unit quaternion space), see e.g. [7,10,12,16]; (b) or more recently, on $\mathbb{R}^{3\times3}$, i.e., deliberately "forgetting" the underlying geometry [2,3,8,14]. Probably the best-known design is the so-called nonlinear complementary filter of [10]; as noticed in [11], it is a special case of so-called invariant observers [5].

In this paper, we propose a new observer of attitude and gyro biases from gyro measurements and (at least) two measurement vectors. It also "forgets" the geometry of SO(3), which allows for uniform global exponential convergence (notice the observer of [10] is only quasi-globally convergent). This observer is an extension of the observer of [14] (which is uniformly globally convergent), itself a modification of the linear cascaded observer of [3] (which is uniformly globally exponentially convergent). The idea of the proof is nevertheless completely different from the approach followed in [3]; it is much more direct, as it relies on a strict, dynamically scaled, Lyapunov function, see [1,9].

2 The Design Model

We consider a moving rigid body subjected to an angular velocity ω. Its orientation matrix $R \in \mathrm{SO}(3)$ is related to the angular velocity by the differential equation

$$\dot{R} = R\omega_\times, \tag{1}$$

where the skew-symmetric matrix $\omega_\times$ is defined by $\omega_\times u := \omega \times u$ whatever the vector $u \in \mathbb{R}^3$.

The rigid body is equipped with a triaxal rate gyro measuring the angular velocity ω, and two additional triaxial sensors (for example accelerometers, magnetometers or sun sensors) providing the measurements of two vectors α and β.

These vectors correspond to the expression in body axes of two known independent vectors α_i and β_i which are constant in inertial axes. In other words,

$$\begin{aligned} \alpha &:= R^T \alpha_i \\ \beta &:= R^T \beta_i. \end{aligned}$$

Since α_i, β_i are constant, we obviously have

$$\begin{aligned} \dot{\alpha} &= \alpha \times \omega \\ \dot{\beta} &= \beta \times \omega. \end{aligned}$$

To take full advantage of the rate gyro, it is wise to take into account that it is biased, hence rather provides the measurement

$$\omega_m := \omega + b,$$

where b is a slowly-varying (for instance with temperature) unknown bias. Since the effect of this bias on attitude estimation may be important, it is worth determining this value. But being not exactly constant, it can not be calibrated in advance and must be estimated online together with the attitude.

Our objective is to design an estimation scheme that can reconstruct online the orientation matrix $R(t)$ and the bias $b(t)$, using (i) the measurements of the gyro and of the two vector sensors; (ii) the knowledge of the constant vectors α_i and β_i. The model on which the design will be based therefore consists of the dynamics

$$\dot{\alpha} = \alpha \times \omega \tag{2}$$
$$\dot{\beta} = \beta \times \omega \tag{3}$$
$$\dot{b} = 0, \tag{4}$$

together with the measurements

$$\omega_m := \omega + b \tag{5}$$
$$\alpha_m := \alpha \tag{6}$$
$$\beta_m := \beta. \tag{7}$$

3 The Observer

We want to show that the state of (2)–(7) can be estimated by the observer

$$\dot{\hat{\alpha}} = \hat{\alpha} \times (\omega_m - \hat{b}) - k_\alpha(\hat{\alpha} - \alpha_m) \tag{8}$$
$$\dot{\hat{\beta}} = \hat{\beta} \times (\omega_m - \hat{b}) - k_\beta(\hat{\beta} - \beta_m) \tag{9}$$
$$\dot{\xi} = l_\alpha(\omega_m - \hat{b}) \times (\hat{\alpha} \times \alpha_m) + l_\beta(\omega_m - \hat{b}) \times (\hat{\beta} \times \beta_m) \tag{10}$$
$$\qquad + l_\alpha k_\alpha \hat{\alpha} \times \alpha_m + l_\beta k_\beta \hat{\beta} \times \beta_m \tag{11}$$
$$\dot{r} = -2\psi_1(r-1) + 2\big(l_\alpha|\alpha_m||\hat{\alpha} - \alpha_m| + l_\beta|\beta_m||\hat{\beta} - \beta_m|\big)r, \tag{12}$$

where

$$\hat{b} := \xi + l_\alpha \hat{\alpha} \times \alpha_m + l_\beta \hat{\beta} \times \beta_m \tag{13}$$

$$k_\alpha := k_1 + r\Big(\frac{1}{2\epsilon} + \frac{l_\alpha^2}{\epsilon_1} r\Big)|\alpha_m|^2 \tag{14}$$

$$k_\beta := k_2 + r\Big(\frac{1}{2\epsilon} + \frac{l_\beta^2}{\epsilon_1} r\Big)|\beta_m|^2. \tag{15}$$

$\hat{\alpha}, \hat{\beta}, \hat{b} \in \mathbb{R}^3$ are the estimates of α, β, b; $\xi \in \mathbb{R}^3$ is the state of the bias observer, and $r \in \mathbb{R}$ is a dynamic scaling variable; the (positive) constants $l_\alpha, l_\beta, \psi_1, k_1, k_2, \epsilon, \epsilon_1$ are tuning gains. Defining the estimation errors as

$$e_\alpha := \hat{\alpha} - \alpha$$
$$e_\beta := \hat{\beta} - \beta$$
$$e_b := \hat{b} - b,$$

the error system reads

$$\dot{e}_\alpha = e_\alpha \times \omega - (\alpha + e_\alpha) \times e_b - k_\alpha(r, \hat{\alpha}) e_\alpha \tag{16}$$

$$\dot{e}_\beta = e_\beta \times \omega - (\beta + e_\beta) \times e_b - k_\beta(r, \hat{\beta}) e_\beta \tag{17}$$

$$\dot{e}_b = (l_\alpha \alpha_\times^2 + l_\beta \beta_\times^2) e_b + l_\alpha e_\alpha \times (\alpha \times e_b) + l_\beta e_\beta \times (\beta \times e_b) \tag{18}$$

$$\dot{r} = -2\psi_1(r-1) + 2\big(l_\alpha |\alpha||e_\alpha| + l_\beta |\beta||e_\beta|\big) r; \tag{19}$$

(18) is obtained thanks to the Jacobi identity $a \times (b \times c) + b \times (c \times a) + c \times (a \times b) = 0$.

The main result is the global exponential convergence of the observer.

Theorem 1. *Assume* $k_1, k_2, \epsilon, \epsilon_1 > 0$, $\psi_1 > \epsilon_1$, *and* l_α, l_β *large enough so that* $-(l_\alpha \alpha_\times^2 + l_\beta \beta_\times^2) > (\psi_1 + \epsilon) I$. *Then the equilibrium point* $(\bar{e}_\alpha, \bar{e}_\beta, \bar{e}_b, \bar{r}) := (0, 0, 0, 1)$ *of the error system* (16)–(19) *is uniformly globally exponentially stable.*

Remark 1 (*see* [4,15]). Since α and β are linearly independent, $-(l_\alpha \alpha_\times^2 + l_\beta \beta_\times^2)$ is a (symmetric) positive definite matrix when $l_\alpha, l_\beta > 0$; moreover, sufficiently large l_α, l_β yield $-(l_\alpha \alpha_\times^2 + l_\beta \beta_\times^2) > \mu I$ whatever the given constant μ.

Proof. First consider the candidate Lyapunov function for the (e_α, e_β)-subsystem

$$V(e_\alpha, e_\beta) := \frac{1}{2}|e_\alpha|^2 + \frac{1}{2}|e_\beta|^2.$$

Its time derivative satisfies

$$\begin{aligned}
\dot{V} &= -\langle e_\alpha, e_\alpha \times \omega \rangle - k_\alpha |e_\alpha|^2 - \langle e_\beta, e_\beta \times \omega \rangle - k_\beta |e_\beta|^2 \\
&\le -k_\alpha |e_\alpha|^2 - k_\beta |e_\beta|^2 + \big(\sqrt{r}|\alpha||e_\alpha|\big)\frac{|e_b|}{\sqrt{r}} + \big(\sqrt{r}|\beta||e_\beta|\big)\frac{|e_b|}{\sqrt{r}} \\
&\le -\Big(k_\alpha - \frac{r|\alpha|^2}{2\epsilon}\Big)|e_\alpha|^2 - \Big(k_\beta - \frac{r|\beta|^2}{2\epsilon}\Big)|e_\beta|^2 + \frac{\epsilon |e_b|^2}{r};
\end{aligned}$$

where we have used $\langle a, a \times b\rangle = 0$ to obtain the first line, and Young's inequality $ab \leq \frac{a^2}{2\epsilon} + \frac{\epsilon b^2}{2}$ to obtain the second line.

Now, the obvious candidate Lyapunov function $V_b(e_b) := \frac{1}{2}|e_b|^2$ for the e_b-subsystem satisfies

$$\begin{aligned}\dot{V}_b &= \langle e_b, (l_\alpha \alpha_\times^2 + l_\beta \beta_\times^2)e_b\rangle + l_\alpha\langle e_b, e_\alpha \times (\alpha \times e_b)\rangle + l_\beta\langle e_b, e_\beta \times (\beta \times e_b)\rangle \\ &\leq -\mu|e_b|^2 + (l_\alpha|\alpha||e_\alpha| + l_\beta|\beta||e_\beta|)|e_b|^2,\end{aligned}$$

where we have used Remark 1. The term $(l_\alpha|\alpha||e_\alpha| + l_\beta|\beta||e_\beta|)|e_b|^2$ happens to be very difficult to dominate with a classical Lyapunov approach. To overcome the problem, we use instead the candidate Lyapunov function

$$\tilde{V}_b(e_b, r) := \frac{1}{2r}|e_b|^2,$$

obtaining by dynamically scaling V_b with r defined by (19). Notice $r(t) \geq 1$ for all positive t as soon as $r(0) \geq 1$. We then have

$$\begin{aligned}\dot{\tilde{V}}_b &:= \frac{\dot{V}_b}{r} - \tilde{V}_b\frac{\dot{r}}{r} \\ &\leq -\mu\frac{|e_b|^2}{r} + (l_\alpha|\alpha||e_\alpha| + l_\beta|\beta||e_\beta|)\frac{|e_b|^2}{r} - \frac{|e_b|^2}{2r}\frac{\dot{r}}{r} \\ &= -(\mu - \psi_1)\frac{|e_b|^2}{r},\end{aligned}$$

where we have used $\frac{r-1}{r} \leq 1$.

We next consider the candidate Lyapunov function for the r-subsystem

$$V_r(r) := \frac{1}{2}(r-1)^2.$$

Its time derivative satisfies

$$\begin{aligned}\dot{V}_r &= -2\psi_1(r-1)^2 + \sqrt{2}(r-1)\sqrt{2}rl_\alpha|\alpha||e_\alpha| + \sqrt{2}(r-1)\sqrt{2}rl_\beta|\beta||e_\beta| \\ &\leq -2(\psi_1 - \epsilon_1)(r-1)^2 + \frac{r^2}{\epsilon_1}(l_\alpha^2|\alpha|^2|e_\alpha||^2 + l_\beta^2|\beta||e_\beta|^2),\end{aligned}$$

where the second line is obtained by Young's inequality.

Finally, consider the complete Lyapunov function

$$W(e_\alpha, e_\beta, e_b, r) := V(e_\alpha, e_\beta) + \tilde{V}_b(e_b, r) + V_r(r).$$

Collecting all the previous findings, its time derivative satisfies

$$\begin{aligned}\dot{W} &\leq -\Big(k_\alpha - \frac{r|\alpha|^2}{2\epsilon}\Big)|e_\alpha|^2 - \Big(k_\beta - \frac{r|\beta|^2}{2\epsilon}\Big)|e_\beta|^2 + \frac{\epsilon|e_b|^2}{r} \\ &\quad - (\mu - \psi_1)\frac{|e_b|^2}{r} \\ &\quad -2\,(\psi_1 - \epsilon_1)(r-1)^2 + \frac{r^2}{\epsilon_1}(l_\alpha^2|\alpha|^2|e_\alpha|^2 + l_\beta^2|\beta||e_\beta|^2) \\ &= -k_1|e_\alpha|^2 - k_2|e_\beta|^2 - (\mu - \psi_1 - \epsilon)\frac{|e_b|^2}{r} - 2(\psi_1 - \epsilon_1)(r-1)^2.\end{aligned}$$

Choosing $k_1, k_2 > 0$, $\psi_1 > \epsilon_1$, and l_α, l_β large enough so that $\mu > \psi_1 + \epsilon$ clearly guarantees the uniform global exponential stability of the equilibrium point $(\bar{e}_\alpha, \bar{e}_m, \frac{\bar{e}_b}{\sqrt{\bar{r}}}, \bar{r}) := (0, 0, 0, 1)$, hence of $(\bar{e}_\alpha, \bar{e}_m, \bar{e}_b, \bar{r}) := (0, 0, 0, 1)$. □

Remark 2. More than two vectors α and β can be used with a direct generalization of the proposed structure.

Remark 3. The observer does not use the knowledge of the constant vectors α_i and β_i. This may be an interesting feature in some applications when those vectors for example are not precisely known and/or (slowly) vary.

We then have the following corollary, which gives an estimate of the true orientation matrix R by using the knowledge of the inertial vectors α_i and β_i. Notice it is considerably simpler than the approach of [3], where the estimated orientation matrix is obtained through an additional observer of dimension 9.

Corollary 1. *Under the assumptions of Theorem 1, the matrix $\tilde{R}$ defined by*

$$\tilde{R}^T := \begin{pmatrix} \frac{\hat{\alpha}}{|\alpha_i|} & \frac{\hat{\alpha}\times\hat{\beta}}{|\alpha_i\times\beta_i|} & \frac{\hat{\alpha}\times(\hat{\alpha}\times\hat{\beta})}{|\alpha_i\times(\alpha_i\times\beta_i)|} \end{pmatrix} \cdot R_i^T$$

$$R_i := \begin{pmatrix} \frac{\alpha_i}{|\alpha_i|} & \frac{\alpha_i\times\beta_i}{|\alpha_i\times\beta_i|} & \frac{\alpha_i\times(\alpha_i\times\beta_i)}{|\alpha_i\times(\alpha_i\times\beta_i)|} \end{pmatrix}$$

exponentially converges to R.

Proof. By Theorem 1, $|e_\alpha(t)| \le C|e_\alpha(0)|e^{-\lambda t}$ and $|e_\beta(t)| \le C|e_\beta(0)|e^{-\lambda t}$ for some $C, \lambda > 0$. Therefore,

$$\begin{aligned} |\hat{\alpha}\times\hat{\beta} - \alpha\times\beta| &= |\alpha\times e_\beta + e_\alpha\times\beta + e_\alpha\times e_\beta| \\ &\le |\alpha||e_\beta| + |\beta||e_\alpha| + |e_\alpha||e_\beta| \\ &\le C|\alpha_i||e_\beta(0)|e^{-\lambda t} + C|\beta_i||e_\alpha(0)|e^{-\lambda t} + C^2|e_\alpha(0)||e_\beta(0)|e^{-\lambda 2t}; \end{aligned}$$

a similar bound is readily obtained for $|\hat{\alpha}\times(\hat{\alpha}\times\hat{\beta}) - \alpha\times(\alpha\times\beta)|$. As a consequence, all the coefficients of the matrix

$$\tilde{R}^T - \begin{pmatrix} \frac{\alpha}{|\alpha_i|} & \frac{\alpha\times\beta}{|\alpha_i\times\beta_i|} & \frac{\alpha\times(\alpha\times\beta)}{|\alpha_i\times(\alpha_i\times\beta_i)|} \end{pmatrix} \cdot R_i^T$$

exponentially converge to 0. The claim follows by noticing

$$\begin{aligned} \begin{pmatrix} \frac{\alpha}{|\alpha_i|} & \frac{\alpha\times\beta}{|\alpha_i\times\beta_i|} & \frac{\alpha\times(\alpha\times\beta)}{|\alpha_i\times(\alpha_i\times\beta_i)|} \end{pmatrix} \cdot R_i^T &= \begin{pmatrix} \frac{R^T\alpha_i}{|\alpha_i|} & \frac{R^T\alpha_i\times R^T\beta_i}{|\alpha_i\times\beta_i|} & \frac{R^T\alpha_i\times(R^T\alpha_i\times R^T\beta_i)}{|\alpha_i\times(\alpha_i\times\beta_i)|} \end{pmatrix} \cdot R_i^T \\ &= R^T R_i R_i^T \\ &= R^T, \end{aligned}$$

where we have used $R^T(u\times v) = R^T u \times R^T v$ since R is a rotation matrix. □

Of course, $\tilde{R}^T$ has no reason to be a rotation matrix (it is only asymptotically so); it is nevertheless the product of a matrix with orthogonal (possibly zero) columns by a rotation matrix. If a bona fide rotation matrix is required at all times, a natural idea is to project $\tilde{R}$ onto the "closest" rotation matrix $\hat{R}$, thanks to a polar decomposition.

References

1. Astolfi, A., Karagiannis, D., Ortega, R.: Nonlinear and Adaptive Control with Applications. Springer, London (2008)
2. Batista, P., Silvestre, C., Oliveira, P.: A GES attitude observer with single vector observations. Automatica **48**(2), 388–395 (2012)
3. Batista, P., Silvestre, C., Oliveira, P.: Globally exponentially stable cascade observers for attitude estimation. Control Eng. Pract. **20**(2), 148–155 (2012)
4. Benziane, L., Benallegue, A., Chitour, Y., Tayebi, A.: Velocity-free attitude stabilization with inertial vector measurements. Int. J. Robust Nonlinear Control **26**, 2478–2493 (2015)
5. Bonnabel, S., Martin, P., Rouchon, P.: Symmetry-preserving observers. IEEE Trans. Autom. Control **53**(11), 2514–2526 (2008)
6. Crassidis, J.L., Markley, F.L., Cheng, Y.: Survey of nonlinear attitude estimation methods. J. Guidance Control Dyn. **30**(1), 12–28 (2007)
7. Grip, H.F., Fossen, T.I., Johansen, T.A., Saberi, A.: Attitude estimation using biased gyro and vector measurements with time-varying reference vectors. IEEE Trans. Autom. Control **57**(5), 1332–1338 (2012)
8. Grip, H.F., Fossen, T.I., Johansen, T.A., Saberi, A.: Globally exponentially stable attitude and gyro bias estimation with application to GNSS/INS integration. Automatica **51**, 158–166 (2015)
9. Karagiannis, D., Sassano, M., Astolfi, A.: Dynamic scaling and observer design with application to adaptive control. Automatica **45**(12), 2883–2889 (2009)
10. Mahony, R., Hamel, T., Pflimlin, J.M.: Nonlinear complementary filters on the special orthogonal group. IEEE Trans. Autom. Control **53**(5), 1203–1218 (2008)
11. Martin, P., Salaün, E.: Invariant observers for attitude and heading estimation from low-cost inertial and magnetic sensors. In: IEEE Conference on Decision and Control, pp. 1039–1045 (2007)
12. Martin, P., Salaün, E.: Design and implementation of a low-cost observer-based attitude and heading reference system. Control Eng. Pract. **18**(7), 712–722 (2010)
13. Martin, P., Salaün, E.: The true role of accelerometer feedback in quadrotor control. In: IEEE International Conference on Robotics and Automation, pp. 1623–1629 (2010)
14. Martin, P., Sarras, I.: A global attitude and gyro bias observer from vector measurements. IFAC-PapersOnLine, vol. 50, IFAC 2017 World Congress (2014)
15. Tayebi, A., Roberts, A., Benallegue, A.: Inertial vector measurements based velocity-free attitude stabilization. IEEE Trans. Autom. Control **58**(11), 2893–2898 (2013)
16. Vasconcelos, J.F., Silvestre, C., Oliveira, P.: A nonlinear observer for rigid body attitude estimation using vector observations. IFAC Proc. Vol. **41**(2), 8599–8604 (2008)
17. Zamani, M., Trumpf, J., Mahony, R., Filtering, N.A.: A comparison study. ArXiv e-prints, arXiv:1502.03990 [cs.SY] (2015)

Geometric Mechanics and Robotics

Geometric Degree of Non Conservativeness

Jean Lerbet[1(✉)], Noël Challamel[2], François Nicot[3], and Félix Darve[4]

[1] IBISC Laboratory, Univ Evry, Paris-Saclay University, Évry, France
jlerbet@gmail.com
[2] LIMATB, South Brittain University, Lorient, France
[3] IRSTEA, Grenoble, France
[4] 3SR, Grenoble Alpes University, Grenoble, France

Abstract. Symplectic structure is powerful especially when it is applied to Hamiltonian systems. We show here how this symplectic structure may define and evaluate an integer index that measures the defect for the system to be Hamiltonian. This defect is called the Geometric Degree of Non Conservativeness of the system. Darboux theorem on differential forms is the key result. Linear and non linear frameworks are investigated.

Keywords: Hamiltonian system · Symplectic geometry · Geometric Degree of Non Conservativeness · Kinematic constraints

1 Position of the Problem

Beyond the hamiltonian framework where external actions (like gravity) and internal actions (like in elasticity) may be described by a potential function, we are concerned here by mechanical systems whose actions are positional but without potential. For external actions, this is the case for example of the so-called follower forces ([1] for example). For internal actions, this is the case of the so called hypoelasticity ([8] for example). One main characteristic of these questions is the loss of symmetry of the stiffness matrix $K(p)$ in the investigated equilibrium configuration and for the load parameter p.

For such systems, the stability issue presents some interesting paradoxical properties. For example, a divergence stable equilibrium configuration can become unstable as the system is subjected to appropriate additional kinematic constraints (see [4,7] for example). This problem and the associate Kinematic Structural Stability concept have been deeply investigated for some years mainly in the linear framework ([3,4] for example). In the present work, we are concerned by the dual question: for such a non conservative system Σ, what is the minimal number of additional kinematic constraints that transform the non conservative system into a conservative one? This minimal number d is called the geometric degree of nonconservativeness (GDNC). The second issue consists in finding the set of appropriate constraints. This issue will be tackled in the framework of discrete mechanics. More precisely, the set of configurations is a n-dimensional manifold $\mathbb{M}$ and the non hamiltonian actions are described by a section ω of

F. Nielsen and F. Barbaresco (Eds.): GSI 2017, LNCS 10589, pp. 355–358, 2017.
https://doi.org/10.1007/978-3-319-68445-1_41

the cotangent bundle $T^*\mathbb{M}$. This one form ω is supposed to be a non closed one form: $\mathbf{d}\omega \neq 0$ where $\mathbf{d}$ is the usual exterior derivative of differential forms. With the differential geometry concepts, the geometric meaning of the GDNC issue is: What is the highest dimension $n-d$ of embedded submanifolds $\mathbb{N}$ of $\mathbb{M}$ such that the “restriction” (in a well defined meaning) $\omega_{\mathbb{N}}$ to $\mathbb{N}$ is closed. We do not tackled in this work the very difficult global issues on $\mathbb{N}$ and, by Poincaré's theorem, the closed form $\omega_{\mathbb{N}}$ will be locally exact.

2 Solution

2.1 Linear Framework

In this subsection, we are concerned by the linearized version of the general problem. A configuration $m_e \in \mathbb{M}$ (we can think to m_e as an equilibrium position) is fixed and a coordinate system $q = (q_1, \ldots, q_n)$ is given. We are looking for solution of the linear GDNC issue at m_e. There is here a real geometric issue to build the linearized counterpart of ω at m_e because it should be obtained by derivative of ω. But there is no connection on $\mathbb{M}$ to make the derivative of ω. We will come back to this problem in the last part. However, as usual, in a coordinate system q, the linearized counterpart of ω is the so-called stiffness matrix $K = K(q_e)$ of the system at m_e whose coordinate system is q_e.

In this framework, the issue is pulled back on the tangent space $T_{m_e}\mathbb{M}$ which will be identified with $\mathbb{R}^n$ thanks to the natural basis of $T_{m_e}\mathbb{M}$ associated with the coordinate system q on $\mathbb{M}$. We indifferently note $E = \mathbb{R}^n$ and E^* its dual space, the vector space of the linear forms on E. Thus, let ϕ the exterior 2-form defined on $E = \mathbb{R}^n$ by:

$$\phi(u, v) = u^T K_a v \tag{1}$$

where K_a is the skew-symmetric part of K. Usual linear algebra says that there is a basis $\mathcal{B} = (e_1, \ldots, e_n)$ of $\mathbb{R}^n$ and a number $r = 2s \leq n$ such that $\phi(e_{2i-1}, e_{2i}) = -\phi(e_{2i}, e_{2i-1}) = 1$ for $i \leq s$ and $\phi(e_i, e_j) = 0$ for the other values of i and j. In the dual basis $(e_1^*, \ldots, e_n^*)$ of $(e_1, \ldots, e_n)$, the form ϕ then reads:

$$\phi = e_1^* \wedge e_2^* + \ldots + e_{2s-1}^* \wedge e_{2s}^* \tag{2}$$

The solution of the linear GDNC issue at m_e is then given by the following:

Proposition 1. *$d = s$ is the GDNC of the mechanical system Σ and a possible set $\mathcal{C} = \{C_1, \ldots, C_s\}$ of linear kinematic constraints making the constrained system $\Sigma_{\mathcal{C}}$ conservative is such that C_i is any in $< e_{2i-1}^*, e_{2i}^* >$ for $i = 1, \ldots, s$.*

In this framework, it is possible to find the set of all such possible constraints. Let then F be the kernel of ϕ. Then $(\mathbb{R}^n/F, \tilde{\phi})$ is a $2s$-dimensional symplectic vector space where $\tilde{\phi}$ is canonically defined by $\tilde{\phi}(\bar{u}, \bar{v}) = \phi(x, y)$ with x (resp, y) any vector of the class $\bar{u}$ (resp. $\bar{v}$).

Proposition 2. *The set of solutions of the GDNC is (isomorphic with) the set of Lagrangian subspaces of $(\mathbb{R}^n/F, \tilde{\phi})$.*

One can find in [6] a concrete construction of this set and in [5] the proof of these results.

2.2 Non Linear Framework

The key of the solution in the nonlinear framework is related to Darboux theorem about the class of 1-form and 2-forms ([2] for example). We suppose now that the 2-form $\mathbf{d}\omega$ is regular on $\mathbb{M}$ meaning that its class r is constant on $\mathbb{M}$. Then here, since the form $\mathbf{d}\omega$ is itself a closed form ($\mathbf{d}^2 = 0$), its class is also equal to its rank and is even: $r = 2s$. s is the unique number such that $(\mathbf{d}\omega)^s \neq 0$ and $(\mathbf{d}\omega)^{s+1} = 0$. We then deduce that $2s \leq n$.

Darboux's theorem gives the local modeling of $\mathbf{d}\omega$ on an open set U of $\mathbb{M}$ and reads:

$$\mathbf{d}\omega = \sum_{k=1}^{s} dy^k \wedge dy^{k+s} \tag{3}$$

where $y^1, \ldots, y^{2s}$ are $2s$ independent functions on U. We then deduce the following

Proposition 3. *Suppose that the class of* $\mathbf{d}\omega$ *is constant at* $m \in \mathbb{M}$ *(namely maximal). The (non linear) GDNC of* Σ *(in a neighborhood of* $m \in \mathbb{M}$ *is then the half* s *of the class* $2s$ *of* $\mathbf{d}\omega$*. The local definition of a submanifold* $\mathbb{N}$ *solution of the problem is given by the family* $f^1 = 0, \ldots, f^s = 0$ *of equations on* $\mathbb{M}$ *where* f^i *is any linear combination (in the vector space on* $\mathbb{R}$ *and not in the modulus on the ring on the functions on* $\mathbb{R}$*) of the above* y^i *and* y^{i+s} *for all* $i = 1, \ldots, s$*.*

3 Open Issues

Two open issues are related to this GDNC issue. The first one concerns the derivative of sections in $T^*\mathbb{M}$. The dual issue is the KISS issue that involves, in a linearized version at m_e, the symmetric part $K_s(q_e)$ of the stiffness matrix $K(q_e)$. It is worth noting that the skew-symmetric aspect $K_a(q_e)$ may be extended to the nonlinear framework through the exterior derivative $\mathbf{d}\omega$ whereas no similar extension is possible for the symmetric part without specify a connection on $\mathbb{M}$. This issue is today partially solved and will be the subject of a forthcoming paper.

The second one concerns the extension to continuum mechanics and infinite dimension spaces. Regarding the dual KISS issue, it is has been performed and will be soon published in an already accepted paper. Regarding the GDNC issue, it remains an interesting challenge because the tools, involved for the finite dimensional solution, are not naturally extendable to the case of infinite dimensional (Hilbert) vector spaces.

References

1. Bolotin, V.V.: Nonconservative Problems of the Theory of Elastic Stability. Pergamon Press, London (1963)
2. Godbillon, C.: Géométrie différentielle et mécanique analytique. Herman, Paris (1969)

3. Lerbet, J., Challamel, N., Nicot, F., Darve, F.: Variational formulation of divergence stability for constrained systems. Appl. Math. Model. doi:10.1016/j.apm.2015.02.052
4. Lerbet, J., Challamel, N., Nicot, F., Darve, F.: Kinematical structural stability. Discr. Contin. Dyn. Syst. Ser. S (DCDS-S) **9**(2), 529–536 (2016). American Institute of Mathematical Sciences (AIMS)
5. Lerbet, J., Challamel, N., Nicot, F., Darve, F.: Geometric degree of nonconservativity: set of solutions for the linear case and extension to the differentiable non linear case. Appl. Math. Modell. (2016). doi:10.1016/j.apm.2016.01.030
6. Souriau, J.M.: Construction explicite de l'indice de Maslov. Applications. In: Janner, A., Janssen, T., Boon, M. (eds.) Group Theoretical Methods in Physics. Lecture Notes in Physics, vol. 50. Springer, Heidelberg (1976). doi:10.1007/3-540-07789-8_13
7. Thompson, J.M.T.: Paradoxical mechanics under fluid flow. Nature **296**(5853), 135–137 (1982)
8. Truesdell, C.: Hypoelasticity. J. Rationa. Mech. Anal. **4**(83–133), 1019–1020 (1955)

A Symplectic Minimum Variational Principle for Dissipative Dynamical Systems

Abdelbacet Oueslati[1], An Danh Nguyen[2], and Géry de Saxcé[1](✉)

[1] Laboratoire de Mécanique de Lille (FRE CNRS 3723), Villeneuve d'ascq, France
{abdelbacet.oueslati,gery.desaxce}@univ-lille1.fr
[2] Institute of General Mechanics, RWTH, Aachen, Germany
andanh@iam.rwth-aachen.de

Abstract. Using the concept of symplectic subdifferential, we propose a modification of the Hamiltonian formalism which can be used for dissipative systems. The formalism is first illustrated through an application of the standard inelasticity in small strains. Some hints concerning possible extensions to non-standard plasticity and finite strains are then given. Finally, we show also how the dissipative transition between macrostates can be viewed as an optimal transportation problem.

Keywords: Symplectic geometry · Convex analysis · Non smooth mechanics

1 Introduction

Realistic dynamical systems considered by engineers and physicists are subjected to energy loss. It may stem from external actions, in which case we call them non conservative. The behaviour of such systems can be represented by Hamilton's least action principle. If the cause is internal, resulting from a broad spectrum of phenomena such as collisions, surface friction, viscosity, plasticity, fracture, damage and so on, we name them dissipative. Hamilton's variational principle failing for such systems, we want to propose another one for them.

Classical dynamics is generally addressed through the world of smooth functions while the mechanics of dissipative systems deals with the one of non smooth functions. Unfortunately, both worlds widely ignore each other. Our aim is laying strong foundations to link both worlds and their corresponding methods.

2 Non Dissipative Systems

Let us consider a dynamical system, which is described by $z = (x, y) \in X \times Y$, where the primal variables x describe the body motion and the dual ones y are the corresponding momenta, both assembled in vectors. X and Y are topological, locally convex, real vector spaces. There is a dual pairing:

$$\langle \cdot, \cdot \rangle : X \times Y \to \mathbb{R}$$

F. Nielsen and F. Barbaresco (Eds.): GSI 2017, LNCS 10589, pp. 359–366, 2017.
https://doi.org/10.1007/978-3-319-68445-1_42

which makes continuous the linear forms $x \mapsto \langle x, y \rangle$ and $y \mapsto \langle x, y \rangle$. The space $X \times Y$ has a natural symplectic form $\omega : (X \times Y)^2 \to \mathbb{R}$ defined by:

$$\omega(z, z') = \langle x, y' \rangle - \langle x', y \rangle$$

For any smooth hamiltonian function $(x, t) \mapsto H(x, t)$, we define the symplectic gradient (or Hamiltonian vector field) by:

$$\dot{z} = XH \quad \Leftrightarrow \quad \forall\, \delta z, \qquad \omega(\dot{z}, \delta z) = \delta H$$

In the particular case $X = Y$, the dual pairing is a scalar product and the space $X \times Y$ is dual with itself, with the duality product:

$$\langle\langle (x, y), (x', y') \rangle\rangle = \langle x, x' \rangle + \langle y, y' \rangle$$

Introducing the linear map $J(x, y) = (-y, x)$ and putting:

$$\omega(z, z') = \langle\langle J(z), z' \rangle\rangle$$

we have $J\,(XH) = D_z H$ that allows to recover the canonical equations governing the motion:

$$\dot{x} = D_y H, \qquad \dot{y} = -D_x H \tag{1}$$

Notice that J makes no sense in the general case when $X \neq Y$.

3 Dissipative Systems

For such systems, the cornerstone hypothesis is to decompose the velocity in the phase space into reversible and irreversible parts:

$$\dot{z} = \dot{z}_R + \dot{z}_I, \quad \dot{z}_R = X\,H, \quad \dot{z}_I = \dot{z} - X\,H$$

the idea being that for a non dissipative system, the irreversible part vanishes and the motion is governed by the canonical equations. Now, it is a crucial turning-point. We will be confronting the tools of the differential geometry to the ones of the non smooth mechanics. We start with a dissipation potential ϕ. It is not differentiable everywhere but convex and lower semicontinuous. This weakened properties allow to model set-valued constitutive laws –currently met in mechanics of disipative materials– through the concept of subdifferential, a set of generalized derivatives (a typical example, the plasticity, will be given at the end of Sect. 5).

We introduce a new subdifferential, called symplectic [2]. Mere sleight of hand: all we have to do is to replace the dual pairing by the symplectic form in the classical definition;

$$\dot{z}_I \in \partial^\omega \phi(\dot{z}) \qquad \Leftrightarrow \qquad \forall \dot{z}', \quad \phi(\dot{z} + \dot{z}') - \phi(\dot{z}) \geq \omega(\dot{z}_I, \dot{z}') \tag{2}$$

From a mechanical viewpoint, it is the constitutive law of the material. Likewise, we define a symplectic conjugate function, by the same sleight of hand in the definition of the Legendre-Fenchel transform:

$$\phi^{*\omega}(\dot{z}_I) = \sup_{\dot{z}} \{\omega(\dot{z}_I, \dot{z}) - \phi(\dot{z})\} \tag{3}$$

satisfying a symplectic Fenchel inequality:

$$\phi(\dot{z}) + \phi^{*\omega}(\dot{z}_I) - \omega(\dot{z}_I, \dot{z}) \geq 0 \tag{4}$$

The equality is reached in the previous relation if and only if the constitutive law (2) is satisfied.

Remarks. Always in the case $X = Y$ where J makes sense, the subdifferential is defined by:

$$\dot{z}_I \in \partial\phi(\dot{z}) \qquad \Leftrightarrow \qquad \forall \dot{z}', \quad \phi(\dot{z} + \dot{z}') - \phi(\dot{z}) \geq \langle\langle \dot{z}_I, \dot{z}' \rangle\rangle$$

Comparing to the definition (2) of the symplectic subdifferential, one has:

$$\dot{z}_I \in \partial^\omega \phi(\dot{z}) \Leftrightarrow J(\dot{z}_I) \in \partial\phi(\dot{z})$$

Recalling the definition of the conjugate function:

$$\phi^*(\dot{z}_I) = \sup_{\dot{z}} \{\langle\langle \dot{z}_I, \dot{z} \rangle\rangle - \phi(\dot{z})\}$$

and comparing to (3), we obtain $\phi^{*\omega}(\dot{z}) = \phi^*(J(\dot{z}))$. Moreover, an interesting fact is that, taking into account the antisymmetry of ω:

$$\langle\langle D_z H, \dot{z} \rangle\rangle = \langle\langle J\,(X\,H), \dot{z} \rangle\rangle = \omega(X\,H, \dot{z}) = \omega(\dot{z}, \dot{z} - X\,H) = \omega(\dot{z}, \dot{z}_I)$$

If we suppose that for all couples $(\dot{z}, \dot{z}_I)$:

$$\phi(\dot{z}) + \phi^{*\omega}(\dot{z}_I) \geq 0$$

the system dissipates for the couples satisfying the constitutive law:

$$\langle\langle D_z H, \dot{z} \rangle\rangle = -\omega(\dot{z}_I, \dot{z}) = -(\phi(z, \dot{z}) + \phi^{*\omega}(z, \dot{z}_I)) \leq 0$$

4 The Symplectic Brezis-Ekeland-Nayroles Principle

The variational formulation can be obtained by integrating the left hand member of (4) on the system evolution. On this ground, we proposed in [3] a symplectic version of the Brezis-Ekeland-Nayroles variational principle:

The natural evolution curve $z : [t_0, t_1] \to X \times Y$ minimizes the functional:

$$\Pi(z) := \int_{t_0}^{t_1} [\phi(\dot{z}) + \phi^{*\omega}(\dot{z} - XH) - \omega(\dot{z} - XH, \dot{z})] \;\mathrm{dt}$$

among all the curves verifying the initial conditions $z(t_0) = z_0$ *and, remarkably, the minimum is zero.*

Observing that $\omega(\dot{z}, \dot{z})$ vanishes and integrating by part, we have also the variant (which is not compulsory):

$$\Pi(z) = \int_{t_0}^{t_1} [\phi(\dot{z}) + \phi^{*\omega}(\dot{z} - XH) - \frac{\partial H}{\partial t}(t, z)] \, \mathrm{d}t + H(t_1, z(t_1)) - H(0, z_0)$$

5 Application to the Standard Plasticity and Viscoplasticity

To illustrate the general formalism and to show how it allows to develop powerful variational principles for dissipative systems within the frame of continuum mechanics, we consider the standard plasticity and viscoplasticity in small deformations based on the additive decomposition of strains into reversible and irreversible strains:

$$\boldsymbol{\varepsilon} = \boldsymbol{\varepsilon}_R + \boldsymbol{\varepsilon}_I$$

where $\boldsymbol{\varepsilon}_I$ is the plastic strain. Let $\Omega \subset \mathbb{R}^n$ be a bounded, open set, with piecewise smooth boundary $\partial\Omega$. As usual, it is divided into two disjoint parts, $\partial\Omega_0$ (called support) where the displacements are imposed and $\partial\Omega_1$ where the surface forces are imposed. The elements of the space X are fields $x = (\boldsymbol{u}, \boldsymbol{\varepsilon}_I) \in U \times E$ where $\boldsymbol{\varepsilon}_I$ is the irreversible strain field and $\boldsymbol{u}$ is a displacement field on the body Ω with trace $\bar{\boldsymbol{u}}$ on $\partial\Omega$. The elements of the corresponding dual space Y are of the form $y = (\boldsymbol{p}, \boldsymbol{\pi})$. Unlike $\boldsymbol{p}$ which is clearly the linear momentum, we do not know at this stage the physical meaning of $\boldsymbol{\pi}$.

The duality between the spaces X and Y has the form:

$$\langle x, y \rangle = \int_{\Omega} (\langle \boldsymbol{u}, \boldsymbol{p} \rangle + \langle \boldsymbol{\varepsilon}_I, \boldsymbol{\pi} \rangle)$$

where the duality products which appear in the integral are finite dimensional duality products on the image of the fields $\boldsymbol{u}, \boldsymbol{p}$ (for our example this means a scalar product on $\mathbb{R}^3$) and on the image of the fields $\boldsymbol{\varepsilon}, \boldsymbol{\pi}$ (in this case this is a scalar product on the space of 3 by 3 symmetric matrices). We denote all these standard dualities by the same $\langle \cdot, \cdot \rangle$ symbols.

The total Hamiltonian of the structure is taken of the integral form:

$$H(t, z) = \int_{\Omega} \left\{ \frac{1}{2\rho} \parallel \boldsymbol{p} \parallel^2 + w(\nabla \boldsymbol{u} - \boldsymbol{\varepsilon}_I) - \boldsymbol{f}(t) \cdot \boldsymbol{u} \right\} - \int_{\partial\Omega_1} \bar{\boldsymbol{f}}(t) \cdot \boldsymbol{u}$$

The first term is the kinetic energy, w is the elastic strain energy, $\boldsymbol{f}$ is the volume force and $\bar{\boldsymbol{f}}$ is the surface force on the part $\partial\Omega_1$ of the boundary, the displacement field being equal to an imposed value $\bar{\boldsymbol{u}}$ on the remaining part $\partial\Omega_0$.

According to (1), its symplectic gradient is:

$$XH = ((D_{\boldsymbol{p}} H, D_{\boldsymbol{\pi}} H), (-D_{\boldsymbol{u}} H, -D_{\boldsymbol{\varepsilon}_I} H))$$

where, introducing as usual the stress field given by the elastic law:

$$\boldsymbol{\sigma} = Dw(\nabla \boldsymbol{u} - \boldsymbol{\varepsilon}_I)$$

$D_{\boldsymbol{u}} H$ is the gradient in the variational sense (from (1) and the integral form of the duality product):

$$D_{\boldsymbol{u}} H = \frac{\partial H}{\partial \boldsymbol{u}} - \nabla \cdot \left(\frac{\partial H}{\partial \nabla \boldsymbol{u}} \right) = -\boldsymbol{f} - \nabla \cdot \boldsymbol{\sigma}$$

and:

$$D_{\bar{\boldsymbol{u}}} H = \boldsymbol{\sigma} \cdot \boldsymbol{n} - \bar{\boldsymbol{f}}$$

Thus one has:

$$\dot{z}_I = \dot{z} - XH = \left(\left(\dot{\boldsymbol{u}} - \frac{\boldsymbol{p}}{\rho}, \dot{\boldsymbol{\varepsilon}}_I \right), (\dot{\boldsymbol{p}} - \boldsymbol{f} - \nabla \cdot \boldsymbol{\sigma}, \dot{\boldsymbol{\pi}} - \boldsymbol{\sigma}) \right)$$

We shall use a dissipation potential which has an integral form:

$$\Phi(z) = \int_{\Omega} \phi(\boldsymbol{p}, \boldsymbol{\pi})$$

and we shall assume that the symplectic Fenchel transform of Φ expresses as the integral of the symplectic Fenchel transform of the dissipation potential density ϕ.

The symplectic Fenchel transform of the function ϕ reads:

$$\phi^{*\omega}(\dot{z}_I) = \sup \left\{ \langle \dot{\boldsymbol{u}}_I, \dot{\boldsymbol{p}}' \rangle + \langle \dot{\boldsymbol{\varepsilon}}_I, \dot{\boldsymbol{\pi}}' \rangle - \langle \dot{\boldsymbol{u}}', \dot{\boldsymbol{p}}_I \rangle - \langle \dot{\boldsymbol{\varepsilon}}'_I, \dot{\boldsymbol{\pi}}_I \rangle - \phi(\dot{z}') : \dot{z}' \in X \times Y \right\}$$

To recover the standard plasticity, we suppose that ϕ is depending explicitly only on $\dot{\boldsymbol{\pi}}$:

$$\phi(\dot{z}) = \varphi(\dot{\boldsymbol{\pi}}) \tag{5}$$

Denoting by χ_K the indicator function of a set K (equal to 0 on K and to $+\infty$ otherwise), we obtain:

$$\phi^{*\omega}(\dot{z}_I) = \chi_{\{\boldsymbol{0}\}}(\dot{\boldsymbol{u}}_I) + \chi_{\{\boldsymbol{0}\}}(\dot{\boldsymbol{p}}_I) + \chi_{\{\boldsymbol{0}\}}(\dot{\boldsymbol{\pi}}_I) + \varphi^*(\dot{\boldsymbol{\varepsilon}}_I)$$

where φ^* is the usual Fenchel transform. In other words, the quantity $\phi^{*\omega}(\dot{z}_I)$ is finite and equal to:

$$\phi^{*\omega}(\dot{z}_I) = \varphi^*(\dot{\boldsymbol{\varepsilon}}_I)$$

if and only if all of the following are true:

(a) $\boldsymbol{p}$ equals the linear momentum

$$\boldsymbol{p} = \rho \dot{\boldsymbol{u}} \tag{6}$$

(b) the balance of linear momentum is satisfied

$$\nabla \cdot \boldsymbol{\sigma} + \boldsymbol{f} = \dot{\boldsymbol{p}} = \rho \ddot{\boldsymbol{u}} \quad \text{on} \quad \Omega, \qquad \boldsymbol{\sigma} \cdot \boldsymbol{n} = \bar{\boldsymbol{f}} \quad \text{on} \quad \partial\Omega_1 \tag{7}$$

(c) and an equality which reveals the meaning of the variable $\boldsymbol{\pi}$:

$$\dot{\boldsymbol{\pi}} = \boldsymbol{\sigma}\,. \tag{8}$$

Eliminating $\dot{\boldsymbol{\pi}}$ by (8), the symplectic Brezis-Ekeland-Nayroles principle applied to standard plasticity states that the evolution curve minimizes:

$$\Pi(z) = \int_{t_0}^{t_1} \left\{ \varphi(\boldsymbol{\sigma}) + \varphi^*(\dot{\boldsymbol{\varepsilon}}_I) - \frac{\partial H}{\partial t}(t, z) \right\} \, \mathrm{d}t + H(t_1, z(t_1)) - H(t_0, z_0)$$

among all curves $z : [t_0, t_1] \rightarrow X \times Y$ such that $z(0) = (x_0, y_0)$, the kinematical conditions on $\partial\Omega_0$, (6) and (7) are satisfied. For instance, in plasticity, the potential φ is the indicator function of the closed convex plastic domain K. The constitutive law $\dot{\boldsymbol{\varepsilon}}_I \in \partial\varphi(\boldsymbol{\sigma})$ reads for $\boldsymbol{\sigma} \in K$:

$$\forall \boldsymbol{\sigma}' \in K, \qquad (\boldsymbol{\sigma}' - \boldsymbol{\sigma}) : \dot{\boldsymbol{\varepsilon}}_I \leq 0 \tag{9}$$

If $\boldsymbol{\sigma}$ is an interior point of K, $\dot{\boldsymbol{\varepsilon}}_I$ vanishes. If $\boldsymbol{\sigma}$ is a boundary point of K, $\dot{\boldsymbol{\varepsilon}}_I$ is a called a subnormal to K at $\boldsymbol{\sigma}$. An important case of interest is the metal plasticity governed by von Mises model for which K is defined as the section:

$$K = \{\boldsymbol{\sigma} \quad \text{such that} \quad f(\boldsymbol{\sigma}) \leq 0\}$$

where f is differentiable on the boundary of K. In this case, if $\boldsymbol{\sigma}$ is a boundary point of K, there exists $\lambda > 0$ such that:

$$\dot{\boldsymbol{\varepsilon}}_I = \lambda \, Df(\boldsymbol{\sigma})$$

that means $\dot{\boldsymbol{\varepsilon}}_I$ is an exterior normal to K. Otherwise, if $\boldsymbol{\sigma}$ does not belong to K, there is no solution to this inequation (9). In short, the previous non smooth constitutive law allows to model the following behavior: below a given stress threshold, there is no plastic deformation then no dissipation; at the threshold, plastic yielding and dissipation occur; over the threshold, no stress state may be reached.

Remark. The assumption that $\boldsymbol{u}, \boldsymbol{\varepsilon}_I$ and $\boldsymbol{p}$ are ignorable in (5) comes down to introduce into the dynamical formalism a "statical" constitutive law:

$$\dot{\boldsymbol{\varepsilon}}_I \in \partial\varphi(\dot{\boldsymbol{\pi}}) = \partial\varphi(\boldsymbol{\sigma})$$

Conversely, the symplectic framework suggests to imagine fully *"dynamical" constitutive laws* of the more general form:

$$(\dot{\boldsymbol{u}}, \dot{\boldsymbol{\varepsilon}}_I) \in \partial^{\omega}\phi\,(\dot{\boldsymbol{p}}, \dot{\boldsymbol{\pi}})$$

6 Extensions to Non Standard Plasticity and Finite Strains

In plasticity and more generally in the mechanics of dissipative materials, some of the constitutive laws are non-associated, *i.e.* cannot be represented by a dissipation potential. A response proposed first in [5] is to introduce a bipotential.

The applications to solid Mechanics are various: Coulomb's friction law, non-associated Drucker-Prager and Cam-Clay models in Soil Mechanics, cyclic Plasticity and Viscoplasticity of metals with the non linear kinematical hardening rule, Lemaitre's damage law (for more details, see reference [4]).

For such constitutive laws, the principle can be easily generalized replacing $\phi(\dot{z}) + \phi^{*\omega}(\dot{z}')$ by a symplectic bipotential $b(\dot{z}, \dot{z}')$:

- separatly convex and lower semicontinuous with respect to $\dot{z}, \dot{z}'$,
- satisfying a cornerstone inequality:

$$\forall \dot{z}, \dot{z}', \quad b(\dot{z}, \dot{z}') \geq \omega(\dot{z}, \dot{z}')$$

extending the symplectic Fenchel inequality (4),

leading to generalize the functional of the symplectic BEN principle:

$$\Pi(z) := \int_{t_0}^{t_1} \left[b(\dot{z}, \dot{z}_I) - \frac{\partial H}{\partial t}(t, z) \right] \, \mathrm{d}t + H(t_1, z(t_1)) - H(t_0, z_0)$$

For the extension to finite strains, we may modify the original framework by working on the tangent bundle and making for instance $\phi = \phi(z, \dot{z})$ [3]. Then the goal is reached in three steps. Firstly, we develop a Lagrangian formalism for the reversible media based on the calculus of variation by jet theory. Next, we propose a corresponding Hamiltonian formalism. Finally, we deduce from it a symplectic minimum principle for dissipative media. This allows, among other things, to get a minimum principle for unstationnary Navier-Stokes models.

7 Dissipative Transition Between Macrostates as an Optimal Transportation Problem

We would like to model a dissipative transition between the macrostates at $t = t_k$ $(k = 0, 1)$. Now, X and Y are separable metric spaces. $X \times Y$ is viewed a the space of microstates z_k with Gibbs probability measure μ_k at $t = t_k$ of density:

$$\mu_k = e^{-(\zeta_k + \beta_k H(z_k))}$$

where β_k is the reciprocal temperature and ζ_k is Planck's (or Massieu's) potential. Let us consider the set of curves from z_0 to z_1:

$$Z(z_0, z_1) = \{ z : [t_0, t_1] \rightarrow X \times Y \quad \text{s.t.} \quad z(t_0) = z_0 \quad \text{and} \quad z(t_1) = z_1 \}$$

We adopt as cost function:

$$c(z_0, z_1) = \inf_{z \in Z(z_0, z_1)} \Pi(z)$$

and suppose it is measurable. It is worth to remark that it is not generally zero. For a measurable map $T : X \times Y \rightarrow X \times Y$, $T(\mu_0)$ denotes the push forward of μ_0 such that for all Borel set B:

$$T(\mu_0)(B) = \mu_0(B)$$

Inspiring from Monge's formulation of the optimal transportation problem, we claim that:

Among all the transport maps T *such that* $\mu_1 = T(\mu_0)$, *the natural one minimizes the functional:*

$$C_{\mu_0}(T) := \int_{X\times Y} c(z_0, T(z_0))\, d\mu_0(z_0)$$

Following Kantorovich's formulation, we consider the set $\Gamma(\mu_0, \mu_1)$ of all probability measures γ on $X\times Y$ with marginals μ_0 on the first factor $X\times Y$ and μ_1 on the second factor $X\times Y$, *i.e.* for all Borel set B on $X\times Y$, $\gamma(B\times(X\times Y)) = \mu_0(B)$ and $\gamma((X\times Y)\times B) = \mu_1(B)$. Hence, we claim that:

The natural probability measure γ *minimizes the functional:*

$$C(\gamma) := \int_{(X\times Y)^2} c(z_0, z_1)\, d\gamma(z_0, z_1)$$

within the set $\Gamma(\mu_0, \mu_1)$.

The advantage of the new formulation is that the latter problem is linear with respect γ while the former one is non linear with respect to T.

The symplectic Wasserstein distance-like function is then defined as the optimal value of the cost C:

$$W_s(\mu_0, \mu_1) := \inf_{\gamma\in\Gamma(\mu_0,\mu_1)} C(\gamma)$$

Aknowledgement. This work was performed thanks to the international cooperation project *Dissipative Dynamical Systems by Geometrical and Variational Methods and Application to Viscoplastic Structures Subjected to Shock Waves* (DDGV) supported by the *Agence Nationale de la Recherche* (ANR) and the *Deutsche Forchungsgemeinschaft* (DFG).

References

1. Brezis, H., Ekeland, I.: Un principe variationnel associé à certaines équations paraboliques. I. Le cas indépendant du temps. C. R. Acad. Sci. Paris Sér. A-B **282**, 971–974 (1976)
2. Buliga, M.: Hamiltonian inclusions with convex dissipation with a view towards applications. Math. Appl. **1**(2), 228–251 (2009)
3. Buliga, M., de Saxcé, G.: A symplectic Brezis-Ekeland-Nayroles principle. Math. Mech. Solid, 1–15 (2016). doi:10.1177/1081286516629532
4. Dang Van, K., de Saxcé, G., Maier, G., Polizzotto, C., Ponter, A., Siemaszko, A., Weichert, D.: In: Weichert, D., Maier, G. (eds.) Inelastic Behaviour of Structures under Variable Repeated Loads. CISM International Centre for Mechanical Sciences, Courses and Lectures, vol. 432. Springer, Vienna (2002)
5. de Saxcé, G., Feng, Z.Q.: New inequation and functional for contact with friction: the implicit standard material approach. Int. J. Mech. Struct. Mach. **19**(3), 301–325 (1991)
6. Nayroles, B.: Deux théorèmes de minimum pour certains systèmes dissipatifs. C. R. Acad. Sci. Paris Sér. A-B **282**, A1035–A1038 (1976)

Torsional Newton-Cartan Geometry

Eric Bergshoeff[1(✉)], Athanasios Chatzistavrakidis[1,2], Luca Romano[1], and Jan Rosseel[3]

[1] Van Swinderen Institute for Particle Physics and Gravity, Nijenborgh 4, 9747 Groningen, AG, The Netherlands
E.A.Bergshoeff@rug.nl, a.chatzistavrakidis@gmail.com, lucaromano2607@gmail.com
[2] Division of Theoretical Physics, Rudjer Bošković Institute, Bijenička 54, 10000 Zagreb, Croatia
[3] Faculty of Physics, University of Vienna, Boltzmanngasse 5, 1090 Vienna, Austria
rosseelj@gmail.com

Abstract. Having in mind applications to Condensed Matter Physics, we perform a null-reduction of General Relativity in $d+1$ spacetime dimensions thereby obtaining an extension to arbitrary torsion of the twistless-torsional Newton-Cartan geometry. We shortly discuss the implementation of the equations of motion.

1 Introduction

Usually, when discussing Newton-Cartan gravity, one defines an absolute time by imposing that the curl of the time-like Vierbein τ_μ vanishes.[1] This condition, sometimes called the zero torsion condition, allows one to solve for τ_μ in terms of a single scalar field $\tau(x)$:

$$\partial_\mu \tau_\nu - \partial_\nu \tau_\mu = 0 \qquad \Rightarrow \qquad \tau_\mu = \partial_\mu \tau. \tag{1}$$

Choosing for this function the time t, i.e. $\tau(x) = t$, defines the absolute time direction:

$$\tau(x) = t \qquad \Rightarrow \qquad \tau_\mu = \delta^0_\mu. \tag{2}$$

The zero-torsion condition (1) is sufficient but not required to obtain a causal behaviour of the theory. A more general condition, that guarantees a time flow orthogonal to Riemannian spacelike leaves, is the so-called hypersurface orthogonality condition:

$$\tau_{ab} \equiv e_a{}^\mu e_b{}^\nu \tau_{\mu\nu} = 0, \qquad\qquad \tau_{\mu\nu} = \partial_{[\mu}\tau_{\nu]}. \tag{3}$$

[1] Most of this presentation applies to any spacetime dimension. We will therefore from now on use the word Vielbein instead of Vierbein and take $\mu = 0, 1, \cdots d-1$.

F. Nielsen and F. Barbaresco (Eds.): GSI 2017, LNCS 10589, pp. 367–374, 2017.
https://doi.org/10.1007/978-3-319-68445-1_43

Here $e_a{}^\mu$, together with τ^μ, are the projective inverses of the spatial and timelike Vielbeine $e_\mu{}^a$ and τ_μ, respectively, with $a = 1, 2, \cdots d-1$. They are defined by the following projective invertibility relations:

$$e^\mu{}_a e_\nu{}^a = \delta^\mu_\nu - \tau^\mu \tau_\nu, \qquad e^\mu{}_a e_\mu{}^b = \delta^a_b,$$

$$\tau^\mu \tau_\mu = 1, \qquad e^\mu{}_a \tau_\mu = 0, \qquad \tau^\mu e_\mu{}^a = 0. \tag{4}$$

The condition (3), also called the twistless torsional condition, was encountered in the context of Lifshitz holography when studying the coupling of Newton-Cartan gravity to the Conformal Field Theory (CFT) at the boundary of spacetime [1]. Twistless-torsional Newton-Cartan geometry has also been applied in a study of the Quantum Hall Effect [2]. It not surprising that the twistless-torsional condition (3) was found in the context of a CFT. The stronger condition (1) simply does not fit within a CFT since it is not invariant under spacetime-dependent dilations $\delta\tau_\mu \sim \Lambda_D(x)\tau_\mu$. Instead, the condition (3) is invariant under spacetime-dependent dilatations due to the relation $e_a{}^\mu \tau_\mu = 0$, see Eq. (4).

One can define a dilatation-covariant torsion as

$$\tau^C_{\mu\nu} \equiv \partial_{[\mu}\tau_{\nu]} - 2b_{[\mu}\tau_{\nu]} = 0, \tag{5}$$

where b_μ transforms as the gauge field of dilatations, i.e. $\delta b_\mu = \partial_\mu \Lambda_D$. Since $\tau^\mu\tau_\mu = 1$ one can use the space-time projection of the equation $\tau^C_{\mu\nu} = 0$ to solve for the spatial components of b_μ:

$$\tau^C_{0a} \equiv \tau^\mu e_a{}^\nu \tau^C_{\mu\nu} = 0 \qquad \Rightarrow \qquad b_a \equiv e_a{}^\mu b_\mu = -\tau_{0a}. \tag{6}$$

This implies that in a conformal theory only the spatial components of the conformal torsion can be non-zero:

$$\tau^C_{ab} \equiv e_a{}^\mu e_b{}^\nu \tau^C_{\mu\nu} = \tau_{ab} \neq 0. \tag{7}$$

At first sight, it seems strange to consider the case of arbitrary torsion since causality is lost in that case. However, in condensed matter applications, one often considers gravity not as a dynamical theory but as background fields that couple to the energy and momentum flux. It was pointed out a long time ago in the seminal paper by Luttinger [3] that to describe thermal transport one needs to consider an auxiliary gravitational field $\psi(x)$ that couples to the energy and is defined by

$$\tau_\mu = e^{\psi(x)}\delta_{\mu,0} \tag{8}$$

corresponding to the case of twistless torsion. Later, it was pointed out that, for describing other properties as well, one also needs to introduce the other components of τ_μ that couple to the energy current. This leads to a full unrestricted τ_μ describing arbitrary torsion [4]. For an earlier discussion on torsional Newton-Cartan geometry, see [5]. For applications of torsion in condensed matter, see [6, 7]. To avoid confusion we will reserve the word 'geometry' if we only consider the background fields and their symmetries whereas we will talk about 'gravity' if also dynamical equations of motion are valid.

In this presentation, we will construct by applying a null reduction of General Relativity, the extension of NC geometry to the case of arbitrary torsion, i.e. $\tau_{\mu\nu} \neq 0$, see Table 1. Null-reductions and Newton-Cartan geometry with or without torsion have been discussed before in [8–11].

The outline of this presentation is as follows. In the next section we will derive NC geometry with arbitrary torsion from an off-shell null reduction, meaning we do not perform a null reduction of the equations of motion, of General Relativity in $d+1$ spacetime dimensions. We point out that performing a null reduction of the equations of motion as well we obtain the equations of motion of NC gravity with *zero* torsion thereby reproducing the result of [8,9]. We comment in the Conclusions on how one could go on-shell keeping arbitrary torsion.

Table 1. Newton-Cartan geometry with torsion.

NC geometry	Geometric constraint
Arbitrary torsion	$\tau_{0a} \neq 0, \tau_{ab} \neq 0$
Twistless torsional	$\tau_{0a} \neq 0, \tau_{ab} = 0$
Zero torsion	$\tau_{0a} = 0, \tau_{ab} = 0$

2 The Null Reduction of General Relativity

One way to obtain NC geometry with arbitrary torsion is by performing a dimensional reduction of General Relativity (GR) from $d+1$ to d spacetime dimensions along a null direction [9]. We show in detail how to perform such a null reduction *off-shell*, i.e. at the level of the transformation rules only. At the end of this section, we point out that, after going on-shell, we obtain the equations of motion of NC gravity with *zero* torsion [8,9].

Our starting point is General Relativity in $d+1$ dimensions in the second order formalism, where the single independent field is the Vielbein $\hat{e}_M{}^A$. Here and in the following, hatted fields are $(d+1)$-dimensional and unhatted ones will denote d-dimensional fields after dimensional reduction. Furthermore, capital indices take $d+1$ values, with M being a curved and A a flat index. The Einstein-Hilbert action in $d+1$ spacetime dimensions is given by

$$S_{\text{GR}}^{(d+1)} = -\frac{1}{2\kappa}\int \mathrm{d}^{d+1}x\, \hat{e}\, \hat{e}^M{}_A \hat{e}^N{}_B \hat{R}_{MN}{}^{AB}\left(\hat{\omega}(\hat{e})\right), \tag{9}$$

where κ is the gravitational constant and $\hat{e}$ is the determinant of the Vielbein. The inverse Vielbein satisfies the usual relations

$$\hat{e}^M{}_A \hat{e}_M{}^B = \delta_A^B, \quad \hat{e}^M{}_A \hat{e}_N{}^A = \delta_N^M. \tag{10}$$

The spin connection is a dependent field, given in terms of the Vielbein as

$$\hat{\omega}_M{}^{BA}(\hat{e}) = 2\hat{e}^{N[A}\partial_{[M}\hat{e}_{N]}{}^{B]} - \hat{e}^{N[A}\hat{e}^{B]P}\hat{e}_{MC}\partial_N\hat{e}_P{}^C, \tag{11}$$

while the curvature tensor is given by

$$\hat{R}_{MN}{}^{AB}\,(\hat{\omega}(\hat{e})) = 2\partial_{[M}\hat{\omega}_{N]}{}^{AB} - 2\hat{\omega}_{[M}{}^{AC}\hat{\omega}_{N]C}{}^{B}. \tag{12}$$

Under infinitesimal general coordinate transformations and local Lorentz transformations, the Vielbein transforms as

$$\delta\hat{e}_M{}^A = \xi^N\partial_N\hat{e}_M{}^A + \partial_M\xi^N\hat{e}_N{}^A + \lambda^A{}_B\hat{e}_M{}^B. \tag{13}$$

In order to dimensionally reduce the transformation rules along a null direction, we assume the existence of a null Killing vector $\xi = \xi^M\partial_M$ for the metric $\hat{g}_{MN} \equiv \hat{e}_M{}^A\hat{e}_N{}^B\eta_{AB}$, i.e.

$$\mathcal{L}_\xi\hat{g}_{MN} = 0 \quad \text{and} \quad \xi^2 = 0. \tag{14}$$

Without loss of generality, we may choose adapted coordinates $x^M = \{x^\mu, v\}$, with μ taking d values, and take the Killing vector to be $\xi = \xi^v\partial_v$. Then the Killing equation implies that the metric is v-independent, i.e. $\partial_v\hat{g}_{MN} = 0$, while the null condition implies the following constraint on the metric: [2]

$$\hat{g}_{vv} = 0. \tag{15}$$

A suitable reduction Ansatz for the Vielbein should be consistent with this constraint on the metric. Such an Ansatz was discussed in [9], and we repeat it below in a formalism suited to our purposes.

First, we split the $(d+1)$-dimensional tangent space indices as $A = \{a, +, -\}$, where the index a is purely spatial and takes $d-1$ values, while $\pm$ denote null directions. Then the Minkowski metric components are $\eta_{ab} = \delta_{ab}$ and $\eta_{+-} = 1$. The reduction Ansatz is specified upon choosing the inverse Vielbein $\hat{e}^M{}_+$ to be proportional to the null Killing vector $\xi = \xi^v\partial_v$. A consistent parametrization is

$$\hat{e}^M{}_A = \begin{array}{c} \\ a \\ - \\ + \end{array}\!\!\begin{array}{c} \begin{array}{cc} \mu & \quad v \end{array} \\ \begin{pmatrix} e^\mu{}_a & e^\mu{}_a m_\mu \\ S\tau^\mu & S\tau^\mu m_\mu \\ 0 & S^{-1} \end{pmatrix} \end{array}. \tag{16}$$

The scalar S is a compensating one and will be gauge-fixed shortly.

Given the expression (16) for the inverse Vielbein, the Vielbein itself is given by

$$\hat{e}_M{}^A = \begin{array}{c} \\ \mu \\ v \end{array}\!\!\begin{array}{c} \begin{array}{ccc} a & \quad - & \quad + \end{array} \\ \begin{pmatrix} e_\mu{}^a & S^{-1}\tau_\mu & -Sm_\mu \\ 0 & 0 & S \end{pmatrix} \end{array}. \tag{17}$$

To avoid confusion, recall that the index a takes one value less than the index μ; thus the above matrices are both square although in block form this is not manifest.

[2] Due to this constraint, we are not allowed to perform the null reduction in the action but only in the transformation rules and equations of motion [9].

Note that the Ansatz (17) has two zeros. The zero in the first column, $\hat{e}_v{}^a = 0$, is due to the fact that we gauge-fixed the Lorentz transformations with parameters $\lambda^a{}_+$. On the other hand, the zero in the second column, $\hat{e}_v{}^- = 0$, is due to the existence of the null Killing vector $\xi = \xi^v \partial_v$:

$$\xi^2 = \xi^v \xi^v \hat{g}_{vv} = 0 \quad \Rightarrow \quad \hat{g}_{vv} = \hat{e}_v{}^A \hat{e}_v{}^B \eta_{AB} = 0 \quad \Rightarrow \quad \hat{e}_v{}^- = 0. \tag{18}$$

We will call $\lambda^a \equiv \lambda^a{}_-$ and $\lambda \equiv \lambda^+{}_+$.

A simple computation reveals that the invertibility relations (10), after substitution of the reduction Ansatz, precisely reproduce the projected invertibility relations (4) provided we identify $\{\tau_\mu \, e_\mu{}^a\}$ as the timelike and spatial Vielbein of NC gravity, respectively. Starting from the transformation rule (13) of the $(d+1)$-dimensional Vielbein, we derive the following transformations of the lower-dimensional fields:

$$\delta\tau_\mu = 0, \tag{19}$$
$$\delta e_\mu{}^a = \lambda^a{}_b e_\mu{}^b + S^{-1}\lambda^a \tau_\mu, \tag{20}$$
$$\delta m_\mu = -\partial_\mu \xi^v - S^{-1}\lambda_a e_\mu{}^a, \tag{21}$$
$$\delta S = \lambda S. \tag{22}$$

Next, fixing the Lorentz transformations with parameter λ by setting $S = 1$ and defining $\sigma := -\xi^v$ we obtain, for arbitrary torsion, the transformation rules

$$\begin{aligned} \delta\tau_\mu &= 0, \\ \delta e_\mu{}^a &= \lambda^a{}_b e_\mu{}^b + \lambda^a \tau_\mu, \\ \delta m_\mu &= \partial_\mu \sigma + \lambda^a e_{\mu a} \end{aligned} \tag{23}$$

of Newton-Cartan geometry in d dimensions provided we identify m_μ as the central charge gauge field associated to the central charge generator of the Bargmann algebra. Note that we have not imposed any constraint on the torsion, i.e. $\tau_{\mu\nu} = \partial_{[\mu}\tau_{\nu]} \neq 0$.

We next consider the null-reduction of the spin-connection given in Eq. (11). Inserting the Vielbein Ansatz (17) with $S = 1$ into Eq. (11) we obtain the following expressions for the torsionful spin-connections:

$$\begin{aligned} \hat{\omega}_\mu{}^{ab}(\hat{e}) &\equiv \omega_\mu{}^{ab}(\tau, e, m) = \mathring{\omega}_\mu{}^{ab}(e, \tau, m) - m_\mu \tau^{ab}, \\ \hat{\omega}_\mu{}^{a+}(\hat{e}) &\equiv \omega_\mu{}^a(\tau, e, m) = \mathring{\omega}_\mu{}^a(e, \tau, m) + m_\mu \tau_0{}^a, \end{aligned} \tag{24}$$

where $\mathring{\omega}_\mu{}^{ab}(e, \tau, m)$ and $\mathring{\omega}_\mu{}^a(e, \tau, m)$ are the torsion-free Newton-Cartan spin connections given by

$$\mathring{\omega}_\mu{}^{ab}(\tau, e, m) = e_{\mu c} e^{\rho a} e^{\sigma b} \partial_{[\rho} e_{\sigma]}{}^c - e^{\nu a} \partial_{[\mu} e_{\nu]}{}^b + e^{\nu b} \partial_{[\mu} e_{\nu]}{}^a - \tau_\mu e^{\rho a} e^{\sigma b} \partial_{[\rho} m_{\sigma]}, \tag{25}$$
$$\mathring{\omega}_\mu{}^a(\tau, e, m) = \tau^\nu \partial_{[\mu} e_{\nu]}{}^a + e_\mu{}^c e^{\rho a} \tau^\sigma \partial_{[\rho} e_{\sigma]c} + e^{\nu a} \partial_{[\mu} m_{\nu]} + \tau_\mu \tau^\rho e^{\sigma a} \partial_{[\rho} m_{\sigma]}. \tag{26}$$

Furthermore, we find that the remaining components of the spin-connections are given by

$$\hat{\omega}_v{}^{ab}(\hat{e}) = \tau^{ab}, \qquad \hat{\omega}_v{}^{a+}(\hat{e}) = -\tau_0{}^a,$$
$$\hat{\omega}_\mu{}^{a-}(\hat{e}) = -\tau_\mu \tau_0{}^a - e_\mu{}^b \tau_b{}^a, \qquad \hat{\omega}_v{}^{a-}(\hat{e}) = 0,$$
$$\hat{\omega}_\mu{}^{-+}(\hat{e}) = -e_\mu{}^b \tau_{0b}, \qquad \hat{\omega}_v{}^{-+}(\hat{e}) = 0. \tag{27}$$

At this point, we have obtained the transformation rules of the independent NC gravitational fields describing NC geometry with arbitrary torsion. Furthermore, we obtained the expressions for the dependent (torsional) spin-connections. The same method cannot be used to obtain the equations of motion of NC gravity with arbitrary torsion [8,9]. A simple argument for this will be given in the next section. Instead, it has been shown [8,9] that reducing the Einstein equations of motion leads to the torsion-less NC equations of motion. More precisely, using flat indices, the equations of motion in the $--, -a$ and ab directions yield the NC equations of motion while those in the $++, +-$ and $+a$ direction constrain the torsion to be zero [8,9]. Since the two sets of equations of motion transform to each other under Galilean boosts, it is not consistent to leave out the second set of equations of motion in the hope of obtaining NC equations of motion with arbitrary torsion. As a final result, we find the following zero torsion NC equations of motion:

$$R_{0a}(G^a) = 0, \qquad R_{c\bar{a}}(J^c{}_b) = 0, \tag{28}$$

where $R(G)$ and $R(J)$ are the covariant curvatures of Galilean boosts and spatial rotations and where in the last equation we collected two field equations into one by using an index $\bar{a} = (a, 0)$.

3 Comments

In this presentation we applied the null-reduction technique to construct the transformation rules corresponding to Newton-Cartan geometry with arbitrary torsion. The null reduction technique has the advantage that the construction is algorithmic and can easily be generalized to other geometries, such as the Schrödinger geometry, as well.

To explain why the on-shell null reduction leads to zero torsion equations of motion, it is convenient to consider the Schrödinger field theory (SFT) [3] that can be associated to the first NC equation of motion of Eq. (28) by adding compensating scalars. To this end, we introduce a complex scalar $\Psi = \varphi\, e^{i\chi}$ that transforms under general coordinate transformations, with parameter $\xi^\mu(x)$, as a scalar and under local dilatations and central charge transformations, with parameters $\lambda_{\rm D}(x)$ and $\sigma(x)$, with weight w and M, respectively:

$$\delta\Psi = \xi^\mu \partial_\mu \Psi + \big(w\lambda_{\rm D} + i{\rm M}\sigma\big)\Psi. \tag{29}$$

[3] SFTs are explained in [12]. The discussion below is partly based upon [12].

We first consider the case of NC geometry with zero torsion. Since the zero torsion NC equations of motion (28) are already invariant under central charge transformations, we consider a *real* scalar φ, i.e. $\chi = 0$. Due to the time/space components of the zero torsion condition, i.e. $\tau_{0a} = 0$, this real scalar must satisfy the constraint $\partial_a \varphi = 0$.[4] At the same time, the first NC equation of motion in Eq. (28) leads to the scalar equation of motion $\partial_0 \partial_0 \varphi = 0$. Given these constraints the second NC equation of motion in Eq. (28) does not lead to additional restrictions. Summarizing, one can show that the two equations just derived form a SFT with the correct Schrödinger symmetries:

$$\text{SFT1}: \qquad \partial_0 \partial_0 \phi = 0, \qquad \partial_a \varphi = 0. \tag{30}$$

We next consider the case of NC geometry with arbitrary torsion. In that case one lacks the second equation of (30) that followed from the torsion constraint $\tau_{0a} = 0$. The real scalar φ now satisfies only the first constraint of (30) which does not constitute a SFT. Instead, one is forced to introduce the second compensating scalar χ, together with a mass parameter M, such that ϕ and χ together form the following SFT:

$$\text{SFT2}: \qquad \partial_0 \partial_0 \varphi - \frac{2}{\text{M}} (\partial_0 \partial_a \varphi) \partial_a \chi + \frac{1}{\text{M}^2} (\partial_a \partial_b \varphi) \partial_a \chi \partial_b \chi = 0. \tag{31}$$

Since χ is the compensating scalar for central charge transformations, this implies that the extension to arbitrary torsion of the first NC equation of motion in (28) cannot be invariant under central charge transformations. Because null-reductions by construction lead to equations of motion that are invariant under central charge transformations, this explains why we did find zero torsion equations of motion in the previous section.

The easiest way to obtain the Newton-Cartan equations of motion with arbitrary torsion would be to couple the SFT2 given in Eq. (31) to Schrödinger gravity with arbitrary torsion that by itself can be obtained by a null-reduction of conformal gravity. Indeed, such a null reduction has been performed and the coupling of SFT2 to Schrödinger gravity can be constructed in three spacetime dimensions [13]. The extension to higher dimensions remains an open question.

As a closing remark, it would be interesting to apply the null reduction technique to supergravity theories. The case of $d = 3$ should lead to a generalization of the off-shell 3d NC supergravity constructed in [14,15] to the case of arbitrary torsion. More interestingly, one can also take $d = 4$ and construct on-shell 4D NC supergravity with zero torsion thereby obtaining, after gauge-fixing, the very first supersymmetric generalization of 4D Newtonian gravity. An intriguing feature of the 3D case is that the Newtonian supergravity theory contains both a Newton potential as well as a dual Newton potential [16] . In analogy to the 3d case, it would be interesting to see which representations of the Newton potential would occur in the 4d case and investigate whether this could have any physical effect.

[4] Note that the space/space components of the zero torsion constraint are already invariant under dilatations.

Acknowledgements. This presentation is based upon work that at the time of writing was in progress [13]. E.A.B. and J.R. gratefully acknowledge support from the Simons Center for Geometry and Physics, Stony Brook University at which some of the research described in this presentation was performed during the workshop *Applied Newton-Cartan Geometry*. E.A.B wishes to thank the University of Vienna for its hospitality.

References

1. Christensen, M.H., Hartong, J., Obers, N.A., Rollier, B.: Torsional Newton-Cartan geometry and Lifshitz holography. Phys. Rev. D **89**, 061901 (2014). doi:10.1103/PhysRevD.89.061901. [arXiv:1311.4794 [hep-th]]
2. Geracie, M., Son, D.T., Wu, C., Wu, S.F.: Spacetime symmetries of the quantum hall effect. Phys. Rev. D **91**, 045030 (2015). doi:10.1103/PhysRevD.91.045030. [arXiv:1407.1252 [cond-mat.mes-hall]]
3. Luttinger, J.M.: Theory of thermal transport coefficients. Phys. Rev. **135**, A1505 (1964). doi:10.1103/PhysRev.135.A1505
4. Gromov, A., Abanov, A.G.: Thermal hall effect and geometry with torsion. Phys. Rev. Lett. **114**, 016802 (2015). doi:10.1103/PhysRevLett.114.016802. [arXiv:1407.2908 [cond-mat.str-el]]
5. Festuccia, G., Hansen, D., Harton, J., Obers, N.A.: Torsional Newton-Cartan geometry from the Noether procedure. Phys. Rev. D **94**(10), 105023 (2016). doi:10.1103/PhysRevD.94.105023. [arXiv:1607.01926 [hep-th]]
6. Geracie, M., Golkar, S., Roberts, M.M.: Hall viscosity, spin density, and torsion. arXiv:1410.2574 [hep-th]
7. Geracie, M., Prabhu, K., Roberts, M.M.: Physical stress, mass, and energy for non-relativistic spinful matter. arXiv:1609.06729 [hep-th]
8. Duval, C., Burdet, G., Kunzle, H.P., Perrin, M.: Bargmann structures and Newton-Cartan theory. Phys. Rev. D **31**, 1841 (1985). doi:10.1103/PhysRevD.31.1841
9. Julia, B., Nicolai, H.: Null Killing vector dimensional reduction and Galilean geometrodynamics. Nucl. Phys. B **439**, 291 (1995). [hep-th/9412002]
10. Jensen, K.: On the coupling of Galilean-invariant field theories to curved spacetime. arXiv:1408.6855 [hep-th]
11. Bekaert, X., Morand, K.: Connections and dynamical trajectories in generalised Newton-Cartan gravity I. An intrinsic view. J. Math. Phys. **57**(2), 022507 (2016)
12. Afshar, H.R., Bergshoeff, E.A., Mehra, A., Parekh, P., Rollier, B.: A Schrödinger approach to Newton-Cartan and Hořava-Lifshitz gravities. JHEP **1604**, 145 (2016). doi:10.1007/JHEP04(2016)145. [arXiv:1512.06277 [hep-th]]
13. Bergshoeff, E., Chatzistavrakidis, A., Romano, L., Rosseel, J.: Null-reductions and Torsion. submitted to JHEP. arXiv:1708.05414 [hep-th]
14. Bergshoeff, E., Rosseel, J., Zojer, T.: Newton-Cartan (super)gravity as a non-relativistic limit. Class. Quant. Grav. **32**(20), 205003 (2015). doi:10.1088/0264-9381/32/20/205003. [arXiv:1505.02095 [hep-th]]
15. Bergshoeff, E., Rosseel, J., Zojer, T.: Newton-Cartan supergravity with torsion and Schrödinger supergravity. JHEP **1511**, 180 (2015). doi:10.1007/JHEP11(2015)180. [arXiv:1509.04527 [hep-th]]
16. Andringa, R., Bergshoeff, E.A., Rosseel, J., Sezgin, E.: 3D Newton-Cartan supergravity. Class. Quant. Grav. **30**, 205005 (2013). doi:10.1088/0264-9381/30/20/205005. [arXiv:1305.6737 [hep-th]]

Self-oscillations of a Vocal Apparatus: A Port-Hamiltonian Formulation

Thomas Hélie[1] and Fabrice Silva[2(✉)]

[1] S3AM Team, UMR STMS 9912, IRCAM-CNRS-UPMC, Paris, France
[2] Aix Marseille Univ., CNRS, Centrale Marseille, LMA, Marseille, France
silva@lma.cnrs-mrs.fr

Abstract. Port Hamiltonian systems (PHS) are open passive systems that fulfil a power balance: they correspond to dynamical systems composed of energy-storing elements, energy-dissipating elements and external ports, endowed with a geometric structure (called Dirac structure) that encodes conservative interconnections. This paper presents a minimal PHS model of the full vocal apparatus. Elementary components are: (a) an ideal subglottal pressure supply, (b) a glottal flow in a mobile channel, (c) vocal-folds, (d) an acoustic resonator reduced to a single mode. Particular attention is paid to the energetic consistency of each component, to passivity and to the conservative interconnection. Simulations are presented. They show the ability of the model to produce a variety of regimes, including self-sustained oscillations. Typical healthy or pathological configuration laryngeal configurations are explored.

1 Motivations

Many physics-based models of the human vocal apparatus were proposed to help understanding the phonation and its pathologies, with a compromise between the complexity introduced in the modelling and the vocal features that can be reproduced by analytical or numerical calculations. Except recent works based on finite elements methods applied to the glottal flow dynamics, most of the models rely on the description of the aerodynamics provided by van den Berg [1] for a glottal flow in static geometries, i.e., that ignores the motion of the vocal folds. Even if enhancements appeared accounting for various effects, they failed to represent correctly the energy exchanges between the flow and the surface of the vocal folds that bounds the glottis.

The port-Hamiltonian approach offers a framework for the modelling, analysis and control of complex system with emphasis on passivity and power balance [2]. A PHS for the classical body-cover model has been recently proposed [3] without connection to a glottal flow nor to a vocal tract, so that no self-oscillations can be produced. The current paper proposes a minimal PHS model of the full vocal apparatus. This power-balanced numerical tool enables the investigation of the various regimes that can be produced by time-domain simulations. Sect. 2 is a reminder on the port-Hamiltonian systems, Sect. 3 is dedicated to the description of the elementary components of the full vocal

F. Nielsen and F. Barbaresco (Eds.): GSI 2017, LNCS 10589, pp. 375–383, 2017.
https://doi.org/10.1007/978-3-319-68445-1_44

apparatus and their interconnection. Sect. 4 presents simulation and numerical results for typical healthy and pathological laryngeal configurations.

2 Port-Hamiltonian Systems

Port-Hamiltonian systems are open passive systems that fulfil a power balance [2,4]. A large class of such finite dimensional systems with input $\mathbf{u}(t) \in \mathbb{U} = \mathbb{R}^P$, output $\mathbf{y}(t) \in \mathbb{Y} = \mathbb{U}$, can be described by a differential algebraic equation

$$\begin{pmatrix} \dot{\mathbf{x}} \\ \mathbf{w} \\ -\mathbf{y} \end{pmatrix} = S(\mathbf{x}, \mathbf{w}) \begin{pmatrix} \nabla_{\mathbf{x}} H \\ \mathbf{z}(\mathbf{w}) \\ \mathbf{u} \end{pmatrix}, \quad \text{with } S = -S^T = \begin{pmatrix} \mathbf{J_x} & -\mathbf{K} & \mathbf{G_x} \\ \mathbf{K}^T & \mathbf{J_w} & \mathbf{G_w} \\ -\mathbf{G_x}^T & -\mathbf{G_w}^T & \mathbf{J_y}, \end{pmatrix} \tag{1}$$

where state $\mathbf{x}(t) \in \mathbb{X} = \mathbb{R}^N$ is associated with energy $E = \mathcal{H}(\mathbf{x}) \geq 0$ and where variables $\mathbf{w}(t) \in \mathbb{W} = \mathbb{R}^Q$ are associated with dissipative constitutive laws $\mathbf{z}$ such that $P_{\text{dis}} = \mathbf{z}(\mathbf{w})^T \mathbf{w} \geq 0$ stands for a dissipated power. Such a system naturally fulfils the power balance $\mathrm{d}E/\mathrm{d}t + P_{\text{dis}} - P_{\text{ext}} = 0$, where the external power is $P_{\text{ext}} = \mathbf{y}^T \mathbf{u}$. This is a straightforward consequence of the skew-symmetry of matrix S, which encodes this geometric structure (Dirac structure, see [2]). Indeed, rewriting Eq. (1) as $B = SA$, it follows that $A^T B = A^T S A = 0$, that is,

$$\nabla_{\mathbf{x}} \mathcal{H}(x)^T \dot{\mathbf{x}} + \mathbf{z}(\mathbf{w})^T \mathbf{w} - \mathbf{u}^T \mathbf{y} = 0 \tag{2}$$

Moreover, connecting several PHS through external ports yields a PHS. This modularity is used in practice, by working on elementary components, separately.

3 Vocal Apparatus

Benefiting from this modularity, the full vocal apparatus is built as the interconnection of the following elementary components: a subglottal pressure supply, two vocal folds, a glottal flow, and an acoustic resonator (see Fig. 1).

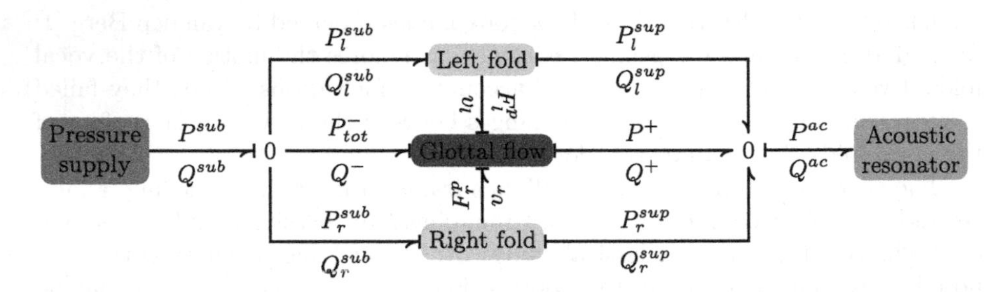

Fig. 1. Components of the vocal apparatus. The interconnection takes place via pairs of effort (P) and flux (Q) variables. The 0 connection expresses the equality of efforts and the division of flux. See Ref. [4] for an introduction to bond graphs.

3.1 The One-Mass Model of Vocal Folds

The left and right vocal folds ($\mathcal{F}_i = \mathcal{L}$ or $\mathcal{R}$ with $i = l$ or r, respectively), are modelled as classical single-d.o.f. oscillators (as in Ref. [5], mass m_i, spring k_i and damping r_i) with a purely elastic cover (as in Ref. [6], spring κ_i). Their dynamics relates the momentum π_i of the mass, and the elongations ξ_i and ζ_i of the body and cover springs, respectively, to the velocity $v_i = \dot{\zeta}_i + \dot{\xi}_i$ of the cover imposed by the glottal flow, and to the transverse resultants of the pressure forces on the upstream (P_i^{sub}) and downstream (P_i^{sup}) faces of the trapezoid-shaped structures (see Fig. 2, left part) :

$$\dot{\pi}_i = -k_i \xi_i - r_i \dot{\xi}_i + \kappa_i \zeta_i - P_i^{sub} S_i^{sub} - P_i^{sup} S_i^{sup}. \tag{3}$$

$F_i^p = -\kappa_i \zeta_i$ is the transverse feedback force opposed by the fold to the flow. The motion of the fold produces the additional flowrates Q_i^{sub} (pumping from the subglottal space, i.e., positive when the fold compresses) and Q_i^{sup} (pulsated into the supraglottal cavity, i.e., positive when the fold inflates).

Port-Hamiltonian modelling of a vocal fold $\mathcal{F}_i$:

$$\mathbf{x}_{\mathcal{F}_i} = \begin{pmatrix} \pi_i \\ \xi_i \\ \zeta_i \end{pmatrix}, \mathbf{u}_{\mathcal{F}_i} = \begin{pmatrix} P_i^{sub} \\ P_i^{sup} \\ v_i \end{pmatrix}, \mathbf{y}_{\mathcal{F}_i} = \begin{pmatrix} -Q_i^{sub} \\ Q_i^{sup} \\ -F_i^p \end{pmatrix}, H_{\mathcal{F}_i} = \frac{1}{2} \mathbf{x}_{\mathcal{F}_i}^T \begin{pmatrix} 1/m_i & & \\ & k_i & \\ & & \kappa_i \end{pmatrix} \mathbf{x}_{\mathcal{F}_i},$$

$$\mathbf{w}_{\mathcal{F}_i} = \dot{\xi}_i, \ \mathbf{z}_{\mathcal{F}_i}(\mathbf{w}_{\mathcal{F}_i}) = r_i \mathbf{w}_{\mathcal{F}_i}, \ \mathbf{J}_{\mathbf{w}}^{\mathcal{F}_i} = 0, \ \mathbf{G}_{\mathbf{w}}^{\mathcal{F}_i} = \mathbb{O}_{1\times 3}, \ \mathbf{J}_{\mathbf{y}}^{\mathcal{F}_i} = \mathbb{O}_{3\times 3},$$

$$\mathbf{J}_{\mathbf{x}}^{\mathcal{F}_i} = \begin{pmatrix} 0 & -1 & 1 \\ 1 & 0 & 0 \\ -1 & 0 & 0 \end{pmatrix}, \mathbf{K}^{\mathcal{F}_i} = \begin{pmatrix} 1 \\ 0 \\ 0 \end{pmatrix}, \text{ and } \mathbf{G}_{\mathbf{x}}^{\mathcal{F}_i} = \begin{pmatrix} -S_i^{sub} & -S_i^{sup} & 0 \\ 0 & 0 & 0 \\ 0 & 0 & 1 \end{pmatrix}.$$

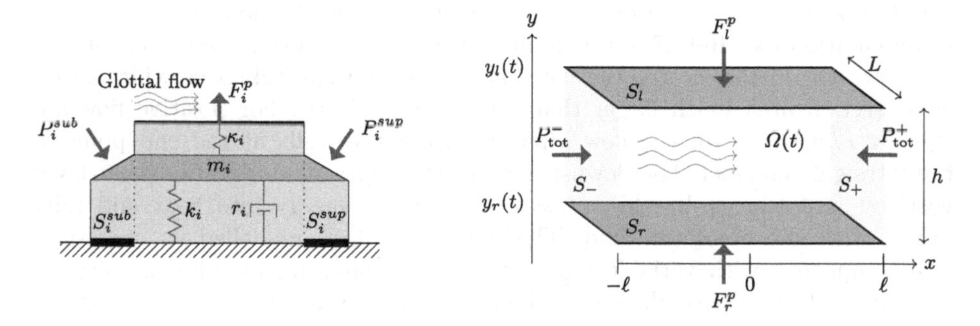

Fig. 2. Left: Schematic of a vocal fold. Right: Schematics of the glottal flow with open boundaries S^- and S^+ and mobile walls S_l and S_r.

3.2 Glottal Flow

We consider a potential incompressible flow of an inviscid fluid of density ρ between two parallel mobile walls located at $y = y_l(t)$ and $y = y_r(t)$, respectively.

The glottis $\mathcal{G}$ has width L, length 2ℓ and height $h = y_l - y_r$, its mid-line being located at $y = y_m = (y_r + y_l)/2$ (see Fig. 2, right part). The simplest kinematics for the fluid velocity $v(x, y)$ obeying the Euler equation

$$\dot{v} + \frac{1}{\rho}\nabla\left(p + \frac{1}{2}\rho|v|^2\right) = 0 \tag{4}$$

and satisfying the normal velocity continuity on the walls is given by:

$$v = \begin{pmatrix} v_x \\ v_y \end{pmatrix} = \begin{pmatrix} v_0 - x\frac{\dot{h}}{h} \\ \dot{y}_m + \frac{\dot{h}}{h}(y - y_m) \end{pmatrix} \quad \forall (x, y) \in \Omega = [-\ell, \ell] \times [y_r, y_l]. \tag{5}$$

The velocity field is thus parametrised by four macroscopic quantities: h, its time derivative $\dot{h}$, and the mean axial and transverse velocities $v_0 = < v_x >_\Omega$ and $\dot{y}_m = < v_y >_\Omega$, respectively. Choosing these quantities as the state allows the exact reduction of the infinite-dimensional problem to a finite-dimension system. The pressure field $p(x, y, t)$ can also be obtained from Eq. (4), as well as the total pressure $p + \frac{1}{2}\rho|v|^2$, but are not expanded here for brevity.

The dynamics for the glottal flow is controlled by the mean total pressures P_{tot}^- and P_{tot}^+ on the open boundaries S^- $(x = -\ell)$ and S^+ $(x = +\ell)$, respectively, and the resultant F_r^p and F_l^p of the pressure forces on the right and left walls, respectively (see Appendix A for the derivation of the equations). The kinetic energy of the fluid on the domain writes as

$$\varepsilon(t) = H_{\mathcal{G}}(\mathbf{x}_{\mathcal{G}}(t)) = \frac{1}{2}\left(m(h)v_0^2 + m(h)\dot{y}_m^2 + m_3(h)\dot{h}^2\right) \tag{6}$$

with the total mass of the fluid $m(h) = 2\rho\ell L h(t)$, and the effective mass for the transverse expansion motion $m_3(h) = m(h)\left(1 + 4\ell^2/h^2\right)/12$. The energy could be written as a function of the momenta to yield a canonical Hamiltonian representation (see Ref. [7] for a similar PHS based on normalised momenta).

Downstream the glottis, the flow enters the supraglottal space which has a cross section area much larger than that of the glottis. For positive flowrate $(Q^+ = Lhv_x(\ell) > 0)$, the flow separates from the walls at the end point of the (straight) channel. The downstream jet then spreads due to the shear-layer vortices until the jet has lost most of its kinetic energy into heat and fully mixed with the quiescent fluid. This phenomenon is modelled as a dissipative component with variable $\mathbf{w}_{\mathcal{G}} = Q^+$ and dissipation function $\mathbf{z}_{\mathcal{G}}(\mathbf{w}_{\mathcal{G}}) = (1/2)\rho(\mathbf{w}_{\mathcal{G}}/Lh)^2\Theta(\mathbf{w}_{\mathcal{G}})$ where Θ is the Heaviside step function. The pressure in the supraglottal space then writes $P^+ = P_{tot}^+ - \mathbf{z}_{\mathcal{G}}$.

Port-Hamiltonian modelling of the glottal flow $\mathcal{G}$:

$$\mathbf{x}_{\mathcal{G}} = \begin{pmatrix} v_0 \\ \dot{y}_m \\ \dot{h} \\ h \end{pmatrix}, \mathbf{u}_{\mathcal{G}} = \begin{pmatrix} P_{tot}^- \\ P^+ \\ F_l^p \\ F_r^p \end{pmatrix}, \mathbf{y}_{\mathcal{G}} = \begin{pmatrix} -Q^- = -Lhv_x(-\ell) \\ +Q^+ = Lhv_x(\ell) \\ -v_l = +\dot{y}_l \\ -v_r = -\dot{y}_r \end{pmatrix},$$

$$H_{\mathcal{G}}(\mathbf{x}_{\mathcal{G}}) = \frac{m(h)}{2}\left(v_0^2 + \dot{y}_m^2\right) + \frac{m_3(h)}{2}\dot{h}^2,$$

$$\mathbf{w}_{\mathcal{G}} = Q^+,\ \mathbf{z}_{\mathcal{G}} = \frac{\rho}{2}\left(\frac{\mathbf{w}_{\mathcal{G}}}{Lh}\right)^2 \Theta(\mathbf{w}_{\mathcal{G}}),$$
$$\mathbf{J}^{\mathcal{G}}_{\mathbf{w}} = \mathbb{O}_{1\times 1},\ \mathbf{G}^{\mathcal{G}}_{\mathbf{w}} = \mathbb{O}_{1\times 4} \text{ and } \mathbf{J}^{\mathcal{G}}_{\mathbf{y}} = \mathbb{O}_{4\times 4},$$
$$\mathbf{J}^{\mathcal{G}}_{\mathbf{x}} = \begin{pmatrix} 0 & 0 & 0 & 0 \\ 0 & 0 & 0 & 0 \\ 0 & 0 & 0 & -\frac{1}{m_3} \\ 0 & 0 & \frac{1}{m_3} & 0 \end{pmatrix}, \mathbf{K}^{\mathcal{G}} = \begin{pmatrix} \frac{Lh}{m} \\ 0 \\ -\frac{L\ell}{m_3} \\ 0 \end{pmatrix}, \text{ and } \mathbf{G}^{\mathcal{G}}_{\mathbf{x}} = \begin{pmatrix} \frac{Lh}{m} & -\frac{Lh}{m} & 0 & 0 \\ 0 & 0 & -\frac{1}{m} & \frac{1}{m} \\ \frac{L\ell}{m_3} & \frac{L\ell}{m_3} & -\frac{1}{2m_3} & -\frac{1}{2m_3} \\ 0 & 0 & 0 & 0 \end{pmatrix}.$$

3.3 Vocal Tract

We assume a modal representation of the input impedance of the vocal tract as seen from the supraglottal cavity, i.e., the supraglottal pressure P^{ac} is defined as the sum of pressure components p_n (for $n = 1, N$, denoted P_n in the Fourier domain) related to the input flowrate Q^{ac} through 2nd order transfer functions:

$$Z_{in}(\omega) = \frac{P^{ac}(\omega)}{Q^{ac}(\omega)} = \sum_{n=1}^{N} \frac{P_n(\omega)}{Q^{ac}(\omega)} = \sum_{n=1}^{N} \frac{j\omega a_n}{\omega_n^2 + jq_n\omega_n\omega - \omega^2} \tag{7}$$

where ω is the angular frequency, ω_n are the modal angular frequencies, q_n are the modal dampings and a_n the modal coefficients. Each mode corresponds to a resonance of the vocal tract, and so to an expected formant in the spectrum of the radiated sound. We follow the convention defined in Ref. [8] for the internal variables of this subsystem.

Port-Hamiltonian modelling of the acoustic resonator $\mathcal{A}$:

$$\mathbf{x}_{\mathcal{A}} = \left(p_1/a_1, \ldots, p_N/a_N, \int_0^t p_1(t')dt', \ldots, \int_0^t p_N(t')dt'\right)^T,$$
$$H_{\mathcal{A}}(\mathbf{x}_{\mathcal{A}}) = \sum_{n=1}^{N} \frac{1}{2}\left(\frac{p_n^2}{a_n} + \frac{\omega_n^2}{a_n}\left(\int_0^t p_n(t')dt'\right)^2\right),$$
$$\mathbf{w}_{\mathcal{A}} = (p_1, \ldots, p_N)^T,\ \mathbf{z}_{\mathcal{A}} = \left(\frac{q_1\omega_1}{a_1}\mathbf{w}_{\mathcal{A}1}, \ldots, \frac{q_N\omega_N}{a_N}\mathbf{w}_{\mathcal{A},N}\right),$$
$$\mathbf{u}_{\mathcal{A}} = (Q^{ac}),\ \mathbf{y}_{\mathcal{A}} = (-P^{ac}),$$
$$\mathbf{J}^{\mathcal{A}}_{\mathbf{x}} = \begin{pmatrix} \mathbb{O}_{N\times N} & -\mathbb{I}_{N\times N} \\ \mathbb{I}_{N\times N} & \mathbb{O}_{N\times N} \end{pmatrix}, \mathbf{K}^{\mathcal{A}} = \begin{pmatrix} \mathbb{I}_{N\times N} \\ \mathbb{O}_{N\times N} \end{pmatrix}, \mathbf{G}^{\mathcal{A}}_{\mathbf{x}} = \begin{pmatrix} 1_N \\ \mathbb{O}_{N\times 1} \end{pmatrix},$$
$$\mathbf{G}^{\mathcal{A}}_{\mathbf{w}} = \mathbb{O}_{N\times 1},\ \mathbf{J}^{\mathcal{A}}_{\mathbf{w}} = \mathbb{O}_{N\times N}, \text{ and } \mathbf{J}^{\mathcal{A}}_{\mathbf{y}} = \mathbb{O}_{1\times 1}.$$

where $\mathbb{I}_{N\times N}$ is the identity matrix of dim $N \times N$, and 1_N is the column vector $N \times 1$ filled with 1.

3.4 Full System

We assume that the lower airways acts as a source able to impose the pressure P^{sub} in the subglottal space of the larynx. The flowrate Q^{sub} coming from this source splits into the flowrate Q^- entering the glottis and the flowrate Q_l^{sub} and Q_r^{sub} pumped by the lower conus elasticus of the left and right vocal folds, respectively, so that $Q^{sub} = Q^- + Q_l^{sub} + Q_r^{sub}$ with $P_{sub} = P_r^{sub} = P_l^{sub} = P_{tot}^-$. Conversely, the flowrate Q^+ sums up with the flowrates Q_l^{sup} and Q_r^{sup} pulsated by the left and right vocal folds, respectively. The resulting flowrate Q^{ac} that enters the acoustic resonator is then $Q^{ac} = Q^+ + Q_l^{sup} + Q_r^{sup}$ with $P^{ac} = P_r^{sup} = P_l^{sup} = P^+$.

The elementary components described above are now put together to assembly the full vocal apparatus. In order to simplify the Dirac structure, the ports of the subsystems have been chosen to be complementary: a port with sink convention is always connected to a port with source convention. As a result, it is trivial to expand the port Hamiltonian modelling of the full system with the following variables, dissipation functions, ports and energy:

$$\mathbf{x} = \begin{pmatrix} \mathbf{x}_\mathcal{R} \\ \mathbf{x}_\mathcal{L} \\ \mathbf{x}_\mathcal{G} \\ \mathbf{x}_\mathcal{A} \end{pmatrix}, \; \mathbf{w} = \begin{pmatrix} \mathbf{w}_\mathcal{R} \\ \mathbf{w}_\mathcal{L} \\ \mathbf{w}_\mathcal{G} \\ \mathbf{w}_\mathcal{A} \end{pmatrix}, \; \mathbf{z} = \begin{pmatrix} \mathbf{z}_\mathcal{R} \\ \mathbf{z}_\mathcal{L} \\ \mathbf{z}_\mathcal{G} \\ \mathbf{z}_\mathcal{A} \end{pmatrix}, \; \mathbf{u} = \left(P^{sub}\right), \; \mathbf{y} = \left(-Q^{sub}\right),$$

$$H(\mathbf{x}) = H_\mathcal{R}(\mathbf{x}_\mathcal{R}) + H_\mathcal{L}(\mathbf{x}_\mathcal{L}) + H_\mathcal{G}(\mathbf{x}_\mathcal{G}) + H_\mathcal{A}(\mathbf{x}_\mathcal{A}).$$

The matrices $\mathbf{J_x}$, $\mathbf{K}$, $\mathbf{G_x}$, $\mathbf{G_w}$, $\mathbf{J_w}$ and $\mathbf{J_y}$ can be obtained using automated generation tools like the PyPHS software [9].

4 Simulations and Results

We here briefly present some preliminary results. In the port-Hamiltonian modelling of the full system, the dissipation variables $\mathbf{w}$ do not explicitly depend on $\mathbf{z}$ (i.e., $\mathbf{J}_w = \mathbb{O}$), so that they can be eliminated leading to a differential realisation that can be numerical integrated (e.g., using the Runge-Kutta 4 scheme). The parameters have the following values: $m_i = 0.2\,\text{g}$, $r_i = 0.05\,\text{kg/s}$, $L = 11\,\text{mm}$, $\ell = 2\,\text{mm}$, $\rho = 1.3\,\text{kg/m}^3$. Due to the sparse data available on the input impedances of vocal tract notably in terms of modal amplitudes a_n, we consider a resonator with a single pole ($N = 1$) with $\omega_n = 2\pi \times 640\,\text{rad/s}$, $q_n = .4$ and $a_n = 1\,\text{M}\Omega$ (from Ref. [10]). The system is driven by a subglottal pressure P^{sub} that increases from 0 to 800 Pa within 20 ms and is then maintained.

In the first simulation, the folds are symmetric ($k_r = k_l = 100\,\text{N/m}$, $\kappa_r = \kappa_l = 3k_r$) and initially separated by a width $h = 1\,\text{mm}$. In such conditions, the folds are pushed away from their rest position (until $h \sim 3\,\text{mm}$), but this equilibrium does not become unstable and the system does not vibrate.

If some adduction is performed bringing the folds closer together ($h = 0.1\,\text{mm}$), the glottis first widens (until $h \sim 2\,\text{mm}$) and the folds then start to vibrate and the acoustic pressure oscillates in the vocal tract (see Fig. 3, top). The sound is stable even if the two folds are slightly mistuned ($k_r = 100\,\text{N/m}$ and $k_l = 97\,\text{N/m}$).

The right fold is then hardened ($k_r = 150\,\text{N/m}$). The system still succeeds to vibrate, but, as visible on Fig. 3 (bottom), the oscillation is supported by the soft left fold at first, and then this latter decays while the hardened right fold starts to vibrate and finally maintains the sound production (even if the oscillations seem intermittent).

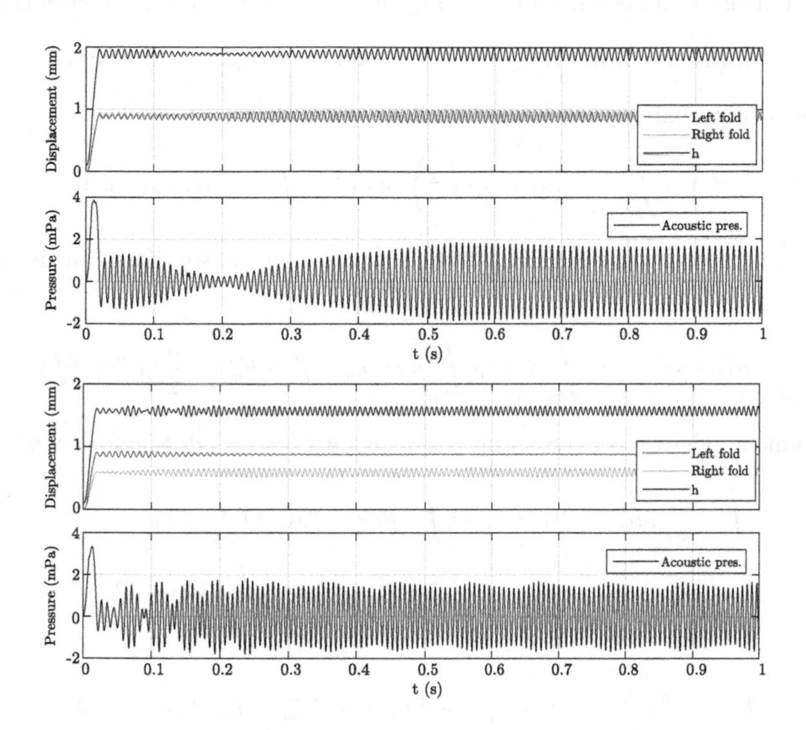

Fig. 3. Adducted (top) and asymmetric (bottom) configurations.

5 Conclusion

To the best knowledge of the authors, this paper proposes the first port-Hamiltonian model of a full vocal apparatus. This ensures passivity and the power balance. Simulations provide a variety of regimes that can be qualitatively related to aphonia (stable equilibrium), phonation (nearly periodic regimes) and dysphonia (irregular oscillations). This preliminary work provides a proof-of-concept for the relevance/interest of the passive and geometric approach.

Further work will be devoted to: (1) analyse regimes and bifurcations of the current model with respect to a few biomechanic parameters, (2) improve the realism of elementary components (separately), (3) account for possible contact between the vocal-folds, and (4) investigate on the synchronisation of coupled asymmetric vocals-folds and explore strategies to treat pathological voices [11].

Acknowledgement. The first author acknowledges the support of the Collaborative Research DFG and ANR project INFIDHEM ANR-16-CE92-0028.

A Dynamics of the glottal flow

The dynamics for the mean velocities can also be derived from the volume integration of the Euler equation (4). Using the gradient theorem, it comes that

$$m(h)\dot{v}_0 = Lh(t)\left(P_{tot}^- - P_{tot}^+\right) \quad \text{and} \quad m(h)\ddot{y}_m = F_r^p - F_l^p. \tag{8}$$

The energy balance for the glottal flow writes down as:

$$\dot{\varepsilon}(t) + \int_{S^- \cup S^+} \left(p + \frac{1}{2}\rho|v|^2\right)(v \cdot n) + \int_{S_l \cup S_r} p\,(v \cdot n) = 0 \tag{9}$$

where n is the outgoing normal. As the normal velocity is uniform on the walls, the last term of the energy balance reduces to

$$\int_{S_l \cup S_r} p\,(v \cdot n) = -\dot{y}_r \int_{S_r} p + \dot{y}_l \int_{S_l} p = \dot{y}_m\left(F_l^p - F_r^p\right) + \frac{\dot{h}}{2}\left(F_r^p + F_l^p\right). \tag{10}$$

The same applies on $S^- \cup S^+$ where $v \cdot n = \pm v_x(x = \pm\ell)$ does not depend on y:

$$\begin{aligned}\int_{S^- \cup S^+} p_{tot}\,(v \cdot n) &= v_x(\ell)\int_{S^+} p_{tot} - v_x(-\ell)\int_{S^-} p_{tot} \\ &= Lv_0\left(P_{tot}^+ - P_{tot}^-\right) - L\ell\frac{\dot{h}}{h}\left(P_{tot}^+ + P_{tot}^-\right). \end{aligned} \tag{11}$$

Thus,

$\dot{\varepsilon} = \dot{y}_m\left(F_r^p - F_l^p\right) - \frac{\dot{h}}{2}\left(F_r^p + F_l^p\right) + Lh(t)v_0\left(P_{tot}^- - P_{tot}^+\right) + L\ell\dot{h}\left(P_{tot}^- + P_{tot}^+\right)$. In the meanwhile, the kinetic energy in Eq. (6) can be derived against time: $\dot{\varepsilon} = m(h)\left(v_0\dot{v}_0 + \dot{y}_m\ddot{y}_m\right) + m_3(h)\dot{h}\ddot{h} + \frac{\partial H}{\partial h}\dot{h}$. The identification of the contribution of the mean axial and transverse velocities (see Eq. (8)) leads to the dynamics of the glottal channel expansion rate :

$$m_3\ddot{h} = L\ell\left(P_{tot}^- + P_{tot}^+\right) - \frac{F_r^p + F_l^p}{2} - \frac{\partial H}{\partial h}. \tag{12}$$

References

1. van den Berg, J.: On the air resistance and the Bernoulli effect of the Human Larynx. J. Acous. Soc. Am. **29**(5), 626–631 (1957)
2. van der Schaft, A., Jeltsema, D.: Port-Hamiltonian Systems Theory: an Introductory Overview. Now Publishers Inc., Hanover (2014)
3. Encina, M., et al.: Vocal fold modeling through the port-Hamiltonian systems approach. In: IEEE Conference on Control Applications (CCA), pp. 1558–1563 (2015)
4. Maschke, B., et al.: An intrinsic Hamiltonian formulation of network dynamics: non-standard Poisson structures and gyrators. J. Frankl. Inst. **329**(5), 923–966 (1992)

5. Flanagan, J., Landgraf, L.: Self-oscillating source for vocal-tract synthesizers. IEEE Trans. Audio Electroacous. **16**(1), 57–64 (1968)
6. Awrejcewicz, J.: Numerical analysis of the oscillations of human vocal cords. Nonlinear Dyn. **2**, 35–52 (1991)
7. Lopes, N., Hélie, T.: Energy balanced model of a jet interacting with a brass player's lip. Acta Acust United Acust. **102**(1), 141–154 (2016)
8. Lopes, N.: Approche passive pour la modélisation, la simulation et l'étude d'un banc de test robotisé pour les instruments de type cuivre. Ph.D. thesis, UPMC, Paris (2016)
9. Falaize, A.: PyPHS: passive modeling and simulation in python. Software. https://afalaize.github.io/pyphs/. last viewed on 21st April 2017
10. Badin, P., Fant, G.: Notes on vocal tract computation. STL-QPSR **25**(2–3), 53–108 (1984)
11. Giovanni, A., et al.: Nonlinear behavior of vocal fold vibration: the role of coupling between the vocal folds. J. Voice **13**(4), 465–476 (1999)

Non Linear Propagation in Reissner Beams: An Integrable System?

Frédéric Hélein[1(✉)], Joël Bensoam[2(✉)], and Pierre Carré[2(✉)]

[1] IMJ-PRG, Institut de Mathématiques de Jussieu - Paris Rive Gauche (UMR 7586), 1 Place I. Stravinsky, 75004 Paris, France
helein@math.jussieu.fr

[2] Ircam, Centre G. Pompidou, CNRS UMR 9912, S3AM, 1 Place I. Stravinsky, 75004 Paris, France
{Bensoam,Pierre.Carre}@ircam.fr

Abstract. In the seventies, Arnold has a geometric approach by considering a dynamical system as a map taking values in an abstract Lie group. As such, he was able to highlight fundamental equivalencies between rigid body motion and fluids dynamic depending on the specific Lie group chosen (group of rotations in the former and group of diffeomorphisms in the latter). Following his idea, nonlinear propagation of waves can also be formalized in their intrinsic qualities by adding space variables independent to time. For a simple one-dimensional acoustical system, it gives rise to the Reissner beam model for which the motion of each different section, labelled by the arc length s, is encoding in the Special Euclidean Lie group $SE(3)$ - a natural choice to describe motion in our 3-dimensional space. It turns out that, fortunately as a map over spacetime, this multi-symplectic approach can be related to the study of harmonic maps for which two dimensional cases can be solved exactly. It allows us to identify, among the family of problems, a particular case where the system is completely integrable. Among almost explicit solutions of this fully nonlinear problem, it is tempting to identify solitons, and to test the known numerical methods on these solutions.

1 Introduction

The Reissner beam is one of the simplest acoustical system that can be treated in the context of mechanics with symmetry. A Lie group is a mathematical construction that handle the symmetry but it is also a manifold on which a motion can take place. As emphasized by Arnold [1], physical motions of symmetric systems governed by the variational principle of least action correspond to geodesic motions on the corresponding group G.

For general problems (wave propagation, field theory), two different geometric approaches are basically available. The first approach, called the "dynamical" approach, uses, as its main ingredient, an infinite dimensional manifold as configuration space (TQ). The reduction techniques developed in the dynamical framework have been studied thoroughly in the literature (see for example [25]

F. Nielsen and F. Barbaresco (Eds.): GSI 2017, LNCS 10589, pp. 384–392, 2017.
https://doi.org/10.1007/978-3-319-68445-1_45

and the references therein cited), but it presents the difficulty to handle geodesic curves in an infinite dimensional function space.

As an alternative, the *covariant* formulation allows to consider a finite dimensional configuration space (the dimension of the symmetry group itself in our case). This can be achieved by increasing the number of independent variables since the validity of the calculus of variations and of the Noether's theorem is not limited to the previous one-variable setting. Although its roots go back to De Donder [26], Weyl [27], Caratheodory [28], after J. M. Souriau in the seventies [29], the classical field theory has been only well understood in the late 20th century (see for example [30] for an extension from symplectic to multisymplectic form). It is therefore not surprising that, in this covariant or jet formulation setting, the geometric constructions needed for reduction have been presented even more recently.

In this context, the multi-symplectic form is obtained from the differential of the Cartan-Poincaré n-form, and is crucial to give rise to an Hamiltonian framework (Lie-Poisson Schouten-Nijenhuis (SN) brackets [31]). The derivation of the conserved quantities from the symmetries of the Lie group is described by a moment map that is no longer a function but must be defined, more generally, as a *Noether's current.* This form is the interior product of the Poincaré-Cartan form by the fundamental vector field of the Lie group and leads to the dynamic equations of the problem. To obtain a well-posed problem, a zero-curvature equation (also know as the Maurer-Cartan equation) must be added to the formulation.

The multi-symplectic approach is developed in the first section through the Reissner's beam model. Inspired from [32], it turns out that, under some assumptions, this system is a completely integrable one. This is done, in the next section, by relating the formulation to the study of harmonic maps (also know under the name of chiral fields in Theoretical Physics and Mathematical Physics) for which the two dimensional case can be solved exactly. In Mathematical Physics this was known for the non linear σ-model since the seventies but it was the Russian school in integrable systems who made an exhaustive study of the principal chiral model (chiral fields with values in a Lie group).

2 Nonlinear Model for Reissner Beam

For the reader convenience, we reproduce below the non-linear Reissner Beam model as it was described in [2].

2.1 Reissner kinematics

A beam of length L, with cross-sectional area A and mass per unit volume ρ is considered. Following the Reissner kinematics, each section of the beam is supposed to be a rigid body. The beam configuration can be described by a position $\mathbf{r}(s,t)$ and a rotation $\mathbf{R}(s,t)$ of each section. The coordinate s corresponds to the position of the section in the reference configuration Σ_0 (see Fig. 1).

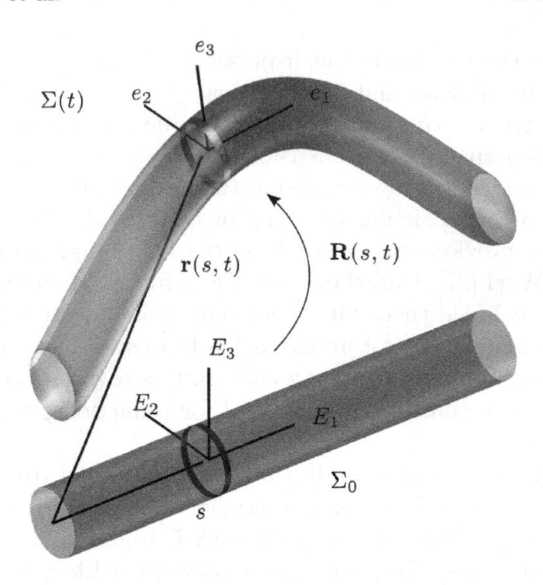

Fig. 1. Reference and current configuration of a beam. Each section, located at position s in the reference configuration Σ_0, is parametrized by a translation $\mathbf{r}(s,t)$ and a rotation $\mathbf{R}(s,t) \in SO_3$ in the current configuration Σ_t.

2.2 Lie Group Configuration Space

Any material point M of the beam which is located at $\mathbf{x}(s,0) = \mathbf{r}(s,0) + \mathbf{w}_0 = s\mathbf{E}_1 + \mathbf{w}_0$ in the reference configuration ($t = 0$) have a new position (at time t) $\mathbf{x}(s,t) = \mathbf{r}(s,t) + \mathbf{R}(s,t)\mathbf{w}_0$. In other words, the current configuration of the beam Σ_t is completely described by a map

$$\begin{pmatrix} \mathbf{x}(s,t) \\ 1 \end{pmatrix} = \underbrace{\begin{pmatrix} \mathbf{R}(s,t) & \mathbf{r}(s,t) \\ 0 & 1 \end{pmatrix}}_{\mathbf{H}(s,t)} \begin{pmatrix} \mathbf{w}_0 \\ 1 \end{pmatrix}, \quad \mathbf{R} \in SO(3), \quad \mathbf{r} \in \mathbf{R}^3, \tag{1}$$

where the matrix $\mathbf{H}(s,t)$ is an element of the Lie group $SE(3) = SO(3) \times \mathbf{R}^3$, where $SO(3)$ is the group of all 3×3 orthogonal matrices with determinant 1 (rotation in R^3). As a consequence, to any motion of the beam a function $\mathbf{H}(s,t)$ of the (scalar) independent variables s and t can be associated. Given some boundary conditions, among all such motions, only a few correspond to physical ones. What are the physical constraints that such motions are subjected to?

In order to formulate those constraints the definition of the Lie algebra is helpful. To every Lie group G, we can associate a Lie algebra $\mathfrak{g}$, whose underlying vector space is the tangent space of G at the identity element, which completely captures the local structure of the group. Concretely, the tangent vectors, $\partial_s\mathbf{H}$ and $\partial_t\mathbf{H}$, to the group $SE(3)$ at the point $\mathbf{H}$, are lifted to the tangent space at the identity e of the group. The definition in general is somewhat technical[1], but in the case of matrix groups this process is simply a multiplication by the inverse matrix $\mathbf{H}^{-1}$. This operation gives rise to definition of two left invariant vector fields in $\mathfrak{g} = \mathfrak{se}(3)$

$$\hat{\epsilon_c}(s,t) = \mathbf{H}^{-1}(s,t)\partial_s\mathbf{H}(s,t) \tag{2}$$

$$\hat{\chi_c}(s,t) = \mathbf{H}^{-1}(s,t)\partial_t\mathbf{H}(s,t), \tag{3}$$

which describe the deformations and the velocities of the beam. Assuming a linear stress-strain relation, those definitions allow to define a reduced Lagrangian by the difference of kinetic and potential energy with

$$E_c(\chi_c) = \int_0^L \frac{1}{2}\chi_c^T \mathbb{J}\chi_c ds, \tag{4}$$

$$E_p(\epsilon_c) = \int_0^L \frac{1}{2}(\epsilon_c - \epsilon_0)^T \mathbb{C}(\epsilon_c - \epsilon_0)ds, \tag{5}$$

where $\hat{\epsilon}_0 = \mathbf{H}^{-1}(s,0)\partial_s\mathbf{H}(s,0)$ correspond to the deformation of the initial configuration and $\mathbb{J}$ and $\mathbb{C}$ are matrix of inertia and Hooke tensor respectively, which are expressed by

$$\mathbb{J} = \begin{pmatrix} \mathbb{J}_r & 0 \\ 0 & \mathbb{J}_d \end{pmatrix}, \quad \mathbb{J}_r = \begin{pmatrix} I_1 & 0 & 0 \\ 0 & I_2 & 0 \\ 0 & 0 & I_3 \end{pmatrix}, \quad \mathbb{J}_d = \begin{pmatrix} m & 0 & 0 \\ 0 & m & 0 \\ 0 & 0 & m \end{pmatrix} \tag{6}$$

$$\mathbb{C} = \begin{pmatrix} \mathbb{C}_r & 0 \\ 0 & \mathbb{C}_d \end{pmatrix}, \quad \mathbb{C}_r = \begin{pmatrix} GI_\rho & 0 & 0 \\ 0 & EI_a & 0 \\ 0 & 0 & EI_a \end{pmatrix}, \quad \mathbb{C}_d = \begin{pmatrix} EA & 0 & 0 \\ 0 & GA & 0 \\ 0 & 0 & GA \end{pmatrix} \tag{7}$$

where $\mathbb{J}_r$, m, I_ρ and I_a are respectively the inertial tensor, the mass, the polar momentum of inertia and the axial moment of inertia of a section, and with E, G and A the Young modulus, shear coefficient and cross-sectional area respectively. The reduced Lagrangian density 2-form yields

$$\ell = l(\chi_L, \epsilon_L)\omega = \frac{1}{2}\left(\chi_L^T \mathbb{J}\chi_L - (\epsilon_L - \epsilon_0)^T \mathbb{C}(\epsilon_L - \epsilon_0)\right) ds \wedge dt.$$

2.3 Equations of Motion

Applying the Hamilton principle to the left invariant Lagrangian l leads to the Euler-Poincaré equation

$$\partial_t \boldsymbol{\pi}_c - ad^*_{\chi_c}\boldsymbol{\pi}_c = \partial_s(\boldsymbol{\sigma}_c - \boldsymbol{\sigma}_0) - ad^*_{\epsilon_c}(\boldsymbol{\sigma}_c - \boldsymbol{\sigma}_0), \tag{8}$$

[1] In the literature, one can find the expression $dL_{g^{-1}}(\dot{g})$ where dL stands for the differential of the left translation L by an element of G defined by

$$\begin{aligned} L_g : G &\rightarrow G \\ h &\rightarrow h \circ g. \end{aligned}$$

where $\boldsymbol{\pi}_c = \mathbb{J}\boldsymbol{\chi}_c$ and $\boldsymbol{\sigma}_c = \mathbb{C}\boldsymbol{\epsilon}_c$, (see for example [4,5] or [6] for details). In order to obtain a well-posed problem, the compatibility condition, obtained by differentiating (2) and (3)

$$\partial_s \boldsymbol{\chi}_c - \partial_t \boldsymbol{\epsilon}_c = ad_{\boldsymbol{\chi}_c} \boldsymbol{\epsilon}_c, \tag{9}$$

must be added to the equation of motion. It should be noted that the operators ad and ad^* in Eq. (8)

$$ad^*_{(\boldsymbol{\omega},\mathbf{v})}(\mathbf{m},\mathbf{p}) = (\mathbf{m}\times\boldsymbol{\omega} + \mathbf{p}\times\mathbf{v}, \mathbf{p}\times\boldsymbol{\omega}) \tag{10}$$
$$ad_{(\boldsymbol{\omega}_1,\mathbf{v}_1)}(\boldsymbol{\omega}_2,\mathbf{v}_2) = (\boldsymbol{\omega}_1\times\boldsymbol{\omega}_2, \boldsymbol{\omega}_1\times\mathbf{v}_2 - \boldsymbol{\omega}_2\times\mathbf{v}_1), \tag{11}$$

depend only on the group $SE(3)$ and not on the choice of the particular "metric" L that has been chosen to described the physical problem [7].

Equations (8) and (9) are written in material (or left invariant) form (c subscript). Spatial (or right invariant) form exist also. In this case, spatial variables (s subscript) are introduced by

$$\hat{\boldsymbol{\epsilon}_s}(s,t) = \partial_s \mathbf{H}(s,t)\mathbf{H}^{-1}(s,t) \tag{12}$$
$$\hat{\boldsymbol{\chi}_s}(s,t) = \partial_t \mathbf{H}(s,t)\mathbf{H}^{-1}(s,t) \tag{13}$$

and (8) leads to the conservation law [19]

$$\partial_t \boldsymbol{\pi}_s = \partial_s(\boldsymbol{\sigma}_s - \boldsymbol{\sigma}_0) \tag{14}$$

where $\boldsymbol{\pi}_s = \mathrm{Ad}^*_{\mathbf{H}^{-1}}\boldsymbol{\pi}_c$ and $\boldsymbol{\sigma}_s = \mathrm{Ad}^*_{\mathbf{H}^{-1}}\boldsymbol{\sigma}_c$. The Ad^* map for $SE(3)$ is

$$\mathrm{Ad}^*_{\mathbf{H}^{-1}}(\mathbf{m},\mathbf{p}) = (\mathbf{R}\mathbf{m} + \mathbf{r}\times\mathbf{R}\mathbf{p}, \mathbf{R}\mathbf{p}). \tag{15}$$

Compatibility condition (9) becomes

$$\partial_s \boldsymbol{\chi}_s - \partial_t \boldsymbol{\epsilon}_s = ad_{\boldsymbol{\epsilon}_s}\boldsymbol{\chi}_s. \tag{16}$$

Equations (8) and (9) (or alternatively (14) and (16)) provide the exact non linear Reissner beam model and can be used to handle the behavior of the beam if the large displacements are taking into account.

Notations and assumptions vary so much in the literature, it is often difficult to recognize this model (see for example [8] for a formulation using quaternions).

3 Comparison with Integrable Systems

3.1 The Zero-Curvature Formulation

I turns out that, under some assumptions, the previous system is a completely integrable one. Our discussion is inspired from [32].

In the following we consider for simplicity the previous system on an infinite 2-dimensional space-time $\mathbb{R}^2$. We also make the hypothesis that the tensors $\mathbb{J}$ and $\mathbb{C}$ are proportional to the identity, so that they can be written $J\mathbb{I}$ and $C\mathbb{I}$

respectively, the Ad operator and the matrix product commutes, and the relation between $\boldsymbol{\pi}_s$ and $\boldsymbol{\chi}_s$ simply becomes :

$$\boldsymbol{\pi}_s = \mathrm{Ad}^*_{\mathbf{H}^{-1}}(\mathbb{J}\mathrm{Ad}_{\mathbf{H}^{-1}}\boldsymbol{\chi}_s) = J\boldsymbol{\chi}_s \tag{17}$$

In the same way we have $\boldsymbol{\sigma}_s = C\boldsymbol{\epsilon}_s$. Without loss of generality, J and C are taken equal to one. Making the hypothesis $\boldsymbol{\sigma}_0 = 0$, the Eqs. (14) and (16) respectively becomes:

$$\begin{cases} \partial_s \boldsymbol{\chi}_s - \partial_t \boldsymbol{\epsilon}_s - [\boldsymbol{\epsilon}_s, \boldsymbol{\chi}_s] = 0 \\ \partial_t \boldsymbol{\chi}_s - \partial_s \boldsymbol{\epsilon}_s = 0 \end{cases} \tag{18}$$

Consider the 1-form $\omega = \boldsymbol{\chi}_s dt + \boldsymbol{\epsilon}_s ds$ and denote $\star\omega := \boldsymbol{\chi}_s ds + \boldsymbol{\epsilon}_s dt$, the previous equations are equivalent to

$$\begin{cases} d\omega - \omega \wedge \omega = 0 \\ d(\star\omega) = 0 \end{cases} \tag{19}$$

We set also $\omega_L := \frac{1}{2}(\omega + \star\omega) = \frac{1}{2}(\boldsymbol{\epsilon}_s + \boldsymbol{\chi}_s)d(s+t)$ (L for *left moving*) and $\omega_R := \frac{1}{2}(\omega - \star\omega) = \frac{1}{2}(\boldsymbol{\epsilon}_s - \boldsymbol{\chi}_s)d(s-t)$ (R for *right moving*) and remark that $\star\omega_L = \omega_L$ and $\star\omega_R = -\omega_R$.

The key in the following is to introduce a so-called *spectral parameter* $\lambda \in (\mathbb{C} \cap \{\infty\}) \setminus \{-1, 1\} = \mathbb{C}P \setminus \{-1, 1\}$ and the following family of connexion forms

$$\omega_\lambda = \frac{\omega_L}{1+\lambda} + \frac{\omega_R}{1-\lambda} = \frac{\omega - \lambda \star \omega}{1-\lambda^2} - . \tag{20}$$

We observe that System (19) is satisfied if and only if

$$d\omega_\lambda - \omega_\lambda \wedge \omega_\lambda = 0, \quad \forall \lambda \in \mathbb{C}P \setminus \{-1, 1\}. \tag{21}$$

This relation is a necessary and sufficient condition for the existence of a family of maps $(H_\lambda)_{\lambda \in \mathbb{C}P \setminus \{\pm 1\}}$ from $\mathbb{R}^2$ to $SE(3)^{\mathbb{C}}$, the complexification of $SE(3)$, such that

$$dH_\lambda = \omega_\lambda H_\lambda \quad \text{on } \mathbb{R}^2. \tag{22}$$

We will assume that $H_\lambda(0,0) = 1$ (the identity element of $SE(3)^{\mathbb{C}}$). Then the solution of (22) is unique. Since ω and $\star\omega$ are real, ω_λ is also real for any $\lambda \in \mathbb{R} \setminus \{\pm 1\}$ and hence H_λ is also real, i.e. takes value in $SE(3)$, for these values of λ. In general H_λ satisfies the *reality condition* $H_{\overline{\lambda}} = \overline{H_\lambda}$, $\forall \lambda \in \mathbb{C}P \setminus \{\pm 1\}$.

3.2 The *undressing* procedure

The family of maps $(H_\lambda)_{\lambda \in \mathbb{C}P \setminus \{\pm 1\}}$ is in correspondence with solutions of linear wave equations, through the following transformation, called the *undressing* procedure.

For that purpose fix some small disks D_{-1} and D_1 in $\mathbb{C}$ centered respectively at -1 and 1 and denote respectively by Γ_{-1} and Γ_1 their boundaries. We temporarily *fix* $(s,t) \in \mathbb{R}^2$, denote $H_\lambda = H_\lambda(s,t)$ and let the variable λ run.

By solving a Riemann-Hibert problem (which admits a solution for H_λ close to the identity), we can find two maps $[\lambda \longmapsto I^L_\lambda]$ and $[\lambda \longmapsto O^L_\lambda]$ on Γ_{-1} with value in $SE(3)^{\mathbb{C}}$ such that:

- $H_\lambda = (I^L_\lambda)^{-1} O^L_\lambda$, $\forall \lambda \in \Gamma_{-1}$;
- I^L_λ can be extended holomorphically *inside* Γ_{-1}, i.e. in D_{-1};
- O^L_λ can be extended holomorphically *outside* Γ_{-1}, i.e. on $\mathbb{C}P \setminus D_{-1}$, and converges to the unity at infinity.

Similarly we can find two maps $[\lambda \longmapsto I^R_\lambda]$ and $[\lambda \longmapsto O^R_\lambda]$ on Γ_1 with value in $SE(3)^{\mathbb{C}}$ such that:

- $H_\lambda = (I^R_\lambda)^{-1} O^R_\lambda$, $\forall \lambda \in \Gamma_1$;
- I^R_λ can be extended holomorphically in D_1;
- O^R_λ can be extended holomorphically on $\mathbb{C}P \setminus D_1$.

Set $o^L_\lambda := dO^L_\lambda \cdot (O^L_\lambda)^{-1}$. We deduce from $O^L_\lambda = I^L_\lambda H_\lambda$ that $o^L_\lambda = dI^L_\lambda (I^L_\lambda)^{-1} + I^L_\lambda \omega_\lambda (I^L_\lambda)^{-1}$ on Γ_{-1}. From its very definition we deduce that o^L_λ can be extended holomorphically on $\mathbb{C}P \setminus D_{-1}$. From $o^L_\lambda = dI^L_\lambda (I^L_\lambda)^{-1} + I^L_\lambda \omega_\lambda (I^L_\lambda)^{-1}$ and by using (20) we deduce that o^L_λ can be extended meromorphically inside D_{-1}, with at most one pole, equal to

$$I^L_{\lambda=-1} \frac{\omega_L}{1+\lambda} (I^L_{\lambda=-1})^{-1}.$$

By using Liouville's theorem and the fact that o^L converges to 0 at infinity we deduce that o^L_λ actually coincides with the latter expression, i.e. has the form

$$o^L_\lambda = \frac{v(s,t)}{1+\lambda} d(s+t). \tag{23}$$

But by the definition of o^L, we know that $do^L_\lambda - o^L_\lambda \wedge o^L_\lambda = 0$. Writing this equation gives us then that $\frac{\partial v}{\partial s} - \frac{\partial v}{\partial t} = 0$, hence $o^L_\lambda = \frac{v(s+t)}{1+\lambda} d(s+t)$.

A similar analysis on Γ_1 gives us that $o^R_\lambda := dO^R_\lambda \cdot (O^R_\lambda)^{-1}$ is of the form $o^R_\lambda = \frac{u(s-t)}{1-\lambda} d(s-t)$.

Hence we constructed from H_λ two maps $(s,t) \longmapsto v(s+t)$ and $(s,t) \longmapsto u(s-t)$ with value in the complexification of the Lie algebra of $SE(3)$ (but which actually satisfy a reality condition) and which are solutions of the linear wave equation (actually v is a left moving solution and u a right moving one).

This construction can be reversed through the so-called *dressing* procedure: starting from the data $(s,t) \longmapsto (u(s-t), v(s+t))$, we build the forms o^L_λ and o^R_λ using the previous expressions. We can integrate these forms for $\lambda \in \Gamma_{\pm 1}$ and we hence get the maps O^L_λ and O^R_λ. We then solve the Riemann–Hilbert problem, consisting of finding a map H_λ on $\Gamma := \Gamma_{-1} \cap \Gamma_1$ and two maps I^L_λ and I^R_λ, defined respectively on Γ_{-1} and Γ_1, such that

- $O^L_\lambda = I^L_\lambda H_\lambda$, $\forall \lambda \in \Gamma_{-1}$;
- $O^R_\lambda = I^R_\lambda H_\lambda$, $\forall \lambda \in \Gamma_1$;
- I^L_λ can be extended holomorphically on D_{-1};

- I^R_λ can be extended holomorphically on D_1;
- H_λ can be extended holomorphically on $\mathbb{C}P \setminus (D_{-1} \cap D_1)$ and converges to 0 at infinity.

Then H_λ will provides us with a solution of (18).

4 Conclusion

A geometrical approach of the dynamic of a Reissner beam has been studied in this article in order to take into account non linear effects due to large displacements. Among the family of problems we identified a particular case where this system is completely integrable. This allows us to find almost explicit solutions of this fully nonlinear problem. Among these solutions, it is tempting to identify solitons, and to test the known numerical methods on these solutions. The existence of such soliton solutions leads also to the question whether soliton solutions exists whenever the tensors $\mathbb{J}$ and $\mathbb{C}$ are less symmetric.

References

1. Arnold, V.: Sur la géométrie différentielle des groupes de Lie de dimension infinie et ses applications à l'hydrodynamique des fluides parfaits. Ann. Inst. Fourier Grenoble **16**, 319–361 (1966)
2. Bensoam, J.: Differential geometry applied to acoustics: non linear propagation in Reissner beams. In: Nielsen, F., Barbaresco, F. (eds.) GSI 2013. LNCS, vol. 8085, pp. 641–649. Springer, Heidelberg (2013). doi:10.1007/978-3-642-40020-9_71
3. Simo, J.: A finite strain beam formulation. The three-dimensional dynamic problem. Part I. Comput. Methods Appl. Mech. Eng. **49**, 55–70 (1985)
4. Roze, D.: Simulation d'une corde avec fortes déformations par les séries de Volterra, Master thesis, Université Pierre et Marie Curie (Paris 6) (2006)
5. Bensoam, J., Roze, D.: Modelling and numerical simulation of strings based on lie groups and algebras. Applications to the nonlinear dynamics of Reissner Beams. In: International Congress on Acoustics, Madrid (2007)
6. Gay-Balmaz, F., Holm, D.D., Ratiu, T.S.: Variational principles for spin systems and the Kirchhoff rod. J. Geometr. Mech. **1**(4), 417–444 (2009)
7. Holm, D.D.: Geometric Mechanics, Part II: Rotating. Translating and Rolling. Imperial College Press, London (2008)
8. Celledoni, E., Safstrom, N.: A Hamiltonian and multi-Hamiltonian formulation of a rod model using quaternions. Comput. Methods Appl. Mech. Eng. **199**, 2813–2819 (2010)
9. Simo, J.C., Tarnow, N., Doblare, M.: Non-linear dynamics of three-dimensional rods: exact energy and momentum conserving algorithms. Int. J. Numer. Methods Eng. **38**(9), 1431–1473 (1995)
10. Leyendecker, S., Betsch, P., Steinmann, P.: Objective energy-momentum conserving integration for the constrained dynamics of geometrically exact beams. Comput. Methods Appl. Mech. Eng. **195**, 2313–2333 (2006)
11. Gams, M., Saje, M., Srpcic, S., Planinc, I.: Finite element dynamic analysis of geometrically exact planar beams. Comput. Struct. **85**, 1409–1419 (2007)

12. Bishop, T.C., Cortez, R., Zhmudsky, O.O.: Investigation of bend and shear waves in a geometrically exact elastic rod model. J. Comput. Phys. **193**, 642–665 (2004)
13. da Fonseca, A.F., de Aguiar, M.A.M.: Solving the boundary value problem for finite Kirchhoff rods. Phys. D **181**, 53–69 (2003)
14. Nizette, M., Goriely, A.: Towards a classification of Euler-Kirchhoff filaments. J. Math. Phys. **40**(6), 2830 (1999)
15. Goriely, A., Tabor, M.: Nonlinear dynamics of filaments II. Nonlinear analysis. Phys. D **105**, 45–61 (1997)
16. Argeri, M., Barone, V., De Lillo, S., Lupoc, G., Sommacal, M.: Elastic rods in life- and material-sciences: a general integrable model. Phys. D **238**, 1031–1049 (2009)
17. Leonard, N.E., Marsden, J.E.: Stability and drift of underwater vehicle dynamics: mechanical systems with rigid motion symmetry. Phys. D **105**(1–3), 130–162 (1997)
18. Holmes, P., Jenkins, J., Leonard, N.E.: Dynamics of the Kirchhoff equations I: coincident centers of gravity and buoyancy. Phys. D **118**, 311–342 (1998)
19. Maddocks, J.H., Dichmann, D.J.: Conservation laws in the dynamics of rods. J. Elast. **34**, 83–96 (1994)
20. do Carmo, M.P.: Differential Geometry of Curves and Surfaces. Prentice-Hall Inc., London (1976)
21. do Carmo, M.P.: Riemannian Geometry. Birkhauser, Boston (1992)
22. Ellis, D.C.P., Gay-Balmaz, F., Holm, D.D., Ratiu, T.S.: Lagrange-Poincaré field equations. J. Geomet. Phys. **61**(11), 2120–2146 (2011)
23. López, M.C., Marsden, J.E.: Covariant and dynamical reduction for principal bundle field theories. Ann. Global Anal. Geomet. **34**, 263–285 (2008)
24. López, M.C., Garcia Perez, P.L.: Multidimensional Euler-Poincaré equations1. In: Differential Geometry and Its Applications, Proceecdings Conference, Opava (Czech Republic), August 27–31, 2001, pp. 383–391 (2001)
25. Marsden, J.E., Ratiu, T.S.: Introduction to Mechanics and Symmetry, 2nd edn. Springer, New York (1999)
26. De Donder, T.: Théorie invariantive du calcul des variations. Gauthier-Villars, Paris (1930)
27. Weyl, H.: Geodesic fields in the calculus of variation for multiple integrals. Ann. Math. **36**, 607–629 (1935)
28. Carathéodory, C.: Calculus of Variations and Partial Differential Equations of the First Order. Chelsea Publishing Company, New York (1999)
29. Souriau, J.-M.: Structure des systèmes dynamiques (1970)
30. Kanatchikov, I.V.: Canonical structure of classical field theory in the polymomentum phase space. Rep. Math. Phys. **41**(1), 49–90 (1998)
31. Schouten, J.A.: Uber Differentialkomitanten zweier kontravarianter Grossen. Proc. K Ned. Akad. Wet. (Amsterdam) **43**, 449–452 (1940)
32. Mañas, M.: The principal chiral model as a completely integrable system. In: Fordy, A.P., Wood, J.C. (eds.) Harmonic Maps and Completely Integrable Systems, pp. 147–173. Viehweg, Springer, Heidelberg (1994)

Quantum Harmonic Analysis and the Positivity of Trace Class Operators; Applications to Quantum Mechanics

Elena Nicola, Maurice de Gosson(✉), and Fabio Nicola

University of Vienna, Vienna, Austria
maurice.de.gosson@univie.ac.at

Abstract. The study of positivity properties of trace class operators is essential in the theory of quantum mechanical density matrices; the latter describe the "mixed states" of quantum mechanics and are essential in information theory. While a general theory for these positivity results is still lacking, we present some new results we have recently obtained and which generalize and extend the well-known conditions given in the 1970s by Kastler, Loupias, and Miracle-Sole, generalizing Bochner's theorem on the Fourier transform of a probability measure. The tools we use are the theory of pseudodifferential operators, symplectic geometry, and Gabor frame theory. We also speculate about some consequences of a possibly varying Planck's constant for the early universe.

1 Introduction

The characterization of positivity properties for trace class operators on $L^2(\mathbb{R}^n)$ is important because of its potential applications to quantum mechanics (positive trace class operators with trace one represent the mixed quantum states). It is also a notoriously difficult part of functional analysis which has been tackled by many authors but there have been few decisive advances since the pioneering work of Kastler [7] and Loupias and Miracle-Sole [8,9]. We begin by reviewing the topic and thereafter state some new results recently obtained by us.

2 The KLM Conditions

Let η be a real parameter. The notion of η*-positivity generalizes the usual notion of positivity:*

Definition 1. *Let $b \in \mathcal{S}(\mathbb{R}^{2n})$. We say that b is of η-positive type if for every integer N the $N \times N$ matrix $\Lambda_{(N)}$ with entries*

$$\Lambda_{jk} = e^{\frac{i\eta}{2}\sigma(z_j,z_k)}b(z_j - z_k)$$

is positive semidefinite for all choices of $(z_1, z_2, ..., z_N) \in (\mathbb{R}^{2n})^N$. (Here σ is the standard symplectic form $\sum_{1\leq j\leq n} dp_j \wedge dx_j$ on $\mathbb{R}^{2n}$).

F. Nielsen and F. Barbaresco (Eds.): GSI 2017, LNCS 10589, pp. 393–398, 2017.
https://doi.org/10.1007/978-3-319-68445-1_46

Notice that $\eta = 0$ corresponds to the ordinary notion of positivity, used in the statement of Bochner's theorem on the Fourier transform of a probability measure [1]. In what follows we denote by $\widehat{A} = \mathrm{Op}_\eta^{\mathrm{W}}(a)$ the η-Weyl operator with symbol a:

$$\langle \widehat{A}\psi, \overline{\phi} \rangle = \langle a, W_\eta(\psi, \phi) \rangle \tag{1}$$

for all $\psi, \phi \in \mathcal{S}(\mathbb{R}^n)$; here $\langle \cdot, \cdot \rangle$ denotes the distributional brackets on $\mathbb{R}^n$ and $\mathbb{R}^{2n}$ (respectively) and $W_\eta(\psi, \phi)$ is the η-cross-Wigner transform defined, for $\psi, \phi \in L^2(\mathbb{R}^n)$, by

$$W_\eta(\psi, \phi)(z) = \left(\tfrac{1}{\pi\eta}\right)^n (R_\eta(z)\psi|\phi)_{L^2}; \tag{2}$$

here $R_\eta(z)$ is the η-parity operator

$$R_\eta(z) = T_\eta(z)R(0)T_\eta(z)^{-1} \tag{3}$$

(with $R(0)\psi(x) = \psi(-x)$) and $T_\eta(z)$ is the Heisenberg–Weyl η-operator

$$T_\eta(z_0)\psi(x) = e^{\frac{i}{\eta}(p_0 x - \frac{1}{2}p_0 x_0)}\psi(x - x_0). \tag{4}$$

We will set $W_\eta(\psi, \psi) = W_\eta\psi$. When $\eta = \hbar > 0$, ($\hbar$ the Planck constant h divided by 2π) we recapture the standard cross-Wigner function $W_\hbar(\psi, \phi)$, simply denoted by $W(\psi, \phi)$.

Using the symplectic Fourier transform of $a \in \mathcal{S}(\mathbb{R}^{2n})$ defined by

$$a_\Diamond(z) = F_\Diamond a(z) = \int_{\mathbb{R}^{2n}} e^{i\sigma(z,z')}a(z')dz'. \tag{5}$$

Kastler [7] proved the following result using the theory of C^*-algebras; we have given in [3] a simpler proof of this result:

Theorem 1. *Let $\widehat{A} = \mathrm{Op}_\eta^{\mathrm{W}}(a)$ be a self-adjoint trace-class operator on $L^2(\mathbb{R}^n)$ (hence a is real). We have $\widehat{A} \geq 0$ if and only the two following conditions hold: (i) $a_\Diamond$ is continuous; (ii) $a_\Diamond$ is of η -positive type.*

Sketch of the proof: assume that $\widehat{A} \geq 0$; there exists a constant $C \in \mathbb{R}$ such that $a = C\sum_j \alpha_j W_\eta \psi_j$ for some family of normalized functions $\psi_j \in L^2(\mathbb{R}^n)$, the coefficients α_j being ≥ 0. Consider now the expression

$$I_N(\psi) = \sum_{1 \leq j,k \leq N} \lambda_j \overline{\lambda_k} e^{-\frac{i}{2\eta}\sigma(z_j,z_k)} F_{\sigma,\eta} W_\eta \psi(z_j - z_k) \geq 0 \tag{6}$$

where $F_{\sigma,\eta}$ is the η-symplectic transform defined by

$$F_{\sigma,\eta}a(z) = a_{\sigma,\eta}(z) = \left(\tfrac{1}{2\pi\eta}\right)^n \int_{\mathbb{R}^{2n}} e^{-\frac{i}{\eta}\sigma(z,z')}a(z')dz; \tag{7}$$

for $\eta \neq 0$ we have $F_{\sigma,\eta}a(z) = (2\pi\eta)^{-n}a_\Diamond(-z/\eta)$ hence the condition "$a_\Diamond$ is of η-positive type" can be restated as: each $N \times N$ matrix $\Lambda'_{(N)}$ with entries

$$\Lambda'_{jk} = e^{-\frac{i}{2\eta}\sigma(z_j,z_k)} F_{\sigma,\eta}a(z_j - z_k)$$

is positive semidefinite for all $(z_1, z_2, ..., z_N) \in (\mathbb{R}^{2n})^N$. We must verify that $I_N(\psi) \geq 0$ but this follows from the observation that

$$I_N(\psi) = \left(\tfrac{1}{2\pi\eta}\right)^n || \sum_{1\leq j\leq N} \lambda_j T_\eta(z_j)\psi||^2_{L^2}. \tag{8}$$

Let us now show that, conversely, the conditions (i) and (ii) imply that $(\widehat{A}\psi|\psi)_{L^2} \geq 0$ for all $\psi \in L^2(\mathbb{R}^n)$; in view of (1) this equivalent to showing that

$$\int_{\mathbb{R}^{2n}} a(z)W_\eta\psi(z)dz \geq 0 \tag{9}$$

for $\psi \in L^2(\mathbb{R}^n)$. Choosing $z_k = 0$ and setting $z_j = z$ in Λ'_{jk} this means that every matrix $(a_{\sigma,\eta}(z))_{1\leq j,k\leq N}$ is positive semidefinite. Setting

$$\Gamma_{jk} = e^{--\frac{i}{2\eta}\sigma(z_j,z_k)}F_{\sigma,\eta}W_\eta\psi(z_j - z_k)$$

the matrix $\Gamma_{(N)} = (\Gamma_{jk})_{1\leq j,k\leq N}$ is also positive semidefinite. Writing

$$M_{jk} = F_{\sigma,\eta}W_\eta\psi(z_j - z_k)a_{\sigma,\eta}(z_j - z_k);$$

one shows using Schur's theorem on the positivity of Hadamard product of the positive semidefinite matrices that the matrix $(M_{jk})_{1\leq j,k\leq N}$ is also positive semidefinite. One then concludes using Bochner's theorem on the Fourier transform of probability measures.

3 A New Positivity Test

The KLM conditions are not easily computable since they involve the verification of a non-countable set of inequalities. The following result replaces the KLM conditions by a *countable* set of conditions:

Theorem 2. *Let $a \in L^2(\mathbb{R}^{2n})$ and $\mathcal{G}(\phi, \Lambda)$ be a Gabor frame for $L^2(\mathbb{R}^n)$. For $(z_\lambda, z_\mu) \in \Lambda \times \Lambda$ set*

$$a_{\lambda,\mu} = \int_{\mathbb{R}^{2n}} e^{-\frac{i}{\eta}\sigma(z,z_\lambda - z_\mu)}a(z)W_\eta\phi(z - \tfrac{1}{2}(z_\lambda + z_\mu))dz. \tag{10}$$

The operator $\widehat{A} = \mathrm{Op}^{\mathrm{W}}_\eta(a)$ is positive semidefinite if and only if for every integer $N \geq 0$ the matrix with entries

$$M_{\lambda,\mu} = e^{-\frac{i}{2\eta}\sigma(z_\lambda,z_\mu)}a_{\lambda,\mu}, |z_\lambda|, |z_\mu| \leq N \tag{11}$$

is positive semidefinite.

Sketch of the proof: Recall that a Gabor frame $\mathcal{G}(\phi, \Lambda)$ in $L^2(\mathbb{R}^n)$ is the datum of a pair (ϕ, Λ) where ϕ ("the window") belongs to a suitable functional space (for instance $L^2(\mathbb{R}^n)$) and $\Lambda \subset \mathbb{R}^{2n}$ is a lattice, and such that there exist constants $a, b > 0$ such that

$$a||\psi||_{L^2}^2 \leq \sum_{z_\lambda \in \Lambda} |(\psi|T(z_\lambda)\phi)|^2 \leq b||\psi||_{L^2}^2$$

for all $\psi \in L^2(\mathbb{R}^n)$, where we set $T(z_\lambda) = T_{1/(2\pi)}(z_\lambda)$. The condition $\widehat{A} \geq 0$ is equivalent to

$$\int_{\mathbb{R}^{2n}} a(z) W_\eta \psi(z) dz \geq 0 \tag{12}$$

for every $\psi \in L^2(\mathbb{R}^n)$. The numbers $a_{\lambda,\mu}$ defined by (10) are the Gabor coefficients with respect to the Gabor system $\mathcal{G}(W_\eta g, \Lambda \times \Lambda)$. Expanding ψ in the frame $\mathcal{G}(\phi, \Lambda)$ we get

$$\psi = \sum_{z_\lambda \in \Lambda} c(z_\lambda) T(z_\lambda)\phi$$

where $c_\lambda = (\psi|T(z_\lambda)\phi)_{L^2}$ and hence

$$\int_{\mathbb{R}^{2n}} a(z) W_\eta \left(\textstyle\sum_{z_\lambda \in \Lambda} c_\lambda T(z_\lambda)\phi\right)(z) dz \geq 0. \tag{13}$$

The claim follows using the relations

$$W_\eta \left(\textstyle\sum_{z_\lambda \in \Lambda} c(z_\lambda) T(z_\lambda)\phi\right) = \textstyle\sum_{z_\lambda \in \Lambda} c_\lambda \overline{c_\mu} W_\eta(T(z_\lambda)\phi, T(z_\mu)\phi);$$

and observing that [5]

$$W_\eta(T(z_\lambda)\phi, T(z_\mu)\phi) = e^{-\frac{i}{2\eta}\sigma(z_\lambda, z_\mu)} e^{-\frac{i}{\eta}\sigma(z, z_\lambda - z_\mu)} W_\eta \phi(z - \tfrac{1}{2}(z_\lambda + z_\mu)).$$

Let us next show that the KLM conditions can be recaptured as a limiting case of the conditions in Theorem 2. To show this claim, we make use of another well-known time-frequency representation: the short-time Fourier transform (STFT). Precisely, for a given function $g \in \mathcal{S}(\mathbb{R}^n) \setminus \{0\}$ (called window), the STFT $V_g f$ of a distribution $f \in \mathcal{S}'(\mathbb{R}^n)$ is defined by

$$V_g f(x, p) = \int_{\mathbb{R}^n} e^{-ip\cdot y} f(y) \overline{g(y - x)}\, dy, \quad (x, p) \in \mathbb{R}^{2n}. \tag{14}$$

Let $\phi_0(x) = (\pi\eta)^{-n/4} e^{-|x|^2/2\eta}$ be the standard Gaussian and $\phi_\nu = T(\nu)\phi_0$, $\nu \in \mathbb{R}^{2n}$. We shall consider the STFT $V_{W\phi_\nu} a$, with window given by the Wigner function $W\phi_\nu$ and symbol a.

Theorem 3. *Let $a \in L^1(\mathbb{R}^n)$ and $z_\lambda, z_\mu \in \mathbb{R}^{2n}$. Setting*

$$M_{\lambda,\mu} = e^{-\frac{i}{2\eta}\sigma(z_\lambda, z_\mu)} a_{\sigma,\eta}(z_\lambda - z_\mu)$$
$$M_{\lambda,\mu}^{\phi_\nu} = e^{-\frac{i}{2\eta}\sigma(z_\lambda, z_\mu)} V_{W\phi_\nu} a(\tfrac{1}{2}(z_\lambda + z_\mu), J(z_\mu - z_\lambda)).$$

we have

$$M_{\lambda,\mu} = \lim_{\varepsilon \to 0+} \sum_{\substack{z_\nu \in \varepsilon\mathbb{Z}^{2n} \\ |z_\nu| < 1/\varepsilon}} \varepsilon^{2n} M_{\lambda,\mu}^{\phi_\nu}.$$

Sketch of the proof: Observe that we can write

$$M_{\lambda,\mu} = e^{-\frac{i}{2\eta}\sigma(z_\lambda,z_\mu)} V_\Phi a(\tfrac{1}{2}(z_\lambda + z_\mu), J(z_\mu - z_\lambda))$$

where $\Phi(z) = 1$ for all $z \in \mathbb{R}^{2n}$ the result follows from the dominated convergence theorem using the limit

$$\lim_{\varepsilon\to 0+} \sum_{\substack{z_\nu \in \varepsilon\mathbb{Z}^{2n} \\ |z_\nu|<1/\varepsilon}} \varepsilon^{2n} W\phi_\nu(z) = 1 \tag{15}$$

for all $z \in \mathbb{R}^{2n}$ and the bound

$$\left| \sum_{\substack{z_\nu \in \varepsilon\mathbb{Z}^{2n} \\ |z_\nu|<1/\varepsilon}} \varepsilon^{2n} W\phi_\nu(z) \right| \leq C. \tag{16}$$

valid for all $z \in \mathbb{R}^{2n}$.

4 Discussion

Positivity questions for operators are notoriously difficult to handle. In the case of trace-class operators not many progresses have been done since the work of Narcowich and his collaborators [12–15] and Bröcker and Werner [2]; also see Dias and Prata [4]. The interest in these questions come from the quantum mechanical problem of characterizing the so-called "mixed states", which are statistical mixtures of well-defined quantum-mechanical states (the "pure states"). Mixed states are mathematically represented by the quantum density operators: such an operator is a self-adjoint positive semidefinite trace class operators with unit trace (on any Hilbert space). While it is usually a rather trivial matter to verify self-adjointness and the trace property, positivity is very delicate – as exemplified by our discussion above –. The importance of this concept has increased since it has been realized by cosmological observations [10,11] that Planck's constant $\eta = h/2\pi$ might very well have been changing it value since the early Universe. If true, this would mean that some quantum states have evolved in classical ones (see our analysis in [6]).

Acknowledgement. Maurice de Gosson has been financed by the grant P27773-N23 of the Austrian Research Foundation FWF.

References

1. Bochner, S.: Lectures on Fourier Integrals. Princeton University Press, Princeton (1959)
2. Bröcker, T., Werner, R.F.: Mixed states with positive Wigner functions. J. Math. Phys. **36**(1), 62–75 (1995)
3. Cordero, E., de Gosson, M., Nicola, F.: On the positivity of trace-class operators, Submitted. ArXiv:1706.06171

4. Dias, N.C., Prata, J.N.: The Narcowich-Wigner spectrum of a pure state. Rep. Math. Phys. **63**(1), 43–54 (2009)
5. de Gosson, M.: Symplectic Methods in Harmonic Analysis and in Mathematical Physics. Birkhäuser, Basel (2011)
6. http://www.sciencedirect.com/science/article/pii/S0375960117307004
7. Kastler, D.: The C^*-Algebras of a free boson field. Commun. Math. Phys. **1**, 14–48 (1965)
8. Loupias, G., Miracle-Sole, S.: C^*-Algèbres des systèmes canoniques, I. Commun. Math. Phys. **2**, 31–48 (1966)
9. Loupias, G., Miracle-Sole, S.: C^*-Algèbres des systèmes canoniques, II. Ann. Inst. Henri Poincaré **6**(1), 39–58 (1967)
10. Murphy, M.T., Webb, J.K., Flambaum, V.V., Dzuba, V.A., Churchill, C.W., Prochaska, J.X., Barrow, J.D., Wolfe, A.M.: Possible evidence for a variable fine-structure constant from QSO absorption lines: motivations, analysis and results. Mon. Not. R. Astron. Soc. **327**, 1208–1222 (2001)
11. Murphy, M.T., Webb, J.K., Flambaum, V.V.: Further evidence for a variable fine-structure constant from Keck/HIRES QSO absorption spectra. Mon. Not. R. Astron. Soc. **345**(2), 609–638 (2003)
12. Narcowich, F.J.: Conditions for the convolution of two Wigner distributions to be itself a Wigner distribution. J. Math. Phys. **29**(9), 2036–2041 (1988)
13. Narcowich, F.J.: Distributions of η-positive type and applications. J. Math. Phys. **30**(11), 2565–2573 (1989)
14. Narcowich, F.J.: Geometry and uncertainty. J. Math. Phys. **31**(2), 354–364 (1990)
15. Narcowich, F.J., O'Connell, R.F.: Necessary and sufficient conditions for a phase-space function to be a Wigner distribution. Phys. Rev. A **34**(1), 1–6 (1986)

Stochastic Geometric Mechanics and Lie Group Thermodynamics

A Variational Formulation for Fluid Dynamics with Irreversible Processes

François Gay-Balmaz[1(✉)] and Hiroaki Yoshimura[2]

[1] CNRS & Ecole Normale Supérieure, LMD, IPSL, 24 Rue Lhomond, 75005 Paris, France
gaybalma@lmd.ens.fr

[2] School of Science and Engineering, Waseda University, Okubo, Shinjuku, Tokyo 169-8555, Japan
yoshimura@waseda.jp

Abstract. In this paper, we present a variational formulation for heat conducting viscous fluids, which extends the Hamilton principle of continuum mechanics to include irreversible processes. This formulation follows from the general variational description of nonequilibrium thermodynamics introduced in [3,4] for discrete and continuum systems. It relies on the concept of thermodynamic displacement. The irreversibility is encoded into a nonlinear nonholonomic constraint given by the expression of the entropy production associated to the irreversible processes involved.

Keywords: Nonequilibrium thermodynamics · Variational formalism · Viscosity · Heat conduction

1 Variational Principle for Discrete Systems

In this section we review the variational formulation for nonequilibrium thermodynamics of discrete (i.e., finite dimensional) systems developed in [3].

1.1 Variational Formulation of Nonequilibrium Thermodynamics of Simple Systems

We shall present the variational formulation by first considering *simple thermodynamic systems* before going into the general setting of the discrete systems. We follow the systematic treatment of thermodynamic systems presented in [12], to which we also refer for the precise statement of the two laws of thermodynamics.

Simple discrete systems. A *discrete thermodynamic system* $\boldsymbol{\Sigma}$ is a collection $\boldsymbol{\Sigma} = \cup_{A=1}^{N} \boldsymbol{\Sigma}_A$ of a finite number of interacting simple thermodynamic systems $\boldsymbol{\Sigma}_A$. By definition, a *simple thermodynamic system* is a macroscopic system for which one (scalar) thermal variable and a finite set of mechanical variables are sufficient to describe entirely the state of the system.

F. Nielsen and F. Barbaresco (Eds.): GSI 2017, LNCS 10589, pp. 401–409, 2017.
https://doi.org/10.1007/978-3-319-68445-1_47

Variational formulation. Let Q be the configuration manifold associated to the mechanical variables of the simple system and denote by TQ and T^*Q its tangent and cotangent bundles. The Lagrangian of a simple thermodynamic system is a function

$$L : TQ \times \mathbb{R} \to \mathbb{R}, \quad (q, v, S) \mapsto L(q, v, S),$$

where $S \in \mathbb{R}$ is the entropy. We assume that the system is subject to exterior and friction forces given by fiber preserving maps $F^{\mathrm{ext}}, F^{\mathrm{fr}} : TQ \times \mathbb{R} \to T^*Q$, and to an external heat power supply $P_H^{\mathrm{ext}}(t)$.

We say that a curve $(q(t), S(t)) \in Q \times \mathbb{R}$, $t \in [t_1, t_2] \subset \mathbb{R}$ is a *solution of the variational formulation of nonequilibrium thermodynamics* if it satisfies the variational condition

$$\delta \int_{t_1}^{t_2} L(q, \dot{q}, S) dt + \int_{t_1}^{t_2} \left\langle F^{\mathrm{ext}}(q, \dot{q}, S), \delta q \right\rangle dt = 0, \quad \text{VARIATIONAL CONDITION} \tag{1}$$

for all variations $\delta q(t)$ and $\delta S(t)$ subject to the constraint

$$\frac{\partial L}{\partial S}(q, \dot{q}, S)\delta S = \left\langle F^{\mathrm{fr}}(q, \dot{q}, S), \delta q \right\rangle, \quad \text{VARIATIONAL CONSTRAINT} \tag{2}$$

with $\delta q(t_1) = \delta(t_2) = 0$, and also if it satisfies the phenomenological constraint

$$\frac{\partial L}{\partial S}(q, \dot{q}, S)\dot{S} = \left\langle F^{\mathrm{fr}}(q, \dot{q}, S), \dot{q} \right\rangle - P_H^{\mathrm{ext}}, \quad \text{PHENOMENOLOGICAL CONSTRAINT} \tag{3}$$

where $\dot{q} = \frac{dq}{dt}$ and $\dot{S} = \frac{dS}{dt}$.

From this variational formulation, we deduce the system of evolution equations for the simple thermodynamic system as

$$\begin{cases} \dfrac{d}{dt}\dfrac{\partial L}{\partial \dot{q}} - \dfrac{\partial L}{\partial q} = F^{\mathrm{fr}} + F^{\mathrm{ext}}, \\ \dfrac{\partial L}{\partial S}\dot{S} = \left\langle F^{\mathrm{fr}}, \dot{q} \right\rangle - P_H^{\mathrm{ext}}. \end{cases} \tag{4}$$

We note that the energy function, defined by $E = \left\langle \frac{\partial L}{\partial \dot{q}}, \dot{q} \right\rangle - L$ verifies $\frac{d}{dt}E = \left\langle F^{\mathrm{ext}}, \dot{q} \right\rangle + P_H^{\mathrm{ext}}$, i.e., the first law of thermodynamics.

Remark 1 (Phenomenological and variational constraints). The explicit expression of the constraint (3) involves phenomenological laws for the friction force F^{fr}, this is why we refer to it as a *phenomenological constraint.* The associated constraint (2) is called a *variational constraint* since it is a condition on the variations to be used in (1). Note that the constraint (3) is nonlinear and also that one passes from the variational constraint to the phenomenological constraint by formally replacing the variations δq, δS by the time derivatives $\dot{q}$, $\dot{S}$. Such a systematic correspondence between the phenomenological and variational constraints still holds for the general discrete systems, as we shall recall below.

For the case of adiabatically closed systems (i.e., $P_H^{\rm ext} = 0$), the evolution Eq. (4) can be geometrically formulated in terms of Dirac structures induced from the phenomenological constraint and from the canonical symplectic form on T^*Q or on $T^*(Q \times \mathbb{R})$, see [6].

1.2 Variational Formulation of Nonequilibrium Thermodynamics of Discrete Systems

Discrete systems. We now consider the case of a discrete system $\boldsymbol{\Sigma} = \cup_{A=1}^N \boldsymbol{\Sigma}_A$, composed of interconnecting simple systems $\boldsymbol{\Sigma}_A$, $A = 1, ..., N$ that can exchange heat and mechanical power, and interact with external heat sources. We follow the description of discrete systems given in [7,12].

The state of the discrete system $\boldsymbol{\Sigma}$ is described by geometric variables $q \in Q_{\boldsymbol{\Sigma}}$ and entropy variables S_A, $A = 1, ..., N$. The Lagrangian is a function

$$L : TQ_{\boldsymbol{\Sigma}} \times \mathbb{R}^N \to \mathbb{R}, \quad (q, \dot{q}, S_1, ..., S_N) \mapsto L(q, \dot{q}, S_1, ..., S_N). \tag{5}$$

We assume that the system is subject to external forces $F^{\rm ext} = \sum_{A=1}^N F^{{\rm ext}\to A} : TQ_{\boldsymbol{\Sigma}} \times \mathbb{R}^N \to T^*Q_{\boldsymbol{\Sigma}}$ and external heat power supply $P_H^{\rm ext} = \sum_{A=1}^N P_H^{{\rm ext}\to A}$.

The friction force associated to system $\boldsymbol{\Sigma}_A$ is $F^{{\rm fr}(A)} : TQ_{\boldsymbol{\Sigma}} \times \mathbb{R}^N \to T^*Q_{\boldsymbol{\Sigma}}$ and we define the total friction force $F^{\rm fr} := \sum_{A=1}^N F^{{\rm fr}(A)}$. The internal heat power exchange between $\boldsymbol{\Sigma}_A$ and $\boldsymbol{\Sigma}_B$ can be described by

$$P_H^{B\to A} = \kappa_{AB}(q, S^A, S^B)(T^B - T^A),$$

where $\kappa_{AB} = \kappa_{BA} \geq 0$ are the heat transfer phenomenological coefficients.

For simplicity, we ignore internal and external matter exchanges in this section. Hence, in particular, the system is closed.

A typical, and historically relevant, example of a discrete (non-simple) system is the *adiabatic piston*. We refer to [8] for a systematic treatment of the adiabatic piston from Stueckelberg's approach.

Variational formulation. Our variational formulation is based on the introduction of new variables, called *thermodynamic displacements*, that allow a systematic inclusion of all the irreversible processes involved in the system. In our case, since we only consider the irreversible processes of mechanical friction and heat conduction, we just need to introduce (in addition to the mechanical displacement q) the *thermal displacements*[1], Γ^A, $A = 1, ..., N$ such that $\dot{\Gamma}^A = T^A$, where Γ^A are monotonically increasing real functions of time t and hence the temperatures T^A of $\boldsymbol{\Sigma}_A$ take positive real values, i.e., $(T^1, ..., T^N) \in \mathbb{R}_+^N$. Each of these variables is accompanied with its dual variable Σ_A whose time rate of change is associated to the entropy production of the simple system $\boldsymbol{\Sigma}_A$.

[1] The notion of thermal displacement was first used by [13] and in the continuum setting by [9]. We refer to the Appendix of [11] for an historical account.

We say that a curve $\left(q(t), S_A(t), \Gamma^A(t), \Sigma_A(t)\right) \in Q_{\boldsymbol{\Sigma}} \times \mathbb{R}^{3N}$, $t \in [t_1, t_2] \subset \mathbb{R}$ is the *solution of the variational formulation of nonequilibrium thermodynamics* if it satisfies the *variational condition*

$$\delta \int_{t_1}^{t_2} \Big[L(q, \dot{q}, S_1, ... S_N) + \sum_{A=1}^{N} (S_A - \Sigma_A) \dot{\Gamma}^A \Big] dt + \int_{t_1}^{t_2} \left\langle F^{\rm ext}, \delta q \right\rangle dt = 0, \quad (6)$$

for all variations $\delta q(t), \delta\Gamma^A(t), \delta\Sigma_A(t)$ subject to the *variational constraint*

$$\frac{\partial L}{\partial S_A} \delta\Sigma_A = \left\langle F^{\rm fr(A)}, \delta q \right\rangle - \sum_{B=1}^{N} \kappa_{AB} (\delta\Gamma^B - \delta\Gamma^A), \quad \text{(no sum on } A) \quad (7)$$

with $\delta q(t_i) = 0$ and $\delta\Gamma(t_i) = 0$, for $i = 1, 2$, and also if it satisfies the nonlinear *phenomenological constraint*

$$\frac{\partial L}{\partial S_A} \dot{\Sigma}_A = \left\langle F^{\rm fr(A)}, \dot{q} \right\rangle - \sum_{B=1}^{N} \kappa_{AB} (\dot{\Gamma}^B - \dot{\Gamma}^A) - P_H^{\rm ext \to A}. \quad \text{(no sum on } A) \quad (8)$$

From this variational formulation, we deduce the system of evolution equations for the discrete thermodynamic system as

$$\begin{cases} \dfrac{d}{dt}\dfrac{\partial L}{\partial \dot{q}} - \dfrac{\partial L}{\partial q} = \displaystyle\sum_{A=1}^{N} F^{\rm fr(A)} + F^{\rm ext}, \\[2ex] \dfrac{\partial L}{\partial S_A} \dot{S}_A = \left\langle F^{\rm fr(A)}, \dot{q} \right\rangle + \displaystyle\sum_{B=1}^{N} \kappa_{AB} \left(\dfrac{\partial L}{\partial S_B} - \dfrac{\partial L}{\partial S_A} \right) - P_H^{\rm ext \to A}, \quad A = 1, ..., N. \end{cases}$$

We refer to [3] for the details regarding the treatment of discrete systems. In a similar way with the situation of simple thermodynamic systems, one passes from the variational constraint (7) to the phenomenological constraint (8) by formally replacing the δ-variations $\delta q, \delta\Sigma_A, \delta\Gamma_A$ by the time derivatives $\dot{q}, \dot{\Sigma}_A, \dot{\Gamma}_A$ (see Remark 1). This is possible thanks to the introduction of the thermodynamic displacements Γ_A.

2 The Heat Conducting Viscous Fluid

We shall now systematically extend to the continuum setting the previous variational formulation by focalising on the case of a heat conducting viscous fluid. We refer to [2,4] for the extension of this approach to the case of diffusion, chemical reaction, and phase changes in fluid dynamics, and to [5] for the variational formulation in terms of the free energy.

Configuration space and geometric setting. We assume that the domain occupied by the fluid is a smooth compact manifold $\mathcal{D}$ with smooth boundary $\partial\mathcal{D}$.

The configuration space is $Q = \mathrm{Diff}_0(\mathcal{D})$, the group of all diffeomorphisms[2] of $\mathcal{D}$ that keep the boundary $\partial\mathcal{D}$ pointwise fixed. This corresponds to no-slip boundary conditions. We assume that the manifold $\mathcal{D}$ is endowed with a Riemannian metric g. The Levi-Civita covariant derivative, the sharp and flat operator associated to g are denoted as ∇^g, $\flat^g : T\mathcal{D} \to T^*\mathcal{D}$, and $\sharp^g : T^*\mathcal{D} \to T\mathcal{D}$.

Given a curve φ_t of diffeomorphisms, starting at the identity at $t = 0$, we denote by $x = \varphi_t(X) = \varphi(t, X) \in \mathcal{D}$ the current position of a fluid particle which at time $t = 0$ is at $X \in \mathcal{D}$. The *mass density* $\varrho(t, X)$ and the *entropy density* $S(t, X)$ in the Lagrangian (or material) description are related to the corresponding quantities $\rho(t, x)$ and $s(t, x)$ in the Eulerian (or spatial) description as

$$\varrho(t,X) = \rho(t,\varphi_t(X))J_{\varphi_t}(X) \quad \text{and} \quad S(t,X) = s(t,\varphi_t(X))J_{\varphi_t}(X), \tag{9}$$

were J_{φ_t} denotes the Jacobian of φ_t relative to the Riemannian metric g, i.e., $\varphi_t^*\mu_g = J_{\varphi_t}\mu_g$, with μ_g the Riemannian volume form.

From the conservation of the total mass, we have $\varrho(t, X) = \varrho_{\mathrm{ref}}(X)$, i.e., the mass density in the material description is time independent. It therefore appears as a parameter in the Lagrangian function and in the variational formulation.

Lagrangian. In a similar way to the case of discrete systems in (5), the Lagrangian in the material description is a map

$$L_{\varrho_{\mathrm{ref}}} : T\,\mathrm{Diff}_0(\mathcal{D}) \times \mathcal{F}(\mathcal{D}) \to \mathbb{R}, \quad (\varphi, \dot{\varphi}, S) \mapsto L_{\varrho_{\mathrm{ref}}}(\varphi, \dot{\varphi}, S),$$

where $T\,\mathrm{Diff}_0(\mathcal{D})$ is the tangent bundle to $\mathrm{Diff}_0(\mathcal{D})$ and $\mathcal{F}(\mathcal{D})$ is a space of real valued functions on $\mathcal{D}$ with a given high enough regularity, so that all the formulas used below are valid. The index notation in $L_{\varrho_{\mathrm{ref}}}$ is used to recall that L depends parametrically on ϱ_{ref}.

Consider a fluid with a given state equation $\varepsilon = \varepsilon(\rho, s)$ where ε is the internal energy density. The Lagrangian is given by

$$\begin{aligned} L_{\varrho_{\mathrm{ref}}}(\varphi,\dot{\varphi},S) &= \int_{\mathcal{D}} \frac{1}{2}\varrho_{\mathrm{ref}}(X)|\dot{\varphi}(X)|_g^2\mu_g(X) - \int_{\mathcal{D}} \varepsilon\left(\frac{\varrho_{\mathrm{ref}}(X)}{J_\varphi(X)}, \frac{S(X)}{J_\varphi(X)}\right) J_\varphi(X)\mu_g(X) \\ &= \int_{\mathcal{D}} \mathfrak{L}(\varphi(X), \dot{\varphi}(X), T_X\varphi, \varrho_{\mathrm{ref}}(X), S(X))\mu_g(X), \end{aligned} \tag{10}$$

where $T_X\varphi : T_X\mathcal{D} \to T_{\varphi(X)}\mathcal{D}$ is the tangent map to φ. The first term of $L_{\varrho_{\mathrm{ref}}}$ represents the total kinetic energy of the fluid, computed with the help of the Riemannian metric g, and the second term represents the total internal energy. The second term is deduced from $\varepsilon(\rho, s)$ by using the relations (9). In the second

[2] In this paper we do not describe the functional analytic setting needed to rigorously work in the framework of infinite dimensional manifolds. For example, one can assume that the diffeomorphisms are of some given Sobolev class, regular enough (at least of class C^1), so that $\mathrm{Diff}_0(\mathcal{D})$ is a smooth infinite dimensional manifold and a topological group with smooth right translation, [1].

line we defined the Lagrangian density $\mathfrak{L}(\varphi, \dot{\varphi}, T\varphi, \varrho_{\rm ref}, S)\mu_g$ as the integrand of the Lagrangian L. The *material temperature* is given by

$$\mathfrak{T} = -\frac{\partial \mathfrak{L}}{\partial S} = \frac{\partial \varepsilon}{\partial s}(\rho, s) \circ \varphi = T \circ \varphi,$$

where T is the *Eulerian temperature*. The derivative of $\mathfrak{L}$ with respect to $T_X\varphi$ is the conservative Piola-Kirchhoff stress tensor

$$\mathbf{P}^{\rm cons} := -\left[\frac{\partial \mathfrak{L}}{\partial T_X\varphi}\right]^{\sharp_g}.$$

Variational formulation in material description. The continuum version of the variational formulation (6)–(8) reads

$$\delta \int_{t_1}^{t_2} \left[L_{\varrho_{\rm ref}}(\varphi, \dot{\varphi}, S) - \int_{\mathcal{D}} (\dot{S} - \dot{\Sigma})\Gamma\mu_g\right] dt = 0, \qquad \text{VARIATIONAL CONDITION} \quad (11)$$

with variational and phenomenological and constraint

$$\dot{\Gamma}\delta\Sigma = (\mathbf{P}^{\rm fr})^{\flat_g} : \nabla^g \delta\varphi - \mathbf{J}_S \cdot \mathbf{d}\delta\Gamma \qquad \text{VARIATIONAL CONSTRAINT} \quad (12)$$

$$\dot{\Gamma}\dot{\Sigma} = (\mathbf{P}^{\rm fr})^{\flat_g} : \nabla^g \dot{\varphi} - \mathbf{J}_S \cdot \mathbf{d}\dot{\Gamma} + \varrho_{\rm ref} R, \quad \text{PHENOMENOLOGICAL CONSTRAINT} \quad (13)$$

where $\mathbf{P}^{\rm fr}(t, X)$ is the friction Piola-Kirchhoff tensor, $\mathbf{J}_S(t, X)$ is the entropy flux density, and $\varrho_{\rm ref}(X)R(t, X)$ is the heat power supply density. In (12) and (13), the double point ":" indicates the contraction with respect to both indices.

In the same way as the case of discrete systems, the introduction of the variables Γ and Σ allows us to propose a variational formulation with a very simple and physically meaningful structure:

- The criticality condition (11) is an extension of the Hamilton critical action principle for fluid dynamics in material representation.

- The nonholonomic constraint (13) is the expression of the power density associated to all the irreversible processes involved (heat transport and viscosity in our case) in the entropy production. This constraint is of phenomenological nature, each of the "thermodynamic forces" being related to the fluxes characterizing an irreversible process via phenomenological laws, see Remark 2 below. The introduction of the variable Γ allows to write this constraint as a sum of force densities acting on velocities, namely $\mathbf{P}^{\rm fr}$ "acting" on $\frac{d}{dt}\varphi$ and $\mathbf{J}_S$ "acting" on $\frac{d}{dt}\Gamma$, resulting in a *power* or *rate of work* density.

- Concerning the virtual constraint (12), the occurrence of the time derivative in (13), also allows us to systematically replace all velocities by "δ-derivatives", i.e., virtual displacements and to formulate the variational constraint as a sum of virtual thermodynamic work densities associated to each of the irreversible processes. It is important to note that this interpretation is possible thanks to the introduction of the variable $\Gamma(t, X)$ whose time derivative is identified with the temperature $\mathfrak{T}(t, X)$:

$$\frac{d}{dt}\Gamma = -\frac{\partial \mathfrak{L}}{\partial S} =: \mathfrak{T},$$

from the stationarity condition associated with the variation δS in the variational principle.

By computing the variations in (11), using the condition $\delta\varphi|_{\partial\mathcal{D}} = 0$ associated to no-slip boundary conditions, and using the variational and phenomenological constraints (12) and (13), we get the following result, see [3] for details.

Proposition 1. *In material representation, the evolution equations for a heat conducting viscous fluid given by*

$$\begin{cases} \varrho_{\text{ref}} \dfrac{D\mathbf{V}}{Dt} = \operatorname{DIV}(\mathbf{P}^{\text{cons}} + \mathbf{P}^{\text{fr}}), \\ \mathfrak{T}(\dot{S} + \operatorname{DIV} \mathbf{J}_S) = (\mathbf{P}^{\text{fr}})^{\flat_g} : \nabla^g \dot{\varphi} - \mathbf{J}_S \cdot \mathbf{d}\mathfrak{T} + \varrho_{\text{ref}} R, \end{cases}$$

with no-slip boundary conditions, follow from the variational formulation for non-equilibrium thermodynamics (11) with variational and phenomenological constraints (12), (13), and with $\delta\Gamma|_{\partial\mathcal{D}} = 0$. If the constraint $\delta\Gamma|_{\partial\mathcal{D}} = 0$ is removed, then it implies $\mathbf{J}_S \cdot \mathbf{n}^{\flat_g} = 0$, where $\mathbf{n}$ is the outward pointing unit vector field on $\partial\mathcal{D}$. If, in addition, $\varrho_{\text{ref}} R = 0$, then the fluid is adiabatically closed.

Variational formulation in spatial description. In terms of the Eulerian velocity $\mathbf{v} = \dot{\varphi} \circ \varphi^{-1}$ and Eulerian variables ρ and s, the Lagrangian (10) reads

$$\ell(\mathbf{v}, \rho, s) = \int_{\mathcal{S}} \frac{1}{2}\rho|\mathbf{v}|_g^2 \mu_g - \int_{\mathcal{S}} \varepsilon(\rho, s)\mu_g.$$

The material variational formulation (11)–(13) induces the spatial variational formulation

$$\delta \int_{t_1}^{t_2} \left[\ell(\mathbf{v}, \rho, s) - \int_{\mathcal{S}} [D_t(s - \sigma)] \, \gamma \, \mu_g \right] dt = 0 \tag{14}$$

with respect to variations

$$\delta\mathbf{v} = \partial_t \boldsymbol{\zeta} + [\mathbf{v}, \boldsymbol{\zeta}], \quad \delta\rho = -\operatorname{div}(\rho\boldsymbol{\zeta}), \quad \delta\gamma, \quad \text{and} \quad \delta\sigma, \tag{15}$$

and subject to the variational and phenomenological constraints

$$D_t \gamma \bar{D}_\delta \sigma = (\boldsymbol{\sigma}^{\text{fr}})^{\flat_g} : \nabla\boldsymbol{\zeta} - \mathbf{j}_S \cdot \mathbf{d} D_\delta \gamma, \tag{16}$$

$$D_t \gamma \bar{D}_t \sigma = (\boldsymbol{\sigma}^{\text{fr}})^{\flat_g} : \nabla\mathbf{v} - \mathbf{j}_S \cdot \mathbf{d} D_t \gamma + \rho r, \tag{17}$$

where γ, σ, $\mathbf{j}_S$, and $\boldsymbol{\sigma}^{\text{fr}}$ are the Eulerian quantities associated to Γ, Σ, $\mathbf{J}_S$, and $\mathbf{P}^{\text{fr}}$, and we used the notations $D_t f = \partial_t f + \mathbf{v} \cdot \mathbf{d} f$, $D_\delta f = \delta f + \boldsymbol{\zeta} \cdot \mathbf{d} f$,

$\bar{D}_t f = \partial_t f + \operatorname{div}(f\mathbf{v})$ and $\bar{D}_\delta f = \delta f + \operatorname{div}(f\boldsymbol{\zeta})$ for the Lagrangian time derivatives and variations of scalar fields and density fields. The first two expressions in (15) are obtained by taking the variations with respect to φ, $\mathbf{v}$, and ρ, of the relations $\mathbf{v} = \dot{\varphi} \circ \varphi^{-1}$ and $\rho = (\varrho_{\rm ref} \circ \varphi^{-1})J_\varphi$ and by defining the vector field $\boldsymbol{\zeta} := \delta\varphi \circ \varphi^{-1}$. These formulas can be directly justified by employing the Euler-Poincaré reduction theory on Lie groups, [10].

By computing the variations in (14), using the condition $\boldsymbol{\zeta}|_{\partial\mathcal{D}} = 0$ associated to no-slip boundary conditions, and using the variational and phenomenological constraints (16) and (17), we get the following result, see [3] for details.

Proposition 2. *In spatial representation, the evolution equations for a viscous heat conducting fluid given by*

$$\begin{cases} \rho(\partial_t \mathbf{v} + \nabla_{\mathbf{v}} \mathbf{v}) = -\operatorname{grad} p + \operatorname{div} \boldsymbol{\sigma}^{\rm fr}, \quad p = \frac{\partial \varepsilon}{\partial \rho}\rho + \frac{\partial \varepsilon}{\partial s} s - \varepsilon \\ \partial_t \rho + \operatorname{div}(\rho \mathbf{v}) = 0 \\ T(\partial_t s + \operatorname{div}(s\mathbf{v}) + \operatorname{div} \mathbf{j}_S) = (\boldsymbol{\sigma}^{\rm fr})^\flat : \nabla \mathbf{v} - \mathbf{j}_S \cdot \mathbf{d}T + \rho r, \quad T = \frac{\partial \varepsilon}{\partial s}, \end{cases} \tag{18}$$

with no-slip boundary conditions, follow from the variational condition for non-equilibrium thermodynamics given in (14)*, with variational and phenomenological constraints* (15), (16), (17)*, and with* $\delta\gamma|_{\partial\mathcal{D}} = 0$. *If the constraint* $\delta\gamma|_{\partial\mathcal{D}} = 0$ *is removed, then it implies* $\mathbf{j}_S \cdot \mathbf{n}^{\flat_g} = 0$. *If, in addition,* $\varrho_{\rm ref} R = 0$*, then the fluid is adiabatically closed.*

Remark 2 (Thermodynamic phenomenology). In order to close the system (18), it is necessary to provide phenomenological expressions of the thermodynamic fluxes in terms of the thermodynamic affinities, compatible with the second law of thermodynamics. In our case, the thermodynamic fluxes are $\boldsymbol{\sigma}^{\rm fr}$ and $\mathbf{j}_s$ and we have the well-known relations:

$$\boldsymbol{\sigma}^{\rm fr} = 2\mu(\operatorname{Def} \mathbf{v})^\sharp + \left(\zeta - \frac{2}{3}\mu\right)(\operatorname{div} \mathbf{v}) g^\sharp \quad \text{and} \quad T\mathbf{j}_s^\flat = -\kappa \mathbf{d}T \quad \text{(Fourier law)},$$

where $\operatorname{Def} \mathbf{v} = \frac{1}{2}(\nabla \mathbf{v} + \nabla \mathbf{v}^{\mathsf{T}})$, $\mu \geq 0$ is the first coefficient of viscosity (shear viscosity), $\zeta \geq 0$ is the second coefficient of viscosity (bulk viscosity), and $\kappa \geq 0$ is the thermal conductivity. Generally, these coefficients depend on ρ and T.

Acknowledgements. F.G.B. is partially supported by the ANR project GEOMFLUID, ANR-14-CE23-0002-01; H.Y. is partially supported by JSPS Grant-in-Aid for Scientific Research (26400408, 16KT0024, 24224004), Waseda University (SR2017K-167), and the MEXT "Top Global University Project".

References

1. Ebin, D.G., Marsden, J.E.: Groups of diffeomorphisms and the motion of an incompressible fluid. Ann. Math. **92**, 102–163 (1970)
2. Gay-Balmaz, F.: A variational derivation of the thermodynamics of a moist atmosphere with irreversible processes (2017). https://arxiv.org/pdf/1701.03921.pdf

3. Gay-Balmaz, F., Yoshimura, H.: A Lagrangian variational formulation for nonequilibrium thermodynamics. Part I: discrete systems. J. Geom. Phys. **111**, 169–193 (2016)
4. Gay-Balmaz, F., Yoshimura, H.: A Lagrangian variational formulation for nonequilibrium thermodynamics. Part II: continuum systems. J. Geom. Phys. **111**, 194–212 (2016)
5. Gay-Balmaz, F., Yoshimura, H.: A free energy Lagrangian variational formulation of the Navier-Stokes-Fourier system. Int. J. Geom. Methods Mod. Phys. (2017, to appear)
6. Gay-Balmaz, F., Yoshimura, H.: Dirac structures in nonequilibrium thermodynamics, preprint (2017). https://arxiv.org/pdf/1704.03935.pdf
7. Gruber, C.: Thermodynamique et Mécanique Statistique, Institut de physique théorique. EPFL (1997)
8. Gruber, C.: Thermodynamics of systems with internal adiabatic constraints: time evolution of the adiabatic piston. Eur. J. Phys. **20**, 259–266 (1999)
9. Green, A.E., Naghdi, P.M.: A re-examination of the basic postulates of thermomechanics. Proc. R. Soc. Lon. Ser. A Math. Phys. Eng. Sci. **432**(1885), 171–194 (1991)
10. Holm, D.D., Marsden, J.E., Ratiu, T.S.: The Euler-Poincaré equations and semidirect products with applications to continuum theories. Adv. Math. **137**, 1–81 (1998)
11. Podio-Guidugli, P.: A virtual power format for thermomechanics. Contin. Mech. Thermodyn. **20**(8), 479–487 (2009)
12. Stueckelberg, E.C.G., Scheurer, P.B.: Thermocinétique phénoménologique galiléenne. Birkhäuser, Basel (1974)
13. von Helmholtz, H.: Studien zur Statik monocyklischer Systeme. Sitzungsberichte der Königlich Preussischen Akademie der Wissenschaften zu Berlin, pp. 159–177 (1884)

Dirac Structures in Nonequilbrium Thermodynamics

Hiroaki Yoshimura[1(✉)] and François Gay-Balmaz[2]

[1] School of Science and Engineering, Waseda University, Okubo, Shinjuku, Tokyo 169-8555, Japan
yoshimura@waseda.jp

[2] CNRS & Ecole Normale Supérieure, LMD, IPSL, 24 Rue Lhomond, 75005 Paris, France
gaybalma@lmd.ens.fr

Abstract. In this paper, we show that the evolution equations for nonequilibrium thermodynamics can be formulated in terms of Dirac structures on the Pontryagin bundle $\mathsf{P} = T\mathsf{Q} \oplus T^*\mathsf{Q}$, where $\mathsf{Q} = Q \times \mathbb{R}$ denotes the thermodynamic configuration manifold. In particular, we extend the use of Dirac structures from the case of linear nonholonomic constraints to the case of nonlinear nonholonomic constraints. Such a nonlinear constraint comes from the entropy production associated with irreversible processes in nonequilibrium thermodynamics. We also develop the induced Dirac structure on $N = T^*Q \times \mathbb{R}$ and the associated Lagrange-Dirac and Hamilton-Dirac dynamical formulations.

Keywords: Nonequilibrium thermodynamics · Dirac structures · Nonlinear constraints · Irreversible processes · Implicit systems

1 Dirac Structures in Thermodynamics

Dirac structures are known as a geometric object that generalizes both (almost) Poisson structures and (pre)symplectic structures on manifolds (see, [2,4]). They were named after Dirac's theory of constraints [3], and various physical systems with constraints such as electric circuits and nonholonomic mechanical systems are shown to be represented in the context of Dirac structures and the associated *implicit Hamiltonian systems* [1,11,12]. On the Lagrangian side, it was shown by [13,14] that the notion of *implicit Lagrangian systems* can be developed in the context of induced Dirac structures, together with its associated variational structure given by the Lagrange-d'Alembert-Pontryagin principle.

1.1 Fundamental Setting for Thermodynamics

Simple Discrete Systems. A *simple thermodynamical system*[1] is a macroscopic system for which one (scalar) thermal variable and a finite set of mechanical variables are sufficient to describe entirely the state of the system. From the

[1] In [9] they are called élément de système (French). We choose to use the English terminology simple system instead of system element.

F. Nielsen and F. Barbaresco (Eds.): GSI 2017, LNCS 10589, pp. 410–417, 2017.
https://doi.org/10.1007/978-3-319-68445-1_48

second law of thermodynamics, we can always choose such a thermal variable as entropy S (see [9]).

In this paper, we focus on the particular case of simple adiabatically closed systems; namely, we assume that there is no exchange of matter and heat with the exterior of the system

Constraints for the Thermodynamics of Simple Systems. Let us consider a simple thermodynamic system with a Lagrangian $L = L(q, v, S) : TQ \times \mathbb{R} \to \mathbb{R}$ and a friction force $F^{\rm fr} : TQ \times \mathbb{R} \to T^*Q$, where Q is a configuration manifold of the mechanical variables q of the system, and $\mathbb{R}$ denotes the space of the thermodynamic variable S. We introduce the *thermodynamic configuration manifold* $\mathsf{Q} := Q \times \mathbb{R}$. Following [5], we define the variational constraint as

$$C_V = \left\{ (q, S, v, W, \delta q, \delta S) \in T\mathsf{Q} \times_{\mathsf{Q}} T\mathsf{Q} \;\middle|\; \frac{\partial L}{\partial S}(q, v, S)\delta S = \left\langle F^{\rm fr}(q, v, S), \delta q \right\rangle \right\},$$

where $(q, S) \in \mathsf{Q}$, $(v, W) \in T_{(q,S)}\mathsf{Q}$, and $(\delta q, \delta S) \in T_{(q,S)}\mathsf{Q}$. Since $\frac{\partial L}{\partial S}(q, v, S) \neq 0$ (temperature is always positive), we obtain that C_V is a submanifold of $T\mathsf{Q} \times_{\mathsf{Q}} T\mathsf{Q}$ of codimension one. For each fixed $(q, S, v, W) \in T\mathsf{Q}$, the annihilator of $C_V(q, S, v, W)$ is given by

$$C_V(q, S, v, W)^{\circ} = \left\{ (q, S, \alpha, \mathcal{T}) \in T^*_{(q,S)}\mathsf{Q} \;\middle|\; \alpha \frac{\partial L}{\partial S}(q, v, S) = -\mathcal{T} F^{\rm fr}(q, v, S) \right\}.$$

The kinematic constraint $C_K \subset T\mathsf{Q}$ is defined from C_V as

$$C_K = \left\{ (q, S, v, W) \in T\mathsf{Q} \;\middle|\; \frac{\partial L}{\partial S}(q, v, S) W = \left\langle F^{\rm fr}(q, v, S), v \right\rangle \right\}.$$

1.2 Dirac Dynamical Systems on the Pontryagin Bundle

Let $\mathsf{P} = T\mathsf{Q} \oplus T^*\mathsf{Q}$ be the Pontryagin bundle over Q. We shall use the notation $x = (q, S, v, W, p, \Lambda)$ for an element of the Pontryagin bundle P. A distribution Δ_{P} on P may be induced from C_V using the projection $\pi_{(\mathsf{P},\mathsf{Q})} : \mathsf{P} \to \mathsf{Q}$, $(q, S, v, W, p, \Lambda) \mapsto (q, S)$ as $\Delta_{\mathsf{P}}(x) := (T_x \pi_{(\mathsf{P},\mathsf{Q})})^{-1}(C_V(q, S, v, W))$. It is locally given by

$$\Delta_{\mathsf{P}}(x) := \left\{ (x, \delta x) \in T\mathsf{P} \;\middle|\; \frac{\partial L}{\partial S}(q, v, S)\delta S = \left\langle F^{\rm fr}(q, v, S), \delta q \right\rangle \right\}.$$

Let $\Omega_{T^*\mathsf{Q}}$ be the canonical symplectic structure on $T^*\mathsf{Q}$. We define the ***induced Dirac structure*** on P from the distribution Δ_{P} and the presymplectic form $\omega_{\mathsf{P}} = \pi^*_{(\mathsf{P},T^*\mathsf{Q})}\Omega_{T^*\mathsf{Q}}$ on P by

$$\begin{aligned} D_{\Delta_{\mathsf{P}}}(x) := \big\{ (v_x, \alpha_x) \in T_x\mathsf{P} \times T^*_x\mathsf{P} \;\mid\; & v_x \in \Delta_{\mathsf{P}}(x) \text{ and} \\ & \langle \alpha_x, w_x \rangle = \omega_{\mathsf{P}}(x)(v_x, w_x) \;\text{ for all }\; w_x \in \Delta_{\mathsf{P}}(x) \big\}. \end{aligned}$$

Using the local expressions $\dot{x} = (\dot{q}, \dot{S}, \dot{v}, \dot{W}, \dot{p}, \dot{\Lambda}) \in T_x\mathsf{P}$, and $\zeta = (\alpha, \mathcal{T}, \beta, \Upsilon, u, \Psi) \in T_x^*\mathsf{P}$, the condition $((x, \dot{x}), (x, \zeta)) \in D_{\Delta_\mathsf{P}}(x)$ is equivalently given by

$$\begin{cases} (\dot{p} + \alpha)\dfrac{\partial L}{\partial S}(q, v, S) = -(\dot{\Lambda} + \mathcal{T})F^{\mathrm{fr}}(q, v, S), \\ \dfrac{\partial L}{\partial S}(q, v, S)\dot{S} = \left\langle F^{\mathrm{fr}}(q, v, S), \dot{q} \right\rangle, \\ \beta = 0, \quad \Upsilon = 0, \quad u = \dot{q}, \quad \Psi = \dot{S}. \end{cases} \tag{1}$$

Dirac Dynamical Formulation on $\mathsf{P} = T^*\mathsf{Q} \oplus T^*\mathsf{Q}$. Let $\mathcal{E} : \mathsf{P} \to \mathbb{R}$ be the generalized energy given by

$$\mathcal{E}(q, S, v, W, p, \Lambda) = \langle p, v \rangle + \Lambda W - L(q, v, S).$$

Using $\mathbf{d}\mathcal{E}(q, S, v, W, p, \Lambda) = \left(q, S, v, W, p, \Lambda, -\frac{\partial L}{\partial q}, -\frac{\partial L}{\partial S}, p - \frac{\partial L}{\partial v}, \Lambda, v, W\right)$ and the condition (1), the Dirac dynamical system

$$((x, \dot{x}), \mathbf{d}\mathcal{E}(x)) \in D_{\Delta_\mathsf{P}}(x)$$

yields the evolution equations of the thermodynamics of simple systems:

$$\begin{cases} \left(\dot{p} - \dfrac{\partial L}{\partial q}(q, v, S)\right)\dfrac{\partial L}{\partial S}(q, v, S) = -\left(\dot{\Lambda} - \dfrac{\partial L}{\partial S}(q, v, S)\right)F^{\mathrm{fr}}(q, v, S), \\ \dfrac{\partial L}{\partial S}(q, v, S)\dot{S} = \left\langle F^{\mathrm{fr}}(q, v, S), \dot{q} \right\rangle, \\ p = \dfrac{\partial L}{\partial v}, \quad \Lambda = 0, \quad v = \dot{q}, \quad W = \dot{S}, \end{cases}$$

which are finally written as

$$\begin{cases} \dfrac{d}{dt}\dfrac{\partial L}{\partial \dot{q}}(q(t), \dot{q}(t), S(t)) - \dfrac{\partial L}{\partial q}(q(t), \dot{q}(t), S(t)) = F^{\mathrm{fr}}(q(t), \dot{q}(t), S(t)), \\ \dfrac{\partial L}{\partial S}(q(t), \dot{q}(t), S(t))\dot{S}(t) = \left\langle F^{\mathrm{fr}}(q(t), \dot{q}(t), S(t)), \dot{q}(t) \right\rangle. \end{cases} \tag{2}$$

These are the evolution equations for the nonequilibrium thermodynamics of simple closed systems, see [5–7].

In the above, the temperature is defined by minus the partial derivative of the Lagrangian with respect to the entropy, namely, $T = -\frac{\partial L}{\partial S}$, which is assumed to be positive. The friction force F^{fr} is dissipative, that is $\left\langle F^{\mathrm{fr}}(q, \dot{q}, S), \dot{q} \right\rangle \leq 0$, for all $(q, \dot{q}, S) \in TQ \times \mathbb{R}$. For the case in which the force is linear in velocity, and in one dimension, we have $F^{\mathrm{fr}}(q, \dot{q}, S) = -\lambda(q, S)\dot{q}$, where $\lambda(q, S) \geq 0$ is the phenomenological coefficient, determined experimentally. The ***internal entropy production*** of the simple system is given by

$$I(t) = -\frac{1}{T}\left\langle F^{\mathrm{fr}}(q, \dot{q}, S), \dot{q} \right\rangle = \frac{1}{T}\lambda(q, S)\dot{q}^2.$$

2 The Lagrange-Dirac Formulation

2.1 Induced Dirac Structures on $N = T^*Q \times \mathbb{R}$.

Here we present the thermodynamic analogue of the Lagrange-Dirac formulation in [13] for nonholonomic mechanics. Namely, we develop the Lagrange-Dirac formulation on $N = T^*Q \times \mathbb{R}$ associated with the induced Dirac structure on N from the variational constraint $C_V \subset TQ \times_{\mathsf{Q}} TQ$ and the canonical symplectic form on T^*Q.

Constraints. We assume that the Lagrangian $L = L(q, v, S) : TQ \times \mathbb{R} \to \mathbb{R}$ of a simple thermodynamic system is *hyperregular with respect to the mechanical variables* (q, v), namely, the map

$$\mathbb{F}L_S : TQ \to T^*Q, \quad (q, v) \mapsto \left(q, \frac{\partial L}{\partial v}(q, v, S) \right)$$

is a diffeomorphism for each fixed $S \in \mathbb{R}$. Given the variational constraint C_V, we can define the constraint $\mathscr{C}_V \subset T^*\mathsf{Q} \times_{\mathsf{Q}} T\mathsf{Q}$ as

$$\mathscr{C}_V(q, S, p, \Lambda) := C_V(q, S, v, W),$$

which can be explicitly described as

$$\mathscr{C}_V(q, S, p, \Lambda) = \left\{ (q, S, \delta q, \delta S) \mid -T(q, p, S)\delta S = \left\langle \mathcal{F}^{\rm fr}(q, p, S), \delta q \right\rangle \right\}.$$

In the above, $T(q, p, S) := -\frac{\partial L}{\partial S}(q, v, S)$ and $\mathcal{F}^{\rm fr}(q, p, S) := F^{\rm fr}(q, v, S)$, in which v is uniquely determined from the condition $\frac{\partial L}{\partial v}(q, v, S) = p$. Since $\mathscr{C}_V$ does not depend on W, we can define from C_V the constraint $\mathscr{C}_V(q, S, p, \Lambda) \in T_{(q,S)}\mathsf{Q}$ and it induces the following distribution on N:

$$\Delta_N(q, S, p) := \left(T_{(q,S,p)} \pi_{(N,\mathsf{Q})} \right)^{-1} \left(\mathscr{C}_V(q, S, p, \Lambda) \right),$$

locally given as

$$\Delta_N(q, S, p) = \left\{ (q, S, p, \delta q, \delta S, \delta p) \in TN \;\middle|\; -T(q, S, p)\delta S = \left\langle \mathcal{F}^{\rm fr}(q, S, p), \delta q \right\rangle \right\}.$$

Using the distribution $\Delta_N(q, S, p)$ and the presymplectic form $\omega_N = \pi^*_{(N,T^*Q)} \Omega_{T^*Q}$, where Ω_{T^*Q} is the canonical symplectic structure, the Dirac structure on N is defined by, for each $n = (q, S, p) \in N$,

$$\begin{aligned} D_{\Delta_N}(n) := \big\{ (v_n, \zeta_n) \in T_nN \times T^*_nN \;\mid\; & v_n \in \Delta_N(n) \text{ and} \\ & \langle \zeta_n, w_n \rangle = \omega_N(n)(v_n, w_n) \;\text{ for all }\; w_n \in \Delta_N(n) \big\}. \end{aligned}$$

Writing locally $(n, \dot{n}) \in TN$ and $(n, \zeta) \in T^*N$, where $\dot{n} = (\dot{q}, \dot{S}, \dot{p})$, and $\zeta = (\alpha, \mathcal{T}, u)$, the condition $\left((n, \dot{n}), (n, \zeta)\right) \in D_{\Delta_N}(n)$ is equivalent to, for each $n = (q, S, p)$,

$$\begin{cases} (\dot{p} + \alpha)T(q, S, p) = \mathcal{T}\mathcal{F}^{\rm fr}(q, S, p), \\ T(q, S, p)\dot{S} = -\left\langle \mathcal{F}^{\rm fr}(q, S, p), \dot{q} \right\rangle, \\ u = \dot{q}. \end{cases} \tag{3}$$

2.2 The Lagrange-Dirac Systems

Recall from [13] that the Dirac differential for a Lagrangian $L : TQ \to \mathbb{R}$ is defined by using the symplectic diffeomorphism $\gamma_Q : T^*TQ \to T^*T^*Q$, locally given by $\gamma_Q(q, v, \alpha, p) = (q, p, -\alpha, v)$, as introduced in [10]. For the case of thermodynamics, we introduce the symplectic diffeomorphism

$$\widehat{\gamma}_Q : T^*(TQ \times \mathbb{R}) \to T^*(T^*Q \times \mathbb{R}), \quad (q, S, v, \alpha, \Lambda, p) \mapsto (q, S, p, -\alpha, -\Lambda, v).$$

Then we define the associated Dirac differential of L as

$$\widehat{\mathbf{d}}_D L(q, S, v) := (\widehat{\gamma}_Q \circ \mathbf{d}L)(q, S, v) = \left(q, S, \frac{\partial L}{\partial v}, -\frac{\partial L}{\partial q}, -\frac{\partial L}{\partial S}, v \right).$$

By this definition and the relations (3), it follows that we have

$$\big((q, S, p, \dot{q}, \dot{S}, \dot{p}), \widehat{\mathbf{d}}_D L(q, S, v)\big) \in D_{\Delta_N}(q, S, p),$$

if and only if

$$\begin{cases} \left(\dot{p} - \dfrac{\partial L}{\partial q}(q, v, S) \right) T(q, p, S) = -\dfrac{\partial L}{\partial S}(q, v, S) \mathcal{F}^{\mathrm{fr}}(q, p, S), \\ T(q, S, p)\dot{S} = -\left\langle \mathcal{F}^{\mathrm{fr}}(q, p, S), \dot{q} \right\rangle, \\ v = \dot{q}, \quad p = \dfrac{\partial L}{\partial v}(q, v, S). \end{cases}$$

The last equality comes from the fact that $(q, S, p, \dot{q}, \dot{S}, \dot{p})$ and $\widehat{\mathbf{d}}_D L(q, S, v)$ both belong to the fibers at $(q, S, p) \in T^*Q \times \mathbb{R}$ and hence we have the following theorem concerning the Lagrange-Dirac formulation for thermodynamics of simple systems.

Theorem 1. *Consider a simple system with a Lagrangian $L = L(q, v, S) : TQ \times \mathbb{R} \to \mathbb{R}$ and a friction force $F^{\mathrm{fr}} : TQ \times \mathbb{R} \to T^*Q$. Assume that L is hyperregular with respect to the mechanical variables (q, v) and define $T(q, p, S)$ and $\mathcal{F}^{\mathrm{fr}}(q, p, S)$ as before. Then the following statements are equivalent:*

- *The curve $(q(t), S(t), v(t), p(t)) \in M$ satisfies the equations*

$$\begin{cases} \left(\dot{p}(t) - \dfrac{\partial L}{\partial q}(q(t), v(t), S(t)) \right) T(q(t), p(t), S(t)) \\ \qquad = -\dfrac{\partial L}{\partial S}(q(t), v(t), S(t)) \mathcal{F}^{\mathrm{fr}}(q(t), p(t), S(t)), \\ T(q(t), v(t), S(t))\dot{S}(t) = -\left\langle \mathcal{F}^{\mathrm{fr}}(q(t), p(t), S(t)), \dot{q}(t) \right\rangle, \\ v(t) = \dot{q}(t), \quad p(t) = \dfrac{\partial L}{\partial v}(q(t), v(t), S(t)). \end{cases} \tag{4}$$

- *The curve $(q(t), S(t), v(t), p(t)) \in M$ satisfies the* ***Lagrange-Dirac system of the simple thermodynamic system***

$$\big((q, S, p, \dot{q}, \dot{S}, \dot{p}), \widehat{\mathbf{d}}_D L(q, S, v)\big) \in D_{\Delta_N}(q, S, p).$$

Moreover, the system (4) is an implicit version of the system of evolution Eq. (2) for the thermodynamics of simple systems.

3 The Hamilton-Dirac Formulation

3.1 Hamilton-Dirac Systems on $N = T^*Q \times \mathbb{R}$

Since we assume that $L : TQ \times \mathbb{R} \to \mathbb{R}$ is *hyperregular with respect to the mechanical variables* (see Sect. 2.1), we can define the *Hamiltonian function* $H : N = T^*Q \times \mathbb{R} \to \mathbb{R}$ by

$$H(q,p,S) = \langle p, \dot{q} \rangle - L(q,\dot{q},S),$$

where $\dot{q}$ is uniquely determined from (q,p,S) by the condition $\frac{\partial L}{\partial \dot{q}}(q,\dot{q},S) = p$.

We shall make use of the same distribution and the same Dirac structure of Sect. 2. In (3) we can directly write the constraint in the Hamiltonian setting in view of $T(q,S,p) = \frac{\partial H}{\partial S}(q,S,p)$. Then, it follows that the ***Hamilton-Dirac system***

$$\left((q,S,p,\dot{q},\dot{S},\dot{p}), \mathbf{d}H(q,S,p)\right) \in D_{\Delta_N}(q,S,p)$$

is equivalent to

$$\begin{cases} \left(\dot{p} + \dfrac{\partial H}{\partial q}(q,S,p)\right)\dfrac{\partial H}{\partial S}(q,S,p) = \dfrac{\partial H}{\partial S}(q,S,p)\mathcal{F}^{\mathrm{fr}}(q,S,p), \\ \dfrac{\partial H}{\partial p} = \dot{q}, \qquad \dfrac{\partial H}{\partial S}(q,S,p)\dot{S} = -\left\langle \mathcal{F}^{\mathrm{fr}}(q,S,p), \dot{q}\right\rangle. \end{cases}$$

We obtain the following theorem.

Theorem 2. *Consider a simple system with a Lagrangian* $L = L(q,v,S) : TQ \times \mathbb{R} \to \mathbb{R}$ *and a friction force* $F^{\mathrm{fr}} : TQ \times \mathbb{R} \to T^*Q$. *Assume that the Lagrangian is hyperregular with respect to the mechanical variables, consider the associated Hamiltonian* $H : T^*Q \times \mathbb{R} \to \mathbb{R}$ *and define* $\mathcal{F}^{\mathrm{fr}}(q,p,S)$ *as before. Then the following statements are equivalent:*

- *The curve* $(q(t),S(t),p(t)) \in N$ *satisfies the equations*

$$\begin{cases} \left(\dot{p}(t) + \dfrac{\partial H}{\partial q}(q(t),p(t),S(t))\right)\dfrac{\partial H}{\partial S}(q(t),p(t),S(t)) \\ \qquad\qquad = \dfrac{\partial H}{\partial S}(q(t),p(t),S(t))\mathcal{F}^{\mathrm{fr}}(q(t),p(t),S(t)), \\ -\dfrac{\partial H}{\partial S}(q(t),p(t),S(t))\dot{S}(t) = \left\langle \mathcal{F}^{\mathrm{fr}}(q(t),p(t),S(t)), \dot{q}(t)\right\rangle, \\ \dfrac{\partial H}{\partial p}(q(t),p(t),S(t)) = \dot{q}(t). \end{cases} \tag{5}$$

- *The curve* $(q(t),S(t),p(t)) \in N$ *satisfies the* **Hamilton-Dirac system**

$$\left((q,S,p,\dot{q},\dot{S},\dot{p}), \mathbf{d}H(q,S,p)\right) \in D_{\Delta_N}(q,S,p).$$

Moreover the system (5), *equivalently written as*

$$\begin{cases} \dot{p}(t) = -\dfrac{\partial H}{\partial q}(q(t), p(t), S(t)) + \mathcal{F}^{\mathrm{fr}}(q(t), p(t), S(t)), \\ \dot{q}(t) = \dfrac{\partial H}{\partial p}(q(t), p(t), S(t)), \\ \dfrac{\partial H}{\partial S}(q(t), p(t), S(t))\dot{S}(t) = -\left\langle \mathcal{F}^{\mathrm{fr}}(q(t), p(t), S(t)), \dot{q}(t) \right\rangle \end{cases} \tag{6}$$

is the Hamiltonian description of the system of evolution equations (2) *for the thermodynamics of simple systems.*

The Hamilton-d'Alembert Principle. To the Hamilton-Dirac formulation is naturally associated a variational structure. In our case, the variational formulation on $N = T^*Q \times \mathbb{R}$ is

$$\delta \int_{t_1}^{t_2} \Big[\langle p, \dot{q} \rangle - H(q, S, p) \Big] dt = 0 \tag{7}$$

for all variations $(\delta q(t), \delta S(t), \delta p(t))$ for the curve $(q(t), S(t), p(t)) \in N$ that satisfy

$$-\frac{\partial H}{\partial S}(q, p, S)\delta S = \left\langle \mathcal{F}^{\mathrm{fr}}(q, p, S), \delta q \right\rangle \tag{8}$$

with $\delta q(t_1) = \delta q(t_2) = 0$, and the curve is subject to the phenomenological constraint

$$-\frac{\partial H}{\partial S}(q, p, S)\dot{S} = \left\langle \mathcal{F}^{\mathrm{fr}}(q, p, S), \dot{q} \right\rangle. \tag{9}$$

The principle (7)–(9) is called the *Hamilton-d'Alembert principle.* From this principle one immediately obtains the system (6).

We refer to [8] for a thorough treatment of Dirac structures in nonequilibrium thermodynamics.

Acknowledgements. F.G.B. is partially supported by the ANR project GEOMFLUID, ANR-14-CE23-0002-01; H.Y. is partially supported by JSPS Grant-in-Aid for Scientific Research (26400408, 16KT0024, 24224004), Waseda University Grant for Special Research Project (2017K-167), and the MEXT "Top Global University Project".

References

1. Bloch, A.M., Crouch, P.E.: Representations of Dirac structures on vector spaces and nonlinear L–C circuits. In: Differential Geometry and Control, Boulder, CO, 1997, vol. 64, pp. 103–117. American Mathematical Society, Providence (1997)
2. Courant, T., Weinstein, A.: Beyond poisson structures. In: Action Hamiltoniennes de Groupes, Troisième théorème de Lie (Lyon, 1986), vol. 27, pp. 39–49. Hermann, Paris (1989)
3. Dirac, P.A.M.: Generalized hamiltonian dynamics. Canad. J. Math. **2**, 129–148 (1950)

4. Dorfman, I.: Dirac Structures and Integrability of Nonlinear Evolution Equations. Nonlinear Science Theory and Applications. Wiley, Chichester (1993)
5. Gay-Balmaz, F., Yoshimura, H.: A Lagrangian variational formulation for nonequilibrium thermodynamics. Part I: discrete systems. J. Geom. Phys. **111**, 169–193 (2016)
6. Gay-Balmaz, F., Yoshimura, H.: A Lagrangian variational formulation for nonequilibrium thermodynamics. Part II: continuum systems. J. Geom. Phys. **111**, 194–212 (2016)
7. Gay-Balmaz, F., Yoshimura, H.: A free energy Lagrangian variational formulation of the Navier-Stokes-Fourier system. Int. J. Geom. Methods Mod. Phys. (2017, to appear)
8. Gay-Balmaz, F., Yoshimura, H.: Dirac structures in nonequilibrium thermodynamics (2017). https://arxiv.org/pdf/1704.03935.pdf
9. Stueckelberg, E.C.G., Scheurer, P.B.: Thermocinétique phénoménologique galiléenne, Birkhäuser (1974)
10. Tulczyjew, W.M.: The Legendre transformation. Ann. Inst. H. Poincaré Sect. A **27**(1), 101–114 (1977)
11. van der Schaft, A.J., Maschke, B.M.: The Hamiltonian formulation of energy conserving physical systems with external ports. Archiv für Elektronik und Übertragungstechnik **49**, 362–371 (1995)
12. van der Schaft, A.J., Maschke, B.M.: Mathematical modelling of constrained Hamiltonian systems. In: Proceedings of the IFAC Symposium on NOLCOS, International Federation of Automatic Control, Tahoe City, CA, pp. 678–683 (1995)
13. Yoshimura, H., Marsden, J.E.: Dirac structures in Lagrangian mechanics. Part I: implicit Lagrangian systems. J. Geom. Phys. **57**, 133–156 (2006)
14. Yoshimura, H., Marsden, J.E.: Dirac structures in Lagrangian mechanics. Part II: Variational structures. J. Geom. Phys. **57**, 209–250 (2006)

About the Definition of Port Variables for Contact Hamiltonian Systems

Bernhard Maschke[1(✉)] and Arjan van der Schaft[2]

[1] Univ Lyon, CNRS, LAGEP UMR 5007, Université Claude Bernard Lyon 1, 43 Boulevard du 11 Novembre 1918, 69100 Villeurbanne, France
bernhard.maschke@univ-lyon1.fr

[2] Johann Bernoulli Institute for Mathematics and Computer Science, University of Groningen, PO Box 407, 9700 AK Groningen, The Netherlands

Abstract. Extending the formulation of reversible thermodynamical transformations to the formulation of irreversible transformations of open thermodynamical systems different classes of nonlinear control systems has been defined in terms of control Hamiltonian systems defined on a contact manifold. In this paper we discuss the relation between the definition of variational control contact systems and the input-output contact systems. We have first given an expression of the variational control contact systems in terms of a nonlinear control systems. Secondly we have shown that the conservative input-output contact systems are a subclass of the contact variational systems with integrable output dynamics.

Keywords: Open irreversible thermodynamic systems · Nonlinear control systems · Hamiltonian systems on contact manifolds

1 Introduction

Extending the formulation of reversible thermodynamical transformations suggested in [11] to the formulation of irreversible transformations of open thermodynamical systems, a class of nonlinear control systems has been defined in terms of control Hamiltonian systems defined on a contact manifold [4,5,7,14]. Their dynamic properties as well as their feedback invariance and stabilization properties have been studied in [2,6,13,15]. An alternative definition, based on a variational formulation has been suggested in [10]. In this paper we shall discuss and compare this definition with the system-theoretic definition suggested in [13].

2 Control Hamiltonian Systems Defined on Contact Manifolds

Since Gibbs' work, it has been established that the Thermodynamic Phase Space is intrinsically defined as a *contact manifold*, that is a differentiable manifold

F. Nielsen and F. Barbaresco (Eds.): GSI 2017, LNCS 10589, pp. 418–424, 2017.
https://doi.org/10.1007/978-3-319-68445-1_49

$\mathcal{M} \ni \tilde{x}$ equipped with a contact form θ. In the sequel we shall denote by $\left(x_0, x, p^\top\right) \in \mathbb{R} \times \mathbb{R}^n \times \mathbb{R}^n$ a set of canonical coordinates[1].

It has also been established that the dynamics of thermodynamic systems subject to reversible and irreversible processes may be formulated in terms of contact Hamiltonian vector fields [4,5,7,11,14].

For *open* thermodynamic systems, a class of nonlinear control systems [12] has been defined, where the drift vector field and the input vector fields are both contact Hamiltonian vector fields [4,5,14]. Deriving from condition on structure preserving state feedback control, the natural output functions have then be defined as the contact Hamiltonian functions defining the input vector field [13]. An alternative definition of control contact systems, derived from a variational formulation, has been suggested in [10].

In this section we shall recall these two different definitions of control Hamiltonian systems and formulate the variational contact systems in terms of nonlinear control systems [12].

2.1 Input - Output Contact Systems [13]

Let us first recall the definition of input-output contact systems.

Definition 1 [13]. *An* input - output contact system *on the contact manifold* $(\mathcal{M}, \theta)$, *with input variable belonging the trivial vector bundle* $F = \mathcal{M} \times \mathbb{R}^m \ni (\tilde{x}, u)$ *over* $\mathcal{M}$ *and output variables being the dual vector bundle* $E = F^* \sim \mathcal{M} \times \mathbb{R}^m \ni (\tilde{x}, y)$, *is defined by the two functions* $K_0 \in C^\infty(\mathcal{M})$, *called the* internal contact Hamiltonian, $K_c \in C^\infty(\mathcal{M})$ *called the* interaction (or control) contact Hamiltonian, *and the state and output equations*

$$\frac{d\tilde{x}}{dt} = X_{K_0} + \sum_{i=1}^{m} X_{K_i} u_i \tag{1}$$

$$y_i = K_i(\tilde{x}) \quad i = 1, \ldots, m \tag{2}$$

where X_{K_0} *and* X_{K_i} *are the contact vector fields*[2] *of* $(\mathcal{M}, \theta)$ *generated by the contact Hamiltonians* K_0 *and* K_i *respectively.*

Note that input - output contact system are the analogue of input-output Hamiltonian systems defined on symplectic manifolds for driven mechanical systems [3,16,17] but extended to contact manifolds.

The models of physical systems such as heat diffusion or the Continuous Stirred Tank Reactor belong to a subclass of contact systems [4,5,14], called *conservative input-output contact systems*.

[1] The reader is referred to the classical textbooks [8, chap. V.] [1, app. 4.].

[2] Recall that a contact vector field X_K generated by the Hamiltonian function $K(\tilde{x})$ is the unique vector field satisfying

$$\begin{aligned} i_X\theta &= K \\ i_X d\theta &= -dK\left(\mathcal{H}(X)\right). \end{aligned} \tag{3}$$

Definition 2 [4]. *A conservative input-output contact system with respect to the Legendre submanifold $\mathcal{L}$ is an input-output contact system with the internal, respectively control, contact Hamiltonians K_0, respectively K_i, satisfying the two conditions:*

(i) they are invariants of the Reeb vector field, satisfying

$$i_E dK_0 = i_E dK_i = 0 \tag{4}$$

(ii) they satisfy the invariance condition

$$K_0\big|_{\mathcal{L}} = 0, \quad K_i\big|_{\mathcal{L}} = 0 \tag{5}$$

2.2 Control Contact System Arizing from a Variational Principle [10]

Arizing from the variational principle defined in [10] a more general class of contact systems has been defined which we briefly recall now.

Definition 3. *A* variational control contact system *[10] on the contact manifold $(\mathcal{M}, \theta)$, is defined by*

(i) the set of output variables *is defined by the vector bundle $E \ni y$ over$\mathcal{M}$ endowed with a (flat) covariant derivative ∇*
(ii) a bundle map $A : T^*\mathcal{M} \to E$ with $A(\theta) = 0$
(ii) the set of conjugated input variables *is the dual bundle $E^* \ni u$ over$\mathcal{M}$*
(iii) the input map *defined by the adjoint bundle map $A^* : E^* \to T\mathcal{M}$*
(iv) a smooth real function $K_0(\tilde{x})$, called internal contact Hamiltonian function

and the dynamical system $\frac{d\tilde{x}}{dt} = X(\tilde{x}, u, y)$ associated with the unique vector field $X(\tilde{x}, u, y)$ satisfying

$$\begin{aligned} i_{(X - A^* u)} d\theta + dK_0 = 0 \\ \theta(X) = i_X \theta = K_0 + \langle u, y \rangle \end{aligned} \tag{6}$$

Let us write the system explicitely in the form of a nonlinear control system. Firstly, notice that the condition $A(\theta) = 0$ is equivalent to

$$\operatorname{im} A^* \subset \ker \theta \tag{7}$$

that is, the image of A^* is contained in the field of contact elements $\ker \theta = \mathcal{C}$ (or *horizontal* with respect to θ).[3] Denoting by X_{K_0} the contact vector field

[3] The tangent bundle $T\mathcal{M}$ may be decomposed into

$$T\mathcal{M} = \ker d\theta \oplus \ker \theta \tag{8}$$

where $\ker d\theta$, called *vertical bundle* , is of rank 1 and is generated by the Reeb vector field and $\ker \theta$, called *horizontal bundle,* is of rank $2n$. Every vector field X on $\mathcal{M}$ may be decomposed in a unique way into

$$X = (i_X \theta) E + (X - (i_X \theta) E) \tag{9}$$

where $(i_X \theta) E \in \ker d\theta$ is vertical and $(X - (i_X \theta) E) = \mathcal{H}(X) \in \ker \theta = \mathcal{C}$ is horizontal with respect to the contact form θ.

generated by the internal contact Hamiltonian K_0 and using the decomposition of the tangent manifold (8), the vector field X defined by (6) becomes

$$\begin{aligned} X(\tilde{x}, u, y) &= \underbrace{(i_X\theta)\,E}_{\in \ker d\theta} + \underbrace{(X-(i_X\theta)\,E)}_{=\mathcal{H}(X)\in\ker\theta=\mathcal{C}} \\ &= (K_0 + \langle u, y\rangle)\,E + \underbrace{\mathcal{H}(X_{K_0}) + A^* u}_{\in\ker\theta=\mathcal{C}} \\ &= \underbrace{X_{K_0}}_{\text{drift contact vect. field}} + \underbrace{\underbrace{\langle u, y\rangle\,E}_{\in\ker d\theta} + \underbrace{A^* u}_{\in\ker\theta=\mathcal{C}}}_{\text{control vector field}} \end{aligned} \tag{10}$$

The second line of (10) shows the decomposition of the control vector field in terms of the vertical component which may be interpreted as the *power balance term* $K_0 + \langle u, y\rangle$ and the horizontal component which, using the tensor $\theta^\sharp$ mapping the semi-basic forms on the contact elements[4] , may be interpreted as a *Hamiltonian control system* defined on the contact elements

$$\theta^\sharp\,(dK_0 - (i_E dK_0)\,\theta) + A^* u \tag{12}$$

Note that these properties are due to the assumption (7).

The third line of (10) shows the decomposition of the control vector field into an *drift contact vector field* X_{K_0} defined by the internal Hamiltonian function K_0 and a *control vector field* decomposed into its vertical and horizontal parts.

The output variable y satisfies a dynamical equation on the output according to [10, p. 786–787]

$$\frac{d}{dt}y = A \circ d\theta\,(X(\tilde{x}, u, y)) \tag{13}$$

Using the expression (10), one obtains

$$\begin{aligned} d\theta\,(X(\tilde{x}, u, y)) &= \imath_{X(\tilde{x}, u, y)} d\theta \\ &= \imath_{X_{K_0}} d\theta + \langle u, y\rangle \underbrace{i_E d\theta}_{=0} + d\theta\,(A^* u) \\ &= [dK_0 - (i_E dK_0)\,\theta] + d\theta\,(A^* u) \end{aligned}$$

Using that $A(\theta) = 0$ hence the dynamics of the output (13) becomes

$$\tfrac{d}{dt}y = A\,([dK_0]) + (A \circ d\theta \circ A^*)\,u$$

[4] Any contact vector fields may be decomposed into

$$X_K = K\,E + \theta^\sharp\,(dK - (i_E dK)\,\theta) \tag{11}$$

where $K\,E$ is the vertical and $\theta^\sharp\,(dK - (i_E dK)\,\theta)$ is the horizontal components of the contact vector field where $\theta^\sharp$ denotes the inverse of the isomorphism $\theta^\flat\big|_{\mathcal{C}}$ from the vector space $\mathcal{C}$ of horizontal vector fields onto the space $\mathcal{F}$ of semi-basic 1-forms induced by the map $\theta^\flat\,(X) = -i_X d\theta$. [8, p. 293].

The Eq. (6) actually define the dynamical equations summarized in the following proposition.

Proposition 1. *The Eq. (6) defining the dynamics of a variational control contact system of definition 3, are equivalent to the dynamical system*

$$\frac{d\tilde{x}}{dt} = X_{K_0} + \langle u, y \rangle E + A^* u \tag{14}$$

$$\frac{dy}{dt} = A([dK_0]) + (A \circ d\theta \circ A^*) u \tag{15}$$

3 Relation Between Variational and Conservative Input-Output Contact Systems

In this section, we shall analyse the relations between conservative input-output contact systems of the definition 2 and the variational control contact systems of the definition 3. We shall give a direct proof that in this case the output dynamics (15) is *integrable* , that is when the output variable y may be expressed as a function of the state variable $\tilde{x}$, as has been stated in [10, Sect. 4.1].

Proposition 2. *The conservative contact input-output system of definition 2 with internal contact Hamiltonian $K_0(\tilde{x})$ and control contact Hamiltonians $-K_i(\tilde{x})$ is a variational control contact system defined in definition 3 with internal contact Hamiltonian $K_0(\tilde{x})$ and bundle map $A : T^*\mathcal{M} \rightarrow \mathbb{R}^n \times \mathcal{M}$ defined by*

$$A(\lambda) = (\langle \lambda, \mathcal{H}(K_i) \rangle)_{i=1,\ldots,m} \tag{16}$$

Proof. Firstly, let us identify the dynamics Eqs. (14) and (1) by decomposing the input contact vector field into its vertical and horizontal part

$$\begin{aligned}
\frac{d\tilde{x}}{dt} &= X_{K_0} - \sum_{i=1}^{m} X_{K_i} u_i \\
&= X_{K_0} - \sum_{i=1}^{m} K_i(\tilde{x}) u_i - \sum_{i=1}^{m} \mathcal{H}(X_{K_i}) u_i
\end{aligned}$$

Comparing this expression with third line of (10), leads to the natural identification of the dual output bundle map $A^*(u) = \sum_{i=1}^{m} \mathcal{H}(X_{K_i}) u_i$ and the outputs $y_i = K_i(\tilde{x})$. The map A^* obviously satisfies the condition (7) and its dual is by definition (16). Let us now check that the defined output indeed satisfies the dynamic Eq. (15). Using that that the functions K_i are invariants of the Reeb vector field : $i_E dK_i = 0$, let us compute the j-th component of $A(dK_i)$[5]

$$A(dK_i)_j = -\langle dK_i, \mathcal{H}(X_{K_i}) \rangle = [K_j, K_i]_\theta \quad i = 0, \ldots, m, \; j = 1, \ldots, m$$

[5] The *Jacobi bracket* $[f, g]_\theta$ of two differentiable functions f and g , defined by $[f, g]_\theta = i_E([X_f, X_g])$ where [,]denotes the Lie bracket on vector fields. We shall use the following identities $[f, g]_\theta = i_{X_f} dg - g\, i_E df = -i_{X_g} df + f\, i_E dg$.

Compute now the control term of the output Eq. (2), using again that that the functions K_i are invariants of the Reeb vector field

$$\begin{aligned}(A \circ d\theta \circ A^*)\, u &= A\,(d\theta\,(A^*\,(u))) \\ &= -A\,(d\theta\,(\textstyle\sum_{i=1}^m u_i\, \mathcal{H}\,(X_{K_i}))) \\ &= -\textstyle\sum_{i=1}^m u_i\, d\theta^\sharp\,(X_i, X_j) \\ &= -\textstyle\sum_{i=1}^m u_i\, \left([K_j,\, K_i]_\theta\right)_{j=1,\dots,m}\end{aligned}$$

Hence the second member of the dynamics (15) of the j-th component of output becomes

$$A\,([dK_0]) + (A \circ d\theta \circ A^*)\, u = [K_j,\, K_0]_\theta - \textstyle\sum_{i=1}^m u_i\, [K_j,\, K_i]_\theta \tag{17}$$

Using that, for functions K_i are invariants of the Reeb vector field $[K_j,\, K_i]_\theta = L_{X_j} K_i$, one obtains

$$A\,([dK_0]) + (A \circ d\theta \circ A^*)\, u = -\frac{dK_j}{dt} \tag{18}$$

Let us firstly notice that the output dynamics has a feedthrough term (depends explicitely on the input variables) which is linear in the Jacobi brackets of the control Hamiltonian functions. This resembles very much the situation for input-output Hamiltonian systems defined in symplectic or Poisson manifolds [9].

Let us discuss the example of integrable system given in [10, Sect. 4.1], for which the control contact Hamiltonians satisfy the conditions that they are in involution with respect to the Jacobi bracket. Indeed a contact manifold may be identified with the 1-jet of some manifold Q, (called *configuration manifold* in [10] and *manifold of independent extensive variables* in the context of Thermodynamics [4]). This 1-jet manifold may be identified with $\mathbb{R} \times T^*Q$ and equiped with the canonical contact structure. As the control Hamiltonian functions are chosen to be function of the configuration manifold only, they are in involution. If $[K_j,\, K_i]_\theta = 0\,, i,\, j = 1, \dots, m$, then the output dynamics (17) does not depend on the control variables. It may be noticed that this condition is not fullfilled for the models of physical systems given in [4,5,14], except for the single input case of course.

4 Conclusion

In this paper we have discussed the relation between the definition of variational control contact systems suggested in [10] and the input-output contact systems defined in [13]. We have first given an expression of the variational control contact systems of [10] in terms of a nonlinear control systems. Secondly we have shown that the conservative input-output contact systems are a subclass of the contact variational systems defined in [10] with integrable output dynamics.

References

1. Arnold, V.I.: Mathematical Methods of Classical Mechanics. Springer, New York (1989). 2 edition, ISBN 0-387-96890-3
2. Bravetti, A., Lopez-Monsalvo, C.S., Nettel, F.: Contact symmetries and Hamiltonian thermodynamics. Ann. Phys. **361**, 377–400 (2015)
3. Brockett, R.W.: Geometric control theory, volume 7 of lie groups: history, frontiers and applications. In: Martin, C., Herman, R. (eds.) Control Theory and Analytical Mechanics, pp. 1–46. Math. Sci. Press, Brookline (1977)
4. Eberard, D., Maschke, B.M., van der Schaft, A.J.: An extension of pseudo-Hamiltonian systems to the thermodynamic space: towards a geometry of non-equilibrium thermodynamics. Rep. Math. Phys. **60**(2), 175–198 (2007)
5. Favache, A., Dochain, D., Maschke, B.M.: An entropy-based formulation of irreversible processes based on contact structures. Chem. Eng. Sci. **65**, 5204–5216 (2010)
6. Favache, A., Dos Santos Martins, V., Maschke, B., Dochain, D.: Some properties of conservative control systems. IEEE trans. Autom. Control **54**(10), 2341–2351 (2009)
7. Grmela, M.: Reciprocity relations in Thermodynamics. Phys. A **309**, 304–328 (2002)
8. Libermann, P., Marle, C.-M.: Symplectic Geometry and Analytical Mechanics. D. Reidel Publishing Company, Dordrecht (1987)
9. Maschke, B.M., van der Schaft, A.J.: Port controlled Hamiltonian systems: modeling origins and system theoretic properties. In: Proceedings of 3rd International IFAC Conference on Nonlinear Systems' Theory and Control, NOLCOS 1992, pp. 282–288, Bordeaux, June 1992
10. Merker, J., Krüger, M.: On a variational principle in Thermodynamics. Continuum Mech. Thermodyn. **25**(6), 779–793 (2013)
11. Mrugała, R.: On a special family of thermodynamic processes and their invariants. Rep. Math. Phys. **46**(3), 461–468 (2000)
12. Nijmeijer, H., van der Schaft, A.J.: Nonlinear Dynamical Control Systems. Springer, New York (1990). 1st edition, ISBN: 0-387-97234-X
13. Ramirez, H., Maschke, B., Sbarbaro, D.: Feedback equivalence of input-output contact systems. Syst. Control Lett. **62**(6), 475–481 (2013)
14. Ramirez, H., Maschke, B., Sbarbaro, D.: Irreversible port-Hamiltonian systems: a general formulation of irreversible processes with application to the CSTR. Chem. Eng. Sci. **89**, 223–234 (2013)
15. Ramirez, H., Maschke, B., Sbarbaro, D.: Partial stabilization of input-output contact systems on a Legendre submanifold. IEEE Trans. Autom. Control **62**(3), 1431–1437 (2017)
16. van der Schaft, A.: Three decades of mathematical system theory. System Theory and Mechanics. LNCIS, vol. 135, pp. 426–452. Springer, Heidelberg (1989). doi:10.1007/BFb0008472
17. van der Schaft, A., Crouch, P.E.: Hamiltonian and self-adjoint control systems. Syst. Control Lett. **8**, 289–295 (1987)

Method of Orbits of Co-Associated Representation in Thermodynamics of the Lie Non-compact Groups

Vitaly Mikheev(✉)

Omsk State Technical University,
644050 Prospekt Mira, 11/8, Omsk, Russian Federation
vvm125@mail.ru

Abstract. A method of the solution of the main problem of homogeneous spaces thermodynamics for non-compact Lie groups is presented in the work. The method originates from formalism of non-commutative Fourier analysis based on method of coadjoint orbits. A formula that allows efficiently evaluate heat kernel and statistic sum on non-compact Lie group is obtained. The algorithm of construction of high temperature heat kernel expansion is also discussed.

Keywords: Heat kernel · Statistic sum · Partition function · Non-commutative integration · Non-compact Lie Groups · High-temperature expansion

1 Introduction

The purpose of that work is to work out the method for solution of the main problem of homogeneous spaces thermodynamics which consists of evaluation of statistic sum (partition function)

$$Z_\beta = \sum_n d_n \exp(-\beta E_n), \tag{1}$$

where d_n is degeneration degree of corresponding E_n. It also may be found as a trace of density matrix (heat kernel)

$$Z_\beta = \int \rho_\beta(x,x) d\mu(x), \quad d\mu(x) = \sqrt{|g|}dx. \tag{2}$$

That problem is important not only because statistic sum and heat kernel are important features of the space and can reveal thermodynamic properties of particles in that manifold [1]. The solution of main problem of homogeneous spaces thermodynamics for arbitrary manifold can move one step further to understanding the problem formulated by Kac "Can we hear the shape of the drum?". In other words we try to understand how can geometry and topology of the space influence spectral properties of Laplace operator on it ([2–4]).

F. Nielsen and F. Barbaresco (Eds.): GSI 2017, LNCS 10589, pp. 425–431, 2017.
https://doi.org/10.1007/978-3-319-68445-1_50

All existing results in that field were related to the compact manifolds or non-compact manifolds of finite volume. There is no algorithm of building heat kernel and statistic sum for arbitrary non-compact manifold because in this case series (1) and integral (2) are divergent since the volume of the manifold is infinite.

Density matrix (heat kernel) is to be found from heat kernel equation (Bloch equation) on homogeneous space with special initial condition

$$\frac{\partial \rho_\beta(x,x')}{\partial \beta} + H(x)\rho_\beta(x,x') = 0, \quad \rho_\beta(x,x')|_{\beta=0} = \delta(x,x'). \tag{3}$$

Solution of Eq. (3) has two problems which can hardly be overcome by existing methods of integration of PDEs, for instance by widely used separation of variables. Firstly one must obtain global solution on entire manifold but separation of variables sufficiently connected with the coordinate system on the manifold and therefor can give only local solutions. Secondly we have to build the solution of Bloch Eq. (3) from the functions which form the solution basis which must satisfy special initial condition chosen as δ - function. That is also a complicated problem.

2 Integration of Heat Kernel Equation on Non-compact Lie Groups

Let's consider Eq. (3) on n–dimensional real Lie group G with operator H being a quadratic function of left-invariant vector fields ξ on the group. That means that H is Laplace operator on a group space with left-invariant Riemann metric

$$H(-i\hbar\xi) = -\hbar^2 G^{ab}\xi_a\xi_b = -\hbar^2\Delta. \tag{4}$$

Solution of Eq. (3) on non-compact Lie group will be obtained using the formalism of non-commutative Fourier analysis on Lie groups based on method of orbits. The method originates from works by Kirillov [6], Souriau [7] and Kostant [8].

For that purpose we induce special irreducible representation of Lie algebra $\mathcal{G}$ (so called λ–representation) on Lagrange submanifold Q to a co-adjoint orbit $\mathcal{O}_\lambda \in \mathcal{G}^*$

$$[l_i(q,\partial_q,\lambda), l_j(q,\partial_q,\lambda)] = C_{ij}^k l_k(q,\partial_q,\lambda). \tag{5}$$

where C_{ij}^k are structural constants of Lie algebra $\mathcal{G}$, and $l_i(q,\partial_q,\lambda)$ - first order differential operators.

It can be shown that any irreducible representation of Lie algebra can be acquired as a certain λ – representation determined by the choice of linear functional $\lambda \in \mathcal{G}^*$. Linear functional $\lambda \equiv \lambda(j)$ where number of parameters j is equal to the number of Casimir functions - index of Lie algebra $\mathcal{G}$. Since that measure $d\mu(\lambda)$ is a spectral measure of Casimir operators on Lie group.

Let's consider representation of the Lie group G in the functional space $C^{\infty}(Q)$ which acts on the functions from that space as follows

$$T_g^{\lambda}\psi(q) = \int D^{\lambda}_{q\overline{q'}}(g)\psi(q')d\mu(q'), \tag{6}$$

and appears to be the lift of λ–representation of Lie algebra to a group

$$l_i(q, \partial_q, \lambda) = \frac{\partial}{\partial g^i} T_g^{\lambda}|_{g=e}. \tag{7}$$

Linear functional λ must be integer, i.e.

$$\int\limits_{\gamma \in H^2(\mathcal{O}_\lambda)} \omega_\lambda = 2\pi i n, \quad n \in Z, \tag{8}$$

where ω_λ is well known Kirillov 2-form on the orbit [6].

Functions $D^{j}_{q\overline{q'}}(g)$ are matrix elements of representation (6) and can be found from equations

$$[\xi_i(g) + \overline{l_i(q', \partial'_q, j)}]D^{\lambda}_{q\overline{q'}}(g) = 0, \quad D^{\lambda}_{q\overline{q'}}(e) = \delta(q, \overline{q'}), \tag{9}$$

here e is identity element of the group.

Functions $D^{\lambda}_{q\overline{q'}}(g)$ perform generalized Fourier transform on Lie group solving the main problem of harmonic analysis [9]. Here J if a manifold of parameters determining covector λ.

$$\varphi(g) = \int\limits_{Q\times Q\times J} \hat{\varphi}_j(q, q')D^{\lambda}_{q\overline{q'}}(g)\, d\mu(q)d\mu(q')d\mu(\lambda). \tag{10}$$

So action of right-invariant and left-invariant vector fields on group goes into action of operators of λ–representation on Lagrange submanifold of the coadjoint orbit [5]

$$\xi_i\varphi(g) \iff l_i(q', \partial'_q, \lambda)\hat{\varphi}_j(q, q'); \quad \eta_i\varphi(g) \iff \overline{l_i(q, \partial_q, \lambda)}\hat{\varphi}_j(q, q'). \tag{11}$$

After transition from the group space to the Lagrange submanifold of the orbit $\mathcal{O}_\lambda$ we have heat kernel equation on coadjoint orbit with smaller number of variables [10]

$$\frac{\partial \mathcal{R}_\beta(q, \overline{\tilde{q}}, j)}{\partial \beta} + H(-i\hbar l)\mathcal{R}_\beta(q, \overline{\tilde{q}}, j) = 0, \quad \mathcal{R}_\beta(q, \overline{\tilde{q}}, j)|_{\beta=0} = \delta(q, \overline{\tilde{q}}), \tag{12}$$

which appears to be ODE and is to be integrated in quadratures if following condition

$$(\dim G - \mathrm{indG})/2 = 1 \tag{13}$$

is satisfied.

In (12) heat kernel $\mathcal{R}_\beta(q,\overline{\tilde{q}},j)$ is connected with $\rho_\beta(g,g')$ on entire space by expression

$$\rho_\beta(g,g') = \int \mathcal{R}_\beta(q,\overline{\tilde{q}},j) D^\lambda_{\overline{q}\tilde{q}}(g'^{-1}g) d\mu(q) d\mu(\tilde{q}) d\mu(\lambda) \tag{14}$$

From solution of (12) we can obtain statistic sum on non-compact Lie group using properties of $D^\lambda_{q\overline{q'}}(g)$

$$Z_\beta = \int\limits_G d\mu(x) \int\limits_{Q\times J} \mathcal{R}_\beta(q,q,j) d\mu(q) d\mu(\lambda) = Vol_G \int\limits_{Q\times J} \mathcal{R}_\beta(q,q,j) d\mu(q) d\mu(\lambda). \tag{15}$$

One can see that integration in (15) over the volume of the manifold goes independently from integration over measure $d\mu(q)$ on coadjoint orbit and spectral measure $d\mu(\lambda)$. So we have opportunity to factorize in statistic sum divergences connected with infinite volume of non-compact space and since that we have following expression for specific statistic sum

$$z_\beta = Z_\beta / Vol_G = \int \mathcal{R}_\beta(q,q,j) d\mu(q) d\mu(\lambda), \tag{16}$$

which is sufficiently finite.

So instead of solution of (2) with n independent variables we solve Eq. (12) with smaller number of variables and get specific statistic sum on non-compact group manifold. Application of statistic sum touches numerous fields of theoretical physics from quantum statistic mechanics and quantum field theory to information theory where traditional physical objects such as entropy find new and quite productive interpretation [11,12].

3 High Temperature Expansion of Heat Kernel on Non-compact Lie Groups

Representation of the partition function and heat kernel itself as a power series (heat kernel expansion) is a significant problem. That expansion in the most general for the homogeneous case is to be written as

$$Z_\beta = \frac{Vol(M)}{(4\pi\beta)^{d/2}} \sum_{n=0}^{\infty} a_n \beta^n. \tag{17}$$

here β is an inverse thermodynamic temperature.

In order to find the coefficients of heat kernel expansion on Lie group is proposed to express the heat kernel as

$$\mathcal{R}_\beta(q,\overline{\tilde{q}},j) = \exp(\frac{i}{\hbar} S_\beta(q,\overline{\tilde{q}},j)), \tag{18}$$

where $S_\beta(q,\overline{\tilde{q}},j)$ is a complex function.

Using regular Fourier transform in respect to the variable $\tilde{q}$

$$\phi(q,p) = \int \phi(q,\tilde{q}) \exp(\frac{ip\tilde{q}}{\hbar}) d\tilde{q},$$

$$\phi(q,\tilde{q}) = \frac{1}{(2\pi\hbar)^{\frac{dim O_\lambda}{2}}} \int \phi(q,p) \exp(-\frac{ipq}{\hbar}) dp,$$

it's possible to pass to the function $\mathcal{R}_\beta(q,p,j)$, which satisfies heat kernel equation

$$\frac{\partial \mathcal{R}_\beta(q,p,j)}{\partial \beta} + \hat{H}(-i\hbar l(q,\partial_q))\mathcal{R}_\beta(q,p,j) = 0. \tag{19}$$

The equation for the function $S_\beta(q,\bar{\tilde{q}},j)$ is

$$\frac{i}{\hbar}\frac{\partial S_\beta(q,p,j)}{\partial \beta} + \exp(-\frac{i}{\hbar}S_\beta(q,p,j))\hat{H}(-i\hbar l(q,\partial_q)) \exp\frac{i}{\hbar}S_\beta(q,p,j)) = 0 \tag{20}$$

with initial condition

$$S_\beta(q,p,j)|_{\beta=0} = pq.$$

To be also represented as a power series

$$S_\beta(q,p,j) = \sum_{n=0}^{\infty} S_n(q,p,j)\beta^n. \tag{21}$$

$\hat{H}(-i\hbar l(q,\partial_q))$ is a second order differential operator and can be represented as

$$\hat{H}(-i\hbar l(q,\partial_q)) = -\hbar^2 G^{ab} l_a(q,\partial_q) l_b(q,\partial_q) = h^{ab}(q)\frac{\partial^2}{\partial q_a \partial q_b} + h^a(q)\frac{\partial}{\partial q_a} + h(q), \tag{22}$$

where coefficients $h^{ab}, h^a, h(q)$ can be easily obtained through the operators of λ — representation (7). Expression (22) using standard notation $\hat{p}_a = i\hbar\frac{\partial}{\partial q_a}$ can be rewritten as

$$\hat{H}(-i\hbar l(q,\partial_q)) = H^{ab}(q)\hat{p}_a\hat{p}_a + H^a(q)\hat{p}_a + H(q). \tag{23}$$

So the Eq. (20) transforms at

$$\frac{i}{h}\sum_{k=0}^{\infty} kS_k\beta^{(k-1)}(q,p,j) + \sum_{k=0}^{\infty} \Theta^{(k)}(q,p,j)\beta^k = 0, \tag{24}$$

with notation

$$\Theta^{(k)}(q,p,j) = -i\hbar H^{ab}S_{k,ab}(q,p,j) + H^{ab}\sum_{m=0}^{k} S_{m,a}(q,p,j)S_{k-m,b}(q,p,j) + H^a S_{k,a}(q,p,j) + H\delta_k^0,$$

Finally we get the recurrent expression to determine coefficients $S_{k+1}(q,p,j)$

$$S_{k+1}(q,p,j) = \frac{i\hbar}{k+1}\Theta^{(k)}(q,p,j). \tag{25}$$

It's obvious that a coefficient corresponding to the first power β in (21) is $H(q,p)$ - a qp-symbol of the hamiltonian $\hat{H}(-i\hbar l(q,\partial_q))$. That allows to get the formula of the first order for the high temperature expansion of partition function

$$z_\beta \approx \frac{1}{(2\pi\hbar)^{\frac{dim\mathcal{O}_\lambda}{2}}}\int \exp(\frac{i(q-\overline{q})}{h} - H(q,p))dpdqdj.$$

Power series of the heat kernel expansion on symplectic sheet to the coadjoint orbit $\mathcal{R}_\beta(q,p,j)$ can be obtained through coefficients $S_k(q,p,j)$ by expression

$$\mathcal{R}_\beta(q,p,j) = \sum_{n=0}^{\infty}\frac{1}{n!}\frac{d}{d\beta^n}\prod_{k=1}^{n+1}\sum_{m=0}^{n+1-k}\frac{((\frac{i}{\hbar}S_k(q,p,j))\beta^k)^m}{m!}|_{\beta=0}\beta^n. \tag{26}$$

High temperature asymptotic of partition function (statistic sum) is to be found by the formula (16), which after inverse Fourier transformations is performed looks as follows

$$z_\beta = \frac{1}{(2\pi\hbar)^{\frac{dim\mathcal{O}_\lambda}{2}}}\int \mathcal{R}_\beta(q,p,j)\exp(-\frac{ip\overline{q}}{\hbar})dpdqdj = \sum_{n=0}^{\infty} z_n\beta^n,$$

so the coefficients z_n of the partition sum expansion are

$$z_n = \frac{1}{(2\pi\hbar)^{\frac{dim\mathcal{O}_\lambda}{2}}}\int \frac{1}{n!}\frac{d}{d\beta^n}\prod_{k=1}^{n+1}\sum_{m=0}^{n+1-k}\frac{((\frac{i}{\hbar}S_k(q,p,j))\beta^k)^m}{m!}|_{\beta=0}e^{(-\frac{ip\overline{q}}{\hbar})}d\mu(p)d\mu(q)d\mu(\lambda), \tag{27}$$

and for the expansion of the heat kernel itself

$$\mathcal{R}_n(q,\overline{q},j) = \frac{1}{(2\pi\hbar)^{\frac{dim\mathcal{O}_\lambda}{2}}}\int \frac{1}{n!}\frac{d}{d\beta^n}\prod_{k=1}^{n+1}\sum_{m=0}^{n+1-k}\frac{((\frac{i}{\hbar}S_k(q,p,j))\beta^k)^m}{m!}|_{\beta=0}e^{-\frac{ip\overline{q}}{\hbar}}d\mu(p). \tag{28}$$

The final result for the heat kernel expansion on the Lie group manifold G is obtained after substitution of coefficients (21) in the formula (14)

$$\rho_n(x,x') = \int \mathcal{R}_n(q,\overline{\tilde{q}},j)D^\lambda_{\overline{q}\tilde{q}}(x'^{-1}x)d\mu(q)d\mu(\tilde{q})d\mu(\lambda). \tag{29}$$

Applications of heat kernel and partition sum are quite useful in many fields of theoretical physics. Among them are worth mentioning problems of quantum field theory and quantum thermodynamics as well as problems of theory of information being considered from a geometric point of view. As an application example of presented method must be mentioned result obtained by the author in [13] for heat kernel on group $E(2)$.

References

1. Hurt, N.: Geometric Quantization in Action. D. Reidel Publishing Company, Dordrecht (1983)
2. Minakshisundaram, S., Pleijel, A.: Some properties of the eigen functions of the laplace operator on riemannian manifolds. Can. J. Math. **1**, 242–256 (1949)
3. Varadhan, S.R.S.: On the behavior of the fundamental solution of the heat equation. Comm. Pure Appl. Math. **20**, 431–455 (1967)
4. Molchanov, S.A.: Diffusion processes and riemannian geometry. Uspekhi Mathem. Nauk. **30**(1), 57 pp. (1975)
5. Shirokov, I.V.: Darboux coordinates on K-orbits and the spectra of Casimir operators on Lie groups. Theoretical and Mathematical Physics **123**(3) (2000)
6. Kirillov, A.A.: Elements of the Theory of Representations. Springer, Heidelberg (1976)
7. Souriau, J.-M.: Structures Des Systmes Dynamiques. Dunod, Paris (1970)
8. Kostant, B.: Quantization and unitary representations. In: Taam, C.T. (ed.) Lectures in Modern Analysis and Applications III. LNM, vol. 170, pp. 87–208. Springer, Heidelberg (1970). doi:10.1007/BFb0079068
9. Barut, A., Razcka, R.: Theory of Group Representations and Applications. World Scientific (1986)
10. Baranovsky, S.P., Mikheyev, V.V., Shirokov, I.V.: Quantum hamiltonian systems on K-orbits: semiclassical spectrum of the asymmetric top. Theor. Mathe. Phys. **129**(1) (2001)
11. Nencka, H., Streater, R.F.: Information geometry for some Lie algebras. Infin. Dimens. Anal. Quantum Probab. Relat. Top. **2**, 441–460 (1999)
12. Barbaresco, F.: Geometric theory of heat from souriau lie groups thermodynamics and koszul hessian geometry. Entropy **18**, 386 (2016)
13. Mikheyev, V.V., Shirokov, I.V.: Application of coadjoint orbits in the thermodynamics of non-compact manifolds. Electron. J. Theor. Phys. **2**(7), 1–10 (2005)

Poly-symplectic Model of Higher Order Souriau Lie Groups Thermodynamics for Small Data Analytics

Frederic Barbaresco(✉)

Advanced Radar Concepts, Thales Air Systems, Paris, France
frederic.barbaresco@thalesgroup.com

Abstract. We introduce poly-symplectic extension of Souriau Lie group Thermodynamics based on higher-order model of statistical physics introduced by R.S. Ingarden. This extended model could be used for small data analytics

Keywords: Higher order thermodynamics · Lie group thermodynamics

1 Preamble

As early as 1966, Souriau applied his theory of geometric mechanics to statistical mechanics, developed in the Chap. 4 of his book "*Structure of Dynamical Systems*" [1, 2], what he called "Lie group thermodynamics". Using Lagrange's viewpoint, in Souriau statistical mechanics, a statistical state is a probability measure on the manifold of motions. Souriau observed that Gibbs equilibrium is not covariant with respect to dynamic groups of Physics. To solve this braking of symmetry, Souriau introduced a new "geometric theory of heat" where the equilibrium states are indexed by a parameter β with values in the Lie algebra of the group, generalizing the Gibbs equilibrium states, where β plays the role of a geometric (Planck) temperature. We will generalize Souriau theory [4, 5] in the framework of higher order thermodynamics as introduced by Ingarden [9–11] for mesoscopic systems. The Gibbs canonical state results from the Maximum Entropy principle when the statistical mean value of energy is supposed to be known. Polish School has studied the maximum entropy inference with higher-order moments of energy (when not only mean values but also statistical moments of higher order of some physical quantities are taken into account). Ingarden in 1992 and Jaworski in 1981 have introduced the concept of second and higher-order temperatures, by assuming a distribution function which includes information not only on the average of the energy but also on higher-order moments, in particular 2nd moment related to fluctuations. This case should be considered in situations where fluctuations are not negligible, such as near phase transitions or critical points, in metastable states in systems with a small number of degrees of freedom. Ingarden idea is that if we can measure more details, such as the first n cumulants of the energy, we can then introduce n high-order temperature, as the Lagrange multipliers when we maximize the Entropy with respect to these values:

F. Nielsen and F. Barbaresco (Eds.): GSI 2017, LNCS 10589, pp. 432–441, 2017.
https://doi.org/10.1007/978-3-319-68445-1_51

$$P_{(\beta_1,\beta_2)} = \frac{1}{Z(\beta_1,\beta_2)} e^{-\beta_1.H-\beta_2(H-U)^2} = e^{\beta_0-\beta_1.H-\beta_2(H-U)^2} \quad (1)$$

Ingarden proposed that if we can measure the second cumulant of the energy (the fluctuation of the energy), the equilibrium state is not the canonical state, but would need two temperatures. Ingarden argues that for a macroscopic system there is very little difference between the two states, and that we would need a mesoscopic or microscopic system to be able to detect the higher temperature. Jaworski [7, 8] has shown that the contribution to the total entropy, arising from the extra information corresponding to the higher-order moments, is *o(N)*, when *N* tends to infinity and *N/V* ratio is constant, with *N* the number of particles and *V* the volume. The main result of Jaworski is that from a purely thermodynamic point of view, the information corresponding to the higher-order moments of extensive physical quantities is not essential and can be neglected in the maximum entropy procedure. Jaworski showed that the maximum entropy inference has a certain stability property with respect to information corresponding to higher order moments of extensive quantities. It can serve as an argument in favor of the maximum entropy method in statistical physics and to understand better why these methods are successful. Streater [3] has prefered to say that the states with generalized temperatures are not in equilibrium, assuming that the final state, at large times, will be the canonical or grand canonical state depending on mixing properties. Streater [3] intends that this occur even for a mesoscopic system, such as a few atoms, adding that his approach is equivalent to Ingarden model if the relaxation time from the state with generalized temperatures to the final equilibrium is very long.

2 Model of Souriau Lie Groups Thermodynamics

In 1970, Souriau [1, 2] introduced the concept of co-adjoint action of a group on its momentum space, based on the orbit method works, that allows to define physical observables like energy, heat and momentum or moment as pure geometrical objects. The moment map is a constant of the motion and is associated to symplectic cohomology. In a first step to establish new foundations of thermodynamics, Souriau has defined a Gibbs canonical ensemble on a symplectic manifold *M* for a Lie group action on *M*. In classical statistical mechanics, a state is given by the solution of Liouville equation on the phase space, the partition function. As symplectic manifolds have a completely continuous measure, invariant by diffeomorphisms, the Liouville measure λ, all statistical states will be the product of the Liouville measure by the scalar function given by the generalized partition function $e^{\Phi(\beta)-\langle\beta,U(\xi)\rangle}$ defined by the energy *U* (defined in the dual of the Lie algebra of this dynamical group) and the geometric temperature β, where Φ is a normalizing constant such the mass of probability is equal to 1, $\Phi(\beta) = -\log \int_M e^{-\langle\beta,U(\xi)\rangle} d\lambda$. Jean-Marie Souriau then generalizes the Gibbs equilibrium state to all symplectic manifolds that have a dynamical group. Souriau has observed that if we apply this theory for non-commutative group (Galileo or Poincaré groups), the symmetry has been broken. For each temperature β, element of the Lie algebra $\mathfrak{g}$, Souriau has introduced a tensor $\widetilde{\Theta}_\beta$, equal to the sum of the cocycle $\widetilde{\Theta}$ and the heat coboundary (with [.,.] Lie bracket):

$$\tilde{\Theta}_\beta(Z_1, Z_2) = \tilde{\Theta}(Z_1, Z_2) + \langle Q, ad_{Z_1}(Z_2) \rangle \tag{2}$$

This tensor $\widetilde{\Theta}_\beta$ has the following properties: $\widetilde{\Theta}(X, Y) = \langle \Theta(X), Y \rangle$ where the map Θ is the symplectic one-cocycle of the Lie algebra $\mathfrak{g}$ with values in $\mathfrak{g}^*$, with $\Theta(X) = T_e\theta(X(e))$ where θ the one-cocycle of the Lie group G. $\widetilde{\Theta}(X, Y)$ is constant on M and the map $\widetilde{\Theta}(X, Y) : \mathfrak{g} \times \mathfrak{g} \to \Re$ is a skew-symmetric bilinear form, and is called the *symplectic two-cocycle of Lie algebra* $\mathfrak{g}$ associated to the *moment map* J, with the following properties:

$$\widetilde{\Theta}(X, Y) = J_{[X,Y]} - \{J_X, J_Y\} \quad \text{with } J \text{ the Moment Map} \tag{3}$$

$$\widetilde{\Theta}([X, Y], Z) + \widetilde{\Theta}([Y, Z], X) + \widetilde{\Theta}([Z, X], Y) = 0 \tag{4}$$

where J_X linear application from $\mathfrak{g}$ to differential function on M: $\mathfrak{g} \to C^\infty(M, R), X \to J_X$ and the associated differentiable application J, called moment (um) map $J : M \to \mathfrak{g}^*, x \mapsto J(x)$ such that $J_X(x) = \langle J(x), X \rangle, X \in \mathfrak{g}$.

The geometric temperature, element of the algebra $\mathfrak{g}$, is in the the kernel of the tensor $\widetilde{\Theta}_\beta$:$\beta \in Ker\widetilde{\Theta}_\beta$ such that $\widetilde{\Theta}_\beta(\beta, \beta) = 0, \quad \forall \beta \in \mathfrak{g}$. The following symmetric tensor $g_\beta([\beta, Z_1], [\beta, Z_2]) = \widetilde{\Theta}_\beta(Z_1, [\beta, Z_2])$, defined on all values of $ad_\beta(.) = [\beta, .]$ is positive definite, and defines extension of classical Fisher metric in Information Geometry (as hessian of the logarithm of partition function):

$$g_\beta([\beta, Z_1], Z_2) = \widetilde{\Theta}_\beta(Z_1, Z_2), \quad \forall Z_1 \in \mathfrak{g}, \quad \forall Z_2 \in \mathrm{Im}\big(ad_\beta(.)\big) \tag{5}$$

with

$$g_\beta(Z_1, Z_2) \geq 0, \quad \forall Z_1, Z_2 \in \mathrm{Im}\big(ad_\beta(.)\big) \tag{6}$$

These equations are universal, because they are not dependent on the symplectic manifold but only on the dynamical group G, the symplectic two-cocycle Θ, the temperature β and the heat Q. Souriau called it "*Lie groups thermodynamics*".

Theorem (Souriau Theorem of Lie Group Thermodynamics). *Let Ω be the largest open proper subset of $\mathfrak{g}$, Lie algebra of G, such that $\int_M e^{-\langle \beta, U(\xi) \rangle} d\lambda$ and $\int_M \xi . e^{-\langle \beta, U(\xi) \rangle} d\lambda$ are convergent integrals, this set Ω is convex and is invariant under every transformation $Ad_g(.)$. Then, the fundamental equations of Lie group thermodynamics are given by the action of the group:*

- *Action of Lie group on Lie algebra:*

$$\beta \rightarrow Ad_g(\beta) \tag{7}$$

- *Characteristic function after Lie group action:*

$$\Phi \rightarrow \Phi - \langle \theta(g^{-1}), \beta \rangle \tag{8}$$

- *Invariance of entropy with respect to action of Lie group:*

$$s \rightarrow s \tag{9}$$

- *Action of Lie group on geometric heat:*

$$Q \rightarrow a(g, Q) = Ad_g^*(Q) + \theta(g) \tag{10}$$

In the framework of Lie group action on a symplectic manifold, equivariance of moment could be studied to prove that there is a unique action *a(.,.)* of the Lie group G on the dual $\mathfrak{g}^*$ of its Lie algebra for which the moment map J is equivariant, that means for each

$$x \in M: J(\Phi_g(x)) = a(g, J(x)) = Ad_g^*(J(x)) + \theta(g) \tag{11}$$

Jean-Louis Koszul has analyzed Souriau model in his book "*Introduction to symplectic geometry*" [6]. Defining classical operation $Ad_s a = sas^{-1}, \quad s \in G, \ a \in \mathfrak{g}, \ ad_a b = [a, b], \quad a \in \mathfrak{g}, \ b \in \mathfrak{g}$ and $Ad_s^* = {}^t Ad_{s^{-1}}, \quad s \in G$ with classical properties $Ad_{\exp a} = \exp(-ad_a), \quad a \in \mathfrak{g}$ or $Ad_{\exp a}^* = \exp{}^t(ad_a), \quad a \in \mathfrak{g}$, we can consider: $x \mapsto sx, x \in M$, $\mu : M \rightarrow \mathfrak{g}^*$, we have $\langle d\mu(v), a \rangle = \omega(ax, v)$.

If we study $\mu \circ s_M - Ad_s^* \circ \mu : M \rightarrow \mathfrak{g}^*$, we have:

$$d\langle Ad_s^* \circ \mu, a \rangle = \langle Ad_s^* d\mu, a \rangle = \langle d\mu, Ad_{s^{-1}} a \rangle$$

$$\langle d\mu(v), Ad_{s^{-1}} a \rangle = \omega(s^{-1} asx, v) = \omega(asx, sv) = \langle d\mu(sv), a \rangle = (d\langle \mu \circ s_M, a \rangle)(v)$$

$d\langle Ad_s^* \circ \mu, a \rangle = d\langle \mu \circ s_M, a \rangle$ and then prove that

$$d\langle \mu \circ s_M - Ad_s^* \circ \mu, a \rangle = 0 \tag{12}$$

If we develop the cocycle given by $\theta_\mu(s) = \mu(sx) - Ad_s^* \mu(x), s \in G$, we can study $\theta_\mu(st) = \theta_\mu(s) - Ad_s^* \theta_\mu(\mathrm{t}), \ s, t \in G$. If we note $\mathrm{c}_\mu(\mathrm{a}, \mathrm{b}) = \langle \mathrm{d}\theta_\mu(a), b \rangle, a, b \in \mathfrak{g}$

$$\theta_\mu(st) = \mu(stx) - Ad_{st}^* \mu(x) = \theta_\mu(s) + Ad_s^* \mu(tx) - Ad_s^* Ad_t^* \mu(x) = \theta_\mu(s) + Ad_s^* \theta_\mu(t)$$

By developing $d\mu(ax) = {}^t ad_a \mu(x) + d\theta_\mu(a), x \in M, a \in \mathfrak{g}$, we obtain:

$$\langle d\mu(ax), b\rangle = \langle \mu(x), [a, b]\rangle + \langle d\theta_\mu(a), b\rangle = \{\langle \mu, a\rangle, \langle \mu, b\rangle\}(x), x \in M, a, b \in \mathfrak{g} \quad (13)$$

We have then $c_\mu(a, b) = \{\langle \mu, a\rangle, \langle \mu, b\rangle\} - \langle \mu, [a, b]\rangle = \langle d\theta_\mu(a), b\rangle, \quad a, b \in \mathfrak{g}$

And the property

$$c_\mu([a, b], c) + c_\mu([b, c], a) + c_\mu([c, a], b) = 0, a, b, c \in \mathfrak{g} \quad (14)$$

If the moment map is transform as

$$\mu' = \mu + \varphi \Rightarrow c_{\mu'}(a, b) = c_\mu(a, b) - \langle \varphi, [a, b]\rangle \quad (15)$$

By considering this action of the group on dual Lie algebra $G \times \mathfrak{g}^* \to \mathfrak{g}^*, (s, \xi) \mapsto s\xi = Ad_s^* \xi + \theta_\mu(s)$

We have the property that

$$\mu(sx) = s\mu(x) = Ad_s^* \mu(x) + \theta_\mu(s), \forall s \in G, x \in M$$

where the cocycle is given by $\theta_\mu(s) = \mu(sx) - Ad_s^* \mu(x)$

We can verify the following properties:

$$G \times \mathfrak{g}^* \to \mathfrak{g}^*, (e, \xi) \mapsto e\xi = Ad_e^* \xi + \theta_\mu(e) = \xi + \mu(x) - \mu(x) = \xi \quad (16)$$

$$\begin{aligned} (s_1 s_2)\xi &= Ad_{s_1 s_2}^* \xi + \theta_\mu(s_1 s_2) = Ad_{s_1}^* Ad_{s_2}^* \xi + \theta_\mu(s_1) + Ad_{s_1}^* \theta_\mu(s_2) \\ (s_1 s_2)\xi &= Ad_{s_1}^* \left(Ad_{s_2}^* \xi + \theta_\mu(s_2) \right) + \theta_\mu(s_1) = s_1(s_2 \xi), \ \forall s_1, s_2 \in G, \xi \in \mathfrak{g}^* \end{aligned} \quad (17)$$

Finally using $c_\mu(a, b) = \{\langle \mu, a\rangle, \langle \mu, b\rangle\} - \langle \mu, [a, b]\rangle = \langle d\theta_\mu(a), b\rangle, \quad a, b \in \mathfrak{g}$:

$$\{\mu^*(a), \mu^*(b)\} = \{\langle \mu, a\rangle, \langle \mu, b\rangle\} = \mu^*([a, b] + c_\mu(a, b)) = \mu^* \{a, b\}_{c_\mu} \quad (18)$$

3 poly-sympectic Higher-Order Lie Groups Thermodynamics

As observed by Jean-Marie Souriau, the Gaussian density is a maximum entropy density of 1st order. This remark is clear if we replace z and (m, R) by ξ and β:

$$p_{(m,R)}(z) = \frac{1}{(2\pi)^{n/2}\det(R)^{1/2}} e^{-\frac{1}{2}(z-m)^T R^{-1}(z-m)} = \frac{1}{(2\pi)^{n/2}\det(R)^{1/2} e^{\frac{1}{2}m^T R^{-1} m}} e^{-\left[-m^T R^{-1} z + \frac{1}{2} z^T R^{-1} z\right]}$$
$$p_{(m,R)}(z) = p_{\hat{\xi}}(\xi) = \frac{1}{Z} e^{-\langle \beta, \xi \rangle} \text{ with } \xi = \begin{bmatrix} z \\ zz^T \end{bmatrix}, \hat{\xi} = \begin{bmatrix} E[z] \\ E[zz^T] \end{bmatrix} = \begin{bmatrix} m \\ R + mm^T \end{bmatrix}$$
$$\text{and } \beta = \begin{bmatrix} -R^{-1}m \\ \frac{1}{2}R^{-1} \end{bmatrix} = \begin{bmatrix} a \\ H \end{bmatrix} \text{ where } \langle \beta, \xi \rangle = a^T z + z^T H z = Tr\left[z a^T + H^T z z^T\right]$$
$$\text{with } \log(Z) = \frac{n}{2}\log(2\pi) + \frac{1}{2}\log\,\det(R) + \frac{1}{2} m^T R^{-1} m \text{ and } S(\hat{\xi}) = \left\langle \hat{\xi}, \beta \right\rangle - \Phi(\beta)$$
$$\hat{\xi} = \Theta(\beta) = \frac{\partial \Phi(\beta)}{\partial \beta} \text{ and } \beta = \Theta^{-1}(\hat{\xi}) \text{ with } \Phi(\beta) = -\log \psi_\Omega(\beta) = -\log \int_{\Omega^*} e^{-\langle \beta, \xi \rangle} d\xi$$
$$Fisher : I(\beta) = \frac{\partial^2 \log \psi_\Omega(\beta)}{\partial \beta^2} = E\left[\frac{\partial \log p_\beta(\xi)}{\partial \beta} \frac{\partial \log p_\beta(\xi)^T}{\partial \beta}\right] = E\left[\left(\xi - \hat{\xi}\right)\left(\xi - \hat{\xi}\right)^T\right] \tag{19}$$

As soon as 1963, R.S. Ingarden has introduced the concept of higher order temperatures for statistical systems such as thermodynamics. In physics, the concept of temperature is connected with the mean value of kinetic energy of molecules in an ideal gas. For a general physical system with interactions among particles (non-ideal gas, liquid or solid), an equilibrium probability distribution is assumed to depend on temperature T as the only statistical parameter of the Gibbs state: $P_\beta(x) = \frac{1}{Z(\beta)} e^{-\beta.H(x)}$ with $\beta = \frac{1}{k_\beta T}$ and $H(x) = H(p, q)$ where p is position, q the mechanical momentum and k_β the Boltzmann constant (a factor to insure that $\beta.H$ is dimensionless). In case of no stochastic interactions between particles (ideal gas), partition function Z is integrable and we obtain Gauss distribution in the momentum space which corresponds to the result of the limit theorem for large N. Boltzmann ideal gas model can fail if the number of particles is not large enough (rnesoscopic systems), and if the interactions between particles are not weak enough. Gibbs hypothesis can also fail if stochastic interactions with the environment are not sufficiently weak. As remarked by R.S. Ingarden, nobody has never observed thermal equilibrium of Gibbs in large and complex systems (Earth's atmosphere, cosmic systems, biological organisms), but only flows, turbulence or pumping, replacing classical approach by the concept of local temperature and thermodynamic flows (thermo-hydrodynamics and non-equilibrium thermodynamics), that is non-coherent with the concept of temperature which is global/intensive by definition and does not depend on position. R.S. Ingarden propose to consider the stationary case by means of the concept of higher order temperatures defined by:

$$P_{(\beta_1,\ldots,\beta_n)}(x) = \frac{1}{Z(\beta_1,\ldots,\beta_n)} e^{-\beta_1.H(x) - \beta_2(H(x)-U)^2 - \ldots - \beta_n(H(x)-U)^n} \tag{20}$$

where $U = E(H)$ is the mean energy introduced to preserve the invariance of the total energy with respect to an arbitrary additive constant, and $\beta_0 = -\log Z(\beta_1,\ldots,\beta_n)$ the normalizing constant. The new constants β_k are said to be β-temperatures of order k.

$H(x)$ is usually a quadratic function of x (for ideal gas only of p, for ideal solid of p and q). The probability distribution is fixed uniquely by all (independent and not contradictory) statistical moments which should be experimentally measured. But if the number of values is too large to make this method practical, we can measure only the lowest moments up to some order (if the higher orders do not change the result to a given accuracy), and to fix the respective β-temperatures as Lagrange multipliers by maximization of entropy of distribution $S = -\int P_{(\beta_1,\ldots,\beta_n)}(x)\log P_{(\beta_1,\ldots,\beta_n)}(x)dx$, with the given moments as additional conditions. R.S. Ingarden observed that the entropy maximization randomizes higher moments in a symmetric way, and it liquidates any possible bias with respect to their special values, and it gives the best estimate to a given accuracy. The values of β can be found by:

$$E(x^k) = \frac{\partial \beta_0}{\partial \beta_k} = \frac{\partial \log Z}{\partial \beta_k} \quad \text{with} \quad E(x^k) = Z^{-1}\int x^k e^{-\sum_{k=1}^{n}\beta_k x^k} dx = \int x^k P_{(\beta_1,\ldots,\beta_n)}(x)dx \tag{21}$$

$$Z = \int e^{-\sum_{k=1}^{n}\beta_k x^k} dx \text{ and the relation: } S = \sum_{k=1}^{n}\beta_k E(x^k) + \log Z = \sum_{k=1}^{n}\beta_k \frac{\partial \beta_0}{\partial \beta_k} - \beta_0 \tag{22}$$

R.S. Ingarden has applied this model for linguistic statistics, assuming the appearance of higher order temperatures since there occur rather strong statistical correlations between phonemes and words as elements of these statistics. He argued his choice observing that in the case of word statistics, the existence of strong correlations is given by grammatical or semantical studies [9]. R.S. Ingarden made the conjecture that his high order thermodynamics is the model of statistically interacting, small systems, and biological living systems, although the calculation/observation are more difficult. We have seen that Souriau has replaced classical Maximum Entropy approach by replacing Lagrange parameters by only one geometric "temperature vector" as element of Lie algebra. In parallel, Ingarden has introduced second and higher order temperature of the Gibbs state that could be extended to Souriau theory of thermodynamics. Ingarden higher order temperatures could be defined in the case when no variation is considered, but when a probability distribution depending on more than one parameter. It has been observed by Ingarden, that Gibbs assumption can fail if the number of components of the sum goes to infinity and the components of the sum are stochastically independent, and if stochastic interactions with the environment are not sufficiently weak. In all these cases, we never observe absolute thermal equilibrium of Gibbs type but only flows or turbulence. Non-equilibrium thermodynamics could be indirectly addressed by means of high order temperatures.

Initiated by Gunther [12, 13] based on n-symplectic model [14, 15], it has been shown that the symplectic structure on the phase space remains true, if we replace the symplectic form by a vector valued form, that is called polysymplectic. This extension defines an action of G over $\mathfrak{g}^* \times \overset{(n)}{\ldots} \times \mathfrak{g}^*$ called n-coadjoint action:

$$\begin{aligned} & Ad_g^{*(n)} : G \times \left(\mathfrak{g}^* \times \overset{(n)}{\ldots} \times \mathfrak{g}^*\right) \to \mathfrak{g}^* \times \overset{(n)}{\ldots} \times \mathfrak{g}^* \\ & g \times \mu_1 \times \ldots \times \mu_n \mapsto Ad_g^{*(n)}(\mu_1, \ldots, \mu_n) = \left(Ad_g^* \mu_1, \ldots, Ad_g^* \mu_n\right) \end{aligned} \tag{23}$$

Let $\mu = (\mu_1, \ldots, \mu_n)$ a poly-momentum, element of $\mathfrak{g}^* \times \overset{(n)}{\ldots} \times \mathfrak{g}^*$, we can define a n-coadjoint orbit $O_\mu = O_{(\mu_1, \ldots, \mu_n)}$ at the point μ, for which the canonical projection $\Pr_k : \mathfrak{g}^* \times \overset{(n)}{\ldots} \times \mathfrak{g}^* \to \mathfrak{g}^*, \quad (\nu_1, \ldots, \nu_n) \mapsto \nu_k$ induces a smooth map between the n-coadjoint orbit O_μ and the coadjoint orbit O_{μ_k}: $\pi_k : O_\mu = O_{(\mu_1, \ldots, \mu_n)} \to O_{\mu_k}$ that is a surjective submersion with $\bigcap_{k=1}^{n} KerT\pi_k = \{0\}$. Extending Souriau approach, equivariance of poly-moment could be studied to prove that there is a unique action *a(.,.)* of the Lie group G on $\mathfrak{g}^* \times \overset{(n)}{\ldots} \times \mathfrak{g}^*$ for which the polymoment map $J^{(n)} = (J^1, \ldots, J^n) : M \to \mathfrak{g}^* \times \overset{(n)}{\ldots} \times \mathfrak{g}^*$ verifies $x \in M$ and $g \in G$:

$$J^{(n)}(\Phi_g(x)) = a(g, J^{(n)}(x)) = Ad_g^{*(n)}\left(J^{(n)}(x)\right) + \theta^{(n)}(g) \tag{24}$$

with $Ad_g^{*(n)}(J^{(n)}(x)) = \left(Ad_g^* J^1, \ldots, Ad_g^* J^n\right)$ and $\theta^{(n)}(g) = \left(\theta^1(g), \ldots, \theta^n(g)\right)$ a poly-symplectic one-cocycle. We can also defined poly-symplectic two-cocycle $\widetilde{\Theta}^{(n)} = \left(\widetilde{\Theta}^1, \ldots, \widetilde{\Theta}^n\right)$ with $\widetilde{\Theta}^k(X, Y) = \langle \Theta^k(X), Y \rangle = J^k_{[X,Y]} - \{J^k_X, J^k_Y\}$ where $\Theta^k(X) = T_e \theta^k(X(e))$. Finally, the poly-symplectic Souriau-Fisher metric is given by:

$$g_\beta([\beta, Z_1], Z_2) = diag\left[\tilde{\Theta}_{\beta_k}(Z_1, Z_2)\right]_k, \forall Z_1 \in g, \forall Z_2 \in \mathrm{Im}\left(ad_\beta(.)\right), \beta = (\beta_1, \ldots, \beta_n) \tag{25}$$

$$\tilde{\Theta}_{\beta_k}(Z_1, Z_2) = -\frac{\partial \Phi(\beta_1, \ldots, \beta_n)}{\partial \beta_k} = \widetilde{\Theta}^k(Z_1, Z_2) + \langle Q_k, ad_{Z_1}(Z_2) \rangle \tag{26}$$

Compared to Souriau model, heat is replaced by previous polysymplectic model:

$$\begin{aligned} Q &= (Q_1, \ldots, Q_n) \in \mathfrak{g}^* \times \overset{(n)}{\ldots} \times \mathfrak{g}^* \text{ with } Q_k = \frac{\partial \Phi(\beta_1, \ldots, \beta_n)}{\partial \beta_k} \\ &= \frac{\int_M U^{\otimes k}(\xi).e^{-\sum_{k=1}^{n} \langle \beta_k, U^{\otimes k}(\xi) \rangle} d\omega}{\int_M e^{-\sum_{k=1}^{n} \langle \beta_k, U^{\otimes k}(\xi) \rangle} d\omega} \end{aligned} \tag{27}$$

with characteristic function:

$$\Phi(\beta_1, \ldots, \beta_n) = -\log \int_M e^{-\sum_{k=1}^{n} \left\langle \beta_k, U^{\otimes k}(\xi) \right\rangle} d\omega \tag{28}$$

We extrapolate Souriau results, who proved in [1, 2] that $\int_M U^{\otimes k}(\xi).e^{-\left\langle \beta_k, U^{\otimes k}(\xi) \right\rangle} d\omega$ is locally normally convergent using multi-linear norm $\left\| U^{\otimes k} \right\| = \underset{U}{Sup} \left\langle E, U \right\rangle^k$ and where $U^{\otimes \mathrm{k}} = U \otimes \overset{(k)}{U} \ldots \otimes U$ is defined as a tensorial product (see [1] and Bourbaki).

Entropy is defined by Legendre transform of Souriau-Massieu characteristic function:

$$S(Q_1, \ldots, Q_n) = \sum_{k=1}^{n} \langle \beta_k, Q_k \rangle - \Phi(\beta_1, \ldots, \beta_n) \text{ where } \beta_k = \frac{\partial S(Q_1, \ldots, Q_n)}{\partial Q_k} \tag{29}$$

The Gibbs density could be then extended with respect to high order temperatures by:

$$p_{Gibbs}(\xi) = e^{\Phi(\beta_1,\ldots,\beta_n) - \sum_{k=1}^{n} \left\langle \beta_k, U^{\otimes k}(\xi) \right\rangle} = \frac{e^{-\sum_{k=1}^{n} \left\langle \beta_k, U^{\otimes k}(\xi) \right\rangle}}{\int_M e^{-\sum_{k=1}^{n} \left\langle \beta_k, U^{\otimes k}(\xi) \right\rangle} d\omega} \tag{30}$$

References

1. Souriau, J.-M.: Structures des Systèmes Dynamiques. Dunod, Paris (1970)
2. Souriau, J.-M.: Mécanique statistique, groupes de Lie et cosmologie, Colloques int. du CNRS numéro 237. Géométrie symplectique et physique mathématique, pp. 59–113 (1974)
3. Nencka, H., Streater, R.F.: Information geometry for some Lie algebras. Infin. Dimens. Anal. Quantum Probab. Relat. Top. **2**, 441–460 (1999)
4. Barbaresco, F.: Geometric theory of heat from Souriau Lie groups thermodynamics and Koszul Hessian geometry. Entropy **18**, 386 (2016)
5. Marle, C.-M.: From tools in symplectic and poisson geometry to J.-M. Souriau's theories of statistical mechanics and thermodynamics. Entropy **18**, 370 (2016)
6. Koszul, J.L.: Introduction to Symplectic Geometry. Science Press, Beijing (1986). (Chinese)
7. Jaworski, W.: Information thermodynamics with the second order temperatures for the simplest classical systems. Acta Phys. Polon. **A60**, 645–659 (1981)
8. Jaworski, W.: Higher-order moments and the maximum entropy inference: the thermodynamical limit approach. J. Phys. A: Math. Gen. **20**, 915–926 (1987)
9. Ingarden, H.S., Meller, J.: Temperatures in linguistics as a model of thermodynamics. Open Sys. Inf. Dyn. **2**, 211–230 (1991)
10. Ingarden, R.S., Nakagomi, T.: The second order extension of the Gibbs state. Open Sys. Inf. Jyn. **1**, 259–268 (1992)

11. Ingarden, R.S., Kossakowski, A., Ohya, M.: Information Dynamics and Open Systems: Classical and Quantum Approach. Springer, Dordrecht (1997). Fundamental Theories of Physics, 86
12. Gunther, C.: The polysymplectic Hamiltonian formalism in field theory and calculus of variations I: the local case. J. Differ. Geom. **25**, 23–53 (1987)
13. Munteanu, F., Rey, A.M., Salgado, M.: The Günther's formalism in classical field theory: momentum map and reduction. J. Math. Phys. **45**(5), 1730–1751 (2004)
14. Awane, A.: k-Symplectic structures. J. Math. Phys. **33**, 4046–4052 (1992)
15. Awane, A., Goze, M.: Pfaffian Systems, k-symplectic Systems. Springer, Dordrecht (2000)

Thermodynamic Equilibrium and Relativity: Killing Vectors and Lie Derivatives

F. Becattini(✉)

University of Florence,
Via G. Sansone 1, 50019 Sesto Fiorentino, Firenze, Italy
becattini@fi.infn.it

Abstract. The main concepts of general relativistic thermodynamics and general relativistic statistical mechanics are reviewed in a quantum framework. The main building block of the proper relativistic extension of classical thermodynamics laws is the four-temperature vector β. The general relativistic thermodynamic equilibrium condition demands β to be a Killing vector field. A remarkable consequence of this condition is that all Lie derivatives of all physical observables along the four-temperature flow vanish.

Keywords: Relativistic thermodynamics · General relativity · Thermodynamic equilibrium

1 Introduction

Relativistic thermodynamics and relativistic statistical mechanics are widespreadly used in advanced research topics: high energy astrophysics, cosmology, and relativistic nuclear collisions. The standard cosmological model views the primordial Universe as a curved manifold with matter content at local thermodynamic equilibrium. Similarly, the matter produced in high energy nuclear collisions is assumed to reach and maintain local thermodynamic equilibrium for a large fraction of its lifetime.

In this paper, we review the basic concepts of thermodynamic equilibrium in a quantum relativistic framework including general relativity. We will follow a rather informal approach leaving mathematical rigour aside for the ease of reading and to better illustrate the reasoning and the methods. We will see that the key role in the extension of thermodynamics to the quantum relativistic realm is played by the inverse temperature or, simply, four-temperature, vector β. This vector field has a precise physical meaning in terms of ideal thermometers, its magnitude being the inverse temperature marked by an ideal thermometer moving along the flow. At global thermodynamic equilibrium, the four-temperature β must be a Killing vector and we will show that, as a consequence, the Lie derivatives of all physical quantities vanish.

F. Nielsen and F. Barbaresco (Eds.): GSI 2017, LNCS 10589, pp. 442–447, 2017.
https://doi.org/10.1007/978-3-319-68445-1_52

Notation

In this paper we use the natural units, with $\hbar = c = k_B = 1$.

The Minkowskian metric tensor is diag$(1, -1, -1, -1)$; for the Levi-Civita symbol we use the convention $\epsilon^{0123} = 1$.

We will use the relativistic notation with repeated indices assumed to be summed over. Quantum operators will be denoted by a large upper hat, e.g. $\widehat{T}$ while unit vectors with a small upper hat, e.g. $\hat{v}$. The stress-energy tensor is assumed to be symmetric with an associated vanishing spin tensor.

2 Density Operator for Local Thermodynamic Equilibrium

The density operator, or density matrix, for local thermodynamic equilibrium in a relativistic framework can be derived by maximizing the Von Neumann entropy with the constraints of fixed energy- time τ. This requires the specification of a space-like hypersurface $\Sigma(\tau)$, where these densities are given, and a continuous set of Lagrange multipliers β and ζ at each point of the hypersurface:

$$-\mathrm{tr}(\widehat{\rho}\log\widehat{\rho})+\int_{\Sigma(\tau)} \mathrm{d}\Sigma n_\mu \left[\left(\langle\widehat{T}^{\mu\nu}(x)\rangle - T^{\mu\nu}(x)\right)\beta_\nu(x) - \left(\langle\widehat{j}^\mu(x)\rangle - j^\mu(x)\right)\zeta(x)\right] \quad (1)$$

where $\mathrm{d}\Sigma$ is the measure of the hypersurface and n the unit vector perpendicular to Σ. The solution is:

$$\widehat{\rho}_{\mathrm{LE}} = \frac{1}{Z_{\mathrm{LE}}} \exp\left[-\int_{\Sigma(\tau)} \mathrm{d}\Sigma n_\mu \left(\widehat{T}^{\mu\nu}(x)\beta_\nu(x) - \zeta(x)\widehat{j}^\mu(x)\right)\right] \quad (2)$$

being:

$$Z_{\mathrm{LE}} = \mathrm{tr}\left(\exp\left[-\int_{\Sigma(\tau)} \mathrm{d}\Sigma n_\mu \left(\widehat{T}^{\mu\nu}(x)\beta_\nu(x) - \zeta(x)\widehat{j}^\mu(x)\right)\right]\right)$$

is the partition function. The operator (2) does depend on the hypersurface Σ, which can be chosen arbitrarily. However, there is a preferential choice which pertains to the definition of local thermodynamic equilibrium itself which relates n to the Lagrange multiplier function β [1].

Starting from the operator (2) and calculating the entropy expression, it is then possible to show that β has a precise physical meaning; its magnitude is the inverse temperature marked by an ideal relativistic thermometer [1], that is an ideal pointlike device which is capable of instantaneously moving with the four-velocity $u = \beta/\sqrt{\beta^2}$ (hence β should be timelike), and marking the temperature $T = 1/\sqrt{\beta^2}$. The four-velocity u is to be interpreted as the four-velocity of the fluid and defines a new frame in relativistic hydrodynamics called β or thermodynamic frame [2].

3 Global Equilibrium Condition

The expression (2) of the density operator is generally covariant and it becomes independent of the hypersurface Σ if - according to the Gauss' theorem - the divergence of the integrand and contribution of the time-like boundary to the integral vanishes. In this case, the operator becomes time-independent and one has thus achieved *global thermodynamic equilibrium.* If the stress-energy tensor is symmetric, the requirement of the vanishing of the divergence implies:

$$\nabla_\mu \beta_\nu + \nabla_\nu \beta_\mu = 0 \tag{3}$$

that is, the four-temperature becomes a Killing vector. This condition was obtained in different fashions. De Groot *et al.* [3] derived it in special relativity from relativistic Boltzmann equation, Souriau [4,5] from a geometric approach. The solution of the above equation in Minkowski spacetime is known:

$$\beta^\mu = b^\mu + \varpi^{\mu\nu} x^\nu$$

where b and ϖ are constants, and includes all known forms of thermodynamic equilibrium: the familiar one with $\varpi = 0$, as well as the rotational and with constant acceleration.

This condition, however, holds in general relativity and may be used to define in the most general fashion the notion of thermodynamic equilibrium in an arbitray space-time. Conversely, the existence of at least one time-like Killing vector field ensures that thermodynamic equilibrium exists because all physical quantities will be stationary along the Killing vector field lines (see next Section). We note that it is not necessary that the Killing vector field is globally time-like; for instance, in Minkowski spacetime the field:

$$\beta = \frac{1}{T_0}(1, \boldsymbol{\omega} \times \mathbf{x})$$

with $\boldsymbol{\omega}$ and T_0 constant, describing equilibrium with rotation [6] is a Killing vector field but it becomes space-like at a radial distance $R = 1/\omega$ from the axis. Similarly, in Schwarzschild space-time the vector field $\partial/\partial t$ is time-like up to the Schwarzschild radius.

4 Thermodynamic Equilibrium and Lie Derivatives

A symmetry transformation Λ in spacetime can be defined as follows [7]:

$$\psi(\Lambda(x)) = D(\Lambda)\psi(x) \tag{4}$$

for every physical field $\psi(x)$ on the tangent hyperplane at x, where $D(\Lambda)$ is a matrix which takes care of the correct transformation of the indices of ψ, whether they are spinorial, vector or tensor. If one has a set of symmetry transformations $\Lambda_\xi(\tau)$ along a vector field ξ described by some real parameter τ, the above equation implies that the Lie derivative along the vector field ξ of ψ vanishes, that is:

$$\mathcal{L}_\xi(\psi) = 0 \tag{5}$$

These notions can be made mathematically rigorous in differential geometry (see e.g. [7]).

Since thermodynamic equilibrium involves stationarity in time, it is reasonable to surmise that a time translation is a symmetry transformation for all physical fields. While this is clear and unambiguous in special relativity, in general relativity there is no preferential frame or set of frames defining a "time". Thus, we need a time-like vector field which is naturally associated to thermodynamic equilibrium such that the Lie derivative of any physical quantity along it vanishes. Such vector field exists and it is just the four-temperature field $\beta(x)$ which meets the above requirement being a Killing vector field.

In order to show that $\mathcal{L}_\beta(\text{anything}) = 0$ two methods are available. The first is to write anything as the mean value of an operator at thermodynamic equilibrium:

$$A = \mathrm{tr}(\widehat{A}\widehat{\rho}) \tag{6}$$

work out all possible dependences of A on the metric tensor, the four-temperature and all of their derivatives and show that all of them have vanishing Lie derivative along β. This method was used in ref. [9] and we refer to that paper for the proof. The second method, which is sketched here, is more elegant, but it requires the use of quantum field theory in curved spacetime, which is a difficult subject for the familiar Hilbert space formalism of flat spacetime getting troublesome. To overcome some of the difficulties, one can take an algebraic approach by maintaining the relations involving quantum field operators in Hilbert space, by interpreting them as elements of a C*-algebra [8]. Particularly, the transformation rule of the Wightman axiom shall be kept:

$$\widehat{U}(\Lambda)\,\widehat{\psi}(x)\,\widehat{U}(\Lambda)^{-1} = D(\Lambda))^{-1}\widehat{\psi}(\Lambda(x)) \tag{7}$$

where $\widehat{U}(\Lambda)$ is the "operator" corresponding to the diffeomorphism Λ in the physical (curved) spacetime and $D(\Lambda)$ its associated matrix (*pull-back* transformation [7]). Also, one can extend the trace operation with its ciclicity property for the calculation of mean values, that is Eq. (6) by suitably redefining the notion of state [8].

With these mathematical tools, we can work out the desired proof. For the sake of simplicity, we will focus on the special case $\zeta = 0$ in Eq. (2), the general case with conserved currents being a straightforward extension. The first step is to define a one-parameter group of diffeomorphisms $\Lambda(\tau)$ along a vector field ξ, with:

$$\frac{\mathrm{d}x^\mu}{\mathrm{d}\tau} = \xi^\mu(\tau)$$

and look for the conditions by which this set defines *symmetry* transformations for the density operator, that is:

$$\widehat{U}(\Lambda(\tau))\,\widehat{\rho}\,\widehat{U}(\Lambda(\tau))^{-1} = \widehat{\rho} \tag{8}$$

If the above condition is met, then, by ciclicity and using (7):

$$
\begin{aligned}
A(x) &= \mathrm{tr}(\widehat{\rho}\widehat{A}(x)) = \mathrm{tr}(\widehat{U}(\Lambda(\tau))\widehat{\rho}\widehat{U}(\Lambda(\tau))^{-1}\widehat{A}(x)) \\
&= \mathrm{tr}(\widehat{\rho}\widehat{U}(\Lambda(\tau))^{-1}\widehat{A}(x)\widehat{U}(\Lambda(\tau))) = D(\Lambda(\tau))\mathrm{tr}(\widehat{\rho}\widehat{A}(\Lambda(\tau)^{-1}(x))) \\
&= D(\Lambda(\tau))A(\Lambda(\tau)^{-1}(x))
\end{aligned}
\tag{9}
$$

for any τ, whence the conclusion:

$$
\mathcal{L}_\xi(A) = 0 \tag{10}
$$

by taking the derivative with respect to τ on both sides follows at once.

Let us now find the conditions which ensure the relation (8). With $\widehat{\rho}$ given by (2), one has:

$$
\begin{aligned}
&\widehat{U}(\Lambda(\tau)) \exp\left[-\int_\Sigma \mathrm{d}\Sigma_\mu \widehat{T}^{\mu\nu}(x)\beta_\nu(x)\right] \widehat{U}(\Lambda(\tau))^{-1} \\
&= \exp\left[-\int_\Sigma \mathrm{d}\Sigma_\mu \widehat{U}(\Lambda(\tau))\widehat{T}^{\mu\nu}(x)\widehat{U}(\Lambda(\tau))^{-1}\beta_\nu(x)\right]
\end{aligned}
$$

One can now apply the Wightman rule (7) to the last expression:

$$
\begin{aligned}
&\exp\left[-\int_\Sigma \mathrm{d}\Sigma n_\mu \widehat{U}(\Lambda(\tau))\widehat{T}^{\mu\nu}(x)\widehat{U}(\Lambda(\tau))^{-1}\beta_\nu(x)\right] \\
&= \exp\left[-\int_\Sigma \mathrm{d}\Sigma n_\mu \Lambda(\tau))^{-1\mu}_{\rho}\Lambda(\tau))^{-1\nu}_{\sigma}\widehat{T}(\Lambda(\tau)(x))^{\rho\sigma}\beta_\nu(x)\right] \\
&= \exp\left[-\int_{\Sigma'(\tau)} \mathrm{d}\Sigma' n'_\rho \Lambda(\tau))^{-1\nu}_{\sigma}\widehat{T}(x')^{\rho\sigma}\beta_\nu(\Lambda(\tau)^{-1}(x'))\right]
\end{aligned}
$$

where, in the last equality we have changed the integration variable to the transformed one, that is $x' = \Lambda(\tau)(x)$ and the integral is now computed over the transformed domain $\Sigma'(\tau)$ with $n' = \Lambda(\tau)(n)$.

One can now make the transformation infinitesimal and expand the last expression to the first order in $\delta\tau$; both the domain and the β field will contribute to the first order expression. It is known that the domain expansion will give rise to:

$$
\int_{\Sigma'} \mathrm{d}\Sigma'_\mu V^\mu \simeq \int_\Sigma \mathrm{d}\Sigma_\mu V^\mu + \delta\tau\frac{1}{2}\int \mathrm{d}\tilde{S}_{\mu\alpha}(V^\mu\xi^\alpha - V^\alpha\xi^\mu) + \delta\tau\int_\Sigma \mathrm{d}\Sigma_\mu \xi^\mu \nabla\cdot V \tag{11}
$$

where $\mathrm{d}\Sigma_\mu \equiv \mathrm{d}\Sigma n_\mu$ and the field V^μ in our case is $\widehat{T}^{\mu\nu}\beta_\nu$. Similarly, it is also well known that:

$$
\Lambda(\tau))^{-1\nu}_{\sigma}\beta_\nu(\Lambda(\tau)^{-1}(x)) \simeq \beta_\sigma(x) - \mathcal{L}_\xi(\beta)_\sigma\delta\tau \tag{12}
$$

where $\mathcal{L}_\xi(\beta)$ is just the Lie derivative of the field β along ξ. If the surface boundary term in Eq. (11) vanishes, one can conclude that if a one-parameter group of diffeomorphisms make up a set of continuous symmetry transformations for $\widehat{\rho}$, then:

$$\int_{\Sigma} \mathrm{d}\Sigma_{\mu} \left(\xi^{\mu} \widehat{T}^{\lambda\nu} \nabla_{\lambda} \beta_{\nu} - \widehat{T}^{\mu\nu} \mathcal{L}_{\xi}(\beta)_{\nu} \right) = 0 \tag{13}$$

A sufficient condition for this is the vanishing of the integrand at any τ, which is the case if β and ξ meet the following requirements:

- β is a Killing vector field fulfilling Eq. (3), so that the first term in the integrand of Eq. (13) vanishes being $\widehat{T}$ symmetric;
- $\xi = \beta$ so that the second term in the integrand of Eq. (13) vanishes as well.

If these conditions are met for any τ, the one-parameter group structure makes it possible to retrace the above reasoning and conclude that the Eq. (8) applies. The above proof can be readily extended to the more general operator involving currents, leading to the conclusion that ζ must be a constant.

5 Consequences

Not only do the mean value of operators have vanishing Lie derivatives along the Killing vector field β, also all of the tensors describing the geometry of the spacetime [9], that is Riemann tensor, its covariant derivatives of any order and combinations thereof. The identification of β with $(1/T)u$ where T is the temperature marked by a thermometer moving with four-velocity u gives β its thermodynamic physical content.

Acknowledgments. The author would like to express its gratitude to F. Barbaresco and G. De Saxcé for pointing out relevant papers and references.

References

1. Becattini, F., Bucciantini, L., Grossi, E., Tinti, L.: Eur. Phys. J. C 75(5), 191 (2015)
2. Jensen, K., Kaminski, M., Kovtun, P., Meyer, R., Ritz, A., Yarom, A.: Phys. Rev. Lett. **109**, 101601 (2012)
3. De Groot, S.R., Van Leeuwen, W.A., Van Weert, C.G.: Relativistic Kinetic Theory. Principles and Applications, Amsterdam, Netherlands: North-holland (1980)
4. Barbaresco, F.: Entropy **18**, 386 (2016). And references therein
5. De, G.: Saxcé. Entropy **18**, 254 (2016). And references therein
6. Landau, L., Lifshitz, E.M.: Statistical Phyisics, vol. 1. Butterworth-Heinemann (2013)
7. Carroll, S.M.: Spacetime and Geometry: An Introduction to General Relativity. Addison-Wesley, San Francisco (2004)
8. Wald, R.M.: General Relativity. Chicago University Press, Chicago (1984)
9. Becattini, F.: Acta Phys. Polon. B **47**, 1819 (2016)

Probability on Riemannian Manifolds

Natural Langevin Dynamics for Neural Networks

Gaétan Marceau-Caron[1(✉)] and Yann Ollivier[2(✉)]

[1] MILA, Université de Montréal, Montréal, Canada
gaetan.marceau.caron@umontreal.ca
[2] CNRS, Université Paris-Saclay, Paris, France
contact@yann-ollivier.org

Abstract. One way to avoid overfitting in machine learning is to use model parameters distributed according to a Bayesian posterior given the data, rather than the maximum likelihood estimator. *Stochastic gradient Langevin dynamics* (SGLD) is one algorithm to approximate such Bayesian posteriors for large models and datasets. SGLD is a standard stochastic gradient descent to which is added a controlled amount of noise, specifically scaled so that the parameter converges in law to the posterior distribution [WT11, TTV16]. The posterior predictive distribution can be approximated by an ensemble of samples from the trajectory.

Choice of the variance of the noise is known to impact the practical behavior of SGLD: for instance, noise should be smaller for sensitive parameter directions. Theoretically, it has been suggested to use the inverse Fisher information matrix of the model as the variance of the noise, since it is also the variance of the Bayesian posterior [PT13, AKW12, GC11]. But the Fisher matrix is costly to compute for large-dimensional models.

Here we use the easily computed Fisher matrix approximations for deep neural networks from [MO16, Oll15]. The resulting *natural Langevin dynamics* combines the advantages of Amari's natural gradient descent and Fisher-preconditioned Langevin dynamics for large neural networks.

Small-scale experiments on MNIST show that Fisher matrix preconditioning brings SGLD close to dropout as a regularizing technique.

Consider a supervised learning problem with a dataset $\mathcal{D} = \{(x_1, y_1), \ldots, (x_N, y_N)\}$ of N input-output pairs, to be modelled by a parametric probabilistic distribution $y_i \sim p_\theta(y|x_i)$ ($x = \varnothing$ amounts to unsupervised learning of y). Defining the log-loss $\ell_\theta(y_i|x_i) := -\ln p_\theta(y_i|x_i)$, the maximum likelihood estimator is the value θ that minimizes $\mathbb{E}_{(x,y)\in\mathcal{D}}\ell_\theta(y|x)$, where $\mathbb{E}_{(x,y)\in\mathcal{D}}$ denotes averaging over the dataset.

Stochastic gradient descent is often used to tackle this minimization problem for large-scale datasets [BL03, Bot10]. This consists in iterating

$$\theta \leftarrow \theta - \eta\, \hat{\mathbb{E}}_{(x,y)\in\mathcal{D}}\, \partial_\theta \ell_\theta(y|x), \tag{1}$$

where η is a step size, ∂_θ denotes the gradient of a function with respect to θ, and $\hat{\mathbb{E}}_{(x,y)\in\mathcal{D}}$ denotes an empirical average of gradients from a random subset of the dataset $\mathcal{D}$ (a minibatch, which may be of size 1).

Estimating the model parameter θ via maximum likelihood, i.e., minimizing the training loss on $\mathcal{D}$, is prone to overfitting. Bayesian methods arguably offer a

F. Nielsen and F. Barbaresco (Eds.): GSI 2017, LNCS 10589, pp. 451–459, 2017.
https://doi.org/10.1007/978-3-319-68445-1_53

protection against overfitting ([Bis06, 3.4], [Mac03, 44.4]; see also [Nea96, Mac92] for Bayesian neural networks). Arguably, the variance of the posterior distribution of θ represents the intrinsic uncertainty on θ given the data, and optimizing θ beyond that point results in overfitting [WT11]; sampling the parameter θ from its Bayesian posterior prevents using a too precisely tuned value.

Stochastic gradient Langevin dynamics (SGLD) [WT11, TTV16] modifies stochastic gradient descent to provide random values of θ that are distributed according to a Bayesian posterior. This is achieved by adding controlled noise to the gradient descent, together with an $O(1/N)$ pull towards a Bayesian prior:

$$\theta \leftarrow \theta - \eta\,\hat{\mathbb{E}}_{(x,y)\in\mathcal{D}}\,\partial_\theta\left(\ell_\theta(y|x) - \frac{1}{N}\ln\alpha(\theta)\right) + \sqrt{\frac{2\eta}{N}}\,\mathcal{N}(0,\mathrm{Id}) \qquad (2)$$

where N is the size of the dataset, $\alpha(\theta)$ is the density of a Bayesian prior on θ, and $\mathcal{N}(0,\mathrm{Id})$ is a random Gaussian vector of size $\dim(\theta)$.[1] The larger N is, the closer SGLD is to simple stochastic gradient descent, as the Bayesian posterior concentrates around a single point. The Bayesian interpretation determines the necessary amount of noise depending on step size and dataset size. SGLD has the same algorithmic complexity as simple stochastic gradient descent.

Thanks to the injected noise, θ does not converge to a single value, but its *distribution* at time t converges to the Bayesian posterior of θ given the data, namely, $\pi(\theta) \propto \alpha(\theta)\prod_{(x,y)\in\mathcal{D}} p_\theta(y|x)$. A formal proof is given in [TTV16, CDC15] for suitably decreasing step sizes; the asymptotically optimal step size is $\eta_k \approx k^{-1/3}$ at step k, thus, larger than the usual Robbins–Monro criterion for stochastic gradient descent. The asymptotic behavior is well understood from [TTV16, CDC15], and [MDM17, DM16] provide sharp non-asymptotic rates in the convex case.

One can then extract information from the distribution of θ. For instance, the Bayesian posterior mean can be approximated by averaging θ over the trajectory. The full Bayesian posterior prediction can be approximated by ensembling [GBC16, 7.12] predictions from several values of θ sampled from the trajectory, though this creates additional computational and memory costs at test time.

We refer to [WT11, TTV16] for a general discussion of SGLD (and other Bayesian methods) for large-scale machine learning.

Practical remarks. For regression problems, the square loss $(y - \hat{y}(\theta))^2$ between observations y and predictions $\hat{y}(\theta)$ must be properly cast as the log-loss of a Gaussian model, $\ell = (y - \hat{y}(\theta))^2/2\sigma^2 + \dim(y)\ln\sigma$ for a proper choice of σ (such as the empirical RMSE). Just using $\sigma^2 = 1$ amounts to using a badly specified error model and will provide a poor Bayesian posterior.

The variance coming from computing gradients on a minibatch from $\mathcal{D}$, $\hat{\mathbb{E}}_{(x,y)\in\mathcal{D}}\partial_\theta\ell_\theta(y|x)$, adds up to the SGLD noise. For small step sizes, $\eta \ll \sqrt{\eta}$, so the SGLD noise dominates. [AKW12] suggest a correction for large η.

[1] Our convention for the step size η differs from [TTV16] by a factor $2/N$, namely, $\delta = \frac{2}{N}\eta$ where δ is the step size in [TTV16, (3)]: this allows for a direct comparison with stochastic gradient descent.

A popular choice of prior $\alpha(\theta)$ is a Gaussian prior $\mathcal{N}(0, \Sigma^2)$; the variance Σ^2 becomes an additional hyperparameter. In line with Bayesian philosophy we also tested the conjugate prior for Gaussian distributions with unknown variance (a mixture of Gaussian priors for all Σ^2), the normal-inverse gamma, with default hyperparameters; empirically, performance comes close enough to the best Σ^2, without having to optimize over Σ^2.

Preconditioning the noise. SGLD as above introduces uniform noise in all parameter directions. This might hurt the optimization process. If performance is more sensitive in certain parameter directions, adapting the noise covariance can largely improve SGLD performance. This requires changing both the noise covariance *and* the gradient step by the same matrix [WT11,GC11,AKW12, LCCC16].

For any positive-definite symmetric matrix C, the *preconditioned SGLD*,

$$\theta \leftarrow \theta - \eta\, C\, \hat{\mathbb{E}}_{(x,y)\in\mathcal{D}}\, \partial_\theta \left(\ell_\theta(y|x) - \frac{1}{N}\ln\alpha(\theta)\right) + \sqrt{\frac{2\eta}{N}}\, C^{1/2}\mathcal{N}(0, \mathrm{Id}) \qquad (3)$$

still converges in law to the Bayesian posterior (it is equivalent to a non-preconditioned Langevin dynamics on $C^{-1/2}\theta$). A diagonal C amounts to having distinct values of the step size η for each parameter direction, both for noise and gradient.

This assumes that C is fixed and does not depend on θ.[2] In practice, this means C should be adapted slowly in the algorithms (hence our use of running averages for C hereafter); the resulting bias is analyzed in [LCCC16, Corollary 2].

[LCCC16] apply preconditioned SGLD to neural networks, with a diagonal preconditioner C taken from the RMSProp optimization scheme, a classical tool to adapt step sizes for each direction of θ.[3]

Langevin preconditioners and information geometry. In order to provide a good or even optimal preconditioner C, it has been suggested to set C to the inverse of the Fisher information matrix [GC11,AKW12,PT13].

The Fisher information matrix $J(\theta)$ at θ, for a model p_θ, is defined by

$$J(\theta) := \mathbb{E}_{(x,y)\in\mathcal{D}}\, \mathbb{E}_{\tilde{y}\sim p_\theta(\tilde{y}|x)} \left[(\partial_\theta \ln p_\theta(\tilde{y}|x))\, (\partial_\theta \ln p_\theta(\tilde{y}|x))^\top\right] \qquad (4)$$

(note that for supervised learning, we fix the distribution of the inputs x from the data but sample y according to the model $p_\theta(y|x)$). Intuitively, the entries of the Fisher matrix represent the sensitivity of the model in each parameter direction.

[2] If $C(\theta)$ depends on θ, the algorithm involves derivatives of $C(\theta)$ with respect to θ [GC11,XSL+14]. In our case (neural networks), these are not readily available.

[3] We could not reproduce the good results from [LCCC16]. Their code contains a bug which produces noise of variance $2\eta/N^2$ instead of $2\eta/N$ in (2), thus greatly suppressing the Langevin noise, and not matching the Bayesian posterior.

Using the inverse Fisher matrix as the SGLD preconditioner C has several theoretical advantages. First, this reduces Langevin noise in sensitive parameter directions (thanks to the Fisher matrix being the average of squared gradients).

Second, since C also affects the gradient term in (3), the gradient part of SGLD becomes Amari's *natural gradient*, known to have theoretically optimal convergence [Ama98]. The resulting algorithm is also insensitive to changes of variables in θ (for small learning rates) and makes sense if θ belongs to a manifold.

Third, the Bayesian posterior variance of the parameter θ is asymptotically proportional to the inverse Fisher information matrix $J(\theta^*)^{-1}$ at the maximum a posteriori θ^* (Bernstein–von Mises theorem [vdV00]). So with Fisher preconditioning, the noise injected in the optimization process has the same shape as the actual noise in the target distribution on θ. Thus, it is tempting to investigate the behavior of SGLD with noise covariance $C \propto J(\theta^*)^{-1}$.

Approximating the Fisher matrix for large models. The Fisher matrix $J(\theta^*)$ can be estimated by replacing the expectation in its definition (4) by an empirical average along the trajectory [AKW12]. This results in Algorithm 3 below.[4]

However, for large-dimensional models such as deep neural networks, the Fisher matrix is too large to be inverted or even stored (it is a full matrix of size $\dim(\theta) \times \dim(\theta)$). So approximation strategies are necessary.

Approximating the Fisher matrix does not invalidate asymptotic convergence of SGLD, since (3) converges to the true Bayesian posterior for any preconditioning matrix C. But the closer C is to the inverse Fisher matrix, the closer SGLD will be to a natural gradient descent, and SGLD noise to the true posterior variance.

One way of building principled approximations of the Fisher matrix is to reason in terms of the associated invariance group. The full Fisher matrix provides invariance under all changes of variables in parameter space θ: optimizing by natural gradient descent over θ or over a reparameterization of θ will yield the same learning trajectories (in the limit of small learning rates). Meanwhile, the Euclidean gradient descent does not have any invariance properties (e.g., inverting black and white in the image inputs of a neural network affects performance). We refer to [Oll15] for further discussion in the context of neural networks.

The diagonal of the Fisher matrix is the most obvious approximation. Its invariance subgroup consists of all rescalings of individual parameter components.

The *quasi-diagonal* approximation of the Fisher matrix [Oll15] is built to retain more invariance properties of the Fisher matrix, at a small computational cost. It provides invariance under all affine transformations of the *activities* of units in a neural network (e.g., shifting or rescaling the inputs, or switching from sigmoid to tanh activation function). The quasi-diagonal approximation maintains the diagonal of the Fisher matrix plus a few well-chosen off-diagonal terms,

[4] The Fisher matrix definition (4) averages over synthetic data $\tilde{y}$ generated by $p_\theta(\tilde{y}|x)$. In practice, using the samples y from the dataset is simpler (the OP variant in Algorithm 3). This can result in significant differences [MO16, Oll15, PB13], even in simple cases.

requiring to store an additional vector of size $\dim(\theta)$. Overall, the resulting algorithmic complexity is of the same order as ordinary backpropagation, thus suitable for large-dimensional models. [Oll15] also provides more complex approximations with a larger invariance group, suited to sparsely connected neural networks.

The resulting *quasi-diagonal natural gradient* can be coded efficiently [MO16]; experimentally, the few extra off-diagonal terms can make a large difference.

Natural Langevin dynamics for neural networks: implementation. Algorithm 1 presents the Langevin dynamics with a generic preconditioner C. For the ordinary SGLD, C would be the identity matrix. The internal setup of a preconditioner decouples from the general implementation of SGLD optimization. A preconditioner C is a matrix object that provides the routines needed by Algorithm 1:

- Multiply a gradient estimate by C: $g \leftarrow Cg$;
- Draw a Gaussian random vector $\xi \sim \mathcal{N}(0, C) = C^{1/2}\mathcal{N}(0, \mathrm{Id})$;
- Update C given recent gradient observations;
- An initialization procedure for C at startup.

We now make these routines explicit for several choices of preconditioner.

The RMSProp preconditioner used in [LCCC16] divides gradients by their recent magnitude: C is diagonal, and for each parameter component i, C_{ii} is the inverse of a root-mean-square average of recent gradients in direction i (Algorithm 2).

Algorithm 3 describes preconditioned SGLD with a preconditioner $C = J^{-1}$ using the full Fisher matrix J at the posterior mean θ^*. This is suitable only for small-dimensional models. The Fisher matrix is obtained as a moving average of rank-one contributions over the trajectory (Algorithm 3). This moving average has the further advantage of smoothing the fluctuations of the parameter θ over the SGLD trajectory, ensuring convergence [AKW12].

Finally we consider SGLD using the *quasi-diagonal* Fisher matrix, the object of the tests in this article, applicable to large-dimensional models.

For a neural network, the parameters are grouped into blocks corresponding to the bias and incoming weights of each neuron, with the bias being the first parameter in a block. The Fisher matrix J is updated as in Algorithm 3, but storing only its diagonal and the first row in each block. Then a Cholesky decomposition $C = AA^\top$ is maintained for the preconditioner C, such that the axioms of the quasi-diagonal approximation are satisfied (Algorithm 4): in each block, A has non-zero entries only on its diagonal and first row, and is built such that $C^{-1} = (A^\top)^{-1}A^{-1}$ has the same first row and diagonal as the Fisher matrix J. The sparse Cholesky decomposition provides the operations of the preconditioner: multiplying by $C = AA^\top$ and sampling from $\mathcal{N}(0, C) = A\mathcal{N}(0, \mathrm{Id})$.

Experiments. We compare empirically four SGLD preconditioners: Euclidean ($C = \mathrm{Id}$, standard SGLD), RMSProp, Diagonal Outer Product (DOP) and Quasi-Diagonal Outer Product (QDOP) on the MNIST dataset. The Euclidean and RMSProp results widely mismatch those from [LCCC16], see footnote 5.

We compare SGLD to Dropout, a standard regularization procedure for neural networks. For SGLD we compare the performance of using a single network set to the posterior mean, and an ensemble of networks sampled from the trajectory (theoretically closer to the true Bayesian posterior, but computationally costlier).

The code for the experiments can be found at https://github.com/gmarceaucaron/natural-langevin-dynamics-for-neural-networks. We use a feedforward ReLU network with two hidden layers of size 400, with the usual $\mathcal{N}(0, 1/\text{fan-in})$ initialization [GBC16]. Inputs are normalized to $[0; 1]$. Step sizes are optimized over $\eta \in \{.001, .01, .1, 1\}$ for Euclidean and $\eta \in \{.0001, .001, .01, .1\}$ for the others, with schedule $\eta \leftarrow \eta/2$ every 10,000 updates [LCCC16]. Minibatch size is 100. The metric decay rate and regularizer are $\gamma_t = 1/\sqrt{t}$ and $\varepsilon = 10^{-4}$. The prior was a Gaussian $\mathcal{N}(0, \sigma^2)$ with $\sigma^2 \in \{0.01, 0.1, 1\}$. The Bayesian posterior ensemble is built by storing every 100-th parameter value of the trajectory after the first 500.

Table 1 shows that SGLD with a quasi-diagonal Fisher matrix preconditioner and Bayesian posterior ensembling outperforms other SGLD settings.

Table 1. Performance on the MNIST test set with a feedforward 400-400 architecture. Hyperparameters were selected based on accuracy on a validation set. The methods are SGD without regularization, Dropout, SGLD ensemble and SGLD posterior mean (PostMean) with a Gaussian prior ($\sigma^2 = 0.1$)

Method	NLL (train)	Accuracy (train)	NLL (test)	Accuracy (test)
SGD	0.0003	100.00	0.0584	98.24
Dropout	0.0006	100.00	0.0519	98.61
Ensemble, Euclidean	0.0357	99.63	0.0726	98.10
Ensemble, RMSProp	0.0415	99.47	0.0742	98.17
Ensemble, DOP	0.0292	99.69	0.0660	98.13
Ensemble, QDOP	0.0229	99.85	0.0591	98.38
PostMean, Euclidean	0.0281	99.12	0.1240	97.16
PostMean, RMSProp	0.0299	99.07	0.1134	97.21
PostMean, DOP	0.0243	99.20	0.1389	97.20
PostMean, QDOP	0.0292	99.60	0.3429	98.14

Bayesian theory favors the use of the full Bayesian posterior at test time, rather than any single parameter value. The results here are consistent with this viewpoint: using a single parameter set to the Bayesian posterior mean offers much poorer performance than either Dropout or a Bayesian posterior ensemble. (Dropout also has a Bayesian inspiration as a mixture of models [SHK+14].) This is also consistent with the generally good performance of ensemble methods.

All other preconditioners perform worse than QDOP or Dropout. In particular, the diagonal Fisher matrix offers no advantage over RMSProp, while the *quasi-diagonal* Fisher matrix does. This is consistent with [MO16] and may vindicate the quasi-diagonal construction via an invariance group viewpoint.

Data: Dataset $\mathcal{D} = \{(x_1, y_1), \ldots, (x_N, y_N)\}$ of size N;
probabilistic model $p_\theta(y|x)$ with log-loss $\ell(y|x) := -\ln p_\theta(y|x)$;
Bayesian prior $\alpha(\theta) = \mathcal{N}(\theta_0, \Sigma_0)$, default: $\theta_0 = 0$;
Learning rate $\eta_t \ll 1$. Preconditioner C (for simple SGLD: $C = \mathrm{Id}$).
Result: Parameter θ whose distribution approximates the Bayesian posterior $\Pr(\theta \mid D, \alpha)$. Approximation $\bar{\theta}$ of the Bayesian posterior mean of θ.
Initialization: $\theta \sim \alpha(\theta)$; $\bar{\theta} \leftarrow \theta_0$; initialize preconditioner;
while *not finished* **do**
- retrieve a data sample x and corresponding target y from $\mathcal{D}$;
- forward x through the network, and compute loss $\ell(y|x)$;
- backpropagate and compute gradient of loss: $g \leftarrow \partial_\theta \ell(y|x)$ (for a minibatch: let g be the *average*, not the sum, of individual gradients);
- incorporate gradient of prior: $g \leftarrow g + \frac{1}{N}\Sigma_0^{-1}(\theta - \theta_0)$;
- update preconditioner C using current sample and gradient g;
- apply preconditioner: $g \leftarrow Cg$;
- sample preconditioned noise: $\xi \sim \mathcal{N}(0, C) = C^{1/2}\mathcal{N}(0, \mathrm{Id})$;
- update parameters: $\theta \leftarrow \theta - \eta_t\, g + \sqrt{(2\eta_t/N)}\,\xi$;
- update posterior mean: $\bar{\theta} \leftarrow (1 - \mu_t)\bar{\theta} + \mu_t \theta$.

end

Algorithm 1. SGLD with a generic preconditioner C. For instance C may be Id (Euclidean SGLD), a diagonal preconditioner such as RMSProp, the inverse of a Fisher matrix approximation...

Data: Preconditioner $C = D^{-1/2}$ with D a diagonal matrix of size $\dim(\theta)$; decay rate γ_t; regularizer $\varepsilon \geq 0$.
Initialization: $D \leftarrow \mathrm{diag}(1)$;
Preconditioner update: $D_{ii} \leftarrow (1 - \gamma_t)D_{ii} + \gamma_t\, g_i^2$ with g_i the components of the gradient of the current sample;
Preconditioner application: $g_i \leftarrow (D_{ii} + \varepsilon)^{-1/2}\, g_i$;
Preconditioned noise: $\xi_i \leftarrow (D_{ii} + \varepsilon)^{-1/4}\mathcal{N}(0, 1)$.

Algorithm 2. RMSProp routines for SGLD, similar to [LCCC16].

Data: Preconditioner $C = J^{-1}$ with J the Fisher matrix; decay rate γ_t; regularizer $\varepsilon \geq 0$.
Initialization: $J \leftarrow \mathrm{diag}(1)$;
Preconditioner update: Synthesize output $\tilde{y} \sim p_\theta(\tilde{y}|x)$ given current model θ and current input x (OP variant: just use $\tilde{y} = y$ from the dataset);
Compute gradient of loss for $\tilde{y}$: $\tilde{v} \leftarrow \partial_\theta \ell(\tilde{y}|x)$;
Update Fisher matrix: $J \leftarrow (1 - \gamma_t)J + \gamma_t \tilde{v}\tilde{v}^\top$;
Preconditioner application: $v \leftarrow (J + \varepsilon\mathrm{Id})^{-1} v$;
Preconditioned noise: $\xi \leftarrow (J + \varepsilon\mathrm{Id})^{-1/2}\mathcal{N}(0, \mathrm{Id})$.

Algorithm 3. Routines for SGLD with full Fisher matrix.

Data: Symmetric positive matrix J of which only the diagonal and first row are known; regularizer $\varepsilon \geq 0$.
Result: Sparse matrix A whose non-zero entries lie only on the diagonal and first row, and such that $(A^\top)^{-1}A^{-1}$ has the same diagonal and first row as $J + \varepsilon \mathrm{Id}$.
$A \leftarrow 0$; $A_{00} \leftarrow \frac{1}{\sqrt{J_{00}+\varepsilon}}$ (Matrix indices start at 0);
$A_{ii} \leftarrow \frac{1}{\sqrt{J_{ii}-(A_{00}J_{0i})^2+\varepsilon}}$ for each index $i \neq 0$;
$A_{0i} \leftarrow -A_{00}^2 A_{ii} J_{0i}$ for each index $i \neq 0$;
return A;

Algorithm 4. Quasi-diagonal Cholesky decomposition.

References

[AKW12] Ahn, S., Korattikara, A., Welling, M.: Bayesian posterior sampling via stochastic gradient Fisher scoring. In: ICML (2012)
[Ama98] Amari, S.: Natural gradient works efficiently in learning. Neural Comput. **10**, 251–276 (1998)
[Bis06] Bishop, C.M.: Pattern Recognition and Machine Learning. Springer, New York (2006)
[BL03] Bottou, L., LeCun, Y.: Large scale online learning. In: NIPS, vol. 30, p. 77 (2003)
[Bot10] Bottou, L.: Large-scale machine learning with stochastic gradient descent. Proceedings of COMPSTAT 2010, pp. 177–186. Springer, Heidelberg (2010)
[CDC15] Chen, C., Ding, N., Carin, L.: On the convergence of stochastic gradient MCMC algorithms with high-order integrators. In: Advances in Neural Information Processing Systems, pp. 2278–2286 (2015)
[DM16] Durmus, A., Moulines, E.: High-dimensional Bayesian inference via the unadjusted Langevin algorithm. arXiv preprint arXiv:1605.01559 (2016)
[GBC16] Goodfellow, I., Bengio, Y., Courville, A.: Deep Learning. MIT Press (2016)
[GC11] Girolami, M., Calderhead, B.: Riemann manifold langevin and hamiltonian monte carlo methods. J. Roy. Stat. Soc. Series B (Statistical Methodology) **73**(2), 123–214 (2011)
[LCCC16] Li, C., Chen, C., Carlson, D.E., Carin, L.: Preconditioned stochastic gradient Langevin dynamics for deep neural networks. In: Schuurmans, D., Wellman, M.P. (eds.) Proceedings of the Thirtieth AAAI Conference on Artificial Intelligence, 12–17 February 2016, Phoenix, Arizona, USA, pp. 1788–1794. AAAI Press (2016)
[Mac92] MacKay, D.J.C.: A practical Bayesian framework for backpropagation networks. Neural Comput. **4**(3), 448–472 (1992)
[Mac03] MacKay, D.J.C.: Information Theory, Inference and Learning Algorithms. Cambridge University Press, Cambridge (2003)
[MDM17] Majewski, S., Durmus, A., Miasojedow, B.: (2017)
[MO16] Marceau-Caron, G., Ollivier, Y.: Practical Riemannian neural networks. arXiv, abs/1602.08007 (2016)
[Nea96] Neal, R.M.: Bayesian Learning for Neural Networks. Springer, New York (1996)
[Oll15] Ollivier, Y.: Riemannian metrics for neural networks I: feedforward networks. Inf. Infer. **4**(2), 108–153 (2015)

[PB13] Pascanu, R., Bengio, Y.: Natural gradient revisited. arXiv, abs/1301.3584 (2013)

[PT13] Patterson, S., Teh, Y.W.: Stochastic gradient Riemannian Langevin dynamics on the probability simplex. In: Advances in Neural Information Processing Systems, pp. 3102–3110 (2013)

[SHK+14] Srivastava, N., Hinton, G.E., Krizhevsky, A., Sutskever, I., Salakhutdinov, R.: Dropout: a simple way to prevent neural networks from overfitting. J. Mach. Learn. Res. **15**(1), 1929–1958 (2014)

[TTV16] Teh, Y.W., Thiery, A.H., Vollmer, S.J.: Consistency and fluctuations for stochastic gradient Langevin dynamics. J. Mach. Learn. Res. **17**(7), 1–33 (2016)

[vdV00] van der Vaart, A.W.: Asymptotic Statistics. Cambridge University Press, Cambridge (2000)

[WT11] Welling, M., Teh, Y.W.: Bayesian learning via stochastic gradient Langevin dynamics. In: Proceedings of the 28th International Conference on Machine Learning (ICML-11), pp. 681–688 (2011)

[XSL+14] Xifara, T., Sherlock, C., Livingstone, S., Byrne, S., Girolami, M.: Langevin diffusions and the Metropolis-adjusted Langevin algorithm. Stat. Probab. Lett. **91**, 14–19 (2014)

3D Insights to Some Divergences for Robust Statistics and Machine Learning

Birgit Roensch[1] and Wolfgang Stummer[1,2](✉)

[1] Department of Mathematics, University of Erlangen–Nürnberg, Cauerstrasse 11, 91058 Erlangen, Germany

[2] Affiliated Faculty Member of the School of Business and Economics, University of Erlangen–Nürnberg, Lange Gasse 20, 90403 Nürnberg, Germany

stummer@math.fau.de

Abstract. Divergences (distances) which measure the similarity respectively proximity between two probability distributions have turned out to be very useful for several different tasks in statistics, machine learning, information theory, etc. Some prominent examples are the Kullback-Leibler information, – for convex functions ϕ – the Csiszar-Ali-Silvey ϕ–divergences CASD, the "classical" (i.e., unscaled) Bregman distances and the more general scaled Bregman distances SBD of [26,27]. By means of 3D plots we show several properties and pitfalls of the geometries of SBDs, also for non-probability distributions; robustness of corresponding minimum-distance-concepts will also be covered. For these investigations, we construct a special SBD subclass which covers both the often used power divergences (of CASD type) as well as their robustness-enhanced extensions with non-convex non-concave ϕ.

Keywords: ϕ–divergences · Power divergences · Scaled Bregman distances · Robustness · Minimum distance estimation

1 Introduction and Results

As exemplary current state of the art, some divergences (distances, (dis)similarity measures, discrepancy measures) between probability distributions have been successfully used for parameter estimation, goodness-of-fit testing, various different machine learning tasks, procedures in information theory, the detection of changes, pattern recognition, etc. Amongst them, let us mention exemplarily that from a (strictly) convex function ϕ one can construct the ϕ–divergences of [1,9], as well as the classical "unscaled" Bregman distances (see e.g. [23]) which also include the density power divergences of [3]. Some comprehensive coverages on their statistical use can e.g. be found in [4,22]. Machine learning applications are e.g. given in [5,8,11,18,21,29–31]. Recently, [27] (cf. also [26]) introduced the concept of scaled Bregman distances (SBD), which cover all the above-mentioned distances as special cases; see [12,13], [15] for some applications of SBD to simultaneous parameter estimation and

F. Nielsen and F. Barbaresco (Eds.): GSI 2017, LNCS 10589, pp. 460–469, 2017.
https://doi.org/10.1007/978-3-319-68445-1_54

goodness-of-fit investigations, and [14] for utilizations in robust change point detections. Notice also [20] for indicating some potential applications of SBD to machine learning tasks, in connection with $v-$conformal divergences; a special sub-setup of the latter was also employed by [19]. In the present paper, we visualize some interesting properties and pitfalls of the SBD geometries induced by the involved divergence balls. To start with, let us assume that the modeled respectively observed random data take values in a state space $\mathscr{Y}$ equipped with a system $\mathscr{A}$ of admissible events ($\sigma-$algebra). On this, let us consider the similarity/proximity of two probability distributions P, Q described by their probability densities $y \mapsto p(y) \geq 0$, $y \mapsto q(y) \geq 0$ via $P[A] = \int_A p(y)\ \mathrm{d}\lambda(y)$, $Q[A] = \int_A q(y)\ \mathrm{d}\lambda(y)$ ($A \in \mathscr{A}$), where λ is a fixed – maybe nonprobability – distribution and one has the normalizations $\int_{\mathscr{Y}} p(y)\ \mathrm{d}\lambda(y) = \int_{\mathscr{Y}} q(y)\ \mathrm{d}\lambda(y) = 1$. The set of all such probability distributions will be denoted by $\mathscr{M}_{\lambda}^{1}$. We also employ the set $\mathscr{M}_{\lambda}$ of all general – maybe nonprobability – distributions ν of the form $\nu[A] = \int_A n(y)\ \mathrm{d}\lambda(y)$ ($A \in \mathscr{A}$) with density $y \mapsto n(y) \geq 0$ satisfying $\int_{\mathscr{Y}} n(y)\ \mathrm{d}\lambda(y) < \infty$. For instance, if λ is the counting distribution (attributing the value 1 to each outcome $y \in \mathscr{Y}$) then $p(\cdot)$, $q(\cdot)$ are (e.g. binomial) probability mass functions and $n(\cdot)$ is a (e.g. unnormalized-histogram-related) general mass function; if λ is the Lebesgue measure on $\mathscr{Y} = \mathbb{R}$, then $p(\cdot)$, $q(\cdot)$ are (e.g. Gaussian) probability density functions and $n(\cdot)$ is a general (possibly unnormalized) density function. In such a context, one can use the general concept of distances (divergences, (dis)similarity measures) between distributions introduced by [27] (see also [15,26]):

Definition 1. *Let $\phi : (0,\infty) \mapsto \mathbb{R}$ be a (for the sake of this paper) strictly convex, differentiable function, continuously extended to $t = 0$. Its derivative is denoted by ϕ'. The Bregman distance of the two probability distributions $P, Q \in \mathscr{M}_{\lambda}^{1}$ scaled by the general distribution $W \in \mathscr{M}_{\lambda}$ (with density w) is defined by*

$$
\begin{aligned}
&0 \leq B_{\phi}\left(P, Q \,||\, W\right) \\
&= \int_{\mathscr{Y}} [\phi(\frac{p(y)}{w(y)}) - \phi(\frac{q(y)}{w(y)}) - \phi'(\frac{q(y)}{w(y)}) \cdot (\frac{p(y)}{w(y)} - \frac{q(y)}{w(y)})]\ \mathrm{d}W(y) \qquad (1) \\
&= \int_{\mathscr{Y}} w(y) \cdot [\phi(\frac{p(y)}{w(y)}) - \phi(\frac{q(y)}{w(y)}) - \phi'(\frac{q(y)}{w(y)}) \cdot (\frac{p(y)}{w(y)} - \frac{q(y)}{w(y)})]\ \mathrm{d}\lambda(y). \qquad (2)
\end{aligned}
$$

To guarantee the existence of the integrals in (1), (2), the zeros of $p(\cdot), q(\cdot), w(\cdot)$ have to be combined by proper conventions. Analogously, we define the scaled Bregman distance $B_{\phi}\left(\mu, \nu \,||\, W\right) \geq 0$ of two general distributions $\mu, \nu \in \mathscr{M}_{\lambda}$ scaled by the general distribution $W \in \mathscr{M}_{\lambda}$, where we additionally assume $\phi(t) \geq 0$.

The papers [26,27] show that if ϕ is (say) strictly convex on $[0,\infty)$, continuous on $(0,\infty)$ with $\phi(1) = 0$, then for the special case $p(y) > 0$, $q(y) = w(y) > 0$ ($y \in \mathscr{Y}$) the scaled Bregman distance (2) becomes

$$
B_{\phi}\left(P, Q \,||\, Q\right) = \int_{\mathscr{Y}} q(y) \cdot \phi(\tfrac{p(y)}{q(y)})\ \mathrm{d}\lambda(y) =: D_{\phi}\left(P, Q\right), \qquad (3)
$$

which is nothing but the well-known $\phi-$divergence between P and Q. The latter has been first studied by [9] as well as [1]; see e.g. also [28] for pitfalls on

ϕ–divergences $D_{\phi}(\mu,\nu)$ between general distributions $\mu,\nu \in \mathscr{M}_{\lambda}$, and [6] for recent applications of the latter for bootstrapping purposes. For "generator" $\phi(t)=\phi_1(t):=t\log t+1-t\geq 0$ $(t>0)$ one ends up with the *Kullback-Leibler KL divergence* $D_{\phi_1}(P,Q)$[1]. The special choice $\phi(t)=\phi_0(t):=-\log t+t-1\geq 0$ leads to the *reversed KL divergence* $D_{\phi_0}(P,Q)=D_{\phi_1}(Q,P)$, and the function $\phi(t)=\phi_{\alpha}(t):=\frac{t^{\alpha}-1}{\alpha(\alpha-1)}-\frac{t-1}{\alpha-1}\geq 0$ $(\alpha\in\mathbb{R}\backslash\{0,1\})$ generates the other *power divergences* $D_{\phi_{\alpha}}(P,Q)$ (cf. [16,24]), where $\alpha=2$ gives the Pearson's chi-square divergence and $\alpha=1/2$ the squared Hellinger distance.

So far, ϕ has been (strictly) convex. However, notice that scaled Bregman distances $B_{\phi}(P,Q\,||\,W)$ can also be used to construct $\widetilde{\phi}$–divergences $D_{\widetilde{\phi}}(P,Q)$ with non-convex non-concave generator $\widetilde{\phi}$ (this new approach contrasts e.g. the construction method of [7,25]). Exemplarily, let $\phi:=\phi_{\alpha}$ and $W:=\widetilde{W}_{\beta,r}(P,Q)$ in terms of the "locally adaptive" scaling density $w(y)=\widetilde{w}_{\beta,r}(p(y),q(y))\geq 0$ defined by the r-th power mean $\widetilde{w}_{\beta,r}(u,v):=(\beta\,u^r+(1-\beta)\,v^r)^{1/r}$, $\beta\in[0,1]$, $r\in\mathbb{R}\backslash\{0\}$, $u\geq 0$, $v\geq 0$. Accordingly, for $\alpha\cdot(\alpha-1)\neq 0$ we derive the corresponding scaled Bregman distance

$$B_{\phi_{\alpha}}(P,Q\,||\,\widetilde{W}_{\beta,r}(P,Q))=\int_{\mathscr{Y}}\frac{\widetilde{w}_{\beta,r}(p(y),q(y))^{1-\alpha}\cdot\{p(y)^{\alpha}+(\alpha-1)\,q(y)^{\alpha}-\alpha\,p(y)\,q(y)^{\alpha-1}\}}{\alpha\cdot(\alpha-1)}\,\mathrm{d}\lambda(y)$$

$$=\int_{\mathscr{Y}}q(y)\cdot\frac{\left(\beta\cdot\left(\frac{p(y)}{q(y)}\right)^r+1-\beta\right)^{(1-\alpha)/r}\cdot\left\{\left(\frac{p(y)}{q(y)}\right)^{\alpha}+\alpha-1-\alpha\cdot\frac{p(y)}{q(y)}\right\}}{\alpha\cdot(\alpha-1)}\,\mathrm{d}\lambda(y)=:D_{\widetilde{\phi}_{\alpha,\beta,r}}(P,Q) \quad (4)$$

$$\text{where the generator}\quad \widetilde{\phi}_{\alpha,\beta,r}(t):=\phi_{\alpha}(t)\cdot(\beta\,t^r+1-\beta)^{(1-\alpha)/r}$$
$$=(\alpha\cdot(\alpha-1))^{-1}\cdot(t^{\alpha}+\alpha-1-\alpha\cdot t)\cdot(\beta\,t^r+1-\beta)^{(1-\alpha)/r}\;>\;0,\quad t>0, \quad (5)$$

can be non-convex non-concave in t; see e.g. Fig. 1(d) which shows $t\mapsto\widetilde{\phi}_{\alpha,\beta,r}(t)$ for $\alpha=7.5$, $\beta=0.05$, $r=7.5$. Analogously, we construct the more general $B_{\phi_{\alpha}}(\mu,\nu\,||\,\widetilde{W}_{\beta,r}(\mu,\nu))=:D_{\widetilde{\phi}_{\alpha,\beta,r}}(\mu,\nu)$ for general distributions μ,ν. The subcase $\beta=0$, $\alpha\cdot(\alpha-1)\neq 0$, leads to the power divergences $D_{\widetilde{\phi}_{\alpha,0,r}}(P,Q)=D_{\phi_{\alpha}}(P,Q)$ where the function $t\mapsto\widetilde{\phi}_{\alpha,0,r}(t)=\phi_{\alpha}(t)$ is strictly convex. We shall see in the RAF discussion in (6) below, that $\beta\neq 0$ opens the gate to enhanced robustness properties; interesting divergence geometries can be achieved, too.

Returning to the general context, with each scaled Bregman divergence $B_{\phi}(\cdot,\cdot\,||\,W)$ one can associate a *divergence-ball* $\mathbb{B}_{\phi}(P,\rho)$ with "center" $P\in\mathscr{M}_{\lambda}^1$ and "radius" $\rho\in]0,\infty[$, defined by $\mathbb{B}_{\phi}(P,\rho):=\{Q\in\mathscr{M}_{\lambda}^1:\ B_{\phi}(P,Q\,||\,W)\leq\rho\}$, whereas the corresponding *divergence-sphere* is given by $\mathbb{S}_{\phi}(P,\rho):=\{Q\in\mathscr{M}_{\lambda}^1:\ B_{\phi}(P,Q\,||\,W)=\rho\}$; see e.g. [10] for a use of some divergence balls with strictly convex generators as a constraint in financial-risk related decisions. Analogously, we define the general-distribution-versions $\mathbb{B}_{\phi}^g(\mu,\rho):=\{\nu\in\mathscr{M}_{\lambda}:\ B_{\phi}(\mu,\nu\,||\,W)\leq\rho\}$ and $\mathbb{S}_{\phi}^g(\mu,R):=\{\nu\in\mathscr{M}_{\lambda}:\ B_{\phi}(\mu,\nu\,||\,W)=\rho\}$. Of course, the "geometry/topology" induced by these divergence balls and

[1] Which is equal to $D_{\check{\phi}_1}(P,Q)$ with $\check{\phi}_1(t):=t\log t\in[-\mathrm{e}^{-1},\infty[$, but generally $D_{\phi_1}(\mu,\nu)\neq D_{\check{\phi}_1}(\mu,\nu)$ where the latter can be negative and thus isn't a distance.

spheres is quite non-obvious. In order to help building up a corresponding intuition, we concretely show several effects in the following, where for the sake of brevity and preparation for the robustness investigations below we confine ourselves to the flexible divergence family $B_{\phi_\alpha}(P, Q \,||\, \widetilde{W}_{\beta,r}(P, Q)) = D_{\widetilde{\phi}_{\alpha,\beta,r}}(P, Q)$ and to $P := P_{\epsilon,\theta_0} = (1-\epsilon)\, Bin(2, \theta_0) + \epsilon\, \delta_2$, where $\epsilon \in]0, 1[$, δ_y denotes Dirac's distribution at y (i.e. $\delta_y[A] = 1$ iff $y \in A$ and $\delta_y[A] = 0$ else), and $Bin(2, \theta_0) =: \widetilde{P}$ is a binomial distribution with parameters $n = 2$ and $\theta_0 \in]0, 1[$ (which amounts to $\mathscr{Y} = \{0, 1, 2\}$, $\widetilde{p}(0) = (1-\theta_0)^2$, $\widetilde{p}(1) = 2\theta_0 \cdot (1-\theta_0)$, $\widetilde{p}(2) = \theta_0^2$). In other words, P_{ϵ,θ_0} is a binomial distribution which is contaminated at the state $y = 2$ with percentage-degree $\epsilon \in]0, 1[$. For the visualization of the divergence spheres $\mathbb{S}_\phi(P_{\epsilon,\theta_0}, \rho)$, all the involved probability distributions (say) P can be – as usual – identified with the 3D column-vectors $P \,\widehat{=}\, (p(0), p(1), p(2))'$ of the corresponding three components of its probability mass function. Thus, each $(p(0), p(1), p(2))'$ lies in the "probability simplex" $\Pi := \{(\pi_1, \pi_2, \pi_3)' \in \mathbb{R}^3 : \pi_1 \geq 0, \pi_2 \geq 0, \pi_3 \geq 0, \pi_1 + \pi_2 + \pi_3 = 1\}$. Analogously, each general distribution (say) ν can be identified with the 3D column-vector $\nu \,\widehat{=}\, (n(0), n(1), n(2))'$ of the corresponding three components of its mass function. Hence, each $(n(0), n(1), n(2))'$ is a point in the first octant $\Sigma := \{(\sigma_1, \sigma_2, \sigma_3)' \in \mathbb{R}^3 : \sigma_1 \geq 0, \sigma_2 \geq 0, \sigma_3 \geq 0\}$.

Of course, data-derived randomness can enter this context – for instance – in the following way: for index $m \in \tau := \mathbb{N}$ let the generation of the m–th data point be represented by the random variable Y_m which takes values in the state space $\mathscr{Y}$. The associated family of random variables $(Y_m, m \in \tau)$ is supposed to be independent and identically distributed (i.i.d.) under the probability distribution P_{ϵ,θ_0}. For each concrete sample $(Y_1, \ldots, Y_N)$ of size N one builds the corresponding (random) empirical distribution $P_N^{emp}[\,\cdot\,] := \frac{1}{N} \cdot \sum_{i=1}^{N} \delta_{Y_i}[\cdot]$ which under the correct model converges (in distribution) to P_{ϵ,θ_0} as the sample size N tends to ∞. Notice that the 3D vector $(p_N^{emp}(0), p_N^{emp}(1), p_N^{emp}(2))'$ of the probability-mass-function components (where $p_N^{emp}(y) := \frac{1}{N} \cdot \#\{i \in \{1, \ldots, N\} : \, Y_i = y\}$) moves randomly in the probability simplex Π as N increases. However, for large N one can (approximately) identify $P_N^{emp} \doteqdot P_{\epsilon,\theta_0}$ which we do in the following.

Within this special context, let us exemplarily explain the following effects:

Effect 1: divergence spheres $\mathbb{S}_\phi(P, \rho)$ can take quite different kinds of shapes, e.g. triangles (with rounded edges), rectangles (with rounded edges), and non-convex non-concave "blobs"; this can even appear with fixed center P $(= P_{\epsilon,\theta_0})$ when only the radius ρ changes; see Fig. 1(a)–(c). As comparative preparation for other effects below, we draw $\theta_0 \mapsto P_{\epsilon,\theta_0}$ as an orange curve, as well as a dark-blue curve which represents the set $\mathscr{C} := \{Bin(2, \theta) : \theta \in]0, 1[\}$ of all binomial distributions, and in Fig. 1(a)–(c) we aim for spheres (in red) which are fully in the green-coloured probability simplex Π (i.e., no need for cutoff on the Π–boundaries) and "on the left of $\mathscr{C}$". Notice that we use viewing angles with "minimal visual distortion". The corresponding, interestingly shaped, "non-simplex-restricted" spheres $\mathbb{S}_\phi^g(P_{\epsilon,\theta_0}, \rho)$ are plotted, too (cf. Fig. 1(e)–(g)).

Effect 2: unlike Euclidean balls, even for fixed radius ρ the divergence spheres $\mathbb{S}_\phi(P,\rho)$ can quite shift their shape as the center P moves in the probability space (e.g., along the orange "contamination" curve $\theta_0 \mapsto P_{\epsilon,\theta_0}$); see Fig. 1(h)–(i).

Effect 3: for fixed center P, increasing the radius ρ may lead to a quite nonlinear growth of the divergence spheres $\mathbb{S}_\phi(P,\rho)$ (with $P = P_{\epsilon,\theta_0}$); see Fig. 1(j)–(l). Notice that the principal shape remains the same (as opposed to Effect 1).

Effect 4: for an i.i.d. sample $(Y_1,\ldots,Y_N)$ under the probability distribution P_0, the corresponding minimum-distance parameter estimator is given by any $\widehat{\theta}$ from the possibly multi-valued set $\arg\min_{\theta\in\Theta} B_\phi\left(P_N^{emp}, Q_\theta \,||\, W\right)$ where $\mathscr{C} := \{Q_\theta \in \mathscr{M}_\lambda^1 : \theta \in \Theta\}$ is a parametric family of probability distributions. At the same time the (distribution of the random) size of $\min_{\theta\in\Theta} B_\phi\left(P_N^{emp}, Q_\theta \,||\, W\right)$ is an indicator for the goodness-of-fit. To visualize some corresponding robustness-concerning geometric effects, we confine ourselves to the above-mentioned contamination context $P_0 = P_{\epsilon,\theta_0} = (1-\epsilon)\, Bin(2,\theta_0) + \epsilon\,\delta_2$ and $\mathscr{C} := \{Bin(2,\theta) : \theta \in]0,1[\}$, and to the special-SBD-subfamily minimization (cf. (4)) $\arg\min_{\theta\in\Theta} D_{\widetilde{\phi}_{\alpha,\beta,r}}\left(P_N^{emp}, Q_\theta\right)$. In fact, for the sake of brevity we only consider in the following the (for large sample sizes N reasonable) deterministic proxy $T(\epsilon) := \arg\min_{\theta\in\Theta} D_{\widetilde{\phi}_{\alpha,\beta,r}}\left(P_{\epsilon,\theta_0}, Q_\theta\right)$, and discuss robustness against contamination in terms of nearness of $T(\epsilon)$ to θ_0 even for "large" contaminations reflected by "large" ϵ; furthermore, we discuss abrupt changes of $\epsilon \mapsto T(\epsilon)$. A formal robustness treatment is given in terms of the RAF below (cf. (6)). As can be seen from Fig. 1(m) for $\alpha = 0.05$, $\beta = 0$ (and thus, $D_{\widetilde{\phi}_{0.05,0,r}}(\cdot,\cdot)$ is the classical 0.05−power divergence independently of $r \neq 0$) and $\theta_0 = 0.08$, the function $\epsilon \mapsto T(\epsilon)$ is quite robust for contamination percentage-degrees $\epsilon \in [0, 0.45]$ (i.e. $T(\epsilon) \approx \theta_0$), but it exhibits a sharp breakdown at $\epsilon \approx 0.46$; this contrasts the non-robust ("uniformly much steeper" but smooth) behaviour in the case $\alpha = 1.0$, $\beta = 0$ of minimum-Kullback-Leibler-divergence estimation which is in one-to-one correspondence with the *maximum likelihood estimation*. A plot which is similar to Fig. 1(m) was first shown by [17] (see also [2]) for the squared Hellinger distance HD $D_{\phi_{0.5}}(\cdot,\cdot)$ – which in our extended framework corresponds to $D_{\widetilde{\phi}_{0.5,0,r}}(\cdot,\cdot)$ – for the larger, *non-visualizable* state space $\mathscr{Y} = \{0,1,\ldots,12\}$ and the contamination $P_0 := (1-\epsilon)\, Bin(12,\frac{1}{2}) + \epsilon\,\delta_{12}$[2]. Because of our low-dimensional state space, we can give further geometric insights to such robustness effects. Indeed, Fig. 1(n) respectively (o) show those spheres $\mathbb{S}_{\phi_{0.05}}(P_{0.45,0.08}, \rho_{min})$ respectively $\mathbb{S}_{\phi_{0.05}}(P_{0.46,0.08}, \widetilde{\rho}_{min})$ which touch the binomial projection set $\mathscr{C}$ (i.e., the dark-blue coloured curve) for the "first time" as the radius ρ grows. The corresponding respective touching points – which represent the resulting estimated probability distributions $Bin(2, T(0.45))$ respectively $Bin(2, T(0.46))$ – are marked as red dots on the blue curve, and are "very far apart". This is also consistent with Fig. 1(p) respectively Fig. 1(q) which show the functions $]0,1[\ni \theta \mapsto D_{\widetilde{\phi}_{0.05,0,r}}(P_{0.45,0.08}, Bin(2,\theta))$ respectively

[2] Also notice that the HD together with $\theta_0 = 0.5$ does not exhibit such an effect for our smaller 3-element-state space, due to the lack of outliers.

$]0,1[\ni \theta \mapsto D_{\widetilde{\phi}_{0.05,0,r}}(P_{0.46,0.08}, Bin(2,\theta))$ where one can see the "global switching between the two local-minimum values". Furthermore, in Fig. 1(n) the red dot lies "robustly close" to the green dot on the dark-blue curve which represents the uncontaminated $Bin(2,\theta_0)$ to be "found out". This contrasts the corresponding behaviour in Fig. 1(o) where the (only 1% higher) contamination-percentage-degree is already in the non-robust range. Additionally, with our new divergence family we can produce similar variants with non-convex divergence spheres (see Figs. 1(r)-(t)) resp. with smoother (non-sharp) breakdown ("smooth-rolling-over of the red dots", see e.g. Figs. 1(u)-(w)). Further, e.g. "cascade-type", transition effects are omitted for the sake of brevity.
Due to the minimization step (in our discrete setup with scalar parameter θ)

$$0 = -\tfrac{\partial}{\partial\theta}\, D_{\phi}\left(P, Q_{\theta}\right) = \; -\textstyle\sum_{x\in\mathscr{X}} \tfrac{\partial}{\partial v}\left(v\cdot\phi\left(\tfrac{p(x)}{v}\right)\right)\Big|_{v=q_{\theta}(x)} \cdot \tfrac{\partial}{\partial\theta} q_{\theta}(x)$$
$$=: \; \textstyle\sum_{x\in\mathscr{X}} a_{\phi}\left(\tfrac{p(x)}{q_{\theta}(x)} - 1\right)\cdot \tfrac{\partial}{\partial\theta} q_{\theta}(x)\;, \qquad \text{with } P = P_N^{emp} \doteqdot P_{\epsilon,\theta_0} \tag{6}$$

the robustness-degree of minimum-distance estimation by $\phi-$divergences $D_{\phi}(\cdot,\cdot)$ can be quantified in terms of the residual adjustment function RAF $a_{\phi}(\delta) := (\delta+1)\cdot\phi'(\delta+1) - \phi(\delta+1)$ with Pearson residual $\delta := \frac{u}{v} - 1 \in [-1,\infty[$ (cf. [2,17]; see also its generalization to density-pair adjustment functions for general scaled Bregman distances given in [15]). More detailed, for both large δ (reflecting outliers) and small δ (reflecting inliers) the RAF $a_{\phi}(\delta)$ should ideally be closer to zero (i.e. more dampening) than that of the Kullback-Leibler (i.e. maximum-likelihood estimation) benchmark $a_{\phi_1}(\delta) = \delta$. Concerning this, for various different $(\alpha,\beta,r)-$constellations our new divergences $D_{\widetilde{\phi}_{\alpha,\beta,r}}(P,Q)$ are *much more robust* against outliers and inliers even than the very-well-performing negative-exponential-divergence NED $D_{\phi_{NED}}(P,Q)$ of [2,17] with $\phi(t) = \phi_{NED}(t) := \exp(1-t) + t - 2$ $(t > 0)$; see Fig. 1(x) for an exemplary demonstration.

Concluding remarks: By means of exemplary 3D plots we have shown some properties and pitfalls of divergence geometries. For this, we have used one special case of scaled Bregman distances SBD – namely power-function-type generators and power-mean-type scalings – which can be represented as $\phi-$divergences with possibly non-convex non-concave generator ϕ; classical Csiszar-Ali-Silvey-type power divergences are covered as a subcase, too. By exploiting the full flexibility of SBD – e.g. those which are not rewriteable as $\phi-$divergence – one can construct further interesting geometric effects. Those contrast principally with the geometric behaviour of the balls constructed from the Bregman divergences with "global NMO-scaling" of [19] defined (in the separable setup) by

$$H(P)\cdot\textstyle\int_{\mathscr{Y}}\left[\phi\left(\tfrac{p(y)}{H(P)}\right) - \phi\left(\tfrac{q(y)}{H(Q)}\right) - \phi'\left(\tfrac{q(y)}{H(Q)}\right)\cdot\left(\tfrac{p(y)}{H(P)} - \tfrac{q(y)}{H(Q)}\right)\right]\,\mathrm{d}\lambda(y) \tag{7}$$

where $H(P) := H\left((p(y))_{y\in\mathscr{Y}}\right)$, $H(Q) = H\left((q(y))_{y\in\mathscr{Y}}\right)$ are real-valued "global" functionals of the (not necessarily probability) density functions $p(\cdot)$, $q(\cdot)$,

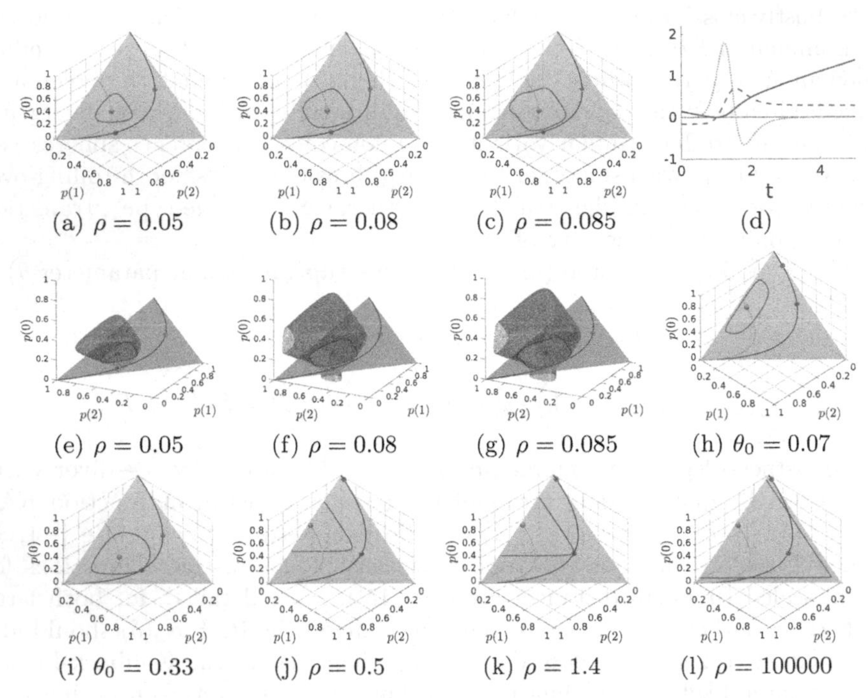

Fig. 1. (a)–(c): divergence spheres $\mathbb{S}_\phi(P_{\epsilon,\theta_0},\rho)$ (in red) for $\phi = \widetilde{\phi}_{\alpha,\beta,r}$ with $\theta_0 = 0.32$, $\epsilon = 0.44$, $\alpha = 7.5$, $\beta = 0.05$, $r = 7.5$ and different radii ρ; the center P_{ϵ,θ_0} is marked as green dot on the orange curve. (d): non-convex non-concave $t \mapsto \widetilde{\phi}_{\alpha,\beta,r}(t)$ with $\alpha = 7.5$, $\beta = 0.05$, $r = 7.5$ (in blue), its first (in magenta) and second derivative (in green). (e)–(g): the to the plots (a)-(c) corresponding $\mathbb{S}^g_\phi(P_{\epsilon,\theta_0},\rho)$ (in different viewing angles) shown as blue surface. (h)-(i): divergence spheres $\mathbb{S}_\phi(P_{\epsilon,\theta_0},\rho)$ for $\phi = \widetilde{\phi}_{\alpha,\beta,r}$ with $\epsilon = 0.44$, $\alpha = 3.35$, $\beta = 0.65$, $r = -6.31$, radius $\rho = 0.2$ and different θ_0. (j)-(l): divergence spheres $\mathbb{S}_\phi(P_{\epsilon,\theta_0},\rho)$ for $\phi = \widetilde{\phi}_{\alpha,\beta,r}$ with $\theta_0 = 0.02$, $\epsilon = 0.44$, $\alpha = 7.5$, $\beta = 0$, arbitrary r (has no effect), and different radii ρ. (m): $\epsilon \mapsto T(\epsilon)$ for $\alpha = 0.05$, $\beta = 0$, arbitrary r (no effect) and $\theta_0 = 0.08$ (dotted line is the KL case $\alpha = 1$, $\beta = 0$); (n)-(o): corresponding "minimizing" (touching) divergence spheres (in red) for $\epsilon = .45$ resp. $\epsilon = .46$; (p)-(q): corresponding $\theta \mapsto D_{\widetilde{\phi}_{0.05,0,r}}(P_{0.45,0.08}, Bin(2,\theta))$ resp. $\theta \mapsto D_{\widetilde{\phi}_{0.05,0,r}}(P_{0.46,0.08}, Bin(2,\theta))$; (r): $\epsilon \mapsto T(\epsilon)$ for $\alpha = 4$, $\beta = 0.35$, $r = 7.5$ and $\theta_0 = 0.08$; (s)-(t): corresponding "minimizing" divergence spheres for $\epsilon = .45$ resp. $\epsilon = .46$; (u): $\epsilon \mapsto T(\epsilon)$ for $\alpha = 2$, $\beta = 0.35$, $r = 7.5$ and $\theta_0 = 0.08$; (v)-(w): corresponding "minimizing" divergence spheres for $\epsilon = .45$ resp. $\epsilon = .46$; (x): residual adjustment functions of KL divergence (in black), negative-exponential divergence (in blue), and of $D_{\widetilde{\phi}_{\alpha,\beta,r}}(\cdot,\cdot)$ (in dotted red) with $\alpha = 10$, $\beta = 0.25$, $r = 10$. (Color figure online)

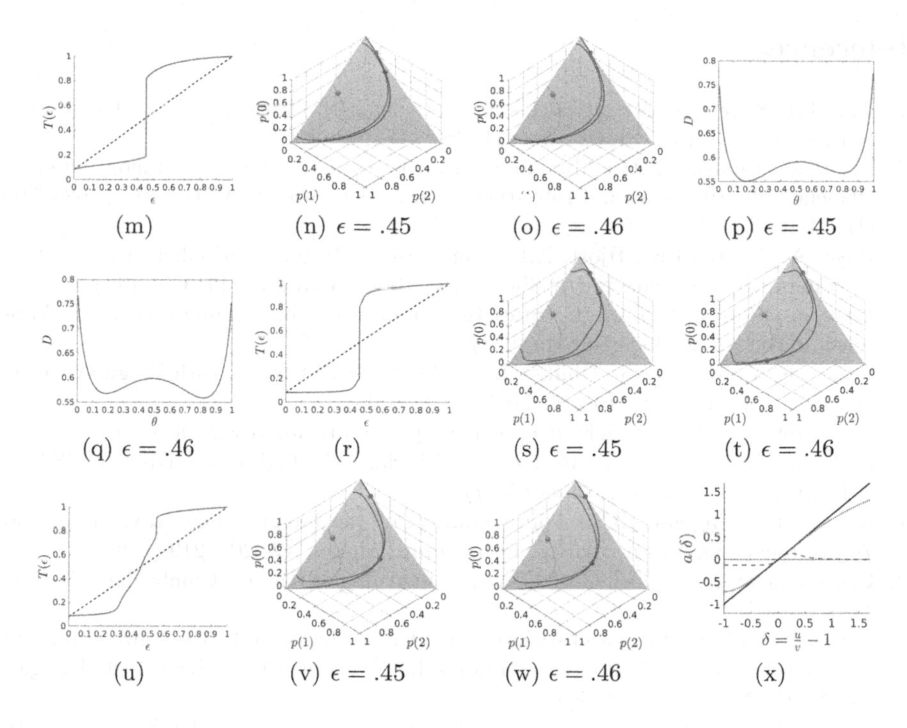

Fig. 1. *(continued)*

e.g. $H(P) := \int_{\mathscr{Y}} h(p(y))\mathrm{d}\lambda(y)$ for some function h. Notice the very substantial difference to SBD, i.e. to the Bregman divergences with the "local SV-scaling" of [26,27] given in (2) (even in the locally adaptive subcase $w(y) = \widetilde{w}_{\beta,r}(p(y), q(y))$). Amongst other things, this difference is reflected by the fact that (under some assumptions) the "NMO-scaled" Bregman divergences can be represented as unscaled Bregman distances with possibly non-convex generator $\overline{\phi}$ (cf. [19]) whereas some "SV-scaled" Bregman divergences can e.g. be represented as Csiszar-Ali-Silvey $\phi-$divergences (which are never unscaled Bregman distances except for KL) with non-convex non-concave generator $\phi := \widetilde{\phi}$, cf. (4). To gain further insights, it would be illuminating to work out closer connections and differences between these two scaling-types – under duality, reparametrization, ambient-space aspects – and to incorporate further, structurally different examples.

Acknowledgement. We are grateful to all three referees for their useful suggestions. W. Stummer thanks A.L. Kißlinger for valuable discussions.

References

1. Ali, M.S., Silvey, D.: A general class of coefficients of divergence of one distribution from another. J. Roy. Statist. Soc. **B–28**, 131–140 (1966)
2. Basu, A., Lindsay, B.G.: Minimum disparity estimation for continuous models: efficiency, distributions and robustness. Ann. Inst. Statist. Math. **46**(4), 683–705 (1994)
3. Basu, A., Harris, I.R., Hjort, N.L., Jones, M.C.: Robust and efficient estimation by minimising a density power divergence. Biometrika **85**, 549–559 (1998)
4. Basu, A., Shioya, H., Park, C.: Statistical Inference: The Minimum Distance Approach. CRC Press, Boca Raton (2011)
5. Banerjee, A., Merugu, S., Dhillon, I.S., Ghosh, J.: Clustering with Bregman divergences. J. Mach. Learn. Res. **6**, 1705–1749 (2005)
6. Broniatowski, M.: A weighted bootstrap procedure for divergence minimization problems. In: Antoch, J., Jureckova, J., Maciak, M., PeSta, M. (eds.) AMISTAT 2015, pp. 1–22. Springer, Cham (2017)
7. Cerone, P., Dragomir, S.S.: Approximation of the integral mean divergence and f−divergence via mean results. Math. Comp. Model. **42**, 207–219 (2005)
8. Cesa-Bianchi, N., Lugosi, G.: Prediction, Learning & Games. Cambridge UP, New York (2006)
9. Csiszar, I.: Eine informationstheoretische Ungleichung und ihre Anwendung auf den Beweis der Ergodizität von Markoffschen Ketten. Publ. Math. Inst. Hungar. Acad. Sci. **A–8**, 85–108 (1963)
10. Csiszar, I., Breuer, T.: Measuring distribution model risk. Mathe. Finance **26**(2), 395–411 (2016)
11. Collins, M., Schapire, R.E., Singer, Y.: Logistic regression, AdaBoost and Bregman distances. Mach. Learn. **48**, 253–285 (2002)
12. Kißlinger, A.-L., Stummer, W.: Some decision procedures based on scaled Bregman distance surfaces. In: Nielsen, F., Barbaresco, F. (eds.) GSI 2013. LNCS, vol. 8085, pp. 479–486. Springer, Heidelberg (2013). doi:10.1007/978-3-642-40020-9_52
13. Kißlinger, A.-L., Stummer, W.: New model search for nonlinear recursive models, regressions and autoregressions. In: Nielsen, F., Barbaresco, F. (eds.) GSI 2015. LNCS, vol. 9389, pp. 693–701. Springer, Cham (2015). doi:10.1007/978-3-319-25040-3_74
14. Kißlinger, A.-L., Stummer, W.: A New Information-Geometric Method of Change Detection. (2015, Preprint)
15. Kißlinger, A.-L., Stummer, W.: Robust statistical engineering by means of scaled Bregman distances. In: Agostinelli, C., Basu, A., Filzmoser, P., Mukherjee, D. (eds.) Recent Advances in Robust Statistics: Theory and Applications, pp. 81–113. Springer, New Delhi (2016). doi:10.1007/978-81-322-3643-6_5
16. Liese, F., Vajda, I.: Convex Statistical Distances. Teubner, Leipzig (1987)
17. Lindsay, B.G.: Efficiency versus robustness: the case for minimum Hellinger distance and related methods. Ann. Statist. **22**(2), 1081–1114 (1994)
18. Murata, N., Takenouchi, T., Kanamori, T., Eguchi, S.: Information geometry of U-boost and Bregman divergence. Neural Comput. **16**(7), 1437–1481 (2004)
19. Nock, R., Menon, A.K., Ong, C.S.: A scaled Bregman theorem with applications. In: Advances in Neural Information Processing Systems 29 (NIPS 2016), pp. 19–27 (2016)
20. Nock, R., Nielsen, F., Amari, S.-I.: On conformal divergences and their population minimizers. IEEE Trans. Inf. Theory **62**(1), 527–538 (2016)

21. Nock, R., Nielsen, F.: Bregman divergences and surrogates for learning. IEEE Trans. Pattern Anal. Mach. Intell. **31**(11), 2048–2059 (2009)
22. Pardo, L.: Statistical Inference Based on Divergence Measures. Chapman H, Boca Raton (2006)
23. Pardo, M.C., Vajda, I.: On asymptotic properties of information-theoretic divergences. IEEE Trans. Inf. Theory **49**(7), 1860–1868 (2003)
24. Read, T.R.C., Cressie, N.A.C.: Goodness-of-Fit Statistics for Discrete Multivariate Data. Springer, New York (1988)
25. Shioya, H., Da-te, T.: A generalisation of Lin divergence and the derivation of a new information divergence measure. Electr. Commun. Japan **78**(7), 34–40 (1995)
26. Stummer, W.: Some Bregman distances between financial diffusion processes. Proc. Appl. Math. Mech. **7**(1), 1050503–1050504 (2007)
27. Stummer, W., Vajda, I.: On Bregman distances and divergences of probability measures. IEEE Trans. Inf.Theory **58**(3), 1277–1288 (2012)
28. Stummer, W., Vajda, I.: On divergences of finite measures and their applicability in statistics and information theory. Statistics **44**, 169–187 (2010)
29. Sugiyama, M., Suzuki, T., Kanamori, T.: Density-ratio matching under the Bregman divergence: a unified framework of density-ratio estimation. Ann. Inst. Stat. Math. **64**, 1009–1044 (2012)
30. Tsuda, K., Rätsch, G., Warmuth, M.: Matrix exponentiated gradient updates for on-line learning and Bregman projection. J. Mach. Learn. Res. **6**, 995–1018 (2005)
31. Wu, L., Hoi, S.C.H., Jin, R., Zhu, J., Yu, N.: Learning Bregman distance functions for semi-supervised clustering. IEEE Trans. Knowl. Data Eng. **24**(3), 478–491 (2012)

A Stochastic Look at Geodesics on the Sphere

Marc Arnaudon[1,2(✉)] and Jean-Claude Zambrini[1,2]

[1] Inst. de Mathématiques de Bordeaux, Univ. de Bordeaux, Bordeaux, France
marc.arnaudon@math.u-bordeaux.fr
[2] Grupo de Física-Matemática Univ. Lisbon, Lisbon, Portugal
jczambrini@fc.ul.pt

Abstract. We describe a method allowing to deform stochastically the completely integrable (deterministic) system of geodesics on the sphere S^2 in a way preserving all its symmetries.

Keywords: Geodesic flow · Stochastic deformation · Integrable systems

1 Introduction

Free diffusions on a sphere S^2 are important case studies in applications, for instance in Biology, Physics, Chemistry, Image processing etc., where they are frequently analysed with computer simulations. However, as for most diffusions on curved spaces, no closed form analytical expressions for their probability densities are available for such simulations. Another way to express the kind of difficulties one faces is to observe that one cannot define Gaussian functions on S^2.

If, instead of free diffusions on S^2 we consider their deterministic counterpart, the classical geodesic flow, a famous integrable system whose complete solution dates back to the 19th century, the situation is much simpler. Indeed, one can use the conservation of angular momentum and energy to foliate the phase space (the cotangent bundle of its configuration space).

We describe here a method allowing to construct free diffusions on S^2 as stochastic deformations of the classical geodesic flow, including a probabilistic counterpart of its conservation laws.

2 Classical Geodesics

The problem of geodesic on the sphere S^2 is a classical example of completely integrable elementary dynamical system [1].

For a unit radius sphere and using spherical coordinates $(q^i) = (\theta, \phi) \in]0, \pi[\times[0, 2\pi]$ where ϕ is the longitude, the Lagrangian L of the system is the scalar defined on the tangent bundle TS^2 of the system by

$$L(\theta, \phi, \dot{\theta}, \dot{\phi}) = (\dot{\theta}^2 + \sin^2\theta\ \dot{\phi}^2) \tag{1}$$

F. Nielsen and F. Barbaresco (Eds.): GSI 2017, LNCS 10589, pp. 470–476, 2017.
https://doi.org/10.1007/978-3-319-68445-1_55

(where $\dot{\theta} = \frac{d\theta}{dt}$ etc. ...), since it coincides with $ds^2 = g_{ij}dq^i dq^j$, here $g = (g_{ij}) = \begin{bmatrix} 1 & 0 \\ 0 & \sin^2\theta \end{bmatrix}$. The Euler-Lagrange equations

$$\frac{d}{dt}\left(\frac{\partial L}{\partial \dot{q}^i}\right) - \frac{\partial L}{\partial q^i} \qquad i = 1, 2 \tag{2}$$

in these coordinates are easily solved. They describe the dynamics of the extremals (here minimal) curves of the action functional

$$S_L[q(\cdot)] = \int_{Q_1}^{Q_2} L(q, \dot{q})dt \tag{3}$$

computed, for instance, between two fixed configurations $Q_1 = (\theta_1, \phi_1)$ and $Q_2 = (\theta_2, \phi_2)$ in the configuration space. Those equations are

$$\ddot{\theta} = \dot{\phi}^2 \sin\theta\cos\theta, \qquad \ddot{\phi} = -2\dot{\theta}\dot{\phi}\ \mathrm{cotg}\ \theta \tag{4}$$

Defining the Hamiltonian $H : T^*S^2 \to \mathbb{R}$ as the Legendre transform of L, we have $H = \frac{1}{2}g^{ij}p_i p_j$, where $p_i = \frac{\partial L}{\partial \dot{q}^i} = g_{ij}\dot{q}^j$ denote the momenta, here

$$H(\theta, \phi, p_\theta, p_\phi) = \frac{1}{2}(p_\theta^2 + \frac{1}{\sin^2\theta}p_\phi^2), \tag{5}$$

with $p_\theta = \dot{\theta}, p_\phi = \sin^2\theta\dot{\phi}$.

It is clear that the energy H is conserved during the evolution. There are three other first integrals for this system, corresponding to the three components of the angular momentum $\mathcal{L}$. They can be expressed as differential operators of the form $X_j^\theta \frac{\partial}{\partial\theta} + X_j^\phi \frac{\partial}{\partial\phi}, j = 1, 2, 3$, namely

$$\mathcal{L}_1 = \sin\phi\frac{\partial}{\partial\theta} + \frac{\cos\phi}{\tan\theta}\frac{\partial}{\partial\phi}$$

$$\mathcal{L}_2 = -\cos\phi\frac{\partial}{\partial\theta} + \frac{\sin\phi}{\tan\theta}\frac{\partial}{\partial\phi}, \qquad \mathcal{L}_3 = -\frac{\partial}{\partial\phi} \tag{6}$$

In geometrical terms, written as $\mathcal{L}_j = (X_j^\theta, X_j^\phi), j = 1, 2, 3$, they are the three Killing vectors for S^2, forming a basis for the Lie algebra of the group of isometries $SO(3)$ of S^2. $\mathcal{L}_3$ corresponds to the conservation of the momentum p_ϕ.

The integrability of this dynamical system relies on the existence of the two first integrals H and p_ϕ. They allow to foliate the phase space by a two-parameter family of two-dimensional tori. Let us recall that the list of first integrals of the system is the statement of Noether's Theorem, according to which the invariance of the Lagrangian L under the local flow of vector field

$$v^{(1)} = X^i(q, t)\frac{\partial}{\partial q^i} + \frac{dX^i}{dt}\frac{\partial}{\partial \dot{q}^i} \tag{7}$$

associated with the group of transformations

$$(q^i, t) \to (Q_\alpha^i = q^i + \alpha X^i(q, t), \tau_\alpha = t + \alpha T(t))$$

for α a real parameter, provides a first integral along extremals of S_L of the form

$$\frac{d}{dt}(X^i p_i - TH) = 0. \tag{8}$$

The coefficients X^i, T must, of course, satisfy some relations between them called "determining equations" of the symmetry group of the system [2]. For instance, for our geodesics on S^2,

$T = 1, X = (X^\theta, X^\phi) = (0, 0)$ corresponds to the conservation of the energy H, and $T = 0, X = (0, -1)$ to the conservation of p_ϕ In fact, the three vectors X_j must satisfy the Killing equations in the (θ, ϕ) coordinates,

$$\nabla^\theta X_j^\phi + \nabla^\phi X_j^\theta = 0, j = 1, 2, 3 \tag{9}$$

where $\nabla^\cdot$ denotes the covariant derivatives.

3 Stochastic Deformation of the Geodesics on the Sphere

Many ways to construct diffusions on S^2 are known. In the spirit of K. Itô [3], we want to deform the above classical dynamical system in a way preserving the essential of its qualitative properties.

Let us start from the backward heat equation for the Laplace-Beltrami "Hamiltonian" operator H (without potentials). in local coordinates (q^i) it can be written $\frac{\partial \eta}{\partial t} = H\eta$, where $g = \det(g_{ij})$ and

$$H = -\frac{1}{2}\Delta_{LB} = -\frac{1}{2\sqrt{\det g}}\frac{\partial}{\partial q^i}\Big(\sqrt{\det g}\; g^{ij}\frac{\partial}{\partial q^j}\Big). \tag{10}$$

A more revealing form in terms of the Christoffel symbols of the Riemannian connection is

$$-\frac{1}{2}\Delta_{LB} = -\frac{g^{ij}}{2}\frac{\partial^2}{\partial q^i \partial q^j} + \frac{1}{2}\Gamma^i_{jk}(q) g^{jk}(q)\frac{\partial}{\partial q^i}. \tag{11}$$

Indeed, the extra first order term, of purely geometric origin, will coincide with the drift of the simplest diffusion on our manifold, the Brownian motion; this was observed by K. Itô, as early as 1962 [3]. In our spherical case, one finds

$$\Gamma^\theta_{jk} g^{jk} = -\text{cotg}\ \theta, \Gamma^\phi_{jk} g^{jk} = 0. \tag{12}$$

Now we shall consider general diffusions z^i on S^2 solving SDEs of the form

$$dz^i(\tau) = (B^i - \frac{1}{2}\Gamma^i_{jk} g^{jk})d\tau + dW^i(\tau), \quad \tau > t \tag{13}$$

for B^i an unspecified vector field, where $dW^i(\tau) = \sigma^i_k d\beta^k(\tau)$ with σ^i_k the square root of g^{ij}, i.e. $g^{ij} = \sigma^i_k \sigma^j_k$, in our case $\sigma = \begin{bmatrix} 1 & 0 \\ 0 & \frac{1}{\sin\theta} \end{bmatrix}$ and β is a two dimensional Wiener process.

Here is the stochastic deformation of the extremality condition for dynamical trajectories in terms of the classical action S_L. It will be convenient to consider S_L as a function of starting configurations q at a time t. For convenience, we shall add a final boundary condition to S_u to S_L.

Let $S_L(q,t)$ be defined now by $S_L(q,t) = -\ln \eta(q,t)$, where $\eta(q,t)$ is a positive solution of the backward heat equation for a (smooth) final boundary condition $S_u(q), u > t$. Let B^i in (13) be adapted to the increasing filtration $\mathcal{P}_\tau$, bounded but otherwise arbitrary. Then

$$S_L(q,t) \leq E_{qt}\{\int_t^u \frac{1}{2} B^i B_i(z(\tau),\tau)d\tau + S_u(z(u))\} \tag{14}$$

where E_{qt} denotes the conditional expectation given $z(t) = q$. The equality holds on the extremal diffusion on S^2, of drift

$$B^i(q,t) = \frac{\partial_i \eta}{\eta}(q,t) = -\nabla^i S_L. \tag{15}$$

This means, on S^2, that S_L minimizes the r.h.s. functional of (14) for the Lagrangian

$$L = \frac{1}{2}[(\frac{\partial_\theta \eta}{\eta})^2 + \sin^2\theta(\frac{\partial_\phi \eta}{\eta})^2] \tag{16}$$

where, manifestly, $(\frac{\partial_\theta \eta}{\eta})$ and $(\frac{\partial_\phi \eta}{\eta})$ plays the roles of $\dot{\theta}$ and $\dot{\phi}$ in the deterministic definition (1).

Let us observe that after the above logarithmic change of variable, it follows from the backward heat equation that the scalar field S_L solves

$$-\frac{\partial S_L}{\partial t} + \frac{1}{2}\|\nabla S_L\|^2 - \frac{1}{2}\nabla^i \nabla_i S_L = 0, \tag{17}$$

with $t < u$ and $S_L(q,u) = S_u(q)$.

This is an Hamilton-Jacobi-Bellman equation, whose relation with heat equations is well known and used in stochastic control [4]. The Laplacian term represents the collective effects of the irregular trajectories $\tau \to z^i(\tau)$ solving (13).

A second order in time dynamical law like (4) requires the definition of the parallel transport of our velocity vector field B^i.

In [3] Itô had already mentioned that there is some freedom of choice in this, involving the Ricci tensor R^i_k on the manifold. One definition is known today in Stochastic Analysis as "Damped parallel transport" [5]. Then the generator of the diffusion z^i acting on a vector field V on S^2 is given by

$$D_t V^i = \frac{\partial V^i}{\partial t} + B^k \nabla_k V^i + \frac{1}{2}(\Delta V)^i \tag{18}$$

where, instead of the Laplace-Beltrami operator, one has now

$$\Delta V^i = \nabla^k \nabla_k V^i + R^i_k V^k \tag{19}$$

When acting on scalar fields φ, D_t reduces to the familiar form

$$D_t\varphi = \frac{\partial \varphi}{\partial t} + B^k \nabla_k \varphi + \frac{1}{2}\nabla^k \nabla_k \varphi \tag{20}$$

When $\varphi = q^k$, $D_t\varphi^k = B^k(z(t), t) = -\nabla^k S_L$, so the r.h.s. Lagrangian of (14) is really $\frac{1}{2}\|D_t z\|^2$, for $\|\cdot\|$ the norm induced by the metric, as it should. For the vector field B^i, we use (17) and the integrability condition $\frac{\partial}{\partial t}\nabla^i S_L = \nabla^i \frac{\partial S_L}{\partial t}$, following from the definition of S_L, to obtain

$$D_t D_t z^i = 0 \tag{21}$$

i.e., the stochastic deformation of both O.D.E.s (4) when $z(t) = (\theta(t), \phi(t))$ solve Eq. (13) namely, in our case,

$$d\theta(t) = \left(\frac{\partial_\theta \eta}{\eta} + \frac{\text{cotg}\theta}{2}\right)dt + dW^\theta(t), \quad d\phi(t) = \frac{1}{\sin^2\theta}\left(\frac{\partial_\phi \eta}{\eta}\right)dt + dW^\phi(t) \tag{22}$$

The bonus of our approach lies in the study of the symmetries of our stochastic system. The symmetry group of the heat equation, in our simple case with constant positive curvature, is generated by differential operators of the form [6]

$$\hat{N} = X^i(q)\nabla_i + T\frac{\partial}{\partial t} + \alpha \tag{23}$$

where T and α are constants, and the X^i are three Killing vectors on (S^2, g). Besides a one dimensional Lie algebra generated by the identity, another one corresponds to $T = 1$ and $X = (X^\theta, X^\phi) = (0, 0)$. This provides the conservation of energy defined here, since $S_L = -\ln \eta$, by $h(\theta(t), \phi(t)) = -\frac{1}{\eta}\frac{\partial \eta}{\partial t}$ or, more explicitly,

$$h = \frac{1}{2}g^{ij}B_iB_j + \frac{1}{2}g^{ij}\frac{\partial}{\partial q^i}B_j - \frac{1}{2}\Gamma^i_{jk}g^{jk}B_i \tag{24}$$

Using (20), one verifies that

$$D_t h(z(t), t) = 0 \tag{25}$$

in other words, h is a martingale of the diffusion $z(t)$ extremal of the Action functional in (14). This is the stochastic deformation of the corresponding classical statement (8) when $X = (0, 0), T = 1$. Analogously, our (deformed) momentum p_ϕ is a martingale. In these conditions, one can define a notion of integrability for stochastic systems (not along Liouville's way, but inspired instead by Jacobi's classical approach) and show that, in this sense, our stochastic problem of geodesics on the sphere is as integrable as its deterministic counterpart. This will be done in [7].

To appreciate better in what sense our approach is a stochastic deformation of the classical problem of geodesics in S^2, replace our metric (g_{ij}) by $\hbar(\sigma_{ij})$ for

σ_{ij} the Riemannian metric, where $\hbar$ is a positive constant, and take into account that our underlying backward heat equation now becomes

$$\frac{\partial \eta}{\partial t} = -\frac{\hbar}{2} \Delta_{LB} \eta \tag{26}$$

then, one verifies easily that, when $\hbar \to 0$, $D_t \to \frac{d}{dt}$, the Lagrangian of (14) reduces to the classical one (1) and the conditional expectation of the action (1) disappears. The Hamilton-Jacobi-Bellman equation (17) reduces to the one of the classical dynamical system and our martingales to its first integrals. In this respect, observe that general (positive) final conditions for Eq. (26) may depend as well on $\hbar$. They provide analogues of Lagrangian submanifolds in the semiclassical limit of Schrödinger equation (Cf Appendix 11 of [13]).

We understand better, now, the role of the future boundary condition S_u in (1): when S_u is constant, the extremal process $z(\cdot)$ coincides with the Brownian motion on S^2 but, of course, in general this is not the case anymore. Stochastic deformation on a Riemannian manifold was treated in [8]. For another approach c.f. [9].

In spite of what was shown here, our approach can be made invariant under time reversal, in the same sense as our underlying classical dynamical system. The reason is that the very same stochastic system can be studied as well with respect to a decreasing filtration and an action functional on the time interval $[s, t]$, with an initial boundary condition $S_s^*(q)$. This relates to the fact that to any classical dynamical systems like ours are associated, in fact, two Hamilton-Jacobi equations adjoint with respect to the time parameter. The same is true after stochastic deformation. So, a time-adjoint heat equation, with initial positive boundary condition, is involved as well. The resulting ("Bernstein reciprocal") diffusions, built from these past and future boundary conditions, are invariant under time reversal on the time interval $[s, u]$. C.f. [10,12].

In particular, Markovian Bernstein processes are uniquely determined from the data of two (stictly positive) probability densities at different times s and u, here on S^2. They solve a "Schrödinger's optimization problem", an aspect very reminiscent of foundational questions of Mass Transportation theory [14]. The close relations between this theory and our method of Stochastic Deformation have been carefully analysed in [11], where many additional references can be found as well.

References

1. Dubrovin, B.A., Krichever, J.M., Novikov, S.P.: Integrable systems I. In: Arnold, V.I., Novikov, S.P. (eds.) Dynamical Systems IV. Encyclopaedia of Mathematical Sciences, vol. 4, pp. 177–332. Springer, Heidelberg (1990)
2. Olver, P.: Applications of Lie Groups to Differential Equations, 2nd edn. Springer, New York (1993)
3. Itô, K.: The Brownian motion and tensor fields on a Riemannian manifold. In: Proceedings of the International Congress Mathematical (Stockholm), pp. 536–539 (1962)

4. Fleming, W.H., Soner, H.M.: Controlled Markov Processes and Viscosity Solutions, 2nd edn. Springer, New York (2005)
5. Malliavin, P.: Stochastic Analysis. Springer, Heidelberg (1997)
6. Kuwabara, R.: On the symmetry algebra of the Schrödinger wave equation. Math. Japonica **22**, 243 (1977)
7. Léonard, C., Zambrini, J.-C.: Stochastic deformation of Jacobi's integrability Theorem in Hamiltonian mechanics. Preparation
8. Zambrini, J.-C.: Probability and quantum symmetries in a riemannian manifold. Prog. Probab. **45** (1999). Birkhäuser
9. Kolsrud, T.: Quantum and classical conserved quantities: martingales, conservation law and constants of motion. In: Benth, F.E., Di Nunno, G., Lindstrøm, T., Øksendal, B., Zhang, T. (eds.) Stochastic Analysis and Applications: Abel Symposium, pp. 461–491. Springer, Heidelberg (2005)
10. Zambrini, J.-C.: The research program of Stochastic Deformation (with a view toward Geometric Mechanics). In: Dalang, R., Dozzi, M., Flandoli, F., Russo, F. (eds.) BirkhäuserStochastic Analysis, A Series of Lectures (2015)
11. Léonard, C.: A survey of the Schrödinger problem and some of its connections with optimal transport. Discrete Contin. Dyn. Syst. A **34**(4) (2014)
12. Léonard, C., Roelly, S., Zambrini, J.-C.: Reciprocal processes. A measure-theoretical point of view. Probab. Surv. **11**, 237–269 (2014)
13. Arnold, V.: Méthodes Mathématiques de la Mécanique Classique. Mir, Moscou (1976)
14. Villani, C.: Optimal Transport, Old and New. Grundlehren der mathematischen Wissenschaften. Springer, Heidelberg (2009)

Constructing Universal, Non-asymptotic Confidence Sets for Intrinsic Means on the Circle

Matthias Glock and Thomas Hotz(✉)

Institut für Mathematik, Technische Universität Ilmenau, 98684 Ilmenau, Germany
{matthias.glock,thomas.hotz}@tu-ilmenau.de

Abstract. We construct confidence sets for the set of intrinsic means on the circle based on i.i.d. data which guarantee coverage of the entire latter set for finite sample sizes without any further distributional assumptions. Simulations demonstrate its applicability even when there are multiple intrinsic means.

Keywords: Circular data · Intrinsic means · Fréchet means · Universal confidence sets

1 Introduction

We are concerned with circular statistics, i.e. with the analysis of data which take values on the unit circle S^1. Such data occur often in practice, e.g. as measurements of wind directions in meteorology or other data with a periodic interpretation like times of the day at which patients are admitted to some hospital unit. Good references for circular statistics which include many more examples are [2,9,11], amongst others.

Here, we will focus on intrinsic means which are Fréchet means with respect to the intrinsic distance on the circle. To be specific, we will henceforth assume that $X, X_1, \ldots, X_n$ (for some sample size $n \in \mathbb{N}$) are independent and identically distributed random variables taking values on the (unit) circle S^1. For convenience, we will think of angular measurements and identify S^1 with $(-\pi, \pi]$, calculating modulo 2π whenever necessary, so that we can treat $X, X_1, \ldots, X_n$ as real-valued.

Of course, the circle is not a vector space so the population (or sample) mean cannot be defined through integration (or averaging). But, following [3] we observe that in a Euclidean space the mean is the unique minimiser of the expected (or summed) squared distances to the random point (or the data). Therefore, given a metric d on S^1, we accordingly define the *set of Fréchet (population) means* to be

$$M = \operatorname*{argmin}_{\mu \in S^1} F(\mu)$$

where $F : S^1 \to [0, \infty)$ is the *Fréchet functional* given by

$$F(\mu) = \mathbf{E}\, d(X, \mu)^2$$

for $\mu \in S^1$, i.e. M is the set of minimisers of F.

F. Nielsen and F. Barbaresco (Eds.): GSI 2017, LNCS 10589, pp. 477–485, 2017.
https://doi.org/10.1007/978-3-319-68445-1_56

There are two popular metrics being used on S^1: if one embeds S^1 as the unit circle in $\mathbb{C}$, $\{\exp(ix) : x \in (-\pi, \pi]\}$, then the *extrinsic* (or chordal) distance is given by $|\exp(ix) - \exp(iy)|$ for $x, y \in (-\pi, \pi]$. On the other hand, there is the *intrinsic* (or arc-length) distance d which is given by $d(x, y) = \min\{|x-y+2\pi k| : k \in \mathbb{Z}\}$ for $x, y \in (-\pi, \pi]$. A comparison between Fréchet means on the circle with respect to these two metrics may be found in [5].

In this article, we are concerned with Fréchet means with respect to the latter, intrinsic distance which are called *intrinsic means*, and we aim to construct a *confidence set* C given $X_1, \ldots, X_n$ which contains the set M of intrinsic means with probability at least $1 - \alpha$ for any pre-specified $\alpha \in (0, 1)$.

The analysis of intrinsic means on the circle is not trivial; the main reason for this is the fact that for any $x \in S^1$, the squared distance to that point, $S^1 \ni \mu \mapsto d(x, \mu)^2$, is everywhere continuously differentiable except at the point x^* "opposite" x which maximises the distance to x, i.e. at the *cut-locus* of x given by $x^* = x+\pi$ for $x \in (-\pi, 0]$ and $x^* = x-\pi$ for $x \in (0, \pi]$. Consequently, F need not be everywhere differentiable. However, F is differentiable at any intrinsic mean $m \in M$ with a vanishing derivative there, $F'(m) = \mathbf{E}2(X - m) = 0$ (calculated modulo 2π), while its cut-locus m^* carries probability measure $\mathbf{P}(X = m^*) = 0$ [10], cf. also [6].

Since intrinsic means are defined as minimisers of the Fréchet functional F, given data $X_1, \ldots, X_n$ it would be natural to consider the minimisers of the *empirical Fréchet functional* $\hat{F}_n : S^1 \to [0, \infty)$ with

$$\hat{F}_n(\mu) = \tfrac{1}{n} \textstyle\sum_{i=1}^{n} d(X_i, \mu)^2 \tag{1}$$

as *M-estimators*, i.e. the so-called *empirical Fréchet means*. Since $\hat{F}_n(\mu)$ converges to $F(\mu)$ almost surely (a.s.) for every $\mu \in S^1$, one might expect the empirical means to be close to the population means, and derive asymptotic confidence sets based on the asymptotic (for $n \to \infty$) behaviour of the empirical means. In fact, one can prove the following result [6,12]: if $M = \{0\}$ (unique population mean), then any measurable selection of empirical Fréchet means $\hat{\mu}_n$ converges a.s. to 0, and if the distribution of X features a continuous Lebesgue density f in a neighbourhood of $0^* = \pi$ with $f(\pi) < \frac{1}{2\pi}$ then $\hat{\mu}_n \xrightarrow{\mathcal{D}} \mathcal{N}\Big(0, \frac{\mathbf{E}X^2}{(1-2\pi f(\pi))^2}\Big)$ while in case $f(\pi) = \frac{1}{2\pi}$ a central limit theorem with a slower rate might hold.

In order to derive asymptotic confidence sets from this central limit theorem, one would need to ensure that M contains only a single point which imposes a restriction on the distribution of X, and that this distribution features a continuous Lebesgue density smaller than $\frac{1}{2\pi}$ at the cut-locus of the intrinsic mean; both conditions would e.g. be fulfilled for distributions with a unimodal density [5]. Then, one could either somehow estimate the asymptotic variance consistently or use a bootstrap approach to obtain asymptotic confidence sets for the unique intrinsic (population) mean, cf. [1] where this has been developed for distributions having no mass at an entire neighbourhood of the cut-locus.

This approach based on the asymptotic distribution of the empirical Fréchet mean has several drawbacks, one being that it guarantees only an approximate coverage probability of $1-\alpha$ for finite sample sizes where the quality of the approximation is usually unknown, and another one being that the assumptions justifying this approach are difficult to check in practice. In particular, judging whether they are fulfilled only by looking at an empirical Fréchet mean may be quite misleading as the following example shows.

Example 1 (equilateral triangle). *Let the distribution of X give equal weight to 3 points forming an equilateral triangle, i.e. set $\mathbf{P}\big(X=-\frac{2}{3}\pi\big)=\mathbf{P}(X=0)=\mathbf{P}\big(X=\frac{2}{3}\pi\big)=\frac{1}{3}$, see Fig. 1(a). It is easy to see that these very points form the set of intrinsic means, i.e. $M=\big\{-\frac{2}{3}\pi,0,\frac{2}{3}\pi\big\}$ in this case, cf. [5].*

For a large sample, however, the empirical measure will comprise 3 different weights with large probability, so that the empirical sample mean will be unique and close to one of the point masses, opposite of which there is no mass at all. Therefore, it will appear as if the assumptions for the central limit theorem are fulfilled though they are not.

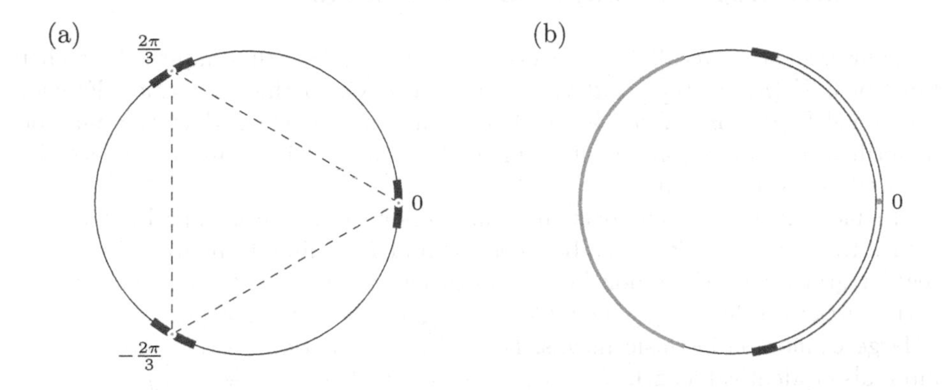

Fig. 1. (a) The distribution in Example 1 gives equal weight to the three (small, red) points which also constitute M (thick, white points); an example confidence set from Sect. 4 based on $n=10,000$ points and $1-\alpha=90\%$ is also shown (thick, blue line) (b) The distribution in Example 2 comprises a point mass (small, red point) and a segment with uniform density (medium, red line) opposite such that M is an entire (thick, white) segment; an example confidence set from Sect. 4 based on $n=10,000$ points and $1-\alpha=90\%$ is also shown (thick, blue line) (Color figure online)

We do not know of any constructions of confidence sets for intrinsic means which are applicable if there is more than one, let alone an entire segment of intrinsic means as in the following example taken from [6, Example 1, case 0b].

Example 2 (point mass with uniform density at the cut locus). *Let the distribution of X comprise a point mass at 0 with weight 0.6 as well as a Lebesgue continuous part with density $\frac{1}{2\pi}\chi_{(-\pi,-0.6\pi]\cup[0.6\pi,\pi]}$ where χ denotes the* *characteristic function of the corresponding segment, see Fig. 1(b). A straightforward calculation shows that the set of intrinsic means is then given by* $M = [-0.4\pi, 0.4\pi]$, *see* [6].

These examples ask for confidence sets which are both *universal*, i.e. they require no distributional assumptions beside the observations being i.i.d., and *non-asymptotic*, i.e. they guarantee coverage of M with probability at least $1-\alpha$ for any finite sample size $n \in \mathbb{N}$. Such confidence sets have been constructed for extrinsic means, i.e. Fréchet means w.r.t. the extrinsic distance on the circle, using geometric considerations for that particular distance in [7,8].

Our construction of such confidence sets for intrinsic means utilises mass concentration inequalities to control both the empirical Fréchet functional $\hat{F}_n$ (Sect. 2) and its derivative (Sect. 3). We then provide simulation results for the two examples above (Sect. 4) before finally discussing the results obtained as well as further research (Sect. 5).

2 Controlling the Empirical Functional

For our first step, recall that at every point $\mu \in S^1$, the empirical Fréchet functional $\hat{F}_n(\mu) = \frac{1}{n}\sum_{i=1}^n d(X_i, \mu)^2$ will be close to the population Fréchet functional $F(\mu) = \mathbf{E}d(X, \mu)^2$ by the law of large numbers; the deviation may be quantified by a mass concentration inequality since S^1 is compact, whence the squared distance is bounded.

In fact, since S^1 is compact and the squared distance is Lipschitz, it will suffice to bound the difference between $\hat{F}_n$ and F at finitely many points on a regular grid (using the union bound) in order to estimate it uniformly on the entire circle; we then may conclude with large probability that points where $\hat{F}_n$ is large cannot be intrinsic means. For this, we partition $(-\pi, \pi]$ into $J \in \mathbb{N}$ intervals of identical length, $I_{j,J} = \left(-\pi + (j-1)\frac{2\pi}{J}, -\pi + j\frac{2\pi}{J}\right]$ for $j = 1, \ldots, J$ whose closure is given by the closed balls with centers $\mu_{j,J} = -\pi + (2j-1)\frac{\pi}{J}$ and radius $\delta_J = \frac{\pi}{J}$.

In order to control the deviation of $\hat{F}_n$ from F at each $\mu_{j,J}$ we employ *Hoeffding's inequality* [4]: if $U_1, \ldots, U_n$, $n \in \mathbb{N}$ are independent random variables taking values in the bounded interval $[a, b]$, $-\infty < a < b < \infty$, then

$$\mathbf{P}\left(|\bar{U}_n - \mathbf{E}\bar{U}_n| \geq t\right) \leq 2\exp\left(-\tfrac{2nt^2}{(b-a)^2}\right) \tag{2}$$

for any $t \in [0, \infty)$ where $\bar{U}_n = \frac{1}{n}\sum_{i=1}^n U_i$.

Now fix some $\mu \in S^1$. Then, since the maximal (intrinsic) distance of two points on the circle is π and $\mathbf{E}\hat{F}_n(\mu) = F(\mu)$, we obtain

$$\mathbf{P}\left(|\hat{F}_n(\mu) - F(\mu)| \geq t\right) \leq 2\exp\left(-\tfrac{2nt^2}{\pi^4}\right)$$

for any $t \in [0, \infty)$. Moreover, for any $\nu, x \in S^1$

$$|d(x,\nu)^2 - d(x,\mu)^2| = \left(d(x,\nu) + d(x,\mu)\right)|d(x,\nu) - d(x,\mu)| \leq 2\pi d(\nu, \mu)$$

by the bound on d and the reverse triangle inequality, so the mapping $S^1 \ni \nu \mapsto d(x,\nu)^2$ is Lipschitz with constant 2π for any fixed $x \in S^1$. This implies that $\hat{F}_n$ and F are also Lipschitz with that very constant, so that $\hat{F}_n - F$ is Lipschitz with constant 4π. Therefore, bounding $|\hat{F}_n(\mu) - F(\mu)|$ for $\mu \in I_{j,J}$ by $|\hat{F}_n(\mu_{j,J}) - F(\mu_{j,J})| + 4\pi\delta_J$, the union bound gives

$$\mathbf{P}\big(\sup_{\mu\in S^1} |\hat{F}_n(\mu) - F(\mu)| \geq t + 4\pi\delta_J\big) \leq 2J\exp\big(-\tfrac{2nt^2}{\pi^4}\big)\,.$$

If we want the right hand side to equal $\beta \in (0,1)$, we have to choose $t = \sqrt{-\frac{\pi^4}{2n}\log\big(\frac{\beta}{2J}\big)}$ which leads to

$$\mathbf{P}\Big(\sup_{\mu\in S^1} |\hat{F}_n(\mu) - F(\mu)| \geq \sqrt{-\tfrac{\pi^4}{2n}\log\big(\tfrac{\beta}{2J}\big)} + \tfrac{4\pi^2}{J}\Big) \leq \beta\,.$$

Since by definition $F(m) = \inf_{\mu\in S^1} F(\mu)$ for any intrinsic mean $m \in M$, the triangle inequality gives

$$\begin{aligned}
\sup_{m\in M} \hat{F}_n(m) - \inf_{\mu\in S^1} \hat{F}_n(\mu) &= \sup_{m\in M} \hat{F}_n(m) - F(m) + \inf_{\mu\in S^1} F(\mu) - \inf_{\mu\in S^1} \hat{F}_n(\mu) \\
&\leq \sup_{m\in M} |\hat{F}_n(m) - F(m)| + \sup_{\mu\in S^1} |\hat{F}_n(\mu) - F(\mu)| \\
&\leq 2\sup_{\mu\in S^1} |\hat{F}_n(\mu) - F(\mu)|\,.
\end{aligned}$$

Thus, choosing $\beta = \frac{\alpha}{2}$, we have obtained our first confidence set for the set M of intrinsic means:

Proposition 1. *Let*

$$C_1 = \Big\{\mu \in S^1 : \hat{F}_n(\mu) < \inf_{\nu\in S^1} \hat{F}_n(\nu) + \Delta_1\Big\} \tag{3}$$

where the critical value $\Delta_1 > 0$ is given by

$$\Delta_1 = \pi^2\Big(\sqrt{-\tfrac{2}{n}\log\big(\tfrac{\alpha}{4J}\big)} + \tfrac{8}{J}\Big)\,.$$

Then we have $\mathbf{P}(C_1 \supseteq M) \geq 1 - \frac{\alpha}{2}$.

Note that $J \in \mathbb{N}$ may be selected in advance by numerical optimisation such that Δ_1 becomes minimal.

Even in the most favourable situation, however, when $\mathbf{P}(X = 0) = 1$, we a.s. have $F(\mu) = \hat{F}_n(\mu) = \mu^2$ for any $\mu \in (-\pi,\pi]$, so that C_1 has Lebesgue measure of the order of $\big(\frac{\log(n\log n)}{n}\big)^{\frac{1}{4}}$ for large n (for J of the order of $\sqrt{n\log n}$) which would give a somewhat slow rate of convergence. This is due to the fact that F itself behaves like a quadratic function at the minimum; this will be improved upon by considering the derivative F' which behaves linearly at the minimum where it vanishes.

Notwithstanding this problem, we observe that $\sup_{\mu\in S^1} |\hat{F}_n(\mu) - F(\mu)|$ and Δ_1 converge to zero in probability when n tends to infinity, which shows that C_1 in a certain sense converges to M in probability, thus ensuring consistency of our approach.

3 Controlling the Derivative

Recall that F is differentiable at any intrinsic mean while $\hat{F}_n$ is differentiable except at points opposite observations which occur opposite an intrinsic mean only with probability 0 [6]. If the derivative of $\hat{F}_n$ exists, i.e. for any $\mu \in S^1$ with $\mu^* \notin \{X_1, \dots, X_n\}$, it is given by

$$\hat{F}_n'(\mu) = \tfrac{2}{n} \textstyle\sum_{i=1}^n [X_i - \mu] \tag{4}$$

where representatives for the X_i, $i = 1, \dots, n$ need to be chosen in $\mathbb{R}$ such that $[X_i - \mu] \in (-\pi, \pi]$. Otherwise, i.e. in case $\mu^* \in \{X_1, \dots, X_n\}$, we simply define $\hat{F}_n'(\mu)$ by (4).

This is utilised as follows: partitioning $(-\pi, \pi]$ into n disjoint intervals $I_{k,n}$, $k = 1, \dots, n$ (using the notation from Sect. 2), let $K = \{k : I_{k,n} \cap M \neq \emptyset\}$ be the set of indices of intervals which contain an intrinsic mean and choose one intrinsic mean $m_k \in M$ for every $k \in K$ whence $M \subseteq \cup_{k \in K} I_{k,n}$. Since $2[X_i - m_k]$ in (4) takes values in $[-2\pi, 2\pi]$, we can employ Hoeffding's inequality (2) again to get

$$\mathbf{P}\big(|\hat{F}_n'(m_k)| \geq t\big) \leq 2 \exp\big(-\tfrac{nt^2}{8\pi^2}\big)$$

for any $k \in K$ where we used $\mathbf{E}\hat{F}_n'(m_k) = 0$. The union bound readily implies

$$\mathbf{P}\big(\exists k \in K : \inf_{\mu \in I_{k,n} \cap M} |\hat{F}_n'(\mu)| \geq t\big) \leq 2|K| \exp\big(-\tfrac{nt^2}{8\pi^2}\big)$$

where $|K|$ is the cardinality of K; choosing $t > 0$ such that the right hand side becomes $\frac{\alpha}{2}$ then gives

$$\mathbf{P}\Big(\exists k \in K : \inf_{\mu \in I_{k,n} \cap M} |\hat{F}_n'(\mu)| \geq \sqrt{-\tfrac{8\pi^2}{n} \log\big(\tfrac{\alpha}{4|K|}\big)}\Big) \leq \tfrac{\alpha}{2}\,. \tag{5}$$

Unfortunately, K is not known in advance, but it can be estimated using the confidence set C_1 for M constructed in Sect. 2: let $\hat{K} = \{k : I_{k,n} \cap C_1 \neq \emptyset\}$. Then, whenever $C_1 \supseteq M$ we have $\hat{K} \supseteq K$, in particular $|\hat{K}| \geq |K|$, and thus $M \subseteq C_1 \cap \cup_{k \in \hat{K}} I_{k,n}$. So, setting

$$C_2 = \cup_{k \in \hat{K} : \inf_{\mu \in I_{k,n} \cap C_1} |\hat{F}_n'(\mu)| < \Delta_2} I_{k,n} \tag{6}$$

with the critical value

$$\Delta_2 = \sqrt{-\tfrac{8\pi^2}{n} \log\big(\tfrac{\alpha}{4|\hat{K}|}\big)}$$

—which will then be larger than the one in (5) based on K— allows to finally construct the desired confidence set:

Proposition 2. *Let $X_1, \dots, X_n$, $n \in \mathbb{N}$ be independent and identically distributed random points on S^1, and let $\alpha \in (0,1)$ be given. Then, the confidence set*

$$C = C_1 \cap C_2 \tag{7}$$

based on the sets C_1 and C_2 constructed in (3) *and* (6) *above, respectively, is a $(1-\alpha)$-confidence set for the set M of intrinsic means, i.e. it fulfills $\mathbf{P}(C \supseteq M) \geq 1 - \alpha$.*

4 Simulations

The construction of the confidence interval C in (7) has been implemented within the statistical software package R [13]. For illustration, we show results for the two examples introduced in Sect. 1.

We simulated Example 1 for sample sizes $n = 10^2, 10^3, 10^4, 10^5, 10^6$ and $1 - \alpha = 90\%$; for each simulation it was checked whether M was covered by C, and the Lebesgue measure of C was computed. This was independently repeated 1,000 times for each sample size. The result of one simulation for $n = 10^4$ is shown in Fig. 1(a). Averages (and standard deviations) of the Lebesgue measure of C computed over the repetitions for each sample size are reported in Table 1.

Table 1. Average Lebesgue measure of C over 1,000 repetitions for different sample sizes (rounded, ± standard deviation) of Example 1

n	Avg. Lebesgue measure (± st. dev.)
10^2	6.25 (±0.07)
10^3	2.76 (±0.01)
10^4	0.94 (±0.00)
10^5	0.32 (±0.00)
10^6	0.11 (±0.00)

As Example 2 leads to numerically more involved calculations, it was simulated only for sample sizes $n = 10^2, 10^3, 10^4, 10^5$ and $1 - \alpha = 90\%$; for each simulation it was checked whether M was covered by C, and the Lebesgue measure of C was computed. This was consequently independently repeated only 100 times for each sample size. Since M has positive Lebesgue measure, averages (and standard deviations) of the Lebesgue measure of $C \setminus M$, i.e. the confidence sets' excess size, computed over the repetitions for each sample size are given in Table 2.

Table 2. Average Lebesgue measure of $C \setminus M$ over 100 repetitions for different sample sizes (rounded, ± standard deviation) of Example 2

n	Avg. Lebesgue measure (± st. dev.)
10^2	2.63 (±0.30)
10^3	0.92 (±0.10)
10^4	0.31 (±0.03)
10^5	0.11 (±0.01)

In both simulation we found that M was covered in all simulations which is to be expected since our use of mass concentration inequalities results in quite conservative confidence sets. This, however, may to a certain extent be the price one has to pay in order to obtain non-asymptotic, universal confidence sets guaranteeing coverage of all of M.

Nonetheless, we observe that the (excess) size of the confidence sets decreases roughly (up to a log-factor) like $n^{-\frac{1}{2}}$ so the second step in our construction had the desired effect to obtain confidence sets of a size usually obtained for M-estimators while the first step was necessary to ensure consistency by also removing local minimisers of F.

5 Discussion and Outlook

We would like to stress again that for both examples asymptotic confidence sets cannot easily be constructed since neither example features a unique intrinsic mean; indeed, to the best of our knowledge, the given construction is the first of a confidence set for M applicable in such situations.

Of course, the given construction may be repeated for more general compact metric spaces as long as one can construct the necessary grids of points at which to control the functionals; carrying this out for other interesting spaces will be left for further research. We also note that the construction may be improved upon by taking the "variance of the estimator" into account; this corresponds here to making usage of the knowledge about F when controlling $\hat{F}_n$.

References

1. Bhattacharya, R.N., Patrangenaru, V.: Large sample theory of intrinsic and extrinsic sample means on manifolds II. Ann. Stat. **33**(3), 1225–1259 (2005)
2. Fisher, N.I.: Statistical Analysis of Circular Data. Cambridge University Press, Cambridge (1993)
3. Fréchet, M.: Les éléments aléatoires de nature quelconque dans un espace distancié. Annales de l'Institut Henri Poincaré **10**(4), 215–310 (1948)
4. Hoeffding, W.: Probability inequalities for sums of bounded random variables. J. Am. Stat. Assoc. **58**(301), 13–30 (1963)
5. Hotz, T.: Extrinsic vs intrinsic means on the circle. In: Nielsen, F., Barbaresco, F. (eds.) GSI 2013. LNCS, vol. 8085, pp. 433–440. Springer, Heidelberg (2013). doi:10.1007/978-3-642-40020-9_47
6. Hotz, T., Huckemann, S.: Intrinsic means on the circle: uniqueness, locus and asymptotics. Ann. Inst. Stat. Math. **67**(1), 177–193 (2015)
7. Hotz, T., Kelma, F., Wieditz, J.: Universal, non-asymptotic confidence sets for circular means. In: Nielsen, F., Barbaresco, F. (eds.) GSI 2015. LNCS, vol. 9389, pp. 635–642. Springer, Cham (2015). doi:10.1007/978-3-319-25040-3_68
8. Hotz, T., Kelma, F., Wieditz, J.: Non-asymptotic confidence sets for circular means. Entropy **18**(10), 375 (2016)
9. Jammalamadaka, S.R., SenGupta, A.: Topics in Circular Statistics. Series on Multivariate Analysis, vol. 5. World Scientific, Singapore (2001)
10. Le, H., Barden, D.: On the measure of the cut locus of a Fréchet mean. Bull. Lond. Math. Soc. **46**(4), 698–708 (2014)

11. Mardia, K.V., Jupp, P.E.: Directional Statistics. Wiley, New York (2000)
12. McKilliam, R.G., Quinn, B.G., Clarkson, I.V.L.: Direction estimation by minimum squared arc length. IEEE Trans. Signal Process. **60**(5), 2115–2124 (2012)
13. R Core Team: R: a language and environment for statistical computing. R Foundation for Statistical Computing, Vienna (2013). http://www.R-project.org/

Noncommutative Geometry and Stochastic Processes

Marco Frasca(✉)

MBDA Italy S.p.A., Via Monte Flavio, 45, 00131 Rome, Italy
marco.frasca@mbda.it
http://marcofrasca.wordpress.com

Abstract. The recent analysis on noncommutative geometry, showing quantization of the volume for the Riemannian manifold entering the geometry, can support a view of quantum mechanics as arising by a stochastic process on it. A class of stochastic processes can be devised, arising as fractional powers of an ordinary Wiener process, that reproduce in a proper way a stochastic process on a noncommutative geometry. These processes are characterized by producing complex values and so, the corresponding Fokker–Planck equation resembles the Schrödinger equation. Indeed, by a direct numerical check, one can recover the kernel of the Schrödinger equation starting by an ordinary Brownian motion. This class of stochastic processes needs a Clifford algebra to exist.

1 Introduction

A comprehension of the link between stochastic processes and quantum mechanics can provide a better understanding of the role of space–time at a quantum gravity level. Indeed, noncommutative geometry, in the way Connes, Chamseddine and Mukhanov provided recently [1,2], seems to fit well the view that a quantized volume yields a link at a deeper level of the connection between stochastic processes and quantum mechanics. This is an important motivation as we could start from a reformulation of quantum mechanics to support or drop proposals to understand quantum gravity and the fabric of space-time.

A deep connection exists between Brownian motion and binomial coefficients. This can be established by recovering the kernel of the heat equation from the binomial distribution for a random walk (Pascal–Tartaglia triangle) and applying the theorem of central limit [3]. When an even smaller step in the random walk is taken a Wiener process is finally approached. So, it is a natural question to ask what would be the analogous of Pascal–Tartaglia triangle in quantum mechanics [4]. This arises naturally by noting the apparent formal similarity between the heat equation and the Schrödinger equation. But this formal analogy is somewhat difficult to understand due to the factor i entering into the Schrödinger equation. An answer to this question hinges on a deep problem not answered yet: Is there a connection between quantum mechanics and stochastic processes? The formal similarity has prompted attempts to answer as in the pioneering work of Edward Nelson [5] and in the subsequent deep analysis by Francesco

F. Nielsen and F. Barbaresco (Eds.): GSI 2017, LNCS 10589, pp. 486–494, 2017.
https://doi.org/10.1007/978-3-319-68445-1_57

Guerra and his group [6]. They dubbed this reformulation of quantum mechanics as "stochastic mechanics". This approach matches directly a Wiener process to the Schrödinger equation passing through a Bohm-like set of hydrodynamic equations and so, it recovers all the drawbacks of Bohm formulation. This view met severe criticisms motivating some researchers to a substantial claim that "no classical stochastic process underlies quantum mechanics" [7], showing contradiction with predictions of quantum mechanics. Subsequent attempts to partially or fully recover this view were proposed with non–Markovian processes [8] or repeated measurements [9–11].

In this paper we will show that a new set of stochastic processes can be devised, starting from noncommutative geometry, that can elucidate such a connection [4,12]. We show how spin is needed also in the non-relativistic limit. These processes are characterized by the presence of a Bernoulli process yielding the values 1 and i, exactly as expected in the volume quantization in noncommutative geometry. In this latter case, it appears that a stochastic process on a quantized manifold is well represented by a fractional power of an ordinary Wiener process when this is properly defined through a technique at discrete time [13]. The kernel of the Schrödinger equation is numerically evaluated through a Brownian motion.

A similar idea to use noncommutative geometry in stochastic processes was proposed in [14] but there it was used to fix univocally the kinetic equation of a real stochastic process.

2 Noncommutative Geometry and Quantization of Volume

A noncommutative geometry, given by the triple $(\mathcal{A}, H, D)$ being $\mathcal{A}$ a set of operators belonging to a *-algebra, H a Hilbert space and D a Dirac operator, implies that the volume of the corresponding Riemann manifold is quantized with two classes of unity of volume $(1, i)$. This has been recently proved by Connes, Chamseddine and Mukhanov [1,2]. The two classes of volume arise from the fact that the Dirac operator should not be limited to Majorana (neutral) states in the Hilbert space and so, we need to associate a charge conjugation operator J to our triple. To complete the characterization of our geometry, we recall that the algebra of Dirac matrices implies a γ^5, the chirality matrix that changes the parity of the states. For an ordinary Riemann manifold, the algebra $\mathcal{A}$ is that of functions and is commuting. Remembering that $[D, a] = i\gamma \cdot \partial a$, and noting that, in four dimensions, x_1, x_2, x_3, x_4 are legal functions of $\mathcal{A}$, it is $[D, x_1][D, x_2][D, x_3][D, x_4] = \gamma^1\gamma^2\gamma^3\gamma^4 = -i\gamma^5$. For generally chosen functions in $\mathcal{A}$, a_0, a_1, a_2, a_3, a_4, $\dots$ a_d, summing over all the possible permutations one has a Jacobian, we can define the chirality operator

$$\gamma = \sum_P (a_0[D, a_1] \dots [D, a_d]). \tag{1}$$

So, in four dimension this gives

$$\gamma = -iJ \cdot \gamma^5 = -i \cdot \det(e)\gamma^5 \tag{2}$$

being J the Jacobian, e^a_μ the vierbein for the Riemann manifold and $\gamma^5 = i\gamma^1\gamma^2\gamma^3\gamma^4$ for $d = 4$, a well-known result. We used the fact that $\det(e) = \sqrt{g}$, being $g_{\mu\nu}$ the metric tensor. So, the definition of the chirality operator is proportional to the factor determining the volume of a Riemannian orientable manifold.

In order to see if a Riemannian manifold can be properly quantized, instead of functions we consider operators Y belonging to an operator algebra $\mathcal{A}'$. These operators have the properties

$$Y^2 = \kappa I \qquad Y^\dagger = \kappa Y. \tag{3}$$

This is a set of compact operators playing the role of coordinates as in the Heisenberg commutation relations. We have to consider two sets of them Y_+ and Y_- as we expect a conjugation of charge operator C to exist such that $CAC^{-1} = Y^\dagger$ for a given operator or complex conjugation for a function. This appears naturally out of a Dirac algebra of gamma matrices. So, a natural way to write down the operators Y is by using an algebra of Dirac matrices Γ^A such that

$$\{\Gamma^A, \Gamma^B\} = 2\delta^{AB}, \qquad (\Gamma^A)^* = \kappa\Gamma^A \tag{4}$$

with $A, B = 1 \ldots d+1$, then

$$Y = \Gamma^A Y^A. \tag{5}$$

We will have two different set of gamma matrices for Y_+ and Y_- that will have independent traces. Using the charge conjugation operator C, we can define a new coordinate

$$Z = 2ECEC^{-1} - I \tag{6}$$

where $E = (1+Y_+)/2 + (1+iY_-)/2$ will project one or the other coordinate. We recognize that the spectrum of Z is in $(1, i)$ given Eq. (3). Now, we generalize our equation for the chirality operator imposing a trace on Γs both for Y_+ and Y_-, normalized to the number of components, and we will have

$$\frac{1}{n!}\langle Z[D, Z] \ldots [D, Z]\rangle = \gamma. \tag{7}$$

where we have introduced the average $\langle \ldots \rangle$ that, in this case, reduces to matrix traces. In order to see the quantization of the volume, let us consider a three dimensional manifold and the sphere $\mathbb{S}^2$. From Eq. (7) one has

$$V_M = \int_M \frac{1}{n!}\langle Z[D, Z] \ldots [D, Z]\rangle d^3x \tag{8}$$

and doing the traces one has

$$V_M = \int_M \left(\frac{1}{2}\epsilon^{\mu\nu}\epsilon_{ABC}Y_+^A\partial_\mu Y_+^B\partial_\nu Y_+^C + \frac{1}{2}\epsilon^{\mu\nu}\epsilon_{ABC}Y_-^A\partial_\mu Y_-^B\partial_\nu Y_-^C\right) d^3x. \tag{9}$$

It is easy to see that this will yield [1,2]

$$\det(e^a_\mu) = \frac{1}{2}\epsilon^{\mu\nu}\epsilon_{ABC}Y^A_+\partial_\mu Y^B_+\partial_\nu Y^C_+ + \frac{1}{2}\epsilon^{\mu\nu}\epsilon_{ABC}Y^A_-\partial_\mu Y^B_-\partial_\nu Y^C_-. \quad (10)$$

The coordinates Y_+ and Y_- belongs to unitary spheres and the Dirac operator has a discrete spectrum being evaluated on a compact manifold, so we are covering all the manifold with a large integer number of these spheres. Thus, the volume is quantized as this condition requires. This can be extended to four dimensions with some more work [1,2].

Differently from an ordinary stochastic process, a Wiener process on a quantized manifold will yield the projection of the spectrum $(1, i)$ of the coordinates on the two kinds of spheres Y_+, Y_-. This will depend on the way a particle moves on the manifold taking into account that the distribution of the two kinds of unitary volumes is absolutely random. One can construct a process Φ such that, against a toss of a coin, one gets 1 or i as outcome, assuming the distribution of the unitary volumes is uniform. This can be written

$$\Phi = \frac{1+B}{2} + i\frac{1-B}{2} \quad (11)$$

with B a Bernoulli process such that $B^2 = I$ producing the value ± 1 depending on the unitary volume hit by the particle and $\Phi^2 = B$. If we want to consider the Brownian motion of the particle on such a manifold we should expect the outcomes to be either Y_+ or Y_-. So, given the set of Γ matrices and the chirality operator γ, the most general form for a stochastic process on the manifold can be written down (summation on A is implied)

$$dY = \Gamma^A \cdot (\kappa_A + \xi_A dX_A \cdot B_A + \zeta_A dt + i\eta_A\gamma^5) \cdot \Phi_A \quad (12)$$

being κ_A, ξ_A, ζ_A, η_A arbitrary coefficients of this linear combination. The Bernoulli processes B_A and the Wiener process dX_A cannot be independent. Rather, the sign arising from the Bernoulli process is the same of that of the corresponding Wiener process. This equation provides the equivalent of the Eq. (3) for the coordinates on the manifold. This is exactly the formula we will obtain for the fractional powers of a Wiener process. It just represents the motion on a quantized Riemannian manifold with two kinds of quanta. Underlying quantum mechanics there appears to be a noncommutative geometry.

3 Powers of Stochastic Processes

We consider an ordinary Wiener W process describing a Brownian motion and define the α-th power of it. We do a proof of existence by construction [13]. A Wiener process W is computed by the cumulative sum of the increments at discrete steps $W_i - W_{i-1}$. Similarly, we will have the process (given $\alpha \in \mathbb{R}^+$) with the formal definition

$$dX = (dW)^\alpha. \quad (13)$$

built through the Euler–Maruyama definition of a stochastic process [15] at discrete times obtained by the cumulative sum

$$X_i = X_{i-1} + (W_i - W_{i-1})^\alpha. \tag{14}$$

as done in computing a Wiener process with $\alpha = 1$. A complete proof of existence of these processes has been shown in [13].

4 "Square Root" Formula and Fokker–Planck Equation

Using Itō calculus to express the "square root" process with more elementary stochastic processes [16], $(dW)^2 = dt$, $dW \cdot dt = 0$, $(dt)^2 = 0$ and $(dW)^\alpha = 0$ for $\alpha > 2$, we could tentatively set

$$dX = (dW)^{\frac{1}{2}} \stackrel{?}{=} \left(\mu_0 + \frac{1}{2\mu_0} dW \cdot \text{sgn}(dW) - \frac{1}{8\mu_0^3} dt\right) \cdot \Phi_{\frac{1}{2}} \tag{15}$$

being $\mu_0 \neq 0$ an arbitrary scale factor and

$$\Phi_{\frac{1}{2}} = \frac{1-i}{2} \text{sgn}(dW) + \frac{1+i}{2} \tag{16}$$

a Bernoulli process equivalent to a coin tossing that has the property $(\Phi_{\frac{1}{2}})^2 = \text{sgn}(dW)$. This process is characterized by the values 1 and i and it is like the Brownian motion went scattering with two different kinds of small pieces of space, each one contributing either 1 or i to the process, randomly. This is the same process seen for the noncommutative geometry in Eq. (11). We have introduced the process $\text{sgn}(dW)$ that yields just the signs of the corresponding Wiener process. Equation (15) is unsatisfactory for a reason, taking the square yields

$$(dX)^2 = \mu_0^2 \text{sgn}(dW) + dW \tag{17}$$

and the original Wiener process is not exactly recovered. We find added a process that has the effect to shift upward the original Brownian motion while retaining the shape. We can fix this problem by using Pauli matrices. Let us consider two Pauli matrices σ_i, σ_k with $i \neq k$ such that $\{\sigma_i, \sigma_k\} = 0$. We can rewrite the above identity as

$$I \cdot dX = I \cdot (dW)^{\frac{1}{2}} = \sigma_i \left(\mu_0 + \frac{1}{2\mu_0} dW \cdot \text{sgn}(dW) - \frac{1}{8\mu_0^3} dt\right) \cdot \Phi_{\frac{1}{2}} + i\sigma_k \mu_0 \cdot \Phi_{\frac{1}{2}} \tag{18}$$

and so, $(dX)^2 = dW$ as it should, after removing the identity matrix on both sides. This idea generalizes easily to higher dimensions using γ matrices. We see that we have recovered a similar stochastic process as in Eq. (12). This will be extended to four dimensions below.

Now, let us consider a more general "square root" process where we assume also a term proportional to dt. We assume implicitly the Pauli matrices simply removing by hand the sgn process at the end of the computation. This forces to take $\mu_0 = 1/2$ when the square is taken, to recover the original stochastic process, and one has

$$dX(t) = [dW(t) + \beta dt]^{\frac{1}{2}} = \left[\frac{1}{2} + dW(t) \cdot \mathrm{sgn}(dW(t)) + (-1 + \beta \mathrm{sgn}(dW(t)))dt\right] \Phi_{\frac{1}{2}}(t). \tag{19}$$

From the Bernoulli process $\Phi_{\frac{1}{2}}(t)$ we can derive

$$\mu = -\frac{1+i}{2} + \beta\frac{1-i}{2} \qquad \sigma^2 = 2D = -\frac{i}{2}. \tag{20}$$

Then, we get a double Fokker–Planck equation for a free particle, being the distribution function $\hat{\psi}$ complex valued,

$$\frac{\partial\hat{\psi}}{\partial t} = \left(-\frac{1+i}{4} + \beta\frac{1-i}{2}\right)\frac{\partial\hat{\psi}}{\partial X} - \frac{i}{4}\frac{\partial^2\hat{\psi}}{\partial X^2}. \tag{21}$$

This should be expected as we have a complex stochastic process and then two Fokker–Planck equations are needed to describe it. We have obtained an equation strongly resembling the Schrödinger equation for a complex distribution function. We can ask at this point if we indeed are recovering quantum mechanics. In the following section we will perform a numerical check of this hypothesis.

5 Recovering the Kernel of the Schrödinger Equation

If really the "square root" process diffuses as a solution of the Schrödinger equation we should be able to recover the corresponding solution for the kernel

$$\hat{\psi} = (4\pi i t)^{-\frac{1}{2}} \exp\left(ix^2/4t\right) \tag{22}$$

sampling the square root process. To see this we note that a Wick rotation, $t \to -it$, turns it into a heat kernel as we get immediately

$$K = (4\pi t)^{-\frac{1}{2}} \exp\left(-x^2/4t\right). \tag{23}$$

A Montecarlo simulation can be easily executed extracting the square root of a Brownian motion and, after a Wick rotation, to show that a heat kernel is obtained. We have generated 10000 paths of Brownian motion and extracted its square root in the way devised in Sect. 3. We have evaluated the corresponding distribution after Wick rotating the results for the square root. The Wick rotation generates real results as it should be expected and a comparison can be performed. The result is given in Fig. 1

The quality of the fit can be evaluated being $\hat{\mu} = 0.007347$ with confidence interval $[0.005916, 0.008778]$, $\hat{\sigma} = 0.730221$ with confidence interval $[0.729210, 0.731234]$ for the heat kernel while one has $\hat{\mu} = 0.000178$ with confidence interval $[-0.002833, 0.003189]$ and $\hat{\sigma} = 1.536228$ with confidence interval $[1.534102, 1.538360]$ for the Schrödinger kernel. Both are centered around 0 and

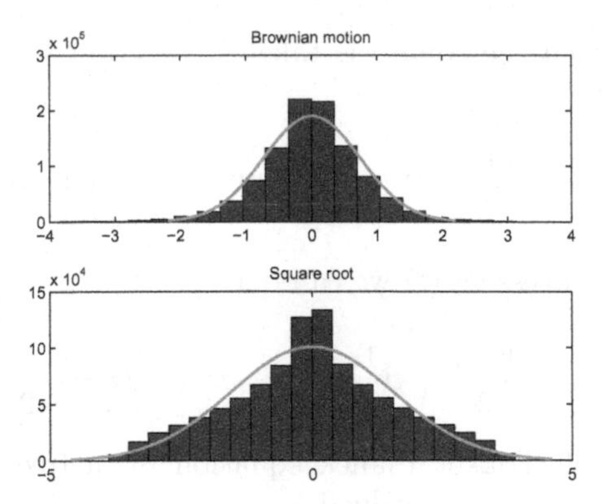

Fig. 1. Comparison between the distributions of the Brownian motion and its square root after a Wick rotation.

there is a factor ~ 2 between standard deviations as expected from Eq. (21). Both the fits are exceedingly good. Having recovered the Schrödinger kernel from Brownian motion with the proper scaling factors in mean and standard deviation, we can conclude that we are doing quantum mechanics: *The "square root" process describes the motion of a quantum particle.* Need for Pauli matrices, as shown in the preceding section, implies that spin cannot be neglected.

6 Square Root and Noncommutative Geometry

We have seen that, in order to extract a sort of square root of a stochastic process, we needed Pauli matrices or, generally speaking, a Clifford algebra. This idea was initially put forward by Dirac to derive his relativistic equation for fermions and the corresponding algebra was proven to exist by construction as it also happens for Pauli matrices. The simplest and non-trivial choice is obtained, as said above, using Pauli matrices $\{\sigma_k \in C\ell_3(\mathbb{C}),\ k = 1, 2, 3\}$ that satisfy

$$\sigma_i^2 = I \qquad \sigma_i\sigma_k = -\sigma_k\sigma_i \qquad i \neq k. \tag{24}$$

This proves to be insufficient to go to dimensions higher than 1+1 for Brownian motion. The more general solution is provided by a Dirac algebra of γ matrices $\{\gamma_k \in C\ell_{1,3}(\mathbb{C}),\ k = 0, 1, 2, 3\}$ such that

$$\gamma_0^2 = I \qquad \gamma_1^2 = \gamma_2^2 = \gamma_3^2 = -I \qquad \gamma_i\gamma_k + \gamma_k\gamma_i = 2\eta_{ik} \tag{25}$$

being η_{ik} the Minkowski metric. In this way one can introduce three different Brownian motions for each spatial coordinates and three different Bernoulli processes for each of them. The definition is now

$$dE = \sum_{k=1}^{3} i\gamma_k \left(\mu_k + \frac{1}{2\mu_k}|dW_k| - \frac{1}{8\mu_k^3}dt\right) \cdot \Phi_{\frac{1}{2}}^{(k)} + \sum_{k=1}^{3} i\gamma_0\gamma_k\mu_k\Phi_{\frac{1}{2}}^{(k)} \tag{26}$$

It is now easy to check that

$$(dE)^2 = I \cdot (dW_1 + dW_2 + dW_3). \tag{27}$$

The Fokker-Planck equations have a solution with 4 components, as now the distribution functions are Dirac spinors. These are given by

$$\frac{\partial \hat{\Psi}}{\partial t} = \sum_{k=1}^{3} \frac{\partial}{\partial X_k} \left(\mu_k \hat{\Psi} \right) - \frac{i}{4} \Delta_2 \hat{\Psi} \tag{28}$$

being $\mu_k = -\frac{1+i}{4} + \beta_k \frac{1-i}{2}$. This implies that, the general formula for the square root process implies immediately spin and antimatter for quantum mechanics that now come out naturally.

7 Conclusions

We have shown the existence of a class of stochastic processes that can support quantum behavior. This formalism could entail a new understanding of quantum mechanics and give serious hints on the properties of space-time for quantum gravity. This yields a deep connection with noncommutative geometry as formulated by Alain Connes through the more recent proposal of space quantization by Connes himself, Chamseddine and Mukhanov. This quantization of volume entails two kinds of quanta implying naturally the unity $(1, i)$ that arises in the "square root" of a Wiener process. Indeed, a general stochastic process for a particle moving on such a quantized volume corresponds to our formula of the "square root" of a stochastic process on a 4-dimensional manifold. Spin appears to be an essential ingredient, already at a formal level, to treat such fractional powers of Brownian motion.

Finally, it should be interesting, and rather straightforward, to generalize this approach to a Dirac equation on a generic manifold. The idea would be to recover also Einstein equations as a fixed point solution to the Fokker-Planck equations as already happens in string theory. Then they would appear as a the result of a thermodynamic system at the equilibrium based on noncommutative geometry. This is left for further study.

I would like to thank Alfonso Farina for giving me the chance to unveil some original points of view on this dusty corner of quantum physics.

References

1. Chamseddine, A.H., Connes, A., Mukhanov, V.: Phys. Rev. Lett. 114(9), 091302 (2015). arXiv:1409.2471, [hep-th]
2. Chamseddine, A.H., Connes, A., Mukhanov, V.: JHEP **1412**, 098 (2014). [arxiv:1411.0977 [hep-th]]
3. Weiss, G.: Aspects and Applications of the Random Walk. North-Holland, Amsterdam (1994)

4. Farina, A., Frasca, M., Sedehi, M.: SIViP **8**, 27 (2014). doi:10.1007/s11760-013-0473-y
5. Nelson, E.: Dynamical Theories of Brownian Motion. Princeton University Press, Princeton (1967)
6. Guerra, F.: Phys. Rept. **77**, 263 (1981)
7. Grabert, H., Hänggi, P., Talkner, P.: Phys. Rev. A **19**, 2440 (1979)
8. Skorobogatov, G.A., Svertilov, S.I.: Phys. Rev. A **58**, 3426 (1998)
9. Blanchard, P., Golin, S., Serva, M.: Phys. Rev. D **34**, 3732 (1986)
10. Wang, M.S., Liang, W.-K.: Phys. Rev. D **48**, 1875 (1993)
11. Blanchard, P., Serva, M.: Phys. Rev. D **51**, 3132 (1995)
12. Frasca, M.: arXiv:1201.5091 [math-ph] (2012). Unpublished
13. Frasca, M., Farina, A.: SIViP **11**, 1365 (2017). doi:10.1007/s11760-017-1094-7
14. Dimakis, A., Tzanakis, C.: J. Phys. A: Math. Gen. **29**, 577 (1996)
15. Higham, D.J.: SIAM Rev. **43**, 525 (2001)
16. Øksendal, B.K.: Stochastic Differential Equations: An Introduction with Applications, p. 44. Springer, Berlin (2003)

Hamilton-Jacobi Theory and Information Geometry

Florio M. Ciaglia[1,2(✉)], Fabio Di Cosmo[1,2], and Giuseppe Marmo[1,2]

[1] Dipartimento di Fisica, Università di Napoli "Federico II",
Via Cinthia Edificio 6, 80126 Napoli, Italy
[2] INFN-Sezione di Napoli, Via Cinthia Edificio 6, 80126 Napoli, Italy
ciaglia@na.infn.it

Abstract. Recently, a method to dynamically define a divergence function D for a given statistical manifold $(\mathcal{M}, g, T)$ by means of the Hamilton-Jacobi theory associated with a suitable Lagrangian function $\mathfrak{L}$ on $T\mathcal{M}$ has been proposed. Here we will review this construction and lay the basis for an inverse problem where we assume the divergence function D to be known and we look for a Lagrangian function $\mathfrak{L}$ for which D is a complete solution of the associated Hamilton-Jacobi theory. To apply these ideas to quantum systems, we have to replace probability distributions with probability amplitudes.

1 Introduction

In the field of information geometry, divergence functions are ubiquitous objects. A divergence function D is a positive semi-definite two-point function defined on $\mathcal{M} \times \mathcal{M}$, where $\mathcal{M}$ is the manifold underlying the statistical model $(\mathcal{M}, g, T)$ under study (see [1–3]), such that $D(m_1, m_2) = 0$ if and only if $m_1 = m_2$. Roughly speaking, the value $D(m_1, m_2)$ is interpreted as a "measure of difference" between the probability distributions parametrized by m_1 and m_2. The exact meaning of this difference depends on the explicit model considered. If we imbed classical probabilities in the space of quantum systems, i.e., we replace probabilities with probability amplitudes, it is still possible to define divergence functions and derive metric tensors for quantum states. For instance, when $\mathcal{M} = \mathcal{P}(\mathcal{H})$ is the space of pure states of a quantum system with Hilbert space $\mathcal{H}$, Wootter has shown (see [21]) that a divergence function D providing a meaningful notion of statistical distance between pure states may be introduced by means of the concepts of distinguishability and statistical fluctuations in the outcomes of measurements. It turns out that this statistical distance coincides with the Riemannian geodesic distance associated with the Fubini-Study metric on the complex projective space. On the other hand, when $\mathcal{M} = \mathcal{P}_n$ is the manifold of positive probability measure on $\chi = \{1, ..., n\}$, and D is the Kullback-Leibler divergence function (see [1–3]), then the meaning of the "difference" between m_1 and m_2 as measured by D is related with the asymptotic estimation theory for an empirical probability distribution extracted from independent samples associated with a given probability distribution (see [1]). One of

F. Nielsen and F. Barbaresco (Eds.): GSI 2017, LNCS 10589, pp. 495–502, 2017.
https://doi.org/10.1007/978-3-319-68445-1_58

the main features of a divergence function D is the possibility to extract from it a metric tensor g, and a skewness tensor T on $\mathcal{M}$ using an algorithm involving iterated derivatives of D and the restriction to the diagonal of $\mathcal{M} \times \mathcal{M}$ (see [1–3]). Given a statistical model $(\mathcal{M}, g, T)$ there is always a divergence function whose associated tensors are precisely g and T (see [17]), and, what is more, there is always an infinite number of such divergence functions. In the context of classical information geometry, all statistical models share the "same" metric g, called the Fisher-Rao metric. This metric arise naturally when we consider $\mathcal{M}$ as immersed in the space $P(\chi)$ of probability distributions on the measure space χ, and, provided some additional requirements on symmetries are satisfied, it is essentially unique (see [3,9]). This means that, once the statistical manifold $\mathcal{M} \subset P(\chi)$ is chosen, all the admissible divergence functions must give back the Fisher-Rao metric g. On the other hand, different admissible divergence functions lead to different third order symmetric tensors T. Quite interestingly, the metric tensor g is no longer unique in the quantum context (see [20]).

In a recent work ([10]), a dynamical approach to divergence functions has been proposed. The main idea is to read a divergence function D, or more generally, a potential function for a given statistical model $(\mathcal{M}, g, T)$, as the Hamilton principal function associated with a suitably defined Lagrangian function $\mathfrak{L}$ on $T\mathcal{M}$ by means of the Hamilton-Jacobi theory (see [7,11]). From this point of view, a divergence function D becomes a dynamical object, that is, the function D is no more thought of as some fixed kinematical function on the double of the manifold of the statistical model, but, rather, it becomes the Hamilton principal function associated with a Lagrangian dynamical system on the tangent bundle of the manifold of the statistical model. In the variational formulation of dynamics [11], the solutions of the equations of motion are expressed as the critical points of the action functional:

$$I(\gamma) = \int_{t_{\text{in}}}^{t_{\text{fin}}} \mathfrak{L}(\gamma, \dot{\gamma})\, \mathrm{d}t\,, \tag{1}$$

where γ are curves on $\mathcal{M}$ with fixed extreme points $m(t_{\text{in}}) = m_{\text{in}}$ and $m(t_{\text{fin}}) = m_{\text{fin}}$, and $\mathfrak{L}$ is the Lagrangian function of the system. In order to avoid technical details, we will always assume that $\mathfrak{L}$ is a regular Lagrangian (see [18]). The evaluation of the action functional on a critical point γ_c gives a two-point function[1]:

$$S(m_{\text{in}}, m_{\text{fin}}) = I(\gamma_c)\,, \tag{2}$$

which is known in the literature as the Hamilton principal function. When a given dynamics admits of alternative Lagrangian description, it is possible to integrate alternative Lagrangians along the same integral curves and get different potential functions. If the determinant of the matrix of the mixed partial derivatives of S is different from zero, then it is possible to prove (see [11]) that S is a complete

[1] In general, this function depends on the additional parameters t_{in} and t_{fin}, however we will always take $t_{\text{in}} = 0$ and $t_{\text{fin}} = 1$.

solution of the Hamilton-Jacobi equation for the dynamics:

$$H\left(x\,,\frac{\partial S}{\partial x}\,,t\right)+\frac{\partial S}{\partial t}=0\,, \tag{3}$$

where H is the Hamiltonian function ([8]) associated with the Lagrangian $\mathfrak{L}$. In this case, $S(m_{\text{in}}\,,m_{\text{fin}})$ is called a complete solution for the Hamilton-Jacobi theory. It turns out that the existence of a complete solution S forces the dynamical system associated with the Lagrangian function $\mathfrak{L}$ to be completely integrable, that is, to adimit $n=dim(\mathcal{M})$ functionally independent constants of the motion which are transversal to the fibre of $T\mathcal{M}$ (see [7]). The main result of [10] is to prove that, given any statistical model $(\mathcal{M}\,,g\,,T)$, the Lagrangian functions:

$$\mathfrak{L}_\alpha=\frac{1}{2}g_{jk}(x)v^j v^k+\frac{\alpha}{6}T_{jkl}(x)v^j v^k v^l\,, \tag{4}$$

labelled by the one-dimensional real parameter α, are such that their associated Hamilton principal functions are potential functions for $(\mathcal{M}\,,g\,,T)$ in the sense that they allow to recover g and T as follows:

$$\left.\frac{\partial^2 S_\alpha}{\partial x^j_{\text{fin}}\partial x^k_{\text{in}}}\right|_{x_{\text{in}}=x_{\text{fin}}}=-g_{jk}(x)\,, \tag{5}$$

$$\left.\frac{\partial^3 S_\alpha}{\partial x^l_{\text{in}}\partial x^k_{\text{in}}\partial x^j_{\text{fin}}}\right|_{x_{\text{in}}=x_{\text{fin}}}-\left.\frac{\partial^3 S_\alpha}{\partial x^l_{\text{fin}}\partial x^k_{\text{fin}}\partial x^j_{\text{in}}}\right|_{x_{\text{in}}=x_{\text{fin}}}=2\alpha T_{jkl}(x)\,. \tag{6}$$

The functions S_α are not in general fair divergence functions because they are not positive-definite. However, the analysis of [10] clearly shows that we may add terms of at least fourth order in the velocities to $\mathfrak{L}_\alpha$ and the resulting Hamilton principal function will be again a potential function for $(\mathcal{M}\,,g\,,T)$. Consequently, we could keep adding terms of higher order in the velocities so that the resulting potential function is actually a divergence function.

In this short contribution we want to formulate an inverse problem for the Hamilton-Jacobi theory focused on some relevant situations in information geometry. Specifically, we ask the following question: Given a fixed divergence function D on $\mathcal{M}\times\mathcal{M}$ generating the statistical model $(\mathcal{M}\,,g\,,T)$, is it possible to find a Lagrangian function $\mathfrak{L}$ on $T\mathcal{M}$ such that D is the Hamilton principal function S of $\mathfrak{L}$? If the answer is yes, then we can analyze the associated dynamical system and its physical interpretation in the context of the adopted model. In the following we will review a case in which the answer exist in full generality, namely, the case of of self-dual statistical manifolds ([3]). An interesting example of such a manifold is given by the space of pure states of quantum mechanics which will be briefly discussed. The possibility to extend this ideas to relevant cases going beyond self-dual statistical manifolds will be addressed in future works.

2 Hamilton-Jacobi, Information Geometry, and the Inverse Problem for Potential Functions

Self-dual statistical manifolds (see [3]) are statistical models for which the symmetric tensor T identically vanishes, so that the only connection available is the self-dual Levi-Civita connection ∇_g associated with the metric g, and a canonical contrast function D_d is given by:

$$D_d(m_{\text{in}}\,, m_{\text{fin}}) = \frac{1}{2}\, d^2(m_{\text{in}}\,, m_{\text{fin}})\,, \tag{7}$$

where $d^2(m_{\text{in}}\,, m_{\text{fin}})$ is the square of the Riemannian geodesic distance associated with the metric g on $\mathcal{M}$. In this particular case, it turns out (see [10]) that the family $\mathfrak{L}_\alpha$ of Lagrangian functions given in Eq. (4) provides a solution to the inverse problem. Indeed, when $T = 0$, the family of Lagrangian functions $\mathfrak{L}_\alpha$ collapses to a single Lagrangian which is the metric Lagrangian $\mathfrak{L}_g = \dfrac{1}{2} g_{jk} v^j v^k$. To prove that $\mathfrak{L}_g$ actually solves the inverse problem for S_d in the case of self-dual manifolds, let us recall that, if the manifold $\mathcal{M}$ is regular enough, the square of Riemannian geodesic distance $d^2(m_{\text{in}}\,, m_{\text{fin}})$ is given by:

$$d^2(m_{\text{in}}\,, m_{\text{fin}}) = \left(\int_0^1 \sqrt{g_{jk}(\gamma_g(t))\,\dot{\gamma_g}^j\,\dot{\gamma_g}^k}\,\mathrm{d}t\right)^2 = \left(\int_0^1 \sqrt{2\mathfrak{L}_g(\gamma_g, \dot{\gamma_g})}\,\mathrm{d}t\right)^2 \tag{8}$$

where γ_g is a geodesics for g with fixed endpoints m_{in} and m_{fin}, and where the square root is introduced in order to ensure the invariance of the distance function under reparametrizations of γ. Geodesics curves are precisely the projection of the integral curves of a vector field Γ on $T\mathcal{M}$ which is the dynamical vector field associated with the Lagrangian function $\mathfrak{L}_g = \frac{1}{2} g_{jk}\, v^j\, v^k$ by means of the Euler-Lagrange equations stemming from the variational principle for the action functional (1). Now, recall that the metric Lagrangian $\mathfrak{L}_g$, as well as all of its functions $F(\mathfrak{L}_g)$ with F analytic, give rise to the same dynamical trajectories ([18]) and are all constants of the motion for this dynamics. Consequently, we can take $\sqrt{2\mathfrak{L}_g(\gamma_g, \dot{\gamma_g})}$ out of the integral in Eq. (8) so that we are left with:

$$D_d(m_{\text{in}}\,, m_{\text{fin}}) = \frac{1}{2} d^2(m_{\text{in}}\,, m_{\text{fin}}) = \mathfrak{L}_g(\gamma_g, \dot{\gamma_g}) =$$

$$= \int_0^1 \mathfrak{L}_g(\gamma_g\,, \dot{\gamma_g})\,\mathrm{d}t = I(\gamma_g) = S(m_{\text{in}}\,, m_{\text{fin}})\,, \tag{9}$$

which means that the canonical divergence function of self-dual manifolds is actually the Hamilton principal function of the metric Lagrangian $\mathfrak{L}_g$ associated with the metric tensor g.

A relevant example of self-dual manifold is given by the space $\mathcal{P}(\mathcal{H})$ of pure states of a quantum system with Hilbert space $\mathcal{H}$. We are here considering probability amplitudes instead of probability distributions. For simplicity, we limit our case to the finite-dimensional case $\mathcal{H} \cong \mathbb{C}^n$. The metric g on $\mathcal{P}(\mathcal{H})$ is the so-called Fubini-Study metric (see [8]). Apart from a constant conformal factor, g is the unique metric on $\mathcal{P}(\mathcal{H})$ which is invariant under the canonical action of the unitary group $\mathcal{U}(n)$ on $\mathcal{P}(\mathcal{H})$. The manifold $\mathcal{P}(\mathcal{H})$ is a homogeneous space for the unitary group, specifically, it is $\mathcal{P}(\mathcal{H}) \cong \mathcal{U}(n)/\mathcal{U}_{\rho_\psi}$, where $\mathcal{U}_{\rho_\psi}$ is the istropy subgroup of the non-negative Hermitean matrix ρ_ψ associated with a pure state ψ with respect to the action $\rho_\psi \mapsto U^\dagger \rho_\psi U$ for which the space of pure states is a homogeneous space of the unitary group. Note that $(\mathcal{P}(\mathcal{H}), g)$ is a Riemannian homogeneous manifold. We may exploit the homogeneous space structure of $\mathcal{P}(\mathcal{H})$ in order to describe the Lagrangian function associated with the metric tensor g by means of a degenerate Lagrangian function on the tangent bundle of the unitary group. This is particularly useful since $\mathcal{U}(n)$ is a Lie group, hence it is parallelizable, and thus a pair of global dual basis $\{X_j\}$ and $\{\theta^j\}$ of, respectively, vector fields and one-forms are available. Let us consider then a fixed positive matrix ρ_ψ associated with a fiducial pure state ψ, and consider the following Lagrangian:

$$\mathfrak{L}(\mathrm{g}, \dot{\mathrm{g}}) = \frac{1}{2} Tr\left([\rho_\psi , \mathrm{g}^{-1}\dot{\mathrm{g}}]^2 \right) = \frac{1}{2} G_{jk} \dot{\theta}^j \dot{\theta}^k , \tag{10}$$

where G_{jk} is a constant matrix, and $\dot{\theta}^j$ is the velocity-like function defined on the tangent space of every Lie group (see [16]). It is clear that $\mathfrak{L}$ is invariant with respect to the tangent lift of the left action of $\mathcal{U}(n)$ on itself and with respect to the tangent lift of the right action of the isotropy subgroup $\mathcal{U}_{\rho_\psi}$. Consequently, $\mathfrak{L}$ is the pullback to $T\mathcal{U}(n)$ of a Lagrangian function $\mathcal{L}$ on $\mathcal{P}(\mathcal{H})$. In order to focus on the main stream of the paper, we will not enter into a full discussion for this dynamical system. We simply state that, using the theory of degenerate Lagrangians ([15]), it is possible to prove that $\mathcal{L}$ is the metric Lagrangian associated with the Fubini-Study metric, and that the dynamical trajectories of the vector field Γ associated with $\mathfrak{L}$ project down onto the geodesics of the Fubini-Study metric on the space of quantum pure states. Specifically, writing $\rho_0 = U_0^\dagger \rho_\psi U_0$, we have:

$$\gamma_{\rho_0, \mathbf{A}}(t) = \mathrm{e}^{-[\rho_\psi, \mathbf{A}]t} \, \rho_0 \, \mathrm{e}^{[\rho_\psi, \mathbf{A}]t}, \tag{11}$$

where $\mathbf{A}$ is a self-adjoint matrix. The dynamical vector field on $T\mathcal{P}(\mathcal{H})$ may be seen as a family of vector fields on $\mathcal{P}(\mathcal{H})$ labelled by the matrix parameter $\mathbf{A}$. Once we select a member of this family, that is we fix $\mathbf{A}$, we are left with a vector field on the space of pure quantum state generating the unitary evolution associated with the Hamiltonian operator $\mathbf{H} = -\imath[\rho_\psi , \mathbf{A}]$. These evolutions have a clear physical meaning, indeed, they represent the dynamical evolution of an isolated quantum system with energy operator $\mathbf{H}$. The Hamilton principal function for $\mathfrak{L}$ is the pullback of the Hamilton principal function associated with the Lagrangian function $\mathcal{L}$ on $T\mathcal{P}(\mathcal{H})$. Writing $\rho_1 = \gamma_{\rho_0, \mathbf{A}}(1)$ we have:

$$S(\rho_0 , \rho_1) = \frac{1}{2} Tr\left([\rho_\psi , [\rho_\psi , \mathbf{A}]]^2 \right) = Tr\left(\rho_\psi \, \mathbf{A} \, [\mathbf{A} , \rho_\psi] \right) . \tag{12}$$

For example, let us consider a two-level quantum system, for which the most general pure state is:

$$\rho = \frac{1}{2}\left(\mathbb{I} + x^j \sigma_j\right) , \tag{13}$$

where $\mathbb{I}$ is the identity matrix, the σ_j's are the Pauli matrices, and $\delta_{jk} x^j x^k = 1$. We take $\rho_\psi = \frac{1}{2}(\mathbb{I} + \sigma_3)$. In this case, the isotropy subgroup $\mathcal{U}(2)_{\rho_\psi}$ is equal to $\mathcal{U}(1) \times \mathcal{U}(1)$, and thus $\mathcal{P}(\mathcal{H})$ is a two-dimensional sphere embedded in the three-dimensional space $\mathbb{R}^3$. The tensor g reads:

$$g = G_{nk} \theta^n \otimes \theta^k = \theta^1 \otimes \theta^1 + \theta^2 \oplus \theta^2 . \tag{14}$$

A direct computation shows that the dynamical trajectories are:

$$\rho(t) = \cos(rt)\rho_0 + \frac{\sin(rt)}{r} [\rho_\psi \, , \mathbf{A}] , \tag{15}$$

where, $r^2 = (A^1)^2 + (A^2)^2$. From this it follows that:

$$[\rho \, , \mathbf{A}] = \frac{1}{\sqrt{1 - (\delta_{jk} \, x_0^j \, x_1^k)^2}} \arccos\left(\delta_{jk} \, x_0^j \, x_1^k\right) \, [\rho_0 \, , \rho_1] , \tag{16}$$

and thus:

$$S(\rho_0 \, , \rho_1) = \arccos^2\left(\delta_{jk} \, x_0^j \, x_1^k\right) . \tag{17}$$

Going back to probability distributions, let us recall a particular case in which the inverse problem formulated here has a positive solution (see [10]). Consider the following family of exponential distributions on $\mathbb{R}^+$ parametrized by $\xi \in \mathbb{R}^+ = \mathcal{M}$:

$$p(x \, , \xi) = \xi \, \mathrm{e}^{-x\xi} \qquad \xi, x > 0 . \tag{18}$$

The Kullback-Leibler divergence function for this model is:

$$D_{KL}(\xi_{\mathrm{in}} \, , \xi_{\mathrm{fin}}) = \int_0^{+\infty} p(x \, , \xi_{\mathrm{in}}) \ln\left(\frac{p(x \, , \xi_{\mathrm{in}})}{p(x \, , \xi_{\mathrm{fin}})}\right) \mathrm{d}x = \ln\left(\frac{\xi_{\mathrm{in}}}{\xi_{\mathrm{fin}}}\right) + \frac{\xi_{\mathrm{fin}}}{\xi_{\mathrm{in}}} - 1 . \tag{19}$$

A direct computation shows that D_{KL} is the Hamilton principal function associated with the Lagrangian function:

$$\mathfrak{L}_{KL}(\xi \, , v) = \mathrm{e}^{\frac{v}{\xi}} - \frac{v}{\xi} - 1 . \tag{20}$$

In this case, it happens that the dynamical system associated with $\mathfrak{L}_{KL}$ and the dynamical system associated with the metric Lagrangian $\mathfrak{L}_g$ of this statistical model are the same, that is, $\mathfrak{L}_{KL}$ and $\mathfrak{L}_g$ are non gauge-equivalent alternative Lagrangians (see [18]).

3 Conclusions

We have seen how the inverse problem for divergence functions in the context of Hamilton-Jacobi theory has a positive answer in the case of self-dual statistical manifolds. In this case, the canonical divergence function $D(m_1, m_2) = \frac{1}{2}d^2(m_1, m_2)$, where $d^2(m_1, m_2)$ is the Riemannian distance, is the Hamilton principal function associated with the metric Lagrangian $\mathfrak{L}_g$. In the case when $\mathcal{M}$ is the space of pure states of a finite-level quantum system, the metric g is the Fubini-study metric and we have seen how to describe the metric Lagrangian $\mathfrak{L}_g$ by means of a degenerate Lagrangian $\mathfrak{L}$ on the unitary group.

In general, both in classical and quantum information geometry, some well-known divergence functions are relative entropies (see [3–5,12,19]), hence, a positive answer to the inverse problem for such divergence functions brings in the possibility of defining dynamical systems associated with relative entropies, and, in accordance with the Hamilton-Jacobi theory, this points to the possibility of looking at relative entropies as generators of canonical transformations. A more thorough analysis of these situations will be presented in future works.

Finally, let us comment on the possible relation of this work with the recent developments in Souriau's Lie group thermodynamic. In this framework, a sort of Hessian metric, called Souriau-Fisher metric, g is defined on a manifold $\mathcal{M}$ by means of a function on $\mathcal{M}$, the so-called Koszul-Vinberg Characteristic function (see [6,13]). It is not possible to compare directly our procedure with the Koszul-Vinberg Characteristic function generating the same statistical structure since the latter is a function defined on $\mathcal{M}$ and not on $\mathcal{M} \times \mathcal{M}$. Moreover, one has to generalize the Hamilton-Jacobi approach along the lines explained in [14], Sect. 6. This generalization amounts to replace $\mathbb{R}$ of the extended formalism with a Lie Group (which could be the Galilei group or Poincarè group). The nontriviality of the second cohomology group for the Galilei group would require to work with suitable central extensions to apply the generalized theory. In Souriau's theory, the so-called Euler-Poincarè equations naturally appear. These equations are equivalent to the equations of motion of a Lagrangian system with symmetries, however, they are defined on the product of the configuration space with the Lie algebra of the group of symmetries of the system rather than on the tangent bundle of the configuration space. Furthermore, they may be derived starting from a variational principle just like Euler-Lagrange equations. Consequently, a possible relation between Hamilton principal function for the action (1) and the Koszul-Vinberg Characteristic function will be possible when the Hamilton-Jacobi theory is generalized to include a Lie group G instead of $\mathbb{R}$.

Acknowledgement. G.M. would like to acknowledge the partial support by the "Excellence Chair Program, Santander-UCIIIM"

References

1. Amari, S.I.: Information Geometry and its Application. Springer, Tokyo (2016)
2. Amari, S.I., Barndorff-Nielsen, O.E., Kass, R.E., Lauritzen, S.L., Rao, C.R.: Differential Geometry in Statistical Inference. Institute of Mathematical Statistics, Hayward (1987)
3. Amari, S.I., Nagaoka, H.: Methods of Information Geometry. American Mathematical Society, Providence (2000)
4. Balian, R.: The entropy-based quantum metric. Entropy **16**(7), 3878–3888 (2014)
5. Balian, R., Alhassid, Y., Reinhardt, H.: Dissipation in many-body systems: a geometric approach based on information theory. Phys. Rep. **131**(1–2), 1–146 (1986)
6. Barbaresco, F.: Geometric theory of heat from souriau lie groups thermodynamics and koszul hessian geometry: applications in information geometry for exponential families. Entropy **18**(11), 386–426 (2016)
7. Cariñena, J.F., Gràcia, X., Marmo, G., Martínez, E., Lecanda, M.C.M., Román-Roy, N.: Geometric hamilton-jacobi theory. Int. J. Geom. Methods Mod. Phys. **03**(07), 1417–1458 (2006)
8. Cariñena, J.F., Ibort, A., Marmo, G., Morandi, G.: Geometry from Dynamics, Classical and Quantum. Springer, Berlin (2015)
9. Cencov, N.N.: Statistical Decision Rules and Optimal Inference. American Mathematical Society, Providence (1982)
10. Ciaglia, F.M., Di Cosmo, F., Felice, D., Mancini, S., Marmo, G., Pérez-Pardo, J.M.: Hamilton-jacobi approach to potential functions in information geometry. J. Mathe. Phys. **58**(6) (2017)
11. Lanczos, C.: The Variational Principles of Mechanics. University of Toronto Press, Toronto (1952)
12. Man'ko, V.I., Marmo, G., Ventriglia, F., Vitale, P.: Metric on the space of quantum states from relative entropy. Tomographic reconstruction. J. Phys. A: Math. Theor. **50**, 302–335 (2016)
13. Marle, C.M.: From tools in symplectic and poisson geometry to J.-M. Souriau's theories of statistical mechanics and thermodynamics. Entropy **18**(10) (2016)
14. Marmo, G., Morandi, G., Mukunda, N.: A geometrical approach to the hamilton-jacobi form of dynamics and its generalizations. La Rivista del Nuovo Cimento **13**, 1–74 (1990)
15. Marmo, G., Mukunda, N., Samuel, J.: Dynamics and symmetry for constrained systems: a geometrical analysis. La Rivista del Nuovo Cimento **6**, 1–62 (1983)
16. Marmo, G., Rubano, C.: Particle Dynamics on Fiber Bundles. Bibliopolis (1988)
17. Matumoto, T.: Any statistical manifold has a contrast function: on the c^3-functions taking the minimum at the diagonal of the product manifold. Hiroshima Mathe. J. **23**(2), 327–332 (1993)
18. Morandi, G., Ferrario, C., Vecchio, G.L., Marmo, G., Rubano, C., Rubano, C.: The inverse problem in the calculus of variations and the geometry of the tangent bundle. Phys. Rept. **188**, 147–284 (1990)
19. Petz, D.: Quantum Information Theory and Quantum Statistics. Springer, Heidelberg (2007)
20. Petz, D., Sudár, C.: Monotone metrics on matrix spaces. Linear Algebra Appl. **244**, 81–96 (1996)
21. Wootters, W.K.: Statistical distance and hilbert space. Phys.Rev. D **23**(2), 357–362 (1981)

Divergence Geometry

Log-Determinant Divergences Between Positive Definite Hilbert-Schmidt Operators

Hà Quang Minh(✉)

Pattern Analysis and Computer Vision (PAVIS), Istituto Italiano di Tecnologia (IIT), Genova, Italy
minh.haquang@iit.it

Abstract. The current work generalizes the author's previous work on the infinite-dimensional Alpha Log-Determinant (Log-Det) divergences and Alpha-Beta Log-Det divergences, defined on the set of positive definite unitized trace class operators on a Hilbert space, to the entire Hilbert manifold of positive definite unitized Hilbert-Schmidt operators. This generalization is carried out via the introduction of the extended Hilbert-Carleman determinant for unitized Hilbert-Schmidt operators, in addition to the previously introduced extended Fredholm determinant for unitized trace class operators. The resulting parametrized family of Alpha-Beta Log-Det divergences is general and contains many divergences between positive definite unitized Hilbert-Schmidt operators as special cases, including the infinite-dimensional generalizations of the affine-invariant Riemannian distance and symmetric Stein divergence.

1 Introduction

The current work is a continuation and generalization of the author's previous work [7,9], which generalizes the finite-dimensional Log-Determinant divergences to the infinite-dimensional setting. We recall that for the convex cone $\mathrm{Sym}^{++}(n)$ of symmetric, positive definite (SPD) matrices of size $n \times n$, $n \in \mathbb{N}$, the Alpha-Beta Log-Determinant (Log-Det) divergence between $A, B \in \mathrm{Sym}^{++}(n)$ is a parametrized family of divergences defined by (see [3])

$$D^{(\alpha,\beta)}(A,B) = \frac{1}{\alpha\beta}\log\det\left[\frac{\alpha(AB^{-1})^{\beta} + \beta(AB^{-1})^{-\alpha}}{\alpha+\beta}\right], \alpha > 0, \beta > 0, \quad (1)$$

along with the limiting cases $(\alpha > 0, \beta = 0), (\alpha = 0, \beta > 0)$, and $(\alpha = 0, \beta = 0)$. This family contains many distance-like functions on $\mathrm{Sym}^{++}(n)$, including

1. The affine-invariant Riemannian distance d_{aiE} [1], corresponding to

$$D^{(0,0)}(A,B) = \frac{1}{2}d^2_{\mathrm{aiE}}(A,B) = \frac{1}{2}||\log(B^{-1/2}AB^{-1/2})||^2_F, \quad (2)$$

where $\log(A)$ denotes the principal logarithm of A and $|| \; ||_F$ denotes the Frobenius norm, with $||A||_F = \sqrt{\mathrm{tr}(A^*A)}$. This is the geodesic distance associated with the affine-invariant Riemannian metric [1,6,10,11].

F. Nielsen and F. Barbaresco (Eds.): GSI 2017, LNCS 10589, pp. 505–513, 2017.
https://doi.org/10.1007/978-3-319-68445-1_59

2. The Alpha Log-Det divergences [2], corresponding to $D^{(\alpha,1-\alpha)}(A,B)$, with

$$D^{(\alpha,1-\alpha)}(A,B) = \frac{1}{\alpha(1-\alpha)}\log\left[\frac{\det[\alpha A + (1-\alpha)B]}{\det(A)^{\alpha}\det(B)^{1-\alpha}}\right], 0 < \alpha < 1, \quad (3)$$

$$D^{(1,0)}(A,B) = \mathrm{tr}(A^{-1}B - I) - \log\det(A^{-1}B), \quad (4)$$

$$D^{(0,1)}(A,B) = \mathrm{tr}(B^{-1}A - I) - \log\det(B^{-1}A). \quad (5)$$

The case $\alpha = 1/2$ gives the symmetric Stein divergence (also called the Jensen-Bregman LogDet divergence), whose square root is a metric on $\mathrm{Sym}^{++}(n)$ [13], with $D^{(1/2,1/2)}(A,B) = 4d^2_{\mathrm{stein}}(A,B) = 4[\log\det(\frac{A+B}{2}) - \frac{1}{2}\log\det(AB)]$.

Previous work. In [9], we generalized the Alpha Log-Det divergences between SPD matrices [2] to the infinite-dimensional Alpha Log-Determinant divergences between positive definite unitized trace class operators on an infinite-dimensional Hilbert space. This is done via the introduction of the extended Fredholm determinant for unitized trace class operators, along with the corresponding generalization of the log-concavity of the determinant for SPD matrices to the infinite-dimensional setting. In [7], we present a formulation for the Alpha-Beta Log-Det divergences between positive definite unitized trace class operators, generalizing the Alpha-Beta Log-Det divergences between SPD matrices as defined by Eq. (1).

Contributions of this work. The current work is a continuation and generalization of [7,9]. In particular, we generalize the Alpha-Beta Log-Det divergences in [7] to the entire Hilbert manifold of positive definite unitized Hilbert-Schmidt operators on an infinite-dimensional Hilbert space. This is done by the introduction of the extended Hilbert-Carleman determinant for unitized Hilbert-Schmidt operators, in addition to the extended Fredholm determinant for unitized trace class operators employed in [7,9]. As in the finite-dimensional setting [3] and in [7,9], the resulting family of divergences is general and admits as special cases many metrics and distance-like functions between positive definite unitized Hilbert-Schmidt operators, including the infinite-dimensional affine-invariant Riemannian distance in [5]. *The proofs for all theorems stated in this paper, along with many other results, are given in the arXiv preprint* [8].

2 Positive Definite Unitized Trace Class and Hilbert-Schmidt Operators

Throughout the paper, we assume that $\mathcal{H}$ is a real separable Hilbert space, with $\dim(\mathcal{H}) = \infty$, unless stated otherwise. Let $\mathcal{L}(\mathcal{H})$ be the Banach space of bounded linear operators on $\mathcal{H}$. Let $\mathrm{Sym}^{++}(\mathcal{H}) \subset \mathcal{L}(\mathcal{H})$ be the set of bounded, self-adjoint, *strictly positive* operators on $\mathcal{H}$, that is $A \in \mathrm{Sym}^{++}(\mathcal{H}) \Longleftrightarrow \langle x, Ax\rangle > 0$ $\forall x \in \mathcal{H}, x \neq 0$. Most importantly, we consider the set $\mathbb{P}(\mathcal{H}) \subset \mathrm{Sym}^{++}(\mathcal{H})$ of self-adjoint, bounded, *positive definite* operators on $\mathcal{H}$, which is defined by

$$A \in \mathbb{P}(\mathcal{H}) \Longleftrightarrow A = A^*, \exists M_A > 0 \text{ such that } \langle x, Ax\rangle \geq M_A||x||^2 \quad \forall x \in \mathcal{H}.$$

We use the notation $A > 0 \Longleftrightarrow A \in \mathbb{P}(\mathcal{H})$. In the following, let $\mathscr{C}_p(\mathcal{H})$ denote the set of pth Schatten class operators on $\mathcal{H}$ (see e.g. [4]), under the norm $|| \ ||_p$, $1 \leq p \leq \infty$, which is defined by

$$\mathscr{C}_p(\mathcal{H}) = \{A \in \mathcal{L}(\mathcal{H}) \ : \ ||A||_p = (\text{tr}|A|^p)^{1/p} < \infty\}, \tag{6}$$

where $|A| = (A^*A)^{1/2}$. The cases we consider in this work are: (i) the space $\mathscr{C}_1(\mathcal{H})$ of trace class operators on $\mathcal{H}$, also denoted by $\text{Tr}(\mathcal{H})$, and (ii) the space $\mathscr{C}_2(\mathcal{H})$ of Hilbert-Schmidt operators on $\mathcal{H}$, also denoted by $\text{HS}(\mathcal{H})$.

Extended (unitized) Trace Class Operators. In [9], we define the set of extended (or unitized) trace class operators on $\mathcal{H}$ to be

$$\text{Tr}_X(\mathcal{H}) = \{A + \gamma I \ : \ A \in \text{Tr}(\mathcal{H}), \gamma \in \mathbb{R}\}. \tag{7}$$

The set $\text{Tr}_X(\mathcal{H})$ becomes a Banach algebra under the *extended trace class norm*

$$||A + \gamma I||_{\text{tr}_X} = ||A||_{\text{tr}} + |\gamma| = \text{tr}|A| + |\gamma|.$$

For $(A + \gamma I) \in \text{Tr}_X(\mathcal{H})$, its *extended trace* is defined to be

$$\text{tr}_X(A + \gamma I) = \text{tr}(A) + \gamma, \quad \text{with} \quad \text{tr}_X(I) = 1. \tag{8}$$

Extended (unitized) Hilbert-Schmidt Operators. In [5], the author considered the following set of extended (unitized) Hilbert-Schmidt operators

$$\text{HS}_X(\mathcal{H}) = \{A + \gamma I \ : A \in \text{HS}(\mathcal{H}), \gamma \in \mathbb{R}\}. \tag{9}$$

The set $\text{HS}_X(\mathcal{H})$ can be equipped with the *extended Hilbert-Schmidt inner product* $\langle \ , \ \rangle_{\text{eHS}}$, defined by

$$\langle A + \gamma I, B + \mu I\rangle_{\text{eHS}} = \langle A, B\rangle_{\text{HS}} + \gamma\mu = \text{tr}(A^*B) + \gamma\mu. \tag{10}$$

along with the associated *extended Hilbert-Schmidt norm*

$$||A + \gamma I||^2_{\text{eHS}} = ||A||^2_{\text{HS}} + \gamma^2 = \text{tr}(A^*A) + \gamma^2, \text{ with } ||I||_{\text{eHS}} = 1. \tag{11}$$

Positive Definite Unitized Trace Class and Hilbert-Schmidt Operators. The set of positive definite unitized trace class operators $\mathscr{PC}_1(\mathcal{H}) \subset \text{Tr}_X(\mathcal{H})$ is defined to be the intersection

$$\mathscr{PC}_1(\mathcal{H}) = \text{Tr}_X(\mathcal{H}) \cap \mathbb{P}(\mathcal{H}) = \{A + \gamma I > 0 : A^* = A, A \in \text{Tr}(\mathcal{H}), \gamma \in \mathbb{R}\}. \tag{12}$$

The set of positive definite unitized Hilbert-Schmidt operators $\mathscr{PC}_2(\mathcal{H}) \subset \text{HS}_X(\mathcal{H})$ is defined to be the intersection

$$\mathscr{PC}_2(\mathcal{H}) = \text{HS}_X(\mathcal{H}) \cap \mathbb{P}(\mathcal{H}) = \{A + \gamma I > 0 : A = A^*, A \in \text{HS}(\mathcal{H}), \gamma \in \mathbb{R}\}. \tag{13}$$

We remark that in [7,9], we use the notations $\text{PTr}(\mathcal{H})$ and $\Sigma(\mathcal{H})$ to denote $\mathscr{PC}_1(\mathcal{H})$ and $\mathscr{PC}_2(\mathcal{H})$, respectively. In the following, we refer to elements of

$\mathscr{PC}_1(\mathcal{H})$ and $\mathscr{PC}_2(\mathcal{H})$ as *positive definite trace class operators* and *positive definite Hilbert-Schmidt operators*, respectively.

In [5], it is shown that the set $\mathscr{PC}_2(\mathcal{H})$ assumes the structure of an infinite-dimensional Hilbert manifold and can be equipped with the following Riemannian metric. For each $P \in \mathscr{PC}_2(\mathcal{H})$, on the tangent space $T_P(\mathscr{PC}_2(\mathcal{H})) \cong \mathcal{H}_{\mathbb{R}} = \{A + \gamma I : A = A^*, A \in \mathrm{HS}(\mathcal{H}), \gamma \in \mathbb{R}\}$, we define the following inner product

$$\langle A + \gamma I, B + \mu I\rangle_P = \langle P^{-1/2}(A + \gamma I)P^{-1/2}, P^{-1/2}(B + \mu I)P^{-1/2}\rangle_{\mathrm{eHS}}.$$

The Riemannian metric given by $\langle \ , \ \rangle_P$ then makes $\mathscr{PC}_2(\mathcal{H})$ an infinite-dimensional Riemannian manifold. Under this Riemannian metric, the geodesic distance between $(A + \gamma I), (B + \mu I) \in \mathscr{PC}_2(\mathcal{H})$ is given by

$$d_{\mathrm{aiHS}}[(A + \gamma I), (B + \mu I)] = || \log[(B + \mu I)^{-1/2}(A + \gamma I)(B + \mu I)^{-1/2}]||_{\mathrm{eHS}}. \quad (14)$$

Aim of this work. In [9], we introduce a parametrized family of divergences, called *Log-Determinant divergences*, between operators in $\mathscr{PC}_1(\mathcal{H})$. In [7], we generalize these to the *Alpha-Beta Log-Determinant divergences* on $\mathscr{PC}_1(\mathcal{H})$, which include the distance d_{aiHS} as a special case. However, these divergences are defined specifically on $\mathscr{PC}_1(\mathcal{H})$, which is a *strict subset* of the set of positive definite Hilbert-Schmidt operators $\mathscr{PC}_2(\mathcal{H})$ when $\dim(\mathcal{H}) = \infty$. In this work, we generalize the divergences in [7,9] to all of $\mathscr{PC}_2(\mathcal{H})$.

3 The Extended Hilbert-Carleman Determinant

We recall that for $A \in \mathrm{Tr}(\mathcal{H})$, the Fredholm determinant is (see e.g. [12])

$$\det(I + A) = \prod_{k=1}^{\infty}(1 + \lambda_k), \quad (15)$$

where $\{\lambda_k\}_{k=1}^{\infty}$ are the eigenvalues of A. To define Log-Determinant divergences between positive definite trace class operators in $\mathscr{PC}_1(\mathcal{H})$, in [9], we generalize the Fredholm determinant to the *extended Fredholm determinant*, which, for an extended trace class operator $(A + \gamma I) \in \mathrm{Tr}_X(\mathcal{H})$, $\gamma \neq 0$, is defined to be

$$\det{}_X(A + \gamma I) = \gamma \det\left(\frac{A}{\gamma} + I\right), \quad \text{when} \quad \dim(\mathcal{H}) = \infty,$$

(we refer to [9] for the derivation leading to this definition). In the case $\dim(\mathcal{H}) < \infty$, we define $\det_X(A + \gamma I) = \det(A + \gamma I)$, the standard matrix determinant.

The extended Fredholm determinant is not sufficient for dealing with positive definite Hilbert-Schmidt operators in $\mathscr{PC}_2(\mathcal{H})$. In order to do so, we introduce the concept of *extended Hilbert-Carleman determinant*. We first recall the Hilbert-Carleman determinant for operators of the form $I + A$, where A is a Hilbert-Schmidt operator. Following [12], $A \in \mathcal{L}(\mathcal{H})$, consider the operator

$$R_n(A) = \left[(I + A)\exp\left(\sum_{k=1}^{n-1}\frac{(-A)^k}{k}\right)\right] - I. \quad (16)$$

If $A \in \mathscr{C}_n(\mathcal{H})$, then $R_n(A) \in \mathscr{C}_1(\mathcal{H})$. Thus the following quantity is well-defined

$$\det_{\mathrm{n}}(I + A) = \det(I + R_n(A)). \quad (17)$$

In particular, for $n = 1$, we obtain $R_1(A) = A$ and thus $\det_1(I+A) = \det(I+A)$. For $n = 2$, we have $R_2(A) = (I + A)\exp(-A) - I$ and thus

$$\det_2(I + A) = \det[(I + A)\exp(-A)]. \quad (18)$$

This is the *Hilbert-Carleman determinant* of $I + A$. For $A \in \mathrm{Tr}(\mathcal{H}) = \mathscr{C}_1(\mathcal{H})$,

$$\det_2(I + A) = \det(I + A)\exp(-\mathrm{tr}(A)). \quad (19)$$

The Hilbert-Carleman determinant $\det_2$ is defined for operators of the form $A + I$, $A \in \mathrm{HS}(\mathcal{H})$, but not for operators of the form $A + \gamma I$, $\gamma > 0, \gamma \neq 1$. In the following, we generalize $\det_2$ to handle these operators. We first have the following generalization of the function $R_2(A) = (I + A)\exp(-A) - I$ above.

Lemma 1. *Assume that* $(A + \gamma I) \in \mathrm{HS}_X(\mathcal{H})$, $\gamma \neq 0$. *Define*

$$R_{2,\gamma}(A) = (A + \gamma I)\exp(-A/\gamma) - \gamma I. \quad (20)$$

Then $R_{2,\gamma}(A) \in \mathrm{Tr}(\mathcal{H})$ *and hence* $R_{2,\gamma}(A)+\gamma I = (A+\gamma I)\exp(-A/\gamma) \in \mathrm{Tr}_X(\mathcal{H})$.

In particular, for $\gamma = 1$, we have $R_{2,1}(A) = R_2(A)$. Motivated by Lemma 1 and the definition of $\det_2$, we arrive at the following generalization of $\det_2$.

Definition 1 (Extended Hilbert-Carleman determinant). *For* $(A+\gamma I) \in \mathrm{HS}_X(\mathcal{H})$, $\gamma \neq 0$, *its extended Hilbert-Carleman determinant is defined to be*

$$\det_{2\mathrm{X}}(A + \gamma I) = \det_{\mathrm{X}}[R_{2,\gamma}(A) + \gamma I] = \det_{\mathrm{X}}[(A + \gamma I)\exp(-A/\gamma)]. \quad (21)$$

If $\gamma = 1$, then we recover the Hilbert-Carleman determinant

$$\det_{2\mathrm{X}}(A + I) = \det[(A + I)\exp(-A)] = \det_2(A + I). \quad (22)$$

If $(A + \gamma I) \in \mathrm{Tr}_X(\mathcal{H})$, then

$$\det_{2\mathrm{X}}(A + \gamma I) = \det_{\mathrm{X}}(A + \gamma I)\exp(-\mathrm{tr}(A)/\gamma). \quad (23)$$

4 Infinite-Dimensional Log-Determinant Divergences Between Positive Definite Hilbert-Schmidt Operators

The following divergence definition, when $\alpha > 0, \beta > 0$, first stated in [7], for $(A+\gamma I), (B+\mu I) \in \mathscr{PC}_1(\mathcal{H})$, is in fact valid for all $(A+\gamma I), (B+\mu I) \in \mathscr{PC}_2(\mathcal{H})$.

Definition 2 (Alpha-Beta Log-Determinant divergences between positive definite Hilbert-Schmidt operators). *Assume that* $\dim(\mathcal{H}) = \infty$. *Let* $\alpha > 0$, $\beta > 0, r \neq 0$ *be fixed. For* $(A+\gamma I), (B+\mu I) \in \mathscr{PC}_2(\mathcal{H})$, *the* (α, β)*-Log-Det divergence* $D_r^{(\alpha,\beta)}[(A+\gamma I), (B+\mu I)]$ *is defined to be*

$$\begin{aligned} & D_r^{(\alpha,\beta)}[(A+\gamma I), (B+\mu I)] \\ &= \frac{1}{\alpha\beta} \log \left[\left(\frac{\gamma}{\mu}\right)^{r(\delta - \frac{\alpha}{\alpha+\beta})} \det{}_X \left(\frac{\alpha(\Lambda + \frac{\gamma}{\mu} I)^{r(1-\delta)} + \beta(\Lambda + \frac{\gamma}{\mu} I)^{-r\delta}}{\alpha+\beta} \right) \right], \end{aligned} \tag{24}$$

where $\Lambda + \frac{\gamma}{\mu} I = (B+\mu I)^{-1/2}(A+\gamma I)(B+\mu I)^{-1/2}$, $\delta = \frac{\alpha\gamma^r}{\alpha\gamma^r + \beta\mu^r}$. *Equivalently,*

$$\begin{aligned} & D_r^{(\alpha,\beta)}[(A+\gamma I), (B+\mu I)] \\ &= \frac{1}{\alpha\beta} \log \left[\left(\frac{\gamma}{\mu}\right)^{r(\delta - \frac{\alpha}{\alpha+\beta})} \det{}_X \left(\frac{\alpha(Z + \frac{\gamma}{\mu} I)^{r(1-\delta)} + \beta(Z + \frac{\gamma}{\mu} I)^{-r\delta}}{\alpha+\beta} \right) \right], \end{aligned} \tag{25}$$

where $Z + \frac{\gamma}{\mu} I = (A+\gamma I)(B+\mu I)^{-1}$.

In Definition 2, the quantity $D_r^{(\alpha,\beta)}[(A+\gamma I), (B+\mu I)]$ is finite $\forall$ $(A+\gamma I), (B+\mu I) \in \mathscr{PC}_2(\mathcal{H})$ by Propositions 2 and 3 in [8]. For the motivation of the factor $(\frac{\gamma}{\mu})^{r(\delta - \frac{\alpha}{\alpha+\beta})}$ in Eqs. (24) and (25), see Theorem 1 in [9].

Finite-Dimensional Case. For $\gamma = \mu$, we have

$$\begin{aligned} & D_r^{(\alpha,\beta)}[(A+\gamma I), (B+\gamma I)] \\ &= \frac{1}{\alpha\beta} \log \det{}_X \left(\frac{\alpha((A+\gamma I)(B+\gamma I)^{-1})^{\frac{r\beta}{\alpha+\beta}} + \beta((A+\gamma I)(B+\gamma I)^{-1})^{-\frac{r\alpha}{\alpha+\beta}}}{\alpha+\beta} \right). \end{aligned}$$

For $A, B \in \mathrm{Sym}^{++}(n)$, we recover Eq. (1) by setting $r = \alpha + \beta$ and $\gamma = 0$.

Limiting Cases. While Definition 2 is stated using the extended Fredholm determinant $\det_X$, the limiting cases $(\alpha > 0, \beta = 0)$ and $(\alpha = 0, \beta > 0)$ both require the concept of the extended Hilbert-Carleman determinant $\det_{2X}$.

Theorem 1 (Limiting case $\alpha > 0, \beta \to 0$). *Let* $\alpha > 0$ *be fixed. Assume that* $r = r(\beta)$ *is smooth, with* $r(0) = r(\beta = 0)$. *Then*

$$\begin{aligned} \lim_{\beta \to 0} D_r^{(\alpha,\beta)}[(A+\gamma I), (B+\mu I)] &= \frac{1}{\alpha^2} \left[\left(\frac{\mu}{\gamma}\right)^{r(0)} - 1 \right] \left(1 + r(0) \log \frac{\mu}{\gamma} \right) \\ &- \frac{1}{\alpha^2} \left(\frac{\mu}{\gamma}\right)^{r(0)} \log \det{}_{2X}([(A+\gamma I)^{-1}(B+\mu I)]^{r(0)}). \end{aligned} \tag{26}$$

Theorem 2 (Limiting case $\alpha \to 0, \beta > 0$). *Let $\beta > 0$ be fixed. Assume that $r = r(\alpha)$ is smooth, with $r(0) = r(\alpha = 0)$. Then*

$$\lim_{\alpha \to 0} D_r^{(\alpha,\beta)}[(A+\gamma I),(B+\mu I)] = \frac{1}{\beta^2}\left[\left(\frac{\gamma}{\mu}\right)^{r(0)} - 1\right]\left(1 + r(0)\log\frac{\gamma}{\mu}\right) \quad (27)$$
$$- \frac{1}{\beta^2}\left(\frac{\gamma}{\mu}\right)^{r(0)} \log\det\nolimits_{2X}([(B+\mu I)^{-1}(A+\gamma I)]^{r(0)}).$$

Motivated by Theorems 1 and 2, the following is our definition of $D_r^{(\alpha,0)}[(A+\gamma I),(B+\mu I)]$ and $D_r^{(0,\beta)}[(A+\gamma I),(B+\mu I)]$, $\alpha > 0, \beta > 0$.

Definition 3 (Limiting cases). *Let $\alpha > 0, \beta > 0, r \neq 0$ be fixed. For $(A+\gamma I),(B+\mu I) \in \mathscr{PC}_2(\mathcal{H})$, $D_r^{(\alpha,0)}[(A+\gamma I),(B+\mu I)]$ is defined to be*

$$D_r^{(\alpha,0)}[(A+\gamma I),(B+\mu I)] = \frac{1}{\alpha^2}[(\frac{\mu}{\gamma})^r - 1](1 + r\log\frac{\mu}{\gamma}) \quad (28)$$
$$- \frac{1}{\alpha^2}(\frac{\mu}{\gamma})^r \log\det\nolimits_{2X}([(A+\gamma I)^{-1}(B+\mu I)]^r).$$

Similarly, the divergence $D_r^{(0,\beta)}[(A+\gamma I),(B+\mu I)]$ is defined to be

$$D_r^{(0,\beta)}[(A+\gamma I),(B+\mu I)] = \frac{1}{\beta^2}[(\frac{\gamma}{\mu})^r - 1](1 + r\log\frac{\gamma}{\mu}) \quad (29)$$
$$- \frac{1}{\beta^2}(\frac{\gamma}{\mu})^r \log\det\nolimits_{2X}([(B+\mu I)^{-1}(A+\gamma I)]^r).$$

The following shows that the square affine-invariant Riemannian distance d^2_{aiHS}, as given in Eq. (14), corresponds to the limiting case $(\alpha = 0, \beta = 0)$.

Theorem 3 (Limiting case $(0,0)$). *Assume that $(A+\gamma I),(B+\mu I) \in \mathscr{PC}_2(\mathcal{H})$. Assume that $r = r(\alpha)$ is smooth, with $r(0) = 0$, $r'(0) \neq 0$, and $r(\alpha) \neq 0$ for $\alpha \neq 0$. Then*

$$\lim_{\alpha \to 0} D_r^{(\alpha,\alpha)}[(A+\gamma I),(B+\mu I)] = \frac{[r'(0)]^2}{8} d^2_{\mathrm{aiHS}}[(A+\gamma I),(B+\mu I)]. \quad (30)$$

In particular, for $r = 2\alpha$,

$$\lim_{\alpha \to 0} D_{2\alpha}^{(\alpha,\alpha)}[(A+\gamma I),(B+\mu I)] = \frac{1}{2} d^2_{\mathrm{aiHS}}[(A+\gamma I),(B+\mu I)]. \quad (31)$$

The divergence $D_0^{(0,0)} = \lim_{\alpha \to 0} D_{2\alpha}^{(\alpha,\alpha)}$ is a member of a parametrized family of symmetric divergences on $\mathscr{PC}_2(\mathcal{H})$, as shown by the following.

Theorem 4 (Symmetric divergences). *The parametrized family $D_{2\alpha}^{(\alpha,\alpha)}[(A+\gamma I),(B+\mu I)]$, $\alpha \geq 0$, is a family of symmetric divergences on $\mathscr{PC}_2(\mathcal{H})$, with $\alpha = 0$ corresponding to the infinite-dimensional affine-invariant Riemannian distance above and $\alpha = 1/2$ corresponding to the infinite-dimensional symmetric Stein divergence, which is given by $\frac{1}{4}D_1^{(1/2,1/2)}[(A+\gamma I),(B+\mu I)]$.*

5 Properties of the Log-Determinant Divergences

Assume in the following that $(A+\gamma I), (B+\mu I) \in \mathscr{PC}_2(\mathcal{H})$.

Theorem 5 (Positivity).

$$D_r^{(\alpha,\beta)}[(A+\gamma I),(B+\mu I)] \geq 0, \tag{32}$$

$$D_r^{(\alpha,\beta)}[(A+\gamma I),(B+\mu I)] = 0 \Longleftrightarrow A = B, \gamma = \mu. \tag{33}$$

Theorem 6 (Dual symmetry).

$$D_r^{(\beta,\alpha)}[(B+\mu I),(A+\gamma I)] = D_r^{(\alpha,\beta)}[(A+\gamma I),(B+\mu I)]. \tag{34}$$

In particular, for $\beta = \alpha$, we have

$$D_r^{(\alpha,\alpha)}[(B+\mu I),(A+\gamma I)] = D_r^{(\alpha,\alpha)}[(A+\gamma I),(B+\mu I)]. \tag{35}$$

Theorem 7 (Dual invariance under inversion).

$$D_r^{(\alpha,\beta)}[(A+\gamma I)^{-1},(B+\mu I)^{-1}] = D_{-r}^{(\alpha,\beta)}[(A+\gamma I),(B+\mu I)] \tag{36}$$

Theorem 8 (Affine invariance). *For any $(A+\gamma I), (B+\mu I) \in \mathscr{PC}_2(\mathcal{H})$ and any invertible $(C+\nu I) \in \mathrm{HS}_X(\mathcal{H})$, $\nu \neq 0$,*

$$D_r^{(\alpha,\beta)}[(C+\nu I)(A+\gamma I)(C+\nu I)^*,(C+\nu I)(B+\mu I)(C+\nu I)^*]$$
$$= D_r^{(\alpha,\beta)}[(A+\gamma I),(B+\mu I)]. \tag{37}$$

Theorem 9 (Invariance under unitary transformations). *For any $(A+\gamma I), (B+\mu I) \in \mathscr{PC}_2(\mathcal{H})$ and any $C \in \mathcal{L}(\mathcal{H})$, with $CC^* = C^*C = I$,*

$$D_r^{(\alpha,\beta)}[C(A+\gamma I)C^*,C(B+\mu I)C^*] = D_r^{(\alpha,\beta)}[(A+\gamma I),(B+\mu I)]. \tag{38}$$

References

1. Bhatia, R.: Positive Definite Matrices. Princeton University Press, Princeton (2007)
2. Chebbi, Z., Moakher, M.: Means of Hermitian positive-definite matrices based on the log-determinant α-divergence function. Linear Algebra Appl. **436**(7), 1872–1889 (2012)
3. Cichocki, A., Cruces, S., Amari, S.: Log-Determinant divergences revisited: Alpha-Beta and Gamma Log-Det divergences. Entropy **17**(5), 2988–3034 (2015)
4. Dunford, N., Schwartz, J.T.: Linear Operators, Part 2: Spectral Theory, Self Adjoint Operators in Hilbert Space. Wiley, New York (1988)
5. Larotonda, G.: Nonpositive curvature: a geometrical approach to Hilbert-Schmidt operators. Differ. Geom. Appl. **25**, 679–700 (2007)
6. Lawson, J.D., Lim, Y.: The geometric mean, matrices, metrics, and more. Am. Math. Monthly **108**(9), 797–812 (2001)

7. Minh, H.Q.: Infinite-dimensional Log-Determinant divergences II: Alpha-Beta divergences. arXiv preprint arXiv:1610.08087v2 (2016)
8. Minh, H.Q.: Infinite-dimensional Log-Determinant divergences between positive definite Hilbert-Schmidt operators. arXiv preprint arXiv:1702.03425 (2017)
9. Minh, H.Q.: Infinite-dimensional Log-Determinant divergences between positive definite trace class operators. Linear Algebra Appl. **528**, 331–383 (2017)
10. Mostow, G.D.: Some new decomposition theorems for semi-simple groups. Mem. Am. Math. Soc. **14**, 31–54 (1955)
11. Pennec, X., Fillard, P., Ayache, N.: A Riemannian framework for tensor computing. Int. J. Comput. Vision **66**(1), 41–66 (2006)
12. Simon, B.: Notes on infinite determinants of Hilbert space operators. Adv. Math. **24**, 244–273 (1977)
13. Sra, S.: A new metric on the manifold of kernel matrices with application to matrix geometric means. In: Advances in Neural Information Processing Systems (NIPS), pp. 144–152 (2012)

Some New Flexibilizations of Bregman Divergences and Their Asymptotics

Wolfgang Stummer[1,2] and Anna-Lena Kißlinger[3(✉)]

[1] Department of Mathematics, University of Erlangen–Nürnberg, Cauerstrasse 11, 91058 Erlangen, Germany
[2] Affiliated Faculty Member of the School of Business and Economics, University of Erlangen–Nürnberg, Lange Gasse 20, 90403 Nürnberg, Germany
[3] Chair of Statistics and Econometrics, University of Erlangen–Nürnberg, Lange Gasse 20, 90403 Nürnberg, Germany
anna-lena.kisslinger@fau.de

Abstract. *Ordinary* Bregman divergences (distances) OBD are widely used in statistics, machine learning, and information theory (see e.g. [5,18]; [4,6,7,14–16,22,23,25]). They can be flexibilized in various different ways. For instance, there are the *Scaled* Bregman divergences SBD of Stummer [20] and Stummer and Vajda [21] which contain both the OBDs as well the Csiszar-Ali-Silvey $\phi-$divergences as special cases. On the other hand, the OBDs are subsumed by the *Total* Bregman divergences of Liu et al. [12,13], Vemuri et al. [24] and the more general Conformal Divergences COD of Nock et al. [17]. The latter authors also indicated the possibility to combine the concepts of SBD and COD, under the name "Conformal Scaled Bregman divergences" CSBD. In this paper, we introduce some new divergences between (non-)probability distributions which particularly cover the corresponding OBD, SBD, COD and CSBD (for separable situations) as special cases. Non-convex generators are employed, too. Moreover, for the case of i.i.d. sampling we derive the asymptotics of a useful new-divergence-based test statistics.

Keywords: Bregman divergences (distances) · Total Bregman divergences · Conformal divergences · Asymptotics of goodness-of-fit divergence

1 Introduction and Results

Let us assume that the modeled respectively observed random data take values in a state space $\mathscr{X}$ (with at least two distinct values), equipped with a system $\mathscr{A}$ of admissible events ($\sigma-$algebra). On this, we want to quantify the divergence (distance, dissimilarity, proximity) $D(P, Q)$ between two probability distributions P, Q[1]. Since the ultimate purposes of a (divergence-based)

[1] our concept can be analogously worked out for non-probability distributions (non-negative measures) P, Q.

F. Nielsen and F. Barbaresco (Eds.): GSI 2017, LNCS 10589, pp. 514–522, 2017.
https://doi.org/10.1007/978-3-319-68445-1_60

statistical inference or machine learning task may vary from case to case, it is of fundamental importance to have at hand a flexible, far-reaching toolbox $\mathscr{D} := \{D_{\phi,M_1,M_2,M_3}(P,Q) : \phi \in \Phi, M_1, M_2, M_3 \in \mathscr{M}\}$ of divergences which allows for goal-oriented situation-based applicability; in the following, we present such a new toolbox, where the flexibility is controlled by various different choices of a "generator" $\phi \in \Phi$, and scalings $M_1, M_2, M_3 \in \mathscr{M}$. In order to achieve this goal, we use the following ingredients: (i) for the class $\mathscr{F}$ of all (measurable) functions from $\mathscr{Y} = (0,\infty)$ to $\overline{\mathbb{R}} := \mathbb{R} \cup \{\infty\} \cup \{-\infty\}$ and for fixed subclass $\mathscr{U} \subset \mathscr{F}$, the divergence-generator family $\Phi = \Phi_{\mathscr{U}}$ is supposed to consist of all functions $\phi \in \mathscr{F}$ which are $\mathscr{U}-$convex and for which the strict $\mathscr{U}-$subdifferential $\partial_{\mathscr{U}}\phi|_{y_0}$ is non-empty for all $y_0 \in \mathscr{Y}$. Typically, the family $\mathscr{U}$ contains ("approximating") functions which are "less complicated" than ϕ. Recall that (see e.g. [19]) a function $u : \mathscr{Y} \mapsto \overline{\mathbb{R}}$ is called a strict $\mathscr{U}-$subgradient of ϕ at a point $y_0 \in \mathscr{Y}$, if $u \in \mathscr{U}$ and $\phi(y) - \phi(y_0) \geq u(y) - u(y_0)$ for all $y \in \mathscr{Y}$ and the last inequality is strict (i.e., $>$) for all $y \neq y_0$; the set of all strict $\mathscr{U}-$subgradients of ϕ at a point $y_0 \in \mathscr{Y}$ is called strict $\mathscr{U}-$subdifferential of ϕ at $y_0 \in \mathscr{Y}$, and is denoted by $\partial_{\mathscr{U}}\phi|_{y_0}$. In case of $\partial_{\mathscr{U}}\phi|_{y_0} \neq \emptyset$ for all $y_0 \in \mathscr{Y}$, a function ϕ is characterized to be $\mathscr{U}-$convex if

$$\phi(y) = \max\{u(y) + c : u \in \mathscr{U}, c \in \mathbb{R}, u(z) + c \leq \phi(z) \text{ for all } z \in \mathscr{Y}\} \text{ for all } y \in \mathscr{Y}, \quad (1)$$

and if furthermore the class $\mathscr{U}$ is invariant under addition of constants (i.e. if $\mathscr{U} + const := \{u + c : u \in \mathscr{U}, c \in \mathbb{R}\} = \mathscr{U}$), then (1) can be further simplified to

$$\phi(y) = \max\{u(y) : u \in \mathscr{U} \text{ and } u(z) \leq \phi(z) \text{ for all } z \in \mathscr{Y}\} \text{ for all } y \in \mathscr{Y}$$

("curved lower envelope" at y). The most prominent special case is the class $\mathscr{U} = \mathscr{U}_{al}$ of all affine-linear functions for which the divergence-generator family $\Phi = \Phi_{\mathscr{U}_{al}}$ is the class of all "usual" strictly convex lower semicontinuous functions on $(0,\infty)$. (ii) As a second group of ingredients, the two probability distributions P, Q are supposed to be described by their probability densities $x \mapsto p(x) \geq 0$, $x \mapsto q(x) \geq 0$ via $P[A] = \int_A p(x)\, \mathrm{d}\lambda(x)$, $Q[A] = \int_A q(x)\, \mathrm{d}\lambda(x)$ $(A \in \mathscr{A})$, where λ is a fixed – maybe nonprobability[2] – distribution and one has the normalizations $\int_{\mathscr{X}} p(x)\, \mathrm{d}\lambda(x) = \int_{\mathscr{X}} q(x)\, \mathrm{d}\lambda(x) = 1$. The set of all such probability distributions will be denoted by $\mathscr{M}_\lambda^1$. We also employ the set $\mathscr{M}_\lambda$ of all general – maybe nonprobability – distributions M of the form $M[A] = \int_A m(x)\, \mathrm{d}\lambda(x)$ $(A \in \mathscr{A})$ with density $x \mapsto m(x) \geq 0$. For instance, in the *discrete setup* where $\mathscr{X} = \mathscr{X}_{count}$ has countably many elements and $\lambda := \lambda_{count}$ is the counting measure (i.e., $\lambda_{count}[\{x\}] = 1$ for all $x \in \mathscr{X}_{count}$) then $p(\cdot)$, $q(\cdot)$ are (e.g. binomial) probability mass functions and $m(\cdot)$ is a (e.g. unnormalized-histogram-related) general mass function. If λ is the Lebesgue measure on $\mathscr{X} = \mathbb{R}$, then $p(\cdot)$, $q(\cdot)$ are (e.g. Gaussian) probability density functions and $m(\cdot)$ is a general (possibly unnormalized) density function. Within such a context, we introduce the following framework of statistical distances:

[2] sigma-finite.

Definition 1. *Let $\phi \in \Phi_{\mathscr{U}}$. Then the divergence (distance) of $P, Q \in \mathscr{M}_\lambda^1$ scaled by $M_1, M_2 \in \mathscr{M}_\lambda$ and aggregated by $M_3 \in \mathscr{M}_\lambda$ is defined by*[3]

$$0 \leq D_{\phi,M_1,M_2,M_3}(P,Q)$$
$$:= \int_{\mathscr{X}} \Big[\inf_{u \in \partial_{\mathscr{U}} \phi \big| \frac{q(x)}{m_2(x)}} \left(\phi\left(\tfrac{p(x)}{m_1(x)}\right) - u\left(\tfrac{p(x)}{m_1(x)}\right) - \phi\left(\tfrac{q(x)}{m_2(x)}\right) + u\left(\tfrac{q(x)}{m_2(x)}\right)\right) \Big] m_3(x) \mathrm{d}\lambda(x). \quad (2)$$

To guarantee the existence of the integrals in (2) (with possibly infinite values), the zeros of p, q, m_1, m_2, m_3 have to be combined by proper conventions (taking into account the limit of $\phi(y)$ at $y = 0$); the full details will appear elsewhere.

Notice that $D_{\phi,M_1,M_2,M_3}(P,Q) \geq 0$, with equality iff $p(x) = \frac{m_1(x)}{m_2(x)} \cdot q(x)$ for all x (in case of absence of zeros). For the special case of the discrete setup $(\mathscr{X}_{count}, \lambda_{count})$, (2) becomes

$$0 \leq D_{\phi,M_1,M_2,M_3}(P,Q)$$
$$:= \sum_{x \in \mathscr{X}} \Big[\inf_{u \in \partial_{\mathscr{U}} \phi \big| \frac{q(x)}{m_2(x)}} \Big(\phi\Big(\tfrac{p(x)}{m_1(x)}\Big) - u\Big(\tfrac{p(x)}{m_1(x)}\Big) - \phi\Big(\tfrac{q(x)}{m_2(x)}\Big) + u\Big(\tfrac{q(x)}{m_2(x)}\Big)\Big)\Big] m_3(x).$$

In the following, we illuminate several special cases, in a "structured" manner: **(I)** Let $\widetilde{\phi}$ be from the class $\Phi := \Phi_{C_1} \subset \Phi_{\mathscr{U}_{al}}$ of functions $\widetilde{\phi} : (0, \infty) \mapsto \mathbb{R}$ which are continuously differentiable with derivative $\widetilde{\phi}'$, strictly convex, continuously extended to $y = 0$, and (say) satisfy $\widetilde{\phi}(1) = 0$. Moreover, let $h : \mathbb{R} \mapsto \mathbb{R}$ be a function which is strictly increasing on the range $\mathscr{R}_{\widetilde{\phi}}$ of $\widetilde{\phi}$ and which satisfies $h(0) = 0$ as well as $h(z) < \inf_{s \in \mathscr{R}_{\widetilde{\phi}}} h(s)$ for all $z \notin \mathscr{R}_{\widetilde{\phi}}$. For generator $\phi(y) := h(\widetilde{\phi}(y))$ we choose $\mathscr{U} = \mathscr{U}_h := \{h(a + b \cdot y) : a \in \mathbb{R}, b \in \mathbb{R}, y \in [0, \infty)\}$ to obtain

$$0 \leq D_{\phi,M_1,M_2,M_3}(P,Q)$$
$$:= \int_{\mathscr{X}} \Big[\phi\Big(\tfrac{p(x)}{m_1(x)}\Big) - h\Big(\widetilde{\phi}\Big(\tfrac{p(x)}{m_1(x)}\Big) + \widetilde{\phi}'\Big(\tfrac{q(x)}{m_2(x)}\Big) \cdot \Big(\tfrac{p(x)}{m_1(x)} - \tfrac{q(x)}{m_2(x)}\Big)\Big)\Big] m_3(x) \mathrm{d}\lambda(x). \quad (3)$$

As a first example, take $\widetilde{\phi}(y) := (y-1)^2/2$ $(y \geq 0)$ with $\mathscr{R}_{\widetilde{\phi}} = [0, \infty)$ and $h(z) := (z-1)^3 + 1$ $(z \in \mathbb{R})$. The generator $\phi(y) := h(\widetilde{\phi}(y)) = (0.5 \cdot y^2 - y - 0.5)^3 + 1$ is a degree-6 polynomial which is neither convex nor concave in the classical sense, and $u_{y_0}(y) := h(\widetilde{\phi}(y_0) + \widetilde{\phi}'(y_0) \cdot (y - y_0)) = (y \cdot y_0 - y - 0.5 \cdot (y_0)^2 - 0.5)^3 + 1 \in \mathscr{U}_h$ is a degree-3 polynomial being a strict $\mathscr{U}_h$−subgradient of ϕ at $y_0 \geq 0$. As a second example, let $\widetilde{\phi} \in \Phi_{C_1}$ have continuous second derivative and h be twice continuously differentiable and strictly convex on $\mathscr{R}_{\widetilde{\phi}}$ with $h'(0) = 1$ (in addition to the above assumptions). Then, $\phi(y) := h(\widetilde{\phi}(y))$ is in Φ_{C_1} having strictly larger curvature than $\widetilde{\phi}$ (except at $y = 1$). Especially, for $h(z) := \exp(z) - 1$ $(z \in \mathbb{R})$ the generator ϕ is basically strictly log-convex and the divergence in (3) becomes

$$\int_{\mathscr{X}} \Big[\exp\big(\widetilde{\phi}(\tfrac{p(x)}{m_1(x)})\big) - \exp\big(\widetilde{\phi}(\tfrac{q(x)}{m_2(x)})\big) \cdot \exp\Big(\widetilde{\phi}'\big(\tfrac{q(x)}{m_2(x)}\big) \cdot \Big(\tfrac{p(x)}{m_1(x)} - \tfrac{q(x)}{m_2(x)}\Big)\Big)\Big] \tfrac{\mathrm{d}\lambda(x)}{1/m_3(x)}. \quad (4)$$

[3] in (2), we can also extend $[\ldots]$ to $G([\ldots])$ for some nonnegative scalar function G satisfying $G(z) = 0$ iff $z = 0$.

(II) If ϕ itself is in the subclass $\varPhi := \varPhi_{C_1} \subset \varPhi_{\mathscr{U}_{al}}$ we obtain from (2)

$$
\begin{aligned}
0 &\leq D_{\phi,M_1,M_2,M_3}(P,Q) \\
&= \int_{\mathscr{X}} \left[\phi\left(\tfrac{p(x)}{m_1(x)}\right) - \phi\left(\tfrac{q(x)}{m_2(x)}\right) - \phi'\left(\tfrac{q(x)}{m_2(x)}\right) \cdot \left(\tfrac{p(x)}{m_1(x)} - \tfrac{q(x)}{m_2(x)}\right)\right] m_3(x)\mathrm{d}\lambda(x). \quad (5)
\end{aligned}
$$

In contrast, if ϕ has a non-differentiable "cusp" at $y_0 = \frac{q(x)}{m_2(x)}$, then one has to take the smaller of the deviations (at $y = \frac{p(x)}{m_1(x)}$) from the right-hand respectively left-hand tangent line at y_0. Notice that in (5) one gets $D_{\phi,M_1,M_2,M_3}(P,Q) = D_{\tilde{\phi},M_1,M_2,M_3}(P,Q)$ for any $\tilde{\phi}(y) := \phi(y) + c_1 + c_2 \cdot y \ \ (y \in (0,\infty))$ with $c_1, c_2 \in \mathbb{R}$. In the subcase $\phi(y) := \exp(\widetilde{\phi}(y)) - 1$ of (I), the divergence in (5) becomes

$$
\int_{\mathscr{X}} \left[\exp(\widetilde{\phi}(\tfrac{p(x)}{m_1(x)})) - \exp(\widetilde{\phi}(\tfrac{q(x)}{m_2(x)})) \cdot \left(1 + \widetilde{\phi}'\left(\tfrac{q(x)}{m_2(x)}\right) \cdot \left(\tfrac{p(x)}{m_1(x)} - \tfrac{q(x)}{m_2(x)}\right)\right)\right] \frac{\mathrm{d}\lambda(x)}{1/m_3(x)}
$$

which is larger than (4) which uses the additional information of log-convexity. This holds analogously for the more general h leading to larger curvature.

(III) By further specializing $\phi \in \varPhi_{C_1}$, $m_1(x) = m_2(x) =: m_\ell(x)$, $m_3(x) = m_\ell(x) \cdot H\big((m_g(x))_{x \in \mathscr{X}}\big)$ for some (measurable) function $m_g : \mathscr{X} \mapsto [0,\infty)$ and some strictly positive scalar functional H thereupon, we deduce from (5)

$$
\begin{aligned}
&0 \leq B_{\phi,M_g,H}\left(P,Q \,|\, M_\ell\right) := D_{\phi,M_1,M_2,M_3}(P,Q) = H\left((m_g(x))_{x\in\mathscr{X}}\right) \cdot \int_{\mathscr{X}} \left[\phi\left(\tfrac{p(x)}{m_\ell(x)}\right)\right. \\
&\left. -\phi\left(\frac{q(x)}{m_\ell(x)}\right) - \phi'\left(\frac{q(x)}{m_\ell(x)}\right) \cdot \left(\frac{p(x)}{m_\ell(x)} - \frac{q(x)}{m_\ell(x)}\right)\right] m_\ell(x)\,\mathrm{d}\lambda(x). \quad (6)
\end{aligned}
$$

The term $H\big((m_g(x))_{x\in\mathscr{X}}\big)$ can be viewed as a "global steepness tuning" multiplier of the generator ϕ, in the sense of $B_{\phi,M_g,H}\left(P,Q \,|\, M_\ell\right) = B_{c\cdot\phi,M_g,}\left(P,Q \,|\, M_\ell\right)$ where $\mathbb{1}$ denotes the functional with constant value 1. This becomes non-trivial for the subcase where the "global" density m_g depends on the probability distributions P,Q of which we want to quantify the distance, e.g. if $M_g = W_g(P,Q)$ in the sense of $m_g(x) = w_g(p(x),q(x)) \geq 0$ for some (measurable) "global scale-connector" $w_g : [0,\infty) \times [0,\infty) \mapsto [0,\infty]$ between the densities $p(x)$ and $q(x)$. Analogously, one can also use "local" scaling distributions of the form $M_\ell = W_\ell(P,Q)$ in the sense that $m_\ell(x) = w_\ell(p(x),q(x)) \geq 0$ (λ–a.a. $x \in \mathscr{X}$) for some "local scale-connector" $w_\ell : [0,\infty) \times [0,\infty) \mapsto [0,\infty]$ between the densities $p(x)$ and $q(x)$ (where w_ℓ is strictly positive on $(0,\infty) \times (0,\infty)$). Accordingly, (6) turns into

$$
\begin{aligned}
&B_{\phi,M_g,H}(P,Q \,|\, M_\ell) = B_{\phi,W_g(P,Q),H}(P,Q \,|\, W_\ell(P,Q)) \\
&= H\big((w_g(p(x),q(x)))_{x\in\mathscr{X}}\big) \cdot \textstyle\int_{\mathscr{X}} \left[\phi\Big(\tfrac{p(x)}{w_\ell(p(x),q(x))}\Big) - \phi\Big(\tfrac{q(x)}{w_\ell(p(x),q(x))}\Big) - \right. \\
&\left.\phi'\Big(\tfrac{q(x)}{w_\ell(p(x),q(x))}\Big) \cdot \Big(\tfrac{p(x)}{w_\ell(p(x),q(x))} - \tfrac{q(x)}{w_\ell(p(x),q(x))}\Big)\right] w_\ell(p(x),q(x))\,\mathrm{d}\lambda(x). \quad (7)
\end{aligned}
$$

In the discrete setup $(\mathscr{X}_{count}, \lambda_{count})$, (7) leads to

$$B_{\phi,M_g,H}(P,Q\,|\,M_\ell) = B_{\phi,W_g(P,Q),H}(P,Q\,|\,W_\ell(P,Q))$$
$$= H\big((w_g(p(x),q(x)))_{x\in\mathscr{X}}\big)\cdot\sum_{x\in\mathscr{X}}\Big[\phi\Big(\tfrac{p(x)}{w_\ell(p(x),q(x))}\Big)-\phi\Big(\tfrac{q(x)}{w_\ell(p(x),q(x))}\Big)-$$
$$\phi'\Big(\tfrac{q(x)}{w_\ell(p(x),q(x))}\Big)\cdot\Big(\tfrac{p(x)}{w_\ell(p(x),q(x))}-\tfrac{q(x)}{w_\ell(p(x),q(x))}\Big)\Big]\cdot w_\ell(p(x),q(x)). \tag{8}$$

Returning to the general setup, from (7) we can extract the following well-known, widely used distances as special subcases of our universal framework:

(IIIa) Ordinary Bregman divergences OBD between probability distributions (see e.g. Pardo and Vajda [18]):

$$\int_{\mathscr{X}}[\phi(p(x))-\phi(q(x))-\phi'(q(x))\cdot(p(x)-q(x))]\,\mathrm{d}\lambda(x)$$
$$= B_{\phi,\,,\,}(P,Q\,|\,\mathbb{I}) = B_{\phi,M_g,H}(P,Q\,|\,M_\ell)$$

where $M_g = \mathbb{I}$, $M_\ell = \mathbb{I}$ means $m_g(x) = 1$, $m_\ell(x) = 1$ for all x.

(IIIb) Csiszar-Ali-Silvey ϕ−divergences CASD (cf. Csiszar [8], Ali and Silvey [3]):

$$\int_{\mathscr{X}}\Big[q(x)\cdot\phi\Big(\tfrac{p(x)}{q(x)}\Big)\Big]\,\mathrm{d}\lambda(x) = B_{\phi,\,,\,}(P,Q\,|\,Q)\,.$$

This includes in particular the Kullback-Leibler information divergence and Pearson's chisquare divergence (see Sect. 2 for explicit formulas).

(IIIc) Scaled Bregman divergences SBD (cf. Stummer [20], Stummer and Vajda [21]):

$$\int_{\mathscr{X}}\Big[\phi\big(\frac{p(x)}{m_\ell(x)}\big)-\phi\big(\frac{q(x)}{m_\ell(x)}\big)-\phi'\big(\frac{q(x)}{m_\ell(x)}\big)\cdot\Big(\frac{p(x)}{m_\ell(x)}-\frac{q(x)}{m_\ell(x)}\Big)\Big]\,m_\ell(x)\,\mathrm{d}\lambda(x)$$
$$= B_{\phi,\,,\,}(P,Q\,|\,M_\ell).$$

The sub-setup $m_\ell(x) = w_\ell(p(x),q(x)) \geq 0$ was used in Kißlinger and Stummer [11] for comprehensive investigations on robustness; see also [9,10].

(IIId) Total Bregman divergences (cf. Liu et al. [12],[13], Vemuri et al. [24]):

$$\frac{1}{\sqrt{1+\int_{\mathscr{X}}(\phi'(q(x)))^2\mathrm{d}\lambda(x)}}\cdot\int_{\mathscr{X}}[\phi(p(x))-\phi(q(x))-\phi'(q(x))\cdot(p(x)-q(x))]\,\mathrm{d}\lambda(x)$$
$$= B_{\phi,M_g^{to},H^{to}}(P,Q\,|\,\mathbb{I})$$

where $M_g^{to} := W_g^{to}(P,Q)$ in the sense of $m_g^{to}(x) := w_g^{to}(p(x),q(x)) := (\phi'(q(x)))^2$, and $H^{to}\big((h(x))_{x\in\mathscr{X}}\big) := \frac{1}{\sqrt{1+\int_{\mathscr{X}}h(x)\,\mathrm{d}\lambda(x)}}$. For example, for the special case of the discrete setup $(\mathscr{X}_{fin}, \lambda_{count})$ where $\mathscr{X} = \mathscr{X}_{fin}$ has only finitely (rather than countably) many elements, Liu et al. [12],[13], Vemuri et al. [24] also deal with non-probability vectors and non-additive aggregations and show that their total Bregman divergences have the advantage to be invariant against certain transformations, e.g. those from the special linear group (matrices whose determinant is equal to 1, for instance rotations).

(IIIe) Conformal divergences:

$$H\big((w_g(q(x)))_{x\in\mathscr{X}}\big)\cdot\int_{\mathscr{X}}[\phi(p(x))-\phi(q(x))-\phi'(q(x))\cdot(p(x)-q(x))]\,\mathrm{d}\lambda(x)$$
$$= B_{\phi,W_g(Q),H}\,(P,Q\,|\,\mathbb{I})\,. \tag{9}$$

For the special case of the finite discrete setup $(\mathscr{X}_{fin},\lambda_{count})$, (9) reduces to the conformal Bregman divergences of Nock et al. [17]; within this $(\mathscr{X}_{fin},\lambda_{count})$ they also consider non-probability vectors and non-additive aggregations.

(IIIf) Scaled conformal divergences:

$$H\left(\left(w_g\left(\tfrac{q(x)}{m_\ell(x)}\right)\right)_{x\in\mathscr{X}}\right)\cdot\int_{\mathscr{X}}\left[\phi\left(\tfrac{p(x)}{m_\ell(x)}\right)-\phi\left(\tfrac{q(x)}{m_\ell(x)}\right)\right.$$
$$\left.-\,\phi'\left(\tfrac{q(x)}{m_\ell(x)}\right)\cdot\left(\tfrac{p(x)}{m_\ell(x)}-\tfrac{q(x)}{m_\ell(x)}\right)\right]m_\ell(x)\,\mathrm{d}\lambda(x)\ = B_{\phi,W_g(Q/M_\ell),H}(P,Q\,|\,M_\ell)\,. \tag{10}$$

In the special finite discrete setup $(\mathscr{X}_{fin},\lambda_{count})$, (10) leads to the scaled conformal Bregman divergences indicated in Nock et al. [17]; within $(\mathscr{X}_{fin},\lambda_{count})$ they also employ non-probability vectors and non-additive aggregations.

(IIIg) Generalized Burbea-Rao divergences with $\beta\in(0,1)$:

$$H\big((m_g(x))_{x\in\mathscr{X}}\big)\cdot\int_{\mathscr{X}}\big[\beta\cdot\phi(p(x))+(1-\beta)\cdot\phi(q(x))$$
$$-\phi(\beta p(x)+(1-\beta)q(x))\,\big]\,\mathrm{d}\lambda(x)\ = B_{\phi_2,M_g,H}\Big(P,Q\,|\,M_\ell^{(\beta,\phi)}\Big)$$

where $\phi_2(y):=(y-1)^2/2$ and $M_\ell^{(\beta,\phi)}{=}W_\ell^{(\beta,\phi)}(P,Q)$ in the sense that $m_\ell^{(\beta,\phi)}(x)$ $=w_\ell^{(\beta,\phi)}(p(x),q(x))$ with $w_\ell^{(\beta,\phi)}(u,v):=\frac{(u-v)^2}{2\cdot(\beta\cdot\phi(u)+(1-\beta)\cdot\phi(v)-\phi(\beta u+(1-\beta)v))}$. In analogy with the considerations in (IIIe) above, one may call the special case $B_{\phi_2,W_g(Q),H}\Big(P,Q\,|\,M_\ell^{(\beta,\phi)}\Big)$ a conformal Burbea-Rao divergence.

To end up Sect. 1, let us mention that there is a well-known interplay between the geometry of parameters for exponential families and divergences, in the setups (IIIa)-(IIIe) (see e.g. [2],[4],[21],[9],[1],[17]). To gain further insights, it would be illuminating to extend this to the context of Definition 1.

2 General Asymptotic Results for the Finite Discrete Case

In this section, we deal with the above-mentioned setup (8) and assume additionally that the function $\phi\in\Phi_{C_1}$ is thrice continuously differentiable on $(0,\infty]$, as well as that all three functions $w_\ell(u,v)$, $w_1(u,v):=\frac{\partial w_\ell}{\partial u}(u,v)$ and $w_{11}(u,v):=\frac{\partial^2 w_\ell}{\partial u^2}(u,v)$ are continuous in all (u,v) of some (maybe tiny) neighbourhood of the diagonal $\{(t,t):t\in(0,1)\}$ (so that the behaviour for $u\approx v$ is technically appropriate). In such a setup, we consider the following context: for $i\in\mathbb{N}$ let the observation of the i−th data point be represented by the random variable X_i which takes values in some finite space $\mathscr{X}:=\{x_1,\ldots,x_s\}$ which has $s:=|\mathscr{X}|\geq 2$ outcomes and thus, we choose the counting distribution $\lambda:=\lambda_{count}$ as reference distribution (i.e., $\lambda_{count}[\{x_k\}]=1$ for all

k). Accordingly, let $X_1, \ldots, X_N$ represent a random sample of independent and identically distributed observations generated from an unknown true distribution $P_{\theta_{true}}$ which is supposed to be a member of a parametric family $\mathscr{P}_\Theta := \{P_\theta \in \mathscr{M}_\lambda^1 : \theta \in \Theta\}$ of hypothetical, potential candidate distributions with probability mass function p_θ. Here, $\Theta \subset \mathbb{R}^\ell$ is a ℓ-dimensional parameter set. Moreover, we denote by $P := P_N^{emp} := \frac{1}{N} \cdot \sum_{i=1}^N \delta_{X_i}[\cdot]$ the empirical distribution for which the probability mass function p_N^{emp} consists of the relative frequencies $p(x) = p_N^{emp}(x) = \frac{1}{N} \cdot \#\{i \in \{1, \ldots, N\} : X_i = x\}$ (i.e. the "histogram entries"). If the sample size N tends to infinity, it is intuitively plausible that the divergence (cf. (8))

$$\begin{aligned}
0 \leq \frac{T_N^\phi(P_N^{emp}, P_\theta)}{2N} &:= B_{\phi, W_g(P_N^{emp}, P_\theta), H}\left(P_N^{emp}, P_\theta \,|\, W_\ell(P_N^{emp}, P_\theta)\right) \\
&= H\left(\left(w_g(p_N^{emp}(x), p_\theta(x))\right)_{x \in \mathscr{X}}\right) \cdot \sum_{x \in \mathscr{X}} \left[\phi\left(\frac{p_N^{emp}(x)}{w_\ell(p_N^{emp}(x), p_\theta(x))}\right) - \phi\left(\frac{p_\theta(x)}{w_\ell(p_N^{emp}(x), p_\theta(x))}\right)\right. \\
&\left. - \phi'\left(\frac{p_\theta(x)}{w_\ell(p_N^{emp}(x), p_\theta(x))}\right) \cdot \left(\frac{p_N^{emp}(x)}{w_\ell(p_N^{emp}(x), p_\theta(x))} - \frac{p_\theta(x)}{w_\ell(p_N^{emp}(x), p_\theta(x))}\right)\right] \cdot w_\ell(p_N^{emp}(x), p_\theta(x)) \\
&=: H\left(\left(w_g(p_N^{emp}(x), p_\theta(x))\right)_{x \in \mathscr{X}}\right) \cdot \zeta_N =: \Upsilon_N \cdot \zeta_N
\end{aligned} \tag{11}$$

between the data-derived empirical distribution P_N^{emp} and the candidate model P_θ converges to zero, provided that we have found the correct model in the sense that P_θ is equal to the true data generating distribution $P_{\theta_{true}}$, and that $H\big((w_g(p_N^{emp}(x), p_\theta(x)))_{x \in \mathscr{X}}\big)$ converges a.s. to a constant $a_\theta > 0$. In the same line of argumentation, $B_{\phi, W_g(P_N^{emp}, P_\theta), H}(P_N^{emp}, P_\theta \,|\, W_\ell(P_N^{emp}, P_\theta))$ becomes close to zero if P_θ is close to $P_{\theta_{true}}$. Notice that (say, for p_N^{emp} and p_θ without zeros) the Kullback-Leibler divergence KL case with $\phi_1(y) := y \log y + 1 - y \geq 0$ $(y > 0)$

$$B_{\phi_1, ,}(P_N^{emp}, P_\theta \,|\, P_\theta) = \sum_{x \in \mathscr{X}} p_\theta(x) \cdot \phi_1\left(\frac{p_N^{emp}(x)}{p_\theta(x)}\right) = \sum_{x \in \mathscr{X}} p_N^{emp}(x) \cdot \log\left(\frac{p_N^{emp}(x)}{p_\theta(x)}\right)$$

is nothing but the (multiple of the) very prominent likelihood ratio test statistics (likelihood disparity); minimizing it over θ produces the maximum likelihood estimate $\widehat{\theta}^{MLE}$. Moreover, by employing $\phi_2(y) := (y-1)^2/2$ the divergence

$$B_{\phi_2, ,}(P_N^{emp}, P_\theta \,|\, P_\theta) = \sum\nolimits_{x \in \mathscr{X}} \frac{(p_N^{emp}(x) - p_\theta(x))^2}{2 p_\theta(x)}$$

represents the (multiple of the) Pearson chi-square test statistics. Concerning the above-mentioned conjectures where the sample size N tends to infinity, in case of $P_{\theta_{true}} = P_\theta$ one can even derive the *limit distribution* of the divergence test statistics $T_N^\phi(P_N^{emp}, P_\theta)$ in quite "universal generality":

Theorem 1. *Under the null hypothesis "H_0: $P_{\theta_{true}} = P_\theta$ with $p_\theta(x) > 0$ for all $x \in \mathscr{X}$" and the existence of a.s.$-\lim_{N \to \infty} H\left((w_g(p_N^{emp}(x), p_\theta(x)))_{x \in \mathscr{X}}\right) =: a_\theta > 0$, the asymptotic distribution (as $N \to \infty$) of*

$$T_N^\phi(P_N^{emp}, P_\theta) = 2N \cdot B_{\phi, W_g(P_N^{emp}, P_\theta), H}\left(P_N^{emp}, P_\theta \,|\, W_\ell(P_N^{emp}, P_\theta)\right)$$

has the following density f_{s^*}[4]*:*

$$f_{s^*}(y;\gamma^{\phi,\theta}) = \frac{y^{\frac{s^*}{2}-1}}{2^{\frac{s^*}{2}}} \sum_{k=0}^{\infty} c_k \cdot \frac{\left(-\frac{y}{2}\right)^k}{\Gamma\left(\frac{s^*}{2}+k\right)}, \qquad y \in [0,\infty[,$$

$$\textit{with } c_0 = \prod_{j=1}^{s^*} \left(\gamma_j^{\phi,\theta}\right)^{-0.5} \textit{ and } c_k = \frac{1}{2k} \sum_{r=0}^{k-1} c_r \sum_{j=1}^{s^*} \left(\gamma_j^{\phi,\theta}\right)^{r-k} \quad (k \in \mathbb{N})$$

where $s^* := rank(\boldsymbol{\Sigma A \Sigma})$ *is the number of the strictly positive eigenvalues* $(\gamma_i^{\phi,\theta})_{i=1,\ldots,s^*}$ *of the matrix* $\boldsymbol{A\Sigma} = (\bar{c}_i \cdot (\delta_{ij} - p_\theta(x_j)))_{i,j=1,\ldots,s}$ *consisting of*

$$\boldsymbol{\Sigma} = (p_\theta(x_i) \cdot (\delta_{ij} - p_\theta(x_j))_{i,j=1,\ldots,s}$$

$$\boldsymbol{A} = \left(\frac{a_\theta \cdot \phi''\left(\frac{p_\theta(x_i)}{w(p_\theta(x_i),p_\theta(x_i))}\right)}{w(p_\theta(x_i),p_\theta(x_i))} \delta_{ij} \right)_{i,j=1,\ldots,s}$$

$$\bar{c}_i = a_\theta \cdot \phi''\left(\frac{p_\theta(x_i)}{w(p_\theta(x_i),p_\theta(x_i))}\right) \cdot \frac{p_\theta(x_i)}{w(p_\theta(x_i),p_\theta(x_i))}.$$

Here we have used Kronecker's delta δ_{ij} *which is 1 iff* $i = j$ *and 0 else.*

In particular, the asymptotic distribution (as $N \to \infty$) of $T_N := T_N^\phi(P_N^{emp}, P_\theta)$ coincides with the distribution of a weighted linear combination of standard-chi-square-distributed random variables where the weights are the $\gamma_i^{\phi,\theta}$ $(i = 1,\ldots,s^*)$. Notice that Theorem 1 extends a theorem of Kißlinger and Stummer [11] who deal with the subcase (IIIc) of scaled Bregman divergences. The proof of the latter can be straightforwardly adapted to verify Theorem 1, due to the representation $T_N = \Upsilon_N \cdot (2N \cdot \zeta_N)$ in (11) and the assumption $a.s.\text{–}\lim_{N\to\infty} \Upsilon_N = a_\theta > 0$. The details will appear elsewhere. Remarkably, within (IIIc) the limit distribution of T_N is even a parameter-free "ordinary" chi-square distribution provided that the condition $w(v,v) = v$ holds for all v (cf. [11]).

Acknowledgement. We are grateful to 3 referees for their useful suggestions.

References

1. Amari, S.-I.: Information Geometry and Its Applications. Springer, Tokyo (2016)
2. Amari, S.-I., Nagaoka, H.: Methods of Information Geometry. Oxford University Press, Oxford (2000)
3. Ali, M.S., Silvey, D.: A general class of coefficients of divergence of one distribution from another. J. Roy. Statist. Soc. **B–28**, 131–140 (1966)
4. Banerjee, A., Merugu, S., Dhillon, I.S., Ghosh, J.: Clustering with Bregman divergences. J. Mach. Learn. Res. **6**, 1705–1749 (2005)
5. Basu, A., Shioya, H., Park, C.: Statistical Inference: The Minimum Distance Approach. CRC Press, Boca Raton (2011)

[4] (with respect to the one-dim. Lebesgue measure).

6. Cesa-Bianchi, N., Lugosi, G.: Prediction, Learning and Games. Cambridge University Press, Cambridge (2006)
7. Collins, M., Schapire, R.E., Singer, Y.: Logistic regression, AdaBoost and Bregman distances. Mach. Learn. **48**, 253–285 (2002)
8. Csiszar, I.: Eine informationstheoretische Ungleichung und ihre Anwendung auf den Beweis der Ergodizität von Markoffschen Ketten. Publ. Math. Inst. Hungar. Acad. Sci. **A–8**, 85–108 (1963)
9. Kißlinger, A.-L., Stummer, W.: Some decision procedures based on scaled Bregman distance surfaces. In: Nielsen, F., Barbaresco, F. (eds.) GSI 2013. LNCS, vol. 8085, pp. 479–486. Springer, Heidelberg (2013). doi:10.1007/978-3-642-40020-9_52
10. Kißlinger, A.-L., Stummer, W.: New model search for nonlinear recursive models, regressions and autoregressions. In: Nielsen, F., Barbaresco, F. (eds.) GSI 2015. LNCS, vol. 9389, pp. 693–701. Springer, Cham (2015). doi:10.1007/978-3-319-25040-3_74
11. Kißlinger, A.-L., Stummer, W.: Robust statistical engineering by means of scaled Bregman distances. In: Agostinelli, C., Basu, A., Filzmoser, P., Mukherjee, D. (eds.) Recent Advances in Robust Statistics: Theory and Applications, pp. 81–113. Springer, New Delhi (2016). doi:10.1007/978-81-322-3643-6_5
12. Liu, M., Vemuri, B.C., Amari, S.-I., Nielsen, F.: Total Bregman divergence and its applications to shape retrieval. In: Proceedings 23rd IEEE CVPR, pp. 3463–3468 (2010)
13. Liu, M., Vemuri, B.C., Amari, S.-I., Nielsen, F.: Shape retrieval using hierarchical total Bregman soft clustering. IEEE Trans. Pattern Anal. Mach. Intell. **34**(12), 2407–2419 (2012)
14. Murata, N., Takenouchi, T., Kanamori, T., Eguchi, S.: Information geometry of U-boost and Bregman divergence. Neural Comput. **16**(7), 1437–1481 (2004)
15. Nock, R., Menon, A.K., Ong, C.S.: A scaled Bregman theorem with applications. In: Advances in Neural Information Processing Systems 29 (NIPS 2016), pp. 19–27 (2016)
16. Nock, R., Nielsen, F.: Bregman divergences and surrogates for learning. IEEE Trans. Pattern Anal. Mach. Intell. **31**(11), 2048–2059 (2009)
17. Nock, R., Nielsen, F., Amari, S.-I.: On conformal divergences and their population minimizers. IEEE Trans. Inf. Theory **62**(1), 527–538 (2016)
18. Pardo, M.C., Vajda, I.: On asymptotic properties of information-theoretic divergences. IEEE Trans. Inf. Theory **49**(7), 1860–1868 (2003)
19. Pallaschke, D., Rolewicz, S.: Foundations of Mathematical Optimization. Kluwer Academic Publishers, Dordrecht (1997)
20. Stummer, W.: Some Bregman distances between financial diffusion processes. Proc. Appl. Math. Mech. **7**(1), 1050503–1050504 (2007)
21. Stummer, W., Vajda, I.: On Bregman distances and divergences of probability measures. IEEE Trans. Inf. Theory **58**(3), 1277–1288 (2012)
22. Sugiyama, M., Suzuki, T., Kanamori, T.: Density-ratio matching under the Bregman divergence: a unified framework of density-ratio estimation. Ann. Inst. Stat. Math. **64**, 1009–1044 (2012)
23. Tsuda, K., Rätsch, G., Warmuth, M.: Matrix exponentiated gradient updates for on-line learning and Bregman projection. J. Mach. Learn. Res. **6**, 995–1018 (2005)
24. Vemuri, B.C., Liu, M., Amari, S.-I., Nielsen, F.: Total Bregman divergence and its applications to DTI analysis. IEEE Trans. Med. Imag. **30**(2), 475–483 (2011)
25. Wu, L., Hoi, S.C.H., Jin, R., Zhu, J., Yu, N.: Learning Bregman distance functions for semi-supervised clustering. IEEE Trans. Knowl. Data Eng. **24**(3), 478–491 (2012)

Quantification of Model Risk: Data Uncertainty

Z. Krajčovičová[1(✉)], P.P. Pérez-Velasco[2], and C. Vázquez[1]

[1] Department of Mathematics, University of A Coruña, A Coruña, Spain
z.krajcovicova@udc.es
[2] Banco Santander, Model Risk Area, Madrid, Spain

Abstract. Worldwide regulations oblige financial institutions to manage and address model risk (MR) as any other type of risk. MR quantification is essential not only in meeting these requirements but also for institution's basic internal operative. In [5] the authors introduce a framework for the quantification of MR based on information geometry. The framework is applicable in great generality and accounts for different sources of MR during the entire lifecycle of a model. The aim of this paper is to extend the framework in [5] by studying its relation with the uncertainty associated to the data used for building the model. We define a metric on the space of samples in order to measure the data intrinsic distance, providing a novel way to probe the data for insight, allowing us to work on the sample space, gain business intuition and access tools such as perturbation methods.

Keywords: Model risk · Calibration · Sampling · Riemannian manifold · Information geometry

1 Introduction

Financial models refer to simplifying mappings of reality that serve a specific purpose by applying mathematical and economic theories to available data. They include mathematical relations, a way to use data or judgment to compute the parameters, and indications on how to apply them to practical issues, refer to [6].

Models focus on specific aspects of reality, degrading or ignoring the rest. Understanding the limitations of the underlying assumptions and their material consequences is essential from a MR perspective. Unrealistic assumptions, poor selection of the model, wrong design, misuse or inadequate knowledge of its usage may expose a financial institution to additional risks. MR refers to potential losses institutions may incur as a consequence of decisions based on the output of models.

In [5] the authors introduce a general framework for the quantification of MR using information geometry, applicable to most modelling techniques currently under usage in financial institutions. This framework copes with relevant aspects of MR management such as usage, performance, mathematical foundations, cal-

F. Nielsen and F. Barbaresco (Eds.): GSI 2017, LNCS 10589, pp. 523–531, 2017.
https://doi.org/10.1007/978-3-319-68445-1_61

ibration or data. Differences between models are determined by the distance under a Riemannian metric.[1]

In this framework, models are represented by particular probability distributions that belong to the set of probability measures $\mathcal{M}$ (the model manifold), available for modelling. We assume that the examined model p_0 is represented by a probability distribution that can be uniquely parametrized using the n–dimensional vector parameter $\theta_0 = (\theta_0^1, \ldots, \theta_0^n)$ and can be described by the probability distribution $p_0 = p(x, \theta_0)$, i.e. $p_0 \in \mathcal{M} = \{p(x, \theta) : \theta \in \Theta \subset \mathbb{R}^n\}$, a differentiable manifold. selecting a particular p_0 is equivalent to fixing a parameter setting $\theta_0 \in \Theta$ and induces $\mathcal{M}$.

A nonlinear weight function K on $\mathcal{M}$ places a relative relevance to every alternative model, and assigns the credibility of the underlying assumptions that would make other models partially or relatively preferable to the nominal one, p_0.[2] Requiring K to be a smooth density function over $\mathcal{M}$ induces a new absolutely continuous probability measure ζ with respect to the Riemannian volume dv defined by

$$\zeta(U) = \int_U d\zeta = \int_U K dv(p), \tag{1}$$

where $U \subseteq \mathcal{M}$ is an open neighborhood around p_0 containing alternate models that are not too far in a sense quantified by the relevance to (missing) properties and limitations of the model (i.e., the uncertainty of the model selection).

The MR measure considers the usage of the model represented by some predefined mapping $f\colon \mathcal{M} \to \mathbb{R}$ with $p \mapsto f(p)$, the output function. MR is then measured as a norm of an appropriate function of the output differences over a weighted Riemannian manifold (K above) endowed with the Fisher–Rao metric and the Levi–Civita connection:

Definition 1. *Let $(\mathcal{F}, \|\cdot\|)$ be a Banach space of measurable functions $f \in \mathcal{F}$ with respect to ζ with f as above. The* model risk *Z of f and p_0 is given by*

$$Z(f, p_0) = \|f - f(p_0)\| \,. \tag{2}$$

This approach is used in [5] to quantify the MR by embedding $\mathcal{M} \hookrightarrow \mathcal{M}'$ where $\mathcal{M}' = \{p(x, \theta) : \theta \in \Theta; \, dim(\theta) \geq dim(\theta_0)\}$ is a Riemannian manifold containing a collection of probability distributions created by varying θ_0, adding properties or considering data and calibration uncertainty (see [5] for more details and examples).

The main objective of this paper is to deepen in the influence of data uncertainty in MR by relating the data uncertainties with the model structure. Pulling

[1] Distance in statistical manifolds has been applied in finance to assess the accuracy of model approximation (for derivatives) or for quantitative comparison of models, e.g. [2,3].

[2] The particular K is depends on the model sensitivity, scenario analysis, relevance of outcomes in decision making, business, intended purpose, or uncertainty of the model foundations.

back the model manifold metric introduces a consistent Riemannian structure on the sample space that allows us to quantify MR, working with samples. In practice it offers a computational alternative, eases application of business intuition and assignment of data uncertainty, as well as insight on MR from the data and the model perspectives.

The rest of the paper is organized as follows: Sect. 2 introduces the concept of input data, and describes the MR that arises from data deficiencies. Also, it is devoted to the sampling and the fitting processes. Section 3 introduces a metric on the sample space, and its Riemannian structure. Quantification of MR in the sample space is proposed in Sect. 4. Section 5 concludes followed by the technical proofs in the Appendix.

2 Sample Space and Fitting Process

Important components of MR associated with the model are flaws in the input data.[3] Issues in financial data that affect consistency and availability of data include backfilling data, availability for only a subset of the cross–section of the subjects, or limited in time.

To assess the uncertainty in the data we move the setting for quantification of MR into a sample space that will represent perturbed inputs and thus alternative models in $\mathcal{M}$ with its own metric. Distance between samples should be defined taking into account the information stored in the samples and their impact on the model usage.

Let $\bar{\sigma} : \mathcal{M} \to \bar{S}$ be a sampling process where $\bar{S}$ is the sample space.[4] Note that a bunch of samples with different elements and varying sizes are associated to the same distributions via $\bar{\sigma}$. Conversely, some distributions may be connected with the same sample via the fitting process (estimation, calibration). Thus, we impose further conditions on $\bar{\sigma}$ (see Eq. 3 below) that will be linked to the fitting process: The sampled data needs to fit back to models in $\mathcal{M}$ and models need to resemble the image in $\bar{S}$.

In general, the fitting process refers to setting the model parameter values so that the behaviour of the model matches features of the measured data in as many dimensions as unknown parameters. Therefore, $\bar{\pi} : \bar{S} \to \mathcal{M}$ associates a probability distribution $p(x, \theta) = \bar{\pi}(x) \in \mathcal{M}$ to the sample $x \in \bar{S}$. We will assume that this map is a smooth immersion, i.e. a local embedding.[5] Besides, we want $\bar{\pi}$ to be smooth in the samples: If a *small* change is applied to x, the change in the parameters should be *relatively small.*

[3] Main categories of financial data: time series, cross–sectional, panel data. Quality and availability deficiencies include errors in definition, lack of historical depth, of critical variables, insufficient sample, migration, proxies, sensitivity to expert judgment or wrong interpretation.

[4] Each sample, i.e. a collection of independent sample points of the probability distribution, is a point in $\bar{S}$. For example, each point in $\bar{S}$ is a particular instance of the portfolio of loans.

[5] For any point $x \in S$, $\exists U \subset S$ of x such that $\bar{\pi} : U \to \mathcal{M}$ is an embedding.

Based on $\bar{\pi}$, we define a quotient space out of $\bar{S}$ with the equivalence classes containing samples that are equally valid with respect to the model manifold $\mathcal{M}$. Different samples, regardless of their sample size, will share the same relevant information of the models for they give rise to the same probability distribution through $\bar{\pi}$.

Definition 2. *Define the equivalence relation* $\sim$ *on the sample space* $\bar{S}$ *by declaring* $x \sim x'$ *if* $p(x, \theta(x)) = p(x', \theta'(x'))$ *where* $\theta(\cdot)$ *is an estimator, i.e. a mapping from the space of random variables to* $\mathcal{M}$. *We denote* $S = \bar{S}/\!\sim = \{[x] : x \in \bar{S}\}$ *the corresponding quotient space and the projection by* $\delta : \bar{S} \to S$.

To guarantee consistency between samples and models, and to ensure uniqueness and compatibility with the sampling process, we require $\bar{\pi}$ and $\bar{\sigma}$ with $Dom(\bar{\sigma}) = \mathcal{M}$ to satisfy

$$\bar{\pi}\bar{\sigma}\bar{\pi} = \bar{\pi}. \tag{3}$$

This in particular implies that not all samples are acceptable. For example, samples with too few elements would not ensure the fitting process to work as inverse of $\bar{\sigma}$.[6]

Proposition 1. *Given the fitting process* $\bar{\pi}$ *as in Definition 2, the equivalence relation* $x \sim x'$ *if* $p(x, \theta(x)) = p(x', \theta'(x'))$ *is well defined.*

Proof. We assume $\bar{\pi} : \bar{S} \to \mathcal{M}$ to be a regular embedding that satisfies Eq. 3. Let $\sim$ be the relation induced by $\bar{\pi}$: $x \sim y \Leftrightarrow \bar{\pi}(x) = p(x, \theta) = p(y, \theta) = \bar{\pi}(y)$. It is easy to check that $\sim$ is reflexive, symmetric and transitive, i.e. an equivalence relation on S.

By contradiction we prove that the equivalence relation is well–defined: Assume that $x \in [y]$ and $x \in [z]$ but $[y] \neq [z]$. Then $\forall y' \in [y], z' \in [z]$ we have $x \sim y'$ and $x \sim z'$. This implies that $\bar{\pi}(x) = \bar{\pi}(y') = p(x, \theta)$ and $\bar{\pi}(x) = \bar{\pi}(z') = p(x, \theta)$, so $\bar{\pi}(y') = \bar{\pi}(z')$. Since $\bar{\pi}\sigma\bar{\pi} = \bar{\pi}$ we have $[y] = [z]$. □

The fitting process $\bar{\pi}$ is invariant under $\sim$ since $\bar{\pi}(x) = \bar{\pi}(x')$ whenever $x \sim x'$, and induces a unique function π on S, such that $\bar{\pi} = \pi \circ \delta$, and similarly for σ.

3 Choice of Metric on the Sample Space

To introduce a measure and a distance intrinsic to the data we define a metric on S. Given Proposition 1 we endow the sample space with the rich structure of the $\mathcal{M}$ by pulling back to S the geometric structure on $\mathcal{M}$.

Let g be the Riemannian metric on $\mathcal{M}$ and $\pi : S \to (\mathcal{M}, g)$ be a smooth immersion (see footnote 7). Then the definition of the pullback of the metric

[6] Many algorithms (Max. Likelihood, Method of Moments, Least Square or Bayesian estimation) and calibration processes satisfy 3 when appropriate restrictions are applied (if needed).

$\pi^* g = h$ acting on two tangent vectors u, v in $T_x S$ is $h(u,v) = (\pi^* g)_x(u,v) = g_{\pi(x)}(\pi_*(u), \pi_*(v))$, where $\pi_* : T_x S \to T_{\pi(x)}\mathcal{M}$ is the tangent (derivative) map.

Similarly, we can define a connection ∇^* on the manifold S as the pullback of the Levi–Civita connection ∇ on $\mathcal{M}$. For $X, Y \in \Gamma(TS)$ vector fields in the tangent bundle, the pullback connection is given by $\nabla^*_X Y = (\pi^* \nabla)_X Y$. The pullback connection exists and is unique, [4], therefore $\pi^* \nabla = \nabla^*$ for the pullback metric h.

Theorem 1. *As above, let π be the fitting process and σ the sampling process such that*

$$\pi \sigma \pi = \pi. \tag{4}$$

Then S becomes a weighted Riemannian manifold through the pullback of π.

Proof: See Appendix.

Since (S, h) and $(\mathcal{M}, g)$ are both Riemannian manifolds and $\pi : S \to \mathcal{M}$ is a smooth immersion, for any $x \in S$ there is an open neighbourhood U of $x \in S$ such that $\pi(x)$ is a submanifold of $\mathcal{M}$ and $\pi\big|_U : U \to \pi(U)$ is a diffeomorphism. Every point $x \in S$ has a neighbourhood U such that $\pi\big|_U$ is an isometry of U onto an open subset of $\mathcal{M}$ and π is a local diffeomorphism. Hence, (S, h) is locally isometric to $(\mathcal{M}, g)$.

In the neighbourhood on which π is an isometry, the probability measure ζ defined on $\mathcal{M}$ given by Eq. 1 can be pulled back to S. The pullback measure of ζ is a Riemannian measure $\pi^*\zeta$ on S with respect to the metric h given by

$$\pi^*\zeta(f) := \zeta(f \circ \sigma), \quad f \in C^\infty(\mathcal{M})$$

Assuming $\mathcal{M}$ being oriented, we can endow S with the pullback orientation via the bundle isomorphism $TS \cong \pi^*(T\mathcal{M})$ over S, and therefore $\int_{\mathcal{M}} \zeta = \int_S \pi^*\zeta$.

Proposition 2. *Let $\mathcal{M}$ be oriented manifold and $\pi\big|_U : U \to \pi(U)$ be an isometry. For any integrable function f on $\pi(U) \subset \mathcal{M}$ we have*

$$\int_{\pi(U)} f d\zeta = \int_U (\pi^* f) d\pi^*\zeta. \tag{5}$$

Proof: See Appendix.

Theorem 1 and Proposition 2 show that in spite of the apparent differences between S and $\mathcal{M}$, one being a space of observations and the other being a statistical manifold, they can be both endowed with the same mathematical structure and become locally equivalent from a geometric point of view.

4 Quantification of Model Risk on the Sample Space

After pulling back the Riemannian structure from $\mathcal{M}$ to S, we can quantify MR directly on S and introduce the sensitivity analysis to different perturbations in the inputs.

The model in previous sections was assumed to be some probability distribution $p \in \mathcal{M}$, or the corresponding class of samples $x \in S$ after the pullback. More likely, a practitioner would define the model as some mapping $f : \mathcal{M} \to \mathbb{R}$ with $p \mapsto f(p)$, i.e. a model outputs some quantity. The associated pullback output function is then given by $F = \pi^* f,\ F : S \to \mathbb{R}$, with F belonging to a Banach space $F \in (\mathcal{F}, ||\cdot||)$ with respect to $\pi^*\zeta$. As an example, for the weighted L^q norms $||F||_q = (\int_S |F|^q d(\pi^*\zeta))^{1/q}, 1 \leq q < \infty$, the weighted Banach space would be $L^q(S, d(\pi^*\zeta)) = \{F : ||F||_q < \infty\}$.

Theorem 2. *Let $\mathcal{M}$ be a model manifold with all alternative models relevant for the quantification of MR. Through σ satisfying Eq. 4 and the projection $\delta : \bar{S} \to S$, S can be endowed with the Riemannian structure via π^*. Letting $(\mathcal{F}, ||\cdot||)$ be a Sobolev space of measurable functions with measure $\pi^*\zeta$, we get an equivalent MR measure Z as in Definition 1 on S, i.e. $F \in \mathcal{F}$, in the neighbourhood U of x_0 where π is a diffeomorphism. The measure is given by $Z(F, x_0) = ||F - F(x_0)||$.*

Proof: See Appendix.

The choice of a specific norm depends among other factors on the purpose of the quantification. Two interesting examples are the L^q and Sobolev norms [1]:

1. $Z^q(F, x_0)$ for $F \in L^q(S, d(\pi^*\zeta))$ is the L^q norm $Z^q(F, x_0) = \|F - F(x_0)\|_q$ defined just before Theorem 2. Every choice of norm provides different information of the model regarding the MR it induces. For instance, the L^1 norm represents the total relative change in the outputs across all relevant sample classes. The L^∞ norm finds the relative worst–case error with respect to p_0, pointing to the sources with largest deviances (using the inverse of the exponential map).
2. $Z^{s,q}(F, x_0), f \in W^{s,q}(S, d(\pi^*\zeta))$, is of interest when the rate of change is relevant:[7] $\|F - F(x_0)\|_{s,q} = (\sum_{|k|\leq s} \int_S \left|\nabla^k \Big(F - F(x_0)\Big)\right|^q d(\pi^*\zeta))^{1/q}$, where ∇ denotes the associated connections on the vector bundles.

5 Further Research

There are many directions for further research, apart from the quantification of MR, both of theoretical and of practical interest. The framework can be applied to sensitivity analysis by using the directional and total derivatives. It is suitable for stress testing (regulatory and planification exercises), for validation of approximations throughout the modeling process, testing model stability, or applied in the MR management. The general methodology can be tailored and made more efficient for specific risks and algorithms. We may also enlarge the neighborhood around the model or adjoin new dimension to $\mathcal{M}$[8] that would

[7] An example can be a derivatives model used not only for pricing but also for hedging.

[8] As an example, consider P&L modeled by a normal distribution $\mathcal{M} = \mathcal{N}(\mu, \sigma)$. To evaluate the impact of relaxing the assumption of symmetry we may introduce skew, and so embed $\mathcal{M}$ into a larger manifold of skew–normal distributions, $\bar{\mathcal{M}} = \{p(x, \mu, \sigma, s) : \mu \in \mathbb{R}, \sigma > 0, s \in \mathbb{R}\}$ where s is the shape parameter. For $s = 0$ we recover the initial normal distribution.

consider missing properties, additional information about model limitations, or wrong underlying assumptions, for verification of model robustness and stability. The framework may be extended using data sub–manifolds in the case of hidden variables and/or incomplete data (suggested by one of the reviewers).

Acknowledgements. The authors express their gratitude to the reviewers for valuable comments. This research has been funded by EU in the HORIZON2020 program, EID "Wakeupcall" (Grant agreement number 643945 inside MSCA-ITN-2014-EID).

6 Appendix

Proof of Theorem 1: π is a smooth immersion that satisfies $\pi\sigma\pi = \pi$. From the definition of pullback, $\forall x \in S$ and $p = \pi(x) \in \mathcal{M}$, and $\forall v_1, v_2 \in T_x S$ tangent vectors, $\pi^* g(x)(v_1, v_2) = g(\pi(x))(D_x\pi \cdot v_1, D_x\pi \cdot v_2)$, so that $\pi^* g$ is symmetric and positive semi–definite for any map π. Thus, for $v \in T_x S$, $v \neq 0$, $\pi^* g$ is a Riemannian metric iff $\pi^* g(x)(v,v) > 0$ iff $D_x\pi \cdot v \neq 0$ (since g is Riemannian) iff $ker\, T_x\pi = 0$. In this case, $\pi : (S, \pi^* g) \to (\mathcal{M}, g)$ is an isometric immersion. Using π we can pull back all the extra structure defined on $\mathcal{M}$ required for the quantification of MR, including the weight function and the Banach space of the output functions. □

Proof of Proposition 2: It suffices to prove Eq. (5) for functions with compact support. As $\pi(U)$ is endowed with a canonical parametrization we only consider f with $supp\, f$ contained in a chart (apply partition of unity otherwise). Let ψ be a chart on $\pi(U)$ with the coordinates $(\theta^1, \ldots, \theta^n)$. We can assume that $\phi = \pi^{-1}(\psi)$ is a chart on S with coordinates $(y^1, \ldots, y^n)$. By pushing forward $(y^1, \ldots, y^n)$ to $\pi(U)$, we can consider $(y^1, \ldots, y^n)$ as new coordinates in ψ. With this identification of ϕ and ψ, π^* becomes the identity. Hence, Eq. 5 amounts to proving that $\pi^*\zeta$ and ζ coincide in ψ. Let g^{θ}_{ij} be the components of the metric g in ψ in coordinates $(\theta^1, \ldots, \theta^n)$, and g^{x}_{kl} the components of the metric g in ψ in coordinates $(y^1, \ldots, y^n)$. Then $g^{x}_{kl} = g^{\theta}_{ij} \cdot \partial\theta^i/\partial y^k \cdot \partial\theta^j/\partial y^l$. Let $\tilde{h}_{kl}$ be the components of the metric h in ϕ in the coordinates $(y^1, \ldots, y^n)$. Since $h = \pi^* g$, we have $\tilde{h}_{kl} = (\pi^* g)_{kl} = g^{\theta}_{ij} \cdot \partial\theta^i/\partial x^k \cdot \partial\theta^j/\partial x^l$, hence $\tilde{h}_{kl} = g^{x}_{kl}$. Since measures $\pi^*\zeta$ and ζ have the same density function, say P, $d\zeta = P\sqrt{\det g^x} dx^1 \cdots dx^n = P\sqrt{\det \tilde{h}} dx^1 \cdots dx^n = d\pi^*\zeta$, which proves the identity of measures ζ and $\pi^*\zeta$. □

Proof of Theorem 2: Recall that Z on $\mathcal{M}$ for p_0 is $Z(f, p_0) = ||f - f(p_0)||$ where $f \in (\mathcal{F}, ||.||)$ is a measurable function belonging to a Banach space with respect to ζ. We want to show that Z is equivalent to $Z(F, x_0) = ||F - F(x_0)||$ defined on S endowed with the pullback structure and measurable functions $F = \pi^* f$ belonging to a Sobolev space with respect to $\pi^*\zeta$. The fitting process π represents a smooth map $\pi : S \to \mathcal{M}$, and so provides a pullback of differential forms from $\mathcal{M}$ to S. Namely, let $D_x\pi$ denote the tangent map of π at $x \in S$. The pullback of any tensor $\omega \in \mathcal{T}^k(\mathcal{M})$, where $\mathcal{T}^k(\mathcal{M})$ denotes the set of all C^∞–covariant tensor fields of order k on $\mathcal{M}$ by π, is defined at $x \in S$ by $(\pi^*\omega)(x) = (D_x\pi)^*(\omega(x))$.

The pullback π^* is a map from $\mathcal{T}^k(\mathcal{M})$ to $\mathcal{T}^k(S)$. The pullback respects exterior products and differentiation:

$$\pi^*(\omega \wedge \eta) = \pi^*\omega \wedge \pi^*\eta, \qquad \pi^*(d\omega) = d(\pi^*\omega), \quad \omega, \eta \in \mathcal{T}(\mathcal{M})$$

Besides being a smooth map, π is an immersion, i.e. for each $p \in \mathcal{M}$, there is a neighborhood U of p such that $\pi\big|U : U \to \pi(U)$ is a diffeomorphism. Since S and $\mathcal{M}$ are Riemannian manifolds with respective volume forms ζ_S and ζ, the tangent map $D_x\pi : T_xS \to T_{\pi(x)}\mathcal{M}$ can be represented by an $n \times n$ matrix Φ independent of the choice of the orthonormal basis. Following [8], the matrix Φ of a local diffeomorphims π has n positive singular values, $\alpha_1(x) \geq \cdots \geq \alpha_n(x) > 0$. Similarly, the inverse map $T_{\pi(x)}(\pi^{-1}) : T_{\pi(x)}\mathcal{M} \to S$ is represented by the inverse matrix Φ^{-1}, whose singular values are the reciprocals of those for Φ, i.e. $\beta_1(\pi(x)) \geq \cdots \geq \beta_n(\pi(x)) > 0$ which satisfies $\beta_i(\pi(x)) = \alpha_{n-i+1}(x)^{-1}$, i.e. $\beta_i = \alpha_{n-i+1}^{-1} \circ \pi^{-1}$, for $i = 1, \ldots, n$. Then the pullback of the volume form on $\mathcal{M}$ is given $\pi^*\zeta = (det\,\Phi)\zeta_S = (\alpha_1 \ldots \alpha_n)\zeta_S$.

Since π is a local isometry on U, the linear map $D_x\pi : T_xS \to T_{\pi(x)}\mathcal{M}$ at each point $x \in U \subset S$ is an orthogonal linear isomorphism and so $D_x\pi$ is invertible [7]. Then, the matrix Φ is orthogonal at every $x \in S$, which implies that the singular values are $\alpha_1 = \cdots = \alpha_n = 1$. So, π preserves the volume, i.e. $\zeta_S = \pi^*\zeta$, and the orientation on a neighbourhood U around each point through the bundle isomorphism $TS \cong \pi^*(T\mathcal{M})$. In [8], the authors provide a general inequality for the L^q–norm of a pullback for an arbitrary k–form on Riemannian manifolds. Given $q, r \in [1, \infty]$ such that $1/q + 1/r = 1$, and some $k = 0, \ldots, n$, suppose that the product $(\alpha_1 \ldots \alpha_{n-k})^{1/q}(\alpha_{n-k+1} \ldots \alpha_n)^{-1/r}$ is uniformly bounded on S. Then, for any smooth k-form $\omega \in L^q\Lambda^k(S)$,

$$\left|\left|(\alpha_1 \ldots \alpha_k)^{1/r}(\alpha_{k+1} \ldots \alpha_n)^{-1/q}\right|\right|_\infty^{-1} ||\omega||_q \leq ||\phi_*\omega||_q \leq \left|\left|(\alpha_1 \ldots \alpha_{n-k})^{1/r}(\alpha_{n-k+1} \ldots \alpha_n)^{-1/p}\right|\right|_\infty ||\omega||_q$$

Similarly, for any $\eta \in L^q\Lambda^k(\mathcal{M})$,

$$\left|\left|(\beta_1 \ldots \beta_k)^{1/r}(\beta_{k+1} \ldots \beta_n)^{-1/q}\right|\right|_\infty^{-1} ||\eta||_q \leq ||\pi^*\eta||_q \leq \left|\left|(\beta_1 \ldots \beta_{n-k})^{1/r}(\beta_{n-k+1} \ldots \beta_n)^{-1/q}\right|\right|_\infty ||\eta||_q$$

For isometry, the singular values are $\alpha_1 = \cdots = \alpha_n = 1$, so that the above stated inequalities reduce to

$$||\omega||_q \leq ||\pi^*\omega||_q \leq ||\omega||_q \text{ and } ||\eta||_q \leq ||\pi^*\eta||_q \leq ||\eta||_q$$

for any $\omega \in L^q\Lambda^k(S)$ and for any $\eta \in L^q\Lambda^k(\mathcal{M})$, respectively. This means that π preserves the L^q norm for all r, and consequently the Sobolev norm since this norm is a finite sum of L^q norms. □

References

1. Adams, R.A.: Sobolev Spaces. Academic Press, New York (1975)
2. Brigo, D., Liinev, J.: On the distributional distance between the Libor and the Swap market models (2002, preprint)

3. Csiszár, I., Breuer, T.: An information geometry problem in mathematical finance. In: Nielsen, F., Barbaresco, F. (eds.) GSI 2015. LNCS, vol. 9389, pp. 435–443. Springer, Cham (2015). doi:10.1007/978-3-319-25040-3_47
4. Elles, J., Lemaire, L.: Selected topics in harmonic maps. Conference Board of the Mathematical Sciences by the American Mathematical Society Providence, Rhode Island (1983)
5. Krajčovičová, Z., Pérez Velasco, P.P., Vázques Cendón, C.: A Novel Approach to Quantification of Model Risk for Practitionners (2017). https://ssrn.com/abstract=2939983
6. Morini, M.: Understanding and Managing Model Risk: A Practical Guide for Quants, Traders and Validators. Wiley, Hoboken (2011)
7. Robbin, J.W.: Extrinsic Differential Geometry, Lecture notes (2008). http://www.math.wisc.edu/~robbin/
8. Stern, A.: L^p change of variables inequalities on manifolds. arXiv preprint arXiv:1004.0401 (2010)

Process Comparison Combining Signal Power Ratio and Jeffrey's Divergence Between Unit-Power Signals

Eric Grivel[1(✉)] and Leo Legrand[1,2]

[1] Bordeaux University - INP Bordeaux ENSEIRB-MATMECA - IMS - UMR CNRS, 5218 Talence, France
eric.grivel@ims-bordeaux.fr

[2] Thales Systèmes Aéroportés, Site de Merignac, France

Abstract. Jeffrey's divergence (JD), the symmetric Kullback-Leibler (KL) divergence, has been used in a wide range of applications. In recent works, it was shown that the JD between probability density functions of k successive samples of autoregressive (AR) and/or moving average (MA) processes can tend to a stationary regime when the number k of variates increases. The asymptotic JD increment, which is the difference between two JDs computed for k and $k-1$ successive variates tending to a finite constant value when k increases, can hence be useful to compare the random processes. However, interpreting the value of the asymptotic JD increment is not an easy task as it depends on too many parameters, i.e. the AR/MA parameters and the driving-process variances. In this paper, we propose to compute the asymptotic JD increment between the processes that have been normalized so that their powers are equal to 1. Analyzing the resulting JD on the one hand and the ratio between the original signal powers on the other hand makes the interpretation easier. Examples are provided to illustrate the relevance of this way to operate with the JD.

1 Introduction

Comparing stochastic processes such as autoregressive and/or moving-average (AR, MA or ARMA) processes can be useful in many applications, from speech processing to biomedical applications, from change detection to process classification. Several ways exist: cepstral distance can be useful, for instance for EEG classification [1]. Power spectrum comparison is also of interest. Several distances have been proposed such as the COSH distance, the log spectral distance or the Itakura-Saito distance. They have been widely used, especially in speech processing [2]. Alternative solutions consist in measuring the dissimilarity between probability density functions (pdf) of data. In this case, divergences or distances can be computed such as the Hellinger distance and the Bhattacharyya divergence. The reader may refer to [3] for a comparative study between them. Metrics in the information geometry can be also seen as dissimilarity measures. The reader

F. Nielsen and F. Barbaresco (Eds.): GSI 2017, LNCS 10589, pp. 532–540, 2017.
https://doi.org/10.1007/978-3-319-68445-1_62

may refer to [4] where the information geometry of AR-model covariance matrices is studied. Alternatively, the practitioner often selects the Kullback-Leibler (KL) divergence [5]. This is probably due to the fact that a 'direct' and 'simple' expression of the KL can be deduced for Gaussian pdf.

- When comparing autoregressive (AR) processes, a recursive way to deduce the Jeffrey's divergence (JD), which is the symmetric Kullback-Leibler (KL) divergence, has been proposed in [6]. The approach has been also extended to classify more than two AR processes in various subsets [7].
- When dealing with the comparison between moving-average (MA) processes, we recently gave the exact analytical expressions of the JD between 1st-order MA processes, for any MA parameter and any number of samples. Moreover, the MA processes can be real or complex, noise-free or disturbed by additive white Gaussian noises. For this purpose, we used the analytical expression of each element of the tridiagonal-correlation-matrix inverse [8].
- Comparing AR and MA processes using the JD has been presented in [9] by taking advantage of the expression of the correlation-matrix inverse [10].

In the above cases, some links with the Rao distance have been drawn [11] when it was possible. It was for instance confirmed that the square of the Rao distance was approximately twice the value of the JD, except when a 1st-order MA process is considered whose zero is close to the unit-circle in the z-plane.

We also concluded that the JD tends to a stationary regime because the JD increment, i.e. the difference between two JDs computed for k and $k-1$ successive variates, tends to a constant value when k increases. This phenomenon was observed for most cases, except for ARMA processes whose zeros are on the unit-circle in the z-plane. In addition, we showed that the asymptotic JD increment was sufficient to compare these random processes. This latter depends on the process parameters. In practice, the comparison can operate with the following steps: given the AR/MA parameters, the asymptotic JD increment is evaluated. The computation cost is smaller and the JD is no longer sensitive to the choice of k.

Nevertheless, interpreting the value of the asymptotic JD increment is not necessarily easy especially because it is a function of all the parameters defining the processes under study. For this reason, we propose to operate differently in this paper. Instead of comparing two stochastic processes by using the asymptotic JD increment, we rather suggest computing the asymptotic JD increment between the processes that have been normalized so that their powers are equal to 1 and looking simultaneously at the ratio between the powers of the two original processes. We will illustrate the benefit of this approach in the following cases: when a 1st-order AR process is compared with a white noise and then when two real 1st-order AR processes are compared. Note that due to the lack of space, other illustrations based on MA processes cannot be presented.

This paper is organized as follows: in Sect. 2, we briefly recall the definitions and properties of the AR processes. In Sect. 3, the expression of the JD is introduced and our contributions are presented. Illustrations are then proposed. In the following, I_k is the identity matrix of size k and Tr the trace of a matrix.

The upper-script ^{T} denotes the transpose. $x_{k_1:k_2} = (x_{k_1}, ..., x_{k_2})$ is the collection of samples from time k_1 to k_2. $l = 1, 2$ is the label of the process under study.

2 Brief Presentation of the AR Processes

Let us consider the lth autoregressive (AR) process with order p. Its nth sample, denoted as x_n^l, is defined as follows:

$$x_n^l = -\sum_{i=1}^{p} a_i^l x_{n-i}^l + u_n^l \tag{1}$$

where the driving process u_n^l is white, Gaussian, zero-mean with variance $\sigma_{u,l}^2$.

These wide-sense stationary processes are characterized by their correlation functions, denoted as $r_{AR,l,\tau}$, with τ the lag. In addition, the Toeplitz covariance matrices are denoted as $Q_{AR,l,k}$ for $l = 1, 2$. Note that for 1st-order AR processes, the correlation function satisfies: $r_{AR,l,\tau} = \frac{(-a_1^l)^{|\tau|}}{1-(a_1^l)^2}\sigma_{u,l}^2$.

In addition, given (1), the AR processes can be seen as the outputs of filters whose inputs are zero-mean white sequences with unit-variances and with transfer functions $H_l(z) = \sigma_{u,l}\frac{1}{\prod_{i=1}^{p}(1-p_i^l z^{-1})}$, where $\{p_i^l\}_{i=1,...,p}$ are the poles. The inverse filters are then defined by the transfer functions $H_l^{-1}(z)$.

3 Jeffrey Divergence Analysis

The Kullback-Leibler (KL) divergence between the joint pdf of k successive values of two random processes, denoted as $p_1(x_{1:k})$ and $p_2(x_{1:k})$, can be evaluated to study the dissimilarities between the processes [12]:

$$KL_k^{(1,2)} = \int_{x_{1:k}} p_1(x_{1:k}) ln\left(\frac{p_1(x_{1:k})}{p_2(x_{1:k})}\right) \mathrm{d}x_{1:k} \tag{2}$$

If the real processes are Gaussian with means $\mu_{1,k}$ and $\mu_{2,k}$ and covariance matrices $Q_{1,k}$ and $Q_{2,k}$, it can be shown that the KL satisfies [13]:

$$KL_k^{(1,2)} = \frac{1}{2}\left[\mathrm{Tr}(Q_{2,k}^{-1}Q_{1,k}) - k - \ln\frac{\det Q_{1,k}}{\det Q_{2,k}} + (\mu_{2,k}-\mu_{1,k})^T Q_{2,k}{}^{-1}(\mu_{2,k}-\mu_{1,k})\right] \tag{3}$$

As the KL is not symmetric, the Jeffrey's divergence (JD) can be preferred:

$$JD_k^{(1,2)} = \frac{1}{2}(KL_k^{(1,2)} + KL_k^{(2,1)}) \tag{4}$$

For zero-mean processes and given (4), the JD can be expressed as follows:

$$JD_k^{(1,2)} = -k + \frac{1}{2}\left[Tr(Q_{2,k}^{-1}Q_{1,k}) + Tr(Q_{1,k}^{-1}Q_{2,k})\right] \tag{5}$$

In the following, our purpose is to study the behavior of the JD when k increases. Therefore, let us introduce the asymptotic JD increment defined by:

$$\Delta JD^{(1,2)} = -1 + \frac{1}{2}\left[\Delta T_{2,1} + \Delta T_{1,2}\right] \tag{6}$$

with the asymptotic increments $T_{l,l'}$ of the trace -$(l,l') = (1,2)$ or $(2,1)$- :

$$\Delta T_{l,l'} = \lim_{k\to+\infty} \left[Tr(Q_{l,k}^{-1}Q_{l',k}) - Tr(Q_{l,k-1}^{-1}Q_{l',k-1})\right] \tag{7}$$

In the next section, we analyze the relevance of JD when comparing processes that are normalized so that their powers are equal to 1.

4 Applications

4.1 JD Between a 1st-Order AR Process and a White Noise

By taking advantage of the inverse of the correlation matrix of a 1st-order AR process [10], the asymptotic JD increment between a 1st-order AR process and a white noise can be expressed as follows:

$$\Delta JD_k{}^{(AR,WN)} = -1 + \frac{1}{2}\left[\Delta T_{WN,AR} + \Delta T_{AR,WN}\right] \tag{8}$$

where

$$\Delta T_{WN,AR} = \frac{\sigma_{u,1}^2}{\sigma_{u,2}^2}\frac{1}{(1-(a_1^1)^2)} \text{ and } \Delta T_{AR,WN} = \frac{\sigma_{u,2}^2}{\sigma_{u,1}^2}(1+(a_1^1)^2) \tag{9}$$

Let us see if one can easily analyze the sensitivity of the JD with respect to the process parameters. For this reason, (8) is first rewritten as follows:

$$\Delta JD^{(AR,WN)} = -1 + \frac{1}{2}\left[\frac{1}{R_u}\frac{1}{(1-(a_1^1)^2)} + R_u(1+(a_1^1)^2)\right] \tag{10}$$

with $R_u = \frac{\sigma_{u,2}^2}{\sigma_{u,1}^2}$. Then, let us express $\sigma_{u,2}^2$ as $\sigma_{u,1}^2 + \delta\sigma_u^2$ and introduce the relative difference between the noise-variances $\Delta\sigma_u^2 = \frac{\delta\sigma_u^2}{\sigma_{u,1}^2}$. This leads to:

$$\Delta JD^{(AR,WN)} = -1 + \frac{1}{2}\left[\frac{1}{1+\Delta\sigma_u^2}\frac{1}{(1-(a_1^1)^2)} + (1+\Delta\sigma_u^2)(1+(a_1^1)^2)\right] \tag{11}$$

In Fig. 1, the asymptotic JD increment between an AR process and a white noise is presented as a function of the AR parameter a_1^1 and the relative difference between the noise-variances $\Delta\sigma_u^2$. This latter varies in the interval $]-1,20]$ with a step equal to 0.03. Only positive values of the AR parameter a_1^1 are considered because $\Delta JD^{(AR,WN)}$ is an even function with respect to the AR parameter in this case. In addition, the AR parameter varies with a step equal to 0.01. Therefore, this illustration makes it possible to present a large set of situations that could happen. Let us now give some comments about Fig. 1:

When the AR parameter is equal to zero, $\Delta JD^{(AR,WN)}$ is only equal to 0 when $\Delta\sigma_u^2$ is equal to 0. Indeed, this amounts to comparing two zero-mean white noises with the same variance. Then, assuming that $\Delta\sigma_u^2$ is still equal to zero, $\Delta JD^{(AR,WN)}$ is all the higher as the power spectral density (PSD) of the AR

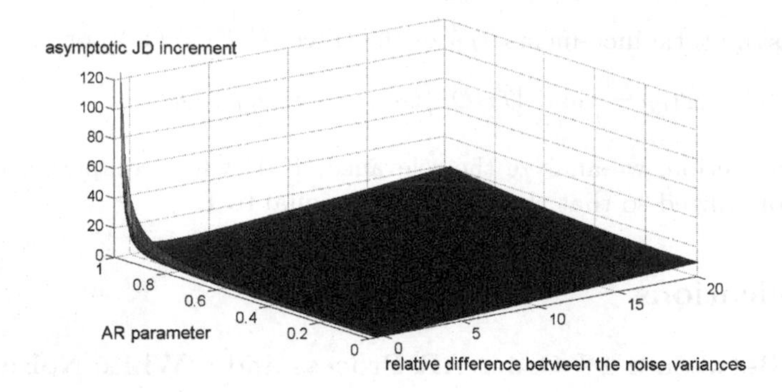

Fig. 1. Asymptotic JD increment as a function of the AR parameter and $\Delta\sigma_u^2$

process is spikier, i.e. the modulus of the AR pole tends to 1. When $\Delta\sigma_u^2$ increases or decreases, the phenomenon remains the same. From a theoretical point of view, this is confirmed by calculating the derivative of $\Delta JD^{(AR,WN)}$ with respect to the AR parameter a_1^1. However, the range of the values taken by $\Delta JD^{(AR,WN)}$ is different. One can also calculate the derivative of $\Delta JD^{(AR,WN)}$ with respect to $\Delta\sigma_u^2$. In this case, one can notice that the minimum value of $\Delta JD^{(AR,WN)}$ is obtained when $\Delta\sigma_u^2 = \sqrt{\frac{1}{1-(a_1^1)^4}} - 1$ and is equal to $\frac{\sqrt{1-(a_1^1)^4}}{1-(a_1^1)^2} - 1$. As a conclusion, when looking at Fig. 1, $\Delta JD^{(AR,WN)}$ takes into account the differences between all the process parameters, but it is not easy to know whether the value that is obtained is due to differences in terms of signal magnitudes and/or spectral shapes. One value of $\Delta JD^{(AR,WN)}$ is related to several situations.

Instead of using this criterion on the processes themselves, we suggest considering the ratio between the random processes on the one hand and the asymptotic increment of the JD between the processes whose powers have been normalized respectively by the square of their process, i.e. $\frac{\sigma_{u,1}}{\sqrt{(1-(a_1^1)^2)}}$ and $\sigma_{u,2}$. It should be noted that in practical cases, the signal powers can be easily estimated from the data. In this case, $\Delta T_{WN,AR}$ is divided by the power of the AR process $\frac{\sigma_{u,1}^2}{(1-(a_1^1)^2)}$ and multiplied by the power of the white noise $\sigma_{u,2}^2$. Similarly, $\Delta T_{AR,WN}$ is divided by the power of the white noise $\sigma_{u,2}^2$ and multiplied by the AR-process power $\frac{\sigma_{u,1}^2}{(1-(a_1^1)^2)}$. Therefore, given (8), it can be easily shown that the asymptotic increment of the JD between the normalized AR process (nAR) and the normalized white noise (nWN) is equal to:

$$\Delta JD_k{}^{(nAR,nWN)} = -1 + \frac{1}{2}\left(1 + \frac{1+(a_1^1)^2}{1-(a_1^1)^2}\right) \tag{12}$$

When a_1^1 is equal to 0, both processes have a PSD that is flat and the asymptotic increment of the JD between the unit-power processes is equal to 0. Meanwhile, one can easily compare the powers of both non-normalized processes. When the

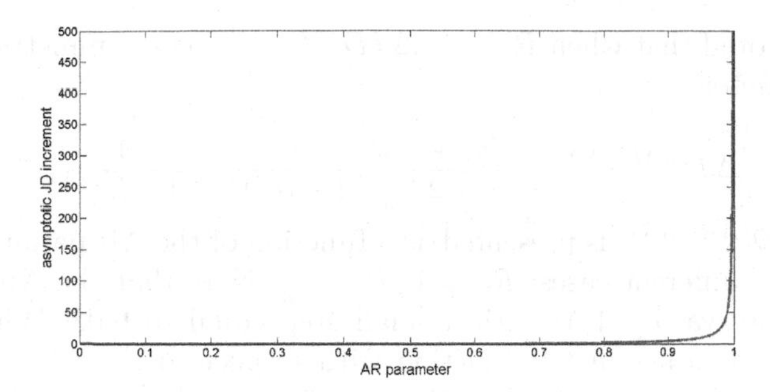

Fig. 2. Asymptotic JD increment between a unit-power AR process and a unit-power white noise as a function of the AR parameter

AR-parameter modulus increases, the PSD of the AR process tends to be spikier and spikier whereas the PSD of the white noise is flat. In this case, the asymptotic JD increment becomes larger and larger. It is also illustrated in Fig. 2.

By removing the influence of the noise variances in the asymptotic JD increment, it is easier to give an interpretation related to the spectral shapes of the processes. Meanwhile, one can compare the powers of both processes.

4.2 JD Between Two 1st-Order AR Processes

In [6], we suggested using a recursive way to deduce the JD two 1st-order AR processes:

$$\Delta JD^{(AR_1,AR_2)} = A + B \tag{13}$$

with

$$\begin{cases} A = -1 + \frac{1}{2}(R_u + \frac{1}{R_u}) \\ B = \frac{(a_1^2 - a_1^1)^2}{2}\left[\frac{1}{1-(a_1^1)^2}\frac{1}{R_u} + \frac{1}{1-(a_1^2)^2}R_u\right] \end{cases} \tag{14}$$

By reorganizing the terms in (14), one has:

$$\begin{aligned} \Delta JD^{(AR_1,AR_2)} &= -1 + \frac{1}{2}\left[\Delta T_{AR_1,AR_2} + \Delta T_{AR_2,AR_1}\right] \\ &= -1 + \frac{1}{2}(R_u \frac{1 - 2a_1^1 a_1^2 + (a_1^1)^2}{1-(a_1^2)^2} + \frac{1}{R_u}\frac{1 - 2a_1^1 a_1^2 + (a_1^2)^2}{1-(a_1^1)^2}) \end{aligned} \tag{15}$$

where

$$\Delta T_{AR_1,AR_2} = R_u \frac{1 - 2a_1^1 a_1^2 + (a_1^1)^2}{1-(a_1^2)^2} \text{ and } \Delta T_{AR_2,AR_1} = \frac{1}{R_u}\frac{1 - 2a_1^1 a_1^2 + (a_1^2)^2}{1-(a_1^1)^2}) \tag{16}$$

It should be noted that when $R_u = 1$, $\Delta JD^{(AR_1,AR_2)}$ is a symmetric function of the AR parameters:

$$\Delta JD^{(AR_1,AR_2)} = \frac{(a_1^2 - a_1^1)^2}{2}\left(\frac{1}{1-(a_1^2)^2} + \frac{1}{1-(a_1^1)^2}\right) \tag{17}$$

In Fig. 3, $\Delta JD^{(AR_1,AR_2)}$ is presented as a function of the AR parameters a_1^1 and a_1^2 and for two different cases: $R_u = 1$, $R_u = \frac{3}{2}$. Note that the AR parameters vary in the interval $]-1,1[$ with a small step equal to 0.03. When $R_u = 1$, $\Delta JD^{(AR_1,AR_2)}$ is a symmetric function with respect to the AR parameters of both processes. Nevertheless, when $R_u \neq 1$, this is no longer true.

To help for interpretation, let us now normalize both processes respectively by $\frac{\sigma_{u,1}}{\sqrt{(1-(a_1^1)^2)}}$ and $\frac{\sigma_{u,2}}{\sqrt{(1-(a_1^2)^2)}}$ so that the process powers become equal to 1. Using (15), it can be easily shown that the asymptotic increment of the JD between the normalized AR processes becomes equal to:

$$\begin{aligned}\Delta JD^{(nAR_1,nAR_2)} &= -1 + \frac{1}{2}\left(\frac{1-2a_1^1a_1^2+(a_1^1)^2}{1-(a_1^1)^2} + \frac{1-2a_1^1a_1^2+(a_1^2)^2}{1-(a_1^2)^2}\right) \\ &= (a_1^1 - a_1^2)\left(\frac{a_1^1}{1-(a_1^1)^2} - \frac{a_1^2}{1-(a_1^2)^2}\right)\end{aligned} \tag{18}$$

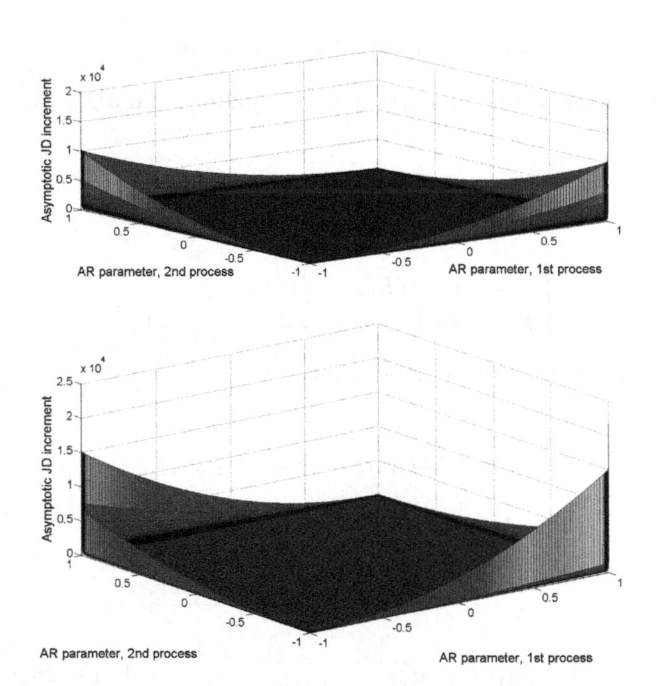

Fig. 3. Asymptotic JD increment between two AR processes as a function of the AR parameters, where $R_u = 1$ and $R_u = 3/2$

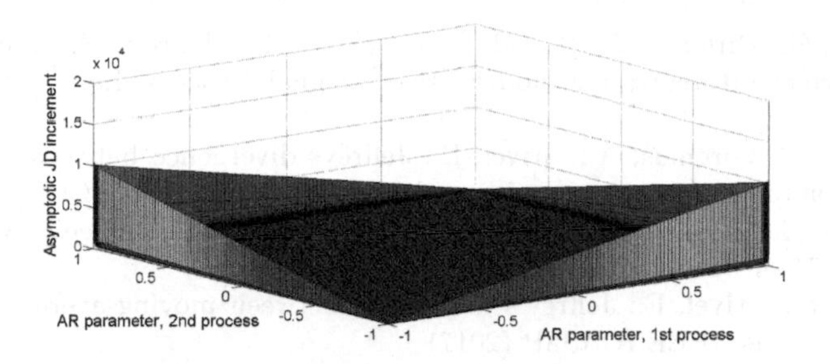

Fig. 4. Asymptotic JD increment between two unit-power AR processes as a function of the AR parameters for any R_u

As depicted in Fig. 4 and according to (18), $\Delta JD^{(nAR_1,nAR_2)}$ is equal to 0 when the AR parameters are the same. In addition, it is a symmetric function with respect to the AR parameters of both processes. Looking at the power ratio at the same time is then a way to clearly see that the processes have the same spectral shapes but their powers are not the same. This could not be pointed out by only looking at the JD between non-normalized AR processes.

5 Conclusions

Interpreting the value of the asymptotic JD increment is not necessarily straightforward because the influences of the process parameters are mixed. To make the interpretation easier, two criteria should rather be taken into account: the process power ratio and the asymptotic increment of the JD between the processes that are preliminary normalized so that their powers are equal to 1.

References

1. Assaleh, K., Al-Nashash, H., Thakor, N.: Spectral subtraction and cepstral distance for enhancing EEG entropy. In: IEEE Proceedings of the Engineering in Medicine and Biology, vol. 3, pp. 2751–2754 (2005)
2. Bobillet, W., Diversi, R., Grivel, E., Guidorzi, R., Najim, M., Soverini, U.: Speech enhancement combining optimal smoothing and errors-in-variables identification of noisy AR processes. IEEE Trans. Signal Proces. **55**, 5564–5578 (2007)
3. Abou-Moustafa, K., Ferrie, F.P.: A note on metric properties for some divergence measures: the Gaussian case. JMLR Workshop and Conference Proceedings, vol. 25, pp. 1–15 (2012)
4. Formont, P., Ovarlez, J., Pascal, F., Vasile, G.: On the extension of the product model in POLSAR processing for unsupervised classification using information geometry of covariance matrices. In: IEEE International Geoscience and Remote Sensing Symposium, pp. 1361–1364 (2011)
5. Murthy, R., Pavlidis, I., Tsiamyrtzis, P.: Touchless monitoring of breathing function. In: IEEE EMBS, pp. 1196–1199 (2004)

6. Magnant, C., Giremus, A., Grivel, E.: On computing Jeffrey's divergence between time-varying autoregressive models. IEEE Signal Process. Lett. **22**(7), 915–919 (2014)
7. Magnant, C., Giremus, A., Grivel, E.: Jeffreys divergence between state models: application to target tracking using multiple models. In: EUSIPCO, pp. 1–5 (2013)
8. Yueh, W.-C.: Explicit inverses of several tridiagonal matrices. Appl. Mathe. E-Notes, 74–83 (2006)
9. Legrand, L., Grivel, E.: Jeffrey's divergence between moving-average and autoregressive models. IEEE ICASSP (2017)
10. Cernuschi-Frias, B.: A derivation of the gohberg-semencul relation [signal analysis]. IEEE Trans. Signal Proces. **39**(1), 190–192 (1991)
11. Rao, C.: Information and the accuracy attainable in the estimation of statistical parameters. Bull. Calcutta Math. Soc. **37**, 81–89 (1945)
12. Kullback, S., Leibler, R.A.: On information and sufficiency. Ann. Math. Stat. **22**(1), 79–86 (1951)
13. Rasmussen, C.E., Williams, C.K.I.: Gaussian Processes for Machine Learning. MIT Press, Cambridge (2006)

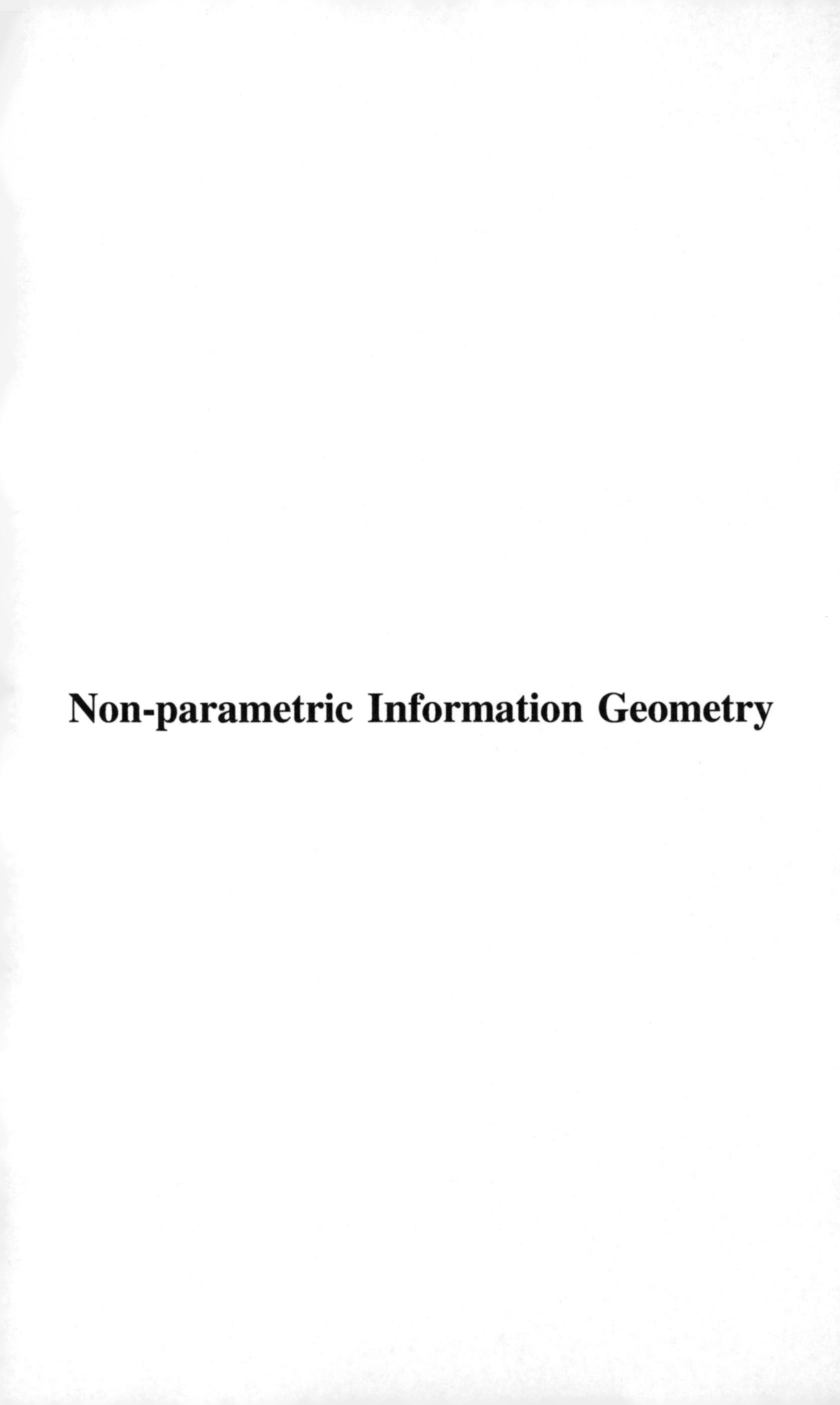

Non-parametric Information Geometry

Itô Stochastic Differential Equations as 2-Jets

John Armstrong[1] and Damiano Brigo[2(✉)]

[1] Department of Mathematics, King's College London, London, UK
[2] Department of Mathematics, Imperial College London,
180 Queen's Gate, London SW7 2AZ, UK
damiano.brigo@imperial.ac.uk

Abstract. We explain how Itô Stochastic Differential Equations on manifolds may be defined as 2-jets of curves and show how this relationship can be interpreted in terms of a convergent numerical scheme. We use jets as a natural language to express geometric properties of SDEs. We explain that the mainstream choice of Fisk-Stratonovich-McShane calculus for stochastic differential geometry is not necessary. We give a new geometric interpretation of the Itô–Stratonovich transformation in terms of the 2-jets of curves induced by consecutive vector flows. We discuss the forward Kolmogorov equation and the backward diffusion operator in geometric terms. In the one-dimensional case we consider percentiles of the solutions of the SDE and their properties. In particular the median of a SDE solution is associated to the drift of the SDE in Stratonovich form for small times.

1 Introduction

This paper is a summary of the preprint by Armstrong and Brigo (2016) [1] and examines the geometry of stochastic differential equations using a coordinate free approach. We suggest for the first time, to the best of our knowledge that SDEs on manifolds can be interpreted as 2-jets of curves driven by Brownian motion, and show convergence of a jet-based numerical scheme.

We will use the language of jets to give geometric expressions for many important concepts that arise in stochastic analysis. These geometric representations are in many ways more elegant than the traditional representations in terms of the coefficients of SDEs. In particular we will give coordinate free formulations of the following: Itô's lemma; the diffusion operators; Itô SDEs on manifolds and Brownian motion on Riemannian manifolds.

We discuss how our formulation is related to the Stratonovich formulation of SDEs. We will prove that sections of the bundle of n-jets of curves in a manifold correspond naturally to n-tuples of vector fields in the manifold. This correspondence shows that Itô calculus and Stratonovich calculus can both be interpreted as simply a choice of coordinate system for the space of sections

Damiano Brigo gratefully acknowledges financial support from the dept. of Mathematics at Imperial College London via the research impulse grant DRI046DB.

F. Nielsen and F. Barbaresco (Eds.): GSI 2017, LNCS 10589, pp. 543–551, 2017.
https://doi.org/10.1007/978-3-319-68445-1_63

of the bundle of n-jets of curves, and provides a new interpretation of the Itô-Stratonovich transformation.

We will also see that the 2-jet defining an Itô SDE can help in studying quantiles of the SDE solution. Moreover we will show that the drift vector of the Stratonovich formulation can be similarly interpreted as a short-time approximation to the median. This observation is related to the coordinate independence of 2-jets and vector fields and the coordinate independence of the notion of median.

Coordinate free formulations of SDEs have been considered before. In particular SDEs have been described either in terms of "second order tangent vectors/diffusors" and Schwartz morphism, see for example [7], or via the Itô bundle of Belopolskaja and Dalecky, see for example [8] or the appendix in [4]. We briefly explore the relationship of the jet approach with earlier approaches in the appendix of [1], but in general we have replaced second order tangent vectors with the more familiar and standard geometric concept of two jets.

We conclude by remarking that our work has numerous applications, including a novel notion of optimal projection for SDEs [2,3] that allows to approximate in a given submanifold the solution of a SDE evolving on a larger space. We already applied the new projection to stochastic filtering in [2].

2 SDEs as 2-jets and Itô-Stratonovich Transformation

Suppose that at every point $x \in \mathbb{R}^n$ we have an associated smooth curve $\gamma_x : \mathbb{R} \to \mathbb{R}^n$ with $\gamma_x(0) = x$. Example: $\gamma^E_{(x_1,x_2)}(t) = (x_1, x_2)+t(-x_2, x_1)+3t^2(x_1, x_2)$. This specific example has zero derivatives with respect to t from the third derivative on. We stop at second order in t in the example since this will be enough to converge to classical stochastic calculus, but our theory is general. We have plotted $t \mapsto \gamma^E_x(t)$ in a grid of possible "centers" $x = (x_1, x_2)$ in Fig. 1.

Given such a γ, a starting point x_0, a Brownian motion W_t and a time step δt we can define a discrete time stochastic process using the following recurrence relation:

$$X_0 := x_0, \quad X_{t+\delta t} := \gamma_{X_t}(W_{t+\delta t} - W_t) \tag{1}$$

In Fig. 1 we have plotted the trajectories of process for γ^E, the starting point $(1, 0)$, a fixed realization of Brownian motion and a number of different time steps. Rather than just plotting a discrete set of points for this discrete time process, we have connected the points using the curves in $\gamma^E_{X_t}$.

As the figure suggests, these discrete time stochastic processes (1) converge in some sense to a limit as the time step tends to zero.

We will use the following notation for the limiting process:

$$\text{Coordinate free SDE:} \quad X_t \smile\!\!\nearrow \gamma_{X_t}(\mathrm{d}W_t), \quad X_0 = x_0. \tag{2}$$

For the time being, let us simply treat Eq. (2) as a short-hand way of saying that Eq. (1) converges in some sense to a limit. We will explore the limit question shortly. Note that it will not converge for arbitrary γ's but for nice γ such as γ^E or more general γ's with sufficiently good regularity.

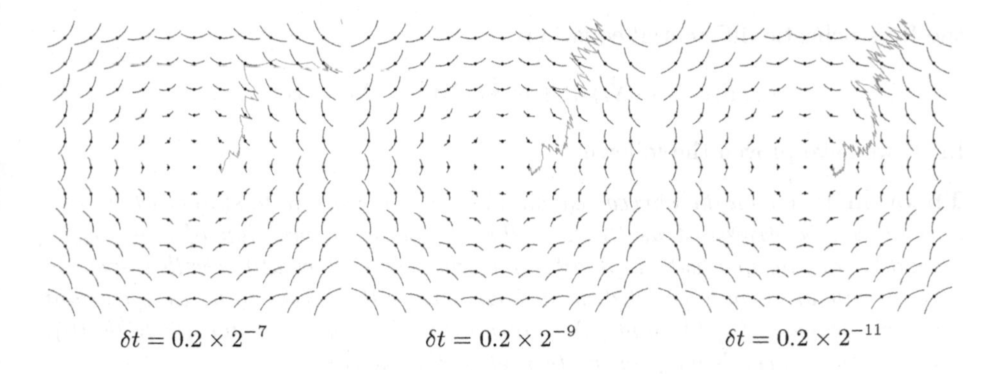

Fig. 1. Discrete time trajectories for γ^E for a fixed W_t and X_0 with different values for δt

An important feature of Eq. (1) is that it makes no reference to the vector space structure of $\mathbb{R}^n$ for our state space X. We could define the same identical scheme in a manifold. We have maintained this in the formal notation used in Eq. (2). By avoiding using the vector space structure on $\mathbb{R}^n$ we will be able to obtain a coordinate free understanding of stochastic differential equations. Now, using $\mathbb{R}^n$ coordinates if we are in an Euclidean space or a coordinate chart if we are in a manifold, consider the (component-wise) Taylor expansion of γ_x. We have:

$$\gamma_x(t) = x + \gamma_x'(0)t + \frac{1}{2}\gamma_x''(0)t^2 + R_x t^3, \quad R_x = \frac{1}{6}\gamma_x'''(\xi), \quad \xi \in [0,t],$$

where $R_x t^3$ is the remainder term in Lagrange form. Substituting this Taylor expansion in our Eq. (1) we obtain

$$\delta X_t = \gamma_{X_t}'(0)\delta W_t + \frac{1}{2}\gamma_{X_t}''(0)(\delta W_t)^2 + R_{X_t}(\delta W_t)^3, \quad X_0 = x_0. \tag{3}$$

Classic strocastic analysis and properties of Brownian motion suggests that we can replace the term $(\delta W_t)^2$ with δt and we can ignore terms of order $(\delta W_t)^3$ and above, see [1] for more details. So we expect that under reasonable conditions, in the chosen coordinate system, the recurrence relation given by (1) and expressed in coordinates by (3) will converge to the same limit as the numerical scheme:

$$\delta \bar{X}_t = \gamma_{\bar{X}_t}'(0)\delta W_t + \frac{1}{2}\gamma_{\bar{X}_t}''(0)\delta t, \quad \bar{X}_0 = x_0. \tag{4}$$

Defining $a(X) := \gamma_X''(0)/2$ and $b(X) := \gamma_X'(0)$ we have that this last equation can be written as

$$\delta \bar{X}_t = a(\bar{X}_t)\delta t + b(\bar{X}_t)\delta W_t. \tag{5}$$

It is well known that this last scheme (Euler scheme) does converge in some appropriate sense to a limit [10] and that this limit is given by the solution to the Itô stochastic differential equation:

$$\mathrm{d}\tilde{X}_t = a(\tilde{X}_t)\,\mathrm{d}t + b(\tilde{X}_t)\mathrm{d}W_t, \quad \tilde{X}_0 = x_0. \tag{6}$$

In [1] we have proven the following

Theorem 1. *Given the curved scheme* (1)*, in any coordinate system where this scheme can be expressed as* (3) *with the coordinate expression of* $h \mapsto \gamma_x(h)$ *smoothly varying in* x *with Lipschitz first and second derivatives with respect to* h*, and with uniformly bounded third derivative with respect to* h*, we have that the three schemes* (3) (4) *and* (5) *converge to the solution of the Itô SDE* (6)*. This solution depends only on the two-jet of the curve* γ*.*

In which sense Eq. (1) and its limit are coordinate free? It is important to note that the coefficients of Eq. (6) only depend upon the first two derivatives of γ. We say that two smooth curves $\gamma : \mathbb{R} \to \mathbb{R}^n$ have the same k-jet ($k \in \mathbb{N}, k > 0$) if they are equal up to order $O(t^k)$ in one (and hence all) coordinate system. The k-jet can then be defined for example as the equivalence class of all curves that are equal up to order $O(t^k)$ in one and hence all coordinate systems. Other definitions are possible, based on operators. Using the jets terminology, we say that the coefficients of Eq. (6) (and (5)) are determined by the 2-jet of the curve γ. Given a curve γ_x, we will write $j_2(\gamma_x)$ for the two jet associated with γ_x. This is formally defined to be the equivalence class of all curves which are equal to γ_x up to $O(t^2)$. Since we stated that, under reasonable regularity conditions, the limit in the symbolic Eq. (2) depends only on the 2-jet of the driving curve, we may rewrite Eq. (2) as:

$$\text{Coordinate-free 2-jet SDE:} \qquad X_t \smile\!\!\nearrow j_2(\gamma_{X_t})(\mathrm{d}W_t), \quad X_0 = x_0. \tag{7}$$

This may be interpreted either as a coordinate free notation for the classical Itô SDE given by Eq. (6) or as a shorthand notation for the limit of the manifestly coordinate-free process given by the discrete time Eq. (1). The reformulation of Itô's lemma in the language of jets we are going to present now shows explicitly that also the first interpretation will be independent of the choice of coordinates. The only issue one needs to consider are the bounds needed to ensure existence of solutions. The details of transferring the theory of existence and uniqueness of solutions of SDEs to manifolds are considered in, for example, [5–7, 9]. As we just mentioned, we can now give an appealing coordinate free version of Itô's formula. Suppose that f is a smooth mapping from $\mathbb{R}^n$ to itself and suppose that X satisfies (1). It follows that $f(X)$ satisfies $(f(X))_{t+\delta t} = (f \circ \gamma_{X_t})(\delta W_t)$. Taking the limit as $\delta t \to 0$ we have:

Lemma 1 (Coordinate free Itô's lemma). *If the process* X_t *satisfies*

$$X_t \smile\!\!\nearrow j_2(\gamma_{X_t})(\mathrm{d}W_t) \quad \textit{then} \quad f(X_t) \quad \textit{satisfies} \quad f(X)_t \smile\!\!\nearrow j_2(f \circ \gamma_{X_t})(\mathrm{d}W_t).$$

We can interpret Itô's lemma geometrically as the statement that the transformation rule for jets under coordinate change is composition of functions.

We now briefly summarize our discussion in [1] generalizing the theory to SDEs driven by d independent Brownian motions W_t^α with $\alpha \in \{1, 2, \dots, d\}$. Consider jets of functions of the form $\gamma_x : \mathbb{R}^d \to \mathbb{R}^n$. Just as before we can consider the candidate 2-jet scheme as a difference equation of the form:

$$X_{t+\delta t} := \gamma_{X_t}\left(\delta W_t^1, \dots, \delta W_t^d\right).$$

Again, the limiting behaviour of such difference equations will only depend upon the 2-jet $j_2(\gamma_x)$ and can be denoted by (7), where it is now understood that dW_t is the vector Brownian motion increment. We obtain a straightforward generalization of Theorem 1, showing that the 2-jet scheme above, in any well-behaving coordinate system, converges in $L^2(\mathbb{P})$ to the classic Itô SDE with the same coefficients (with Einstein summation convention):

$$\mathrm{d}X_t^i = \frac{1}{2}\partial_\alpha \partial_\beta \gamma^i \mathrm{d}W_t^\alpha \mathrm{d}W_t^\beta + \partial_\alpha \gamma^i \,\mathrm{d}W_t^\alpha = \frac{1}{2}\partial_\alpha \partial_\beta \gamma^i g_E^{\alpha\beta} dt + \partial_\alpha \gamma^i \,\mathrm{d}W_t^\alpha \tag{8}$$

Here x^α are the standard $\mathbb{R}^d$ orthonormal coordinates. Our equation should be interpreted with the convention that $\mathrm{d}W_t^\alpha \mathrm{d}W_t^\beta = g_E^{\alpha\beta} \mathrm{d}t$ where g_E is equal to 1 if α equals β and 0 otherwise. We choose to write g_E instead of using a Kronecker δ because one might want to choose non orthonormal $\mathbb{R}^d$ coordinates and so it is useful to notice that g_E represents the symmetric 2-form defining the Euclidean metric on $\mathbb{R}^d$. Using a Kronecker δ would incorrectly suggest that this term transforms as an endomorphism rather than as a symmetric 2-form.

Guided by the above discussion, we introduce the following

Definition 1 (Coordinate free Itô SDEs driven by Brownian motion). *A Ito SDE on a manifold M is a section of the bundle of 2-jets of maps $\mathbb{R}^d \to M$ together with d Brownian motions W_t^i, $i = 1, \dots, d$.*

Take now $f : M \to \mathbb{R}$. We can define a differential operator acting on *functions* in terms of a 2-jet associated with γ_x as follows.

Definition 2 (Backward diffusion operator via 2-jets). *The Backward diffusion operator for the Itô SDE defined by the 2-jet associated with the map γ_x is defined on suitable functions f as*

$$\mathcal{L}_{\gamma_x} f := \frac{1}{2}\Delta_E(f \circ \gamma_x) = \frac{1}{2}\partial_\alpha \partial_\beta (f \circ \gamma_x) g_E^{\alpha\beta}. \tag{9}$$

Here Δ_E is the Laplacian defined on $\mathbb{R}^d$.

$\mathcal{L}_{\gamma_x}$ acts on functions defined on M. In [1] we further express the forward Kolmogorov or Fokker-Planck equation in terms of 2-jets, highlighting the coordinate-free interpretation of the backward and forward diffusion operators. Furthermore, we see that both the Itô SDE (8) and the backward diffusion operator use only part of the information contained in the 2-jet: only the diagonal terms of $\partial_\alpha \partial_\beta \gamma^i$ (those with $\alpha = \beta$) influence the SDE and even for these terms it is only their average value that is important. The same consideration applies to the backward diffusion operator. This motivates our definition of weakly equivalent and strongly equivalent 2-jets given in [1], where weak equivalence is defined between γ^1 and γ^2 if $\mathcal{L}_{\gamma^1_x} = \mathcal{L}_{\gamma^2_x}$, while strong equivalence requires, in addition, the same 1-jets: $j_1(\gamma^1) = j_1(\gamma^2)$. With strong equivalence, given the same realization of the driving Brownian motions W^α_t, the solutions of the SDEs will be almost surely the same (under reasonable assumptions to ensure pathwise uniqueness). When the 2-jets are weakly equivalent, the transition probability distributions resulting from the dynamics of the related SDEs are the same even though the dynamics may be different for any specific realisation of the Brownian motions. For this reason one can define a diffusion process on a manifold as a smooth selection of a second order linear operator $\mathcal{L}$ at each point that determines the transition of densities. A diffusion can be realised locally as an SDE, but not necessarily globally.

Recall that the top order term of a quasi linear differential operator is called its symbol. In the case of a second order quasi linear differential operator D which maps $\mathbb{R}$-valued functions to $\mathbb{R}$-valued functions, the symbol defines a section of S^2T, the bundle of symmetric tensor products of tangent vectors, which we will call g_D. In local coordinates, if the top order term of D is $Df = a^{ij}\partial_i\partial_j f +$ lower order, then g_D is given by $g_D(X_i, X_j) = a^{ij}X_iX_j$. We are using the letter g to denote the symbol for a second order operator because, in the event that g is positive definite and $d = \dim M$, g defines a Riemannian metric on M. In these circumstances we will say that the SDE/diffusion is *non-singular.* Thus we can associate a canonical Riemannian metric $g_{\mathcal{L}}$ to any non-singular SDE/diffusion.

Definition 3. *A non-singular diffusion on a manifold M is called a Riemannian Brownian motion if $\mathcal{L}(f) = \frac{1}{2}\Delta_{g_{\mathcal{L}}}(f)$.*

Note that given a Riemannian metric h on M there is a unique Riemannian Brownian motion (up to diffusion equivalence) with $g_{\mathcal{L}} = h$. This is easily checked with a coordinate calculation.

This completes our definitions of the key concepts in stochastic differential geometry and indicates some of the important connections between stochastic differential equations, Riemannian manifolds, second order linear elliptic operators and harmonic maps. We emphasize that all our definitions are coordinate free and we have worked exclusively with Itô calculus. However, it is more conventional to perform stochastic differential geometry using Stratonovich calculus. The justification usually given for this is that Stratonovich calculus obeys the usual chain rule so the coefficients of Stratonovich SDEs can be interpreted as vector fields. For example one can immediately see if the trajectories to a Stratonovich SDE almost surely lie on particular submanifold by testing if the coefficients of the SDE are all tangent to the manifold. We would argue that the corresponding test for Itô SDEs is also perfectly simple and intuitive: one checks whether the 2-jets lie in the manifold. As is well known, the probabilistic properties of the Stratonovich integral are not as nice as the properties of the Itô integral, and the use of Stratonovich calculus on manifolds comes at a price. Our results allows us to retain the probabilistic advantages of Itô calculus while working directly with geometry. However, we will now further clarify the relationship of our jets approach with Stratonovich calculus.

For simplicity, let us assume for a moment that the SDE driver is one dimensional Brownian motion. Thus to define an SDE on a manifold, one must choose a 2-jet of a curve at each point of the manifold. One way to specify a k-jet of a curve at every point in a neighbourhood is to first choose a chart for the neighbourhood and then consider curves of the form $\gamma_x(t) = x + \sum_{i=1}^{k} a_i(x)t^i$ where $a_i : \mathbb{R}^n \to \mathbb{R}^n$. As we have already seen in (1), these coefficient functions a_i depend upon the choice of chart in a relatively complex way. For example for 2-jets the coefficient functions are not vectors but instead transform according to Itô's lemma. We will call this the *standard representation* for a family of k-jets.

An alternative way to specify the k-jet of a curve at every point is to choose k vector fields $A_1, \ldots, A_k$ on the manifold. One can then define $\Phi^t_{A_i}$ to be the vector flow associated with the vector field A_i. This allows one to define curves at each point x as $\gamma_x(t) = \Phi^{t^k}_{A_k}(\Phi^{t^{k-1}}_{A_{k-1}}(\ldots(\Phi^t_{A_1}(x))\ldots))$ where t^k denotes the k-th power of t. We will call this the *vector representation* for a family of k-jets. It is not immediately clear that all k-jets of curves can be written in this way. In [1] we prove that this is indeed the case. The standard and vector representations simply give us two different coordinate systems for the infinite dimensional space of families of k-jets. Let us apply this to SDEs seen as 2-jets.

Lemma 2. *Suppose that a family of 2-jets of curves is given in the vector representation as $\gamma_x(t) = \Phi^{t^2}_A(\Phi^t_B(x))$ for vector fields A and B. Choose a coordinate chart and let A^i, B^i be the components of the vector fields in this chart. Then the corresponding standard representation for the family of 2-jets is (see [1] for a proof) $\gamma_x(t) = x + a(x)t^2 + b(x)t$ with $a^i = A^i + \frac{1}{2}\frac{\partial B^i}{\partial x^j}B^j$, $b^i = B^i$.*

As we have already discussed, the standard representation of a 2-jet corresponds to conventional *Itô calculus.* What we have just demonstrated is that the vector representation of a 2-jet corresponds to *Fisk–Stratonovich–McShane calculus* [11] (Stratonovich from now on). Moreover we have given a geometric interpretation of how the coordinate free notion of a 2-jet of a curve is related to the vector fields defining a Stratonovich SDE, and of the Itô-Stratonovich transformation. A much richer discussion on Itô and Stratonovich SDEs, in relation to our result above, and on why the Itô Stratonovich transformation is not enough to work with SDE on manifolds is presented in [1].

We conclude with a new interpretation of the drift of one-dimensional Stratonovich SDEs as median. This is part of more general results on quantiles of SDEs solutions that are given in full detail in [1]. We begin by noticing that the definition of the α-percentile depends only upon the ordering of $\mathbb{R}$ and not its vector space structure. As a result, for continuous monotonic f and X with connected state space, the median of $f(X)$ is equal to f applied to the median of X. If f is strictly increasing, the analogous result holds for the α percentile. This has the implication that the trajectory of the α-percentile of an $\mathbb{R}$ valued stochastic process is invariant under smooth monotonic coordinate changes of $\mathbb{R}$. In other words, percentiles have a coordinate free interpretation.

The mean does not. This raises the question of how the trajectories of the percentiles can be related to the coefficients of the stochastic differential equation. In [1] we calculate this relationship. We summarize the main results proven in that paper for the one-dimensional SDE with non-vanishing diffusion term b:

$$\mathrm{d}X_t = a(X_t, t)\,\mathrm{d}t + b(X_t, t)\mathrm{d}W_t, \qquad X_0 = x_0. \tag{10}$$

Theorem 2. *For small* t, *the* α*-th percentile of solutions to* (10) *is given by:*

$$x_0 + b_0\sqrt{t}\Phi^{-1}(\alpha) + \left(a_0 - \frac{1}{2}b_0 b_0'(1 - \Phi^{-1}(\alpha)^2)\right)t + O(t^{3/2}) \tag{11}$$

so long as the coefficients of (10) *are smooth, the diffusion coefficient* b *never vanishes, and further regularity conditions specified in [1] hold. In this formula* a_0 *and* b_0 *denote the values of* $a(x_0, 0)$ *and* $b(x_0, 0)$ *respectively. In particular, the median process is a straight line up to* $O(t^{\frac{3}{2}})$ *with tangent given by the drift of the Stratonovich version of the Itô SDE* (10). *The* $\Phi(1)$ *and* $\Phi(-1)$ *percentiles correspond up to* $O(t^{\frac{3}{2}})$ *to the curves* $\gamma_{X_0}(\pm\sqrt{t})$ *where* γ_{X_0} *is any representative of the* 2*-jet that defines the SDE in Itô form.*

The theorem above has given us the median as a special case, and a link between the median and the Stratonovich version of the SDE. By contrast the mean process has tangent given by the drift of the Itô SDE as the Itô integral is a martingale. In [1] we derive the mode interval equations too and discuss their relationship with median and mean.

References

1. Armstrong, J., Brigo, D.: Coordinate free stochastic differential equations as jets (2016). http://arxiv.org/abs/1602.03931
2. Armstrong, J., Brigo, D.: Optimal approximation of SDEs on submanifolds: the Ito-vector and Ito-jet projections (2016). http://arxiv.org/abs/1610.03887
3. Armstrong, J., Brigo, D.: Extrinsic projection of Itô SDEs on submanifolds with applications to non-linear filtering. In: Nielsen, F., Critchley, F., Dodson, C.T.J. (eds.) Computational Information Geometry. SCT, pp. 101–120. Springer, Cham (2017). doi:10.1007/978-3-319-47058-0_5
4. Brzeźniak, Z., Elworthy, K.D.: Stochastic differential equations on Banach manifolds. Methods Funct. Anal. Topology **6**(1), 43–84 (2000)
5. Elworthy, K.D.: Stochastic Differential Equations on Manifolds. Cambridge University Press, Cambridge, New York (1982)
6. Elworthy, D.: Geometric aspects of diffusions on manifolds. In: Hennequin, P.-L. (ed.) École d'Été de Probabilités de Saint-Flour XV–XVII, 1985–87. LNM, vol. 1362, pp. 277–425. Springer, Heidelberg (1988). doi:10.1007/BFb0086183
7. Emery, M.: Stochastic Calculus in Manifolds. Springer, Heidelberg (1989)
8. Gliklikh, Y.E.: Global and Stochastic Analysis with Applications to Mathematical Physics. Theoretical and Mathematical Physics. Springer, London (2011)

9. Itô, K.: Stochastic differential equations in a differentiable manifold. Nagoya Math. J. **1**, 35–47 (1950)
10. Kloeden, P.E., Platen, E.: Numerical solution of stochastic differential equations. Applications of mathematics. Springer, New York (1999). Third printing
11. Stratonovich, R.L.: A new representation for stochastic integrals and equations. SIAM J. Control **4**(2), 362–371 (1966)

The Cramér-Rao Inequality on Singular Statistical Models

Hông Vân Lê[1(✉)], Jürgen Jost[2], and Lorenz Schwachhöfer[3]

[1] Mathematical Institute of ASCR, Zitna 25, 11567 Praha, Czech Republic
hvle@math.cas.cz

[2] Max-Planck-Institut Für Mathematik in den Naturwissenschaften, Inselstrasse 22, 04103 Leipzig, Germany
jost@mis.mpg.de

[3] Technische Universität Dortmund, Vogelpothsweg 87, 44221 Dortmund, Germany
Lorenz.Schwachhoefer@math.uni-dortmund.de

Abstract. We introduce the notions of essential tangent space and reduced Fisher metric and extend the classical Cramér-Rao inequality to 2-integrable (possibly singular) statistical models for general φ-estimators, where φ is a V-valued feature function and V is a topological vector space. We show the existence of a φ-efficient estimator on strictly singular statistical models associated with a finite sample space and on a class of infinite dimensional exponential models that have been discovered by Fukumizu. We conclude that our general Cramér-Rao inequality is optimal.

1 k-integrable Parametrized Measure Models and the Reduced Fisher Metric

In this section we recall the notion of a *k-integrable parametrized measure model* (Definitions 1, 3). Then we give a characterization of k-integrability (Theorem 1), which is important for later deriving the classical Cramér-Rao inequalities from our general Cramér-Rao inequality. Finally we introduce the notion of *essential tangent space* of a 2-integrable parametrized measure model (Definition 4) and the related notion of *reduced Fisher metric.*

Notations. For a measurable space Ω and a finite measure μ_0 on Ω we denote

$$\begin{aligned} \mathcal{P}(\Omega) &:= \{\mu \ : \ \mu \text{ a probability measure on } \Omega\}, \\ \mathcal{M}(\Omega) &:= \{\mu \ : \ \mu \text{ a finite measure on } \Omega\}, \\ \mathcal{S}(\Omega) &:= \{\mu \ : \ \mu \text{ a signed finite measure on } \Omega\}, \\ \mathcal{S}(\Omega, \mu_0) &= \{\mu = \phi\, \mu_0 \ : \ \phi \in L^1(\Omega, \mu_0)\}. \end{aligned}$$

Hông Vân Lê—Speaker, partially supported by RVO: 6798584.

F. Nielsen and F. Barbaresco (Eds.): GSI 2017, LNCS 10589, pp. 552–560, 2017.
https://doi.org/10.1007/978-3-319-68445-1_64

Definition 1 ([AJLS2016b, Definition 4.1]). *Let Ω be a measurable space.*

1. *A* parametrized measure model *is a triple $(M, \Omega, \mathbf{p})$ where M is a (finite or infinite dimensional) Banach manifold and $\mathbf{p} : M \to \mathcal{M}(\Omega) \subset \mathcal{S}(\Omega)$ is a Frechét-C^1-map, which we shall call simply a C^1-map.*
2. *The triple $(M, \Omega, \mathbf{p})$ is called a* statistical model *if it consists only of probability measures, i.e., such that the image of $\mathbf{p}$ is contained in $\mathcal{P}(\Omega)$.*
3. *We call such a model* dominated by μ_0 *if the image of $\mathbf{p}$ is contained in $\mathcal{S}(\Omega, \mu_0)$. In this case, we use the notation $(M, \Omega, \mu_0, \mathbf{p})$ for this model.*

Let $(M, \Omega, \mathbf{p})$ be a parametrized measure model. It follows from [AJLS2016b, Proposition 2.1] that for all $\xi \in M$ the differential $d_\xi \mathbf{p}(V)$ is dominated by $\mathbf{p}(\xi)$. Hence *the logarithmic derivative of $\mathbf{p}$ at ξ in direction V* [AJLS2016b, (4.2)]

$$\partial_V \log \mathbf{p}(\xi) := \frac{d\{d_\xi \mathbf{p}(V)\}}{d\mathbf{p}(\xi)} \tag{1}$$

is an element in $L^1(\Omega, \mathbf{p}(\xi))$. If measures $\mathbf{p}(\xi)$, $\xi \in M$, are dominated by μ_0, we also write

$$\mathbf{p}(\xi) = p(\xi) \cdot \mu_0 \text{ for some } p(\xi) \in L^1(\Omega, \mathbf{p}_0). \tag{2}$$

Definition 2 ([AJLS2016b, Definition 4.2]). *We say that a parametrized model $(M, \Omega, \mu_0, \mathbf{p})$ has a* regular density function *if the density function $p : \Omega \times M \to \mathbb{R}$ satisfying (2) can be chosen such that for all $V \in T_\xi M$ the partial derivative $\partial_V p(.; \xi)$ exists and lies in $L^1(\Omega, \mu_0)$ for some fixed μ_0.*

If the model has a positive regular density function, we have

$$\partial_V \log \mathbf{p}(\xi) = \partial_V \log p. \tag{3}$$

Next we recall the notion of k-integrability. On the set $\mathcal{M}(\Omega)$ we define the preordering $\mu_1 \leq \mu_2$ if μ_2 dominates μ_1. Then $(\mathcal{M}(\Omega), \leq)$ is a directed set, meaning that for any pair $\mu_1, \mu_2 \in \mathcal{M}(\Omega)$ there is a $\mu_0 \in \mathcal{M}(\Omega)$ dominating both of them (e.g. $\mu_0 := \mu_1 + \mu_2$).

For fixed $r \in (0, 1]$ and measures $\mu_1 \leq \mu_2$ on Ω we define the linear embedding

$$\imath_{\mu_2}^{\mu_1} : L^{1/r}(\Omega, \mu_1) \longrightarrow L^{1/r}(\Omega, \mu_2), \qquad \phi \longmapsto \phi \left(\frac{d\mu_1}{d\mu_2}\right)^r.$$

Observe that

$$\begin{aligned} \|\imath_{\mu_2}^{\mu_1}(\phi)\|_{1/r} &= \left|\int_\Omega |\imath_{\mu_2}^{\mu_1}(\phi)|^{1/r} \, d\mu_2\right|^r = \left|\int_\Omega |\phi|^{1/r} \frac{d\mu_1}{d\mu_2} \, d\mu_2\right|^r \\ &= \left|\int_\Omega |\phi|^{1/r} \, d\mu_1\right|^r = \|\phi\|_{1/r}. \end{aligned} \tag{4}$$

It has been proved that $\imath_{\mu_2}^{\mu_1}$ is an isometry [AJLS2016b, (2.6)]. Moreover, $\imath_{\mu_2}^{\mu_1}\imath_{\mu_3}^{\mu_2} = \imath_{\mu_3}^{\mu_1}$ whenever $\mu_1 \le \mu_2 \le \mu_3$. Then we define the *space of r-th roots of measures on Ω* to be the directed limit over the directed set $(\mathcal{M}(\Omega), \le)$

$$\mathcal{S}^r(\Omega) := \underrightarrow{\lim}\, L^{1/r}(\Omega, \mu). \tag{5}$$

By [AJLS2016b, (2.9)] the space $\mathcal{S}^r(\Omega)$ is a Banach space provided with the norm $||\phi||_{1/r}$ defined in (4).

Denote the equivalence class of $\phi \in L^{1/r}(\Omega, \mu)$ by $\phi\mu^r$, so that $\mu^r \in \mathcal{S}^r(\Omega)$ is the equivalence class represented by $1 \in L^{1/r}(\Omega, \mu)$.

In [AJLS2016b, Proposition 2.2], for $r \in (0, 1]$ and $0 < k \le 1/r$ we defined a map

$$\tilde{\pi}^k : \mathcal{S}^r(\Omega) \to \mathcal{S}^{rk}(\Omega),\ \phi \cdot \mu^r \mapsto \mathrm{sign}(\phi)|\phi|^{\mathrm{k}}\mu^{\mathrm{rk}}.$$

For $1 \le k \le 1/r$ the map $\tilde{\pi}^k$ is a C^1-map between Banach spaces [AJLS2016b, (2.13)]. Using the same analogy, we set [AJLS2016b, (4.3)]

$$\mathbf{p}^{1/k} := \tilde{\pi}^{1/k} \circ \mathbf{p} : M \to \mathcal{S}^{1/k}(\Omega) \tag{6}$$

and

$$d_\xi \mathbf{p}^{1/k}(V) := \frac{1}{k}\partial_V \log \mathbf{p}(\xi)\ \mathbf{p}^{1/k}(\xi) \in \mathcal{S}^{1/k}(\Omega, \mathbf{p}(\xi)). \tag{7}$$

Definition 3 ([JLS2017a, Definition 2.6]). *A parametrized measure model $(M, \Omega, \mathbf{p})$ is called* k-integrable, *if the map $\mathbf{p}^{1/k}$ from (6) is a Fréchet-C^1-map.*

The k-integrability of parametrized measure models can be characterized in different ways.

Theorem 1 ([JLS2017a, Theorem 2.7]). *Let $(M, \Omega, \mathbf{p})$ be a parametrized measure model. Then the model is k-integrable if and only if the map*

$$V \longmapsto \|d\mathbf{p}^{1/k}(V)\|_k < \infty \tag{8}$$

defined on TM is continuous.

Thus, $(M, \Omega, \mathbf{p})$ is k-integrable if and only if the map $d\mathbf{p}^{1/k} : M \to \mathcal{S}^{1/k}(\Omega)$ from (7) is well defined (i.e., $\partial_V \log \mathbf{p}(\xi) \in L^k(\Omega, \mathbf{p}(\xi))$) and continuous. In particular, the definition of k-integrability in Definition 3 above is equivalent to that in [AJLS2016b, Definition 4.4] and [AJLS2015, Definition 2.4].

Remark 1. 1. The Fisher metric $\mathfrak{g}$ on a 2-integrable parametrized measure model $(M, \Omega, \mathbf{p})$ is defined as follows for $v, w \in T_\xi M$

$$\mathfrak{g}_\xi(v, w) := \langle \partial_v \log \mathbf{p}; \partial_w \log \mathbf{p}\rangle_{L^2(\Omega, \mathbf{p}(\xi))} = \langle d\mathbf{p}^{1/2}(v); d\mathbf{p}^{1/2}(w)\rangle_{\mathcal{S}^{1/2}(\Omega)} \tag{9}$$

2. The standard notion of a statistical model always assumes that it is dominated by some measure and has a positive regular density function (e.g. [Borovkov1998, p. 140, 147], [BKRW1998, p. 23], [AN2000, Sect. 2.1], [AJLS2015, Definition 2.4]). In fact, the definition of a parametrized measure model or statistical model in [AJLS2015, Definition 2.4] is equivalent to a parametrized measure model or statistical model with a positive regular density function in the sense of Definition 2, see also [AJLS2016] for detailed discussion.

Let $(M, \Omega, \mathbf{p})$ be a 2-integrable parametrized measure model. Formula (9) shows that the kernel of the Fisher metric $\mathfrak{g}$ at $\xi \in M$ coincides with the kernel of the map $\Lambda_\xi : T_\xi M \to L^2(\Omega, \mathbf{p}(\xi))$, $V \mapsto \partial_V(\log \mathbf{p})$. In other words, the degeneracy of the Fisher metric $\mathfrak{g}$ is caused by the non-effectiveness of the parametrisation of the family $\mathbf{p}(\xi)$ by the map $\mathbf{p}$. The tangent cone $T_{\mathbf{p}(\xi)}\mathbf{p}(M)$ of the image $\mathbf{p}(M) \subset \mathcal{S}(\Omega)$ is isomorphic to the quotient $T_\xi M/\ker \Lambda_x$. This motivates the following

Definition 4. ([JLS2017a, Definition 2.9]). *The quotient $\hat{T}_\xi M := T_\xi M/\ker \Lambda_\xi$ will be called the essential tangent space of M at ξ.*

Clearly the Fisher metric $\mathfrak{g}$ descends to a non-degenerated metric $\hat{\mathfrak{g}}$ on $\hat{T}M$, which we shall call the *reduced Fisher metric.*

Denote by $\hat{T}^{\hat{\mathfrak{g}}}M$ the fiberwise completion of $\hat{T}M$ wrt the reduced Fisher metric $\hat{\mathfrak{g}}$. Its inverse $\hat{\mathfrak{g}}^{-1}$ is a well-defined quadratic form on the fibers of the dual bundle $\hat{T}^{*,\hat{\mathfrak{g}}^{-1}}M$, which we can therefore identify with $\hat{T}^{\hat{\mathfrak{g}}}M$.

2 The General Cramér-Rao Inequality

In this section we assume that $(M, \Omega, \mathbf{p})$ is a 2-integrable measure model. We introduce the notion of a regular function on a measure space Ω (Definition 5), state a rule of differentiation under integral sign (Proposition 1) and derive a general Cramér-Rao inequality (Theorem 2).

We set for $k \in \mathbb{N}^+$

$$L^k_M(\Omega) := \{f \in L^k(\Omega, \mathbf{p}(\xi)) \text{ for all } \xi \in M\}.$$

Definition 5. *Let $(M, \Omega, \mathbf{p})$ be a parametrized measure model. We call an element $f \in L^k_M(\Omega)$* regular *if the function $\xi \mapsto \|f\|_{L^k(\Omega,\mathbf{p}(\xi))}$ is locally bounded, i.e. if for all $\xi_0 \in M$*

$$\limsup_{\xi \to \xi_0} \|f\|_{L^k(\Omega,\mathbf{p}(\xi))} < \infty.$$

The regularity of a function f is important for the validity of differentiation under the integral sign.

Proposition 1. *Let $k, k' > 1$ be dual indices, i.e. $k^{-1} + k'^{-1} = 1$, and let $(M, \Omega, \mathbf{p})$ be a k'-integrable parametrized measure model. If $f \in L^k_M(\Omega)$ is regular, then the map*

$$M \longrightarrow \mathbb{R}, \qquad \xi \longmapsto \mathbb{E}_{\mathbf{p}(\xi)}(f) = \int_\Omega f\, d\mathbf{p}(\xi) \tag{10}$$

is Gatêaux-differentiable, and for $X \in TM$ the Gâteaux-derivative is

$$\partial_X \mathbb{E}_{\mathbf{p}(\xi)}(f) = \mathbb{E}_{\mathbf{p}(\xi)}(f\, \partial_X \log \mathbf{p}(\xi)) = \int_\Omega f\, \partial_X \log \mathbf{p}(\xi)\, d\mathbf{p}(\xi). \tag{11}$$

Let V be a topological vector space over the real field $\mathbb{R}$, possibly infinite dimensional. We denote by V^M the vector space of all V-valued functions on M. A V-valued function φ will stand for the coordinate functions on M, or in general, a feature of M (cf. [BKRW1998]). Let V^* denote the dual space of V. Later, for $l \in V^*$ we denote the composition $l \circ \varphi$ by φ^l. This should be considered as the l-th coordinate of φ.

Assume that $(M, \Omega, \mathbf{p})$ is a 2-integrable parametrized measure model. A Gateaux-differentiable function f on M whose differential df vanishes on $\ker d\mathbf{p} \subset TP$ will be called a *visible function*.

Recall that an *estimator* is a map $\hat{\sigma} : \Omega \to M$. If $k, k' > 1$ are dual indices, i.e., $k^{-1} + k'^{-1} = 1$, and given a k'-integrable parametrized measure model $(M, \Omega, \mathbf{p})$ and a function $\varphi \in V^M$, we define

$$L^k_\varphi(M, \Omega) := \{\hat{\sigma} : \Omega \to M \mid \varphi^l \circ \hat{\sigma} \in L^k_M(\Omega) \text{ for all } l \in V^*\}.$$

We call an estimator $\hat{\sigma} \in L^k_\varphi(M, \Omega)$ *φ-regular* if $\varphi^l \circ \hat{\sigma} \in L^k_M(\Omega)$ is regular for all $l \in V^*$.

Any $\hat{\sigma} \in L^k_\varphi(M, \Omega)$ induces a V^{**}-valued function $\varphi_{\hat{\sigma}}$ on M by computing the expectation of the composition $\varphi \circ \hat{\sigma}$ as follows

$$\langle \varphi_{\hat{\sigma}}(\xi), l \rangle := \mathbb{E}_{\mathbf{p}(\xi)}(\varphi^l \circ \hat{\sigma}) = \int_\Omega \varphi^l \circ \hat{\sigma}\, d\mathbf{p}(\xi) \tag{12}$$

for any $l \in V^*$. If $\hat{\sigma} \in L^k_\varphi(M, \Omega)$ is φ-regular, then Proposition 1 immediately implies that $\varphi_{\hat{\sigma}} : M \to V^{**}$ is visible with Gâteaux-derivative

$$\langle \partial_X \varphi_{\hat{\sigma}}(\xi), l \rangle = \int_\Omega \varphi^l \circ \hat{\sigma} \cdot \partial_X \log \mathbf{p}(\xi)\, \mathbf{p}(\xi). \tag{13}$$

Let $pr : TM \to \hat{T}M$ denote the natural projection.

Definition 6. ([JLS2017a, Definition 3.8]). *A section $\xi \mapsto \nabla_{\hat{\mathfrak{g}}} f(\xi) \in \hat{T}^{\hat{\mathfrak{g}}}_\xi M$ will be called the generalized Fisher gradient of a visible function f, if for all $X \in T_\xi M$ we have*

$$df(X) = \hat{\mathfrak{g}}(pr(X), \nabla_{\hat{\mathfrak{g}}} f).$$

If the generalized gradient belongs to $\hat{T}M$ we will call it the Fisher gradient.

We set (cf. [Le2016])

$$\mathcal{L}_1^k(\Omega) := \{(f, \mu) | \mu \in \mathcal{M}(\Omega) \text{ and } f \in L^k(\Omega, \mu)\}.$$

For a map $\mathbf{p} : P \to \mathcal{M}(\Omega)$ we denote by $\mathbf{p}^*(\mathcal{L}_1^k(\Omega))$ the pull-back "fibration" (also called the fiber product) $P \times_{\mathcal{M}(\Omega)} \mathcal{L}_1^k(\Omega)$.

Definition 7. ([JLS2017a, Definition 3.10]). *Let h be a visible function on M. A section*

$$M \to \mathbf{p}^*(\mathcal{L}_1^2(\Omega)),\ \xi \mapsto \nabla h_\xi \in L^2(\Omega, \mathbf{p}(\xi)),$$

is called a pre-gradient of h, if for all $\xi \in M$ and $X \in T_\xi M$ we have

$$dh(X) = \mathbb{E}_{\mathbf{p}(\xi)}((\partial_X \log \mathbf{p}) \cdot \nabla h_\xi).$$

Proposition 2. ([JLS2017a, Proposition 3.12]).

1. *Let $(M, \Omega, \mathbf{p})$ be a 2-integrable measure model and $f \in L_M^2(\Omega, V)$ is a regular function. Then the section of the pullback fibration $\mathbf{p}^*(\mathcal{L}_1^2(\Omega))$ defined by $\xi \mapsto f \in L^2(\Omega, \mathbf{p}(\xi))$ is a pre-gradient of the visible function $E_{\mathbf{p}(\xi)}(f)$.*
2. *Let $(P, \Omega, \mathbf{p})$ be a 2-integrable statistical model and $f \in L_P^2(\Omega, V)$. Then the section of the pullback fibration $\mathbf{p}^*(\mathcal{L}_1^2(\Omega))$ defined by $\xi \mapsto f - \mathbb{E}_{\mathbf{p}(\xi)}(f) \in L^2(\Omega, \mathbf{p}(\xi))$ is a pre-gradient of the visible function $E_{\mathbf{p}(\xi)}(f)$.*

For an estimator $\hat{\sigma} \in L_\varphi^2(P, \Omega)$ we define the *variance of $\hat{\sigma}$ w.r.t. φ* to be the quadratic form $V_{\mathbf{p}(\xi)}^\varphi[\hat{\sigma}]$ on V^* such that for all $l, k \in V^*$ we have [JLS2017a, (4.3)]

$$V_{\mathbf{p}(\xi)}^\varphi[\hat{\sigma}](l, k) := E_{\mathbf{p}(\xi)}[(\varphi^l \circ \hat{\sigma} - E_{\mathbf{p}(\xi)}(\varphi^l \circ \hat{\sigma})) \cdot (\varphi^k \circ \hat{\sigma} - E_{\mathbf{p}(\xi)}(\varphi^k \circ \hat{\sigma}))]. \quad (14)$$

We regard $||d\varphi_{\hat{\sigma}}^l||_{\hat{\mathfrak{g}}^{-1}}^2(\xi)$ as a quadratic form on V^* and denote the latter one by $(\hat{\mathfrak{g}}_{\hat{\sigma}}^\varphi)^{-1}(\xi)$, i.e.

$$(\hat{\mathfrak{g}}_{\hat{\sigma}}^\varphi)^{-1}(\xi)(l, k) := \langle d\varphi_{\hat{\sigma}}^l, d\varphi_{\hat{\sigma}}^k \rangle_{\hat{\mathfrak{g}}^{-1}}(\xi).$$

Theorem 2 (General Cramér-Rao inequality) ([JLS2017a, Theorem 4.4]). *Let $(P, \Omega, \mathbf{p})$ be a 2-integrable statistical model, φ a V-valued function on P and $\hat{\sigma} \in L_\varphi^2(P, \Omega)$ a φ-regular estimator. Then the difference $V_{\mathbf{p}(\xi)}^\varphi[\hat{\sigma}] - (\hat{\mathfrak{g}}_{\hat{\sigma}}^\varphi)^{-1}(\xi)$ is a positive semi-definite quadratic form on V^* for any $\xi \in P$.*

Remark 2. Assume that V is finite dimensional and φ is a coordinate mapping. Then $\mathfrak{g} = \hat{\mathfrak{g}}$, $d\varphi^l = d\xi^l$, and abbreviating $b_{\hat{\sigma}}^\varphi$ as b, we write

$$(\mathfrak{g}_{\hat{\sigma}}^\varphi)^{-1}(\xi)(l, k) = \langle \sum_i (\frac{\partial \xi^l}{\partial \xi^i} + \frac{\partial b^l}{\partial \xi^i}) d\xi^i, \sum_j (\frac{\partial \xi^k}{\partial \xi^j} + \frac{\partial b^k}{\partial \xi^j}) d\xi^j \rangle_{\mathfrak{g}^{-1}}(\xi). \quad (15)$$

Let $D(\xi)$ be the linear transformation of V whose matrix coordinates are

$$D(\xi)_k^l := \frac{\partial b^l}{\partial \xi^k}.$$

Using (15) we rewrite the Cramér-Rao inequality in Theorem 2 as follows

$$V_\xi[\hat{\sigma}] \geq (E + D(\xi))\mathfrak{g}^{-1}(\xi)(E + D(\xi))^T. \quad (16)$$

The inequality (16) coincides with the Cramér-Rao inequality in [Borovkov1998, Theorem 1.A, p. 147]. By Theorem 1, the condition (R) in [Borovkov1998, p. 140, 147] for the validity of the Cramér-Rao inequality is essentially equivalent to the 2-integrability of the (finite dimensional) statistical model with positive density function under consideration, more precisely Borokov ignores/excludes the points $x \in \Omega$ where the density function vanishes for computing the Fisher metric. Borovkov also uses the φ-regularity assumption, written as $\mathbb{E}_\theta((\theta^*)^2) < c < \infty$ for $\theta \in \Theta$, see also [Borovkov1998, Lemma 1, p. 141] for a more precise formulation. Classical versions of Cramér-Rao inequalities, as in e.g. [CT2006], [AN2000], are special cases of the Cramér-Rao inequality in [Borovkov1998]. We refer the reader to [JLS2017a] for comparison of our Cramér-Rao inequality with more recent Cramér-Rao inequalities in parametric statistics.

3 Optimality of the General Cramér-Rao Inequality

To investigate the optimality of our general Cramér-Rao inequality we introduce the following

Definition 8 ([JLS2017b]). *Assume that φ is a V-valued function on P, where $(P, \Omega, \mathbf{p})$ is a 2-integrable statistical model. A φ-regular estimator $\hat{\sigma} \in L^2_\varphi(P, \Omega)$ will be called φ-efficient, if $V^\varphi_{\mathbf{p}(\xi)} = (\hat{\mathfrak{g}}^\varphi_{\hat{\sigma}})^{-1}(\xi)$ for all $\xi \in P$.*

If a statistical model $(P, \Omega, \mathbf{p})$ admits a φ-efficient estimator, the Cramér-Rao inequality is optimal on $(P, \Omega, \mathbf{p})$.

Example 1. Assume that $(P \subset \mathbb{R}^n, \Omega \subset \mathbb{R}^n, \mathbf{p})$ is a minimal full regular exponential family, $\varphi : P \to \mathbb{R}^n$ - the canonical embedding $P \to \mathbb{R}^n$, and $\hat{\sigma} : \Omega \to P$ - the mean value parametrization. Then it is well known that $\hat{\sigma}$ is an unbiased φ-efficient estimator, see e.g. [Brown1986, Theorem 3.6, p. 74]. Let S be a submanifold in P and $f : P' \to P$ is a blowing-up of P along S, i.e. f is a smooth surjective map such that $\ker df$ is non -trivial exactly at $f^{-1}(S)$. Then $(P', \Omega, \mathbf{p} \circ f)$ is a strictly singular statistical model which admits an unbiased φ-efficient estimator, since $(P, \Omega, \mathbf{p})$ admits unbiased φ-efficient estimator.

Example 2. Let Ω_n be a finite set of n elements. Let $A : \Omega_n \to \mathbb{R}^d_+$ be a map, where $d \leq m - 1$. We define an exponential family $P^A(\cdot|\theta) \subset \mathcal{M}(\Omega_m)$ with parameter θ in $\mathbb{R}^d$ as follows.

$$P^A(x|\theta) = Z_A(\theta) \cdot \exp\langle\theta, A(x)\rangle, \text{ for } \theta \in \mathbb{R}^d, \text{ and } x \in \Omega_m. \quad (17)$$

Here $Z_A(\theta)$ is the normalizing factor such that $P^A(\cdot|\theta) \cdot \mu_0$ is a probability measure, where μ_0 is the counting measure on Ω_m: $\mu_0(x_i) = 1$ for $x_i \in \Omega_m$.

Denote $A^l(x) := \langle l, A(x)\rangle$ for $l \in (\mathbb{R}^d)^*$. We set

$$\hat{\sigma} : \Omega_n \to \mathbb{R}^d,\ x \mapsto \log A(x) := (\log A^1(x), \cdots, \log A^d(x)),$$

$$\varphi : \mathbb{R}^d \to \mathbb{R}^d_+ \subset \mathbb{R}^d,\ \theta \mapsto \exp\theta.$$

Then $\hat{\sigma}$ is a (possibly biased) φ-efficient estimator [JLS2017b]. Using blowing-up, we obtain strictly singular statistical models admitting (possibly biased) φ-efficient estimators.

In [Fukumizu2009] Fukumizu constructed a large class of infinite dimensional exponential families using reproducing kernel Hilbert spaces (RKHS). Assume that Ω is a topological space and μ is a Borel probability measure such that $sppt(\mu) = \Omega$. Let $k : \Omega \times \Omega \to \mathbb{R}$ be a continuous positive definite kernel on Ω. It is known that for a positive definite kernel k on Ω there exists a unique RKHS $\mathcal{H}_k$ such that

1. $\mathcal{H}_k$ consists of functions on Ω,
2. Functions of the form $\sum_{i=1}^m a_i k(\cdot, x_i)$ are dense in $\mathcal{H}_k$,
3. For all $f \in \mathcal{H}_k$ we have $\langle f, k(\cdot, x)\rangle = f(x)$ for all $x \in \Omega$,
4. $\mathcal{H}_k$ contains the constant functions $c_{|\Omega}$, $c \in \mathbb{R}$.

For a given positive definite kernel k on Ω we set

$$\hat{k} : \Omega \to \mathcal{H}_k,\ \hat{k}(x) := k(\cdot, x).$$

Theorem 3 ([JLS2017b]). *Assume that Ω is a complete topological space and μ is a Borel probability measure with $sppt(\mu) = \Omega$. Suppose that a kernel k on Ω is bounded and satisfies the following relation whenever $x, y \in \Omega$*

$$\hat{k}(x) - \hat{k}(y) = c_{|\Omega} \in \mathcal{H}_k \Longrightarrow c_{|\Omega} = 0 \in \mathcal{H}_k. \tag{18}$$

Let

$$\mathcal{P}_\mu := \{f \in L^1(\Omega, \mu) \cap C^0(\Omega) |\ f > 0 \text{ and } \int_\Omega f d\mu = 1\}.$$

Set

$$\mathbf{p} : \mathcal{P}_\mu \to \mathcal{M}(\Omega),\ f \mapsto f \cdot \mu_0.$$

Then there exists a map $\varphi : \mathcal{P}_\mu \to \mathcal{H}_k$ such that $(\mathcal{P}_\mu, \Omega, \mathbf{p})$ admits a φ-efficient estimator.

References

[AJLS2015] Ay, N., Jost, J., Lê, H.V., Schwachhöfer, L.: Information geometry and sufficient statistics. Probab. Theory Relat. Fields **162**, 327–364 (2015)

[AJLS2016] Ay, N., Jost, J., Lê, H.V., Schwachhöfer, L.: Information Geometry. Springer (2017). Ergebnisse der Mathematik und ihrer Grenzgebiete

[AJLS2016b] Ay, N., Jost, J., Lê, H.V., Schwachhöfer, L.: Parametrized measure models. (accepted for Bernoulli Journal). arXiv:1510.07305

[AN2000] Amari, S., Nagaoka, H.: Methods of Information Geometry. Translations of Mathematical Monographs, vol. 191. American Mathematical Society (2000)

[BKRW1998] Bickel, P., Klaassen, C.A.J., Ritov, Y., Wellner, J.A.: Efficient and Adaptive Estimation for Semiparametric Models. Springer, New York (1998)

[Borovkov1998] Borovkov, A.A.: Mathematical Statistics. Gordon and Breach Science Publishers (1998)

[Brown1986] Brown, L.D.: Fundamentals of Statistical Families with Applications in Statistical Decision Theory, Lecture Notes-Monograph Series, vol. 9. IMS (1986)

[CT2006] Cover, T.M., Thomas, J.A.: Elements of Information Theory, 2nd edn. Wiley, Hoboken (2006)

[Fukumizu2009] Fukumizu, K.: Exponential manifold by reproducing kernel Hilbert spaces. In: Gibilisco, P., Riccomagno, E., Rogantin, M.-P., Winn, H. (eds.) Algebraic and Geometric methods in Statistics, pp. 291–306. Cambridge University Press (2009)

[JLS2017a] Jost, J., Lê, H.V., Schwachhöfer, L.: The Cramér-Rao inequality on singular statistical models I. arXiv:1703.09403

[JLS2017b] Jost, J., Lê, H.V., Schwachhöfer, L.: The Cramér-Rao inequality on singular statistical models II (2017, preprint)

[Le2016] Lê, H.V.: The uniqueness of the Fisher metric as information metric. AISM **69**, 879–896 (2017)

Information Geometry of Predictor Functions in a Regression Model

Shinto Eguchi[1,2(✉)] and Katsuhiro Omae[2]

[1] The Institute of Statistical Mathematics, Tachikawa, Tokyo 190-8562, Japan
eguchi@ism.ac.jp

[2] Department of Statistical Science, The Graduate University for Advanced Studies, Tachikawa, Tokyo 190-8562, Japan

Abstract. We discuss an information-geometric framework for a regression model, in which the regression function accompanies with the predictor function and the conditional density function. We introduce the e-geodesic and m-geodesic on the space of all predictor functions, of which the pair leads to the Pythagorean identity for a right triangle spinned by the two geodesics. Further, a statistical modeling to combine predictor functions in a nonlinear fashion is discussed by generalized average, and in particular, we observe the flexible property of the log-exp average.

1 Introducton

We discuss a framework of information geometry for a regression model to be compatible with that for a probability density model. The framework consists of three spaces of the regression, predictor and conditional density functions which are connected in a one-to-one correspondence. In this way, the framework of the regression analysis is more complicated than that only of a density function model. The key to build the new framework is to keep compatibility with the information geometry established in the density function model such that the dualistic pair of e-geodesic and m-geodesics plays a central role under the information metric, cf. [1].

Let X be a p-dimensional explanatory vector and Y a response variable, in which our interests are focused on an association of X with Y. Thus we write the regression function by

$$\mu(x) = \mathbb{E}(Y|X = x). \tag{1}$$

A major goal in the regression analysis is to make an inference on the regression function, which is described by the conditional density function. The predictor function models the conditional density function. We adopt the formulation of the generalized linear model (GLM) to formulate more precisely the relation among the regression, predictor and conditional density functions. GLM is a standard model for the statistical regression analysis, which gives comprehensive unification of Gaussian, Poisson, Gamma, logistic regression models and so forth, cf [4]. We begin with a predictor function $f(x)$ rather than the regression function

F. Nielsen and F. Barbaresco (Eds.): GSI 2017, LNCS 10589, pp. 561–568, 2017.
https://doi.org/10.1007/978-3-319-68445-1_65

$\mu(x)$. Because $\mu(x)$ is sometimes confined to be in a finite range as in a binary regression, which implies that it is difficult to directly model $\mu(x)$ by a parameter such as a linear model. Let $\mathcal{F}$ be the space of all predictor functions $f(x)$'s which satisfy mild regularity conditions for the smoothness. In the formation of GLM, $f(x)$ and $\mu(x)$ are connected by a one-to-one function ℓ, called the mean link function such that $\mu(x) = \ell(f(x))$. The conditional density function of Y given $X = x$ is assumed by

$$p(y|x, f) = \exp\{y\vartheta(f(x)) - \psi(\vartheta(f(x)))\}, \tag{2}$$

where $f \in \mathcal{F}$ and $\vartheta(f) = (\partial\psi/\partial\theta)^{-1}(\ell(f))$, called canonical link function. Note that the canonical and mean parameters θ and μ are connected by $\mu = (\partial\psi/\partial\theta)(\theta)$, which is a basic property in the exponential model (2). Typically, if (2) is a Bernoulli distribution, then a logistic model is led to as $\mu(x) = 1/(1+\exp\{-f(x)\})$, or equivalently $f(x) = \log\{\mu(x)/(1-\mu(x))\}$. In the standard framework we write a linear predictor function by

$$f(x) = \beta^\top x + \alpha \tag{3}$$

with the slope vector β and intercept α, in which the conditional density function is reduced to a parametric model. Thus, the logistic model is written by

$$\mathbb{E}(Y|X = x, \alpha, \beta) = \frac{1}{1 + \exp\{-(\beta^\top x + \alpha)\}} \tag{4}$$

via the mean link function. In practice, the model density function (2) is often added to a dispersion parameter in order to give a reasonable fitting to data with overdispersion, however we omit this description for notational simplicity. Based on this formulation, GLM has been expended to the hierarchy Bayesian GLM and generalized additive model, see [4] for recent developments.

Unless we consider such a parametric form (3), then the conditional density function is in a semiparametric model

$$\mathcal{P} = \{p(y|x, f) : f \in \mathcal{F}\}, \tag{5}$$

where f is a nonparametric component in the exponential density model. We give an information-geometric framework for $\mathcal{F}$ in association with this semiparametric model $\mathcal{P}$. In reality, we like to explore more flexible form than the linear predictor function (3). Let f_0 and f_1 be in $\mathcal{F}$ and ϕ a strictly increasing function defined on $\mathbb{R}$. Then we introduce a one-parameter family,

$$f_t^{(\phi)}(x) = \phi^{-1}((1-t)\phi(f_0(x)) + t\phi(f_1(x))) \tag{6}$$

for $t \in [0, 1]$. We call $f_t^{(\phi)}$ ϕ-geodesic connecting between f_0 and f_1. It is noted that, if f_0 and f_1 are both linear predictor functions as in (3) and ϕ is an identity function, or $\phi(f) = f$, then $f_t^{(\phi)}(x)$ is also a linear predictor function because

$$f_t^{(\phi)}(x) = \{(1-t)\beta_0 + t\beta_1\}^\top + (1-t)\alpha_0 + t\alpha_1$$

for $f_a(x) = \beta_a^\top x + \alpha_a$ with $a = 0, 1$.

The ϕ-geodesic induces to a one-parameter density family $p(y|x, f_t^{(\phi)})$ in the space $\mathcal{P}$, which connects $p(y|x, f_0)$ with $p(y|x, f_1)$. Specifically, if we take the canonical link function $\vartheta(f)$ as $\phi(f)$, then

$$f_t^{(\vartheta)}(x) = \vartheta^{-1}((1-t)\vartheta(f_0(x)) + t\vartheta(f_1(x))), \tag{7}$$

which induces to a conditional density function

$$p(y|x, f_t^{(\vartheta)}) = \exp\{y\theta_t(x) - \psi(\theta_t(x))\}, \tag{8}$$

where $\theta_t(x) = (1-t)\vartheta(f_0(x)) + t\vartheta(f_1(x))$. This is nothing but the e-geodesic connecting between $p(y|x, f_0)$ and $p(y|x, f_1)$ in $\mathcal{P}$ because we observe that

$$p(y|x, f_t^{(\vartheta)}) = z_t p_0(y|x)^{1-t} p_1(y|x)^t, \tag{9}$$

where z_t is a normalizing factor and $p_a(y|x) = p(y|x, f_a)$ for $a = 0, 1$. Back to the standard case of GLM with the canonical link function. Then we have a linear model $\theta(x) = \beta^\top x + \alpha$. If $f_0(x)$ and $f_1(x)$ is in the linear model, then $f_t^{(\vartheta)}(x)$ is also in the linear model with $\theta_t(x) = \{(1-t)\beta_0 + t\beta_1\}^\top x + (1-t)\alpha_0 + t\alpha_1$ for $\vartheta(f_a(x)) = \beta_a^\top x + \alpha_a$ $(a = 0, 1)$.

Alternatively, we can consider a connection between $\mathcal{F}$ and the space of regression function, say $\mathcal{R}$. If we take the mean link function $\ell(f)$ as $\phi(f)$, then

$$f_t^{(\ell)}(x) = \ell^{-1}((1-t)\ell(f_0(x)) + t\ell(f_1(x))), \tag{10}$$

which leads to the mixture geodesic

$$\mu_t^{(\ell)}(x) = (1-t)\mu_0(x) + t\mu_1(x) \tag{11}$$

in $\mathcal{R}$, where $\mu_a(x) = \ell(f_a(x))$ for $a = 0, 1$. Here we implicitly assume that $\mu_t^{(\ell)}(x) = (1-t)\mu_0(x) + t\mu_1(x)$ belongs to $\mathcal{R}$ for any $f_0(x)$ and $f_1(x)$ of $\mathcal{F}$. Through the discussion as given above, the canonical link function induces to the exponential geodesic in $\mathcal{P}$; the mean link function induces to the mixture geodesic in $\mathcal{R}$, see Fig. 1. Henceforth, we refer to (7) and (10) as e-geodesic and m-geodesic in $\mathcal{F}$, respectively.

We next discuss a triangle associated with three points in $\mathcal{R}$. Let $C_1 = \{f_t : t \in [0,1]\}$ and $C_2 = \{g_s : s \in [0,1]\}$ be curves intersecting when $(s,t) = (1,1)$, so that $g_1 = f_1$, say f. Then C_1 and C_2 are said to orthogonally intersects at f if

$$\mathbb{E}\left\{\frac{\partial}{\partial t}\ell(f_t(X))\frac{\partial}{\partial s}\vartheta(g_s(X))\right\}\Big|_{(s,t)=(1,1)} = 0. \tag{12}$$

The orthogonality is induced from that for curves of density functions defined by the information metric, see [1]. We like to consider a divergence measure between f and g be in $\mathcal{F}$. Thus we define

$$D_{\mathrm{KL}}(f, g) = \mathbb{E}\{\widetilde{D}_{\mathrm{KL}}(p(\cdot|X, f), p(\cdot|X, g))\}, \tag{13}$$

where $\widetilde{D}_{\mathrm{KL}}$ is the Kullback-Leibler divergence in $\mathcal{P}$. Thus, we can write

$$D_{\mathrm{KL}}(f, g) = \mathbb{E}[\ell(f(X))\{\vartheta(f(X)) - \vartheta(g(X))\} - \psi(\vartheta(f(X))) + \psi(\vartheta(g(X)))].$$

Hence, we can discuss the Pythagorean identity on $\mathcal{F}$ as in the space of density functions.

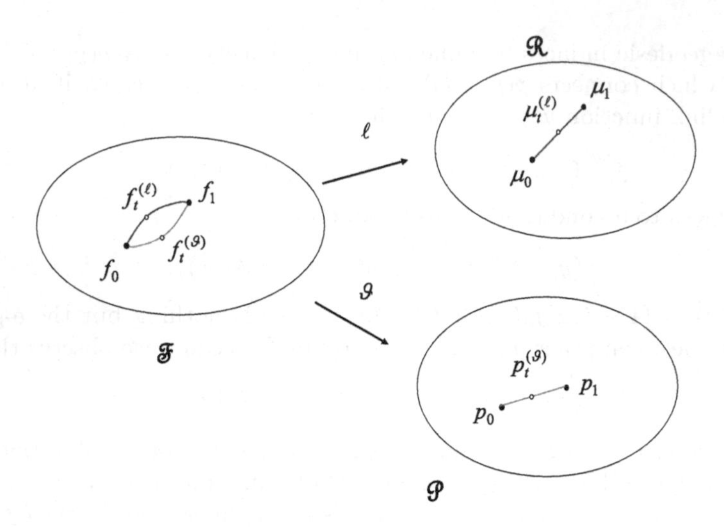

Fig. 1. The e-gedesic and m-geodesic in $\mathcal{F}$ are injectively connected with the e-gedesic in $\mathcal{R}$ and the m-geodesic in $\mathcal{P}$, respectively.

Proposition 1. *Let f, g and h be in $\mathcal{F}$. Consider two curves: one is the m-geodesic connecting between f and g as*

$$C^{(\mathrm{m})} = \{f_t^{(\ell)}(x) := \ell^{-1}((1-t)\ell(f(x)) + t\ell(g(x))) : t \in [0,1]\} \tag{14}$$

and the other is the e-geodesic connecting between h and g as

$$C^{(\mathrm{e})} = \{h_s^{(\vartheta)}(x) := \vartheta^{-1}((1-s)\vartheta(h(x)) + s\vartheta(g(x))) : s \in [0,1]\}. \tag{15}$$

Then the triangle induced by vertices f, g and h satisfies a Pythagorean identity

$$D_{\mathrm{KL}}(f,g) + D_{\mathrm{KL}}(g,h) = D_{\mathrm{KL}}(f,h) \tag{16}$$

if and only if the curves defined in (14) and (15) orthogonally intersects at g.

Proof is easily confirmed by a fact that

$$\mathbb{E}\Big\{\frac{\partial}{\partial t}\ell(f_t^{(\ell)}(X))\frac{\partial}{\partial s}\vartheta(h_s^{(\vartheta)}(X))\Big\}\Big|_{s=t=1} = D_{\mathrm{KL}}(f,h) - \{D_{\mathrm{KL}}(f,g) + D_{\mathrm{KL}}(g,h)\}.$$

We note that the orthogonality means the further identity:

$$D_{\mathrm{KL}}(f_t^{(\ell)},g) + D_{\mathrm{KL}}(g,h_s^{(\vartheta)}) = D_{\mathrm{KL}}(f_t^{(\ell)},h_s^{(\vartheta)}) \tag{17}$$

for any $(t,s) \in [0,1]^2$.

In accordance with Proposition 1 the information geometry for the predictor space $\mathcal{F}$ is induced from that for the density space $\mathcal{P}$ by way of that for the regression space $\mathcal{R}$, in which the distribution assumption (2) plays a central role on the induction from $\mathcal{P}$ and $\mathcal{R}$ to $\mathcal{F}$. If a regression function (1) is degenerated, or constant in x, then any predictor function should be so, which means that both $\mathcal{R}$ and $\mathcal{F}$ are singletons, and $\mathcal{P}$ is just a one-parameter exponential family such as a Gaussian, Bernoulli and Poisson distributions. In effect, $\mathcal{R}$ should be sufficiently a rich space so that the richness can cover a flexible association of the explanatory vector X with the response variable Y.

2 Log-Exp Means

We have discussed that a geometric property associated with the pair of e-geodesic (7) and m-geodesic (10) as specific ϕ-geodesics in $\mathcal{F}$. There is another potential expansion of ϕ-geodesics which enables to flexibly modeling nonlinear predictor functions in $\mathcal{F}$.

Let $f_k(x)$ be a predictor function for $k = 1, \cdots, K$. Then we propose a generalized mean as

$$f_{\tau,\pi}^{(\phi)}(x) = \frac{1}{\tau}\phi^{-1}\Big(\sum_{k=1}^{K} \pi_k \phi\big(\tau f_k(x)\big)\Big) \tag{18}$$

for a generator function ϕ assumed to be strictly increasing, where τ is a shape parameter and π_k's are proportions of the k-th predictor, so that $\sum_{k=1}^{K} \pi_k = 1$. Two properties of $f_{\tau,\pi}^{(\phi)}(x)$ are observed as follows.

Proposition 2. *Let $f_{\tau,\pi}^{(\phi)}(x)$ be a generalized mean defined in (18). Then, the following property holds:*

$$\min_{1\le k\le K} f_k(x) \le f_{\tau,\pi}^{(\phi)}(x) \le \max_{1\le k\le K} f_k(x) \tag{19}$$

for any τ, and $\lim_{\tau\to 0} f_{\tau,\pi}^{(\phi)}(x) = \sum_{k-1}^{K} \pi_k f_k(x)$.

We focus on another typical example of ϕ-geodesic than e-geodesic and m-geodesic taking a choice as $\phi = \exp$ such that

$$f_{\tau,\pi}^{(\exp)}(x) = \frac{1}{\tau}\log\Big\{\sum_{k=1}^{K} \pi_k \exp(\tau f_k(x))\Big\}, \tag{20}$$

we call log-exp mean, cf. the geometric mean for density functions discussed in [3]. We observe specific behaviors with respect to τ as follows:

$$\lim_{\tau\to-\infty} f_{\tau,\pi}^{(\exp)}(x) = \min_{1\le k\le K} f_k(x) \text{ and } \lim_{\tau\to\infty} f_{\tau,\pi}^{(\exp)}(x) = \max_{1\le k\le K} f_k(x). \tag{21}$$

This suggests that the combined predictor function $f_{\tau,\pi}^{(\exp)}(x)$ attains the two bounds observed in (19) when τ goes to $-\infty$ or ∞. In this way $f_{\tau,\pi}^{(\exp)}(x)$ can express flexible performance of the prediction via an appropriate selection for the tuning parameter τ.

3 Statistical Model of Log-Exp Means

We discuss a practical application of ϕ-geodesics focusing on the log-exp mean. Typical candidates for K predictors are linear predictor functions, however it would suffer from model identification as in a Gaussian mixture model. Further, it is difficult to get any reasonable understanding for the association between X and Y because we have got K different slope vectors and intercepts for the explanatory vector. This problem is closely related with that of the multilayer perceptron, in which the causality is confounding by several links among the multilayer components. As a solution of this problem we will discuss a parsimonious model in the following subsection.

3.1 Parsimonious Model

We consider a parsimonious modeling for combing linear predictor functions to keep the model identifiability as follows. Assume that the explanatory vector x is partitioned into sub-vectors $x_{(k)}$ with $k = 1, \cdots, K$ in unsupervised learning manner, for example, K-means for an empirical dataset, so that

$$x = (x_{(1)}, \cdots, x_{(K)}). \tag{22}$$

Then we employ $f_k(x) = \beta_{(k)}^\top x_{(k)} + \alpha$ for $k, 1 \le k \le K$ such that

$$f_\tau(x, \beta, \pi, \alpha) = \frac{1}{\tau} \log \Big[\sum_{k=1}^{K} \pi_k \exp\{\tau(\beta_{(k)}^\top x_{(k)} + \alpha)\} \Big], \tag{23}$$

where $\beta = (\beta_{(1)}, \cdots, \beta_{(K)})$, see [7] for detailed discussion. We note that α is the threshold of the integrated predictor as

$$f_\tau(x, \beta, \pi, \alpha) = f_\tau(x, \beta, \pi, 0) + \alpha, \tag{24}$$

cf. [5,6] for the log-sum-exp trick. We have a reasonable understanding similar to the linear model (3) since

$$\sum_{k=1}^{K} \beta_{(k)}^\top x_{(k)} = \beta^\top x. \tag{25}$$

Thus, the model (23) connects among K linear predictor functions in a flexibly nonlinear manner including linear combination as discussed in Proposition 2. In effect, we can assume for the proportions π_k's to be known because π_k's are estimated in the preprocessing by the unsupervised learning like K-means. Then we remark that the dimension of the model (23) is exactly equal to the linear predictor model (3). Subsequently, we will take another parametrization as

$$f_\tau(x, \beta, \gamma) = \frac{1}{\tau} \log \Big[\sum_{k=1}^{K} \exp\{\tau(\beta_{(k)}^\top x_{(k)} + \gamma_k)\} \Big], \tag{26}$$

where $\gamma_k = \alpha + \tau^{-1} \log \pi_k$.

We investigate a local behavior of (26). The gradient vectors are

$$\frac{\partial}{\partial \beta_{(k)}} f_\tau(x,\beta,\gamma) = w_{\tau k}(x,\beta,\gamma) x_{(k)}, \tag{27}$$

$$\frac{\partial}{\partial \gamma_k} f_\tau(x,\beta,\gamma) = w_{\tau k}(x,\beta,\gamma), \tag{28}$$

where

$$w_{\tau k}(x,\beta,\gamma) = \frac{\exp\{\tau(\beta_{(k)}^\top x_{(k)} + \gamma_k)\}}{\sum_{\ell=1}^K \exp\{\tau(\beta_{(\ell)}^\top x_{(\ell)} + \gamma_\ell)\}}. \tag{29}$$

We remark that

$$\lim_{\tau\to\infty} w_{\tau k}(x,\beta,\gamma) = \begin{cases} 1 & \text{if } k = k_{\max} \\ 0 & \text{otherwise} \end{cases} \tag{30}$$

where $k_{\max} = \operatorname{argmax}_{1\le k\le K} \beta_k^\top x_{(k)} + \gamma_k$. On the other hand, if τ gos to $-\infty$, then $w_{\tau k}(x,\beta,\gamma)$ converges to a weight vector degenerated at the index to minimize $\beta_k^\top x_{(k)} + \gamma_k$ with respect to k. Taking a limit of τ into 0, $w_{\tau k}(x,\beta,\gamma)$ becomes the weight π_k.

We discuss to incorporate the model (26) with the generalized linear model (2). Let $\mathcal{D} = \{(x_i, y_i) : 1 \le i \le n\}$ be a data set. Then the log-likelihood function is written by

$$L_{\mathcal{D}}(\beta,\gamma) = \sum_{i=1}^n \{y_i \vartheta(f_\tau(x_i,\beta,\gamma))) - \psi(\vartheta(f_\tau(x_i,\beta,\gamma)))\}. \tag{31}$$

The gradient vector is given in a weighted form as

$$\frac{\partial}{\partial \beta_{(k)}} L_{\mathcal{D}}(\beta,\gamma) = \sum_{i=1}^n \widetilde{w}_\tau(x_i,\beta,\gamma) x_{(k)} \{y_i - \ell(f_\tau(x_i,\beta,\gamma))\}, \tag{32}$$

$$\frac{\partial}{\partial \gamma} L_{\mathcal{D}}(\beta,\gamma) = \sum_{i=1}^n \widetilde{w}_\tau(x_i,\beta,\gamma) \{y_i - \ell(f_\tau(x_i,\beta,\gamma))\}, \tag{33}$$

where

$$\widetilde{w}_\tau(x_i,\beta,\gamma) = \frac{\ell'(f_\tau(x_i,\beta,\gamma)) w_{\tau k}(x_i,\beta,\gamma)}{\operatorname{var}(\vartheta(f_\tau(x_i,\beta,\gamma)))} \tag{34}$$

with $\operatorname{var}(\theta) = (\partial^2/\partial\theta^2)\psi(\theta)$. Hence the MLE for (β,γ) can be obtained by a gradient-type algorithm, or the Fisher score method in a straightforward manner.

4 Discussion

The framework of information geometry for a regression model utilizes the close relation among three function spaces $\mathcal{F}$, $\mathcal{R}$ and $\mathcal{P}$ with one-to-one correspondence, where the e-geodesic and m-geodesic on $\mathcal{F}$ are induced from those in $\mathcal{P}$

and $\mathcal{R}$, respectively. The information metric on $\mathcal{P}$ is naturally translated into the space $\mathcal{F}$ taking the marginal expctation. Consider a parametric model

$$\mathcal{M} = \{f_\omega(x) : \omega \in \Omega\} \tag{35}$$

embedded in $\mathcal{F}$ with a parameter vector ω. Then the e-connection and m-connection on $\mathcal{M}$ are induced from those on the model

$$\widetilde{\mathcal{M}} = \{p(y|x, f_\omega) : \omega \in \Omega\} \tag{36}$$

embedded in $\mathcal{P}$. If we consider a divergence D than D_{KL}, then another pair of connections on M is associated, see [2] for detailed formulae.

The discussion in this paper strongly depends on the assumption for the conditional density function $p(y|x, f)$ as in (2) in accordance with GLM formulation. If $p(y|x, f)$ does not belong to such an exponential model but another type of model, then the framework of the information geometry should be adapted. For example, if $p(y|x, f)$ is in a deformed exponential model, then the geometry is suggested by the deformation. However, the structure is still valid including dually flatness and Pythagorean relation associated with the canonical divergence.

References

1. Amari, S., Nagaoka, H.: Methods of Information Geometry. Oxford University Press, Oxford (2000)
2. Eguchi, S.: Geometry of minimum contrast. Hiroshima Math. J. **22**, 631–647 (1992)
3. Eguchi, S., Komori, O.: Path connectedness on a space of probability density functions. In: Nielsen, F., Barbaresco, F. (eds.) GSI 2015. LNCS, vol. 9389, pp. 615–624. Springer, Cham (2015). doi:10.1007/978-3-319-25040-3_66
4. Hastie, T., Tibshirani, R., Friedman, J.: The Elements of Statistical Learning. Springer, New York (2009)
5. Nielsen, F., Sun, K.: Guaranteed bounds on the Kullback-Leibler divergence of univariate mixtures using piecewise log-sum-exp inequalities. arXiv preprint arXiv:1606.05850 (2016)
6. Murphy, K.: Naive Bayes classifiers. University of British Columbia (2006)
7. Omae, K., Komori, O., Eguchi, S.: Quasi-linear score for capturing heterogeneous structure in biomarkers. BMC Bioinform. **18**(1), 308 (2017)

Translations in the Exponential Orlicz Space with Gaussian Weight

Giovanni Pistone(✉)

de Castro Statistics, Collegio Carlo Alberto, Torino, Italy
giovanni.pistone@carloalberto.org
http://www.giannidiorestino.it

Abstract. We study the continuity of space translations on non-parametric exponential families based on the exponential Orlicz space with Gaussian reference density.

1 Introduction

On the Gaussian probability space $(\mathbb{R}^n, \mathcal{B}, M \cdot \ell)$, M being the standard Gaussian density and ℓ the Lebesgue measure, we consider densities of the form $\mathrm{e}_M(U) = \exp\left(U - K_M(U)\right) \cdot M$, where U belongs to the exponential Orlicz space $L^{(\cosh -1)}(M)$, $\mathbb{E}_M[U] = 0$, and $K_M(U)$ is constant [7,8]. An application to the homogeneous Boltzmann equation has been discussed in [5].

The main limitation of the standard version of Information Geometry is its inability to deal with the structure of the sample space as it provides a geometry of the "parameter space" only. As a first step to overcome that limitation, we want to study the effect of a space translation τ_h, $h \in \mathbb{R}^n$, on the exponential probability density $\mathrm{e}_M(U)$. Such a model has independent interest and, moreover, we expect such a study to convey informations about the case where the density $\mathrm{e}_M(U)$ admits directional derivatives.

The present note is devoted to the detailed discussion of the some results concerning the translation model that have been announced at the IGAIA IV Conference, Liblice CZ on June 2016. All results are given in Sect. 2, in particular the continuity result in Proposition 4. The final Sect. 3 gives some pointers to further research work to be published elsewhere.

2 Gauss-Orlicz Spaces and Translations

The *exponential space* $L^{(\cosh -1)}(M)$ and the *mixture space* $L^{(\cosh -1)_*}(M)$ are the Orlicz spaces associated the Young functions $(\cosh -1)$ and its convex conjugate $(\cosh -1)_*$, respectively [6]. They are both Banach spaces and the second one has the Δ_2-property, because of the inequality

$$(\cosh -1)_*(ay) \leq \max(1, a^2)(\cosh -1)_*(y), \quad a, y \in \mathbb{R}.$$

F. Nielsen and F. Barbaresco (Eds.): GSI 2017, LNCS 10589, pp. 569–576, 2017.
https://doi.org/10.1007/978-3-319-68445-1_66

The closed unit balls are

$$\left\{f \middle| \int \phi(f(x))\, M(x) dx \leq 1\right\}$$

with $\phi = \cosh - 1$ and $\phi = (\cosh - 1)_*$, respectively. Convergence to 0 in norm of a sequence g_n, $n \in \mathbb{N}$ holds if, and only if, for all $\rho > 0$ one has

$$\limsup_{n \to \infty} \int \phi(\rho g_n(x))\, M(x) dx \leq 1.$$

If $1 < a < \infty$, the following inclusions hold

$$L^\infty(M) \hookrightarrow L^{(\cosh - 1)}(M) \hookrightarrow L^a(M) \hookrightarrow L^{(\cosh - 1)_*}(M) \hookrightarrow L^1(M),$$

and the restrictions to the ball $\Omega_R = \{x \in \mathbb{R}^n | |x| < R\}$,

$$L^{(\cosh - 1)}(M) \to L^a(\Omega_R), \quad L^{(\cosh - 1)_*}(M) \to L^1(\Omega_R),$$

are continuous.

The exponential space $L^{(\cosh - 1)}(M)$ contains all functions $f \in C^2(\mathbb{R}^n; \mathbb{R})$ whose Hessian is uniformly bounded in operator's norm. In particular, it contains all polynomials with degree up to 2, hence all functions which are bounded by such a polynomial. The mixture space $L^{(\cosh - 1)_*}(M)$ contains all random variables $f \colon \mathbb{R}^d \to \mathbb{R}$ which are bounded by a polynomial, in particular, all polynomials.

Let us review those properties of the exponential function on the space $L^{(\cosh - 1)}(M)$ that justify our definition of *non-parametric exponential* model as the set of densities $\mathrm{e}_M(U) = \exp(U - K_M(U)) \cdot M$, where U has zero M-expectation and belongs to the interior $\mathcal{S}_M$ of the proper domain of the partition functional $Z_M(U) = \mathbb{E}_M\left[\mathrm{e}^U\right]$.

Proposition 1. *1. The functionals Z_M and $K_M = \log Z_M$ are both convex.*
2. The proper domain of both Z_M and K_M contains the open unit ball of $L^{(\cosh - 1)}(M)$, hence its interior $\mathcal{S}_M$ is nonempty.
3. The functions Z_M and K_M are both Fréchet differentiable on $\mathcal{S}_M$.

Proof. Statements 1–3 above are all well known. Nevertheless, we give the proof of the differentiability. We have

$$0 \leq \exp(U + H) - \exp(U) - \exp(U)\, H = \int_0^1 (1 - s) \exp(U + sH)\, H^2\, ds.$$

For all $U, U + H \in \mathcal{S}_M$, choose $\alpha > 1$ such that $\alpha U \in \mathcal{S}_M$. We have

$$0 \leq Z_M(U+H) - Z_M(U) - \mathbb{E}_M\left[\exp(U)\, H\right] = \int_0^1 (1-s) \mathbb{E}_M\left[\exp(U + sH)\, H^2\right] ds,$$

where the derivative term $H \mapsto \mathbb{E}_M\left[\exp(U)\, H\right]$ is continuous at U because

$$\begin{aligned} \left|\mathbb{E}_M\left[\exp(U)\, H\right]\right| &\leq \mathbb{E}_M\left[\exp(\alpha U)\right]^{1/\alpha} \mathbb{E}_M\left[|H|^{\alpha/(\alpha-1)}\right]^{(\alpha-1)/\alpha} \leq \\ &\quad \text{const} \times \mathbb{E}_M\left[\exp(\alpha U)\right]^{1/\alpha} \|H\|_{L^{(\cosh - 1)}(M)}. \end{aligned}$$

The remainder term is bounded by

$$|Z_M(U+H) - Z_M(U) - \mathbb{E}_M[\exp(U)H]| = \int_0^1 (1-s)\mathbb{E}_M\left[\exp(U+sH)H^2\right] ds \leq$$

$$\mathbb{E}_M\left[e^{\alpha U}\right]^{1/\alpha} \int_0^1 (1-s)\mathbb{E}_M\left[\exp\left(s\frac{\alpha}{\alpha-1}H\right)H^{2\frac{\alpha}{\alpha-1}}\right]^{(\alpha-1)/\alpha} ds \leq$$

$$\text{const} \times \mathbb{E}_M\left[H^{4\frac{\alpha}{\alpha-1}}\right]^{(\alpha-1)/2\alpha} \int_0^1 (1-s)\mathbb{E}_M\left[\exp\left(s\frac{2\alpha}{\alpha-1}H\right)\right]^{(\alpha-1)/2\alpha} ds.$$

We have

$$\mathbb{E}_M\left[\exp\left(s\frac{2\alpha}{\alpha-1}H\right)\right] \leq 2\left(\mathbb{E}_M\left[(\cosh-1)\left(s\frac{2\alpha}{\alpha-1}H\right)+1\right]\right) \leq 4$$

if $\|H\|_{L^{(\cosh-1)}(M)} \leq (\alpha-1)/2\alpha$. Under this condition, we have

$$|Z_M(U+H) - Z_M(U) - \mathbb{E}_M[\exp(U)H]| \leq \text{const} \times \|H\|^2_{L^{4\alpha/(\alpha-1)}(M)} \leq \text{const} \times \|H\|^2_{L^{(\cosh-1)}(M)},$$

where the constant depends on U. □

The space $L^{(\cosh-1)}(M)$ is neither separable nor reflexive. However, we have the following density property for the bounded point-wise convergence. The proof uses a form of the Monotone-Class argument [3, 22.3]. Let $C_c(\mathbb{R}^n)$ and $C_c^\infty(\mathbb{R}^n)$ respectively denote the space of continuous real functions with compact support and its sub-space of infinitely-differentiable functions.

Proposition 2. *For each $f \in L^{(\cosh-1)}(M)$ there exists a nonnegative function $h \in L^{(\cosh-1)}(M)$ and a sequence $f_n \in C_c^\infty(\mathbb{R}^n)$ with $|f_n| \leq h$, $n = 1, 2, \ldots$, such that $\lim_{n\to\infty} f_n = f$ a.e. As a consequence, $C_c^\infty(\mathbb{R}^n)$ is weakly dense in $L^{(\cosh-1)}(M)$.*

Proof. Before starting the proof, let us note that $L^{(\cosh-1)}(M)$ is stable under bounded a.e. convergence. Assume $f_n, h \in L^{(\cosh-1)}(M)$ with $|f_n| \leq h$, $n = 1, 2, \ldots$ and $\lim_{n\to\infty} f_n = f$ a.e. By definition of $h \in L^{(\cosh-1)}(M)$, for $\alpha = \|h\|^{-1}_{L^{(\cosh-1)}(M)}$ we have the bound $\mathbb{E}_M[(\cosh-1)(\alpha h)] \leq 1$. The sequence of functions $(\cosh-1)(\alpha f_n)$, $n = 1, 2, \ldots$, is a.e. convergent to $(\cosh-1)(\alpha f)$ and it is bounded by the integrable function $(\cosh-1)(\alpha h)$. The inequality $\mathbb{E}_M[(\cosh-1)(\alpha f)] \leq 1$ follows now by dominated convergence and is equivalent to $\|f\|_{L^{(\cosh-1)}(M)} \leq \|h\|_{L^{(\cosh-1)}(M)}$. By taking a converging sequences (f_n) in $C_c^\infty(\mathbb{R}^n)$ we see that the condition in the proposition is sufficient. Conversely, let $\mathcal{L}$ be the set of all functions $f \in L^{(\cosh-1)}(M)$ such that there exists a sequence $(f_n)_{n\in\mathbb{N}}$ in $C_c(\mathbb{R}^n)$ which is dominated by a function $h \in L^{(\cosh-1)}(M)$ and converges to f point-wise. The set $\mathcal{L}$ contains the constant functions and $C_c(\mathbb{R}^n)$ itself. The set $\mathcal{L}$ is a vector space: if $f^1, f^2 \in \mathcal{L}$ and both $f_n^1 \to f^1$ a.s. with $\left|f_n^1\right| \leq h^1$ and $f_n^2 \to f^2$ point-wise with $\left|h_n^2\right| \leq h^2$, then $\alpha_1 f_n^1 + \alpha_2 f_n^2 \to \alpha_1 f^1 + \alpha_2 f^2$ point-wise with $\left|\alpha_1 f_n^1 + \alpha_2 f_n^2\right| \leq |\alpha_1| h^1 + |\alpha_2| h^2$. Moreover, $\mathcal{L}$ is closed under the min operation: if $f^1, f^2 \in \mathcal{L}$, with both $f_n^1 \to f^1$ with $\left|g_n^1\right| \leq h^1$ and $f_n^2 \to f^2$ with $\left|g_n^2\right| \leq h^2$, then $f_n^1 \wedge f_n^2 \to f_1 \wedge f_2$ and $\left|f_n^1 \wedge f_n^2\right| \leq h^1 \wedge h^2 \in L^{(\cosh -1)}(M)$. $\mathcal{L}$ is closed for the maximum too, because $f^1 \vee f^2 = -\left((-f^1) \wedge (-f^2)\right)$. We come now to the application of the Monotone-Class argument. As $\mathbf{1}_{f>a} = ((f-a) \vee 0) \wedge 1 \in \mathcal{L}$, each element of $\mathcal{L}$ is the point-wise limit of linear combinations of indicator functions in $\mathcal{L}$. Consider the class $\mathcal{C}$ of sets whose indicator belongs to $\mathcal{L}$. $\mathcal{C}$ is a σ-algebra because of the closure properties of $\mathcal{L}$ and contains all open bounded rectangles of $\mathbb{R}^n$ because they are all of the form $\{f > 1\}$ for some $f \in C_c(\mathbb{R}^n)$. Hence $\mathcal{C}$ is the Borel σ-algebra and $\mathcal{L}$ is the set of Borel functions which are bounded by an element of $L^{(\cosh -1)}(M)$, namely $\mathcal{L} = L^{(\cosh -1)}(M)$. To conclude, note that each $g \in C_c(\mathbb{R}^n)$ is the uniform limit of a sequence in $C_c^\infty(\mathbb{R}^n)$. The last statement is proved by bounded convergence. □

Let us discuss some consequences of this result. Let be given $u \in \mathcal{S}_M$ and consider the exponential family $p(t) = \exp(tu - K_M(tu)) \cdot M$, $t \in]-1,1[$. From Proposition 2 we get a sequence $(f_n)_{n\in\mathbb{N}}$ in $C_c^\infty(\mathbb{R}^n)$ and a bound $h \in L^{(\cosh -1)}(M)$ such that $f_n \to u$ point-wise and $|f_n|, |u| \leq h$. As $\mathcal{S}_M$ is open and contains 0, we have $\alpha h \in \mathcal{S}_M$ for some $0 < \alpha < 1$. For each $t \in]-\alpha, \alpha[$, $\exp(tf_n) \to \exp(tu)$ point-wise and $\exp(tf) \leq \exp(\alpha h)$ with $\mathbb{E}_M[\mathbb{E}(\alpha h)] < \infty$. It follows that $K_M(tf_n) \to K(tu)$, so that we have the point-wise convergence of the density $p_n(t) = \exp(tf_n - K_M(tf_n)) \cdot M$ to the density $p(t)$. By Scheffé's lemma, the convergence holds in $L^1(\mathbb{R}^n)$. In particular, for each $\phi \in C_c^\infty(\mathbb{R}^n)$, we have the convergence

$$\int \partial_i\phi(x) p_n(x;t)\, dx \to \int \partial_i\phi(x) p(x;t)\, dx, \quad n \to \infty.$$

for all t small enough. By computing the derivatives, we have

$$\int \partial_i\phi(x) p_n(x;t)\, dx = -\int \phi(x)\partial_i\left(e^{tf_n(x) - K_M(tf_n)} M(x)\right) dx = \int \phi(x)\left(x_i - t\partial_i f_n(x)\right) p_n(x;t)\, dx,$$

that is,

$$(X_i - t\partial_i f_n)\, p_n(t) \to -\partial_i p(t)$$

in the sense of (Schwartz) distributions. It would be of interest to discuss the possibility of the stronger convergence of $p_n(t)$ in $L^{(\cosh -1)_*}(M)$, but we do follow this development here.

The norm convergence of the point-wise bounded approximation will not hold in general. Consider the following example. The function $f(x) = |x|^2$ belongs in $L^{(\cosh -1)}(M)$, but for the tails $f_R(x) = (|x| > R)\,|x|^2$ we have

$$\int (\cosh -1)(\epsilon^{-1} f_R(x))\, M(x) dx \geq \frac{1}{2}\int_{|x|>R} e^{\epsilon^{-1}|x|^2}\, M(x) dx = +\infty, \quad \text{if } \epsilon \leq 2,$$

hence there is no convergence to 0. However, the truncation of $f(x) = |x|$ does converge. This, together with Proposition 2, suggests the following variation of the classical definition of Orlicz class.

Definition 1. *The* exponential class, $C_c^{(\cosh -1)}(M)$, *is the closure of* $C_c^\infty(\mathbb{R}^n)$ *in the space* $L^{(\cosh -1)}(M)$.

Proposition 3. *Assume* $f \in L^{(\cosh -1)}(M)$ *and write* $f_R(x) = f(x)(|x| > R)$. *The following conditions are equivalent:*

1. *The real function* $\rho \mapsto \int (\cosh -1)(\rho f(x))\, M(x) dx$ *is finite for all* $\rho > 0$.
2. f *is the limit in* $L^{(\cosh -1)}(M)$*-norm of a sequence of bounded functions.*
3. $f \in C_c^{(\cosh -1)}(M)$.

Proof. (1) $\Leftrightarrow$ (2) This is well known, but we give a proof for sake of clarity. We can assume $f \geq 0$ and consider the sequence of bounded functions $f_n = f \wedge n$, $n = 1, 2, \ldots$. We have for all $\rho > 0$ that $\lim_{n\to\infty}(\cosh -1)(\rho(f - f_n)) = 0$ point-wise and $(\cosh -1)(\rho(f - f_n))M \leq (\cosh -1)(\rho(f)M$ which is integrable by assumption. Hence

$$0 \leq \limsup_{n\to\infty} \int (\cosh -1)(\rho(f(x) - f_n(x)))M(x)\, dx \leq$$

$$\int \limsup_{n\to\infty} (\cosh -1)(\rho(f(x) - f_n(x)))M(x)\, dx = 0,$$

which in turn implies $\lim_{n\to\infty} \|f - f_n\|_{L^{(\cosh -1)}(M)} = 0$. Conversely, observe first that we have from the convexity of $(\cosh -1)$ that

$$2(\cosh -1)(\rho(x + y)) \leq (\cosh -1)(2\rho x) + (\cosh -1)(2\rho y).$$

It follows that, for all $\rho > 0$ and $n = 1, 2, \ldots$, we have

$$2 \int (\cosh -1)(\rho f(x))M(x)\, dx \leq$$

$$\int (\cosh -1)(2\rho(f(x) - f_n(x)))M(x)\, dx + \int (\cosh -1)(2\rho f_n(x))M(x)\, dx,$$

where the $\limsup_{n\to\infty}$ of the first term of the RHS is bounded by 1 because of the assumption of strong convergence, while the second term is bounded by $(\cosh -1)(2\rho n)$. Hence the LHS is finite for all $\rho > 0$.

(2) $\Rightarrow$ (3) Assume first f bounded and use Proposition 2 to find a point-wise approximation $f_n \in C_0(\mathbb{R}^n)$, $n \in \mathbb{N}$, of f together with a dominating function $|f_n(x)| \leq h(x)$, $h \in L^{(\cosh -1)}(M)$. As f is actually bounded, we can assume h to be equal to the constant bounding f. We have $\lim_{n\to\infty}(\cosh -1)(\rho(f - f_n)) = 0$ point-wise, and $(\cosh -1)(\rho(f - f_n)) \leq (\cosh -1)(2\rho h)$. By dominated convergence we have $\lim_{n\to\infty} \int (\cosh -1)(\rho(f(x) - f_n(x)))M(x)\, dx = 0$ for all $\rho > 0$, which implies the convergence $\lim_{n\to\infty} \|f - f_n\|_{L^{(\cosh -1)}(M)} = 0$. Because of (2), we have the desired result.

(3) ⇒ (2) Obvious from $C_c(\mathbb{R}^n) \subset L^\infty(M)$. □

We discuss now properties of translation operators in a form adapted to the exponential space $L^{(\cosh -1)}(M)$. Define $\tau_h f(x) = f(x-h)$, $h \in \mathbb{R}^n$.

Proposition 4 (Translation by a vector)

1. *For each $h \in \mathbb{R}^n$, the mapping $f \mapsto \tau_h f$ is linear from $L^{(\cosh -1)}(M)$ to itself and $\|\tau_h f\|_{L^{(\cosh -1)}(M)} \leq 2\|f\|_{L^{(\cosh -1)}(M)}$ if $|h| \leq \sqrt{\log 2}$.*
2. *The transpose of τ_h is defined on $L^{(\cosh -1)_*}(M)$ by $\langle \tau_h f, g\rangle_M = \langle f, \tau_h^* g\rangle_M$, $f \in L^{(\cosh -1)}(M)$, and is given by $\tau_h^* g(x) = \mathrm{e}^{-h\cdot x + |h|^2/2}\tau_{-h}g(x)$. For the dual norm, the bound $\|\tau_h^* g\|_{L^{(\cosh -1)}(M)^*} \leq 2\|g\|_{L^{(\cosh -1)}(M)^*}$ holds if $|h| \leq \sqrt{\log 2}$.*
3. *If $f \in C_c^{(\cosh -1)}(M)$ then $\tau_h f \in C_c^{(\cosh -1)}(M)$, $h \in \mathbb{R}^n$ and the mapping $\mathbb{R}^n \colon h \mapsto \tau_h f$ is continuous in $L^{(\cosh -1)}(M)$.*

Proof. 1. Let us first prove that $\tau_h f \in L^{(\cosh -1)}(M)$. It is enough to consider the case $\|f\|_{L^{(\cosh -1)}(M)} \leq 1$. For each $\rho > 0$, with $\Phi = \cosh -1$, we have

$$\int \Phi(\rho\tau_h f(x))\, M(x)dx = \mathrm{e}^{-\frac{1}{2}|h|^2} \int \mathrm{e}^{-z\cdot h}\Phi(\rho f(z))\, M(z)dz,$$

hence, using the elementary inequality $\Phi(u)^2 \leq \Phi(2u)/2$, we obtain

$$\int \Phi(\rho\tau_h f(x))\, M(x)dx \leq \mathrm{e}^{-\frac{1}{2}|h|^2}\left(\int \mathrm{e}^{-2z\cdot h}\, M(z)dz\right)^{\frac{1}{2}}\left(\int \Phi^2(\rho f(z))\, M(z)dz\right)^{\frac{1}{2}} \leq \frac{1}{\sqrt{2}}\mathrm{e}^{\frac{|h|^2}{2}}\left(\int \Phi(2\rho f(z))M(z)\, dz\right)^{\frac{1}{2}}.$$

Take $\rho = 1/2$ to get $\mathbb{E}_M\left[\Phi\left(\tau_h \frac{1}{2} f(x)\right)\right] \leq \mathrm{e}^{\frac{|h|^2}{2}}/\sqrt{2}$, which in particular implies $f \in L^{(\cosh -1)}(M)$. Moreover, $\|\tau_h f\|_{L^{(\cosh -1)}(M)} \leq 2$ if $\mathrm{e}^{\frac{|h|^2}{2}} \leq \sqrt{2}$.

2. The computation of τ_h^* is

$$\begin{aligned}\langle \tau_h f, g\rangle_M &= \int f(x-h)g(x)\, M(x)dx = \int f(x)g(x+h)M(x+h)\, dx \\ &= \int f(x)\mathrm{e}^{-h\cdot x - \frac{|h|^2}{2}}\tau_{-h}g(x)\, M(x)dx = \langle f, \tau_h^* g\rangle_M.\end{aligned}$$

If $|h| \leq \sqrt{\log 2}$,

$$\begin{aligned}\|\tau_h^* g\|_{(L^{(\cosh -1)}(M))^*} &= \sup\left\{\langle \tau_h f, g\rangle_M \middle| \|f\|_{L^{(\cosh -1)}(M)} \leq 1\right\} \leq \\ &\sup\left\{\|\tau_h f\|_{L^{(\cosh -1)}(M)}\|g\|_{(L^{(\cosh -1)}(M))^*} \middle| \|f\|_{L^{(\cosh -1)}(M)} \leq 1\right\} \leq \\ &2\|g\|_{(L^{(\cosh -1)}(M))^*}.\end{aligned}$$

3. For each $\rho > 0$ we have found that

$$\mathbb{E}_M\left[\Phi(\rho\tau_h f)\right] \leq \frac{1}{\sqrt{2}}\mathrm{e}^{\frac{|h|^2}{2}}\left(\int \Phi(2\rho f(z))M(z)\,dz\right)^{\frac{1}{2}}$$

where the right-end-side if finite for all ρ if $f \in C_c^{(\cosh -1)}(M)$. It follows that $\tau_h f \in C_c^{(\cosh -1)}(M)$. Recall that $f \in C_c(\mathbb{R}^n)$, implies $\tau_h f \in C_c(\mathbb{R}^n)$ and $\lim_{h\to 0}\tau_h f = f$ in the uniform topology. Let f_n be a sequence in $C_c(\mathbb{R}^n)$ that converges to f in $L^{(\cosh -1)}(M)$-norm. Let $|h| \leq \sqrt{\log 2}$ and let A be positive and $\Phi(A) = 1$.

$$\begin{aligned}
&\|\tau_h f - f\|_{L^{(\cosh -1)}(M)} = \\
&\qquad \|\tau_h(f - f_n) + (\tau_h f_n - f_n) - (f - f_n)\|_{L^{(\cosh -1)}(M)} \leq \\
&\|\tau_h(f - f_n)\|_{L^{(\cosh -1)}(M)} + \|\tau_h f_n - f_n\|_{L^{(\cosh -1)}(M)} + \|f - f_n\|_{L^{(\cosh -1)}(M)} \leq \\
&\quad 2\|f - f_n\|_{L^{(\cosh -1)}(M)} + A^{-1}\|\tau_h f_n - f_n\|_\infty + \|f - f_n\|_{L^{(\cosh -1)}(M)} \leq \\
&\qquad\qquad 3\|f - f_n\|_{L^{(\cosh -1)}(M)} + A^{-1}\|\tau_h f_n - f_n\|_\infty,
\end{aligned}$$

which implies the desired limit at 0. The continuity at a generic point follows from the continuity at 0 and the semigroup property,

$$\lim_{k\to h}\|\tau_k f - \tau_h f\|_{L^{(\cosh -1)}(M)} = \lim_{k-h\to 0}\|\tau_{k-h}(\tau_h f) - \tau_h f\|_{L^{(\cosh -1)}(M)} = 0.$$

□

We conclude by giving, without proof, the corresponding result for a translation by a probability measure μ, namely $\tau_\mu f(x) = \int f(x-y)\mu(dy)$. We denote by $\mathcal{P}_{\mathrm{e}}$ the set of probability measures μ such that $h \mapsto \mathrm{e}^{\frac{1}{2}|h|^2}$ is integrable for example, μ could be a normal with variance $\sigma^2 I$ and $\sigma^2 < 1$, or μ could have a bounded support.

Proposition 5 (Translation by a probability). *Let $\mu \in \mathcal{P}_{\mathrm{e}}$*

1. *The mapping $f \mapsto \tau_\mu f$ is linear and bounded from $L^{(\cosh -1)}(M)$ to itself. If, moreover, $\int \mathrm{e}^{|h|^2/2}\,\mu(dh) \leq \sqrt{2}$, then its norm is bounded by 2.*
2. *If $f \in C_c^{(\cosh -1)}(M)$ then $\tau_\mu f \in C_c^{(\cosh -1)}(M)$. The mapping $\mathcal{P}_{\mathrm{e}} \colon \mu \mapsto \tau_\mu f$ is continuous at δ_0 from the weak convergence to the $L^{(\cosh -1)}(M)$ norm.*

We can use the previous proposition to show the existence of sequences of mollifiers. A bump function is a non-negative function ω in $C_c^\infty(\mathbb{R}^n)$ such that $\int \omega(x)\,dx = 1$. It follows that $\int \lambda^{-n}\omega(\lambda^{-1}x)\,dx = 1$, $\lambda > 0$ and the family of mollifiers $\omega_\lambda(dx) = \lambda^{-n}\omega(\lambda^{-1}x)dx$ converges weakly to the Dirac mass at 0 as $\lambda \downarrow 0$, so that for all $f \in C_c^{(\cosh -1)}(M)$, the translations $\tau_{\omega_\lambda} f \in C_c^\infty(\mathbb{R}^n)$ and convergence to f in $L^{(\cosh -1)}(M)$ holds for $\lambda \to 0$.

3 Conclusions

We have discussed the density for the bounded point-wise convergence of the space of smooth functions $C_c^\infty(\mathbb{R}^n)$ in the exponential Orlicz space with Gaussian weight $L^{(\cosh -1)}(M)$. The exponential Orlicz class $C_c^{(\cosh -1)}(M)$ has been defined as the norm closure of the space of smooth functions. The continuity of translations holds in the latter space.

The continuity of translation is the first step in the study of differentiability in the exponential Gauss-Orlicz space. The aim is to apply non-parametric exponential models to the study of Hyvärinen divergence [4,5] and the projection problem for evolution equations [1,2]. A preliminary version of the Gauss-Orlicz-Sobolev theory has been published in the second part of [5].

Acknowledgments. The author thanks Bertrand Lods (Università di Torino and Collegio Carlo Alberto, Moncalieri) for his comments and acknowledges the support of de Castro Statistics and Collegio Carlo Alberto, Moncalieri. He is a member of GNAMPA-INDAM.

References

1. Brigo, D., Hanzon, B., Le Gland, F.: Approximate nonlinear filtering by projection on exponential manifolds of densities. Bernoulli **5**(3), 495–534 (1999)
2. Brigo, D., Pistone, G.: Dimensionality reduction for measure valued evolution equations in statistical manifolds. In: Nielsen, F., Critchley, F., Dodson, C.T.J. (eds.) Computational Information Geometry. SCT, pp. 217–265. Springer, Cham (2017). doi:10.1007/978-3-319-47058-0_10
3. Dellacherie, C., Meyer, P.A.: Probabilités et potentiel. Chapitres I à IV. Hermann, Paris (1975). Édition entièrment refondue
4. Hyvärinen, A.: Estimation of non-normalized statistical models by score matching. J. Mach. Learn. Res. **6**, 695–709 (2005)
5. Lods, B., Pistone, G.: Information geometry formalism for the spatially homogeneous Boltzmann equation. Entropy **17**(6), 4323–4363 (2015)
6. Musielak, J.: Orlicz Spaces and Modular Spaces. LNM, vol. 1034. Springer, Heidelberg (1983)
7. Pistone, G.: Nonparametric information geometry. In: Nielsen, F., Barbaresco, F. (eds.) GSI 2013. LNCS, vol. 8085, pp. 5–36. Springer, Heidelberg (2013). doi:10.1007/978-3-642-40020-9_3
8. Pistone, G., Sempi, C.: An infinite-dimensional geometric structure on the space of all the probability measures equivalent to a given one. Ann. Statist. **23**(5), 1543–1561 (1995)

On Mixture and Exponential Connection by Open Arcs

Marina Santacroce, Paola Siri(✉), and Barbara Trivellato

Dipartimento di Scienze Matematiche "G.L. Lagrange", Politecnico di Torino, Corso Duca degli Abruzzi 24, 10129 Torino, Italy
{marina.santacroce,paola.siri,barbara.trivellato}@polito.it

Abstract. Results on mixture and exponential connections by open arcs are revised and used to prove additional duality properties of statistical models.

Keywords: Exponential models · Mixture models · Orlicz spaces · Kullback-Leibler divergence · Dual systems

1 Introduction

In this paper we review some results on mixture and exponential connections by arc and their relation to Orlicz spaces. These results are essentially contained in our previous works, as well as in papers by Pistone and different coauthors.

We use some of them in order to prove a new theorem concerning the duality between statistical exponential models and Lebesgue spaces. Moreover, the notions of connection by mixture and exponential arcs, as well as divergence finiteness between two densities, are presented here in a unified framework.

The geometry of statistical models started with the paper of Rao [12] and has been described in its modern formulation by Amari [1,2] and Amari and Nagaoka [3]. Until the nineties, the theory was developed only in the parametric case. The first rigorous infinite dimensional extension has been formulated by Pistone and Sempi [11]. In that paper, using the Orlicz space associated to an exponentially growing Young function, the set of positive densities is endowed with a structure of exponential Banach manifold.

More recently, different authors have generalized this structure replacing the exponential function with a new class of functions, called deformed exponentials (see, e.g., Vigelis and Cavalcante [16]). However, the connection to open arcs has not been investigated yet.

The geometry of nonparametric exponential models and its analytical properties in the topology of Orlicz spaces have been also studied in subsequent works, such as Cena and Pistone [6] and Santacroce, Siri and Trivellato [14,15], among others.

In the exponential framework, the starting point is the notion of *maximal exponential model* centered at a given positive density p, introduced by Pistone

F. Nielsen and F. Barbaresco (Eds.): GSI 2017, LNCS 10589, pp. 577–584, 2017.
https://doi.org/10.1007/978-3-319-68445-1_67

and Sempi [11]. One of the main result in Cena and Pistone [6] states that any density belonging to the maximal exponential model centered at p is connected by an *open* exponential arc to p and viceversa (by *open*, we essentially mean that the two densities are not the extremal points of the arc). Further upgrades of these statements have been proved in Santacroce, Siri and Trivellato [14,15]. In [14], the equivalence between the equality of the maximal exponential models centered at two (connected) densities p and q and the equality of the Orlicz spaces referred to the same densities is proved. In [15], another additional equivalent condition, involving transport mappings, is given. Moreover, in the last paper, it is also shown that exponential connection by arc is stable with respect to projections and that projected densities belong to suitable sub-models.

The manifold setting of exponential models, introduced in Pistone and Sempi [11], turns out to be well-suited for applications in physics as some recent papers show (see, e.g., Lods and Pistone [9]). On the other hand, statistical exponential models built on Orlicz spaces have been exploited in several fields, such as differential geometry, algebraic statistics, information theory and, very recently, in mathematical finance (see Santacroce, Siri and Trivellato [15]).

In a large branch of mathematical finance convex duality is strongly used to tackle portfolio optimization problems. In particular, the duality between Orlicz spaces has been receiving a growing attention (see [4] among the others). In the last section of this paper we prove a general duality result involving the vector space generated by the maximal exponential model which could be well suited for a financial framework.

2 Mixture and Exponential Arcs

Let $(\mathcal{X}, \mathcal{F}, \mu)$ be a fixed probability space and denote with $\mathcal{P}$ the set of all densities which are positive μ-a.s. and with $\mathbb{E}_p$ the expectation with respect to $p d\mu$, for each fixed $p \in \mathcal{P}$.

Let us consider the Young function $\Phi_1(x) = \cosh(x) - 1$, equivalent to the more commonly used $\Phi_2(x) = e^{|x|} - |x| - 1$.
Its conjugate function is $\Psi_1(y) = \int_0^y \sinh^{-1}(t)dt$, which, in its turn, is equivalent to $\Psi_2(y) = (1 + |y|)\log(1 + |y|) - |y|$.

Given $p \in \mathcal{P}$, we consider the Orlicz space associated to Φ_1, defined by

$$L^{\Phi_1}(p) = \{u \text{ measurable } : \exists\, \alpha > 0 \ s.t.\ \mathbb{E}_p(\Phi_1(\alpha u)) < +\infty\}. \tag{1}$$

Recall that $L^{\Phi_1}(p)$ is a Banach space when endowed with the *Luxembourg norm*

$$\|u\|_{\Phi_1,p} = \inf\left\{k > 0 \ : \ \mathbb{E}_p\left(\Phi_1\left(\frac{u}{k}\right)\right) \leq 1\right\}. \tag{2}$$

Finally, it is worth to note the following chain of inclusions:

$$L^\infty(p) \subseteq L^{\Phi_1}(p) \subseteq L^a(p) \subseteq L^{\psi_1}(p) \subseteq L^1(p),\ a > 1.$$

Definition 1. *$p, q \in \mathcal{P}$ are connected by an open exponential arc if there exists an open interval $I \supset [0,1]$ such that one the following equivalent relations is satisfied:*

1. $p(\theta) \propto p^{(1-\theta)} q^{\theta} \in \mathcal{P}$, $\forall \theta \in I$;
2. $p(\theta) \propto e^{\theta u} p \in \mathcal{P}$, $\forall \theta \in I$, *where* $u \in L^{\Phi_1}(p)$ *and* $p(0) = p,\ p(1) = q$.

Observe that connection by open exponential arcs is an equivalence relation.

Let us consider the cumulant generating functional map defined on $L_0^{\Phi_1}(p) = \{u \in L^{\Phi_1}(p) : \mathbb{E}_p(u) = 0\}$, by the relation $K_p(u) = \log \mathbb{E}_p(e^u)$. We recall from Pistone and Sempi [11] that K_p is a positive convex and lower semicontinuous function, vanishing at zero, and that the interior of its proper domain, denoted here by $\overset{\circ}{\operatorname{dom}} K_p$, is a non empty convex set.

For every density $p \in \mathcal{P}$, we define the maximal exponential model at p as

$$\mathcal{E}(p) = \left\{ q = e^{u - K_p(u)} p : u \in \overset{\circ}{\operatorname{dom}} K_p \right\} \subseteq \mathcal{P}. \tag{3}$$

We now state one of the central results of [6,14,15], which gives equivalent conditions to open exponential connection by arcs, in a complete version, containing all the recent improvements.

Theorem 1. *(Portmanteau Theorem)*
Let $p, q \in \mathcal{P}$. The following statements are equivalent.

(i) $q \in \mathcal{E}(p)$;
(ii) q *is connected to* p *by an open exponential arc;*
(iii) $\mathcal{E}(p) = \mathcal{E}(q)$;
(iv) $\log \frac{q}{p} \in L^{\Phi_1}(p) \cap L^{\Phi_1}(q)$;
(v) $L^{\Phi_1}(p) = L^{\Phi_1}(q)$;
(vi) $\frac{q}{p} \in L^{1+\varepsilon}(p)$ *and* $\frac{p}{q} \in L^{1+\varepsilon}(q)$, *for some* $\varepsilon > 0$;
(vii) the mixture transport mapping

$$\begin{aligned} {}^m\mathbb{U}_p^q : L^{\Psi_1}(p) &\longrightarrow L^{\Psi_1}(q) \\ v &\mapsto \frac{p}{q} v \end{aligned} \tag{4}$$

is an isomorphism of Banach spaces.

The equivalence of conditions $(i) \div (iv)$ is proved in Cena and Pistone [6]. Statements (v) and (vi) have been added by Santacroce, Siri and Trivellato [14], while statement (vii) by Santacroce, Siri and Trivellato [15]. It is worth noting that, among all conditions of Portmanteau Theorem, (v) and (vi) are the most useful from a practical point of view: the first one allows to switch from one Orlicz space to the other at one's convenience, while the second one permits to work with Lebesgue spaces. On the other hand condition (vii), involving the mixture transport mapping, could be a useful tool in physics applications of exponential models, as the recent research on the subject demonstrates (see, e.g. Pistone [10], Lods and Pistone [9], Brigo and Pistone [5]). In these applications, finiteness of Kullback-Leibler divergence, implied from Portmanteau Theorem, is a desirable property.

Corollary 1. *If $q \in \mathcal{E}(p)$, then the Kullback-Leibler divergences $D(q\|p)$ and $D(p\|q)$ are both finite.*

The converse of this corollary does not hold, as the counterexamples in Santacroce, Siri and Trivellato [14,15] show.

In the following we introduce mixture connection between densities and study its relation with exponential arcs.

Definition 2. *We say that two densities $p, q \in \mathcal{P}$ are connected by an open mixture arc if there exists an open interval $I \supset [0,1]$ such that $p(\theta) = (1-\theta)p+\theta q$ belongs to $\mathcal{P}$, for every $\theta \in I$.*

The connection by open mixture arcs is an equivalence relation as well as in the exponential case.

Given $p \in \mathcal{P}$, we denote by $\mathcal{M}(p)$ the set of all densities $q \in \mathcal{P}$ which are connected to p by an open mixture arc.

Theorem 2. *Let $p, q \in \mathcal{P}$. The following statements are equivalent.*

(i) $q \in \mathcal{M}(p)$;
(ii) $\mathcal{M}(p) = \mathcal{M}(q)$;
(iii) $\frac{q}{p}, \frac{p}{q} \in L^\infty$.

The previous theorem is the counterpart of Portmanteau Theorem for open mixture arcs (Santacroce, Siri and Trivellato [14]).

From Theorems 1 and 2, it immediately follows that $\mathcal{M}(p) \subseteq \mathcal{E}(p)$, while, in general, the other inclusion does not hold. A counterexample is given in Santacroce, Siri and Trivellato [14].

Moreover, in the same paper $\mathcal{E}(p)$ and $\mathcal{M}(p)$ are proved to be convex.

The following proposition restates some of the previous results concerning densities either connected by open mixture or exponential arcs or with finite relative divergence. Assuming a different perspective, a new condition is expressed in term of the ratios $\frac{q}{p}$ and $\frac{p}{q}$ which have to belong to L^∞ or to its closure with respect to a suitable topology.

Proposition 1. *Let $p, q \in \mathcal{P}$. The following statements are true.*

(i) $q \in \mathcal{M}(p)$ *if and only if* $\frac{q}{p}, \frac{p}{q} \in L^\infty$;
(ii) $q \in \mathcal{E}(p)$ *if and only if* $\frac{q}{p} \in L^{1+\epsilon}(p) = \overline{L^\infty}^{\Phi_{1+\epsilon},p}$ *and* $\frac{p}{q} \in L^{1+\epsilon}(q) = \overline{L^\infty}^{\Phi_{1+\epsilon},q}$, *for some* $\epsilon > 0$, *where* $\Phi_{1+\epsilon}(x) = x^{1+\epsilon}$;
(iii) $D(q\|p) < +\infty$ *and* $D(p\|q) < +\infty$ *if and only if* $\frac{q}{p} \in L^{\Psi_1}(p) = \overline{L^\infty}^{\Psi_1,p}$ *and* $\frac{p}{q} \in L^{\Psi_1}(q) = \overline{L^\infty}^{\Psi_1,q}$.

Proof. Since (i) and (ii) have been already discussed, we consider only condition (iii). To prove it, we just need to observe that $D(q\|p) < +\infty$ if and only if $\frac{q}{p} \in L^{\Psi_1}(p)$ (Cena and Pistone [6]) and, that simple functions are dense in $L^{\Psi_1}(p)$ (Rao and Ren [13]).

The next theorem states a closure result concerning densities belonging to the open mixture model.

Theorem 3. *For any $p \in \mathcal{P}$ the open mixture model $\mathcal{M}(p)$ is $L^1(\mu)$-dense in the non negative densities $\mathcal{P}_{\geq}$, that is $\overline{\mathcal{M}(p)} = \mathcal{P}_{\geq}$, where the overline denotes the closure in the $L^1(\mu)$-topology.*

(See Santacroce, Siri and Trivellato [14] for the proof.)

Remark 1. Since $\mathcal{M}(p) \subseteq \mathcal{E}(p)$, we immediately deduce that also $\mathcal{E}(p)$ is $L^1(\mu)$-dense in $\mathcal{P}_{\geq}$. The last result was already proved, by different arguments, in Imparato and Trivellato [8].

As a consequence, the positive densities with finite Kullback-Leibler divergence with respect to any $p \in \mathcal{P}$ is $L^1(\mu)$-dense in the set of all densities $\mathcal{P}_{\geq}$. This also corresponds to the choice $\varphi(x) = x(\log(x))^+$ in the following result, proved in [14].

Proposition 2. *Assume $\varphi : (0, +\infty) \to (0, +\infty)$ is a continuous function. Then the set*

$$\mathcal{P}_\varphi = \left\{ q \in \mathcal{P} : \mathbb{E}_p \left(\varphi \left(\frac{q}{p} \right) \right) < +\infty \right\}$$

is $L^1(\mu)$-dense in $\mathcal{P}_{\geq}$.

3 Dual Systems

In this paragraph we show a new duality result concerning the linear space generated by the maximal exponential model.

Let $p \in \mathcal{P}$ and define

$$U = \bigcap_{q \in \mathcal{E}(p)} L^1(q), \qquad V = \mathrm{Lin}\{\mathcal{E}(p)\}.$$

Proposition 3. *It holds*

(i) $L^{\Phi_1}(p) \subseteq U$
(ii) $V \subseteq pL^{\Psi_1}(p)$.

Proof. In order to prove (i) it is sufficient to observe that if $u \in L^{\Phi_1}(p)$, by (v) of Portmanteau Theorem, $u \in L^{\Phi_1}(q) \subseteq L^1(q)$ for any $q \in \mathcal{E}(p)$. With regard to (ii), we consider $v \in V$. Since $v = \sum_{i \in F} \alpha_i q_i$, with F a finite set, $\alpha_i \in \mathbb{R}$ and $q_i \in \mathcal{E}(p)$, we have that

$$\frac{v}{p} = \sum_{i \in F} \alpha_i \frac{q_i}{p}.$$

From Corollary 1 and of Proposition 1 (iii) we get $\frac{q_i}{p} \in L^{\Psi_1}(p)$ and the conclusion follows.

The next result is our main contribution and shows that U and V are dual spaces with the duality given by the bilinear map $(u,v) \to \langle u,v\rangle = \mathbb{E}_\mu(uv)$ and that the dual system is separated in both U and V (see Grothendieck [7] for a standard reference on general dual systems). Therefore, if we endow U and V with the weak topologies $\sigma(U,V)$ and $\sigma(V,U)$, respectively, they become locally convex Hausdorff topological vector spaces.

Theorem 4. *The map*

$$\begin{aligned}\langle\cdot,\cdot\rangle : U\times V &\longrightarrow \mathbb{R}\\ (u,v) &\longmapsto \langle u,v\rangle = \mathbb{E}_\mu(uv)\end{aligned}$$

is a well-defined bilinear form. Moreover, the two separation axioms are satisfied

(a.1) $\langle u,v\rangle = 0 \quad \forall u\in U \implies v = 0$ *μ-a.s.*
(a.2) $\langle u,v\rangle = 0 \quad \forall v\in V \implies u = 0$ *μ-a.s..*

Proof. The map $\langle\cdot,\cdot\rangle$ is clearly well-defined by the definitions of U and V, and its bilinearity trivially follows from the linearity of the expectation.

We first show statement $(a.1)$ holds. We consider $v\in V$ such that $\langle u,v\rangle = 0$ $\forall u\in U$. Since, $\forall A\in\mathcal{F}$, $\mathbb{1}_A\in U$, we immediately get $\langle \mathbb{1}_A, v\rangle = \mathbb{E}_\mu(\mathbb{1}_A v) = 0$ and, therefore, $v = 0$, μ-a.s..
With regard to statement $(a.2)$, let us suppose $u\in U$ such that $\langle u,v\rangle = 0$ $\forall v\in V$. By definition (3) of maximal exponential model, if $v\in\mathcal{E}(p)$, then $v = e^{w-K_p(w)}p$, with $w\in \overset{\circ}{\text{dom}}\, K_p$, from which $\langle u,v\rangle = e^{-K_p(w)}\mathbb{E}_p(ue^w)$.

Then the hypothesis, restricted to $\mathcal{E}(p)$, becomes

$$\mathbb{E}_p(ue^w) = 0, \quad \forall w\in \overset{\circ}{\text{dom}}\, K_p, \tag{5}$$

from which we will deduce $u = 0$ μ-a.s..

In order to do this, we define $A = \{u > 0\}$, which we suppose not negligible, without loss of generality. Let $\bar{w} = c\mathbb{1}_A + d\mathbb{1}_{A^c}$ where the two constants $c\neq d$ are chosen in order to have $\mathbb{E}_p(\bar{w}) = 0$. We check now that $\pm\bar{w}$ belong to $\overset{\circ}{\text{dom}}\, K_p$, which is equivalent to $\pm\bar{w}$ belong to $L_0^{\Phi_1}(p)$ and $\mathbb{E}_p(e^{(1+\epsilon)(\pm\bar{w})}) < +\infty$ for some $\epsilon > 0$.

In fact, it holds the stronger result

$$\mathbb{E}_p(e^{\alpha\bar{w}}) = e^{\alpha c}\int_A p d\mu + e^{\alpha d}\int_{A^c} p d\mu < +\infty,$$

for any $\alpha\in\mathbb{R}$.

From condition (5) applied to $\bar{w}$, that is $\mathbb{E}_p(ue^{\bar{w}}) = 0$, we deduce that $\mathbb{E}_p(u\mathbb{1}_{A^c}) = -e^{d-c}\mathbb{E}_p(u\mathbb{1}_A)$.

Similarly, from $\mathbb{E}_p(ue^{-\bar{w}}) = 0$, we have that $\mathbb{E}_p(u\mathbb{1}_{A^c}) = -e^{c-d}\mathbb{E}_p(u\mathbb{1}_A)$.

Since $c\neq d$, it follows that $\mathbb{E}_p(u\mathbb{1}_A) = \mathbb{E}_p(u\mathbb{1}_{A^c}) = 0$ and, therefore, $u = 0$ μ-a.s..

Remark 2. Note that $\mathcal{E}(p)$ is obviously contained in $V \cap \mathcal{P}$. The next example shows that the inclusion is strict.

Example 1. Let $\mathcal{X} = (2, \infty)$, endowed with the probability measure μ whose Radon-Nikodym derivative with respect to the Lebesgue measure is $\frac{1}{kx(\log x)^2}$ ($k > 0$ normalizing constant).

Define the densities p, q_1 and $q_2 \in \mathcal{P}$ where $q_1(x) = p(x) = 1$ and $q_2(x) = \frac{x-1}{cx}$ ($c > 0$ normalizing constant).

Since

$$\mathbb{E}_p(q_2^{1+\epsilon}) = \mathbb{E}_\mu(q_2^{1+\epsilon}) = \int_2^\infty \left(\frac{x-1}{cx}\right)^{1+\epsilon} \frac{1}{kx(\log x)^2}\, dx < \infty$$

and

$$\mathbb{E}_p(q_2^{-\epsilon}) = \mathbb{E}_\mu(q_2^{-\epsilon}) = \int_2^\infty \left(\frac{cx}{x-1}\right)^{\epsilon} \frac{1}{kx(\log x)^2}\, dx < \infty$$

we deduce that $q_2 \in \mathcal{E}(p)$.

Define now $q \in V$ by

$$q(x) := \frac{1}{1-c} q_1(x) + \frac{c}{c-1} q_2(x) = \frac{1}{1-c} + \frac{c}{c-1}\frac{x-1}{cx} = \frac{1}{(1-c)x}, \qquad \forall x \in \mathcal{X}.$$

Let us observe that

$$c = \int_2^\infty \frac{x-1}{x}\, d\mu(x) = \int_2^\infty \frac{x-1}{kx^2(\log x)^2} dx = 1 - \int_2^\infty \frac{1}{kx^2(\log x)^2}\, dx,$$

which implies $c \in (0, 1)$ and thus $q > 0$. Since $1/(1-c) + c/(c-1) = 1$, we immediately get $q \in \mathcal{P}$.

On the other hand, since for every $\epsilon > 0$, we get

$$\mathbb{E}_p(q^{-\epsilon}) = \mathbb{E}_\mu(q^{-\epsilon}) = (1-c)^\epsilon \int_2^\infty \frac{x^\epsilon}{kx(\log x)^2}\, dx = \infty,$$

we infer that $q \notin \mathcal{E}(p)$.

4 Conclusions

In the paper, we review several results contained in our previous works, sometimes presenting them under different perspectives. We prove an original result in Theorem 4, where a duality is stated between the intersection of $L^1(q)$, when q ranges in a maximal exponential model, and the linear space generated by the exponential model itself. The search for dual systems represents the preliminary step to formulate minimax results, which are fundamental instruments for solving utility maximization problems through convex analysis. Thus, the duality of Theorem 4 could be used in portfolio optimization when allowing for more general set of strategies than the ones considered in the literature.

References

1. Amari, S.: Differential geometry of curved exponential families-curvatures and information loss. Ann. Stat. **10**, 357–385 (1982)
2. Amari, S.: Differential-Geometrical Methods in Statistics. Lecture Notes in Statistics, vol. 28. Springer, New York (1985)
3. Amari, S., Nagaoka, H.: Methods of information geometry. In: Translations of Mathematical Monographs, vol. 191. American Mathematical Society, Providence, RI. Oxford University Press, Oxford (2000)
4. Biagini, S., Frittelli, M.: A unified framework for utility maximization problems: an Orlicz space approach. Ann. Appl. Probab. **18**(3), 929–966 (2008)
5. Brigo, D., Pistone, G.: Projection based dimensionality reduction for measure valued evolution equations in statistical manifolds (2016). arXiv:1601.04189v3
6. Cena, A., Pistone, G.: Exponential statistical manifold. AISM **59**, 27–56 (2007)
7. Grothendieck, A.: Topological Vector Spaces. Gordon & Breach Science Publishers, London (1973)
8. Imparato, D., Trivellato, B.: Geometry of extended exponential models. In: Algebraic and Geometric Methods in Statistics, pp. 307–326 (2009)
9. Lods, B., Pistone, G.: Information geometry formalism for the spatially homogeneous Boltzmann equation. Entropy **17**, 4323–4363 (2015)
10. Pistone, G.: Examples of the application of nonparametric information geometry to statistical physics. Entropy **15**, 4042–4065 (2013)
11. Pistone, G., Sempi, C.: An infinite-dimensional geometric structure on the space of all the probability measures equivalent to a given one. Ann. Stat. **23**(5), 1543–1561 (1995)
12. Rao, C.R.: Information and accuracy attainable in the estimation of statistical parameters. Bull. Calcutta Math. Soc. **37**, 81–91 (1945)
13. Rao, M.M., Ren, Z.D.: Theory of Orlicz Spaces. Marcel Dekker Inc., New York (1991)
14. Santacroce, M., Siri, P., Trivellato, B.: New results on mixture and exponential models by Orlicz spaces. Bernoulli **22**(3), 1431–1447 (2016)
15. Santacroce, M., Siri, P., Trivellato, B.: Exponential models by Orlicz spaces and Applications. Submitted (2017)
16. Vigelis, R.F., Cavalcante, C.C.: On φ-families of probability distributions. J. Theor. Probab. **26**(3), 870–884 (2013)

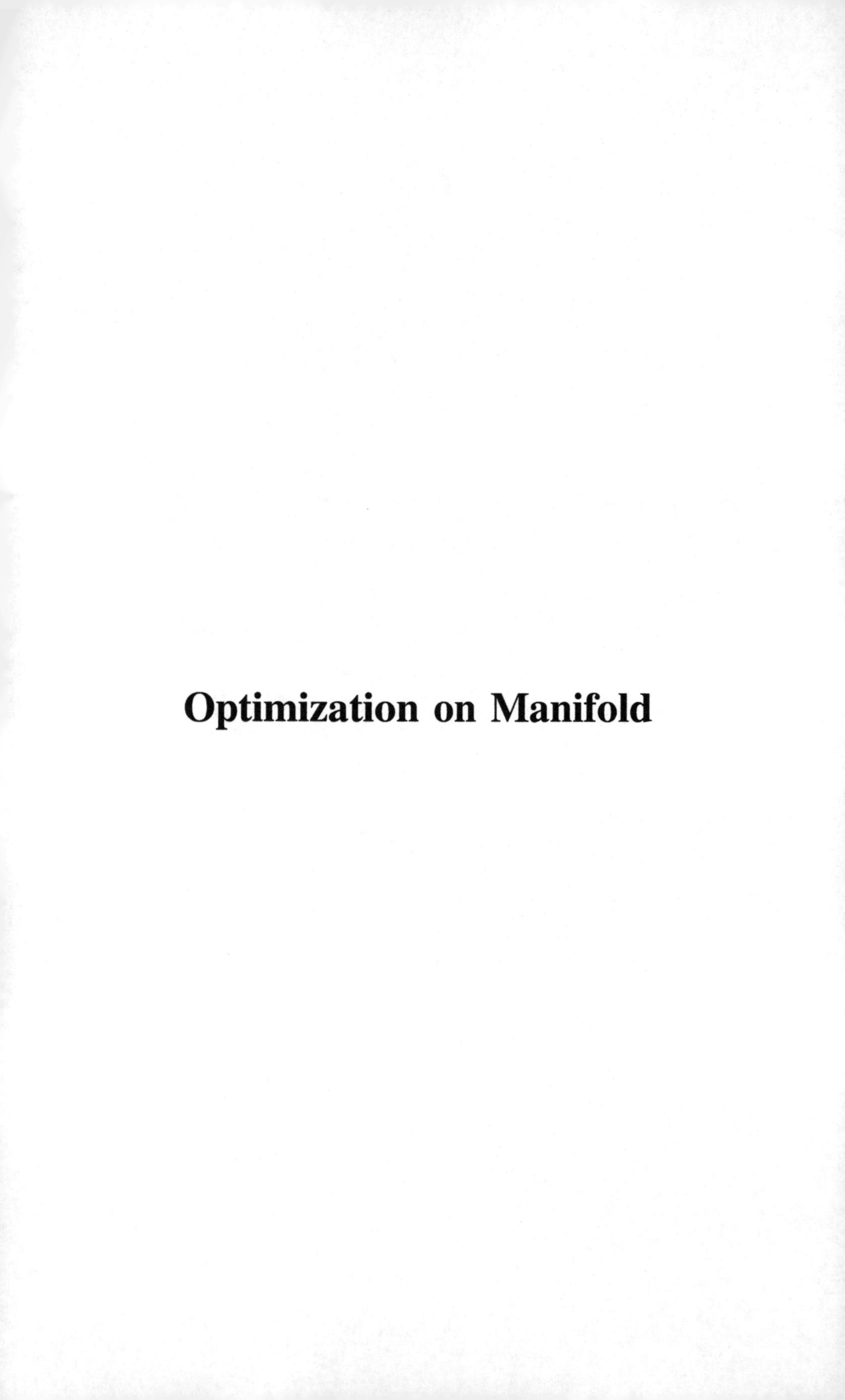

Optimization on Manifold

Applying Backward Nested Subspace Inference to Tori and Polyspheres

Benjamin Eltzner(✉) and Stephan Huckemann

Felix-Bernstein-Institute for Mathematical Statistics in the Biosciences, University of Goettingen, Goettingen, Germany
beltzne@uni-goettingen.de

Abstract. For data with non-Euclidean geometric structure, hypothesis testing is challenging because most statistical tools, for example principal component analysis (PCA), are specific for linear data with a Euclidean structure. In the last 15 years, the subject has advanced through the emerging development of central limit theorems, first for generalizations of means, then also for geodesics and more generally for lower dimensional subspaces. Notably, there are data spaces, even geometrically very benign, such as the torus, where this approach is statistically not feasible, unless the geometry is changed, to that of a sphere, say. This geometry is statistically so benign that nestedness of Euclidean PCA, which is usually not given for the above general approaches, is also naturally given through principal nested great spheres (PNGS) and even more flexible than Euclidean PCA through principal nested (small) spheres (PNS). In this contribution we illustrate applications of bootstrap two-sample tests for the torus and its higher dimensional generalizations, polyspheres.

1 Introduction

For the characterization of complex biological and medical data, non-Euclidean spaces often present the only satisfactory representation, say, for shape and multi-directional information, which often play crucial roles in biological and medical applications. As computation power surges, such data spaces have moved into the scope of statistical analysis in recent years. We will concentrate here on tori of arbitrary dimension and products of hyperspheres of any dimensions, which we call *polyspheres*. Polyspheres occur in shape description by skeletal representations (S-reps), cf. [10], while RNA backbones described by dihedral angles are represented in torus data spaces [2].

Determining differences of populations at given confidence level based on empirical quantities of two samples lies at the heart of statistics. A typical statistic by which one may distinguish two data sets is their mean value, but also their variances, and in multivariate statistics covariances play a crucial part. In this context, in the case of a Euclidean data space, principal component analysis (PCA) has proven to be a valuable tool. However, for data on non-Euclidean

Acknowledging the Niedersachsen Vorab of the Volkswagen Foundation.

F. Nielsen and F. Barbaresco (Eds.): GSI 2017, LNCS 10589, pp. 587–594, 2017.
https://doi.org/10.1007/978-3-319-68445-1_68

spaces, like shape data or directional data, PCA is not applicable, as it relies on the linearity of the underlying data space.

The very obvious approach to data on non-Euclidean manifolds consists of applying linear statistics to a tangent space projection at a sample mean, via the exponential map, say. However, such an approach disregards topology and curvature, for tori and polyspheres, say, and may lead to a misrepresentation of data variance. To preserve geometry and topology, approaches relying on geodesics, like geodesic [4,5] and horizontal PCA [11], were developed. However, like the intrinsic mean, also principal geodesics may fail to lie close to the data. As an alternative that aims at higher data fidelity, backward nested families of descriptors (BNFD) have been established. A prime example are principal nested spheres (PNS) proposed by [8]. Recently, we have established asymptotics of general nested descriptors in [7].

In polyspheres and tori, also the above presented intrinsic approaches face severe problems, because almost all geodesics wind around indefinitely, so that geodesic-based PCA-like methods approximate any data arbitrarily well. And the very low-dimensional symmetry group allows only for a low dimensional backward nested approach (cf. discussion in [6]). To overcome these limitations, we have introduced a deformation method for polyspheres in [1] and more recently have detailed the intricacies of a similar deformation method for tori in [2].

These deformation methods deform polyspheres and tori into a single high dimensional sphere which naturally comes with an even richer set of canonical subspaces than Euclidean PCA. While PNS was originally proposed for dimension reduction for skeletal representations, where it was only applied to the individual spheres of the polysphere, we apply PNS to entire high dimensional polyspheres, after deformation. Relying on the asymptotic theory and the two-sample test proposed in [7], we perform two-sample tests in the setting of PNS on deformed polyspheres in application to S-reps discrimination and on tori in application to RNA discrimination.

2 Theoretical Framework

2.1 Principal Nested Spheres as Backward Nested Subspaces

We begin by demonstrating that principal nested great spheres (PNGS) as presented in [8] fit the setting of BNFDs as introduced in [7].

The data space of PNGS is $Q = \mathbb{S}^D$. The parameter spaces are Grassmannians, where $O(m, n)$ denote Stiefel manifolds and $O(n)$ the orthogonal group

$$\begin{aligned} P_j = Gr(j+1, D+1) &= O(D-j, D+1)/O(D-j) \\ &= O(D+1)/(O(D-j) \times O(j+1)) \end{aligned}$$

because a j-dimensional subsphere of $\mathbb{S}^j \subset \mathbb{S}^D \subset \mathbb{R}^{D+1}$ is defined by intersecting with a $j+1$-dimensional linear subspace $\mathbb{R}^{j+1} \subset \mathbb{R}^{D+1}$ and the space of these linear subspaces is precisely the Grassmannian.

For any $p^j = [v] \in P_j$ using any representative $v \in O(D-j, D+1)$, we define

$$S_{p^j} := \left\{[v'] = [v, v_{D-j+1}] \middle| v^T v_{D-j+1} = 0\right\} \subset P_{j-1}\,.$$

Further, for any $p^{j-1} = [v'] \in S_{p^j}$ with any representative $v' \in O(D+1-j, D+1)$ we define a map, using the vector $y \in \mathbb{R}^{D+1}$ for $y \in p^j \subset \mathbb{S}^D \subset \mathbb{R}^{D+1}$

$$\rho_{p^j} : p^j \times S_p \to [0, \infty) \qquad \rho_{p^j}(y, p^{j-1}) = \arccos\left(y^T (I_{D+1} - v'v'^T) y\right)\,. \tag{1}$$

Finally, we define projections using the above notation

$$\pi_{p^j, p^{j-1}} : p^j \to p^{j-1}, \quad q \mapsto y = \frac{(I_{D+1} - v_{D-j+1} v_{D-j+1}^T) q}{\|(I_{D+1} - v_{D-j+1} v_{D-j+1}^T) q\|}\,. \tag{2}$$

For $j \in \{1, \dots, D-2\}$ a family

$$f = \{p^j, \dots, p^{D-1}\}, \text{ with } p^{k-1} \in S_{p^k}, k = j+1, \dots, D$$

is then a *backward nested family of descriptors* (BNFD) in the sense introduced by [7] and the asymptotic theory is applicable correspondingly.

2.2 Sausage Transformation

We now explain the specific properties of the torus deformation which we introduced in [2] in some detail. We rely on that publication for the following.

Let $T^D = (\mathbb{S}^1)^{\times D}$ be the D-dimensional unit torus and $\mathbb{S}^D = \{x \in \mathbb{R}^{D+1} : \|x\| = 1\}$ the D-dimensional unit sphere, $D \in \mathbb{N}$. The definition of the deformation map $P : T^D \longrightarrow \mathbb{S}^D$ defined in this section is based on comparing squared Riemannian line elements. For $\psi_k \in \mathbb{S}^1$, the squared line element of T^D is given by the squared Euclidean line element

$$ds^2_{T^D} = \sum_{k=1}^{D} d\psi_k^2\,.$$

For $\mathbb{S}^D$, in polar coordinates $\phi_k \in (0, \pi)$ for $k = 1, \dots, D-1$ and $\phi_D \in \mathbb{S}^1$, whose relation to embedding Euclidean coordinates x_k is given by

$$x_1 = \cos\phi_1, \quad \forall 2 \le k \le D : x_k = \left(\prod_{j=1}^{k-1} \sin\phi_j\right) \cos\phi_k, \quad x_{D+1} = \left(\prod_{j=1}^{D} \sin\phi_j\right),$$

the spherical squared line element is given by

$$ds^2_{\mathbb{S}^D} = d\phi_1^2 + \sum_{k=2}^{D} \left(\prod_{j=1}^{k-1} \sin^2\phi_j\right) d\phi_k^2\,. \tag{3}$$

In fact, this squared line element is not defined for the full sphere since $\phi_k \in (0, \pi)$ $(k = 1, \ldots, D-1)$, i.e. the coordinate singularities of $\phi_k = 0, \pi$ are excluded. Coordinate singularities appear because a hypersphere cannot be covered by a single chart. In our construction, real singularities of P will coincide with the coordinate singularities at $\phi_k = 0, \pi$, resulting in a self-gluing as explained below.

Following colloquial usage, we use the term *distortion* for the effects of deformation in the following. The line element (3) features a factor $\prod_{j=1}^{k-1} \sin^2 \phi_j$ in front of the squared line element $d\phi_k^2$. Thus, no distortion at all occurs for ϕ_1, i.e. distance along this angle corresponds directly to spherical distance. Conversely, the distance along ϕ_D is distorted by all other angles. For this reason, we will refer to ϕ_D as the *innermost angle* and to ϕ_1 as the *outermost angle* in the following. The amount of distortion can be qualitatively described as follows:

Remark 1. *Near the equatorial great circle given by $\phi_k = \frac{\pi}{2}$ $(k = 1, \ldots, D-1)$ the squared line element ds^2 is nearly Euclidean. Distortions occur whenever leaving the equatorial great circle. More precisely, distortions are higher when angles ϕ_k with low values of the index k (outer angles) are close to zero or π, than when angles ϕ_k with high values of the index k (inner angles) are close to zero or π.*

Definition 2 (Torus to Sphere Deformation). *With a data-driven permutation p of $\{1, \ldots, D\}$, data-driven central angles μ_k $(k = 1, \ldots, D)$ and data-driven scalings α_k, all of which are described below, set*

$$\phi_k = \frac{\pi}{2} + \alpha_{p(k)}(\psi_{p(k)} - \mu_{p(k)}), \quad k = 1, \ldots, D \tag{4}$$

where $p(k)$ is the index k permuted by p and the difference $(\psi_{p(k)} - \mu_{p(k)})$ is taken modulo 2π such that it is in the range $(-\pi, \pi]$.

For a thorough discussion of the free parameters, we refer to [2]. Here we choose the scalings $\alpha_{k'} = 1$ $(k' = 1, \ldots, D-1)$ and $\alpha_D = 1$, $k' = p(k)$. The reference $\mu_{k'}$ is chosen as follows. Let $\psi_{k',\mathrm{gap}}$ the center of the largest gap between neighboring $\psi_{k'}$ values of data points, then use its antipodal point $\mu_{k'} = \psi^*_{k',\mathrm{gap}}$. The choice of the permutation p_k is driven by analyses of the *data spread*

$$\sigma_k^2 = \sum_{i=1}^{n} (\psi_{k,i} - \mu_k)^2, \quad k = 1, \ldots, D \tag{5}$$

for each angle, where $\psi_{k,i} \in \mathbb{S}^1$ are the torus data and n is the number of data points on T^D. The angles are ordered by increasing data spread, such that $\sigma^2_{p(1)}$ is minimal and $\sigma^2_{p(D)}$ is maximal. In view of Remark 1, the change of distances between data points caused by the deformation factors $\sin^2 \phi_j$ in Eq. (3) is minimized. Figure 1 illustrates the case $D = 2$.

In the following, we give a brief overview of the resulting deformation. Due to periodicity on the torus, $\psi_k = 0$ is identified with $\psi_k = 2\pi$ for all $k = 1, \ldots, D$. In contrast, for all angles ϕ_k $(k = 1, \ldots, D-1)$, $\phi_k = 0$ denotes

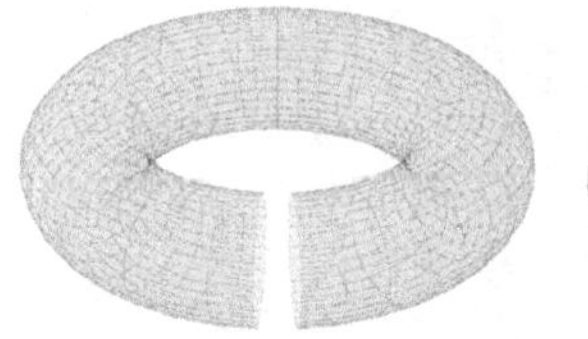

(a) *Cutting open along a circle, giving two circles.*

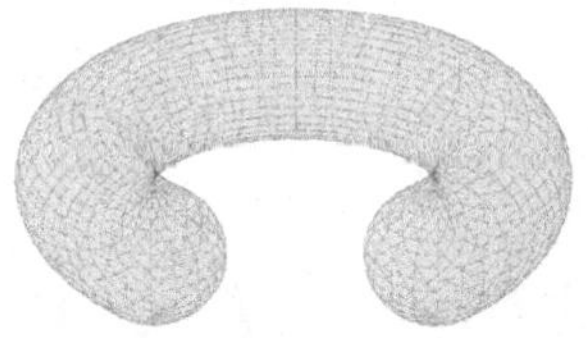

(b) *Separately collapsing each circle to a point.*

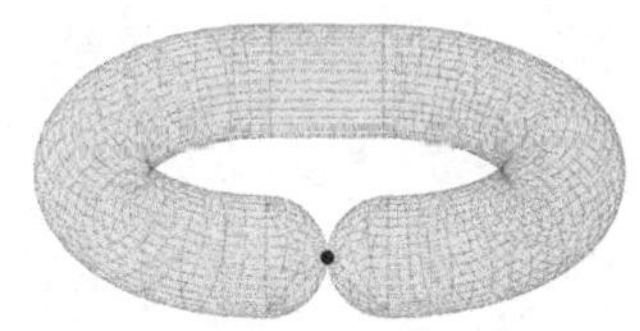

(c) *Identifying the points restores the cyclic topology.*

Fig. 1. Self-gluing of T^2: From a donut to a sausage. These operations are only topological.

spherical locations different from $\phi_k = \pi$. For a representation respecting the torus' topology, however, it is necessary to identify these locations accordingly. Due to the spherical geometry, each of those regions is of dimension $D - j - 1$, in which all angles vary except for j of the $\phi_1, \ldots, \phi_{D-1}$ which are set to fixed values in $\{0, \pi\}$. In the topology of the torus, all those regions with a specific choice of fixed angles are identified with one-another. In particular, there are $2(D-1)$ such regions of highest dimension $D-2$ on the sphere (where only one angle is fixed to 0 or π), two of which are pairwise identified in the topology of the torus. In fact, in the topology of the torus, each of these $D - 1$ regions of highest dimension $D-2$ itself carries the topology of a torus of dimension $D-2$, each glued to each other torus along a subtorus of dimension $D - 3$, and so on. Thus the *self-gluing* of S^D giving the topology of T^D can be iteratively achieved along a topological subsphere of dimension $D - 2$ which is suitably divided into $2(D-1)$ regions that are pairwise identified by way of a torus, sharing common boundaries which correspond to lower dimensional tori.

3 Applications

We apply the two-sample test put forth in [7] in two examples. The first example compares two data sets of S-reps by applying PNGS on a deformed polysphere. The second example is a cross-test between four RNA data sets, two tRNA samples and two rRNA samples, where we apply PNGS on a deformed torus.

3.1 Discriminating S-Reps

The two S-rep data sets we compare are data sets of a toy model S-rep from [9], each consisting of 66 spokes of unit length. Sample I corresponds to shapes of an ellipsoid which is bent along its longest axis. Sample II corresponds to shapes of an ellipsoid which is bent along and twisted around its longest axis. Since the polysphere is mapped to an $\mathbb{S}^{132}$ but either of the two samples only consists of 30 data, we proceed as follows.

The samples are pooled and mapped jointly to an $\mathbb{S}^{132}$ using the deformation outlined in our earlier work [1], then we perform PNGS until we are left with

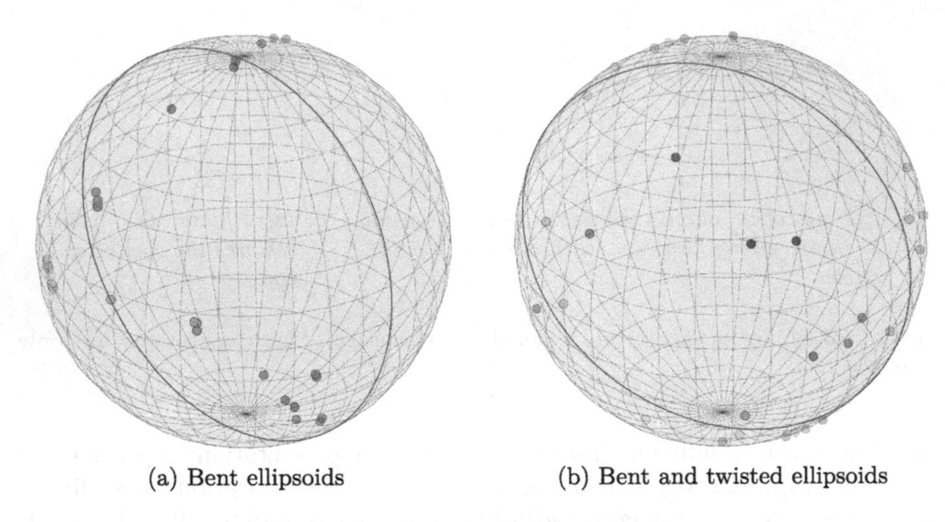

Fig. 2. $\mathbb{S}^2$ PNGS projections of the two S-rep samples compared here with PNGS $\mathbb{S}^1$. The PNGS up to $\mathbb{S}^5$ was done jointly for the data sets and from that point PNGS was conducted separately.

$30 + 30$ data points on an $\mathbb{S}^5$. Finally, we apply our hypothesis test to PNGS on the two data sets.

As determined in earlier work [1], the data variation is essentially at most two-dimensional for these samples. Therefore, we focus on the comparison of the subspaces up to dimension $d = 2$. The test clearly rejects with a p-value below the accuracy of 10^{-3} for the nested mean. The p-value for the $\mathbb{S}^1$ is 0.644 and the value for the $\mathbb{S}^2$ is 1.0, so these tests clearly do not reject.

The fact that the test does not reject for the $\mathbb{S}^2$ suggests that the two-dimensional projections of the S-rep data after deformation are compatible. The $\mathbb{S}^1$ fit for Sample II has very large variance, therefore the test does not reject. The test for the nested mean rejects with 99% confidence and thus shows that the two samples can clearly be distinguished.

3.2 Discriminating RNA

The RNA data sets we consider are sets of geometries of single residues which are characterized in terms of 7 dihedral angles as displayed in Fig. 3.

The data sets considered here are subsets of the data set of 7544 residue backbones used in [2,12]. We remove residues further than 50° in torus distance from their nearest neighbor as outliers, leading to a set of 7544 residues. From these we then pick the 1105 residues from tRNA structures and the 3989 residues stemming from the two large rRNA data sets (with PDB codes 1s72 and 1xmq). We jointly map all these data from T^7 to $\mathbb{S}^7$ as elaborated above in Sect. 2. From either of the two data sets on $\mathbb{S}^7$ we then sample two disjoint random subsets of 500 residues each. We call these samples tRNA1, tRNA2, rRNA1, and rRNA2, and we apply a PNGS hypothesis test on each pair of samples.

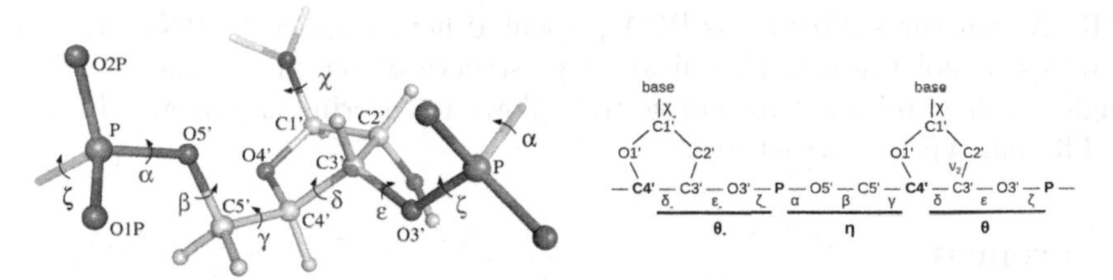

(a) *3D structure of an RNA residue.* (b) *2D scheme of an RNA residue.*

Fig. 3. Part of an RNA backbone. Dihedral angles (Greek letters) are defined by three bonds, the central bond carries the label; pseudo-torsion angles (bold Greek letters) are defined by the pseudo-bonds between bold printed atoms (Fig. 3b). The subscript "–" denotes angles of the neighboring residue. Figure 3a is reproduced from [3].

Table 1. Bootstrap PNGS p-values for pairs of similar samples. These tests are used as control. We use $B = 1000$ bootstrap samples for all tests.

Dimension	0	1	2	3	4	5	6
tRNA1 vs tRNA2	0.168	1.0	1.0	1.0	0.969	0.544	0.5
rRNA1 vs rRNA2	0.668	1.0	1.0	1.0	0.628	0.280	0.838

Table 2. Bootstrap PNGS p-values for pairs of different samples. These tests distinguish rRNA from tRNA. We use $B = 1000$ bootstrap samples for all tests.

Dimension	0	1	2	3	4	5	6
tRNA1 vs rRNA1	0.003	1.0	0.932	0.999	0.134	0.008	0.070
tRNA1 vs rRNA2	0.001	1.0	0.690	0.537	0.267	0.048	0.021
tRNA2 vs rRNA1	< 0.001	1.0	1.0	0.985	0.405	0.007	0.415
tRNA2 vs rRNA2	< 0.001	0.999	0.999	0.378	0.244	0.029	0.377

The tRNAs are short chains with similar shapes, while the large ribosomal complexes have a much more varied structure. As a consequence, we expect tRNAs and rRNAs to feature different distributions of residue geometries. Furthermore, we certainly expect visible differences in nested means. We therefore focus on the nested mean p-values, although we list all p-values for completeness in Tables 1 and 2.

The test results clearly conform with our expectation. The control tests displayed in Table 1 do not reject, which is expected for two random samples from a larger common set. However, the null hypothesis of equal nested mean is always rejected at 99% confidence level for the tests displayed in Table 2. This shows that rRNA and tRNA feature different distributions of residue geometries.

The development of dimension reduction methods for RNA backbone shapes is crucial for improving backbone fits to X-ray spectroscopy data. To improve fits it is important to know which structural elements prevail for which types of RNA structures. Our torus-PCA presented here, applied to RNA structures, provides a tool towards classification of structures on the smallest scale of a single residue and our two-sample test allows uncovering significant differences in different types of structures.

References

1. Eltzner, B., Jung, S., Huckemann, S.: Dimension reduction on polyspheres with application to skeletal representations. In: Nielsen, F., Barbaresco, F. (eds.) GSI 2015. LNCS, vol. 9389, pp. 22–29. Springer, Cham (2015). doi:10.1007/978-3-319-25040-3_3
2. Eltzner, B., Huckemann, S.F., Mardia, K.V.: Deformed torus PCA with applications to RNA structure (2015). arXiv:1511.04993
3. Frellsen, J., Moltke, I., Thiim, M., Mardia, K.V., Ferkinghoff-Borg, J., Hamelryck, T.: A probabilistic model of RNA conformational space. PLoS Comput. Biol. **5**(6), e1000406 (2009)
4. Huckemann, S., Hotz, T., Munk, A.: Intrinsic shape analysis: Geodesic principal component analysis for Riemannian manifolds modulo Lie group actions (with discussion). Stat. Sin. **20**(1), 1–100 (2010)
5. Huckemann, S., Ziezold, H.: Principal component analysis for Riemannian manifolds with an application to triangular shape spaces. Adv. Appl. Probab. (SGSA) **38**(2), 299–319 (2006)
6. Huckemann, S.F., Eltzner, B.: Polysphere PCA with applications. In: Proceedings of the 33th LASR Workshop, pp. 51–55. Leeds University Press (2015). http://www1.maths.leeds.ac.uk/statistics/workshop/lasr2015/Proceedings15.pdf
7. Huckemann, S.F., Eltzner, B.: Backward nested descriptors asymptotics with inference on stem cell differentiation (2017). arXiv:1609.00814
8. Jung, S., Dryden, I., Marron, J.: Analysis of principal nested spheres. Submitted to Biometrika (2010)
9. Schulz, J., Jung, S., Huckemann, S., Pierrynowski, M., Marron, J., Pizer, S.M.: Analysis of rotational deformations from directional data. J. Comput. Graph. Stat. **24**(2), 539–560 (2015)
10. Siddiqi, K., Pizer, S.: Medial Representations: Mathematics, Algorithms and Applications. Springer, Heidelberg (2008)
11. Sommer, S.: Horizontal dimensionality reduction and iterated frame bundle development. In: Nielsen, F., Barbaresco, F. (eds.) GSI 2013. LNCS, vol. 8085, pp. 76–83. Springer, Heidelberg (2013). doi:10.1007/978-3-642-40020-9_7
12. Wadley, L.M., Keating, K.S., Duarte, C.M., Pyle, A.M.: Evaluating and learning from RNA pseudotorsional space Quantitative validation of a reduced representation for RNAstructure. J. Mol. Biol. **372**(4), 942–957 (2007). http://www.sciencedirect.com/science/article/pii/S0022283607008509

Fast Method to Fit a $\mathcal{C}^1$ Piecewise-Bézier Function to Manifold-Valued Data Points: How Suboptimal is the Curve Obtained on the Sphere $\mathbb{S}^2$?

Pierre-Yves Gousenbourger(✉), Laurent Jacques, and P.-A. Absil

ICTEAM Institute, Université catholique de Louvain,
1348 Louvain-la-Neuve, Belgium
pierre-yves.gousenbourger@uclouvain.be

Abstract. We propose an analysis of the quality of the fitting method proposed in [7]. This method fits smooth paths to manifold-valued data points using $\mathcal{C}^1$ piecewise-Bézier functions. This method is based on the principle of minimizing an objective function composed of a data-attachment term and a regularization term chosen as the mean squared acceleration of the path. However, the method strikes a tradeoff between speed and accuracy by following a strategy that is guaranteed to yield the optimal curve only when the manifold is linear. In this paper, we focus on the sphere $\mathbb{S}^2$. We compare the quality of the path returned by the algorithms from [7] with the path obtained by minimizing, over the same search space of $\mathcal{C}^1$ piecewise-Bézier curves, a finite-difference approximation of the objective function by means of a derivative-free manifold-based optimization method.

Keywords: Path fitting on Riemannian manifolds · Bézier functions · Optimization on manifolds

1 Introduction

We consider the problem of fitting an univariate C^1 piecewise-Bézier curve to manifold-valued data points. This problem is motivated by several applications in engineering and the sciences, such as projection-based model order reduction of dynamical systems that depend on one parameter [10]. In that case, the data points are projectors from the full state space to the reduced state space and hence belong to a Grassmann manifold. In a recent paper, Gousenbourger *et al.* [7] illustrated the benefits of this approach by estimating wind fields: the task required to fit a curve to a set of data points belonging to the manifold of $p \times p$ positive semidefinite (PSD) matrices of rank r. We also mention the case of image denoising, as in Bergmann *et al.* [3], where one seeks a two-parameter function fitting an image with manifold-valued pixels, or the blood vessels tracking in the eyes in Sanguinetti *et al.* [12] as an application to the sub-Riemannian manifold SE(2).

F. Nielsen and F. Barbaresco (Eds.): GSI 2017, LNCS 10589, pp. 595–603, 2017.
https://doi.org/10.1007/978-3-319-68445-1_69

Fitting and interpolation on manifolds has been an active research topic in the past few years. For instance Samir *et al.* [11] proposed a fitting method where the search space is infinite-dimensional. In that paper, the fitting curve $\mathfrak{B}$ is discretized with a small stepsize and the objective function is minimized with a manifold-valued gradient descent. An application in image processing can be found in Su *et al.* [13]. In Absil *et al.* [1] (interpolation) or more recently in Gousenbourger *et al.* [7] (fitting), the search space is restricted to a finite dimensional space of C^1 piecewise-Bézier functions.We also mention Machado *et al.* [8] for the specific case of the sphere.

The method proposed in [7] seeks a C^1 piecewise-Bézier curve as in [2]. It also considers a smoothing objective function—a roughness penalty and a data-fitting term—as in [11]. This approach has several advantages. With respect to [2], interpolation is replaced by smoothing, which is more apt for noisy data. Compared to [11], *(i)* it reduces the space complexity (instead of being discretized, the solution curve is represented by only a few Bézier control points on the manifold) and *(ii)* it provides a very simple algorithm that only requires two objects on the manifold: the Riemannian exponential and the Riemannian logarithm. However, the proposed approach tends to be suboptimal for two reasons. First, the search space is restricted to C^1 piecewise-Bézier curves; and second, the proposed computational method ensures optimality (within the restricted search space) only if the manifold is flat.

The study of this second drawback is the subject of this paper: in particular, we aim to evaluate the quality of the fitting curve obtained with the method developed in [7] compared to a more accurate solution obtained with a more general (but also slower) optimization tool (like, for instance, Manopt [5]).

The paper is organized as follows. We first recall some generalities on Bézier curves and introduce the composite Bézier curve $\mathfrak{B}$ we would like to fit to data points (Sect. 2). In Sect. 3, we summarize the method from [7] and then introduce a more accurate (but also less efficient) method based on a discretization. We also look for an acceptable discretization stepsize. Finally, we present results on the sphere $\mathbb{S}^2$ in Sect. 4

2 Notations and Framework

We consider the case in which the data points $\{d_0, \dots, d_n\} \subset \mathcal{M}$ take values on a manifold $\mathcal{M}$ and are associated with measurement parameters $t_0 \leq t_1 \leq \dots \leq t_n$. For simplicity, we will let $t_i = i$, $i = 0, \dots, n$. We seek a composite Bézier curve $\mathfrak{B} : \mathbb{R} \to \mathcal{M}$ such that $\mathfrak{B}(t_i) \simeq d_i$, $i = 0, \dots, n$. We note $T_a\mathcal{M}$ the (Euclidean) tangent space to $\mathcal{M}$ at $a \in \mathcal{M}$; $T\mathcal{M} = \cup_a T_a\mathcal{M}$ the tangent bundle to $\mathcal{M}$; $\langle \cdot, \cdot \rangle_a$, the inner product in the tangent space at a and from which we deduce the norm of $v \in T_a\mathcal{M}$, $\|v\|_{\mathcal{M}} = \langle v, v \rangle_a$; $\exp_a(\cdot) : T_a\mathcal{M} \to \mathcal{M} : v \mapsto b = \exp_a(v)$, the Riemannian exponential; $\log_a(\cdot) : \mathcal{M} \to T_a\mathcal{M} : b \mapsto v = \log_a(b)$, the Riemannian logarithm which can be viewed as the inverse Riemannian exponential. We also introduce the notation $\gamma_{a,b}(t)$ for the shortest geodesic between $a = \gamma_{a,b}(0)$ and $b = \gamma_{a,b}(1)$. We assume throughout that we can compute these objects.

2.1 Preliminaries on Bézier Curves

We first consider the trivial case where $\mathcal{M} = \mathbb{R}^r$ to define the Bézier curve. A *Bézier curve of degree* $K \in \mathbb{N}$ is a function β parametrized by $K+1$ *control points* $\{b_0, \ldots, b_K\} \subset \mathbb{R}^r$ taking the form

$$\beta_K(\cdot; b_0, \ldots, b_K) : [0,1] \to \mathbb{R}^r, t \mapsto \textstyle\sum_{j=0}^K b_j B_{jK}(t),$$

where $B_{jK}(t) = \binom{K}{j} t^j (1-t)^{K-j}$ are the Bernstein basis polynomials (also called binomial functions) [6]. The first control point and the last one are interpolated by construction while the position of the other control points models the shape of the curve. More specifically, the quadratic and cubic Bézier curves are respectively

$$\beta_2(t; b_0, b_1, b_2) = b_0(1-t)^2 + 2b_1(1-t)t + b_2 t^2 \tag{1}$$

$$\beta_3(t; b_0, b_1, b_2, b_3) = b_0(1-t)^3 + 3b_1(1-t)^2 t + 3b_2(1-t)t^2 + b_3 t^3 \tag{2}$$

One well-known way to generalize Bézier curves to a Riemannian manifold $\mathcal{M}$ is via the De Casteljau algorithm. This algorithm, generalized to manifolds by Popiel and Noakes [9, Sect. 2], only requires the Riemannian exponential and logarithm and conserves the interpolation property of the first and last control points.

2.2 Composite Bézier Function on Manifolds

We now consider a general manifold $\mathcal{M}$. As illustrated in Fig. 1, the composite Bézier function $\mathfrak{B} \in \mathcal{M}$ is a $\mathcal{C}^1$ composition of n Bézier curves, *i.e.*,

$$\mathfrak{B} : [0,n] \to \mathcal{M}, \; t \mapsto \beta^i(t-i) \text{ on } [i, i+1], \quad i = 0, \ldots, n-1,$$

where β^i defines a piece of $\mathfrak{B}$ associated to the endpoints $\{p_i, p_{i+1}\} \subset \mathcal{M}$. The control points of the $(i-1)^{\text{th}}$ and i^{th} piece of $\mathfrak{B}$ defined on the left and right of p_i are noted $\{b_i^-, b_i^+\} \subset \mathcal{M}$, $i = 1, \ldots, n-1$. The first and last segments of $\mathfrak{B}$ are quadratic Bézier curves respectively noted $\beta^0(t) = \beta_2(t; p_0, b_1^-, p_1)$ and $\beta^{n-1}(t) = \beta_2(t; p_{n-1}, b_{n-1}^+, p_n)$. All the other segments are cubic and denoted by $\beta^i(t) = \beta_3(t; p_i, b_i^+, b_{i+1}^-, p_{i+1})$. Note the use of the superscript to refer to the ith segment of $\mathfrak{B}$ while the subscript refers to the degree of the Bézier curve.

The continuity of $\mathfrak{B}$ is trivial as $\mathfrak{B}(i) = \beta^i(i) = \beta^{i-1}(i) = p_i$. Differentiability is ensured by taking $p_1 = \text{av}[(b_1^-, b_1^+), (\frac{2}{5}, \frac{3}{5})]$, $p_i = \text{av}[(b_i^-, b_i^+), (\frac{1}{2}, \frac{1}{2})]$ $(i = 2, \ldots, n-2)$ and $p_{n-1} = \text{av}[(b_{n-1}^-, b_{n-1}^+), (\frac{3}{5}, \frac{2}{5})]$, where $\text{av}[(x,y), (1-\alpha, \alpha)] = \exp_x(\alpha \log_x(y))$ stands for the convex combination of $x, y \in \mathcal{M}$ with weight $\alpha \in [0,1]$. A proof of these properties can be found in [1].

As stated in the introduction, we would ideally like $\mathfrak{B}$ to minimize its mean square acceleration and its fidelity to data points. Specifically,

$$\min_{p_i, b_i^+, b_i^-} f(p_i, b_i^+, b_i^-) = \min_{p_i, b_i^+, b_i^-} \underbrace{\sum_{i=0}^{n-1} \int_0^1 \|\ddot{\beta}^i(t)\|_{\mathcal{M}}^2 \mathrm{d}t}_{\text{"mean square acceleration"}} + \lambda \underbrace{\sum_{i=0}^{n} \mathrm{d}^2(p_i, d_i)}_{\text{"fidelity"}}, \tag{3}$$

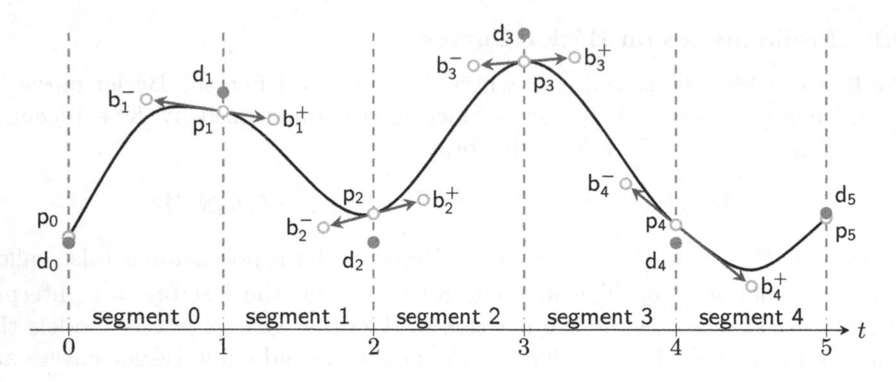

Fig. 1. Schematic representation of the composite Bézier function $\mathfrak{B}(t)$: the data points d_i are represented in red; the circled green ones are control points. The first and last Bézier segments are quadratic Bézier functions while all the other segments are cubic Bézier functions. (Color figure online)

where $\ddot{g}(t)$ stands for the temporal covariant second derivative of $g(t)$, under continuity and differentiability constraints. The parameter $\lambda > 0$ adjusts the balance between data fidelity and the "smoothness" of $\mathfrak{B}$. This balance tends to the interpolation problem from [2] when $\lambda \to \infty$.

3 Methods

In this section, we first summarize the method from [7] which is a generalization of optimality conditions holding only when $\mathcal{M} = \mathbb{R}^r$. This generalization holds for any manifold $\mathcal{M}$ if it is possible to compute the exponential map and the logarithm map. In a second time, we introduce a version $\tilde{f}_{\Delta\tau}$ of the objective f (Eq. (3)) obtained by discretizing the time domain of the mean square acceleration term with a step size $\Delta\tau$. We determine experimentally $\Delta\tau$ for which $\tilde{f}_{\Delta\tau}$ is a sufficiently good approximation of f, *i.e.*, the relative error between $\tilde{f}_{\Delta\tau}$ and f is small. Then, in Sect. 4, we will compare the solution from [7] to the minimizer of $\tilde{f}_{\Delta\tau}$.

3.1 Summary of the optimality conditions from [7]

In [7], the problem (3) is not directly addressed on a manifold $\mathcal{M}$. The (suboptimal) solution is obtained in two steps.

Step 1. The problem is considered on $\mathcal{M} = \mathbb{R}^r$ where $\mathrm{d}(\cdot,\cdot)$ and $\|.\|_{\mathcal{M}}$ are the classical Euclidean distance and norm, respectively. Hence (3) is a quadratic function in the $2n$ variables p_0, $(b_i^-, b_i^+)_{i=1}^{n-1}$ and p_n. Therefore, its optimality conditions take the form of a linear system $(A_0 + \lambda A_1)\mathbf{x} = \lambda C\mathbf{d}$, where

$A_0, A_1 \in \mathbb{R}^{2n \times 2n}$ and $C \in \mathbb{R}^{2n \times n+1}$ are matrices of coefficients, where $\mathbf{x} = [x_0, x_1, \ldots, x_{2n-1}]^T := [p_0, b_1^-, b_1^+, \ldots, b_{n-1}^+, p_n]^T \in \mathbb{R}^{2n \times r}$ contains the $2n$ optimization variables, and where $\mathbf{d} := [d_0, \ldots, d_n]^T \in \mathbb{R}^{n+1 \times r}$ contains the data points. The solution reads $\mathbf{x} = Q(\lambda)\mathbf{d}$, or

$$x_j = \sum_{l=0}^{n} q_{jl}(\lambda) d_l, \tag{4}$$

with $Q(\lambda) \in \mathbb{R}^{2n \times n+1}$, a matrix of coefficients depending on λ.

Step 2. Because (3) is invariant to translation on $\mathbb{R}^r$, the conditions (4) can be generalized to any Riemannian manifold. Indeed, $x_j - d_j^\star = \sum_{l=0}^{n} q_{jl}(\lambda)(d_l - d_j^\star)$, by translation with respect to a reference point $d_j^\star$. The generalization arises by interpreting the Euclidean difference as a logarithm map on a general manifold $\mathcal{M}$. Thus, a simple and natural way to generalize (4) to $\mathcal{M}$ is

$$x_j = \exp_{d_j^\star}\left(\sum_{l=0}^{n} q_{jl}(\lambda) \log_{d_j^\star}(d_l)\right). \tag{5}$$

By default, $d_j^\star := d_i$ when x_j is one of the control points b_i^-, b_i^+ or p_i.

Finally, the curve $\mathfrak{B}$ is reconstructed using the De Casteljau algorithm (as mentioned in Sect. 2).

3.2 Discretization of the Mean Square Acceleration on Manifolds

In comparison to Sect. 3.1, we here solve (3) directly on an arbitrary manifold $\mathcal{M}$. However, there is no simple expression of the Bézier curves β^i on $\mathcal{M}$, which means that it is not possible to express its mean squared acceleration in general. To overcome this difficulty, we replace f by a version $\tilde{f}_{\Delta\tau}$ where the acceleration of the curves is approached by a Riemannian second order finite difference (generalized with the log map from the Euclidean finite differences as in [4]), and the integration is replaced by a classical trapezoidal rule. The new objective function $\tilde{f}_{\Delta\tau}(p_i, b_i^+, b_i^-)$ now reads

$$\sum_{k=1}^{M-1} \Delta\tau \left\| \frac{\log_{\mathfrak{B}(t_k)}(\mathfrak{B}(t_{k-1})) + \log_{\mathfrak{B}(t_k)}(\mathfrak{B}(t_{k+1}))}{\Delta\tau^2} \right\|_{\mathcal{M}}^2 + \lambda \sum_{i=0}^{n} \mathrm{d}^2(p_i, d_i), \tag{6}$$

where $\Delta\tau = \frac{n}{M}$. As there is also no general expression of the Riemannian gradient of $\tilde{f}_{\Delta\tau}$ with respect to p_0, $(b_i^-, b_i^+)_{i=1}^{n-1}$ and p_n, we solve this problem with a Riemannian derivative-free optimization method, like the Particle Swarm Optimization algorithm provided in Manopt [5].

As there is no exact solution of (3) on a general Riemannian manifold, there is also no way to determine with precision the stepsize $\Delta\tau$ for which $\tilde{f}_{\Delta\tau}$ is close enough to f on $\mathcal{M}$. To overcome this, we determine an acceptable $\Delta\tau$ on the Euclidean space and then use this stepsize to optimize (6) on $\mathcal{M}$. This behavior is illustrated in Fig. 2. We can see that a stepsize of $\Delta\tau \simeq 10^{-2}$ is already acceptable on the Euclidean space for $\tilde{f}_{\Delta\tau}$ to approach f with a relative error of less than 1%.

We will now use this stepsize to compare x^*, the solution (5) from Sect. 3.1, to $\tilde{x}^*$, the solution obtained with (6).

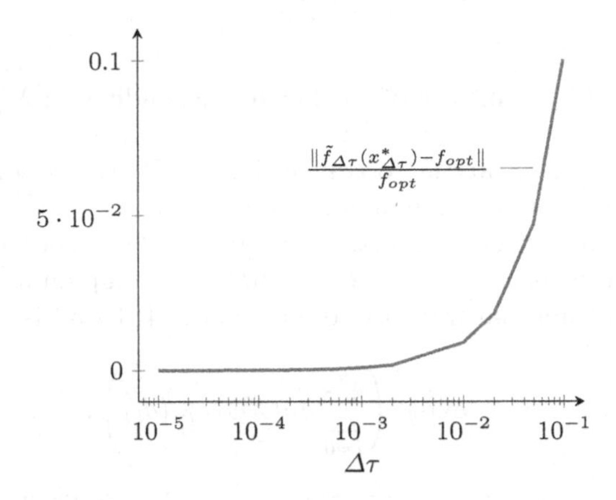

Fig. 2. On the Euclidean space, the continuous objective function (3) is approached by its discretized version (6) when $\Delta\tau$ tends to be small. A stepsize of $\Delta\tau = 10^{-2}$ already leads to a relative error of less than 1% on a random set of data points

4 Results

In this section, we evaluate the quality of the method from [7] (Sect. 3.1) on the sphere $\mathbb{S}^2$. To do so, we compare its solution x^* with the solution $\tilde{x}^*$ obtained by optimizing the discretized version of the objective function given in equation (6). This comparison is easily extendable to other manifolds provided that the log and exp map can be computed.

Data points and error evaluation. Consider the points $a = [0, 0, 1]^T$ and $b = [1, 0, 0]^T$ and the geodesic $\gamma_{a,b} : [0, 1] \to \mathcal{M} : t \mapsto \gamma_{a,b}(t)$. We construct $S = 20$ sets (indexed by m) of $n \in \{3, \ldots, 10\}$ data points $(\hat{d}_i^m)_{i=1}^n$, $m = 1, \ldots, S$ aligned and equispaced on the geodesic $\gamma_{a,b}(t)$ and then slightly disturbed with a noise η. Specifically,

$$\hat{d}_i^m = \frac{d_i^m + \eta}{\|d_i^m + \eta\|}, \quad i = 1, \ldots, n, \quad m = 1, \ldots, S,$$

such that $d_i^m = \gamma_{a,b}(\frac{i-1}{n-1})$, and $\eta \sim \mathcal{N}(0, (0.1)^2)$, as shown on Fig. 3, left.

For each set m and each number of data points n, we compute $x^*_{m,n} \in \mathbb{S}^{2\times 2n}$, the solution from [7] given by Eq. (5), and $\tilde{x}^*_{m,n} \in (\mathbb{S}^2)^{2n}$, the solution to the problem (6) with a discretization stepsize $\Delta\tau = 10^{-2}$. We evaluate the distance $\varepsilon_{m,n}$ of the objective value obtained with $x^*_{m,n}$ and $\tilde{x}^*_{m,n}$ in (6) as

$$\varepsilon_{m,n} = \frac{\tilde{f}_{\Delta\tau}(\tilde{x}^*_{m,n}) - \tilde{f}_{\Delta\tau}(x^*_{m,n})}{\tilde{f}_{\Delta\tau}(\tilde{x}^*_{m,n})}, \tag{7}$$

with $\Delta\tau = 10^{-4}$.

Note that two different stepsizes are used to evaluate the relative distance ε: a larger one ($\Delta\tau = 10^{-2}$) to compute $\tilde{x}^*$ and another one ($\Delta\tau = 10^{-4}$) to evaluate the quality of the solutions. We chose a larger stepsize in the minimization because solving (6) with a derivative free algorithm becomes less and less tractable when $\Delta\tau$ decreases. However, $\tilde{f}_{\Delta\tau}$ approaches the actual manifold-valued objective function (3) when $\Delta\tau$ is small. Thus, we used a finer stepsize the evaluate the quality of x^* in (7).

Results. On Fig. 3 (right), we represent the mean $\mathcal{E}(n)$ and the standard deviation of the distances $(\varepsilon_{m,n})_{m=1}^{S}$, for each number n of data points. We can observe that the fast algorithm from [7] returns results close to the optimum in the case of this geodesic-like proof of concept, even if still slightly suboptimal (relative error of about 1% of the cost $\tilde{f}$). Indeed, this proof of concept might be too easy as data points are chosen close to a geodesic and Fig. 3 (right) could be so good only in this case. However, finding a solution to the discretized problem with the particle-swarm optimization is less and less tractable for n growing and $\Delta\tau$ decreasing. This is why the main advantage of [7] is its efficiency to compute an acceptable solution to (3) in a very short computation time.

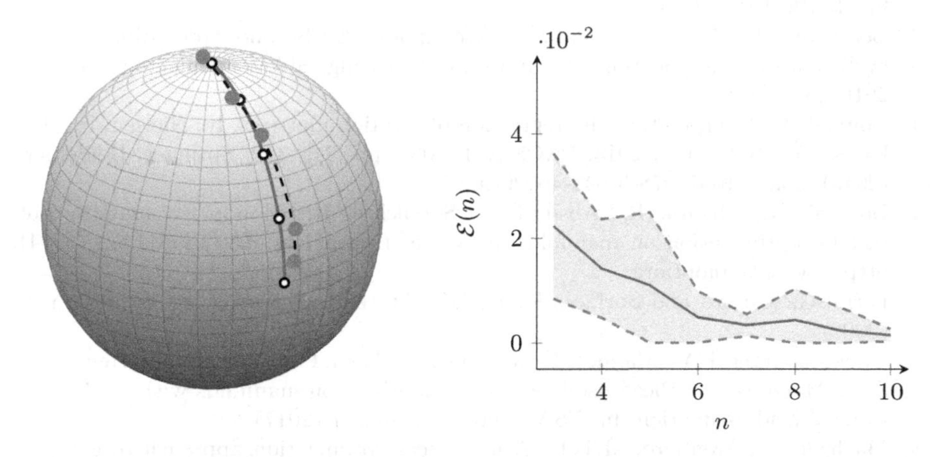

Fig. 3. *Left* - the data points (red) are a noisy version of points (black circles) aligned on a geodesic (blue line). The Bézier curve computed via [7] based on the data points (red) is in dashed line. *Right* - the fast algorithm from [7] returns solutions close to optimum. The relative error $\mathcal{E}(n)$ is about 1% (solid) with a standard deviation (dashed) of 2%. (Color figure online)

5 Future Work

The seeked goal in this paper was to evaluate the suboptimality of the fitting curve computed by the fast algorithm from [7]. We showed as a proof of concept that the method proposed in [7] approaches $\tilde{x}^*$ with a very satisfactory small relative error of 1% of the cost $\tilde{f}$ on the sphere $\mathbb{S}^2$, when the data points lie close to a geodesic.

Different pieces of work can be considered for the future. For instance, it may be worth considering a more advanced configuration of the data points to evaluate better the limits of the method. Estimating a theoretical upper bound on $|f(x_{\text{opt}}) - f(x^*)|$, where x_{opt} is the actual (and not numerical) solution of (3) is also left for future work. Furthermore, using a derivative-free optimization tool appeared to be time-consuming: a gradient-based approach could be investigated, exploiting the iterative structure of the De Casteljau algorithm to approach the gradient of a general Bézier curve on $\mathcal{M}$.

References

1. Absil, P.A., Gousenbourger, P.Y., Striewski, P., Wirth, B.: Differentiable piecewise-Bézier surfaces on riemannian manifolds. SIAM J. Imaging Sci. **9**(4), 1788–1828 (2016)
2. Arnould, A., Gousenbourger, P.-Y., Samir, C., Absil, P.-A., Canis, M.: Fitting smooth paths on riemannian manifolds: endometrial surface reconstruction and preoperative mri-based navigation. In: Nielsen, F., Barbaresco, F. (eds.) GSI 2015. LNCS, vol. 9389, pp. 491–498. Springer, Cham (2015). doi:10.1007/978-3-319-25040-3_53
3. Bergmann, R., Laus, F., Steidl, G., Weinmann, A.: Second order differences of cyclic data and applications in variational denoising. SIAM J. Imaging Sci. **7**(4), 2916–2953 (2014)
4. Boumal, N.: Interpolation and regression of rotation matrices. In: Nielsen, F., Barbaresco, F. (eds.) GSI 2013. LNCS, vol. 8085, pp. 345–352. Springer, Heidelberg (2013). doi:10.1007/978-3-642-40020-9_37
5. Boumal, N., Mishra, B., Absil, P.A., Sepulchre, R.: Manopt, a matlab toolbox for optimization on manifolds. J. Mach. Learn. Res. **15**, 1455–1459 (2014). http://www.manopt.org
6. Farin, G.: Curves and Surfaces for CAGD, 5th edn. Academic Press, New York (2002)
7. Gousenbourger, P.Y., Massart, E., Musolas, A., Absil, P.A., Jacques, L., Hendrickx, J.M., Marzouk, Y.: Piecewise-Bézier C^1 smoothing on manifolds with application to wind field estimation. In: ESANN2017, to appear (2017)
8. Machado, L., Monteiro, M.T.T.: A numerical optimization approach to generate smoothing spherical splines. J. Geom. Phys. **111**, 71–81 (2017)
9. Popiel, T., Noakes, L.: Bézier curves and $\mathcal{C}^2$ interpolation in Riemannian manifolds. J. Approximation Theory **148**(2), 111–127 (2007)
10. Pyta, L., Abel, D.: Interpolatory galerkin models for the navier-stokes-equations. IFAC Pap. Online **49**(8), 204–209 (2016)
11. Samir, C., Absil, P.A., Srivastava, A., Klassen, E.: A gradient-descent method for curve fitting on Riemannian manifolds. Found. Comput. Math. **12**, 49–73 (2012)

12. Sanguinetti, G., Bekkers, E., Duits, R., Janssen, M.H.J., Mashtakov, A., Mirebeau, J.-M.: Sub-riemannian fast marching in SE(2). Progress in Pattern Recognition, Image Analysis, Computer Vision, and Applications. LNCS, vol. 9423, pp. 366–374. Springer, Cham (2015). doi:10.1007/978-3-319-25751-8_44
13. Su, J., Dryden, I., Klassen, E., Le, H., Srivastava, A.: Fitting smoothing splines to time-indexed, noisy points on nonlinear manifolds. Image Vis. Comput. **30**(67), 428–442 (2012)

Nonlocal Inpainting of Manifold-Valued Data on Finite Weighted Graphs

Ronny Bergmann[1(✉)] and Daniel Tenbrinck[2]

[1] Fachbereich Mathematik, Technische Universität Kaiserslautern, 67663 Kaiserslautern, Germany
bergmann@mathematik.uni-kl.de
[2] Institute for Computational and Applied Mathematics, Westfälische Wilhelms-Universität Münster, 48149 Münster, Germany
daniel.tenbrinck@uni-muenster.de

Abstract. Recently, there has been a strong ambition to translate models and algorithms from traditional image processing to non-Euclidean domains, e.g., to manifold-valued data. While the task of denoising has been extensively studied in the last years, there was rarely an attempt to perform image inpainting on manifold-valued data. In this paper we present a nonlocal inpainting method for manifold-valued data given on a finite weighted graph. We introduce a new graph infinity-Laplace operator based on the idea of discrete minimizing Lipschitz extensions, which we use to formulate the inpainting problem as PDE on the graph. Furthermore, we derive an explicit numerical solving scheme, which we evaluate on two classes of synthetic manifold-valued images.

1 Introduction

Variational methods and partial differential equations (PDEs) play a key role for both modeling and solving image processing tasks. When processing real world data certain information might be missing due to structural artifacts, occlusions, or damaged measurement devices. Reconstruction of missing image information is known as inpainting task and there exist various variational models to perform inpainting. One successful method from the literature is based on a discretization of the ∞-Laplace operator [7]. This idea has been adapted to finite weighted graphs in [10]. The graph model enables to perform local and nonlocal inpainting within the same framework based on the chosen graph construction. Nonlocal inpainting has the advantage of preserving structural features by using all available information in the given image instead of only local neighborhood values.

With the technological progress in modern data sensors there is an emerging field of processing non-Euclidean data. We concentrate our discussion in the following on manifold-valued data, i.e., each data value lies on a Riemannian manifold. Real examples for manifold-valued images are interferometric synthetic aperture radar (InSAR) imaging [14], where the measured phase-valued data may be noisy and/or incomplete. Sphere-valued data appears, e.g., in directional analysis [13]. Another application is diffusion tensor imaging (DT-MRI) [3],

F. Nielsen and F. Barbaresco (Eds.): GSI 2017, LNCS 10589, pp. 604–612, 2017.
https://doi.org/10.1007/978-3-319-68445-1_70

where the diffusion tensors can be represented as 3×3 symmetric positive definite matrices, which also constitute a manifold. For such data, there were several variational methods and algorithms proposed to perform image processing tasks, see e.g., [2,4,9,12,17]. Recently, the authors generalized the graph p-Laplacian for manifold-valued data, $1 \leq p < \infty$, in [5] and derived an explicit as well as an semi-implicit iteration scheme for computing solutions to related partial difference equations on graphs as mimetic approximation of continuous PDEs. While the previous work concentrated on denoising, the present work deals with the task of image inpainting of manifold-valued data. For this, we extend the already defined family of manifold-valued graph p-Laplacians by a new operator, namely the graph ∞-Laplacian for manifold valued data. We derive an explicit numerical scheme to solve the corresponding PDE and illustrate its capabilities by performing nonlocal inpainting of synthetic manifold-valued data.

The remainder of this paper is organized as follows: In Sect. 2 we introduce the necessary notations of Riemannian manifolds, finite weighted graphs, and manifold-valued vertex functions. In Sect. 3 we introduce a new graph ∞-Laplace operator for manifold-valued data based on the idea of discrete minimizing Lipschitz extensions. Furthermore, we derive an explicit numerical scheme to solve the corresponding PDE $\Delta_\infty f = 0$ with suitable boundary conditions. In Sect. 4 we apply the proposed method to inpainting of synthetic manifold-valued images. Finally, Sect. 5 concludes the paper.

2 Preliminaries

In this section we first introduce the needed theory and notations on Riemannian manifolds in Sect. 2.1 and introduce finite weighted graphs in Sect. 2.2. We then combine both concepts to introduce vertex functions and tangential edge functions needed for the remainder of this paper in Sect. 2.3. For further details we refer to [5].

2.1 Riemannian Manifolds

For a detailed introduction to functions on Riemannian manifolds we refer to, e.g., [1,11]. The values of the given data lie in a *complete, connected, m-dimensional Riemannian manifold* $\mathcal{M}$ with Riemannian metric $\langle\cdot,\cdot\rangle_x \colon \mathrm{T}_x\mathcal{M} \times \mathrm{T}_x\mathcal{M} \to \mathbb{R}$, where $\mathrm{T}_x\mathcal{M}$ is the tangent space at $x \in \mathcal{M}$. In every tangent space $\mathrm{T}_x\mathcal{M}$ the metric induces a norm, which we denote by $\|\cdot\|_x$. The disjoint union of all tangent spaces is called the *tangent bundle* $\mathrm{T}\mathcal{M} := \dot\cup_{x\in\mathcal{M}} \mathrm{T}_x\mathcal{M}$. Two points $x, y \in \mathcal{M}$ can be joined by a (not necessarily unique) shortest curve $\gamma_{\widehat{x,y}} \colon [0, L] \to \mathcal{M}$, where L is its length. This generalizes the idea of shortest paths from the Euclidean space $\mathcal{M} = \mathbb{R}^m$, i.e., straight lines, to a manifold and induces the geodesic distance denoted $d_{\mathcal{M}} \colon \mathcal{M} \times \mathcal{M} \to \mathbb{R}^+$. A curve γ can be reparametrized such that derivative vector field $\dot\gamma(t) := \frac{d}{dt}\gamma(t) \in T_{\gamma(t)}\mathcal{M}$ has constant norm, i.e., $\|\dot\gamma_{\widehat{x,y}}(t)\|_{\gamma_{\widehat{x,y}}(t)} = 1$, $t \in [0, L]$. The corresponding curve then has unit speed. We employ another notation of a geodesic, namely $\gamma_{x,\xi}$, $\xi \in \mathrm{T}_x\mathcal{M}$, $x \in \mathcal{M}$, which denotes the locally unique geodesic fulfilling $\gamma_{x,\xi}(0) = x$ and $\dot{\gamma}_{x,\xi}(0) = \xi \in \mathrm{T}_x\mathcal{M}$. This is unique due to the Hopf–Rinow Theorem, cf. [11, Theorem 1.7.1]. We further introduce the *exponential map* $\exp_x\colon \mathrm{T}_x\mathcal{M} \to \mathcal{M}$ as $\exp_x(\xi) = \gamma_{x,\xi}(1)$. Let $r_x \in \mathbb{R}^+$ denote the injectivity radius, i.e., the largest radius such that $\exp_x$ is injective for all ξ with $\|\xi\|_x < r_x$. Furthermore, let

$$\mathcal{D}_x := \left\{ y \in \mathcal{M} : y = \exp_x \xi, \text{ for some } \xi \in T_x\mathcal{M} \text{ with } \|\xi\|_x < r_x \right\}.$$

Then the inverse map $\log_x : \mathcal{D}_x \to \mathrm{T}_x\mathcal{M}$ is called the *logarithmic map* and maps a point $y = \gamma_{x,\xi}(1) \in \mathcal{D}_x$ to ξ.

2.2 Finite Weighted Graphs

Finite weighted graphs allow to model relations between arbitrary discrete data. Both local and nonlocal methods can be unified within the graph framework by using different graph construction methods: spatial vicinity for local methods like finite differences, and feature similarity for nonlocal methods. A finite weighted graph $G = (V, E, w)$ consists of a finite set of indices $V = \{1, \dots, n\}$, $n \in \mathbb{N}$, denoting the vertices, a set of directed edges $E \subset V \times V$ connecting a subset of vertices, and a nonnegative weight function $w\colon E \to \mathbb{R}^+$ defined on the edges of a graph. For an edge $(u, v) \in E$, $u, v \in V$ the node u is the start node, while v is the end node. We also denote this relationship by $v \sim u$. Furthermore, the weight function w can be extended to all $V \times V$ by setting $w(u, v) = 0$ when $v \not\sim u$. The neighborhood $\mathcal{N}(u) := \{v \in V : v \sim u\}$ is the set of adjacent nodes.

2.3 Manifold-Valued Vertex Functions and Tangential Edge Functions

The functions of main interest in this work are manifold-valued vertex functions, which are defined as

$$f\colon V \to \mathcal{M}, \quad u \mapsto f(u),$$

The range of the vertex function f is the Riemannian manifold $\mathcal{M}$. We denote the set of admissible vertex functions by $\mathcal{H}(V; \mathcal{M}) := \{f\colon V \to \mathcal{M}\}$. This set can be equipped with a metric given by

$$d_{\mathcal{H}(V;\mathcal{M})}(f, g) := \left(\sum_{u \in V} d^2_{\mathcal{M}}(f(u), g(u)) \right)^{\frac{1}{2}}, \qquad f, g \in \mathcal{H}(V; \mathcal{M}).$$

Furthermore, we need the notion of a tangential vertex function. The space $\mathcal{H}(V; \mathrm{T}\mathcal{M})$ consists of all functions $H\colon V \to \mathrm{T}\mathcal{M}$, i.e., for each $u \in V$ there exists a value $H(u) \in \mathrm{T}_x\mathcal{M}$ for some $x \in \mathcal{M}$.

3 Methods

In this section we generalize the ∞-Laplacian from the real-valued, continuous case to the manifold-valued setting on graphs. We discuss discretizations of the ∞-Laplacian both for real-valued functions on bounded open sets and on graphs in Sect. 3.1. We generalize these to manifold-valued functions on graphs in Sect. 3.2 and state a corresponding numerical scheme in Sect. 3.3.

3.1 Discretizations of the ∞-Laplace Operator

Let $\Omega \subset \mathbb{R}^d$ be a bounded, open set and let $f\colon \Omega \to \mathbb{R}$ be a smooth function. Following [8] the infinity Laplacian of f at $x \in \Omega$ can be defined as

$$\Delta_\infty f(x) \;=\; ((\nabla f)^{\mathrm{T}} \Delta f \nabla f)(x) \;=\; \sum_{j=1}^{d}\sum_{k=1}^{d} \frac{\partial f}{\partial x_j}\frac{\partial f}{\partial x_k}\frac{\partial^2 f}{\partial x_j x_k}(x). \tag{1}$$

As discussed above, this operator is not only interesting in theory, but also has applications in image processing [7], e.g., for image interpolation and inpainting. Oberman discussed in [16] different possibilities for a consistent discretization scheme of the infinity Laplacian defined in (1). One basic observation is that the operator can be well approximated by the maximum and minimum values of the function in a local ε-ball neighborhood, i.e.,

$$\Delta_\infty f(x) \;=\; \frac{1}{\varepsilon^2}\left(\min_{y\in B_\varepsilon(x)} f(y) + \max_{y\in B_\varepsilon(x)} f(y) - 2f(x)\right) \;+\; \mathcal{O}(\varepsilon^2). \tag{2}$$

The approximation in (2) has inspired Elmoataz et al. [10] to propose a definition of a discrete graph ∞-Laplacian operator for real-valued vertex functions, i.e.,

$$\begin{aligned}\Delta_\infty f(u) = \max_{v\sim u} &|\max(\sqrt{w(u,v)}(f(v)-f(u)),0)| \\ &- \max_{v\sim u} |\min(\sqrt{w(u,v)}(f(v)-f(u)),0)|\end{aligned} \tag{3}$$

Furthermore, Oberman uses in [16] the well-known relationship between solutions of the homogeneous infinity Laplace equation $-\Delta_\infty f = 0$ and absolutely minimizing Lipschitz extensions to derive a numerical scheme based on the idea of minimizing the discrete Lipschitz constant in a neighborhood.

3.2 The Graph ∞-Laplacian for Manifold-Valued Data

Instead of following the approach proposed by Elmoataz et al. in [10] we propose a new graph ∞-Laplace operator for manifold valued functions based on the idea of computing discrete minimal Lipschitz extensions, i.e., for a vertex function $f \in \mathcal{H}(V;\mathcal{M})$ we define the graph ∞-Laplacian for manifold valued data $\Delta_\infty\colon \mathcal{H}(V;\mathcal{M}) \to H(V;T\mathcal{M})$ in a vertex $u \in V$ as

$$\Delta_\infty f(u) \;:=\; \frac{\left(\sqrt{w(u,v_1^*)}\log_{f(u)} f(v_1^*) \;+\; \sqrt{w(u,v_2^*)}\log_{f(u)} f(v_2^*)\right)}{\sqrt{w(u,v_1^*)} + \sqrt{w(u,v_2^*)}} \tag{4}$$

for which the designated neighbors $v_1^*, v_2^* \in \mathcal{N}(u)$ are characterized by maximizing the discrete Lipschitz constant in the local tangential plane $T_{f(u)}\mathcal{M}$ among all neighbors, i.e.,

$$(v_1^*, v_2^*) = \operatorname*{arg\,max}_{(v_1,v_2)\in\mathcal{N}^2(u)} \left\| \sqrt{w(u,v_1)} \log_{f(u)} f(v_1) - \sqrt{w(u,v_2)} \log_{f(u)} f(v_2) \right\|_{f(u)}$$

By means of the proposed operator in (4) we are interested in solving discrete interpolation problems on graphs for manifold-valued data. Let $U \subset V$ be a subset of vertices of the finite weighted graph $G = (V, E, w)$ and let $f: V/U \to \mathcal{M}$ be a given vertex function on the complement of U. The interpolation task now consists in computing values of f on $\mathcal{M}$ for vertices $u \in U$ in which f is unknown. For this we solve the following PDE on a graph based on the proposed operator in (4) with given boundary conditions:

$$\begin{cases} \Delta_\infty f(u) = 0 & \text{for all } u \in U, \\ f(u) = g(u) & \text{for all } u \in V/U. \end{cases} \tag{5}$$

3.3 Numerical Iteration Scheme

In order to numerically solve the PDE in (5) on a finite weighted graph we introduce an artificial time variable t and derive a related parabolic PDE, i.e.,

$$\begin{cases} \frac{\partial f}{\partial t}(u,t) = \Delta_\infty f(u,t) & \text{for all } u \in U,\ t \in (0,\infty), \\ f(u,0) = f_0(u) & \text{for all } u \in U, \\ f(u,t) = g(u,t) & \text{for all } u \in V/U, t \in [0,\infty). \end{cases} \tag{6}$$

We propose an explicit Euler time discretization scheme with sufficiently small time step size $\tau > 0$ to iteratively solve (6). Note that we proposed a similar explicit scheme for the computation of solutions of the graph p-Laplacian operator for manifold-valued data in [5]. Using the notation $f_k(u) := f(u, k\tau)$, i.e. we discretize the time t in steps $k\tau$, $k \in \mathbb{N}$, an update for the vertex function f can be computed by

$$f_{k+1}(u) = \exp_{f_k(u)}\big(\tau \Delta_\infty f_k(u)\big), \tag{7}$$

for which the graph ∞-Laplacian is defined in (4) above.

4 Numerical Examples

We first describe our graph construction and details of our inpainting algorithm in Sect. 4.1. We then consider two synthetic examples of manifold-valued data in Sect. 4.2, namely directional image data and an image consisting of symmetric positive matrices of size 2×2.

4.1 Graph Construction and Inpainting Algorithm

We construct a nonlocal graph from a given manifold-valued image $f \in \mathcal{M}^{m,n}$ as follows: we consider patches $q_{i,j} \in \mathcal{M}^{2p+1,2p+1}$ of size $2p+1$ around each pixel $(i,j), i \in \{1,\dots,m\}, j \in \{1,\dots,n\}$ with periodic boundary conditions. We denote for two patches $q_{i,j}$ and $q_{i',j'}$ the set $I \subset \{1,\dots,2p+1\} \times \{1,\dots,2p+1\}$ as pixels that are known in both patches and compute $d_{i,j,i',j'} := \frac{1}{|I|} d_{\mathcal{H}(I;\mathcal{M})}(q_{i,j}, q_{i',j'})$. For $(i,j) \in \{1,\dots,n\} \times \{1,\dots,m\}$ we introduce edges to the pixels (i',j') with the k smallest patch distances. We define the weight function as $w(u,v) = \mathrm{e}^{-d^2_{i,j,i',j'}/\sigma^2}$, where $u = (i,j), v = (i',j') \in \mathcal{G}$ with $v \sim u$. For computational efficiency we further introduce a search window size $r \in \mathbb{N}$, i.e., similar patches are considered within a window of size $2r+1$ around (i,j) around $i,j)$, a construction that is for example also used for non-local means [6].

We then solve the iterative scheme (6) with $\tau = \frac{1}{10}$ in (7) on all pixels (i,j) that where initially unknown. We start with all border pixels, i.e., unknown pixels with at least one known local neighbor pixel. We stop our scheme if the relative change between two iterations falls below $\epsilon = 10^{-7}$ or after a maximum of $t = 1000$ iterations. We then add all now border pixel to the active set we solve the equation on and reinitialize our iterative scheme (7). We iterate this algorithm until all unknown pixel have been border pixel and are hence now known. Our algorithm is implemented in MathWorks MATLAB employing the Manifold-valued Image Restoration Toolbox (MVIRT)[1].

4.2 Inpainting of Directional Data

We investigate the presented algorithm for artificial manifold-valued data: first, let $\mathcal{M} = \mathbb{S}^2$ be the unit sphere, i.e., our data items are directions in $\mathbb{R}^3$. Its data items are drawn as small three-dimensional arrows color-encoded by elevation, i.e., the south pole is blue (dark), the north pole yellow (bright). The periodicity of the data is slightly obstructed by two vertical and horizontal discontinuities of jump height $\frac{\pi}{16}$ dividing the image into nine parts. The input data is given by the lossy data, cf. Fig. 1a). We set the search window to $r = 32$, i.e., global comparison of patches in the graph construction from Sect. 4.1. Using $k = 25$ most similar patches and a patch radius of $p = 12$, the iterative scheme (7) yields Fig. 1b). The proposed methods finds a reasonable interpolation in the missing pixels.

4.3 Inpainting of Symmetric Positive Definite Matrices

As a second example we consider an image of symmetric positive definite (s.p.d.) matrices from $\mathbb{R}^{2\times 2}$, i.e., $\mathcal{M} = \mathcal{P}(2)$. These can be illustrated as ellipses using their eigenvectors as main axes and their eigenvalues as their lengths and the geodesic anisotropy index [15] in the hue colormap. The input data of 64×64

[1] open source, http://www.mathematik.uni-kl.de/imagepro/members/bergmann/mvirt/.

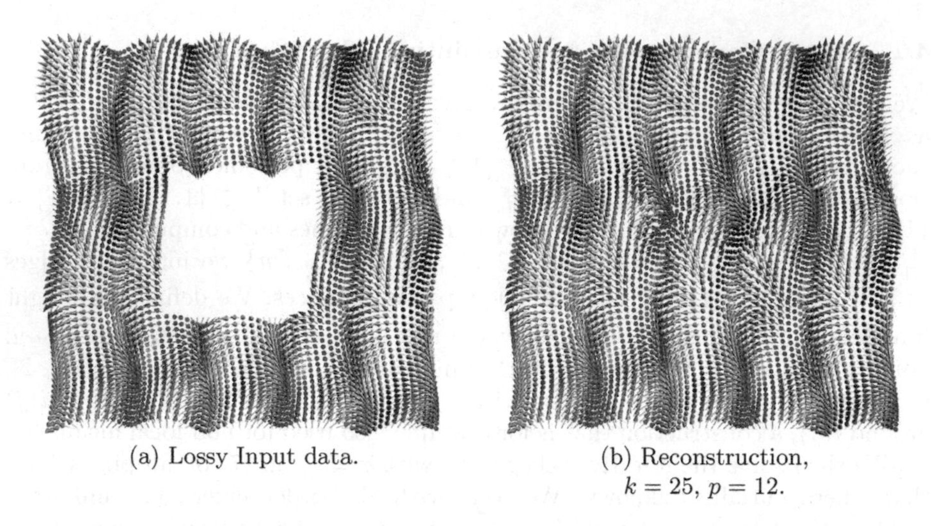

(a) Lossy Input data.

(b) Reconstruction, $k = 25$, $p = 12$.

Fig. 1. Reconstruction of directional data $f \in (\mathbb{S}^2)^{64\times 64}$ from lossy given data, i.e. in (a) the original data is shown with missing data (the center). Using the $k = 25$ most similar patches and a patch radius of $p = 12$ leads to a reasonable reconstruction in (b).

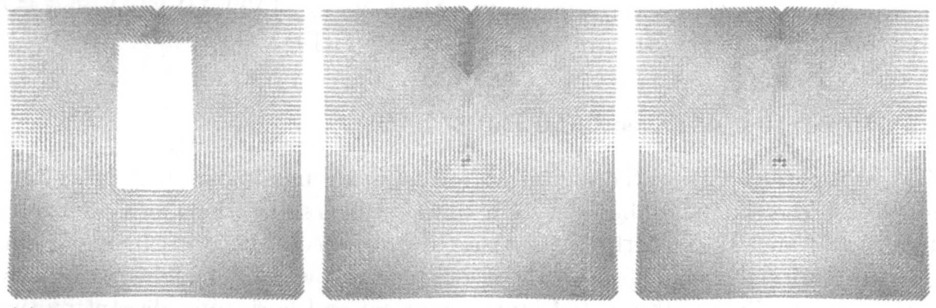

(a) Input data with missing rectangle.

(b) Reconstruction, $k = 5$, $p = 6$.

(c) Reconstruction, $k = 25$, $p = 6$.

Fig. 2. Reconstruction of s.p.d. matrices $f \in (\mathcal{P}(2))^{64\times 64}$ of (a) a lossy image. Increasing the number of neighbors from $k = 5$ in (b) to $k = 25$ in (c) broadens the center feature and smoothens the discontinuity in the center.

pixel is missing a rectangular area, cf. Fig. 2a). We set again $r = 32$ for a global comparison. Choosing $k = 5$, $p = 6$ yields a first inpainting result shown in Fig. 2b) which preserves the discontinuity line in the center and introduces a red area within the center bottom circular structure. Increasing the nonlocal neighborhood to $k = 25$, cf. in Fig. 2c) broadens the red center feature and the discontinuity gets smoothed along the center vertical line.

5 Conclusion

In this paper we introduced the graph ∞-Laplacian operator for manifold-valued functions by generalizing a reformulation of the ∞-Laplacian for real-valued functions and using discrete minimizing Lipschitz extensions. To the best of our knowledge, this generalization induced by our definition is even new for the vector-valued ∞-Laplacian on images. This case is included within this framework by setting e.g. $\mathcal{M} = \mathbb{R}^3$ for color images. We further derived an explicit numerical scheme to solve the related parabolic PDE on a finite weighted graph. First numerical examples using a nonlocal graph construction with patch-based similarity measures demonstrate the capabilities and performance of the inpainting algorithm applied to manifold-valued images.

Despite an analytic investigation of the convergence of the presented scheme, future work includes further development of numerical algorithms, as well as properties of the ∞-Laplacian for manifold-valued vertex functions on graphs.

References

1. Absil, P.A., Mahony, R., Sepulchre, R.: Optimization Algorithms on Matrix Manifolds. Princeton University Press, Princeton and Oxford (2008)
2. Bačák, M., Bergmann, R., Steidl, G., Weinmann, A.: A second order non-smooth variational model for restoring manifold-valued images. SIAM J. Sci. Comput. **38**(1), A567–A597 (2016)
3. Basser, P., Mattiello, J., LeBihan, D.: MR diffusion tensor spectroscopy and imaging. Biophys. J. **66**, 259–267 (1994)
4. Bergmann, R., Persch, J., Steidl, G.: A parallel Douglas-Rachford algorithm for minimizing ROF-like functionals on images with values in symmetric hadamard manifolds. SIAM J. Imaging Sci. **9**(4), 901–937 (2016)
5. Bergmann, R., Tenbrinck, D.: A graph framework for manifold-valued data (2017). arXiv preprint https://arxiv.org/abs/1702.05293
6. Buades, A., Coll, B., Morel, J.M.: A non-local algorithm for image denoising. In: Proceedings of the IEEE CVPR 2005, vol. 2, pp. 60–65 (2005)
7. Caselles, V., Morel, J.M., Sbert, C.: An axiomatic approach to image interpolation. Trans. Image Process. **7**(3), 376–386 (1998)
8. Crandall, M., Evans, L., Gariepy, R.: Optimal lipschitz extensions and the infinity laplacian. Calc. Var. Partial Differ. Equ. **13**(2), 123–139 (2001)
9. Cremers, D., Strekalovskiy, E.: Total cyclic variation and generalizations. J. Math. Imaging Vis. **47**(3), 258–277 (2013)
10. Elmoataz, A., Desquesnes, X., Lezoray, O.: Non-local morphological PDEs and p-Laplacian equation on graphs with applications in image processing and machine learning. IEEE J. Sel. Topics Signal Process. **6**(7), 764–779 (2012)
11. Jost, J.: Riemannian Geometry and Geometric Analysis. Springer, Heidelberg (2011)
12. Lellmann, J., Strekalovskiy, E., Koetter, S., Cremers, D.: Total variation regularization for functions with values in a manifold. In: Proceedings of the IEEE ICCV 2013, pp. 2944–2951 (2013)
13. Mardia, K.V., Jupp, P.E.: Directional Statistics. Wiley, Chichester (2000)

14. Massonnet, D., Feigl, K.L.: Radar interferometry and its application to changes in the earth's surface. Rev. Geophys. **36**(4), 441–500 (1998)
15. Moakher, M., Batchelor, P.G.: Symmetric positive-definite matrices: From geometry to applications and visualization. In: Weickert, J., Hagen, H. (eds.) Visualization and Processing of Tensor Fields, pp. 285–298. Springer, Heidelberg (2006)
16. Oberman, A.M.: A convergent difference scheme for the infinity Laplacian: Construction of absolutely minimizing lipschitz extensions. Math. Comput. **74**(251), 1217–1230 (2004)
17. Weinmann, A., Demaret, L., Storath, M.: Total variation regularization for manifold-valued data. SIAM J. Imaging Sci. **7**(4), 2226–2257 (2014)

Affine-Invariant Orders on the Set of Positive-Definite Matrices

Cyrus Mostajeran(✉) and Rodolphe Sepulchre

Department of Engineering, University of Cambridge, Cambridge CB2 1PZ, UK
csm54@cam.ac.uk

Abstract. We introduce a family of orders on the set S_n^+ of positive-definite matrices of dimension n derived from the homogeneous geometry of S_n^+ induced by the natural transitive action of the general linear group $GL(n)$. The orders are induced by affine-invariant cone fields, which arise naturally from a local analysis of the orders that are compatible with the homogeneous structure of S_n^+. We then revisit the well-known Löwner-Heinz theorem and provide an extension of this classical result derived using differential positivity with respect to affine-invariant cone fields.

1 Introduction

The question of how one can order the elements of a space in a consistent and well-defined manner is of fundamental importance to many areas of applied mathematics, including the theory of monotone functions and matrix means in which the notion of order plays a defining role [1,6,8,9]. These concepts play an important role in a wide variety of applications across information geometry where one is interested in performing statistical analysis on sets of matrices. In such applications, the choice of order relation is often taken for granted. This choice, however, is of crucial significance since a function that is not monotone with respect to one order, may be monotone with respect to another, in which case powerful results from monotonicity theory would become relevant.

In this paper, we outline an approach to systematically generate orders on homogeneous spaces, which form a class of nonlinear spaces that are ubiquitous in many applications in information engineering and control theory. A homogeneous space is a manifold on which a Lie group acts transitively, in the sense that any point on the manifold can be mapped onto any other point by an element of a group of transformations that act on the space. The geometry of homogeneous spaces, coupled with the observation that cone fields induce conal orders on continuous spaces [7], forms the basis for the approach taken in this paper. The aim is to systematically generate cone fields that are invariant with respect to the homogeneous geometry, thereby defining families of conal orders built upon the underlying symmetries of the space.

The focus of this paper is on ordering the elements of the set of symmetric positive-definite matrices S_n^+ of dimension n. Positive definite matrices arise in numerous applications, including as covariance matrices in statistics and computer vision, as variables in convex and semidefinite programming, as unknowns

F. Nielsen and F. Barbaresco (Eds.): GSI 2017, LNCS 10589, pp. 613–620, 2017.
https://doi.org/10.1007/978-3-319-68445-1_71

in fundamental problems in systems and control theory, as kernels in machine learning, and as diffusion tensors in medical imaging. The space S_n^+ forms a smooth manifold that can be viewed as a homogeneous space admitting a transitive action by the general linear group $GL(n)$, which endows the space with an affine-invariant geometry as reviewed in Sect. 2. In Sect. 3, this geometry is used to construct affine-invariant cone fields and new partial orders on S_n^+. In Sect. 4, we discuss how differential positivity [5] can be used to study and characterize monotonicity on S_n^+ with respect to the invariant orders introduced in this paper. We also state a generalized version of the celebrated Löwner-Heinz theorem [6,9] of operator monotonicity theory derived using this approach.

2 Homogeneous Geometry of S_n^+

The set S_n^+ of symmetric positive definite matrices of dimension n has the structure of a homogeneous space with a transitive $GL(n)$-action. This follows by noting that any $\Sigma \in S_n^+$ admits a Cholesky decomposition $\Sigma = AA^T$ for some $A \in GL(n)$. The Cauchy polar decomposition of the invertible matrix A yields a unique decomposition $A = PQ$ of A into an orthogonal matrix $Q \in O(n)$ and a symmetric positive definite matrix $P \in S_+^n$. Now note that if Σ has Cholesky decomposition $\Sigma = AA^T$ and A has a Cauchy polar decomposition $A = PQ$, then $\Sigma = PQQ^TP = P^2$. That is, Σ is invariant with respect to the orthogonal part Q of the polar decomposition. Therefore, we can identify any $\Sigma \in S_n^+$ with the equivalence class $[\Sigma^{1/2}] = \Sigma^{1/2} \cdot O(n)$ in the quotient space $GL(n)/O(n)$, where $\Sigma^{1/2}$ denotes the unique positive definite square root of Σ. That is,

$$S_n^+ \cong GL(n)/O(n). \tag{1}$$

The identification in (1) can also be made by noting the transitive action of $GL(n)$ on S_n^+ defined by

$$\tau_A : \Sigma \mapsto A\Sigma A^T \quad \forall A \in GL(n),\ \forall \Sigma \in S_n^+. \tag{2}$$

This action is said to be *almost effective* in the sense that $\pm I$ are the only elements of $GL(n)$ that fix every $\Sigma \in S_n^+$. The isotropy group of this action at $\Sigma = I$ is precisely $O(n)$, since $\tau_Q : I \mapsto QIQ^T = I$ if and only if $Q \in O(n)$. Once again, if $\Sigma \in S_n^+$ has Cholesky decomposition $\Sigma = AA^T$ and A has polar decomposition $A = PQ$, then $\tau_A(I) = AIA^T = P^2 = \Sigma$.

A homogeneous space G/H is said to be reductive if there exists a subspace $\mathfrak{m}$ of the Lie algebra $\mathfrak{g}$ of G such that $\mathfrak{g} = \mathfrak{h} \oplus \mathfrak{m}$ and $\mathrm{Ad}(H)\mathfrak{m} \subseteq \mathfrak{m}$. Recall that the Lie algebra $\mathfrak{gl}(n)$ of $GL(n)$ consists of the set $\mathbb{R}^{n\times n}$ of all real $n \times n$ matrices equipped with the Lie bracket $[X, Y] = XY - YX$, while the Lie algebra of $O(n)$ is $\mathfrak{o}(n) = \{X \in \mathbb{R}^{n\times n} : X^T = -X\}$. Since any matrix $X \in \mathbb{R}^{n\times n}$ has a unique decomposition $X = \frac{1}{2}(X - X^T) + \frac{1}{2}(X + X^T)$, as a sum of an antisymmetric part and a symmetric part, we have $\mathfrak{gl}(n) = \mathfrak{o} \oplus \mathfrak{m}$, where $\mathfrak{m} = \{X \in \mathbb{R}^{n\times n} : X^T = X\}$. Furthermore, since $\mathrm{Ad}_Q(S) = QSQ^{-1} = QSQ^T$ is a symmetric matrix for each $S \in \mathfrak{m}$, we have $\mathrm{Ad}_{O(n)} = \{QSQ^{-1} : Q \in O(n),\ S \in \mathfrak{m}\} \subseteq \mathfrak{m}$. Hence,

$S_n^+ = GL(n)/O(n)$ is indeed a reductive homogeneous space with reductive decomposition $\mathfrak{gl}(n) = \mathfrak{o}(n) \oplus \mathfrak{m}$.

The tangent space $T_o S_n^+$ of S_n^+ at the base-point $o = [I] = I \cdot O(n)$ is identified with $\mathfrak{m}$. For each $\Sigma \in S_n^+$, the action $\tau_{\Sigma^{1/2}} : S_n^+ \to S_n^+$ induces the vector space isomorphism $d\tau_{\Sigma^{1/2}}|_I : T_I S_n^+ \to T_\Sigma S_n^+$ given by

$$d\tau_{\Sigma^{1/2}}\big|_I X = \Sigma^{1/2} X \Sigma^{1/2}, \quad \forall X \in \mathfrak{m}. \tag{3}$$

The map (3) can be used to extend structures defined in $T_o S_n^+$ to structures defined on the tangent bundle TS_n^+ through affine-invariance, provided that the structures in $T_o S_n^+$ are $\mathrm{Ad}_{O(n)}$-invariant. The $\mathrm{Ad}_{O(n)}$-invariance is required to ensure that the extension to TS_n^+ is unique and thus well-defined. For instance, any homogeneous Riemannian metric on $S_n^+ \cong GL(n)/O(n)$ is determined by an $\mathrm{Ad}_{O(n)}$-invariant inner product on $\mathfrak{m}$. Any such inner product induces a norm that is rotationally invariant and so can only depend on the scalar invariants $\mathrm{tr}(X^k)$ where $k \geq 1$ and $X \in \mathfrak{m}$. Moreover, as the inner product is a quadratic function, $\|X\|^2$ must be a linear combination of $(\mathrm{tr}(X))^2$ and $\mathrm{tr}(X^2)$. Thus, any $\mathrm{Ad}_{O(n)}$-invariant inner product on $\mathfrak{m}$ must be a scalar multiple of

$$\langle X, Y \rangle_{\mathfrak{m}} = \mathrm{tr}(XY) + \mu \mathrm{tr}(X)\mathrm{tr}(Y), \tag{4}$$

where μ is a scalar parameter with $\mu > -1/n$ to ensure positive-definiteness [12]. Therefore, the corresponding affine-invariant Riemannian metrics are generated by (3) and given by

$$\begin{aligned} \langle X, Y \rangle_\Sigma &= \langle \Sigma^{-1/2} X \Sigma^{-1/2}, \Sigma^{-1/2} Y \Sigma^{-1/2} \rangle_{\mathfrak{m}} \\ &= \mathrm{tr}(\Sigma^{-1} X \Sigma^{-1} Y) + \mu \mathrm{tr}(\Sigma^{-1} X)\mathrm{tr}(\Sigma^{-1} Y), \end{aligned} \tag{5}$$

for $\Sigma \in S_n^+$ and $X, Y \in T_\Sigma S_n^+$. In the case $\mu = 0$, (5) yields the most commonly used 'natural' Riemannian metric on S_n^+, which corresponds to the Fisher information metric for the multivariate normal distribution [4,13], and has been widely used in applications such as tensor computing in medical imaging.

3 Affine-Invariant Orders

3.1 Affine-Invariant Cone Fields

A cone field $\mathcal{K}$ on S_n^+ smoothly assigns a cone $\mathcal{K}(\Sigma) \subset T_\Sigma S_n^+$ to each point $\Sigma \in S_n^+$. We say that $\mathcal{K}$ is affine-invariant or homogeneous with respect to the quotient geometry $S_n^+ \cong GL(n)/O(n)$ if

$$\left(d\tau_{\Sigma_2^{1/2} \Sigma_1^{-1/2}}\big|_{\Sigma_1} \right) \mathcal{K}(\Sigma_1) = \mathcal{K}(\Sigma_2), \tag{6}$$

for all $\Sigma_1, \Sigma_2 \in S_n^+$. The procedure we will use for constructing affine-invariant cone fields on S_n^+ is similar to the approach taken for generating the affine-invariant Riemannian metrics in Sect. 2. We begin by defining a cone $\mathcal{K}(I)$ at I that is $\mathrm{Ad}_{O(n)}$-invariant:

$$X \in \mathcal{K}(I) \Leftrightarrow \mathrm{Ad}_Q X = d\tau_Q X = QXQ^T \in \mathcal{K}(I), \quad \forall Q \in O(n). \tag{7}$$

Using such a cone, we generate a cone field via

$$\mathcal{K}(\Sigma) = d\tau_{\Sigma^{1/2}}\big|_I \mathcal{K}(I) = \{X \in T_\Sigma S_n^+ : \Sigma^{-1/2} X \Sigma^{-1/2} \in \mathcal{K}(I)\}. \tag{8}$$

The $\mathrm{Ad}_{O(n)}$-invariance condition (7) is satisfied if $\mathcal{K}(I)$ has a spectral characterization; that is, we can check to see if any given $X \in T_I S_n^+ \cong \mathfrak{m}$ lies in $\mathcal{K}(I)$ using only properties of X that are characterized by its spectrum. For instance, $\mathrm{tr}(X)$ and $\mathrm{tr}(X^2)$ are both properties of X that are spectrally characterized and indeed $\mathrm{Ad}_{O(n)}$-invariant. Furthermore, quadratic $\mathrm{Ad}_{O(n)}$-invariant cones are defined by inequalities on suitable linear combinations of $(\mathrm{tr}(X))^2$ and $\mathrm{tr}(X^2)$.

Proposition 1. *For any choice of parameter $\mu \in (0, n)$, the set*

$$\mathcal{K}(I) = \{X \in T_I S_n^+ : (tr(X))^2 - \mu tr(X^2) \geq 0,\ tr(X) \geq 0\}, \tag{9}$$

defines an $Ad_{O(n)}$-invariant cone in $T_I S_n^+ = \{X \in \mathbb{R}^{n\times n} : X^T = X\}$.

Proof. $\mathrm{Ad}_{O(n)}$-invariance is clear since $\mathrm{tr}(X^2) = \mathrm{tr}(QXQ^TQXQ^T)$ and $\mathrm{tr}(X) = \mathrm{tr}(QXQ^T)$ for all $Q \in O(n)$. To prove that (9) is a cone, first note that $0 \in \mathcal{K}(I)$ and for $\lambda > 0$, $X \in \mathcal{K}(I)$, we have $\lambda X \in \mathcal{K}(I)$ since $\mathrm{tr}(\lambda X) = \lambda \mathrm{tr}(X) \geq 0$ and

$$(\mathrm{tr}(\lambda X))^2 - \mu \mathrm{tr}((\lambda X)^2) = \lambda^2[(\mathrm{tr}(X))^2 - \mu \mathrm{tr}(X^2)] \geq 0. \tag{10}$$

To show convexity, let $X_1, X_2 \in \mathcal{K}(I)$. Now $\mathrm{tr}(X_1 + X_2) = \mathrm{tr}(X_1) + \mathrm{tr}(X_2) \geq 0$, and

$$\begin{aligned}(\mathrm{tr}(X_1 + X_2))^2 - \mu \mathrm{tr}((X_1 + X_2)^2) &= [(\mathrm{tr}(X_1))^2 - \mu \mathrm{tr}(X_1^2)] \\ + \quad [(\mathrm{tr}(X_2))^2 - \mu \mathrm{tr}(X_2^2)] &+ 2[\mathrm{tr}(X_1)\mathrm{tr}(X_2) - \mu \mathrm{tr}(X_1 X_2)] \geq 0,\end{aligned} \tag{11}$$

since $\mathrm{tr}(X_1X_2) \leq (\mathrm{tr}(X_1^2))^{\frac{1}{2}}(\mathrm{tr}(X_2^2))^{\frac{1}{2}} \leq \frac{1}{\sqrt{\mu}}\mathrm{tr}(X_1)\frac{1}{\sqrt{\mu}}\mathrm{tr}(X_2)$, where the first inequality follows by Cauchy-Schwarz. Finally, we need to show that $\mathcal{K}(I)$ is pointed. If $X \in \mathcal{K}(I)$ and $-X \in \mathcal{K}(I)$, then $\mathrm{tr}(-X) = -\mathrm{tr}(X) = 0$. Thus, $(\mathrm{tr}(X))^2 - \mu \mathrm{tr}(X^2) = -\mu \mathrm{tr}(X^2) \geq 0$, which is possible if and only if all of the eigenvalues of X are zero; i.e., if and only if X = 0. □

The parameter μ controls the opening angle of the cone. If $\mu = 0$, then (9) defines the half-space $\mathrm{tr}(X) \geq 0$. As μ increases, the opening angle of the cone becomes smaller and for $\mu = n$ (9) collapses to a ray. For any fixed $\mu \in (0, n)$, we obtain a unique well-defined affine-invariant cone field given by

$$\mathcal{K}(\Sigma) = \{X \in T_\Sigma S_n^+ : (\mathrm{tr}(\Sigma^{-1}X))^2 - \mu \mathrm{tr}(\Sigma^{-1}X\Sigma^{-1}X) \geq 0,\ \mathrm{tr}(\Sigma^{-1}X) \geq 0\}. \tag{12}$$

It should be noted that of course not all $\mathrm{Ad}_{O(n)}$-invariant cones at I are quadratic. Indeed, it is possible to construct polyhedral $\mathrm{Ad}_{O(n)}$-invariant cones that arise as the intersections of a collection of spectrally defined half-spaces in $T_I S_n^+$. The clearest example of such a construction is the cone of positive semi-definite matrices in $T_I S_n^+$, which of course itself has a spectral characterization $\mathcal{K}(I) = \{X \in T_I S_n^+ : \lambda_i(X) \geq 0,\ i = 1, \ldots, n\}$, where $(\lambda_i(X))$ denote the n real eigenvalues of the symmetric matrix X.

3.2 Visualization of Affine-Invariant Cone Fields on S_2^+

It is well-known that the set of positive semidefinite matrices of dimension n forms a cone in the space of symmetric $n \times n$ matrices. Moreover, S_n^+ forms the interior of this cone. A concrete visualization of this identification can be made in the $n = 2$ case, as shown in Fig. 1. The set S_2^+ can be identified with the interior of $K = \{(x, y, z) \in \mathbb{R}^3 : z^2 - x^2 - y^2 \geq 0,\ z \geq 0\}$, through the map $\phi : S_2^+ \to K$ given by

$$\phi : \begin{pmatrix} a & b \\ b & c \end{pmatrix} \mapsto (x, y, z) = \left(\sqrt{2}b, \frac{1}{\sqrt{2}}(a - c), \frac{1}{\sqrt{2}}(a + c) \right). \tag{13}$$

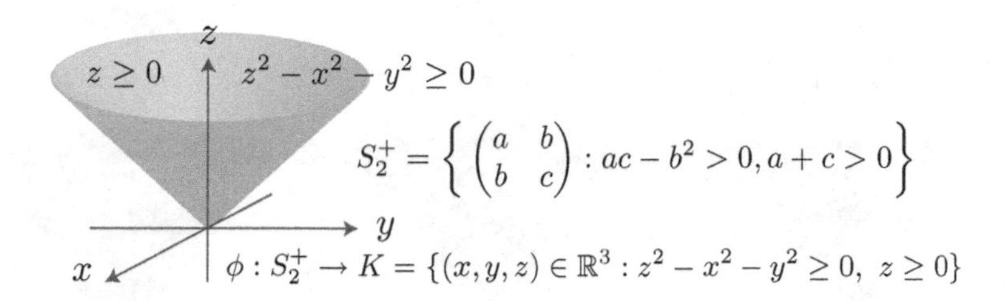

z
$z \geq 0$
$z^2 - x^2 - y^2 \geq 0$
$S_2^+ = \left\{ \begin{pmatrix} a & b \\ b & c \end{pmatrix} : ac - b^2 > 0, a + c > 0 \right\}$
y
x
$\phi : S_2^+ \to K = \{(x, y, z) \in \mathbb{R}^3 : z^2 - x^2 - y^2 \geq 0,\ z \geq 0\}$

Fig. 1. Identification of S_2^+ with the interior of the closed, convex, pointed cone $K = \{(x, y, z) \in \mathbb{R}^3 : z^2 - x^2 - y^2 \geq 0,\ z \geq 0\}$ in $\mathbb{R}^3$.

Inverting ϕ, we find that $a = \frac{1}{\sqrt{2}}(z+y)$, $b = \frac{1}{\sqrt{2}}x$, $c = \frac{1}{\sqrt{2}}(z-y)$. Note that the point $(x, y, z) = (0, 0, \sqrt{2})$ corresponds to the identity matrix $I \in S_2^+$. We seek to arrive at a visual representation of the affine-invariant cone fields generated from the $\mathrm{Ad}_{O(n)}$-invariant cones (9) for different choices of the parameter μ. The defining inequalities $\mathrm{tr}(X) \geq 0$ and $(\mathrm{tr}(X))^2 - \mu \mathrm{tr}(X^2) \geq 0$ in $T_I S_2^+$ take the forms

$$\delta z \geq 0, \quad \text{and} \quad \left(\frac{2}{\mu} - 1 \right) \delta z^2 - \delta x^2 - \delta y^2 \geq 0, \tag{14}$$

respectively, where $(\delta x, \delta y, \delta z) \in T_{(0,0,\sqrt{2})} K \cong T_I S_2^+$. Clearly the translation-invariant cone fields generated from this cone are given by the same equations as in (14) for $(\delta x, \delta y, \delta z) \in T_{(x,y,z)} K \cong T_{\phi^{-1}(x,y,z)} S_2^+$.

To obtain the affine-invariant cone fields, note that at $\Sigma = \phi^{-1}(x, y, z) \in S_2^+$, the inequality $\mathrm{tr}(\Sigma^{-1} X) \geq 0$ takes the form

$$\mathrm{tr} \left[\begin{pmatrix} c & -b \\ -b & a \end{pmatrix} \begin{pmatrix} \delta a & \delta b \\ \delta b & \delta c \end{pmatrix} \right] = c\ \delta a - 2b\ \delta b + a\ \delta c \geq 0 \tag{15}$$

$$\Leftrightarrow\ z\ \delta z - x\ \delta x - y\ \delta y \geq 0. \tag{16}$$

Similarly, the inequality $(\mathrm{tr}(\Sigma^{-1} X))^2 - \mu \mathrm{tr}(\Sigma^{-1} X \Sigma^{-1} X) \geq 0$ is given by

$$\begin{aligned} 2(x\ \delta x + y\ \delta y - z\ \delta z)^2 - \mu \big[&(z^2 + x^2 - y^2) \delta x^2 + (z^2 - x^2 - y^2) \delta y^2 \\ &+ (x^2 + y^2 + z^2) \delta z^2 + 4xy\ \delta x \delta y - 4xz\ \delta x \delta z - 4yz\ \delta y \delta z \big] \geq 0, \end{aligned} \tag{17}$$

where $(\delta x, \delta y, \delta z) \in T_{(x,y,z)}K \cong T_\Sigma S_2^+$. In the case $\mu = 1$, this reduces to $(\frac{2}{\mu} - 1)\delta z^2 - \delta x^2 - \delta y^2 \geq 0$. That is, for $\mu = 1$ the quadratic cone field generated by affine-invariance coincides with the corresponding translation-invariant cone field. Generally, however, affine-invariant and translation-invariant cone fields do not agree, as depicted in Fig. 2. Each of the different cone fields in Fig. 2 induces a distinct partial order on S_n^+.

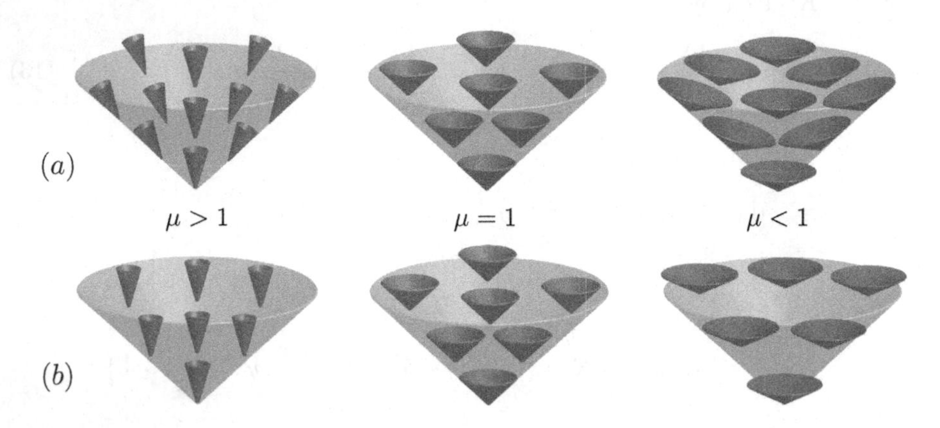

Fig. 2. Cone fields on S_2^+: (a) Quadratic affine-invariant cone fields for different choices of the parameter $\mu \in (0, 2)$. (b) The corresponding translation-invariant cone fields.

3.3 The Löwner Order

The Löwner order is the partial order $\geq_L$ on S_n^+ defined by

$$A \geq_L B \quad \Leftrightarrow \quad A - B \geq_L O, \tag{18}$$

where the inequality on the right denotes that $A - B$ is positive semidefinite [2]. The definition in (18) is based on translations and the 'flat' geometry of S_n^+. It is clear that the Löwner order is translation invariant in the sense that $A \geq_L B$ implies that $A + C \geq_L B + C$ for all $A, B, C \in S_n^+$. From the perspective of conal orders, the Löwner order is the partial order induced by the cone field generated by translations of the cone of positive semidefinite matrices at $T_I S_n^+$.

In the previous section, we gave an explicit construction showing that the cone field generated through translations of the cone of positive semidefinite matrices at $T_I S_n^+$ coincides with the cone field generated through affine-invariance in the $n = 2$ case. We will now show that this is a general result which holds for all n. First note that the cone at $T_I S_n^+$ can be expressed as

$$\mathcal{K}(I) = \{X \in T_I S_n^+ : u^T X u \geq 0 \; \forall u \in \mathbb{R}^n, \; u^T X u = 0 \Rightarrow u = 0\}, \tag{19}$$

and the resulting translation-invariant cone field is simply given by

$$\mathcal{K}_T(\Sigma) = \{X \in T_\Sigma S_n^+ : u^T X u \geq 0 \; \forall u \in \mathbb{R}^n, \; u^T X u = 0 \Rightarrow u = 0\}. \tag{20}$$

The corresponding affine-invariant cone field is given by

$$\mathcal{K}_A(\Sigma) = \{X \in T_\Sigma S_n^+ : u^T \Sigma^{-1/2} X \Sigma^{-1/2} u \geq 0 \; \forall u \in \mathbb{R}^n, \\ u^T \Sigma^{-1/2} X \Sigma^{-1/2} u = 0 \Rightarrow u = 0\}, \quad (21)$$

which is seen to be equal to $\mathcal{K}_T$ by introducing the invertible transformation $\bar{u} = \Sigma^{-1/2} u$ in (21). Thus we see that the Löwner order enjoys the special status of being both affine-invariant and translation-invariant, even though its classical definition is based on the 'flat' or translational geometry on S_n^+.

4 Monotonicity on S_n^+

Let f be a map of S_n^+ into itself. We say that f is *monotone* with respect to a partial order $\geq$ on S_n^+ if $f(\Sigma_1) \geq f(\Sigma_2)$ whenever $\Sigma_1 \geq \Sigma_2$. Such functions were introduced by Löwner in his seminal paper [9] on operator monotone functions. Since then operator monotone functions have been studied extensively and found applications to many fields including electrical engineering, network theory, and quantum information theory [3,10]. One of the most fundamental results in operator theory is the Löwner-Heinz theorem [6,9] stated below.

Theorem 1 (Löwner-Heinz). *If $\Sigma_1 \geq_L \Sigma_2$ in S_n^+ and $r \in [0,1]$, then*

$$\Sigma_1^r \geq_L \Sigma_2^r. \quad (22)$$

Furthermore, if $n \geq 2$ and $r > 1$, then $\Sigma_1 \geq_L \Sigma_2 \not\Rightarrow \Sigma_1^r \geq_L \Sigma_2^r$.

There are several different proofs of the Löwner-Heinz theorem. See [2,6,9, 11], for instance. Most of these proofs are based on analytic methods, such as integral representations from complex analysis. Instead we employ a geometric approach to study monotonicity based on a differential analysis of the system. One of the advantages of such an approach is that it is immediately applicable to all of the conal orders considered in this paper, while providing a deeper geometric insight into the behavior of the map under consideration. Recall that a smooth map $f : S_n^+ \to S_n^+$ is said to be differentially positive with respect to a cone field $\mathcal{K}$ on S_n^+ if

$$\delta\Sigma \in \mathcal{K}(\Sigma) \quad \Rightarrow \quad df|_\Sigma(\delta\Sigma) \in \mathcal{K}(f(\Sigma)), \quad (23)$$

where $df|_\Sigma : T_\Sigma S_n^+ \to T_{f(\Sigma)} S_n^+$ denotes the differential of f at Σ. Assuming that $\geq_\mathcal{K}$ is a partial order induced by $\mathcal{K}$, then f is monotone with respect to $\geq_\mathcal{K}$ if and only if it is differentially positive with respect to $\mathcal{K}$. Applying this to the family of affine-invariant cone fields in (12), we arrive at the following extension to the Löwner-Heinz theorem.

Theorem 2 (Generalized Löwner-Heinz). *For any of the quadratic affine-invariant orders (12) parameterized by μ, and $r \in [0,1]$, the map $f(\Sigma) = \Sigma^r$ is monotone on S_n^+.*

This result suggests that the monotonicity of the map $f : \Sigma \mapsto \Sigma^r$ for $r \in (0,1)$ is intimately connected to the affine-invariant geometry of S_n^+ and not its translational geometry. The proof of Theorem 2 has been omitted from this abstract due to length limitations. A more detailed treatment of the topics discussed here, alongside new results and a proof of Theorem 2 will be provided in a subsequent journal paper.

5 Conclusion

The choice of partial order is a key part of studying monotonicity of functions that is often taken for granted. Invariant cone fields provide a geometric approach to systematically construct 'natural' orders by connecting the geometry of the state space to the search for orders. Coupled with differential positivity, invariant cone fields provide an insightful and powerful method for studying monotonicity, as shown in the case of S_n^+.

References

1. Ando, T.: Concavity of certain maps on positive definite matrices and applications to hadamard products. Linear Algebra Appl. **26**, 203–241 (1979)
2. Bhatia, R.: Positive Definite Matrices. Princeton University Press, Princeton (2007)
3. Bhatia, R.: Matrix Analysis, vol. 169. Springer, New York (2013)
4. Burbea, J., Rao, C.: Entropy differential metric, distance and divergence measures in probability spaces: A unified approach. J. Multivar. Anal. **12**(4), 575–596 (1982)
5. Forni, F., Sepulchre, R.: Differentially positive systems. IEEE Trans. Autom. Control **61**(2), 346–359 (2016)
6. Heinz, E.: Beiträge zur störungstheorie der spektralzerlegung. Math. Ann. **123**, 415–438 (1951)
7. Hilgert, J., Hofmann, K.H., Lawson, J.: Lie Groups, Convex Cones, and Semigroups. Oxford University Press, Oxford (1989)
8. Kubo, F., Ando, T.: Means of positive linear operators. Math. Ann. **246**, 205–224 (1979)
9. Löwner, K.: Über monotone matrixfunktionen. Math. Z. **38**, 177–216 (1934)
10. Nielsen, M.A., Chuang, I.: Quantum computation and quantum information. Am. J. Phys. **70**, 558 (2002)
11. Pedersen, G.K.: Some operator monotone functions. Proc. Am. Math. Soc. **36**(1), 309–310 (1972)
12. Pennec, X.: Statistical computing on manifolds for computational anatomy. Ph.D. thesis, Université Nice Sophia Antipolis (2006)
13. Skovgaard, L.T.: A Riemannian geometry of the multivariate normal model. Scand. J. Stat. **11**(4), 211–223 (1984)

Geodesic Least Squares Regression on the Gaussian Manifold with an Application in Astrophysics

Geert Verdoolaege[1,2(✉)]

[1] Department of Applied Physics, Ghent University, 9000 Ghent, Belgium
geert.verdoolaege@ugent.be

[2] Laboratory for Plasma Physics – Royal Military Academy (LPP–ERM/KMS), 1000 Brussels, Belgium

Abstract. We present a new regression method called *geodesic least squares* (GLS), which is particularly robust against data and model uncertainty. It is based on minimization of the Rao geodesic distance on a probabilistic manifold. We apply GLS to Tully-Fisher scaling of the total baryonic mass vs. the rotation velocity in disk galaxies and we show the excellent robustness properties of GLS for estimating the coefficients and the tightness of the scaling.

Keywords: Robust regression · Geodesic least squares · Rao geodesic distance · Tully-Fisher scaling

1 Introduction

Many natural phenomena can be described by means of scaling laws, often in the form of a power law, e.g. in astrophysics, fluid and plasma dynamics, biology, geology, climatology and finance. However, in many application fields relatively simple or outdated statistical techniques are frequently used to estimate power laws. In the vast majority of cases, ordinary least squares (OLS) is applied to estimate the exponents (coefficients) of the power law on a logarithmic scale, despite its often poor performance in all but the simplest regression problems. Indeed, in more realistic settings, particularly when the goal is extrapolation of the scaling law, robustness is at least as important a quality compared to goodness-of-fit. This can become an issue in the presence of model uncertainty, heterogeneous data, atypical measurements (outliers) and skewed likelihoods [1].

Astrophysical data are often relatively complex from the statistical perspective and it has long been recognized that various assumptions of ordinary least squares regression are not valid in many applications in the field. Accordingly, several techniques from the domains of frequentist statistics and Bayesian probability theory have been applied to address the shortcomings of OLS. However, presently most techniques are designed to address one or a few shortcomings of OLS, but not all. In addition, judicious application of these techniques may

F. Nielsen and F. Barbaresco (Eds.): GSI 2017, LNCS 10589, pp. 621–628, 2017.
https://doi.org/10.1007/978-3-319-68445-1_72

require considerable expertise from the practitioner in statistics or probability theory, which can be an issue in various physics-centered application fields. Presently, in many application domains there is a need for a robust general-purpose regression technique for estimating scaling laws.

For these reasons we have developed a new, robust regression method that is simple to implement, called *geodesic least squares regression* (GLS). It is based on minimization of the Rao geodesic distance between, on the one hand, the probability distribution of the response variable predicted by the regression model, and, on the other hand, a more data-driven distribution model of the response variable. GLS has recently been tested and applied in the field of magnetic confinement fusion [2,3], showing its enhanced robustness over various traditional methods.

In this contribution, we apply GLS regression to estimate a key scaling law in astrophysics: the baryonic Tully-Fisher relation. This is a remarkably tight relation between the total baryonic mass of disk galaxies and their rotational velocity, of great practical and theoretical significance in astrophysics and cosmology.

2 Geodesic Least Squares Regression

2.1 Principles of GLS

We here provide a brief overview of the GLS regression method. A more detailed description can be found in [1]. Implicitly, GLS performs regression on a probabilistic manifold characterized by the Fisher information. However, it is not directly based on a manifold regression technique like geodesic regression [4], where the relation between a manifold-valued response variable and a scalar predictor variable is modeled as a geodesic curve on the manifold. Rather, the idea behind GLS is to consider two different proposals for the distribution of a real-valued response variable y, conditional on the real-valued predictor variables, all of which can be affected by uncertainty. On the one hand, there is the distribution that one would expect if all assumptions were correct regarding the deterministic component of the regression model (regression function) and the stochastic component. We call this the *modeled distribution*. On the other hand, we try to capture the distribution of y by relying as little as possible on the model assumptions, and much more on the actual measurements of y. For this we will use the term *observed distribution*. In this sense, GLS is similar to minimum distance estimation (MDE), where the Hellinger distance is a popular similarity measure [5], but there are several differences. First and foremost, GLS calculates the geodesic distance between each *individual* pair of modeled and observed distributions of the response variable. This often corresponds to an individual measurement point, together with an estimate of its error bar, provided by the experimentalist. The error bar estimate may have been obtained from previous experiments, or from a time series obtained at fixed (or stationary) values of the predictor variables. As such, each single data point is replaced by a probability density function describing the distribution of the response variable under fixed measured values of the predictor variables. In contrast, MDE usually considers a distance between a kernel density estimate of the distribution of residuals on the one hand, and the parametric model on the other hand, but based on the entire data sample. Secondly, we explicitly model all parameters of the modeled distribution, similar to the idea behind the link function in the generalized linear model. In the present work this will be accomplished by explicitly modeling both the mean and standard deviation of the Gaussian modeled distribution. A final difference is that we use the Rao geodesic distance as a similarity measure.

2.2 The GLS Algorithm

We start from a parametric multiple regression model between m predictor variables ξ_j ($j = 1, \ldots, m$) and a single response variable η, all assumed to be infinitely precise. For n realizations of these variables, the regression model can be written as follows:

$$\eta_i = f(\xi_{i1}, \ldots, \xi_{im}, \beta_1, \ldots, \beta_p) \equiv f(\{\xi_{ij}\}, \{\beta_k\}), \qquad \forall i = 1, \ldots, n. \tag{1}$$

Here, f is the regression model function, in general nonlinear and characterized by p parameters β_k ($k = 1, \ldots, p$). In regression analysis within the astronomy community, it is customary to add a noise variable to the idealized relation (1). This so-called *intrinsic scatter* serves to model the intrinsic uncertainty on the theoretical relation, i.e. uncertainty not related to the measurement process. We take another route for capturing model uncertainty, however.

In any realistic situation, we have no access to the quantities ξ_{ij} and η_i. Instead, a series of noisy measurements x_{ij}, resp. y_i is acquired for the predictor and response variables:

$$\begin{aligned} y_i &= \eta_i + \epsilon_{y,i}, \qquad & \epsilon_{y,i} &\sim \mathcal{N}\left(0, \sigma_{y,i}^2\right), \\ x_{ij} &= \xi_{ij} + \epsilon_{x,ij}, \qquad & \epsilon_{x,ij} &\sim \mathcal{N}\left(0, \sigma_{x,ij}^2\right). \end{aligned}$$

We have assumed independent Gaussian noise, but this can be generalized to any distribution. Also, in general the standard deviations are different for each point. For instance, in many real-world situations, such as the one discussed in this paper, there is a constant relative error on the measurements, so the standard deviation can be modeled to be proportional to the measurement itself.

Under this model, the distribution of the variable y, conditional on measured values x_{ij} of the m predictor variables (fixed i), as well as the parameters β_k, is given by

$$p_{\text{mod}}(y|\{x_{ij}\}, \{\beta_k\}) = \frac{1}{\sqrt{2\pi}\sigma_{\text{mod},i}} \exp\left\{-\frac{1}{2}\frac{\left[y_i - f\left(\{x_{ij}\}, \{\beta_k\}\right)\right]^2}{\sigma_{\text{mod},i}^2}\right\}. \tag{2}$$

This is the modeled distribution, where we suppose that estimates of the standard deviations $\sigma_{x,ij}$ and $\sigma_{y,i}$ are available. The uncertainty on the predictor variables propagates through the function f and adds to the conditional uncertainty on the response variable, determined by $\sigma_{\text{mod},i}$. We use standard Gaussian error propagation theory as a practical solution for this purpose. For example, referring to $f\big(\{x_{ij}\},\{\beta_k\}\big)$ as the modeled mean $\mu_{\text{mod},i}$, for a linear model we have (with relabeled β_k):

$$\mu_{\text{mod},i} \equiv \beta_0 + \beta_1 x_{i1} + \ldots + \beta_m x_{im},$$
$$\sigma^2_{\text{mod},i} \equiv \sigma^2_{y,i} + \beta_1^2\sigma^2_{x,i1} + \ldots + \beta_m^2\sigma^2_{x,im}.$$

Relying on the maximum likelihood method, one would proceed to estimate the parameters β_k by maximizing (2), or, under the assumption of symmetry of the likelihood distribution and homoscedasticity, by minimizing the sum of squared differences (Euclidean distances) between each measured y_i and predicted $\mu_{\text{mod},i}$. However, this assumes that the model is exact, specifically that $\sigma_{\text{mod},i}$ is the only source of data variability. In order to take into account additional uncertainty sources, in particular model uncertainty, we therefore also consider the observed distribution of y, relying on as few assumptions as possible regarding the regression model. Specifically, we replace each data point y_i by a distribution $p_{\text{obs}}(y|y_i)$. In the context of the GLM, this is known as the *saturated model*. In the present application, we choose again the normal distribution, but centered on each data point: $\mathcal{N}\big(y_i,\sigma^2_{\text{obs},i}\big)$, where $\sigma_{\text{obs},i}$ is to be estimated from the data. The extra parameters $\sigma_{\text{obs},i}$ give the method added flexibility, since they are not *a priori* required to equal $\sigma_{\text{mod},i}$. As a result, GLS is less sensitive to incorrect model assumptions. Choosing a Gaussian form for both the modeled and observed distribution offers a computational advantage, since the corresponding expression for the GD has a closed form [6]. Also, in principle, $\sigma_{\text{obs},i}$ can be different for each point, although in practice it is clear that we will need to introduce some sort of regularization to render the model identifiable. In this paper we either assume $\sigma_{\text{obs},i}$ a constant s_{obs}, or proportional to the response variable, $\sigma_{\text{obs},i} = r_{\text{obs}}|\bar{y}_i|$. The parameters s_{obs} or r_{obs} have to be estimated from the data. More complicated (parametrized) relations between $\sigma_{\text{obs},i}$ and the response variable or other data would be possible too, but one should be careful not to put too many restrictions on p_{obs}, thereby defeating its purpose.

GLS now proceeds by minimizing the total GD between, on the one hand, the joint observed distribution of the n realizations of the variable y and, on the other hand, the joint modeled distribution. Owing to the independence assumption in this example, we can write this in terms of products of the corresponding marginal distributions (including all dependencies and with γ_{obs} either s_{obs} or r_{obs}):

$$\begin{aligned}
&\left\{\hat{\beta}_k,\hat{\gamma}_{\text{obs}}\right\}\\
&= \underset{\beta_k,\gamma_{\text{obs}}\in\mathbb{R}}{\text{argmin}}\ \text{GD}^2\left[\prod_{i=1}^n p_{\text{obs}}\left(y|y_i,\gamma_{\text{obs}}\right),\prod_{i=1}^n p_{\text{mod}}\left(y|\{x_{ij}\},\{\beta_k\},\sigma_{y_i},\{\sigma_{x_{ij}}\}\right)\right]\\
&= \underset{\beta_k,\gamma_{\text{obs}}\in\mathbb{R}}{\text{argmin}}\ \sum_{n=1}^n \text{GD}^2\Big[p_{\text{obs}}\left(y|y_i,\gamma_{\text{obs}}\right),p_{\text{mod}}\left(y|\{x_{ij}\},\{\beta_k\},\sigma_{y_i},\{\sigma_{x_{ij}}\}\right)\Big]. \qquad (3)
\end{aligned}$$

Note that the parameters β_k occur both in the mean and the variance of the modeled distribution. The last equality in (3) entails a considerable simplification, owing to the property that the squared GD between products of distributions can be written as the sum of squared GDs between the corresponding factors [6]. Hence, the optimization procedure involves, on the level of each measurement, matching not only y_i with $\mu_{\text{mod},i}$, but also $\sigma_{\text{obs},i}$ with $\sigma_{\text{mod},i}$, in a way dictated by the geometry of the likelihood distribution. As will be shown in the experiments, the result is that GLS is relatively insensitive to uncertainties in both the stochastic and deterministic components of the regression model. The same quality renders the method also robust against outliers.

In the experiments below, we employed a classic active-set algorithm to carry out the optimization [7]. Furthermore, presently the GLS method does not directly offer confidence (or credible) intervals on the estimated quantities. Future work will address this issue in more detail, but for now error estimates were derived by a bootstrap procedure. The bootstrapping involved creating, from the measured data set, 100 artificial data sets of the same size, by resampling with replacement. The regression analysis was then carried out on each of the data sets and the mean and standard deviation, over all data sets, of each estimated regression parameter were used as estimates of the parameter and its error bar, respectively. This scheme typically results in rather conservative error bars, which could possibly be narrowed down using more sophisticated methods.

Incidentally, forcing $\sigma_{\text{obs},i} \equiv \sigma_{\text{mod},i}$ in (3), $\forall\, i$, would take us back to standard maximum likelihood estimation, since the Rao GD between two Gaussians p_1 and p_2 with means y_i, resp. $f(\{x_{ij}\}, \{\beta_k\})$, but with identical standard deviations σ_i (fixed along the geodesic path), is precisely the Mahalanobis distance [8]:

$$\text{GD}(p_1, p_2) = \frac{\left|y_i - f(\{x_{ij}\}, \{\beta_k\})\right|}{\sigma_i}.$$

3 Application of GLS to Tully-Fisher Scaling

3.1 The Baryonic Tully-Fisher Relation

The baryonic Tully-Fisher relation (BTFR) between the total (stellar + gaseous) baryonic mass M_b of disk galaxies and their rotational velocity V_f is of fundamental importance in astrophysics and cosmology [9]. It is a remarkably simple and tight empirical relation of the form

$$M_b = \beta_0 V_f^{\beta_1}. \tag{4}$$

The BTFR serves as a tool for determining cosmic distances, provides constraints on galaxy formation and evolution models, and serves as a test for the Lambda cold dark matter paradigm (ΛCDM) in cosmology. In this scaling problem, we use data from 47 gas-rich galaxies, as plotted in Fig. 1 and detailed in [9]. The data also contain estimates of the observational errors, which we treat here as a single standard deviation. Figure 2 shows a scatter plot of $\sigma_{\text{mod},i}$, which is

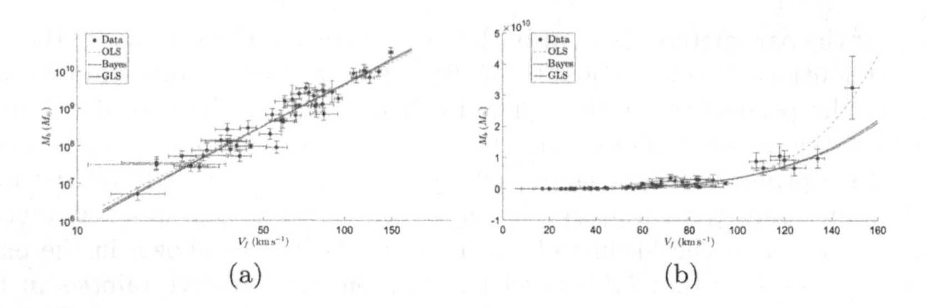

Fig. 1. Baryonic mass M_b vs. rotation velocity V_f for 47 gas-rich galaxies and the fitted BTFR using OLS, Bayesian inference and GLS. (a) On the logarithmic scale and (b) on the original scale.

Table 1. Regression estimates for the BTFR parameters using loglinear and nonlinear OLS, Bayesian inference and GLS.

Loglinear	$\hat{\beta}_0$	$\hat{\beta}_1$	Nonlinear	$\hat{\beta}_0$	$\hat{\beta}_1$
OLS	310	3.56	OLS	0.063	5.37
Bayes	160	3.72	Bayes	91	3.80
GLS	110	3.81	GLS	79	3.83

almost entirely determined by σ_{M_b}, vs. M_b for the 47 galaxies in the database. This suggests a measurement error on the response variable proportional to M_b, about 38%, i.e. a constant error bar on the logarithmic scale.

3.2 Regression Analysis

Owing to the power law character of most scaling laws, they are often estimated by linear regression on a logarithmic scale. However, it is known that this may lead to unreliable estimates, as the logarithm (heavily) distorts the distribution of the data [1]. This is in particular the case when the estimation is done using simple OLS or when there are outliers in the data. In contrast, we will show that GLS regression produces consistent results on both the logarithmic and original scales, demonstrating its robustness.

In view of the proportional error on M_b, the observed standard deviation in GLS is modeled here as $\sigma_{\text{obs},i} = r_{\text{obs}} M_b$, with r_{obs} an unknown scale factor to be estimated from the data using the optimization routine.

We compare the results of GLS regression with OLS and a standard Bayesian method. For the latter we choose the likelihood given in (2), allowing uncertainty on the standard deviation through a scale factor with a Jeffreys prior. We also use uninformative prior distributions for the regression parameters.

The scalings obtained using the various methods are shown in Fig. 1a for the case of linear regression on the logarithmic scale, and in Fig. 1b for power-law regression on the original scale. The coefficient estimates are given in Table 1. It is clear that GLS yields estimates that are much more consistent compared

Table 2. Average regression estimates and 95% confidence intervals for the BTFR using loglinear and nonlinear OLS, Bayesian inference and GLS, obtained from 100 bootstrap samples.

Loglinear	$\hat{\beta}_0$	$\hat{\beta}_1$
OLS	360 ± 220	3.57 ± 0.15
Bayes	220 ± 220	3.72 ± 0.19
GLS	140 ± 82	3.80 ± 0.16
Nonlinear	$\hat{\beta}_0$	$\hat{\beta}_1$
OLS	$(3.6 \pm 6.2) \times 10^3$	4.56 ± 1.19
Bayes	130 ± 160	3.80 ± 0.21
GLS	390 ± 280	3.85 ± 0.18

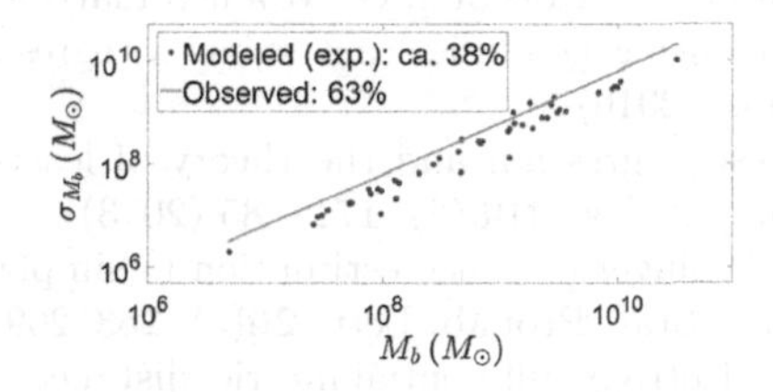

Fig. 2. Plot of σ_{M_b} ($\approx \sigma_{\text{mod}}$) and $r_{\text{obs}} M_b$ ($= \sigma_{\text{obs}}$) vs. M_b, as estimated by GLS.

to OLS. In particular, whereas the data point corresponding to the largest V_f and M_b does not have the characteristics of an outlier on the logarithmic scale, it may be considered as such on the original scale. The nonlinear OLS estimate for the exponent β_1 is heavily influenced by this point, causing the discrepancy with the estimate on the logarithmic scale. Furthermore, the results of GLS are verified by the Bayesian method.

Next, 100 bootstrap samples were created from the data, yielding average parameter estimates and 95% confidence intervals on the basis of the OLS and GLS results, shown in Table 2. Again, the enhanced robustness of GLS compared to OLS stands out.

Finally, Fig. 2 shows the plot of $\sigma_{\text{obs}} = r_{\text{obs}} M_b$ vs. M_{b}, with the scale factor r_{obs} (observed relative error) amounting to 63%. This is considerably larger than the value of 38% predicted by the model, possibly indicating that the scatter on the scaling law is not due to measurement error alone. This will be an important area of further investigation, as it may provide evidence for the ΛCDM vs. MOND cosmological models.

4 Conclusion

We have introduced geodesic least squares, a versatile and robust regression method based on regression between probability distributions. Part of the strength of the method is its simplicity, allowing straightforward application by users in various application fields, without the need for parameter tuning. We have applied GLS to baryonic Tully-Fisher scaling, thereby demonstrating the robustness of the method and providing an alternative means for testing cosmological models based on the estimated intrinsic scatter.

References

1. Verdoolaege, G.: A new robust regression method based on minimization of geodesic distances on a probabilistic manifold: Application to power laws. Entropy **17**(7), 4602–4626 (2015)
2. Verdoolaege, G., Noterdaeme, J.-M.: Robust scaling in fusion science: case study for the L-H power threshold. Nucl. Fusion 55(11), 113019, (19 pp.) (2015)
3. Verdoolaege, G., Shabbir, A., Hornung, G.: Robust analysis of trends in noisy tokamak confinement data using geodesic least squares regression. Rev. Sci. Instrum. 87(11), 11D422, (3 pp.) (2016)
4. Fletcher, P.T.: Geodesic regression and the theory of least squares on riemannian manifolds. Int. J. Comput. Vis. **105**(2), 171–185 (2013)
5. Pak, R.J.: Minimum hellinger distance estimation in simple regression models; distribution and efficiency. Stat. Probab. Lett. **26**(3), 263–269 (1996)
6. Burbea, J., Rao, C.R.: Entropy differential metric, distance and divergence measures in probability spaces: a unified approach. J. Multivar. Anal. **12**(4), 575–596 (1982)
7. Gill, P.E., Murray, W., Wright, M.H.: Numerical Linear Algebra and Optimization, vol. 1. Addison Wesley, Boston (1991)
8. Rao, C.R.: Differential metrics in probability spaces. In: Differential Geometry in Statistical Inference. Institute of Mathematical Statistics, Hayward, CA (1987)
9. McGaugh, S.S.: The baryonic Tully-Fisher relation of gas-rich galaxies as a test of ΛCDM and MOND. Astron. J. 143(2), 40, (15 pp.) (2012)

Computational Information Geometry

Warped Metrics for Location-Scale Models

Salem Said(✉) and Yannick Berthoumieu

Laboratoire IMS (CNRS - UMR 5218), Université de Bordeaux, Bordeaux, France
{salem.said,yannick.berthoumieu}@ims-bordeaux.fr

Abstract. This paper argues that a class of Riemannian metrics, called warped metrics, plays a fundamental role in statistical problems involving location-scale models. The paper reports three new results: (i) the Rao-Fisher metric of any location-scale model is a warped metric, provided that this model satisfies a natural invariance condition, (ii) the analytic expression of the sectional curvature of this metric, (iii) the exact analytic solution of the geodesic equation of this metric. The paper applies these new results to several examples of interest, where it shows that warped metrics turn location-scale models into complete Riemannian manifolds of negative sectional curvature. This is a very suitable situation for developing algorithms which solve problems of classification and on-line estimation. Thus, by revealing the connection between warped metrics and location-scale models, the present paper paves the way to the introduction of new efficient statistical algorithms.

Keywords: Rao-fisher metric · Warped metric · Location-scale model · Sectional curvature · Geodesic equation

1 Introduction: Definition and Two Examples

This paper argues that a class of Riemannian metrics, called warped metrics, is natural and useful to statistical problems involving location-scale models. A warped metric is defined as follows [1]. Let M be a Riemannian manifold with Riemannian metric ds^2_M. Consider the manifold $\mathcal{M} = M \times (0\,,\infty)$, equipped with the Riemannian metric,

$$ds^2(z) = I_0(\sigma)\, d\sigma^2 \, + \, I_1(\sigma)\, ds^2_M(\bar{x}) \tag{1}$$

where each $z \in \mathcal{M}$ is a couple $(\bar{x}\,,\sigma)$ with $\bar{x} \in M$ and $\sigma \in (0\,,\infty)$. The Riemannian metric (1) is called a warped metric on $\mathcal{M}$. The functions I_0 and I_1 have strictly positive values and are part of the definition of this metric.

The main claim of this paper is that warped metrics arise naturally as Rao-Fisher metrics for a variety of location-scale models. Here, to begin, two examples of this claim are given. Example 1 is classic, while Example 2, to our knowledge, is new in the literature. As of now, the reader is advised to think of $\mathcal{M}$ as a statistical manifold, where $\bar{x}$ is a location parameter and σ is either a scale parameter or a concentration parameter.

F. Nielsen and F. Barbaresco (Eds.): GSI 2017, LNCS 10589, pp. 631–638, 2017.
https://doi.org/10.1007/978-3-319-68445-1_73

Example 1 (univariate normal model): Let $M = \mathbb{R}$, with $ds^2_M(\bar{x}) = d\bar{x}^2$ the canonical metric of $\mathbb{R}$. If each $z = (\bar{x}\,,\sigma)$ in $\mathcal{M}$ is identified with the univariate normal density of mean $\bar{x}$ and standard deviation σ, then the resulting Rao-Fisher metric on $\mathcal{M}$ is given by [2]

$$ds^2(z) = \sigma^{-2}\,d\sigma^2 + \frac{1}{2}\sigma^{-2}\,d\bar{x}^2 \qquad (2)$$

Example 2 (von Mises-Fisher model): let $M = S^2$, the unit sphere with $ds^2_M = d\theta^2$ its canonical metric induced from $\mathbb{R}^3$. Identify $z = (\bar{x}\,,\sigma)$ in $\mathcal{M}$ with the von Mises-Fisher density of mean direction $\bar{x}$ and concentration parameter σ [3]. The resulting Rao-Fisher metric on $\mathcal{M}$ is given by

$$ds^2(z) = \left(\sigma^{-2} - \sinh^{-2}\sigma\right)d\sigma^2 + \left(\sigma\coth\sigma - 1\right)d\theta^2(\bar{x}) \qquad (3)$$

Remark a: note that σ is a scale parameter in Example 1, but a concentration parameter in Example 2. Accordingly, at $\sigma = 0$, the metric (2) becomes infinite, while the metric (3) remains finite and degenerates to $ds^2(z)\big|_{\sigma=0} = (1/3)\,d\sigma^2$. Thus, (3) gives a Riemannian metric on the larger Riemannian manifold $\hat{\mathcal{M}} = \mathbb{R}^3$, which contains $\mathcal{M}$, obtained by considering σ as a radial coordinate and $\sigma = 0$ as the origin of $\mathbb{R}^3$. ■

2 A General Theorem: From Rao-Fisher to Warped Metrics

Examples 1 and 2 of the previous section are special cases of Theorem 1, given here. To state this theorem, let (M, ds^2_M) be an irreducible Riemannian homogeneous space, under the action of a group of isometries G [4]. Denote by $g \cdot x$ the action of $g \in G$ on $x \in M$. Then, assume each $z = (\bar{x}\,,\sigma)$ in $\mathcal{M}$ can be identified uniquely and regularly with a probability density $p(x|\,z) = p(x|\,\bar{x}\,,\sigma)$ on M, with respect to the Riemannian volume element, such that the following property is verified,

$$p(g\cdot x|\,g\cdot\bar{x}\,,\sigma) = p(x|\,\bar{x}\,,\sigma)\;\; g \in G \qquad (4)$$

The densities $p(x|\,\bar{x}\,,\sigma)$ form a statistical model on M, where $\bar{x}$ is a location parameter and σ can be chosen as either a scale or a concentration parameter, (roughly, a scale parameter is the inverse of a concentration parameter).

In the statement of Theorem 1, $\ell(z) = \log p(x|\,z)$ and $\nabla_{\bar{x}}\,\ell(z)$ denotes the Riemannian gradient vector field of $\ell(z)$, with respect to $\bar{x} \in M$. Moreover, $\|\nabla_{\bar{x}}\ell(z)\|$ denotes the length of this vector field, as measured by the metric ds^2_M.

Theorem 1 (warped metrics). *The Rao-Fisher metric of the statistical model* $\{\,p(x|\,z)\,;\, z \in \mathcal{M}\,\}$ *is a warped metric of the form (1), defined by*

$$I_0(\sigma) = \mathbb{E}_z\left(\,\partial_\sigma\ell(z)\,\right)^2 \;\; I_1(\sigma) = \mathbb{E}_z\,\|\nabla_{\bar{x}}\ell(z)\|^2\,/\dim M \qquad (5)$$

where $\mathbb{E}_z$ *denotes expectation with respect to* $p(x|\,z)$. *Due to property (4), the two expectations appearing in (5) do not depend on the parameter* $\bar{x}$, *so* I_0 *and* I_1 *are well-defined functions of* σ .

Remark b: the proof of Theorem 1 cannot be given here, due to lack of space. It relies strongly on the assumption that the Riemannian homogeneous space M is irreducible. In particular, this allows the application of Schur's lemma, from the theory of group representations [5]. To say that M is an irreducible Riemannian homogeneous space means that the following property is verified: if $K_{\bar{x}}$ is the stabiliser in G of $\bar{x} \in M$, then the isotropy representation $k \mapsto dk|_{\bar{x}}$ is an irreducible representation of $K_{\bar{x}}$ in the tangent space $T_{\bar{x}}M$. ■

Remark c: if the assumption that M is irreducible is relaxed, then Theorem 1 generalises to a similar statement, involving so-called multiply warped metrics. Roughly, this is because a homogeneous space which is not irreducible, may still decompose into a direct product of irreducible homogeneous spaces [4]. ■

Remark d: statistical models on M which verify (4) often arise under an exponential form,

$$p(x|\,\bar{x}\,,\sigma) \;=\; \exp\left(\,\eta \cdot D(x\,,\bar{x}) - \psi(\eta)\,\right) \tag{6}$$

where $\eta = \eta(\sigma)$ is a natural parameter, and $\psi(\eta)$ is the cumulant generating function of the statistic $D(x\,,\bar{x})$. Then, for assumption (4) to hold, it is necessary and sufficient that

$$D(g\cdot x\,,\,g\cdot\bar{x}) \;=\; D(x\,,\bar{x}) \tag{7}$$

Both Examples 1 and 2 are of the form (6), as is Example 3, in the following section, which deals with the Riemannian Gaussian model [6,7]. ■

3 Curvature Equations and the Extrinsic Geometry of M

For each $\sigma \in (0\,,\infty)$, there is an embedding of M into $\mathcal{M}$, as the surface $M \times \{\sigma\}$. This embedding yields an extrinsic geometry of M, given by the first and second fundamental forms [8].

The first fundamental form is the restriction of the metric ds^2 of $\mathcal{M}$ to the tangent bundle of M. This will be denoted $ds^2_M(x|\,\sigma)$ for $x \in M$. It is clear from (1) that

$$ds^2_M(x|\,\sigma) \;=\; I_1(\sigma)\, ds^2_M(x) \tag{8}$$

This extrinsic Riemannian metric on M is a scaled version of its intrinsic metric ds^2_M . It induces an extrinsic Riemannian distance given by

$$d^{\,2}(x\,,y|\,\sigma) \;=\; I_1(\sigma)\, d^{\,2}(x\,,y)\;\; x\,,\, y\, \in M \tag{9}$$

where $d(x\,,y)$ is the intrinsic Riemannian distance, induced by the metric ds^2_M .

The extrinsic distance (9) is a generalisation of the famous Mahalanobis distance. In fact, replacing in Example 1 yields the classical expression of the Mahalanobis distance $d^{\,2}(x\,,y) = |x-y|^2/2\sigma^2$. The significance of this distance can be visualised as follows: if σ is a dispersion parameter, the extrinsic distance between two otherwise fixed points $x, y \in M$ will decrease as σ increases, as if the space M were contracting, (for a concentration parameter, there is an expansion, rather than a contraction).

The second fundamental form is given by the tangent component of the covariant derivative of the unit normal to the surface $M \times \{\sigma\}$. This unit normal is ∂_r where r is the vertical distance coordinate, given by $dr/d\sigma = I_0^{\frac{1}{2}}(\sigma)$. Using Koszul's formula [9], it is possible to express the second fundamental form,

$$S(v) = \frac{1}{2}\left(\partial_r I_1 \Big/ I_1\right) v \tag{10}$$

for any v tangent to M. Knowledge of the second fundamental form is valuable, as it yields the relationship between extrinsic and intrinsic curvatures of M.

Proposition 1 (curvature equations). *Let $K^{\mathcal{M}}$ and K^M denote the sectional curvatures of $\mathcal{M}$ and M. The following are true*

$$K^{\mathcal{M}}(u, v) = \left(1 \Big/ I_1\right) K^M(u, v) - \frac{1}{4}\left(\partial_r I_1 \Big/ I_1\right)^2 \tag{11}$$

$$K^{\mathcal{M}}(u, \partial_r) = -\left(\partial_r^2 I_1^{\frac{1}{2}} \Big/ I_1^{\frac{1}{2}}\right) \tag{12}$$

for any linearly independent u, v tangent to M.

Remark e: here, Eq. (11) is the Gauss curvature equation. Roughly, it shows that embedding M into $\mathcal{M}$ adds negative curvature. Eq. (12) is the mixed curvature equation. If the intrinsic sectional curvature K^M is negative, then (11) and (12) show that the sectional curvature $K^{\mathcal{M}}$ of $\mathcal{M}$ is negative if and only if $I_1^{\frac{1}{2}}$ is a convex function of the vertical distance r. ■

Return to example 1: here, $M = \mathbb{R}$ is one-dimensional, so the Gauss equation (11) does not provide any information. The mixed curvature equation gives the curvature of the two-dimensional manifold $\mathcal{M}$. In this equation, $\partial_r = \sigma\,\partial_\sigma$, and it follows that

$$K^{\mathcal{M}}(u, \partial_r) = -1 \tag{13}$$

so $\mathcal{M}$ has constant negative curvature. In fact, it was observed long ago that the metric (2) is essentially the Poincaré half-plane metric [2]. ■

Return to example 2: in Example 2, $M = S^2$ so $K^M \equiv 1$ is constant. It follows from the Gauss equation that each sphere $S^2 \times \{\sigma\}$ has constant extrinsic curvature, equal to

$$K^{\mathcal{M}}\big|_\sigma = \left(1 \Big/ I_1\right) - \frac{1}{4}\left(\partial_r I_1 \Big/ I_1\right)^2 \tag{14}$$

Upon replacing the expressions of I_1 and ∂_r based on (3), this is found to be strictly negative for $\sigma > 0$,

$$K^{\mathcal{M}}\big|_{\sigma} < 0 \ \text{ for } \sigma > 0 \tag{15}$$

Thus, the Rao-Fisher metric (3) induces a negative extrinsic curvature on each spherical surface $S^2 \times \{\sigma\}$. In fact, by studying the mixed curvature equation (12), it is seen the whole manifold $\mathcal{M}$ equipped with the Rao-Fisher metric (3) is a manifold of negative sectional curvature. ∎

Example 3 (Riemannian Gaussian model): A Riemannian Gaussian distribution may be defined on any Riemannian symmetric space M of non-positive curvature. It is given by the probability density with respect to Riemannian volume

$$p(x|\,\bar{x}\,,\sigma) = Z^{-1}(\sigma)\,\exp\left[\,-\frac{d^2(x\,,\bar{x})}{2\sigma^2}\,\right] \tag{16}$$

where the normalising constant $Z(\sigma)$ admits a general expression, which was given in [7]. If M is an irreducible Riemannian symmetric space, then Theorem 1 above applies to the Riemannian Gaussian model (16), leading to a warped metric with

$$I_0(\sigma) \;=\; \psi''(\eta)\;I_1(\sigma) \;=\; 4\eta^2\,\psi'(\eta)\Big/\dim M \tag{17}$$

where $\eta = -1/2\sigma^2$ and $\psi(\eta) = \log\, Z(\sigma)$. The result of Eq. (17) is here published for the first time. Consider now the special case where M is the hyperbolic plane. The analytic expression of I_0 and I_1 can be found from (17) using

$$Z(\sigma) \;=\; \text{Const.}\;\sigma\,\times\,e^{\sigma^2/4}\,\times\,\text{erf}(\sigma/2) \tag{18}$$

which was derived in [6]. Here, erf denotes the error function. Then, replacing (17) in the curvature equations (11) and (12) yields the same result as for Example 2: the manifold $\mathcal{M}$ equipped with the Rao-Fisher metric (17) is a manifold of negative sectional curvature. ∎

Remark f (a conjecture): based on the three examples just considered, it seems reasonable to conjecture that warped metrics arising from Theorem 1 will always lead to manifolds $\mathcal{M}$ of negative sectional curvature. ∎

4 Solution of the Geodesic Equation: Conservation Laws

If the assumptions of Theorem 1 are slightly strengthened, then an analytic solution of the geodesic equation of the Riemannian metric (1) on $\mathcal{M}$ can be obtained, by virtue of the existence of a sufficient number of conservation laws. To state this precisely, let $\langle\cdot,\cdot\rangle_M$ and $\langle\cdot,\cdot\rangle_{\mathcal{M}}$ denote respectively the scalar products defined by the metrics ds^2_M and ds^2.

Two kinds of conservation laws hold along any affinely parameterised geodesic curve $\gamma(t)$ in $\mathcal{M}$, with respect to the metric ds^2. These are conservation of energy and conservation of moments [10]. If the geodesic $\gamma(t)$ is expressed as a couple $(\sigma(t), x(t))$ where $\sigma(t) > 0$ and $x(t) \in M$, then the energy of this geodesic is

$$E = I_0(\sigma)\,\dot{\sigma}^2 + I_1(\sigma)\,\|\dot{x}\|^2 \tag{19}$$

where the dot denotes differentiation with respect to t, and $\|\dot{x}\|$ the Riemannian length of $\dot{x}$ as measured by the metric ds^2_M.

On the other hand, if ξ is any element of the Lie algebra of the group of isometries G acting on M, the corresponding moment of the geodesic $\gamma(t)$ is

$$J(\xi) = I_1(\sigma)\,\langle \dot{x}, X_\xi \rangle_M \tag{20}$$

where X_ξ is the vector field on M given by $X_\xi(x) = \frac{d}{dt}\big|_{t=0}\, e^{t\xi}\cdot x$. The equation of the geodesic $\gamma(t)$ is given as follows.

Proposition 2 (conservation laws and geodesics). *For any geodesic $\gamma(t)$, its energy E and its moment $J(\xi)$ for any ξ are conserved quantities, remaining constant along this geodesic. If M is an irreducible Riemannian symmetric space, the equation of the geodesic $\gamma(t)$ is the following,*

$$x(t) = \mathrm{Exp}_{x(0)}\left[\left(\int_0^t \frac{I_1(\sigma(0))}{I_1(\sigma(s))}ds\right)\dot{x}(0)\right] \tag{21}$$

$$t = \pm\int_{\sigma(0)}^{\sigma(t)} \frac{I_0^{\frac{1}{2}}(\sigma)\,d\sigma}{\sqrt{E - V(\sigma)}} \tag{22}$$

where Exp *denotes the Riemannian exponential mapping of the metric ds^2_M on M, and $V(\sigma)$ is the function $V(\sigma) = J_0 \times I_1(\sigma(0))\big/I_1(\sigma)$, with $J_0 = I_1(\sigma(0))\,\|\dot{x}(0)\|^2$.*

Remark g: under the assumption that M is an irreducible Riemannian symmetric space, the second part of Proposition 2, stating the equations of $x(t)$ and $\sigma(t)$ is a corollary of the first part, stating the conservation of energy and moment. The proof, as usual not given due to lack of space, relies on a technique of lifting the geodesic equation to the Lie algebra of the group of isometries G. ■

Remark h: Here, Eq. (21) states that $x(t)$ describes a geodesic curve in the space M, with respect to the metric ds^2_M, at a variable speed equal to $I_1(\sigma(0))\big/I_1(\sigma(t))$. Equation (22) states that $\sigma(t)$ describes the one-dimensional motion of a particle of energy E and mass $2I_0(\sigma)$, in a potential field $V(\sigma)$. ■

Remark i (completeness of $\mathcal{M}$): From Eq. (22) it is possible to see that any geodesic $\gamma(t)$ in $\mathcal{M}$ is defined for all $t > 0$, if and only if the following conditions are verified

$$\int_0 I_0^{\frac{1}{2}}(\sigma)\,d\sigma = \infty \quad \int^\infty I_0^{\frac{1}{2}}(\sigma)\,d\sigma = \infty \tag{23}$$

where the missing integration bounds are arbitrary. The first condition ensures that $\gamma(t)$ may not escape to $\sigma = 0$ within a finite time, while the second condition ensures the same for $\sigma = \infty$. The two conditions (23), taken together, are necessary and sufficient for $\mathcal{M}$ to be a complete Riemannian manifold. ■

Return to Example 2: for the von Mises-Fisher model of Example 2, the second condition in (23) is verified, but not the first. Therefore, a geodesic $\gamma(t)$ in $\mathcal{M}$ may escape to $\sigma = 0$ within a finite time. However, $\gamma(t)$ is also a geodesic in the larger manifold $\hat{\mathcal{M}} = \mathbb{R}^3$, which contains $\sigma = 0$ as its origin. If $\gamma(t)$ arrives at $\sigma = 0$ at some finite time, it will just go through this point and immediately return to $\mathcal{M}$. In fact, $\hat{\mathcal{M}}$ is a complete Riemannian manifold which has $\mathcal{M}$ as an isometrically embedded submanifold. ■

5 The Road to Applications: Classification and Estimation

The theoretical results of the previous chapters have established that warped metrics are natural statistical objects arising in connection with location-scale models, which are invariant under some group action. Precisely, Theorem 1 has stated that warped metrics appear as Rao-Fisher metrics for all location-scale models which verify the group invariance condition (4).

Analytical knowledge of the Rao-Fisher metric of a statistical model is potentially useful to many applications. In particular, to problems of classification and efficient on-line estimation. However, in order for such applications to be realised, it is necessary for the Rao-Fisher metric to be well-behaved. Propositions 2 and 3 in the above seem to indicate such a good behavior for warped metrics on location-scale models.

Indeed, as conjectured in Remark f, the curvature equations of Proposition 2 would indicate that the sectional curvature of these warped metrics is always negative. Then, if the conditions for completeness, given in Remark i based on Proposition 3, are verified, the location-scale models equipped with these warped metrics appear as complete Riemannian manifolds of negative curvature. This is a favourable scenario, (which at least holds for the von Mises-Fisher model of Example 2), under which many algorithms can be implemented.

For classification problems, it becomes straightforward to find the analytic expression of Rao's Riemannian distance, and to compute Riemannian centres of mass, whose existence and uniqueness will be guaranteed. These form the building blocks of many classification methodologies.

For efficient on-line estimation, Amari's natural gradient algorithm turns out to be identical to the stochastic Riemannian gradient algorithm, defined using the Rao-Fisher metric. Then, analytical knowledge of the Rao-Fisher metric, (which is here a warped metric), and of its completeness and curvature properties, yields an elegant formulation of the natural gradient algorithm, and a geometrical means of proving its efficiency and understanding its convergence properties.

References

1. Petersen, P.: Riemannian Geometry, 2nd edn. Springer, New York (2006)
2. Atkinson, C., Mitchell, A.: Rao's distance measure. Sankhya Ser. A **43**, 345–365 (1981)
3. Mardia, K.V., Jupp, P.E.: Directional Statistics. Wiley, New York (2000)
4. Kobayashi, S., Nomizu, K.: Foundations of Differential Geometry, Volume II. Wiley, New York (1969)
5. Chevalley, C.: Theory of Lie Groups, Volume I. Princeton University Press, Princeton (1946)
6. Said, S., Bombrun, L., Berthoumieu, Y., Manton, J.H.: Riemannian gaussian distributions on the space of symmetric positive definite matrices. IEEE Trans. Inf. Theory **63**(4), 2153–2170 (2016)
7. Said, S., Hajri, H., Bombrun, L., Vemuri, B.C.: Gaussian distributions on Riemannian symmetric spaces: statistical learning with structured covariance matrices (under review). IEEE Trans. Inf. Theory (2017)
8. Do Carmo, M.P.: Riemannian Geometry. Birkhauser, Basel (1992)
9. Helgason, S.: Differential Geometry, Lie Groups, and Symmetric Spaces. American Mathematical Society, Providence (2001)
10. Gallot, S., Hulin, D., Lafontaine, J.: Riemannian Geometry. Springer, Heidelberg (2004)

Bregman Divergences from Comparative Convexity

Frank Nielsen[1,2](✉) and Richard Nock[3,4,5]

[1] École Polytechnique, Palaiseau, France
Frank.Nielsen@acm.org
[2] Sony CSL, Tokyo, Japan
[3] Data61, Sydney, Australia
Richard.Nock@data61.csiro.au
[4] Australian National University, Canberra, Australia
[5] University of Sydney, Camperdown, Australia

Abstract. Comparative convexity is a generalization of ordinary convexity based on abstract means instead of arithmetic means. We define and study the Bregman divergences with respect to comparative convexity. As an example, we consider the convexity induced by quasi-arithmetic means, report explicit formulas, and show that those Bregman divergences are equivalent to conformal ordinary Bregman divergences on monotone embeddings.

Keywords: Convexity · Regular mean · Quasi-arithmetic weighted mean · Skew Jensen divergence · Bregman divergence · Conformal divergence

1 Introduction: Convexity, Jensen and Bregman Divergences

Convexity allows one to define classes of *dissimilarity measures* parameterized by functional generators. Let $F : \mathcal{X} \to \mathbb{R}$ be a real-valued function. Burbea and Rao [8] studied the *Jensen difference* for $F \in \mathcal{C}$ as such a family of dissimilarities:

$$J_F(p,q):=\frac{F(p)+F(q)}{2} - F\left(\frac{p+q}{2}\right). \tag{1}$$

A dissimilarity $D(p,q)$ is *proper* iff $D(p,q) \geq 0$ with equality iff $p = q$. It follows from the strict midpoint convex property of F that J_F is proper. Nowadays, these Jensen differences are commonly called *Jensen Divergences* (JD), where a *divergence* is a *smooth* dissimilarity measure inducing a dual geometry [2]. One can further define the proper *skewed Jensen divergences* for $\alpha \in (0,1)$, see [15,20]:

$$J_{F,\alpha}(p:q):=(1-\alpha)F(p)+\alpha F(q) - F((1-\alpha)p+\alpha q), \tag{2}$$

F. Nielsen and F. Barbaresco (Eds.): GSI 2017, LNCS 10589, pp. 639–647, 2017.
https://doi.org/10.1007/978-3-319-68445-1_74

with $J_{F,\alpha}(q:p) = J_{F,1-\alpha}(p:q)$. The ":" notation emphasizes the fact that the dissimilarity may be asymmetric. Another popular class of dissimilarities are the Bregman Divergences [6] (BDs):

$$B_F(p:q) := F(p) - F(q) - (p-q)^\top \nabla F(q), \tag{3}$$

where ∇F denotes the gradient of F. Let $J'_{F,\alpha}(p:q) := \frac{1}{\alpha(1-\alpha)} J_{F,\alpha}(p:q)$ denote the scaled skew JDs. Then it was proved that BDs can be obtained as limit cases of skew JDs [15,20]:

$$B_F(p:q) = \lim_{\alpha \to 0^+} J'_{F,\alpha}(p:q), \quad B_F(q:p) = \lim_{\alpha \to 1^-} J'_{F,1-\alpha}(p:q).$$

2 Jensen and Bregman Divergences with Comparative Convexity

2.1 Comparative Convexity

The branch of *comparative convexity* [14] studies classes $\mathcal{C}_{M,N}$ of (M,N)-strictly convex functions F that satisfies the following *generalized strict midpoint convex* inequality:

$$F \in \mathcal{C}_{M,N} \Leftrightarrow F(M(p,q)) < N(F(p), F(q)), \quad \forall p, q \in \mathcal{X}, \tag{4}$$

where M and N are two abstract means defined on the domain $\mathcal{X}$ and codomain $\mathbb{R}$, respectively. When $M = N = A$, the arithmetic mean, we recover the ordinary convexity.

An *abstract mean* $M(p,q)$ aggregates two values to produce an intermediate quantity that satisfies the *innerness property* [7]: $\min\{x,y\} \leq M(x,y) \leq \max\{x,y\}$. There are many families of means. For example, the family of *power means* P_δ (Hölder means [10]) is defined by: $P_\delta(x,y) = \left(\frac{x^\delta + y^\delta}{2}\right)^{\frac{1}{\delta}}$. The arithmetic, harmonic and quadratic means are obtained for $\delta = 1$, $\delta = -1$, and $\delta = 2$, respectively. To get a continuous family of power means for $\delta \in \mathbb{R}$, we define for $\delta = 0$, $P_0(x,y) = \sqrt{xy}$, the geometric mean. Notice that power means satisfy the innerness property, and include in the limit cases the minimum and maximum values: $\lim_{\delta \to -\infty} P_\delta(x,y) = \min\{x,y\}$ and $\lim_{\delta \to \infty} P_\delta(x,y) = \max\{x,y\}$. Moreover, the power means are ordered, $P_\delta(x,y) \leq P_{\delta'}(x,y)$ for $\delta' \geq \delta$, a property generalizing the well-known inequality of arithmetic and geometric means [7].

There are many ways to define *parametric* family of means [7]: For example, let us cite the Stolarksy, Lehmer and Gini means, with the Gini means including the power means. Means can also be parameterized by monotone functions: Let us cite the quasi-arithmetic means [9,11,13], the Lagrange means [4], the Cauchy means [12], etc.

2.2 Generalized Skew Jensen Divergences

We shall introduce *univariate* divergences for $\mathcal{X} \subset \mathbb{R}$ in the remainder. Multivariate divergences for $\mathcal{X} \subset \mathbb{R}^d$ can be built from univariate divergences componentwise.

Definition 1 (Comparative Convexity Jensen Divergence). *The Comparative Convexity Jensen Divergence (ccJD) is defined for a midpoint (M, N)-strictly convex function $F : I \subset \mathbb{R} \rightarrow \mathbb{R}$ by:*

$$J_F^{M,N}(p,q):=N(F(p),F(q))-F(M(p,q)) \tag{5}$$

It follows from the strict midpoint (M, N)-convexity that the ccJDs are proper: $J_F^{M,N}(p,q) \geq 0$ with equality iff $p = q$.

To define generalized skew Jensen divergences, we need (i) to consider *weighted means*, and (ii) to ensure that the divergence is proper. This restrict weighted means to be *regular*:

Definition 2 (Regular mean). *A mean M is said* regular *if it is (i) symmetric ($M(p,q) = M(q,p)$), (ii) continuous, (iii) increasing in each variable, and (iv) homogeneous ($M(\lambda p, \lambda q) = \lambda M(p,q), \forall \lambda > 0$).*

Power means are regular: They belong to a broader family of regular means, the quasi-arithmetic means. A *quasi-arithmetic mean* is defined for a continuous and strictly increasing function $f : I \subset \mathbb{R} \rightarrow J \subset \mathbb{R}$ as:

$$M_f(p,q):=f^{-1}\left(\frac{f(p)+f(q)}{2}\right). \tag{6}$$

These means are also called Kolmogorov-Nagumo-de Finetti means [9, 11,13]. By choosing $f(x) = x$, $f(x) = \log x$ or $f(x) = \frac{1}{x}$, we obtain the Pythagorean arithmetic, geometric, and harmonic (power) means, respectively. A quasi-arithmetic weighted mean is defined by $M_f(p,q;1-\alpha,\alpha):=f^{-1}\left((1-\alpha)f(p)+\alpha f(q)\right)$ for $\alpha \in [0,1]$. Let $M_\alpha(p,q):=M(p,q;1-\alpha,\alpha)$ denote a shortcut for a weighted regular mean.

A *continuous* function F satisfying the (M, N)-midpoint convex property for regular means M and N is (M, N)*-convex* (Theorem A of [14]):

$$N_\alpha(F(p),F(q)) \geq F(M_\alpha(p,q)), \forall p,q \in \mathcal{X}, \forall \alpha \in [0,1]. \tag{7}$$

Thus we can define a proper divergence for a strictly (M, N)-convex function when considering regular weighted means:

Definition 3 (Comparative Convexity skew Jensen Divergence). *The Comparative Convexity skew α-Jensen Divergence (ccsJD) is defined for a strictly (M, N)-convex function $F \in \mathcal{C}_{M,N} : I \rightarrow \mathbb{R}$ by:*

$$J_{F,\alpha}^{M,N}(p:q):=N_\alpha(F(p),F(q))-F(M_\alpha(p,q)), \tag{8}$$

where M and N are regular weighted means, and $\alpha \in (0,1)$.

For regular weighted means, we have $J_{F,\alpha}^{M,N}(q,p) = J_{F,1-\alpha}^{M,N}(p:q)$ since the weighted means satisfy $M_\alpha(p,q) = M_{1-\alpha}(q,p)$. This generalized ccsJD can be extended to a positively weighted set of values by defining a notion of *diversity* [8] as:

Definition 4 (Comparative Convexity Jensen Diversity Index). *Let* $\{(w_i,x_i)\}_{i=1}^n$ *be a set of* n *positive weighted values so that* $\sum_{i=1}^n w_i = 1$. *Then the* Jensen diversity index *with respect to the strict* (M,N)*-convexity of a function* F *for regular weighted means is:*

$$J_F^{M,N}(x_1,\ldots,x_n;w_1,\ldots,w_n):=N(\{(F(x_i),w_i)\}_i) - F(M(\{(x_i,w_i)\}_i)).$$

When both means M and N are set to the arithmetic mean, this diversity index has also been called the *Bregman information* [3] in the context of k-means clustering.

2.3 Generalized Bregman Divergences

By analogy to the ordinary setting, let us define the (M,N)-Bregman divergence as the limit case of a scaled skew (M,N)-ccsJDs. Let $J'^{M,N}_{F,\alpha}(p:q) = \frac{1}{\alpha(1-\alpha)} J_{F,\alpha}^{M,N}(p:q)$.

Definition 5 ((M, N)-Bregman divergence). *For regular weighted means* M *and* N, *the* (M,N)*-Bregman divergence is defined for a strictly* (M,N)*-convex function* $F: I \to \mathbb{R}$ *by*

$$B_F^{M,N}(p:q) := \lim_{\alpha\to 1^-} J'^{M,N}_{F,\alpha}(p:q). \tag{9}$$

It follows from the symmetry $J'_{F,\alpha}(p:q) = J'_{F,1-\alpha}(q:p)$ that we get the *reverse Bregman divergence* as: $B_F^{M,N}(q:p) = \lim_{\alpha\to 0^+} J'^{M,N}_{F,\alpha}(p:q)$.

Note that a generalization of Bregman divergences has also been studied by Petz [18] to get generalized quantum relative entropies when considering the arithmetic weighted means: Petz defined the Bregman divergence between two points p and q of a convex set C sitting in a Banach space for a given function $F: C \to \mathcal{B}(\mathcal{H})$ (Banach space induced by a Hilbert space $\mathcal{H}$) as: $B_F(p:q):=F(p) - F(q) - \lim_{\alpha\to 0^+}\frac{1}{\alpha}(F(q+\alpha(p-q)) - F(q))$. This last equation can be rewritten in our framework as $B_F(p:q) = \lim_{\alpha\to 1^-}\frac{1}{1-\alpha}J_{F,\alpha}^{A,A}(p,q)$.

In general, we need to prove that (i) when the limits exists, (ii) the (M,N)-Bregman divergences are proper: $B_F^{M,N}(q:p) \geq 0$ with equality iff $p=q$.

3 Quasi-arithmetic Bregman Divergences

Let us report direct formulas for the generalized Bregman divergences defined with respect to comparative convexity when using regular quasi-arithmetic means. Let ρ and τ be two continuous differentiable functions defining the quasi-arithmetic means M_ρ and M_τ, respectively.

3.1 A Direct Formula

By definition, a function $F \in \mathcal{C}_{\rho,\tau}$ is (ρ,τ)-convex iff $M_\tau(F(p),F(q))) \geq F(M_\rho(p,q))$. This (ρ,τ)-midpoint convexity property with the continuity of F yields the more general definition of (ρ,τ)-convexity $M_{\tau,\alpha}(F(p),F(q))) \geq F(M_{\rho,\alpha}(p,q)),\quad \alpha \in [0,1]$. Let us study the generalized Bregman Divergences $B_F^{\rho,\tau}$ obtained when taking the limit:

$$B_F^{\rho,\tau}(q:p) := \lim_{\alpha \to 0} \frac{M_{\tau,\alpha}(F(p),F(q))) - F\left(M_{\rho,\alpha}(p,q)\right)}{\alpha(1-\alpha)}. \tag{10}$$

for $M_{\rho,\alpha}$ and $M_{\tau,\alpha}$ two quasi-arithmetic weighted means obtained for continuous and monotonic functions ρ and τ, respectively.

We state the generalized Bregman divergence formula obtained with respect to quasi-arithmetic comparative convexity:

Theorem 1 (Quasi-arithmetic Bregman divergences). *Let $F : I \subset \mathbb{R} \to \mathbb{R}$ be a real-valued strictly (ρ,τ)-convex function defined on an interval I for two strictly monotone and differentiable functions ρ and τ. The Quasi-Arithmetic Bregman divergence (QABD) induced by the comparative convexity is:*

$$\begin{aligned} B_F^{\rho,\tau}(p:q) &= \frac{\tau(F(p)) - \tau(F(q))}{\tau'(F(q))} - \frac{\rho(p)-\rho(q)}{\rho'(q)} F'(q), \\ &= \kappa_\tau(F(q):F(p)) - \kappa_\rho(q:p)F'(q), \end{aligned} \tag{11}$$

where

$$\kappa_\gamma(x:y) = \frac{\gamma(y)-\gamma(x)}{\gamma'(x)}. \tag{12}$$

Proof. By taking the first-order Taylor expansion of $\tau^{-1}(x)$ at x_0, we get $\tau^{-1}(x) \simeq_{x_0} \tau^{-1}(x_0) + (x-x_0)(\tau^{-1})'(x_0)$. Using the property of the derivative of an inverse function, $(\tau^{-1})'(x) = \frac{1}{(\tau'(\tau^{-1})(x))}$, it follows that the first-order Taylor expansion of $\tau^{-1}(x)$ is $\tau^{-1}(x) \simeq \tau^{-1}(x_0) + (x-x_0)\frac{1}{(\tau'(\tau^{-1})(x_0))}$. Plugging $x_0 = \tau(p)$ and $x = \tau(p) + \alpha(\tau(q) - \tau(p))$, we get a first-order approximation of the weighted quasi-arithmetic mean M_τ when $\alpha \to 0$:

$$M_\alpha(p,q) \simeq p + \frac{\alpha(\tau(q)-\tau(p))}{\tau'(p)}. \tag{13}$$

For example, when $\tau(x) = x$ (*i.e.*, arithmetic mean), we have $A_\alpha(p,q) \simeq p + \alpha(q-p)$, when $\tau(x) = \log x$ (*i.e.*, geometric mean), we obtain $G_\alpha(p,q) \simeq p + \alpha p \log \frac{q}{p}$, and when $\tau(x) = \frac{1}{x}$ (*i.e.*, harmonic mean) we get $H_\alpha(p,q) \simeq p + \alpha(p - \frac{p^2}{q})$. For the regular power means, we have $P_\alpha(p,q) \simeq p + \alpha\frac{q^\delta - p^\delta}{\delta p^{\delta-1}}$. These are first-order weighted mean approximations obtained for small values of α.

Now, consider the comparative convexity skewed Jensen Divergence defined by $J_{F,\alpha}^{\tau,\rho}(p:q) = (M_{\tau,\alpha}(F(p),F(q)) - F(M_{\rho,\alpha}(p,q)))$, and apply a first-order

Taylor expansion to get $F(M_{\tau,\alpha}(p,q))) \simeq F\left(p + \frac{\alpha(\tau(q)-\tau(p))}{\tau'(p)}\right) \simeq F(p) + \frac{\alpha(\tau(q)-\tau(p))}{\tau'(p)}F'(p)$. Thus it follows that the Bregman divergence for quasi-arithmetic comparative convexity is $B_F^{\rho,\tau}(q:p) = \lim_{\alpha\to 0} J'_{\tau,\rho,\alpha}(p:q) = \frac{\tau(F(q))-\tau(F(p))}{\tau'(F(p))} - \frac{\rho(q)-\rho(p)}{\rho'(p)}F'(p)$, and the reverse Bregman divergence $B_F^{\rho,\tau}(p:q) = \lim_{\alpha\to 1}\frac{1}{\alpha(1-\alpha)}J_\alpha^{\tau,\rho}(p:q) = \lim_{\alpha\to 0}\frac{1}{\alpha(1-\alpha)}J_\alpha^{\tau,\rho}(q:p)$.

Since power means are regular quasi-arithmetic means, we get the following family of *power mean Bregman divergences*:

Corollary 1 (Power Mean Bregman Divergences). *For $\delta_1, \delta_2 \in \mathbb{R}\backslash\{0\}$ with $F \in \mathcal{C}_{P_{\delta_1},P_{\delta_2}}$, we have the family of Power Mean Bregman Divergences (PMBDs):*

$$B_F^{\delta_1,\delta_2}(p:q) = \frac{F^{\delta_2}(p) - F^{\delta_2}(q)}{\delta_2 F^{\delta_2-1}(q)} - \frac{p^{\delta_1} - q^{\delta_1}}{\delta_1 q^{\delta_1-1}}F'(q) \tag{14}$$

A sanity check for $\delta_1 = \delta_2 = 1$ let us recover the ordinary Bregman divergence.

3.2 Quasi-arithmetic Bregman Divergences Are Proper

Appendix A proves that a function $F \in \mathcal{C}_{\rho,\tau}$ iff $G = F_{\rho,\tau} = \tau \circ F \circ \rho^{-1} \in \mathcal{C}$. We still need to prove that QABDs are proper: $B_F^{\rho,\tau}(p:q) \geq 0$ with equality iff $p = q$. Defining the ordinary Bregman divergence on the convex generator $G(x) = \tau(F(\rho^{-1}(x)))$ for a (ρ,τ)-convex function with $G'(x) = \tau(F(\rho^{-1}(x)))' = \frac{1}{(\rho'(\rho^{-1})(x))}F'(\rho^{-1}(x))\tau'(F(\rho^{-1}(x)))$, we get an ordinary Bregman divergence that is, in general, *different* from the generalized quasi-arithmetic Bregman divergence $B_F^{\rho,\tau}$: $B_G(p:q) \neq B_F^{\rho,\tau}(p:q)$ with:

$$B_G(p:q) = \tau(F(\rho^{-1}(p))) - \tau(F(\rho^{-1}(q))) - (p-q)\frac{F'(\rho^{-1}(q))\tau'(F(\rho^{-1}(q)))}{(\rho'(\rho^{-1})(q))}$$

A sanity check shows that $B_G(p:q) = B_F^{\rho,\tau}(p:q)$ when $\rho(x) = \tau(x) = x$ (since we have the derivatives $\rho'(x) = \tau'(x) = 1$).

Let us notice the following remarkable identity:

$$B_F^{\rho,\tau}(p:q) = \frac{1}{\tau'(F(q))}B_G(\rho(p):\rho(q)). \tag{15}$$

This identity allows us to prove that QABDs are proper divergences.

Theorem 2 (QABDs are proper). *The quasi-arithmetic Bregman divergences are proper divergences.*

Proof.. B_G is a proper ordinary BD, $\tau' > 0$ a positive function since τ is a strictly increasing function, and $\rho(p) = \rho(q)$ iff $p = q$ since ρ is strictly monotonous. It follows that $\frac{1}{\tau'(F(q))}B_G(\rho(p):\rho(q)) \geq 0$ with equality iff $p = q$.

3.3 Conformal Bregman Divergences on Monotone Embeddings

A closer look at Eq. 15 allows one to interpret the QABDs $B_F^{\rho,\tau}(p:q)$ as conformal divergences. A conformal divergence [2,16,17] $D_\kappa(p:q)$ of a divergence $D(p:q)$ is defined by a positive conformal factor function κ as follows: $D_\kappa(p:q) = \kappa(q)D(p:q)$. An example of Bregman conformal divergence is the total Bregman divergence [19] with $\kappa(q) = \frac{1}{\sqrt{1+\|\nabla F(q)\|^2}}$.

Property 1 (QABDs as conformal BDs). The quasi-arithmetic Bregman divergence $B_F^{\rho,\tau}(p:q)$ amounts to compute an ordinary Bregman conformal divergence in the ρ-embedded space:

$$B_F^{\rho,\tau}(p:q) = \kappa(\rho(q))B_G(\rho(p):\rho(q)), \tag{16}$$

with conformal factor $\kappa(x) = \frac{1}{\tau'(F(\rho^{-1}(x)))} > 0$.

4 Concluding Remarks

We have introduced generalized (M,N)-Bregman divergences as limit of scaled skew (M,N)-Jensen divergences for regular M and N means. Regular means include power means, quasi-arithmetic means, Stolarsky means, etc. But not all means are regular: For example, the Lehmer mean $L_2(x,y) = \frac{x^2+y^2}{x+y}$ is not increasing and therefore not regular. We reported closed-form expression for quasi-arithmetic (ρ,τ)-Bregman divergences, prove that those divergences are proper, and show that they can be interpreted as conformal ordinary Bregman divergences on a monotone embedding [21]. This latter observation further let us extend usual Bregman divergence results to quasi-arithmetic Bregman divergences (*eg.*, conformal Bregman k-means [19], conformal Bregman Voronoi diagrams [5]).

A Quasi-arithmetic to Ordinary Convexity Criterion

Lemma 1 ((ρ,τ)-convexity $\leftrightarrow$ ordinary convexity [1]**).** *Let $\rho: I \to \mathbb{R}$ and $\tau: J \to \mathbb{R}$ be two continuous and strictly monotone real-valued functions with τ increasing, then function $F: I \to J$ is (ρ,τ)-convex iff function $G = F_{\rho,\tau} = \tau \circ F \circ \rho^{-1}$ is (ordinary) convex on $\rho(I)$.*

Proof. Let us rewrite the (ρ,τ)-convexity midpoint inequality as follows:

$$F(M_\rho(x,y)) \leq M_\tau(F(x),F(y)),$$

$$F\left(\rho^{-1}\left(\frac{\rho(x)+\rho(y)}{2}\right)\right) \leq \tau^{-1}\left(\frac{\tau(F(x))+\tau(F(y))}{2}\right),$$

Since τ is strictly increasing, we have:

$$(\tau \circ F \circ \rho^{-1})\left(\frac{\rho(x)+\rho(y)}{2}\right) \leq \frac{(\tau \circ F)(x)+(\tau \circ F)(y)}{2}. \tag{17}$$

Let $u = \rho(x)$ and $v = \rho(y)$ so that $x = \rho^{-1}(u)$ and $y = \rho^{-1}(v)$ (with $u, v \in \rho(I)$). Then it comes that:

$$(\tau \circ F \circ \rho^{-1})\left(\frac{u+v}{2}\right) \leq \frac{(\tau \circ F \circ \rho^{-1})(u) + (\tau \circ F \circ \rho^{-1})(v)}{2}. \quad (18)$$

This last inequality is precisely the ordinary midpoint convexity inequality for function $G = F_{\rho,\tau} = \tau \circ F \circ \rho^{-1}$. Thus a function F is (ρ, τ)-convex iff $G = \tau \circ F \circ \rho^{-1}$ is ordinary convex, and vice-versa.

References

1. Aczél, J.: A generalization of the notion of convex functions. Det Kongelige Norske Videnskabers Selskabs Forhandlinger, Trondheim 19(24), 87–90 (1947)
2. Amari, S.: Information Geometry and Its Applications. Applied Mathematical Sciences. Springer, Tokyo (2016)
3. Banerjee, A., Merugu, S., Dhillon, I.S., Ghosh, J.: Clustering with bregman divergences. J. Mach. Learn. Res. **6**, 1705–1749 (2005)
4. Berrone, L.R., Moro, J.: Lagrangian means. Aequationes Math. **55**(3), 217–226 (1998)
5. Boissonnat, J.-D., Nielsen, F., Nock, R.: Bregman Voronoi diagrams. Discrete Comput. Geom. **44**(2), 281–307 (2010)
6. Bregman, L.M.: The relaxation method of finding the common point of convex sets and its application to the solution of problems in convex programming. USSR Comput. Math. Math. Phys. **7**(3), 200–217 (1967)
7. Bullen, P.S., Mitrinovic, D.S., Vasic, M.: Means and Their Inequalities, vol. 31. Springer, New York (2013)
8. Burbea, J., Rao, C.: On the convexity of some divergence measures based on entropy functions. IEEE Trans. Inf. Theory **28**(3), 489–495 (1982)
9. De Finetti, B.: Sul concetto di media 3, 369–396 (1931). Istituto italiano degli attuari
10. Holder, O.L.: Über einen Mittelwertssatz. Nachr. Akad. Wiss. Gottingen Math.-Phys. Kl, pp. 38–47 (1889)
11. Kolmogorov, A.N.: Sur la notion de moyenne. Acad. Naz. Lincei Mem. Cl. Sci. His. Mat. Natur. Sez. 12, 388–391 (1930)
12. Matkowski, J.: On weighted extensions of Cauchy's means. J. Math. Anal. Appl. **319**(1), 215–227 (2006)
13. Nagumo, M.: Über eine klasse der mittelwerte. Jpn. J. Math. Trans. Abstracts **7**, 71–79 (1930). The Mathematical Society of Japan
14. Niculescu, C.P., Persson, L.-E.: Convex Functions and Their Applications: A Contemporary Approach. Springer, New York (2006)
15. Nielsen, F., Boltz, S.: The Burbea-Rao and Bhattacharyya centroids. IEEE Trans. Inf. Theory **57**(8), 5455–5466 (2011)
16. Nielsen, F., Nock, R.: Total jensen divergences: definition, properties and clustering. In: IEEE International Conference on Acoustics, Speech and Signal Processing (ICASSP), pp. 2016–2020. IEEE (2015)
17. Nock, R., Nielsen, F., Amari, S.: On conformal divergences and their population minimizers. IEEE Trans. Inf. Theory **62**(1), 527–538 (2016)

18. Petz, D.: Bregman divergence as relative operator entropy. Acta Mathematica Hung. **116**(1–2), 127–131 (2007)
19. Vemuri, B.C., Liu, M., Amari, S., Nielsen, F.: Total bregman divergence and its applications to DTI analysis. IEEE Trans. Med. Imaging **30**(2), 475–483 (2011)
20. Zhang, J.: Divergence function, duality, and convex analysis. Neural Comput. **16**(1), 159–195 (2004)
21. Zhang, J.: Nonparametric information geometry: From divergence function to referential-representational biduality on statistical manifolds. Entropy **15**(12), 5384–5418 (2013)

Weighted Closed Form Expressions Based on Escort Distributions for Rényi Entropy Rates of Markov Chains

Philippe Regnault[1(✉)], Valérie Girardin[2], and Loïck Lhote[3]

[1] Laboratoire de Mathématiques de Reims, EA 4535, Université de Reims Champagne-Ardenne, BP 1039, 51687 Reims Cedex 2, France
philippe.regnault@univ-reims.fr

[2] Laboratoire de Mathématiques Nicolas Oresme UMR CNRS 6139, Université de Caen Normandie,BP 5186, 14032 Caen, France
valerie.girardin@unicaen.fr

[3] ENSICAEN, GREYC, UMR CNRS 6072, Université de Caen Normandie, BP 5186, 14032 Caen, France
loick.lhote@unicaen.fr

Abstract. For Markov chains with transition probabilities p_{ij}, the Shannon entropy rate is well-known to be equal to the sum of the $-\sum_j p_{ij} \log p_{ij}$ weighted by the stationary distribution. This expression derives from the chain rule specific to Shannon entropy. For an ergodic Markov chain, the stationary distribution is the limit of the marginal distributions of the chain.

Here a weighted expression for the Rényi entropy rate is derived from a chain rule, that involves escort distributions of the chain, specific to Rényi entropy. Precisely, the rate is equal to the logarithm of a weighted sum of the $-\sum_j p_{ij}^s$, where s is the parameter of Rényi entropy. The weighting distribution is the normalized left eigenvector of the maximum eigenvalue of the matrix (p_{ij}^s). This distribution, that plays the role of the stationary distribution for Shannon entropy, is shown to be the limit of marginals of escort distributions of the chain.

1 Introduction

Originally, the concept of entropy $\mathbb{S}(m)$ of a probability measure m on a finite or denumerable set has been introduced by Shannon, by setting

$$\mathbb{S}(m) = -\sum_{i\in E} m_i \log m_i.$$

It constitutes a measure of information – or uncertainty – of the distribution m in the sense that it is minimal when m is a Dirac measure and, for a finite set E, maximal if m is uniform. Apart from information theory to which it gave birth, it is an essential tool in all scientific fields involving probability measures,

F. Nielsen and F. Barbaresco (Eds.): GSI 2017, LNCS 10589, pp. 648–656, 2017.
https://doi.org/10.1007/978-3-319-68445-1_75

such as signal processing, computer science, probability and statistics. A large toolbox has been developed; see e.g., [3].

Due to the diversity of applications and constraints specific to each domain, alternative measures of information have been developed since. Rényi entropy functionals, depending on a parameter $s > 0$, are an important instance of such measures. They are defined by $\mathbb{R}_1(m) = \mathbb{S}(m)$ and

$$\mathbb{R}_s(m) = \frac{1}{1-s} \log \left(\sum_{i \in E} m_i^s \right), \quad s \neq 1.$$

A related toolbox has been developed too; see [13]. Further, the Rényi entropy of a random variable is the entropy of its distribution. Thus, $\mathbb{R}_s(\mathbf{X}_0^n) = \mathbb{R}_s(\mathbf{m}_0^n)$ for any random vector $\mathbf{X}_0^n = (X_0, \dots, X_{n-1})$ with distribution $\mathbf{m}_0^n$. For a random sequence $\mathbf{X} = (X_n)_{n \geq 0}$, the usual notion of averaging per time unit leads – when the limit exists – to the Rényi entropy rate

$$\mathbb{H}_s(\mathbf{X}) = \lim_{n \to \infty} \frac{1}{n} \mathbb{R}_s(\mathbf{X}_0^n),$$

The entropy rate $\mathbb{H} = \mathbb{H}_1$ associated to Shannon entropy was originally defined in [15] for an ergodic – homogeneous aperiodic irreducible – Markov chain with a finite state space E as the sum of the entropies of the transition distributions $p_i = (p_{ij})_{j \in E}$ weighted by the probability of occurrence of each state $i \in E$ according to the stationary distribution π of the chain, namely

$$\mathbb{H}(\mathbf{X}) = -\sum_{i \in E} \pi_i \sum_{j \in E} p_{ij} \log p_{ij}, \tag{1}$$

where $p_{ij} = \mathbb{P}(X_n = j | X_{n-1} = i)$. This expression, which has a direct interpretation in terms of dynamics of the chain, has allowed for a complete toolbox to be developed around Shannon entropy rate, including probability and statistical results. The convergence of $\frac{1}{n}\mathbb{S}(\mathbf{X}_0^n)$ to (1) is proven in [15] for ergodic finite chains. The proof, based on the chain rule specific to Shannon entropy, has since been extended to the denumerable case, and then under hypotheses weakened in many directions; see [6].

For ergodic Markov chains, the Shannon entropy rate is also known to be given by

$$\mathbb{H}(\mathbf{X}) = \lambda'(1), \tag{2}$$

where $\lambda(s)$ denotes the dominant eigenvalue of the perturbated matrix

$$P(s) = (p_{ij}^s)_{(i,j) \in E^2}, \quad s > 0. \tag{3}$$

The Rényi entropy rate for $s > 0$ of finite state space ergodic Markov chains is shown in [12] to be well-defined and given by

$$\mathbb{H}_s(\mathbf{X}) = (1-s)^{-1} \log \lambda(s), \quad s \neq 1. \tag{4}$$

An attempt to extend this expression to denumerable Markov chains has been proposed in [10], based on positive operator theory. Unfortunately, some necessary conditions on the involved operator are not checked by the authors, making their proof inaccurate. We refer to [5] for a correct statement and proof of (4) for denumerable Markov chains. Still, no closed form weighted expression similar to (1) exists when $s \neq 1$. The aim of the present paper is to fill this gap. Precisely, we will show that, under technical assumptions to be specified,

$$\mathbb{H}_s(\mathbf{X}) = \frac{1}{1-s} \log \left[\sum_{i \in E} \mathsf{u}_i(s) \sum_{j \in E} p_{ij}^s \right], \tag{5}$$

where $\mathsf{u}(s)$ denotes the normalized left eigenvector of $P(s)$ associated to the dominant eigenvalue $\lambda(s)$. The proof will be based on a functional identity satisfied by Rényi entropy. This identity, strongly related to the escort distributions of the marginals of the chain, is seemingly closed to the chain rule satisfied by Shannon entropy.

Escort distributions have initially been introduced in [2]. The escort distributions of a probability distribution $m = (m_i)_{i \in E}$ on a finite or denumerable set E are $m^{*s} = (m_i^{*s})_{i \in E}$, for $s \in \mathbb{R}$, with

$$m_i^{*s} = \frac{m_i^s}{\sum_{j \in E} m_j^s}, \quad i \in E,$$

whenever $\sum_{j \in E} m_j^s$ is finite. They provide a tool for zooming at different parts of m, or for adapting m to constraints through their ability to scan its structure. They also constitute the geodesics of information geometry with respect to an affine connection naturally induced by the Kullback-Leibler divergence; see [1,4, 16], and also [8] for an application to hypothesis testing. The escort distributions of a random variable are the escort distributions of its distribution, in short escorts. The chain rule for Rényi entropy is based on escorts. Further, we will show that the eigenvector $\mathsf{u}(s)$ is closely linked to the asymptotic behavior of the escorts of the marginals of the Markov chain.

The paper is organized as follows. Since they constitute the keystone of the present paper, the chain rules for Shannon and Rényi entropies are recalled in Sect. 2. In Sect. 3, the asymptotic behavior of the sequence $(P(s)^n)_n$ together with the existence and uniqueness of $\mathsf{u}(s)$, are considered. In Sect. 4, links between $\mathsf{u}(s)$ and the asymptotic behavior of escorts of marginal distributions of the Markov chain are highlighted. Finally, the weighted expression for Rényi entropy (5) is derived in Sect. 5.

2 Chain Rules for Shannon and Rényi entropies

The convergence of the normalized marginal entropies $\mathbb{S}(X_0^n)/n$ to (1) has been derived in [15, Theorem 5], from the so-called chain rule satisfied by Shannon entropy

$$\mathbb{S}(X, Y) = \mathbb{S}(X) + \mathbb{S}(Y|X), \tag{6}$$

for any pair of random variables X, Y taking values in finite or denumerable sets E and F. Here $\mathbb{S}(Y|X)$ denotes the so-called conditional entropy of Y given X, i.e., the entropy of the distribution of Y conditional to X weighted by the distribution m_X of X,

$$\mathbb{S}(Y|X) = -\sum_{i\in E} m_X(i) \sum_{j\in F} \mathbb{P}(Y=j|X=i)\log \mathbb{P}(Y=j|X=i).$$

Weighting by escort distributions of X instead of the distribution of X itself yields a sort of chain rule for Rényi entropy; see the axiomatics in [11]. Precisely, supposing that $\sum_{j\in E}\mathbb{P}(X=j)^s$ is finite,

$$\mathbb{R}_s(X,Y) = \mathbb{R}_s(X) + \mathbb{R}_s(Y|X), \tag{7}$$

with

$$\mathbb{R}_s(Y|X) = \frac{1}{1-s}\log\left[\sum_{i\in E} m_X^{*s}(i)\sum_{j\in F}\mathbb{P}(Y=j|X=i)^s\right].$$

Therefore, adapting the original proof of the weighted expression of Shannon entropy rate in order to obtain a similar expression for Rényi entropy rate requires to study the escorts of the marginals of the Markov chains. The main tool in this aim is the perturbation $P(s)$ given by (3) of the transition matrix $P = (p_{ij})_{(i,j)\in E^2}$.

3 The Eigenvectors of the Perturbated Transition Matrix

For finite ergodic Markov chains, the matrix $P(s)$ is positive, irreducible and aperiodic so that the Perron-Frobenius theorem applies. A unique positive dominant – with maximal modulus – eigenvalue $\lambda(s)$ exists, with a spectral gap, namely

$$|\lambda(s)| > \sup\{|\lambda| : \lambda \text{ eigenvalue of } P(s),\ \lambda \neq \lambda(s)\}.$$

Left and right positive eigenvectors associated to $\lambda(s)$ for $P(s)$ exist, say $\mathsf{u}(s)$ and $\mathsf{v}(s)$ such that

$${}^t\mathsf{u}(s).P(s) = \lambda(s){}^t\mathsf{u}(s) \quad \text{and} \quad P(s).\mathsf{v}(s) = \lambda(s)\mathsf{v}(s). \tag{8}$$

Moreover they can be normalized so that $\sum_{i\in E}\mathsf{u}_i(s) = 1$ and $\sum_{i\in E}\mathsf{u}_i(s)\mathsf{v}_i(s) = 1$.

For denumerable ergodic Markov chains, a positive eigenvalue with maximal modulus $\lambda(s)$ again generally exists for $P(s)$. Still, a sequence of eigenvalues converging to $\lambda(s)$ may exist, hindering the existence of a spectral gap. Additional assumptions are thus required. Sufficient conditions for the operator $u \mapsto {}^tu.P(s)$ to be compact on $\ell_1 = \{u = (u_i)_{i\in E} : \sum_{i\in E}|u_i| < \infty\}$, and hence for the existence of a spectral gap, are proposed in the following proposition whose proof can be found in [5]. Moreover, the asymptotic behavior of $P(s)^n$ is also derived.

Proposition 1. *Let* $\mathbf{X} = (X_n)_{n\in\mathbb{N}}$ *be an ergodic Markov chain with finite or denumerable state space* E*, transition matrix* $P = (p_{ij})_{(i,j)\in E^2}$ *and initial distribution* $\mu = (\mu_i)_{i\in E}$ *satisfying the following assumptions:*

C1. $\sup_{(i,j)\in E^2} p_{ij} < 1$;
C2. $\exists \sigma_0 < 1$ *such that* $\forall s > \sigma_0$*, both* $\sup_{i\in E}\sum_{j\in E} p_{ij}^s$ *and* $\sum_{i\in E}\mu_i^s$ *are finite;*
C3. $\forall \varepsilon > 0$ *and* $\forall s > \sigma_0$*,* $\exists A \subset E$ *finite such that* $\sup_{i\in E}\sum_{j\in E\setminus A} p_{ij}^s < \varepsilon$.

Then, for $s > \sigma_0$*, the perturbed matrix* $P(s)$ *defined by (3) has a unique positive eigenvalue* $\lambda(s)$ *with maximum modulus and a spectral gap. Positive left and right eigenvectors* $\mathsf{u}(s)$ *and* $\mathsf{v}(s)$ *exist for* $\lambda(s)$*, and can be chosen such that* $\sum_{i\in E}\mathsf{u}_i(s) = 1$ *and* $\sum_{i\in E}\mathsf{u}_i(s)\mathsf{v}_i(s) = 1$.
Moreover, for all $s > \sigma_0$*, a positive real number* $\rho(s) < 1$ *exists such that*

$$P(s)^n = \lambda(s)^n C(s) + O_{\ell_1}((\rho(s)\lambda(s))^n), \quad n \in \mathbb{N}^*, \tag{9}$$

where $C(s) = \mathsf{v}(s).{}^t\mathsf{u}(s)$.

Proposition 1 clearly induces the following property

$$\sum_{i_0^n\in E^n} m_0^n(i_0)^s = {}^t\mu^s.P(s)^{n-1}.\mathbf{1} = \lambda(s)^{n-1}c(s) + O((\rho(s)\lambda(s))^{n-1}), \quad n \in \mathbb{N}^*, \tag{10}$$

where $\mu^s = (\mu_i^s)_{i\in E}$ and $c(s) = {}^t\mu^s.C(s).\mathbf{1} = {}^t\mu^s.\mathsf{v}(s)$. This key property is called the quasi-power property in [7]. It is naturally satisfied for any finite ergodic Markov chain, since its transition matrix and initial distribution obviously satisfy conditions C1 to C3 with $\sigma_0 = -\infty$. In the denumerable case, condition C1 ensures mixing between the states of the chain, avoiding some very probable paths to asymptotically capture the chain. Condition C2 insures that ${}^tu.P(s) \in \ell_1$ for any $u \in \ell_1$. Condition C3 forbids the probability mass to escape to infinity.

4 Escort Distributions and Left Eigenvectors

Up to our knowledge, escort distributions of random vectors or sequences have never been studied in the literature. Clearly, we do not aim here at studying them in a comprehensive way. We will establish a few properties linked to ergodic Markov chains, interesting in their own and that will finally yield the weighted closed form expression of Rényi entropy rates.

First, the escort distribution of a random vector with independent components is equal to the product of the escort distributions of its components. In mathematical words, for n independent random variables $\mathbf{X}_0^n = (X_0, \dots, X_{n-1})$,

$$\mathbf{m}^{*s}_{\mathbf{X}_0^n} = \bigotimes_{k=0}^{n-1} m^{*s}_{X_k}, \tag{11}$$

for all $s > 0$ such that the escort distributions exist. If, in addition, X_0^n are independent and identically distributed (i.i.d.) with common distribution m, then

$\mathbf{m}^{*s}_{\mathbf{X}_0^n} = (m^{*s})^{\otimes n}$. Such a random vector can be viewed as the n first components of an i.i.d. sequence and hence, as a Markov chain $\mathbf{X}$ whose transition matrix P has all rows equal to m. Then, provided that $\sum_{i\in E} m_i^s$ is finite, its perturbation $P(s)$ has only two eigenvalues, first 0, and second $\lambda(s) = \sum_{i\in E} m_i^s$ with multiplicity 1. Easy computation yield ${}^t m^{*s}.P(s) = \lambda(s){}^t m^{*s}$, meaning that the escort distribution m^{*s} of m is equal to the normalized left eigenvector $\mathsf{u}(s)$ of $P(s)$.

For dependent random sequences, in particular Markov chains, (11) is not true anymore. Still, the escorts of the marginals of the chain converge to $\mathsf{u}(s)$.

Proposition 2. *Let the assumptions of Proposition 1 be satisfied. Let* $\mathsf{u}(s)$ *be defined by* (8)*. For any* $n \in \mathbb{N}^*$*, let* $\mathbf{m}_{\mathbf{X}_0^n}$ *denote the distribution of* $\mathbf{X}_0^n$ *and* $\mathbf{m}^{*s}_{\mathbf{X}_0^n}$ *its escorts.*

Then, the n*-th marginal of* $\mathbf{m}^{*s}_{\mathbf{X}_0^n}$ *converges to* $\mathsf{u}(s)$ *as* n *goes to infinity.*

Proof. Thanks to C2, $\sum_{i\in E} \mathbf{m}^{*s}_{\mathbf{X}_0^n} = {}^t\mu^s.P(s)^{n-1}.\mathbf{1}$ is finite and the escort $\mathbf{m}^{*s}_{\mathbf{X}_0^n}$ is well-defined for $s > \sigma_0$. Let $\text{margin}_n(\mathbf{m}^{*s}_{\mathbf{X}_0^n})$ denote its n-th marginal distribution. We have

$$ {}^t\text{margin}_n(\mathbf{m}^{*s}_{\mathbf{X}_0^n}) = \frac{{}^t\mu^s.P(s)^{n-1}}{{}^t\mu^s.P(s)^{n-1}.\mathbf{1}}. $$

Thanks to (9),

$$ {}^t\mu^s.P(s)^{n-1} = \lambda(s)^{n-1}c(s){}^t\mathsf{u}(s) + O_{\ell_1}((\rho(s)\lambda(s))^{n-1}), $$

where $c(s) = {}^t\mu^s.\mathsf{v}(s)$. Together with (10), this implies that

$$ {}^t\text{margin}_n(\mathbf{m}^{*s}_{\mathbf{X}_0^n}) = \frac{\lambda(s)^{n-1}c(s){}^t\mathsf{u}(s) + O_{\ell_1}((\rho(s)\lambda(s))^{n-1})}{\lambda(s)^{n-1}c(s) + O((\rho(s)\lambda(s))^{n-1})} = {}^t\mathsf{u}(s) + o_{\ell_1}(1), $$

and the result follows. □

Note that no clear relationship exists between the limit $\mathsf{u}(s)$ of the n-th marginal of the escort distribution and the escort distribution of the asymptotic distribution of the chain. First of all, they are not equal, and hence, limit and escort are not commuting operators in general.

5 Weighted Expression for Rényi Entropy Rates

We can now state and prove the main result of the paper.

Theorem 1. *Under the assumptions of Proposition 1, the Rényi entropy rate of the ergodic Markov chain* $\mathbf{X}$ *is*

$$ \mathbb{H}_s(\mathbf{X}) = \frac{1}{1-s}\log\left[\sum_{i\in E}\mathsf{u}_i(s)\sum_{j\in E}p_{ij}^s\right], \quad s > \sigma_0, s \neq 1, \tag{12} $$

$$ \mathbb{H}_1(\mathbf{X}) = -\sum_{i\in E}\pi_i\sum_{j\in E}p_{ij}\log p_{ij}. \tag{13} $$

Proof. Let $s > \sigma_0$, with $s \neq 1$. Applying (7) recursively yields

$$\mathbb{R}_s(\mathbf{X}_0^n) = \mathbb{R}_s(X_{n-1}|\mathbf{X}_0^{n-1}) + \mathbb{R}_s(\mathbf{X}_0^{n-1}) = \mathbb{R}_s(X_0) + \sum_{k=1}^{n-1} \mathbb{R}_s(X_k|\mathbf{X}_0^k). \quad (14)$$

Iterated use of the Markov property yields

$$\begin{aligned}\mathbb{R}_s(X_k|\mathbf{X}_0^k) &= \frac{1}{1-s} \log \left[\sum_{i_0^k \in E^k} \mathbf{m}^{*s}_{\mathbf{X}_0^k}(i_0^k) \sum_{i_k \in E} p^s_{i_{k-1} i_k} \right] \\ &= \frac{1}{1-s} \log \left(\sum_{i \in E} \left[\sum_{i_0^{k-1} \in E^{k-1}} \mathbf{m}^{*s}_{\mathbf{X}_0^k}(i_0^{k-1}, i) \right] \sum_{j \in E} p^s_{ij} \right).\end{aligned}$$

Since the sum $\sum_{i_0^{k-1}} \mathbf{m}^{*s}_{X_0^k}(i_0^{k-1}, i)$ is the k–th marginal distribution of $\mathbf{m}^{*s}_{\mathbf{X}_0^k}$, Proposition 2 applies to show that it converges to $\mathsf{u}_i(s)$. Finally,

$$\mathbb{R}_s(X_k|\mathbf{X}_0^k) \xrightarrow[k \to \infty]{} \frac{1}{1-s} \log \left[\sum_{i \in E} \mathsf{u}_i(s) \sum_{j \in E} p^s_{ij} \right],$$

so that (14) yields

$$\frac{1}{n} \mathbb{R}_s(\mathbf{X}_0^n) \xrightarrow[n \to \infty]{} \frac{1}{1-s} \log \left[\sum_{i \in E} \mathsf{u}_i(s) \sum_{j \in E} p^s_{ij} \right].$$

For $s = 1$, $\mathbb{R}_1(\mathbf{X}_0^n) = \mathbb{S}(\mathbf{X}_0^n)$ and hence (13) is classical. □

An obvious analogy exists between (12) and (13). The left eigenvector $\mathsf{u}(s)$ plays in (12) the role of the stationary distribution π of the chain in (13); if $s = 1$, then $\mathsf{u}(1) = \pi$ is the stationary distribution of $\mathbf{X}$. The stationary distribution π is the asymptotic limit of the marginal distribution of the chain, while Proposition 2 provides an analogous asymptotic interpretation of $\mathsf{u}(s)$.

Further, thanks to normalization, $\sum_{i \in E} \mathsf{u}'_i(s) = 0$, so that (12) tends to (13) as s tends to 1, proving that $\mathbb{H}_s$ is a continuous function of $s > 0$.

Note that if $\mathbf{X}$ is an i.i.d. sequence with common distribution m, then all n-th marginals of $\mathbf{m}^{*s}_{\mathbf{X}_0^n}$ are $m^{*s} = \mathsf{u}(s)$; see (11). Thus, $\mathbb{R}_s(X_n|\mathbf{X}_0^{n-1})$ does not depend on n and equals

$$\mathbb{R}_s(X_n|\mathbf{X}_0^{n-1}) = \frac{1}{1-s} \log \left(\sum_{i \in E} m_i^{*s} \sum_{j \in E} m_j^s \right) = \mathbb{R}_s(m).$$

This expresses the well-known additivity of Rényi entropy for independent variables. Therefore, the Rényi entropy rate of any i.i.d. sequence reduces to the Rényi entropy of the common distribution, namely $\mathbb{H}_s(\mathbf{X}) = \mathbb{R}_s(m)$.

6 General Entropy Functionals

A huge class of entropy functionals is defined by setting

$$\mathbb{S}_{h,\varphi}(m) = h(\sum_{i \in E} \varphi(m_i)).$$

In all cases of interest, $\varphi : [0,1] \to \mathbb{R}_+$ and $h : \mathbb{R}^+ \to \mathbb{R}$, with either φ concave and h increasing or φ convex and h decreasing. Moreover h and φ behave such that $\mathbb{S}_{h,\varphi}$ takes only non-negative values, and $h(z)$ is finite for all $z \in \mathbb{R}_+$. See [14], and also [5].

Rényi entropy is the unique (h,φ)-entropy satisfying (7) as well as Shannon entropy is the unique one satisfying the chain rule (6); see [11]. Up to our knowledge, no tractable functional identity of this sort exists for the others, making this approach hard to extend.

An alternative consists in studying the asymptotic behavior of the entropy $\mathbb{S}_{h,\varphi}(\mathbf{m}_0^n)$ of the marginals of the chain through analytic combinatoric techniques and then derive the related entropy rate. So doing, [5] proves that the entropy rates $\mathbb{S}_{h,\varphi}(\mathbf{X}) = \lim_{n\to\infty} \frac{1}{n}\mathbb{S}_{h,\varphi}(\mathbf{X}_0^n)$ of ergodic Markov chains $\mathbf{X}$, are degenerated (equal 0 or $\pm\infty$), for all (h,φ)-entropy functionals except Shannon and Rényi. Indeed, the normalizing term n appears to be badly adapted to other (h,φ)-entropy functionals. Through rescaling by a suitable normalizing sequence, [7] defines non-degenerated entropy rates $\mathbb{S}_{h,\varphi,r}(\mathbf{X}) = \lim_{n\to\infty} \frac{1}{r_n}\mathbb{S}_{h,\varphi}(\mathbf{X}_0^n)$. Closed form expressions are obtained for ergodic Markov chains as functions of $\lambda(s)$ or $\lambda'(s)$, thus extending (2) and (4). In the continuation of [7], weighted closed form expressions similar to (12) and (13) are to be derived for all (h,φ)-entropy functionals in [9].

Obtaining weighted closed form expressions for divergences between Markov chains would be a relevant further step, as well as the statement of properties in geometry of information linked to generalized entropy and divergence rates.

References

1. Amari, S., Nagaoka, H.: Methods of Information Geometry. Oxford University Press, Oxford (2000)
2. Beck, C., Schlogl, F.: Thermodynamics of Chaotic Systems. Cambridge University Press, Cambridge, Great Britain (1993)
3. Cover, L., Thomas, J.: Elements of Information Theory. Wiley Series in Telecommunications, New York (1991)
4. Csiszár, I.: I-Divergence geometry of probability distributions and minimization problems. Ann. Probab. **3**, 141–158 (1975)
5. Ciuperca, G., Girardin, V., Lhote, L.: Computation and estimation of generalized entropy rates for denumerable markov chains. IEEE Trans. Inf. Theory **57**, 4026–4034 (2011)
6. Girardin, V.: On the different extensions of the ergodic theorem of information theory. In: Baeza-Yates, R., Glaz, J., Gzyl, H., Hüsler, J., Palacios, J.L. (eds.) Recent Advances in Applied Probability, pp. 163–179. Springer, San Francisco (2005)

7. Girardin, V., Lhote, L.: Rescaling entropy and divergence rates. IEEE Trans. Inf. Theory **61**, 5868–5882 (2015)
8. Girardin, V., Regnault, P.: Escort distributions minimizing the Kullback-Leibler divergence for a large deviations principle and tests of entropy level. Ann. Inst. Stat. Math. **68**, 439–468 (2016)
9. Girardin, V., Lhote, L., Regnault, P.: Different closed-form expressions for generalized entropy rates of Markov chains. Work in Progress
10. Golshani, L., Pasha, E., Yari, G.: Some properties of Rényi entropy and Rényi entropy rate. Inf. Sci. **179**, 2426–2433 (2009)
11. Jizba, P., Arimitsu, T.: Generalized statistics: yet another generalization. Physica A **340**, 110–116 (2004)
12. Rached, Z., Alajaji, F., Campbell, L.L.: Rényi's divergence and entropy rates for finite alphabet Markov sources. IEEE Trans. Inf. Theory **47**, 1553–1561 (2001)
13. Rényi, A.: Probability Theory. North Holland, Amsterdam (1970)
14. Salicrú, M., Menéndez, M.L., Morales, D., Pardo, L.: Asymptotic distribution of (h, φ)-entropies. Commun. Stat. Theory Methods 22, 2015–2031 (1993)
15. Shannon, C.: A mathematical theory of communication. Bell Syst. Tech. J. 27, 379–423, 623–656 (1948)
16. Sgarro, A.: An informational divergence geometry for stochastic matrices. Calcolo **15**, 41–49 (1978)

On the Error Exponent of a Random Tensor with Orthonormal Factor Matrices

Rémy Boyer[1(✉)] and Frank Nielsen[2]

[1] L2S, Department of Signals and Statistics, University of Paris-Sud, Orsay, France
remy.boyer@l2s.centralesupelec.fr
[2] École Polytechnique, LIX, Palaiseau, France
Frank.Nielsen@acm.org

Abstract. In signal processing, the detection error probability of a random quantity is a fundamental and often difficult problem. In this work, we assume that we observe under the alternative hypothesis a noisy tensor admitting a Tucker decomposition parametrized by a set of orthonormal factor matrices and a random core tensor of interest with fixed multilinear ranks. The detection of the random entries of the core tensor is hard to study since an analytic expression of the error probability is not tractable. To cope with this difficulty, the Chernoff Upper Bound (CUB) on the error probability is studied for this tensor-based detection problem. The tightest CUB is obtained for the minimal error exponent value, denoted by $s^\star$, that requires a costly numerical optimization algorithm. An alternative strategy to upper bound the error probability is to consider the Bhattacharyya Upper Bound (BUB) by prescribing $s^\star = 1/2$. In this case, the costly numerical optimization step is avoided but no guarantee exists on the tightness optimality of the BUB. In this work, a simple analytical expression of $s^\star$ is provided with respect to the Signal to Noise Ratio (SNR). Associated to a compact expression of the CUB, an easily tractable expression of the tightest CUB is provided and studied. A main conclusion of this work is that the BUB is the tightest bound at low SNRs but this property is no longer true at higher SNRs.

1 Introduction

The theory of tensor decomposition is an important research topic (see for instance [1,2]). They are useful to extract relevant information confined into a small dimensional subspaces from a massive volume of measurements while reducing the computational cost. In the context where the measurements are naturally modeled according to more than two axes of variations, *i.e.*, in the case of tensors, the problem of obtaining a *low rank approximation* faces a number of practical and fundamental difficulties. Indeed, even if some aspects of the

Frank Nielsen—This research was partially supported by Labex DigiCosme (project ANR-11-LABEX-0045-DIGICOSME) operated by ANR as part of the program "Investissement d'Avenir" Idex Paris-Saclay (ANR-11-IDEX-0003-02).

F. Nielsen and F. Barbaresco (Eds.): GSI 2017, LNCS 10589, pp. 657–664, 2017.
https://doi.org/10.1007/978-3-319-68445-1_76

tensor algebra can be considered as mature, several "obvious" algebraic concepts in the matrix case such as decomposition uniqueness, rank, or the notions of singular and eigen-values remain active and challenging research areas [3]. The Tucker decomposition [4] and the HOSVD (High-Order SVD) [5] are two popular decompositions being an alternative to the Canonical Polyadic decomposition [6]. In this case, the notion of tensorial rank is no longer relevant and an alternative rank definition is used. Specifically, it is standard to use the multilinear ranks defined as the set of strictly positive integers $\{R_1, R_2, R_3\}$ where R_p is the usual rank (in the matrix sense) of the p-th mode or unfolding matrix. Its practical construction is non-iterative and optimal in the sense of the Eckart-Young theorem at each mode level. This approach is interesting because it can be computed in real time [7] or adaptively [8]. Unfortunately, it is shown that the fixed (multilinear) rank tensor based on this procedure is generally suboptimal in the Fröbenius norm sense [5]. In other words, there does not exist a generalization of the Eckart-Young theorem for tensor of order strictly greater than two. Despite of this theoretical singularity, we focus our effort in the detection performance of a given multilinear rank tensor following the Tucker model with orthonormal factor matrices, leading to the HOSVD. It is important to note that the detection theory for tensors is an under-studied research topic. To the best of our knowledge, only the publication [9] tackles this problem in the context of RADAR multidimensional data detection. A major difference with this publication is that their analysis is dedicated to the performance of a low rank detection after matched filtering. More specifically, the goal is to decompose a 3-order tensor $\boldsymbol{\mathcal{X}}$ of size $N_1 \times N_2 \times N_3$ into a core tensor denoted by $\boldsymbol{\mathcal{S}}$ of size $R_1 \times R_2 \times R_3$ and into three rank-R_p orthonormal factor matrices $\{\boldsymbol{\Phi}_1, \boldsymbol{\Phi}_2, \boldsymbol{\Phi}_3\}$ each of size $N_p \times R_p$ with $R_p < N_p$, $\forall p$. For zero-mean independent Gaussian core and noise tensors, a key discriminative parameter is the Signal to Noise Ratio defined by $\text{SNR} = \sigma_{\mathbf{s}}^2/\sigma^2$ where $\sigma_{\mathbf{s}}^2$ and σ^2 are the variances of the vectorized core and noise tensors, respectively. The *binary hypothesis test* can be described under the null hypothesis $\mathcal{H}_0 : \text{SNR} = 0$ (*i.e.*, only the noise is present) and the alternative hypothesis $\mathcal{H}_1 : \text{SNR} \neq 0$ (*i.e.*, there exists a signal of interest). First note that the exact derivation of the analytical expression of the error probability is not tractable in an analytical way even in the matrix case [10]. To tackle this problem, we adopt an information-geometric characterization of the detection performance [11,12].

2 Chernoff Information Framework

2.1 The Bayes' Detection Theory

Let $\Pr(\mathcal{H}_i)$ be the *a priori* hypothesis probability with $\Pr(\mathcal{H}_0) + \Pr(\mathcal{H}_1) = 1$. Let $p_i(\mathbf{y}) = p(\mathbf{y}|\mathcal{H}_i)$ and $\Pr(\mathcal{H}_i|\mathbf{y})$ be the i-th *conditional* and the *posterior* probabilities, respectively. The Bayes' detection rule chooses the hypothesis $\mathcal{H}_i$ associated with the largest posterior probability $\Pr(\mathcal{H}_i|\mathbf{y})$. Introduce the indicator hypothesis function according to

$$\phi(\mathbf{y}) \sim \text{Bernou}(\alpha),$$

where Bernou(α) stands for the Bernoulli distribution of success probability $\alpha = \Pr(\phi(\mathbf{y}) = 1) = \Pr(\mathcal{H}_1)$. Function $\phi(\mathbf{y})$ is defined on $\mathcal{X} \to \{0,1\}$ where $\mathcal{X}$ is the data-set enjoying the following decomposition $\mathcal{X} = \mathcal{X}_0 \cup \mathcal{X}_1$ where $\mathcal{X}_0 = \{\mathbf{y} : \phi(\mathbf{y}) = 0\} = \mathcal{X} \setminus \mathcal{X}_1$ and

$$\begin{aligned}\mathcal{X}_1 &= \{\mathbf{y} : \phi(\mathbf{y}) = 1\} \\ &= \left\{\mathbf{y} : \Omega(\mathbf{y}) = \log \frac{\Pr(\mathcal{H}_1|\mathbf{y})}{\Pr(\mathcal{H}_0|\mathbf{y})} > 0\right\} \\ &= \left\{\mathbf{y} : \Lambda(\mathbf{y}) = \log \frac{p_1(\mathbf{y})}{p_0(\mathbf{y})} > \log \tau\right\}\end{aligned}$$

in which $\tau = \frac{1-\alpha}{\alpha}$, $\Omega(\mathbf{y})$ is the log posterior-odds ratio and $\Lambda(\mathbf{y})$ is the log-likelihood ratio. The average error probability is defined as

$$P_e^{(N)} = \mathbb{E}\{\Pr(\text{Error}|\mathbf{y})\}, \tag{1}$$

with

$$\Pr(\text{Error}|\mathbf{y}) = \begin{cases} \Pr(\mathcal{H}_0|\mathbf{y}) \; if \; \mathbf{y} \in \mathcal{X}_1, \\ \Pr(\mathcal{H}_1|\mathbf{y}) \; if \; \mathbf{y} \in \mathcal{X}_0. \end{cases}$$

2.2 Chernoff Upper Bound (CUB)

Using the fact that $\min\{a, b\} \le a^s b^{1-s}$ with $s \in (0,1)$ and $a, b > 0$ in Eq. (1), the minimal error probability is upper bounded as follows

$$P_e^{(N)} \le \frac{\alpha}{\tau^s} \cdot \int_{\mathcal{X}} p_0(\mathbf{y})^{1-s} p_1(\mathbf{y})^s d\mathbf{y} = \frac{\alpha}{\tau^s} \cdot \exp[-\mu_s], \tag{2}$$

where the Chernoff s-divergence is defined according to

$$\mu_s = -\log M_{\Lambda(\mathbf{y}|\mathcal{H}_0)}(s),$$

with $M_{\Lambda(\mathbf{y}|\mathcal{H}_0)}(s)$ the M oment Generating Function (mgf) of the log-likelihood ratio under the null hypothesis.

The Chernoff Upper Bound (CUB) of the error probability is given by

$$P_e^{(N)} \le \frac{\alpha}{\tau^s} \cdot \exp[-\mu_s] = \alpha \cdot \exp[-r_s] \tag{3}$$

where

$$r_s = \mu_s + s \log \tau$$

is the exponential decay rate of the CUB. Assume that an optimal value of s denoted by $s^\star \in (0,1)$ exists then the tightest CUB verifies

$$P_e^{(N)} \le \alpha \cdot \exp[-r_{s^\star}] < \alpha \cdot \exp[-r_s].$$

The above condition is equivalent to maximize the exponential decay rate, *i.e.*,

$$s^\star = \arg \max_{s \in (0,1)} r_s. \tag{4}$$

Finally using Eqs. (3) and (4), we obtain the Chernoff Upper Bound. The Bhattacharyya Upper Bound (BUB) is obtained by Eq. (3) by fixing $s = 1/2$ instead of solving Eq. (4). The two bounds verify the following inequality relation:

$$P_e^{(N)} \leq \alpha \cdot \exp[-r_{s^\star}] \leq \alpha \cdot \exp\left[-r_{\frac{1}{2}}\right].$$

Note that in many encountered problems, the two hypothesis are assumed to be equi-probable, *i.e.*, $\alpha = 1/2$ and $\tau = 1$. Then the exponential decay rate is in this scenario given by the Chernoff s-divergence since $r_s = \mu_s$ and the tightest CUB is

$$P_e^{(N)} \leq \frac{1}{2} \exp[-\mu_{s^\star}].$$

3 Tensor Detection with Orthonormal Factors

3.1 Binary Hypothesis Test Formulation for Random Tensors

Tucker Model with Orthonormal Factors. Assume that the multidimensional measurements follow a noisy 3-order tensor of size $N_1 \times N_2 \times N_3$ given by

$$\boldsymbol{\mathcal{Y}} = \boldsymbol{\mathcal{X}} + \boldsymbol{\mathcal{N}}$$

where $\boldsymbol{\mathcal{N}}$ is the $N_1 \times N_2 \times N_3$ is the noise tensor where each entry is centered *i.i.d.* Gaussian, *i.e.* $[\boldsymbol{\mathcal{N}}]_{n_1,n_2,n_3} \sim \mathcal{N}(0, \sigma^2)$ and

$$\boldsymbol{\mathcal{X}} = \boldsymbol{\mathcal{S}} \times_1 \boldsymbol{\Phi}_1 \times_2 \boldsymbol{\Phi}_2 \times_3 \boldsymbol{\Phi}_3$$

is the $N_1 \times N_2 \times N_3$ "data" tensor following a Tucker model of (R_1, R_2, R_3)-multilinear rank. Matrices $\{\boldsymbol{\Phi}_1, \boldsymbol{\Phi}_2, \boldsymbol{\Phi}_3\}$ are the three orthonormal factors each of size $N_p \times R_p$ with $N_p > R_p$. These factors are for instance involved in the Higher-Order SVD (HOSVD) [5] with $\boldsymbol{\Phi}_p^T \boldsymbol{\Phi}_p = \mathbf{I}_{R_p}$ and $\boldsymbol{\Pi}_p = \boldsymbol{\Phi}_p \boldsymbol{\Phi}_p^T$ a $N_p \times N_p$ orthogonal projector on the range space of $\boldsymbol{\Phi}_p$. The $R_1 \times R_2 \times R_3$ core tensor is given by

$$\boldsymbol{\mathcal{S}} = \boldsymbol{\mathcal{X}} \times_1 \boldsymbol{\Phi}_1^T \times_2 \boldsymbol{\Phi}_2^T \times_3 \boldsymbol{\Phi}_3^T. \tag{5}$$

Formulating the Detection Test. We assume that each entry of the core tensor is centered *i.i.d.* Gaussian, *i.e.* $[\boldsymbol{\mathcal{S}}]_{r_1,r_2,r_3} \sim \mathcal{N}(0, \sigma_{\mathbf{s}}^2)$. Let $\mathbf{Y}_n$ be the n-th frontal $N_1 \times N_2$ slab of the 3-order tensor $\boldsymbol{\mathcal{Y}}$, the vectorized tensor expression is defined according to

$$\mathbf{y} = \left[(\text{vec}\mathbf{Y}_1)^T \ldots (\text{vec}\mathbf{Y}_{N_3})^T\right]^T = \mathbf{x} + \mathbf{n} \in \mathbb{R}^{N \times 1}$$

where $N = N_1 \cdot N_2 \cdot N_3$, $\mathbf{n}$ is the vectorization of the noise tensor $\boldsymbol{\mathcal{N}}$ and

$$\mathbf{x} = \boldsymbol{\Phi}\mathbf{s}$$

with $\mathbf{s}$ the vectorization of the core tensor $\boldsymbol{\mathcal{S}}$ and

$$\boldsymbol{\Phi} = \boldsymbol{\Phi}_3 \otimes \boldsymbol{\Phi}_2 \otimes \boldsymbol{\Phi}_1$$

is a $N \times R$ structured matrix with $R = R_1 \cdot R_2 \cdot R_3$.

In this framework, the associated equi-probable binary hypothesis test for the detection of the random signal, $\mathbf{s}$, is

$$\begin{cases} \mathcal{H}_0 : \mathbf{y}|\boldsymbol{\Phi}, \sigma^2 \sim \mathcal{N}\left(\mathbf{0}, \boldsymbol{\Sigma}_0 = \sigma^2 \mathbf{I}_N\right), \\ \mathcal{H}_1 : \mathbf{y}|\boldsymbol{\Phi}, \sigma^2 \sim \mathcal{N}\Big(\mathbf{0}, \boldsymbol{\Sigma}_1 = \sigma^2 \left(\mathrm{SNR} \cdot \boldsymbol{\Pi} + \mathbf{I}_N\right)\Big) \end{cases}$$

where $\mathrm{SNR} = \sigma_{\mathbf{s}}^2/\sigma^2$ is the signal to noise ratio and $\boldsymbol{\Pi} = \boldsymbol{\Pi}_3 \otimes \boldsymbol{\Pi}_2 \otimes \boldsymbol{\Pi}_1$ is an orthogonal projector. The performance of the detector of interest is quite often difficult to determine analytically [10]. As a consequence, we adopt the methodology of the CUB to upper bound it.

Geometry of the Expected Log-Likelihood Ratio. Consider $p(\mathbf{y}|\hat{\mathcal{H}}) = \mathcal{N}(\mathbf{0}, \boldsymbol{\Sigma})$ associated to the estimated hypothesis $\hat{\mathcal{H}}$. The expected log-likelihood ratio over $\mathbf{y}|\hat{\mathcal{H}}$ is given by

$$\begin{aligned} \mathbb{E}_{\mathbf{y}|\hat{\mathcal{H}}} \Lambda(\mathbf{y}) &= \int_{\mathcal{X}} p(\mathbf{y}|\hat{\mathcal{H}}) \log \frac{p_1(\mathbf{y})}{p_0(\mathbf{y})} d\mathbf{y} \\ &= \int_{\mathcal{X}} p(\mathbf{y}|\hat{\mathcal{H}}) \log \left[\frac{p(\mathbf{y}|\hat{\mathcal{H}})}{p_0(\mathbf{y})} \cdot \left(\frac{p(\mathbf{y}|\hat{\mathcal{H}})}{p_1(\mathbf{y})} \right)^{-1} \right] d\mathbf{y} \\ &= \mathcal{KL}(\hat{\mathcal{H}}, \mathcal{H}_0) - \mathcal{KL}(\hat{\mathcal{H}}, \mathcal{H}_1) \end{aligned}$$

where the Kullback-Leibler pseudo-distances are

$$\mathcal{KL}(\hat{\mathcal{H}}, \mathcal{H}_0) = \int_{\mathcal{X}} p(\mathbf{y}|\hat{\mathcal{H}}) \log \frac{p(\mathbf{y}|\hat{\mathcal{H}})}{p_0(\mathbf{y})} d\mathbf{y}, \quad \mathcal{KL}(\hat{\mathcal{H}}, \mathcal{H}_1) = \int_{\mathcal{X}} p(\mathbf{y}|\hat{\mathcal{H}}) \log \frac{p(\mathbf{y}|\hat{\mathcal{H}})}{p_1(\mathbf{y})} d\mathbf{y}.$$

The corresponding data-space for hypothesis $\mathcal{H}_1$ is

$$\mathcal{X}_1 = \{\mathbf{y} : \Lambda(\mathbf{y}) > \tau'\}$$

with

$$\begin{aligned} \Lambda(\mathbf{y}) &= \mathbf{y}^T (\boldsymbol{\Sigma}_0^{-1} - \boldsymbol{\Sigma}_1^{-1}) \mathbf{y} = \frac{1}{\sigma^2} \mathbf{y}^T \boldsymbol{\Phi} \left(\boldsymbol{\Phi}^T \boldsymbol{\Phi} + \mathrm{SNR} \cdot \mathbf{I} \right)^{-1} \boldsymbol{\Phi}^T \mathbf{y} \\ \tau' &= \log \frac{\det(\boldsymbol{\Sigma}_0)}{\det(\boldsymbol{\Sigma}_1})) = -\log\det\left(\mathrm{SNR} \cdot \boldsymbol{\Pi} + \mathbf{I}_N\right) \end{aligned}$$

where det(·) stands for the determinant. Thus, the alternative hypothesis is selected, *i.e.*, $\hat{\mathcal{H}} = \mathcal{H}_1$ if

$$\mathbb{E}_{\mathbf{y}|\hat{\mathcal{H}}} \Lambda(\mathbf{y}) = \frac{1}{\sigma^2} \mathrm{Tr} \left\{ \left(\boldsymbol{\Phi}^T \boldsymbol{\Phi} + \mathrm{SNR} \cdot \mathbf{I} \right)^{-1} \boldsymbol{\Phi}^T \boldsymbol{\Sigma} \boldsymbol{\Phi} \right\} > \tau'$$

or equivalently

$$\mathcal{KL}(\hat{\mathcal{H}}, \mathcal{H}_0) > \tau' + \mathcal{KL}(\hat{\mathcal{H}}, \mathcal{H}_1).$$

4 Tightest CUB

4.1 Derivation of the Bound

Theorem 1. *Let $c = R/N < 1$. The Chernoff s-divergence for the above test is given by*

$$\mu_s = \frac{c}{2} \left((1-s) \cdot \log(\mathrm{SNR} + 1) - \log\left(\mathrm{SNR} \cdot (1-s) + 1\right) \right).$$

Proof. According to [13], the Chernoff s-divergence for the above test is given by

$$\mu_s = \frac{1-s}{2N} \log\det\left(\mathrm{SNR} \cdot \boldsymbol{\Pi} + \mathbf{I}\right) - \frac{1}{2N} \log\det\left(\mathrm{SNR} \cdot (1-s)\boldsymbol{\Pi} + \mathbf{I}\right).$$

Using $\lambda\{\boldsymbol{\Pi}\} = \{\underbrace{1, \dots, 1}_{R}, \underbrace{0, \dots, 0}_{N-R}\}$ in the above expression yields Theorem 1.

Theorem 2. *1. The Chernoff s-divergence is a strictly convex function and admits an unique minimizer given by*

$$s^\star = \frac{1}{\mathrm{SNR}} \left(1 + \mathrm{SNR} - \frac{1}{\psi(\mathrm{SNR})} \right) \tag{6}$$

where $\psi(\mathrm{SNR}) = \frac{\log(\mathrm{SNR}+1)}{\mathrm{SNR}}$.

2. The tightest CUB for the (R_1, R_2, R_3)-multilinear rank orthonormal Tucker decomposition of Eq. (5) *is given by*

$$\mu_{s^\star} = \frac{c}{2} \Big(1 - \psi(\mathrm{SNR}) + \log\psi(\mathrm{SNR}) \Big).$$

Proof. The proof is straightforward and thus omitted due to the lack of space.

4.2 Analysis in Typical Limit Regimes

We can identify the two following limit scenarii:

- At low SNR, the tightest divergence, denoted by $\mu_{s^\star}$, coincides with the divergence $\mu_{1/2}$ associated with the BUB. Indeed, the optimal value in (6) admits a second-order approximation for SNR $\ll 1$ according to

$$s^\star \approx 1 + \frac{1}{\mathrm{SNR}} \left(1 - \left(1 + \frac{\mathrm{SNR}}{2} \right) \right) = \frac{1}{2}.$$

– At contrary for SNR $\gg 1$, we have $s^\star \approx 1$. So, the BUB is a loose bound in this regime.

To illustrate our analysis, on Fig. 1, the optimal s value obtained thanks to a numerical optimization of Eq. (4) using the divergence given in Theorem 1 and the analytical solution reported in Eq. (6) are plotted. We can check that the predicted analytical optimal s-value is in agreement with the approximated numerical one. We also verify the s-value in the low and high SNR regimes. In particular, for high SNRs, the optimal value is far from 1/2.

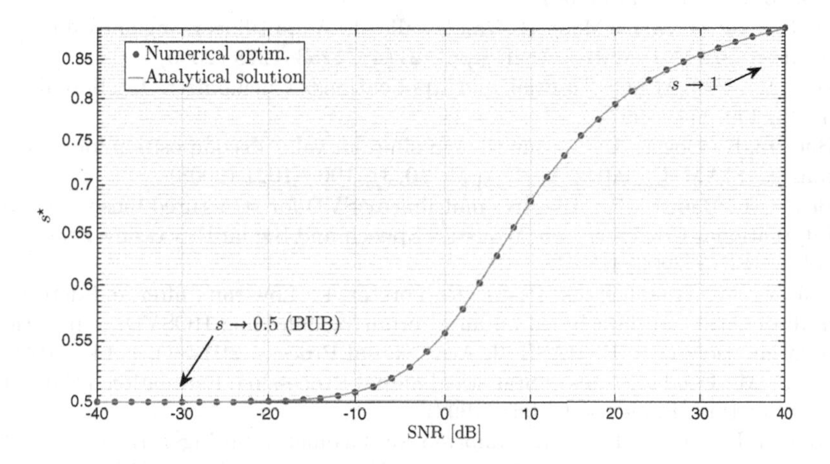

Fig. 1. Optimal s-value *vs* SNR in dB for a (3, 3, 3)-multilinear rank tensor $\boldsymbol{\mathcal{X}}$ of size $4 \times 4 \times 4$: The exact analytical formula is in full agreement with the numerical approximation scheme.

5 Conclusion

Performance detection in terms of the minimal Bayes' error probability for multidimensional measurements is a fundamental problem at the heart of many challenging applications. Interestingly, this tensor detection problem has received little attention so far. In this work, we derived analytically a tightest upper bound on the minimal Bayes' error probability for the detection of a random core tensor denoted by $\boldsymbol{\mathcal{S}}$ given a $N_1 \times N_2 \times N_3$ noisy observation tensor $\boldsymbol{\mathcal{X}}$ following an orthonormal Tucker model with a (R_1, R_2, R_3)-multilinear rank with $R_p < N_p$, $\forall p$. In particular, we showed that the tightest upper bound in the high SNR regime is not the Bhattacharyya upper bound.

References

1. Comon, P.: Tensors: A brief introduction. IEEE Signal Process. Mag. **31**(3), 44–53 (2014)
2. Cichocki, A., Mandic, D., De Lathauwer, L., Zhou, G., Zhao, Q., Caiafa, C., Phan, H.A.: Tensor decompositions for signal processing applications: From two-way to multiway component analysis. IEEE Signal Process. Mag. **32**(2), 145–163 (2015)
3. Chang, K., Qi, L., Zhang, T.: A survey on the spectral theory of nonnegative tensors. Numer. Linear Algebra Appl. **20**(6), 891–912 (2013)
4. Tucker, L.R.: Some mathematical notes on three-mode factor analysis. Psychometrika **31**(3), 279–311 (1966)
5. De Lathauwer, L., De Moor, B., Vandewalle, J.: A multilinear singular value decomposition. SIAM J. Matrix Anal. Appl. **21**(4), 1253–1278 (2000)
6. Bro, R.: PARAFAC: Tutorial and applications. Chemometr. Intell. Lab. Syst. **38**(2), 149–171 (1997)
7. Badeau, R., Boyer, R.: Fast multilinear singular value decomposition for structured tensors. SIAM J. Matrix Anal. Appl. **30**(3), 1008–1021 (2008)
8. Boyer, R., Badeau, R.: Adaptive multilinear SVD for structured tensors. In: IEEE International Conference on Acoustics Speech and Signal Processing Proceedings, vol. 3. IEEE (2006) III-III
9. Boizard, M., Ginolhac, G., Pascal, F., Forster, P.: Low-rank filter and detector for multidimensional data based on an alternative unfolding HOSVD: application to polarimetric STAP. EURASIP J. Adv. Signal Process. **2014**(1), 1–14 (2014)
10. Kay, S.M.: Fundamentals of Statistical Signal Processing: Estimation Theory. PTR Prentice-Hall, Englewood Cliffs (1993)
11. Nielsen, F., Bhatia, R.: Matrix Information Geometry. Springer, Heidelberg (2013)
12. Cover, T.M., Thomas, J.A.: Elements of Information Theory. Wiley, New York (2012)
13. Boyer, R., Nielsen, F.: Information geometry metric for random signal detection in large random sensing systems. In: ICASSP (2017)

Coordinate-Wise Transformation and Stein-Type Densities

Tomonari Sei(✉)

The University of Tokyo, 7-3-1 Hongo, Bunkyo-ku, Tokyo 113-8656, Japan
sei@mist.i.u-tokyo.ac.jp

Abstract. A Stein-type density function is defined as a stationary point of the free-energy functional over a fiber that consists of probability densities obtained by coordinate-wise transformations of a given density. It is shown that under some conditions there exists a unique Stein-type density in each fiber. An application to rating is discussed.

Keywords: Coordinate-wise transformation · Copositivity · Copula · Free-energy functional · Optimal transport · Positive dependence · Stein-type density

1 Introduction

Let $V(\boldsymbol{x})$ be a continuous function of $\boldsymbol{x} \in \mathbb{R}^d$. Consider a minimization problem of the free-energy functional

$$\mathcal{E}(p) = \mathcal{E}_V(p) = \int p(\boldsymbol{x}) \log p(\boldsymbol{x}) \mathrm{d}\boldsymbol{x} + \int p(\boldsymbol{x}) V(\boldsymbol{x}) \mathrm{d}\boldsymbol{x} \tag{1}$$

over a restricted set of probability density functions $p(\boldsymbol{x})$ on $\mathbb{R}^d$. The functional $\mathcal{E}(p)$ is, as is well known, the Lagrange function for maximizing entropy under given moment $\int p(\boldsymbol{x}) V(\boldsymbol{x}) \mathrm{d}\boldsymbol{x}$. Equation (1) is also discussed in the theory of optimal transport (e.g. [7,15]).

We first recall the solution of the unconstrained problem.

Lemma 1. *Let $Z = \int e^{-V(\boldsymbol{x})} \mathrm{d}\boldsymbol{x}$. If $Z < \infty$, then the functional $\mathcal{E}$ is minimized at $q(\boldsymbol{x}) = e^{-V(\boldsymbol{x})}/Z$. If $Z = \infty$, then $\mathcal{E}$ is not bounded from below.*

Proof. If $Z < \infty$, we have $\mathcal{E}(p) = \mathrm{KL}(p, q) - \log Z$, where $\mathrm{KL}(p, q)$ denotes the Kullback-Leibler divergence. Therefore $\mathcal{E}(p)$ is minimized at $p = q$. If $Z = \infty$, then $\mathcal{E}(p_n) \to -\infty$ as $n \to \infty$ for $p_n(\boldsymbol{x}) \propto e^{-V(\boldsymbol{x})} I_{[-n,n]^d}(\boldsymbol{x})$. □

Now consider a set of probability densities obtained by coordinate-wise transformations of a given density $p_0(\boldsymbol{x})$. Here a coordinate-wise transformation means

$$T(\boldsymbol{x}) = (T_1(x_1), \ldots, T_d(x_d)), \quad T_i'(x_i) > 0. \tag{2}$$

F. Nielsen and F. Barbaresco (Eds.): GSI 2017, LNCS 10589, pp. 665–672, 2017.
https://doi.org/10.1007/978-3-319-68445-1_77

If p_0 is pushed forward by T, then the resultant probability density is

$$p(\boldsymbol{x}) = (T_\sharp p_0)(\boldsymbol{x}) = p_0(T_1^{-1}(x_1), \ldots, T_d^{-1}(x_d)) \prod_{i=1}^{d} (T_i^{-1})'(x_i).$$

We will call the set $\{p = T_\sharp p_0 \mid T \text{ satisfies (2)}\}$ *a fiber*. It is widely known that each fiber has a unique copula density, which plays an important role in dependence modeling (e.g. [9]).

Our problem is to minimize $\mathcal{E}$ over the fiber, even if $Z = \int e^{-V(\boldsymbol{x})} \mathrm{d}\boldsymbol{x} = \infty$. In Theorem 1, it is shown that the stationary condition of $\mathcal{E}$ over the fiber is

$$\int f(x_i) \partial_i V(\boldsymbol{x}) p(\boldsymbol{x}) \mathrm{d}\boldsymbol{x} = \int f'(x_i) p(\boldsymbol{x}) \mathrm{d}\boldsymbol{x}, \quad \forall f \in \mathrm{C}^1(\mathbb{R}). \tag{3}$$

This equation is applied to a rating problem in Sect. 5. The following definition is a generalization of that in [13].

Definition 1 (Stein-type density). *A d-dimensional probability density function $p(\boldsymbol{x})$ is called* a Stein-type density *with respect to V if it satisfies (3).*

The Stein-type density is named after the Stein identity (see e.g. [2,14])

$$\int f(x) x \phi(x) \mathrm{d}x = \int f'(x) \phi(x) \mathrm{d}x,$$

that characterizes the standard normal density function $\phi(x) = e^{-x^2/2}/\sqrt{2\pi}$. The Stein identity corresponds to $d = 1$ and $V(x) = x^2/2$ in (3).

The remainder of the present paper is organized as follows. In Sect. 2, we describe the unique existence theorem on the constrained minimization problem, where the proof of existence is more challenging. In Sect. 3, we provide sufficient conditions for existence. In Sect. 4, examples of Stein-type densities are shown. In Sect. 5, we briefly discuss an application to rating. Throughout the paper, we assume that the density functions are continuous and positive over $\mathbb{R}^d$. Otherwise, more careful treatment is necessary.

2 Main Result

Suppose that $V(\boldsymbol{x})$ satisfies the following condition:

$$V(\boldsymbol{x}) = \psi(x_1 + \cdots + x_d), \quad \psi: \text{non-negative, convex}, \quad \lim_{x \to \pm\infty} \psi(x) = \infty. \tag{4}$$

The last condition is called coercive. For example, $V(\boldsymbol{x}) = |x_1 + \cdots + x_d|$ satisfies the condition. Note that $Z = \int e^{-V(\boldsymbol{x})} \mathrm{d}\boldsymbol{x} = \infty$ when $d \geq 2$.

As a referee pointed out, the restriction that V is a function of the sum can be relaxed. However we assumed it to make the description simpler.

Under Eq. (4), we can restrict the domain of $\mathcal{E}$ to

$$\mathcal{P} = \left\{ p \mid \int x_1 p_1(x_1) \mathrm{d}x_1 = \cdots = \int x_d p_d(x_d) \mathrm{d}x_d \right\} \tag{5}$$

without loss of generality, where p_i denotes the i-th marginal of p. Indeed, $\mathcal{E}(p)$ is invariant under the translation $x_i \mapsto x_i + a_i$ for any constants a_i with $\sum_i a_i = 0$. Note that the translation is a coordinate-wise transformation.

For each $p \in \mathcal{P}$, let $\mathcal{T}_{\rm cw}(p)$ be the set of coordinate-wise transformations such that $T_\sharp p \in \mathcal{P}$. Then *the p-fiber* is defined by

$$\mathcal{F}_p = \{T_\sharp p \mid T \in \mathcal{T}_{\rm cw}(p)\}.$$

If p is not specified explicitly, we call it *a fiber*. The space $\mathcal{P}$ is the disjoint union of fibers. In the context of optimal transport, each fiber is a totally-geodesic subspace of the L^2-Wasserstein space (e.g. [15]).

Stein-type densities are characterized by the following theorem. It holds for any convex function $V(\boldsymbol{x})$ without the restriction (4).

Theorem 1 (Characterization). *Let V be a convex function on $\mathbb{R}^d$. Then the following two conditions on $p \in \mathcal{P}$ are equivalent to each other:*

(i) p is Stein-type,
(ii) the functional $\mathcal{E}$ restricted to the p-fiber is minimized at p.

Proof. The proof relies on McCann's displacement convexity [7]. For each p,

$$\begin{aligned}\mathcal{E}(T_\sharp p) &= \int (T_\sharp p)(\boldsymbol{x}) \log (T_\sharp p)(\boldsymbol{x}) \mathrm{d}\boldsymbol{x} + \int (T_\sharp p)(\boldsymbol{x}) V(\boldsymbol{x}) \mathrm{d}\boldsymbol{x} \\ &= \int p(\boldsymbol{x}) \log \frac{1}{\prod_i T_i'(x_i)} \mathrm{d}\boldsymbol{x} + \int p(\boldsymbol{x}) V(T(\boldsymbol{x})) \mathrm{d}\boldsymbol{x},\end{aligned}$$

that is a convex functional of T. Thus it suffices to check the stationary condition. Consider a coordinate-wise transformation $T^t(\boldsymbol{x}) = \boldsymbol{x} + t f(x_i) \boldsymbol{e}_i$ parameterized by t, where f is any function and $\boldsymbol{e}_i$ is the i-th unit vector. Then we have

$$\frac{d}{dt}\mathcal{E}((T^t)_\sharp p)\Big|_{t=0} = -\int p_i(x_i) f'(x_i) \mathrm{d}x_i + \int p(\boldsymbol{x}) f(x_i) \{\partial_i V(\boldsymbol{x})\} \mathrm{d}\boldsymbol{x}.$$

Then the stationary condition $(d/dt)\mathcal{E}((T^t)_\sharp p)|_{t=0} = 0$ is equivalent to (3). □

To state the main result, we define additional symbols. For each $p \in \mathcal{P}$, denote the product of the marginal densities of p by $p^\perp(\boldsymbol{x}) = \prod_{i=1}^d p_i(x_i)$.

Definition 2 (Copositivity). *For each $p \in \mathcal{P}$, define*

$$\beta(p) = \beta_V(p) = \inf_{T \in \mathcal{T}_{\rm cw}(p)} \frac{\int V(T(\boldsymbol{x})) p(\boldsymbol{x}) \mathrm{d}\boldsymbol{x}}{\int V(T(\boldsymbol{x})) p^\perp(\boldsymbol{x}) \mathrm{d}\boldsymbol{x}}.$$

If $\beta(p) > 0$, p is called copositive *with respect to V.*

It is shown that $\beta(p)$ takes a common value in each fiber and that $\beta(p) \in [0,1]$. Sufficient conditions for copositivity are discussed in Sect. 3. On the other hand, there is a positive density function that is not copositive [13].

The following theorem is our main result. The result for $\psi(x) = x^2/2$ is proved in [13]. In that case, the theorem is interpreted as a non-linear analogue of the diagonal scaling theorem on matrices established by [6].

Theorem 2 (Existence and uniqueness). *Let V satisfy Eq. (4). Assume that $p_0 \in \mathcal{P}$ is copositive. Then there exists a unique Stein-type density in the p_0-fiber.*

Proof. The uniqueness is a consequence of the displacement convexity used in the proof of Theorem 1, where strict convexity holds under the restriction (5).

Now prove the existence in line with [13]. By Theorem 1, it is enough to show that the functional $\mathcal{E}|_{\mathcal{F}_{p_0}}$ restricted to the p_0-fiber has its minimum. For that purpose, we prove that $\mathcal{E}|_{\mathcal{F}_{p_0}}$ is bounded from below, and each sublevel set $\{p \in \mathcal{F}_{p_0} \mid \mathcal{E}(p) \leq M\}$ is tight (refer to [7] for details). Note that $\mathcal{E}$ itself is not bounded from below.

Let $p = T_\sharp p_0$ with $T \in \mathcal{T}_{\mathrm{cw}}(p_0)$. The assumption on copositivity of p_0 implies

$$\int V(\boldsymbol{x})p(\boldsymbol{x})\mathrm{d}\boldsymbol{x} \geq \beta \int V(\boldsymbol{x})p^\perp(\boldsymbol{x})\mathrm{d}\boldsymbol{x},$$

where $\beta = \beta(p) = \beta(p_0) > 0$. Hence we obtain

$$\begin{aligned}
\mathcal{E}(p) &= \int p(\boldsymbol{x}) \log \frac{p(\boldsymbol{x})}{p^\perp(\boldsymbol{x})}\mathrm{d}\boldsymbol{x} + \int p(\boldsymbol{x}) \log p^\perp(\boldsymbol{x})\mathrm{d}\boldsymbol{x} + \int V(\boldsymbol{x})p(\boldsymbol{x})\mathrm{d}\boldsymbol{x} \\
&\geq \int p(\boldsymbol{x}) \log p^\perp(\boldsymbol{x})\mathrm{d}\boldsymbol{x} + \beta \int V(\boldsymbol{x})p^\perp(\boldsymbol{x})\mathrm{d}\boldsymbol{x} \\
&= \sum_{i=1}^d \int p_i(x_i) \log p_i(x_i)\mathrm{d}x_i + \beta \int V(\boldsymbol{x})p^\perp(\boldsymbol{x})\mathrm{d}\boldsymbol{x}.
\end{aligned}$$

Therefore the problem is essentially reduced to the independent case $p = p^\perp$.

By using Jensen's inequality for $V(\boldsymbol{x}) = \psi(\sum_i x_i)$, we have

$$\int V(\boldsymbol{x})p^\perp(\boldsymbol{x})\mathrm{d}\boldsymbol{x} \geq \int \psi(x_i + (d-1)c)p_i(x_i)\mathrm{d}x_i$$

for each i, where $c = c(p) = \int x_j p_j(x_j)\mathrm{d}x_j$ does not depend on the index j due to the definition of $\mathcal{P}$. By combining these results, we obtain

$$\mathcal{E}(p) \geq \sum_{i=1}^d \left\{ \int p_i(x_i) \log p_i(x_i)\mathrm{d}x_i + \frac{\beta}{d} \int \psi(x_i + (d-1)c)p_i(x_i)\mathrm{d}x_i \right\}.$$

Define a probability density

$$q(x) = \frac{e^{-\frac{\beta}{2d}\psi(x+(d-1)c)}}{A}, \quad A = \int e^{-\frac{\beta}{2d}\psi(x)}\mathrm{d}x < \infty,$$

to obtain

$$\begin{aligned}
&\int p_i(x_i) \log p_i(x_i)\mathrm{d}x_i + \frac{\beta}{d} \int \psi(x_i + (d-1)c)p_i(x_i)\mathrm{d}x_i \\
&= \int p_i(x_i) \log \frac{p_i(x_i)}{q(x_i)}\mathrm{d}x_i - \log A + \frac{\beta}{2d} \int \psi(x_i + (d-1)c)p_i(x_i)\mathrm{d}x_i \\
&\geq -\log A + \frac{\beta}{2d} \int \psi(x_i + (d-1)c)p_i(x_i)\mathrm{d}x_i.
\end{aligned}$$

Then, putting $\alpha = -d \log A$, we have

$$\mathcal{E}(p) \geq \alpha + \frac{\beta}{2d} \sum_{i=1}^{d} \int \psi(x_i + (d-1)c) p_i(x_i) \mathrm{d}x_i \tag{6}$$

$$\geq \alpha + \frac{\beta}{2} \psi(dc) \quad \text{(Jensen's inequality)} \tag{7}$$

$$\geq \alpha. \tag{8}$$

By (8), $\mathcal{E}|_{\mathcal{F}_{p_0}}$ is bounded from below. Define the sublevel set $\mathcal{P}_M = \{p \in \mathcal{F}_{p_0} \mid \mathcal{E}(p) \leq M\}$ for a fixed M. Equation (7) implies that, if $p \in \mathcal{P}_M$, then $\psi(dc)$ is bounded from above and therefore c is contained in a bounded interval $[-C, C]$. Now a function defined by

$$\psi_*(x) = \min_{c \in [-C,C]} \psi(x + (d-1)c)$$

is coercive. In addition, the inequality (6) shows that $\int \psi_*(x_i) p_i(x_i) \mathrm{d}x_i$ is bounded from above. Therefore we deduce that $\mathcal{P}_M$ is tight. This completes the proof. □

For an independent density $p(\boldsymbol{x}) = \prod_{i=1}^{d} p_i(x_i)$, it is obvious that $\beta(p) = 1$ from the definition. Therefore we have the following corollary. This is interpreted as a variational Bayes method [1] except that $\int e^{-V(\boldsymbol{x})} \mathrm{d}\boldsymbol{x} = \infty$.

Corollary 1. *Let V satisfy (4). Then there is a unique independent Stein-type density.*

3 Sufficient Conditions for Copositivity

In this section, we discuss sufficient conditions for copositivity (see Definition 2). The notion of positive dependence plays a central role.

A function $f : \mathbb{R}^d \to \mathbb{R}$ is called *super-modular* if for any $\boldsymbol{x}, \boldsymbol{y} \in \mathbb{R}^d$,

$$f(\boldsymbol{x} \vee \boldsymbol{y}) + f(\boldsymbol{x} \wedge \boldsymbol{y}) \geq f(\boldsymbol{x}) + f(\boldsymbol{y}).$$

Here $\boldsymbol{x} \vee \boldsymbol{y}$ and $\boldsymbol{x} \wedge \boldsymbol{y}$ are coordinate-wise maximum and minimum, respectively. For smooth functions f, the super-modularity is equivalent to

$$\frac{\partial^2 f}{\partial x_i \partial x_j} \geq 0, \quad i \neq j.$$

The following lemma is straightforward.

Lemma 2. *Let $V(\boldsymbol{x}) = \psi(x_1 + \cdots + x_d)$ with a convex function ψ. Then for any coordinate-wise transformation $T(\boldsymbol{x})$, the composite function $V(T(\boldsymbol{x}))$ is super-modular.*

Although there are a number of variants of positive dependence, we use only three of them. Refer to [11] for further information.

Definition 3 (Positive dependence). *Let $p(\boldsymbol{x})$ be a probability density function on $\mathbb{R}^d$. Then*

1. *$p(\boldsymbol{x})$ is called* MTP$_2$ (multivariate totally positive of order 2) *if*
$$p(\boldsymbol{x} \vee \boldsymbol{y})p(\boldsymbol{x} \wedge \boldsymbol{y}) \geq p(\boldsymbol{x})p(\boldsymbol{y})$$
for all $\boldsymbol{x}$ and $\boldsymbol{y}$ in $\mathbb{R}^d$. In other words, $\log p(\boldsymbol{x})$ is super-modular.
2. *$p(\boldsymbol{x})$ is said to be* associated *if*
$$\int \phi(\boldsymbol{x})\psi(\boldsymbol{x})p(\boldsymbol{x})\mathrm{d}\boldsymbol{x} \geq \int \phi(\boldsymbol{x})p(\boldsymbol{x})\mathrm{d}\boldsymbol{x} \int \psi(\boldsymbol{x})p(\boldsymbol{x})\mathrm{d}\boldsymbol{x}$$
for any increasing functions $\phi, \psi : \mathbb{R}^d \to \mathbb{R}$.
3. *$p(\boldsymbol{x})$ is called* PSMD (positive super-modular dependent) *if*
$$\int \phi(\boldsymbol{x})p(\boldsymbol{x})\mathrm{d}\boldsymbol{x} \geq \int \phi(\boldsymbol{x})p^{\perp}(\boldsymbol{x})\mathrm{d}\boldsymbol{x}$$
for any super-modular function ϕ. Recall that $p^{\perp}(\boldsymbol{x}) = \prod_i p_i(x_i)$.

These variants of positive dependence have the following implications. The first implication is called the FKG inequality [5].

Lemma 3 ([3,5])**.** *MTP$_2$ $\Rightarrow$ associated $\Rightarrow$ PSMD.*

MTP$_2$ is relatively easy to confirm whereas association is interpretable. A Gaussian distribution is MTP$_2$ (resp. PSMD) if and only if all the partial correlation coefficients (resp. all the correlation coefficients) are non-negative [8,10]. Graphical models with the MTP$_2$ property are discussed in [4].

PSMD meets our purpose as follows.

Theorem 3. *Let $V(\boldsymbol{x})$ satisfy (4). If p is PSMD, then p is copositive.*

Proof. For any $T \in \mathcal{T}_{\mathrm{cw}}(p)$, Lemma 2 implies that $V(T(\boldsymbol{x}))$ is super-modular. Therefore if p is PSMD, then
$$\int V(T(\boldsymbol{x}))p(\boldsymbol{x})\mathrm{d}\boldsymbol{x} \geq \int V(T(\boldsymbol{x}))p^{\perp}(\boldsymbol{x})\mathrm{d}\boldsymbol{x},$$
which means $\beta(p) = 1$. □

We also provide two other sufficient conditions for copositivity. Each of them holds for any non-negative $V(\boldsymbol{x})$, and its proof is straightforward.

Lemma 4. *If $p_0(\boldsymbol{x})$ is copositive and there exist constants $M, \delta > 0$ such that $\delta \leq p(\boldsymbol{x})/p_0(\boldsymbol{x}) \leq M$ for all $\boldsymbol{x}$, then $p(\boldsymbol{x})$ is also copositive.*

Lemma 5. *If there exists $\delta > 0$ such that $p(\boldsymbol{x})/p^{\perp}(\boldsymbol{x}) \geq \delta$ for all $\boldsymbol{x}$, then $p(\boldsymbol{x})$ is copositive. This condition holds if and only if the copula density (e.g. [9]) corresponding to $p(\boldsymbol{x})$ is greater than or equal to the constant δ.*

4 Examples of Stein-Type Densities

Let $V(\boldsymbol{x}) = \psi(x_1 + \cdots + x_d)$. First, by integral-by-parts formula, Eq. (3) is equivalent to

$$\left(\sum_i \partial_i p(\boldsymbol{x})\right) + \psi'(x_1 + \cdots + x_d)p(\boldsymbol{x}) = r(\boldsymbol{x}), \quad \int_{\mathbb{R}^{d-1}} r(\boldsymbol{x})\mathrm{d}\boldsymbol{x}_{-i} = 0, \tag{9}$$

When $r(\boldsymbol{x})$ is given, the partial differential equation (9) is solved by the characteristic curve method. In particular, if $r(\boldsymbol{x}) = 0$, the general solution is

$$p(\boldsymbol{x}) = \frac{1}{Z} e^{-\psi(x_1+\cdots+x_d)} q(\boldsymbol{Q}^\top \boldsymbol{x}), \tag{10}$$

where Z is the normalizing constant, $\boldsymbol{Q}$ is a matrix such that $(\mathbf{1}/\sqrt{d}, \boldsymbol{Q})$ is an orthogonal matrix, and q is an arbitrary $(d-1)$-dimensional density function.

However, for given $p_0 \in \mathcal{P}$, it is generally difficult to find a Stein-type density that belongs to the p_0-fiber. The case for $\psi(x) = x^2/2$ is investigated in [13]. Here we give another example.

Example 1. Let $d = 2$ and $V(x_1, x_2) = |x_1 + x_2|$. Then the independent Stein-type density has the marginals

$$p_i(x_i) = \frac{1}{4\cosh^2(x_i/2)} = \frac{e^{x_i}}{(e^{x_i}+1)^2}, \quad i = 1, 2.$$

This is a logistic distribution. One can directly confirm (9).

5 Application to Rating

Suppose that a d-dimensional density $p(\boldsymbol{x})$ denotes a distribution of students' marks on d subjects. We make a rule to determine the general score of each student. An answer is given as follows.

Fix a convex function ψ. Typically $\psi(x) = x^2/2$ or $\psi(x) = |x|$. As long as p is copositive, Theorem 2 implies that there exists a coordinate-wise transformation T such that $T_\sharp p$ is Stein-type. In particular, we obtain

$$\int p(\boldsymbol{x}) f(x_i) \psi'(T_1(x_1) + \cdots + T_d(x_d))\mathrm{d}\boldsymbol{x} > 0 \tag{11}$$

for any increasing function f. Then we can use $T_1(x_1) + \cdots + T_d(x_d)$ as the general score of $\boldsymbol{x}$. Refer to [12] for relevant information.

Example 2. Let $\psi(x) = |x|$. Consider a probability density function

$$p(x_1, x_2) = \frac{0.1}{4} e^{-|x_1+0.1x_2|}\{e^{-|x_1-0.1x_2-1|} + e^{-|x_1-0.1x_2+1|}\}.$$

A map $T(x_1, x_2) = (x_1, 0.1x_2)$ attains the Stein-type density due to (10). The following contingency tables show that, under the law p, the signs of x_1 and x_1+x_2 have negative correlation whereas those of x_1 and $x_1+0.1x_2$ have positive.

	$x_1 + x_2 < 0$	$x_1 + x_2 > 0$
$x_1 < 0$	0.198	0.302
$x_1 > 0$	0.302	0.198

	$x_1 + 0.1x_2 < 0$	$x_1 + 0.1x_2 > 0$
$x_1 < 0$	0.342	0.158
$x_1 > 0$	0.158	0.342

Indeed (11) implies that the sign of $x_i - a$ for any i and $a \in \mathbb{R}$ has positive correlation with the sign of the general score.

As referees pointed out, the general score depends on the choice of the potential ψ. A natural choice would be $\psi(x) = x^2/2$ because then the scores $x_1, \ldots, x_d$ are transformed into normal quantiles if they are independent. On the other hand, if one concerns the "passing point" of a particular grading, the potential $\psi(x) = |x|$ seems more preferable as seen in Example 2.

Acknowledgements. The author is grateful to three anonymous referees for their constructive comments. This work was supported by JSPS KAKENHI Grant Numbers JP26108003 and JP17K00044.

References

1. Bishop, C.M.: Pattern Recognition and Machine Learning. Springer, Heidelberg (2006)
2. Chen, L.H.Y., Goldstein, L., Shao, Q.: Normal Approximation by Stein's Method. Springer, Heidelberg (2011)
3. Christofides, T.C., Vaggelatou, E.: A connection between supermodular ordering and positive/negative association. J. Multivar. Anal. **88**, 138–151 (2004)
4. Fallat, S., Lauritzen, S., Sadeghi, K., Uhler, C., Wermuth, N., Zwiernik, P.: Total positivity in Markov structures. Ann. Statist. **45**(3), 1152–1184 (2017)
5. Fortuin, C.M., Kasteleyn, P.W., Ginibre, J.: Correlation inequalities on some partially ordered sets. Comm. Math. Phys. **22**, 89–103 (1971)
6. Marshall, A.W., Olkin, I.: Scaling of matrices to achieve specified row and column sums. Numer. Math. **12**, 83–90 (1968)
7. McCann, R.J.: A convexity principle for interacting gases. Adv. Math. **128**, 153–179 (1997)
8. Müller, A., Stoyan, D.: Comparison Methods for Stochastic Models and Risks. Wiley, New York (2002)
9. Nelsen, R.B.: An Introduction to Copulas, 2nd edn. Springer, Heidelberg (2006)
10. Rüschendorf, L.: Characterization of dependence concepts in normal distributions. Ann. Inst. Statist. Math. **33**, 347–359 (1981)
11. Rüschendorf, L.: Mathematical Risk Analysis. Springer, Heidelberg (2013)
12. Sei, T.: An objective general index for multivariate ordered data. J. Multivar. Anal. **147**, 247–264 (2016)
13. Sei, T.: Coordinate-wise transformation of probability distributions to achieve a Stein-type identity, Technical Report METR2017-04. The University of Tokyo, Department of Mathematical Engineering and Information Physics (2017)
14. Stein, C.: A bound for the error in the normal approximation to the distribution of a sum of dependent random variables. In: Proceedings of the Sixth Berkeley Symposium on Mathematical Statistics and Probability, vol. 2, pp. 583–602 (1972)
15. Villani, C.: Topics in Optimal Transportation. American Mathematical Society, Providence (2003)

Probability Density Estimation

Riemannian Online Algorithms for Estimating Mixture Model Parameters

Paolo Zanini[1,2], Salem Said[1,2(✉)], Yannick Berthoumieu[1,2], Marco Congedo[1,2], and Christian Jutten[1,2]

[1] Laboratoire IMS (CNRS - UMR 5218), Université de Bordeaux, Bordeaux, France
{salem.said,yannick.berthoumieu}@ims-bordeaux.fr
[2] Gipsa-lab (CNRS - UMR 5216), Université de Grenoble, Grenoble, France
{paolo.zanini,marco.congedo,christian.jutten}@gipsa-lab.fr

Abstract. This paper introduces a novel algorithm for the online estimate of the Riemannian mixture model parameters. This new approach counts on Riemannian geometry concepts to extend the well-known Titterington approach for the online estimate of mixture model parameters in the Euclidean case to the Riemannian manifolds. Here, Riemannian mixtures in the Riemannian manifold of Symmetric Positive Definite (SPD) matrices are analyzed in details, even if the method is well suited for other manifolds.

Keywords: Riemannian mixture estimation · Information geometry · Online EM algorithm

1 Introduction

Information theory and Riemannian geometry have been widely developed in the recent years in a lot of different applications. In particular, Symmetric Positive Definite (SPD) matrices have been deeply studied through Riemannian geometry tools. Indeed, the space $\mathcal{P}_m$ of $m \times m$ SPD matrices can be equipped with a Riemannian metric. This metric, usually called Rao-Fisher or affine-invariant metric, gives it the structure of a Riemannian manifold (specifically a homogeneous space of non-positive curvature). SPD matrices are of great interest in several applications, like diffusion tensor imaging, brain-computer interface, radar signal processing, mechanics, computer vision and image processing [1–5]. Hence, it is very useful to develop statistical tools to analyze objects living in the manifold $\mathcal{P}_m$. In this paper we focus on the study of Mixtures of Riemannian Gaussian distributions, as defined in [6]. They have been successfully used to define probabilistic classifiers in the classification of texture images [7] or Electroencephalography (EEG) data [8]. In these examples mixtures parameters are estimated through suitable EM algorithms for Riemannian manifolds. In this paper we consider a particular situation, that is the observations are observed one at a time. Hence, an online estimation of the parameters is needed. Following the Titterington's approach [9], we derive a novel approach for the online estimate of parameters of Riemannian Mixture distributions.

F. Nielsen and F. Barbaresco (Eds.): GSI 2017, LNCS 10589, pp. 675–683, 2017.
https://doi.org/10.1007/978-3-319-68445-1_78

The paper is structured as follows. In Sect. 2 we describe the Riemannian Gaussian Mixture Model. In Sect. 3, we introduce the reference methods for online estimate of mixture parameters in the Euclidean case, and we describe in details our approach for the Riemannian framework. For lack of space, some equation's proofs will be omitted. Then, in Sect. 4, we present some simulations to validate the proposed method. Finally we conclude with some remarks and future perspectives in Sect. 5.

2 Riemannian Gaussian Mixture Model

We consider a Riemannian Gaussian Mixture model $g(x;\theta) = \sum_{k=1}^{K} \omega_k p(x;\psi_k)$, with the constraint $\sum_{k=1}^{K} \omega_k = 1$. Here $p(x;\psi_k)$ is the Riemannian Gaussian distribution studied in [6], defined as $p(x;\psi_k) = \frac{1}{\zeta(\sigma_k)} \exp\left(-\frac{d_R^2(x,\overline{x}_k)}{2\sigma_k^2}\right)$, where x is a SPD matrix, $\overline{x}_k$ is still a SPD matrix representing the center of mass of the kth component of the mixture, σ_k is a positive number representing the dispersion parameter of the kth mixture component, $\zeta(\sigma_k)$ is the normalization factor, and $d_R(\cdot,\cdot)$ is the Riemannian distance induced by the metric on $\mathcal{P}_m$. $g(x;\theta)$ is also called incomplete likelihood. In the typical mixture model approach, indeed, we consider some latent variables Z_i, categorical variables over $\{1,...,K\}$ with parameters $\{\omega_k\}_{k=1}^{K}$, assuming $X_i|Z_i = k \sim p(\cdot,\psi_k)$. Thus, the complete likelihood is defined as $f(x,z;\theta) = \sum_{k=1}^{K} \omega_k p(x;\psi_k)\delta_{z,k}$, where $\delta_{z,k} = 1$ if $z = k$ and 0 otherwise. We deal here with the problem to estimate the model parameters, gathered in the vector $\theta = [\omega_1, \overline{x}_1, \sigma_1, ..., \omega_K, \overline{x}_K, \sigma_K]$. Usually, given a set of N i.i.d. observations $\chi = \{x_i\}_{i=1}^{N}$, we look for $\widehat{\theta}_N^{MLE}$, that is the MLE of θ, i.e. the maximizer of the log-likelihood $l(\theta;\chi) = \frac{1}{N}\sum_{i=1}^{N} \log \sum_{k=1}^{K} \omega_k p(x_i;\psi_k)$.

To obtain $\widehat{\theta}_N^{MLE}$, EM or stochastic EM approaches are used, based on the complete dataset $\chi_c = \{(x_i, z_i)\}_{i=1}^{N}$, with the unobserved variables Z_i. In this case, average complete log-likelihood can be written:

$$l_c(\theta;\chi_c) = \frac{1}{N}\sum_{i=1}^{N} \log \prod_{k=1}^{K} (\omega_k p(x_i;\psi_k))^{\delta_{z_i,k}} = \frac{1}{N}\sum_{i=1}^{N}\sum_{k=1}^{K} \delta_{z_i,k} \log(\omega_k p(x_i;\psi_k)). \quad (1)$$

Here we consider a different situation, that is the dataset χ is not available entirely, rather the observations are observed one at a time. In this situation online estimation algorithms are needed.

3 Online Estimation

In the Euclidean case, reference algorithms are the Titterington's algorithm, introduced in [9], and the Cappé-Moulines's algorithm presented in [10].

We focus here on Titterington's approach. In classic EM algorithms, the Expectation step consists in computing $Q(\theta;\widehat{\theta}^{(r)},\chi) = E_{\widehat{\theta}^{(r)}}[l_c(\theta;\chi_c)|\chi]$, and then, in the Maximization step, in maximizing Q over θ. These steps are performed iteratively and at each iteration r an estimate $\widehat{\theta}^{(r)}$ of θ is obtained exploiting the whole dataset. In the online framework, instead, the current estimate will be indicated by $\widehat{\theta}^{(N)}$, since in this setting, once $x_1, x_2, ..., x_N$ are observed we want to update our estimate for a new observation x_{N+1}. Titterington approach corresponds to the direct optimization of $Q(\theta; \widehat{\theta}^{(N)}, \chi)$ using a Newton algorithm:

$$\widehat{\theta}^{(N+1)} = \widehat{\theta}^{(N)} + \gamma^{(N+1)} I_c^{-1}(\widehat{\theta}^{(N)}) u(x_{N+1}; \widehat{\theta}^{(N)}), \tag{2}$$

where $\{\gamma^{(N)}\}_N$ is a decreasing sequence, the Hessian of Q is approximated by the Fisher Information matrix I_c for the complete data $I_c^{-1}(\widehat{\theta}^{(N)}) = -E_{\widehat{\theta}^{(N)}}[\frac{\log f(x,z;\theta)}{\partial\theta\partial\theta^T}]$, and the score $u(x_{N+1}; \widehat{\theta}^{(N)})$ is defined as $u(x_{N+1}; \widehat{\theta}^{(N)}) = \nabla_{\widehat{\theta}^{(N)}} \log g(x_{N+1}; \widehat{\theta}^{(N)}) = E_{\widehat{\theta}^{(N)}}[\nabla_{\widehat{\theta}^{(N)}} \log f(x_{N+1}; \widehat{\theta}^{(N)}) | x_{N+1}]$ (where last equality is presented in [10]).

Geometrically speaking, Tittetington algorithm consists in modifying the current estimate $\widehat{\theta}^{(N+1)}$ adding the term $\xi^{(N+1)} = \gamma^{(N+1)} I_c^{-1}(\widehat{\theta}^{(N)}) u(x_{N+1}; \widehat{\theta}^{(N)})$. If we want to consider parameters belonging to Riemannian manifolds, we have to suitably modify the update rule. Furthermore, even in the classical framework, Titterington update does not necessarily constraint the estimates to be in the parameters space. For instance, the weights could be assume negative values. The approach we are going to introduce solves this problem, and furthermore is suitable for Riemannian Mixtures.

We modify the update rule, exploiting the Exponential map. That is:

$$\widehat{\theta}^{(N+1)} = \mathrm{Exp}_{\widehat{\theta}^{(N)}}(\xi^{(N+1)}), \tag{3}$$

where our parameters become $\theta_k = [s_k, \overline{x}_k, \eta_k]$. Specifically, $s_k^2 = w_k \rightarrow s = [s_1, ..., s_K] \in \mathbb{S}^{K-1}$ (i.e., the sphere), $\overline{x}_k \in P(m)$ and $\eta_k = -\frac{1}{2\sigma_k^2} < 0$.

Actually we are not forced to choose the exponential map, in the update formula (3), but we can consider any retraction operator. Thus, we can generalize (3) in $\widehat{\theta}^{(N+1)} = \mathcal{R}_{\widehat{\theta}^{(N)}}(\xi^{(N+1)})$.

In order to develop a suitable update rule, we have to define $I(\theta)$ and the score $u()$ in the manifold, noting that every parameter belongs to a different manifold. Firstly we note that the Fisher Information matrix $I(\theta)$ can be written as:

$$I(\theta) = \begin{pmatrix} I(s) & & \\ & I(\overline{x}) & \\ & & I(\eta) \end{pmatrix}.$$

Now we can analyze separately the update rule for s, $\overline{x}$, and η. Since they belong to different manifold the exponential map (or the retraction) will be different, but the philosophy of the algorithm is still the same.

For the update of weights s_k, the Riemannian manifold considered is the sphere $\mathbb{S}^{K-1}$, and, given a point $s \in \mathbb{S}^{K-1}$, the tangent space $T_s\mathbb{S}^{K-1}$ is identified as $T_s\mathbb{S}^{K-1} = \{\xi \in \mathbb{R}^K : \xi^T s = 0\}$. We can write the complete log-likelihood only in terms of s: $l(x,z;s) = \log f(x,z;s) = \sum_{k=1}^{K} \log s_k^2 \delta_{z,k}$. We start by evaluating $I(s)$, that will be a $K \times K$ matrix of the quadratic form

$$I_s(u,w) = E[\langle u, v(z,s)\rangle \langle v(z,s), w\rangle], \quad (4)$$

for u, w elements of the tangent space in s, and $v(z,s)$ is the Riemannian gradient, defined as $v(z,s) = \frac{\partial l}{\partial s} - \left(\frac{\partial l}{\partial s}, s\right) s$. In this case we obtain $\frac{\partial l}{\partial s_k} = 2\frac{\delta_{z,k}}{s_k} \rightarrow v(z,s_k) = 2\left(\frac{\delta_{z,k}}{s_k} - s_k\right)$. It is easy to see that the matrix of the quadratic form has elements

$$I_{kl}(s) = E[v_k(z,s)v_l(z,s)] = E\left[4\left(\frac{\delta_{z,k}}{s_k} - s_k\right)\left(\frac{\delta_{z,l}}{s_l} - s_l\right)\right]$$
$$= E\left[4\left(\frac{\delta_{z,k}\delta_{z,l}}{s_k s_l} - \frac{s_l}{s_k}\delta_{z,k} - \frac{s_k}{s_l}\delta_{z,l} + s_k s_l\right)\right] = 4(\delta_{kl} - s_k s_l - s_k s_l + s_k s_l) = 4(\delta_{kl} - s_k s_l).$$

Thus, the Fisher Information matrix $I(s)$ applied to an element ξ of the tangent space results to be $I(s)\xi = 4\xi$, hence $I(s)$ corresponds to 4 times the identity matrix. Thus, if we consider update rule (3), we have $\xi^{(N+1)} = \frac{\gamma^{(N+1)}}{4} u(x_{N+1}; \widehat{\theta}^{(N)})$. We have to evaluate $u(x_{N+1}; \widehat{\theta}^{(N)})$. We proceed as follows:

$$u_k(x_{N+1}; \widehat{\theta}^{(N)}) = E[v_k(z,s)|x_{N+1}] = E\left[2\left(\frac{\delta_{z,k}}{s_k} - s_k\right)|x_{N+1}\right] = 2\left(\frac{h_k(x_{N+1}; \widehat{\theta}^{(N)})}{s_k} - s_k\right),$$

where $h_k(x_{N+1}; \widehat{\theta}^{(N)}) \propto s_k^2 p(x_{N+1}; \widehat{\theta}_k^{(N)})$. Thus we obtain

$$\widehat{s}^{(N+1)} = \mathrm{Exp}_{\widehat{s}^{(N)}}\left(\frac{\gamma^{(N+1)}}{2}\left(\frac{h_1(x_{N+1}; \widehat{\theta}^{(N)})}{\widehat{s}_1^{(N)}} - \widehat{s}_1^{(N)}, \ldots, \frac{h_K(x_{N+1}; \widehat{\theta}^{(N)})}{\widehat{s}_K^{(N)}} - \widehat{s}_K^{(N)}\right)\right)$$
$$= \mathrm{Exp}_{\widehat{s}^{(N)}}\left(\xi^{(N+1)}\right). \quad (5)$$

Considering the classical exponential map on the sphere (i.e., the geodesic), the update rule (5) becomes

$$\widehat{s}_k^{(N+1)} = \widehat{s}_k^{(N)} \cos(\|\xi^{(N+1)}\|) + \frac{\frac{\gamma^{(N+1)}}{2}\left(\frac{h_k}{\widehat{s}_k^{(N)}} - \widehat{s}_k^{(N)}\right)}{\|\xi^{(N+1)}\|} \sin(\|\xi^{(N+1)}\|). \quad (6)$$

Actually, as anticipated before, we are not forced to used the exponential map, but we can consider other retractions. In particular, on the sphere, we could consider the "projection" retraction $\mathcal{R}_x(\xi) = \frac{x+\xi}{\|x+\xi\|}$, deriving update rule accordingly.

For the update of barycenters $\overline{x}_k$ we have, for every barycenter $\overline{x}_k, k = 1, \ldots, K$, an element of $\mathcal{P}_m$, the Riemannian manifold of $m \times m$ SPD matrices. Thus, we derive the update rule for a single k.

First of all we have to derive expression (4). But this expression is true only for irreducible manifolds, as the sphere. In the case of $\mathcal{P}_m$ we have to introduce some theoretical results. Let $\mathcal{M}$ a symmetric space of negative curvature (like $\mathcal{P}_m$), it can be expressed as a product $\mathcal{M} = \mathcal{M}_1 \times \cdots \times \mathcal{M}_R$, where each $\mathcal{M}_r$

is an irreducible space [11]. Now let $\overline{x}$ an element of $\mathcal{M}$, and v, w elements of the tangent space $T_{\overline{x}}\mathcal{M}$. We can write $\overline{x} = (\overline{x}_1, ..., \overline{x}_R)$, $v = (v_1, ..., v_R)$ and $w = (w_1, ..., w_R)$. We can generalize (4) by the following expression:

$$I_{\overline{x}}(u, w) = \sum_{r=1}^{R} E[\langle u_r, v_r(\overline{x}_r)\rangle_{\overline{x}} \langle v_r(\overline{x}_r), w_r\rangle_{\overline{x}}], \tag{7}$$

with $v_r(\overline{x}_r) = \nabla_{\overline{x}} l(\overline{x})$ being the Riemannian score. In our case $\mathcal{P}_m = \mathbb{R} \times \mathcal{SP}_m$, where $\mathcal{SP}_m$ represents the manifold of SPD matrices with unitary determinant, while $\mathbb{R}$ takes into account the part relative to the determinant. Thus, if $x \in \mathcal{P}_m$, we can consider the isomorphism $\phi(x) = (x_1, x_2)$ with $x_1 = \log \det x \in \mathbb{R}$ and $x_2 = e^{-x_1/m} x \in \mathcal{SP}_m$, $(\det x_2 = 1)$. The idea is to use the procedure adopted to derive $\widehat{s}^{(N+1)}$, for each component of $\widehat{\overline{x}}_k^{(N+1)}$. Specifically we proceed as follows:

- we derive $I(\overline{x}_k)$ through formula (7), with components I_r.
- we derive the Riemannian score $u(x_{N+1}; \widehat{\theta}^{(N)}) = E\left[v(x_{N+1}, z_{N+1}; \widehat{\overline{x}}_k^{(N)}, \widehat{\sigma}_k^{(N)}) | x_{N+1}\right]$, with components u_r.
- for each component $r = 1, 2$ we evaluate $\xi_r^{(N+1)} = \gamma^{(N+1)} I_r^{-1} u_r$
- we update each component $\left(\widehat{\overline{x}}_k^{(N+1)}\right)_r = \text{Exp}_{\left(\widehat{\overline{x}}_k^{(N)}\right)_r}\left(\xi_r^{(N+1)}\right)$ and we could use $\phi^{-1}(\cdot)$ to derive $\widehat{\overline{x}}_k^{(N+1)}$ if needed.

We start deriving $I(\overline{x}_k)$ for the complete model (see [12] for some derivations):

$$\begin{aligned} I_{\overline{x}_k}(u, w) &= E\left[\langle u, v(x, z; \overline{x}_k, \sigma_k)\rangle \langle v(x, z; \overline{x}_k, \sigma_k), w\rangle\right] = E\left[\frac{\delta_{z,k}}{\sigma_k^4} \langle u, \text{Log}_{\overline{x}_k} x\rangle \langle \text{Log}_{\overline{x}_k} x, w\rangle\right] = \\ &= E\left[\frac{\delta_{z,k}}{\sigma_k^4} I(u, w)\right] = \frac{\omega_k}{\sigma_k^4} \sum_{r=1}^{2} \frac{\psi_r'(\eta_k)}{dim(\mathcal{M}_r)} \langle u_r, w_r\rangle_{(\overline{x}_k)_r}, \end{aligned} \tag{8}$$

where $\psi(\eta_k) = \log \zeta$ as a function of $\eta_k = -\frac{1}{2\sigma_k^2}$, and we have the result introduced in [13] that says that if $x \in \mathcal{M}$ is distributed with a Riemannian Gaussian distribution on $\mathcal{M}$, x_r is distributed as a Riemannian Gaussian distribution on $\mathcal{M}_r$ and $\zeta(\sigma_k) = \prod_{r=1}^{R} \zeta_r(\sigma_k)$. In our case $\zeta_1(\sigma_k) = \sqrt{2\pi m \sigma_k^2}$ $(\psi_1(\eta_k) = \frac{1}{2}\log(-\frac{\pi m}{\eta_k}))$, and then we obtain $\zeta_2(\sigma_k) = \frac{\zeta(\sigma_k)}{\zeta_1(\sigma_k)}$ easily, since $\zeta(\sigma_k)$ has been derived in [6,8]. From (8), we observe that for both components $r = 1, 2$ the Fisher Information matrix is proportional to the identity matrix with a coefficient $\frac{\omega_k}{\sigma_k^4} \frac{\psi_r'(\eta_k)}{dim(\mathcal{M}_r)}$.

We derive now the Riemannian score $u(x_{N+1}; \widehat{\theta}_k^{(N)}) \in T_{\widehat{\overline{x}}_k^{(N)}} P(m)$:

$$u(x_{N+1}; \widehat{\theta}_k^{(N)}) = E\left[v(x, z; \widehat{\overline{x}}_k^{(N)}, \widehat{\sigma}_k^{(N)}) | x_{N+1}\right] = \frac{h_k(x_{N+1}; \widehat{\theta}^{(N)})}{\widehat{\sigma}_k^{2(N)}} \text{Log}_{\widehat{\overline{x}}_k^{(N)}} x_{N+1}.$$

In order to find u_1 and u_2 we have simply to apply the Logarithmic map of Riemannian manifold $\mathcal{M}_1$ and $\mathcal{M}_2$, which in our case are $\mathbb{R}$ and $\mathcal{SP}_m$, respectively, to the component 1 and 2 of x_{N+1} and $\widehat{\overline{x}}_k^{(N)}$:

$$u_1 = \frac{h_k(x_{N+1}; \widehat{\theta}^{(N)})}{\widehat{\sigma}_k^{2(N)}} \left((\widehat{\overline{x}}_k^{(N)})_1 - (x_{N+1})_1 \right)$$

$$u_2 = \frac{h_k(x_{N+1}; \widehat{\theta}^{(N)})}{\widehat{\sigma}_k^{2(N)}} \left(\widehat{\overline{x}}_k^{(N)}\right)_2^{1/2} \log \left(\left(\widehat{\overline{x}}_k^{(N)}\right)_2^{-1/2} (x_{N+1})_2 \left(\widehat{\overline{x}}_k^{(N)}\right)_2^{-1/2} \right) \left(\widehat{\overline{x}}_k^{(N)}\right)_2^{1/2}$$

Expliciting $\psi_r'(\eta_k)$, specifically $\psi_1'(\eta_k) = -\frac{1}{2\eta_k} = \sigma_k^2$ and $\psi_2'(\eta_k) = \psi'(\eta_k) + \frac{1}{2\eta_k}$, we can easily apply the Fisher Information matrix to u_r. In this way we can derive $\xi_1^{(N+1)} = \gamma^{(N+1)} I_1^{-1}(\widehat{\theta}^{(N)}) u_1$ and $\xi_2^{(N+1)} = \gamma^{(N+1)} I_2^{-1}(\widehat{\theta}^{(N)}) u_2$. We are now able to obtain the update rules through the respective exponential maps:

$$\left(\widehat{\overline{x}}_k^{(N+1)}\right)_1 = \left(\widehat{\overline{x}}_k^{(N)}\right)_1 - \xi_1^{(N+1)} \tag{9}$$

$$\left(\widehat{\overline{x}}_k^{(N+1)}\right)_2 = \left(\widehat{\overline{x}}_k^{(N)}\right)_2^{1/2} \exp \left(\left(\widehat{\overline{x}}_k^{(N)}\right)_2^{-1/2} \xi_2^{(N+1)} \left(\widehat{\overline{x}}_k^{(N)}\right)_2^{-1/2} \right) \left(\widehat{\overline{x}}_k^{(N)}\right)_2^{1/2} \tag{10}$$

For the update of dispersion parameters σ_k, we consider $\eta_k = -\frac{1}{2\sigma_k^2}$. Thus, we consider a real parameter, and then our calculus will be done in the classical Euclidean framework. First of all we have $l(x, z; \eta_k) = \log f(x, z; \eta_k) = \sum_{k=1}^{K} \delta_{z,k} \left(-\psi(\eta_k) + \eta_k d_R^2(x, \overline{x}_k) \right)$. Thus, we can derive $v(x, z; \eta_k) = \frac{\partial l}{\partial \eta_k} = \delta_{z,k}(-\psi'(\eta_k) + d_R^2(x, \overline{x}_k))$. Knowing that $I(\eta_k) = \omega_k \psi''(\eta_k)$, we can evaluate the score:

$$u(x_{N+1}; \widehat{\theta}^{(N)}) = E[v(x, z; \eta_k) | x_{N+1}] = h_k(x_{N+1}; \widehat{\theta}^{(N)}) \left(d_R^2 \left(x_{N+1}, \widehat{\overline{x}}_k^{(N)} \right) - \psi'(\widehat{\eta}_k^{(N)}) \right). \tag{11}$$

Hence we can obtain the updated formula for the dispersion parameter

$$\widehat{\eta}_k^{(N+1)} = \widehat{\eta}_k^{(N)} + \gamma^{(N+1)} \frac{h_k(x_{N+1}; \widehat{\theta}^{(N)})}{\widehat{\omega}_k^{(N)} \psi''(\widehat{\eta}_k^{(N)})} \left(d_R^2 \left(x_{N+1}, \widehat{\overline{x}}_k^{(N)} \right) - \psi'(\widehat{\eta}_k^{(N)}) \right), \tag{12}$$

and, obviously $\widehat{\sigma}_k^{2\,(N+1)} = -\frac{1}{2\widehat{\eta}_k^{(N+1)}}$.

4 Simulations

We consider here two simulation frameworks to test the algorithm described in this paper.

The first framework corresponds to the easiest case. Indeed we consider only one mixture component (i.e., $K = 1$). Thus, this corresponds to a simple online mean and dispersion parameter estimate for a Riemannian Gaussian sample.

We consider matrices in $\mathcal{P}_3$ and we analyze three different simulations corresponding to three different value of the barycenter $\overline{x}_1$:

$$\overline{x}_1 = \begin{pmatrix} 1 & 0 & 0 \\ 0 & 1 & 0 \\ 0 & 0 & 1 \end{pmatrix} \; ; \; \overline{x}_1 = \begin{pmatrix} 1 & 0.8 & 0.64 \\ 0.8 & 1 & 0.8 \\ 0.64 & 0.8 & 1 \end{pmatrix} \; ; \; \overline{x}_1 = \begin{pmatrix} 1 & 0.3 & 0.09 \\ 0.3 & 1 & 0.3 \\ 0.09 & 0.3 & 1 \end{pmatrix}$$

The value of dispersion parameter σ is taken equal to 0.1 for the three simulations. We analyze different initial estimates $\widehat{\theta}_{in}$, closer to the true values at the beginning, and further at the end. We focus only on the barycenter, while the initial estimate for σ corresponds to the true value. We consider two different initial values for each simulation. Specifically for case (a), $d_R(\overline{x}_1, \widehat{\overline{x}}_1^{(0)})$ is lower, varying between 0.11 and 0.14. For case (b) it is greater, varying between 1.03 and 1.16. For every simulation we generate $N_{rep} = 100$ samples, each one of $N = 100$ observations. Thus at the end we obtain N_{rep} different estimates $(\widehat{\overline{x}}_{1r}, \widehat{\sigma}_r)$ for every simulation and we can evaluate the mean m and standard deviation s of the error, where the error is measured as the Riemannian distance between $\widehat{\overline{x}}_{1r}$ and $\overline{x}_1$ for the barycenter, and as $|\sigma - \widehat{\sigma}|$ for the dispersion parameter. The results are summarized in Table 1.

Table 1. Mean and standard deviation of the error for the first framework

Simulation	$m_{\overline{x}_1}$	$s_{\overline{x}_1}$	m_σ	s_σ
1	0.0308	0.0092	0.0097	0.0556
2	0.0309	0.0098	0.0117	0.0570
3	0.0308	0.0096	0.0047	0.0051

In the second framework we consider the mixture case, in particular $K = 2$. The true weight are 0.4 and 0.6, while $\sigma_1 = \sigma_2 = 0.1$. The true barycenters are:

$$\overline{x}_1 = \begin{pmatrix} 1 & 0 & 0 \\ 0 & 1 & 0 \\ 0 & 0 & 1 \end{pmatrix} \quad \overline{x}_2 = \begin{pmatrix} 1 & 0.7 & 0.49 \\ 0.7 & 1 & 0.7 \\ 0.49 & 0.7 & 1 \end{pmatrix}$$

We make the initial estimates varying from the true barycenters to some SPD different from the true ones. In particular we analyze three cases. Case (a), where $d_R(\overline{x}_1, \widehat{\overline{x}}_1^{(0)}) = d_R(\overline{x}_2, \widehat{\overline{x}}_2^{(0)}) = 0$; case (b), where $d_R(\overline{x}_1, \widehat{\overline{x}}_1^{(0)}) = 0.2$ and $d_R(\overline{x}_2, \widehat{\overline{x}}_2^{(0)}) = 0.26$; case (c), where $d_R(\overline{x}_1, \widehat{\overline{x}}_1^{(0)}) = d_R(\overline{x}_2, \widehat{\overline{x}}_2^{(0)}) = 0.99$. The results obtained are shown in Table 2. In both frameworks it is clear that we can obtain very good results when starting close to the real parameter values, while the goodness of the estimates becomes weaker as the starting points are further from real values.

Table 2. Mean and standard deviation of the error for the second framework

	m_w	s_w	$m_{\overline{x}_1}$	$s_{\overline{x}_1}$	m_{σ_1}	s_{σ_1}	$m_{\overline{x}_2}$	$s_{\overline{x}_2}$	m_{σ_2}	s_{σ_2}
Case a	0.059	0.077	0.078	0.078	0.142	0.172	0.051	0.050	0.071	0.241
Case b	0.089	0.114	0.119	0.136	0.379	0.400	0.100	0.109	0.265	0.325
Case c	0.515	0.090	1.035	0.215	0.455	0.230	0.812	0.292	0.184	0.323

5 Conclusion

This paper has addressed the problem of the online estimate of mixture model parameters in the Riemannian framework. In particular we dealt with the case of mixtures of Gaussian distributions in the Riemannian manifold of SPD matrices. Starting from a classical approach proposed by Titterington for the Euclidean case, we extend the algorithm to the Riemannian case. The key point was that to look at the innovation part in the step-wise algorithm as an exponential map, or a retraction, in the manifold. Furthermore, an important contribution was that to consider Information Fisher matrix in the Riemannian manifold, in order to implement the Newton algorithm. Finally, we presented some first simulations to validate the proposed method. We can state that, when the starting point of the algorithm is close to the real parameters, we are able to estimate the parameters very accurately. The simulation results suggested us the next future work needed, that is to investigate on the starting point influence in the algorithm, to find some ways to improve convergence towords the good optimum. Another perspective is to apply this algorithm on some real dataset where online estimation is needed.

References

1. Pennec, X., Fillard, P., Ayache, N.: A riemannian framework for tensor computing. Int. J. Comput. Vis. **66**(1), 41–66 (2006)
2. Barachant, A., Bonnet, S., Congedo, M., Jutten, C.: Multiclass brain-computer interface classification by riemannian geometry. IEEE Trans. Biomed. Eng. **59**(4), 920–928 (2012)
3. Arnaudon, M., Barbaresco, F., Yang, L.: Riemannian medians and means with applications to Radar signal processing. IEEE J. Sel. Topics Signal Process. **7**(4), 595–604 (2013)
4. Tuzel, O., Porikli, F., Meer, P.: Pedestrian detection via classification on Riemannian manifolds. IEEE Trans. Pattern Anal. Mach. Intell. **30**(10), 1713–1727 (2008)
5. Dong, G., Kuang, G.: Target recognition in SAR images via classification on riemannian manifolds. IEEE Geosci. Remote Sens. Lett. **21**(1), 199–203 (2015)
6. Said, S., Bombrun, L., Berthoumieu, Y., Manton, J.H.: Riemannian gaussian distributions on the space of covariance matrices. IEEE Trans. Inf. Theory **63**(4), 2153–2170 (2017)
7. Said, S., Bombrun, L., Berthoumieu, Y.: Texture classification using Rao's distance: An EM algorithm on the Poincaré half plane. In: International Conference on Image Processing (ICIP) (2015)

8. Zanini, P., Congedo, M., Jutten, C., Said, S., Berthomieu, Y.: Parameters estimate of riemannian gaussian distribution in the manifold of covariance matrices. In: IEEE Sensor Array and Multichannel Signal Processing Workshop (IEEE SAM 2016) (2016)
9. Titterington, D.: Recursive parameter estimation using incomplete data. J. Royal Stat. Soc. Ser. B (Stat. Methodologies) **46**(2), 257–267 (1984)
10. Cappé, O., Moulines, E.: Online EM algorithm for latent data models. J. Roy. Stat. Soc. Ser. B (Stat. Methodologies) 593–613 (2009)
11. Helgason, S.: Differential Geometry, Lie Groups, and Symmetric Space, vol. 34. American Mathematical Society, Providence (2012)
12. Said, S., Berthoumieu, Y.: Warped metrics for location-scale models (2017). arXiv:1702.07118v1
13. Said, S., Hajri, H., Bombrun, L., Vemuri, B.: Gaussian distributions on riemannian symmetric spaces: statistical learning with structured covariance matrices (2016). arXiv:1607.06929v1

Density Estimation for Compound Cox Processes on Hyperspheres

Florent Chatelain[1], Nicolas Le Bihan[1(✉)], and Jonathan H. Manton[2]

[1] Univ. Grenoble Alpes, CNRS, Grenoble INP, GIPSA-lab (UMR 5216), Grenoble, France
nicolas.le-bihan@gipsa-lab.grenoble-inp.fr

[2] Department of Electrical and Electronic Engineering, The University of Melbourne, Melbourne, Australia

Abstract. Cox multiple scattering processes on hyperspheres are a class of doubly stochastic Poisson processes that can be used to describe scattering phenomenon in Physics (optics, micro-waves, acoustics, etc.). In this article, we present an EM (Expectation Maximization) technique to estimate the concentration parameter of a Compound Cox process with values on hyperspheres. The proposed algorithm is based on an approximation formula for multiconvolution of von Mises Fisher densities on spheres of any dimension.

Keywords: Multiple scattering processes · Hyperspheres · von Mises Fisher distribution · Compound Cox processes · Characteristic function · Parametric estimation · Expectation maximization

1 Introduction

We consider isotropic multiple scattering processes taking values on hyperspheres $\mathcal{S}^{m-1}$. The elements of $\mathcal{S}^{m-1}$ are vectors of unit length in $\mathbb{R}^m$, *i.e.* $\mathcal{S}^{m-1} = \left\{\boldsymbol{x} \in \mathbb{R}^m; \left(\sum_{i=1}^m x_i^2\right)^{1/2} = 1\right\}$ with x_i, $i = 1, \ldots, m$ the components of $\boldsymbol{x}$. Multiple scattering phenomena occur in many areas of Physics, and the studied model here consists of elastic multiple scattering in any dimension. This means that the description of the wave/particle encountering multiple scattering can be performed by modeling of the direction of propagation only (no energy loss). In such situation, one can consider that the trajectory of one particle is simply a random walk on an hypersphere.

In this work, we introduce a Compound Cox process model for multiple scattering in possibly dynamically varying media, and propose an estimation algorithm for the concentration parameter of the scatterers density. This density models the way a heterogeneity scatters the particle/wave, and is based on von Mises Fisher distribution (vMF). The concentration parameter of the vMF is thus a parameter of interest to describe a random medium and we propose an Expectation Maximisation procedure to infer this parameter, given a set of observations at a time t of the output vector in $\mathcal{S}^{m-1}$ of the multiple scattering process.

F. Nielsen and F. Barbaresco (Eds.): GSI 2017, LNCS 10589, pp. 684–691, 2017.
https://doi.org/10.1007/978-3-319-68445-1_79

The model used in this work was introduced originally in [1]. The estimation technique here has computational advantages compared to the one proposed in [2] and is an alternate to the one studied in [3].

2 von Mises Fisher Random Walk on $\mathcal{S}^{m-1}$

A multiple scattering process is made up of an infinite number of contributions from random walks of finite different lengths. In this section, we give the *pdf* and characteristic function expressions for a random walk on hypersphere $\mathcal{S}^{m-1}$. These results are detailed in [1], and summarized here for use in Sect. 4.

An *isotropic* random walk on $\mathcal{S}^{m-1}$ consists of a sequence of vectors starting at $\boldsymbol{x}_0$, with element after k steps given by:

$$\boldsymbol{x}_k = \mathbf{R}_k \boldsymbol{x}_{k-1} = \mathbf{R}_k \mathbf{R}_{k-1} \dots \mathbf{R}_1 \boldsymbol{x}_0 \tag{1}$$

where $\mathbf{R}_i$, $i = 1, \dots, k$ is the rotation associated with the ith step. The sequence of unit vectors $\boldsymbol{x}_0, \dots, \boldsymbol{x}_k$ is *isotropic* and obeys the Markov property condition. This means that the conditional *pdf* $f(\boldsymbol{x}_k | \boldsymbol{x}_{k-1}, \dots, \boldsymbol{x}_0)$ is given by:

$$f(\boldsymbol{x}_k | \boldsymbol{x}_{k-1}, \dots, \boldsymbol{x}_0) = g_{k,k-1}(\boldsymbol{x}_{k-1}^T \boldsymbol{x}_k) \tag{2}$$

which is only a function of the cosine of the angle between $\boldsymbol{x}_{k-1}$ and $\boldsymbol{x}_k$.

In directional statistics, the von Mises Fisher (vMF) distribution is amongst the most popular [4], thanks to its similarity with the normal distribution on the real line. Here, we consider the vMF distribution on $\mathcal{S}^{m-1}$, denoted $M_m(\boldsymbol{\mu}, \kappa)$, with *pdf* given by:

$$f(\boldsymbol{x}; \boldsymbol{\mu}, \kappa) = \frac{\kappa^{m/2-1}}{(2\pi)^{m/2} I_{m/2-1}(\kappa)} e^{\kappa \boldsymbol{\mu}^T \boldsymbol{x}}, \tag{3}$$

where $I_\nu(\cdot)$ is the modified Bessel function [5, p. 374], $\boldsymbol{\mu} \in \mathbb{S}^{m-1}$ is the mean direction and $\kappa \geq 0$ is the *concentration parameter*: the larger the value of κ, the more concentrated is the distribution about the mean direction $\boldsymbol{\mu}$. The *vMF random walk* with concentration parameter κ is thus defined like:

$$\boldsymbol{x}_k | \boldsymbol{x}_{k-1} \sim M_m(\boldsymbol{x}_{k-1}, \kappa),$$

Indeed, this random walk is unimodal with mode $\boldsymbol{\mu} \equiv \boldsymbol{x}_0$ as demonstrated in [1]. In order to provide an expression of the *pdf* of $\boldsymbol{x}_n$, the position of the random walker after n steps on the hypersphere, we first introduce an approximation formula for the multiple convolution of vMF *pdf* over $\mathcal{S}^{m-1}$.

2.1 Convolution of vMF Pdfs

Consider a *vMF* isotropic n-step random walk of over $\mathcal{S}^{m-1}$, then:

$$f_n(\boldsymbol{x}_n; \boldsymbol{\mu}) = (g_{n,n-1} \star \dots \star g_{1,0})\ (\boldsymbol{x}_n) \tag{4}$$

where $f(\boldsymbol{x}_k|\boldsymbol{x}_{k-1}) = g_{k,k-1}(\boldsymbol{x}_k^T \boldsymbol{x}_{k-1})$ for $k = 1, \ldots, n$, $\boldsymbol{x}_0 \equiv \boldsymbol{\mu}$ is the initial direction and where $\star$ represents the convolution over the double coset $SO(m-1)\backslash SO(m)/SO(m-1)$ [1]. The *pdf* $f_n(\boldsymbol{x}_n; \boldsymbol{\mu})$ is unimodal (see [1] for proof and details) with mode $\boldsymbol{\mu}$ and rotationally invariant with respect to $\boldsymbol{\mu}$. As it is well known in directional statistics [4], the vMF distribution is not stable by convolution. Here, we make use of the approximation introduced in [1, Theorem 4.1] for the multiconvolution of vMF *pdf* s with identical mean direction $\boldsymbol{\mu} \in \mathcal{S}^{m-1}$ and concentration parameter κ. In the *high concentration asymptotic* case of large κ and small n, *i.e.* $n/\kappa \to 0$, a n-step vMF random walk $\boldsymbol{x}_n$ is distributed as $M_m(\boldsymbol{\mu}, \tilde{\kappa}_n)$ with:

$$\tilde{\kappa}_n = \frac{\kappa - 1/2}{n} + 1/2 \tag{5}$$

the equivalent concentration parameter. As a consequence, it is possible to approximate the Fourier series (*i.e.* the characteristic function) of $f_n(\boldsymbol{x}_n; \boldsymbol{\mu})$ based on the multiconvolution n-step random walk vMF. First, recall that a vMF *pdf* $f \in L^1(SO(m-1)\backslash SO(m)/SO(m-1), \mathbb{R})$ can be written as:

$$f(\boldsymbol{x}; \boldsymbol{\mu}, \kappa) = \sum_{\ell \geq 0} \beta_{m,\ell} \hat{f}_\ell(\kappa) P_\ell(\boldsymbol{\mu}^T \boldsymbol{x}) \tag{6}$$

where the normalisation constant $\beta_{m,\ell}$ is given by:

$$\beta_{m,\ell} = \frac{1}{\omega_{m-1}} \frac{(2\ell + m - 2)\Gamma(\ell + m - 2)}{\ell! \Gamma(m-1)}$$

for all $\ell \geq 0$ with $\omega_{m-1} = 2\frac{\pi^{m/2}}{\Gamma(m/2)}$ the area of the $(m-1)$-dimensional sphere $\mathbb{S}^{m-1}$. The basis elements $P_\ell(\boldsymbol{\mu}^T \boldsymbol{x})$ are the Legendre polynomials of order ℓ in dimension m, taken at $\boldsymbol{x}$ and with respect to $\boldsymbol{\mu}$, the symmetry axis of f. The coefficients $\hat{f}_\ell(\kappa)$ are given by [6]:

$$\hat{f}_\ell(\kappa) = \mathbb{E}\left[P_\ell(\boldsymbol{\mu}^T \boldsymbol{x})\right] = \frac{I_{\ell+\nu}(\kappa)}{I_\nu(\kappa)} \tag{7}$$

with $\nu = m/2 - 1$ for $\kappa > 0$ and $\ell \geq 0$.

In the case of a isotropic n-step multiconvolution, the coefficients of the Fourier series of the *pdf* given in (4), denoted $\widehat{f_\ell^{\otimes n}}$ can be approximated by:

$$\widehat{f_\ell^{\otimes n}} = \tilde{f}_\ell^n + O\left(\left(\frac{n}{\kappa}\right)^3\right) \tag{8}$$

as $n/\kappa \to 0$, where $\tilde{f}_\ell^n$ are the Legendre coefficients of asymptotic distribution $M_m(\boldsymbol{\mu}, \tilde{\kappa}_n)$ given in (5).

Thus the Fourier series of a n-step random walk can be expressed using the asymptotic approximation introduced in this Section. In the sequel, we will make use of the *high concentration asymptotic* approximation to derive an expression for the *pdf* of a compound Cox process on $\mathcal{S}^{m-1}$.

3 vMF Multiple Scattering Process on $\mathcal{S}^{m-1}$

The study of multiple scattering processes on manifolds has been originally motivated by their usage in describing the behaviour of waves/particles propagating through a random medium [1,7]. Here, we consider the case of a multiple scattering process occurring in $\mathbb{R}^m$, where the direction of propagation of the wave/particle is a unit vector $\boldsymbol{x} \in \mathcal{S}^{m-1}$. Now assuming that during the propagation, the wave/particle encounters a random number of scatterers, then the direction of propagation after a time t, denoted $\boldsymbol{x}_t$, consists of a mixture of weighted n-steps random walks with $n = 0, \ldots, \infty$ and with weights being the probability of having n scattering events during the elapsed time t, *i.e.* $\mathbb{P}(N(t) = n)$. The process $N(t)$ is called the counting process. When the time between two scattering events follows an exponential law, $\boldsymbol{x}_t$ is a Compound Poisson process with weights equal to $e^{-\lambda t}(\lambda t)^n/n!$ and λ the Poisson intensity parameter. The parameter λ can be related to the mean free path of the random medium [7]. In the sequel, we consider the more general case where the Poisson process $N(t)$ is no more homogeneous, *i.e.* its intensity parameter is a random process $\Lambda(t)$. $N(t)$ is then called a Cox process [8]. It is well known [9] that it that case, the probability $\mathbb{P}(N(t) = n)$ takes the form:

$$\mathbb{P}(N(t) = n) = \mathcal{P}_n\left[f_{\Lambda(t)}\right] = \int_0^{\infty} \frac{e^{-\lambda t}(\lambda t)^n}{n!} f_{\Lambda(t)}(\lambda_t) d\lambda_t \tag{9}$$

where $\mathcal{P}_n\left[f_{\Lambda(t)}\right]$ is the *Poisson transform* of the *pdf* $f_{\Lambda(t)}$. It is then possible to give the expression of the *pdf* of $\boldsymbol{x}_t$ in the case where the n-steps random walks are vMFs. Using the high concentration approximation introduced in (8), for a given intensity distribution of $\Lambda(t)$, in the limit of large κ, one gets:

$$f(\boldsymbol{x}_t; \mu, \kappa) \simeq \mathcal{P}_0\left[f_{\Lambda(t)}\right] \delta_{\boldsymbol{\mu}}(\boldsymbol{x}_t) + \sum_{n \geq 1} \mathcal{P}_n\left[f_{\Lambda(t)}\right] f(\boldsymbol{x}_t; \mu, \tilde{\kappa}_n) \tag{10}$$

where $f(\cdot\,; \mu, \tilde{\kappa}_n)$ is the vMF *pdf* of the n-step random walk, *i.e.* $M_m(\boldsymbol{\mu}, \tilde{\kappa}_n)$[1]. Note that this expression is valid if there exists $a > 0$ such that $\mathbb{E}[\Lambda(t)^a] < +\infty$. In practical applications, it is required that only a small number of scattering events that weakly pertubate the direction of propagation occurred, for the random medium to be identified. The high concentration approximation used here fits exactly this framework. The high concentration approximation in (10) will be used in Sect. 4 for estimation purpose.

3.1 Compound Cox Process with Gamma Intensity Distribution

Thus for a given $t > 0$, the intensity of the Poisson process is now assumed to be Gamma distributed, i.e. $\Lambda(t) \sim \mathcal{G}(r_t, p)$ where $r_t > 0$ is a fixed and known

[1] Here we make use of a notation abuse by expressing $f(\boldsymbol{x}_t; \mu, \kappa)$ as a sum of dirac measure and a *pdf*. It has to be understood the following way: when $\boldsymbol{x} = \boldsymbol{\mu}$, it equals the dirac mass, and for other cases it equals the density function.

shape parameter and $p > 0$ is an unknown scale parameter. This means that the density of the mixing process reads

$$f_{\Lambda(t)}(x) = \frac{x^{r_t - 1}}{\Gamma(r_t) p^{r_t}} e^{-\frac{x}{p}}, \tag{11}$$

for all $x \geq 0$. In this case the counting process $N(t)$ for the scattering events obeys a negative binomial $\mathcal{NB}(r_t, q)$ distribution with stopping-time r_t and success probability $q = p/(p+1)$:

$$\Pr(N(t) = n) = \mathcal{P}_n[f_{\Lambda_t}] = \frac{\Gamma(n + r_t)}{n!\Gamma(r_t)} \frac{p^k}{(p+1)^{n+r_t}}, \tag{12}$$

for all $n \in \mathbb{N}$.

In this case the high concentration distribution given in (10) is still valid and the Poisson transform masses are given in (12).

4 Estimation

The distribution of the multiple scattering process yields a likelihood function which is a product of Fourier series. This function is quite expensive to calculate. In such situations, approximate Bayesian computation (ABC) is a popular likelihood-free method that can be used to perform Bayesian inference, as shown in [2], at a price of a high computational cost.

However the asymptotic expression (10) allows us to model the multiple scattering process as a (infinite) mixture model of vMF distributions. As a consequence, it becomes quite straightforward to apply classical procedures to estimate the parameters of this multiple scattering process. Such procedures include expectation-maximization (EM) algorithm, or Gibbs sampling in a Bayesian framework. One benefit of these standard estimation methods is the possibility to reduce significantly the computational burden w.r.t. likelihood-free methods such as ABC.

4.1 EM Algorithm

In this work, we derive an EM algorithm in order to estimate the parameters of vMF compound Cox process when the intensity of the Poisson process is Gamma distributed. Observations are assumed to be made at a given time t. Let $\boldsymbol{x} = (\boldsymbol{x}_1, \ldots \boldsymbol{x}_N)$ be a sample of N independent observations in $\mathcal{S}^{m-1}$ from a vMF multiple scattering process with known initial direction $\boldsymbol{\mu}$ and whose intensity $\Lambda(t)$ is Gamma distributed as defined in (11). The vector of the unknown parameter of the multiple scattering process to be estimated is $\boldsymbol{\theta} = (p, \kappa)$, where p is related to the intensity of the Poisson process, and κ is the concentration parameter of the vMF distribution given in (3). In order to derive the EM equations, we can introduce the following binary latent variables for all $1 \leq i \leq N$, $n \geq 0$,

$$z_{i,n} = \begin{cases} 1 & \text{if } N(t) = n \text{ for the sample } \boldsymbol{x}_i, \\ 0 & \text{otherwise.} \end{cases}$$

Then the log-likelihood of the complete data $(\boldsymbol{x}, \boldsymbol{z})$, where $\boldsymbol{z}$ is the set of all the z_{in}, can be (approximately) computed based on the high concentration distribution (10) as

$$\ell(\boldsymbol{\theta}; \boldsymbol{x}, \boldsymbol{z}) \simeq \sum_{i=1}^{N} \sum_{n \geq 0} z_{in} \log \left[\mathcal{P}_n \left[f_{\Lambda_t}\right] f(\boldsymbol{x}_i; \boldsymbol{\mu}, \tilde{\kappa}_n)\right],$$

with by convention $f(\boldsymbol{x}_i; \boldsymbol{\mu}, \tilde{\kappa}_0) = \delta_{\boldsymbol{\mu}}(\boldsymbol{x}_i)$.

E Step. Given $\boldsymbol{x}$ and a current estimate of the parameter vector $\boldsymbol{\theta}^{(\alpha)} = \left(p^{(\alpha)}, \kappa^{(\alpha)}\right)$, the conditional expected value of the log-likelihood reads

$$Q\left(\boldsymbol{\theta} | \boldsymbol{\theta}^{(\alpha)}\right) = \sum_{i=1}^{N} \sum_{n \geq 0} t_{i,n} \log \left[\mathcal{P}_n \left[f_{\Lambda_t}\right] f(\boldsymbol{x}_i; \boldsymbol{\mu}, \tilde{\kappa}_n)\right], \tag{13}$$

where $t_{i,n} = E\left[z_{i,n} | \boldsymbol{x}, \boldsymbol{\theta}^{(\alpha)}\right] = \Pr\left(z_{i,n} = 1 | \boldsymbol{x}, \boldsymbol{\theta}^{(\alpha)}\right)$ for all $1 \leq i \leq N$, $n \geq 0$. Using Bayes rule, it comes that

$$t_{i,n} = \frac{\mathcal{P}_n \left[f_{\Lambda_t}\right] f(\boldsymbol{x}_i; \boldsymbol{\mu}, \tilde{\kappa}_n^{(\alpha)})}{\sum_{l \geq 0} \mathcal{P}_l \left[f_{\Lambda_t}\right] f(\boldsymbol{x}_i; \boldsymbol{\mu}, \tilde{\kappa}_l^{(\alpha)})}, \tag{14}$$

where $\tilde{\kappa}_n^{(\alpha)} = \frac{\kappa^{(\alpha)} - 1/2}{n} + 1/2$ as defined in (5) and $\mathcal{P}_n \left[f_{\Lambda_t}\right]$ is obtained by replacing the scale parameter p by its current value $p^{(\alpha)}$ in (12).

M Step. In order to maximize $Q\left(\boldsymbol{\theta} | \boldsymbol{\theta}^{(\alpha)}\right)$, it is useful to note that this objective function is separable w.r.t. the two parameters p and κ and can thus be maximized independently.

The scale parameter p is obtained by maximizing $p \mapsto \sum_{i=1}^{N} \sum_{n \geq 0} t_{i,n} \log \mathcal{P}_n \left[f_{\Lambda_t}\right]$, where the negative binomial masses $\mathcal{P}_n \left[f_{\Lambda_t}\right]$ are given in (12). This has the same form as a weighted maximum likelihood estimator for a negative binomial distribution. Moreover the negative binomial distribution $\mathcal{NB}(r_t, p)$ with fixed stopping-time parameter r_t describes a natural exponential family with mean parameter equal to $r_t p$. Thus it comes directly that

$$p^{(\alpha+1)} = \frac{1}{r_t} \frac{1}{N} \sum_{i=1}^{N} \sum_{n \geq 0} n t_{i,n}. \tag{15}$$

The concentration parameter κ is obtained by maximizing

$$\kappa \mapsto Q\left(\boldsymbol{\theta} | \boldsymbol{\theta}^{(\alpha)}\right) = \sum_{i=1}^{N} \sum_{n \geq 1} t_{i,n} \left(\log c_m \left(\tilde{\kappa}_n\right) + \tilde{\kappa}_n \boldsymbol{\mu}^T \boldsymbol{x}_i\right) + \text{constant},$$

where $c_m(\kappa) = \frac{\kappa^{m/2-1}}{(2\pi)^{m/2} I_{m/2-1}(\kappa)}$ is the vMF normalizing constant, and $\tilde{\kappa}_n$ depends on κ as defined in (5). By differentiating this function w.r.t. κ, we obtain the following score function

$$v(\kappa) = \sum_{i=1}^{N} \sum_{n \geq 1} t_{i,n} \frac{-A_m(\tilde{\kappa}_n) + \boldsymbol{\mu}^T \boldsymbol{x}_i}{n},$$

where $A_m(\kappa) = \frac{-c'_m(\kappa)}{c_m(\kappa)} = \frac{I_{m/2}(\kappa)}{I_{m/2-1}(\kappa)}$. For a fixed direction $\boldsymbol{\mu}$ the distribution of the scalar $\boldsymbol{\mu}^T \boldsymbol{y}$, where $\boldsymbol{y}$ is vMF distributed as $M_m(\boldsymbol{\mu}, \kappa)$, describes a natural exponential family on the set $\kappa > 0$. Thus the associated log-likelihood is strictly concave w.r.t. κ and $v(\kappa)$ is monotonically decreasing as a positively weighted sum of monotonically decreasing function. This shows that the value that maximizes $\kappa \mapsto Q\left(\boldsymbol{\theta}|\boldsymbol{\theta}^{(\alpha)}\right)$ is the unique zero of $v(\kappa)$. Unfortunately there is no tractable closed-form expression of this zero. However, similarly to maximum likelihood procedures for vMF distributions [10], it is possible to perform one Newton's iteration to improve the objective criterion $Q\left(\boldsymbol{\theta}|\boldsymbol{\theta}^{(\alpha)}\right)$. Indeed such improvement ensures, as a standard property of EM algorithm, that the marginal likelihood of the observations $\boldsymbol{x}$ is improved. This yields the following update rule

$$\kappa^{(\alpha+1)} = \kappa^{(\alpha)} - \frac{v(\kappa^{(\alpha)})}{v'(\kappa^{(\alpha)})}, \tag{16}$$

where $v'(\kappa) = -\sum_{n \geq 1} \frac{A'_m(\tilde{\kappa}_n)}{n^2} \sum_{i=1}^{N} t_{i,n}$ is the derivative of the score function $v(\kappa)$. Moreover $A'_m(\kappa)$ can be efficiently computed using $A'_m(\kappa) = 1 - A_m(\kappa)^2 - \frac{m-1}{\kappa} A_m(\kappa)$ as shown in [4].

Note finally that the EM algorithm can be initialized by using method of moment estimates as defined in [3].

4.2 Simulation Results

Several simulations have been conducted on synthetic data to evaluate the statistical performances of the estimates given by the EM procedure.

Note that in practice, to compute the EM equations and thus to implement the EM algorithm, we can truncate the infinite series in (14), (15) and (16) to compute the probabilities $t_{i,n}$ and the updated estimates. This can be done by introducing a tolerance parameter $\epsilon > 0$ and the associated index upper bound $L_\epsilon = \arg\min_{L \in \mathbb{N}} \sum_{l=0}^{L} \mathcal{P}_l \left[f_{\Lambda_t} \right] > 1 - \epsilon$. Thus it comes that

$$t_{i,n} \approx \frac{\mathcal{P}_n \left[f_{\Lambda_t} \right] f(\boldsymbol{x}_i; \boldsymbol{\mu}, \tilde{\kappa}_n^{(\alpha)})}{\sum_{l=0}^{L_\epsilon} \mathcal{P}_l \left[f_{\Lambda_t} \right] f(\boldsymbol{x}_i; \boldsymbol{\mu}, \tilde{\kappa}_l^{(\alpha)})}, \tag{17}$$

for $0 \leq n \leq L_\epsilon$, and $t_{i,n} \approx 0$ for $n > L_\epsilon$.

Table 1. Performances of EM estimates $(\hat{p}, \hat{\kappa})$ as a function of the true parameters p and κ, from 1000 Monte-Carlo runs, for a vMF multiple scattering process with $\mathcal{G}(r_t, p)$ mixing distribution ($r_t = 1$). Sample size is $N = 1000$, and samples belong to $\mathcal{S}^2$ ($m = 3$).

p		1		2		3		5		10	
		$\widehat{p}$	$\widehat{\kappa}$	$\widehat{p}$	$\widehat{\kappa}$	$\widehat{p}$	$\widehat{\kappa}$	$\widehat{p}$	$\widehat{\kappa}$	$\widehat{p}$	$\widehat{\kappa}$
$\kappa = 20$	bias	0.011	0.275	0.025	0.447	0.043	0.695	0.141	1.386	0.330	2.996
	stdev	0.059	1.288	0.121	1.281	0.195	1.344	0.327	1.313	0.752	1.496
$\kappa = 50$	bias	0.007	0.425	0.022	0.624	0.042	0.975	0.151	1.914	0.437	3.433
	stdev	0.059	3.011	0.125	3.063	0.193	3.241	0.337	3.564	0.752	3.966
$\kappa = 100$	bias	0.007	0.502	0.020	0.898	0.041	1.168	0.129	2.743	0.446	5.097
	stdev	0.060	6.297	0.124	6.198	0.190	6.660	0.336	6.971	0.756	8.132
$\kappa = 200$	bias	0.007	1.477	0.021	2.178	0.042	2.792	0.140	5.519	0.484	9.486
	stdev	0.058	12.28	0.123	12.55	0.199	13.54	0.319	13.80	0.749	15.58

The biases and the standard deviations of EM estimates are reported in Table 1. These numerical results underline the accuracy of the EM estimators based on the high concentration approximation even when the ratio κ/p is not very large, e.g. $\kappa/p \geq 2$.

References

1. Le Bihan, N., Chatelain, F., Manton, J.: Isotropic multiple scattering processes on hyperspheres. IEEE Trans. Inf. Theory **62**, 5740–5752 (2016)
2. Chatelain, F., Le Bihan, N., Manton, J.: Parameter estimation for multiple scattering process on the sphere. In: IEEE International Conference on Acoustics, Speech and Signal Processing (ICASSP) (2015)
3. Chatelain, F., Le Bihan, N.: von-mises fisher approximation of multiple scattering process on the hypershpere. In: IEEE International Conference on Acoustics, Speech and Signal Processing (ICASSP) (2013)
4. Mardia, K., Jupp, P.: Directional Statistics. Wiley, New York (2000)
5. Abramowitz, M., Stegun, I.A.: Handbook of Mathematical Functions: With Formulas, Graphs, and Mathematical Tables. Dover Publications, New York (1972)
6. Kent, J.: Limiting behaviour of the von Mises-Fisher distribution. Math. Proc. Cambridge Philos. Soc. **84**, 531–536 (1978)
7. Le Bihan, N., Margerin, L.: Nonparametric estimation of the heterogeneity of a random medium using compound poisson process modeling of wave multiple scattering. Phys. Rev. E **80**, 016601 (2009)
8. Lefebvre, M.: Applied Stochastic Processes. Springer, New York (2006)
9. Saleh, B.: Photoelectron Statistics. Springer, Heidelberg (1978)
10. Sra, S.: A short note on parameter approximation for von mises-fisher distributions. Comput. Stat. **27**, 177–190 (2012)

Maximum Likelihood Estimators on Manifolds

Hatem Hajri[1(✉)], Salem Said[2], and Yannick Berthoumieu[2]

[1] Institut Vedecom, 77 rue des Chantiers, Versailles, France
hatem.hajri@vedecom.fr

[2] Laboratoire IMS (CNRS - UMR 5218), Université de Bordeaux, Bordeaux, France
{salem.said,yannick.berthoumieu}@ims-bordeaux.fr

Abstract. Maximum likelihood estimator (MLE) is a well known estimator in statistics. The popularity of this estimator stems from its asymptotic and universal properties. While asymptotic properties of MLEs on Euclidean spaces attracted a lot of interest, their studies on manifolds are still insufficient. The present paper aims to give a unified study of the subject. Its contributions are twofold. First it proposes a framework of asymptotic results for MLEs on manifolds: consistency, asymptotic normality and asymptotic efficiency. Second, it extends popular testing problems on manifolds. Some examples are discussed.

Keywords: Maximum likelihood estimator · Consistency · Asymptotic normality · Asymptotic efficiency of MLE · Statistical tests on manifolds

1 Introduction

Density estimation on manifolds has many applications in signal and image processing. To give some examples of situations, one can mention

Covariance matrices: In recent works [1–5], new distributions called Gaussian and Laplace distributions on manifolds of covariance matrices (positive definite, Hermitian, Toeplitz, Block Toeplitz...) are introduced. Estimation of parameters of these distributions has led to various applications (image classification, EEG data analysis, etc.).

Stiefel and Grassmann manifolds: These manifolds are used in various applications such as pattern recognition [6–8] and shape analysis [9]. Among the most studied density functions on these manifolds, one finds the Langevin, Bingham and Gaussian distributions [10]. In [6–8], maximum likelihood estimations of the Langevin and Gaussian distributions are applied for tasks of activity recognition and video-based face recognition.

Lie groups: Lie groups arise in various problems of signal and image processing such as localization, tracking [11,12] and medical image processing [13]. In [13], maximum likelihood estimation of new distributions on Lie groups, called Gaussian distributions, is performed and applications are given in medical image processing. The recent work [4] proposes new Gaussian distributions on Lie

F. Nielsen and F. Barbaresco (Eds.): GSI 2017, LNCS 10589, pp. 692–700, 2017.
https://doi.org/10.1007/978-3-319-68445-1_80

groups and a complete program, based on MLE, to learn data on Lie groups using these distributions.

The present paper is structured as follows. Section 2 focuses on consistency of MLE on general metric spaces. Section 3 discusses asymptotic normality and asymptotic efficiency of MLE on manifolds. Finally Sect. 4 presents some hypothesis tests on manifolds.

2 Consistency

In this section it is shown that, under suitable conditions, MLEs on general metric spaces are consistent estimators. The result given here may not be optimal. However, in addition to its simple form, it is applicable to several examples of distributions on manifolds as discussed below.

Let (Θ, d) denote a metric space and let $\mathcal{M}$ be a measurable space with μ a positive measure on it. Consider $(\mathbb{P}_\theta)_{\theta\in\Theta}$ a family of distributions on $\mathcal{M}$ such that $\mathbb{P}_\theta(dx) = f(x,\theta)\mu(dx)$ and $f > 0$.

If $x_1, \cdots, x_n$ are independent random samples from $\mathbb{P}_{\theta_0}$, a maximum likelihood estimator is any $\hat{\theta}_n$ which solves

$$\max_\theta L_n(\theta) = L_n(\hat{\theta}_n) \text{ where } L_n(\theta) = \frac{1}{n}\sum_{i=1}^{n} \log f(x_i, \theta)$$

The main result of this section is Theorem 1 below. The notation $\mathbb{E}_\theta[g(x)]$ stands for $\int_{\mathcal{M}} g(y) f(y,\theta)\mu(dy)$.

Theorem 1. *Assume the following assumptions hold for some $\theta_0 \in \Theta$.*

(1) For all x, $f(x,\theta)$ is continuous with respect to θ.
(2) $\mathbb{E}_{\theta_0}[|\log f(x,\theta)|] < \infty$ for all θ, $L(\theta) = \mathbb{E}_{\theta_0}[\log f(x,\theta)]$ is continuous on Θ and uniquely maximized at θ_0.
(3) For all compact K of Θ,

$$Q(\delta) := \mathbb{E}_{\theta_0}[\sup\{|\log f(x,\theta) - \log f(x,\theta')| : \theta, \theta' \in K, d(\theta,\theta') \leq \delta\}]$$

satisfies $\lim_{\delta\to 0} Q(\delta) = 0$.
Let $x_1, \cdots, x_n, \cdots$ be independent random samples of $\mathbb{P}_{\theta_0}$. For every compact K of Θ, the following convergence holds in probability

$$\lim_{n\to\infty} \sup_{\theta\in K} |L_n(\theta) - L(\theta)| = 0$$

Assume moreover
(4) There exists a compact $K_0 \subset \Theta$ containing θ_0 such that

$$\mathbb{E}_{\theta_0}[|\sup\{\log f(x,\theta) : \theta \in K_0^c\}|] < \infty$$

and

$$\mathbb{E}_{\theta_0}[\sup\{\log f(x,\theta) : \theta \in K_0^c\}] < L(\theta_0)$$

Then, whenever $\hat{\theta}_n$ exists and is unique for all n, it satisfies $\hat{\theta}_n$ converges to θ_0 in probability.

Proof. Since L is a deterministic function, it is enough to prove, for every compact K,

(i) Convergence of finite dimensional distributions: $(L_n(\theta_1), \cdots, L_n(\theta_p))$ weakly converges to $(L(\theta_1), \cdots, L(\theta_p))$ for any $\theta_1, \cdots, \theta_p \in K$.
(ii) Tightness criterion: for all $\varepsilon > 0$,

$$\lim_{\delta \to 0} \limsup_{n \to \infty} \mathbb{P}\Big(\sup_{\theta, \theta' \in K, d(\theta, \theta') < \delta} |L_n(\theta) - L_n(\theta')| > \varepsilon \Big) = 0$$

Fact (i) is a consequence of the first assumption in (2) and the strong law of large numbers (SLLN). For (ii), set $F = \{(\theta, \theta') \in K^2, d(\theta, \theta') < \delta\}$ and note

$$\mathbb{P}\big(\sup_F |L_n(\theta) - L_n(\theta')| > \varepsilon \big) \leq \mathbb{P}(Q_n(\delta) > \varepsilon)$$

where $Q_n(\delta) = \frac{1}{n}\sum_{i=1}^n \sup_F |\log f(x_i, \theta) - \log f(x_i, \theta')|$. By assumption (3), there exists $\delta_0 > 0$ such that $Q(\delta) \leq Q(\delta_0) < \varepsilon$ for all $\delta \leq \delta_0$. An application of the SLLN shows that, for all $\delta \leq \delta_0$, $\lim_n Q_n(\delta) = Q(\delta)$ and consequently

$$\limsup_{n \to \infty} \mathbb{P}(Q_n(\delta) > \varepsilon) = \limsup_{n \to \infty} \mathbb{P}(Q_n(\delta) - Q(\delta) > \varepsilon - Q(\delta)) = 0$$

This proves fact (ii). Assume (4) holds. The bound

$$\mathbb{P}(\hat{\theta}_n \notin K_0) \leq \mathbb{P}(\sup_{K_0^c} L_n(\theta) > \sup_{K_0} L_n(\theta)) \leq \mathbb{P}(\sup_{K_0^c} L_n(\theta) > L_n(\theta_0))$$

and the inequality $\sup_{\theta \in K_0^c} L_n(\theta) \leq \frac{1}{n}\sum_{i=1}^n \sup_{\theta \in K_0^c} \log f(x_i, \theta)$ give

$$\mathbb{P}(\hat{\theta}_n \notin K_0) \leq \mathbb{P}\left(\frac{1}{n} \sum_{i=1}^n \sup_{\theta \in K_0^c} \log f(x_i, \theta) > L_n(\theta_0) \right)$$

By the SLLN, $\limsup_n \mathbb{P}(\hat{\theta}_n \notin K_0) \leq 1_{\{\mathbb{E}_{\theta_0}[\sup_{\theta \in K_0^c} \log f(x, \theta)] \geq L(\theta_0)\}} = 0$. With $K_0(\varepsilon) := \{\theta \in K_0 : d(\theta, \theta_0) \geq \varepsilon\}$, one has

$$\mathbb{P}(d(\hat{\theta}_n, \theta_0) \geq \varepsilon) \leq \mathbb{P}(\hat{\theta}_n \in K_0(\varepsilon)) + \mathbb{P}(\hat{\theta}_n \notin K_0)$$

where $\mathbb{P}(\hat{\theta}_n \in K_0(\varepsilon)) \leq \mathbb{P}(\sup_{K_0(\varepsilon)} L_n > L_n(\theta_0))$. Since L_n converges to L uniformly in probability on $K_0(\varepsilon)$, $\sup_{K_0(\varepsilon)} L_n$ converges in probability to $\sup_{K_0(\varepsilon)} L$ and so $\limsup_n \mathbb{P}(d(\hat{\theta}_n, \theta_0) \geq \varepsilon) = 0$ using assumption (2).

2.1 Some Examples

In the following some distributions which satisfy assumptions of Theorem 1 are given. More examples will be discussed in a forthcoming paper.

(i) Gaussian and Laplace distributions on $\mathcal{P}_m$. Let $\Theta = \mathcal{M} = \mathcal{P}_m$ be the Riemannian manifold of symmetric positive definite matrices of size $m \times m$ equipped with Rao-Fisher metric and its Riemannian distance d called Rao's distance. The Gaussian distribution on $\mathcal{P}_m$ as introduced in [1] has density with respect to the Riemannian volume given by $f(x,\theta) = \frac{1}{Z_m(\sigma)} \exp\left(-\frac{d^2(x,\theta)}{2\sigma^2}\right)$ where $\sigma > 0$ and $Z_m(\sigma) > 0$ is a normalizing factor only depending on σ.

Points (1) and (3) in Theorem 1 are easy to verify. Point (2) is proved in Proposition 9 [1]. To check (4), define $O = \{\theta : d(\theta,\theta_0) > \varepsilon\}$ and note

$$\mathbb{E}_{\theta_0}[\sup_{O}(-d^2(x,\theta))] \leq \mathbb{E}_{\theta_0}[\sup_{O}(-d^2(x,\theta))1_{2d(x,\theta_0)\leq\varepsilon-1}] \tag{1}$$

By the triangle inequality $-d^2(x,\theta) \leq -d(x,\theta_0)^2 + 2d(\theta,\theta_0)d(x,\theta_0) - d^2(\theta,\theta_0)$ and consequently (1) is smaller than

$$\mathbb{E}_{\theta_0}[\sup_{O}(2d(\theta,\theta_0)d(x,\theta_0) - d^2(\theta,\theta_0))1_{2d(x,\theta_0)\leq\varepsilon-1}]$$

But if $2d(x,\theta_0) \leq \varepsilon - 1$ and $d(\theta,\theta_0) > \varepsilon$,

$$2d(\theta,\theta_0)d(x,\theta_0) - d^2(\theta,\theta_0) < d(\theta,\theta_0)(\varepsilon - 1 - \varepsilon) < -\varepsilon$$

Finally (1) $\leq -\varepsilon$ and this gives (4) since $K_0 = O^c$ is compact.

Let $x_1, \cdots, x_n, \cdots, ...$ be independent samples of $f(\cdot,\theta_0)$. The MLE based on these samples is the Riemannian mean $\hat{\theta}_n = \text{argmin}_\theta \sum_{i=1}^n d^2(x_i,\theta)$. Existence and uniqueness of $\hat{\theta}_n$ follow from [14]. Theorem 1 shows the convergence of $\hat{\theta}_n$ to θ_0. This convergence was proved in [1] using results of [15] on convergence of empirical barycenters.

(ii) Gaussian and Laplace distributions on symmetric spaces. Gaussian distributions can be defined more generally on Riemannian symmetric spaces [4]. MLEs of these distributions are consistent estimators [4]. This can be recovered by applying Theorem 1 as for $\mathcal{P}_m$. In the same way, it can be checked that Laplace distributions on $\mathcal{P}_m$ [2] and symmetric spaces satisfy assumptions of Theorem 1 and consequently their estimators are also consistent. Notice, for Laplace distributions, MLE coincides with the Riemannian median $\hat{\theta}_n = \text{argmin}_\theta \sum_{i=1}^n d(x_i,\theta)$.

3 Asymptotic Normality and Asymptotic Efficiency of the MLE

Let Θ be a smooth manifold with dimension p equipped with an affine connection ∇ and an arbitrary distance d. Consider $\mathcal{M}$ a measurable space equipped with a positive measure μ and $(\mathbb{P}_\theta)_{\theta\in\Theta}$ a family of distributions on $\mathcal{M}$ such that $\mathbb{P}_\theta(dx) = f(x,\theta)\mu(dx)$ and $f > 0$.

Consider the following generalization of estimating functions [16].

Definition 1. *An estimating form is a function $\omega : \mathcal{M} \times \Theta \longrightarrow T^*\Theta$ such that for all $(x,\theta) \in \mathcal{M} \times \Theta$, $\omega(x,\theta) \in T^*_\theta\Theta$ and $\mathbb{E}_\theta[\omega(x,\theta)] = 0$ or equivalently $\mathbb{E}_\theta[\omega(x,\theta)X_\theta] = 0$ for all $X_\theta \in T_\theta\Theta$.*

Assume $l(x,\theta) = log(f(x,\theta))$ is smooth in θ and satisfies appropriate integrability conditions, then differentiating with respect to θ, the identity $\int_{\mathcal{M}} f(x,\theta)\mu(dx) = 1$, one finds $\omega(x,\theta) = dl(x,\theta)$ is an estimating form.

The main result of this section is the following

Theorem 2. *Let $\omega : \mathcal{M} \times \Theta \longrightarrow T^*\Theta$ be an estimating form. Fix $\theta_0 \in \Theta$ and let $(x_n)_{n\geq 1}$ be independent samples of $\mathbb{P}_{\theta_0}$. Assume*

(i) There exist $(\hat{\theta}_N)_{N\geq 1}$ such that $\sum_{n=1}^N \omega(x_n, \hat{\theta}_N) = 0$ for all N and $\hat{\theta}_N$ converges in probability to θ_0.

(ii) For all $u, v \in T_{\theta_0}\Theta$, $\mathbb{E}_{\theta_0}[|\nabla\omega(x,\theta_0)(u,v)|] < \infty$ and there exists $(e_a)_{a=1,\cdots,p}$ a basis of $T_{\theta_0}\Theta$ such that the matrix A with entries $A_{a,b} = \mathbb{E}_{\theta_0}[\nabla\omega(x,\theta_0)(e_a,e_b)]$ is invertible.

(iii) The function $R(\delta) =$

$$\mathbb{E}_{\theta_0}[\sup_{t\in[0,1],\overline{\theta}\in B(\theta_0,\delta)} |\nabla\omega(x,\gamma(t))(e_a(t),e_b(t)) - \nabla\omega(x,\theta_0)(e_a,e_b)|]$$

satisfies $\lim_{\delta\to 0} R(\delta) = 0$ where $(e_a, a = 1\cdots,p)$ is a basis of $T_{\theta_0}\Theta$ as in (ii) and $e_a(t), t \in [0,1]$ is the parallel transport of e_a along γ the unique geodesic joining θ_0 and $\bar{\theta}$.

Let $Log_\theta(\hat{\theta}_N) = \sum_{a=1}^p \Delta_a e_a$ be the decomposition of $Log_\theta(\hat{\theta}_N)$ in the basis $(e_a)_{a=1,\cdots,p}$. The following convergence holds in distribution as $N \longrightarrow \infty$

$$\sqrt{N}(\Delta_1,\cdots,\Delta_p)^T \Rightarrow \mathcal{N}(0,(A^\dagger)^{-1}\Gamma A^{-1})$$

where Γ is the matrix with entries $\Gamma_{a,b} = \mathbb{E}_{\theta_0}[\omega(x,\theta_0)e_a.\omega(x,\theta_0)e_b]$.

Proof. Take V a small neighborhood of θ_0 and let $\gamma : [0,1] \longrightarrow V$ be the unique geodesic contained in V such that $\gamma(0) = \theta_0$ and $\gamma(1) = \hat{\theta}_N$. Let $(e_a, a = 1\cdots,p)$ be a basis of $T_{\theta_0}\Theta$ as in (ii) and define $e_a(t), t \in [0,1]$ as the parallel transport of e_a along γ: $\frac{De_a(t)}{dt} = 0,\ t \in [0,1],\ e_a(0) = e_a$ where D is the covariant derivative along γ. Introduce

$$\omega_N(\theta) = \sum_{n=1}^N \omega(x_n,\theta) \text{ and } F_a(t) = \omega_N(\gamma(t))(e_a(t))$$

By Taylor formula, there exists $c_a \in [0,1]$ such that

$$F_a(1) = F_a(0) + F_a'(c_a) \tag{2}$$

Note $F_a(1) = 0, F_a(0) = \omega_N(\theta_0)(e_a)$ and $F_a'(t) = (\nabla\omega_N)(\gamma'(t),e_a(t)) = \sum_b \Delta_b(\nabla\omega_N)(e_b(t),e_a(t))$. In particular, $F_a'(0) = \sum_b \Delta_b(\nabla\omega_N)(e_b,e_a)$. Dividing (2) by $\sqrt{N}$, gives

$$-\frac{1}{\sqrt{N}}\omega_N(\theta_0)(e_a) = \frac{1}{\sqrt{N}}\sum_b \Delta_b(\nabla\omega_N)(e_b(c_a),e_a(c_a)) \tag{3}$$

Define $Y^N = \left(-\frac{1}{\sqrt{N}}\omega_N(\theta_0)(e_1), \cdots, -\frac{1}{\sqrt{N}}\omega_N(\theta_0)(e_p)\right)^\dagger$ and let A_N be the matrix with entries $A_N(a,b) = \frac{1}{N}(\nabla\omega_N)(e_a(c_a), e_b(c_a))$. Then (3) writes as $Y^N = (A_N)^\dagger(\sqrt{N}\Delta_1, \cdots, \sqrt{N}\Delta_p)^\dagger$. Since $\mathbb{E}_{\theta_0}[\omega(x,\theta_0)] = 0$, by the central limit theorem, Y^N converges in distribution to a multivariate normal distribution with mean 0 and covariance Γ. Note

$$A^N_{a,b} = \frac{1}{N}(\nabla\omega_N)(e_a, e_b) + R^N_{a,b}$$

where $R^N_{a,b} = \frac{1}{N}(\nabla\omega_N)(e_a(c_a), e_b(c_a)) - \frac{1}{N}(\nabla\omega_N)(e_a, e_b)$. By the SLLN and assumption (ii), the matrix B_N with entries $B_N(a,b) = \frac{1}{N}(\nabla\omega_N)(e_a, e_b)$ converges almost surely to the matrix A. Note $|R^N_{a,b}|$ is bounded by

$$\frac{1}{N}\sum_{n=1}^{N} \sup_{t\in[0,1]} \sup_{\overline{\theta}\in B(\theta_0,\delta)} |\nabla\omega(x_n, \gamma(t))(e_a(t), e_b(t)) - \nabla\omega(x_n, \theta_0)(e_a, e_b)|$$

By the SLLN, for δ small enough, the right-hand side converges to $R(\delta)$ defined in (iii). The convergence in probability of $\hat{\theta}_N$ to θ_0 and assumption (iii) show that $R^N_{a,b} \to 0$ in probability and so A_N converges in probability to A. By Slutsky lemma $((A_N^\dagger)^{-1}, Y_N)$ converges in distribution to $((A^\dagger)^{-1}, \mathcal{N}(0,\Gamma))$ and so $(A_N^\dagger)^{-1}Y_N$ converges in distribution to $(A^\dagger)^{-1}\mathcal{N}(0,\Gamma) = \mathcal{N}(0, (A^\dagger)^{-1}\Gamma A^{-1})$.

Remark 1 on $\omega = dl$. For ω an estimating form, one has $\mathbb{E}_\theta[\omega(x,\theta)] = 0$. Taking the covariant derivative, one gets $\mathbb{E}_\theta[dl(U)\omega(V)] = -\mathbb{E}_\theta[\nabla\omega(U,V)]$ for all vector fields U, V. When $\omega = dl$, this writes $\mathbb{E}_\theta[\omega(U)\omega(V)] = -\mathbb{E}_\theta[\nabla\omega(U,V)]$. In particular $\Gamma = \mathbb{E}_{\theta_0}[dl \otimes dl(e_a, e_b)] = -A$ and $A^\dagger = A = \mathbb{E}_{\theta_0}[\nabla(dl)(e_a, e_b)] = \mathbb{E}_{\theta_0}[\nabla^2 l(e_a, e_b)]$ where ∇^2 is the Hessian of l. The limit matrix is therefore equal to Fisher information matrix $\Gamma^{-1} = -A^{-1}$. This yields the following corollary.

Corollary 1. *Assume $\Theta = (M, g)$ is a Riemannian manifold and let d be the Riemannian distance on Θ. Assume $\omega = dl$ satisfies the assumptions of Theorem 2 where ∇ is the Levi-Civita connection on Θ. The following convergence holds in distribution as $N \to \infty$.*

$$Nd^2(\hat{\theta}_N, \theta_0) \Rightarrow \sum_{i=1}^{p} X_i^2$$

where $X = (X_1, \cdots, X_p)^T$ is a random variable with law $\mathcal{N}(0, I^{-1})$ with $I(a,b) = \mathbb{E}_{\theta_0}[\nabla^2 l(e_a, e_b)]$.

The next proposition is concerned with asymptotic efficiency of MLE. It states that the lower asymptotic variance for estimating forms satisfying Theorem 2 is attained for $\omega_0 = dl$.

Take ω an estimating from and consider the matrices E, F, G, H with entries $E_{a,b} = \mathbb{E}_{\theta_0}[dl(\theta_0, x)e_a dl(\theta_0, x)e_b]$, $F_{a,b} = \mathbb{E}_{\theta_0}[dl(\theta_0, x)e_a\omega(\theta_0, x)e_b] = -A_{a,b}$, $G_{a,b} = F_{b,a}$, $H_{a,b} = \mathbb{E}_{\theta_0}[\omega(\theta_0, x)e_a\omega(\theta_0, x)e_b] = \Gamma_{a,b}$. Recall E^{-1} is the limit distribution when $\omega_0 = dl$. Note $M = \begin{pmatrix} E & F \\ G & H \end{pmatrix}$ is symmetric. When $\omega = dl$, it is furthermore positive but not definite.

Proposition 1. *If M is positive definite, then $E^{-1} < (A^\dagger)^{-1}\Gamma A^{-1}$.*

Proof. Since M is symmetric positive definite, the same also holds for its inverse. By Schur inversion lemma, $E - FH^{-1}G$ is symmetric positive definite. That is $E > FH^{-1}G$ or equivalently $E^{-1} < (A^\dagger)^{-1}\Gamma A^{-1}$.

Remark 2. As an example, it can be checked that Theorem 2 is satisfied by $\omega = dl$ of the Gaussian and Laplace distributions discussed in paragraph Sect. 2.1. For the Gaussian distribution on $\mathcal{P}_m$, this result is proved in [1]. More examples will be given in a future paper.

Remark 3 on Cramér-Rao lower bound. Assume Θ is a Riemannian manifold and $\hat{\theta}_n$ defined in Theorem 2 (i) is unbiased: $\mathbb{E}[\mathrm{Log}_{\theta_0}(\hat{\theta}_n)] = 0$. Consider $(e_1, \cdots, e_p)$ an orthonormal basis of $T_{\theta_0}\Theta$ and denote by $a = (a_1, \cdots, a_p)$ the coordinates in this basis of $\mathrm{Log}_{\theta_0}(\hat{\theta}_n)$. Smith [17] gave an intrinsic Cramér-Rao lower bound for the covariance $C(\theta_0) = \mathbb{E}[aa^T]$ as follows

$$\mathcal{C} \geq \mathcal{F}^{-1} + \text{curvature terms} \tag{4}$$

where $\mathcal{F} = (\mathcal{F}_{i,j} = \mathbb{E}[dL(\theta_0)e_i dL(\theta_0)e_j], i,j \in [1,p])$ is Fisher information matrix and $L(\theta) = \sum_{i=1}^{N} \log f(x_i, \theta)$. Define $\mathcal{L}$ the matrix with entries $\mathcal{L}_{i,j} = \mathbb{E}[dl(\theta_0)e_i dl(\theta_0)e_j]$ where $l(\theta) = \log f(x_1, \theta)$. By multiplying (4) by $\sqrt{n}$, one gets, with $y = \sqrt{n}a$,

$$\mathbb{E}[yy^T] \geq \mathcal{L}^{-1} + n \times \text{curvature terms}$$

It can be checked that as $n \to \infty$, $n \times$ curvature terms $\to 0$. Recall y converges in distribution to $\mathcal{N}(0, (A^\dagger)^{-1}\Gamma A^{-1})$. Assume it is possible to interchange limit and integral, from Theorem 2 one deduces $(A^\dagger)^{-1}\Gamma A^{-1} \geq \mathcal{L}^{-1}$ which is similar to Proposition 1.

4 Statistical Tests

Asymptotic properties of MLE have led to another fundamental subject in statistics which is testing. In the following, some popular tests on Euclidean spaces are generalized to manifolds.

Let $\Theta, \mathcal{M}$ and f be as in the beginning of the previous section.

Wald and score tests. Given $x_1, \cdots, x_n$ independent samples of $f(., \theta)$ where θ is unknown, consider the test $H_0 : \theta = \theta_0$. Define the Wald test statistic for H_0 by

$$Q_W = n(\Delta_1, \cdots, \Delta_p) I(\theta_0) (\Delta_1, \cdots, \Delta_p)^T$$

where $I(\theta_0)$ is Fisher matrix with entries $I(\theta_0)(a,b) = -\mathbb{E}_{\theta_0}[\nabla^2 l(e_a, e_b)]$ and $\Delta_1, \cdots, \Delta_p$, $(e_a)_{a=1:p}$ are defined as in Theorem 2.

Continuing with the same notations, the score test is based on the statistic

$$Q_S = U(\theta_0)^T I(\theta_0) U(\theta_0)$$

where $U(\theta_0) = (U_1(\theta_0), \cdots, U_p(\theta_0))$, $(U_a(\theta_0))_{a=1:p}$ are the coordinates of $\nabla_{\theta_0} l(\theta_0, X)$ in the basis $(e_a)_{a=1:p}$ and $l(\theta, X) = \sum_{i=1}^{n} \log(f(x_i, \theta))$.

Theorem 3. *Assume $\omega = dl$ satisfies conditions of Theorem 2. Then, under $H_0 : \theta = \theta_0$, Q_W (respectively Q_S) converges in distribution to a χ^2 distribution with $p = dim(\Theta)$ degrees of freedom. In particular, Wald test (resp. the score test) rejects H_0 when Q_W (resp. Q_S) is larger than a chi-square percentile.*

Because of the lack of space, the proof of this theorem will be published in a future paper. One can also consider a generalization of Wilks test to manifolds. An extension of this test to the manifold $\mathcal{P}_m$ appeared in [1].

References

1. Said, S., et al.: Riemannian Gaussian distributions on the space of symmetric positive definite matrices. IEEE Trans. Inf. Theory **63**, 2153–2170 (2017)
2. Hajri, H., et al.: Riemannian Laplace distribution on the space of symmetric positive definite matrices. Entropy **18**, 98 (2016)
3. Hajri, H., et al.: A geometric learning approach on the space of complex covariance matrices. In: ICASSP 2017 (2017)
4. Said, S., et al.: Gaussian distributions on Riemannian symmetric spaces: statistical learning with structured covariance matrices. IEEE Trans. Inf. Theory (2017)
5. Zanini, P., et al.: Parameters estimate of Riemannian Gaussian distribution in the manifold of covariance matrices. In: IEEE Sensor Array. Rio de Janeiro (2016)
6. Turaga, P.K., et al.: Statistical analysis on Stiefel and Grassmann manifolds with applications in computer vision. In: IEEE Computer Society (2008)
7. Aggarwal, G., et al.: A system identification approach for video-based face recognition. In: ICPR, vol. 4, pp. 175–178. IEEE Computer Society (2004)
8. Turaga, P.K., et al.: Statistical computations on Grassmann and Stiefel manifolds for image and video-based recognition. IEEE Trans. Pattern Anal. Mach. Intell. **33**(11), 2273–2286 (2011)
9. Kendall, D.G.: Shape manifolds, Procrustean metrics, and complex projective spaces. Bull. London Math. Soc. **16**, 81–121 (1984)
10. Chikuse, Y.: Statistics on Special Manifolds. Lecture Notes in Statistics, vol. 174. Springer, New York (2003)
11. Kwon, J., et al.: A geometric particle filter for template-based visual tracking. IEEE Trans. Pattern Anal. Mach. Intell. **36**(4), 625–643 (2014)
12. Trumpf, J., et al.: Analysis of non-linear attitude observers for time-varying reference measurements. IEEE Trans. Autom. Control **57**(11), 2789–2800 (2012)
13. Fletcher, P.T., et al.: Gaussian distributions on Lie groups and their application to statistical shape analysis. In: 18th International Conference on Information Processing in Medical Imaging, UK (2003)

14. Afsari, B.: Riemannian L^p center of mass: existence, uniqueness and convexity. Proc. Am. Math. Soc. **139**(2), 655–673 (2011)
15. Bhattacharya, R., Patrangenaru, V.: Large sample theory of intrinsic and extrinsic sample means on manifolds. Ann. Stat. **3**(1), 1–29 (2003)
16. Heyde, C.C.: Quasi-Likelihood and Its Application: A General Approach to Optimal Parameter Estimation. Springer, New York (1997)
17. Smith, S.T.: Covariance, subspace, and intrinsic Cramér-Rao bounds. IEEE Trans. Signal Process. **53**(5), 1610–1630 (2005)

Von Mises-Like Probability Density Functions on Surfaces

Florence Nicol and Stéphane Puechmorel(✉)

Université Fédérale de Toulouse, Laboratoire ENAC,
7 Avenue Edouard Belin, 31055 Toulouse, France
{florence.nicol,stephane.puechmorel}@enac.fr

Abstract. Directional densities were introduced in the pioneering work of von Mises, with the definition of a rotationally invariant probability distribution on the circle. It was further generalized to more complex objects like the torus or the hyperbolic space. The purpose of the present work is to give a construction of equivalent objects on surfaces with genus larger than or equal to 2, for which an hyperbolic structure exists. Although the directional densities on the torus were introduced by several authors and are closely related to the original von Mises distribution, allowing more than one hole is challenging as one cannot simply add more angular coordinates. The approach taken here is to use a wrapping as in the case of the circular wrapped Gaussian density, but with a summation taken over all the elements of the group that realizes the surface as a quotient of the hyperbolic plane.

Keywords: Directional densities · Hyperbolic geometry · von Mises probability distributions

1 Introduction

Estimating probability densities by the means of kernels is a basic procedure in non-parametric statistics. For finite dimensional vector spaces, the choice of the kernel bandwidth is the critical point, while the kernel itself is not as important. When dealing with manifolds, it is no longer the case, since the kernel must be well defined as a density on the manifold itself. A classical case arises when data of interest belong to a unit sphere, that yields the von Mises-Fischer distribution. It relies on the embedding of the unit sphere $\mathbb{S}^{d-1}$ in $\mathbb{R}^d$ to build a kernel that depends on the inner product $\langle x, y \rangle$ of the radial unit vectors associated to a couple of points (x, y) of $\mathbb{S}^{d-1}$. It is obviously invariant by rotation, as applying an isometry will not change the inner product. Since $\langle x, y \rangle$ is also $\cos\theta$, with θ the angle between x and y, it can be seen as a function of the geodetic distance $d(x, y)$ on the sphere. Finally, it has the maximum entropy property, which makes it similar to the normal distribution (in fact, the normal distribution is a limiting case of the von Mises-Fischer, the other being the uniform distribution). Spherical distributions have numerous applications in statistics, as many data

F. Nielsen and F. Barbaresco (Eds.): GSI 2017, LNCS 10589, pp. 701–708, 2017.
https://doi.org/10.1007/978-3-319-68445-1_81

can be interpreted as directions in $\mathbb{R}^d$. Attempts was made to generalize them to the d-dimensional torus [6] to obtain multivariate von Mises distributions. Here, the approach taken is somewhat different, since the primary goal was to build a density that is only invariant coordinate-wise. The existence of an angular coordinate system on the d-dimensional torus is the basis of the construction: rotation invariance in each coordinate is gained just by using angle differences in the overall distribution. An interpretation using the geodesic distance on the embedded surface is no longer possible. While not strictly compliant with the geometer's view of a torus density, the multivariate von Mises-Fischer distribution has still a geometrical interpretation. If we restrict our attention to the two-dimensional case, $\mathbb{T}^2$ admits a flat structure, with universal covering space $\mathbb{R}^2$ and fundamental domain a rectangle. It can be obtain as the quotient of $\mathbb{R}^2$ by a group G generated by two translations, respectively parallel to the x and y axis. The most natural density for such an object will be a wrapped $\mathbb{R}^2$ heat kernel, namely a sum of the form $p(x, y, t) = \sum_{g \in G} k(x, gy, t)$ with k the heat kernel on $\mathbb{R}^2$ and x the point at which the density is centered. It turns out that, due to the commutativity of the two translations and the particular shape of the $\mathbb{R}^2$ heat kernel, it boils down to a product of wrapped normal densities. Recalling that the one-dimensional von Mises distribution is an approximation of the wrapped normal and is rotation invariant, one can think of the multivariate von Mises density as a product of two densities invariant by the respective actions of the two translation cyclic groups.

The purpose of the present work is to introduce a class of probability distributions on orientable surfaces of genus larger than 1 and endowed with an hyperbolic structure, that may be used as kernels for non-parametric density estimates or to generate random data on such surfaces. By analogy with the multivariate von Mises distribution on the d-dimensional torus, the proposed density will approximate the wrapped heat kernel on the surface in the limit of time parameter going to 0. Furthermore, invariance of the density with respect to the action of a primitive element, similar to rotation invariance, will be enforced.

The overall procedure will closely mimic the construction of the multivariate von Mises distribution, starting with a representation of the surface as a quotient of the hyperbolic space $\mathbb{H}^2$ by a group G of hyperbolic isometries. The heat kernel is obtained readily as a wrapped sum of heat kernels on $\mathbb{H}^2$ over the elements of G. Using the property that the centralizer of an hyperbolic element is an infinite cyclic group with generator a primitive element, the wrapped kernel can be written in such a way that an equivalent to a wrapped one dimensional kernel appears.

2 Directional Densities: A Brief Survey

Directional densities are roughly speaking probability distributions depending on angular parameters. One of the most commonly used is the von Mises-Fischer on the unit sphere $\mathbb{S}^{d-1}$ of $\mathbb{R}^d$, that depend on two parameters $\mu \in \mathbb{S}^{d-1}$ and $\kappa > 0$, respectively called the mean and concentration. Its value at a point $x \in \mathbb{S}^{d-1}$ is given by:

$$p(x; \mu, \kappa) = \frac{\kappa^{d/2-1}}{(2\pi)^{d/2} I_{d/2-1}(\kappa)} \exp\left(\kappa \langle \mu, x \rangle\right), \tag{1}$$

where I_k stands for the modified Bessel function of order k and x, μ are given as unit vectors in $\mathbb{R}^d$. It enjoys many properties, like infinite divisibility [4] and maximal entropy [5]. It has been generalized to other Riemannian manifolds like the d-dimensional torus $\mathbb{T}^d$, on which it becomes the multivariate von Mises distribution [6]:

$$p(\theta; \mu, \kappa, \Lambda) \propto \exp\left(\langle \kappa, c(\theta, \mu) \rangle + \frac{1}{2} s(\theta, \mu)^T \Lambda\, s(\theta, \mu)\right), \tag{2}$$

where θ, μ are d-dimensional vectors of angles, κ is a d-dimensional vector of positive real numbers and Λ is a $d \times d$ symmetric, positive definite matrix describing the covariance between the angular parameters. The terms $c(\theta, \mu), s(\theta, \mu)$ occurring in the expression are given by:

$$c(\theta, \mu)^T = (\cos(\theta_1 - \mu_1), \dots, \cos(\theta_d - \mu_d)), \tag{3}$$
$$s(\theta, \mu)^T = (\sin(\theta_1 - \mu_1), \dots, \sin(\theta_d - \mu_d)). \tag{4}$$

Another generalization is made in [1] and, with a different approach in [3], to the hyperbolic d-dimensional space $\mathbb{H}^d$. Following the later, the starting point is the hyperbolic Brownian motion defined as a diffusion on $\mathbb{H}^d$ with infinitesimal generator:

$$\frac{x_d^2}{2} \left(\sum_{i=1}^{d} \frac{\partial^2}{\partial x_i^2} \right) - \frac{(d-2) x_d}{2} \frac{\partial}{\partial x_d}, \tag{5}$$

where all the coordinates are given in the half-space model of $\mathbb{H}^d$:

$$\mathbb{H}^d = \{x_1, \dots, x_d \colon x_i \in \mathbb{R},\, i = 1, \dots, d-1,\, x_d \in \mathbb{R}^+\}.$$

In the sequel, only the case $d = 2$ will be considered, as the primary object of interest are surfaces. It is convenient to represent the half-space model of $\mathbb{H}^2$ in $\mathbb{C}$, with $z = x_1 + i x_2$. The 2-dimensional hyperboloid embedded in $\mathbb{R}^3$ associated with $\mathbb{H}^2$ is given by:

$$\{(x_1, x_2, x_3) \colon x_1^2 + x_2^2 - x_3^2 = -1\}. \tag{6}$$

It admits hyperbolic coordinates:

$$x_1 = \sinh(r)\cos(\theta), x_2 = \sinh(r)\sin(\theta), x_3 = \cosh(r) \tag{7}$$

that transforms to the unit disk model as:

$$u = \frac{\sinh(r)\cos(\theta)}{1 + \cosh(r)},\ v = \frac{\sinh(r)\sin(\theta)}{1 + \cosh(r)} \tag{8}$$

where θ and r are the angular and radius coordinates. Finally, using a complex representation $z = u+iv$ and the Möbius mapping $z \rightarrow i(1-z)/(1+z)$, it comes the expression of the half-plane coordinates:

$$x = \frac{\sinh(r)\sin(\theta)}{\cosh(r) + \sinh(r)\cos(\theta)},\ y = \frac{1}{\cosh(r) + \sinh(r)\cos(\theta)}. \tag{9}$$

The hyperbolic Von Mises distribution is then defined, for a given $r > 0$, as the density of the first exit on the circle of center i and radius r of the hyperbolic Brownian motion starting at i. Its expression is given in [3] as:

$$p_{vm}(r,\theta) = \frac{1}{2\pi P^0_{-\nu}(\cosh(r))}\left(\cosh(r) + \sinh(r)\cos(\theta)\right)^{-\nu} \tag{10}$$

where $P^0_{-\nu}$ is the Legendre function of the first kind with parameters $0, -\nu$, that acts as a normalizing constant to get a true probability density. The parameter ν is similar to the concentration used in the classical Von Mises distribution.

3 Closed Geodesics and Wrapping

A classical circular density is the wrapped (centered) Gaussian distribution:

$$p_{wg}(\theta;\sigma) = \frac{1}{\sqrt{2\pi}\sigma}\sum_{k\in\mathbb{Z}}\exp(-(\theta+2k\pi)/(2\sigma^2)) \tag{11}$$

It is clearly a periodic distribution with period 2π, with σ acting as an inverse concentration parameter. Von Mises densities can approximate the circular wrapped Gaussian density quite well when the concentration is large enough. It worth notice that the wrapped Gaussian can be seen as a circular heat kernel $K(\theta,t) = p_{wg}(\theta,\sigma^2/2)$, with the angular parameter θ being interpreted as a distance on the unit circle. The starting point for defining an equivalent of the von Mises distributions on surfaces is the wrapping formula given above. First of all, only compact orientable surfaces of genus g larger that 1 will be considered, as the case of the sphere or the torus is already covered. It is a classical result in hyperbolic geometry that such a surface M can be endowed with an hyperbolic structure that is obtained as the quotient of the hyperbolic plane by a group G of hyperbolic isometries. Any non-trivial element of G is conjugate to an isometry of hyperbolic type and in turn to a scaling acting as: $z \mapsto q^2 z, q^2 \neq 0, 1$. It is quite interesting to note that this is exactly the case considered in [3] for the definition of the Brownian motion with drift in $\mathbb{H}^2$, the drift component being an hyperbolic isometry.

The first possible definition of a directional density on an orientable surface of genus $g > 1$ will be to use $k_{\mathbb{H}^2}(x,y,t)$, the heat kernel on $\mathbb{H}^2$, as an analogous of the Gaussian heat kernel, and to consider its wrapping over all possible elements in G. The expression of $k_{\mathbb{H}^2}$ is given by [8]:

$$K_{\mathbb{H}^2}(x,y,t) = \frac{\sqrt{2}e^{-\frac{t}{4}}}{(4\pi t)^{3/2}}\int_{d(x,y)}^{\infty}\frac{se^{-\frac{s^2}{4t}}}{\sqrt{\cosh s - \cosh d(x,y)}}ds \tag{12}$$

with $d(x, y)$ the hyperbolic distance between x, y. Please note that $K_{\mathbb{H}^2}(x, y, t)$ can be written using the hyperbolic distance only as:

$$K_{\mathbb{H}^2}(x, y, t) = k_{\mathbb{H}^2}(d(x, y), t)$$

The group G admits an hyperbolic polygon with $4g$ sides as fundamental region in $\mathbb{H}^2$. For any $g \in G$, its length is defined to be $l(g) = \inf_x d(x, gx)$, or using the conjugacy class of g: $l(g) = \inf_x d(x, kgk^{-1})$ where k runs over G. Elements of G with non zero length are conjugate to hyperbolic elements in $SL(2, \mathbb{R})$ (elliptic and parabolic ones are associated to rotations and translation in $\mathbb{H}^2$ so that the length can be made arbitrary small), and are thus conjugate to a scaling $x \mapsto \lambda^2 x$. Furthermore, a conjugacy class represents a free homotopy class of closed curves, that contains a unique minimal geodesic whose length is $l(g)$, where g is a representative element.

M can be identified with the quotient $\mathbb{H}^2/G$ so that one can define a wrapped heat kernel on M by the formula:

$$K_M \colon (x, y, t) \in M^2 \times \mathbb{R}^+ \mapsto \sum_{g \in G} K_{\mathbb{H}^2}(x, gy, t) \tag{13}$$

K_M is clearly invariant by the left action of G and is symmetric since:

$$\begin{aligned} K_M(x, y, t) &= \sum_{g \in G} k_{\mathbb{H}^2}(d(x, gy), t) && (14) \\ &= \sum_{g \in G} k_{\mathbb{H}^2}(d(g^{-1}x, y), t) = \sum_{g \in G} k_{\mathbb{H}^2}(d(y, g^{-1}x), t) && (15) \\ &= K_M(y, x, t) && (16) \end{aligned}$$

Finally, primitive elements in G (i.e. those $p \in G$ that cannot be written as a non trivial power of another element) play a central role in the sum defining K_M. For p a primitive element, let G_p denote its centralizer in G. The conjugacy classes in G are all of the form $gp^n g^{-1}, g \in G/G_p$ with p primitive and $n \in \mathbb{Z}$. The wrapped kernel can then be rewritten as:

$$K_M(x, y, t) = \sum_{p} \sum_{g \in G/G_p} \sum_{n \in \mathbb{Z}} K_{\mathbb{H}^2}(gx, p^n gy, t) \tag{17}$$

where p runs through the primitive elements of G.

It indicates that the kernel K_M can be understood as a sum of elementary wrapped kernels associated to primitive elements, namely those $\tilde{k}_p$ defined by:

$$\tilde{k}_p(x, y, t) = \sum_{n \in \mathbb{Z}} K_{\mathbb{H}^2}(x, p^n y, t) \tag{18}$$

with p primitive. Finally, p being hyperbolic, it is conjugate to a scaling, so it is enough to consider kernels of the form:

$$\tilde{k}_p(x, y, t) = \sum_{n \in \mathbb{Z}} K_{\mathbb{H}^2}(x, (\lambda^2)^n y, t) \tag{19}$$

with $\lambda > 1$ a real number. To each primitive element p, a simple closed minimal geodesic loop is associated, which projects onto the axis of the hyperbolic transformation p. In the Poincaré half-plane model, such a loop unwraps onto the segment of the imaginary axis that lies between i and $i\lambda^2$. It is easily seen that the action of the elements $p^n, n \in \mathbb{Z}$ will give rise to a tiling of the positive imaginary axis with segments of the form $[\lambda^{2n}, \lambda^{2(n+1)}[$. This representation allows a simple interpretation of the elementary wrapped kernels $\tilde{k}_p$, where the wrapping is understood as a winding.

4 Von Mises Like Distributions

The wrapped kernels are most natural from the viewpoint of surfaces as quotient spaces, since the group action appears directly within the definition. However, it quite difficult to use them for the purpose of density estimation: even with truncated expansions, it requires quite a huge amount of computation. In the case of circular data, the usual von Mises distribution behaves much like the wrapped Gaussian, but does not involves a summation. The same is true for the generalized multivariate von Mises (2). As mentioned in the introduction, it is invariant under the action of the two generating translations. Using the same principle, a distribution invariant under the action of a primitive element will be used in place of the wrapped sum defining the elementary kernels (19).

Since any primitive element is conjugate to a scaling λ^2, it is natural to seek after a distribution on a simple hyperbolic surface that is obtained from the quotient of the hyperbolic half-plane by the cyclic group ξ_λ generated by λ^2. A fundamental domain for its action is the subset of $\mathbb{C}$ defined as:

$$\{z = x + iy, x \geq 0, 1 \leq y < \lambda^2\}$$

In the quotient, the upper line $y = \lambda^2$ will be identified with the lower line $y = 1$, yielding an hyperbolic cylinder.

It is convenient to use an exponential coordinate system in order to allow for a simple invariant expression. For a given $z = x + iy$ in the hyperbolic half plane, let $x = ue^v, y = e^v$. The hyperbolic distance between two elements $z_1 = (u_1, v_1), z_2 = (u_2, v_2)$ is given by:

$$\begin{aligned}\cosh d(z_1, z_2) &= 1 + \frac{|z_1 - z_2|^2}{2\Im z_1 \Im z_2} \\ &= 1 + (u_1 e^{\frac{v_1 - v_2}{2}} - u_2 e^{\frac{v_2 - v_1}{2}}) + 2\sinh^2\left(\frac{v_1 - v_2}{2}\right) \\ &= (u_1 e^{\frac{v_1 - v_2}{2}} - u_2 e^{\frac{v_2 - v_1}{2}}) + \cosh(v_1 - v_2)\end{aligned}$$

If the second term in the sum is considered only, it remains $|v_1 - v_2|$, which is similar to the angle difference in the case of the multivariate von Mises density. This fact can be made explicit when considering points (y_1, y_2) located on the y axis only. In such a case, the hyperbolic distance between them is easily seen to be $l = |\log(y_2/y_1)|$. Letting $L = 2\log(\lambda)$, we have the following result:

Theorem 1. *The wrapped kernel* $\tilde{k}_p(y_1, y_2, t) = \tilde{k}p(l, t)$ *is periodic of period* L, *with Fourier coefficient of order* p *given by:*

$$a_p(t) = \sqrt{\frac{\pi}{2}} \frac{1}{L} e^{-t/8} \int_0^\infty v^{-1/2} K_{\frac{ip2\pi}{L}}(v)\theta_v(t)dv$$

where K *is the modified bessel function and* θ_v *is:*

$$\theta_v(t) = \frac{v}{\sqrt{2\pi^3 t}} \int_0^\infty e^{(\pi^2 - b^2)/2t} e^{-v \cosh b} \sinh(b) \sin(\pi b/t) db$$

Proof (Sketch). The starting point is the hyperbolic heat kernel representation given in [7]:

$$k(r, t) = \frac{e^{-t/8}}{\sqrt{2\pi}} \int_0^\infty v^{-1/2} \exp(-v \cosh r)\theta_v(t) dv$$

By wrapping it, it appears the periodic function:

$$\sum_{n \in \mathbb{Z}} e^{-v \cosh(l + nL)}$$

whose fourier coefficients can be expressed using the modified bessel function $K_{\frac{ip2\pi}{L}}(v)$. The technical part is the use of asymptotics of K to legitimate the summation.

The fourier series expansion of the wrapped kernel has quickly decreasing coefficients, so that it is well approximated by the first term. Please note that the wrapped distance may be interpreted as an angular distance on the circle after scaling by $2\pi/l$, and due to the previous remark, it gives rise to a standard one dimensional (circular) von Mises kernel depending on the wrapped hyperbolic distance between the two points.

Unlike the flat torus case, the hyperbolic translations induced by primitive elements will not commute, so that summation is much more intricate. The way the computation must be performed is currently under study to obtain if possible the most tractable expression. However, it can be organized in such a way to make appear the multivariate von Mises distribution (2) on the angular parameters associated to primitive elements, the correlation matrix Λ in the expression being related to the commutation relations between the corresponding primitive elements.

5 Conclusion and Future Work

The extension of directional statistics to surfaces of genus larger than 1, endowed with an hyperbolic structure, can be performed using a special kind of angular parameters. Writing the surface as the quotient of the hyperbolic space by a group G of hyperbolic isometry, one can construct an invariant kernel by summing over the translates of an element. Using the primitive elements of G, the summation can be split in such a way that the innermost sum can be understood as a wrapped kernel on an hyperbolic cylinder. This allows to replace it by a standard directional kernel depending on an angular parameter. Proceeding further in the sum, the summation over the elements of the conjugacy classes will yield a multivariate directional density, with correlated components.

The practical computation of such a density is still difficult due to the fact that all the possible combinations between primitive elements must be considered. An approximation can be made by neglecting those words in G involving more that a given number of terms: this makes senses as the kernels considered must decay very fast. An important part of the future developments will be dedicated to the computational aspect as it is one of the key points for being able to use the densities on real data.

The second aspect that needs to be addressed is the shape of the kernel itself. Due to the fact that important parameters may be reduced to angles, the initial approach was to use the already available multivariate von Mises. Since it is known that the choice of the kernel is of secondary importance in classical settings, the same may be expected here. However, if one wants to get some extra properties, like maximum entropy, an especially tailored distribution will be needed. The possibility of defining it using the Abel transform [2]of kernels on $\mathbb{R}$ is currently investigated.

References

1. Barndorff-Nielsen, O.: Hyperbolic distributions and distributions on hyperbolae. Scand. J. Stat. **5**(3), 151–157 (1978)
2. Beerends, R.J.: An introduction to the abel transform. In: Miniconference on Harmonic Analysis, Canberra, Australia, pp. 21–33 (1987). Centre for Mathematical Analysis, The Australian National University
3. Gruet, J.-C.: A note on hyperbolic von mises distributions. Bernoulli **6**(6), 1007–1020 (2000)
4. Kent, J.T.: The infinite divisibility of the von mises-fisher distribution for all values of the parameter in all dimensions. Proc. London Mathe. Soc. **s3–35**(2), 359–384 (1977)
5. Mardia, K.V.: Statistics of directional data. J. Roy. Stat. Soc.: Ser. B (Methodol.) **37**(3), 349–393 (1975)
6. Mardia, K.V., Hughes, G., Taylor, C.C., Singh, H.: A multivariate von mises distribution with applications to bioinformatics. Canadian J. Stat. **36**(1), 99–109 (2008)
7. Matsumoto, H.: Closed form formulae for the heat kernels and the green functions for the laplacians on the symmetric spaces of rank one. Bull. des Sci. Mathématiques **125**(6), 553–581 (2001)
8. McKean, H.P.: An upper bound to the spectrum of δ on a manifold of negative curvature. J. Differential Geom. **4**(3), 359–366 (1970)

Riemannian Gaussian Distributions on the Space of Positive-Definite Quaternion Matrices

Salem Said[1], Nicolas Le Bihan[2(✉)], and Jonathan H. Manton[3]

[1] Laboratoire IMS (CNRS - UMR 5218), Paris, France
[2] Univ. Grenoble Alpes, CNRS, Grenoble INP, GIPSA-lab (UMR 5216), Grenoble, France
nicolas.le-bihan@gipsa-lab.inpg.fr
[3] Department of Electrical and Electronic Engineering, The University of Melbourne, Melbourne, Australia

Abstract. Recently, Riemannian Gaussian distributions were defined on spaces of positive-definite real and complex matrices. The present paper extends this definition to the space of positive-definite quaternion matrices. In order to do so, it develops the Riemannian geometry of the space of positive-definite quaternion matrices, which is shown to be a Riemannian symmetric space of non-positive curvature. The paper gives original formulae for the Riemannian metric of this space, its geodesics, and distance function. Then, it develops the theory of Riemannian Gaussian distributions, including the exact expression of their probability density, their sampling algorithm and statistical inference.

Keywords: Riemannian gaussian distribution · Quaternion · Positive-definite matrix · Symplectic group · Riemannian barycentre

1 Introduction

The Riemannian geometry of the spaces $\mathcal{P}_n$ and $\mathcal{H}_n$, respectively of $n \times n$ positive-definite real and complex matrices, is well-known to the information science community [1,2]. These spaces have the property of being Riemannian symmetric spaces of non-positive curvature [3,4],

$$\mathcal{P}_n = \mathrm{GL}(n,\mathbb{R})/\mathrm{O}(n) \qquad \mathcal{H}_n = \mathrm{GL}(n,\mathbb{C})/\mathrm{U}(n)$$

where $\mathrm{GL}(n,\mathbb{R})$ and $\mathrm{GL}(n,\mathbb{C})$ denote the real and complex linear groups, and $\mathrm{O}(n)$ and $\mathrm{U}(n)$ the orthogonal and unitary groups. Using this property, Riemannian Gaussian distributions were recently introduced on $\mathcal{P}_n$ and $\mathcal{H}_n$ [5,6]. The present paper introduces the Riemannian geometry of the space $\mathcal{Q}_n$ of $n \times n$ positive-definite quaternion matrices, which is also a Riemannian symmetric space of non-positive curvature [4],

$$\mathcal{Q}_n = \mathrm{GL}(n,\mathbb{H})/\mathrm{Sp}(n)$$

F. Nielsen and F. Barbaresco (Eds.): GSI 2017, LNCS 10589, pp. 709–716, 2017.
https://doi.org/10.1007/978-3-319-68445-1_82

where $\mathrm{GL}(n,\mathbb{H})$ denotes the quaternion linear group, and $\mathrm{Sp}(n)$ the compact symplectic group. It then studies Riemannian Gaussian distributions on $\mathcal{Q}_n$. The main results are the following: Proposition 1 gives the Riemannian metric of the space $\mathcal{Q}_n$, Proposition 2 expresses this metric in terms of polar coordinates on the space $\mathcal{Q}_n$, Proposition 3 uses Proposition 2 to compute the moment generating function of a Riemannian Gaussian distribution on $\mathcal{Q}_n$, and Propositions 4 and 5 describe the sampling algorithm and maximum likelihood estimation of Riemannian Gaussian distributions on $\mathcal{Q}_n$. Motivation for studying matrices from $\mathcal{Q}_n$ comes from their potential use in multidimensional bivariate signal processing [7].

2 Quaternion Matrices, GL($\mathbb{H}$) and Sp(n)

Recall the non-commutative division algebra of quaternions, denoted $\mathbb{H}$, is made up of elements $q = q_0 + q_1\,\mathrm{i} + q_2\,\mathrm{j} + q_3\,\mathrm{k}$ where $q_0, q_1, q_2, q_3 \in \mathbb{R}$, and the imaginary units $\mathrm{i},\mathrm{j},\mathrm{k}$ satisfy the relations [8]

$$\mathrm{i}^2 = \mathrm{j}^2 = \mathrm{k}^2 = \mathrm{ijk} = -1 \tag{1}$$

The real part of q is $\mathrm{Re}(q) = q_0$, its conjugate is $\bar{q} = q_0 - q_1\,\mathrm{i} - q_2\,\mathrm{j} - q_3\,\mathrm{k}$ and its squared norm is $|q|^2 = q\bar{q}$. The multiplicative inverse of $q \neq 0$ is given by $q^{-1} = \bar{q}/|q|^2$.

The set $M_n(\mathbb{H})$ consists of $n \times n$ quaternion matrices A [9]. These are arrays $A = (A_{ij}\,;\, i,j = 1,\ldots,n)$ where $A_{ij} \in \mathbb{H}$. The product $C = AB$ of $A, B \in M_n(\mathbb{H})$ is the element of $M_n(\mathbb{H})$ with

$$C_{ij} = \sum_{l=1}^{n} A_{il} B_{lj} \tag{2}$$

A quaternion matrix A is said invertible if it has a multiplicative inverse A^{-1} with $AA^{-1} = A^{-1}A = I$ where I is the identity matrix. The conjugate-transpose of A is $A^\dagger$ which is a quaternion matrix with $A^\dagger_{ij} = \bar{A}_{ji}$.

The rules for computing with quaternion matrices are quite different from the rules for computing with real or complex matrices [9]. For example, in general, $\mathrm{tr}(AB) \neq \mathrm{tr}(BA)$, and $(AB)^T \neq B^T A^T$ where ^{T} denotes the transpose. For the results in this paper, only the following rules are needed [9],

$$(AB)^{-1} = B^{-1}A^{-1} \quad (AB)^\dagger = B^\dagger A^\dagger \quad \mathrm{Re}\,\mathrm{tr}(AB) = \mathrm{Re}\,\mathrm{tr}(BA) \tag{3}$$

$\mathrm{GL}(n,\mathbb{H})$ consists of the set of invertible quaternion matrices $A \in M_n(\mathbb{H})$. The subset of $A \in \mathrm{GL}(n,\mathbb{H})$ such that $A^{-1} = A^\dagger$ is denoted $\mathrm{Sp}(n) \subset \mathrm{GL}(n,\mathbb{H})$.

It follows from (3) that $\mathrm{GL}(n,\mathbb{H})$ and $\mathrm{Sp}(n)$ are groups under the operation of matrix multiplication, defined by (2). However, one has more. Both these groups are real Lie groups. Usually, $\mathrm{GL}(n,\mathbb{H})$ is called the quaternion linear group, and $\mathrm{Sp}(n)$ the compact symplectic group. In fact, $\mathrm{Sp}(n)$ is a compact connected Lie subgroup of $\mathrm{GL}(n,\mathbb{H})$ [10].

The Lie algebras of these two Lie groups are given by

$$\mathfrak{gl}(n,\mathbb{H}) = M_n(\mathbb{H}) \qquad \mathfrak{sp}(n) = \left\{ X \in \mathfrak{gl}(n,\mathbb{H}) \,|\, X + X^\dagger = 0 \right\} \tag{4}$$

with the bracket operation $[X,Y] = XY - YX$. The Lie group exponential is identical to the quaternion matrix exponential

$$\exp(X) = \sum_{m \geq 0} \frac{X^m}{m!} \quad X \in \mathfrak{gl}(n,\mathbb{H}) \tag{5}$$

For $A \in \mathrm{GL}(n,\mathbb{H})$ and $X \in \mathfrak{gl}(n,\mathbb{H})$, let $\mathrm{Ad}(A) \cdot X = AXA^{-1}$. Then,

$$A \exp(X) A^{-1} = \exp(\,\mathrm{Ad}(A) \cdot X\,) \tag{6}$$

as can be seen from (5).

3 The Space $\mathcal{Q}_n$ and its Riemannian metric

The space $\mathcal{Q}_n$ consists of all quaternion matrices $S \in M_n(\mathbb{H})$ which verify $S = S^\dagger$ and

$$\sum_{i,j=1}^{n} \bar{x}_i \, S_{ij} \, x_j > 0 \quad \text{for all non-zero } (x_1, \ldots, x_n) \in \mathbb{H}^n \tag{7}$$

In other words, $\mathcal{Q}_n$ is the space of positive-definite quaternion matrices. Note that, due to the condition $S = S^\dagger$, the sum in (7) is a real number.

Define now the action of $\mathrm{GL}(n,\mathbb{H})$ on $\mathcal{Q}_n$ by $A \cdot S = ASA^\dagger$ for $A \in \mathrm{GL}(n,\mathbb{H})$ and $S \in \mathcal{Q}_n$. This is a left action, and is moreover transitive. Indeed [9], each $S \in \mathcal{Q}_n$ can be diagonalized by some $K \in \mathrm{Sp}(n)$,

$$S = K \exp(R) K^{-1} = \exp(\,\mathrm{Ad}(K) \cdot R\,) \quad ; R \text{ real diagonal matrix} \tag{8}$$

where the second equality follows from (6). Thus, each $S \in \mathcal{Q}_n$ can be written $S = AA^\dagger$ for some $A \in \mathrm{GL}(n,\mathbb{H})$, which is the same as $S = A \cdot I$.

For $A \in \mathrm{GL}(n,\mathbb{H})$, note that $A \cdot I = I$ iff $AA^\dagger = I$, which means that $A \in \mathrm{Sp}(n)$. Therefore, as a homogeneous space under the left action of $\mathrm{GL}(n,\mathbb{H})$,

$$\mathcal{Q}_n = \mathrm{GL}(n,\mathbb{H}) \,/\, \mathrm{Sp}(n) \tag{9}$$

The space $\mathcal{Q}_n$ is a real differentiable manifold. In fact, if $\mathfrak{p}_n$ is the real vector space of $X \in \mathfrak{gl}(n,\mathbb{H})$ such that $X = X^\dagger$, then it can be shown $\mathcal{Q}_n$ is an open subset of $\mathfrak{p}_n$. Therefore, $\mathcal{Q}_n$ is a manifold, and for each $S \in \mathcal{Q}_n$ the tangent space $T_S\mathcal{Q}_n$ may be identified with $\mathfrak{p}_n$. Moreover, $\mathcal{Q}_n$ can be equipped with a Riemannian metric as follows.

Define on $\mathfrak{gl}(n,\mathbb{H})$ the $\mathrm{Sp}(n)$-invariant scalar product

$$\langle X | Y \rangle = \mathrm{Re}\,\mathrm{tr}(XY^\dagger) \quad X, Y \in \mathfrak{gl}(n,\mathbb{H}) \tag{10}$$

For u, v in $T_S\mathcal{Q}_n \simeq \mathfrak{p}_n$, let

$$(u,v)_S = \left\langle\, (A^{-1})u(A^{-1})^\dagger \,|\, (A^{-1})v(A^{-1})^\dagger \,\right\rangle \tag{11}$$

where A is any element of $\mathrm{GL}(n,\mathbb{H})$ such that $S = A \cdot I$.

Proposition 1 (Riemannian metric)

(i) For each $S \in \mathcal{Q}_n$, formula (11) defines a scalar product on $T_S\mathcal{Q}_n \simeq \mathfrak{p}_n$, which is independent of the choice of A.
(ii) Moreover,

$$(u, v)_S = \operatorname{Re}\operatorname{tr}\left(S^{-1}u\,S^{-1}v\right) \tag{12}$$

which yields a Riemannian metric on $\mathcal{Q}_n$.
(iii) This Riemannian metric is invariant under the action of $\mathrm{GL}(n, \mathbb{H})$ *on $\mathcal{Q}_n$.*

The proof of Proposition 1 only requires the fact that (11) is a scalar product on $\mathfrak{p}_n$, and application of the rules (3). It is here omitted for lack of space.

4 The Metric in Polar Coordinates

In order to provide analytic expressions in Sects. 5 and 6, we now introduce the expression of the Riemannian metric (12) in terms of polar coordinates. For $S \in \mathcal{Q}_n$, the polar coordinates of S are the pair (R, K) appearing in the decomposition (8). It is an abuse of language to call them coordinates, as they are not unique. However, this terminology is natural and used quite often in the literature [5,6].

The expression of the metric (12) in terms of the polar coordinates (R, K) is here given in Proposition 2. This requires the following notation. For $i, j = 1, \ldots, n$, let θ_{ij} be the quaternion-valued differential form on $\mathrm{Sp}(n)$,

$$\theta_{ij}(K) = \sum_{l=1}^{n} K^{\dagger}_{il}\, dK_{lj} \tag{13}$$

Note that, by differentiating the identity $K^{\dagger}K = I$, it follows that $\theta_{ij} = -\bar{\theta}_{ji}$. Proposition 2 expresses the length element corresponding to the Riemannian metric (12).

Proposition 2 (the metric in polar coordinates). *In terms of the polar coordinates (R, K), the length element corresponding to the Riemannian metric (12) is given by,*

$$ds^2(R, K) = \sum_{i=1}^{n} dr_i^2 + 8 \sum_{i<j} \sinh^2\left(|r_i - r_j|/2\right) |\theta_{ij}|^2 \tag{14}$$

where r_i denote the diagonal elements of the matrix R.

The proof of this proposition cannot be given here, due to lack of space.

Proposition 2 is valuable to understanding the Riemannian geometry of the space $\mathcal{Q}_n$. Precisely, it can be used to infer, with almost no calculation, the expressions of geodesics and of distance, on this space. Indeed, it becomes clear from (14) that the shortest curve connecting the identity $I \in \mathcal{Q}_n$ to a diagonal (and therefore real) element $a \in \mathcal{Q}_n$, is given by $t \mapsto a^t$ for $t \in [0, 1]$. Using this simple result, and the fact that the metric (12) is invariant under the action of $\mathrm{GL}(n,\mathbb{H})$ on $\mathcal{Q}_n$, the equation of the minimising geodesic curve $\gamma(t)$ connecting two elements $S, Q \in \mathcal{Q}_n$ can be obtained,

$$\gamma(t) = S^{\frac{1}{2}} \left(S^{-\frac{1}{2}} Q S^{-\frac{1}{2}} \right)^t S^{\frac{1}{2}} \tag{15}$$

Accordingly, the distance between S and Q is

$$d(S,Q) = \left\| \log \left(S^{-\frac{1}{2}} Q S^{-\frac{1}{2}} \right) \right\| \tag{16}$$

where $\|\cdot\|$ is the norm corresponding to the scalar product (10).

In (15) and (16) matrix functions, such as elevation to a power and logarithm, are computed via the decomposition (8), where the functions are applied to the diagonal matrix $\exp(R)$.

5 Riemannian Gaussian Distributions on $\mathcal{Q}_n$

It is possible to define Riemannian Gaussian distributions on any Riemannian symmetric space of non-positive curvature [6]. This is indeed the case of the space $\mathcal{Q}_n$, as can be seen from its representation (9) as a quotient space, by consulting the tables which classify irreducible Riemannian symmetric spaces of type III [4].

Accordingly, it is possible to define Riemannian Gaussian distributions on $\mathcal{Q}_n$. Precisely, a Riemannian Gaussian distribution on $\mathcal{Q}_n$ with Riemannian barycentre $\breve{S} \in \mathcal{Q}_n$ and dispersion parameter $\sigma > 0$ has the following probability density

$$p(S|\,\breve{S},\sigma) = \frac{1}{Z(\sigma)} \exp\left[-\frac{d^2(S,\,\breve{S})}{2\sigma^2} \right] \tag{17}$$

with respect to the Riemannian volume element of $\mathcal{Q}_n$, here denoted dv. In this probability density, $d(S,\,\breve{S})$ is the Riemannian distance given by (16).

The first step to understanding this definition is computing the normalising constant $Z(\sigma)$. This is given by the integral,

$$Z(\sigma) = \int_{\mathcal{Q}_n} \exp\left[-\frac{d^2(S,\,\breve{S})}{2\sigma^2} \right] dv(S) \tag{18}$$

As shows in [6], this does not depend on $\breve{S}$, and therefore it is possible to take $\breve{S} = I$. From the decomposition (8) and formula (16), it follows that

$$d^2(S,I) = \sum_{i=1}^{n} r_i^2 \tag{19}$$

Given this simple expression, it seems reasonable to pursue the computation of the integral (18) in polar coordinates. This is achieved in the following Proposition 3. For the statement, write the quaternion-valued differential form θ_{ij} of (13) as $\theta_{ij} = \theta^a_{ij} + \theta^b_{ij}\,\mathrm{i} + \theta^c_{ij}\,\mathrm{j} + \theta^d_{ij}\,\mathrm{k}$ where $\theta^a_{ij}, \theta^b_{ij}, \theta^c_{ij}, \theta^d_{ij}$ are real-valued.

Proposition 3 (normalising constant)

(i) In terms of the polar coordinates (R, K), the Riemannian volume element $dv(S)$ corresponding to the Riemannian metric (12) is given by

$$dv(R, K) = 8^{n(n-1)} \prod_{i<j} \sinh^4 (|r_i - r_j|/2) \prod_{i=1}^{n} dr_i \bigwedge_{i<j} \theta^a_{ij} \bigwedge_{i<j} \theta^b_{ij} \bigwedge_{i<j} \theta^c_{ij} \bigwedge_{i<j} \theta^d_{ij} \tag{20}$$

(ii) The integral $Z(\sigma)$ appearing in (18) is given by

$$Z(\sigma) = \text{Const.} \times \int_{\mathbb{R}^n} \exp\left(-\frac{1}{2\sigma^2} \sum_{i=1}^{n} r_i^2 \right) \prod_{i<j} \sinh^4 (|r_i - r_j|/2) \prod_{i=1}^{n} dr_i \tag{21}$$

This proposition is a corollary of Proposition 2. Formula (20) is a straightforward consequence of formula (14). Furthermore, (21) is an immediate application of (19) and (20).

6 Sampling and Inference

The present section describes two aspects of Riemannian Gaussian distributions on $\mathcal{Q}_n$: (i) sampling from these distributions, (ii) maximum likelihood estimation of these distributions.

The first of these aspects is given in Proposition 4 below. This relies on the use of polar coordinates (R, K) which appear in the decomposition (8).

Proposition 4 (Gaussian distribution in polar coordinates). *Let K and r be independent random variables, with their values in $\mathrm{Sp}(n)$ and $\mathbb{R}^n$ respectively. Assume K is uniformly distributed on $\mathrm{Sp}(n)$, and r has the following probability density, with respect to the Lebesgue measure on $\mathbb{R}^n$,*

$$p(r_1, \ldots, r_n) \propto \exp\left(-\frac{1}{2\sigma^2} \sum_{i=1}^{n} r_i^2 \right) \prod_{i<j} \sinh^4 (|r_i - r_j|/2) \tag{22}$$

If S is given by (8), where the matrix R has diagonal elements r_i, then S has a Riemannian Gaussian distribution (17) with Riemannian barycentre $\breve{S} = I$ and dispersion parameter σ. Moreover, for any $\breve{S} \in \mathcal{Q}_n$ and $A \in \mathrm{GL}(n, \mathbb{H})$ such that $A \cdot I = \breve{S}$, if $Q = A \cdot S$ then Q has Riemannian Gaussian distribution with Riemannian barycentre $\breve{S}$ and dispersion parameter σ.

Proposition 4 provides a sampling algorithm for Riemannian Gaussian distributions on $\mathcal{Q}_n$. Indeed, the proposition states that in order to obtain Q with Riemannian Gaussian distribution of barycentre $\breve{S}$ and dispersion σ, it is enough to know how to sample S from a Riemannian Gaussian distribution with barycentre I. In turn, this is done using polar coordinates, through decomposition (8).

In this decomposition, K must be sampled from a uniform distribution on $\mathrm{Sp}(n)$, and R with diagonal elements r_i from the multivariate density (22). Sampling from a uniform distribution on $\mathrm{Sp}(n)$ can be achieved as follows : let Z be an $n \times n$ quaternion matrix whose elements are independent normal proper quaternion random variables [11], and write $Z = KP$ for the polar decomposition of Z [9]. Then, K has a uniform distribution on $\mathrm{Sp}(n)$. On the other hand, sampling from the multivariate density (22) can be carried out using a Metropolis-Hastings algorithm, which is included in most statistical software [12].

Consider now maximum likelihood estimation of Riemannian Gaussian distributions. This is given by the following Proposition 5. This proposition brings out the important role of the function $Z(\sigma)$ defined by (18) and (21). Precisely, this is the moment generating function of the Riemannian Gaussian distribution (17). If $\eta = -1/2\sigma^2$ and $\psi(\eta) = \log Z(\sigma)$, then $\psi(\eta)$ is a strictly convex function, which is the cumulant generating function of the distribution (17).

Proposition 5 (Maximum likelihood estimation). *Let $S_1, \ldots, S_N$ be independent samples from a Riemannian Gaussian distribution with density (17). Based on these samples, the maximum likelihood estimate of $\breve{S}$ is the sample Riemannian barycentre $\hat{S}_N$,*

$$\hat{S}_N = \mathrm{argmin}_{S \in \mathcal{Q}_n} \sum_{i=1}^{N} d^2(S_i, S) \tag{23}$$

where the distance $d(S_i, S)$ is given by (16). Moreover, the maximum likelihood estimate of $\eta = -1/2\sigma^2$ is $\hat{\eta}_N$,

$$\hat{\eta}_N = (\psi')^{-1}\left(\frac{1}{N}\sum_{i=1}^{N} d^2(S_i, \hat{S}_N)\right) \tag{24}$$

where $(\psi')^{-1}$ is the reciprocal function of ψ', the derivative of ψ.

Proposition 5 indicates how the maximum likelihood estimates $\hat{S}_N$ and $\hat{\eta}_N$ can be computed. First, $\hat{S}_N$ is the sample Riemannian barycentre of $S_1, \ldots, S_N$. Its existence and uniqueness are guaranteed by the fact that $\mathcal{Q}_N$ is a Riemannian manifold of non-positive curvature. In practice, it can be computed using a Riemannian gradient descent algorithm [13,14]. Once $\hat{S}_N$ has been obtained, $\hat{\eta}_N$ is found by direct application of (24). This only requires knowledge of the cumulant generating function $\psi(\eta)$, which can be tabulated using the Monte Carlo method of [15].

References

1. Pennec, X.: Intrinsic statistics on Riemannian manifolds: basic tools for geometric measurements. J. Math. Imaging Vis. **25**(1), 127–154 (2006)
2. Chebbi, Z., Moakher, M.: Means of Hermitian positive-definite matrices based on the log-determinant alpha-divergence function. Linear Algebra Appl. **436**(7), 1872–1889 (2012)

3. Helgason, S.: Differential Geometry, Lie Groups, and Symmetric Spaces. American Mathematical Society, Providence (2001)
4. Besse, A.L.: Einstein Manifolds, 1st edn. Springer, Heidelberg (2007)
5. Said, S., Bombrun, L., Berthoumieu, Y., Manton, J.H.: Riemannian Gaussian distributions on the space of symmetric positive definite matrices (accepted). IEEE Trans. Inf. Theory **63**, 2153–2170 (2016)
6. Said, S., Hajri, H., Bombrun, L., Vemuri, B.C.: Gaussian distributions on Riemannian symmetric spaces : statistical learning with structured covariance matrices (under review). IEEE Trans. Inf. Theory (2017, under review)
7. Flamant, J., Le Bihan, N., Chainais, P.: Time-frequency analysis of bivariate signals (under review). Applied and Computational Harmonic Analysis (2017)
8. Conway, J.H., Smith, D.A.: On Quaternions and Octonions, their Geometry, Arithmetic and Symmetry. CRC Press, Boca Raton (2003)
9. Zhang, F.: Quaternions and matrices of quaternions. Linear Algebra Appl. **251**, 21–57 (1997)
10. Kirillov, A.: An Introduction to Lie Groups and Lie Algebras. Cambridge University Press, Cambridge (2008)
11. Le Bihan, N.: The geometry of proper quaternion random variables. Signal Processing (2017, to appear)
12. Robert, C.P., Casella, G.: Monte Carlo Statistical Methods. Springer, New York (2004). doi:10.1007/978-1-4757-4145-2
13. Manton, J.H.: A globally convergent numerical algorithm for computing the centre of mass on compact Lie groups. In: ICARCV 2004 8th Control, Automation, Robotics and Vision Conference, 2004, vol. 3, pp. 2211–2216, December 2004
14. Manton, J.H.: A framework for generalising the Newton method and other iterative methods from Euclidean space to manifolds. Numer. Math. **129**(1), 91–125 (2014)
15. Zanini, P., Said, S., Congedo, M., Berthoumieu, Y., Jutten, C.: Parameter estimates of Riemannian Gaussian distributions in the manifold of covariance matrices. In: Sensor Array and Multichannel Signal Processsing Workshop (SAM) (2016)

A Family of Anisotropic Distributions on the Hyperbolic Plane

Emmanuel Chevallier(✉)

Weizmann Institute of Science, Rehovot, Israel
emmanuelchevallier1@gmail.com

Abstract. Most of the parametric families of distributions on manifold are constituted of radial distributions. The main reason is that quantifying the anisotropy of a distribution on a manifold is not as straightforward as in vector spaces and usually leads to numerical computations. Based on a simple definition of the covariance on manifolds, this paper presents a way of constructing anisotropic distributions on the hyperbolic space whose covariance matrices are explicitly known. The approach remains valid on every manifold homeomorphic to vector spaces.

1 Introduction

Probability density estimation on Riemannian manifolds is the subject of several recent studies. The different approaches can be separated into two categories, the parametric and non-parametric ones. The context of Riemannian manifolds brings difficulties of two kinds. Firstly, the theoretical results about distributions and the convergence of estimators known for random variables valued in $\mathbb{R}^n$ have to be adapted to the case of random variables valued in Riemannian manifolds, see [1–3,8,9,12–15]. Secondly, the construction of probability distribution and of density estimators should require a reasonable amount of computational complexity, see [8,12,13,16–18]. A generalization of the Gaussian distribution on manifolds was proposed in [8]. Although the expression of the proposed law is hard to compute on general manifolds, expressions of radial Gaussians on symmetric spaces can be found in [12–14]. On isotropic spaces, an isotropic density is simply a radial density. The anisotropy of a density can be evaluated with the notion of covariance proposed in [8].

In this paper, we are interested in the construction of anisotropic distributions on the hyperbolic space. The problem of anisotropic normal distributions on manifold have been addressed in [19] through anisotropic diffusion. The construction is valid on arbitrary manifolds but requires important computations. The hyperbolic space is a very particular Riemannian manifold: it is at the same time isotropic and diffeomorphic to a vector space. These two specificities significantly ease the construction of probability distributions and probability density estimators. Generally, it is difficult to control the covariance of a distribution on a Riemannian manifold, e.g. the covariance of the Gaussian law proposed in [8]. We propose a simple way of constructing distributions whose covariance is fully

F. Nielsen and F. Barbaresco (Eds.): GSI 2017, LNCS 10589, pp. 717–724, 2017.
https://doi.org/10.1007/978-3-319-68445-1_83

controlled. The method is derived from the density kernel proposed by [1]. These distributions can be used in the non parametric kernel density estimator but also to design mixture models for parametric density estimation.

The paper is organised as follows. Section 2 is a very brief introduction to the hyperbolic plane. Section 3 reviews some general facts about probabilities on Riemannian manifolds. Section 4 describes how to built anisotropic density functions on the hyperbolic space.

2 The Hyperbolic Space

The hyperbolic geometry results of a modification of the fifth Euclid's postulate on parallel lines. In two dimensions, given an line D and a point $p \notin D$, the hyperbolic geometry is an example where there are at least two lines going through p, which do not intersect D. Let us consider the open unit disk of the Euclidean plane endowed with the Riemannian metric:

$$ds^2_{\mathbb{D}} = 4\frac{dx^2 + dy^2}{(1 - x^2 - y^2)^2} \tag{1}$$

where x and y are the Cartesian coordinates. The unit disk $\mathbb{D}$ endowed with $ds_{\mathbb{D}}$ is called the Poincaré disk and is a model of the two-dimensional hyperbolic geometry. The construction is generalized to higher dimensions. Let ISO be the isometry group of $\mathbb{D}$. It can be shown that:

- $\mathbb{D}$ is homogeneous: $\forall p, q \in \mathbb{D}, \exists \phi \in ISO, \phi(p) = q$, points are indistinguishable.
- $\mathbb{D}$ is isotropic: for any couple of geodesics γ_1 and γ_2 going through a point $p \in \mathbb{D}$, there exists $\phi \in ISO$ such that $\phi(p) = p$ and $\phi(\gamma_1) = \gamma_2$. In other words, directions are indistinguishable.
- the Riemannian exponential applications are bijective.
- $\mathbb{D}$ has a constant negative curvature.

Let x denote the coordinates of elements of $T_p\mathbb{D}$ in an orthogonal basis. x is mapped to a point on $\mathbb{D}$ by the Riemannian exponential application noted exp_p and form thus a chart of $\mathbb{D}$. This chart is called an exponential chart at the point p.

Given a reference point p the point of polar coordinates (r, α) of the hyperbolic space is defined as the point at distance r of p on the geodesic with initial direction $\alpha \in \mathbb{S}^1$. Since the hyperbolic space is isotropic, the expression of the metric in polar coordinates only depends on r,

$$ds^2 = dr^2 + \sinh(r)^2 d\alpha^2, \tag{2}$$

see [10, 11].

3 Distributions on $\mathbb{D}$

3.1 Densities

The metric of a Riemannian manifold provides a measure of volumes *vol*. In a chart, if G is the matrix of the metric, the density of *vol* with respect to the Lebesgue measure of the chart is

$$\frac{dvol}{dLeb} = |det(\sqrt{G})|$$

where $\sqrt{G}$ is the matrix square root of G. Let μ be a measure on $\mathcal{M}$. If μ has a density f with respect to the Lebesgue measure of a chart, then the density with respect to the Riemannian volume measure is given by

$$\frac{d\mu}{dvol} = \frac{d\mu}{dLeb}\frac{dLeb}{dvol} = \frac{1}{|det(\sqrt{G})|}f. \tag{3}$$

3.2 Intrinsic Means

Given a distribution μ, the variance at p be defined by

$$\sigma^2(p) = \int_{\mathbb{D}} d(p,.)^2 d\mu.$$

When the variance is finite everywhere, its minima are called mean points. The hyperbolic space is a Cartan-Hadamar manifold, that is to say it is complete, simply connected and of negative curvature. On Cartan-Hadamar manifolds, when the variance is finite everywhere, the mean exists and is unique, see [8] corollary 2. It is achieved at p such that

$$\int_{T_p\mathbb{D}} x d\tilde{\mu} = 0,$$

where $\tilde{\mu}$ is the image of the measure μ by the inverse of the exponential application at p.

3.3 Covariance on Manifold

The covariance of a random vector is the matrix formed by the covariance of its coordinates. In a vector space the coordinates of a vector are given in terms of projection on the corresponding axis. On a Riemannian manifold the notions of projection on a geodesic usually do not lead to explicit expressions. Even if it does not conserve all the properties of the covariance of vectors, when possible, the simplest generalisation to manifolds is to take the Euclidean covariance after lifting the distribution on a tangent space by the inverse of the exponential map, see [8]. Since on the hyperbolic space the exponential application a bijection, it is always possible to lift distributions on tangent spaces. Given a distribution μ and a orthogonal basis of $T_p\mathbb{D}$, the covariance at $p \in \mathbb{D}$ is thus defined as

$$\Sigma_p(\mu) = \int_{T_p\mathbb{D}} xx^t d\tilde{\mu}$$

This definition of covariance was used to define a notion of principal geodesic analysis on manifolds in [20]. It can be noted that the covariance at the point p is a point in $T\mathbb{D} \otimes T\mathbb{D}$.

4 Constructing Anisotropic Distributions

The author of [8] proposes a generalization of Gaussian distributions on manifolds as the distribution that maximizes the entropy given its barycenter and covariance. This generalization leads to a density of the form,

$$N_{(p,\Gamma)}(exp_p(x)) = k.\exp\left(-\frac{x^t\Gamma x}{2}\right)$$

Given p and the covariance matrix Σ_p, the main difficulties are to obtain expressions of the normalizing factor k and of the concentration matrix Γ. Since hyperbolic space is homogenous, k and Γ only depend on the matrix Σ_p. The expression of k and Γ when Σ_p is a (positive) multiple of the identity matrix can be found in [12]. However, it is difficult to obtain these relations when Σ_p is not diagonal.

It might be interesting to define parametric families of distributions whose means and covariances can easily be controlled, even if they do not verify the same statistical properties as the Gaussian distributions. Let $K : \mathbb{R}_+ \rightarrow \mathbb{R}_+$ be a function such that,

i. $\int_{\mathbb{R}^2} K(\|y\|)\, dy = 1$
ii. $\int_{\mathbb{R}^2} \|y\|^2 K(\|y\|)\, dy = 2$

Given Γ a symmetric positive definite matrix, we have then

$$\int_{\mathbb{R}^2} \frac{1}{\sqrt{det(\Gamma)}} K(\sqrt{x^t\Gamma^{-1}x})dx = 1.$$

Let $\overline{p}$ be a point in $\mathbb{D}$. Set an orthonormal basis of the tangent space $T_{\overline{p}}\mathbb{D}$ and consider the distribution $\nu_{\overline{p},\Gamma}$ on $T_{\overline{p}}\mathbb{D}$ whose density with respect to the Lebesgue measure of $T_{\overline{p}}\mathbb{D}$ is given by $\frac{1}{\sqrt{det(\Sigma)}}K(\sqrt{x^t\Gamma^{-1}x})$, where x and Γ are expressed in the reference basis. Let $\mu_{\overline{p},\Gamma} = exp_{\overline{p}*}(\nu_{\overline{p},\Gamma})$ be the pushforward measure of $\nu_{\overline{p},\Gamma}$ by the Riemannian exponential at $\overline{p}$.

Theorem 1. *$\overline{p}$ is the unique mean of $\mu_{\overline{p},\Gamma}$.*

Proof. It can be checked that $\mu_{\overline{p},\Gamma}$ has a finite variance everywhere. Moreover,

$$\int_{T_{\overline{p}}\mathbb{D}} \sqrt{\Gamma^{-1}}x \frac{1}{\sqrt{\det \Gamma}} K(\sqrt{x^t \Gamma^{-1} x})\, dx = 0.$$

The integrability of the function can be deduced from i and ii and the nullity from its symmetry. Therefore according to Sect. 3.2 $\overline{p}$ is the unique mean of $\mu_{\overline{p},\Gamma}$.

Theorem 2. *The covariance $\Sigma_{\overline{p}}$ of $\mu_{\overline{p},\Gamma}$ at $\overline{p}$ and the concentration matrix Γ are equal.*

Proof. In the reference basis, making use of ii with the change of variables $y = \sqrt{\Gamma^{-1}}x$

$$\begin{aligned}
\Sigma_{\overline{p}} &= \int_{\mathbb{R}^2} xx^t \frac{1}{\sqrt{\det(\Gamma)}} K(\sqrt{x^t \Gamma^{-1} x}) dx \\
&= \Gamma^{1/2} \int_{\mathbb{R}^2} yy^t K(\sqrt{y^t y}) dy \Gamma^{1/2} \\
&= \Gamma^{1/2} \left(\int_{\mathbb{R}} \int_0^{2\pi} r^2 \begin{pmatrix} \cos(\theta) \\ sin(\theta) \end{pmatrix} \begin{pmatrix} \cos(\theta) \\ sin(\theta) \end{pmatrix}^t K(r) r dr d\theta \right) \Gamma^{1/2} \\
&= \Gamma^{1/2} \left(\frac{1}{2} \int_{\mathbb{R}} r^2 I K(r) 2\pi r dr \right) \Gamma^{1/2} \\
&= \Gamma^{1/2} I \left(\frac{1}{2} \int_{\mathbb{R}^2} \|y\|^2 K(\|y\|)\, dy \right) \Gamma^{1/2} \\
&= \Gamma.
\end{aligned}$$

The tangent space $T_{\overline{p}}\mathbb{D}$ endowed with the reference basis provides a parametrization of the hyperbolic space. By definition, the density of $\mu_{\overline{p},\Gamma}$ in this parametrization is given by $\frac{1}{\sqrt{det(\Sigma)}} K(\sqrt{x^t \Sigma^{-1} x})$. In order to obtain the density with respect to the Riemannian measure this term should be multiplied by the density of the Lebesgue measure of the parametrization with respect to the Riemannian measure, see Eq. 3. In an adapted orthonormal basis of $T_{\overline{p}}\mathbb{D}$, Eq. 2 leads to the following expression of the matrix of the metric,

$$G = \begin{pmatrix} 1 & 0 \\ 0 & \frac{\sinh(r)^2}{r^2} \end{pmatrix}.$$

Thus,

$$det(\sqrt{G}) = \frac{\sinh(r)}{r}.$$

Equation 3 leads to the density ratio,

$$\frac{dx}{dvol}(x) = \frac{||x||}{\sinh(||x||)},$$

where dx is the Lebesgue measure induced by the reference basis. Recall that in this parametrization, the Euclidean norm of x is the distance between $exp_{\overline{p}}(x)$ and $\overline{p}$. The density of $\mu_{\overline{p},\Gamma}$ with respect to the Riemannian measure is given by

$$f(exp_{\overline{p}}(x)) = \frac{||x||}{\sinh(||x||)\sqrt{det(\Sigma)}} K\left(\sqrt{x^t \Sigma^{-1} x}\right).$$

Figure 1 shows the level lines when K is Gaussian.

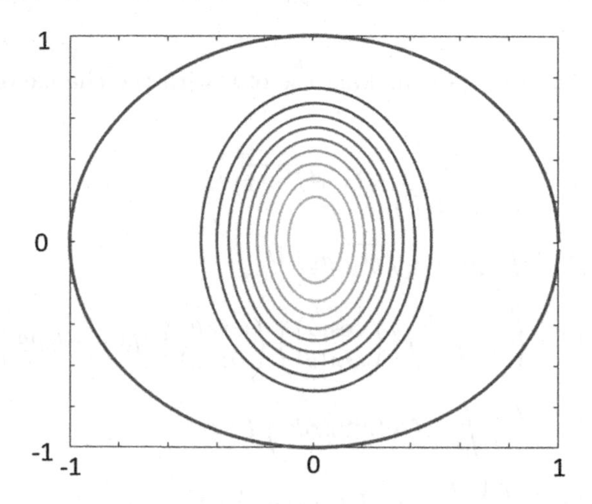

Fig. 1. In this example $K(x) = \frac{1}{\sqrt{2\pi}} e^{-x^2}$ and Σ has 1 and $\frac{1}{4}$ as eigenvalues. The level lines of the corresponding density f are flattened circle but are not ellipses.

5 Estimating the Mean and the Covariance

Let the function K and the distribution $\mu_{\overline{p},\Gamma}$ be as defined in Sect. 4. Given a set of draws drawn from this distribution it is important to have estimators of the two parameters: the mean and the covariance. In order to estimate the unknown parameters $(\overline{p}, \Sigma_{\overline{p}})$ given a set of independent samples $(p_1, .., p_n)$, it is usual to try to maximize the likelihood function. The log-likelihood of a set of samples is defined as

$$\mathcal{L}(p_1, .., p_n; (\hat{p}, \hat{\Sigma})) = \sum_i \log\left(\frac{||x_i||}{\sinh(||x_i||)\sqrt{det(\hat{\Sigma})}} K\left(\sqrt{x_i^t \hat{\Sigma}^{-1} x_i}\right)\right)$$

$$= \sum_i \log\left(\frac{||x_i||}{\sinh(||x_i||)\sqrt{det(\hat{\Sigma})}}\right) + log\left(K\left(\sqrt{x_i^t \hat{\Sigma}^{-1} x_i}\right)\right).$$

The major difficulty is that it is not possible to optimize the mean and the covariance separately. Thus there might not be explicit expressions of the maximum likelihood. However, the mean and the covariance have natural estimators. It is already known that the empirical barycenter is a strongly consistent estimator of the barycenter, see [21] Theorem 2.3.

Given an estimate of the barycenter, it is possible to compute the empirical covariance in the corresponding tangent plane,

$$\hat{\Sigma}_{\hat{p}} = \frac{1}{N}\sum x_i x_i^t \tag{4}$$

Using a similar construction as the Sasakian metric, see [22], the vector bundle $T\mathbb{D}\otimes T\mathbb{D}$ can be endowed with a Riemannian metric. Although we do not prove it in this paper, we are convinced that almost surely

$$d((\hat{p}, \hat{\Sigma}_{\hat{p}}), (\overline{p}, \Sigma)) \underset{n\to+\infty}{\longrightarrow} 0,$$

where d is the Riemannian distance on $T\mathbb{D}\otimes T\mathbb{D}$.

6 Conclusion

In this paper we proposed a set of parametric families of anisotropic distributions on the hyperbolic plane. The main interest of these distributions is that the covariance matrix and concentration matrix are equal. The empirical mean and covariance provide thus simple estimators of the parameters of the distribution. Working with anisotropic distributions is expected to reduce the number of distributions used in mixture models and thus to reduce the computational complexity of the parameter estimation of the mixture models. On the one hand, our future work will focus on deriving convergence rates of the estimation of the covariance. On the other hand, we will study the use of these distributions in problems of radar signal classification.

References

1. Pelletier, B.: Kernel density estimation on Riemannian manifolds. Stat. Probab. Lett. **73**, 297–304 (2005)
2. Hendriks, H.: Nonparametric estimation of a probability density on a Riemannian manifold using Fourier expansions. Ann. Stat. **18**, 832–849 (1990)
3. Huckemann, S., Kim, P., Koo, J., Munk, A.: Mobius deconvolution on the hyperbolic plan with application to impedance density estimation. Ann. Stat. **38**, 2465–2498 (2010)
4. Barbaresco, F.: Robust statistical radar processing in Fréchet metric space: OS-HDR-CFAR and OS-STAP processing in siegel homogeneous bounded domains. In: Proceedings of the 2011 12th International Radar Symposium (IRS), Leipzig, Germany, 7–9 September 2011

5. Barbaresco, F.: Information geometry of covariance matrix: Cartan-Siegel homogeneous bounded domains, Mostow/Berger fibration and Fréchet median. In: Nielsen, F., Bhatia, R. (eds.) Matrix Information Geometry, pp. 199–256. Springer, Heidelberg (2012). doi:10.1007/978-3-642-30232-9_9
6. Barbaresco, F.: Information geometry manifold of Toeplitz Hermitian positive definite covariance matrices: Mostow/Berger fibration and Berezin quantization of Cartan-Siegel domains. Int. J. Emerg. Trends Signal Process. (IJETSP) **1**, 1–87 (2013)
7. Cannon, J.W., Floyd, W.J., Kenyon, R., Parry, W.R.: Hyperbolic Geometry, vol. 31. MSRI Publications, Cambridge (1997)
8. Pennec, X.: Intrinsic statistics on Riemannian manifolds: basic tools for geometric measurements. J. Math. Imaging Vis. **25**, 127 (2006)
9. Kim, P.T., Richards, D.St.P: Deconvolution density estimation on the space of positive definite symmetric matrices. In: Nonparametric Statistics and Mixture Models, pp. 147–168 (2011)
10. Grigoryan, A.: Heat Kernel and Analysis on Manifolds, vol. 47. American Mathematical Soc., Providence (2012)
11. Anker, J.-P., Ostellari, P.: The heat kernel on noncompact symmetric spaces. In: Lie Groups and Symmetric Spaces. AMS Transl. Ser. 2, vol. 210, pp. 27–46 (2003)
12. Said, S., Bombrun, L., Berthoumieu, Y.: New Riemannian priors on the univariate normal model. Entropy **16**(7), 4015–4031 (2014)
13. Said, S., Bombrun, L., Berthoumieu, Y., Jonathan Manton, J.: Riemannian Gaussian distributions on the space of symmetric positive definite matrices. arXiv:1507.01760 [math.ST] (2015)
14. Said, S., Hajri, H., Bombrun, L., Vemuri, B.C.: Gaussian distributions on Riemannian symmetric spaces: statistical learning with structured covariance matrices. arXiv:1607.06929 [math.ST] (2016)
15. Le Bihan, N., Flamant, J., Manton, J.H.: Density estimation on the rotation group using diffusive wavelets. arXiv:1512.06023 (2015)
16. Chevallier, E., Barbaresco, F., Angulo, J.: Probability density estimation on the hyperbolic space applied to radar processing. In: Nielsen, F., Barbaresco, F. (eds.) GSI 2015. LNCS, vol. 9389, pp. 753–761. Springer, Cham (2015). doi:10.1007/978-3-319-25040-3_80
17. Chevallier, E., Forget, T., Barbaresco, F., Angulo, J.: Kernel density estimation on the siegel space with an application to radar processing. Entropy **18**(11), 396 (2016)
18. Chevallier, E., Kalunga, E., Angulo, J.: Kernel density estimation on spaces of Gaussian distributions and symmetric positive definite matrices. SIAM J. Imaging Sci. **10**(1), 191–215 (2017)
19. Sommer, S.: Anisotropic distributions on manifolds: template estimation and most probable paths. In: Ourselin, S., Alexander, D.C., Westin, C.-F., Cardoso, M.J. (eds.) IPMI 2015. LNCS, vol. 9123, pp. 193–204. Springer, Cham (2015). doi:10.1007/978-3-319-19992-4_15
20. Fletcher, P.T., Lu, C., Pizer, S.M., Joshi, S.: Principal geodesic analysis for the study of nonlinear statistics of shape. IEEE Trans. Med. Imaging **23**(8), 995–1005 (2004)
21. Bhattacharya, R., Patrangenaru, V.: Large sample theory of intrinsic and extrinsic sample means on manifolds. I. Ann. Stat. **31**, 1–29 (2003)
22. Kappos, E.: Natural metric on tangent bundles. Master's thesis (2001)

Session Geometry of Tensor-Valued Data

A Riemanian Approach to Blob Detection in Manifold-Valued Images

Aleksei Shestov(✉) and Mikhail Kumskov

Faculty of Mechanics and Mathematics, Lomonosov Moscow State University, GSP-1, 1 Leninskiye Gory, Main Building, 119991 Moscow, Russia
shestov.msu@gmail.com, mkumskov@gmail.com

Abstract. This paper is devoted to the problem of blob detection in manifold-valued images. Our solution is based on new definitions of blob response functions. We define the blob response functions by means of curvatures of an image graph, considered as a submanifold. We call the proposed framework Riemannian blob detection. We prove that our approach can be viewed as a generalization of the grayscale blob detection technique. An expression of the Riemannian blob response functions through the image Hessian is derived. We provide experiments for the case of vector-valued images on 2D surfaces: the proposed framework is tested on the task of chemical compounds classification.

Keywords: Blob detection · Image processing · Manifold-valued images · Vector-valued images · Differential geometry

1 Introduction

Blob detection [1] is a widely used method of keypoints detection in grayscale images. Informally speaking, blob detection aims to find ellipse-like regions of different sizes with similar intensity inside. Blobs are sought as local extremums of a blob response function. Several color blob detection algorithms were proposed in [2,3]. Blob detection has applications in 3D face recognition, object recognition, panorama stitching, 3D scene modeling, tracking, action recognition, medical images processing, etc.

Our goal is to propose a blob detection framework for the general setting of an image being a map between Riemannian manifolds. Our approach is based on a definition of blob response functions by means of image graph curvatures. Furthermore, we derive the expression of Riemannian blob response functions through image Hessian. This expression shows that Riemannian blob detection coincides with the classical blob detection framework for the grayscale case. Also this expression provides a more convenient way to calculate Riemannian blob response functions for vector- and manifold-valued images.

Research of connections between image processing methods and image graph geometry is of its own interest. This research helps deeply understand traditional methods, provides insights and gives natural generalizations of classical methods

F. Nielsen and F. Barbaresco (Eds.): GSI 2017, LNCS 10589, pp. 727–735, 2017.
https://doi.org/10.1007/978-3-319-68445-1_84

to vector-valued and manifold-valued images [4–6]. Connections between the blob response functions and image graph curvatures were mentioned in papers [7,8]. Our work is the first to accurately analyze this question in the general setting.

Contributions:

1. We are the first to provide a blob detection framework for the general setting of an image being a map between manifolds. This framework can be viewed as a generalization of grayscale blob detection. Our framework provides blob response functions for the previously uncovered problems: blob detection in color images on manifold domain and blob detection in manifold-valued images (both on Euclidian and manifold domains).
2. We are the first to analyze connections between the blob response functions and curvatures of image graph both for Euclidian and manifold domains.
3. The experiments on the task of chemical compounds classification show the effectiveness of our approach for the case of vector-valued images on 2d surfaces.

2 The Problem Introduction

Blob detection was firstly proposed for grayscale images on 2D Euclidian domain [1]. In [9] blob detection was generalized to 2D surfaces. Several approaches to generalization of blob detection to color case were proposed in [2,3]. However, these approaches are based on global or local conversion of a color image to the grayscale, so they can't be used for manifold-valued images.

Consider a grayscale image $I(x) : X \to \mathbb{R}$ on a smooth 2-dimensional manifold X. The blob detection framework by [9] is as follows:

1. Calculate the scale-space $L(x,t) : X \times \mathbb{R}_+ \to \mathbb{R}$. $L(x,t)$ is the solution of the heat equation on the surface $\partial_t L(x,t) = -\Delta_{\mathrm{LB}} L(x,t), L(x,0) = I(x)$, where Δ_{LB} is the Laplace-Beltrami operator;
2. Choose a blob response function and calculate it:

$$\textit{the determinant blob response} : BR_{\mathrm{det}}(x,t) = \det H_L(x,t) \text{ or} \tag{1}$$

$$\textit{the trace blob response} : BR_{\mathrm{tr}}(x,t) = \operatorname{tr} H_L(x,t), \tag{2}$$

   where H_L is the Hessian of $L(x,t)$ as a function of x with fixed t;
3. Find blobs centers and scales as $C = \{(x,t) = \arg\min_{x,t} \tilde{BR}(x,t) \text{ or } (x,t) = \arg\max_{x,t} \tilde{BR}(x,t)\}$, where $\tilde{BR} = t\, BR_{\mathrm{tr}}$ or $\tilde{BR} = t^2\, BR_{\mathrm{det}}$. Find the blobs radii as $s = \sqrt{2t}$.

For the general case of a map between manifolds $I(x) : X \to Y$ the Hessian is the covariant differential of the differential: $H_L = \nabla \mathrm{d}L, H_L \in T^*X \otimes T^*X \otimes TY$. Consider the straightforward generalization of the blob detection stages:

1. Scale-space calculation. $L(x,t)$ is calculated as the solution of $\partial_t L(x,t) = -\operatorname{tr} H_L(x,t), L(x,0) = I(x)$. Methods of manifold-valued PDEs solution for different cases are discussed in the papers [5,10,11] and others. These methods are out of scope of our work.
2. Blob response calculation. The determinant blob response $BR_{\text{det}} = \det H_L$ is not defined.
3. Blobs centers calculation. We can't find maximums or minimums of the trace blob response because it is not scalar-valued: $BR_{\text{tr}} = \operatorname{tr} H_L \in TY$.

We see that there is no straightforward generalization of the blob response functions to the manifold-valued case. How can the problem of blob response generalization be solved? Our key ideas are the following:

1. Consider the image graph Gr as a submanifold embedded in $X \times Y$. The grayscale and manifold-valued cases differ only by a co-dimension of the embedding. Then a formulation of the blob response through notions defined for all co-dimensions will give an immediate generalization to the manifold-valued case.
2. What notions to use? The scalar and the mean curvatures are defined for all co-dimensions and are close to the determinant and the trace of the image Hessian respectively if tangent planes to Gr and to X are "close".

3 The Proposed Method

3.1 Used Notations

All functions and manifolds here and further are considered to be smooth. Consider m- and n-dimensional manifolds X and Y. Denote the $(n+m)$-dimensional manifold $X \times Y$ as E. Consider the isometric embeddings $i_x(y) = id(x,y) : Y \to E, i_y(x) = id(x,y) : X \to E$. Further we identify X (resp. Y) and related notions with $i_y(X)$ (resp. $i_x(Y)$). The letters i,j,k,l (resp. α,β,γ) are used as indices for notions related to X (resp. Y). The set $\{e_i\}$ (resp. $\{e_\alpha\}$) is an orthonormal basis of T_xX (resp. T_yY).

For a map $f(x) : X \to Y$ its graph Gr_f is an n-dimensional manifold embedded in E. Denote the Hessian of f as H_f. Let μY, $\mu \in \mathbb{R}^+$, be the manifold Y with the metric μG_Y. For a map $f : X \to Y$ denote $\mu f : X \to \mu Y$.

We analyze a manifold-valued image $I(x) : X \to Y$. Denote $L(x,t) : X \times \mathbb{R}_+ \to \mathbb{R}$ the solution of the heat equation $\partial_t L(x,t) = -\Delta_{\text{LB}} L(x,t), L(x,0) = I(x)$, where Δ_{LB} is the Laplace-Beltrami operator.

For a manifold N and its submanifold M denote the mean curvature of M as h_M^N, its scalar curvature as r_M, an exponential map from T_mM to N as $\exp_M^N$.

Subscripts and superscripts are omitted when they are clear from a context. The definitions of used differential geometric notions can be found in textbooks [12].

3.2 Main Definitions and Theorems

Definition 1. *The scalar blob response is defined as:*

$$BR_{\text{scalar}} = \lim_{\mu \to 0} \frac{1}{\mu^2} \left(r_{Gr_{\mu L}} - r_{\exp^{X \times \mu Y}_{Gr_{\mu L}}} \right),$$

the mean blob response is defined as:

$$BR_{\text{mean}} = \lim_{\mu \to 0} \frac{1}{\mu} h^{X \times \mu Y}_{Gr_{\mu L}}.$$

The next theorem connects BR_{scalar} and BR_{mean} with the scale-space Hessian. The obtained expression provides a more convenient way for calculation of the Riemannian blob response functions.

Theorem 1. *Let $H_{ij} = H_L(e_i, e_j)$, $H^\alpha(,) = \langle H_L(,), e_\alpha \rangle_Y$. Then*

$$BR_{\text{scalar}} = \sum_{i,j=1}^{n} \left(\langle H_{ij}, H_{ji} \rangle_Y - \langle H_{ii}, H_{jj} \rangle_Y \right),$$

$$BR_{\text{mean}} = \|(\operatorname{tr} H^1, \ldots, \operatorname{tr} H^m)\|_Y.$$

The next corollary from Theorem 1 states that for the grayscale case Riemannian blob detection coincides with usual blob detection. This corollary allows to consider our method as a generalization of grayscale blob detection.

Corollary 1. *Let $dim(X) = 2$. Then the scalar blob response is equal to the determinant blob response (1):*

$$BR_{\text{scalar}} = BR_{\text{det}},$$

the mean blob response is equal to the trace blob response (2):

$$BR_{\text{mean}} = BR_{\text{tr}}.$$

3.3 Proof of the Theorem 1

Additional Notations. Consider maps $y = f(x) : X \to Y$, $\tilde{f}(x) : X \to E$, $\tilde{f}(x) = (x, f(x)) = \tilde{y}$. $\{e_i^{'} = d\tilde{f}(e_i)\}$ is a basis (not orthonormal) of $T_{\tilde{y}} Gr_f$, $\{e_\alpha^{'} : (e_\alpha^{'}, e_i^{'})_E = 0 \, \forall i \, \forall \alpha\}$ is a basis of $T_{\tilde{y}}(Gr_f)^\perp$. Then $\{e_i^{'}, e_\alpha^{'}\}$ is a basis of $T_{\tilde{y}} E$.

For a manifold M denote its metric as $g(,)_M$ or $\langle , \rangle_M$, the Levi-Civita connection as ∇^M, a connection on a vector bundle $\mathcal{E}$ over M as $\nabla^{\mathcal{E}}$.

Denote as P_V (resp. P_V^U) an orthogonal (resp. along a subspace U) projection on a subspace V.

Some minor formal details of the proofs are omitted due to the space constraints.

Proposition 1. *Let $f : X \to Y$, $u, v \in T_xX$. Then $H_f(u,v) = \nabla_v^{f^*TY}\mathrm{d}f(u)$ $-\mathrm{d}f(\nabla_v^X u)$. If f is injective then $H_f(u,v) = \nabla_{\mathrm{d}f(v)}^Y \mathrm{d}f(u) - \mathrm{d}f(\nabla_v^X u)$.*

Proof. Consider the Hessian H_f as $H_f : TX \otimes TX \to f^*TY$, then $H_f(u,v) = \sum_{\alpha=1}^m \nabla_v(\mathrm{d}f(u,e_\alpha))e_\alpha$. We apply the Leibniz rule to this expression and obtain the first statement. Recall that if f is injective then $\mathrm{d}f$ is an isomorphism between f^*TY and TY. This gives the second statement. □

Lemma 1. *Let $u, v \in T_xX$. Let $\nabla^{\tilde{f}(X)}$ be the connection on Gr_f induced by the isomorphism $\tilde{f}$. Let II be the second fundamental form of the submanifold Gr_f of E with respect to the connection $\nabla^{\tilde{f}(X)}$. Then $H_{\tilde{f}}(u,v) = II(\mathrm{d}\tilde{f}(u), \mathrm{d}\tilde{f}(v))$.*

Proof. As $\tilde{f}$ is injective, by Proposition 1: $H_{\tilde{f}}(u,v) = \nabla_{\mathrm{d}\tilde{f}(v)}^Y \mathrm{d}\tilde{f}(u) - \mathrm{d}\tilde{f}(\nabla_v^X u)$ $= \nabla_{\mathrm{d}\tilde{f}(v)}^Y \mathrm{d}\tilde{f}(u) - \nabla_{\mathrm{d}\tilde{f}(v)}^{\tilde{f}(X)} \mathrm{d}\tilde{f}(u) = II(\mathrm{d}u, \mathrm{d}v)$. □

Proposition 2. *Let $u, v \in T_xX$, $\mathbf{0} \in T_xX$, $H_f(u,v) \in T_{f(x)}Y$. Then $H_{\tilde{f}}(u,v)$ $= (\mathbf{0}, H_f(u,v))$.*

Proof. By Proposition 1: $H_{\tilde{f}}(u,v) = \nabla_v^{\tilde{f}^*(X\times Y)}\mathrm{d}\tilde{f}(u) - \mathrm{d}\tilde{f}(\nabla_v^X u)$.
Recall that $\nabla_{(u_1,u_2)}^{X\times Y}(v_1,v_2) = (\nabla_{u_1}^X v_1, \nabla_{u_2}^Y v_2)$. Then
$\nabla_v^{\tilde{f}^*(X\times Y)}(\mathrm{d}i_y, \mathrm{d}f)(u) - (\mathrm{d}i_y(\nabla_v^X u), \mathrm{d}f(\nabla_v^X u))$
$= (\nabla_v^X(\mathrm{d}i_y u), \nabla_{\mathrm{d}f(v)}^Y(\mathrm{d}fu)) - (\nabla_v^X u, \mathrm{d}f(\nabla_v^X u))$
$= (\mathbf{0}, \nabla_{\mathrm{d}f(v)}^Y(\mathrm{d}fu) - \mathrm{d}f(\nabla_v^X u)) = (\mathbf{0}, H_f(u,v))$. □

Proposition 3. *Let $II_{\tilde{f}}$ be the second fundamental form of the submanifold Gr_f of E with respect to the connection $\nabla^{\tilde{f}(X)}$ and II_E be the second fundamental form with respect to the connection ∇^{Gr_f} induced by ∇^E. Let $u, v \in T_{\tilde{y}}Gr_f$. Then $II_E(u,v) = \mathrm{P}_{T_{\tilde{y}}Gr_f^\perp} II_{\tilde{f}}(u,v)$.*

Proof. By properties of a second fundamental form of a normalized manifold: $II_E(u,v) = \mathrm{P}_{T_{\tilde{y}}Gr_f^\perp}\nabla_u^E v$ and $\exists N \subset T_{\tilde{y}}E : II_{\tilde{f}}(u,v) = \mathrm{P}_N^{T_{\tilde{y}}Gr_f}\nabla_u^E v$. Then by simple operations with vectors we obtain the lemma proposition.

Lemma 2. $II_{Gr_f}(e_i^{'}, e_j^{'}) = \sum_{\alpha,\beta=1}^m H_{ij}^\alpha g^{'\alpha\beta} e_\beta^{'}$.

Proof. $\mathrm{P}_{T_{\tilde{y}}Gr_f^\perp} e_\alpha = g^{'\alpha\beta}e_\beta^{'}$. Then $II_{Gr_f}(e_i^{'}, e_j^{'})$ = (from Proposition 3 and Lemma 1) = $\mathrm{P}_{T_{\tilde{y}}Gr_f^\perp} H_{\tilde{f}}(e_i, e_j)$ = (by Proposition 2) $\mathrm{P}_{T_{\tilde{y}}Gr_f^\perp} H_{fij}^\alpha e_\alpha$ = $\sum_{\alpha,\beta=1}^m H_{ij}^\alpha g^{'\alpha\beta} e_\beta^{'}$. □

Proposition 4. *Consider $\{e_i^{'}\}$ as a basis of $T_{\tilde{y}}Gr_f$. $\mathrm{d}F$ is the matrix of $\mathrm{d}f$ in the basis $\{e_i, e_\alpha\}$ and E is the $n \times n$ unit matrix.*

Then: the induced metric $g_{ij}^{'}$ on Gr_f has the matrix $E + \mathrm{d}F^T\,\mathrm{d}F$; the induced metric on $Gr_{\mu f}$ has the matrix $E + \mu^2\mathrm{d}F^T\,\mathrm{d}F$; the covariant induced metric on $Gr_{\mu f}$ has the matrix $E - \mu^2\mathrm{d}F^T\mathrm{d}F + o(\mu^2)$; $H_{\mu f} = \mu H_f$.

Lemma 3. $\lim_{\mu\to 0}\frac{1}{\mu^2}\left(r_{Gr_{\mu f}} - r_{\exp^{X\times\mu Y}_{Gr_{\mu f}}}\right) = \sum_{i,j=1}^{n}\Big(\langle H_{ij}, H_{ji}\rangle_Y - \langle H_{ii}, H_{jj}\rangle_Y\Big)$.

Proof. Write the Gauss equation, the scalar curvature definition and apply Lemma 2: $r_{Gr_f} - r_{\exp^E_{Gr_f}} = g^{'ik}g^{'jl}\sum_{\alpha,\beta=1}^{m} g^{'\alpha\beta}\Big(H^\alpha_{ik}H^\alpha_{jl} - H^\alpha_{il}H^\alpha_{jk}\Big)$. Then substitute μf for f as μ tends to 0, apply Proposition 4 and obtain the needed equality. □

Lemma 4. $\lim_{\mu\to 0}\frac{1}{\mu}h^{X\times\mu Y}_{Gr_{\mu f}} = \|(\operatorname{tr} H^1, \ldots, \operatorname{tr} H^m)\|_Y$.

Proof. $h^E_{Gr_f}{}^2 =$ (by Lemma 2) $\sum_{\gamma,\delta=1}^{m} g^{'ii}g^{'ii}g^{'}_{\alpha\beta}g^{'\alpha\gamma}H^\gamma_{fii}g^{'\beta\delta}H^\delta_{fii}$. For μf: $\lim_{\mu\to 0}\frac{1}{\mu}h^{X\times\mu Y}_{Gr_{\mu f}} =$ (by Prop. 4) $\lim_{\mu\to 0}\Big(\sum_{i,\alpha}H^\alpha_{fii}H^\alpha_{fii} + o(1)\Big)^{\frac{1}{2}} = \|(\operatorname{tr} H^1, \ldots, \operatorname{tr} H^m)\|_Y$ □

Theorem 1 follows from Lemmas 3 and 4. The formulation of Theorem 1 is obtained by substitution of f with L.

4 The Experiments

Experimental Setup. We apply our blob detection framework to a chemical compounds classification problem, called also the QSAR problem [13]. The task is to distinguish active and non-active compounds using their structure. Each compound is represented by a triangulated molecular surface [14] and several physico-chemical and geometrical properties on the surface. So an input data element can be modeled as a 2-dimensional manifold X with a vector-valued function $f(x) : X \to \mathbb{R}^m$. We use the following properties: the electrostatic and the steric potentials, the Gaussian and the mean curvatures. These properties are calculated in each triangulation vertex.

Implementation. We use Riemannian blob detection for the construction of descriptor vectors. The procedure is the following:

1. Detect blobs by our method in each compound surface;
2. Form pairs of blobs on each surface;
3. Transform the blobs pairs into vectors of fixed length by using the bag of words approach [15].

The Riemannian blob response functions are calculated for each triangulation vertex v. The procedure is the following:

1. Find the directional derivatives $\partial_{z_j}L_i$ by the finite differences approximation, where z_j are the directions from v to its neighbour vertices.
2. Find the differential $\mathrm{d}L = (\mathrm{d}L_i)$ by solving the overdetermined linear system $\mathrm{d}L(Z) = \partial_{z_j}L_i$, Z is a matrix which columns are vectors z_j.
3. Find the covariant derivatives of the differential in the neighbour directions, i.e. find $\nabla^X_{z_j}\mathrm{d}L$ for each j as by $\nabla^X_{z_j}\mathrm{d}L = \mathrm{P}_{T_xX}(\nabla^{\mathbb{R}^3}_{z_j}\mathrm{d}L)$. $\nabla^{\mathbb{R}^3}_{z_j}\mathrm{d}L$ are found by the finite differences approximation.

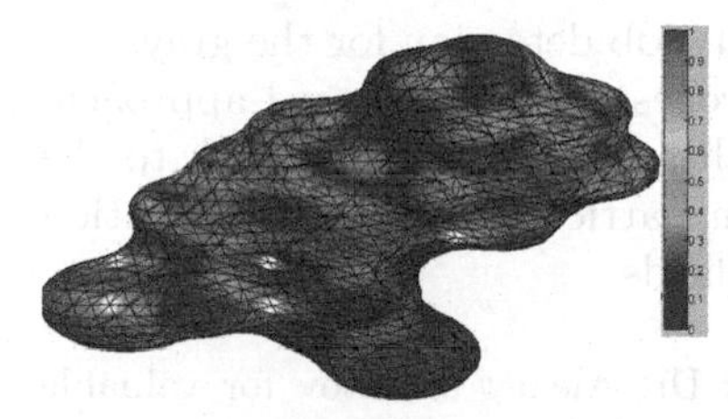

Fig. 1. A molecular surface with BR_{scalar} on it and found centers (denoted by white color) of blobs of radii 3.

Table 1. The results: the cross-validation of the models, based on feature vectors built by the blob detection methods.

	BR_{scalar}	Naive	BR_{mean}	Adapt. [3]
glik	1.0	0.954	0.975	1.0
pirim	0.99	0.96	0.97	0.98
sesq	1.0	0.98	0.976	1.0
bzr	0.992	0.971	0.975	0.983
er_lit	0.98	0.961	0.956	0.98
cox2	0.991	0.967	0.985	0.986

4. Find the covariant differential $\nabla^X \mathrm{d}L$ by solving the overdetermined linear system $\nabla^X \mathrm{d}L(Z) = \nabla^X_{z_j} \mathrm{d}L$, Z is a matrix which columns are vectors z_j.

$\nabla^X \mathrm{d}L = \{H^\alpha_{ij}\}$ is obtained. Calculate $BR_{\text{scalar}}(x,t) = \sum_{\alpha=1}^m \det H^\alpha$, $BR_{\text{mean}}(x,t) = \| \operatorname{tr} H^\alpha \|$.

The Results. An example of the algorithm result is presented in Fig. 1. We compare the prediction models built on the base of the following blob detection methods:

1. Riemannian blob detection with BR_{scalar} as a blob response function;
2. A naive method of applying blob detection to each channel separately;
3. Riemannian blob detection with BR_{mean} as a blob response function. It coincides with the method [2], adapted to the case of 2D surface;
4. The method of adaptive neighbourhood projection [3]. It is adapted by us to the case of 2D surface.

The feature reduction SVM [17] is used for construction of the prediction model. The cross-validation functional [16] is used as an index of the performance quality. The test data is the following: 3 datasets (bzr, er_lit, cox2) from [18], 3 datasets (glik, pirim, sesq) from Russian Oncology Science Center. The results are presented in Table 1.

Riemannian blob detection with BR_{scalar} as a blob response function is the best performing method. This shows the effectiveness of our approach. This particular method for vector-valued functions on 2D surfaces wasn't presented in the literature before.

5 Conclusion and Future Work

We propose the Riemannian framework for blob detection in manifold-valued images. This framework is based on the definition of the blob response functions by means of the image graph curvatures. Our approach gives new methods for the uncovered problems and coincides with classical blob detection for the grayscale case. The experiments results show the effectiveness of the proposed approach. The next direction for the research is a generalization of our framework to the case of sections of non-trivial fiber bundles. In particular, such generalization will cover an important case of tangent vector fields.

Acknowledgments. The authors want to thank Dr. Alexey Malistov for valuable discussions and for a help with the article editing.

References

1. Lindeberg, T.: Feature detection with automatic scale selection. Int. J. Comput. Vision **30**(2), 79–116 (1998)
2. Khanina, N.A., Semeikina, E.V., Yurin, D.V.: Scale-space color blob and ridge detection. Pattern Recogn. Image Anal. **22**(1), 221–227 (2012)
3. Smirnov, P., Semenov, P., Lyakh, M., Chun, A., Gusev, D., Redkin, A., Srinivasan, S.: GRoM – Generalized robust multichannel feature detector. In: 2011 IEEE International Conference on Signal and Image Processing Applications, pp. 585–590, November 2011
4. Saucan, E., Wolansky, G., Appleboim, E., Zeevi, Y.Y.: Combinatorial ricci curvature and Laplacians for image processing. In: 2nd International Congress on Image and Signal Processing, CISP 2009, pp. 1–6 (2009)
5. Sochen, N., Kimmel, R., Malladi, R.: A general framework for low level vision. IEEE Trans. Image Process. **7**(3), 310–318 (1998)
6. Batard, T., Berthier, M.: Spinor Fourier transform for image processing. IEEE J. Sel. Topics Sig. Process. **7**(4), 605–613 (2013)
7. Ferraz, L., Binefa, X.: A sparse curvature-based detector of affine invariant blobs. Comput. Vis. Image Underst. **116**(4), 524–537 (2012)
8. Ferraz, L., Binefa, X.: A scale invariant interest point detector for discriminative blob detection. In: Araujo, H., Mendonça, A.M., Pinho, A.J., Torres, M.I. (eds.) IbPRIA 2009. LNCS, vol. 5524, pp. 233–240. Springer, Heidelberg (2009). doi:10.1007/978-3-642-02172-5_31
9. Zaharescu, A., Boyer, E., Varanasi, K., Horaud, R.: Surface feature detection and description with applications to mesh matching. In: Computer Vision and Pattern Recognition, 2009, pp. 373–380, June 2009
10. Mmoli, F., Sapiro, G., Osher, S.: Solving variational problems and partial differential equations mapping into general target manifolds. J. Comput. Phys. **195**(1), 263–292 (2004)
11. Tschumperle, D., Deriche, R.: Diffusion tensor regularization with constraints preservation. In: Proceedings of the 2001 IEEE Computer Society Conference on Computer Vision and Pattern Recognition, CVPR 2001, Vol. 1, pp. I–I (2001)
12. Spivak, M.: Comprehensive introduction to differential geometry, vol. IV A (1981)
13. Baskin, I., Varnek, I.: Fragment descriptors in SAR/QSAR/QSPR studies, molecular similarity analysis and in virtual screening. In: Chemoinformatic Approaches to Virtual Screening, pp. 1–43 (2008)
14. Connolly, M.L.: Analytical molecular surface calculation. J. Appl. Crystallogr. **16**(5), 548–558 (1983)
15. Csurka, G., Dance, C., Fan, L., Willamowski, J., Bray, C.: Visual categorization with bags of keypoints. In: Workshop on Statistical Learning in Computer Vision, ECCV, vol. 1, no. 1–2, pp. 1–22, May 2004

16. Kohavi, R.: A study of cross-validation and bootstrap for accuracy estimation and model selection. In: IJCAI, vol. 14, no. 2, pp. 1137–1145, August 1995
17. Weston, J., Mukherjee, S., Chapelle, O., Pontil, M., Poggio, T., Vapnik, V.: Feature selection for SVMs. In: Proceedings of the 13th International Conference on Neural Information Processing Systems, pp. 647–653, January 2000
18. Sutherland, J.J., OBrien, L.A., Weaver, D.F.: Spline-fitting with a genetic algorithm: a method for developing classification structure-activity relationships. J. Chem. Inf. Comput. Sci. **43**, 1906–1915 (2003)

Co-occurrence Matrix of Covariance Matrices: A Novel Coding Model for the Classification of Texture Images

Ioana Ilea[1,2], Lionel Bombrun[1(✉)], Salem Said[1], and Yannick Berthoumieu[1]

[1] Laboratoire IMS, Groupe Signal et Image, Université de Bordeaux, Bordeaux, France
{ioana.ilea,lionel.bombrun,salem.said, yannick.berthoumieu}@ims-bordeaux.fr

[2] Technical University of Cluj-Napoca, Cluj-Napoca, Romania
ioana.ilea@com.utcluj.ro

Abstract. This paper introduces a novel local model for the classification of covariance matrices: the co-occurrence matrix of covariance matrices. Contrary to state-of-the-art models (BoRW, R-VLAD and RFV), this local model exploits the spatial distribution of the patches. Starting from the generative mixture model of Riemannian Gaussian distributions, we introduce this local model. An experiment on texture image classification is then conducted on the VisTex and Outex_TC000_13 databases to evaluate its potential.

Keywords: Co-occurrence matrix · Riemannian Gaussian distributions · Classification · Covariance matrix

1 Introduction

Material image classification from texture contents is to assign one or more category labels to an image. It is one of the most fundamental problems in a wide range of applications such as industrial inspection [1], image retrieval [2], medical imaging [3,4], remote sensing [5,6], object recognition, and facial recognition [7–9]. In the general framework of image classification, feature coding techniques for bag-of-features methodologies have proven their efficiency in the recent literature. From a given feature space, bag-of-features techniques consist of first generating a codebook composed by a finite set of codewords, also called dictionary, followed by a coding step which associate to each image an activation map.

In the context of texture analysis, recent works [10–14] proposed compact and discriminative representations from localized structured descriptors in the form of region covariances, i.e. symmetric positive definite (SPD) matrices or local covariance matrices (LCM). Considering the intrinsic Riemannian geometry properties of the SPD matrix space, this paper aims at providing a competitive study of different coding techniques based on LCM codewords for texture classification.

F. Nielsen and F. Barbaresco (Eds.): GSI 2017, LNCS 10589, pp. 736–744, 2017.
https://doi.org/10.1007/978-3-319-68445-1_85

The paper is structured as follows. Section 2 introduces the general workflow for the classification based on local descriptors. Then, Sects. 3 and 4 focuses on two of its main steps, namely the dictionary learning and the coding steps. Finally, Sect. 5 presents an experiment on texture images databases to evaluate the potential of the proposed coding model.

2 General Workflow

Figure 1 presents the general workflow for the classification methods based on local features.

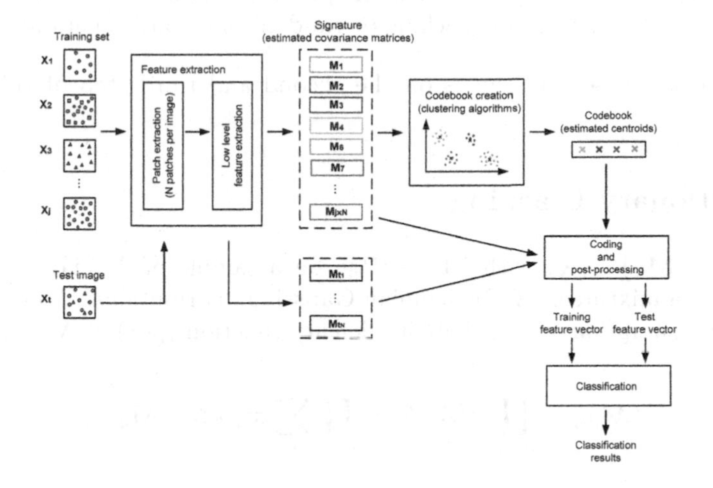

Fig. 1. Classification workflow for local features based methods.

1. During the first step (called feature extraction), some low level features are computed from each element in the database. These descriptors are often computed on patches and as a result, a set of feature vectors (or signature) is obtained for each element in the database. These features can be covariance matrices characterizing for example the color or spatial dependencies.
2. The second step consists in the codebook creation. For that, a clustering algorithm such as the k-means or expectation maximization (EM) one is applied on the training set. By using these algorithms, the set is partitioned into a predefined number of clusters, each of them being described by parameters, such as the cluster's centroid, the dispersion and the associated weight. These estimated parameters are called codewords and are grouped in a codebook.
3. The third step is the coding stage. During this step, each signature set is projected onto the codebook space. For that, various approaches have been proposed in the literature for features being covariance matrices such as the bag of Riemannian words model (BoRW) [12], the Riemannian vectors of

locally aggregated descriptors (R-VLAD) [13] and the Riemannian Fisher vectors (RFV) [14]. Inspired by the concept of gray-level co-occurrences matrices (GLCM), the main contribution of the paper is to propose a novel coding approach which exploits the spatial arrangement between the extracted covariance matrices.
4. After the coding step, a post-processing step is classically applied, consisting in two possible normalizations, namely the ℓ_2 [15] and power normalizations [16]. These post-processing are respectively used to minimize the influence of the background information on the image signature and to correct the independence assumption made on the patches.
5. For the final classification stage, the test image is labeled to the class of the most similar training observation. In practice, classifiers such as k-nearest neighbors, support vector machine or random forest are generally employed.

The next two sections focus on the second and third step of this general workflow.

3 Dictionary Learning

Let $\mathcal{M} = \{\mathbf{M}_n\}_{n=1:N}$, with $\mathbf{M}_n \in \mathcal{P}_m$, be a sample of N i.i.d observations modeled as a mixture of K Riemannian Gaussian distributions. Under the independence assumption, the probability density function (pdf) of $\mathcal{M}$ is given by:

$$p(\mathcal{M}|\theta) = \prod_{n=1}^{N} p(\mathbf{M}_n|\theta) = \prod_{n=1}^{N}\sum_{k=1}^{K} \varpi_k p(\mathbf{M}_n|\bar{\mathbf{M}}_k, \sigma_k), \tag{1}$$

where $p(\mathbf{M}_n|\bar{\mathbf{M}}_k, \sigma_k)$ is the Riemannian Gaussian density (RGD) defined on the manifold $\mathcal{P}_m$ of $m \times m$ real, symmetric and positive definite matrices [17]. The pdf of the RGD, with respect to the Riemannian volume element, has been introduced in [17] as:

$$p(\mathbf{M}|\bar{\mathbf{M}}, \sigma) = \frac{1}{Z(\sigma)} \exp\Big\{-\frac{d^2(\mathbf{M}, \bar{\mathbf{M}})}{2\sigma^2}\Big\}, \tag{2}$$

where $Z(\sigma)$ is the normalization factor independent of the centroid $\bar{\mathbf{M}}$ and $d(\cdot)$ is the Riemannian distance given by $d(\mathbf{M}_1, \mathbf{M}_2) = \big[\sum_i (\ln \lambda_i)^2\big]^{\frac{1}{2}}$, with λ_i, $i = 1, \ldots, m$ being the eigenvalues of $\mathbf{M}_1^{-1}\mathbf{M}_2$.

The codebook is hence composed by the K codewords which are the distribution parameters of each component in the mixture model defined in (1), *i.e.* the mixture weight ϖ_k, the centroid $\bar{\mathbf{M}}_k$ and the dispersion parameter σ_k. In practice, the parameters of the mixture model are estimated by considering an intrinsic k-means algorithm or an EM algorithm. For more information on the implementation of the EM algorithm, the interested reader is referred to [18].

In the experimental part, in order to ensure that each class is represented by a set of codewords, a within-class strategy is adopted to estimate the codebook.

This means that a mixture model is learned for each class in the training set, and the final codebook is obtained by concatenating each codewords (from all the classes).

Once the codebook is created, a coding step is used to encode each image in the database. For that, different strategies can be adopted such as the bag of Riemannian words (BoRW) [12], the Riemannian vectors of locally aggregated descriptors (R-VLAD) [13], the Riemannian Fisher vectors (RFV) [14] and the Co-occurrences of covariances (CoC). The next section describes each of these strategies.

4 Coding Step

Let $\mathcal{M} = \{\mathbf{M}_n\}_{n=1:N}$, with $\mathbf{M}_n \in \mathcal{P}_m$, be a sample of N i.i.d covariance matrices. The aim of the coding step is to project this set $\mathcal{M}$ onto the codebook elements.

4.1 Bag of Riemannian Words (BoRW)

The bag of words (BoW) models is probably one of the most conventional methods used to encode an image. This approach is used in a wide variety of applications in computer vision and signal and image processing. But, when features are living in a non-Euclidean space such as the Riemannian manifold $\mathcal{P}_m$ of $m \times m$ covariance matrices, this model should be readapted. For that the so-called bag of Riemannian words (BoRW) [12] and log-Euclidean bag of words (LE-BoW) [19] models have been introduced.

In these models, the data space is partitioned in K Voronoï regions by maximizing the corresponding pdf. Then, each observation $\mathbf{M}_n$ is assigned to the cluster k, $k = 1, \dots, K$ according to:

$$\arg\max_k \varpi_k \, p(\mathbf{M}_n | \bar{\mathbf{M}}_k, \sigma_k), \tag{3}$$

where $p(\mathbf{M}_n | \bar{\mathbf{M}}_k, \sigma_k)$ is the RGD pdf given in (2). In practice, the homoscedasticity assumption is generally considered (*i.e.* $\sigma_k = \sigma \; \forall k \in [1, K]$) and the codewords are assumed to be equiprobable (*i.e.* $\varpi_k = 1/K$). Further on, for each image in the dataset, its signature is determined by computing the histogram of the number of occurrences of each codeword.

The BoRW model is a simple but effective method. Nevertheless, it suffer from a major drawback, it only counts the number of local descriptors assigned to each Voronoï region. In order to increase the classification performances, some authors have proposed some models which include second order statistics. This is the case for the R-VLAD and RFV models which are presented next.

4.2 Riemannian Vectors of Locally Aggregated Descriptors (R-VLAD)

The Riemannian version of the VLAD descriptors, called Riemannian Vectors of Locally Aggregated Descriptors (R-VLAD), has been developed in [13]. For each cluster c_k, $k \in [1, K]$, a vector containing the differences between the cluster's centroid $\bar{\mathbf{M}}_k$ and each element $\mathbf{M}_i$ in that cluster is computed. Next, the sum of differences concerning each cluster c_k is determined:

$$\mathbf{v}_k = \sum_{\mathbf{M}_i \in c_k} \text{Log}_{\bar{\mathbf{M}}_k} \mathbf{M}_i, \tag{4}$$

where Log(·) is the Riemannian logarithm mapping [20]. This model assumes two hypotheses:

- an hard assignment scheme, this means that each observation $\mathbf{M}_i$ belongs only to one cluster c_k.
- the homoscedasticity assumption, that is $\sigma_k = \sigma$, $\forall k = 1, \ldots, K$

In order to relax these two assumptions, the Riemannian Fisher vectors model has been introduced in [14].

4.3 Riemannian Fisher Vectors (RFV)

Starting from the generative model introduced in (1), the RFV model is obtained by computing the derivative of the log-likelihood of the mixture model with respect to the distribution parameters [14].

$$\frac{\partial \log p(\mathcal{M}|\theta)}{\partial \bar{\mathbf{M}}_k} = \sum_{n=1}^{N} \gamma_k(\mathbf{M}_n)\ \sigma_k^{-2}\ \text{Log}_{\bar{\mathbf{M}}_k}(\mathbf{M}_n), \tag{5}$$

$$\frac{\partial \log p(\mathcal{M}|\theta)}{\partial \sigma_k} = \sum_{n=1}^{N} \gamma_k(\mathbf{M}_n) \Big\{ - \frac{Z'(\sigma_k)}{Z(\sigma_k)} + \frac{d^2(\mathbf{M}_n, \bar{\mathbf{M}}_k)}{\sigma_k^3} \Big\}, \tag{6}$$

$$\frac{\partial \log p(\mathcal{M}|\theta)}{\partial \alpha_k} = \sum_{n=1}^{N} \left[\gamma_k(\mathbf{M}_n) - \varpi_k \right], \tag{7}$$

where $Z'(\sigma)$ is the derivative of the normalizing factor $Z(\sigma)$ with respect to the dispersion parameter σ. The term $\gamma_k(\mathbf{M}_n)$ corresponds to the contribution of each observation $\mathbf{M}_n$ to the cluster c_k, it is defined by:

$$\gamma_k(\mathbf{M}_n) = \frac{\varpi_k\ p(\mathbf{M}_n|\bar{\mathbf{M}}_k, \sigma_k)}{\sum_{j=1}^{K} \varpi_j\ p(\mathbf{M}_n|\bar{\mathbf{M}}_j, \sigma_j)}, \tag{8}$$

Note that the following parametrization of the weights in the mixture model is used in order to ensure the positivity and sum to one constraints of the weights

$$\varpi_k = \frac{\exp(\alpha_k)}{\sum_{j=1}^{K} \exp(\alpha_j)}. \tag{9}$$

As explained in [14], R-VLAD features are a particular case of RFV features. They are retrieved from the RFV when only the derivative with respect to $\bar{\mathbf{M}}_k$ is considered (5) and when the two hypotheses recalled in Sect. 4.2 are assumed.

4.4 Co-occurrences of Covariances (CoC)

These three models (BoRW, R-VLAD and RFV) have shown promising results, but all of these methods do not exploit one main characteristic: the spatial distribution of the patches. Inspired by the concept of GLCM to texture analysis, we introduce a novel coding approach: the co-occurrences of covariances (CoC).

For a dictionary of K codewords, the $K \times K$ co-occurrence matrix of covariance matrices for an image I describes the spatial interactions between the covariance matrices $\mathbf{M}_n$ computed on patches separated from a distance (Δ_x, Δ_y). The element $\mathbf{C}_{\Delta_x,\Delta_y}(k,l)$ of this co-occurrence matrix contains the number of times a covariance matrix which belongs to the codeword l occurs in the neighborhood $\mathcal{N}_{\Delta_x,\Delta_y}$ a covariance matrix which belongs to the codeword k:

$$\mathbf{C}_{\Delta_x,\Delta_y}(k,l) = \sum_{\mathbf{M}_n \in \mathcal{M}} \sum_{\mathbf{M}_p \in \mathcal{N}_{\Delta_x,\Delta_y}(\mathbf{M}_n)} \begin{cases} 1 \text{ if } \mathbf{M}_n \in c_k \text{ and } \mathbf{M}_p \in c_l \\ 0 \text{ otherwise.} \end{cases} \tag{10}$$

Once the co-occurrence matrices are computed, the proximity between two CoC $\mathbf{C}^1$ and $\mathbf{C}^2$ is computed as their intersection by:

$$\sum_{k=1}^{K} \sum_{l=1}^{K} min\left(\mathbf{C}^1_{\Delta_x,\Delta_y}(k,l),\ \mathbf{C}^2_{\Delta_x,\Delta_y}(k,l)\right) \tag{11}$$

This similarity measure is then used in the classification procedure.

5 Application to Texture Image Classification

In this section, we present an application to texture image classification. The aim of this part is to evaluate the potential of the four coding models presented in Sect. 4: BoRW, R-VLAD, RFV and CoC.

For this experiment, two databases are considered: the VisTex [21] database and the Outex_TC000_13 [22] database. The VisTex database is composed by 40 texture classes. Each class is represented by a set of 64 images of size 64×64 pixels. The Outex_TC000_13 database contains 68 texture classes, where each class is represented by a set of 20 images of size 128×128 pixels. For both databases, the feature extraction and classification steps shown in Fig. 1 are similar. We consider the same protocol as the one presented in [14]. First, covariance matrices are computed on sliding patches of size 15×15 pixels. These covariance matrices describe the interaction between the image intensities $I(x, y)$ and the norms of the first and second order derivatives of $I(x, y)$ in both directions x and y [10]. Then, once the images are encoded with one of the four presented model (BoRW, R-VLAD, RFV or CoC), an SVM classifier with a Gaussian kernel is used for the final classification step. In practice, the dispersion parameter of this kernel is optimized by using a cross validation procedure on the training set.

Table 1 presents the classification results in term of overall accuracy obtained on the VisTex and Outex TC000_13 databases for the four coding models. For the RFV model, the contribution of each parameter (centroid, dispersion, weight) is analyzed. For example, the row "RFV: ϖ" shows the classification accuracy when only the derivatives with respect to the weights are considered to calculate the RFV (see (7)), ... For the CoC model, an 8-neighborhood with a displacement of two pixels between the patches is considered. As observed in Table 1, the best classification results are observed for the proposed CoC model which exploits the spatial distribution of the patches. A significant gain of about 1% is observed on both VisTex and Outex TC000_13 databases compared to other state-of-the-art coding models (BoRW, R-VLAD and RFV).

Table 1. Classification results on the VisTex and Outex databases in terms of overall accuracy.

Method	VisTex	Outex TC000_13
BoRW [12]	86.87 ± 1.56	83.86 ± 1.41
R-VLAD [13]	87.91 ± 0.74	83.13 ± 1.50
RFV: ϖ [14]	89.42 ± 0.63	84.97 ± 0.87
RFV: σ [14]	79.32 ± 1.38	76.75 ± 1.48
RFV: $\bar{\mathbf{M}}$ [14]	87.77 ± 0.84	84.20 ± 0.65
RFV: σ, ϖ [14]	82.13 ± 1.19	79.35 ± 1.39
RFV: $\bar{\mathbf{M}}, \varpi$ [14]	88.73 ± 0.89	84.57 ± 0.54
RFV: $\bar{\mathbf{M}}, \sigma$ [14]	89.43 ± 0.79	84.01 ± 0.65
RFV: $\bar{\mathbf{M}}, \sigma, \varpi$ [14]	89.80 ± 0.57	84.22 ± 0.62
CoC	$\mathbf{91.08 \pm 0.61}$	$\mathbf{85.19 \pm 0.97}$

6 Conclusion

This paper has introduced a novel local model for image classification on the manifold of covariance matrices. Based on the concept of co-occurrence matrices, this local model exploits the spatial distribution of the patches, allowing to improve the classification performances compared to standard coding models (BoRW, R-VLAD and RFV).

Further works will concern the extension of such coding model to fuzzy co-occurrence matrices [23].

References

1. Liu, C., Sharan, L., Adelson, E.H., Rosenholtz, R.: Exploring features in a Bayesian framework for material recognition. In: CVPR, pp. 239–246. IEEE Computer Society (2010)
2. Hiremath, P., Pujari, J.: Content based image retrieval using color, texture and shape features. In: 2012 18th International Conference on Advanced Computing and Communications (ADCOM), pp. 780–784 (2007)
3. de Luis-García, R., Westin, C.F., Alberola-López, C.: Gaussian mixtures on tensor fields for segmentation: applications to medical imaging. Comput. Med. Imaging Graph. **35**(1), 16–30 (2011)
4. Cirujeda, P., Cid, Y.D., Müller, H., Rubin, D.L., Aguilera, T.A., Loo, B.W., Diehn, M., Binefa, X., Depeursinge, A.: A 3-D Riesz-covariance texture model for prediction of nodule recurrence in Lung CT. IEEE Trans. Med. Imaging **35**(12), 2620–2630 (2016)
5. Zhu, C., Yang, X.: Study of remote sensing image texture analysis and classification using wavelet. Int. J. Remote Sens. **19**(16), 3197–3203 (1998)
6. Regniers, O., Bombrun, L., Lafon, V., Germain, C.: Supervised classification of very high resolution optical images using wavelet-based textural features. IEEE Trans. Geosci. Remote Sens. **54**(6), 3722–3735 (2016)
7. Tan, X., Triggs, B.: Enhanced local texture feature sets for face recognition under difficult lighting conditions. In: Zhou, S.K., Zhao, W., Tang, X., Gong, S. (eds.) AMFG 2007. LNCS, vol. 4778, pp. 168–182. Springer, Heidelberg (2007). doi:10.1007/978-3-540-75690-3_13
8. Vu, N.S., Dee, H.M., Caplier, A.: Face recognition using the POEM descriptor. Pattern Recogn. **45**(7), 2478–2488 (2012)
9. Nguyen, T.P., Vu, N., Manzanera, A.: Statistical binary patterns for rotational invariant texture classification. Neurocomputing **173**, 1565–1577 (2016)
10. Tuzel, O., Porikli, F., Meer, P.: Region covariance: a fast descriptor for detection and classification. In: Leonardis, A., Bischof, H., Pinz, A. (eds.) ECCV 2006. LNCS, vol. 3952, pp. 589–600. Springer, Heidelberg (2006). doi:10.1007/11744047_45
11. Jayasumana, S., Hartley, R.I., Salzmann, M., Li, H., Harandi, M.T.: Kernel methods on the Riemannian manifold of symmetric positive definite matrices. In: IEEE CVPR, pp. 73–80 (2013)
12. Faraki, M., Harandi, M.T., Wiliem, A., Lovell, B.C.: Fisher tensors for classifying human epithelial cells. Pattern Recogn. **47**(7), 2348–2359 (2014)
13. Faraki, M., Harandi, M.T., Porikli, F.: More about VLAD: a leap from Euclidean to Riemannian manifolds. In: IEEE CVPR, pp. 4951–4960, June 2015
14. Ilea, I., Bombrun, L., Germain, C., Terebes, R., Borda, M., Berthoumieu, Y.: Texture image classification with Riemannian Fisher vectors. In: IEEE ICIP, pp. 3543–3547 (2016)
15. Perronnin, F., Sánchez, J., Mensink, T.: Improving the Fisher kernel for large-scale image classification. In: Daniilidis, K., Maragos, P., Paragios, N. (eds.) ECCV 2010. LNCS, vol. 6314, pp. 143–156. Springer, Heidelberg (2010). doi:10.1007/978-3-642-15561-1_11
16. Perronnin, F., Liu, Y., Sánchez, J., Poirier, H.: Large-scale image retrieval with compressed Fisher vectors. In: The Twenty-Third IEEE Conference on Computer Vision and Pattern Recognition, San Francisco, CA, USA, pp. 3384–3391 (2010)
17. Said, S., Bombrun, L., Berthoumieu, Y., Manton, J.H.: Riemannian Gaussian distributions on the space of symmetric positive definite matrices. IEEE Trans. Inf. Theory **63**(4), 2153–2170 (2017)

18. Said, S., Bombrun, L., Berthoumieu, Y.: Texture classification using Rao's distance on the space of covariance matrices. In: Geometric Science of Information (2015)
19. Faraki, M., Palhang, M., Sanderson, C.: Log-Euclidean bag of words for human action recognition. IET Comput. Vision **9**(3), 331–339 (2015)
20. Higham, N.J.: Functions of Matrices: Theory and Computation. Society for Industrial and Applied Mathematics, Philadelphia (2008)
21. Vision Texture Database. MIT Vision and Modeling Group. http://vismod.media.mit.edu/pub/VisTex
22. Outex Texture Database. Center for Machine Vision Research of the University of Oulu. http://www.outex.oulu.fi/index.php?page=classification
23. Ledoux, A., Losson, O., Macaire, L.: Texture classification with fuzzy color co-occurrence matrices. In: IEEE ICIP, pp. 1429–1433, September 2015

Positive Signal Spaces and the Mehler-Fock Transform

Reiner Lenz(✉)

Linköping University, 58183 Linköping, Sweden
reiner.lenz@liu.se

Abstract. Eigenvector expansions and perspective projections are used to decompose a space of positive functions into a product of a half-axis and a solid unit ball. This is then used to construct a conical coordinate system where one component measures the distance to the origin, a radial measure of the distance to the axis and a unit vector describing the position on the surface of the ball. A Lorentz group is selected as symmetry group of the unit ball which leads to the Mehler-Fock transform as the Fourier transform of functions depending an the radial coordinate only. The theoretical results are used to study statistical properties of edge magnitudes computed from databases of image patches. The constructed radial values are independent of the orientation of the incoming light distribution (since edge-magnitudes are used), they are independent of global intensity changes (because of the perspective projection) and they characterize the second order statistical moment properties of the image patches. Using a large database of images of natural scenes it is shown that the generalized extreme value distribution provides a good statistical model of the radial components. Finally, the visual properties of textures are characterized using the Mehler-Fock transform of the probability density function of the generalized extreme value distribution.

1 Introduction and Overview

Many applications are based on measurements which have only positive values. Length, weight and age/duration are some common examples. In the following we will consider a theoretical framework in which a signal s is an element in a Hilbert space H. We also have a number of subspaces $\mathrm{H}_k, (k = 1, \dots K)$ and c_k is the length of the projection of s in H_k. Being a length measurement (and assuming that the signal is never perfect) we assume that for all k the c_k are positive. In the following the space H is the finite-dimensional vector space of gray value distributions on a collections of pixels. The subspaces H_k are given by the irreducible representations of the underlying symmetry group of the pixel grid. Descriptions of group representations can be found in [2–4,8,9,13]. Applying principal component analysis (PCA) and using the Perron-Frobenius theorem one can introduce a conical coordinate system in which the first coordinate corresponds to the first eigenvector coefficient. Using perspective projection along this

F. Nielsen and F. Barbaresco (Eds.): GSI 2017, LNCS 10589, pp. 745–753, 2017.
https://doi.org/10.1007/978-3-319-68445-1_86

axis shows that the remaining coordinates describe a solid unit ball. The angular part of the new system can be analyzed with the help of spherical techniques like expansion in surface harmonics whereas the radial part can be studied with the help of the Mehler-Fock-Transform (MFT) which corresponds to the Fourier transform in the case where the underlying transformation group is the group of hyperbolic rotations. Three-dimensional color spaces (such as the common RGB-system) have the same structure. Here the main axis of the cone describes the gray colors and the projection the intensity. The solid ball is the unit disk where the radius measures saturation and the angle corresponds to hue.

In the second contribution of this paper the conical framework is used to investigate statistical properties of image patches from calibrated images taken in a habitat similar to the one in which the human eye developed. The construction leads to texture descriptors which are orientation invariant (due to the usage of the edge-magnitudes) and independent of overall intensity changes (because the perspective projection). They are also independent of the distributions between the higher order eigenvectors since only the radial variation in the unit ball is analyzed. The results show that the family of generalized-extreme-value distributions (GEV's) provide a class of probability distributions that give good fitting results for these measurements.

In the third contribution the group of hyperbolic rotations is used as natural class of transformations acting on the radial variable. This is a one-parameter subgroup of the Lorentz group and the corresponding Fourier transform in this variable is the Mehler-Fock transform. The study of the relation between the visual properties of a patch and the properties of the MFT are the topic of the final experiments.

2 Conical Structure of Positive Signals

Fourier related techniques (like the continuous Fourier transform (CFT), the discrete (DFT) and the fast Fourier transform (FFT)) are some of the most powerful signal processing tools. For images whose pixels are located on a square grid the corresponding symmetry group is the dihedral group D(4) and the Fourier transform are the dihedral filters. A special class of dihedral filters corresponds to the traditional edge detectors in low-level image processing. These edge filters come in pairs and correspond to the two-dimensional irreducible representations of the dihedral group. The magnitude of a pair of such filter results corresponds to the projection length mentioned above. In the following the space of all gray value distributions on a 5×5 window is used. Its dimension is 25, the six edge detector pairs span a 12-dimensional subspace and define six edge magnitudes. Application of the dihedral filtering results thus for every pixel $\mathbf{x}$ in a six-dimensional vector $\mathbf{f}(\mathbf{x})$ with non-negative elements (the value zero requires all underlying pixel values to be zero).

Next the 6-D (local) mean vector $\mathbf{m}(\mathbf{x})$ and the 6×6 matrix $\mathbf{C}(\mathbf{x})$ of (local) second order moments of $\mathbf{f}(\mathbf{x})$ are computed. The matrix $\mathbf{C}(\mathbf{x})$ is symmetric, positive-definite and positive and the Perron-Frobenius theorem shows that the eigenvector $\mathbf{b}_0(\mathbf{x})$ belonging to the highest eigenvalue has only positive entries. At pixel $\mathbf{x}$ a new coordinate system $\mathbf{b}_0(\mathbf{x}), \mathbf{b}_1(\mathbf{x}), \ldots, \mathbf{b}_5(\mathbf{x})$ is defined spanned by the eigenvectors of $\mathbf{C}(\mathbf{x})$ (ordered by the value of the eigenvalues). This gives the expansion of the vector $\mathbf{f}(\mathbf{x}) = \sum_{k=0}^{5} \langle \mathbf{f}(\mathbf{x}), \mathbf{b}_k(\mathbf{x}) \rangle \mathbf{b}_k(\mathbf{x}) = \sum_{k=0}^{5} c_k(\mathbf{x}) \, \mathbf{b}_k(\mathbf{x})$, where $\langle .,. \rangle$ denotes the scalar product.

The vectors $\mathbf{f}(\mathbf{x})$ and $\mathbf{b}_0(\mathbf{x})$ have both positive entries and their scalar product is therefore positive and the ratios $q_k(\mathbf{x}) = \frac{c_k(\mathbf{x})}{c_0(\mathbf{x})}, k = 1, 2$ are well-defined. Eigenvectors are orthogonal and the eigenvectors $\mathbf{b}_k(\mathbf{x}), k > 0$ must have negative elements. In general one can expect that $c_0(\mathbf{x}) > 0$ will be *"big"* and $|c_k(\mathbf{x})|, k > 0$ will be *"small"*. Furthermore, the values $q_k(\mathbf{x})$ are invariant under scaling of the original pixel values. If $q(\mathbf{x})$ denotes the vector containing these ratios then it is assumed that its norm is bounded by some value R independent of $\mathbf{x}$. This is the case for all examples used later. Therefore polar coordinates $(\rho(\mathbf{x}), \theta(\mathbf{x}))$ with $(q_1(\mathbf{x}), q_2(\mathbf{x})) = (R\rho(\mathbf{x}) \cos\theta(\mathbf{x}), R\rho(\mathbf{x}) \sin\theta(\mathbf{x}))$ can be used.

The result describes $\mathbf{f}(\mathbf{x})$ in the coordinate system $(c_0(\mathbf{x}), \rho(\mathbf{x}), \theta(\mathbf{x}))$ where $c_0(\mathbf{x})$ measures the projection along the mean-direction, $\rho(\mathbf{x})$ is the distance from the mean-vector in the space spanned by the second and third eigenvector and $\theta(\mathbf{x})$ depends on the relation between the second and third eigenvector. In human color perception this corresponds to a characterization in terms of intensity $c_0(\mathbf{x})$, saturation $\rho(\mathbf{x})$ and hue $\theta(\mathbf{x})$ (more information can be found in [10,11]). This construction is not limited to the analysis of the first three eigenvector coefficients. Using more then three components only leads to the replacement of the angular variable by spherical coordinates.

3 The Action of the Group SU(1, 1)

In a group theoretical context the groups used so far are: the group D(4) related to the sensor array. The scaling group $\mathbb{R}^+$ acting by multiplication on $c_0(\mathbf{x})$ and the group SU(1, 1) acting on the points on the unit disk. The details of this constructions are as follows: points on the unit disk are complex variables z and the group SU(1, 1) is defined as the 2×2 matrices with complex elements:

$$\mathrm{SU}(1,1) = \left\{ \mathbf{M} = \begin{pmatrix} a & b \\ \overline{b} & \overline{a} \end{pmatrix}, a, b \in \mathbb{C}, |a|^2 - |b|^2 = 1 \right\} \tag{1}$$

The group operation is the usual matrix multiplication and the group acts as a transformation group on the open unit disk $\mathcal{D}$ (consisting of all points $z \in \mathbb{C}$ with $|z| < 1$) as the Möbius transforms:

$$(\mathbf{M}, z) = \left(\begin{pmatrix} a & b \\ \overline{b} & \overline{a} \end{pmatrix}, z \right) \mapsto \mathbf{M}\langle z \rangle = \frac{az + b}{\overline{b}z + \overline{a}}, z \in \mathcal{D} \tag{2}$$

with $(\mathbf{M}_1\mathbf{M}_2)\langle z \rangle = \mathbf{M}_1 \langle \mathbf{M}_2 \langle z \rangle \rangle$ for all matrices and all points. The notation $\mathbf{M}$ will be used when the group elements are represented as matrices. Otherwise the symbol g is used. An ordinary three-dimensional rotation can be written as a product of three rotations around the coordinate axes. A similar decomposition holds also for the group SU(1, 1) . Denoting the three parameters by ψ, τ, φ the decomposition is given as:

$$g(\varphi,\tau,\psi) = g(\varphi,0,0)g(0,\tau,0)g(0,0,\psi) = \begin{pmatrix} \cosh\frac{\tau}{2}e^{i(\varphi+\psi)/2} & \sinh\frac{\tau}{2}e^{i(\varphi-\psi)/2} \\ \sinh\frac{\tau}{2}e^{-i(\varphi-\psi)/2} & \cosh\frac{\tau}{2}e^{-i(\varphi+\psi)/2} \end{pmatrix} \tag{3}$$

Introducing the two one-parameter subgroups

$$K = \{g(\varphi,0,0) = \begin{pmatrix} e^{i\varphi/2} & 0 \\ 0 & e^{-i\varphi/2} \end{pmatrix} : -2\pi \le \varphi < 2\pi\} \tag{4}$$

$$A = \{g(0,\tau,0) = \begin{pmatrix} \cosh\frac{\tau}{2} & \sinh\frac{\tau}{2} \\ \sinh\frac{\tau}{2} & \cosh\frac{\tau}{2} \end{pmatrix} : \tau \in \mathbb{R}\}. \tag{5}$$

shows that $g(\varphi,0,0), g(0,0,\psi) \in K$ and $g(0,\tau,0) \in A$. This is known as the Cartan or the polar decomposition of the group and the ψ, τ, φ are the Cartan coordinates.

The elements in K are rotations and leave the origin fixed. For a general element in SU(1, 1) the Cartan decomposition gives:

$$g(\varphi,\tau,\psi)\langle 0\rangle = g(\varphi,0,0)g(0,\tau,0)g(0,0,\psi)\langle 0\rangle = \tanh\frac{\tau}{2}e^{i\varphi}.$$

This shows that $\mathcal{D} = \mathrm{SU}(1,1)/K$ and functions on the unit disk are functions on the group that are independent of the last argument of the Cartan decomposition.

Next consider two points on the unit disk, corresponding to group elements $g = g(\varphi_0,\tau_0,0)$ and $h = g(\varphi_1,\tau_1,0)$. The difference between these two elements is given by $h^{-1}g$ with its own decomposition $h^{-1}g = g(\varphi,\tau,0)$ with $g(\varphi,\tau,0) = g(\varphi_1,\tau_1,0)^{-1}g(\varphi_0,\tau_0,0) = g(0,-\tau_1,0)g(\varphi_1-\varphi_0,\tau_0,0)$. In [17], (vol. 1, p. 271) the following relation between the parameters of the three group elements is derived:

$$\cosh\tau = \cosh\tau_1\cosh\tau_0 + \sinh\tau_1\sinh\tau_0\cos(\varphi_1-\varphi_0). \tag{6}$$

For SU(1, 1) the role of the exponential function is played by the associated Legendre functions (zonal or Mehler functions, [17], p. 324) of order m and degree $\alpha = -1/2 + i\kappa$. They are defined as (see Eq.(7)):

$$\mathfrak{P}_\alpha^m(\cosh\tau) = \frac{1}{2\pi}\frac{\Gamma(\alpha+m+1)}{\Gamma(\alpha+1)}\int_0^{2\pi}(\sinh\tau\cos\theta+\cosh\tau)^\alpha\, e^{im\theta}\,d\theta \tag{7}$$

and satisfy the addition formula ([17], p. 327)

$$\mathfrak{P}_\alpha(\cosh\tau_l\cosh\tau_0 + \sinh\tau_l\sinh\tau_0\cos\theta) = \sum_{m\in\mathbb{Z}}\mathfrak{P}_\alpha^{-m}(\cosh\tau_l)\,\mathfrak{P}_\alpha^m(\cosh\tau_0)\,e^{-im\theta}. \tag{8}$$

The transform for SU(1, 1) corresponding to the Fourier transform is the Mehler-Fock transform. The following theorem shows that a large class of functions are combinations of the associated Legendre functions:

Theorem 1 (Mehler-Fock Transform; MFT). *For a function* k *defined on the interval* $[1,\infty)$ *define its transform* c *as:*

$$c(\kappa) = \int_0^\infty \mathrm{k}(\cosh\tau)\mathfrak{P}_{-1/2+i\kappa}(\cosh\tau)\sinh\tau\, d\tau \tag{9}$$

Then k *can be recovered by the inverse transform:*

$$\mathrm{k}(\cosh\tau) = \int_0^\infty \kappa\tanh(\pi\kappa)\mathfrak{P}_{-1/2+i\kappa}(\cosh\tau)c(\kappa)\, d\kappa \tag{10}$$

Details about the transform, special cases and its applications can be found in [14], Sect. 7.6, [1,7,15,17]. The MFT also preserves the scalar product (Parseval relation): using the parametrization $x = \cosh\tau$ and defining $c_n(\kappa) = \int_1^\infty f_n(x)\mathfrak{P}_{1/2+i\kappa}(x)\, dx$ gives (see [14] (7.6.16) and (7.7.1)):

$$\int_0^\infty c_1(\kappa)c_2(\kappa)\kappa\tanh(\pi\kappa)\, d\kappa = \int_1^\infty f_1(x)f_2(x)\, dx \tag{11}$$

One possible application of the MFT in statistics is the application as a tool to study parametric probability distributions. Using its connection to group convolutions it can also be used to simplify kernel density estimators of densities defined on the positive half-axis. In the following it will be used to investigate generalized-extreme-value distribution based models of natural image statistics. Kernel-density estimators will be discussed elsewhere.

4 Implementation

An implementation of the transform has to take into account at least three problems: (1) the computation of the associated Legendre functions, (2) numerical evaluation of infinite integrals and (3) sampling schemes in the signal and transform domain. The definition of the associated Legendre functions using Eq. (7) is not useful, instead the relations to other special functions can be used (see [6,12]). In the following the LegendreP function in Mathematica is used. Computation of the infinite integrals involved is difficult in general. In the experiments the density functions of the generalized-extreme-value distribution are used. For these functions the quantiles are known and the integration domain selected extends to the maximum value of the 0.99 quantiles for all distributions in the database. Sampling is another factor to be taken into account. Since the main underlying structure is the one-parameter group A a linear sampling scheme in the signal and the transform domain is chosen. Integrals are computed as scalar products of vectors. The results were compared to the application of numerical integration methods (NIntegrate in Mathematica) which confirmed the reliability of the results for the functions involved in this study. In all the numerical results reported in the following a matrix of size 101×301 was used where 301 sampling points were used to sample the pdfs and 101 points sampling the transform domain. The range of the transform domain was determined manually.

5 The Botswana Dataset

Statistical properties of images are of interest in the study of biological vision systems and technical applications such as image restoration and image classification and compression (see [5] for a review of natural image statistics research).

The following experiments make use of a large database of natural images[1] and described in [16]). Some of the main characteristics of the database are the following: it contains 5677 calibrated natural images, collected in a single environment: a savanna habitat in Botswana which is thought to be similar to the environment where the human eye evolved. The images in the database are organized in 103 folders (albums) containing images of a common theme. There are 61 albums characterized as scenes and 42 albums showing objects. The original images were taken with a Nikon D70 camera which was calibrated to take into account the optical and signal processing properties of the camera. The images in the LMS format were downloaded. They represent the input to the long/medium/short wavelength sensitive cones in the human retina and the M-channel which is most important for lightness perception was used.

The raw images are filtered with the six edge-detection filter pairs and the results combined in the six dimensional vector with the edge-magnitude values. Then the outer product is computed at every pixel and the (matrix valued) image is filtered with a Gaussian filter kernel of size 15×15. The result is a matrix valued image with the empirical second-order moments. Next the (normalized) projection variables $\rho(\mathbf{x})$ are computed and converted to hyperbolic form. For every image the 128×128 patch at the center of the image is selected. Pixels with zero-valued first eigenvectors (produced by constant 5×5 patches) are ignored and only patches with more than 500 valid measurements were evaluated. This resulted in 5560 patches used in the following. The quality of the fitting was measured with the adjusted R-Squared value and the lowest values found was 0.9623. The mean value over all 5560 patches was 0.998.

As illustration an example with an R-Squared value nearest to the mean value of 0.998 is selected. In Fig. 1 the original scene (cd20A, DSC-0055), the selected patch, the image of the group parameters in the projection, the histogram and the GEV-fitting and finally the MFT of the distribution are shown.

The next three figures illustrate the effect of an operation similar to low-pass filtering in traditional signal processing. Here the first 10 MFT coefficients were extracted. Then the scalar product (see Eq. 11) in the transform domain was used to compute the length of the projection of a patch in this MFT-region. The patches were then ordered and in Fig. 2 the patches, the values of the projections and three distributions are shown. The left and the middle column show 3×5 patches. The patches on top are the five patches with lowest projection value in the MFT-bands, the patches in the middle correspond to positions 251 to 255 and the lowest five are number 4001 to 4005. The pixel values in these images are all scaled such that the 0.9 quantile of the values in all patches has maximum gray value. The left column (Fig. 2(a)) with the original patches

[1] Available at http://tofu.psych.upenn.edu/~upennidb.

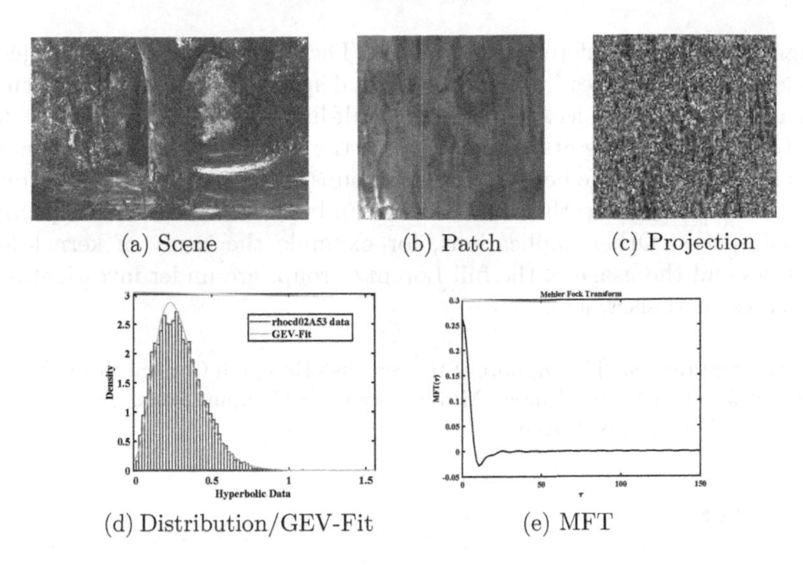

(a) Scene (b) Patch (c) Projection

(d) Distribution/GEV-Fit (e) MFT

Fig. 1. Scene (cd02A; DSC-0055-LMS)

show that the projection eliminates global intensity changes: Some parts of the patches are almost black and others have maximum gray value. The patches in the center of the figure (Fig. 2(b)) show that low contributions to the lower MFT channels correspond to low filter values (high probability of low edge magnitude responses) and the patches at the top are therefore more homogeneous than the others. The plot in the right column (Fig. 2(c)) shows the histograms and the fitted GEV-distributions for the three left-most patches. This illustrates the strength of the GEV-approach since the shapes of the three distributions vary widely and the locations of the distributions indicate the increasing value of the projection values of the three images.

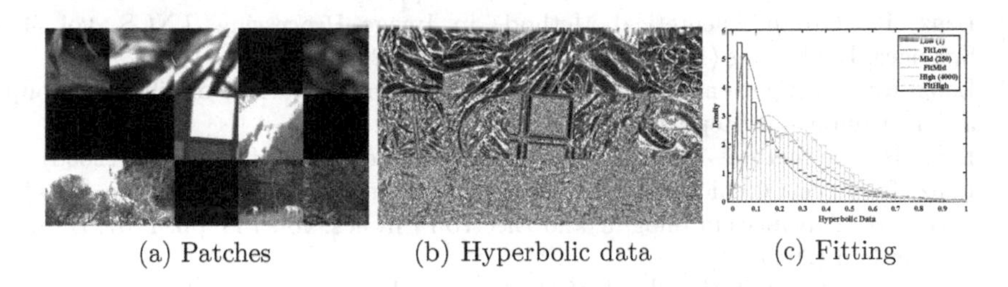

(a) Patches (b) Hyperbolic data (c) Fitting

Fig. 2. Bandpass filtering

6 Summary and Conclusions

Symmetry-based methods entered the study of positive signals at several levels: first the representation theory of the symmetry group of the sensor is used to split the signal space into different components. Then the lengths of the projections in the different subspaces form another signal space of positive signals which in turn can be represented as a product of a half-line and a unit ball. The half-axis is related to scaling operations modeled by the scaling group and the unit-ball can be analyzed with the help of Lorentz groups. The description here is only an illustration of how group theoretical tools can be used to analyze the properties of signal spaces. Other applications, for example the study of kernel-density estimators and the usage of the full Lorentz group, are under investigation and will be presented elsewhere.

Acknowledgements. The support of the Swedish Research Council through a framework grant for the project “Energy Minimization for Computational Cameras” (2014-6227) is gratefully acknowledged.

References

1. Bateman, H., Erdélyi, A., States, U. (eds.): Tables of Integral Transforms, vol. 2. McGraw-Hill, New York (1954)
2. Chirikjian, G., Kyatkin, A.: Harmonic Analysis for Engineers and Applied Scientists. Dover Publications, Mineola (2016)
3. Diaconis, P.: Group representations in probability and statistics. In: Gupta, S.S. (ed.) Institute of Mathematical Statistics Lecture Notes - Monograph Series, vol. 11, Institute of Mathematical Statistics, Hayward, California (1988)
4. Fässler, A., Stiefel, E.: Group Theoretical Methods and Their Applications. Birkhäuser, New York (1992)
5. Gerhard, H.E., Theis, L., Bethge, M.: Modeling natural image statistics. In: Biologically-Inspired Computer Vision, Fundamentals and Applications. Wiley (2015)
6. Gil, A., Segura, J., Temme, N.M.: Computing the conical function $p^{\mu}_{-1/2+i\tau}(x)$. SIAM J. Sci. Comput. **31**(3), 1716–1741 (2009)
7. Lebedev, N.N.: Special Functions and Their Applications. Dover, New York (1972)
8. Lenz, R.: Group Theoretical Methods in Image Processing. LNCS, vol. 413. Springer, Heidelberg (1990). doi:10.1007/3-540-52290-5
9. Lenz, R.: Investigation of receptive fields using representations of dihedral groups. J. Vis. Comm. Im. Repr. **6**(3), 209–227 (1995)
10. Lenz, R.: Lie methods for color robot vision. Robotica **26**(4), 453–464 (2008)
11. Lenz, R.: Spectral color spaces: their structure and transformations. In: Hawkes, P.W. (ed.) Advances in Imaging and Electron Physics, vol. 138, pp. 1–67. Elsevier, Amsterdam (2005)
12. Olver, F.W.J.: NIST Handbook of Mathematical Functions. Cambridge University Press, New York (2010)
13. Serre, J.: Linear Representations of Finite Groups. Springer, New York (2012). Softcover. ISBN 10: 1468494600 ISBN 13: 9781468494600
14. Sneddon, I.N.: The Use of Integral Transforms. McGraw-Hill, New York (1972)
15. Terras, A.: Harmonic Analysis on Symmetric Spaces and Applications I. Springer, New York (1985). doi:10.1007/978-1-4612-5128-6

16. Tkacik, G., Garrigan, P., Ratliff, C., Milcinski, G., Klein, J., Seyfarth, L., Sterling, P., Brainard, D., Balasubramanian, V.: Natural images from the birthplace of the human eye. PLoS ONE **6**(6), e20409 (2011)
17. Vilenkin, N.J., Klimyk, A.U.: Representation of Lie Groups and Special Functions: Volume 1: Simplest Lie Groups, Special Functions and Integral Transforms. In: Springer Science & Business Media, Mathematics, 6 Dec 2012, 612 pages. Springer, Dordrecht (2012)

Bounding the Convergence Time of Local Probabilistic Evolution

Simon Apers[1(✉)], Alain Sarlette[1,2], and Francesco Ticozzi[3,4]

[1] Department of Electronics and Information Systems, Ghent University, Ghent, Belgium
simon.apers@ugent.be
[2] QUANTIC Lab, INRIA Paris, Paris, France
alain.sarlette@inria.fr
[3] Dipartimento di Ingegneria dell'Informazione, Università di Padova, Padua, Italy
ticozzi@dei.unipd.it
[4] Department of Physics and Astronomy, Dartmouth College, Hanover, NH 03755, USA

Abstract. Isoperimetric inequalities form a very intuitive yet powerful characterization of the connectedness of a state space, that has proven successful in obtaining convergence bounds. Since the seventies they form an essential tool in differential geometry [1,2], graph theory [3,4] and Markov chain analysis [5,6,8]. In this paper we use isoperimetric inequalities to construct a bound on the convergence time of any local probabilistic evolution that leaves its limit distribution invariant. We illustrate how this general result leads to new bounds on convergence times beyond the explicit Markovian setting, among others on quantum dynamics.

This paper is concerned with the discrete-time spreading of a distribution along the edges of a graph. In essence we establish that even by exploiting global information about the graph and allowing a very general use of this information, this spreading can still not be accelerated beyond the conductance bound. Before providing more ample context, we start with a motivating example ascribed to Eugenio Calabi, but which came to our attention through the seminal 1969 paper by Jeff Cheeger [1]. Whereas the original example concerns differential geometry, we will apply it to a graph setting.

Consider a locality structure (discrete geometry) prescribed by the "dumbbell" graph family $K_n - K_n$ shown in Fig. 1, consisting of two complete graphs over n nodes, connected by a single edge. The diameter of this graph, being the "longest shortest path" between any two nodes, is three. However, a random walk over this graph converging to the uniform distribution has an expected convergence time in $O(n^2)$. This convergence time can be improved with a "global design" but without violating locality of the evolution, by adding some memory to the walker. In Fig. 1, the system designer has superimposed a cycle (dashed line) over the dumbbell graph. By adding subnodes that allow to conditionally select different subflows through the graph (formally we "lift" the walk [9]), the walker can be restricted to walk along this cycle. Using this cycle, we can

F. Nielsen and F. Barbaresco (Eds.): GSI 2017, LNCS 10589, pp. 754–762, 2017.
https://doi.org/10.1007/978-3-319-68445-1_87

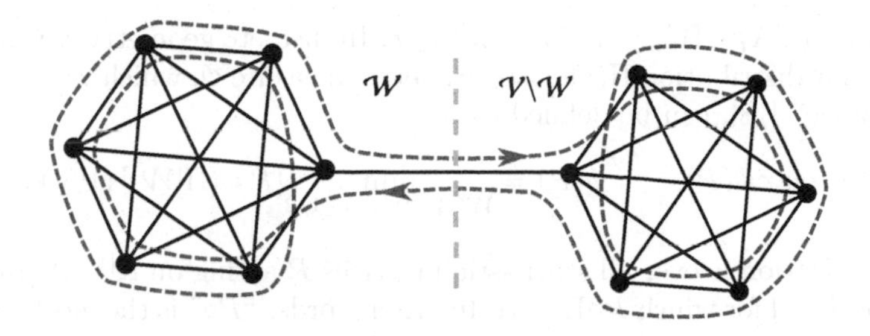

Fig. 1. (solid line) Dumbbell graph $K_n - K_n$ for $n = 6$. (dashed line) superimposed cycle of length $4n$ in a construction towards faster mixing.

impose a strategy by Diaconis, Holmes and Neal [9,16] to efficiently speed-up mixing over this cycle: *let the walker cycle in the same direction with a probability* $1-1/n$, *and switch direction with probability* $1/n$. This way the walk will mix over the graph in $O(n)$, i.e. quadratically faster than the original random walk. But this is still order n times slower than the diameter. Nevertheless, we show in our paper that this improvement is the best possible for any local probabilistic process that leaves the target distribution invariant. So mixing in diameter time may be possible, but not without loosening any of these constraints.

1 Problem Description and Main Result

Consider a graph $\mathcal{G}$ with nodes $\mathcal{V}$ and edges $\mathcal{E} \subseteq \mathcal{V} \times \mathcal{V}$. We use the convention that $(i,i) \in \mathcal{E}\ \forall i \in \mathcal{V}$. We define "states" X as probability distributions over $\mathcal{V}$. Given an initial state X_0, some system "$\rightarrow$" propagates it over t time steps as $X_0 \xrightarrow{t} X_t$. For a subset $\mathcal{W} \subseteq \mathcal{V}$ and a state X, we define $X(\mathcal{W})$ the probability of $\mathcal{W}$ according to X, and $X|\mathcal{W}$ as the state X conditioned on being in $\mathcal{W}$. We call $\mathcal{N}(\mathcal{W})$ the neighborhood of $\mathcal{W} \subseteq \mathcal{V}$, i.e., the nodes outside $\mathcal{W}$ that have an edge going to $\mathcal{W}$. We impose the following fundamental properties.

- **linear initialization:** $X_0 \xrightarrow{t} X_t,\ \tilde{X}_0 \xrightarrow{t} \tilde{X}_t \Rightarrow pX_0+(1\text{-}p)\tilde{X}_0 \xrightarrow{t} pX_t+(1\text{-}p)\tilde{X}_t$
- **locality:** $\forall X_0, t \geq 0, \mathcal{W} \subseteq \mathcal{V}:\ \ X_{t+1}(\mathcal{W}) \leq X_t(\mathcal{W}) + X_t(\mathcal{N}(\mathcal{W}))$
- **invariance:** $X_0 \xrightarrow{t\to\infty} \Pi\ \forall X_0\ \ \Rightarrow\ \ \Pi \xrightarrow{t} \Pi$

The last property states that the unique steady state distribution of the system must be invariant as an initial condition. The second property expresses that probability weight can only flow along an edge at each time, without referring to details of the system mapping "$\rightarrow$". The first property is natural as the input is a probability distribution. The point however is that the general process "$\rightarrow$" may e.g. contain hidden states, and we here impose a linear initialization with the hidden states as well (see example below).

Our theorem presents a bound on the convergence of a system "$\rightarrow$" that obeys these conditions towards its steady state Π. Explicitly, let τ be a time step such that $\|X_t - \Pi\|_1 \leq 1/2$ for all $t \geq \tau$. In discrete geometry, given a graph $\mathcal{G}$ and a limit distribution Π, the isoperimetric measure Φ, which we also call the "conductance" [18], can be defined as:

$$\Phi = \max_P \Phi(P), \qquad \Phi(P) = \min_{\mathcal{W} \subseteq \mathcal{V} : \Pi(\mathcal{W}) \leq 1/2} [P \circ (\Pi|\mathcal{W})] (\mathcal{V} \setminus \mathcal{W}).$$

The maximization is over all stochastic matrices P acting on $\mathbb{R}^{|\mathcal{V}|}$ that obey the locality of $\mathcal{G}$ and for which $P \circ \Pi = \Pi$. In other words, "$P\circ$" is the most basic type of system "$\rightarrow$" satisfying our requirements: it is time-invariant and memoryless ("Markov"). If Π is the uniform distribution, then Φ is upper bounded by the edge expansion of $\mathcal{G}$, which is $1/n$ for the dumbbell graph. We establish the following "conductance bound" for *any* more complicated system.

Theorem 1. *If a system is linear, local and invariant, then* $\tau \geq 1/(8\Phi)$.

So for the dumbbell graph with Π the uniform distribution, we find $\tau \geq n/8$ for any linear, local and invariant system.

Mixing on graph structures has drawn much interest for sampling algorithms, see e.g. perfect matching [15], or the Metropolis-Hastings algorithm used a lot in statistical mechanics. Bounds similar to Thm.1 have originally been proven by Cheeger [1] and Buser [2] in a differential geometry setting, and by Fiedler [4], Dodziuk [8] and Alon [3] in a discrete geometry and graph setting. In Markov chain analysis, early uses trace back to Aldous [5], Lawler and Sokal [6] and Mihail [7]. More recent examples are by Chen, Lovász and Pak [9] who used a similar bound to prove that a restricted class of extended Markov chains called "lifted Markov chains" can at most quadratically accelerate convergence, and by Aharonov et al. [10] to (loosely) bound the convergence speed of certain quantum processes.

Our result allows to improve known mixing bounds, e.g. for quantum processes, and to generalize bounds beyond usual Markov chain settings, e.g. by including nonlinear decision rules. Examples are briefly discussed after the proof.

2 Proof

Our proof essentially comes down to two steps:

- *locality implies particular simulability:* the locality condition implies that the dynamics can always be described using (time- and state-dependent) local stochastic matrices. This is not entirely trivial in such generality.
- *bound for extended Markov chains:* we rather straightforwardly combine these matrices in an extended Markov chain model, for which we can prove the bound along standard lines.

2.1 Locality Implies Simulability

A stochastic matrix P is local if the system $X_0 \rightarrow X_t = P^t \circ X_0$ is local. It is not hard to check that this coincides with the traditional definition, where locality means $P_{i,j} = 0$ whenever $(i, j) \notin \mathcal{E}$. The following Lemma kind of proves the converse. Its proof is inspired by a related result of Scott Aaronson [17], establishing the lemma for quantum systems whose evolution is governed by a local unitary matrix.

Lemma 1. *If "$\rightarrow$" is a local system, then for every pair (X_0, t) with $t > 0$ there exists a local stochastic matrix $P(t, X_0)$ such that $X_0 \xrightarrow{t} X_t = P(t, X_0) \circ X_{t-1}$.*

Proof. Call $Y = X_{t-1}$ and $Z = X_t$. We make a digression to *flows over capacitated networks* [11] and consider the one shown in Fig. 2. The network consists of a source node s, a sink node t, and two copies of the graph nodes $\mathcal{V}$ and $\mathcal{V}'$. Node s is connected to any node $i \in \mathcal{V}$ with capacity $Y(i)$, any node $i \in \mathcal{V}$ is connected with capacity 1 to any node $j \in \mathcal{V}'$ iff $(i, j) \in \mathcal{E}$ (else the nodes are not connected), and any node $j \in \mathcal{V}'$ is connected to node t with capacity $Z(j)$. If this network can route a steady flow of value 1 from node s to node t, then the fraction of $Y(i)$ that is routed from $i \in \mathcal{V}$ towards $j \in \mathcal{V}'$ directly defines $P_{j,i}(t, X_0)$, as $Z(j) = \sum_{i \in \mathcal{V}} P_{j,i}(t, X_0) Y(i)$ and so $P(t, X_0) \circ Y = Z$.

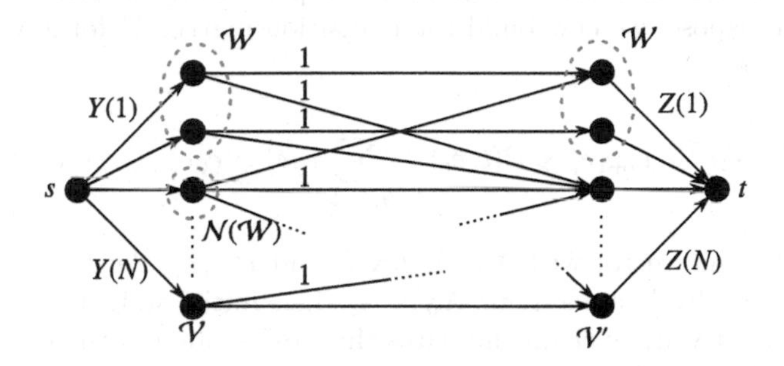

Fig. 2. Capacitated network construction used in Lemma 1.

The max-flow-min-cut theorem [11] states that the maximum steady flow which can be routed from node s to node t is equal to the minimum cut value of the graph, where a cut value is the sum of the capacities of a set of edges that disconnects s from t.

It is clear that cutting all edges arriving at t disconnects the graph, with a cut value of 1, whereas cutting any middle edge between $\mathcal{V}$ and $\mathcal{V}'$ gives a cut value ≥ 1. So the minimum cut need involve no such middle edge. Let us try to not cut the edges from $\mathcal{W} \subseteq \mathcal{V}'$ to t. To block any flow from s to t while keeping all middle edges, we must then cut the edges from s to all the $l \in \mathcal{V}$ which have an edge to $\mathcal{W}$. This corresponds to all $l \in \mathcal{W} \cup \mathcal{N}(\mathcal{W})$. The value of this cut is thus

$$1 - \sum_{j \in \mathcal{W}} Z(j) + \sum_{j \in \mathcal{W}} Y(j) + \sum_{j \in \mathcal{N}(\mathcal{W})} Y(j) .$$

Recalling that $Y = X_{t-1}$ and $Z = X_t$, locality imposes

$$\sum_{j \in \mathcal{W}} Z(j) \leq \sum_{j \in \mathcal{W}} Y(j) + \sum_{j \in \mathcal{N}(\mathcal{W})} Y(j) ,$$

from which follows that the minimum cut value is ≥ 1. According to the previous arguments, this concludes the proof. □

2.2 Bound for "extended" Markov Chains

On the basis of these $P(t, X_0)$, we show how to construct a local Markov chain with at most twice the convergence time τ of our original system "$\rightarrow$". Thereto, we first define a closely related system

$$\overset{t}{\rightsquigarrow} \quad \equiv \quad \text{iterate } \overset{\tau}{\rightarrow} \quad (\text{namely floor}(t/\tau) \text{ times, plus } t \bmod \tau \text{ steps of it}) .$$

By construction, "$\rightsquigarrow$" has the same convergence time τ as "$\rightarrow$", it has the same limit and it obeys the same locality and invariance conditions. We will now build a standard, time-invariant Markov chain that simulates the system "$\rightsquigarrow$".

To this end we first extend our state space: the original node set $\mathcal{V}$ is lifted to $\hat{\mathcal{V}} = (\mathcal{V}; \{0, \ldots, \tau - 1\}; \mathcal{V})$. From the perspective of a random walker, the first item contains its starting position, the second item a clock variable, and the last item its current position. In matrix form, with $\otimes$ representing the Kronecker product and $\cdot^\dagger$ the transpose, we now build the transition matrix M for a Markov chain on $\hat{\mathcal{V}}$ as follows:

$$M = \sum_{i \in \mathcal{V}} \sum_{t=0}^{\tau-2} e_i e_i^\dagger \otimes e_{t+1} e_t^\dagger \otimes P(t, e_i) + \sum_{i,j \in \mathcal{V}} e_j e_i^\dagger \otimes e_0 e_{\tau-1}^\dagger \otimes e_j e_j^\dagger P(\tau - 1, e_i).$$

Here e_i is the unit vector with 1 at index i, and $P(t, e_i)$ denotes the transition matrix obtained by Lemma 1 for $X_0 = e_i$, i.e., initial weight concentrated at node $i \in \mathcal{V}$. This Markov chain simulates the "$\rightsquigarrow$" system in the following sense: when we locally initialize it in the state

$$v[X_0] = \sum_{i \in V} X_0(i) e_i \otimes e_0 \otimes e_i \, ,$$

the distribution over the subsets $(\mathcal{V}; \{0, \ldots, \tau - 1\}; i)$ of the resulting state $M^t v[X_0]$ at time t exactly corresponds to X_t resulting from $X_0 \rightsquigarrow X_t$.

A priori our Markov chain only simulates "$\rightsquigarrow$" for special initial states of the form $v[X_0]$ over $\hat{\mathcal{V}}$. The following lemma shows that in fact when starting from an arbitrary distribution over $\hat{\mathcal{V}}$, it takes at most twice the time to converge to $\|X_t - \Pi\|_1 \leq 1/2$ (over sets as just mentioned).

Lemma 2. *If $\rightarrow$ has a convergence time τ, then the Markov chain M on $\hat{\mathcal{V}}$ has a convergence time at most 2τ over the subsets $\{ (\mathcal{V}; \{0, \ldots, \tau - 1\}; i) : i \in \mathcal{V} \}$.*

Proof. Consider an arbitrary initial state $e_i \otimes e_T \otimes e_k$ for the Markov chain M. After $\tau - T$ steps, this state will necessarily have evolved to one of the special initial states of the form $v[X_0]$, for some X_0. By construction, the distribution of this state over the subsets $\{ (\mathcal{V}; \{0, \ldots, \tau - 1\}; i) : i \in \mathcal{V} \}$ will then simulate the evolution $X_0 \overset{t}{\rightsquigarrow} X_t$, which converges to $\|X_t - \Pi\|_1 \leq 1/2$ for all $t \geq \tau$. Note indeed that $\rightsquigarrow$ and $\rightarrow$ are equivalent over the first τ time steps, and furthermore by invariance, iterating $\rightarrow$ via $\rightsquigarrow$ will never increase $\|X_t - \Pi\|_1$. Hence the Markov chain will have converged after $\tau - T + \tau \leq 2\tau$ steps at most. We have thus proved the convergence time for initial states with all weight concentrated on one element of $\hat{\mathcal{V}}$. By linearity, this also proves the convergence time for arbitrary initial states. □

The last element of the proof is a lower bound on τ for standard Markov chains. It essentially follows from a result by Chen, Lovász and Pak [9], stating that a class of extended Markov chains called "lifted Markov chains" that converge over $\hat{\mathcal{V}}$ can converge at best in order $1/\Phi$, with Φ the conductance of the original graph. Our Markov chain M does not exactly fit into this framework, because it is periodic on $\hat{\mathcal{V}}$ and only its projection onto $\mathcal{V}$ via the subsets of Lemma (2) will converge. The proof can however be adapted to this case. Due to space constraints we must refer the reader to [20] for a detailed proof, and we here only provide the statement:

Lemma 3. *The convergence time of M over the sets $\{(\mathcal{V}; \{0, \ldots, \tau-1\}; i) : i \in \mathcal{V}\}$ is lower bounded by $1/(4\Phi)$.*

By combining Lemmas 2 and 3, we obtain that 2τ must be larger than $1/(4\Phi)$ and hence τ larger than $1/(8\Phi)$, as stated in the main theorem.

3 Examples

We now discuss a few examples to illustrate the generality of our result. Note that the mathematical result is not restricted to cases where X_t represents a probability distribution. It can apply to any situation where X_t remains positive, bounded and preserves the sum of its components. Such dynamics can appear in flow dynamics and e.g. average consensus algorithms for weight distribution [13]. In such settings our result might suggest how e.g. relaxing the linearity constraint is necessary for beating the conductance bound.

3.1 Time-Inhomogeneous Markov Chains and Cesaro Mixing

The bound clearly includes time-varying Markov chains (that satisfy invariance), as appear in the proof. Practical examples of such processes can be found in [21], and in [22] for card shuffling. The difficulty to analyze the convergence time of such processes is explicitly stated. Our paper thus provides a clear bound on the maximal achievable acceleration by exploiting the time-inhomogeneity degree of freedom in mixing algorithms.

Cesaro mixing [19] using a stochastic matrix P is defined by the system $X_0 \xrightarrow{t} X_t = \frac{1}{t+1} \sum_{k=0}^{t} P^k X_0$. There appears to be no obvious way to write this as a Markov chain. However, one can show that Cesaro mixing satisfies our assumptions, so Thm.1 allows to directly bound the mixing time of such processes.

3.2 Processes with State-Dependent or Nonlocal Decision Rules

The locality condition concerns how much probability weight is transferred at each step, but not how the decision about this transfer is taken. The latter is constrained by linearity. In this sense, our result directly bounds any attempts at adapting fast converging nonlinear algorithms e.g. from consensus, towards truly probabilistic Markov chains where linearity is natural. Consider a nonlinear update rule from consensus, like:

$$X_{t+1} = P(Z_t)\, X_t\,.$$

In [14], Z_t is a static function of the weight differences on the respective links, e.g. the weight associated to link (i,j) in $P(Z)$ is a function of $X_t(i) - X_t(j)$. Our framework would even admit Z_t being a dynamic function of X, possibly nonlinear, taking values in any space, and *would not even require that it is based on local values of X only:* e.g. the weight associated to link (i,j) in $P(Z)$ might be a function of some $X_t(k)$ where node k is totally elsewhere in the graph.

Such update rule is in general not linear in X_0, but one may attempt to adapt it in this sense in the hope of designing e.g. stochastic automata that improve mixing over standard Markov chains. For instance, one might imagine a system that distinguishes, in memory, each part of X_t that has started from a different node at X_0. Once this is done, we are free to choose the evolution (possibly nonlinear, nonlocal) for each of these X_0-indexed parts, postulating that the full X_t consists of their linear combination. One might thus wonder whether such heuristic approach could lead to faster mixing on e.g. the dumbbell graph. Our result implies that — provided also invariance is required — such acceleration attempts are all limited to the conductance bound.

3.3 Finite-Time Convergence

Consider the following algebraic problem, related to finite-time convergence [23] and the inverse eigenvalue problem [24]:

> What is the minimal number of symmetric stochastic matrices over a graph $\mathcal{G}$ whose product has all but one eigenvalue equal to zero?

From Theorem 1 it follows that this number is bounded by $1/(8\Phi)$. To see this, note that a set of local, symmetric, stochastic matrices $\{P(l), 1 \leq l \leq T\}$ over a graph $\mathcal{G}$ defines a linear and local system by $X_0 \xrightarrow{t} X_t = (\Pi_{l=1}^{t} P(l))\, X_0$ for $t \leq T$. The system is also invariant as the matrix product leaves the all-ones vector $\mathbf{1}$ invariant: $\mathbf{1} \xrightarrow{t} \mathbf{1}$. If the product $\Pi_{l=1}^{T} P(l)$ has all but one eigenvalue equal to zero, the remaining eigenvalue necessarily being 1 with eigenvector $\mathbf{1}$, then necessarily the system has converged: $X_0 \xrightarrow{T} \mathbf{1}/\|X_0\|_1, \forall X_0$ and so $\tau \leq T$. By Theorem 1 the convergence of any linear, local and invariant system is bounded, specifically stating that $T \geq 1/(8\Phi)$.

3.4 Quantum Walks

The convergence properties of quantum processes spreading over localized state spaces play a role both in physics (e.g. transport of excitations in photosynthesis [12]) and in quantum computation (e.g. quantum random walks [10]).

A discrete-time quantum walk is (although to our knowledge this has never been said explicitly) the generalization of a lifted walk, by keeping coherences among the node options. Denote by ρ the quantum state, i.e. a positive definite "density matrix" with trace one, whose diagonal represents probabilities over $\hat{\mathcal{V}} = \{(i, z)\}$ where $i \in \mathcal{V}$ is a graph node and z is a possible auxiliary degree of freedom (see introductory example [16], coined quantum walks [10], or Sect. 2.1). A general quantum walk follows:

$$\rho_{t+1} = \Psi \circ \rho_t$$

where Ψ is a completely positive trace-preserving map; most popular is the unitary quantum walk, where $\Psi \circ \rho_t = U \rho_t U^\dagger$, with U a unitary matrix satisfying the locality of the graph $\mathcal{G}$, exactly as P does for a Markov chain and M does in Sect. 2.1. If ρ_t would remain diagonal, this would correspond exactly to a lifted Markov chain [9]. Authors have been wondering for some time whether the additional information contained in the off-diagonal elements of ρ_t ("quantum coherences") might allow faster mixing. In [10] a conductance bound is given for *unitary* quantum walks and within a factor of the graph degree; but such factor becomes dominant for e.g. the dumbbell graph.

As will be further worked out in a future publication, quantum walks do satisfy the conditions of this paper, for most reasonable initializations. Then our result improves the bound of [10], both by generalizing it to non-unitary walks and by getting rid of the degree-dependent factor. Quantum walks indeed satisfy locality, including the hidden (complex) variables representing coherences. Linearity trivially holds, except if one allows to initialize the walk with nonlocal coherences already. In other words, since ρ_0 would necessarily be block-diagonal when all the initial weight is concentrated on a single node, introducing off-diagonal initial blocks when starting with a distribution over nodes would break linearity — and by Thm. 1, this would be necessary to potentially beat the conductance bound.

References

1. Cheeger, J.: A lower bound for the smallest eigenvalue of the Laplacian. In: Proceedings of the Princeton Conference in Honor of Professor S. Bochner (1969)
2. Buser, P.: A note on the isoperimetric constant. Annales Scientifiques de l'Ecole Normale Supérieure **15**(2), 213–230 (1982). Buser, P.: Ueber den ersten Eigenwert des Laplace-Operators auf kompakten Flächen. Commentarii Mathematici Helvetici **54**(1), 477–493 (1979)
3. Alon, N., Milman, V.D.: λ_1, isoperimetric inequalities for graphs, and superconcentrators. J. Combin. Theory Ser. B **38**(1), 73–88 (1985). Alon, N.: Eigenvalues and expanders. Combinatorica **6**(2), 83–96 (1986)

4. Fiedler, M.: Algebraic connectivity of graphs. Czech. Math. J. **23**(2), 298–305 (1973)
5. Aldous, D.: On the Markov chain simulation method for uniform combinatorial distributions and simulated annealing. Probab. Eng. Inf. Sci. **1**(1), 33–46 (1987)
6. Lawler, G.F., Sokal, A.D.: Bounds on the L^2 spectrum for Markov chains and Markov processes: a generalization of Cheeger's inequality. Trans. Am. Math. Soc. **309**(2), 557–580 (1988)
7. Mihail, M.: Conductance and convergence of Markov chains-a combinatorial treatment of expanders. In: IEEE Annual Symposium on Foundations of Computer Science (1989)
8. Dodziuk, J.: Difference equations, isoperimetric inequality and transience of certain random walks. Trans. Am. Math. Soc. **284**(2), 787–794 (1984)
9. Chen, F., Lovász, L., Pak, I.: Lifting Markov chains to speed up mixing. In: Proceedings of the Thirty-First Annual ACM Symposium on Theory of Computing. ACM (1999)
10. Aharonov, D., et al.: Quantum walks on graphs. In: Proceedings of the Thirty-Third Annual ACM Symposium on Theory of Computing. ACM (2001)
11. Ford, L.R., Fulkerson, D.R.: Maximal flow through a network. Can. J. Math. **8**(3), 399–404 (1956)
12. Mohseni, M., et al.: Environment-assisted quantum walks in photosynthetic energy transfer. J. Chem. Phys. **129**(17), 11B603 (2008)
13. Tsitsiklis, J.N., Athans, M.: Problems in decentralized decision making and computation. Ph.D. thesis, MIT (1984)
14. Murray, R., Saber, R.O.: Consensus protocols for networks of dynamic agents. In: Proceedings of the American Control Conference (2003)
15. Jerrum, M., Sinclair, A.: Approximating the permanent. SIAM J. Comput. **18**(6), 1149–1178 (1989)
16. Diaconis, P., Holmes, S., Neal, R.M.: Analysis of a nonreversible Markov chain sampler. Ann. Appl. Probab. 726–752 (2000)
17. Aaronson, S.: Quantum computing and hidden variables. Phys. Rev. A **71**(3), 032325 (2005)
18. Aldous, D., Fill, J.: Reversible Markov chains and random walks on graphs (2002)
19. Levin, D.A., Peres, Y., Wilmer, E.L.: Markov Chains and Mixing Times. American Mathematical Soc. (2009)
20. Apers, S., Sarlette, A., Ticozzi, F.: Lifting Markov chains to mix faster: limits and opportunities. IEEE Trans. Inf. Theory (2017, in preparation)
21. Saloff-Coste, L., Zúniga, J.: Convergence of some time inhomogeneous Markov chains via spectral techniques. Stochast. Process. Appl. **117**(8), 961–979 (2007). Touri, B., Nedić, A.: Alternative characterization of ergodicity for doubly stochastic chains. In: Proceedings of the IEEE Conference on Decision and Control (2011)
22. Mossel, E., Peres, Y., Sinclair, A.: Shuffling by semi-random transpositions. In: Proceedings of the IEEE Symposium on Foundations of Computer Science (2004). Diaconis, P., Ram, A.: Analysis of systematic scan Metropolis algorithms using Iwahori-Hecke algebra techniques. Stanford University, Department of Statistics (2000)
23. Hendrickx, J.M., et al.: Graph diameter, eigenvalues, and minimum-time consensus. Automatica **50**(2), 635–640 (2014). Hendrickx, J.M., Shi, G., Johansson, K.H.: Finite-time consensus using stochastic matrices with positive diagonals. IEEE Trans. Autom. Control **60**(4), 1070–1073 (2015)
24. Hogben, L.: Spectral graph theory and the inverse eigenvalue problem of a graph. Electron. J. Linear Algebra **14**(1), 3 (2005)

Inductive Means and Sequences Applied to Online Classification of EEG

Estelle M. Massart[1(✉)] and Sylvain Chevallier[2]

[1] ICTEAM Institute, Université catholique de Louvain, Louvain-la-Neuve, Belgium
estelle.massart@uclouvain.be
[2] LISV, Université de Versailles Saint-Quentin, Versailles, France

Abstract. The translation of brain activity into user command, through Brain-Computer Interfaces (BCI), is a very active topic in machine learning and signal processing. As commercial applications and out-of-the-lab solutions are proposed, there is an increased pressure to provide online algorithms and real-time implementations. Electroencephalography (EEG) systems offer lightweight and wearable solutions, at the expense of signal quality. Approaches based on covariance matrices have demonstrated good robustness to noise and provide a suitable representation for classification tasks, relying on advances in Riemannian geometry. We propose to equip the minimum distance to mean (MDM) classifier with a new family of means, based on the inductive mean, for block-online classification tasks and to embed the inductive mean in an incremental learning algorithm for online classification of EEG.

1 Introduction

Real-time recording and decoding of brain signals allow to control a large variety of systems, such as wheelchairs, exoskeletons, robotic arms or other types of Brain-Computer Interface (BCI) devices [3]. With electroencephalography (EEG), the brain signal is recorded at the surface of the head (on the scalp), offering a simple setup that does not require surgery as it is the case for invasive recording methods. The signal quality of EEG is lower than with invasive methods and the recording is very sensitive to noise, nonetheless possible applications offer promising results [11]. As technologies and signal processing techniques are more and more mature, out-of-the-lab applications and commercial systems are the focus of growing interests [3]. These applications and systems rely on a small number of electrodes for recording and low-cost hardware for signal processing. Thus the denoising and classification algorithms should work online and with a reasonable computational load. One of the most challenging issues with EEG-based BCI is to harness the individual variability of brain signals, which could change from hour-to-hour for a user and are highly variable from one user to the other.

Among all the methods considered in the literature for EEG signal processing, the ones relying on covariance matrices were shown numerically to achieve

F. Nielsen and F. Barbaresco (Eds.): GSI 2017, LNCS 10589, pp. 763–770, 2017.
https://doi.org/10.1007/978-3-319-68445-1_88

good performances [12]. In this approach, a portion of the EEG signal is represented by a covariance matrix, whose elements correspond to the covariance of the signals recorded with different electrodes, possibly filtered around different frequencies. The fact that covariance matrices belong to a non-Euclidean space – the manifold of symmetric positive definite (SPD) matrices – calls for efficient classifiers adapted to that geometry.

In this paper, we work with the Minimum Distance to Mean (MDM) classifier, initially proposed in [2]. This classifier assigns covariance matrices to the class with the closest mean. The classification results were shown to depend heavily on the mean and distance definition used, and many possibilities were compared in [5]. In the following we will distinguish the offline setting, where the classifier's parameters are selected and evaluated using all available data, the block-online setting, where the classifier is parametrized on a first batch of data (usually the beginning of a session) and evaluated on another batch of data (the rest of the session), and the online setting, where there is no data available beforehand from the user and the classifier is assessed directly on new data. We equip here this classifier with a new family of means based on the so-called inductive mean, which has the main advantage of being computed incrementally, a key property when working in an online setting. This property was already used in [4] for k-means clustering. We show numerically that the use of these new means achieves a classification accuracy in a block-online framework comparable to the most accurate nonparametric mean: the Riemannian barycenter with respect to the affine-invariant metric (less than 1% of difference on average), while their computation cost is lower. We also propose a variant of the online classification algorithm proposed in [6]. In our algorithm, the means of the classes are adapted online, following an incremental learning scheme. Starting from classes learned with other users, the goal is to enable the algorithm to progressively fit with the observed data of a new user.

The paper is organized as follows. Section 2 is devoted to block-online classification: we define the MDM classifier and the family of means we use, and compare numerically the classification results with other state-of-the-art methods. In Sect. 3, we present our incremental learning algorithm for online classification.

2 Offline and Block-Online Classification of EEG

The proposed approaches are applied on steady-state visual evoked potentials (SSVEP), that is brain responses to visual stimuli, but are valid on other kinds of BCI stimuli. In a SSVEP experiment, blinking LEDs are placed at different locations in the visual field of a user. The LEDs are blinking at F different frequencies ($\text{freq}_1, \ldots, \text{freq}_F$). The subject is either asked to focus on one specific blinking LED (with a known frequency) or to focus on a location without LED (resting state). The blinking LED elicit induced oscillations in the brain, which are visible in the EEG. The goal is to determine based on the EEG if the user is focusing on a blinking LED and if so, on which one.

We summarize in Algorithm 1 the block-online classification method proposed in [5]. Each time that the user is asked to focus on a stimulus, the portion of the EEG recording following the cue onset (the time at which the user was instructed to focus on the blinking LED) is first transformed into a covariance matrix and then classified using the MDM classifier. The means of the classes are estimated beforehand, based on a collection of labelled data, according to the offline training scheme detailed in Algorithm 2.

Algorithm 1. Block-online classification - MDM algorithm

Inputs: $\bar{\Sigma}^{(k)}$, the mean of the class k, for $k = 1, \ldots, K$ (obtained using Algorithm 2) and an unlabelled EEG trial $X \in \mathbb{R}^{C \times N}$ (with C the number of electrodes and N the number of time samples).

Output: $\hat{k}$, the predicted label of X.

1: Compute $\hat{\Sigma}$, an estimate of the covariance matrix of X (see Sect. 2.1).

2: Define the class label associated to trial X as $\hat{k} = \text{argmin}_{k=1,\ldots,K}\, \delta(\hat{\Sigma}, \bar{\Sigma}^{(k)})$, where $\delta(\Sigma_1, \Sigma_2) = ||\Sigma_1^{-1/2} \Sigma_2 \Sigma_1^{-1/2}||_{\text{F}}$ is the Riemannian distance between Σ_1 and Σ_2.

Algorithm 2. Offline training

Inputs : $X_i \in \mathbb{R}^{C \times N}$, for $i = 1, \ldots, l$, a set of labelled EEG trials, and $\mathcal{I}(k)$, $k = 1, \ldots, K$, the set of indices of trials belonging to class k.

Output: $\bar{\Sigma}^{(k)}$, the mean of the class k, for $k = 1, \ldots, K$.

1: Compute $\hat{\Sigma}_i$, an estimate of the covariance matrix of X_i, for $i = 1, \ldots, l$ (see Sect. 2.1).

2: **For** $k = 1 : K$ **do**

3: Compute the center of class $\bar{\Sigma}^{(k)} = \mu(\{\hat{\Sigma}_i | i \in \mathcal{I}(k)\})$ (see Sect. 2.2).

2.1 Estimation of Covariance Matrices

Algorithms 1 and 2 require to estimate the covariance matrix of an EEG trial $X \in \mathbb{R}^{C \times N}$, where N is the number of time samples and C the number of electrodes. The signal X is first band-pass filtered around the F frequencies used in the experiment, to yield an extended signal as follows:

$$X \in \mathbb{R}^{C \times N} \to X_{\text{Ext}} = \left[X_{\text{freq}_1}^T, \ldots, X_{\text{freq}_F}^T\right]^T \in \mathbb{R}^{FC \times N}.$$

The covariance matrix $\Sigma \in \mathbb{P}_{CF}$ of the signal X_{Ext}, with $\mathbb{P}_{CF}$ the set of SPD matrices of size $CF \times CF$, is then estimated using the Schäfer estimator [10]. We refer the reader to [6] for more information regarding the choice of the estimator.

2.2 Inductive Means and Sequences

The training of the classifier also relies on the definition of a mean μ on the set of SPD matrices. Several means were already considered in [5]. Among the non-parametric means, the Riemannian barycenter with respect to the affine-invariant metric was shown numerically to provide the most accurate classification results (we will use the shortcut “Riemannian barycenter” in the rest of the paper, implying that we work here with the affine-invariant metric). However, its computation is rather costly. To remedy this problem, another family of means was proposed in [8]. These means are based on the inductive mean (see [9]).

The inductive mean of a set of SPD matrices $\Sigma_1, \dots, \Sigma_l \in \mathbb{P}_{CF}$ is defined as:

$$M^{\text{Ind}}(\Sigma_1, \dots, \Sigma_l) = \left(\left(\left(\Sigma_1 \#_{\frac{1}{2}} \Sigma_2 \right) \#_{\frac{1}{3}} \Sigma_3 \right) \dots \#_{\frac{1}{l}} \Sigma_l \right), \tag{1}$$

where $A \#_s B = A^{\frac{1}{2}} (A^{-\frac{1}{2}} B A^{-\frac{1}{2}})^s A^{\frac{1}{2}}$, with $s \in [0, 1]$, is the (unique) point located on the geodesic from A to B, at a distance $s\delta(A, B)$ of A.

If all the matrices pairwise commute, then the Riemannian barycenter and the inductive mean coincide. Otherwise, the inductive mean looses the property of invariance under permutation: in general, $M^{\text{Ind}}(\Sigma_1, \dots, \Sigma_l) \neq M^{\text{Ind}}(\Sigma_{\pi(1)}, \dots, \Sigma_{\pi(l)})$, where π is a permutation of $(1, \dots, l)$. Moreover, in [8], the authors illustrate numerically that the inductive mean $M^{\text{Ind}}(\Sigma_1, \dots, \Sigma_l)$ tends to overemphasize the last data points (i.e., Σ_l, Σ_{l-1}, ...). To remedy this, they developed an inductive sequence $(X_j^{\text{Ind}})_{j=1,2,\dots}$, i.e., an extension of the inductive mean in which each element X_{jl}^{Ind}, with $j = 1, 2, \dots$ and l the total number of matrices, is defined as:

$$X_{jl}^{\text{Ind}} = M^{\text{Ind}} \left(\pi \left(\underbrace{\Sigma_1, \Sigma_2, \dots, \Sigma_l, \dots, \Sigma_1, \Sigma_2, \dots, \Sigma_l}_{j \times l \text{ elements}} \right) \right) \tag{2}$$

where π is a shuffling operator. The sequence $(X_j^{\text{Ind}})_{j=1,2,\dots}$ converges to the Riemannian barycenter, and the shuffling improves the convergence rate by reducing the bias mentioned above.

2.3 Experimental Results for Block-Online Classification

Table 1 compares block-online classification accuracy and computation times for several mean definitions. Our validation is performed on the same datasets as in [5]. These datasets were obtained in a SSVEP experiment with three frequencies (13, 17, or 21 Hz). This is thus a classification task with four classes (one for each frequency and one for the resting class). For each subject, the recorded session is made of several batches (from 2 to 5), one batch consisting in 32 trials (i.e., the responses to 32 stimuli, 8 for each class). As in [5], we used, for each subject, the last batch as validation set and all other batches as training set. We refer the reader to [5] for more detail regarding the experimental protocol.

We compare inductive means with the Euclidean mean, the Log-Euclidean mean and the Riemannian barycenter (estimated using a steepest descent algorithm). For comparison, we also provide results obtained with a state-of-the-art method not based on covariance matrices: the SVM algorithm with CCA filtering used in [5]. The last row of Table 1 presents the average performances obtained with the different means. It indicates that the inductive mean is a nice trade-off between the Log-Euclidean mean, which is cheaper to compute but

Table 1. Performances obtained for block-online classification. Each row of the table corresponds to one subject. The computation times recorded are the average times needed to compute the mean of the covariance matrices of a given class in the training set of the subject. They are larger for the subjects 7, 10 and 12 since the number of matrices to average were bigger for those subjects (their training sets were made of respectively 2, 3 and 4 batches instead of one for the other subjects).

	CCA + SVM [7]	MDM											
		Euclidean		LogEuclid.		Riem. Baryc.		Ind. Mean		Ind. Seq. X_{2l}^{Ind}		Ind. Seq. X_{5l}^{Ind}	
	acc(%)	acc(%)	t(ms)	acc(%)	t(ms)	acc(%)	t(ms)	acc(%)	t(ms)	acc(%)	t(ms)	acc(%)	T(ms)
S1	54.68	53.12	0.6	71.88	15	**73.44**	74	70.31	15	**73.44**	20	**73.44**	54
S2	37.50	43.75	0.5	78.12	16	**79.69**	77	78.12	11	78.12	19	78.12	43
S3	**89.06**	67.19	0.7	85.94	18	85.94	72	85.94	16	85.94	23	85.94	60
S4	79.69	54.69	0.5	84.38	13	**87.50**	68	**87.50**	11	**87.50**	22	**87.50**	42
S5	50.00	37.50	0.6	62.50	16	**68.75**	63	67.19	11	**68.75**	21	**68.75**	44
S6	**87.50**	34.38	0.5	84.38	18	85.94	69	84.38	11	85.94	19	85.94	49
S7	77.08	60.42	0.9	87.50	29	88.54	131	**89.58**	27	**89.58**	39	**89.58**	99
S8	73.44	67.19	0.7	90.62	18	**92.19**	71	**92.19**	16	**92.19**	24	**92.19**	46
S9	60.94	57.81	0.6	**70.31**	13	**70.31**	69	**70.31**	15	**70.31**	23	**70.31**	56
S10	67.97	38.28	1.2	75.00	43	**80.47**	179	78.91	38	78.91	67	**80.47**	137
S11	**71.88**	48.44	0.8	60.94	18	65.62	82	64.06	16	64.06	22	65.62	47
S12	95.63	71.25	1.5	96.25	58	**96.88**	216	**96.88**	49	**96.88**	86	**96.88**	198
Avg	70.45	52.83	0.8	78.98	23	81.27	97	80.45	20	80.97	32	81.23	73

also less accurate, and the Riemannian barycenter, which is more accurate but considerably more costly. Inductive sequences improve further the accuracy, but become also more costly.

Observe finally that Algorithms 1 and 2 are only suitable for block-online classification, and require to know the cue onsets, i.e., the time at which the stimuli are applied to the subject. Those requirements will be relaxed in the online classification approach presented in the next section.

3 Online Classification Using Inductive Means

In most cases, cue onsets are not available. The goal is then to detect parts of the EEG signal corresponding with a high probability to a given stimulus. Based on the incremental definition of the inductive mean, we propose a variant of the online classification algorithm detailed in [6]. Indeed, conversely to most other means, including the Riemannian barycenter, the inductive mean of $N+1$ data points can be easily computed from the inductive mean of N points:

$$M^{\text{Ind}}(\Sigma_1, \ldots, \Sigma_{N+1}) = M^{\text{Ind}}(\Sigma_1, \ldots, \Sigma_N) \#_{\frac{1}{N+1}} \Sigma_{N+1}.$$

It is then possible to update the means of the classes 'on-the-fly' in the classification algorithm. The complete classification scheme is presented in Algorithm 3. It works as follows. The algorithm scans the EEG signal, considering successive frames of size w, the starting times of two successive frames being separated by Δn samples. The covariance matrix of the current frame is estimated and classified using the MDM classifier. The most recurrent class among the last D ones is considered to be the current class. If the confidence in this decision is high enough (see Sect. 3.1), the class is returned and the mean of the class is updated. Otherwise, the algorithm moves immediately to the next frame. It is of tremendous importance to avoid that possible misclassifications move the means of the classes in an erroneous direction. To this aim, we added in Algorithm 3 a filtering step, following similar ideas as in the Riemannian potato [1]. The mean of the class is updated at most once per trial (i.e., per different stimulus), in the direction of the 'best' covariance matrix scanned in the trial.

Algorithm 3. Online classification

Inputs : $\bar{\Sigma}^{(k)} \in \mathbb{P}_{FC}$, the mean of class k, for all class $k = 1, \ldots, K$ (offline training, or default initialisation based on data available from other subjects), $\bar{d}_k$ the average distance between the training matrices belonging to class k and $\bar{\Sigma}^{(k)}$, a EEG recording $\mathcal{X}(n) \in \mathbb{R}^C$, $n = 0, \ldots, N$, hyperparameters w, Δn, D, s.
Output: Classification decisions $\hat{k}(n)$.

1: Initialisation: $\Sigma^{\text{best}} = \bar{\Sigma}^{(1)}$, $d^{\text{best}} = \infty$, $k_{\text{cur}} = -1$, $\hat{k}(n) = -1\ \forall n$ (default value, meaning no decision).
2: **For** $d = 0, \ldots, \lfloor \frac{N-w}{\Delta n} \rfloor$ **do**
3: $X_d := \mathcal{X}(d\Delta n, \ldots, d\Delta n + w)$
4: Compute $\hat{\Sigma}_d$, an estimate of the covariance matrix of X_d, and classify it:

$$k_d^* := \underset{k=1,\ldots,K}{\operatorname{argmin}}\ \delta(\hat{\Sigma}_d, \bar{\Sigma}^{(k)}).$$

5: **If** $d \geq D$ **then** find most recurrent class among D last classifications:

$$\bar{k} := \underset{k=1,\ldots,K}{\operatorname{argmax}}\ \rho(k) \quad \text{with} \quad \rho(k) := \frac{\#\{k_j^* = k\}_{j=d,d-1,\ldots,d-D+1}}{D}.$$

6: Evaluate confidence criterion C (see Sect. 3.1)
7: **If** C = true **then**
8: **If** $k_{\text{cur}} > 0$ and $\bar{k} \neq k_{\text{cur}}$ (we left previous class) and $d^{\text{best}} \leq \bar{d}_{k_{\text{cur}}}$ **then** update previous class:
9: $\bar{\Sigma}^{(k_{\text{cur}})} := \bar{\Sigma}^{(k_{\text{cur}})} \#_{\frac{\alpha}{s+\alpha}} \hat{\Sigma}^{\text{best}}$ with $\alpha := 1 - \frac{\delta(\bar{\Sigma}^{(k_{\text{cur}})}, \hat{\Sigma}^{\text{best}})}{\bar{d}_{k_{\text{cur}}}}$
10: $s := s + 1$, $\Sigma^{\text{best}} := \hat{\Sigma}_d$, $d^{\text{best}} := \delta(\bar{\Sigma}^{(\bar{k})}, \hat{\Sigma}_d)$
11: **elseif** $\delta(\bar{\Sigma}^{(\bar{k})}, \hat{\Sigma}_d) \leq d^{\text{best}}$, improve current estimates:
12: $\Sigma^{\text{best}} := \hat{\Sigma}_d$, $d^{\text{best}} := \delta(\bar{\Sigma}^{(\bar{k})}, \hat{\Sigma}_d)$
13: $k_{\text{cur}} := \bar{k}$
14: $\hat{k}(n) := \bar{k}$ for $n \in [d\Delta n, d\Delta n + w]$

3.1 Confidence Criterion

Similarly as in [6], a confidence criterion is used in Algorithm 3 to discard unreliable classifications. Two conditions have to be encountered for this criterion to be satisfied. The first one verifies that the current classification decision is consistent with previous classifications: the class $\bar{k}$ should have been chosen among the D previous classes with a proportion larger than or equal to given threshold ϑ, i.e.,

$$\rho(\bar{k}) \geq \vartheta\ . \tag{3}$$

The second condition is related to the displacement of the covariance matrices: those should be in the direction of the mean of the class. Otherwise, we might expect that a new stimulus has been applied, and that the covariance matrices is moving away from the mean of the old class, to get closer to the mean of the new class. Hence, the relative distances to means should on average decrease on the last D frames:

$$\sum_{j=d-D+2}^{d} (\delta_{\bar{k}}^{\text{rel}}(j) - \delta_{\bar{k}}^{\text{rel}}(j-1)) \leq 0 \quad \text{with} \quad \delta_{\bar{k}}^{\text{rel}}(d) = \frac{\delta(\hat{\Sigma}_d, \bar{\Sigma}^{(\bar{k})})}{\sum_{k=1}^{K} \delta(\hat{\Sigma}_d, \bar{\Sigma}^{(k)})}. \tag{4}$$

If conditions (3) and (4) are satisfied, the confidence criterion is satisfied, i.e., $C = \text{true}$, otherwise $C = \text{false}$.

3.2 Numerical Results for Online Classification

The main interest of Algorithm 3 is that it allows to progressively update the user's means of the classes. To illustrate this, we used EEG batches from the three first subjects to initialize the centers of the classes and we run Algorithm 3 to perform classification on all the batches of the other users. In Fig. 1, we compare the results obtained using Algorithm 3 with those obtained when the means of the classes are not updated, i.e. removing lines 10 to 12 in Algorithm 3, for the two subjects with the highest number of batches available, that is 5 for subject 12 and 4 for subject 10. Hyperparameters were set empirically to $w = 2.6$s, $\Delta n = 0.2$s, $D = 5$, $s = 8$, $\vartheta = 0.7$. For subject 12, the classification accuracy improves with the batches, compared to the version with frozen means of the classes. However, this is not the case for subject 10: despite the use of the filtering step, some misclassification resulted in the displacement of the mean of one

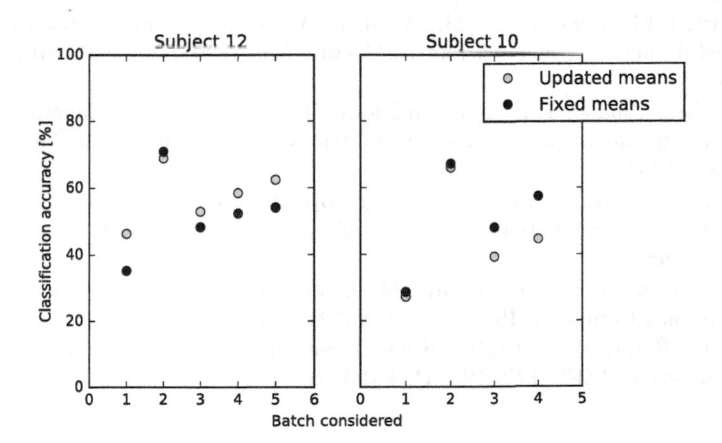

Fig. 1. Classification results of Algorithm 3 on the two subjects for which the largest number of recordings are available.

class in an erroneous direction, which alters subsequent classification decisions. Unfortunately, the low number of recordings per subject makes it difficult to obtain a reliable measure of the performance of our online algorithm. Further work should therefore aim at assessing the performance of the algorithm on larger datasets. Other filtering strategies can also be investigated for Algorithm 3, as well as the influence of the hyperparameters on the classification results.

References

1. Barachant, A., Andreev, A., Congedo, M.: The Riemannian Potato: an automatic and adaptive artifact detection method for online experiments using Riemannian geometry. In: TOBI Workshop lV, pp. 19–20 (2013)
2. Barachant, A., Bonnet, S., Congedo, M., Jutten, C.: Riemannian geometry applied to BCI classification. In: Vigneron, V., Zarzoso, V., Moreau, E., Gribonval, R., Vincent, E. (eds.) LVA/ICA 2010. LNCS, vol. 6365, pp. 629–636. Springer, Heidelberg (2010). doi:10.1007/978-3-642-15995-4_78
3. Clerc, M., Bougrain, L., Lotte, F.: Brain-Computer Interfaces 2: Technology and Applications. Wiley, London (2016)
4. Ho, J., Cheng, G., Salehian, H., Vemuri, B.: Recursive Karcher expectation estimators and geometric law of large numbers. In: Artificial Intelligence and Statistics, pp. 325–332 (2013)
5. Kalunga, E.K., Chevallier, S., Barthélemy, Q., Djouani, K., Hamam, Y., Monacelli, E.: From euclidean to riemannian means: information geometry for SSVEP classification. In: Nielsen, F., Barbaresco, F. (eds.) GSI 2015. LNCS, vol. 9389, pp. 595–604. Springer, Cham (2015). doi:10.1007/978-3-319-25040-3_64
6. Kalunga, E.K., Chevallier, S., Barthélemy, Q., Djouani, K., Monacelli, E., Hamam, Y.: Online SSVEP-based BCI using Riemannian geometry. Neurocomputing **191**, 55–68 (2016)
7. Kalunga, E.K., Djouani, K., Hamam, Y., Chevallier, S., Monacelli, E.: SSVEP enhancement based on canonical correlation analysis to improve BCI performances. In: Africon, IEEE (2013)
8. Massart, E.M., Hendrickx, J.M., Absil, P.-A.: Matrix geometric means based on shuffled inductive sequences. Linear Algebra Appl. (2017). doi:10.1016/j.laa.2017.05.036
9. Sagae, M., Tanabe, K.: Upper and lower bounds for the arithmetic-geometric-harmonic means of positive definite matrices. Linear Multilinear Algebra **37**(4), 279–282 (1994)
10. Schäferm, J., Strimmer, K.: A shrinkage approach to large-scale covariance matrix estimation and implications for functional genomics. Stat. Appl. Genet. Mol. Biol. **4**(1), 32 (2005)
11. van Erp, J., Lotte, F., Tangermann, M.: Brain-computer interfaces: beyond medical applications. Computer **45**(4), 26–34 (2012)
12. Yger, F., Berar, M., Lotte, F.: Riemannian approaches in brain-computer interfaces: a review. IEEE TNSRE **PP**, 1 (2016)

Geodesic Methods with Constraints

Vessel Tracking via Sub-Riemannian Geodesics on the Projective Line Bundle

Erik J. Bekkers[1], Remco Duits[1(✉)], Alexey Mashtakov[2(✉)], and Yuri Sachkov[2]

[1] Department of Mathematics and Computer Science,
Eindhoven University of Technology, Eindhoven, The Netherlands
{E.J.Bekkers,R.Duits}@tue.nl

[2] Control Processes Research Center,
Program Systems Institute of RAS, Pereslavl-Zalessky, Russia
alexey.mashtakov@gmail.com, yusachkov@gmail.com

Abstract. We study a data-driven sub-Riemannian (SR) curve optimization model for connecting local orientations in orientation lifts of images. Our model lives on the projective line bundle $\mathbb{R}^2 \times P^1$, with $P^1 = S^1/_\sim$ with identification of antipodal points. It extends previous cortical models for contour perception on $\mathbb{R}^2 \times P^1$ to the data-driven case. We provide a complete (mainly numerical) analysis of the dynamics of the 1st Maxwell-set with growing radii of SR-spheres, revealing the cut-locus. Furthermore, a comparison of the cusp-surface in $\mathbb{R}^2 \times P^1$ to its counterpart in $\mathbb{R}^2 \times S^1$ of a previous model, reveals a general and strong reduction of cusps in spatial projections of geodesics. Numerical solutions of the model are obtained by a single wavefront propagation method relying on a simple extension of existing anisotropic fast-marching or iterative morphological scale space methods. Experiments show that the projective line bundle structure greatly reduces the presence of cusps. Another advantage of including $\mathbb{R}^2 \times P^1$ instead of $\mathbb{R}^2 \times S^1$ in the wavefront propagation is reduction of computational time.

Keywords: Sub-Riemannian geodesic · Tracking · Projective line bundle

1 Introduction

In image analysis extraction of salient curves such as blood vessels, is often tackled by first lifting the image data to a new representation defined on the higher dimensional space of positions and directions, followed by a geodesic

Joint main authors. The ERC is gratefully acknowledged for financial support (ERC-StG nr. 335555). Sections 1, 2 of the paper are written by R. Duits and A. Mashtakov, Sect. 3 is written by A. Mashtakov, Yu. Sachkov and R. Duits, Sects. 4, 6 are written by R. Duits and A. Mashtakov, and Sect. 5 is written by E.J. Bekkers. The work of A. Mashtakov and Yu. Sachkov is supported by the Russian Science Foundation under grant 17-11-01387 and performed in Ailamazyan Program Systems Institute of Russian Academy of Sciences.

F. Nielsen and F. Barbaresco (Eds.): GSI 2017, LNCS 10589, pp. 773–781, 2017.
https://doi.org/10.1007/978-3-319-68445-1_89

tracking [1–3] in this lifted space [4–6]. Benefits of such approaches are that one can generically deal with complex structures such as crossings [4,6,7], bifurcations [8], and low-contrast [5,6,9], while accounting for contextual alignment of local orientations [5,6]. The latter can be done in the same way as in cortical models of visual perception of lines [10–13], namely via sub-Riemannian (SR) geometry on the combined space of positions and orientations. In these cortical models, it is sometimes stressed [12] that one should work in a projective line bundle $\mathbb{R}^2 \times P^1$ with a partition of equivalence classes $P^1 := S^1/_\sim$ with $\mathbf{n}_1 \sim \mathbf{n}_2 \Leftrightarrow \mathbf{n}_1 = \pm\mathbf{n}_2$. Furthermore, in the statistics of line co-occurrences in retinal images the same projective line bundle structure is crucial [14]. Also, for many image analysis applications the orientation of an elongated structure is a well defined characteristic of a salient curve in an image, in contrast to an artificially imposed direction.

At first sight the effect of the identification of antipodal points might seem minor as the minimizing SR geodesic between two elements in $\mathbb{R}^2 \times P^1$ is obtained by the minimum of the two minimizing SR geodesics in $\mathbb{R}^2 \times S^1$ that arise (twice) by flipping the directions of the boundary conditions. However, this appearance is deceptive, it has a rather serious impact on geometric notions such as (1) the 1st Maxwell set (where two distinct geodesics with equal length meet for the first positive time), (2) the cut-locus (where a geodesic looses optimality), (3) the cusp-surface (where spatial projections of SR geodesics show a cusp). Besides an analysis of the geometric consequences in Sects. 2, 3 and 4, we show that the projective line bundle provides a better tracking with much less cusps in Sect. 5.

2 The Projective Line Bundle Model

The projective line bundle $\mathrm{PT}(\mathbb{R}^2)$ is a quotient of Lie group SE(2), and one can define a sub-Riemannian structure (SR) on it. The group $\mathrm{SE}(2) = \mathbb{R}^2 \rtimes SO(2)$ of planar roto-translations is identified with the coupled space of positions and orientations $\mathbb{R}^2 \times S^1$, and for each $g = (x, y, \theta) \in \mathbb{R}^2 \times S^1 \cong \mathrm{SE}(2)$ one has

$$L_g g' = g \odot g' = (x' \cos\theta + y' \sin\theta + x, -x' \sin\theta + y' \cos\theta + y, \theta' + \theta). \quad (1)$$

Via the push-forward $(L_g)_*$ one gets the left-invariant frame $\{\mathcal{A}_1, \mathcal{A}_2, \mathcal{A}_3\}$ from the Lie-algebra basis $\{A_1, A_2, A_3\} = \{\partial_x|_e, \partial_\theta|_e, \partial_y|_e\}$ at the unity $e = (0, 0, 0)$:

$$\mathcal{A}_1 = \cos\theta\, \partial_x + \sin\theta\, \partial_y, \quad \mathcal{A}_2 = \partial_\theta, \quad \mathcal{A}_3 = -\sin\theta\, \partial_x + \cos\theta\, \partial_y.$$

Let $\mathcal{C} : \mathrm{SE}(2) \to \mathbb{R}^+$ denote a smooth cost function strictly bounded from below. The SR-problem on SE(2) is to find a Lipschizian curve $\gamma : [0, T] \to \mathrm{SE}(2)$, s.t.

$$\begin{aligned} \dot{\gamma}(t) &= u^1(t)\, \mathcal{A}_1|_{\gamma(t)} + u^2(t)\, \mathcal{A}_2|_{\gamma(t)}, \qquad \gamma(0) = g_0, \quad \gamma(T) = g_1, \\ l(\gamma(\cdot)) &:= \int_0^T \mathcal{C}(\gamma(t)) \sqrt{\xi^2 |u^1(t)|^2 + |u^2(t)|^2}\, \mathrm{d}t \to \min, \end{aligned} \quad (2)$$

with controls $u^1, u^2 : [0, T] \to \mathbb{R}$ are in $L^\infty[0, T]$, boundary points g_0, g_1 are given, $\xi > 0$ is constant, and terminal time $T > 0$ is free.

Thanks to reparametrization invariance the SR distance can be defined as

$$d(g_0, g_1) = \min_{\substack{\gamma \in \mathrm{Lip}([0,1], \mathrm{SE}(2)), \\ \dot\gamma \in \Delta|_\gamma,\ \gamma(0)=g_0, \gamma(1)=g_1}} \int_0^1 \sqrt{\mathcal{G}_{\gamma(\tau)}(\dot\gamma(\tau), \dot\gamma(\tau))}\, \mathrm{d}\tau, \tag{3}$$

with $\mathcal{G}_{\gamma(\tau)}(\dot\gamma(\tau), \dot\gamma(\tau)) = \mathcal{C}^2(\gamma(\tau))\left(\xi^2|u^1(\tau T)|^2 + |u^2(\tau T)|^2\right)$, $\tau = \frac{t}{T} \in [0,1]$, and $\Delta := \mathrm{span}\{\mathcal{A}_1, \mathcal{A}_2\}$ with dual $\Delta^* = \mathrm{span}\{\cos\theta\, \mathrm{d}x + \sin\theta\, \mathrm{d}y, \mathrm{d}\theta\}$. The projective line bundle $\mathrm{PT}(\mathbb{R}^2)$ is a quotient $\mathrm{PT}(\mathbb{R}^2) = \mathrm{SE}(2)/_\sim$ with identification $(x, y, \theta) \sim (x, y, \theta+\pi)$. The SR distance in $\mathrm{PT}(\mathbb{R}^2) \cong \mathbb{R}^2 \times P^1 = \mathbb{R}^2 \times \mathbb{R}/\{\pi\mathbb{Z}\}$ is

$$\begin{aligned}\overline{d}(q_0, q_1) &:= \min\{d(g_0, g_1),\, d(g_0 \odot (0,0,\pi), g_1 \odot (0,0,\pi)), \\ &\qquad d(g_0, g_1 \odot (0,0,\pi)),\, d(g_0 \odot (0,0,\pi), g_1)\} \\ &= \min\{\, d(g_0, g_1),\, d(g_0 \odot (0,0,\pi), g_1)\}\end{aligned} \tag{4}$$

for all $q_i = (x_i, y_i, \theta_i) \in \mathrm{PT}(\mathbb{R}^2)$, $g_i = q_i = (x_i, y_i, \theta_i) \in \mathrm{SE}(2), i \in \{0,1\}$. Equation (4) is due to $\gamma^*_{g_0 \to g_1}(\tau) = \gamma^*_{\tilde g_1 \to \tilde g_0}(1-\tau)$, with $\tilde g_i := g_i \odot (0,0,\pi)$, with $\gamma^*_{g_0 \to g_1}$ a minimizing geodesic from $g_0 = (\mathbf{x}_0, \theta_0)$ to $g_1 = (\mathbf{x}_1, \theta_1)$, and **has 2 consequences:**

(1) One can account for the $\mathrm{PT}(\mathbb{R}^2)$ structure in the building of the distance function before tracking takes place, cf. Proposition 1 below.
(2) It affects cut-locus, the first Maxwell set (Propositions 2 and 3), and cusps (Proposition 4).

We apply a Riemannian limit [8, Theorem 2] where $\overline{d}$ is approximated by Riemannian metric $\overline{d}^\epsilon$ induced by $\mathcal{G}^\epsilon_q(\dot q, \dot q) := \mathcal{G}_q(\dot q, \dot q) + \frac{\mathcal{C}^2(q)\,\xi^2}{\epsilon^2}\left|-\dot x \sin\theta + \dot y \cos\theta\right|^2$ for $\dot q = (\dot x, \dot y, \dot\theta), q = (x, y, \theta), 0 < \epsilon \ll 1$, and use SR gradient $\mathcal{G}_q^{-1}\mathrm{d}W(q) := \mathcal{G}_q^{-1} P_{\Delta^*}\mathrm{d}W(q) = \frac{\mathcal{A}_1 W(q)}{\xi^2 \mathcal{C}^2(q)}\mathcal{A}_1|_q + \frac{\mathcal{A}_2 W(q)}{\mathcal{C}^2(q)}\mathcal{A}_2|_q$ for steepest descent on $W = \overline{d}(\cdot, e)$.

Proposition 1. *Let $q \neq e$ be chosen such that there exists a unique minimizing geodesic $\gamma^*_\epsilon : [0,1] \to \mathrm{PT}(\mathbb{R}^2)$ of $\overline{d}^\epsilon(q, e)$ for $\epsilon > 0$ sufficiently small, that does not contain conjugate points (i.e. the differential of the exponential map of the Hamiltonian system is non-degenerate along γ^*_ϵ, cf. [15]). Then $\tau \mapsto \overline{d}(e, \gamma^*_0(\tau))$ is smooth and $\gamma^*_0(\tau)$ is given by $\gamma^*_0(\tau) = \gamma^*_b(1-\tau)$ with*

$$\begin{cases} \dot\gamma^*_b(\tau) = -W(q)\,(\mathcal{G}^{-1}_{\gamma^*_b(\tau)}\mathrm{d}W)(\gamma^*_b(\tau)), & \tau \in [0,1] \\ \gamma^*_b(0) = q, \end{cases} \tag{5}$$

with $W(q)$ the viscosity solution of the following boundary value problem:

$$\begin{cases} \mathcal{G}_q\left(\mathcal{G}_q^{-1}\mathrm{d}W(q), \mathcal{G}_q^{-1}\mathrm{d}W(q)\right) = 1 \ \textit{for}\ q \neq e, \\ W(x, y, \pi) = W(x, y, 0), \ \textit{for all}\ (x, y) \in \mathbb{R}^2, \\ W(0,0,0) = W(0,0,\pi) = 0. \end{cases} \tag{6}$$

Proof. By [8, Theorems 2 and 4], (extending [7, Theorem 3.2] to non-uniform cost) we get minimizing SR geodesics in SE(2) by intrinsic gradient descent on

W. The 2nd condition in (6) is due to $P^1 = S^1/_\sim$, the 3rd is due to (4). When applying [8, Theorem 4] we need differentiability of the SR distance. As our assumptions exclude conjugate and Maxwell-points, this holds by [16, Theorem 11.15]. □

At least for $\epsilon = 0$ and $\mathcal{C} = 1$ the assumption in Proposition 1 on conjugate points is obsolete by [17] and [7, Theorem 3.2, Appendix D].

3 Analysis of Maxwell Sets for $\mathcal{C} = 1$

A *sub-Riemannian sphere* is a set of points equidistant from e. A sphere of radius R centred at e is given by $\mathcal{S}(R) = \{q \in \mathrm{PT}(\mathbb{R}^2) \,|\, \overline{d}(e,q) = R\}$. We define (the first) *Maxwell point* as a point in $\mathrm{PT}(\mathbb{R}^2)$ connected to e by multiple SR length minimizers. I.e. its *multiplicity* is >1. All Maxwell points form a *Maxwell set*:

$$\begin{aligned}\mathcal{M} = \{q \in \mathrm{PT}(\mathbb{R}^2) \,|\, \exists\, \gamma^1, \gamma^2 \in \mathrm{Lip}([0,1], \mathrm{PT}(\mathbb{R}^2)),\ \text{s. t. } \dot{\gamma}^i \in \Delta|_{\gamma^i},\\ \gamma^i(0) = e,\ \gamma^i(1) = q,\ \text{ for } i = 1,2, \text{ and } \gamma^1 \neq \gamma^2,\ l(\gamma^1) = l(\gamma^2) = \overline{d}(e,q)\}.\end{aligned}$$

The set $\mathcal{M}$ is a stratified manifold $\mathcal{M} = \bigcup_i \mathcal{M}_i$. We aim for maximal dimension strata: $\dim(\mathcal{M}_i) = 2$.

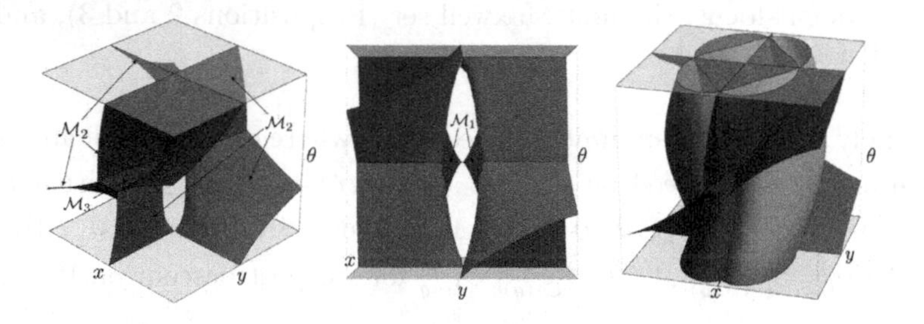

Fig. 1. Maxwell set and its intersection (right image) with the SR sphere in Fig. 2. The folds on the Green surface are in $\mathcal{M}_1$, the intersections of the Green surface with Red and Blue surface are in $\mathcal{M}_2$, the intersection of the Red and Blue surface is in $\mathcal{M}_3$. (Color figure online)

Proposition 2. *Let* $W(q) = \overline{d}(e,q)$ *and let* $W^{\mathrm{SE}(2)}(g) = d(e,g)$. *The Maxwell set* $\mathcal{M}$ *is given by* $\mathcal{M} = \bigcup_{i=1}^3 \mathcal{M}_i$, *see Fig. 1, where*

- $\mathcal{M}_1$ *is a part of local component of Maxwell set* $\mathrm{Exp}(\mathrm{MAX}^2)$ *in* SE(2), *see [18, Theorem 5.2], restricted by the condition* $t_1^{\mathrm{MAX}} = W(\gamma(t_1^{\mathrm{MAX}}))$;
- $\mathcal{M}_2$ *is given by* $W^{\mathrm{SE}(2)}(g) = W^{\mathrm{SE}(2)}(g \odot (0,0,\pi))$;
- $\mathcal{M}_3$ *is a part of global component of Maxwell set* $\mathrm{Exp}(\mathrm{MAX}^5)$ *in* SE(2), *see [18, Theorem 5.2], restricted by the condition* $t_1^{\mathrm{MAX}} = W(\gamma(t_1^{\mathrm{MAX}}))$.

Proof. There are two possible reasons for $\mathrm{PT}(\mathbb{R}^2) \ni q = g/_{\sim}$ be a Maxwell point: (1) if g is a Maxwell point in SE(2), s.t. $W^{\mathrm{SE}(2)}(g) = W(q)$ (i.e. $W^{\mathrm{SE}(2)}(g) \leq W^{\mathrm{SE}(2)}(g \odot (0,0,\pi))$); (2) if q is a (new) Maxwell point induced by the quotient (i.e. q is a root of $W^{\mathrm{SE}(2)}(g) = W^{\mathrm{SE}(2)}(g \odot (0,0,\pi))$). Strata $\mathcal{M}_1$, $\mathcal{M}_3$ follow from $\mathrm{Exp}(\mathrm{MAX}^2)$, $\mathrm{Exp}(\mathrm{MAX}^5)$ [18], while $\mathcal{M}_2$ is induced by $P^1 = S^1/_{\sim}$. Set $\mathcal{M}_3$ is in $\theta = 0$, as $\mathrm{Exp}(\mathrm{MAX}^5)$ is in $\theta = \pi$, which is now identified with $\theta = 0$.

Proposition 3. *The maximal multiplicity ν of a Maxwell point on a SR sphere depends on its radius R. Denote $\mathcal{M}^R = \mathcal{M} \cap \mathcal{S}(R)$ and $\mathcal{M}_i^R = \mathcal{M}_i \cap \mathcal{S}(R)$. One has the following development of Maxwell set as R increases, see Figs. 2 and 3:*

1. *if $0 < R < \frac{\pi}{2}$ then $\mathcal{S}(R)$ is homeomorphic to S^2 and it coincides with SR sphere in* SE(2), *$\mathcal{M}^R = \mathcal{M}_1^R$ and $\nu = 2$;*
2. *if $R = \frac{\pi}{2}$ then $\mathcal{S}(R)$ is homeomorphic to S^2 glued at one point, $\mathcal{M}^R = \mathcal{M}_1^R \cup \mathcal{M}_2^R$, $\mathcal{M}_1^R \cap \mathcal{M}_2^R = \emptyset$, and $\nu = 2$;*
3. *if $\frac{\pi}{2} < R < \overline{R}$ then $\mathcal{S}(R)$ is homeomorphic to T^2, $\mathcal{M}^R = \mathcal{M}_1^R \cup \mathcal{M}_2^R$, $\mathcal{M}_1^R \cap \mathcal{M}_2^R = \emptyset$ and $\nu = 2$;*
4. *if $R = \overline{R} \approx \frac{17}{18}\pi$ then $\mathcal{S}(R)$ is homeomorphic to T^2, $\mathcal{M}^R = \mathcal{M}_1^R \cup \mathcal{M}_2^R$, and $\mathcal{M}_1^R$ intersects $\mathcal{M}_2^R$ at four (conjugate) points, $\nu = 2$;*
5. *if $\overline{R} < R < \tilde{R}$ then $\mathcal{S}(R)$ is homeomorphic to T^2, $\mathcal{M}^R = \mathcal{M}_1^R \cup \mathcal{M}_2^R$, and $\mathcal{M}_1^R$ intersects $\mathcal{M}_2^R$ at four points, where $\nu = 3$;*
6. *if $R = \tilde{R} \approx \frac{10}{9}\pi$ then $\mathcal{S}(R)$ is homeomorphic to T^2, $\mathcal{M} = \mathcal{M}_1^R \cup \mathcal{M}_2^R \cup \mathcal{M}_3^R$, $\mathcal{M}_1^R = \mathcal{M}_3^R$, and $\mathcal{M}_2^R$ intersects $\mathcal{M}_1^R$ at two points, where $\nu = 4$;*
7. *if $R > \tilde{R}$ then $\mathcal{S}(R)$ is homeomorphic to T^2, $\mathcal{M}^R = \mathcal{M}_2^R \cup \mathcal{M}_3^R$ and $\mathcal{M}_2^R$ intersects $\mathcal{M}_3^R$ at four points, where $\nu = 3$.*

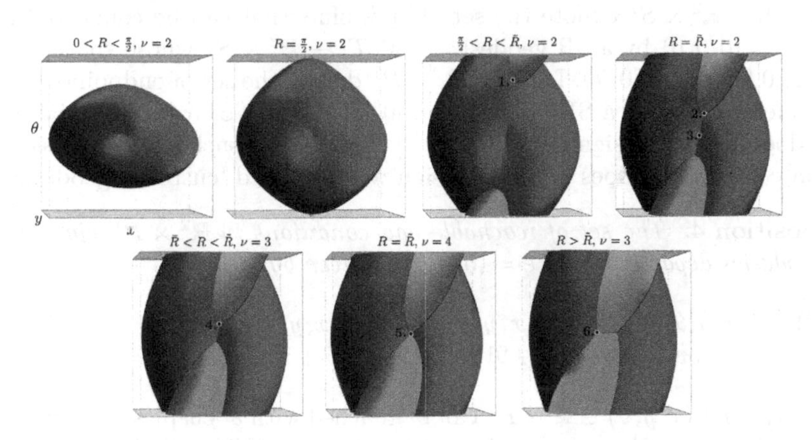

Fig. 2. Evolution of the 1st Maxwell set as the radius R of the SR-spheres increases.

Remark 1. Results in [19, Sect. 4] imply that $\tilde{R}$ can be computed from the system:

$$\tilde{R}/2 = K(k_1) = k_2\, p_1(k_2), \quad \frac{K(k_1) - E(k_1)}{k_1\sqrt{1-k_2^2}} = \frac{p_1(k_2) - \mathrm{E}(p_1(k_2), k_2)}{\mathrm{dn}(p_1(k_2), k_2)}, \tag{7}$$

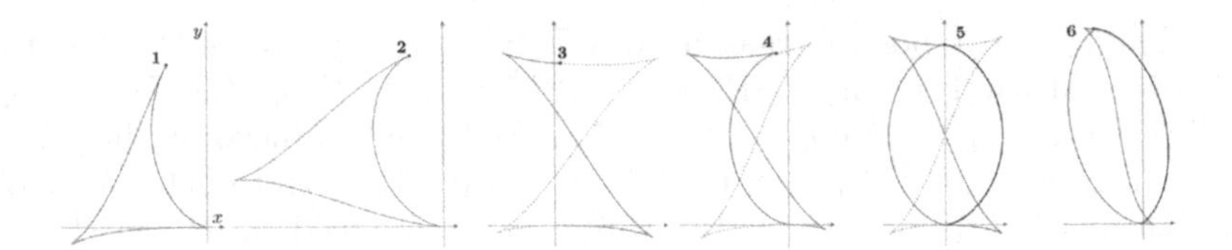

Fig. 3. SR length minimizers ending at the points indicated at Fig. 2.

where $K(k)$ and $E(k)$ are complete elliptic integrals of the 1st and 2nd kind; $\mathrm{E}(u,k) = E(\mathrm{am}(u,k),k)$, while $E(v,k)$ is the incomplete elliptic integral of the 2nd kind and $\mathrm{am}(u,k)$ is the Jacobian amplitude; $p_1(k)$ is the first positive root of $\mathrm{cn}(p,k)(\mathrm{E}(p,k)-p) - \mathrm{dn}(p,k)\mathrm{sn}(p,k) = 0$; and $\mathrm{sn}(p,k)$, $\mathrm{cn}(p,k)$, $\mathrm{dn}(p,k)$ are Jacobian elliptic functions. Solving (7), we get $\tilde{R} \approx 1.11545\,\pi \approx 10/9\,\pi$. Radius $\overline{R}$ is s.t. $\mathcal{S}(\bar{R})$ hits the 1st conjugate set and can be computed as well (see item 2 of Fig. 3).

4 Set of Reachable End Conditions by Cuspless Geodesics

A *cusp point* $\mathbf{x}(t_0)$ on a spatial projection of a (SR) geodesic $t \mapsto (\mathbf{x}(t), \theta(t))$ in $\mathbb{R}^2 \times S^1$ is a point where the only spatial control switches sign, i.e. $u^1(t_0) := \dot{x}(t_0)\cos\theta(t_0) + \dot{y}(t_0)\sin\theta(t_0) = 0$ and $(u^1)'(t_0) \neq 0$. In fact, the 2nd condition $(u^1)'(t_0) \neq 0$ is obsolete [8, Appendix C]. The next proposition shows that the occurrence of cusps is greatly reduced in $\mathbb{R}^2 \times P^1$.

Let $\mathfrak{R} \subset \mathbb{R}^2 \times S^1$ denote the set of endpoints that can be connected to the origin $e = (0,0,0)$ by a SR geodesic $\gamma : [0,T] \to \mathbb{R}^2 \times S^1$ whose spatial control $u^1(t) > 0$ for all $t \in [0,T]$. Let $\tilde{\mathfrak{R}} \subset \mathbb{R}^2 \times P^1$ denote the set of endpoints that can be connected to e by a SR geodesic $\gamma : [0,T] \to \mathbb{R}^2 \times S^1$ whose spatial control $u^1(t)$ does not switch sign for all $t \in [0,T]$. Henceforth, such a SR geodesic whose spatial control $u^1(\cdot)$ does not switch sign will be called 'cuspless' geodesic.

Proposition 4. *The set of reachable end-conditions in $\mathbb{R}^2 \times P^1$ via 'cuspless' SR geodesics departing from $e = (0,0,0)$ is given by*

$$\begin{aligned}\tilde{\mathfrak{R}} = \{(x,y,\theta) \in \mathrm{PT}(\mathbb{R}^2) | &(x,y,\theta) \in \mathfrak{R} \;\; or \; (x,y,\theta+\pi) \in \mathfrak{R} \\ &or \; (-x,y,-\theta) \in \mathfrak{R} \; or \; (-x,y,-\theta+\pi) \in \mathfrak{R} \; or \; x=y=0\}.\end{aligned} \tag{8}$$

Proof. A point $(x,y,\theta) \in \mathbb{R}^2 \times P^1$ can be reached with a 'cuspless' SR geodesic if (1) $(x,y,\theta) \in \mathbb{R}^2 \rtimes S^1$ can be reached with a 'cuspless' SR geodesic in SE(2) or (2) if $(-x,y,-\theta)$ can be reached with a 'cuspless' SR geodesic in SE(2). Recall from [20, Theorem 7] that $(x,y,\theta) \in \mathfrak{R} \Rightarrow (x \geq 0$ and $(x,y) \neq (0,0))$. If $x \geq 0$ and $(x,y) \neq (0,0)$, the first option holds if $(x,y,\theta) \in \mathfrak{R}$, and the second option holds if $(x,y,\theta+\pi) \in \mathfrak{R}$. If $x < 0$, the endpoint can only be reached by a 'cuspless' SR geodesic in SE(2) with a negative spatial control function $u^1 < 0$. Here we rely on symmetry $(x,y,\theta) \mapsto (-x,y,-\theta) \Rightarrow (x(t),y(t),\theta(t)) \mapsto (-x(t),y(t),-\theta(t)))$ that holds for SR geodesics $(x(\cdot),y(\cdot),\theta(\cdot))$ in SE(2). For the control u^1 in (2), this symmetry implies $u^1(t) \mapsto -u^1(t)$. By [20, Theorem 10] one has $(x, y, \theta) \in \mathfrak{R} \Rightarrow (x, y, \theta + \pi) \notin \mathfrak{R}$, and points with $x = y = 0$ are not in $\mathfrak{R}$ [20, Remark 5.5] so all 'or' conditions in (8) are exclusive. $\square$

Set $\mathfrak{R}$ yields a single cone field of reachable angles in $x > 0$, see [20, Fig. 14, Theorem 9]. By Proposition 4, set $\tilde{\mathfrak{R}}$ is a union of 2 such cone fields that is also reflected to $x < 0$.

5 Practical Advantages in Vessel Tracking

Distance $W(q)$ can be numerically obtained by solving the eikonal PDE of Eq. (6) via similar approaches as was previously done for the SE(2) case. E.g., via an iterative upwind scheme [7], or a fast marching (FM) solver [21] in which case the SR metric tensor is approximated by an anisotropic Riemannian metric tensor [22]. A gradient descent (cf. Eq. (5)) on W then provides the SR geodesics.

We construct the cost function $\mathcal{C}$ in the same way as in [7]: (1) a retinal image is lifted via the orientation score transform using cake wavelets [23]; (2) vessels are enhanced via left-invariant Gaussian derivatives using $\mathcal{A}_3$; (3) a cost function is constructed via $\mathcal{C} = \frac{1}{1+\lambda \mathcal{V}^p}$, with $\mathcal{V}$ the max-normalized vessel enhanced orientation score, and with λ and p respectively a "cost-strength" and contrast parameter. We use the same data and settings ($\lambda = 100$, $p = 3$ and $\xi = 0.01$) as in [7], and perform vessel tracking on 235 vessel segments. For the results on all retinal image patches, see http://erikbekkers.bitbucket.io/PTR2.html.

Figure 4 shows the results on three different vessel segments with comparison between SR geodesics in SE(2) and PT($\mathbb{R}^2$). As expected, with the PT($\mathbb{R}^2$) model we always obtain the SE(2) geodesic with minimum SR length (cf. Eq. (4)). This has the advantage that overall we encounter less cusps in the tracking. Additionally, the PT($\mathbb{R}^2$) model is approximately four times faster since now we only have to consider half of the domain $\mathbb{R}^2 \times S^1$, and by Proposition 1 we only run once (instead of twice). The average computation time via FM for constructing W with the SE(2) model for 180×140 pixel patches is 14.4 s, whereas for the PT($\mathbb{R}^2$) model this is only 3.4 s. The rightmost image in Fig. 4 shows an exceptional case in which the reversed boundary condition (red arrow) is preferred as this leads to a geodesic with only one cusp instead of two.

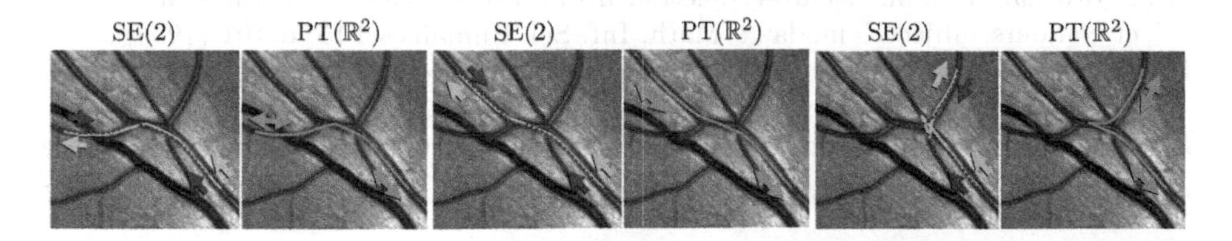

Fig. 4. Data-adaptive SR geodesics in SE(2) (in green and red-dashed) compared to SR geodesics in PT($\mathbb{R}^2$) (in blue). For the SE(2) case we specify antipodal boundary conditions since the correct initial and end directions are not known a priori. (Color figure online)

6 Conclusion

We have shown the effect of including the projective line bundle structure SR in optimal geodesic tracking (Proposition 1), in SR geometry (Proposition 2), and in Maxwell-stratification (Proposition 3), and in the occurrence of cusps in spatially projected geodesics (Proposition 4). It supports our experiments that show benefits of including such a projective line bundle structure: A better vessel tracking algorithm with a reduction of cusps and computation time. As the cusp-free model without reverse gear [8] also benefits [8, Fig. 12] from $PT(\mathbb{R}^2)$-structure, we leave the Maxwell stratification of this combined model for future work.

References

1. Peyré, G., Péchaud, M., Keriven, R., Cohen, L.D.: Geodesic methods in computer vision and graphics. Found. Trends Comput. Graph. Vis. **5**(34), 197–397 (2010)
2. Caselles, V., Kimmel, R., Sapiro, G.: Geodesic active contours. IJCV **22**(1), 61–79 (1997)
3. Cohen, L., Kimmel, R.: Global minimum for active contour models. IJCV **24**(1), 57–78 (1997)
4. Péchaud, M., Keriven, R., Peyré, G.: Extraction of tubular structures over an orientation domain. In: IEEE Conference on CVPR, pp. 336–342 (2009)
5. Bekkers, E.J.: Retinal Image Analysis using Sub-Riemannian Geometry in SE(2), Ph.D. thesis, Eindhoven University of Technology, Biomedical Engineering (2017)
6. Chen, D.: New Minimal Path Models for Tubular Structure Extraction and Image Segmentation, Ph.D. thesis, Universite Paris Dauphine, PSL Research Univ. (2016)
7. Bekkers, E.J., Duits, R., Mashtakov, A., Sanguinetti, G.: A PDE approach to data-driven sub-riemannian geodesics in SE(2). SIAM-SIIMS **8**(4), 2740–2770 (2015)
8. Duits, R., Meesters, S., Mirebeau, J., Portegies, J.: Optimal paths for variants of the 2d, 3d Reeds-Shepp. car with applications in image analysis (arXiv: 1612.06137) (2017)
9. Zhang, J., Dashtbozorg, B., Bekkers, E., Pluim, J., Duits, R., ter Haar Romeny, B.: Robust retinal vessel segmentation via locally adaptive derivative frames in orientation scores. IEEE TMI **35**(12), 2631–2644 (2016)
10. Citti, G., Sarti, A.: A cortical based model of perceptual completion in the roto-translation space. JMIV **24**(3), 307–326 (2006)
11. Petitot, J.: Vers une Neuro-gèométrie: fibrations corticales, structures de contact et contours subjectifs modaux. Math. Inf. Sci. Humaines **145**, 5–101 (1999)
12. Boscain, U., Duits, R., Rossi, F., Sachkov, Y.: Curve cuspless reconstruction via sub-Riemannian geometry. ESAIM: COCV **20**, 748–770 (2014)
13. Mashtakov, A.P., Ardentov, A.A., Sachkov, Y.L.: Parallel algorithm and software for image inpainting via sub-Riemannian minimizers on the group of rototranslations. NMTMA **6**(1), 95–115 (2013)
14. Abbasi-Sureshjani, S., Zhang, J., Duits, R., ter Haar Romeny, B.: Retrieving challenging vessel connections in retinal images by line co-occurence statistics. Biol. Cybern. **111**(3), 237–247 (2017)
15. Agrachev, A.A., Sachkov, Y.L.: Control Theory from the Geometric Viewpoint. Springer, New York (2004)

16. Agrachev, A.A., Barilari, D., Boscain, U.: Introduction to Riemannian and Sub-Riemannian Geometry from the Hamiltonian Viewpoint, preprint SISSA 09/2012/M, 20 November 2016
17. Sachkov, Y.L.: Conjugate and cut time in the sub-Riemannian problem on the group of motions of a plane. ESAIM: COCV **16**(4), 1018–1039 (2009)
18. Moiseev, I., Sachkov, Y.L.: Maxwell strata in sub-Riemannian problem on the group of motions of a plane. ESAIM: COCV **16**(2), 380–399 (2010)
19. Sachkov, Y.L.: Cut locus and optimal synthesis in the sub-Riemannian problem on the group of motions of a plane. ESAIM: COCV **17**(2), 293–321 (2011)
20. Duits, R., Boscain, U., Rossi, F., Sachkov, Y.: Association fields via cuspless sub-Riemannian geodesics in SE(2). JMIV **49**(2), 384–417 (2014)
21. Mirebeau, J.-M.: Anisotropic fast-marching on cartesian grids using lattice basis reduction. SIAM J. Num. Anal. **52**(4), 1573–1599 (2014)
22. Sanguinetti, G., Bekkers, E., Duits, R., Janssen, M.H.J., Mashtakov, A., Mirebeau, J.-M.: Sub-riemannian fast marching in SE(2). Progress in Pattern Recognition, Image Analysis, Computer Vision, and Applications. LNCS, vol. 9423, pp. 366–374. Springer, Cham (2015). doi:10.1007/978-3-319-25751-8_44
23. Duits, R., Felsberg, M., Granlund, G., ter Haar Romeny, B.M.: Image analysis and reconstruction using a wavelet transform constructed from a reducible representation of the Euclidean motion group. IJCV **72**(1), 79–102 (2007)

Anisotropic Edge-Based Balloon Eikonal Active Contours

Da Chen(✉) and Laurent D. Cohen

CEREMADE, CNRS,
University Paris Dauphine, PSL Research University, UMR 7534,
75016 Paris, France
chenda@ceremade.dauphine.fr

Abstract. In this paper, we propose a new edge-based active contour model for image segmentation and curve evolution by an asymmetric Finsler metric and the corresponding minimal paths. We consider the edge anisotropy information and the balloon force term to build a Finsler metric comprising of a symmetric quartic term and an asymmetric linear term. Unlike the traditional geodesic active contour model where the curve evolution is carried out by the level set framework, we search for a more robust optimal curve by solving an Eikonal partial differential equation (PDE) associated to the Finsler metrics. Moreover, we present an interactive way for geodesics extraction and closed contour evolution. Compared to the level set-based geodesic active contour model, our model is more robust to spurious edges, and also more efficient in numerical solution.

1 Introduction

Active contours model or the snakes model [1] was proposed by Kass *et al.* for boundary detection. The basic idea is to extract a sequence of time-dependent curves to minimize the curve-based energy where the limit of these curves denotes the boundary of an object. The snakes energy involves a potential function P such that $E_{\text{snake}}(\gamma) = \int_0^1 \Big(\eta_1\|\gamma'(v)\|^2 + \eta_2\|\gamma''\|^2 + P(\gamma(v))\Big)dv$, where η_1 and η_2 are two constants. A curve $\gamma \in H^2([0,1],\Omega)$ lies at an open domain $\Omega \subset \mathbb{R}^2$ with H^2 is a Sobolev space. The terms $\|\gamma'\|$ and $\|\gamma''\|$ are respective the first- and second-order derivatives of the path γ. In the past decades, a series of approaches have been devoted to overcome the drawbacks of the snakes model [1] such as the initialization sensitivity and the dependence of the parameterization.

The geodesic active contours (GAC) model [2,3] reformulated the snakes energy E_{snake} and removed the second-order derivative $\|\gamma''\|$ from E_{snake}. The GAC model leads to important theoretical results. However, in its basic formulation, the geodesic metric is actually an isotropic Riemannian metric which cannot take into account the curve orientation. In [4,5], the authors extended the isotropic metric to the anisotropic case and the Finslerian case. The curve evolution is originally carried out based on the level set framework [6] and the

F. Nielsen and F. Barbaresco (Eds.): GSI 2017, LNCS 10589, pp. 782–790, 2017.
https://doi.org/10.1007/978-3-319-68445-1_90

Euler-Lagrange equation. Such a curve evolutional strategy costs expensive computation time and known to be sensitive to noise and spurious edges due to the numerous undesired local minimums. Cohen and Kimmel [7] proposed an efficiently minimal path model, which can be naturally used for open curve detection. For object segmentation, more efforts [8–10] have been devoted to for closed contours detection which are used to delineate object boundaries.

In this paper, we propose a new curve evolution scheme based on the Eikonal interpretation framework of a general regional active contour energy [10]. The main contribution lies at the construction of a Finsler metric induced from the balloon force [11] and the anisotropic edge information. In contrast to [10], our method mainly depends on the anisotropic edge saliency information and balloon force, which is insensitive to gray levels inhomogeneities.

2 Background on Minimal Path and Eikonal PDE

Let $\Im([0,1],\Omega)$ be the collection of all Lipschitz continuous curves $\gamma:[0,1]\to\Omega$. We denote by S_2^+ the collection of 2×2 symmetric positive definite matrices. A norm $\|\mathbf{u}\|_M$ is defined by $\sqrt{\langle \mathbf{u},\, M\mathbf{u}\rangle}$, where $M\in S_2^+$.

Cohen and Kimmel [7] proposed an Eikonal PDE-based method to globally minimize the following geodesic energy $\mathcal{L}_{\text{Iso}}(\gamma):=\int_0^1(\mathcal{P}(\gamma(t))+\epsilon)\|\gamma'(t)\|\,dt$ with $\epsilon>0$ a constant used for minimal geodesic regularization. The geodesic distance map $\mathcal{U}$ associated to a source $\mathbf{s}$ is defined as $\mathcal{U}_{\mathbf{s}}(\mathbf{x}):=\min\{\mathcal{L}_{\text{Iso}}(\gamma);\gamma\in\Im([0,1],\Omega)\}$, which is the viscosity solution to the isotropic Eikonal PDE [7]

$$\|\nabla\mathcal{U}(\mathbf{x})\|=\mathcal{P}(\mathbf{x})+\epsilon,\quad \forall\mathbf{x}\in\Omega\backslash\{\mathbf{s}\},\quad \text{and}\quad \mathcal{U}(\mathbf{s})=0. \tag{1}$$

A general Finsler metric $\mathcal{F}:\Omega\times\mathbb{R}^2\to\mathbb{R}^+$ is a positive, 1-homogeneous, and potentially asymmetric function [4,12], based on which the curve length associated to the Finsler metric $\mathcal{F}$ is defined by $\mathcal{L}_{\text{F}}(\gamma):=\int_0^1\mathcal{F}(\gamma(t),\gamma'(t))\,dt$. The Finsler Eikonal PDE [12–15] associated to $\mathcal{L}_{\text{F}}$ can be expressed by

$$\sup_{\|\mathbf{v}\|=1}\frac{\langle\nabla\mathcal{U}(\mathbf{x}),\mathbf{v}\rangle}{\mathcal{F}(\mathbf{x},\mathbf{v})}=1,\quad \forall\mathbf{x}\in\Omega\backslash\{\mathbf{s}\},\forall\mathbf{u}\in\mathbb{R}^2,\quad \text{and}\quad \mathcal{U}(\mathbf{s})=0. \tag{2}$$

We consider the Randers metric [16], a special Finsler metric with the form of

$$\mathcal{F}(\mathbf{x},\mathbf{u})=\|\mathbf{u}\|_{\mathcal{M}(\mathbf{x})}+\langle\boldsymbol{\omega}(\mathbf{x}),\,\mathbf{u}\rangle,\quad \forall\mathbf{x}\in\Omega,\ \forall\mathbf{u}\in\mathbb{R}^2, \tag{3}$$

where $\mathcal{M}:\Omega\to S_2^+$ is a positive symmetric definite tensor field and $\omega:\Omega\to\mathbb{R}^2$ is a vector field. The tensor field $\mathcal{M}$ and the vector field ω should satisfy

$$\langle\boldsymbol{\omega}(\mathbf{x}),\mathcal{M}^{-1}(\mathbf{x})\,\boldsymbol{\omega}(\mathbf{x})\rangle<1,\quad \forall\mathbf{x}\in\Omega, \tag{4}$$

to ensure the positivity of $\mathcal{F}$ [13,16].

3 Edge-Based Balloon Eikonal Active Contours

Geodesic interpretation of an edge-based balloon energy. Let χ_B be the characteristic function of a region $B \subset \Omega$. The balloon force [11] was designed as external force for active contours by minimizing the region-based term [17]

$$E_{\text{balloon}}(\chi_B) = \int_{\Omega} \chi_B(\mathbf{x})d\mathbf{x} = \int_B d\mathbf{x}, \tag{5}$$

where $B \subset \mathbb{R}^2$ is the interior region of a close path $\gamma_B \in \Im([0,1],\Omega)$.

A complete edge-based active contour energy $\mathcal{E}$ can be defined by the summation of an anisotropic edge-based term and a balloon term

$$\mathcal{E}(\gamma_B) = \int_0^1 \|\gamma_B'(v)\|_{\mathcal{M}_{\text{e}}(\gamma_A(v))}\, dv + \alpha\, E_{\text{balloon}}(\chi_B), \tag{6}$$

where $\mathcal{M}_{\text{e}} : \Omega \to S_2^+$ is an edge-based tensor field and $\alpha < 0$ is a constant.

Let $\mathfrak{g} \subset \Omega$ be a fixed shape and let $U_{\mathfrak{g}}$ be a tubular neighbourhood of a curve $\gamma_{\mathfrak{g}}$ such that $U_{\mathfrak{g}} := \{\mathbf{x} \in \Omega;\, \min_{v\in[0,1]} \|\mathbf{x} - \gamma_{\mathfrak{g}}(v)\| < \mathrm{r}\}$ where $r \in \mathbb{R}^+$ is a constant. Denoting by $\mathfrak{g}' = \mathfrak{g}\backslash U_{\mathfrak{g}}$ that is entirely determined by $U_{\mathfrak{g}}$[1]. We define an admissible shape set $\Phi(U_{\mathfrak{g}}) := \{B \subset \Omega; \gamma_B \in \Im([0,1], U_{\mathfrak{g}}), \mathfrak{g}' \subset B\}$.

In the course of curve evolution, let γ_{B_k} $(k > 0)$ be the resulting curve in the k-th step. We note U_k as the tubular neighbour of γ_{B_k}. Our goal is to find an optimal curve $\gamma_{B_{k+1}}$ such that $B_{k+1} \in \Phi(U_k)$. This can be done by solving

$$\inf_{A\in\Phi(U_k)} \mathcal{E}(\gamma_A) = \inf_{A\in\Phi(U_k)} \left\{ \int_0^1 \|\gamma_A'(v)\|_{\mathcal{M}_{\text{e}}(\gamma_A(v))} dv + \alpha \int_A d\mathbf{x} \right\}. \tag{7}$$

For any shape $A \in \Phi(U_k)$, one has the following equations

$$\begin{aligned}\mathcal{E}(\gamma_A) &= \int_0^1 \|\gamma_A'(v)\|_{\mathcal{M}_{\text{e}}(\gamma_A(v))}\, dv + \alpha \int_A d\mathbf{x} \\ &= \int_0^1 \|\gamma_A'(v)\|_{\mathcal{M}_{\text{e}}(\gamma_A(v))}\, dv + \alpha \int_A \chi_{U_k}(\mathbf{x})\, d\mathbf{x} + C_{B_k},\end{aligned} \tag{8}$$

where $C_{B_k} = \alpha \int_{A\backslash U_k} d\mathbf{x}$ is a constant associated to the shape B_k. We consider a vector field $\boldsymbol{a}_k$ that satisfies the following divergence equation

$$\operatorname{div} \boldsymbol{a}_k = \alpha\, \chi_{U_k} \tag{9}$$

[1] This is because $\mathfrak{g}'$ is the bounded connected component of $\Omega\backslash U_{\mathfrak{g}}$.

and suppose that $\boldsymbol{\omega}_k(\mathbf{x}) = R\,\boldsymbol{a}_k(\mathbf{x})$, $\forall \mathbf{x} \in \Omega$, where R is the clockwise rotation matrix with angle $\pi/2$. We rewrite Eq. (8) by removing the constant C_{B_k}

$$\begin{aligned}
&\int_0^1 \|\gamma_A'(v)\|_{\mathcal{M}_e(\gamma_A(v))} dv + \alpha \int_A \chi_{U_k}(\mathbf{x})\, d\mathbf{x} \\
&= \int_0^1 \|\gamma_A'(v)\|_{\mathcal{M}_e(\gamma_A(v))} dv + \int_0^1 \langle \boldsymbol{a}_k(\gamma_A(v)), \mathcal{N}(v)\rangle \|\gamma_A'(v)\| dv \qquad (10)\\
&= \int_0^1 \|\gamma_A'(v)\|_{\mathcal{M}_e(\gamma_A(v))} dv + \int_0^1 \langle \boldsymbol{\omega}_k(\gamma_A(v)), \mathcal{T}(v)\rangle \|\gamma_A'(t)\| dt \\
&= \int_0^1 \mathcal{F}_k(\gamma_A(v), \gamma_A'(v)) dv, \qquad (11)
\end{aligned}$$

where $\mathcal{T} = R\mathcal{N}$ is the *clockwise* tangent of γ with the reality that $\gamma' = \mathcal{T}\|\gamma'\|$. The function $\mathcal{F}_k$ is defined by

$$\mathcal{F}_k(\mathbf{x}, \mathbf{u}) = \|\mathbf{u}\|_{\mathcal{M}_e(\mathbf{x})} + \langle \boldsymbol{\omega}_k(\mathbf{x}), \mathbf{u}\rangle. \qquad (12)$$

For a given shape B_k, the problem (7) is equivalent to

$$\inf_{A \in \Phi(U_k)} \int_0^1 \mathcal{F}_k(\gamma_A(v), \gamma_A'(v)) dv, \qquad (13)$$

where the shape A is the interior region of the path γ_A. Note that the formulation (11) was first used in [10] for geodesic energy interpretation of a general region-based energy. Here we use it to convert the balloon force energy to a geodesic energy by a Finsler metic $\mathcal{F}_k$. The crucial point for the curve length energy (11) is the construction of the vector $\boldsymbol{\omega}_k$ in Eq. (9). As discussed in [10], we solve the following PDE-constrained problem

$$\min\left\{\int_{U_k} \|\boldsymbol{\omega}_k(\mathbf{x})\|^2 d\mathbf{x}\right\}, \quad s.t. \quad \mathrm{div}\boldsymbol{\omega}_k = \alpha\,\chi_{U_k}, \qquad (14)$$

in an *optimization-then-discretization* manner to obtain the vector field $\boldsymbol{\omega}_k$.

A new robust Finsler Metric. The tensor field $\mathcal{M}_e$ can be expressed by its eigenvalues λ_i and eigenvectors $\boldsymbol{\nu}_i$ such that $\mathcal{M}_e(\cdot) = \sum_i \lambda_i(\cdot)\boldsymbol{\nu}_i(\cdot)\boldsymbol{\nu}_i^{\mathrm{T}}(\cdot)$ following that $1 \leq \lambda_1(\cdot) \leq \lambda_2(\cdot)$. The eigenvalues λ_i are computed according to the Frobenius norm of the gradient $\nabla(G_\sigma * \mathbf{I})$ of a color image $\mathbf{I} : \Omega \to \mathbb{R}^3$, where $\nabla(G_\sigma * \mathbf{I})$ is a *Jacobian matrix* with size 2×3 and G_σ is a Gaussian filter with variance σ. Letting g be the Frobenius norm of the gradient $\nabla(G_\sigma * \mathbf{I})$, one has

$$\lambda_1(\cdot) = \exp\left(\beta_1\left(\|g\|_\infty - g(\cdot)\right)\right), \quad \lambda_2(\cdot) = \exp\left(\beta_2\, g(\cdot)\right)\lambda_1(\cdot),$$

where β_1 and β_2 are two positive constants. The vector $\boldsymbol{\nu}_1(\cdot)$ is the eigenvector of $\nabla(G_\sigma * \mathbf{I})(\cdot)$ corresponding to the smaller eigenvalue. Thus $\boldsymbol{\nu}_1(\mathbf{x})$ is collinear to the edge orientation at $\mathbf{x}$. The vector $\boldsymbol{\nu}_2(\cdot)$ is the remaining eigenvector of $\nabla(G_\sigma * \mathbf{I})(\cdot)$. In this case, β_2 controls the anisotropy of the tensor filed $\mathcal{M}_e$.

If $\mathbf{x}$ is far from the boundaries, one has $\lambda_1(\mathbf{x}) \approx \lambda_2(\mathbf{x}) \gg 1$, leading to an approximately isotropic tensor $\mathcal{M}_{\mathrm{e}}(\mathbf{x})$ and high metric cost at $\mathbf{x}$. In contrast, if $\mathbf{x}$ is near a boundary, one has $\lambda_2(\mathbf{x}) \gg 1$ and $\lambda_1(\mathbf{x}) \approx 1$ corresponding to an highly anisotropic tensor $\mathcal{M}_{\mathrm{e}}(\mathbf{x})$.

To obey the positive constraint (4), we should ensure that $\inf_{\mathbf{x}} \|\boldsymbol{\omega}_k(\mathbf{x})\| < 1$. We make use of a non-linear map to construct a new vector field $\tilde{\boldsymbol{\omega}}$ such that

$$\tilde{\boldsymbol{\omega}}_k(\mathbf{x}) = \big(1 - \exp(-\tilde{\alpha}\, \|\boldsymbol{\omega}_k(\mathbf{x})\|)\big)\, \boldsymbol{\omega}_k(\mathbf{x})/\|\boldsymbol{\omega}_k(\mathbf{x})\|, \tag{15}$$

where $\tilde{\alpha}$ is a positive parameter. The Finsler metric $\mathcal{F}_k$ in Eq. (12) thus becomes

$$\tilde{\mathcal{F}}_k(\cdot, \mathbf{u}) = \|\mathbf{u}\|_{\mathcal{M}_{\mathrm{e}}(\cdot)} + \langle \tilde{\boldsymbol{\omega}}(\cdot), \mathbf{u} \rangle. \tag{16}$$

Since the balloon force is only used to drive the curves outward, the use of the reconstructed Finsler metric $\tilde{\mathcal{F}}$ will not modify the goal that $\mathcal{F}$ services for [10].

Interactive Segmentation. A curve concatenation operator can be defined by

$$\Gamma(v) = (\Gamma_1 \uplus \Gamma_2)(v) = \begin{cases} \Gamma_1(2v), & \text{if } 0 \le v < \frac{1}{2}, \\ \Gamma_2(2v-1), & \text{if } \frac{1}{2} \le v < 1, \end{cases}$$

where $\Gamma, \Gamma_1, \Gamma_2 \in \Im([0,1], \Omega)$ are clockwise paths.

Considering a collection $\{\mathbf{p}_i\}_{i \le m}$ of m ($m \ge 3$) user-provided control points distributed in a *clockwise* order along an object boundary. We aim to search for a closed contour to delineate the target object boundary. This can be done by concatenating a set of minimal paths associated to the metric $\mathcal{F}_k$, each of which links a pair of successive landmark points $\{\mathbf{p}_i, \mathbf{p}_{i+1}\}$. In Fig. 1a, we show three control points denoted by red dots. During the curve evolution, in the k-th iteration, we denote by $\mathcal{C}_{i,k}$ the paths between each pair of successive control points $\mathbf{p}_i$ and $\mathbf{p}_{i+1}$ for $i < m$, and by $\mathcal{C}_{m,k}$ the path linking $\mathbf{p}_m$ to $\mathbf{p}_1$. A closed contour γ_{B_k}, indicating the exterior boundary of the shape B_k, can be concatenated by $\gamma_{B_k} = \biguplus_{i=1}^{m} \mathcal{C}_{i,k}$. Let U_k be the tubular neighbourhood of γ_{B_k}. One can identify a subregion $\Re_i \subset U_k$ for each path $\mathcal{C}_{i,k}$

$$\Re_i := \{\mathbf{x} \in U; d(\mathbf{x}, \mathcal{C}_{i,k}) < d(\mathbf{x}, \mathcal{C}_{j,k}),\, \forall j \ne i\} \cup \{\mathcal{C}_{i,k}(0),\, \mathcal{C}_{i,k}(1)\}.$$

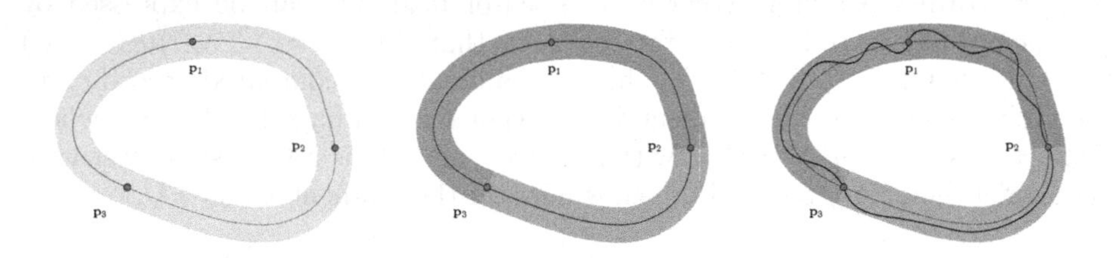

Fig. 1. Illustration for the procedure of interactive image segmentation. **Column 1** Control points $\mathbf{p}_i$ (red dots) and tubular neighbourhood (gray region). **Column 2** Separated tubular subregions. **Column 3** Extracted minimal paths (solid black curves) between successive control points. (Color figure online)

In Fig. 1b, we illustrated each subregion $\Re_i$ by different colours.

Within each region $\Re_i$, we take $\mathbf{p}_i$ as the source point to compute the geodesic distance map $\mathcal{U}_{\mathbf{p}_i}$ with respect to the metric $\mathcal{F}$ via the solution to the Eikonal PDE (1). Then a minimal path $\mathcal{C}_{i,k+1}$ is obtained by using $\mathcal{U}_{\mathbf{p}_i}$ and the gradient descent ODE (2). The desired closed contour $\gamma_{B_{k+1}}$ can be concatenated by

$$\gamma_{B_{k+1}} = \Cup_{i=1}^{m} \mathcal{C}_{i,k+1}, \tag{17}$$

and the shape B_{k+1} can be simply identified as the interior region of $\gamma_{B_{k+1}}$. Once we obtain $\gamma_{B_{k+1}}$, the vector field $\tilde{\omega}_{k+1}$ and metrics $\tilde{\mathcal{F}}_{k+1}$ can be updated using Eqs. (15) and (16), respectively. We illustrate the course of the interactive segmentation in Fig. 2, where the proposed model can converge to the desired object boundary in only 4 steps. The curve evolution can be terminated when the Hausdorff distance between two curves γ_{B_k} and $\gamma_{B_{k+1}}$ is small enough.

Remark. The path $\mathcal{C}_{i,k}$ is actually a globally minimizing curve in the domain $\Re_i$ with respect to the Finsler metric $\tilde{\mathcal{F}}_k$, which leads the proposed method to be insensitive to spurious edges and noise. Moreover, the definition of $\Re_i$ guarantees the extracted closed contour $\gamma_{B_{k+1}}$ (see Eq. (17)) to be a simple curve since each pair of subregions $\Re_i$ and $\Re_j$ has only one intersection point.

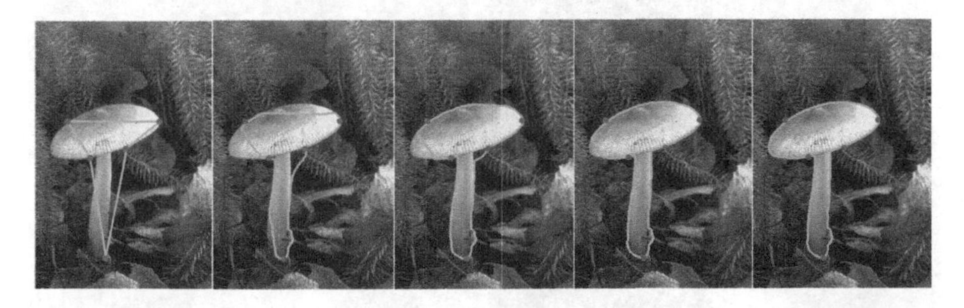

Fig. 2. Evolution course of the interactive image segmentation scheme. **Column 1** Initialization. Red dots are the control points. **Columns 2–5** Segmentation results from the first iteration to the fourth iteration. (Color figure online)

4 Experimental Results

In Fig. 3, we show the curve evolution results by setting the tensor field $\mathcal{M}_{\mathrm{e}} \equiv \mathcal{I}_d$, where $\mathcal{I}_d$ is the identity matrix. These contours (blue curves) inflate outward in the course of the curve evolution due to the balloon force (negative value of α). In this experiment the control points $\mathbf{p}_i$ (red dots) have been fixed. Moreover, as an option, these control points can be resampled in each iteration (for details we refer to [10]). In this case, the contours (blue curves) will tend to appear as a circle and will expand indefinitely since there is no edges to stop the evolution.

Fig. 3. Curve evolution with $\mathcal{M}_e \equiv \mathbf{I}_d$. **Column 1** Control points and initial contour. **Columns 2-3** Evolution results on different iterations. (Color figure online)

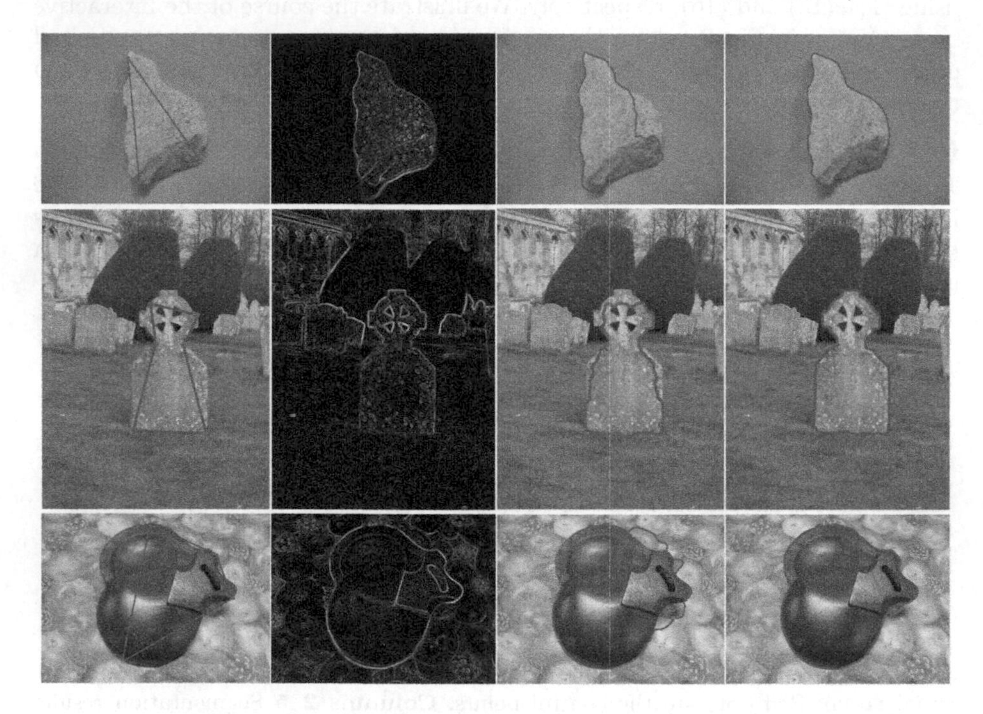

Fig. 4. Image segmentation results. **Column 1** Initializations. Red dots are user-specified control points. **Column 2** Edge saliency map. **Column 3** Segmentation from GAC model. **Column 4** Segmentation from the proposed model. (Color figure online)

We compare our method to the GAC model [2]. The gradient flow of the GAC model with respect to a level set function[2] ψ can be expressed by

$$\psi_t = \|\nabla\psi\|\mathrm{div}(f\,\nabla\psi/\|\nabla\psi\|) + c\,f\,\|\nabla\psi\|, \tag{18}$$

where $f(\cdot) = \exp(-\beta_2 g(\cdot))$ and g is defined as the Frobenius norm of the gradient $\nabla(G_\sigma * \mathbf{I})$. The term $c\,f\|\nabla\psi\|$ with $c < 0$ services as the adaptive balloon force

[2] We use the distance preserving method [18] to avoid level set reinitialization.

such that the curves will go outward in the flatten region where the edge indicator $f(\cdot) \gg 0$. In columns 3 and 4 of Fig. 4, we show the comparison results of the GAC model and our method, where the corresponding initializations are illustrated in column 1. We also show the edge saliency map in column 2. One can see that the proposed model can successfully catch the desired boundaries. In each tubular subregions $\Re_i$, our method can find the robust and globally (w.r.t $\Re_i$) minimizing curve. In the column 3 of the GAC results, some portions of the contours leak outside the boundaries due to the constant c for the adaptive balloon force in Eq. (18). At the same time, some parts of the contours fall into unexpected local minimums that are inside the objects. We can claim that compared to the GAC model, the main advantages of the proposed method are the robust optimality and the use of the user-specified control points.

5 Conclusion

In this paper, we propose a new edge-based active contour model based on the Finsler Eikonal PDE. The basic idea is to convert the balloon regional term as a curve energy via an asymmetric Finsler metric including the anisotropic edge information. The proposed model is able to blend the benefits from the global optimality of minimal path framework, the efficiency of the fast marching method and the user intervention. Experiments show that our model indeed obtains promising results.

References

1. Kass, M., Witkin, A., Terzopoulos, D.: Snakes: active contour models. IJCV **1**(4), 321–331 (1988)
2. Caselles, V., Kimmel, R., Sapiro, G.: Geodesic active contours. IJCV **22**(1), 61–79 (1997)
3. Yezzi, A., Kichenassamy, S., Kumar, A., Olver, P., Tannenbaum, A.: A geometric snake model for segmentation of medical imagery. TMI **16**(2), 199–209 (1997)
4. Melonakos, J., Pichon, E., Angenent, S., Tannenbaum, A.: Finsler active contours. TPAMI **30**(3), 412–423 (2008)
5. Jbabdi, S., Bellec, P., et al.: Accurate anisotropic fast marching for diffusion-based geodesic tractography. JBI **2008**, 2 (2008)
6. Osher, S., Sethian, J.A.: Fronts propagating with curvature-dependent speed: algorithms based on Hamilton-Jacobi formulations. JCP **79**(1), 12–49 (1988)
7. Cohen, L.D., Kimmel, R.: Global minimum for active contour models: a minimal path approach. IJCV **24**(1), 57–78 (1997)
8. Appia, V., Yezzi, A.: Active geodesics: region-based active contour segmentation with a global edge-based constraint. In: Proceedings of ICCV 2011, pp. 1975–1980 (2011)
9. Mille, J., Bougleux, S., Cohen, L.: Combination of piecewise-geodesic paths for interactive segmentation. IJCV **112**(1), 1–22 (2014)
10. Chen, D., Mirebeau, J.M., Cohen, L.D.: Finsler geodesic evolution model for region-based active contours. In: Proceedings of BMVC 2016 (2016)

11. Cohen, L.D.: On active contour models and balloons. CVGIP Image Underst. **53**(2), 211–218 (1991)
12. Chen, D., Mirebeau, J.M., Cohen, L.D.: Global minimum for a Finsler elastica minimal path approach. IJCV **122**(3), 458–483 (2017)
13. Mirebeau, J.M.: Efficient fast marching with Finsler metrics. Numer. Math. **126**(3), 515–557 (2014)
14. Sethian, J.A., Vladimirsky, A.: Ordered upwind methods for static Hamilton-Jacobi equations: theory and algorithms. SINUM **41**(1), 325–363 (2003)
15. Chen, D., Cohen, L.D.: Fast asymmetric fronts propagation for image segmentation (2017, preprint)
16. Randers, G.: On an asymmetrical metric in the four-space of general relativity. Phys. Rev. **59**(2), 195 (1941)
17. Chan, T.F., Vese, L.A.: Active contours without edges. TIP **10**(2), 266–277 (2001)
18. Li, C., Xu, C., et al.: Distance regularized level set evolution and its application to image segmentation. TIP **19**(12), 3243–3254 (2010)

Automatic Differentiation of Non-holonomic Fast Marching for Computing Most Threatening Trajectories Under Sensors Surveillance

Jean-Marie Mirebeau[1(✉)] and Johann Dreo[2]

[1] University Paris-Sud, CNRS, University Paris-Saclay, Orsay, France
jean-marie.mirebeau@math.u-psud.fr
[2] THALES Research & Technology, Palaiseau, France
johann.dreo@thalesgroup.com

Abstract. We consider a two player game, where a first player has to install a surveillance system within an admissible region. The second player needs to enter the monitored area, visit a target region, and then leave the area, while minimizing his overall probability of detection. Both players know the target region, and the second player knows the surveillance installation details. Optimal trajectories for the second player are computed using a recently developed variant of the fast marching algorithm, which takes into account curvature constraints modeling the second player vehicle maneuverability. The surveillance system optimization leverages a reverse-mode semi-automatic differentiation procedure, estimating the gradient of the value function related to the sensor location in time $\mathcal{O}(N \ln N)$.

Keywords: Anisotropic fast-marching · Motion planning · Sensors placement · Game theory · Optimization

1 Introduction

This paper presents a proof of concept numerical implementation of a motion planning algorithm related to a two player game. A first player selects, within an admissible class Ξ, an integral cost function on paths, which takes into account their position, orientation, and possibly curvature. The second player selects a path, within an admissible class Γ, with prescribed endpoints and an intermediate keypoint. The players objective is respectively to maximize and minimize the path cost

$$\mathfrak{C}(\Xi, \Gamma) := \sup_{\xi \in \Xi} \inf_{\gamma \in \Gamma} \mathfrak{C}(\xi, \gamma), \quad \text{where } \mathfrak{C}(\xi, \gamma) := \int_0^{T(\gamma)} \mathcal{C}_\xi(\gamma(t), \gamma'(t), \gamma''(t))\, dt, \tag{1}$$

This work has been supported by the 662107-SWARMs-ECSEL-2014-1 European project.

F. Nielsen and F. Barbaresco (Eds.): GSI 2017, LNCS 10589, pp. 791–800, 2017.
https://doi.org/10.1007/978-3-319-68445-1_91

where the path γ is parametrized at unit Euclidean speed, and the final time $T(\gamma)$ is free. From a game theoretic point of view, this is a non-cooperative zero-sum game, where player Ξ has no information and player Γ has full information over the opponent's strategy.

The game (1) typically models a surveillance problem [17], and $\exp(-\mathfrak{C}(\Xi,\Gamma))$ is the probability for player Γ to visit a prescribed keypoint without being detected by player Ξ. For instance player Ξ is responsible for the installation of radar [1] or sonar detection systems [17], and would like to prevent vehicles sent by player Γ from spying on some objectives without being detected.

The dependence of the cost $\mathcal{C}_\xi$ w.r.t. the path tangent $\gamma'(t)$ models the variation of a measure of how detectable the target is (radar cross section, directivity index, etc.) w.r.t. the relative positions and orientations of the target and sensor. The dependence of $\mathcal{C}_\xi$ on the path curvature $\gamma''(t)$ models the airplane maneuverability constraints, such as the need to slow down in tight turns [9], or even a hard bound on the path curvature [8].

Strode [17] has shown the interplay of motion planning and game theory in a similar setting, on a multistatic sonar network use case, but using isotropic graph-based path planning. The same year, Barbaresco [2] used fast-marching for computing threatening paths toward a single radar, but without taking into account curvature constraints and without considering a game setting.

The main contributions of this paper are as follows:

1. *Anisotropy and curvature penalization:* Strategy optimization for player Γ is an optimal motion planning problem, with a known cost function. This is addressed by numerically solving a generalized eikonal PDE posed on a two or three dimensional domain, and which is strongly anisotropic in the presence of a curvature penalty and a detection measurement that depends on orientation. A Fast-Marching algorithm, relying on recent adaptive stencils constructions, based on tools from lattice geometry, is used for that purpose [9,12,13]. In contrast, the classical fast marching method [14] used in [5] is limited to cost functions $\mathcal{C}_\xi(\gamma(t))$ independent of the path orientation $\gamma'(t)$ and curvature $\gamma''(t)$.
2. *Gradient computation for sensors placement:* Strategy optimization for player Ξ is typically a non-convex problem, to which various strategies can be applied, yet gradient information w.r.t. the variable $\xi \in \Xi$ is usually of help. For that purpose, we implement efficient differentiation algorithms, forward and reverse, for estimating the gradient of the value function of player Ξ

$$\nabla_\xi \mathfrak{C}(\xi,\Gamma), \qquad \text{where } \mathfrak{C}(\xi,\Gamma) := \inf_{\gamma\in\Gamma} \mathfrak{C}(\xi,\gamma). \tag{2}$$

Reverse mode differentiation reduced the computation cost of $\nabla_\xi \mathfrak{C}(\xi,\Gamma)$ from $\mathcal{O}(N^2)$, as used in [5], to $\mathcal{O}(N \ln N)$, where N denotes the number of discretization points of the domain. As a result, we can reproduce examples from [5] with computation times reduced by several orders of magnitude, and address complex three dimensional problems.

Due to space constraints, this paper is focused on problem modeling and numerical experiments, rather than on mathematical aspects of wellposedness and convergence analysis. Free and open source codes for reproducing (part of) the presented numerical experiments are available on the first author's webpage[1].

2 Mathematical Background of Trajectory Optimization

We describe in this section the PDE formalism, based on generalized eikonal equations, used to compute the value function $\min_{\gamma \in \Gamma} \mathfrak{C}(\xi, \gamma)$ of the second player, where ξ is known and fixed. Their discretization is discussed in Sect. 3. We distinguish two cases, depending on whether the path local cost function $\mathcal{C}_\xi(x, \dot{x}, \ddot{x})$ appearing in (1) depends on the last entry $\ddot{x}$, i.e. on path curvature.

2.1 Curvature Independent Cost

Let $\Omega \subset \mathbb{E} := \mathbb{R}^2$ be a bounded domain, and let the source set Υ and target set Θ be disjoint subsets of Ω. For each $x \in \Omega$, let Γ_x denote the set of all paths $\gamma \in C^1([0,T], \Omega)$, where $T = T(\gamma)$ is free, such that $\gamma(0) \in \Upsilon$, $\gamma(T) = x$ and $\forall t \in [0,T]$, $\|\gamma'(t)\| = 1$. The problem description states that the first player needs to go from Υ to Θ and back, hence its set of strategies is $\Gamma = \bigcup_{x \in \Theta} \Gamma_x^+ \times \Gamma_x^-$, where $\Gamma_x^+ = \Gamma_x^- := \Gamma_x$, and

$$\mathfrak{C}(\xi, \Gamma) = \inf_{x \in \Theta} u_\xi^+(x) + u_\xi^-(x), \qquad \text{where } u_\xi^\pm(x) := \inf_{\gamma \in \Gamma_x^\pm} \mathfrak{C}^\pm(\xi, \gamma). \tag{3}$$

Here and below, the symbol "±" must be successively replaced with "+" and then "−". We denoted by $\mathfrak{C}^\pm$ the path cost defined in terms of the local cost $\mathcal{C}_\xi(x, \pm\dot{x})$. In practice though, we only consider symmetric local costs, obeying $\mathcal{C}_\xi(x, \dot{x}) = \mathcal{C}_\xi(x, -\dot{x})$, hence the forward and return paths are identical and we denote $u_\xi := u_\xi^+ = u_\xi^-$. Define the 1-homogenous metric $\mathcal{F}_\xi : \Omega \times \mathbb{E} \to [0, \infty]$, the Lagrangian $\mathcal{L}_\xi$ and the Hamiltonian $\mathcal{H}_\xi$ by

$$\mathcal{F}_\xi(x, \dot{x}) := \|\dot{x}\| \mathcal{C}_\xi(x, \dot{x}/\|\dot{x}\|), \quad \mathcal{L}_\xi := \frac{1}{2}\mathcal{F}_\xi^2, \quad \mathcal{H}_\xi(x, \hat{x}) := \sup_{\dot{x} \in \mathbb{E}} \langle \hat{x}, \dot{x} \rangle - \mathcal{L}_\xi(x, \dot{x}).$$

Here and below, symbols denoting tangent vectors are distinguished with a "dot", e.g. $\dot{x}$, and co-vectors with a "hat", e.g. $\hat{x}$. Under mild assumptions [3], the function $u_\xi : \Omega \to \mathbb{R}$ is the unique viscosity solution to a generalized eikonal equation $\forall x \in \Omega \setminus \Upsilon$, $\mathcal{H}_\xi(x, \nabla_x u_\xi(x)) = 1/2$, $\forall x \in \Upsilon$, $u_\xi(x) = 0$, with out-flow boundary conditions on $\partial\Omega$. The discretization of this PDE is discussed in Sect. 3. We limit in practice our attention to Isotropic costs $\mathcal{C}_\xi(x)$, and Riemannian costs $\mathcal{C}_\xi(x, \dot{x}) = \sqrt{\langle \dot{x}, M_\xi(x) \dot{x} \rangle}$ where $M_\xi(x)$ is symmetric positive definite, for which efficient numerical strategies have been developed [11,12].

[1] github.com/Mirebeau/HamiltonianFastMarching.

2.2 Curvature Dependent Cost

Let $\Omega \subset \mathbb{R}^2 \times \mathbb{S}^1$ be a bounded domain, within the three dimensional space of all positions and orientations. As before, let $\Upsilon, \Theta \subset \Omega$. For all $x \in \Omega$ let $\Gamma_x^\pm$ be the collection of all $\gamma \in C^2([0,T],\Omega)$, such that $\eta := (\gamma, \pm\gamma')$ satisfies $\eta(0) \in \Upsilon$, $\eta(T) = x$ and $\forall t \in [0,T]$, $\|\gamma'(t)\| = 1$. Since the first player needs to go from Υ to Θ and back, its set of strategies is $\Gamma = \bigcup_{x\in\Theta} \Gamma_x^+ \times \Gamma_x^-$. Equation (3) holds, where $\mathfrak{C}^\pm$ denotes the path cost defined in terms of the local cost $\mathcal{C}_\xi(p, \pm\dot{p}, \ddot{p})$.

Consider the 1-homogeneous metric $\mathcal{F}_\xi^\pm : \mathrm{T}\Omega \to [0,\infty]$, defined on the tangent bundle to $\Omega \subset \mathbb{R}^2 \times \mathbb{S}^1$ by

$$\mathcal{F}_\xi^\pm((p,n),(\dot{p},\dot{n})) := \begin{cases} +\infty & \text{if } \dot{p} \neq \|\dot{p}\|n, \\ \|\dot{p}\|\mathcal{C}_\xi(p,n,\pm\dot{n}/\|\dot{p}\|) & \text{else,} \end{cases}$$

where $p \in \mathbb{R}^2$, $n \in \mathbb{S}^1$ is a unit vector, and the tangent vector satisfies $\dot{p} \in \mathbb{R}^2$, $\dot{n} \perp n$. This choice is motivated by the fact that $\int_0^T \mathcal{F}_\xi^\pm(\eta(t),\eta'(t))dt$ is finite iff $\eta : [0,T] \to \Omega$ is of the form $(\gamma, \pm\gamma')$, and then it equals $\int_0^T \mathcal{C}_\xi(\gamma(t), \pm\gamma'(t), \gamma''(t))dt$.

Introducing the Lagrangian $\mathcal{L}_\xi^\pm = \frac{1}{2}(\mathcal{F}_\xi^\pm)^2$ on $\mathrm{T}\Omega$, and its Legendre-Fenchel dual the Hamiltonian $\mathcal{H}_\xi^\pm$, one can again under mild assumptions characterize $u_\xi^\pm$ as the unique viscosity solution to the generalized eikonal PDE $\mathcal{H}_\xi^\pm(x, \nabla u_\xi^\pm(x)) = 1/2$ with appropriate boundary conditions [3]. In practice, we choose cost functions of the form $\mathcal{C}_\xi(p,\dot{p},\ddot{p}) = \mathcal{C}_\xi^\circ(p,\dot{p})\mathcal{C}_\star(|\ddot{p}|)$, where $\mathcal{C}_\star$ is the Reeds-Shepp car or Dubins car [8] curvature penalty, with respective labels $\star =$ RS and D, namely

$$\mathcal{C}_{\mathrm{RS}}(\kappa) := \sqrt{1+\rho^2\kappa^2}, \qquad \mathcal{C}_{\mathrm{D}}(\kappa) := \begin{cases} 1 & \text{if } |\rho\kappa| \leq 1, \\ +\infty & \text{otherwise,} \end{cases}$$

where $\rho > 0$ is a parameter which has the dimension of a curvature radius. The Dubins car can only follow paths which curvature radius is $\leq \rho$, whereas the Reeds-Shepp car (in the sense of [9] and without reverse gear), can rotate into place if needed. The Hamiltonian then has the explicit expression $\mathcal{H}((p,n),(\hat{p},\hat{n})) = \frac{1}{2}\mathcal{C}_\xi^0(p,n)^{-2}\mathcal{H}_*(n,(\hat{p},\hat{n}))$ where $\mathcal{H}_{\mathrm{RS}} = \frac{1}{2}(\langle\hat{p},n\rangle_+^2 + \|\hat{n}/\rho\|^2)$ and $\mathcal{H}_{\mathrm{D}} = \frac{1}{2}\max\{0, \langle\hat{p},n\rangle + \|\hat{n}/\rho\|\}^2$.

3 Discretization of Generalized Eikonal Equations

We construct a discrete domain X by intersecting the computational domain with an orthogonal grid of scale $h > 0$: $X = \Omega \cap (h\mathbb{Z})^d$, where $d = 2$ for the curvature independent models, and $d = 3$ for the other models which are posed on $\mathbb{R}^2 \times (\mathbb{R}/2\pi\mathbb{Z})$ —using the angular parametrization $\mathbb{S}^1 \cong \mathbb{R}/2\pi\mathbb{Z}$ (in the latter periodic case, $2\pi/h$ must be an integer). We design weights $c_\xi(x,y)$, $x,y \in X$ such that for any tangent vector $\dot{x}$ at $x \in \Omega$ one has

$$\mathcal{H}_\xi(x,\dot{x}) \approx h^{-2} \sum_{y\in X} c_\xi^2(x,y)\langle x-y, \dot{x}\rangle_+^2, \tag{4}$$

where $a_+ := \max\{0, a\}$ (expression (4) is typical, although some models require a slight generalization). The weights $c_\xi(x,y)$ are non-zero for only few $(x,y) \in X$ at distance $\|x-y\| = \mathcal{O}(h)$. Their construction exploits the additive structure of the discretization grid X and relies on techniques from lattice geometry [16], see [9,12,13] for details. The generalized eikonal PDE $\mathcal{H}_\xi(x, \nabla_x u_\xi(x)) = 1/2$, which solution $u_\xi(x)$ should be regarded as a distance map, is discretized as

$$\sum_{y \in X} c_\xi^2(x,y)(U_\xi(x) - U_\xi(y))_+^2 = h^2/2, \tag{5}$$

with adequate boundary conditions. The solution $U_\varepsilon : X \to \mathbb{R}$ to this system of equations is computed in a single pass with $\mathcal{O}(N \ln N)$ complexity [14], using a variant of the Fast-Marching algorithm. This is possible since the l.h.s. of (5) is a *non-decreasing function* of the *positive parts* of the finite differences $(U_\xi(x) - U_\xi(y))_{y \in X}$. Note that the eikonal PDE discretization (5), based on upwind finite differences, differs from the semi-Lagrangian approach [15], which can also be solved in a single pass but is usually less efficient due to the large cardinality and radius of its stencils. Image segmentation techniques relying on the numerical solutions to anisotropic eikonal PDEs were proposed in [6] using Riemannian metrics, and in [4,7] based on the reversible Reeds-Shepp car and Euler elastica curvature penalized models respectively. However these early works rely on non-causal discretizations, which have super-linear complexity $\mathcal{O}(N^{1+1/d})$ where the unspecified constant is large for strongly anisotropic and non-uniform metrics. This alternative approach yields (much) longer solve times, incompatible our application - where strongly anisotropic three dimensional eikonal PDEs are solved as part of an inner loop of an optimization procedure.

To be able to use the gradient to solve the problem (1), we need to differentiate the cost $\mathfrak{C}(\xi, \Gamma)$ w.r.t. the first player strategy $\xi \in \Xi$. In view of (3), this only requires the sensitivity of the discrete solution values $U_\xi(x_*)$ at the few points $x_* \in X \cap \Theta$, w.r.t to variations in the weights $c_\xi(x,y)$, $x, y \in X$. For that purpose we differentiate (5) w.r.t. ξ at an arbitrary point $x \in X \setminus \Upsilon$, and obtain

$$\sum_{y \in X} \omega_\xi(x,y)\left(\mathrm{d}U_\xi(x) - \mathrm{d}U_\xi(y) + (U_\xi(x) - U_\xi(y))\, \mathrm{d} \ln c_\xi(x,y)\right) = 0,$$

where $\omega_\xi(x,y) := c_\xi^2(x,y)(U_\xi(x) - U_\xi(y))_+$. Therefore

$$\mathrm{d}U_\xi(x) = \sum_{y \in X} \alpha_\xi(x,y) \mathrm{d}U_\xi(y) + \sum_{y \in X} \beta_\xi(x,y) \mathrm{d}c_\xi(x,y), \tag{6}$$

where $\alpha_\xi(x,y) := \omega_\xi(x,y)/\sum_y \omega_\xi(x,y)$, and $\beta_\xi(x,y) := \alpha_\xi(x,y)/c_\xi(x,y)$. We first choose $x = x_*$ in (6), and then recursively eliminate the terms $\mathrm{d}U_\xi(y)$ by applying the same formula at these points, except for points in the source set $y \in \Upsilon$ for which one uses the explicit expression $\mathrm{d}U_\xi(y) = 0$ (since $U_\xi(y) = 0$ is in this case independent of ξ). This procedure terminates: indeed, whenever $\mathrm{d}U_\xi(x)$ depends on $\mathrm{d}U_\xi(y)$ in (6), one has $\alpha_\xi(x,y) > 0$, thus $\omega(x,y) > 0$, hence $U_\xi(x) > U_\xi(y)$. It is closely related to automatic differentiation by reverse accumulation [10], and has the modest complexity $\mathcal{O}(N)$.

4 Numerical Results

The chosen physical domain R is the rectangle $[0,2]\times[0,1]$ minus some obstacles, as illustrated on Fig. 1. Source point is $(0.2, 0.5)$ and target keypoint $(1.8, 0.5)$. The computational domain is thus $\Omega = R$ for curvature independent models and $\Omega = R \times \mathbb{S}^1$ for curvature dependent models, which is discretized on a 180×89 or $180 \times 89 \times 60$ grid.

No intervention from the first player. The cost function is $\mathcal{C}_\xi(p, \dot{p}, \ddot{p}) = \mathcal{C}_*(|\ddot{p}|)$, where $\mathcal{C}_*(\kappa)$ is respectively 1, $\sqrt{1+\rho^2\kappa^2}$ and (1 if $\rho\kappa \le 1$, otherwise $+\infty$), with $\rho := 0.3$. The differences between the three models are apparent: the curvature independent model uses the same path forward and back; the Reeds-Shepp car spreads some curvature along the way but still makes an angle at the target point; the Dubins car maintains the radius of curvature below the bound ρ, and its trajectory is a succession of straight and circular segments. A referee notes that following an optimal trajectory for the Dubins model is dangerous in practice, since any small deviation is typically impossible to correct locally, and may drive into an obstacle; these trajectories are also easier to detect due to the large circular arc motions.

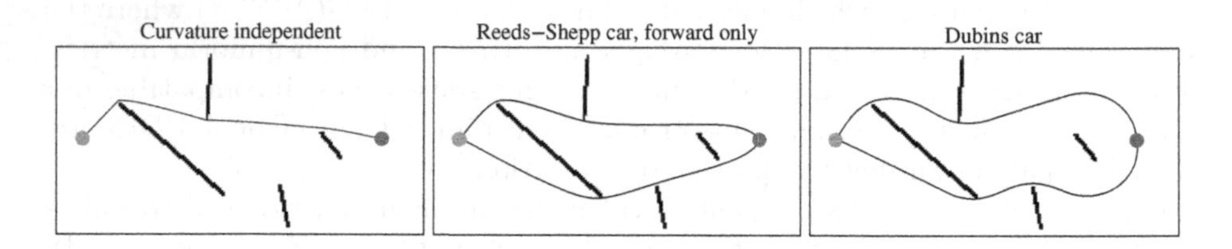

Fig. 1. Shortest path from the blue point (left) to the red keypoint (right) and back. (Color figure online)

Next we study three games where player one aims to detect player two along its way from the source set Υ to the target Θ and back, using different means. If the first player does not intervene, see Fig. 1, or if its strategy is not optimized, see Fig. 3, then there is typically a unique optimal path (optimal loop in our games) for player two. In contrast, an interesting qualitative property of the *optimal* strategy $\xi \in \Xi$ for the first player is that it has a large number of optimal responses from player two, see Fig. 4, in some cases even a continuum, see Fig. 2 (bottom) and [5]. This is typical of two player games.

Fresh paint based detection. In this toy model, see Fig. 2, the first player spreads some fresh paint over the domain, and the second player is regarded as detected if he comes back covered in it from his visit to the keypoint. The cost function is $\mathcal{C}_\xi(p, \dot{p}, \ddot{p}) = \xi(p)\mathcal{C}_*(|\ddot{p}|)$, where $\xi : R \to \mathbb{R}_+$ is the fresh paint density, decided by the first player, and $\mathcal{C}_*(\kappa)$ is as above. For wellposedness, we impose upper

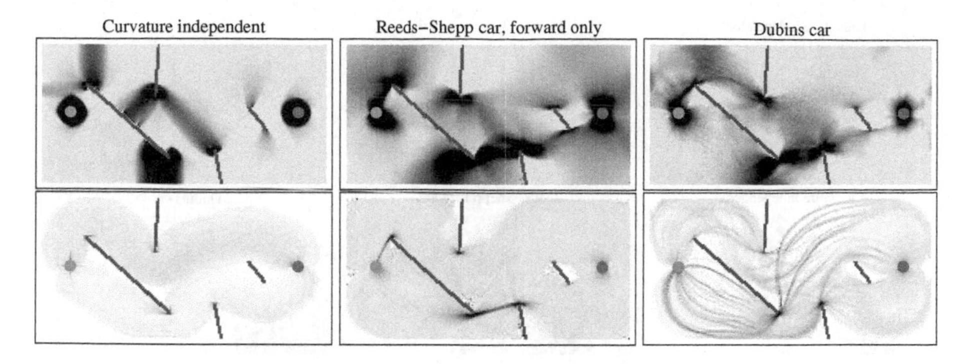

Fig. 2. Top: Optimal distribution of paint, to mark a path from the blue point (left) to the red keypoint (right) and back. Bottom: Geodesic density at the optimal paint distribution. (Color figure online)

and lower bounds on the paint density, namely $0.1 \leq \xi(p) \leq 1$, and subtract the paint supply cost $\int_R \xi(p)\mathrm{d}p$ to (1). The main interest of this specific game, also considered in [5], is that $\mathfrak{C}(\xi, \Gamma)$ is concave w.r.t. $\xi \in \Xi$. The observed optimal strategy for player Ξ is in the curvature independent case to make some "fences" of paint between close obstacles, and in the curvature penalized models to deposit paint at the edges of obstacles, as well as along specific circular arcs for the Dubins model.

Visual detection. The first player places some cameras, e.g. with 360-degree field of view and mounted at the ceiling, which efficiency at detecting the second player decreases with distance and is blocked by obstacles, see Fig. 3. The cost function is

$$\mathcal{C}_\xi(p, \dot{p}, \ddot{p}) = \mathcal{C}_*(\kappa) \sum_{\substack{q \in \xi \\ [p,q] \subset R}} \frac{1}{\|q - p\|^2}, \tag{7}$$

where $\xi \in \Xi$ is a subset of R with prescribed cardinality, two in our experiments. The green arrows on Fig. 3 originate from the current (non optimal) camera position, and point in the direction of greatest growth $\nabla \mathfrak{C}(\xi, \Gamma)$ for the first player objective function.

Radar based detection. The first player places some radars on the domain $R = [0, 2] \times [0, 1]$, here devoid of obstacles, and the second player has to fly by undetected. The cost function is

$$\mathcal{C}_\xi(p, \dot{p}, \ddot{p}) = \mathcal{C}_*(|\ddot{p}|) \sqrt{\sum_{q \in \xi} \frac{\langle \dot{p}, n_{\boldsymbol{pq}} \rangle^2 + \delta^2 \langle \dot{p}, n_{\boldsymbol{pq}}^\perp \rangle}{\|p - q\|^4}} \tag{8}$$

where $n_{\boldsymbol{pq}} := (q - p)/\|q - p\|$. The first player strategy ξ contains the positions of three radars, constrained to lie in the subdomain $[0.4, 1.6] \times [0, 1]$. The parameter

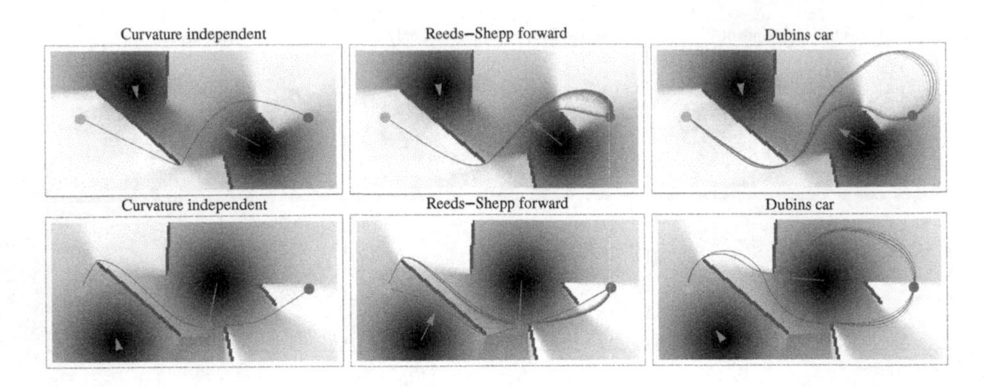

Fig. 3. Field of view of the cameras (black gradients), optimal furtive paths (red lines), local direction of improvement of the camera position (green arrows). (Color figure online)

δ is set to 1 for an isotropic radar cross section (RCS), or to 0.2 for an anisotropic RCS. In the latter case a plane showing its side to radar is five times less likely to be detected than a plane showing its nose or back, at the same position. Green arrows on Fig. 4 point from the original position to the (locally) optimized position for player Ξ. At this position, several paths are optimal for player Γ, shown in red on Fig. 4.

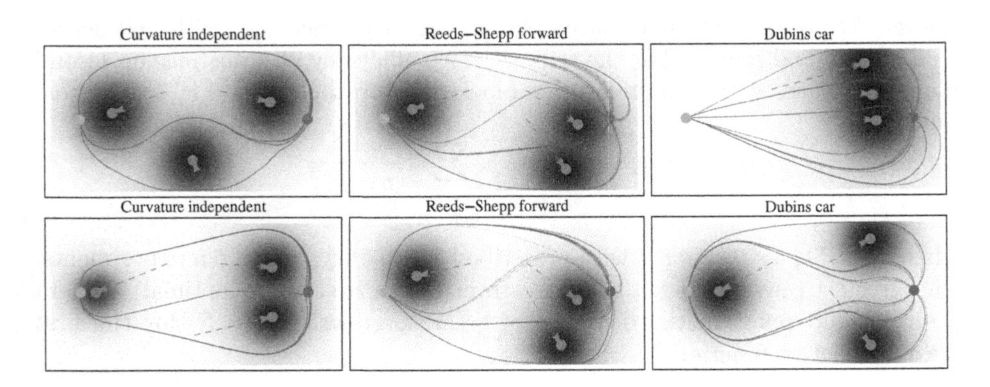

Fig. 4. Optimal radar placement with an isotropic (top) or anisotropic (bottom) radar cross section.

Computational cost. On a standard Laptop computer (2.7 Ghz, 16 GB ram), optimizing the second player objective, by solving a generalized eikonal equation, takes ≈1 s in the curvature dependent case, and ≈ 60 times less in the curvature independent case thanks to the absence of angular discretization of the domain. Optimizing the first player objective takes ≈100 *L-BFGS* iterations, each one taking at most 8 s. For the stability of the minimization procedure, the problems considered were slightly regularized by the use of soft-minimum functions and by "blurring" the target keypoint over the 3×3 box of adjacent pixels.

5 Conclusion

We have modeled a motion planning problem that minimize an anisotropic probability of detection, taking into account navigation constraints while computing the gradient of the value function related to the sensors location. This model is thus useful for surveillance applications modeled as a two-player zero-sum game involving a target that tries to avoid detection.

References

1. Barbaresco, F., Monnier, B.: Minimal geodesics bundles by active contours: radar application for computation of most threathening trajectories areas & corridors. In: 10th European Signal Processing Conference, Tampere, pp. 1–4 (2000)
2. Barbaresco, F.: Computation of most threatening radar trajectories areas and corridors based on fast-marching & Level Sets. In: IEEE Symposium on Computational Intelligence for Security and Defense Applications (CISDA), Paris, pp. 51–58 (2011)
3. Bardi, M., Capuzzo-Dolcetta, I.: Optimal Control and Viscosity Solutions of Hamilton-Jacobi-Bellman Equations. Bikhauser, Bosto (1997)
4. Bekkers, E.J., Duits, R., Mashtakov, A., Sanguinetti, G.R.: Data-driven sub-riemannian geodesics in SE(2). In: Aujol, J.-F., Nikolova, M., Papadakis, N. (eds.) SSVM 2015. LNCS, vol. 9087, pp. 613–625. Springer, Cham (2015). doi:10.1007/978-3-319-18461-6_49
5. Benmansour, F., Carlier, G., Peyré, G., Santambrogio, F.: Derivatives with respect to metrics and applications: subgradient marching algorithm. Numer. Math. **116**(3), 357–381 (2010)
6. Benmansour, F., Cohen, L.: Tubular structure segmentation based on minimal path method and anisotropic enhancement. Int. J. Comput. Vis. **92**(2), 192–210 (2011)
7. Chen, D., Mirebeau, J.-M., Cohen, L.D.: A new finsler minimal path model with curvature penalization for image segmentation and closed contour detection. In: Proceedings of CVPR 2016, Las Vegas, USA (2016)
8. Dubins, L.E.: On curves of minimal length with a constraint on average curvature, and with prescribed initial and terminal positions and tangents. Amer. J. Math. **79**, 497–516 (1957)
9. Duits, R., Meesters, S.P.L., Mirebeau, J.-M., Portegies, J.M.: Optimal Paths for Variants of the 2D and 3D Reeds-Shepp. Car with Applications in Image Analysis (Preprint available on arXiv)
10. Griewank, A., Walther, A.: Evaluating Derivatives: Principles and Techniques of Algorithmic Differentiation. Society for Industrial and Applied Mathematics, Philadelphia (2008)
11. Mirebeau, J.-M.: Anisotropic Fast-Marching on cartesian grids using Lattice Basis Reduction. SIAM J. Numer. Anal. **52**(4), 1573–1599 (2014)
12. Mirebeau, J.-M.: Anisotropic fast-marching on cartesian grids using Voronois first reduction of quadratic forms (preprint available on HAL)

13. Mirebeau, J.-M.: Fast Marching methods for Curvature Penalized Shortest Paths (preprint available on HAL)
14. Rouy, E., Tourin, A.: A viscosity solutions approach to shape-from-shading. SIAM J. Numer. Anal. **29**(3), 867–884 (1992)
15. Sethian, J., Vladimirsky, A.: Ordered upwind methods for static Hamilton-Jacobi equations. Proc. Natl. Acad. Sci. **98**(20), 11069–11074 (2001)
16. Schürmann, A.: Computational geometry of positive definite quadratic forms, University Lecture Series (2009)
17. Strode, C.; Optimising multistatic sensor locations using path planning and game theory. In: IEEE Symposium on Computational Intelligence for Security and Defense Applications (CISDA), Paris, pp. 9–16 (2011)

On the Existence of Paths Connecting Probability Distributions

Rui F. Vigelis[1(✉)], Luiza H.F. de Andrade[2], and Charles C. Cavalcante[3]

[1] Computer Engineering, Campus Sobral, Federal University of Ceará, Sobral, CE, Brazil
rfvigelis@ufc.br

[2] Center of Exact and Natural Sciences, Federal Rural University of the Semi-arid Region, Mossoró, RN, Brazil
luizafelix@ufersa.edu.br

[3] Department of Teleinformatics Engineering, Wireless Telecommunication Research Group, Federal University of Ceará, Fortaleza, CE, Brazil
charles@ufc.br

Abstract. We introduce a class of paths defined in terms of two deformed exponential functions. Exponential paths correspond to a special case of this class of paths. Then we give necessary and sufficient conditions for any two probability distributions being path connected.

1 Introduction

In Non-parametric Information Geometry, many geometric structures can be defined in terms of paths connecting probability distributions. It is shown in [1,5,6] that two probability distributions are connected by an open exponential path (or exponential arc) if and only if they belong to the same exponential family. Exponential paths are the auto-parallel curves w.r.t. the exponential connection. Using deformed exponential functions, we can define an analogue version of exponential paths. A *deformed exponential* $\varphi\colon \mathbb{R} \to [0, \infty)$ is a convex function such that $\lim_{u\to-\infty} \varphi(u) = 0$ and $\lim_{u\to\infty} \varphi(u) = \infty$. Eguchi and Komori in [3] introduced and investigated a class of paths defined in terms of deformed exponential functions. In the present paper we extend the definition of paths given in [3], and then we show equivalent conditions for any two probability distributions being path connected.

Throughout the text, (T, Σ, μ) denotes the σ-finite measure space on which probability distributions (or probability density functions) are defined. All probability distributions are assumed to have positive density w.r.t. the underlying measure μ. In other words, they belong to the collection $\mathcal{P}_\mu = \{p \in L^0 : \int_T p d\mu = 1 \text{ and } p > 0\}$, where L^0 is the space of all real-valued, measurable functions on T, with equality μ-a.e.

Let us fix a positive, measurable function $u_0\colon T \to (0, \infty)$. Given two probability distributions p and q in $\mathcal{P}_\mu$, a *φ_1/φ_2-path* (or *φ_1/φ_2-arc*) is a curve

F. Nielsen and F. Barbaresco (Eds.): GSI 2017, LNCS 10589, pp. 801–808, 2017.
https://doi.org/10.1007/978-3-319-68445-1_92

in $\mathcal{P}_\mu$ defined by $\alpha \mapsto \varphi_1(\alpha\varphi_2^{-1}(p) + (1-\alpha)\varphi_2^{-1}(q) + \kappa(\alpha)u_0)$. The constant $\kappa(\alpha) := \kappa(\alpha; p, q) \in \mathbb{R}$ is introduced so that

$$\int_T \varphi_1(\alpha\varphi_2^{-1}(p) + (1-\alpha)\varphi_2^{-1}(q) + \kappa(\alpha)u_0)d\mu = 1. \tag{1}$$

We use the word "φ-path" in the place of "φ/φ-path" (i.e., if $\varphi_1 = \varphi_2 = \varphi$). The case where $\varphi_1 = \varphi_2 = \varphi$ and $u_0 = 1$ was analyzed by Eguchi and Komori in [3]. Exponential paths correspond to $\varphi_1(\cdot)$ and $\varphi_2(\cdot)$ equal to $\exp(\cdot)$, and $u_0 = 1$. A φ_1/φ_2-path can be seen as a φ_1-path connecting $\varphi_1(\varphi_2^{-1}(p) + \kappa(1)u_0)$ and $\varphi_1(\varphi_2^{-1}(q) + \kappa(0)u_0)$. Unless $\varphi_1 = \varphi_2 = \varphi$, a φ_1/φ_2-path does not connect p and q. We can use $\kappa(\alpha)$ to define the divergence

$$\mathcal{D}^{(\alpha)}(p \parallel q) = -\frac{1}{\alpha}\kappa(0) - \frac{1}{1-\alpha}\kappa(1) + \frac{1}{\alpha(1-\alpha)}\kappa(\alpha).$$

This divergence for $\varphi_1 = \varphi_2 = \varphi$ is related to a generalization of Rényi divergence, which was introduced by the authors in [2]. If $\varphi_1(\cdot)$ and $\varphi_2(\cdot)$ are equal to $\exp(\cdot)$, and $u_0 = 1$, then $\mathcal{D}^{(\alpha)}(\cdot \parallel \cdot)$ reduces to Rényi divergence.

The main goal of this notes is to give necessary an sufficient conditions for the existence of $\kappa(\alpha)$ in (1) for every $p, q \in \mathcal{P}_\mu$ and $\alpha \in [0,1]$.

Proposition 1. *Assume that the measure μ is non-atomic. Let $\varphi_1, \varphi_2 \colon \mathbb{R} \to (0,\infty)$ be two positive, deformed exponential functions, and let $u_0 \colon T \to (0,\infty)$ be a positive, measurable function. Fix any $\alpha \in (0,1)$. For every pair of probability distributions p and q in $\mathcal{P}_\mu$, there exists a constant $\kappa(\alpha) := \kappa(\alpha; p, q)$ satisfying* (1) *if, and only if,*

$$\int_T \varphi_1(c + \lambda u_0)d\mu < \infty, \qquad \textit{for all } \lambda \geq 0, \tag{2}$$

for each measurable function $c \colon T \to \mathbb{R}$ satisfying $\int_T \varphi_2(c)d\mu < \infty$.

A proof of this proposition is shown in the next section. Using some results involved in the proof of Proposition 1, we give an equivalent criterion for the existence of u_0 satisfying condition (2) for $\varphi_1 = \varphi_2 = \varphi$. As consequence, there may exist functions $\varphi_1 = \varphi_2 = \varphi$ for which we cannot find u_0 satisfying (2), a result which was shown in [2] (Example 2).

2 Results

We begin by showing an equivalent criterion for condition (2).

Proposition 2. *Two deformed exponential functions $\varphi_1, \varphi_2 \colon \mathbb{R} \to [0,\infty)$ and a measurable function $u_0 \colon T \to (0,\infty)$ satisfy condition* (2) *if, and only if, for each $\lambda > 0$, we can find $\alpha \in (0,1)$ and a measurable function $c \colon T \to \mathbb{R} \cup \{-\infty\}$ such that $\int_T \varphi_1(c)d\mu < \infty$ and*

$$\alpha\varphi_1(u) \leq \varphi_2(u - \lambda u_0(t)), \qquad \textit{for all } u \geq c(t), \tag{3}$$

for μ-a.e. $t \in T$.

The proof of Proposition 2 requires a preliminary result.

Lemma 1. *Suppose that, for each $\lambda > 0$, we cannot find $\alpha \in (0,1)$ and a measurable function $c\colon T \to \mathbb{R} \cup \{-\infty\}$ such that $\int_T \varphi_1(c)d\mu < \infty$ and*

$$\alpha\varphi_1(u) \le \varphi_2(u - \lambda u_0(t)), \qquad \text{for all } u \ge c(t). \tag{4}$$

Then there exist sequences $\{\lambda_n\}$, $\{c_n\}$ and $\{A_n\}$ of positive numbers $\lambda_n \downarrow 0$, measurable functions, and pairwise disjoint, measurable sets, respectively, such that

$$\int_{A_n} \varphi_1(c_n)d\mu = 1 \quad \text{and} \quad \int_{A_n} \varphi_2(c_n - \lambda_n u_0)d\mu \le 2^{-n}, \quad \text{for all } n \ge 1. \tag{5}$$

Proof. Let $\{\lambda'_m\}$ be a sequence of positive numbers $\lambda'_m \downarrow 0$. For each $m \ge 1$, we define the function

$$f_m(t) = \sup\{u \in \mathbb{R} : 2^{-m}\varphi_1(u) > \varphi_2(u - \lambda'_m u_0(t))\},$$

where we use the convention $\sup\emptyset = -\infty$. We will verify that f_m is measurable. For each rational number r, define the measurable sets

$$E_{m,r} = \{t \in T : 2^{-m}\varphi_1(r) > \varphi_2(r - \lambda'_m u_0(t))\}$$

and the simple functions $u_{m,r} = r\chi_{E_{m,r}}$. Let $\{r_i\}$ be an enumeration of the rational numbers. For each $m, k \ge 1$, consider the non-negative, simple functions $v_{m,k} = \max_{1\le i\le k} u_{m,r_i}$. Moreover, denote $B_{m,k} = \bigcup_{i=1}^k E_{m,r_i}$. By the continuity of $\varphi_1(\cdot)$ and $\varphi_2(\cdot)$, it follows that $v_{m,k}\chi_{B_{m,k}} \uparrow f_m$ as $k \to \infty$, which shows that f_m is measurable. Since (4) is not satisfied, we have that $\int_T \varphi_1(f_m)d\mu = \infty$ for all $m \ge 1$. In virtue of the Monotone Convergence Theorem, for each $m \ge 1$, we can find some $k_m \ge 1$ such that the function $v_m = v_{m,k_m}$ and the set $B_m = B_{m,k_m}$ satisfy $\int_{B_m} \varphi_1(v_m)d\mu \ge 2^m$. Clearly, we have that $\varphi_1(v_m)\chi_{B_m} < \infty$ and $2^{-m}\varphi_1(v_m)\chi_{B_m} \ge \varphi_2(v_m - \lambda'_m u_0)\chi_{B_m}$. By Lemma 8.3 in [4], there exist an increasing sequence $\{m_n\}$ of indices and a sequence $\{A_n\}$ of pairwise disjoint, measurable sets such that $\int_{A_n} \varphi_1(v_{m_n})d\mu = 1$. Clearly, $\int_{A_n} \varphi_2(v_{m_n} - \lambda'_{m_n} u_0)d\mu \le 2^{-m_n}$. Denoting $\lambda_n = \lambda'_{m_n}$, $c_n = v_{m_n}$, we obtain (5).

Proof (Proposition 2). Assume that $\varphi_1(\cdot)$, $\varphi_2(\cdot)$ and u_0 satisfy condition (2). Suppose that expression (3) does not hold. Let $\{\lambda_n\}$, $\{c_n\}$ and $\{A_n\}$ be as stated in Lemma 1. Then we define $c = c_0\chi_{T\setminus A} + \sum_{n=1}^\infty (c_n - \lambda_n u_0)\chi_{A_n}$, where $A = \bigcup_{n=1}^\infty A_n$ and $c_0\colon T \to \mathbb{R}$ is any measurable function such that $\int_{T\setminus A} \varphi_2(c_0)d\mu < \infty$. In view of (5), we have

$$\begin{aligned}\int_T \varphi_2(c)d\mu &= \int_{T\setminus A} \varphi_2(c_0)d\mu + \sum_{n=1}^\infty \int_{A_n} \varphi_2(c_n - \lambda_n u_0)d\mu \\ &\le \int_{T\setminus A} \varphi_2(c_0)d\mu + \sum_{n=1}^\infty 2^{-n} < \infty.\end{aligned}$$

Given any $\lambda > 0$, we take $n_0 \geq 1$ such that $\lambda \geq \lambda_n$ for all $n \geq n_0$. Then we can write

$$\begin{aligned}\int_T \varphi_1(c + \lambda u_0) d\mu &\geq \sum_{n=n_0}^{\infty} \int_{A_n} \varphi_1(c_n + (\lambda - \lambda_n) u_0) d\mu \\ &\geq \sum_{n=n_0}^{\infty} \int_{A_n} \varphi_1(c_n) d\mu = \sum_{n=1}^{\infty} 1 = \infty. \end{aligned} \tag{6}$$

which is a contradiction to condition (2).

Conversely, suppose that expression (3) holds for a given $\lambda > 0$. Let $\widetilde{c}\colon T \to \mathbb{R}$ be any measurable function satisfying $\int_T \varphi_2(\widetilde{c}) d\mu < \infty$. Denote $A = \{t : \widetilde{c}(t) + \lambda u_0 \geq c(t)\}$. We use inequality (3) to write

$$\begin{aligned}\alpha \int_T \varphi_1(\widetilde{c} + \lambda u_0) d\mu &\leq \alpha \int_A \varphi_1(\widetilde{c} + \lambda u_0) d\mu + \alpha \int_{T \setminus A} \varphi_1(c) d\mu \\ &\leq \int_A \varphi_2(\widetilde{c}) d\mu + \int_{T \setminus A} \varphi_2(c - \lambda u_0) d\mu < \infty. \end{aligned}$$

Thus, condition (2) follows.

Before we give a proof of Proposition 1, we show the following technical result.

Lemma 2. *Let $\varphi_1, \varphi_2\colon \mathbb{R} \to (0, \infty)$ be positive, deformed exponential functions, and $\widetilde{c}\colon T \to \mathbb{R}$ a measurable function such that $\int_A \varphi_i(\widetilde{c}) d\mu < 1$, for $i = 1, 2$, where A and $B = T \setminus A$ are measurable sets such that $\mu(A) > 0$ and $\mu(B) > 0$. Fix any $\alpha \in (0, 1)$. Then we can find measurable functions $b_1, b_2\colon T \to \mathbb{R}$ for which $p = \varphi_2(c_1)$ and $q = \varphi_2(c_2)$ are in $\mathcal{P}_\mu$, where $c_1 = \widetilde{c}\chi_A + b_1\chi_B$ and $c_2 = \widetilde{c}\chi_A + b_2\chi_B$, and*

$$\int_T \varphi_1(\alpha \varphi_2^{-1}(p) + (1 - \alpha) \varphi_2^{-1}(q)) < 1. \tag{7}$$

In addition, we assume $b_1\chi_B \neq b_2\chi_B$.

Proof. Let $\{B_n\}$ be a sequence of measurable sets such that $B = \bigcup_{n=1}^{\infty} B_n$ and $0 < \mu(B_n) < \infty$. For each $n \geq 1$, we select measurable sets C_n and D_n such that $B_n = C_n \cup D_n$ and $\mu(C_n) = \mu(D_n) = \frac{1}{2}\mu(B_n)$. Let $\{\gamma_n^{(1)}\}$ and $\{\gamma_n^{(2)}\}$ be sequences of positive numbers satisfying

$$\sum_{n=1}^{\infty} \gamma_n^{(1)} < 1 - \int_A \varphi_1(\widetilde{c}) d\mu, \quad \text{and} \quad \sum_{n=1}^{\infty} \gamma_n^{(2)} = 1 - \int_A \varphi_2(\widetilde{c}) d\mu.$$

Then we take $\beta_n \in \mathbb{R}$ and $\theta_n > 0$ such that

$$\varphi_2(\beta_n) + \varphi_2(-\theta_n) = 2 \frac{\gamma_n^{(2)}}{\mu(B_n)} \tag{8}$$

and

$$\varphi_1(\alpha\beta_n - (1-\alpha)\theta_n) + \varphi_1(-\alpha\theta_n + (1-\alpha)\beta_n) \leq 2\frac{\gamma_n^{(1)}}{\mu(B_n)}. \quad (9)$$

Numbers β_n and θ_n satisfying (8) and (9) exist because $\varphi_1(\cdot)$ and $\varphi_2(\cdot)$ are positive, and $\beta_n < \varphi_2^{-1}(2\gamma_n^{(2)}/\mu(B_n))$. Let us define

$$b_1 = \sum_{n=1}^{\infty} \beta_n \chi_{C_n} - \theta_n \chi_{D_n}$$

and

$$b_2 = \sum_{n=1}^{\infty} -\theta_n \chi_{C_n} + \beta_n \chi_{D_n}.$$

From these choices, it follows that

$$\begin{aligned}
\int_B \varphi_2(b_1) d\mu &= \sum_{n=1}^{\infty} \varphi_2(\beta_n)\mu(C_n) + \varphi_2(-\theta_n)\mu(D_n) \\
&= \sum_{n=1}^{\infty} [\varphi_2(\beta_n) + \varphi_2(-\theta_n)] \frac{\mu(B_n)}{2} \\
&= \sum_{n=1}^{\infty} \gamma_n^{(2)} = 1 - \int_A \varphi_1(\widetilde{c}) d\mu,
\end{aligned}$$

which implies that $\int_T \varphi_2(c_1) = 1$, where $c_1 = \widetilde{c}\chi_A + b_1\chi_B$. Similarly, we have that $\int_T \varphi_2(c_2) = 1$, where $c_2 = \widetilde{c}\chi_A + b_2\chi_B$. On the other hand, we can write

$$\begin{aligned}
&\int_B \varphi_1(\alpha b_1 + (1-\alpha) b_2) \\
&\quad = \sum_{n=1}^{\infty} \varphi_1(\alpha\beta_n - (1-\alpha)\theta_n)\mu(C_n) + \varphi_1(-\alpha\theta_n + (1-\alpha)\beta_n)\mu(D_n) \\
&\quad = \sum_{n=1}^{\infty} [\varphi_1(\alpha\beta_n - (1-\alpha)\theta_n) + \varphi_1(-\alpha\theta_n + (1-\alpha)\beta_n)] \frac{\mu(B_n)}{2} \\
&\quad \leq \sum_{n=1}^{\infty} \gamma_n^{(1)} < 1 - \int_A \varphi_1(\widetilde{c}) d\mu,
\end{aligned}$$

from which expression (7) follows.

Finally we can present a proof of Proposition 1.

Proof (Proposition 1). Because $\varphi_2(\cdot)$ is convex, it follows that $\int_T \varphi_2(c) d\mu < \infty$, where $c = \alpha\varphi_2^{-1}(p) + (1-\alpha)\varphi_2^{-1}(q)$. Condition (2) along with the Monotone Convergence Theorem and the continuity of $\varphi_1(\cdot)$ implies the existence and uniqueness of $\kappa(\alpha)$.

Conversely, assume the existence of $\kappa(\alpha)$ in (1) for every $p, q \in \mathcal{P}_\mu$. We begin by showing that

$$\int_T \varphi_1(c - \lambda u_0) d\mu < \infty, \qquad \text{for all } \lambda \geq 0, \tag{10}$$

for every measurable function $c\colon T \to \mathbb{R}$ such that $\int_T \varphi_2(c) d\mu < \infty$. If expression (10) does not hold, then for some measurable function $c\colon T \to \mathbb{R}$ with $\int_T \varphi_2(c) d\mu < \infty$, and some $\lambda_0 \geq 0$, we have

$$\begin{cases} \displaystyle\int_T \varphi_1(c - \lambda u_0) d\mu < \infty, & \text{for } \lambda_0 \leq \lambda, \\ \displaystyle\int_T \varphi_1(c - \lambda u_0) d\mu = \infty, & \text{for } 0 \leq \lambda < \lambda_0, \end{cases} \tag{11}$$

or

$$\begin{cases} \displaystyle\int_T \varphi_1(c - \lambda u_0) d\mu < \infty, & \text{for } \lambda_0 < \lambda, \\ \displaystyle\int_T \varphi_1(c - \lambda u_0) d\mu = \infty, & \text{for } 0 \leq \lambda \leq \lambda_0. \end{cases} \tag{12}$$

Notice that expression (11) with $\lambda_0 = 0$ corresponds to (10). So in (11) we assume that $\lambda_0 > 0$. Let $\{T_n\}$ be a sequence of non-decreasing, measurable sets with $0 < \mu(T_n) < \mu(T)$ and $\mu(T \setminus \bigcup_{n=1}^\infty T_n) = 0$. Define $E_n = T_n \cap \{c - \lambda_0 u_0 \leq n\}$, for each $n \geq 1$. Clearly, the sequence $\{E_n\}$ is non-decreasing and satisfies $\mu(E_n) < \infty$ and $\mu(T \setminus \bigcup_{n=1}^\infty E_n) = 0$.

If expression (11) is satisfied for $\lambda_0 > 0$, we select a sufficiently large $n_0 \geq 1$ such that $\int_{T \setminus E_{n_0}} \varphi_i(c - \lambda_0 u_0) d\mu < 1$, for $i = 1, 2$. Denote $A := T \setminus E_{n_0}$ and $B := E_{n_0}$. According to Lemma 2, we can find measurable functions for which $p = \varphi_2(c_1)$ and $q = \varphi_2(c_2)$ are in $\mathcal{P}_\mu$, where $c_1 = (c - \lambda_0 u_0)\chi_A + b_1 \chi_B +$ and $c_2 = (c - \lambda_0 u_0)\chi_A + b_2 \chi_B$, and inequality (7) is satisfied. For any $\lambda > 0$, we can write

$$\begin{aligned} \int_T \varphi_1(\alpha \varphi_2^{-1}(p) + (1 - \alpha)\varphi_2^{-1}(q) + \lambda u_0) &\geq \int_B \varphi_1(c - (\lambda_0 - \lambda)u_0) d\mu \\ = \int_T \varphi_1(c - (\lambda_0 - \lambda)u_0) d\mu &- \int_{A_{n_0}} \varphi_1(c - (\lambda_0 - \lambda)u_0) d\mu = \infty. \end{aligned}$$

By this expression and inequality (7), we conclude that the constant $\kappa(\alpha)$ as defined by (1) cannot be found.

Now suppose that (12) is satisfied. Let $\{\lambda_n\}$ be a sequence in (λ_0, ∞) such that $\lambda_n \downarrow \lambda_0$. We define inductively an increasing sequence $\{k_n\} \subseteq \mathbb{N}$ as follows. Choose $k_0 \geq 1$ such that $\int_{T \setminus E_{k_0}} \varphi_1(c - \lambda_1 u_0) d\mu \leq 2^{-2}$. Given k_{n-1} we select some $k_n > k_{n-1}$ such that

$$\int_{E_{k_n} \setminus E_{k_{n-1}}} \varphi_1(c - \lambda_0 u_0) d\mu \geq 1$$

and

$$\int_{T \setminus E_{k_n}} \varphi_1(c - \lambda_{n+1} u_0) d\mu \leq 2^{-(n+2)}.$$

Let us denote $A_n = E_{k_n} \setminus E_{k_{n-1}}$ for $n \geq 1$. Notice that the sets A_n are pairwise disjoint. Take $n_0 > 1$ such that $\int_A \varphi_2(c) d\mu < 1$, where $A = \bigcup_{n=n_0}^{\infty} A_n$. Now we define $\widetilde{c} = \sum_{n=n_0}^{\infty} (c - \lambda_n u_0) \chi_{A_n}$. As a result of these choices, it follows that

$$\int_A \varphi_1(\widetilde{c}) d\mu = \sum_{n=n_0}^{\infty} \int_{A_n} \varphi(c - \lambda_n u_0) d\mu \leq \sum_{n=n_0}^{\infty} 2^{-n_0} < 1$$

and

$$\int_A \varphi_2(\widetilde{c}) d\mu < \int_A \varphi_2(c) d\mu < 1.$$

Denote $B = T \setminus A$. In view of Lemma 2, there exist measurable functions $b_1, b_2 \colon T \to \mathbb{R}$ such that $p = \varphi_2(c_1)$ and $q = \varphi_2(c_2)$ are in $\mathcal{P}_\mu$, where $c_1 = \widetilde{c}\chi_A + b_1 \chi_B$ and $c_2 = \widetilde{c}\chi_A + b_2 \chi_B$, and inequality (7) is satisfied. Consequently, if the constant $\kappa(\alpha)$ as defined in (1) exists, then $k(\alpha) > 0$. Fixed arbitrary $\lambda > 0$, we take $n_1 \geq n_0$ such that $\lambda_n - \lambda \leq \lambda_0$ for all $n \geq n_1$. Observing that $\int_{A_n} \varphi_1(c - \lambda_0 u_0) d\mu \geq 1$, we can write

$$\begin{aligned}\int_T \varphi_1(\alpha \varphi_2^{-1}(p) + (1-\alpha)\varphi_2^{-1}(q) + \lambda u_0) d\mu &\geq \int_A \varphi_1(\widetilde{c} + \lambda u_0) d\mu \\ &\geq \sum_{n=n_1}^{\infty} \int_{A_n} \varphi_1(c - (\lambda_n - \lambda) u_0) d\mu \geq \sum_{n=n_1}^{\infty} 1 = \infty,\end{aligned}$$

which shows that $\kappa(\alpha)$ cannot be found.

Suppose that condition (2) is not satisfied. By Proposition 2 and Lemma 1, we can find sequences $\{\lambda_n\}$, $\{c_n\}$ and $\{A_n\}$ of positive numbers $\lambda_n \downarrow 0$, measurable functions, and pairwise disjoint, measurable sets, respectively, such that

$$\int_{A_n} \varphi_1(c_n) d\mu = 1 \quad \text{and} \quad \int_{A_n} \varphi_2(c_n - \lambda_n u_0) d\mu \leq 2^{-n}, \quad \text{for all } n \geq 1.$$

By expression (10), we can conclude that $\sum_{n=n_0}^{\infty} \int_{A_n} \varphi_1(c_n - \lambda_n u_0) d\mu < \infty$. Then we can take some $n_0 > 1$ for which the function $\widetilde{c} = \sum_{n=n_0}^{\infty} (c_n - \lambda_n u_0) \chi_{A_n}$ satisfies $\int_A \varphi_i(\widetilde{c}) d\mu < 1$, for $i = 1, 2$, where $A = \bigcup_{n=n_0}^{\infty} A_n$. Let us denote $B = T \setminus A$. From (2), there exist measurable functions $b_1, b_2 \colon T \to \mathbb{R}$ such that $p = \varphi_2(c_1)$ and $q = \varphi_2(c_2)$ belong to $\mathcal{P}_\mu$, where $c_1 = \widetilde{c}\chi_A + b_1 \chi_B$ and $c_2 = \widetilde{c}\chi_A + b_2 \chi_B$, and inequality (7) holds. Given any $\lambda > 0$, take $n_1 \geq n_0$ such that $\lambda \geq \lambda_n$ for all $n \geq n_1$. Then we can write

$$\begin{aligned}\int_T \varphi_1(\alpha \varphi_2^{-1}(p) + (1-\alpha)\varphi_2^{-1}(q) + \lambda u_0) d\mu &\geq \int_A \varphi_1(\widetilde{c} + \lambda u_0) d\mu \\ &\geq \sum_{n=n_1}^{\infty} \int_{A_n} \varphi_1(c_n + (\lambda - \lambda_n) u_0) d\mu \geq \sum_{n=n_1}^{\infty} \int_{A_n} \varphi_1(c_n) d\mu = \infty.\end{aligned}$$

This expression and inequality (7) imply that the constant $\kappa(\alpha)$ as defined by (1) cannot be found. Therefore, condition (2) have to be satisfied.

The next result is a consequence of Proposition 2.

Proposition 3. *Let $\varphi\colon \mathbb{R} \to [0,\infty)$ be a deformed exponential. Then we can find a measurable function $u_0\colon \mathbb{R} \to (0,\infty)$ for which condition* (2) *holds for $\varphi_1 = \varphi_2 = \varphi$ if, and only if,*

$$\limsup_{u\to\infty} \frac{\varphi(u)}{\varphi(u-\lambda_0)} < \infty, \qquad \textit{for some } \lambda_0 > 0. \tag{13}$$

Proof. By Proposition 2 we can conclude that the existence of u_0 implies (13). Conversely, assume that expression (13) holds for some $\lambda_0 > 0$. In this case, there exists $M \in (1,\infty)$ and $\overline{c} \in \mathbb{R}$ such that $\frac{\varphi(u)}{\varphi(u-\lambda_0)} \leq M$ for all $u \geq \overline{c}$. Let $\{\lambda_n\}$ be any sequence in $(0,\lambda_0]$ such that $\lambda_n \downarrow 0$. For each $n \geq 1$, define

$$c_n = \sup\{u \in \mathbb{R} : \alpha\varphi(u) > \varphi(u-\lambda_n)\}, \tag{14}$$

where $\alpha = 1/M$ and we adopt the convention $\sup\emptyset = -\infty$. From the choice of $\{\lambda_n\}$ and α, it follows that $-\infty \leq c_n \leq \overline{c}$. We claim that $\varphi(c_n) \downarrow 0$. If the sequence $\{c_n\}$ converges to some $c > -\infty$, the equality $\alpha\varphi(c_n) = \varphi(c_n - \lambda_n)$ implies $\alpha\varphi(c) = \varphi(c)$ and then $\varphi(c) = 0$. In the case $c_n \downarrow -\infty$, it is clear that $\varphi(c_n) \downarrow 0$. Let $\{T_k\}$ be a sequence of pairwise disjoint, measurable sets with $\mu(T_k) < \infty$ and $\mu(T \setminus \bigcup_{k=1}^{\infty} T_k) = 0$. Thus we can select a sub-sequence $\{c_{n_k}\}$ such that $\sum_{k=1}^{\infty} \varphi(c_{n_k})\mu(T_k) < \infty$. Let us define $c = \sum_{k=1}^{\infty} c_{n_k}\chi_{T_k}$ and $u_0 = \sum_{k=1}^{\infty} \lambda_{n_k}\chi_{T_k}$. From (14) it follows that $\alpha\varphi(u) \leq \varphi(u - u_0(t))$, for all $u \geq c(t)$. Proposition 2 implies that $\varphi(\cdot)$ and u_0 satisfy condition (2).

References

1. Cena, A., Pistone, G.: Exponential statistical manifold. Ann. Inst. Statist. Math. **59**(1), 27–56 (2007)
2. de Souza, D.C., Vigelis, R.F., Cavalcante, C.C.: Geometry induced by a generalization of Rényi divergence. Entropy **18**(11), 407 (2016)
3. Eguchi, S., Komori, O.: Path connectedness on a space of probability density functions. In: Nielsen, F., Barbaresco, F. (eds.) GSI 2015. LNCS, vol. 9389, pp. 615–624. Springer, Cham (2015). doi:10.1007/978-3-319-25040-3_66
4. Musielak, J.: Orlicz Spaces and Modular Spaces. LNM, vol. 1034. Springer, Heidelberg (1983). doi:10.1007/BFb0072210
5. Giovanni Pistone and Maria Piera Rogantin: The exponential statistical manifold: mean parameters, orthogonality and space transformations. Bernoulli **5**(4), 721–760 (1999)
6. Santacroce, M., Siri, P., Trivellato, B.: New results on mixture and exponential models by Orlicz spaces. Bernoulli **22**(3), 1431–1447 (2016)

Classification of Totally Umbilical CR-Statistical Submanifolds in Holomorphic Statistical Manifolds with Constant Holomorphic Curvature

Michel Nguiffo Boyom[1], Aliya Naaz Siddiqui[2(✉)], Wan Ainun Mior Othman[3], and Mohammad Hasan Shahid[2]

[1] IMAG, Alexander Grothendieck Research Institute, Université of Montpellier, Montpellier, France
nguiffo.boyom@gmail.com

[2] Department of Mathematics, Jamia Millia Islamia University, New Delhi 110025, India
aliyanaazsiddiqui9@gmail.com, hasan_jmi@yahoo.com

[3] Institute of Mathematical Sciences, University of Malaya, Kuala Lumpur, Malaysia
wanainun@um.edu.my

Abstract. In 1985, Amari [1] introduced an interesting manifold, i.e., statistical manifold in the context of information geometry. The geometry of such manifolds includes the notion of dual connections, called conjugate connections in affine geometry, it is closely related to affine geometry. A statistical structure is a generalization of a Hessian one, it connects Hessian geometry.

In the present paper, we study CR-statistical submanifolds in holomorphic statistical manifolds. Some results on totally umbilical CR-statistical submanifolds with respect to $\overline{\nabla}$ and $\overline{\nabla}^*$ in holomorphic statistical manifolds with constant holomorphic curvature are obtained.

Keywords: CR-statistical submanifolds · Holomorphic statistical manifolds · Totally umbilical submanifolds

1 Introduction

In 1978, A. Bejancu [3] introduced the notion of a CR-submanifold of Kaehler manifolds with complex structure $\mathcal{J}$. CR submanifolds arise as a natural generalization of both holomorphic and totally real submanifolds in complex geometry. Since then such submanifolds have been investigated extensively by many geometers and many interesting results were obtained. On the other hand, statistical manifolds are abstract generalizations of statistical models. Even if a statistical manifold is treated as a purely geometric object, however, the motivation for the definitions is inspired from statistical models. Geometry of statistical manifolds lies at the confluence of some research areas such as information geometry,

F. Nielsen and F. Barbaresco (Eds.): GSI 2017, LNCS 10589, pp. 809–817, 2017.
https://doi.org/10.1007/978-3-319-68445-1_93

affine differential geometry, and Hessian geometry. Beyond expectations, statistical manifolds are familiar to geometers and many interesting results were obtained [2,10,11,14]. In 2004, Kurose [9] defined their complex version, i.e., holomorphic statistical manifolds. It is natural for geometers to try to build the submanifold theory and the complex manifold theory of statistical manifolds. Recently, Furuhata and Hasegawa [7] studied CR-statistical submanifold theory in holomorphic statistical manifolds. Motivated by their work, we wish to give some more results on CR-statistical submanifolds of holomorphic statistical manifolds, which are new objects originating from information geometry.

Our work is structured as follows: Sect. 2 is devoted to preliminaries. Section 3 deals with some basic results in holomorphic statistical manifolds. In Sect. 4, we give complete classification of totally umbilical CR-statistical submanifolds with respect to $\overline{\nabla}$ and $\overline{\nabla}^*$ in holomorphic statistical manifolds with constant holomorphic curvature.

2 Preliminaries

This section is fully devoted to a brief review of several fundamental notions, formulas and some definitions which are required later.

Definition 1 [7]. *A statistical manifold is a Riemannian manifold* $(\overline{M}, g)$ *of dimension* $(n+k)$*, endowed with a pair of torsion-free affine connections* $\overline{\nabla}$ *and* $\overline{\nabla}^*$ *satisfying*

$$Zg(X,Y) = g(\overline{\nabla}_Z X, Y) + g(X, \overline{\nabla}^*_Z Y)$$

for any $X, Y, Z \in \Gamma(T\overline{M})$*. It is denoted by* $(\overline{M}, g, \overline{\nabla}, \overline{\nabla}^*)$*. The connections* $\overline{\nabla}$ *and* $\overline{\nabla}^*$ *are called dual connections on* $\overline{M}$ *and it is easily shown that* $(\overline{\nabla}^*)^* = \overline{\nabla}$*. If* $(\overline{\nabla}, g)$ *is a statistical structure on* $\overline{M}$*, then* $(\overline{\nabla}^*, g)$ *is also a statistical structure.*

Let $(\overline{M}, g)$ be a Riemannian manifold and M a submanifold of $\overline{M}$. If (M, ∇, g) is a statistical manifold, then we call (M, ∇, g) a statistical submanifold of $(\overline{M}, g)$, where ∇ is an affine connection on M and the Riemannian metric for M and $\overline{M}$ is denoted by the same symbol g. Let $\overline{\nabla}$ be an affine connection on $\overline{M}$. If $(\overline{M}, g, \overline{\nabla})$ is a statistical manifold and M a submanifold of $\overline{M}$, then (M, ∇, g) is also a statistical manifold with the induced connection ∇ and induced metric g.

In the geometry of Riemannian submanifolds [15], the fundamental equations are the Gauss and Weingarten formulas and the equations of Gauss, Codazzi and Ricci. In our case, for any $X, Y \in \Gamma(TM)$ and $\mathcal{V} \in \Gamma(T^{\perp}M)$, Gauss and Weingarten formulas are, respectively, defined by [7]

$$\left.\begin{aligned} \overline{\nabla}_X Y &= \nabla_X Y + \sigma(X,Y), & \overline{\nabla}^*_X Y &= \nabla^*_X Y + \sigma^*(X,Y), \\ \overline{\nabla}_X \mathcal{V} &= -\mathcal{A}_{\mathcal{V}}(X) + D_X \mathcal{V}, & \overline{\nabla}^*_X \mathcal{V} &= -\mathcal{A}^*_{\mathcal{V}}(X) + D^*_X \mathcal{V}, \end{aligned}\right\} \tag{1}$$

where $\overline{\nabla}$ and $\overline{\nabla}^*$ (respectively, ∇ and ∇^*) are the dual connections on $\overline{M}$ (respectively, on M), σ and σ^* are symmetric and bilinear, called the imbedding curvature tensor of M in $\overline{M}$ for $\overline{\nabla}$ and the imbedding curvature tensor of M in $\overline{M}$ for $\overline{\nabla}^*$, respectively. Since σ and σ^* are bilinear, we have the linear transformations $\mathcal{A}_{\mathcal{V}}$ and $\mathcal{A}^*_{\mathcal{V}}$, defined by [7]

$$\left.\begin{aligned} g(\sigma(X,Y),\mathcal{V}) &= g(\mathcal{A}^*_{\mathcal{V}}(X),Y), \\ g(\sigma^*(X,Y),\mathcal{V}) &= g(\mathcal{A}_{\mathcal{V}}(X),Y) \end{aligned}\right\} \tag{2}$$

for any $X, Y \in \Gamma(TM)$ and $\mathcal{V} \in \Gamma(T^{\perp}M)$.

Let $\overline{R}$ and R be the curvature tensor fields of $\overline{\nabla}$ and ∇, respectively. The corresponding Gauss, Codazzi and Ricci equations, respectively, are given by [7]

$$\left.\begin{aligned} \overline{R}(X,Y,Z,W) =& R(X,Y,Z,W) + g(\sigma(X,Z),\sigma^*(Y,W)) \\ & - g(\sigma^*(X,W),\sigma(Y,Z)), \\ \overline{R}(X,Y,Z,\mathcal{N}) =& g((\overline{\nabla}_X\sigma)(Y,Z),\mathcal{N}) - g((\overline{\nabla}_Y\sigma)(X,Z),\mathcal{N}), \\ \overline{R}(X,Y,\mathcal{N},Z) =& g((\overline{\nabla}_Y\mathcal{A})_{\mathcal{N}}X,Z) - g((\overline{\nabla}_X\mathcal{A})_{\mathcal{N}}Y,Z), \\ \overline{R}(X,Y,\mathcal{N},\mathcal{V}) =& R^{\perp}(X,Y,\mathcal{N},\mathcal{V}) + g(\sigma(Y,\mathcal{A}_{\mathcal{N}}X,\mathcal{V}) - g(\sigma(X,\mathcal{A}_{\mathcal{N}}Y,\mathcal{V}) \end{aligned}\right\} \tag{3}$$

for any $X, Y, Z, W \in \Gamma(TM)$ and $\mathcal{N}, \mathcal{V} \in \Gamma(T^{\perp}M)$.

Similarly, $\overline{R}$ and R are the curvature tensor fields of $\overline{\nabla}$ and ∇, respectively and duals of all equations in (3) can be obtained for the connections $\overline{\nabla}^*$ and ∇^* [7].

Definition 2 [7,8]. *Let M be a submanifold of a statistical manifold $\overline{M}$. Then M is said to be a*

(A) *totally geodesic with respect to $\overline{\nabla}$ if $\sigma = 0$.*
(A*) *totally geodesic with respect to $\overline{\nabla}^*$ if $\sigma^* = 0$.*
(B) *totally tangentially umbilical with respect to $\overline{\nabla}$ if $\sigma(X,Y) = g(X,Y)\mathcal{H}$ for any $X, Y \in \Gamma(TM)$. Here $\mathcal{H}$ is the mean curvature vector of M in $\overline{M}$ for $\overline{\nabla}$.*
(B*) *totally tangentially umbilical with respect to $\overline{\nabla}^*$ if $\sigma^*(X,Y) = g(X,Y)\mathcal{H}^*$ for any $X, Y \in \Gamma(TM)$. Here $\mathcal{H}^*$ is the mean curvature vector of M in $\overline{M}$ for $\overline{\nabla}^*$.*
(C) *totally normally umbilical with respect to $\overline{\nabla}$ if $\mathcal{A}_{\mathcal{N}}X = g(\mathcal{H},\mathcal{N})X$ for any $X \in \Gamma(TM)$ and $\mathcal{N} \in \Gamma(T^{\perp}M)$.*
(C*) *totally normally umbilical with respect to $\overline{\nabla}^*$ if $\mathcal{A}^*_{\mathcal{N}}X = g(\mathcal{H}^*,\mathcal{N})X$ for any $X \in \Gamma(TM)$ and $\mathcal{N} \in \Gamma(T^{\perp}M)$.*

Definition 3 [7]. *Let $(\overline{M},\mathcal{J},g)$ be a Kähler manifold and $\overline{\nabla}$ be an affine connection on $\overline{M}$. Then $(\overline{M},\overline{\nabla},g,\mathcal{J})$ is said to be a holomorphic statistical manifold if $(\overline{M},\overline{\nabla},g)$ is a statistical manifold and a 2–form ω on $\overline{M}$, given by*

$$\omega(X,Y) = g(X,\mathcal{J}Y)$$

for any $X, Y \in \Gamma(T\overline{M})$, is a $\overline{\nabla}$–parallel, i.e., $\overline{\nabla}\omega = 0$.

Let $(\overline{M}, \mathcal{J}, g)$ be a holomorphic statistical manifold. Then [7]

$$\overline{\nabla}_X(\mathcal{J}Y) = \mathcal{J}\overline{\nabla}^*_X Y \tag{4}$$

for any $X, Y \in \Gamma(TM)$, where $\overline{\nabla}^*$ is the dual connection of $\overline{\nabla}$ with respect to g.

Remark 1. A holomorphic statistical manifold $(\overline{M}, \overline{\nabla}, g, \mathcal{J})$ is nothing but a special Kähler manifold if $\overline{\nabla}$ is flat.

For any vector field $X \in \Gamma(TM)$ and $V \in \Gamma(T^{\perp}M)$, respectively, we put [15]

$$\mathcal{J}X = GX + LX \quad and \quad \mathcal{J}V = CV + BV, \tag{5}$$

where $GX = tan(\mathcal{J}X)$, $LX = nor(\mathcal{J}X)$, $CV = tan(\mathcal{J}V)$ and $BV = nor(\mathcal{J}V)$. It is easy to see that [15]

$$g(GX, Y) = -g(X, GY), \quad g(BV, \mathcal{N}) = -g(V, B\mathcal{N}) \quad and \quad g(LX, \mathcal{N}) = -g(X, C\mathcal{N}) \tag{6}$$

for any $X, Y \in \Gamma(TM)$ and $V, \mathcal{N} \in \Gamma(T^{\perp}M)$.

Definition 4 [7]. *A holomorphic statistical manifold $\overline{M}$ of constant holomorphic curvature $k \in \mathbb{R}$ is said to be a holomorphic statistical space form $\overline{M}(k)$ if the following curvature equation holds*

$$\begin{aligned} \overline{R}(X, Y)Z = \frac{k}{4}\{&g(Y, Z)X - g(X, Z)Y + g(\mathcal{J}Y, Z)\mathcal{J}X \\ &-g(\mathcal{J}X, Z)\mathcal{J}Y + 2g(X, \mathcal{J}Y)\mathcal{J}Z\}. \end{aligned} \tag{7}$$

for any $X, Y, Z \in \Gamma(T\overline{M})$.

The statistical version of definition of CR-submanifold as follows:

Definition 5 [7]. *A statistical submanifold M is called a CR-statistical submanifold in a holomorphic statistical manifold $\overline{M}$ of dimension $2m \geq 4$ if M is CR-submanifold in $\overline{M}$, i.e., there exists a differentiable distribution $\mathcal{D} : x \to \mathcal{D}_x \subseteq T_xM$ on M satisfying the following conditions:*

(A) *$\mathcal{D}$ is holomorphic, i.e., $\mathcal{J}\mathcal{D}_x = \mathcal{D}_x \subset T_xM$ for each $x \in M$, and*

(B) *the complementary orthogonal distribution $\mathcal{D}^{\perp} : x \to \mathcal{D}^{\perp}_x \subseteq T_xM$ is totally real, i.e., $\mathcal{J}\mathcal{D}^{\perp}_x \subset T^{\perp}_xM$ for each $x \in M$.*

Remark 2 [7]. If $\mathcal{D} \neq 0$ and $\mathcal{D}^{\perp} \neq 0$, then M is said to be *proper*.

Remark 3 [7]. CR-statistical submanifolds are characterized by the condition $LG = 0$.

Definition 6 [7]. *A statistical submanifold M of a holomorphic statistical manifold $\overline{M}$ is called holomorphic ($L = 0$ and $C = 0$) if the almost complex structure $\mathcal{J}$ of $\overline{M}$ carries each tangent space of M into itself whereas it is said to be totally real ($G = 0$) if the almost complex structure $\mathcal{J}$ of $\overline{M}$ carries each tangent space of M into its corresponding normal space.*

For a CR-statistical submanifold M we shall denote by μ the orthogonal complementary subbundle of $\mathcal{J}\mathcal{D}^{\perp}$ in $T^{\perp}M$, we have [7]

$$T^{\perp}M = \mathcal{J}\mathcal{D}^{\perp} \oplus \mu. \tag{8}$$

3 Some Basic Results in Holomorphic Statistical Manifolds

In this section, we propose some basic results which are based on $D_X LY = L\nabla^*_X Y$:

Theorem 1. *Let M be a statistical submanifold of a holomorphic statistical manifold $\overline{M}$. Then $D_X LY = L\nabla^*_X Y$ holds if and only if $\mathcal{A}^*_{\mathcal{N}} GY = -\mathcal{A}_{B\mathcal{N}} Y$ for any $X, Y \in \Gamma(TM)$ and $\mathcal{N} \in \Gamma(T^{\perp}M)$.*

Proof. From Lemma 5(7.40) of [7], we have

$$\sigma(X, GY) + D_X LY = L\nabla^*_X Y + B\sigma^*(X, Y).$$

In the light of (6) and (2), we get

$$g(\mathcal{A}^*_{\mathcal{N}} GY, X) = -g(\mathcal{A}_{B\mathcal{N}} Y, X)$$

for any $X, Y \in \Gamma(TM)$ and $\mathcal{N} \in \Gamma(T^{\perp}M)$. This proves our theorem.

The following theorem shall be required to prove some results in the next section:

Theorem 2. *Let M be a statistical submanifold of a holomorphic statistical manifold $\overline{M}$. If $D_X LY = L\nabla^*_X Y$ holds, then the curvature tensor R^* and the normal curvature tensor $R^{\perp}$ satisfy $LR^*(X, Y)Z = R^{\perp}(X, Y)LZ$ for any $X, Y, Z \in \Gamma(TM)$.*

Proof. Since, we have assumed that $D_X(LY) = L\nabla^*_X Y$ for any $X, Y \in \Gamma(TM)$. Therefore, we derive the following:

$$\begin{aligned} LR^*(X, Y)Z &= L\big(\nabla^*_X \nabla^*_Y Z - \nabla^*_Y \nabla^*_X Z - \nabla^*_{[X,Y]} Z\big) \\ &= D_X D_Y LZ - D_Y D_X LZ - D_{[X,Y]} LZ \\ &= R^{\perp}(X, Y)LZ \end{aligned}$$

for any $X, Y, Z \in \Gamma(TM)$. This proves our assertion.

4 Classification of Totally Umbilical CR-Statistical Submanifolds

Chen [5] studied totally umbilical submanifolds in the case of spaces of constant curvature. Also, Chen and Ogiue [6] considered such immersions in complex space forms. Blair and Vanhecke [4] considered in Sasakian space forms. In 2002, Kurose [8] studied totally tangentially umbilical and totally normally umbilical in statistical manifolds. In this section, we study a special class of CR-statistical submanifolds which is totally umbilical CR-statistical submanifolds with respect to $\overline{\nabla}$ and $\overline{\nabla}^*$ in holomorphic statistical manifolds.

Theorem 3. *Let M be a CR-statistical submanifold in a holomorphic statistical manifold $\overline{M}$. If M is totally umbilical with respect to $\overline{\nabla}$ and $\overline{\nabla}^*$ such that $\mathcal{J}\mathcal{H}^* \in \Gamma(\mu)$, then we have either*

(A) *M is a totally geodesic with respect to $\overline{\nabla}^*$, or*
(B) *$dim\ \mathcal{D} \geq 2$.*

Proof. For any $X, Y \in \Gamma(\mathcal{D})$, we have

$$\overline{\nabla}_X \mathcal{J}Y = \overline{\nabla}_X GY.$$

By our assumption, last relation takes the following form:

$$\mathcal{J}\nabla^*_X Y + g(Y, X)\mathcal{J}\mathcal{H}^* = \nabla_X GY + g(X, GY)\mathcal{H}. \tag{9}$$

Taking inner product with $\mathcal{J}\mathcal{H}^*$ on both sides of (9), we obtain

$$g(Y, X)||\mathcal{H}^*||^2 = g(X, GY)g(\mathcal{J}\mathcal{H}^*, \mathcal{H}). \tag{10}$$

Interchanging the role of X and Y in above equation, we get

$$g(X, Y)||\mathcal{H}^*||^2 = g(Y, GX)g(\mathcal{J}\mathcal{H}^*, \mathcal{H}). \tag{11}$$

Combining both Eqs. (10) and (11), we find that

$$g(X, Y)||\mathcal{H}^*||^2 = 0. \tag{12}$$

From (12), we conclude that $\mathcal{H}^* = 0$ or $g(X, Y)$ for any $X, Y \in \Gamma(\mathcal{D})$ which shows that M is totally geodesic with respect to $\overline{\nabla}^*$ or dim $\mathcal{D} \geq 2$, respectively. This completes the proof of the theorem.

Theorem 4. *Let M be a CR-statistical submanifold in a holomorphic statistical manifold $\overline{M}$. If M is totally umbilical with respect to $\overline{\nabla}$ and $\overline{\nabla}^*$, then, for any $X \in \Gamma(\mathcal{D}^\perp)$, we have*

(A) *$\mathcal{D}_X \mathcal{H} \in \mathcal{J}\mathcal{D}^\perp$, or*
(B) *$\mathcal{D}_X \mathcal{H} = 0$,*

(A*) $\mathcal{D}^*_X\mathcal{H}^* \in \mathcal{J}\mathcal{D}^\perp$, *or*
(B*) $\mathcal{D}^*_X\mathcal{H}^* = 0$.

Proof. For any $X \in \Gamma(\mathcal{D}^\perp)$ and $Y, Z \in \Gamma(\mathcal{D})$, we have the following [7]:

$$\mathcal{J}\overline{R}^*(X,Y)Z = \overline{R}(X,Y)\mathcal{J}Z$$
$$G\overline{R}^*(X,Y)Z + L\overline{R}^*(X,Y)Z = g(Y,GZ)D_X\mathcal{H}, \tag{13}$$

where we have used Codazzi equation for a totally tangentially umbilical submanifold. Taking inner product on both sides of (13) with $\mathcal{N} \in \Gamma(\mu)$ and putting $Y = GZ$, we arrive at

$$||Z||^2 g(D_X\mathcal{H}, \mathcal{N}) = 0. \tag{14}$$

Similarly, we can easily obtain dual of (14), i.e.,

$$||Z||^2 g(D^*_X\mathcal{H}^*, \mathcal{N}) = 0, \tag{15}$$

where we have used Codazzi equation for a totally normally umbilical submanifold. From (14), we conclude that $D_X\mathcal{H} \in \mathcal{J}\mathcal{D}^\perp$ or $D_X\mathcal{H} = 0$. And (15) gives $D^*_X\mathcal{H}^* \in \mathcal{J}\mathcal{D}^\perp$ or $D^*_X\mathcal{H}^* = 0$. This completes the proof of the theorem.

For CR-statistical submanifolds in holomorphic statistical space forms, we have the followings:

Theorem 5. *Let M be a CR-statistical submanifold in a holomorphic statistical space form $\overline{M}(k)$. If M is totally umbilical with respect to $\overline{\nabla}$ and $\overline{\nabla}^*$ such that $\mathcal{D}_X\mathcal{H} = 0$ and $\mathcal{D}^*_X\mathcal{H}^* = 0$ for any $X \in \Gamma(\mathcal{D}^\perp)$, then*

(A) $k = 0$, *or*
(B) *dim* $\mathcal{D} \geq 2$, *or*
(C) *H and H^* are perpendicular to $J\mathcal{D}^\perp$, or*
(D) *M is totally geodesic with respect to $\overline{\nabla}$ and $\overline{\nabla}^*$.*

Proof. From Proposition 3(7.33) of [7] and (7), we have

$$(\overline{\nabla}_X\sigma)(Y,Z) - (\overline{\nabla}_Y\sigma)(X,Z) + (\overline{\nabla}^*_X\sigma^*)(Y,Z) - (\overline{\nabla}^*_Y\sigma^*)(X,Z)$$
$$= 2\left\{\frac{k}{4}[g(GY,Z)LX - g(GX,Z)LY + 2g(X,GY)LZ]\right\} \tag{16}$$

for any $X, Y, Z \in \Gamma(TM)$. Now we evaluate (16) for $Z = GW$, $W, Y \in \Gamma(\mathcal{D})$ and $X \in \Gamma(\mathcal{D}^\perp)$, we get

$$(\overline{\nabla}_X\sigma)(Y,GW) - (\overline{\nabla}_Y\sigma)(X,GW) + (\overline{\nabla}^*_X\sigma^*)(Y,GW) - (\overline{\nabla}^*_Y\sigma^*)(X,GW)$$
$$= \frac{k}{4}\Big[g(GY,GW)LX\Big].$$

By virtue of Codazzi equation for a totally umbilical with respect to $\overline{\nabla}$ and $\overline{\nabla}^*$, we arrive at $\frac{k}{4}\Big[g(Y,W)LX\Big] = 0$. If we take inner product with H (respectively, H^*) on both sides of above relation, then we conclude that $k = 0$ or dim $\mathcal{D} \geq 2$ or both H and H^* are perpendicular to $J\mathcal{D}^\perp$ or M is totally geodesic with respect to $\overline{\nabla}$ and $\overline{\nabla}^*$. This completes our proof.

Theorem 6. *Let M be a proper CR-statistical submanifold in a holomorphic statistical space form $\overline{M}(k)$. If M is totally umbilical with respect to $\overline{\nabla}$ and $\overline{\nabla}^*$ such that $D_X LY = L\nabla^*_X Y$ for any $Y \in \Gamma(\mathcal{D}^\perp)$ and $X \in \Gamma(\mathcal{D})$, then we have*

(A) $k = 0$, *or*
(B) $dim\ \mathcal{D}^\perp \geq 2$.

Proof. From Theorem 2, 4th equation of (3) and its dual, we can easily get $\frac{k}{4}\left[g(Y,Z)GX\right] = 0$ for any $Y, Z \in \Gamma(\mathcal{D}^\perp)$ and $X \in \Gamma(\mathcal{D})$. Thus, we get $k = 0$ or dim $\mathcal{D}^\perp \geq 2$. This completes our proof.

An immediate consequence of Theorem 6 as follows:

Corollary 1. *Let M be a proper CR-statistical submanifold in a holomorphic statistical space form $\overline{M}(k)$. If M is totally geodesic with respect to $\overline{\nabla}$ and $\overline{\nabla}^*$ such that $D_X LY = L\nabla^*_X Y$ for any $Y \in \Gamma(\mathcal{D}^\perp)$ and $X \in \Gamma(\mathcal{D})$, then we have*

(A) $k = 0$, *or*
(B) $dim\ \mathcal{D}^\perp \geq 2$.

Acknowledgment. The authors are grateful to the referee for his/her valuable comments and suggestions.

References

1. Amari, S.: Differential Geometric Methods in Statistics. Lecture Notes in Statistics, vol. 28. Springer, New York (1985)
2. Aydin, M.E., Mihai, A., Mihai, I.: Some inequalities on submanifolds in statistical manifolds of constant curvature. Filomat **29**(3), 465–477 (2015)
3. Bejancu, A.: CR submanifolds of a Kaehler manifold I. Proc. Am. MAth. Soc. **69**, 135–142 (1978)
4. Blair, D.E., Vanhecke, L.: Umbilical submanifolds of sasakian space forms. J. Differ. Geom. **13**(2), 273–278 (1978)
5. Chen, B.Y.: Geometry of Submanifolds. Marcel Dekker, New York (1973)
6. Chen, B.Y., Ogiue, K.: On totally real submanifolds. Trans. Amer. Math. Soc. **193**, 257–266 (1974)
7. Furuhata, H., Hasegawa, I.: Submanifold theory in holomorphic statistical manifolds. In: Dragomir, S., Shahid, M.H., Al-Solamy, F.R. (eds.) Geometry of Cauchy-Riemann Submanifolds, pp. 179–215. Springer, Singapore (2016). doi:10.1007/978-981-10-0916-7_7
8. Kurose, T.: Conformal-projective geometry of statistical manifolds. Interdisc. Inf. Sci. **8**(1), 89–100 (2002)
9. Kurose, T.: Geometry of statistical manifolds. In: Mathematics in the 21st Century, Nihon-hyouron-sha, pp. 34–43 (2004). (in Japanese)
10. Nguiffo Boyom, M.: Foliations-webs-hessian geometry-information geometry-entropy and cohomology. Entropy **18**, 433 (2016). doi:10.3390/e18120433

11. Boyom, M.N., Jamali, M., Shahid, M.H.: Multiply CR-warped product statistical submanifolds of a holomorphic statistical space form. In: Nielsen, F., Barbaresco, F. (eds.) GSI 2015. LNCS, vol. 9389, pp. 257–268. Springer, Cham (2015). doi:10.1007/978-3-319-25040-3_29
12. Nomizu, K.: Generalized central spheres and the notion of spheres in Riemannian geometry. Tohoku Math. J. **25**, 129–137 (1973)
13. Nomizu, K., Yano, K.: On circles and spheres in Riemannian geometry. Math. Ann. **210**(2), 163–170 (1974)
14. Siddiqui, A.N., Al-Solamy, F.R., Shahid, M.H.: On CR-statistical submanifolds of holomorphic statistical manifolds. Communicated
15. Yano, K., Kon, M.: Structures on Manifolds. Worlds Scientific, Singapore (1984)

Applications of Distance Geometry

An Approach to Dynamical Distance Geometry

Antonio Mucherino[1(✉)] and Douglas S. Gonçalves[2]

[1] IRISA, University of Rennes 1, Rennes, France
antonio.mucherino@irisa.fr
[2] Department of Mathematics, CFM, Federal University of Santa Catarina, Florianópolis, Brazil
douglas@mtm.ufsc.br

Abstract. We introduce the dynamical distance geometry problem (dynDGP), where vertices of a given simple weighted undirected graph are to be embedded at different times t. Solutions to the dynDGP can be seen as motions of a given set of objects. In this work, we focus our attention on a class of instances where motion inter-frame distances are not available, and reduce the problem of embedding every motion frame as a static distance geometry problem. Some preliminary computational experiments are presented.

1 Introduction

Given a simple weighted undirected graph G, the Distance Geometry Problem (DGP) asks whether there exists an embedding of the graph into a Euclidean space $\mathbb{R}^K$ so that the distances between embedded vertices correspond to the weights assigned to the edges of G [10]. There is a growing interest in this problem, as it is shown by the increasing number of books and journal collections on this topic (see for example [11,12], other publications are currently under production). However, in most of the published material on this topic, the DGP is presented as a static problem, and there is only a little mention to its potential extension to dynamical applications. This paper presents a preliminary step toward the study of dynamical DGPs.

We focus our attention on the class of problems where the instances can be represented by employing a graph G whose vertex set $V \times T$ is the product of two sets: a set V representing predefined objects, and a set $T \subset \mathbb{N}$ of temporal values. The vertex $\{v, t\} \in V \times T$ represents a given object v at a certain instant t (in the following, we will use the notation v_t for the vertex $\{v, t\}$ of G). The edge set E contains edges $\{u_q, v_t\}$, whose weights provide information about the distance between two vertices u and v at times q and t, respectively. In this context, a possible embedding for G is one of the possible *motions* for the objects in V that are compatible with the distance constraints in G.

Let $G = (V \times T, E, d)$ be therefore a simple weighted undirected graph representing an instance of the *dynamical* DGP (dynDGP). The function

$$d : \{u_q, v_t\} \in E \longrightarrow (\delta, \pi) \in \mathbb{R}_+ \times \mathbb{R}_+,$$

F. Nielsen and F. Barbaresco (Eds.): GSI 2017, LNCS 10589, pp. 821–829, 2017.
https://doi.org/10.1007/978-3-319-68445-1_94

assigns pairs of nonnegative real values to every pair of vertices of $V \times T$ belonging to the edge set of G. The nonnegative value δ is the distance between u and v at times q and t, respectively, while π is a nonnegative value representing the "importance" of the distances δ (higher values indicate higher importance). We also refer to π as the *priority level* of the distance δ.

From a given instance of the dynDGP, we can extract some meaningful sub-instances. For example, the subgraph $G[\{v\} \times T]$ corresponds to the trajectory of one v_t for different times t. Moreover, the subgraph $G[V \times \{t\}]$, induced by the set product between V and only one temporal value, gives a "classical" DGP instance for a fixed time t. In this context, a possible embedding of $G[V \times \{t\}]$ is a candidate *frame* at time t of a motion which is an embedding for G. Let E_t be the edge set of the induced subgraph $G[V \times \{t\}]$. The set $\hat{E} = E \setminus \bigcup_{t \in T} E_t$ contains all edges that relate vertices at different times t. In this work, we will make the assumption that $\hat{E} = \emptyset$. As a consequence, with a little abuse of notation, we will use the notation δ^t_{uv} for indicating the distance between u and v at time t. We will use a similar notation π^t_{uv} for the priorities of such distances. We will omit the time t when not relevant.

Our approach to this class of the dynDGP is to solve, for every time $t \in T$, a classical DGP on the sub-instance $G[V \times \{t\}]$. For solving the sub-instances $G[V \times \{t\}]$, we consider the optimization problem proposed in [6], which solves DGPs where priorities are associated to the available distances. We tackle this optimization problem with a non-monotone spectral gradient method [1,2,16], and obtain a fluid motion even in absence of distances in the subset $\hat{E}$.

The rest of the paper is organized as follows. We will provide the details of the optimization problem to be solved for every sub-instance $G[V \times \{t\}]$ in Sect. 2, while Sect. 3 will present the non-monotone spectral gradient method. In Sect. 4, we will propose some computational experiments on a set of motions from which we derived instances of the dynDGP such that $\hat{E} = \emptyset$. Finally, Sect. 5 concludes the paper with a short discussion on some applications that can be modeled by a dynDGP.

2 An Optimization Problem with Priorities on Distances

Let $G = (V \times T, E, d)$ be an instance of the dynDGP such that $\hat{E} = \emptyset$. In these hypotheses, there are no distances in E that relate vertices u_q and v_t such that the times q and t are different (see Introduction). Therefore, we can split our dynDGP instance in a sequence of static DGPs represented by the subgraphs $G[V \times \{t\}]$, for each time t, without losing any information. In order to guarantee a smooth variation for the positions assigned to the vertices v_t, we consider, for increasing time values t, the solution found when solving the sub-instance $G[V \times \{t-1\}]$ as an initial approximation for the sub-instance $G[V \times \{t\}]$.

As already described in the Introduction, we suppose that our graphs G have two kinds of weights assigned to the edges. The weight δ^t_{uv} is a numerical approximation of the distance between the vertices u_t and v_t at the same time

t; the weight π^t_{uv} represents the priority on the distance δ^t_{uv}. In fact, since our distances δ^t_{uv} may be not precise, and we have no a priori information on the error they may carry, it is fundamental to coupled them with the priority levels π^t_{uv}. This way, even if the distance matrix $[\delta_{uv}]$ is not a Euclidean distance matrix (EDM), we can seek an embedding where distances having a higher priority are privileged. This approach is also justified by some of the potential applications of the dynDGP (see Conclusions).

For solving every sub-instance $G[V \times \{t\}]$, we consider the optimization problem proposed by Glunt et al. in [6]. Let $n = |V \times \{t\}| = |V|$ and recall that E_t is the edge set of the induced subgraph $G[V \times \{t\}]$. At each time t, we seek a set of positions $\{x^t_1, x^t_2, \ldots, x^t_n\} \in \mathbb{R}^K$ for the vertices in $V \times \{t\}$, which minimizes the following objective function:

$$\sigma(X) = \frac{1}{2} \sum_{\{u,v\} \in E_t} \pi_{uv} (\|x_u - x_v\| - \delta_{uv})^2,$$

where $X = [x_1\, x_2 \ldots x_n]^T \in \mathbb{R}^{n \times K}$ is a matrix representation of the set of positions for the vertices, with x_v a column vector. Notice that we omitted the time t because it is supposed to be fixed in every subgraph. The variables defined in the following and derived from X are supposed to inherit the same matrix structure.

As shown in [5], the function $\sigma(X)$ is differentiable at X if and only if $\|x_u - x_v\| > 0$ for all $\{u, v\} \in E_t$ such that $\pi_{uv}\, \delta_{uv} > 0$. In such a case, the gradient can be written as $\nabla \sigma(X) = 2(WX - B(X)X)$, where the matrix $W = [w_{uv}]$ is defined as

$$w_{uv} = \begin{cases} -\pi_{uv}, & \text{if}\, u \neq v, \\ \sum_{w \neq u} \pi_{uw} & \text{otherwise,} \end{cases}$$

and $B(X) = [b_{uv}(X)]$ is a function of X and is defined as

$$b_{uv}(X) = \begin{cases} -\dfrac{\pi_{uv}\, \delta_{uv}}{\|x_u - x_v\|}, & \text{if}\, u \neq v \text{ and } \|x_u - x_v\| > 0, \\ 0, & \text{if}\, u \neq v \text{ and } \|x_u - x_v\| = 0, \\ -\sum_{w \neq u} b_{uw}(X), & \text{otherwise.} \end{cases}$$

In [6], Glunt et al. have proposed a pure spectral gradient method for the solution of this optimization problem. This method does not require line search, but convergence was established only for strictly convex quadratic functions. In order to ensure global convergence from arbitrary starting points, we consider the non-monotone line search strategy initially proposed by Grippo, Lampariello and Lucidi in [7] and subsequently by Zhang and Hager in [16]. Although the gradient $\nabla \sigma(X)$ may not be continuous when $\|x_u - x_v\| = 0$, we did not observe numerical instabilities in our computational experiments within the required tolerance in the stopping criteria.

3 A Non-monotone Spectral Gradient Method

For a general unconstrained optimization problem, where f is a continuously differentiable objective function, iterative methods produce a sequence of points $\{X_k\}_{k \in \mathbb{N}}$ where each X_{k+1} is generated from X_k by applying the formula $X_{k+1} = X_k + \alpha_k D_k$, where D_k gives the search direction in the domain of the objective function, and α_k is the step that is performed from X_k in the direction D_k. Since $-\nabla f(X_k)$ provides the direction of steepest decrease of the objective function from X_k, it is common that D_k is proportional to $-\nabla f(X_k)$. When f is twice continuously differentiable, Newton-type methods use directions of the form $D_k = -H_k^{-1} \nabla f(X_k)$, where H_k is an approximation for the Hessian of f [14].

Since the work of Barzilai and Borwein [1], the use of spectral gradient methods have been employed with success in large scale optimization [3]. In such methods, $H_k = \mu_k I$, for $\mu_k > 0$, where I is the identity matrix. As proposed in [1,6], we adopt $\mu_k = \langle Y_{k-1}, S_{k-1} \rangle / \langle S_{k-1}, S_{k-1} \rangle$, where $Y_{k-1} = \nabla f(X_k) - \nabla f(X_{k-1})$, $S_{k-1} = X_k - X_{k-1}$ and $\langle A, B \rangle$ denotes the inner product, computed as the trace of $B^\top A$.

The choice of the step α_k is crucial in a line search: when it is too short, further improvements may be possible on the objective function value; when it is too long, the best objective function values may be missed during the step. In monotone searches, it is imposed that $f(X_{k+1}) < f(X_k)$, for every k. However, it seems reasonable to combine a non-monotone line search strategy with a spectral gradient method, because in the latter the objective function does not generally decrease monotonically.

For the optimization problem presented in Sect. 2, for every sub-instance $G[V \times \{t\}]$, we employed the spectral gradient method with the non-monotone line search method proposed by Zhang and Hager [16]. The main steps are sketched in Algorithm 1. For this particular line search, global convergence was proved for smooth and non-convex functions, and R-linear convergence was proved for strongly convex functions.

In Algorithm 1, we have used the classical choice for μ_k with safeguards $\mu_{min} = 10^{-4}$ and $\mu_{max} = 10^4$ in order to guarantee a positive bounded sequence $\{\mu_k\}$. The iterations are stopped when $\|\nabla \sigma(X_k)\| < \varepsilon = 10^{-8}$. At line 11, the algorithm attempts to perform an initial step $\alpha = 1$. While the non-monotone Armijo condition is not satisfied (with $\gamma = 10^{-4}$), the value of α is divided by 2. Following the non-monotone line search in [16], by setting $\eta_k = 0$ one obtains a classical monotone line-search whereas $\eta_k = 1$ (used in the our experiments) implies a non-monotone line search where C_k corresponds to the average of objective function values over all previous iterations.

Concerning the starting point, for $t = 1$, we randomly select the vertex positions for X_0 and center this realization around the origin. For all other $t > 0$, X_0 is set to the solution obtained for the sub-instance $G[V \times \{t-1\}]$. One interesting remark about Algorithm 1 is that if X_0 is centered, then all iterates are centered due to the forms of $B(X)$ and W [6]. For more information about the algorithm, the reader is referred to the publications cited above.

Algorithm 1. The non-monotone spectral gradient method.

1: NonmonotoneSpectralGradient (X_0, $\mu_{\min}$, $\mu_{\max}$, ε, γ, η_k)
2: **set** $k = 0$, $Q_0 = 1$, $C_0 = \sigma(X_0)$;
3: **while** ($\|\nabla\sigma(X)\| > \varepsilon$) **do**
4:     **evaluate** $\sigma(X_k)$ and $\nabla\sigma(X_k)$;
5:     **if** ($k = 0$) **then**
6:         **set** $D_0 = -\nabla\sigma(X_0)$; **go** to line 11;
7:     **end if**
8:     **let** $Y_{k-1} = \nabla\sigma(X_k) - \nabla\sigma(X_{k-1})$; **let** $S_{k-1} = X_k - X_{k-1}$;
9:     **let** $\mu_k = \min\left(\mu_{\max}, \max\left(\mu_{\min}, \frac{\langle Y_{k-1}, S_{k-1}\rangle}{\langle S_{k-1}, S_{k-1}\rangle}\right)\right)$;
10:     **let** $D_k = -\frac{1}{\mu_k}\nabla\sigma(X_k)$;
11:     **let** $\alpha_k = 1$;
12:     **while** ($\sigma(X_k + \alpha_k D_k) > C_k + \gamma\,\alpha_k\langle\nabla\sigma(X_k), D_k\rangle$) **do**
13:         **let** $\alpha_k = \alpha_k/2$;
14:     **end while**
15:     **let** $X_{k+1} = X_k + \alpha_k D_k$; **let** $Q_{k+1} = \eta_k Q_k + 1$; **let** $C_{k+1} = (\eta_k Q_k C_k + \sigma(X_{k+1}))/Q_{k+1}$;
16:     **let** $k = k + 1$;
17: **end while**

4 Computational Experiments

We present some computational experiments on a set of artificially generated instances of the dynDGP. All codes were written in Matlab 2016b and the experiments were carried out on an Intel Core 2 Duo @ 2.4 GHz with 2 GB RAM, running Mac OS X.

We mainly focus on two kinds of motions, both defined in a two-dimensional Euclidean space. The *cascade* motion is an animation consisting of 8 points initially placed at the top of a $[0,1] \times [0,1]$ domain and subsequently moving downwards in the direction of the x axis in a cascade fashion. Two of such points are fixed during the motion, in order to avoid to define frames with almost constant inter-point distances. We refer to the second motion that we consider as the *black-hole* motion. Only one vertex remains steady in the center of a $[0,1] \times [0,1]$ domain, while the others, initially located around the boarders of the domain, tend to come closer and closer to the steady vertex. All considered motions are composed by 100 frames. The reader can make reference to the first and third column in Fig. 1 for having a better understanding of the described motions.

The two motions were initially generated by defining the trajectories of the coordinates of their 8 vertices in the two-dimensional space. From these trajectories, we generated two instances of the dynDGP such that the corresponding graph G has an edge set E for which $\hat{E} = \emptyset$. We were able to solve the inverse problem, i.e. the one of reconstructing the trajectories from the distance information in G, by solving the optimization problem presented in Sect. 2, with the method described in Sect. 3. The quality of the found solutions is measured with the function

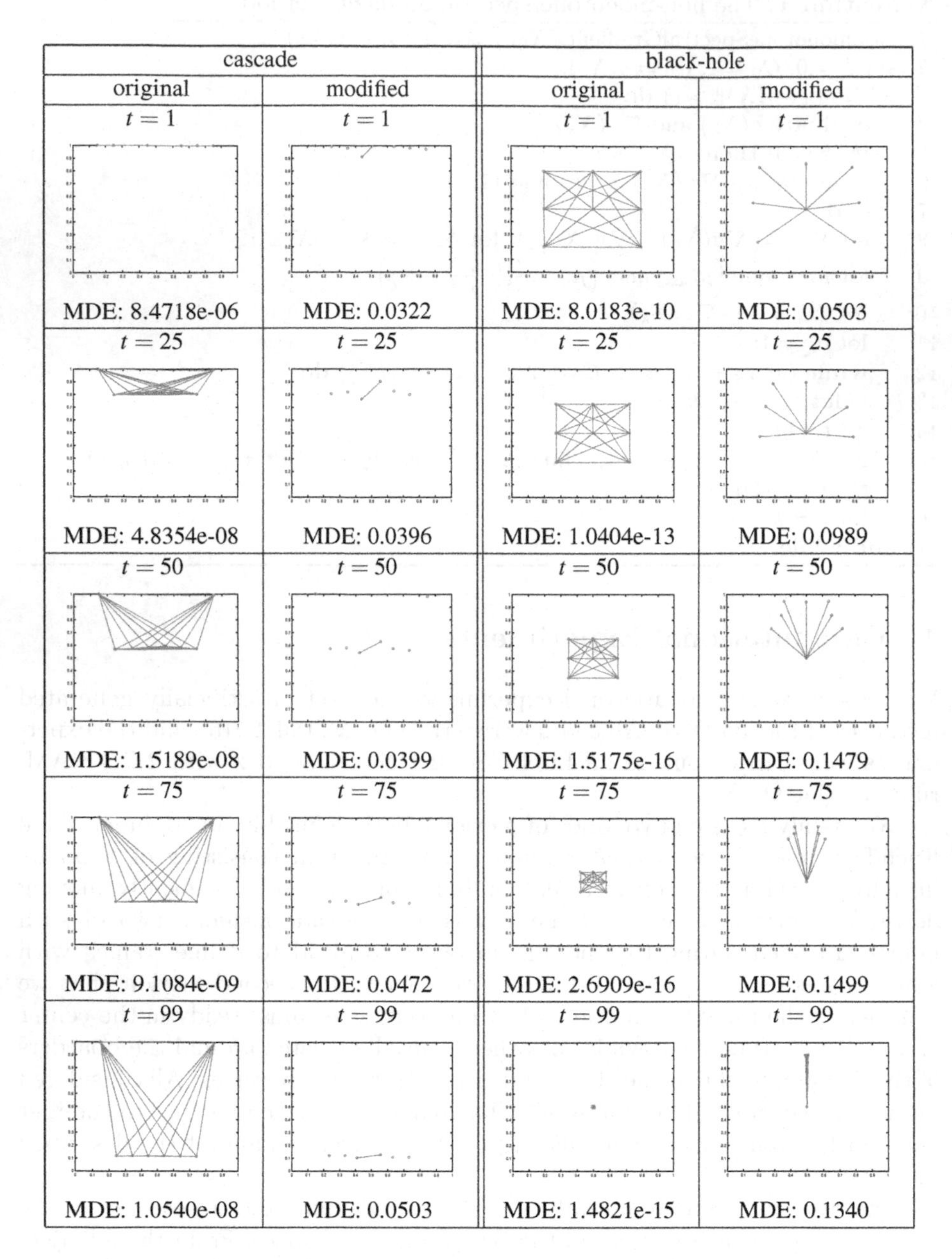

cascade
black-hole
original
modified
original
modified
t = 1
MDE: 8.4718e-06
MDE: 0.0322
MDE: 8.0183e-10
MDE: 0.0503
t = 25
MDE: 4.8354e-08
MDE: 0.0396
MDE: 1.0404e-13
MDE: 0.0989
t = 50
MDE: 1.5189e-08
MDE: 0.0399
MDE: 1.5175e-16
MDE: 0.1479
t = 75
MDE: 9.1084e-09
MDE: 0.0472
MDE: 2.6909e-16
MDE: 0.1499
t = 99
MDE: 1.0540e-08
MDE: 0.0503
MDE: 1.4821e-15
MDE: 0.1340

Fig. 1. Some selected frames of the obtained motions, for both instances cascade and black-hole, without and with some modifications on the original distances ($\tau = 3.0$ and $d_0 = 0.4$). The segments mark the distances with highest priorities. When no distances are modified, they all have the same priority.

$$MDE(X) = \sum_{\{u,v\} \in E_t} \frac{|\,||x_u - x_v|| - \delta_{uv}\,|}{\delta_{uv}},$$

which has maximal values $4 \cdot 10^{-6}$ and $8 \cdot 10^{-10}$, respectively, for the cascade and black-hole motions (see Fig. 1, first and third columns).

We also consider instances of the dynDGP with modified distances. Given a graph G obtained from a known motion, we wish to control this motion by imposing a new set of distance constraints. In our cascade motion, we impose that the distance δ_{uv}^t between two particular vertices increases by a factor $\tau \in [0, 3]$ during the motion ($\delta_{uv}^t \leftarrow \tau\, \delta_{uv}^t$, for all t). In the black-hole motion, we impose instead that the distance δ_{uv}^t between the steady vertex and any other is always greater than a given threshold $d_0 > 0$ ($\delta_{uv}^t \leftarrow \max(d_0, \delta_{uv}^t)$).

In both cases, these modifications on the original distances concern a small percentage of distances. Therefore, the optimization process may tend to optimize the original and "genuine" distances while ignoring the new ones, because they would only increase a relatively small number of terms in the objective function. The use of the weights π_{uv}^t on the distances δ_{uv}^t becomes therefore fundamental: in our experiments with modified distances, we give two levels of priorities to the distances. Higher priority is given to the modified distances, i.e. $\pi_{uv}^t = 1$ (π_{uv}^t is instead set to 0.5 when the corresponding distance value is not modified).

Table 1 shows the smallest and the largest values for the MDE function for all frames t composing the obtained motions. This table shows an increase of the MDE values when the values of τ and d_0 are larger: this was expected because the perturbation on the original distances becomes more and more significant. It is important to remark that the MDE values may become smaller when assigning no priorities to the distances, but then, as mentioned above, the obtained motions may not satisfy the newly introduced distance constraints. In this context, the analysis of the motion by a viewer, rather than the measure of a quality index such as the MDE, is often preferred for validating the results (see Fig. 1, second and forth columns).

Table 1. Computational experiments with different values for τ and d_0, for both instances cascade and black-hole. The smallest and the largest MDE values over the frames at time t of the obtained motions are reported.

Cascade			Black-hole		
τ	MDE_{min}	MDE_{max}	d_0	MDE_{min}	MDE_{max}
1.0	2.4035e-09	4.5095e-06	0.0	1.0637e-16	8.0183e-10
1.4	0.0113	0.0608	0.1	8.0183e-10	0.1518
1.8	0.0245	0.0416	0.2	8.0183e-10	0.1518
2.0	0.0315	0.0504	0.3	1.7766e-08	0.1518
3.0	0.0639	0.1016	0.4	0.0503	0.1518

5 Conclusions

We introduced the dynDGP where an embedding represents a motion of a given set of objects that satisfies the distance constraints in the graph G. We focused our attention on problems where no inter-frame distance information is given, and proposed to formulate an optimization problem for the identification of every frame of the motion, where the distance values are coupled with priority levels.

This work is motivated by various emerging real-life applications. In character animation, for example, motions are generally simulated by modifying pre-recorded motions [13], and an attempt to perform this modification through distance constraints was already proposed in [9]. In air-traffic control, the positions of a set of flying airplanes needs to be predicted so that some distance constraints are satisfied, which are defined for guaranteeing collision avoidance (see for example the recent work in [15]). Similarly, in crowd simulations, one is interested in simulating a crowd motion in different situations, where distances between pairs of pedestrians can be estimated and exploited for the simulations [4].

The formulation of these applications as a dynDGP, as well as the applicability of the proposed method for the computation of the motions, will be the subject of future research. Notice that our assumption $\hat{E} = \emptyset$ may not be feasible for all these applications: our method may need to be extended and adapted to the different situations.

Acknowledgments. This work was partially supported by an INS2I-CNRS 2016 "PEPS" project. The authors are thankful to Franck Multon and Ludovic Hoyet for the fruitful discussions.

References

1. Barzilai, J., Borwein, J.: Two-point step size gradient methods. IMA J. Num. Anal. **8**, 141–148 (1988)
2. Birgin, E.G., Martínez, J.M., Raydan, M.: Nonmonotone spectral projected gradient methods on convex sets. SIAM J. Optim. **10**, 1196–1211 (2000)
3. Birgin, E.G., Martínez, J.M.: Large-scale active-set box-constrained optimization method with spectral projected gradients. Comput. Optim. Appl. **23**, 101–125 (2002)
4. Bruneau, J., Dutra, T.B., Pettré, J.: Following behaviors: a model for computing following distances. Transp. Res. Procedia **2**, 424–429 (2014)
5. de Leeuw, J.: Convergence of the majorization method for multidimensional scaling. J. Classif. **5**, 163–180 (1988)
6. Glunt, W., Hayden, T.L., Raydan, M.: Molecular conformations from distance matrices. J. Comput. Chem. **14**(1), 114–120 (1993)
7. Grippo, L., Lampariello, F., Lucidi, S.: A nonmonotone line search technique for Newton's method. SIAM J. Num. Anal. **23**, 707–716 (1986)
8. Grippo, L., Lampariello, F., Lucidi, S.: A truncated newton method with nonmonotone line search for uncontrained optimization. J. Optim. Theory Appl. **60**, 401–419 (1989)

9. LeNaour, T., Courty, N., Gibet, S.: Cinématique guidée par les distances. Rev. Electron. Francophone Inform. Graph. Assoc. Franaise Inform. Graph. **6**(1), 15–25 (2012)
10. Liberti, L., Lavor, C., Maculan, N., Mucherino, A.: Euclidean distance geometry and applications. SIAM Rev. **56**(1), 3–69 (2014)
11. Mucherino, A., de Freitas, R., Lavor, C.: Distance geometry and applications. Discrete Appl. Math. **197**, 1–144 (2015). Special Issue
12. Mucherino, A., Lavor, C., Liberti, L., Maculan, N. (eds.) Distance Geometry: Theory, Methods and Applications, 410 p., Springer, Heidelberg (2013)
13. Multon, F., France, L., Cani-Gascuel, M.P., Debunne, G.: Computer animation of human walking: a survey. J. Vis. Comput. Anim. **10**(1), 39–54 (1999)
14. Nocedal, J., Wright, S.J.: Numerical Optimization. Springer Series in Operations Research and Financial Engineering. Springer, New York (2006)
15. Omer, J.: A space-discretized mixed-integer linear model for air-conflict resolution with speed and heading maneuvers. Comput. Oper. Res. **58**, 75–86 (2015)
16. Zhang, H., Hager, W.W.: A nonmonotone line search technique and its applications to unconstrained optimization. SIAM J. Optim. **14**(4), 1043–1056 (2004)

Distance Geometry in Linearizable Norms

Claudia D'Ambrosio(✉) and Leo Liberti

LIX CNRS (UMR7161), École Polytechnique, 91128 Palaiseau, France
{dambrosio,liberti}@lix.polytechnique.fr

Abstract. Distance Geometry puts the concept of distance at its center. The basic problem in distance geometry could be described as drawing an edge-weighted undirected graph in $\mathbb{R}^K$ for some given K such that the positions for adjacent vertices have distance which is equal to the corresponding edge weight. There appears to be a lack of exact methods in this field using any other norm but ℓ_2. In this paper we move some first steps using the ℓ_1 and ℓ_∞ norms: we discuss worst-case complexity, propose mixed-integer linear programming formulations, and sketch a few heuristic ideas.

Keywords: Distance geometry · Norms · Mathematical programming

1 Introduction

We discuss the following basic problem in Distance Geometry (DG)

> Distance Geometry Problem (DGP). Given an integer $K > 0$ and a simple, edge-weighted, undirected graph $G = (V, E, d)$, where $d : E \to \mathbb{R}_+$, determine whether there exists a *realization function* $x : V \to \mathbb{R}^K$ such that:
> $$\forall \{i, j\} \in E \quad \|x_i - x_j\| = d_{ij}, \qquad (1)$$

where $\|\cdot\|$ is either the ℓ_1 or the ℓ_∞ norm. We assume all along that, without loss of generality, G is connected, otherwise it suffices to realize the disconnected components independently.

Most existing work concerning the DGP focuses on the ℓ_2 (or Euclidean) norm, the only exception being [4] on ℓ_∞ norm DGPs. In this paper, we move some steps forward in the direction of the ℓ_1 and ℓ_∞ norms, which we call *linearizable norms*, since their unit spheres are polyhedral.

DG in the ℓ_2 norm recently received a lot of attention [7,14] due to its widespread use in engineering and science applications, such as, for example, finding the structure of proteins from Nuclear Magnetic Resonance (NMR) interatomic distances [20] and many others. It was shown in [18] that the DGP (with the Euclidean norm and $K = 1$) is **NP**-hard, by reduction from Partition.

L. Liberti—Partly supported by the ANR "Bip:Bip" project under contract ANR-10-BINF-0003.

F. Nielsen and F. Barbaresco (Eds.): GSI 2017, LNCS 10589, pp. 830–837, 2017.
https://doi.org/10.1007/978-3-319-68445-1_95

The limited attention to DG in other norms stems from a scarcity of applications. Yet, recently, we were made aware of applications for both of the linearizable norms. The DGP with the ℓ_1 norm arises in the positioning of mobile sensors in urban areas [2]. The DGP with the ℓ_∞ or with the ℓ_1 can be used in order to fill a hypercube with a pre-fixed number of "well-distributed" points, which is relevant in the design of experiments [5,19].

The rest of this paper is organized as follows. We recall some notions relating to linearizable norms in Sect. 2. We propose two new formulations for linearizable norms in Sect. 3, and prove that the DGP in linearizable norms is **NP**-complete for any K. Lastly, we sketch some new ideas for solving the linearizable norm DGP using heuristics in Sect. 4.

2 Known Results for ℓ_1 and ℓ_∞ Norms

Complexity: Since the $\ell_1, \ell_2, \ell_\infty$ norms coincide for $K = 1$, the reduction from the PARTITION problem given in [18] shows that the DGP is **NP**-complete for these three norms in $K = 1$. Since a realization in one dimension can be embedded isometrically in any number of dimensions, this shows by inclusion that the DGP is **NP**-hard for these three norms. We show in Theorem 1 that the ℓ_1 and ℓ_∞ norm variants are also **NP**-complete for $K > 1$. This strikes a remarkable difference with the Euclidean norm DGP, for which the status of membership in **NP** is currently unknown [1].

Isometric embeddings: For the ℓ_∞ norm, the isometric embedding problem can be solved in a very elegant way [10]: any finite metric (X, d) with $X = \{x_1, \ldots, x_n\}$ can be embedded in $\mathbb{R}^n$ using the ℓ_∞ norm by means of the *Fréchet embedding*: $\forall i, j \leq n \quad T(x_i) = (d(x_i, x_1), \ldots, d(x_i, x_n))$. The proof is short and to the point: for any $i, j \leq n$ we have:

$$\|T(x_i) - T(x_j)\|_\infty = \max_{k \leq n} |d(x_i, x_k) - d(x_j, x_k)| \leq \max_{k \leq n} d(x_i, x_j) = d(x_i, x_j)$$

by the triangle inequality on the given metric d. Moreover, the maximum over k of $|d(x_i, x_k) - d(x_j, x_k)|$ is obviously achieved when $k \in \{i, j\}$, in which case $|d(x_i, x_k) - d(x_j, x_k)| = d(x_i, x_j)$. So we have $\|T(x_i) - T(x_j)\|_\infty = d(x_i, x_j)$ as claimed. We remark that (X, d) need not be given explicitly: the distance matrix is enough. We also remark that a Fréchet embedding can be constructed for any given square symmetric matrix A: if the Fréchet embedding of A is infeasible w.r.t. A, it means that A is not a valid distance matrix.

For the ℓ_1 norm no such general result is known. It is known that ℓ_2 metric spaces consisting of n points can be embedded in a vector space of $O(n)$ dimensions in ℓ_1 norm almost isometrically [16, Sect. 2.5] (the "almost" refers to a multiplicative distortion measure of the form: $(1 - \varepsilon)\|x\|_2 \leq \|T(x)\|_1 \leq (1 + \varepsilon)\|x\|_2$ for some $\varepsilon \in (0, 1)$).

3 MILP Formulations for Linearizable Norms

The DGP is a nonlinear feasibility problem. As such, it can be modelled by means of Mathematical Programming (MP), which is a formal language for describing optimization problems. When the norm is linearizable, it is possible to replace nonlinear functions by piecewise linear forms, which are routinely modelled in MP using binary variables and linear forms. This yields MPs of the Mixed-Integer Linear Programming (MILP) class. A convenient feature of MILP is that solution technology is very advanced — MILP solvers are currently at the forefront for their generality and empirical efficiency. To the best of our knowledge, no MILP formulations have ever been proposed for the DGP in linearizable norms. We give here two MILP formulations for ℓ_1 and ℓ_∞ norms.

We first re-write Eq. (1) as follows: $\min\limits_x \sum\limits_{\{i,j\}\in E} |\, \|x_i - x_j\|_p - d_{ij}\,|$, for $p \in \{1,\infty\}$. Obviously, even if the unconstrained optimization problem above always has a feasible solution, it has global optimal value zero if and only if the global optimum is a solution of Eq. (1). The MILP formulations below can be solved using any off-the-shelf MILP solver, such as, e.g., CPLEX [13].

The ℓ_1 norm. For $p = 1$ we write:

$$\min_x \sum_{\{i,j\}\in E} \left| \sum_{k\le K} |x_{ik} - x_{jk}| - d_{ij} \right|. \tag{2}$$

The MILP reformulation we propose is the following:

$$\left.\begin{array}{rr} \min\limits_{x,s,t,z} & \sum\limits_{\{i,j\}\in E} s_{ij} \\ \forall \{i,j\}\in E & -s_{ij} \le \sum\limits_{k\le K} (t^+_{ijk} + t^-_{ijk}) - d_{ij} \le s_{ij} \\ \forall k\le K, \{i,j\}\in E & t^+_{ijk} - t^-_{ijk} = x_{ik} - x_{jk} \\ \forall k\le K, \{i,j\}\in E & t^+_{ijk} \le d_{ij} z_{ijk} \\ \forall k\le K, \{i,j\}\in E & t^-_{ijk} \le d_{ij}(1 - z_{ijk}) \\ \forall k\le K & \sum\limits_{i\in V} x_{ik} = 0 \\ \forall \{i,j\}\in E & s_{ij} \in [0, d_{ij}] \\ \forall k\le K, \{i,j\}\in E & t^+_{ijk}, t^-_{ijk} \in [0, d_{ij}] \\ \forall k\le K, \{i,j\}\in E & z_{ijk} \in \{0,1\}. \end{array}\right\} \tag{3}$$

Additional variables $s_{ij} \ge 0$, for each $\{i,j\} \in E$, are non-negative decision variables $s_{ij} \ge 0$ that represent the outermost absolute value of (2) thanks to the well known fact that $\min |f|$ is equivalent to $\min\limits_{\hat{f}\ge 0} \hat{f}$ subject to $-\hat{f} \le f \le \hat{f}$. Moreover, additional slack and surplus variables t^+, t^- were introduced to reformulate the innermost absolute value terms. In order to do this, they are subject to complementarity constraints $t^+_{ijk} t^-_{ijk} = 0$, $\forall k \le K, \{i,j\} \in E$, that in (3) were linearized by adding binary variables z and the two sets of standard "big-M" constraints that links the t and the z variables.

Note that, as each realization can be translated at will, constraint $\sum_{i \in V} x_i = 0$ can be safely added. It means that we can, without loss of generality, impose that the barycenter is zero this is not necessary but it is useful in practice, see [6] for detailed empirical results. Again for practical efficiency, we let $U = \sum_{\{i,j\} \in E} d_{ij}$ and use it to bound x, so that for each $i \in V$ and $k \leq K$, we have $x_{ik} \in [-U, U]$.

Proposition 1. *Equation* (3) *is a valid formulation for the DGP using the* ℓ_1 *norm.*

The ℓ_∞ norm. For $p = \infty$ we write: $\min_x \sum_{\{i,j\} \in E} \left| \max_{k \leq K} |x_{ik} - x_{jk}| - d_{ij} \right|$.

The MILP reformulation that we propose is the following:

$$\left.\begin{array}{rr} \min\limits_{x,w,z,s,t} & \sum\limits_{\{i,j\} \in E} s_{ij} \\ \forall \{i,j\} \in E, k \in K & t^+_{ijk} + t^-_{ijk} - d_{ij} \leq s_{ij} \\ \forall \{i,j\} \in E, k \in K & t^+_{ijk} + t^-_{ijk} + s_{ij} \geq d_{ij} w_{ijk} \\ \forall \{i,j\} \in E & \sum\limits_{k \leq K} w_{ijk} \geq 1 \\ \forall k \leq K, \{i,j\} \in E & t^+_{ijk} - t^-_{ijk} = x_{ik} - x_{jk} \\ \forall k \leq K, \{i,j\} \in E & t^+_{ijk} \leq d_{ij} z_{ijk} \\ \forall k \leq K, \{i,j\} \in E & t^-_{ijk} \leq d_{ij}(1 - z_{ijk}) \\ \forall k \leq K & \sum\limits_{i \in V} x_{ik} = 0 \\ \forall \{i,j\} \in E & s_{ij} \in [0, d_{ij}] \\ \forall k \leq K, \{i,j\} \in E & t^+_{ijk}, t^-_{ijk} \in [0, d_{ij}] \\ \forall k \leq K, \{i,j\} \in E & w_{ijk}, z_{ijk} \in \{0,1\} \end{array}\right\} \quad (4)$$

where the outermost and the innermost absolute values are reformulated as before and the barycenter constraint is added.

At this point similarities with the ℓ_1 norm stop. Note that the first three set of constraints model the linearization of constraints $-s_{ij} \leq \max_{k \leq K}(t^+_{ijk} + t^-_{ijk}) - d_{ij} \leq s_{ij}, \forall \{i,j\} \in E$. The center and right-hand-side can be reformulated as the first set of constraints. The left-hand-side and the center is more complicated as it corresponds to a non-convex constraint. In order to linearize it, we need to add auxiliary binary variables w and the second and third set of constraints, which express the fact that *at least one out of K components satisfies the constraint* (that component being the maximum, of course).

Proposition 2. *Equation* (4) *is a valid formulation for the DGP using the* ℓ_∞ *norm.*

The DGP with linearizable norms is in NP. Our MILP formulations make it easy to prove the following.

Theorem 1. *The DGP in linearizable norms is* **NP***-complete.*

Proof. It was already remarked in Sect. 2 that the DGP in linearizable norms is **NP**-hard by reduction from PARTITION to the DGP in $K = 1$, where the three norm $\ell_1, \ell_2, \ell_\infty$ coincide. Now by Propositions 1 and 2 we know that certificates to the DGP in linearizable norms are rational for rational input, since the MP formulations in (3) and (4) can be solved by a Branch-and-Bound (BB) algorithm, each node of which involves the solution of a Linear Program (LP), which is known to be in **NP**. Simple BB implementations will continue branching until the incumbent, found by solving the LP at some node, is proven optimal. So the certificate (solution) provided by the BB is polynomially sized, as claimed. □

3.1 Computational Results

Both formulations (Eqs. (3) and (4)) belong to the MILP class, and can be solved by several existing MILP solvers. We employ CPLEX 12.6.2 [13] on a MacBook Pro mid-2015 running Darwin 15.5.0 on a (virtual) quad-core i7 at 3.1 GHz with 16 GB RAM.

We generated a set of (feasible) random DGP instances in both norms as follows: for each cardinality value $n = |V|$ of the vertex set of G ranging over $\{10, 15, 20, 25, 30, 35, 40\}$ we sampled n points in the plane, bounded by a box. We then added a Hamiltonian cycle to the edge set, in order to guarantee connectedness. Lastly, we add the remaining edges with an Erdős-Rényi generation process [8] having probability s ranging over $\{0.1, 0.2, 0.3, 0.5, 0.8\}$. This yields 35 instances per norm $p \in \{1, \infty\}$.

We deployed CPLEX on these 70 instances with a time limit of 10 min enforced on the "wall clock", meaning the actual elapsed time. We used this measure instead of the user time since CPLEX exploits all four processor cores, which means that the system time taken for parallel execution tasks is essential. Since the running time is limited, we do not always find feasible solutions. We evaluate the error by employing two well known measures: the scaled Mean Distance Error (MDE) and the scaled Largest Distance Error (LDE).

$$\mathsf{MDE}(x) = \frac{1}{|E|} \sum_{\{i,j\} \in E} \frac{|\|x_i - x_j\|_p - d_{ij}|}{d_{ij}}, \quad \mathsf{LDE}(x) = \max_{\{i,j\} \in E} \frac{|\|x_i - x_j\|_p - d_{ij}|}{d_{ij}}.$$

Intuitively, the scaled MDE gives an idea of the percentual average discrepancy of the given realization from the given data. The scaled LDE gives an idea of the percentual average worst error over all edges. Obviously, the LDE is generally higher than the MDE.

The detailed results are not reported for lack of space. For each instance we computed scaled MDE and LDE scores, and compared the wall clock CPU time taken by CPLEX running on Eqs. (3) and (4).

From these results, it appears clear that DGP instances for linearizable norms can be solved in practice up to at least $n = 40$, though the densest instances are more difficult with the ℓ_∞ norm. A good ℓ_1 norm solution (with MDE and

LDE smaller than 10^{-4}) has been obtained using CPLEX on Eq. (3) for a 50 vertex instance with $s = 0.8$ in just over 304 s of wall clock time, but no solution was obtained within 10 min of wall clock time for a 60 vertex instance and $s = 0.8$. This seems to indicate the need for heuristic methods (discussed below) to address larger instance sizes.

4 Heuristic Ideas

In this section we provide some ideas to design heuristic methods for the DGP problem with ℓ_∞ and ℓ_1 norm and report computational results.

4.1 Solving the DMCP in the ℓ_∞ Norm

A problem related to DGP, the Distance Matrix Completion Problem (DMCP), asks whether a partially defined matrix can be completed to a (full) distance matrix in a specified norm. Although a solution of the DMCP is a distance matrix, finding the missing distance values usually entails finding a realization that is consistent with the given values: this realization is then used to compute the missing entries in the given partial matrix. In this sense, realizations provide certificates for both DGP and DMCP. In the DMCP, however, differently from the DGP, the dimensionality K of the embedding space is not given — realizations into $\mathbb{R}^K$ for any $K > 0$ will provide a feasibility certificate for DMCP instances.

In the following we propose a heuristic for DMCP that will be the starting point for a heuristic for DGP. Our first proposal concerns the exploitation of the Fréchet embedding to "approximately solve" the DMCP (rather than the DGP) in the ℓ_∞ norm. Our algorithm works as follows:

1. let A' be the $n \times n$ weighted adjacency matrix of G, where off-diagonal zeros are to be considered as "unspecified entries"
2. complete A' to a full symmetric matrix A using the Floyd-Warshall all-shortest-paths algorithm [17]
3. output the realization $x : V \to \mathbb{R}^n$ given by the Fréchet embedding $x_i = A_i$ for each $i \in V$.

By "approximately solve" we mean that the output realization x is generally not going to be a valid certificate for the given DMCP instance, but that its MDE and LDE error measures are hopefully going to be low enough. The worst-case complexity of this heuristic is dominated by the Floyd-Warshall algorithm, which is generally $O(n^3)$. Experimentally, we found that this heuristic is useless for sparse graphs (where the reconstruction of A' has the highest chances of being wrong), but is both fast and successful for dense graphs, notably those with $s = 0.8$, which represented the hardest instances in the tests. This heuristic took 2.42 s to find *all* realizations for 17 random graphs obtained with $s = 0.8$ and n varying in $\{5 + 5\ell \mid 1 \leq \ell \leq 17\}$. The cumulative scaled MDE over all instances is 0.019, with cumulative scaled LDE at 9.61 mostly due to one bad outlier. Overall, it found MDE and LDE smaller than 10^{-4} for 11 over 17 instances. In conclusion, it is both fast and effective.

4.2 Solving the DGP in the ℓ_∞ Norm

The second idea turns the above heuristic into a method for solving DGP in ℓ_∞ norm for a given dimensionality K: it consists in selecting the K columns from the realization $x \in \mathbb{R}^{n\times n}$ obtained in Sect. 4.1 which best match the given distances. For this purpose, we solve the following MILP (based on Eq. (4)):

$$\left.\begin{array}{rr} \min\limits_{w,y,s} & \sum\limits_{\{i,j\}\in E} s_{ij} \\ \forall \{i,j\}\in E, k\le n & |x_{ik}-x_{jk}|y_k - s_{ij} \le d_{ij} \\ \forall \{i,j\}\in E, k\le n & |x_{ik}-x_{jk}|y_k + s_{ij} \ge d_{ij}w_{ijk} \\ \forall \{i,j\}\in E & \sum\limits_{k\le K} w_{ijk} \ge 1 \\ & \sum\limits_{k\le n} y_k = K \\ \forall \{i,j\}\in E & s_{ij} \in [0,d_{ij}] \\ \forall k\le K, \{i,j\}\in E & w_{ijk}, y_k \in \{0,1\}. \end{array}\right\} \quad (5)$$

In Eq. (5), note that x are no longer decision variables, but constants (output of the Fréchet heuristic for the DMCP). The new decision variables $y \in \{0,1\}^n$ decide whether coordinate k should be in the realization in $\mathbb{R}^K$ or not. Since solving MILP takes exponential time, and we would like the heuristic to be fast, we set a 120 s time limit on CPLEX. For instances up to $n = 70$ with s fixed at 0.8, the average scaled MDE error is non-negligible but still acceptable (around 0.18) while the average scaled LDE is disappointingly close to 0.9. However, this heuristic allows us to solve instances which the MP formulations of Sect. 3 cannot solve.

4.3 Solving the DGP in the ℓ_1 Norm

At first, we tested a Variable Neighbourhood Search (VNS) [12] type algorithm with neighbourhoods defined by centers given by infeasible realization x', and radii given by the maximum number of coordinates of x' that are allowed to change during a BB search. This constraint is enforced by a Local Branching (LB) [9] mechanism added to Eq. (3). The centers are computed using the probabilistic constructive heuristic proposed in [15], which extends the concept of Fréchet embeddings to sets. Unfortunately, this idea yielded very poor results, both quality-wise and in terms of CPU time. We only report it to prevent other researchers from pursuing this same direction.

The second idea we tested is based on alternating solutions between two continuous reformulations of Eq. (2), denoted $\mathcal{A}$ (Eq. (2) with $|\cdot|$ replaced by $\sqrt{(\cdot)^2+\varepsilon}$) and $\mathcal{B}$ ((3) with nonlinear complementarity constraints): we randomly sample an infeasible realization as a starting point for $\mathcal{A}$, then use $\mathcal{A}$'s solution as a starting point to $\mathcal{B}$. We repeat this loop, updating the "best solution found so far", until a CPU time-based termination condition (set to 600s) becomes true. We used SNOPT [11] to solve $\mathcal{A}$ and IPOPT [3] to solve $\mathcal{B}$. The results, obtained on 85 instances with $n \in \{10, 15, 20, \ldots, 90\}$ of any density

$s \in \{0.1, 0.2, 0.3, 0.5, 0.8\}$, count 13 failures (generally instances with high density where SNOPT had convergence issues) and 8 feasible realizations (scaled MDE/LDE scores $< 10^{-4}$). If we consider only the results for the instances ended with no failures, overall, this heuristic displayed average scaled MDE and LDE of 0.13 and 1.16.

References

1. Beeker, N., Gaubert, S., Glusa, C., Liberti, L.: Is the distance geometry problem in NP? In: Mucherino, A., Lavor, C., Liberti, L., Maculan, N. (eds.) Distance Geometry: Theory, Methods, and Applications. Springer, New York (2013)
2. Chiu, W.-Y., Chen, B.-S.: Mobile positioning problem in Manhattan-like urban areas: Uniqueness of solution, optimal deployment of BSs, and fuzzy implementation. IEEE Trans. Sig. Process. **57**(12), 4918–4929 (2009)
3. COIN-OR. Introduction to IPOPT: A tutorial for downloading, installing, and using IPOPT (2006)
4. Crippen, G.M.: An alternative approach to distance geometry using L∞ distances. Discrete Appl. Math. **197**, 20–26 (2015). Distance Geometry and Applications
5. D'Ambrosio, C., Nannicini, G., Sartor, G.: MILP models for the selection of a small set of well-distributed points. Oper. Res. Lett. **45**, 46–52 (2017)
6. D'Ambrosio, C., Vu, K., Lavor, C., Liberti, L., Maculan, N.: New error measures and methods for realizing protein graphs from distance data. Discrete Comput. Geom. **57**(2), 371–418 (2017)
7. Dokmanić, I., Parhizkar, R., Ranieri, J., Vetterli, M.: Euclidean distance matrices: essential theory, algorithms and applications. IEEE Sig. Process. Mag. **1053–5888**, 12–30 (2015)
8. Erdős, P., Rényi, A.: On random graphs i. Publ. Math. (Debrecen) **6**, 290–297 (1959)
9. Fischetti, M., Lodi, A.: Local branching. Math. Program. **98**, 23–37 (2005)
10. Fréchet, M.: Sur quelques points du calcul fonctionnel. Rend. Circ. Mat. Palermo **22**, 1–74 (1906)
11. Gill, P.E.: User's guide for SNOPT version 7.2. Systems Optimization Laboratory. Stanford University, California (2006)
12. Hansen, P., Mladenović, N.: Variable neighbourhood search: principles and applications. Eur. J. Oper. Res. **130**, 449–467 (2001)
13. IBM: ILOG CPLEX 12.6 User's Manual. IBM (2014)
14. Liberti, L., Lavor, C., Maculan, N., Mucherino, A.: Euclidean distance geometry and applications. SIAM Rev. **56**(1), 3–69 (2014)
15. Matousek, J.: On the distortion required for embedding finite metric spaces into normed spaces. Isr. J. Math. **93**, 333–344 (1996)
16. Matoušek, J.: Lecture notes on metric embeddings. Technical report, ETH Zürich (2013)
17. Mehlhorn, K., Sanders, P.: Algorithms and Data Structures. Springer, Berlin (2008)
18. Saxe, J.: Embeddability of weighted graphs in k-space is strongly NP-hard. In: Proceedings of 17th Allerton Conference in Communications, Control and Computing, pp. 480–489 (1979)
19. Vu, K., D'Ambrosio, C., Hamadi, Y., Liberti, L.: Surrogate-based methods for black-box optimization. Int. Trans. Oper. Res. **24**(3), 393–424 (2017)
20. Wüthrich, K.: Protein structure determination in solution by nuclear magnetic resonance spectroscopy. Science **243**, 45–50 (1989)

Self-similar Geometry for Ad-Hoc Wireless Networks: Hyperfractals

Philippe Jacquet(✉) and Dalia Popescu(✉)

Bell Labs, Nokia, Paris, France
{philippe.jacquet,dalia-georgiana.herculea}@nokia-bell-labs.com

Abstract. In this work we study a Poisson patterns of fixed and mobile nodes distributed on straight lines designed for 2D urban wireless networks. The particularity of the model is that, in addition to capturing the irregularity and variability of the network topology, it exploits self-similarity, a characteristic of urban wireless networks. The pattern obeys to "Hyperfractal" measures which show scaling properties corresponding to an apparent dimension larger than 2. The hyperfractal pattern is best suitable for capturing the traffic over the streets and highways in a city. The scaling effect depends on the hyperfractal dimensions. Assuming radio propagation limited to streets, we prove results on the scaling of routing metrics and connectivity graph.

1 Introduction

The modeling of topology of *ad hoc* wireless networks makes extensive use of stochastic geometry. Uniform Poisson spatial models have been successfully applied to the analysis of wireless networks which exhibit a high degree of randomness [1,2]. Other modeling of networks such as lattice structures like Manhattan grid [6] are often used for their high degree of regularity.

We recently introduced new models based on fractal repartition [3,4] which are proven to model successfully an environment displaying self-similarity characteristics. Results have shown, [4], that a limit of the capacity in a network with a non-collaborative protocol is inversely proportional to the fractal dimension of the support map of the terminals. In the model of [4], the nodes have locations defined as a Poisson shot inside a fractal subset, for example a Cantor set.

By definition, a fractal set has a dimension smaller than the Euclidean dimension of the embedding vector space; it can be arbitrary smaller. In this work we study a recently introduced model [5], which we named "Hyperfractal", for the *ad hoc* urban wireless networks. This model is focused on the self-similarity of the topology and captures the irregularity and variability of the nodes distribution. The hyperfractal model is a Poisson shot model which has support a *measure* with scaling properties instead of a pure fractal set. It is a kind of generalization of fractal Poisson shot models, and in our cases it will have a dimension that is *larger* than the dimension of the underlying Euclidean space and this dimension can be arbitrarily large. This holds for every case of our urban traffic models.

F. Nielsen and F. Barbaresco (Eds.): GSI 2017, LNCS 10589, pp. 838–846, 2017.
https://doi.org/10.1007/978-3-319-68445-1_96

A novel aspect is that the radio model now comprises specific phenomenons such as the "urban canyon" propagation effect, a characteristic property of metropolitan environments with tall or medium sized buildings.

Using insights from stochastic geometry and fractal geometry, we study scaling laws of information routing metrics and we prove by numerical analysis and simulations the accuracy of our expressions.

The papers is organized as follows. Section 2 reminds the newly introduced geometrical model and its basic properties. Section 3 gives results on the connectivity graph and information routing metrics that are validated through simulations in Sect. 4.

2 System Model and Geometry

Let us briefly remind the new model we introduced in a recent work [5] and its basic properties. The map is assumed to be the unit square and it is divided into a grid of streets similar to a Manhattan grid but with an infinite resolution. The horizontal (resp vertical streets) streets have abscissas (resp. ordinates) which are integer multiple of inverse power of two. The number of binary digits after the coma minus indicates the level of the street, starting with the street with abscissas (resp ordinate) 1/2 being at level 0. Notice that these two streets make a central cross.

This model can be modified and generalized by taking integer multiple of inverse power of any other number called the street-*arity*, the street-arity could change with the levels, the central cross could be an initial grid of main streets, *etc.*. Figure 1 shows a map of Indianapolis as an example. It could also model the pattern of older cities in the ancient world. In this case, the model would display a similar hierarchical street distribution but plugged into a more chaotic geometric pattern instead that of the grid pattern.

2.1 Hyperfractal Mobile Nodes Distribution

The process of assigning points to the streets is performed recursively, in iterations, similar to the process for obtaining the Cantor Dust.

The two streets of level 0 form a central cross which splits the map in exactly 4 quadrants. We denote by probability p' the probability that the mobile node is located on the cross according to a uniform distribution and q' the complementary probability. With probability $q'/4$, it is located in one the four quadrants. The assignation procedure recursively continues and it stops when the mobile node is assigned to a cross of a level $m \geq 0$. A cross of level m consists of two intersecting segments of streets of level m. An example of a decreasing density in the street assignment process performed in $L = 4$ steps is given in Fig. 2.

Taking the unit density for the initial map, the density of mobile nodes in a quadrant is $q'/4$. Let μ_H be the density of mobile nodes assigned on a street of level H. It satisfies:

$$\mu_H = (p'/2)(q'/2)^H \tag{1}$$

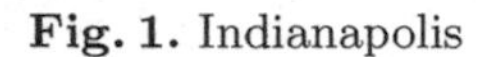

Fig. 1. Indianapolis

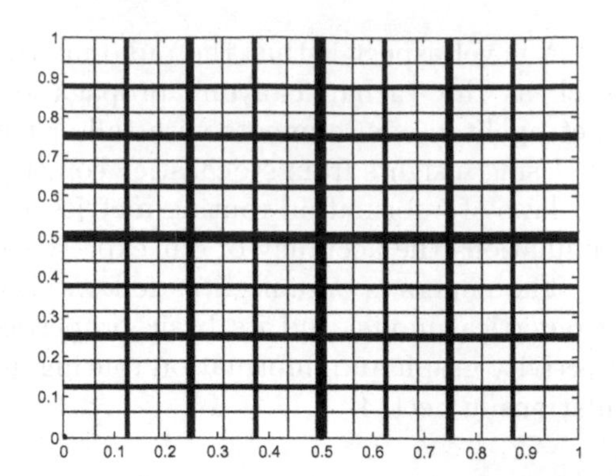

Fig. 2. Hyperfractal support

The measure (understood in the Lebesgue meaning) which represents the actual density of mobile nodes in the map has strong scaling properties. The most important one is that the map as a whole is identically reproduced in each of the four quadrants but with a weight $q'/4$ instead of 1. Thus the measure has a structure which recalls the structure of a fractal set, such as the Cantor map. A crucial difference lies in the fact that its dimension, d_m, is in fact *greater* than 2, the dimension of the underlying Euclidean space. Indeed, considering the map in only half of its length consists into considering the same map but with a reduced weight by a factor $q'/4$. One obtains:

$$\left(\frac{1}{2}\right)^{d_m} = q'/4 \qquad \text{thus} \qquad d_m = \frac{\log(\frac{4}{q'})}{\log 2} > 2 \tag{2}$$

This property can only be explained via the concept of measure. Notice that when $p' \to 0$ then $d_m \to 2$ and the measure tends to the uniform measure in the unit square (weak convergence).

2.2 Canyon Effect and Relays

Due to the presence of buildings, the radio wave can hardly propagate beyond the streets borders. Therefore, we adopt the *canyon* propagation model where the signal emitted by a mobile node propagates only on the axis where it stands on. Considering the given construction process, the probability that a mobile node is placed in an intersection tends to zero and mobiles positioned on two different streets will never be able to communicate. Therefore, one needs to add relays in some street crossings in order to guarantee connectivity and packet delivery.

We again make use of a hyperfractal process to select the intersections where a relay will be placed. We denote by p a fixed probability and $q = 1 - p$ the complementary probability. A run for selecting a street crossing requires two processes: the in-quadrant process and the in-segment process. The selection starts with the in-quadrant process as follows. (i) With probability p^2, the selection is the central crossing of the two streets of level 0; (ii) with probability $p(q/2)$, the relay is placed in one of the four street segments of level 0 starting at this point: North, South, West or East, and the process continues on the segment with the in-segment process. Otherwise, (iii) with probability $(q/2)^2$, the relay is placed in one of the four quadrants delimited by the central cross and the in-quadrant process recursively continues.

The process of placing the relays is illustrated in Fig. 3. We perform M independent runs of selection. If one crossing is selected several times (*e.g.* the central crossing), only one relay will be installed in the respective crossing. This reduction will mean that the number of actually placed relays will be much smaller than M.

Some basics results following the construction process and shown in [5] are as following. The relay placement is hyperfractal with dimension d_r:

$$d_r = 2\frac{\log(2/q)}{\log 2}. \tag{3}$$

Let $p(H, V)$ be the probability that the run selects a crossing of two streets, one horizontal street of level H and one vertical street of level V. There are 2^{H+V} of such crossings. We have:

$$p(H, V) = p^2(q/2)^{H+V}. \tag{4}$$

Thus the probability that such crossing is selected to host a relay is $1 - (1 - p(H, V))^M$ which is equivalent to $1 - \exp(-Mp(H, V))$ when M is large. From now, we assume that the number of crossing selection run is a Poisson random variable of mean ρ, and the probability that a crossing hosts a relay is now exactly $1 - \exp(-Mp(H, V))$.

The average number of relays on a streets of level H is denoted by $L_H(\rho)$ and satisfies the identity:

$$L_H(\rho) = \sum_{V\geq 0} 2^V (1 - \exp(-p(H, V)\rho)). \tag{5}$$

The average total number of relays in the city, $R(\rho)$, has the expression:

$$R(\rho) = \sum_{H,V\geq 0} 2^{H+V} (1 - \exp(-p(H, V)\rho)) \tag{6}$$

Furthermore:

$$R(\rho) = O(\rho^{2/d_r} \log \rho) \tag{7}$$

A complete Hyperfractal map containing both mobile nodes and relays is presented in Fig. 4.

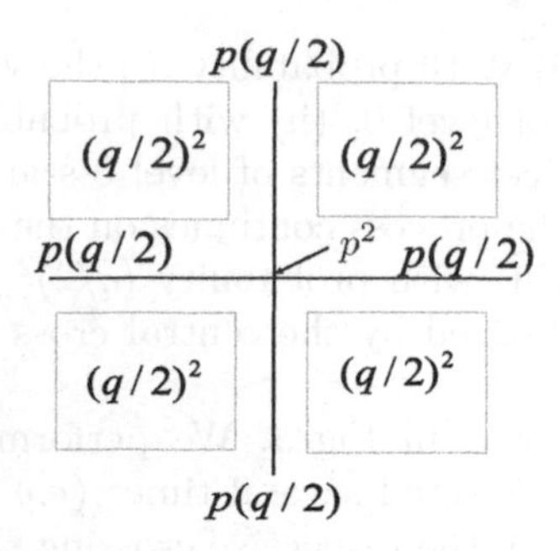

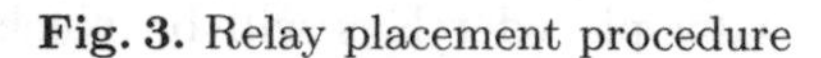
Fig. 3. Relay placement procedure

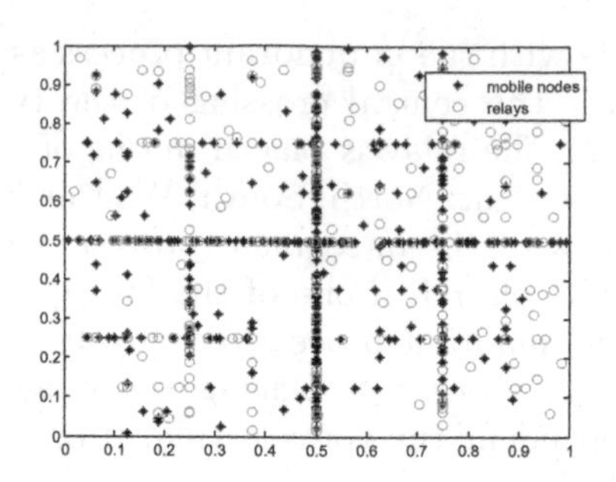

Fig. 4. Mobiles and relays

3 Routing

The routing table will be computed according to a minimum cost path over a cost matrix $[t_{ij}]$ where t_{ij} represents the cost of directly transmitting a packet from node i to node j. The min cost path from node i to node j which optimizes the relaying nodes (either mobile nodes or fixed relays) is denoted m_{ij} and satisfies:

$$m_{ij} = \min_k \{m_{ik} + t_{kj}\}, \ \forall(i, j), \tag{8}$$

Furthermore, we study here the nearest neighbor routing scenario considering the canyon effect. In this strategy the next relay is always a next neighbor on an axis, *i.e.* there exist no other nodes between the transmitter and the receiver. Thus

$$\begin{cases} t_{ij} = 1 & \text{if nodes } i \text{ and } j \text{ are aligned} \\ & \text{and } \nexists k \text{ such that } d(i,j) = d(i,k) + d(k,j) \\ t_{ij} = \infty & \text{otherwise} \end{cases}$$

Due to the canyon effect some nodes can be disconnected from the rest of the network, several connected components may appear and some routes may not exist. In the case node i and node j cannot communicate $m_{ij} = \infty$. Therefore, a very important implication in the routing process has the connectivity of the hyperfractal. Next, we study this connectivity by looking at the properties of the giant component.

3.1 Giant Component

We restrict our analysis to the giant component of the network. Following the construction process which assigns a relay in the central cross with high probability, the giant component will be formed around this relay with coordinates $[\frac{1}{2}, \frac{1}{2}]$.

Theorem 1. *The fraction of mobile nodes in the giant component tends to* 1 *when* $\rho \to \infty$ *and the average number of mobile nodes outside the giant component is* $O(N\rho^{-2(d_m-2)/d_r})$ *when* $\rho \to \infty$.

Remark: For a configuration where $d_m - 2 > d_r/2$, the average number of mobile nodes outside the giant component tends to zero when $\rho = O(N)$.

Let us now sketch a proof that will show the validity of Theorem 1.

Proof. Given a horizontal line of level H, the probability that the line is connected to the vertical segment in the central cross is $1 - e^{-\rho p^2 (q/2)^H}$.

On each line of level H the density of mobiles is $(p'/2)(q'/2)^H$.

Furthermore, there are 2^H of such lines intersecting each of the lines forming the central cross. We multiply by 2 and obtain $g(\rho)$, the cumulated density of lines connected to the central cross with a single relay:

$$g(\rho) = 2 \sum_{H \geq 1} 2^H (p'/2)(q'/2)^H (1 - e^{-\rho p^2 (q/2)^H}) \tag{9}$$

The quantity $g(\rho)$ is a lower bound of the fraction of mobile nodes connected to the central cross. It is indeed a lower bound as a line can be connected to the central cross via a sequence of relays, while above we consider the lines which are connected via a single relay. The fraction of nodes connected to the central nodes including those nodes in the central relay (which are in density $2p$) is therefore lower bounded by the quantity $2p + g(\rho)$.

Let $E(\rho, N)$ be the average number of mobile nodes outside the giant component. We have $E(\rho, N) \leq Ne(\rho)$ represented with an harmonic sum:

$$e(\rho) = 1 - 2p - g(\rho) = 2 \sum_{H \geq 1} 2^H (p'/2)(q'/2)^H e^{-\rho p^2 (q/2)^H} \tag{10}$$

The Mellin Transform $e^*(s)$ of $e(\rho)$ is:

$$e^*(s) = \int_0^\infty e(\rho)\rho^{s-1} d\rho = \Gamma(s) 2 \sum_{H \geq 1} 2^H (\frac{p'}{2})(\frac{q'}{2})^H (p^2 (q/2)^H)^{-s} = \frac{p'q'(\frac{p^2 q}{2})^{-s}}{1 - q'(\frac{q}{2})^{-s}} \Gamma(s) \tag{11}$$

The Mellin transform is defined for $R(s) > 0$ and has s a simple pole at $s_0 = \frac{\log(1/q')}{\log(2/q)} = \frac{-2(d_m - 2)}{d_r}$.

Using the inverse Mellin transform $e(\rho) = \frac{1}{2i\pi} \int_{c-i\infty}^{c+i\infty} e^*(s)\rho^{-s} ds$ for any given number c within the definition domain of $e^*(s)$, following the similar analysis as with the expressions used for mobile fractal dimension from Eq. 2 and relay fractal dimension from Eq. 3, one gets $e(\rho) = O(\rho^{-s_0})$ which proves the theorem.

Notice than when $s_0 > 1$ we have $E(\rho, N) \to 0$. Therefore, the connectivity graph tends to contain all the nodes.

3.2 Routing on the Giant Component

Let us now remind shortly a result on routing from [5] performed on the giant component. In the context of nearest neighbor routing strategy, the following holds:

Theorem 2. *The average number of hops in a Hyperfractal of N mobile nodes and hyperfractal dimensions of mobile nodes and relays d_m and d_r is:*

$$D_N = O\left(N^{1-\frac{2}{(1+1/d_m)d_r}}\right) \tag{12}$$

Proof. Mobile node mH on a line of level H sends a packet to mV, on a line of level V as in Fig. 5(a) [5]. The dominant case is $V = H = 0$. As lines H and V have high density of population, the packet will be diverted by following a vertical line of level $x > 0$ with a much lower density. A similar phenomenon happens towards mobile mV. We will consider [5] that $x = y$.

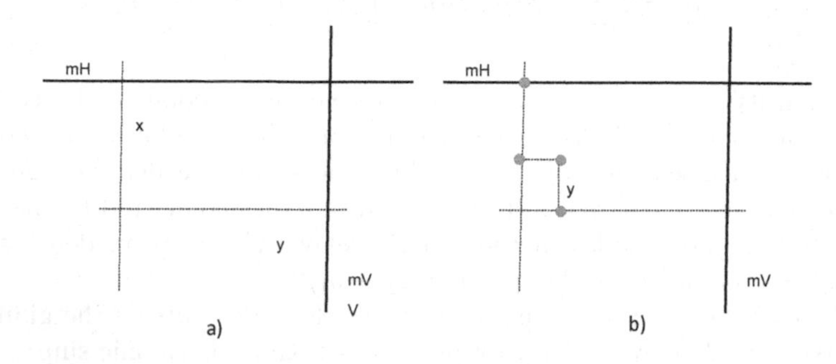

Fig. 5. Routing in a Hyperfractal (a) intermediate levels x and y, (b)extra intermediate levels

For changing direction in the route, it is mandatory that a relay exists at the crossing. Let $L(a, b)$ be the average distance between a random mobile node on a street of level a to the first relay to a street of level b. Every crossing between streets of level a and b is independent and holds a relay with probability $1 - \exp(-\rho p(a, b))$. Since such crossings are regularly spaced by interval 2^{-a} we get:

$$L(a, b) \leq \frac{2^{-a}}{1 - \exp(-\rho p(a, b))}. \tag{13}$$

Let us assume that the two streets of level x have a relay at their intersection. In this case, the average number of traversed nodes is upper bounded by $2N\mu_0 L(x, 0) + 2N\mu_x$.

In [5], the authors showed that the probability that there exists a valid relay at level x street intersection is very low, therefore an intermediate level (see Fig. 5b) has to be added. Given the probabilities of existence of relays and the densities of lines computed according to the logic described in Sect. 2, the order of the number of nodes that the packet hops on towards its destination is obtained to be as [5]: The minimized value of the number of hops is, thus:

$$D_N = O(N\rho^{-\frac{2}{(1+1/d_m)d_r}}) \tag{14}$$

4 Numerical Results

The configuration studies is chosen as to validate the constraint given by Theorem 1, $d_m - 2 > \frac{d_r}{2}$, $d_m = d_r = 3$, $N = \rho = 200, 300, 400, 500, 800, 1200, 1600$ nodes.

Figure 6 illustrates the variation of fraction of mobile nodes that are not included in the giant component. As claimed by Theorem 1, the fraction decreases with the increase of number of mobiles. Furthermore, the actual number of mobiles comprised in the giant component nodes follows the scaling law $O(N^{\frac{1}{3}})$ as shown in Fig. 7.

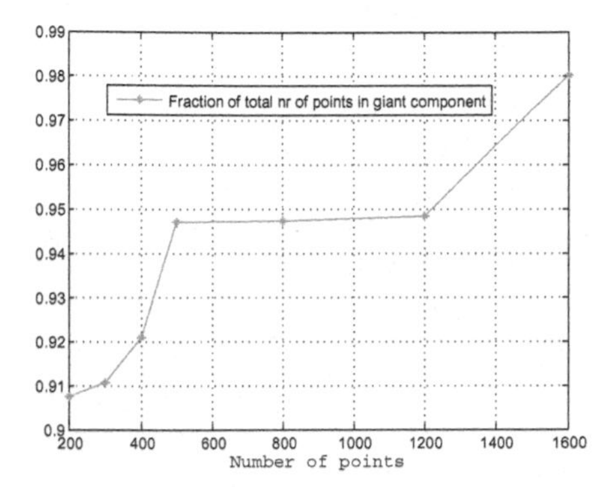

Fig. 6. Fraction of points in the giant component

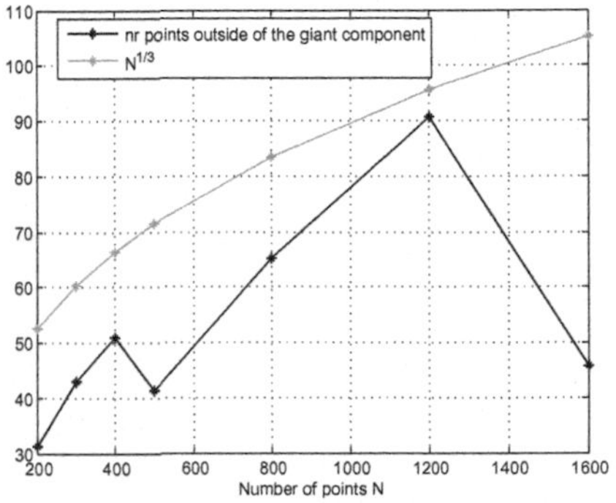

Fig. 7. Number of points outside the giant component

5 Conclusion

This work studies routing properties and scaling of a recently introduced geometrical model for wireless ad-hoc networks, called "Hyperfractal". This model which is best fit for urban wireless networks as it captures not only the irregularity and variability of the node configuration but also the self-similarity of the topology. We showed here that the connectivity properties of the Hyperfractal exhibits good properties, supporting the Hyperfractal as a new topology for urban ad-hoc wireless networks.

References

1. Blaszczyszyn, B., Muhlethaler, P.: Random linear multihop relaying in a general field of interferers using spatial aloha. IEEE TWC, July 2015
2. Herculea, D., Chen, C.S., Haddad, M., Capdevielle, V.: Straight: stochastic geometry and user history based mobility estimation. In: HotPOST (2016)

3. Jacquet, P.: Capacity of simple multiple-input-single-output wireless networks over uniform or fractal maps. In: MASCOTS, August 2013
4. Jacquet, P.: Optimized outage capacity in random wireless networks in uniform and fractal maps. In: ISIT, pp. 166–170, June 2015
5. Jacquet, P., Popescu, D.: Self-similarity in urban wireless networks: hyperfractals. In: Workshop on Spatial Stochastic Models for Wireless Networks (SpaSWiN)
6. Karabacak, M., et al.: Mobility performance of macrocell-assisted small cells in manhattan model. In: VTC, May 2014

Information Distances in Stochastic Resolution Analysis

Radmila Pribić(✉)

Sensors Advanced Developments, Thales Nederland, Delft, The Netherlands
Radmila.Pribic@nl.thalesgroup.com

Abstract. A stochastic approach to resolution is explored that uses information distances computed from the geometry of data models characterized by the Fisher information in cases with spatial-temporal measurements for multiple parameters. Stochastic resolution includes probability of resolution at signal-to-noise ratio (SNR) and separation of targets. The probability of resolution is assessed by exploiting different information distances in likelihood ratios. Taking SNR into account is especially relevant in compressive sensing (CS) due to its fewer measurements. Our stochastic resolution is also compared with actual resolution from sparse-signal processing that is nowadays a major part of any CS sensor. Results demonstrate the suitability of the proposed analysis due to its ability to include crucial impacts on the performance guarantees: array configuration or sensor design, SNR, separation and probability of resolution.

Keywords: Resolution · Information geometry · Compressive sensing · Radar

1 Introduction

Resolution is primarily described by the minimum distance between two objects that still can be resolved (e.g. [1]). Stochastic resolution has been introduced [2] by including the Cramér-Rao bound (CRB). The stochastic approach was extended with the probability of resolution at a given separation and signal-to-noise ratio (SNR) obtained via an asymptotic generalized likelihood ratio (GLR) test based on Euclidean distances [3]. Information resolution have also been explored with an arbitrary test [4]. For completeness of the stochastic approach, information geometry (IG, [5–9]) and compressive sensing (CS, [10, 11]) are combined due to their focus on information content [12–15]. In [13, 14], the Fisher-Rao information distance is recognized in the asymptotic GLR. In [15], links to other information distances and tighter resolution bounds have been obtained that we expand here to multiple parameters.

In the IG-based resolution analysis, the Fisher information metric (FIM) is employed for computing resolution bounds or information resolution. The stochastic analysis is crucial when using fewer measurements what is typical for compressive data acquisition in the front-end of a CS sensor (e.g. [10]). In the back-end, the analysis provides metrics for the high-resolution performance of sparse-signal processing (SSP). In radar, SSP can be seen as a model-based refinement of existing processing (e.g. [16]). Despite substantial CS research during last decade (e.g. [10, 11]), complete

F. Nielsen and F. Barbaresco (Eds.): GSI 2017, LNCS 10589, pp. 847–855, 2017.
https://doi.org/10.1007/978-3-319-68445-1_97

guarantees of CS-radar resolution performance have not been developed yet. Both IG and CS can improve existing sensors due to their focus on the information content in measurements rather than the sensing bandwidth only.

The resolution ability is primarily given by the Rayleigh distance determined by the sensing bandwidth from sensor design. In array processing, this deterministic resolution relies on the sensor wavelength and the array size. The stochastic resolution analysis includes also SNR available from the data acquisition. Moreover, besides the sensor design, it also involves targets of interest with their SNR and separation. In [13, 14], we assess the probability of resolution at a given SNR and separation by applying the Fisher-Rao information distances in the asymptotic distribution of the GLR.

In this paper and in [15], we extend the stochastic resolution analysis with tighter resolution bounds obtained via an LR test with different types of information distances. In Sect. 2, modeling of radar measurements and their SSP are summarized. In Sect. 3, our stochastic resolution analysis is explained and expanded to multiple parameters. In Sect. 4, numerical results from the stochastic resolution analysis and from SSP are compared. In the end, conclusions are drawn and future work indicated.

2 Data Modelling and SSP

The data models in array processing and in SSP needed in Sect. 3 are given here.

In a linear array (LA) of M_s antenna elements, a measurement y_i at time t from a single point-target (in a far field) at an array-element position μ_i measured in half-wavelength units (and centered, i.e. $\sum_i \mu_i = 0$), $i = 1, 2, \ldots, M_s$, is given by:

$$y_i = \alpha e^{j\mu_i \mathrm{u}} + z_i = \alpha_0 a_i(\mathrm{u}, 0) + z_i = m_i(\mathrm{u}) + z_i = m_i(\boldsymbol{\theta}) + z_i \tag{1}$$

where α is a target echo at time t, u is an angle parameter (containing the target azimuth φ, $\mathrm{u} = \pi \sin\varphi$), $a_i(\mathrm{u}, 0)$ is the model function (at zero Doppler) and z_i is complex-Gaussian receiver noise with zero mean and variance γ, $z_i \sim CN(0, \gamma)$. The target echo α is assumed to have constant nonrandom amplitude α_0 (so-called SW0, [17]), and thus, the input SNR equal to $|\alpha_0|^2/\gamma$. Thus, $y_i \sim CN(m_i(\boldsymbol{\theta}), \gamma)$, $\boldsymbol{\theta} = [\mathrm{u}]$.

Doppler v is included by modeling α in y_i from (1) at times t_l (normalized by sampling interval Δt) from an LA of size M_t (during coherent processing), as follows:

$$y_{i,l} = \alpha_0 e^{j(\mu_i \mathrm{u} + t_l \mathrm{v})} + z_{i,l} = \alpha_0 a_{i,l}(\mathrm{u}, \mathrm{v}) + z_{i,l} = m_{i,l}(\boldsymbol{\theta}) + z_{i,l} \tag{2}$$

where v is a frequency shift (from the narrowband assumption of the Doppler effect), $\mathrm{v} = 4\pi f v_0 \Delta t / c$ with f, v_0 and c being the carrier frequency, constant target velocity and the propagation speed, respectively. The noise $z_{i,l}$ and the echo α_0 are as in (1).

In SSP, data in (1) or (2) form a vector $\boldsymbol{y}$, $\boldsymbol{y} \in \mathbb{C}^M$, $\boldsymbol{y} \sim CN(\mathbf{m}(\boldsymbol{\theta}), \gamma \boldsymbol{I}_M)$, given by:

$$\boldsymbol{y} = \boldsymbol{A}\boldsymbol{x} + \boldsymbol{z} = \mathbf{m}(\boldsymbol{\theta}) + \boldsymbol{z} \tag{3}$$

where $\boldsymbol{A} \in \mathbb{C}^{M \mathrm{x} N}$ is a sensing matrix over the M spatial and temporal observations, $M = M_s M_t$, and for N parameter pairs $(\mathrm{u}_n, \mathrm{v}_n)$, $n = 1, 2, \ldots, N$, $\boldsymbol{x} \in \mathbb{C}^N$ is a sparse angle-Doppler profile of multiple targets, and $z \in \mathbb{C}^M$ is the receiver-noise vector, $z \sim CN(0, \gamma \boldsymbol{I}_M)$. The usual SSP, e.g. LASSO [18], applies as:

$$\boldsymbol{x}_{\text{ssp}} = \arg\min_x \|\boldsymbol{y} - \boldsymbol{A}\boldsymbol{x}\|^2 + \eta \|\boldsymbol{x}\|_1 \tag{4}$$

where the l_1-norm $\|\boldsymbol{x}\|_1$ promotes sparsity, the l_2-norm $\|\boldsymbol{y} - \boldsymbol{A}\boldsymbol{x}\|$ minimizes the errors, and a parameter η regulates the two tasks. In radar, the parameter η is closely related to the detection threshold (e.g. [16, 19, 21]). SSP in (4) relies on incoherence of $\boldsymbol{A}$ and sparsity, i.e. only K nonzeros (or targets) in $\boldsymbol{x}$, $K < M \le N$ (e.g. [10]). The mutual coherence $\kappa(\boldsymbol{A})$ is an incoherence measure, $\kappa(\boldsymbol{A}) = \max_{i,j,i \neq j} |\boldsymbol{a}_i^{\mathrm{H}} \boldsymbol{a}_j|$ where $\boldsymbol{a}_n$ is an n th column of $\boldsymbol{A}$, $\|\boldsymbol{a}_n\| = 1, n = 1, 2, \ldots, N$.

In radar signal processing, a sensing matrix $\boldsymbol{A}$ is intrinsically deterministic and its incoherence is also intrinsically strong because of the physics of radar sensing. Moreover, the model matrices for different parameters are mutually incoherent by their physical nature. Namely, typical radar data models employ shifts in time (for range), shifts in frequency (for Doppler) and shifts in phase (for angles) that are correctly isolated by the underlying physics.

In array processing, the sensing matrix $\boldsymbol{A}$ from (3) is an IFFT matrix, i.e. $\kappa(\boldsymbol{A}) = 0$ when $M = N$. Accordingly, with a uniform LA of size M_s, the grid cell Δu is $2\pi/M_s$ large. With regular sampling over a coherent processing interval of length T_{CPI} (each Δt), the grid cell Δv is $2\pi/T_{\text{CPI}}$ large. Such a grid-cell size can be called the Nyquist grid-cell size. Fewer measurements or smaller grid cells make $\kappa(\boldsymbol{A})$ increase.

3 Stochastic Resolution Analysis

The stochastic resolution analysis is presented gradually by introducing our approach in a simpler LA case as in [15], and completing it with a joint angle-Doppler case.

In an LA whose measurement y_i is modeled at position μ_i as in (1), the FIM $G(\text{u})$ for unknown u and Gaussian pdf $p(\boldsymbol{y}|\text{u})$ of $\boldsymbol{y}$ can be written as (e.g. [14, 20]):

$$G(\text{u}) = -\text{E}\left[\frac{\partial^2 \ln p(\boldsymbol{y}|\text{u})}{\partial \text{u}^2}\right] = 2\frac{|\alpha_0|^2}{\gamma}\left\|\frac{\partial \boldsymbol{a}(\text{u})}{\partial u}\right\|^2 = 2\text{SNR}\|\boldsymbol{\mu}\|^2 = G_{\text{u}} \tag{5}$$

In joint LAs whose measurement $y_{i,l}$ is modeled at position μ_i and at time t_l, as in (2), the FIM $\boldsymbol{G}(\boldsymbol{\theta})$, $\boldsymbol{\theta} = [\text{u v}]^{\mathrm{T}}$, $G_{i,l} \equiv 2\frac{|\alpha|^2}{\gamma} Re\left\{\left[\frac{\partial \boldsymbol{a}(\boldsymbol{\theta})}{\partial \theta_i}\right]^H \frac{\partial \boldsymbol{a}(\boldsymbol{\theta})}{\partial \theta_l}\right\}$, is given by (e.g. [14]):

$$\boldsymbol{G}\left(\begin{bmatrix}\text{u}\\ \text{v}\end{bmatrix}\right) = 2\frac{|\alpha_0|^2}{\gamma}\begin{bmatrix} M_{\text{t}} G_{\text{u}} & G_{\text{u,v}} \\ G_{\text{v,u}} & M_{\text{s}} G_{\text{v}} \end{bmatrix} = 2\text{SNR}\begin{bmatrix} M_{\text{t}}\|\boldsymbol{\mu}\|^2 & \sum_{i,l} \mu_i t_l \\ \sum_{i,l} \mu_i t_l & M_{\text{s}}\|\boldsymbol{t}\|^2 \end{bmatrix} = \boldsymbol{G}_{\text{u,v}} \tag{6}$$

where $\sum_{i,l} \mu_i t_l$ is zero because the positions μ_i and the times t_l are centered (to restrict the influence of positions to the FIM), i.e. $\sum_i \mu_i = 0$ and $\sum_l t_l = 0$, respectively.

The FIM $\boldsymbol{G}(\boldsymbol{\theta})$ is typically applied in the accuracy analysis to the accuracy bounds such as e.g. so-called Cramer-Rao bound (CRB) of the mean squared error (MSE) of an unbiased estimator $\hat{\theta}$ of $\boldsymbol{\theta}$, i.e. $\mathrm{MSE}\left(\hat{\theta}\right) \geq \mathrm{CRB}(\boldsymbol{\theta}) = [\boldsymbol{G}(\boldsymbol{\theta})]^{-1}$ (e.g. [19]).

In addition, we have been exploring how the FIM $\boldsymbol{G}(\boldsymbol{\theta})$ can also be used to compute resolution bounds based on information distances between $p(\boldsymbol{y}|\boldsymbol{\theta})$ and $p(\boldsymbol{y}|\boldsymbol{\theta}+d\boldsymbol{\theta})$ when the parameters $\boldsymbol{\theta}$ change a bit by $d\boldsymbol{\theta}$ (e.g. [13–15]). Among many information distances available in IG [6], we start with the Fisher-Rao information distance ([5], FRID) because it is directly related to the Riemann distance and the basic infinitesimal Fisher-Rao metric ds, $ds^2 = d\boldsymbol{\theta}^T\boldsymbol{G}(\boldsymbol{\theta})d\boldsymbol{\theta}$. Moreover, most information (pseudo-)distances from IG such as e.g. the Kullback-Leibler (KL) divergence and Bhattacharyya (BT) distance, reduce to simple functions of ds and of FRID, especially, in the case of Gaussian-distributed measurements.

A FRID $d_{\boldsymbol{\theta}}$ (related to $\boldsymbol{\theta}$) is a geodesic computed from the integrals of ds over possible curves of integration on the statistical manifold, given by:

$$d_{\boldsymbol{\theta}} \equiv \min_{\vartheta(t)} \int_0^1 \sqrt{\dot{\boldsymbol{\vartheta}}(t)^T G(\boldsymbol{\vartheta}(t))\dot{\boldsymbol{\vartheta}}(t)dt}$$

where $\boldsymbol{\vartheta}(t)$ are coordinates of an integration path parameterized by t, $t \in [01]$. The coordinates are computed from the geodesic equations [5] written as:

$$\ddot{\vartheta}_k(t) + \sum\nolimits_{i,j} C_{ij}^k \dot{\vartheta}_i(t)\dot{\vartheta}_j(t) = 0, \quad C_{ij}^k = \sum\nolimits_l G^{kl}\left(\frac{\partial G_{i,l}}{\partial \theta_j} + \frac{\partial G_{j,l}}{\partial \theta_i} + \frac{\partial G_{i,j}}{\partial \theta_k}\right)$$

where C_{ij}^k is a connection coefficient and G^{kl} an element from the inverse of the FIM $\boldsymbol{G}(\boldsymbol{\theta})$, $i, j, k, l = 1, 2, \ldots \dim(\boldsymbol{\theta})$.

In the LA case from (5), $\dim(\boldsymbol{\theta}) = 1$ and the integration curve is a line given by: $\vartheta(t) = \delta\mathrm{u}t + \mathrm{u}$, $t \in [01]$, because the connection coefficient C_{11}^1 equals zero as $G(\mathrm{u})$ is constant w.r.t. u, $G(\mathrm{u}) \equiv G_{\mathrm{u}}$, i.e. $\ddot{\vartheta}(t) = 0$. Accordingly, the FRID d_{u} between $p(\boldsymbol{y}|\mathrm{u})$ and $p(\boldsymbol{y}|\mathrm{u}+\delta\mathrm{u})$ on the 1D statistical manifold is given by:

$$d_{\mathrm{u}} = \int_0^1 \sqrt{G_{\mathrm{u}}}\delta\mathrm{u}\; dt = \sqrt{G_{\mathrm{u}}}\delta\mathrm{u} \; = \sqrt{2\mathrm{SNR}}\|\boldsymbol{\mu}\delta\mathrm{u}\|. \tag{7}$$

In the joint case from (6), $\boldsymbol{G}(\boldsymbol{\theta})$ is also constant w.r.t. $\boldsymbol{\theta}$, $\boldsymbol{G}(\boldsymbol{\theta}) \equiv \boldsymbol{G}_{\mathrm{u,v}}$, i.e. $\ddot{\boldsymbol{\vartheta}}(t) = \boldsymbol{0}$, so that the FRID $d_{\mathrm{u,v}}$ between $p(\boldsymbol{y}|\mathrm{u},\mathrm{v})$ and $p(\boldsymbol{y}|\mathrm{u}+\delta\mathrm{u}, \mathrm{v}+\delta\mathrm{v})$ is given by:

$$d_{\mathrm{u,v}} = \sqrt{[\delta\mathrm{u}\;\delta\mathrm{v}]\boldsymbol{G}_{\mathrm{u,v}}[\delta\mathrm{u}\;\delta\mathrm{v}]^{\mathrm{T}}} = \sqrt{2\mathrm{SNR}\left(M_{\mathrm{t}}\|\boldsymbol{\mu}\delta\mathrm{u}\|^2 + M_{\mathrm{s}}\|\boldsymbol{t}\delta\mathrm{v}\|^2\right)} \tag{8}$$

The FRID $d_{\mathrm{u.v}}$ contains separations $\delta\mathrm{u}$ and $\delta\mathrm{v}$ in a weighted form rather than directly as in the norm: $\|\delta\boldsymbol{\theta}\| \equiv \sqrt{\delta\mathrm{u}^2 + \delta\mathrm{v}^2}$. This weighting makes a FRID, e.g. here d_{u} or $d_{\mathrm{u.v}}$, have no unit because each $\delta\theta$ is actually normalized with the Rayleigh distance.

Next we derive the resolution bounds (or the information resolution) from the FRID $d_{\boldsymbol{\theta}}$ as the probability that two targets can be resolved at a separation $\delta\boldsymbol{\theta}$ and an SNR.

In some early work on IG [5], Rao proposed a test of a resolution hypothesis $H_1 : \delta\boldsymbol{\theta} \neq 0$ and its alternative $H_0 : \delta\boldsymbol{\theta} = \mathbf{0}$, with a test statistic w from $N(0,1)$, $w = d/\sqrt{\text{CRLB}[d]}$, by using a FRID d as the distance between populations $p(\mathbf{y}|\boldsymbol{\theta})$ and $p(\mathbf{y}|\boldsymbol{\theta}+\delta\boldsymbol{\theta})$. The test with w can only suggest a true H_1 when H_0 is rejected.

In our resolution analysis [13–15], we quantify the resolution potential when there are two close targets, i.e. we prefer assessing $H_1 : \delta\boldsymbol{\theta} \neq \mathbf{0}$. Hence, in [13–15], we tested the same hypotheses via the likelihood ratio (LR), $\text{LR} = p(\mathbf{y}|\boldsymbol{\theta},\boldsymbol{\theta}+\delta\boldsymbol{\theta})/p(\mathbf{y}|\boldsymbol{\theta})$, first at the maximum likelihood (ML) estimate $\delta\hat{\theta}_{\text{ml}}$, i.e. so-called the generalized LR (GLR). A statistic ln GLR is asymptotically χ^2-distributed with a number of degrees of freedom equal to $\dim(\boldsymbol{\theta})$, and the parameter ε, given by ([21] in 6):

$$\ln \text{GLR} \overset{a}{\rightarrow} \ln \text{GLR}_a = \delta\hat{\boldsymbol{\theta}}_{\text{ML}}^T G\left(\boldsymbol{\theta}+\delta\hat{\boldsymbol{\theta}}_{\text{ML}}\right)\delta\hat{\boldsymbol{\theta}}_{\text{ML}} \sim \chi^2_{\varepsilon,\dim(\boldsymbol{\theta})} \tag{9}$$

In [13, 14], ε from (9), $\varepsilon = \delta\boldsymbol{\theta}^T G(\boldsymbol{\theta}+\delta\boldsymbol{\theta})\delta\boldsymbol{\theta} \equiv \delta\boldsymbol{\theta}^T G_{\boldsymbol{\theta}}\delta\boldsymbol{\theta}$ (as in 6 from [21]), is linked to the FRID $d_{\boldsymbol{\theta}}$ from (7) or (8) as: $\varepsilon = d_{\boldsymbol{\theta}}^2$. To assess the probability of resolution $P_{\text{res,FRID}}$, the statistic $\ln \text{GLR}_a$ is tested as follows:

$$P_{\text{res,FRID}} = \mathcal{P}\{\ln \text{GLR}_a > \rho | H_1\}, \quad \ln \text{GLR}_a \sim \chi^2_{\varepsilon,\dim(\boldsymbol{\theta})}. \tag{10}$$

where ρ is a threshold from the inverse central χ^2-distribution (under H_0) at a probability of false alarms P_{fa}, $\rho = \chi^{2,\text{inv}}_{0,\dim(\boldsymbol{\theta})}(P_{\text{fa}})$.

In [14], we had noticed that $d_{\boldsymbol{\theta}}$ and the resolution bounds from (10) were too optimistic. Therefore, we started investigating another FRID $d_{\mathbf{m}(\boldsymbol{\theta})}$ related to the mean $\mathbf{m}(\boldsymbol{\theta})$ of measurements. In [15], the FRID $d_{\mathbf{m}(\theta)}$ with a single θ as in (1) gives tighter resolution bounds because radar models are nonlinear w.r.t. θ, while FRID d_θ can be interpreted as the Taylor expansion. Here we expand the analysis to multiple parameters in $\boldsymbol{\theta}$ as required in (2). The FRID $d_{\mathbf{m}(\boldsymbol{\theta})}$ between $CN(\mathbf{m}(\boldsymbol{\theta}), \gamma \boldsymbol{I}_M)$ and $CN(\mathbf{m}(\boldsymbol{\theta}+\delta\boldsymbol{\theta}), \gamma \boldsymbol{I}_M)$ with different means, $\delta\mathbf{m} = \mathbf{m}(\boldsymbol{\theta}+\delta\boldsymbol{\theta}) - \mathbf{m}(\boldsymbol{\theta})$, but the same covariance matrix is equal to the Mahalanobis distance [5], as follows:

$$d_{\mathbf{m}(\boldsymbol{\theta})} = \sqrt{\delta\mathbf{m}^H \boldsymbol{G}_{\mathbf{m}} \delta\mathbf{m}} = \|\delta\mathbf{m}\|/\sqrt{\gamma/2} \qquad \neq d_{\boldsymbol{\theta}} = \sqrt{\delta\boldsymbol{\theta}^H \boldsymbol{G}_{\boldsymbol{\theta}} \delta\boldsymbol{\theta}} \tag{11}$$

where $\boldsymbol{G}_{\mathbf{m}}$ is the FIM about the mean, $\boldsymbol{G}_{\mathbf{m}} = 2\boldsymbol{I}_M/\gamma$. The FRID $d_{\mathbf{m}(\boldsymbol{\theta})}$ would be equal to $d_{\boldsymbol{\theta}}$ only if $\mathbf{m}(\boldsymbol{\theta})$ would be affine transformation of $\boldsymbol{\theta}$ what is not a case in radar. However, the basic infinitesimal metric $ds(\boldsymbol{\theta})$ is always invariant to different parameterization. E.g., if $\text{m} = \text{m}(\theta)$, $ds^2(\text{m}) = ds^2(\theta)$ because $G_\theta = |dm/d\theta|^2 G_{\text{m}}$. Regarding links to CS, we can note that the cross-correlations from the coherence $\kappa(\boldsymbol{A})$ make $d_{\mathbf{m}(\boldsymbol{\theta})}$ decrease, $d_{\mathbf{m}(\boldsymbol{\theta})} = \|\delta\mathbf{m}/\|\sqrt{\gamma/2} = \sqrt{2\text{SNR}(1 - Re\{\boldsymbol{a}^{\text{H}}(\boldsymbol{\theta})\boldsymbol{a}(\boldsymbol{\theta}+\delta\boldsymbol{\theta})\})}$.

Next we explored the LR test at the true separation $\delta\boldsymbol{\theta}$ with the equivalent hypotheses being aimed for detecting $\delta\mathbf{m}$, expressed as [15]:

$$\begin{aligned} H_0 &: \boldsymbol{y} = 2\mathbf{m}(\boldsymbol{\theta}) + \boldsymbol{z} = \boldsymbol{y}_0 \\ H_1 &: \boldsymbol{y} = \mathbf{m}(\boldsymbol{\theta}) + \mathbf{m}(\boldsymbol{\theta} + \delta\boldsymbol{\theta}) + \boldsymbol{z} = \boldsymbol{y}_0 + \delta\mathbf{m} \end{aligned} \tag{12}$$

where measurements $\boldsymbol{y}$ contain responses from two point-targets separated by $\delta\boldsymbol{\theta}$.

The test statistic ln LR is derived and linked to $d_{\mathbf{m}(\boldsymbol{\theta})}$, as follows:

$$\ln \mathrm{LR} = 2Re\{[\mathbf{y} - 2\mathbf{m}(\boldsymbol{\theta})]^H \delta\mathbf{m}\}/\gamma \sim N(d^2_{\mathbf{m}(\boldsymbol{\theta})}, d^2_{\mathbf{m}(\boldsymbol{\theta})}) \tag{13}$$

as Gaussian distributed with mean and variance equal to $d^2_{\mathbf{m}(\boldsymbol{\theta})}$. Moreover, ln LR can be tested directly, with no need for an asymptotic ln GLR. Note also that we end up at a normal distribution, $\ln \mathrm{LR}/d_{\mathbf{m}(\boldsymbol{\theta})} \sim N\left(d_{\mathbf{m}(\boldsymbol{\theta})}, 1\right)$, like in a test from [5], $w \sim N(0,1)$, but via different tests and different test statistics.

The test statistic ln LR from (13), is tested against a threshold ρ_{m} at P_{fa}, $\rho_{\mathrm{m}} = N^{-1}\left(0, d^2_{\mathbf{m}(\boldsymbol{\theta})}, P_{\mathrm{fa}}\right)$, to assess the probability of resolution $P_{\mathrm{res,FRIDm}}$ given by:

$$P_{\mathrm{res,FRIDm}} = \mathcal{P}\{\ln \mathrm{LR} > \rho_{\mathrm{m}} | H_1\} \ \ln \mathrm{LR} \sim N\left(d^2_{\mathbf{m}(\boldsymbol{\theta})}, d^2_{\mathbf{m}(\boldsymbol{\theta})}\right) \tag{14}$$

In cases with Gaussian-distributed measurements $\boldsymbol{y}$, we can extend (14) with other information (pseudo-) distances such as the KL divergence d_{KL} and BT distance d_{BT}, because they are directly related to $d_{\mathbf{m}(\boldsymbol{\theta})}$. Moreover, d_{KL} and d_{BT} between $p(\boldsymbol{y}|\boldsymbol{\theta})$ and $p(\boldsymbol{y}|\boldsymbol{\theta} + d\boldsymbol{\theta})$ are also related to the LR, and can be easily derived as follows:

$$d_{\mathrm{KL}} = \mathrm{E}_{H_1}[\ln \mathrm{LR}] = d^2_{\mathbf{m}(\boldsymbol{\theta})} \text{ and } d_{\mathrm{BT}} = -\ln \mathrm{E}_{H_0}\left[\sqrt{\mathrm{LR}}\right] = d^2_{\mathbf{m}(\boldsymbol{\theta})}/4 \tag{15}$$

Finally, the FRID-based probabilities $P_{\mathrm{res,FRID}}$ and $P_{\mathrm{res,FRIDm}}$ are compared with the resolution from SSP whose probability $P_{\mathrm{res,SSP}}$ is assessed numerically from $\boldsymbol{x}_{\mathrm{SSP}}$ in (4) for two point-targets in cells i and j, $i \neq j$, by:

$$P_{\mathrm{res,SSP}} = \mathcal{P}\left\{\left(x_{\mathrm{SSP},i} \neq 0\right) \wedge \left(x_{\mathrm{SSP},j} \neq 0\right) | H_1\right\} \tag{16}$$

where η is related to P_{fa} as: $\eta^2 = -\gamma \ln P_{\mathrm{fa}} \equiv \chi^{2,\mathrm{inv}}_{0,2}(P_{\mathrm{fa}})/2$ (e.g. [16, 19, 21]).

4 Numerical Results

The stochastic resolution analysis presented in Sect. 3 is demonstrated with numerical tests with spatial and temporal measurements for joint angle-Doppler processing of two close equal point-targets at different SNRs. Measurements are acquired from joint uniform LAs (ULAs) of size M_{s} in space and of size M_{t} in time. For the sake of simplicity, the sizes are equal, $M_{\mathrm{s}} = M_{\mathrm{t}}$, and moreover, $\|\boldsymbol{\mu}\|^2 = \|\boldsymbol{t}\|^2$.

The measurements $\boldsymbol{y}$ from (3) contain responses from two point-targets separated by δu or δv, $\boldsymbol{y} = \alpha\boldsymbol{a}(\mathrm{u},\mathrm{v}) + \alpha\boldsymbol{a}(\mathrm{u}+\delta\mathrm{u},\mathrm{v}+\delta\mathrm{v}) + \boldsymbol{z}$. The angle-Doppler estimation grid of size N is Nyquist, i.e. without up-sampling, $N = M_s M_t$. The targets are placed in the middle of the estimation grid, and separated in u or v by the IFFT bin size, i.e. δu equals $\frac{2\pi}{M_s}$ or δv equals $\frac{2\pi}{M_t}$. The amplitude α_0 is nonrandom, $|\alpha_0|^2 = \gamma$ SNR. The noise variance γ is kept constant, $\gamma = 1$. For the thresholds ρ, ρ_m and η in (10), (14) and (16), respectively, P_{fa} is set to 0.000001 (as typically low in radar).

The FRID-based probabilities $P_{\mathrm{res,FRID}}$ and $P_{\mathrm{res,FRIDm}}$ from (10) and (14) are assessed by computing the true ε and $d_{\mathbf{m}(\mathbf{0})}$ from (9) and (11), respectively. The probability $P_{\mathrm{res,SSP}}$ from (16) is assessed numerically from a sufficient number of noise runs. The SSP optimization from (4) is performed with yall1 [22].

In Fig. 1, the different information (pseudo-)distances d_{u}, $d_{\mathbf{m}(\mathrm{u})}$, and d_{KL} and d_{BT} from (8), (11) and (15), respectively, are shown in the same test case. While d_{u} grows linearly with δu, $d_{\mathbf{m}(\mathrm{u})}$ reaches its maximum quite fast (at around 1.4 Nyquist cell) and stays around that value despite the growing separation δu. The same tendency holds also for d_{KL} and d_{BT} (as given in (15), d_{KL} and d_{BT} are squared versions of $d_{\mathbf{m}(\mathrm{u})}$).

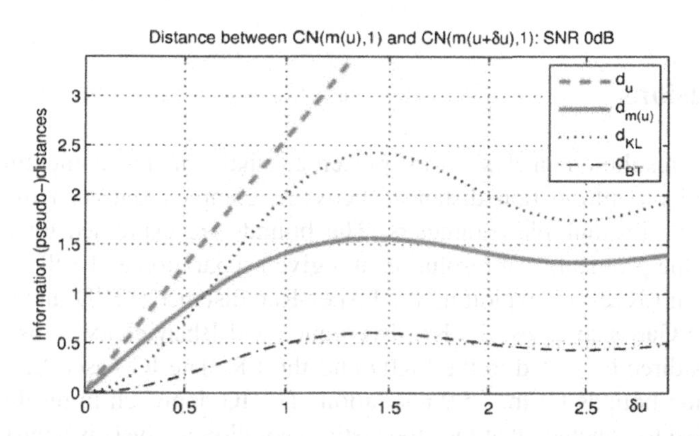

Fig. 1. Information distances from (8), (11) and (15) in a test case from (1) with a uniform linear array of size equal to 10, and two point-targets separated by δu (expressed in Nyquist cells).

In Fig. 2, the resolution probabilities are shown in the same test case from angle-Doppler processing. The SSP resolution $P_{\mathrm{res,SSP}}$ is quite far from the resolution bounds given by the FRID-based probabilities $P_{\mathrm{res,FRID}}$ and $P_{\mathrm{res,FRIDm}}$. In addition, at a bit larger separation (Fig. 2, right), $P_{\mathrm{res,FRID}}$ increases clearly while $P_{\mathrm{res,FRIDm}}$ as well as $P_{\mathrm{res,SSP}}$ remain realistic, i.e. nearly the same. This behavior agrees with related information distances such as d_{u} and $d_{\mathrm{m(u)}}$ in Fig. 1.

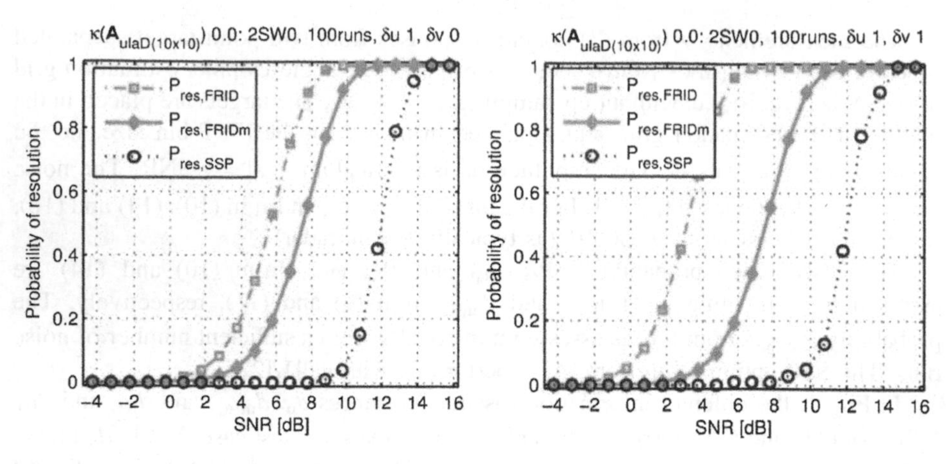

Fig. 2. Resolution bounds $P_{\mathrm{res,FRID}}$ and $P_{\mathrm{res,FRIDm}}$ (green squares and diamonds) from (10) and (14), respectively, and SSP resolution $P_{\mathrm{res,SSP}}$ from (16) (blue circles) in a test case with joint uniform LAs of equal size in space and time, $M_{\mathrm{s}} = M_{\mathrm{t}} = 10$, and two point-targets separated in angle u by δu equal to one, and in Doppler v by δv equal to zero (left) and one (right). (Color figure online)

5 Conclusions

A stochastic resolution analysis was presented that enables computing resolution bounds based on information distances between complex-Gaussian distributions of measurements with multiple parameters. The bounds are expressed for an array configuration as the probability of resolution at a given separation and SNR of targets, and assessed via an LR test by exploiting the Fisher-Rao distance (FRID) in the distribution of the LR. In Gaussian cases, the KL divergence and Bhattacharyya distance are also applicable as directly related to the FRID and the LR. The IG-based LR test provides tight resolution bounds for the SSP resolution. The IG approach to resolution analysis enables us also to conclude that the stochastic resolution analysis is appropriate in (but not limited to) radar because of the sensitivity to the crucial features in the performance guarantees: array configuration or sensor design as well as SNR, separation and probability of resolution.

In future work, this stochastic resolution analysis based on IG is being further interpreted and applied to fewer measurements from sparse sensing in the front-end and to all radar parameters: range, Doppler and angles. In addition, incorporating the geometry of the Laplacian manifold related to sparsity, and also links to the importance of information distances in the information theory are being explored.

References

1. den Dekker, A.J., van den Bos, A.: Resolution: a survey. J. Opt. Soc. Am. A **14**(3), 547–557 (1997)
2. Smith, S.T.: Statistical resolution limits and the complexified CR bounds. IEEE Trans. SP **53** (5), 1597–1609 (2005)
3. Liu, Z., Nehorai, A.: Statistical angular resolution limit for point sources. IEEE Trans. SP **55** (11), 5521–5527 (2007)
4. Cheng, Y., Wang, X., Caelli, T., Li, X., Moran, B.: On information resolution of radar systems. IEEE Trans. AES **48**(4), 3084–3102 (2012)
5. Rao, C.R.: Information and the accuracy attainable in the estimation of statistical parameters. Bull. Calcutta Math. Soc. **37**, 81–89 (1945)
6. Amari, S.: Information geometry of statistical inference - an overview. In: IEEE ITW (2002)
7. Nielsen, F.: CRLB and Information Geometry (2013). https://arxiv.org/abs/1301.3578
8. Arnaudon, M., Barbaresco, F., Yang, L.: Riemannian medians and means with applications to radar signal processing. IEEE J. Sel. Top. SP **7**(4), 595–604 (2013)
9. Brigo, D., Hanzon, B., LeGland, F.: A differential geometric approach to nonlinear filtering: the projection filter. IEEE Trans. Autom. Control **43**(2), 247–252 (1998)
10. Donoho, D.: Compressed sensing. IEEE Trans. IT **52**(4), 1289–1306 (2005)
11. Candès, E., Fernandez-Granda, C.: Towards a mathematical theory of super-resolution. Commun. Pure Appl. Math. **67**(7), 906–956 (2014). Willey
12. Coutino, M., Pribić, R., Leus, G.: Direction of arrival estimation based on information geometry. In: IEEE ICASSP (2016)
13. Pribić, R., Coutino, M., Leus, G.: Stochastic resolution analysis of co-prime arrays in radar. In: IEEE SSP (2016)
14. Pribić, R.: Stochastic resolution analysis via a GLR test in radar. In: IEEE CoSeRa (2016)
15. Pribić, R., Leus, G.: Information distances for radar resolution analysis. In: Submitted to IEEE CAMSAP (2017)
16. Pribić, R., Kyriakides, I.: Design of SSP in radar systems. In: IEEE ICASSP (2014)
17. Cook, C.E., Bernfeld, M.: Radar Signals; an Introduction to Theory and Application. Academic Press, Cambridge (1967)
18. Tibshirani, R.: Regression shrinkage and selection via the lasso. J. Roy. Stat. Soc. Ser. B **58** (1), 267–288 (1996)
19. Fuchs, J.J.: The generalized likelihood ratio test and the sparse representations approach. In: Elmoataz, A., Lezoray, O., Nouboud, F., Mammass, D., Meunier, J. (eds.) ICISP 2010. LNCS, vol. 6134, pp. 245–253. Springer, Heidelberg (2010). doi:10.1007/978-3-642-13681-8_29
20. Van Trees, H.L.: Optimum Array Processing. Wiley, Hoboken (2002)
21. Kay, S.M.: Fundamentals of Statistical Signal Processing Volume II: Detection Theory. Prentice Hall, Upper Saddle River (1998)
22. YALL1: your algorithms for L1. http://yall1.blogs.rice.edu/

k-Means Clustering with Hölder Divergences

Frank Nielsen[1,2(✉)], Ke Sun[3], and Stéphane Marchand-Maillet[4]

[1] École Polytechnique, Palaiseau, France
nielsen@lix.polytechnique.fr
[2] Sony Computer Science Laboratories Inc., Tokyo, Japan
[3] King Abdullah University of Science and Technology (KAUST), Thuwal, Saudi Arabia
[4] University of Geneva, Geneva, Switzerland

Abstract. We introduced two novel classes of Hölder divergences and Hölder pseudo-divergences that are both invariant to rescaling, and that both encapsulate the Cauchy-Schwarz divergence and the skew Bhattacharyya divergences. We review the elementary concepts of those parametric divergences, and perform a clustering analysis on two synthetic datasets. It is shown experimentally that the symmetrized Hölder divergences consistently outperform significantly the Cauchy-Schwarz divergence in clustering tasks.

1 Introduction

To build dissimilarity measures between p and q in a common domain, one can use bi-parametric inequalities (Mitrinovic et al. 2013) $\mathrm{lhs}(p,q) \leq \mathrm{rhs}(p,q)$, and measure the inequality tightness. When $\mathrm{lhs}(p,q) > 0$, a dissimilarity can be constructed by the log-ratio gap:

$$D(p:q) = -\log\left(\frac{\mathrm{lhs}(p,q)}{\mathrm{rhs}(p,q)}\right) = \log\left(\frac{\mathrm{rhs}(p,q)}{\mathrm{lhs}(p,q)}\right) \geq 0. \tag{1}$$

Notice that this divergence construction allows one to consider the equivalence class of scaled inequalities: $\lambda \times \mathrm{lhs}(p,q) \leq \lambda \times \mathrm{rhs}(p,q), \forall \lambda > 0$. Following this divergence construction principle, we defined Hölder divergences based on the Hölder's inequality, and presented the basic properties of this divergence family (Nielsen et al. 2017). In this paper, we further extend the empirical clustering study with respect to Hölder divergences, and show that symmetrized Hölder divergences consistently outperform significantly the Cauchy-Schwarz divergence (Hasanbelliu et al. 2014). We build Hölder divergences that are invariant by rescaling: These divergences D are called projective divergences and satisfy the property $D(\lambda p : \lambda' q) = D(p:q), \forall \lambda, \lambda' > 0$.

The term "Hölder divergence" was coined previously based on the definition of the Hölder score (Kanamori et al. 2014; Kanamori 2014): The score-induced Hölder divergence $D(p:q)$ is a proper gap divergence that yields a scale-invariant divergence. A key difference with our work is that this score-induced divergence

F. Nielsen and F. Barbaresco (Eds.): GSI 2017, LNCS 10589, pp. 856–863, 2017.
https://doi.org/10.1007/978-3-319-68445-1_98

is not projective and does not include the Cauchy-Schwarz (CS) divergence, while our definition is projective and includes the CS divergence.

This paper is organized as follows: Sect. 2 reviews the definition of Hölder pseudo divergence and Hölder proper divergence. Section 3 gives algorithms for clustering based on Hölder divergences, and presents the experimental clustering results. Section 4 concludes this work.

2 Hölder Divergences: Definitions and Properties

Let $(\mathcal{X}, \mathcal{F}, \mu)$ be a measurable space where μ is the Lebesgue measure, and let $L^\gamma(\mathcal{X}, \mu)$ denote the space of functions with their γ-th power of absolute value Lebesgue integrable, for any $\gamma > 0$. When $\gamma \geq 1$, this is a Lebesgue space but we consider the wider scope of $\gamma > 0$ in this work. Hölder's inequality states that $\|pq\|_1 \leq \|p\|_\alpha \|q\|_\beta$ for conjugate exponents $\alpha > 0$ and $\beta > 0$ (satisfying $\frac{1}{\alpha} + \frac{1}{\beta} = 1$), $p \in L^\alpha(\mathcal{X}, \mu)$ and $q \in L^\beta(\mathcal{X}, \mu)$. Let $p(x) \in L^{\alpha\sigma}(\mathcal{X}, \mu)$ and $q(x) \in L^{\beta\tau}(\mathcal{X}, \mu)$ be positive measures where $\sigma > 0$ and $\tau > 0$ are prescribed parameters. We define (Nielsen et al. 2017) a tri-parametric family of divergences as follows:

Definition 1 (Hölder pseudo-divergence). *The Hölder pseudo-divergence (HPD) between $p(x)$ and $q(x)$ is the log-ratio-gap:*

$$D^{\mathrm{H}}_{\alpha,\sigma,\tau}(p : q) := -\log \left(\frac{\int_{\mathcal{X}} p(x)^\sigma q(x)^\tau \mathrm{d}x}{\left(\int_{\mathcal{X}} p(x)^{\alpha\sigma} \mathrm{d}x\right)^{\frac{1}{\alpha}} \left(\int_{\mathcal{X}} q(x)^{\beta\tau} \mathrm{d}x\right)^{\frac{1}{\beta}}} \right).$$

The non-negativeness follows straightforwardly from Hölder's inequality (1889). However the symmetry, the triangle-inequality, and the law of indiscernibles (self distance equals to zero) are not satisfied for HPDs. Therefore these dissimilarity measures are said to belong to a broader class of "pseudo-divergences".

In order to have a proper divergence with the law of the indiscernibles, we note that the equality $D^{\mathrm{H}}_{\alpha,\sigma,\tau}(p : q) = 0$ holds if and only if $p(x)^{\alpha\sigma} \propto q(x)^{\beta\tau}$ (almost everywhere). To make this equality condition to be $p(x) = q(x)$ (ae.) for probability distributions, we take $\gamma := \alpha\sigma = \beta\tau$.

Let $p(x)$ and $q(x)$ be positive measures in $L^\gamma(\mathcal{X}, \mu)$ for a prescribed scalar value $\gamma > 0$. Let $\alpha, \beta > 0$ be conjugate exponents. We define (Nielsen et al. 2017) a bi-parametric divergence family, which is a sub-family of HPD that satisfies both non-negativeness and law of the indiscernibles as follows:

Definition 2 (Proper Hölder divergence). *The proper Hölder divergence (HD) between two densities $p(x)$ and $q(x)$ is:*

$$D^{\mathrm{H}}_{\alpha,\gamma}(p : q) = D^{\mathrm{H}}_{\alpha,\frac{\gamma}{\alpha},\frac{\gamma}{\beta}}(p : q) := -\log \left(\frac{\int_{\mathcal{X}} p(x)^{\gamma/\alpha} q(x)^{\gamma/\beta} \mathrm{d}x}{(\int_{\mathcal{X}} p(x)^\gamma \mathrm{d}x)^{1/\alpha} (\int_{\mathcal{X}} q(x)^\gamma \mathrm{d}x)^{1/\beta}} \right).$$

Notice that D^{H} is used to denote both HPD and HD. One has to check the number of subscripts to distinguish between these two pseudo and proper cases.

An important fact about Hölder divergences is that they encapsulate both the Cauchy-Schwarz divergence and the one-parameter family of skew Bhattacharyya divergences (Nielsen and Boltz 2011). In the definition of HD, setting $\alpha = \beta = \gamma = 2$ yields the CS divergence:

$$D^{\mathrm{H}}_{2,2}(p:q) = D^{\mathrm{H}}_{2,1,1}(p:q) = \mathrm{CS}(p:q) := -\log\left(\frac{\int_{\mathcal{X}} p(x)q(x)\mathrm{d}x}{\left(\int_{\mathcal{X}} p(x)^2\mathrm{d}x\right)^{\frac{1}{2}}\left(\int_{\mathcal{X}} q(x)^2\mathrm{d}x\right)^{\frac{1}{2}}}\right).$$

In the definition of HD, setting $\gamma = 1$ yields the skew Bhattacharyya divergences:

$$D^{\mathrm{H}}_{\alpha,1}(p:q) = D^{\mathrm{H}}_{\alpha,\frac{1}{\alpha},\frac{1}{\beta}}(p:q) = -\log\int_{\mathcal{X}} p(x)^{1/\alpha}q(x)^{1/\beta}\mathrm{d}x := B_{1/\alpha}(p:q).$$

It is easy to check from Definition 1 that the HPD is a projective divergence which is invariant to scaling of its parameter densities:

$$D^{\mathrm{H}}_{\alpha,\sigma,\tau}(\lambda p:\lambda' q) = D^{\mathrm{H}}_{\alpha,\sigma,\tau}(p:q) \quad (\forall\lambda,\lambda' > 0).$$

Figure 1 illustrates the relationships between those divergence families.

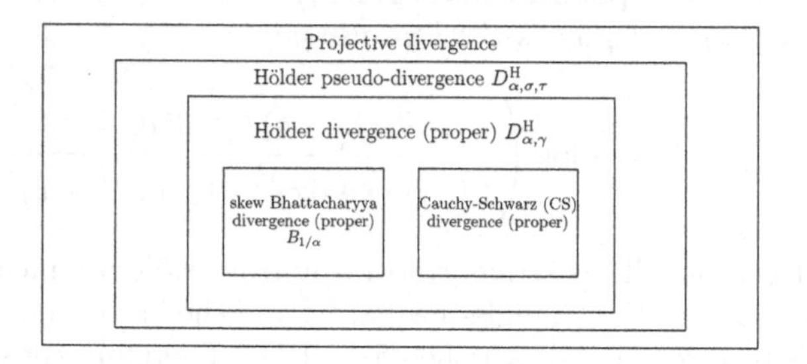

Fig. 1. Hölder proper divergence (bi-parametric) and Hölder improper pseudo-divergence (tri-parametric) encompass the Cauchy-Schwarz divergence and the skew Bhattacharyya divergences.

By definition, the HPD is asymmetric and satisfies the reference duality:

$$D^{\mathrm{H}}_{\alpha,\sigma,\tau}(p:q) = D^{\mathrm{H}}_{\beta,\tau,\sigma}(q:p),$$

for conjugate exponents α and β. Similarly, the HD satisfies:

$$D^{\mathrm{H}}_{\alpha,\gamma}(p:q) = D^{\mathrm{H}}_{\beta,\gamma}(q:p).$$

The HPD and the HD admit closed-form formulas for exponential family distributions. For example, consider $p(x) = \exp(\theta_p^\top t(x) - F(\theta_p))$ and $q(x) = \exp(\theta_q^\top t(x) - F(\theta_q))$, where $t(x)$ is a vector of sufficient statistics, and $F(\theta)$ is the convex cumulant generating function. Then from straightforward derivations, the symmetrized Hölder divergence is:

$$\begin{aligned} S^{\mathrm{H}}_{\alpha,\gamma}(p:q) &:= \frac{1}{2}\left(D^{\mathrm{H}}_{\alpha,\gamma}(p:q) + D^{\mathrm{H}}_{\alpha,\gamma}(q:p)\right) \\ &= \frac{1}{2}\left[F(\gamma\theta_p) + F(\gamma\theta_q) - F\left(\frac{\gamma}{\alpha}\theta_p + \frac{\gamma}{\beta}\theta_q\right) - F\left(\frac{\gamma}{\beta}\theta_p + \frac{\gamma}{\alpha}\theta_q\right)\right]. \end{aligned}$$

This $S^{\mathrm{H}}_{\alpha,\gamma}$ has the key advantage that its centroid can be solved efficiently using the concave-convex procedure (CCCP) (Nielsen and Boltz 2011). Consider a set of fixed densities $\{\theta_1, \cdots, \theta_n\}$ with positive weights $\{w_1, \cdots, w_n\}$ ($\sum_{i=1}^{n} w_i = 1$) of the same exponential family. The symmetrized HD centroid with respect to $\alpha, \gamma > 0$ is defined as:

$$\begin{aligned} O_{\alpha,\gamma} &:= \arg\min_{\theta} \sum_{i=1}^{n} w_i S^{\mathrm{H}}_{\alpha,\gamma}(\theta_i : \theta) \\ &= \arg\min_{\theta} \sum_{i=1}^{n} w_i \left[F(\gamma\theta) - F\left(\frac{\gamma}{\alpha}\theta_i + \frac{\gamma}{\beta}\theta\right) - F\left(\frac{\gamma}{\beta}\theta_i + \frac{\gamma}{\alpha}\theta\right)\right]. \end{aligned} \tag{2}$$

Because $F(\theta)$ is a strictly convex function, the energy function to be minimized in Eq. (2) is the difference between two convex functions. Setting the derivatives to zero, we get the CCCP iterations given by:

$$O^{t+1}_{\alpha,\gamma} = \frac{1}{\gamma}(\nabla F)^{-1}\left[\sum_{i=1}^{n} w_i \left(\frac{1}{\beta}\nabla F\left(\frac{\gamma}{\alpha}\theta_i + \frac{\gamma}{\beta}O^t_{\alpha,\gamma}\right) + \frac{1}{\alpha}\nabla F\left(\frac{\gamma}{\beta}\theta_i + \frac{\gamma}{\alpha}O^t_{\alpha,\gamma}\right)\right)\right],$$

where ∇F and $(\nabla F)^{-1}$ are forward and reverse transformations between the natural parameters and the dual expectation parameters, respectively.

3 Clustering Based on Symmetric Hölder Divergences

Given a set of densities $\{p_1, \cdots, p_n\}$, we can perform variational k-means (Nielsen and Nock 2015) clustering based on $S^{\mathrm{H}}_{\alpha,\gamma}$. The cost function is the Hölder information:

$$E := \sum_{i=1}^{n} S^{\mathrm{H}}_{\alpha,\gamma}(p_i : O_{l_i}), \tag{3}$$

where $O_1, \cdots, O_L$ are the cluster centers and $l_i \in \{1, \cdots, L\}$ is the cluster label of p_i. Algorithm 1 presents a revision of the clustering algorithm given in (Nielsen et al. 2017) with k-means++ initialization (Arthur and Vassilvitskii 2007).

We investigate two different datasets. The first (Nielsen et al. 2017) consists of n random 2D Gaussians with two or three clusters. In the first cluster, the mean of each Gaussian $G(\mu, \Sigma)$ has the prior distribution $\mu \sim G((-2,0), I)$; the covariance matrix is obtained by first generating $\sigma_1 \sim \Gamma(7, 0.01)$, $\sigma_2 \sim \Gamma(7, 0.003)$, where Γ means a gamma distribution with prescribed shape and scale, then

Algorithm 1. Hölder variational k-means.

Input: $p_1, \cdots, p_n$; number of clusters L; $1 < \alpha \leq 2$; $\gamma > 0$
Output: A clustering scheme assigning each p_i to a label in $\{1, \cdots, L\}$

1 Randomly pick one center $O_1 \in \{p_i\}_{i=1}^n$, then sequentially pick O_k $(2 \leq k \leq L)$ with probability proportional to $\min_{j=1}^{k-1} S^{\mathrm{H}}_{\alpha,\gamma}(p_i : O_j)$
2 **while** *not converged* **do**
3 **for** $i = 1, \ldots, n$ **do**
4 Assign $l_i = \arg\min_l S^{\mathrm{H}}_{\alpha,\gamma}(p_i : O_l)$
5 **for** $l = 1, \ldots, L$ **do**
`/* Carry CCCP iterations until the current center improves the former cluster Hölder information */`
6 Compute the centroid $O_l = \arg\min_O \sum_{i:l_i=l} S^{\mathrm{H}}_{\alpha,\gamma}(p_i : O)$;
7 **return** $\{l_i\}_{i=1}^n$;

rotating the covariance matrix $\mathrm{diag}(\sigma_1, \sigma_2)$ so that the resulting Gaussian has a "radial direction" with respect to the center $(-2, 0)$. The second and third clusters are similar to the first cluster with the only difference being that their μ's are centered around $(2, 0)$ and $(0, 2\sqrt{3})$, respectively.

The second dataset consists of multinomial distributions in Δ_9, the 9D probability simplex. The dataset presents two or three clusters. For each cluster, we first pick a random center $(c_0, \cdots, c_d)$ based on the uniform distribution in Δ_9. Then we randomly generate a distribution $(p_0, \cdots, p_d)$ based on $p_i = \frac{\exp(\log c_i + \sigma\epsilon_i)}{\sum_{i=0}^{d}\exp(\log c_i + \sigma\epsilon_i)}$, where $\sigma > 0$ is a noise level parameter, and each ϵ^i follows independently a standard Gaussian distribution. We repeat generating random samples for each cluster center, and make sure that different clusters have almost the same number of samples.

Our clustering algorithm involves two additional hyper-parameters γ and α as compared with standard k-means clustering. Therefore it is meaningful to study how these two hyper-parameters affect the performance. We extend the experiments reported previously (Nielsen et al. 2017) (where $\alpha = \gamma$ is applied for simplicity) with a grid of α and γ values. Notice that the reference duality gives $S^{\mathrm{H}}_{\alpha,\gamma} = S^{\mathrm{H}}_{\beta,\gamma}$ for conjugate exponents α and β. Therefore we assume $1 < \alpha \leq 2$ without loss of generality. If we choose $\alpha = \gamma = 2$, then $S^{\mathrm{H}}_{\alpha,\gamma}$ becomes the CS divergence, and Algorithm 1 reduces to traditional CS clustering.

We performed clustering experiments by setting the number of clusters $k \in \{2, 3\}$ and setting the sample size $n \in \{50, 100\}$. Tables 1 and 2 show the clustering accuracy measured by the Normalized Mutual Information (NMI). The large variance of the clustering accuracy is because different runs are based on different random datasets. We see that the symmetric Hölder divergence can give *strikingly better* clustering as compared to CS clustering. An empirical range of well-performed parameter values is given by $\gamma \in [0.5, 1.5]$ and $\alpha \in (1, 2]$. In practice, one has to setup a configuration grid and apply cross-validation to find the best α and γ values.

Table 1. Performance in NMI (mean±std) when clustering 2D Gaussians based on 1000 independent runs for each configuration. Bold numbers indicate the best obtained performance. The boxed numbers are given by the Cauchy-Schwarz (CS) clustering.

(a) $k = 2$; $n = 50$

	$\alpha = 1.01$	$\alpha = 1.2$	$\alpha = 1.4$	$\alpha = 1.6$	$\alpha = 1.8$	$\alpha = 2$
$\gamma = 0.25$	0.91 ± 0.10	0.89 ± 0.13	0.86 ± 0.14	0.85 ± 0.14	0.85 ± 0.15	0.85 ± 0.16
$\gamma = 0.5$	$\mathbf{0.92 \pm 0.09}$	0.86 ± 0.16	0.84 ± 0.17	0.84 ± 0.17	0.82 ± 0.19	0.82 ± 0.17
$\gamma = 0.75$	$\mathbf{0.92 \pm 0.10}$	0.85 ± 0.16	0.83 ± 0.17	0.82 ± 0.18	0.82 ± 0.19	0.81 ± 0.18
$\gamma = 1$	$\mathbf{0.92 \pm 0.10}$	0.84 ± 0.18	0.81 ± 0.20	0.82 ± 0.18	0.82 ± 0.20	0.81 ± 0.20
$\gamma = 1.5$	$\mathbf{0.92 \pm 0.10}$	0.82 ± 0.18	0.80 ± 0.20	0.81 ± 0.19	0.81 ± 0.19	0.80 ± 0.21
$\gamma = 2$	$\mathbf{0.92 \pm 0.10}$	0.81 ± 0.20	0.82 ± 0.19	0.80 ± 0.21	0.80 ± 0.20	$\boxed{0.81 \pm 0.20}$

(b) $k = 2$; $n = 100$

	$\alpha = 1.01$	$\alpha = 1.2$	$\alpha = 1.4$	$\alpha = 1.6$	$\alpha = 1.8$	$\alpha = 2$
$\gamma = 0.25$	0.91 ± 0.07	0.88 ± 0.08	0.87 ± 0.09	0.86 ± 0.10	0.86 ± 0.10	0.86 ± 0.10
$\gamma = 0.5$	0.91 ± 0.07	0.87 ± 0.12	0.85 ± 0.11	0.85 ± 0.13	0.84 ± 0.14	0.84 ± 0.12
$\gamma = 0.75$	0.91 ± 0.07	0.86 ± 0.11	0.84 ± 0.13	0.84 ± 0.13	0.84 ± 0.14	0.84 ± 0.14
$\gamma = 1$	$\mathbf{0.92 \pm 0.07}$	0.86 ± 0.12	0.83 ± 0.15	0.83 ± 0.13	0.83 ± 0.14	0.84 ± 0.12
$\gamma = 1.5$	$\mathbf{0.92 \pm 0.07}$	0.84 ± 0.14	0.83 ± 0.14	0.83 ± 0.15	0.83 ± 0.14	0.83 ± 0.13
$\gamma = 2$	0.91 ± 0.08	0.84 ± 0.14	0.82 ± 0.15	0.83 ± 0.14	0.83 ± 0.14	$\boxed{0.83 \pm 0.14}$

(c) $k = 3$; $n = 50$

	$\alpha = 1.01$	$\alpha = 1.2$	$\alpha = 1.4$	$\alpha = 1.6$	$\alpha = 1.8$	$\alpha = 2$
$\gamma = 0.25$	0.88 ± 0.12	0.83 ± 0.14	0.81 ± 0.15	0.80 ± 0.15	0.79 ± 0.14	0.80 ± 0.15
$\gamma = 0.5$	0.88 ± 0.12	0.80 ± 0.15	0.77 ± 0.16	0.77 ± 0.15	0.77 ± 0.15	0.76 ± 0.16
$\gamma = 0.75$	$\mathbf{0.89 \pm 0.12}$	0.80 ± 0.14	0.77 ± 0.15	0.76 ± 0.16	0.75 ± 0.15	0.76 ± 0.16
$\gamma = 1$	0.88 ± 0.12	0.78 ± 0.15	0.76 ± 0.16	0.75 ± 0.16	0.75 ± 0.16	0.76 ± 0.15
$\gamma = 1.5$	0.88 ± 0.13	0.76 ± 0.16	0.76 ± 0.16	0.76 ± 0.15	0.76 ± 0.16	0.76 ± 0.16
$\gamma = 2$	0.88 ± 0.12	0.76 ± 0.16	0.75 ± 0.16	0.74 ± 0.16	0.75 ± 0.16	$\boxed{0.76 \pm 0.16}$

(d) $k = 3$; $n = 100$

	$\alpha = 1.01$	$\alpha = 1.2$	$\alpha = 1.4$	$\alpha = 1.6$	$\alpha = 1.8$	$\alpha = 2$
$\gamma = 0.25$	$\mathbf{0.89 \pm 0.08}$	0.84 ± 0.11	0.82 ± 0.12	0.82 ± 0.11	0.82 ± 0.11	0.82 ± 0.12
$\gamma = 0.5$	$\mathbf{0.89 \pm 0.08}$	0.83 ± 0.11	0.81 ± 0.12	0.79 ± 0.12	0.78 ± 0.14	0.79 ± 0.13
$\gamma = 0.75$	$\mathbf{0.89 \pm 0.09}$	0.81 ± 0.12	0.79 ± 0.13	0.78 ± 0.13	0.77 ± 0.14	0.78 ± 0.14
$\gamma = 1$	0.88 ± 0.10	0.80 ± 0.12	0.78 ± 0.14	0.78 ± 0.14	0.78 ± 0.13	0.78 ± 0.13
$\gamma = 1.5$	$\mathbf{0.89 \pm 0.09}$	0.78 ± 0.13	0.77 ± 0.14	0.77 ± 0.14	0.76 ± 0.14	0.77 ± 0.14
$\gamma = 2$	$\mathbf{0.89 \pm 0.09}$	0.78 ± 0.13	0.77 ± 0.13	0.77 ± 0.14	0.77 ± 0.13	$\boxed{0.78 \pm 0.13}$

Table 2. Performance in NMI (mean±std) when clustering multinomial distributions in Δ_9 based on 1000 independent runs for each configuration. Bold numbers indicate the best obtained performance. The boxed numbers are given by the Cauchy-Schwarz (CS) clustering.

(a) $k = 2$; $n = 50$

	$\alpha = 1.01$	$\alpha = 1.2$	$\alpha = 1.4$	$\alpha = 1.6$	$\alpha = 1.8$	$\alpha = 2$
$\gamma = 0.25$	$\mathbf{0.93 \pm 0.14}$	$\mathbf{0.93 \pm 0.15}$	$\mathbf{0.93 \pm 0.13}$	$\mathbf{0.93 \pm 0.15}$	0.92 ± 0.16	$\mathbf{0.93 \pm 0.13}$
$\gamma = 0.5$	0.91 ± 0.16	0.92 ± 0.15	0.90 ± 0.18	0.91 ± 0.17	0.91 ± 0.16	0.91 ± 0.16
$\gamma = 0.75$	0.87 ± 0.20	0.86 ± 0.21	0.87 ± 0.20	0.87 ± 0.20	0.88 ± 0.19	0.88 ± 0.19
$\gamma = 1$	0.83 ± 0.23	0.83 ± 0.23	0.83 ± 0.23	0.82 ± 0.24	0.81 ± 0.23	0.82 ± 0.23
$\gamma = 1.5$	0.75 ± 0.26	0.71 ± 0.28	0.72 ± 0.27	0.70 ± 0.28	0.71 ± 0.28	0.71 ± 0.28
$\gamma = 2$	0.68 ± 0.28	0.65 ± 0.29	0.64 ± 0.29	0.62 ± 0.29	0.62 ± 0.30	$\boxed{0.61 \pm 0.30}$

(b) $k = 2$; $n = 100$

	$\alpha = 1.01$	$\alpha = 1.2$	$\alpha = 1.4$	$\alpha = 1.6$	$\alpha = 1.8$	$\alpha = 2$
$\gamma = 0.25$	0.93 ± 0.12	0.93 ± 0.12	0.93 ± 0.11	0.93 ± 0.12	$\mathbf{0.94 \pm 0.11}$	0.93 ± 0.12
$\gamma = 0.5$	0.92 ± 0.14	0.91 ± 0.14	0.92 ± 0.13	0.91 ± 0.15	0.92 ± 0.14	0.91 ± 0.14
$\gamma = 0.75$	0.89 ± 0.16	0.88 ± 0.16	0.89 ± 0.17	0.89 ± 0.16	0.88 ± 0.16	0.89 ± 0.15
$\gamma = 1$	0.83 ± 0.20	0.84 ± 0.19	0.84 ± 0.19	0.83 ± 0.19	0.84 ± 0.19	0.84 ± 0.19
$\gamma = 1.5$	0.77 ± 0.24	0.74 ± 0.25	0.74 ± 0.23	0.73 ± 0.24	0.74 ± 0.23	0.74 ± 0.24
$\gamma = 2$	0.70 ± 0.26	0.67 ± 0.26	0.65 ± 0.27	0.64 ± 0.27	0.63 ± 0.27	$\boxed{0.63 \pm 0.27}$

(c) $k = 3$; $n = 50$

	$\alpha = 1.01$	$\alpha = 1.2$	$\alpha = 1.4$	$\alpha = 1.6$	$\alpha = 1.8$	$\alpha = 2$
$\gamma = 0.25$	$\mathbf{0.87 \pm 0.16}$	$\mathbf{0.87 \pm 0.16}$	$\mathbf{0.87 \pm 0.16}$	$\mathbf{0.87 \pm 0.15}$	$\mathbf{0.87 \pm 0.16}$	0.86 ± 0.16
$\gamma = 0.5$	0.84 ± 0.17	0.84 ± 0.17	0.84 ± 0.17	0.83 ± 0.17	0.84 ± 0.17	0.84 ± 0.18
$\gamma = 0.75$	0.80 ± 0.18	0.79 ± 0.18	0.79 ± 0.18	0.78 ± 0.19	0.79 ± 0.18	0.78 ± 0.19
$\gamma = 1$	0.73 ± 0.20	0.72 ± 0.20	0.73 ± 0.20	0.73 ± 0.20	0.72 ± 0.20	0.71 ± 0.21
$\gamma = 1.5$	0.65 ± 0.21	0.63 ± 0.20	0.61 ± 0.19	0.59 ± 0.20	0.59 ± 0.20	0.60 ± 0.20
$\gamma = 2$	0.57 ± 0.20	0.55 ± 0.20	0.53 ± 0.19	0.51 ± 0.18	0.52 ± 0.18	$\boxed{0.51 \pm 0.18}$

(d) $k = 3$; $n = 100$

	$\alpha = 1.01$	$\alpha = 1.2$	$\alpha = 1.4$	$\alpha = 1.6$	$\alpha = 1.8$	$\alpha = 2$
$\gamma = 0.25$	$\mathbf{0.90 \pm 0.13}$	0.88 ± 0.14	0.88 ± 0.14	0.88 ± 0.13	0.89 ± 0.13	0.89 ± 0.12
$\gamma = 0.5$	0.87 ± 0.14	0.86 ± 0.14	0.86 ± 0.15	0.86 ± 0.14	0.86 ± 0.14	0.86 ± 0.14
$\gamma = 0.75$	0.82 ± 0.16	0.82 ± 0.17	0.83 ± 0.15	0.82 ± 0.16	0.82 ± 0.16	0.82 ± 0.16
$\gamma = 1$	0.77 ± 0.18	0.77 ± 0.17	0.77 ± 0.18	0.75 ± 0.18	0.76 ± 0.18	0.76 ± 0.18
$\gamma = 1.5$	0.66 ± 0.19	0.63 ± 0.19	0.64 ± 0.19	0.63 ± 0.19	0.63 ± 0.19	0.63 ± 0.19
$\gamma = 2$	0.57 ± 0.18	0.56 ± 0.18	0.54 ± 0.18	0.53 ± 0.18	0.53 ± 0.19	$\boxed{0.53 \pm 0.18}$

This hints that one should use instead the general Hölder divergence to replace CS in similar clustering applications (Hasanbelliu et al. 2014; Rami et al. 2016). Although one faces the problem of tuning the parameters α and γ, Hölder divergences can potentially give much better results.

4 Conclusion

We experimentally confirmed the usefulness of the novel parametric Hölder classes of statistical divergences and pseudo-divergences. In general one should use Hölder clustering instead of Cauchy-Schwarz clustering to get much better results. These new concepts can open up new insights and applications in statistics and information sciences.

Reproducible source code is available online at:

https://www.lix.polytechnique.fr/~nielsen/HPD/.

References

Mitrinovic, D.S., Pecaric, J., Fink, A.M.: Classical and New Inequalities in Analysis, vol. 61. Springer Science & Business Media, New York (2013)

Kanamori, T., Fujisawa, H.: Affine invariant divergences associated with proper composite scoring rules and their applications. Bernoulli **20**, 2278–2304 (2014)

Kanamori, T.: Scale-invariant divergences for density functions. Entropy **16**, 2611–2628 (2014)

Arthur, D., Vassilvitskii, S.: k-means++: the advantages of careful seeding. In: ACM-SIAM Symposium on Discrete Algorithms, pp. 1027–1035 (2007)

Nielsen, F., Sun, K., Marchand-Maillet, S.: On Hölder projective divergences. Entropy **19**, 122 (2017)

Holder, O.L.: Über einen Mittelwertssatz. Nachr. Akad. Wiss. Gottingen Math. Phys. Kl. **44**, 38–47 (1889)

Nielsen, F., Boltz, S.: The Burbea-Rao and Bhattacharyya centroids. IEEE Trans. Inf. Theory **57**, 5455–5466 (2011)

Nielsen, F., Nock, R.: Total Jensen divergences: definition, properties and clustering. In: Proceedings of the 2015 IEEE International Conference on Acoustics, Speech and Signal Processing (ICASSP), South Brisbane, Queensland, Australia, 19–24 April 2015, pp. 2016–2020 (2015)

Hasanbelliu, E., Giraldo, L.S., Principe, J.C.: Information theoretic shape matching. IEEE Trans. Pattern Anal. Mach. Intell. **36**, 2436–2451 (2014)

Rami, H., Belmerhnia, L., Drissi El Maliani, A., El Hassouni, M.: Texture retrieval using mixtures of generalized gaussian distribution and cauchy-schwarz divergence in wavelet domain. Image Commun. **42**, 45–58 (2016)

Real Hypersurfaces in the Complex Quadric with Certain Condition of Normal Jacobi Operator

Imsoon Jeong[1], Gyu Jong Kim[2], and Young Jin Suh[2(✉)]

[1] Pai Chai University, Daejeon 35345, Republic of Korea
[2] Kyungpook National University, Daegu 41566, Republic of Korea
yjsuh@knu.ac.kr

Abstract. We introduce the notion of normal Jacobi operator of Codazzi type for real hypersurfaces in the complex quadric Q^m. The normal Jacobi operator of Codazzi type implies that the unit normal vector field N becomes $\mathfrak{A}$-principal or $\mathfrak{A}$-isotropic. Then according to each case, we give a non-existence theorem of real hypersurfaces in Q^m with normal Jacobi operator of Codazzi type.

Keywords: Normal Jacobi operator of Codazzi type · $\mathfrak{A}$-isotropic · $\mathfrak{A}$-principal · Kähler structure · Complex conjugation · Complex quadric

1 Introduction

In the complex projective space $\mathbb{C}P^{m+1}$ and the quaternionic projective space $\mathbb{Q}P^{m+1}$ some classifications related to the Ricci tensor and the structure Jacobi operator were investigated by Kimura [10,11], Pérez and Suh [17–19], Pérez and Santos [14], and Pérez, Santos and Suh [15,16], respectively. Some examples of Hermitian symmetric space of rank 2 are $G_2(\mathbb{C}^{m+2}) = SU_{m+2}/S(U_2U_m)$ and $G_2^*(\mathbb{C}^{m+2}) = SU_{2,m}/S(U_2U_m)$, which are said to be complex two-plane Grassmannians and complex hyperbolic two-plane Grassmannians, respectively (see [23–25]). These are viewed as Hermitian symmetric spaces and quaternionic Kähler symmetric spaces equipped with the Kähler structure J and the quaternionic Kähler structure $\mathfrak{J}$.

The classification problems of real hypersurfaces in the complex two-plane Grassmannian $SU_{m+2}/S(U_2U_m)$ with certain geometric conditions were mainly investigated by Jeong et al. [6], Jeong, Machado, Pérez and Suh [7,8] and Suh [23–25], where the classification of commuting and parallel Jacobi operator, contact hypersurfaces, parallel Ricci tensor, and harmonic curvature for real hypersurfaces in complex two-plane Grassmannians were extensively studied. Moreover, in [26] we have asserted that the Reeb flow on a real hypersurface in $SU_{2,m}/S(U_2U_m)$ is isometric if and only if M is an open part of a tube around a totally geodesic $SU_{2,m-1}/S(U_2U_{m-1}) \subset SU_{2,m}/S(U_2U_m)$.

As another kind of Hermitian symmetric space with rank 2 of compact type different from the above ones, we can give the example of complex quadric

F. Nielsen and F. Barbaresco (Eds.): GSI 2017, LNCS 10589, pp. 864–871, 2017.
https://doi.org/10.1007/978-3-319-68445-1_99

$Q^m = SO_{m+2}/SO_mSO_2$, which is a complex hypersurface in complex projective space $\mathbb{C}P^{m+1}$ (see Klein [9], and Smyth [21]). The complex quadric can also be regarded as a kind of real Grassmann manifold of compact type with rank 2 (see Kobayashi and Nomizu [12]). Accordingly, the complex quadric admits two important geometric structures, a complex conjugation structure A and a Kähler structure J, which anti-commute with each other, that is, $AJ = -JA$. Then for $m \geq 2$ the triple (Q^m, J, g) is a Hermitian symmetric space of compact type with rank 2 and its maximal sectional curvature is equal to 4 (see Klein [9] and Reckziegel [20]).

In addition to the complex structure J there is another distinguished geometric structure on Q^m, namely a parallel rank two vector bundle $\mathfrak{A}$ which contains an S^1-bundle of real structures, that is, complex conjugations A on the tangent spaces of Q^m. This geometric structure determines a maximal $\mathfrak{A}$-invariant subbundle $\mathcal{Q}$ of the tangent bundle TM of a real hypersurface M in Q^m as follows:

$$\mathcal{Q} = \{X \in T_zM \,|\, AX \in T_zM \quad \text{for all} \quad A \in \mathfrak{A}\}.$$

Moreover, the derivative of the complex conjugation A on Q^m is defined by

$$(\bar{\nabla}_X A)Y = q(X)JAY$$

for any vector fields X and Y on M and q denotes a certain 1-form defined on M.

Recall that a nonzero tangent vector $W \in T_{[z]}Q^m$ is called singular if it is tangent to more than one maximal flat in Q^m. There are two types of *singular* tangent vectors for the complex quadric Q^m (see Berndt and Suh [2] and Reckziegel [20]):

- If there exists a conjugation $A \in \mathfrak{A}$ such that $W \in V(A)$(the $(+1)$-eigenspace of a conjugation A in $T_{[z]}Q^m$), then W is singular. Such a singular tangent vector is called $\mathfrak{A}$-principal.
- If there exists a conjugation $A \in \mathfrak{A}$ and orthonormal vectors $X, Y \in V(A)$ such that $W/||W|| = (X + JY)/\sqrt{2}$, then W is singular. Such a singular tangent vector is called $\mathfrak{A}$-isotropic.

When we consider a real hypersurface M in the complex quadric Q^m, under the assumption of some geometric properties the unit normal vector field N of M in Q^m can be either $\mathfrak{A}$-isotropic or $\mathfrak{A}$-principal (see [27, 28]). In the first case, where N is $\mathfrak{A}$-isotropic, Suh has shown in [27] that M is locally congruent to a tube over a totally geodesic $\mathbb{C}P^k$ in Q^{2k}. In the second case, when the unit normal N is $\mathfrak{A}$-principal, he proved that a contact hypersurface M in Q^m is locally congruent to a tube over a totally geodesic and totally real submanifold S^m in Q^m (see [28]).

Jacobi fields along geodesics of a given Riemannian manifold $\bar{M}$ satisfy a well known differential equation. Naturally the classical differential equation inspires the so-called *Jacobi operator*. That is, if $\bar{R}$ is the curvature operator of $\bar{M}$, the Jacobi operator with respect to X at $z \in \bar{M}$, is defined by

$$(\bar{R}_X Y)(z) = (\bar{R}(Y, X)X)(z)$$

for any $Y \in T_z\bar{M}$. Then $\bar{R}_X$ becomes a symmetric endomorphism of the tangent bundle $T\bar{M}$ of $\bar{M}$, that is, $\bar{R}_X \in \mathrm{End}(T_z\bar{M})$. Clearly, each tangent vector field X to $\bar{M}$ provides a Jacobi operator with respect to X (see Pérez and Santos [14], and Pérez, Santos and Suh [15,16]). From such a view point, for a real hypersurface M in Q^m with unit normal vector field N the *normal Jacobi operator* $\bar{R}_N$ is defined by

$$\bar{R}_N = \bar{R}(\cdot, N)N \in \mathrm{End}\,(T_zM), \quad z \in M,$$

where $\bar{R}$ denotes the curvature tensor of Q^m. Of course, the normal Jacobi operator $\bar{R}_N$ is a symmetric endomorphism of T_zM (see Jeong, Kim and Suh [6], Jeong, Machado, Pérez and Suh [7,8]). We introduce the notion of parallelism with respect to the normal Jacobi operator $\bar{R}_N$ of M in $\bar{M}$. It is defined by $(\nabla_X \bar{R}_N)Y = 0$ for all tangent vector fields X and Y on M. This has a geometric meaning that the eigenspaces of $\bar{R}_N$ are parallel, that is, invariant under any parallel displacements along any curves on M in $\bar{M}$. Using this notion, specially, in [6] they gave a non-existence theorem for Hopf hypersurfaces with parallel normal Jacobi operator in $G_2(\mathbb{C}^{m+2})$. Moreover, Suh [30] gave a non-existence theorem for the case of a Hopf hypersurface with parallel normal Jacobi operator in Q^m. Here M is called Hopf if the Reeb vector field ξ defined by $\xi = -JN$ is principal, that is, $S\xi = \alpha\xi$, where S is the shape operator of M associated with the unit normal N.

As a generalized notion of parallel normal Jacobi operator, in this paper we want to introduce the definition of the *normal Jacobi operator of Codazzi type* from the view point of exterior derivative (see Besse [2], and Derdzinski and Shen [3]). Let E be a vector bundle over a manifold $\bar{M}$. For any section ω of $\bigwedge^p \bar{M} \bigotimes E$ the exterior derivative $d^\nabla \omega$ is the section of $\bigwedge^{p+1} \bar{M} \bigotimes E$ such that for $X_0, \cdots, X_p$ in $T_z\bar{M}$, $z \in \bar{M}$, extended to vector fields $\tilde{X}_0, \cdots, \tilde{X}_p$ in a neighborhood as follows:

$$\begin{aligned}(d^\nabla \omega)(X_0, \cdots, X_p) &= \sum\nolimits_i (-1)^i \nabla_{X_i} (\omega(\tilde{X}_0, \cdots, \hat{\tilde{X}}_i, \cdots, \tilde{X}_p)) \\ &\quad + \sum\nolimits_{i \neq j} (-1)^{i+j} \omega([\tilde{X}_i, \tilde{X}_j], \tilde{X}_0, \cdots, \hat{\tilde{X}}_i, \cdots, \hat{\tilde{X}}_j, \cdots, \tilde{X}_p).\end{aligned}$$

In case of the normal Jacobi operator $\bar{R}_N$, then the exterior derivative of the normal Jacobi operator gives

$$\begin{aligned}(d^\nabla \bar{R}_N)(X, Y) &= \nabla_X(\bar{R}_N Y) - \nabla_Y(\bar{R}_N X) - \bar{R}_N([X, Y]) \\ &= (\nabla_X \bar{R}_N)Y - (\nabla_Y \bar{R}_N)X.\end{aligned}$$

Now we apply above equation to $\bar{R}_N$, which is a tensor field of type $(1, 1)$ on a real hypersurface M in $\bar{M}$. The normal Jacobi operator $\bar{R}_N$ of M in $\bar{M}$ is said to be of *Codazzi type* if the normal Jacobi operator $\bar{R}_N$ satisfies $d^\nabla \bar{R}_N = 0$, that is, the normal Jacobi operator is closed with respect to the exterior derivative d^∇ related to the induced connection ∇. Then by the above formula we have the following

$$(\nabla_X \bar{R}_N)Y = (\nabla_Y \bar{R}_N)X$$

for any $X, Y \in T_zM$, $z \in M$. Related to this definition, Machado, Pérez, Jeong, and Suh considered the case of a real hypersurface in $G_2(\mathbb{C}^{m+2})$ and gave a non-existence theorem as follows.

Remark 1. There does not exist any connected Hopf real hypersurface in complex two-plane Grassmannians $G_2(\mathbb{C}^{m+2})$, $m \geq 3$, whose normal Jacobi operator is of Codazzi type if the distribution $\mathcal{D}$ or the $\mathcal{D}^{\perp}$-component of the Reeb vector field is invariant under the shape operator (see [13]).

Remark 2. Many geometers have studied the various tensor of Codazzi type (1,1), for example the shape operator, the structure Jacobi operator $R_\xi := R(\cdot, \xi)\xi \in \mathrm{End}(TM)$, Ricci tensor etc., on a real hypersurface M in $G_2(\mathbb{C}^{m+2})$ (see [4,22,25] and so on). In particular, in [22] the third author deal with the parallelism for the shape operator on M. But we know that the proofs contain for the case of Codazzi type with respect to the shape operator.

On the other hand, the Ricci operator Ric of M in Q^m is said to be of *harmonic curvature* if the Ricci operator Ric satisfies

$$(\nabla_X \mathrm{Ric})Y = (\nabla_Y \mathrm{Ric})X$$

for any $X, Y \in T_zM$, $z \in M$. In the study of real hypersurfaces in the complex quadric Q^m we considered some notions of parallel Ricci tensor or more generally, harmonic curvature, that is, $\nabla \mathrm{Ric} = 0$ or $(\nabla_X \mathrm{Ric})Y = (\nabla_Y \mathrm{Ric})X$ respectively(see Suh [28,29]). But from the assumptions of Ricci parallel or harmonic curvature, it was difficult for us to derive the fact that either the unit normal N is $\mathfrak{A}$-isotropic or $\mathfrak{A}$-principal. So in [1,28] and [29] we gave a classification with the further assumption of $\mathfrak{A}$-isotropic. But fortunately, when we consider the normal Jacobi operator $\bar{R}_N$ of Codazzi type, first we can assert that the unit normal vector field N becomes either $\mathfrak{A}$-isotropic or $\mathfrak{A}$-principal.

Theorem 1. *Let M be a Hopf real hypersurface in Q^m, $m \geq 3$, with normal Jacobi operator of Codazzi type. Then the unit normal vector field N is singular, that is, N is $\mathfrak{A}$-isotropic or $\mathfrak{A}$-principal.* □

Then motivated by such a result, next we give a non-existence theorem for Hopf hypersurfaces in the complex quadric Q^m with normal Jacobi operator of Codazzi type as follows:

Theorem 2. *There does not exist any Hopf real hypersurface in Q^m, $m \geq 3$ with normal Jacobi operator of Codazzi type.* □

2 Outline of Proofs

It is well known that complex quadric $Q^m = SO_{m+2}/SO_mSO_2$ is a compact Kählelr manifold with the metric and complex structure induced from $\mathbb{C}P^{m+1}$ with constant holomorphic sectional curvature 4. Moreover it becomes a

Hermitian symmetric space of rank 2 being equipped with both a complex structure J and a real structure (or complex conjugation) A satisfying $AJ = -JA$ and $\mathrm{Tr}A = \mathrm{Tr}JA = 0$.

Hence by the equation of Gauss, the curvature tensor $\bar{R}(X,Y)Z$ for a complex hypersurface Q^m in $\mathbb{C}P^{m+1}$ induced from the curvature tensor $\widetilde{R}$ of $\mathbb{C}P^{m+1}$ can be described in terms of the complex structure J and the complex conjugation $A \in \mathfrak{A}$ as follows:

$$\begin{aligned}\bar{R}(X,Y)Z &= g(Y,Z)X - g(X,Z)Y + g(JY,Z)JX - g(JX,Z)JY \\ &\quad - 2g(JX,Y)JZ + g(AY,Z)AX - g(AX,Z)AY \\ &\quad + g(JAY,Z)JAX - g(JAX,Z)JAY \\ &\quad + g(SY,Z)SX - g(SX,Z)SY\end{aligned} \tag{1}$$

for any $X,Y,Z \in T_zQ^m$, $z \in Q^m$.

From the induced almost contact metric structure (ϕ,ξ,η,g) of M let us put

$$\begin{aligned}JX &= \phi X + \eta(X)N, \quad JN = -\xi \\ AX &= BX + \rho(X)N\end{aligned}$$

for any vector field $X \in T_zQ^m$, $z \in M$, $\rho(X) = g(AX,N)$, where BX and $\rho(X)N$ respectively denote the tangential and normal component of the vector field AX. Moreover, it follows that $A\xi = B\xi + \rho(\xi)N$ and $\rho(\xi) = g(A\xi,N) = 0$. Then it follows that

$$AN = AJ\xi = -JA\xi = -JB\xi = -(\phi B\xi + \eta(B\xi)N).$$

By virtue of the equation of Gauss and (1), the normal Jacobi operator $\bar{R}_N$ of M can be described in terms of the complex structure J and the complex conjugations $A \in \mathfrak{A}$ as follows:

$$\bar{R}_N(Y) = Y + 3\eta(Y)\xi + g(AN,N)BY + g(AY,N)\phi A\xi - g(AY,\xi)A\xi$$

for any $Y \in T_zM$, $z \in M$. Now the derivative of the normal Jacobi operator $\bar{R}_N$ is given by

$$(\nabla_X\bar{R}_N)Y = \nabla_X(\bar{R}_N(Y)) - \bar{R}_N(\nabla_XY),$$

where ∇ denotes the Riemannian connection defined on M in Q^m.

Now let us consider the following formula for the connection $\bar{\nabla}$ on Q^m and any tangent vector fields X and Y on M

$$\begin{aligned}\bar{\nabla}_X(AY) - A\nabla_XY &= (\bar{\nabla}_XA)Y + A\bar{\nabla}Y - A\nabla_XY \\ &= q(X)JAY + A\sigma(X,Y) \\ &= q(X)JAY + g(SX,Y)AN,\end{aligned}$$

where q denotes a certain 1-form defined on M in the introduction, and $\sigma(X,Y)$ denotes the normal part of $\bar{\nabla}_XY$. From this, together with normal Jacobi operator of Codazzi type for any tangent vector fields X and Y on M in Q^m, it follows that

$$\begin{aligned}
0 &= (\nabla_X \bar{R}_N)Y - (\nabla_Y \bar{R}_N)X \\
&= 3\{g(\phi SX, Y) - g(\phi SY, X)\}\xi + 3\{\eta(Y)\phi SX - \eta(X)\phi SY\} \\
&\quad - 2g(ASX, N)AY + 2g(ASY, N)AX \\
&\quad + g(AN, N)\{q(X)JAY - q(Y)JAX\} \\
&\quad - \{q(X)g(JAY, N) - q(Y)g(JAX, N)\}AN \\
&\quad + g(AY, SX)AN - g(AX, SY)AN \\
&\quad - g(AY, N)\{(\bar{\nabla}_X A)N + A\bar{\nabla}_X N\} \\
&\quad + g(AX, N)\{(\bar{\nabla}_Y A)N + A\bar{\nabla}_Y N\} \\
&\quad - \{g(\bar{\nabla}_X(AY) - A\nabla_X Y, \xi) - g(\bar{\nabla}_Y(AX) - A\nabla_Y X, \xi)\}A\xi \\
&\quad - \{g(AY, \phi SX + \sigma(X, \xi)) - g(AX, \phi SY + \sigma(Y, \xi))\}A\xi \\
&\quad - g(AY, \xi)\{(\bar{\nabla}_X A)\xi + A\bar{\nabla}_X \xi\} \\
&\quad + g(AX, \xi)\{(\bar{\nabla}_Y A)\xi + A\bar{\nabla}_Y \xi\}.
\end{aligned} \tag{2}$$

Using this equation, we can prove following lemma.

Lemma 1. *Let M be a Hopf real hypersurface in Q^m, $m \geq 3$, with normal Jacobi operator of Codazzi type. Then the unit normal vector field N is either $\mathfrak{A}$-isotropic or $q(\xi) = -g(AN, N)q(A\xi)$.*

By Lemma 1, we can assert the following

Lemma 2. *Let M be a Hopf real hypersurface in Q^m, $m \geq 3$, with normal Jacobi operator of Codazzi type. Then the unit normal vector field N is singular, that is, N is either $\mathfrak{A}$-isotropic or $\mathfrak{A}$-principal.*

Proof. When $g(AN, N) = 0$, we have proved that the unit normal N is $\mathfrak{A}$-isotropic. Now let us consider the case that $g(AN, N) \neq 0$. Then by Lemma 1, we know that

$$\alpha g(AN, N)^2 = 0.$$

Here, if the Reeb function $\alpha \neq 0$, then $g(AN, N) = 0$ also gives that the unit normal vector field N is $\mathfrak{A}$-isotropic.

When the Reeb function α is vanishing, that is, the equation

$$Y\alpha = (\xi\alpha)\eta(Y) - 2g(\xi, AN)g(Y, A\xi) + 2g(Y, AN)g(\xi, A\xi)$$

becomes

$$g(Y, (AN)^T)g(\xi, A\xi) = 0,$$

where we have used $g(\xi, AN) = 0$ and the vector field $(AN)^T$ denotes the tangential component of the vector field AN.

Since in the second case we have assumed that N is not $\mathfrak{A}$-isotropic, we know $g(\xi, A\xi) \neq 0$. So it follows that $(AN)^T = 0$. This means that

$$AN = (AN)^T + g(AN, N)N = g(AN, N)N.$$

Then it implies that

$$N = A^2 N = g(AN, N)AN = g^2(AN, N)N.$$

This gives that $g(AN, N) = \pm 1$. So $g(AN, N) = \cot 2t = \pm 1$ implies $t = 0$ or $\frac{\pi}{2}$. Accordingly, we should have $t = 0$, that is, $g(AN, N) = 1$. Then we can take the unit normal N such that $AN = N$. So the unit normal vector field N becomes $\mathfrak{A}$-principal, that is, $AN = N$. □

Then motivated by such a result, we give a non-existence for real hypersurfaces in the complex quadric Q^m with normal Jacobi operator of Codazzi type as Theorem 1 given in Sect. 1 (see [5]).

Usually, normal Jacobi operator of Codazzi type is a generalization of parallel normal Jacobi operator $\bar{R}_N$ of M in Q^m, that is, $\nabla_X \bar{R}_N = 0$ for any tangent vector field X on M. The parallelism of normal Jacobi operator has a geometric meaning that every eigen spaces of the normal Jacobi operator $\bar{R}_N$ are parallel along any direction on M in Q^m. Then naturally, by Theorem 1 we give the following:

Corollary 1. [30] *There do not exist any Hopf real hypersurfaces in Q^m, $m \geq 3$ with parallel normal Jacobi operator.*

Acknowledgments. This work was supported by grant Proj. NRF-2015-R1A2A1A-01002459 from National Research Foundation of Korea and the first author NRF-2017-R1A2B-4005317.

References

1. Berndt, J., Suh, Y.J.: Real hypersurfaces with isometric Reeb flow in complex quadric. Int. J. Math. **24**, 1350050 (2013). (18 pages)
2. Besse, A.L.: Einstein Manifolds. Springer, Heidelberg (2008). doi:10.1007/978-3-540-74311-8
3. Derdzinski, A., Shen, C.L.: Codazzi tensor fields, curvature and Pontryagin forms. Proc. London Math. Soc. **47**, 15–26 (1980)
4. Jeong, I., Lee, H., Suh, Y.J.: Real hypersurfaces in complex two-plane Grassmannians whose structure Jacobi operator is of Codazzi type. Acta Math. Hungar. **125**, 141–160 (2009)
5. Jeong, I., Kim G.J., Suh, Y.J.: Real hypersurfaces in the complex quadric with normal Jacobi operator of Codazzi type. Bull. Malays. Math. Sci. Soc. (to appear). doi:10.1007/s40840-017-0485-9
6. Jeong, I., Kim, H.J., Suh, Y.J.: Real hypersurfaces in complex two-plane Grassmannians with parallel normal Jacobi operator. Publ. Math. Debrecen **76**, 203–218 (2010)
7. Jeong, I., Machado, C.J.G., Pérez, J.D., Suh, Y.J.: Real hypersurfaces in complex two-plane Grassmannians with $\mathfrak{D}^\perp$-parallel structure Jacobi operator. Int. J. Math. **22**, 655–673 (2011)
8. Jeong, I., Machado, C.J.G., Pérez, J.D., Suh, Y.J.: $\mathcal{D}$-parallelism of normal and structure Jacobi operators for hypersurfaces in complex two-plane Grassmannians. Ann. Mat. Pura. Appl. **193**, 591–608 (2014)

9. Klein, S.: Totally geodesic submanifolds in the complex quadric. Differ. Geom. Appl. **26**, 79–96 (2008)
10. Kimura, M.: Real hypersurfaces and complex submanifolds in complex projective space. Trans. Amer. Math. Soc. **296**, 137–149 (1986)
11. Kimura, M.: Some real hypersurfaces of a complex projective space. Saitama Math. J. **5**, 1–5 (1987)
12. Kobayashi, S., Nomizu, K.: Foundations of Differential Geometry. Vol. II. A Wiley-Interscience Publication, Wiley Classics Library Edition (1996)
13. Machado, C.J.G., Pérez, J.D., Jeong, I., Suh, Y.J.: Real hypersurfaces in complex two-plane Grassmannians whose normal Jacobi opertor is of Codazzi type. Cent. Eur. J. Math. **9**, 578–582 (2011)
14. Pérez, J.D., Santos, F.G.: Real hypersurfaces in complex projective space with recurrent structure Jacobi operator. Differential Geom. Appl. **26**, 218–223 (2008)
15. Pérez, J.D., Santos, F.G., Suh, Y.J.: Real hypersurfaces in complex projective space whose structure Jacobi operator is Lie ξ-parallel. Differ. Geom. Appl. **22**, 181–188 (2005)
16. Pérez, J.D., Santos, F.G., Suh, Y.J.: Real hypersurfaces in complex projective space whose structure Jacobi operator is $\mathcal{D}$-parallel. Bull. Belg. Math. Soc. **13**, 459–469 (2006)
17. Pérez, J.D., Suh, Y.J.: Real hypersurfaces of quaternionic projective space satisfying $\nabla_{U_i} R = 0$. Differ. Geom. Appl. **7**, 211–217 (1997)
18. Pérez, J.D., Suh, Y.J.: Certain conditions on the Ricci tensor of real hypersurfaces in quaternionic projective space. Acta Math. Hungar. **91**, 343–356 (2001)
19. Pérez, J.D., Suh, Y.J.: The Ricci tensor of real hypersurfaces in complex two-plane Grassmannians. J. Korean Math. Soc. **44**, 211–235 (2007)
20. Reckziegel, H.: On the geometry of the complex quadric. In: Geometry and Topology of Submanifolds VIII. World Scientific Publishing, River Edge. pp. 302–315 (1995)
21. Smyth, B.: Differential geometry of complex hypersurfaces. Ann. Math. **85**, 246–266 (1967)
22. Suh, Y.J.: Real hypersurfaces in complex two-plane Grassmannians with parallel shape operator. Bull. Austral. Math. Soc. **68**, 493–502 (2003)
23. Suh, Y.J.: Real hypersurfaces of type B in complex two-plane Grassmannians. Monatsh. Math. **147**, 337–355 (2006)
24. Suh, Y.J.: Real hypersurfaces in complex two-plane Grassmannians with parallel Ricci tensor. Proc. Roy. Soc. Edinburgh Sect. A. **142**, 1309–1324 (2012)
25. Suh, Y.J.: Real hypersurfaces in complex two-plane Grassmannians with harmonic curvature. J. Math. Pures Appl. **100**, 16–33 (2013)
26. Suh, Y.J.: Hypersurfaces with isometric Reeb flow in complex hyperbolic two-plane Grassmannians. Adv. Appl. Math. **50**, 645–659 (2013)
27. Suh, Y.J.: Real hypersurfaces in the complex quadric with Reeb parallel shape operator. Internat. J. Math. 25, 1450059 (2014). 17 pp
28. Suh, Y.J.: Real hypersurfaces in the complex quadric with parallel Ricci tensor. Adv. Math. **281**, 886–905 (2015)
29. Suh, Y.J.: Real hypersurfaces in the complex quadric with harmonic curvature. J. Math. Pures Appl. **106**, 393–410 (2016)
30. Suh, Y.J.: Real hypersurfaces in the complex quadric with parallel normal Jacobi operator. Math. Nachr. **290**, 442–451 (2017)

Newton's Equation on Diffeomorphisms and Densities

Boris Khesin[1], Gerard Misiołek[2], and Klas Modin[3(✉)]

[1] Department of Mathematics, University of Toronto, Toronto, Canada
khesin@math.toronto.edu

[2] Department of Mathematics, University of Notre Dame, Notre Dame, USA
gmisiole@nd.edu

[3] Mathematical Sciences, Chalmers and University of Gothenburg, Gothenburg, Sweden
klas.modin@chalmers.se

Abstract. We develop a geometric framework for Newton-type equations on the infinite-dimensional configuration space of probability densities. It can be viewed as a second order analogue of the "Otto calculus" framework for gradient flow equations. Namely, for an n-dimensional manifold M we derive Newton's equations on the group of diffeomorphisms $\mathrm{Diff}(M)$ and the space of smooth probability densities $\mathrm{Dens}(M)$, as well as describe the Hamiltonian reduction relating them. For example, the compressible Euler equations are obtained by a Poisson reduction of Newton's equation on $\mathrm{Diff}(M)$ with the symmetry group of volume-preserving diffeomorphisms, while the Hamilton–Jacobi equation of fluid mechanics corresponds to potential solutions. We also prove that the Madelung transform between Schrödinger-type and Newton's equations is a symplectomorphism between the corresponding phase spaces $T^*\mathrm{Dens}(M)$ and $PL^2(M, C)$. This improves on the previous symplectic submersion result of von Renesse [1]. Furthermore, we prove that the Madelung transform is a Kähler map provided that the space of densities is equipped with the (prolonged) Fisher–Rao information metric and describe its dynamical applications. This geometric setting for the Madelung transform sheds light on the relation between the classical Fisher–Rao metric and its quantum counterpart, the Bures metric. In addition to compressible Euler, Hamilton–Jacobi, and linear and nonlinear Schrödinger equations, the framework for Newton equations encapsulates Burgers' inviscid equation, shallow water equations, two-component and μ-Hunter–Saxton equations, the Klein–Gordon equation, and infinite-dimensional Neumann problems.

Keywords: Newton's equation · Wasserstein distance · Fisher–Rao metric · Madelung transform · Compressible Euler equations

Reference

1. von Renesse, M.K.: An optimal transport view of Schrödinger's equation. Canad. Math. Bull **55**, 858–869 (2012)

F. Nielsen and F. Barbaresco (Eds.): GSI 2017, LNCS 10589, p. 873, 2017.
https://doi.org/10.1007/978-3-319-68445-1

Author Index

Printed in the United States
By Bookmasters